T0142193

Neural Networks and Statistical Learning

Ke-Lin Du · M. N. S. Swamy

Neural Networks and Statistical Learning

Second Edition

Ke-Lin Du
Department of Electrical
and Computer Engineering
Concordia University
Montreal, QC, Canada

Xonlink Inc.
Hangzhou, China

M. N. S. Swamy
Department of Electrical
and Computer Engineering
Concordia University
Montreal, QC, Canada

ISBN 978-1-4471-7454-7 ISBN 978-1-4471-7452-3 (eBook)
https://doi.org/10.1007/978-1-4471-7452-3

This Springer imprint is published by the registered company Springer-Verlag London Ltd. part of Springer Nature.
The registered company address is: The Campus, 4 Crinan Street, London, N1 9XW, United Kingdom

To Falong Xing and Jie Zeng
—Ke-Lin Du

To my teachers and my students
—M. N. S. Swamy

Preface to the Second Edition

Since the publication of the first edition in December 2013, the rapid rise of deep learning and AI has resulted in a wave of research activities and numerous new results. During the past few years, there have been several breakthroughs in deep learning and AI. At the same time, research and application of big data are widespread. Machine learning has become the brains behind big data.

In such a setting, this book has become one of the best sellers of Springer books. Under the suggestion of Anthony Doyle at Springer London Ltd., we decided to publish this second edition.

In this second edition, we will add six new chapters to the first edition:

- Chapter 3 focuses on computation learning theory. Part of its content is split from Chap. 2 of the first edition.
- Chapter 18 introduces compressed sensing and sparse coding. In this approach, a datum is represented as a linear combination of basis functions, and the coefficients are assumed to be sparse.
- Chapter 19 deals with matrix completion. Recovery of a data matrix from a subset of its entries is an extension of compressed sensing and sparse approximation.
- Chapter 23 introduces the Boltzmann machine. Part of its content is split from Chap. 19 of the first edition.
- Chapter 24 describes deep learning and deep neural networks. Deep learning is the state-of-the-art approach to solving complex problems.
- Chapter 31 introduces big data, cloud computing, and Internet of Things. These topics go hand in hand. Machine learning functions as the major tool for data analytics.

We also update each chapter by including major contributions published in the past 6 years.

The authors wish to thank Jie Zeng (Richcon MC, Inc., China), Biaobiao Zhang (Yuantu Internet Technology Ltd., China), Li Yu (Zhejiang University of Technology, China), Zhijiang Xu (Zhejiang University of Technology, China), and Renwang Li (Zhejiang Sci-Tech University, China) for their help during the preparation of this second edition.

Hangzhou, China/Montreal, Canada Ke-Lin Du
Montreal, Canada M. N. S. Swamy
May 2019

Preface to the First Edition

The human brain, consisting of nearly 10^{11} neurons, is the center of human intelligence. Human intelligence has been simulated in various ways. Artificial intelligence (AI) pursues exact logical reasoning based on symbol manipulation. Fuzzy logics model the highly uncertain behavior of decision-making. Neural networks model the highly nonlinear infrastructure of brain networks. Evolutionary computation models the evolution of intelligence. Chaos theory models the highly nonlinear and chaotic behaviors of human intelligence.

Soft computing is an evolving collection of methodologies for the representation of ambiguity in human thinking; it exploits the tolerance for imprecision and uncertainty, approximate reasoning, and partial truth in order to achieve tractability, robustness, and low-cost solutions. The major methodologies of soft computing are fuzzy logic, neural networks, and evolutionary computation.

Conventional model-based data processing methods require experts' knowledge for the modeling of a system. Neural network methods provide a model-free, adaptive, fault-tolerant, parallel, and distributed processing solution. A neural network is a black box that directly learns the internal relations of an unknown system, without guessing functions for describing cause-and-effect relationships. The neural network approach is a basic methodology of information processing. Neural network models may be used for function approximation, classification, nonlinear mapping, associative memory, vector quantization, optimization, feature extraction, clustering, and approximate inference. Neural networks have wide applications in almost all areas of science and engineering.

Fuzzy logic provides a means for treating uncertainty and computing with words. This mimics human recognition, which skillfully copes with uncertainty. Fuzzy systems are conventionally created from explicit knowledge expressed in the form of fuzzy rules, which are designed based on experts' experience. A fuzzy system can explain its action by fuzzy rules. Neurofuzzy systems, as a synergy of fuzzy logic and neural networks, possess both learning and knowledge representation capabilities.

This book is our attempt to bring together the major advances in neural networks and machine learning, and to explain them in a statistical framework. While some mathematical details are needed, we emphasize the practical aspects of the models and methods rather than the theoretical details. To us, neural networks are merely some statistical methods that can be represented by graphs and networks. They can iteratively adjust the network parameters. As a statistical model, a neural network can learn the probability density function from the given samples, and then predict, by generalization according to the learnt statistics, outputs for new samples that are not included in the learning sample set.

The neural network approach is a general statistical computational paradigm. Neural network research solves two problems: the direct problem and the inverse problem. The direct problem employs computer and engineering techniques to model biological neural systems of the human brain. This problem is investigated by cognitive scientists and can be useful in neuropsychiatry and neurophysiology. The inverse problem simulates biological neural systems for their problem-solving capabilities for application in scientific or engineering fields. Engineering and computer scientists have conducted an extensive investigation in this area. This book concentrates mainly on the inverse problem, although the two areas often shed light on each other. The biological and psychological plausibility of the neural network models have not been seriously treated in this book, though some background material is discussed.

This book is intended to be used as a textbook for advanced undergraduate and graduate students in engineering, science, computer science, business, arts, and medicine. It is also a good reference book for scientists, researchers, and practitioners in a wide variety of fields, and assumes no previous knowledge of neural network or machine learning concepts.

This book is divided into 25 chapters and 2 appendices. It contains almost all the major neural network models and statistical learning approaches. We also give an introduction to fuzzy sets and logic, and neurofuzzy models. Hardware implementations of the models are discussed. Two chapters are dedicated to the applications of neural network and statistical learning approaches to biometrics/bioinformatics and data mining. Finally, in the appendices, some mathematical preliminaries are given, and benchmarks for validating all kinds of neural network methods and some web resources are provided.

First and foremost, we would like to thank the supporting staff from Springer London, especially Anthony Doyle and Grace Quinn for their enthusiastic and professional support throughout the period of manuscript preparation.

K.-L. Du also wishes to thank Jiabin Lu (Guangdong University of Technology, China), Jie Zeng (Richcon MC, Inc., China), Biaobiao Zhang and Hui Wang (Enjoyor, Inc., China), and many of his graduate students, including Na Shou, Shengfeng Yu, Lusha Han, Xiaolan Shen, Yuanyuan Chen, and Xiaoling Wang (Zhejiang University of Technology, China) for their consistent assistance.

In addition, we should mention at least the following names for their help: Omer Morgul (Bilkent University, Turkey), Yanwu Zhang (Monterey Bay Aquarium Research Institute, USA), Chi Sing Leung (City University of Hong Kong,

Hong Kong), M. Omair Ahmad and Jianfeng Gu (Concordia University, Canada), Li Yu, Limin Meng, Jingyu Hua, Zhijiang Xu, and Luping Fang (Zhejiang University of Technology, China), Yuxing Dai (Wenzhou University, China), and Renwang Li (Zhejiang Sci-Tech University, China). Last, but not least, we would like to thank our families for their support and understanding during the course of writing this book.

A book of this length is certain to have some errors and omissions. Feedback is welcome via email at kldu@ieee.org or swamy@encs.concordia.ca. Due to restriction on the length of this book, we have placed two appendices, namely, Mathematical preliminaries, and Benchmarks and resources, on the website of this book. MATLAB code for the worked examples is also downloadable from the website of this book.

Hangzhou, China
Montreal, Canada
April 2013

Ke-Lin Du
M. N. S. Swamy

Contents

Abbreviations

A/D	Analog-to-digital
adaline	Adaptive linear element
AI	Artificial intelligence
AIC	Akaike information criterion
ALA	Adaptive learning algorithm
ANFIS	Adaptive-network-based fuzzy inference system
AOSVR	Accurate online SVR
APCA	Asymmetric PCA
APEX	Adaptive principal components extraction
API	Application programming interface
ART	Adaptive resonance theory
ASIC	Application-specific integrated circuit
ASSOM	Adaptive-subspace SOM
BAM	Bidirectional associative memory
BFGS	Broyden–Fletcher–Goldfarb–Shanno
BIC	Bayesian information criterion
BIRCH	Balanced iterative reducing and clustering using hierarchies
BP	Backpropagation
BPTT	Backpropagation through time
BSB	Brain-states-in-a-box
BSS	Blind source separation
CBIR	Content-based image retrieval
CCA	Canonical correlation analysis
CCCP	Constrained concave-convex procedure
cdf	Cumulative distribution function
CEM	Classification EM
CG	Conjugate gradient
CMAC	Cerebellar model articulation controller
COP	Combinatorial optimization problem
CORDIC	Coordinate rotation digital computer

CoSaMP	Compressive sampling matching pursuit
CPT	Conditional probability table
CPU	Central processing units
CURE	Clustering using representation
DBSCAN	Density-based spatial clustering of applications with noise
DCS	Dynamic cell structures
DCT	Discrete cosine transform
DFP	Davidon–Fletcher–Powell
DFT	Discrete Fourier Transform
ECG	Electrocardiogram
ECOC	Error-correcting output code
EEG	Electroencephalogram
EKF	Extended Kalman filtering
ELM	Extreme learning machine
EM	Expectation–maximization
ERM	Empirical risk minimization
E-step	Expectation step
ETF	Elementary transcendental function
EVD	Eigenvalue decomposition
FCM	Fuzzy *C*-means
FFT	Fast Fourier Transform
FIR	Finite impulse response
fMRI	Functional magnetic resonance imaging
FPGA	Field-programmable gate array
FSCL	Frequency-sensitive competitive learning
GAP-RBF	Growing and pruning algorithm for RBF
GCS	Growing cell structures
GHA	Generalized Hebbian algorithm
GLVQ-F	Generalized LVQ family algorithms
GNG	Growing neural gas
GSO	Gram–Schmidt orthonormal
HWO	Hidden weight optimization
HyFIS	Hybrid neural fuzzy inference system
ICA	Independent component analysis
IHT	Iterative hard thresholding
iid	Independently drawn and identically distributed
IoT	Internet of Things
KKT	Karush–Kuhn–Tucker
LASSO	Least absolute selection and shrinkage operator
LBG	Linde–Buzo–Gray
LDA	Linear discriminant analysis
LM	Levenberg–Marquardt
LMAM	LM with adaptive momentum
LMI	Linear matrix inequality
LMS	Least mean squares

LMSE	Least mean squared error
LMSER	Least mean square error reconstruction
LP	Linear programming
LS	Least squares
LSI	Latent semantic indexing
LTG	Linear threshold gate
LVQ	Learning vector quantization
MAD	Median of the absolute deviation
MAP	Maximum a posteriori
MCA	Minor component analysis
MDL	Minimum description length
MEG	Magnetoencephalogram
MFCC	Mel frequency cepstral coefficient
MIMD	Multiple instruction multiple data
MKL	Multiple kernel learning
ML	Maximum likelihood
MLP	Multilayer perceptron
MSA	Minor subspace analysis
MSE	Mean squared error
MST	Minimum spanning tree
M-step	Maximization step
NARX	Nonlinear autoregressive with exogenous input
NEFCLASS	Neurofuzzy classification
NEFCON	Neurofuzzy controller
NEFLVQ	Non-Euclidean FLVQ
NEFPROX	Neuronfuzzy function approximation
NIC	Novel information criterion
k-NN	*k*-nearest neighbor
NOVEL	Nonlinear optimization via external lead
OBD	Optimal brain damage
OBS	Optimal brain surgeon
OLAP	Online analytical processing
OLS	Orthogonal least squares
OMP	Orthogonal matching pursuit
OWO	Output weight optimization
PAC	Probably approximately correct
PAST	Projection approximation subspace tracking
PASTd	PAST with deflation
PCA	Principal component analysis
PCM	Possibilistic *C*-means
pdf	Probability density function
PSA	Principal subspace analysis
QP	Quadratic programming
QR-cp	QR with column pivoting
RAN	Resource-allocating network

RBF	Radial basis function
RBM	Restricted Boltzmann machine
ReLU	Rectified linear unit
RIC	Restricted isometry constant
RIP	Restricted isometry property
RLS	Recursive least squares
RPCCL	Rival penalized controlled competitive learning
RPCL	Rival penalized competitive learning
Rprop	Resilient propagation
RTRL	Real-time recurrent learning
RVM	Relevance vector machine
SDP	Semidefinite programs
SIMD	Single instruction, multiple data
SLA	Subspace learning algorithm
SMO	Sequential minimal optimization
SOM	Self-organization maps
SPMD	Single program multiple data
SRM	Structural risk minimization
SVD	Singular value decomposition
SVDD	Support vector data description
SVM	Support vector machine
SVR	Support vector regression
TDNN	Time-delay neural network
TDRL	Time-dependent recurrent learning
TLMS	Total least mean squares
TLS	Total least squares
TREAT	Trust-region-based error aggregated training
TRUST	Terminal repeller unconstrained subenergy tunneling
TSK	Takagi–Sugeno–Kang
TSP	Traveling salesman problem
VC	Vapnik–Chervonenkis
VLSI	Very large-scale integrated
WINC	Weighted information criterion
k-WTA	*k*-winners-take-all
WTA	Winner-takes-all
XML	eXtensible markup language

Chapter 1
Introduction

1.1 Major Events in Machine Learning Research

The discipline of neural networks models the human brain. The average human brain consists of nearly 10^{11} neurons of various types, with each neuron connecting to up to tens of thousands synapses. As such, neural network models are also called *connectionist models*. Information processing is mainly in the cerebral cortex, the outer layer of the brain. Cognitive functions, including language, abstract reasoning, and learning and memory, represent the most complex brain operations to define in terms of neural mechanisms.

In the 1940s, McCulloch and Pitts [39] found that a neuron can be modeled as a simple threshold device to perform logic function. In 1949, Hebb [24] proposed the Hebbian rule to describe how learning affects the synaptics between two neurons. In 1952, based upon the physical properties of cell membranes and the ion currents passing through transmembrane proteins, Hodgkin and Huxley [26] incorporated the neural phenomena such as neuronal firing and action potential propagation into a set of evolution equations, yielding quantitatively accurate spikes and thresholds. This work brought Hodgkin and Huxley a Nobel Prize in 1963. In the late 1950s and early 1960s, Rosenblatt [45] proposed the perceptron model, and Widrow and Hoff [54] proposed the adaline (adaptive linear element) model, trained with a least mean squares (LMS) method.

In 1969, Minsky and Papert [40] proved mathematically that the perceptron cannot be used for complex logic function. This substantially waned the interest in the field of neural networks. During the same period, the adaline model as well as its multilayer version called the madaline was successfully used in many problems; however, they cannot solve linearly inseparable problems due to the use of linear activation function.

In the 1970s, Grossberg [22, 23], von der Malsburg [52], and Fukushima [19] conducted pioneering work on competitive learning and self-organization, inspired from the connection patterns found in the visual cortex. Fukushima proposed his cognitron [19] and neocognitron models [20, 21], under the competitive learning paradigm. Neocognitron, inspired by the primary visual cortex, is a hierarchical multi-layered

K.-L. Du and M. N. S. Swamy, *Neural Networks and Statistical Learning*,
https://doi.org/10.1007/978-1-4471-7452-3_1

neural network specially designed for robust visual pattern recognition. Several linear associative memory models were also proposed in that period [33]. In 1982, Kohonen proposed the self-organization map (SOM) [34]. SOM adaptively transforms incoming signal patterns of arbitrary dimensions into one- or two-dimensional discrete maps in a topologically ordered fashion. Grossberg and Carpenter [10, 23] proposed the adaptive resonance theory (ART) model in the mid-1980s. ART model, also based on competitive learning, is recurrent and self-organizing.

Hopfield model introduced in 1982 [28] ushered in the modern era of neural network research. The model works at the system level rather than at a single neuron level. It is a recurrent neural network working with the Hebbian rule. This network can be used as an associative memory for information storage and for solving optimization problems. The Boltzmann machine [1] was introduced in 1985 as an extension to the Hopfield network by incorporating stochastic neurons. Boltzmann learning is based on a method called *simulated annealing* [31]. In 1987, Kosko proposed the adaptive bidirectional associative memory (BAM) [35]. The Hamming network proposed by Lippman in the mid-1980s [37] is based on competitive learning, and is the most straightforward associative memory. In 1988, Chua and Yang [11] extended the Hopfield model by proposing cellular neural network model. The cellular network is a dynamical network model and is particularly suitable for two-dimensional signal processing and VLSI implementation.

The most prominent landmark in neural network research is backpropagation (BP) learning algorithm proposed for the multilayer perceptron (MLP) model in 1986 by Rumelhart et al. [47]. Later on, the BP algorithm was discovered to have already been invented in 1974 by Werbos [53]. In 1988, Broomhead and Lowe proposed the radial basis function (RBF) network model [7]. Both the MLP and the RBF network are universal approximators.

In 1982, Oja proposed the principal component analysis (PCA) network for classical statistical analysis [41]. In 1994, Common proposed independent component analysis (ICA) [12]. ICA is a generalization of PCA, and it is usually used for feature extraction and blind source separation (BSS). Since then, many neural network algorithms for classical statistical methods, such as Fisher's linear discriminant analysis (LDA), canonical correlation analysis (CCA), and factor analysis, have been proposed.

In 1985, Pearl introduced the Bayesian network model [43]. Bayesian network is the best known graphical model in AI. It possesses the characteristic of being both a statistical and a knowledge representation formalism. It establishes the foundation for inference of modern AI.

Another landmark in the machine learning and neural network communities is the support vector machine (SVM) proposed by Vapnik et al. in the early 1990s [51]. SVM is based on the statistical learning theory and is particularly useful for classification with small sample sizes. SVM has been used for classification, regression, and clustering. Thanks to its successful application in SVM, the kernel method has aroused wide interest.

Around 2004, Candes et al. [9], and Donoho [13] proved that given the knowledge about the sparsity of a signal, the signal may be reconstructed with even fewer samples than the Nyquist–Shannon sampling theorem requires. This idea is the basis of compressed sensing. Based on this, a lot of sparse coding or sparse recovery methods have been proposed. In 2009, Candes and Recht [8] extended sparse recovery to a data matrix from a sampling of its entries, leading to matrix completion.

Since 2006, the breakthrough in deep learning [5, 25] has propelled the AI industry to a worldwide enthusiasm. Deep learning is based on greedy layerwise unsupervised pretraining of each layer of features.

In addition to neural networks, fuzzy logic and evolutionary computation are two other major soft computing paradigms. Soft computing is a computing framework that can tolerate imprecision and uncertainty instead of depending on exact mathematical computations. Fuzzy logic [55] can incorporate the human knowledge into a system by means of fuzzy rules. Evolutionary computation [27, 48] originates from Darwin's theory of natural selection, and can optimize in a domain that is difficult to solve by other means. These techniques are now widely used to enhance the interpretability of the neural networks or to select optimum architecture and parameters of neural networks.

In summary, the brain is a dynamic information processing system that evolves its structure and functionality in time through information processing at different hierarchical levels: quantum, molecular (genetic), single neuron, ensemble of neurons, cognitive, and evolutionary [30]:

- At a quantum level, particles, that constitute every molecule, move continuously, being in several states at the same time that are characterized by probability, phase, frequency, and energy. These states can change following the principles of quantum mechanics.
- At a molecular level, RNA and protein molecules evolve in a cell and interact in a continuous way, based on the stored information in the DNA and on external factors, and affect the functioning of a cell (neuron).
- At the level of a single neuron, the internal information processes and the external stimuli change the synapses and cause the neuron to produce a signal to be transferred to other neurons.
- At the level of neuronal ensembles, all neurons operate together as a function of the ensemble through continuous learning.
- At the level of the whole brain, cognitive processes take place in a life-long incremental multiple task/multiple modalities learning mode, such as language and reasoning, and global information processes are manifested, such as consciousness.
- At the level of a population of individuals, species evolve through evolution via changing the genetic DNA code.

Building computational models that integrate principles from different information levels may be efficient for solving complex problems. These models are called *integrative connectionist learning systems* [30]. Information processes at different levels in the information hierarchy interact and influence each other.

1.2 Neurons

Among the 10^{11} neurons in the human brain, about 10^{10} are in the cortex. The cortex is the outer mantle of cells surrounding the central structures, e.g., brainstem and thalamus. Cortical thickness varies mostly between 2 and 3 mm in the human, and is folded with an average surface area is about 2,200 cm^2 [56].

The neuron, or nerve cell, is the fundamental anatomical and functional unit of the nervous system including the brain. A neuron is an extension of the simple cell with two types of appendages: multiple dendrites and an axon. A neuron possesses all the internal features of a regular cell. A neuron has four components: the dendrites, the soma (cell body), the axon, and the synapse. A soma contains a cell nucleus. Dendrites branch into a bushy network around the cell to receive input from other neurons, whereas the axon stretches out for a long distance, typically a centimeter and as far as a meter in extreme cases. The axon is an output channel to other neurons; it branches into strands and substrands to connect to the dendrites and cell bodies of other neurons. The connecting junction is called a synapse. Each cortical neuron receives 10^4–10^5 synaptic connections, with most inputs coming from distant neurons. Thus connections in the cortex are said to exhibit long-range excitation and short-range inhibition.

A neuron receives signals from other neurons through its soma and dendrites, integrates them, and sends output signals to other neurons through its axon. The dendrites receive signals from several neighborhood neurons and pass these onto the cell body, and are processed therein and the resulting signal is transferred through an axon. A schematic diagram shown in Fig. 1.1.

Like any other cell, neurons have a membrane potential, that is, an electric potential difference between the intracellular and extracellular compartments, caused by the different densities of sodium (Na) and potassium (K). Neuronal membrane is

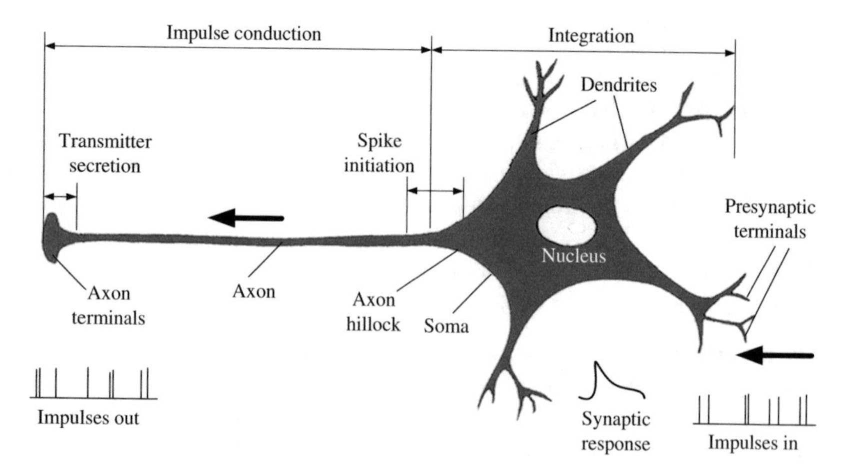

Fig. 1.1 Schematic drawing of a prototypical neuron

endowed with relatively selective ionic channels that allow some specific ions to cross the membrane. The cell membrane has an electrical resting potential of −70 mV, which is maintained by pumping positive ions (Na^+) out of the cell. Unlike an ordinary cell, the neuron is excitable. Because of inputs from the dendrites, the cell may not be able to maintain the −70 mV resting potential, resulting in an action potential that is a pulse transmitted down the axon. Signals are propagated from neuron to neuron by a complicated electrochemical reaction. Chemical transmitter substances pass the synapses and enter the dendrite, changing the electrical potential of the cell body. When the potential is above a threshold, an electrical pulse or action potential is sent along the axon. After releasing the pulse, the neuron returns to its resting potential. The action potential causes a release of certain biochemical agents for transmitting messages are to the dendrites of nearby neurons. These biochemical transmitters may have either an excitatory or inhibitory effect on neighboring neurons. A synapse that increases the potential is excitatory, whereas a synapse that decreases it is inhibitory.

Synaptic connections exhibit plasticity—long-term changes in the strength of connections in response to the pattern of stimulation. Neurons also form new connections with other neurons, and sometimes entire collections of neurons can migrate from one place to another. These mechanisms are thought to form the basis for learning in the brain. Synaptic plasticity is a basic biological mechanism underlying learning and memory. Two types of plasticity are Hebbian plasticity [24] for positive feedback instability, and compensatory homeostatic plasticity for stabilizing neural activity. Inspired by this, a large number of learning rules, specifying how activity and training experience change synaptic efficacies, have been advanced.

Physiological neurons are typically either excitatory or inhibitory, but not both. Dale's Principle states that neurons typically release the same transmitters at all branches of their axons [15]. Although there are exceptions, most neurons are exclusively excitatory or inhibitory most of the time. Any mixture of excitatory and inhibitory functional connections could be realized by a purely excitatory projection in parallel with a two-synapse projection through an inhibitory population [42]. This works well with ratios of excitatory and inhibitory neurons that are realistic for the neocortex [42]. Mixed excitatory and inhibitory functional connections can be realized in networks that are dominated by inhibition, such as those of the basal ganglia [49].

1.2.1 McCulloch–Pitts Neuron Model

A neuron is a basic processing unit in a neural network. It is a node that processes all fan-in from other nodes and generates an output according to a transfer function called the *activation function*. The activation function represents a linear or nonlinear mapping from the input to the output and is denoted by $\phi(\cdot)$. The variable synapses is modeled by weights. The McCulloch–Pitts neuron model [39], which employs the sigmoidal activation function, was inspired biologically.

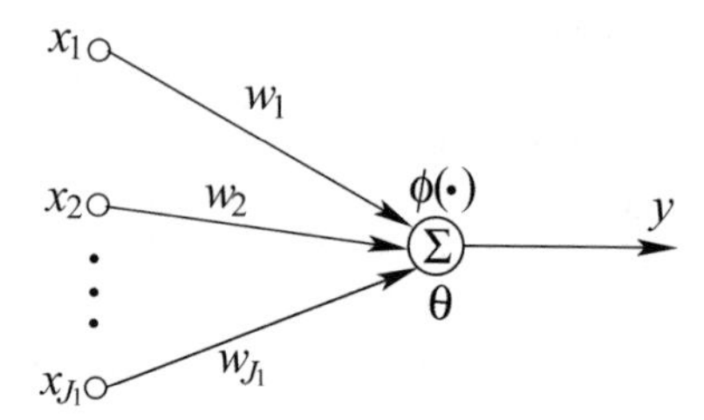

Fig. 1.2 The mathematical model of McCulloch–Pitts neuron

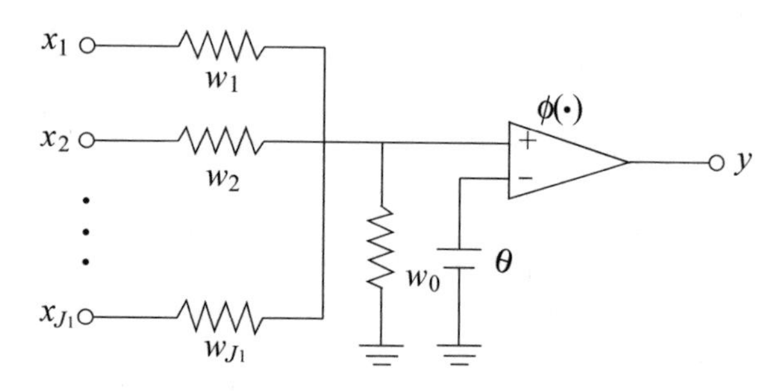

Fig. 1.3 VLSI model of a neuron

Figure 1.2 illustrates the simple McCulloch–Pitts neuron model. The output of the neuron is given by

$$net = \sum_{i=1}^{J_1} w_i x_i - \theta = \boldsymbol{w}^T \boldsymbol{x} - \theta, \tag{1.1}$$

$$y = \phi(net), \tag{1.2}$$

where x_i is the ith input, w_i is the link weight from the ith input, $\boldsymbol{w} = (w_1, \ldots, w_{J_1})^T$, $\boldsymbol{x} = (x_1, \ldots, x_{J_1})^T$, θ is a threshold or bias, and J_1 is the number of inputs. The activation function $\phi(\cdot)$ is usually some continuous or discontinuous function, mapping the real numbers into the interval $(-1, 1)$ or $(0, 1)$.

Neural networks are suitable for VLSI circuit implementations. The analog approach is extremely attractive in terms of size, power, and speed. A neuron can be realized with a simple amplifier and the synapse is realized with a resistor. Memristor is a two-terminal passive circuit element that acts as a variable resistor whose value can be varied by varying the current passing through it [2]. The circuit of a neuron is given in Fig. 1.3. Since weights from the circuits can only be positive, an inverter can be applied to the input voltage so as to realize a negative synaptic weight.

By Kirchhoff's current law, the output voltage of the neuron is derived as

$$y = \phi\left(\frac{\sum_{i=1}^{J_1} w_i x_i}{\sum_{i=0}^{J_1} w_i} - \theta\right), \tag{1.3}$$

where x_i is the ith input voltage, w_i is the conductance of the ith resistor, θ the bias voltage, and $\phi(\cdot)$ is the transfer function of the amplifier. The bias voltage of a neuron in a VLSI circuit is caused by device mismatches, and is difficult to control.

The McCulloch–Pitts neuron model is known as the classical perceptron model, and it is used in most neural network models, including the MLP and the Hopfield network. Many other neural networks are also based on the McCulloch–Pitts neuron model, but use other activation functions. For example, the adaline [54] and the SOM [34] use linear activation functions, and the RBF network adopts a radial basis function (RBF).

1.2.2 Spiking Neuron Models

Many of the intrinsic properties seen within the brain were not included in the classical perceptron, limiting their functionality and use to linear discrimination tasks. A single classical perceptron is not capable of solving nonlinear problems, such as the XOR problem. The brain functions as a spatiotemporal information processing machine. Spiking neuron and spiking neural network models mimic the spiking activity of neurons in the brain when processing information.

Spiking neurons tend to gather in functional groups firing together during strict time intervals, also referred to as *events*. Spiking neural networks exchange information in the form of precisely timed events called *spikes*. A spike train is a sequence of stereotyped events generated at regular or irregular intervals. The form of the spike does not carry any information, and but the number and the timing of spikes do.

In nature, a spike is represented by a short pulse of voltage, typically having an amplitude of about 100 mV and a duration of 1–2 ms. The activation function is a differential equation that tries to model the dynamic properties of a biological neuron in terms of spikes. The spikes are sparse in time and space, and event-driven. With bio-plausible local learning rules, it is easier to build low-power, neuromorphic hardware for spiking neural networks. This results in a much higher number of patterns stored in a model and more flexible processing.

During training of a spiking neural network, the weight of the synapse is modified according to the timing difference between the presynaptic spike and the postsynaptic spike. This synaptic plasticity is called *spike-timing-dependent plasticity (STDP)* [38]. STDP is a set of Hebbian learning rules firmly based on biological evidence. In a biological system, a neuron integrates the excitatory postsynaptic current, which is produced by presynaptic stimulus, to change the voltage of its soma. If the soma voltage is larger than a defined threshold, an action potential (spike) is produced.

Hodgkin–Huxley model [26] incorporates the principal neurobiological properties of a neuron in order to understand phenomena such as the action potential. It was obtained from empirical investigation of the physiological properties of the squid axon into a dynamical system framework. The model is a set of conductance-based coupled ordinary differential equations, incorporating sodium (Na), potassium (K), and chloride (Cl) ion flows through their respective channels. These equations are

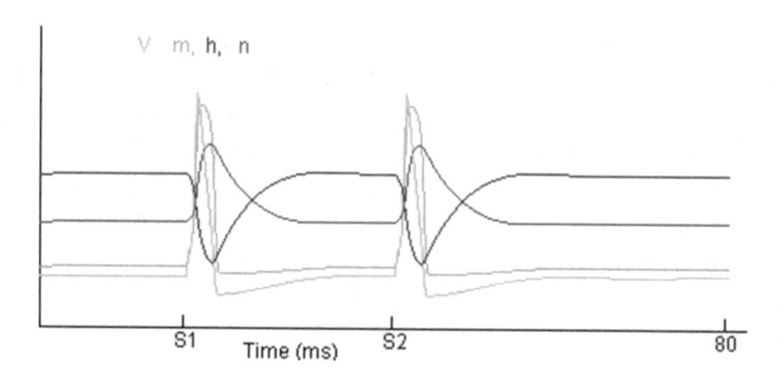

Fig. 1.4 Parameters of the Hodgkin–Huxley model for a neuron

based upon the physical properties of cell membranes and the ion currents passing through transmembrane proteins. Chloride channel conductances are static (not voltage dependent) and hence leaky.

According to the Hodgkin–Huxley model, the dynamics of the membrane potential $V(t)$ of the neuron can be described by

$$C\frac{dV}{dt} = -g_{Na}m^3h(V - V_{Na}) - g_K n^4(V - V_K) - g_L(V - V_L) + I(t), \quad (1.4)$$

where the first three terms on the right-hand side correspond to the potassium, sodium, and leakage currents, respectively, and $g_{Na} = 120\,\text{mS/cm}^2$, $g_K = 36\,\text{mS/cm}^2$, and $g_L = 0.3\,\text{mS/cm}^2$ are the maximal conductances of sodium, potassium, and leakage, respectively. The membrane capacitance $C = 1\,\text{mF/cm}^2$; $V_{Na} = 50\,\text{mV}$, $V_K = -77\,\text{mV}$, and $V_L = -54.4\,\text{mV}$ are the reversal potentials of sodium, potassium, and leakage currents, respectively. $I(t)$ is the injected current. The stochastic gating variables n, m and h represent the activation term of the potassium channel, the activation term, and the inactivation term of the sodium channel, respectively. The factors n^4 and m^3h are the mean portions of the open potassium and sodium ion channels within the membrane patch. To take into account the channel noise, m, h, and n obey the Langevin equations. When the stimuli S1 and S2 occur at 15 ms and 40 ms of 80 ms, the simulated results for V, m, h, and n are plotted in Fig. 1.4; this figure was generated by a Java applet (http://thevirtualheart.org/HHindex.html).

Integrate-and-fire neuron [50], FitzHugh–Nagumo neuron [17], and Izhikevich model [29] are simplified versions of Hodgkin–Huxley neuron model. Both the integrate-and-fire and FitzHugh–Nagumo neurons model key features of biological neurons such as the membrane potential, excitatory postsynaptic potential, and inhibitory postsynaptic potential. A single neuron incorporating these key features has a higher dimension to the information it processes in terms of its membrane threshold, firing rate, and postsynaptic potential, than a classical perceptron.

Integrate-and-fire model is derived from Hodgkin–Huxley model but neglects the shape of the potential actions. This model assumes that all potential actions are uniform but differ in the time of occurrence. As a result, all the spikes have the same characteristics such as shape, width, and amplitude. The leaky integrate-and-fire model decays the membrane potential of the neuron over time if no potentials reach the neuron. Integrate-and-fire neuron model, whose output is binary on a short timescale, either fires an action potential or does not. A spike train $s \in \mathcal{S}(\mathcal{T})$ is a sequence of ordered spike times $s = \{t_m \in \mathcal{T} : m = 1, \ldots, N\}$ corresponding to the time instants in the interval $\mathcal{T} = [0, T]$ at which a neuron fires.

FitzHugh–Nagumo model is a simplified version of Hodgkin–Huxley model which models in a detailed manner the activation and deactivation dynamics of a spiking neuron.

Izhikevich model [29] is a simple model, which has only nine dimensionless parameters. The spiking model can reproduce various intrinsic firing patterns based on different values of the parameters.

Each neuron maintains an internal membrane potential, which is a function of input spikes, associated synaptic weights, current membrane potential, and a constant membrane potential leakage coefficient. A neuron fires (emits a spike to all connected synapses/neurons) when its membrane potential exceeds its firing threshold value.

1.3 Neural Networks

A neural network is characterized by the network architecture, node characteristics, and learning rules.

Architecture

The network architecture is represented by the connection weight matrix $\mathbf{W} = [w_{ij}]$, where w_{ij} denotes the connection weight from node i to node j. When $w_{ij} = 0$, there is no connection from node i to node j. By setting some w_{ij}'s to zero, different network topologies can be realized. Neural networks can be grossly classified into feedforward neural networks, recurrent neural networks, and their hybrids.

Popular network topologies are fully connected layered feedforward networks, recurrent networks, lattice networks, layered feedforward networks with lateral connections, and cellular networks, as shown in Fig. 1.5.

- In a feedforward network, the connections between neurons are in one direction. A feedforward network is usually arranged in the form of layers. In such a layered feedforward network, there is no connection between the neurons in the same layer, and there is no feedback between layers. In a fully connected layered feedforward network, every node in any layer is connected to every node in its adjacent forward layer. The MLP and the RBF network are fully connected layered feedforward networks.
- In a recurrent network, there exists at least one feedback connection. The Hopfield model and the Boltzmann machine are two examples of recurrent networks.

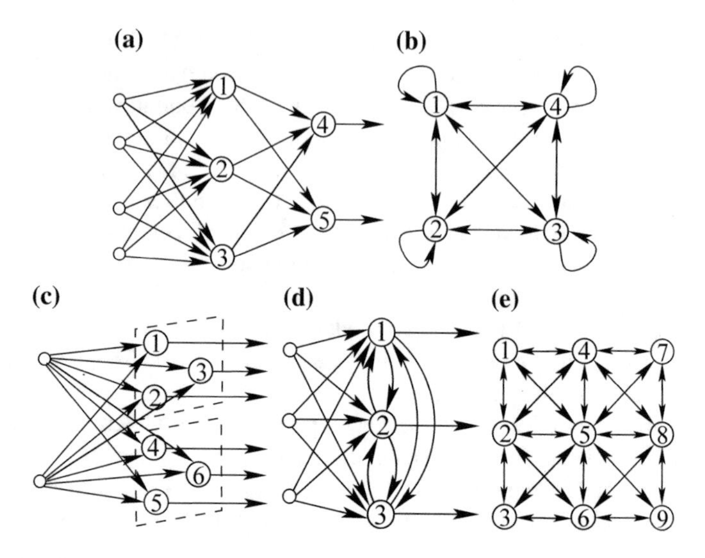

Fig. 1.5 Architecture of neural networks. **a** Layered feedforward network. **b** Recurrent network. **c** Two-dimensional lattice network. **d** Layered feedforward network with lateral connections. **e** Cellular network. The big numbered circles stand for neurons and the small ones for input nodes

- A lattice network consists of one-, two-, or higher dimensional array of neurons. Each array has a corresponding set of input nodes. The Kohonen network [34] uses a one- or two-dimensional lattice architecture.
- A layered feedforward network with lateral connections has lateral connections between the units at the same layer of its layered feedforward network architecture. A competitive learning network has a two-layered network of such an architecture. The feedforward connections are excitatory, while the lateral connections in the same layer are inhibitive. Some PCA networks using the Hebbian/anti-Hebbian learning rules [46] also employ this kind of network topology.
- A cellular network consists of regularly spaced neurons, called *cells*, which communicate only with the neurons in its immediate neighborhood. Adjacent cells are connected by mutual interconnections. Each cell is excited by its own signals and by signals flowing from its adjacent cells [11].

In this book, we use the notation J_1-J_2-$\cdots$-J_M to represent a neural network with a layered architecture of M layers, where J_i is the number of nodes in the ith layer. Notice that the input layer is counted as layer 1 and nodes at this layer are not neurons. Layer M is the output layer.

Operation

The operation of neural networks is divided into two stages: learning (training) and generalization (recalling). Network training is typically accomplished by using examples, and network parameters are adapted using a learning algorithm, in an online or offline manner. Once the network is trained to accomplish the desired performance, the learning process is terminated and it can then be used directly to replace the complex system dynamics. The trained network can be used to operate in a static manner: to emulate an unknown dynamics or nonlinear relationship.

For real-time applications, a neural network is required to have a constant processing delay regardless of the number of input nodes and to have a minimum number of layers. As the number of input nodes increases, the size of the network layers should grow at the same rate without additional layers.

Adaptive neural networks are a class of neural networks that do not need to be trained by providing a training pattern set. They can learn when they are performing. For adaptive neural networks, unsupervised learning methods are usually used. For example, the Hopfield model uses a generalized Hebbian learning rule for implementation as associative memory. Any time a pattern is presented to it, the Hopfield network always updates the connection weights. After the network is trained with standard patterns and is prepared for generalization, the learning capability should be disabled; otherwise, when an incomplete or noisy pattern is presented to the network, it will search the closest matching, meanwhile, the memorized pattern is replaced by this new pattern.

Reinforcement learning is also naturally adaptive, where the environment is treated as a teacher. Supervised learning is not adaptive in nature.

Properties

One of the main functions of the brain is prediction. Neural networks are biologically motivated. Each neuron is a computational node, which represents a nonlinear function. Neural networks possess the following advantages [14]:

- **Adaptive learning**: They can adapt themselves by changing the network parameters in a surrounding environment.
- **Generalization**: A trained neural network has superior generalization capability.
- **General-purpose nonlinear nature**: They perform like a black box.
- **Self-organizing**: Some neural networks such as the SOM [34] and competitive learning-based neural networks have a self-organization property.
- **Massive parallelism and simple VLSI implementations**: Each basic processing unit usually has a uniform property. This parallel structure allows for highly parallel software and hardware implementations.
- **Robustness and fault tolerance**: A neural network can easily handle imprecise, fuzzy, noisy, and probabilistic information. It is a distributed information system, where information is stored in the whole network in a distributed manner by the network structure such as $\mathbf{W}$. Thus, the overall performance does not degrade significantly when the information at some nodes is lost or some connections in the network are damaged. The network repairs itself, and thus possesses a fault-tolerant capability.

Applications

Neural networks can be treated as a general statistical tool for almost all disciplines of science and engineering. The applications can be in modeling and system identification, classification, pattern recognition, optimization, control, industrial application, communications, signal processing, image analysis, bioinformatics, and data mining. Pattern recognition is central to biological and artificial intelligence; it is a complete process that gathers the observations, extracts features from the observations, and classifies or describes the observations. Pattern recognition is one of the most fundamental applications of neural networks. More specific, some neural network models have the following functions:

- **Function approximation**: This capability is generally used for modeling and system identification, regression, and prediction, control, signal processing, pattern recognition and classification, and associative memory. Image restoration is also a function approximation problem. The MLP and RBF networks are universal approximators for nonlinear functions. Some recurrent networks are universal approximators of dynamical systems. Prediction is an open-loop problem while control is a closed-loop problem.
- **Classification**: Classification is the most fundamental application of neural networks. Classification can be based on the function approximation capability of neural networks.
- **Clustering and vector quantization**: Clustering groups together similar objects, based on some distance measure. Unlike in classification problems, the class membership of a pattern is not known a priori. Vector quantization is similar to clustering. Clustering is a fundamental tool for data analysis. It finds wide applications in pattern recognition, feature extraction, vector quantization, image segmentation, bioinformatics, and data mining. Clustering is a classical method for the prototype selection of kernel-based neural networks such as the RBF network, and is most useful for neurofuzzy systems.
- **Associative memory**: An association is an input–output pair. Associative memory, also known as *content-addressable memory*, is a memory organization that accesses memory by its content instead of its address. It picks up a desirable match from all stored prototypes, when an incomplete or corrupted sample is presented. Associative memories are useful for pattern recognition, pattern association, or pattern completion.
- **Optimization**: Some neural network models, such as the Hopfield model and the Boltzmann machine, can be used to solve combinatorial optimization problems (COPs).
- **Feature extraction and information compression**: Coding and information compression is an essential task in the transmission and storage of speech, audio, image, video, and other information. PCA, ICA, and vector quantization can achieve the objective of feature extraction and information compression.

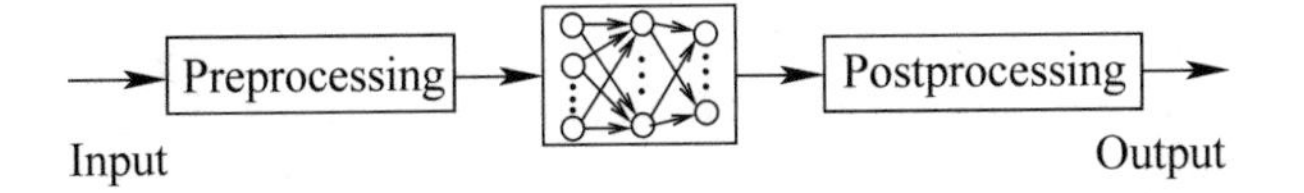

Fig. 1.6 Architecture of a neural network processor

1.4 Neural Network Processors

A typical architecture of a neural network processor is illustrated in Fig. 1.6. It is composed of three components: input preprocessing, a neural network for performing inversion, and output postprocessing. Input preprocessing is used to remove redundant and/or irrelevant information in order to achieve a small network and to reduce the dimensionality of the signal parameter space, thus improving the generalization capability of the network. Postprocessing the output of the network generates the desired information.

Preprocessing is to transform the raw data into a new representation before being presented to a neural network. If the input data is preprocessed at the training stage, accordingly at the generalizing stage, the input data also needs to be preprocessed before being passed on to the neural network. Similarly, if the output data is preprocessed at the training stage, the network output at the generalization stage is also required to be postprocessed to generate the target output corresponding to the raw output patterns.

For high-dimensional data, dimensionality reduction is the key to cope with the curse of dimensionality. Preprocessing has a significant influence on the generalization performance of a neural network. This process removes the redundancy in the input space and reduces the space of the input data, thus usually resulting in a reduction in the amount or the dimensionality of the input data. This helps to alleviate the problem of the curse of dimensionality. A network with preprocessed inputs may be constrained by a smaller dataset, and thus one needs only to train a small network, which also achieves a better generalization capability.

Preprocessing usually takes the form of linear or nonlinear transformation of the raw input data to generate input data for the network. It can also be based on the prior knowledge of the network architecture, or the problem itself. When preprocessing removes redundant information in the input data, it also results in a loss of information. Thus, preprocessing should retain as much relevant information as possible.

The data are sometimes called *features*, and preprocessing for input raw data can be either feature selection or feature extraction. Feature selection concentrates on selecting from the original set of features a smaller subset of salient features, while feature extraction is to combine the original features in such a way as to produce a new reduced set of salient features. The raw data may be orders of magnitude in range, and linear scaling and data whitening of the raw data are usually employed as a preprocessing step.

When some examples of the raw data suffer from missing components, one simple treatment is to discard those examples from the dataset. This, however, is applicable only when the data is abundant, the percentage of examples with missing components is small, and the mechanism for loss of data is independent of the data itself. However, this may lead to a biased subset of the data. The missing values should be replaced according to various criteria. For function approximation problems, one can represent any variable with a missing value as a regression over the other variables using the available data, and then find the missing value by interpolating the regression function. For density estimation problems, the ML solution to problems with the missing data can be found by applying an expectation–maximization (EM) algorithm.

Feature extraction reduces the dimension of the features by orthogonal transforms. The extracted features do not have any physical meaning. In comparison, feature selection decreases the size of the feature set or reduces the dimension of the features by discarding the raw information according to a criterion.

Feature Selection

Feature selection is to select the best subset or the best subspace of the features out of the original set, since irrelevant features degrade the performance. A criterion is required to evaluate each subset of the features so that an optimum subset can be selected. The selection criterion should be the same as that for assessing the complete system, such as the MSE criterion for function approximation and the misclassification rate for classification. Theoretically, the global optimum subset of the features can only be selected by an exhaustive search of all the possible subsets of the features.

Feature selection algorithms can be categorized as either filter or wrapper approaches. During the process of feature selection, the generalization ability of a subset of features needs to be estimated. This type of feature selection is called a wrapper method [32]. The problem of searching the best r variables is solved by means of a greedy algorithm based on backward selection [32]. The filter approach basically preselects the features, and then applies the selected feature subset to the clustering algorithm. Filter-based greedy algorithms using the sequential selection of the feature with the best criterion value are computationally more efficient than wrappers. In general, the wrapper method outperforms the filter method, but at the expense of training a large number of classifiers.

Some nonexhaustive search methods such as the branch and bound procedure, sequential forward selection, and sequential backward elimination are discussed in [6]. Usually, backward selection is slower, but is more stable in selecting optimal features than forward selection. Backward selection starts from all the features and deletes one feature at a time, which deteriorates the selection criterion the least, until the selection criterion reaches a specified value. In contrast, forward selection starts from an empty set of features and adds one feature at a time that improves the selection criterion the most.

Mutual information-based feature selection is a common method for feature selection [4, 16]. The mutual information measures the arbitrary dependence between random variables, whereas linear relations, such as the correlation-based methods, are prone to mistakes. By calculating the mutual information, the importance levels of the features are ranked based on their ability to maximize the evaluation criterion. Relevant inputs are found by estimating the mutual information between the inputs and the desired outputs. The normalized mutual information feature selection [16] does not require a user-defined parameter.

Feature Extraction

Feature extraction is usually conducted by using orthogonal transforms, though the Gram–Schmidt orthonormalization (GSO) is more suitable for feature selection. This is due to the fact that the physically meaningless features in the Gram–Schmidt space can be linked back to the same number of variables of the measurement space, resulting in no dimensionality reduction. In situations where the features are used for pattern understanding and analysis, the GSO transform provides a good option.

The advantage of employing an orthogonal transform is that the correlations among the candidate features are decomposed so that the significance of the individual features can be evaluated independently. PCA is a well-known orthogonal transform. Taking all the data into account, PCA computes vectors that have the largest variance associated with them. The generated PCA features may not have clear physical meanings. Dimensionality reduction is achieved by dropping the variables with insignificant variance. Projection pursuit [18] is a general approach to feature extraction, which extracts features by repeatedly choosing projection vectors and then orthogonalizing.

PCA is often used to select inputs, but it is not always useful, since the variance of a signal is not always related to the importance of the variable, for example, for non-Gaussian signals. An improvement on PCA is provided by nonlinear generalizations of PCA, which extend the ability of PCA to incorporate nonlinear relationships in the data. ICA can extract the statistically independent components from the input dataset. It is to estimate the mutual information between the signals by adjusting the estimated matrix to give outputs that are maximally independent [3]. The dimensions to remove are those that are independent of the output. LDA searches for those vectors in the underlying space that best discriminate among the classes (rather than those that best describe the data). In [36], the proposed scheme for linear feature extraction in classification is based on the maximization of the mutual information between the features extracted and the classes.

For time/frequency-continuous signal systems such as speech-recognition systems, the fixed time–frequency resolution FFT power spectrum, and the multiresolution discrete wavelet transform and wavelet packets are usually used for feature extraction. The features used are chosen from the Fourier or wavelet coefficients having high energy. The cepstrum and its time derivative remain the most commonly used feature set [44]. These features are calculated by taking the discrete cosine transform (DCT) of the logarithm of the energy at the output of a Mel filter and are commonly called *Mel frequency cepstral coefficients (MFCC)*. In order to have the temporal information, the first- and second-time derivatives of the MFCC are taken.

1.5 Scope of the Book

This book contains 30 chapters and 2 appendices:

- Chapter 2 describes some fundamental topics on neural networks and machine learning.
- Chapter 3 deals with computational learning theory.
- Chapter 4 is dedicated to the perceptron.
- The MLP is the topic of Chaps. 5 and 6. The MLP with BP learning is introduced in Chap. 5, and structural optimization of the MLP is also described in this chapter.
- The MLP with second-order learning is introduced in Chap. 6.
- Chapter 7 treats the Hopfield model, its application for solving COPs, simulation annealing, chaotic neural networks, and cellular networks.
- Chapter 8 describes associative memory models and algorithms.
- Chapters 9 and 10 are dedicated to clustering. Chapter 9 introduces Kohonen networks, ART networks, C-means, subtractive, and fuzzy clustering.
- Chapter 10 provides an introduction to many advanced topics in clustering.
- In Chap. 11, we offer an elaboration on the RBF network model.
- Chapter 12 introduces the learning of general recurrent networks.
- Chapter 13 deals with PCA networks and algorithms. The minor component analysis (MCA), crosscorrelation PCA networks, generalized eigenvalue decomposition (EVD), and CCA are also introduced in this chapter.
- Nonnegative matrix factorization (NMF) is introduced in Chap. 14.
- ICA and BSS are explicated in Chap. 15.
- Discriminant analysis is described in Chap. 16.
- Reinforcement learning is introduced in Chap. 17.
- Compressed sensing and sparse coding are described in Chap. 18.
- Matrix completion and recovery are related to sparse coding. They are treated in Chap. 19.
- Kernel methods other than SVMs are introduced in Chap. 20.
- SVMs are given a detailed introduction in Chap. 21.
- Probabilistic and Bayesian networks are introduced in Chap. 22. Many topics such as EM algorithms, hidden Markov model (HMM), and sampling (Monte Carlo) methods are treated in this framework.
- Boltzmann machines are introduced in Chap. 23.
- Chapter 24 gives an account of deep learning and deep neural networks.
- Ensemble learning is addressed in Chap. 25.
- Fuzzy sets and logic are introduced in Chap. 26.
- Neurofuzzy models are described in Chap. 27. Transformations between fuzzy logic and neural networks are also discussed.
- Implementation of neural networks in hardware is treated in Chap. 28.

- In Chap. 29, we give an introduction to neural network applications to biometrics and bioinformatics.
- Data mining, as well as the application of neural networks to the field, is introduced in Chap. 30.
- Big data, cloud computing, and Internet of Things are elaborated in Chap. 31.
- Mathematical preliminaries are included in Appendix A.
- Some benchmarks and resources are included in Appendix B.

Examples and exercises are included in most of the chapters.

Problems

1.1 List the major differences between the neural network approach and classical information processing approaches.

1.2 Formulate a McCulloch–Pitts neuron for four variables: white blood count, systolic blood pressure, diastolic blood pressure, and pH of the blood.

1.3 Derive Eq. (1.3) from Fig. 1.3.

References

1. Ackley, D. H., Hinton, G. E., & Sejnowski, T. J. (1985). A learning algorithm for Boltzmann machines. *Cognitive Science*, *9*, 147–169.
2. Adhikari, S. P., Yang, C., Kim, H., & Chua, L. O. (2012). Memristor bridge synapse-based neural network and its learning. *IEEE Transactions on Neural Networks and Learning Systems*, *23*(9), 1426–1435.
3. Back, A. D., & Trappenberg, T. P. (2001). Selecting inputs for modeling using normalized higher order statistics and independent component analysis. *IEEE Transactions on Neural Networks*, *12*(3), 612–617.
4. Battiti, R. (1994). Using mutual information for selecting features in supervised neural net learning. *IEEE Transactions on Neural Networks*, *5*(4), 537–550.
5. Bengio, Y., Lamblin, P., Popovici, D., & Larochelle, H. (2007). Greedy layer-wise training of deep networks. In B. Schlkopf, J. Platt, & T. Hofmann (Eds.), *Advances in neural information processing systems* (Vol. 19, pp. 153–160). Cambridge, MA: MIT Press.
6. Bishop, C. M. (1995). *Neural networks for pattern recognition*. New York: Oxford Press.
7. Broomhead, D. S., & Lowe, D. (1988). Multivariable functional interpolation and adaptive networks. *Complex Systems*, *2*, 321–355.
8. Candes, E. J., & Recht, B. (2009). Exact matrix completion via convex optimization. *Foundations of Computational Mathematics*, *9*(6), 717–772.
9. Candes, E. J., Romberg, J. K., & Tao, T. (2006). Stable signal recovery from incomplete and inaccurate measurements. *Communications on Pure and Applied Mathematics*, *59*(8), 1207–1223.
10. Carpenter, G. A., & Grossberg, S. (1987). A massively parallel architecture for a self-organizing neural pattern recognition machine. *Computer Vision, Graphics, and Image Processing*, *37*, 54–115.

11. Chua, L. O., & Yang, L. (1988). Cellular neural network: I. Theory; II. Applications. *IEEE Transactions on Circuits and Systems*, *35*, 1257–1290.
12. Comon, P. (1994). Independent component analysis—A new concept? *Signal Processing*, *36*(3), 287–314.
13. Donoho, D. L. (2006). Compressed sensing. *IEEE Transactions on Information Theory*, *52*(4), 1289–1306.
14. Du, K.-L., & Swamy, M. N. S. (2006). *Neural networks in a softcomputing framework*. London: Springer.
15. Eccles, J. (1976). From electrical to chemical transmission in the central nervous system. *Notes and Records of the Royal Society of London*, *30*(2), 219–230.
16. Estevez, P. A., Tesmer, M., Perez, C. A., & Zurada, J. M. (2009). Normalized mutual information feature selection. *IEEE Transactions on Neural Networks*, *20*(2), 189–201.
17. FitzHugh, R. (1961). Impulses and physiological states in theoretical models of nerve membrane. *Biophysical Journal*, *1*, 445–466.
18. Friedman, J. H., & Tukey, J. W. (1974). A projection pursuit algorithm for exploratory data analysis. *IEEE Transactions on Computers*, *23*(9), 881–889.
19. Fukushima, K. (1975). Cognition: A self-organizing multulayered neural network. *Biological Cybernetics*, *20*, 121–136.
20. Fukushima, K. (1980). Neocognitron: A self-organizing neural network model for a mechanism of pattern recognition unaffected by shift in position. *Biological Cybernetics*, *36*, 193–202.
21. Fukushima, K. (2011). Increasing robustness against background noise: Visual pattern recognition by a neocognitron. *Neural Networks*, *24*(7), 767–778.
22. Grossberg, S. (1972). Neural expectation: Cerebellar and retinal analogues of cells fired by unlearnable and learnable pattern classes. *Kybernetik*, *10*, 49–57.
23. Grossberg, S. (1976). Adaptive pattern classification and universal recording: I. Parallel development and coding of neural feature detectors; II. Feedback, expectation, olfaction, and illusions. *Biological Cybernetics*, *23*, 121–134 & 187–202.
24. Hebb, D. O. (1949). *The organization of behavior*. New York: Wiley.
25. Hinton, G. E., & Salakhutdinov, R. R. (2006). Reducing the dimensionality of data with neural networks. *Science*, *313*(5786), 504–507.
26. Hodgkin, A. L., & Huxley, A. F. (1952). A quantitative description of ion currents and its applications to conductance and excitation in nerve membranes. *Journal of Physiology*, *117*, 500–544.
27. Holland, J. (1975). *Adaptation in natural and artificial systems*. Ann Arbor, MI: University of Michigan Press.
28. Hopfield, J. J. (1982). Neural networks and physical systems with emergent collective computational abilities. *Proceedings of the National Academy of Sciences of the United States of America*, *79*, 2554–2558.
29. Izhikevich, E. M. (2003). Simple model of spiking neurons. *IEEE Transactions on Neural Networks*, *14*(6), 1569–1572.
30. Kasabov, N. (2009). Integrative connectionist learning systems inspired by nature: Current models, future trends and challenges. *Natural Computing*, *8*, 199–218.
31. Kirkpatrick, S., Gelatt, C. D, Jr., & Vecchi, M. P. (1983). Optimization by simulated annealing. *Science*, *220*, 671–680.
32. Kohavi, R., & John, G. H. (1997). Wrappers for feature subset selection. *Artificial Intelligence*, *97*, 273–324.
33. Kohonen, T. (1972). Correlation matrix memories. *IEEE Transactions on Computers*, *21*, 353–359.
34. Kohonen, T. (1982). Self-organized formation of topologically correct feature maps. *Biological Cybernetics*, *43*, 59–69.
35. Kosko, B. (1987). Adaptive bidirectional associative memories. *Applied Optics*, *26*, 4947–4960.
36. Leiva-Murillo, J. M., & Artes-Rodriguez, A. (2007). Maximization of mutual information for supervised linear feature extraction. *IEEE Transactions on Neural Networks*, *18*(5), 1433–1441.

37. Lippman, R. P. (1987). An introduction to computing with neural nets. *IEEE ASSP Magazine*, *4*(2), 4–22.
38. Markram, H., Lubke, J., Frotscher, M., & Sakmann, B. (1997). Regulation of synaptic efficacy by coincidence of postsynaptic APs and EPSPs. *Science*, *275*(5297), 213–215.
39. McCulloch, W. S., & Pitts, W. (1943). A logical calculus of the ideas immanent in nervous activity. *Bulletin of Mathematical Biology*, *5*, 115–133.
40. Minsky, M. L., & Papert, S. (1969). *Perceptrons*. Cambridge, MA: MIT Press.
41. Oja, E. (1982). A simplified neuron model as a principal component analyzer. *Journal of Mathematical Biology*, *15*, 267–273.
42. Parisien, C., Anderson, C. H., & Eliasmith, C. (2008). Solving the problem of negative synaptic weights in cortical models. *Neural Computation*, *20*(6), 1473–1494.
43. Pearl, J. (1988). *Probabilistic reasoning in intelligent systems: Networks of plausible inference*. San Mateo, CA: Morgan Kaufmann.
44. Picone, J. (1993). Signal modeling techniques in speech recognition. *Proceedings of the IEEE*, *81*(9), 1215–1247.
45. Rosenblatt, R. (1962). *Principles of neurodynamics*. New York: Spartan Books.
46. Rubner, J., & Tavan, P. (1989). A self-organizing network for principal-component analysis. *Europhysics Letters*, *10*, 693–698.
47. Rumelhart, D. E., Hinton, G. E., & Williams, R. J. (1986). Learning internal representations by error propagation. In D. E. Rumelhart & J. L. McClelland (Eds.), *Parallel distributed processing: Explorations in the microstructure of cognition, 1: Foundation* (pp. 318–362). Cambridge, MA: MIT Press.
48. Schwefel, H. P. (1981). *Numerical optimization of computer models*. Chichester: Wiley.
49. Tripp, B., & Eliasmith, C. (2016). Function approximation in inhibitory networks. *Neural Networks*, *77*, 95–106.
50. Tuckwell, H. C. (1988). *Introduction to theoretical neurobiology*. Cambridge, UK: Cambridge University Press.
51. Vapnik, V. N. (1998). *Statistical learning theory*. New York: Wiley.
52. von der Malsburg, C. (1973). Self-organizing of orientation sensitive cells in the striata cortex. *Kybernetik*, *14*, 85–100.
53. Werbos, P. J. (1974). Beyond regressions: New tools for prediction and analysis in the behavioral sciences. Ph.D. thesis, Harvard University, Cambridge, MA.
54. Widrow, B., & Hoff, M. E. (1960). Adaptive switching circuits. *Convention Record of IRE Eastern Electronic Show & Convention (WESCON1960)*, *4*, 96–104.
55. Zadeh, L. A. (1965). Fuzzy sets. *Information and Control*, *8*, 338–353.
56. Zilles, K. (1990). Cortex. In G. Pixinos (Ed.), *The human nervous system*. New York: Academic Press.

Chapter 2
Fundamentals of Machine Learning

2.1 Learning and Inference Methods

Learning is of utmost importance for survival and evolution of any living organism. It is a fundamental capability of neural networks. Learning rules are algorithms for finding suitable weights $\mathbf{W}$ and/or other network parameters. Learning of a neural network can be viewed as a nonlinear optimization problem for finding a set of network parameters that minimize the cost function for given examples. This kind of parameter estimation is also called a *learning* or *training algorithm*.

Neural networks are usually trained by epoch. An epoch is a complete run when all the training examples are presented to the network and are processed using the learning algorithm only once. After learning, a neural network represents a complex relationship, and possesses the ability for generalization. To control a learning process, a criterion is defined to decide the time for terminating the process. The complexity of an algorithm is usually denoted as $O(m)$, indicating that the order of number of floating-point operations is m.

In logic and statistical inference, transduction is reasoning from observed, specific (training) cases to specific (test) cases. In contrast, induction is reasoning from observed training cases to general rules, which are then applied to the test cases. Machine learning falls into two broad classes: inductive learning or transductive learning. Inductive learning pursues the standard goal in machine learning, which is to accurately classify the entire input space. In contrast, transductive learning focuses on a predefined target set of unlabeled data, the goal being to label the specific target set.

Inductive inference estimates the model function based on the relation of data to the entire hypothesis space, and uses this model to forecast output values for examples beyond the training set. Many machine learning methods fall into this category, including SVMs, neural network models, and neuro fuzzy models.

Transductive learning, also known as transductive inference, attempts to predict exclusive model functions on specific test cases by using additional observations on the training dataset in relation to the new cases [114].

K.-L. Du and M. N. S. Swamy, *Neural Networks and Statistical Learning*,
https://doi.org/10.1007/978-1-4471-7452-3_2

2.1.1 *Scientific Reasoning*

Scientific reasoning is basically classified as deduction, induction, and abduction, of which deduction is necessarily true, induction plausibly true, and abduction hypothetically true.

Deductive reasoning starts from a cause to deduce the consequence or effects. Deduction examines the premises, which are a rule and a case under the rule, and the result is a necessary truth. Inductive reasoning allows us to deduce possible causes from the consequence. A rule is concluded from a case and a result. The conclusion may not be necessarily but only plausibly true. For abductive reasoning, a case is concluded from a rule and a result (an observation). The conclusion is hypothetically plausible.

Syllogistic is a theory of deductive reasoning. In a syllogistic inference, the conclusion is derived from the major and minor premises. Reasoning may arise from a series of syllogistic inferences, through transitive closure (deduction), generalization (induction), and experimental verification (abduction). The three types of reasoning are related and their truth perceptions can be explained as different degrees of belief [95]. The source of this difference can be found in the conversion of a premise required by syllogistic processing.

Deductive Reasoning

In deductive reasoning (top-down logic), also deductive logic, a conclusion is reached reductively by applying general rules that hold over the entirety of a closed domain of discourse, narrowing the range under consideration until only the conclusion(s) is left. The conclusion of a deductive argument is certain.

The law of detachment, also known as *affirming the antecedent* and *modus ponens* (Latin for "the way that affirms by affirming"), is the first form of deductive reasoning. In propositional logic, *modus ponens* or implication elimination is a rule of inference. It can be summarized as "P implies Q ($P \rightarrow Q$)" and "if P is asserted to be true (P), then Q must be true (Q)."

The law of syllogism takes two conditional statements and forms a conclusion by combining the hypothesis of one statement with the conclusion of another: If $P \rightarrow Q$ and $Q \rightarrow R$, then $P \rightarrow R$.

In propositional logic, *modus tollens* (Latin for "the way that denies by denying") or *denying the consequent* is a valid argument form and a rule of inference. It is an application of the general truth that if a statement is true, then so is its contrapositive.

The law of contrapositive states that, in a conditional, if the conclusion is false, then the hypothesis must be false also: $P \rightarrow Q$. If $\neg Q$, then $\neg P$.

Every use of modus tollens can be converted to a use of modus ponens and one use of transposition to the premise which is a material implication. Likewise, every use of modus ponens can be converted to a use of modus tollens and transposition.

Inductive Reasoning

In inductive reasoning (bottom-up logic), the conclusion is reached by generalizing or extrapolating from specific cases to general rules, i.e., there is epistemic uncertainty. However, inductive reasoning is not the same as induction used in mathematical proofs—mathematical induction is actually a form of deductive reasoning.

Inductive learning is a special class of the supervised learning techniques, where given a set of $\{\boldsymbol{x}_i, f(\boldsymbol{x}_i)\}$ pairs, we determine a hypothesis $h(\boldsymbol{x}_i)$ such that $h(\boldsymbol{x}_i) \approx f(\boldsymbol{x}_i), \forall i$. In inductive learning, given many positive and negative instances of a problem the learner has to form a concept that supports most of the positive but no negative instances. This requires a number of training instances to form a concept in inductive learning.

Abductive Reasoning

Abductive reasoning (also called abduction, abductive inference, or retroduction) is a form of logical inference which starts with an observation, then seeks to find the simplest and most likely explanation by provisionally adopting a hypothesis. Every possible consequence of the hypothesis can be verified experimentally. In abductive reasoning, the premises do not guarantee the conclusion. One can understand abductive reasoning as inference to the best explanation.

The scientific method of answering a question is illustrated in Fig. 2.1. A question can be answered by a three-phase process [111]. In the first phase (*Question in search of answers*), the analyst guesses answers (hypotheses). In the second phase (*Hypothesis in search of evidence*), each hypothesis is verified to discover evidence that supports that hypothesis. In the third phase (*Evidentiary assessment of hypotheses*), the probability of each hypothesis is assessed based on the discovered evidence. New

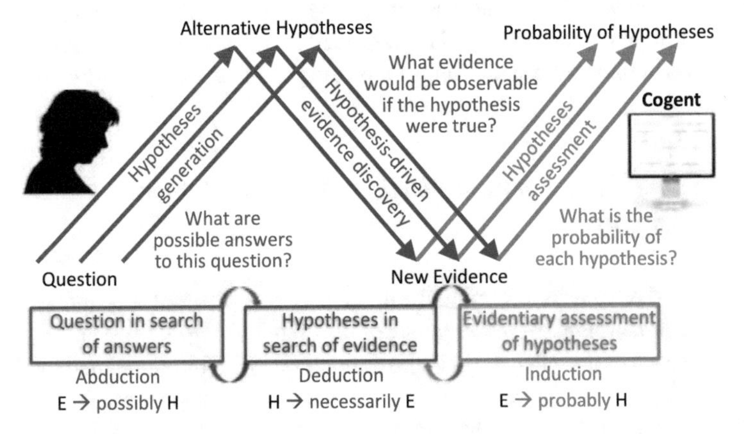

Fig. 2.1 The reasoning process of answering a question, where Cogent is the system. From [111]

evidence discovered may lead to new answers to the question, which may further lead to the discovery of new evidence and revised assessments of probability. The process integrates abductive, deductive, and inductive reasoning. Hypotheses generation involves abductive reasoning, which shows that something is possibly true. Hypothesis-driven evidence discovery involves deductive reasoning, which shows that something is necessarily true. Hypotheses assessment involves inductive reasoning, which shows that something is probably true.

Analogical Reasoning

Unlike inductive learning, *analogical learning* can be accomplished from a single example; for instance, given a training instance of plural of `fungus` as `fungi`, one can determine the plural of `bacilus`: `bacillus -> bacilli`.

Case-Based Reasoning

Case-based reasoning and knowledge generalization are the two main approaches leveraging past experiences. The main concern of case-based reasoning lies in the retrieval of the relevant case(s) to aid learning in the target task, among a collection of past experiences (cases).

Ontologies

Ontologies are formal systems that assign data objects to classes and that relate classes to other classes. Ontologies are unrestrained classifications. An object is permitted to be a direct subclass of more than one class. The simplest form of ontology is classification, in which each class is limited to one parent class. In an ontology, the assignment of an object to a class is determined by rules. Every class, subclass, and superclass is defined by rules, which can be programmed into the software.

2.1.2 *Supervised, Unsupervised, and Reinforcement Learnings*

Learning methods are conventionally divided into supervised, unsupervised, and reinforcement learning; these schemes are illustrated in Fig. 2.2. $\boldsymbol{x}_p$ and $\boldsymbol{y}_p$ are the input and output of the pth pattern in the training set, $\hat{\boldsymbol{y}}_p$ is the neural network output for the pth input, and E is an error function. From a statistical viewpoint, unsupervised learning learns the pdf of the training set, $p(\boldsymbol{x})$, while supervised learning learns about the pdf of $p(\boldsymbol{y}|\boldsymbol{x})$. Supervised learning is widely used in classification, approx-

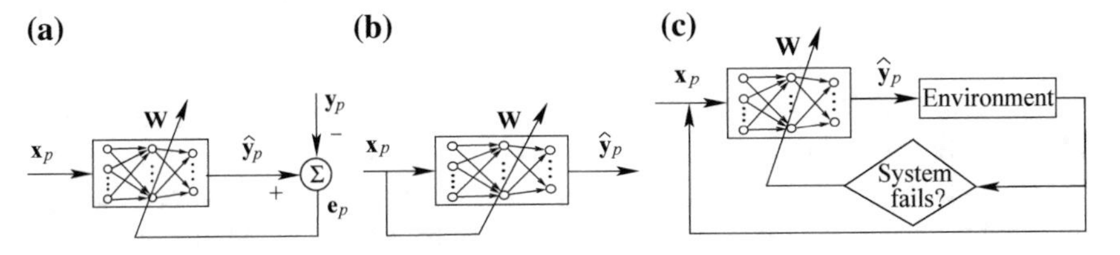

Fig. 2.2 Learning methods. **a** Supervised learning. $\boldsymbol{e}_p = \hat{\boldsymbol{y}}_p - \boldsymbol{y}_p$. **b** Unsupervised learning. **c** Reinforcement learning

imation, control, modeling and identification, signal processing, and optimization. Unsupervised learning schemes are mainly used for clustering, vector quantization, feature extraction, signal coding, and data analysis. Reinforcement learning is usually used in control and artificial intelligence.

Supervised Learning

The task of supervised learning is to learn a function $f : \mathcal{X} \to \mathcal{Y}$ from a given data set $\{(\boldsymbol{x}_i, y_i)|i = 1, \ldots, N\}$, where $\boldsymbol{x}_i \in \mathcal{X}$ is an instance and $y_i \in \mathcal{Y}$ is the known label of $\boldsymbol{x}_i$.

Supervised learning adjusts network parameters by a direct comparison between the actual network output and the desired output. Supervised learning is a closed-loop feedback system, where the error is the feedback signal. The error measure, which shows the difference between the network output and the output from the training samples, is used to guide the learning process. The error measure is usually defined by the mean squared error (MSE)

$$E = \frac{1}{N} \sum_{p=1}^{N} \left\| \boldsymbol{y}_p - \hat{\boldsymbol{y}}_p \right\|^2, \tag{2.1}$$

where N is the number of pattern pairs in the sample set, $\boldsymbol{y}_p$ is the output part of the pth pattern pair, and $\hat{\boldsymbol{y}}_p$ is the network output corresponding to the pattern pair p. The error E is calculated anew after each epoch. The learning process is terminated when E is sufficiently small or a failure criterion is met.

To decrease E toward zero, a gradient-descent procedure is usually applied. The gradient-descent method always converges to a local minimum in a neighborhood of the initial solution of network parameters. The LMS and BP algorithms are the two most popular gradient-descent-based algorithms. Second-order methods are based on the computation of the Hessian matrix.

Unsupervised Learning

Unsupervised learning involves no target values. It tries to autoassociate information from the inputs with an intrinsic reduction of data dimensionality or total amount of input data. Unsupervised learning is solely based on the correlations among the input data, and is used to find the significant patterns or features in the input data without the help of a teacher. Unsupervised learning is particularly suitable for biological learning in that it does not rely on a teacher and it uses intuitive primitives like neural competition and cooperation.

A criterion is needed to terminate the learning process. Without a stopping criterion, a learning process continues even when a pattern, which does not belong to the training patterns set, is presented to the network. The network is adapted according to a constantly changing environment. Hebbian learning, competitive learning, and the SOM are the three well-known unsupervised learning approaches. Generally speaking, unsupervised learning is slow to settle into stable conditions.

In Hebbian learning, learning is a purely local phenomenon, involving only two neurons and a synapse. The synaptic weight change is proportional to the correlation between the pre- and postsynaptic signals. Many neural networks for PCA and associative memory are based on Hebbian learning. In competitive learning, the output neurons of a neural network compete for the right to respond. The SOM is also based on competitive learning. Competitive learning is directly related to clustering. The Boltzmann machine uses a stochastic training technique known as simulated annealing, which can be treated as a special type of unsupervised learning based on the inherent property of a physical system.

Reinforcement Learning

Reinforcement learning is a class of computational algorithms that specifies how an artificial agent (e.g., a real or simulated robot) can learn to select actions in order to maximize the cumulative expected reward [9]. This computed difference, termed reward prediction error, has been shown to correlate very well with the phasic activity of dopamine-releasing neurons projecting from the substantia nigra in nonhuman primates [96].

Reinforcement learning is a special case of supervised learning, where the exact desired output is unknown. The teacher supplies only feedback about the success or failure of an answer. This is cognitively more plausible than supervised learning since a fully specified correct answer might not always be available to the learner or even the teacher. It is based only on the information as to whether or not the actual output is close to the estimate. Reinforcement learning is a learning procedure that *rewards* the neural network for its *good* output result and *punishes* it for the *bad* output result. Explicit computation of derivatives is not required. This, however, presents a slower learning process. For a control system, if the controller still works properly after an input, the output is judged as *good*; otherwise, it is considered as *bad*. The evaluation of the binary output, called *external reinforcement*, is used as the error signal.

2.1.3 Semi-supervised Learning and Active Learning

In many machine learning applications, such as bioinformatics, web and text mining, text categorization, database marketing, spam detection, face recognition, and video-indexing, abundant amounts of unlabeled data can be cheaply and automatically collected. However, manual labeling is often slow, expensive, and error-prone. When only a small number of labeled samples are available, unlabeled samples could be used to prevent the performance degradation due to overfitting.

The goal of semi-supervised learning is to employ a large collection of unlabeled data jointly with a few labeled examples for improving generalization performance. Some semi-supervised learning methods are based on some assumptions that relate the probability $P(x)$ to the conditional distribution $P(Y = 1|X = x)$. Semi-supervised learning is related to the problem of transductive learning. Two typical semi-supervised learning approaches are learning with the cluster assumption [114] and learning with the manifold assumption [14]. The cluster assumption requires that data within the same cluster are more likely to have the same label. The most prominent example is the transductive SVM [114].

Universum data are given a set of unlabeled examples and do not belong to either class of the classification problem of interest. Contradiction happens when two functions in the same equivalence class have different signed outputs on a sample from the Universum. Universum learning is conceptually different from semi-supervised learning or transduction [114], because the Universum data is not from the same distribution as the labeled training data. Universum learning implements a trade-off between explaining training samples (using large margin hyperplanes) and maximizing the number of contradictions (on the Universum).

In active learning, or so-called *pool-based active learning*, the labels of data points are initially hidden, and the learner must pay for each label he wishes to be revealed. The goal of active learning is to actively select the most informative examples for manual labeling in these learning tasks, that is, designing input signals for optimal generalization [37]. Based on conditional expectation of the generalization error, a pool-based active learning method effectively copes with model misspecification by weighting training samples according to their importance [108]. Reinforcement learning can be regarded as a form of active learning. At this point, a query mechanism proactively asks for the labels of some of the unlabeled data.

Examples of situations in which active learning can be employed are web searching, email filtering, and relevance feedback for a database or website. The first two examples involve induction. The goal is to create a classifier that works well on unseen future instances. The third situation is an example of transduction [114]. The learner's performance is assessed on the remaining instances in the database rather than a totally independent test set.

The query-by-committee algorithm [39] is an active learning algorithm for classification, which uses a prior distribution over hypotheses. In this algorithm, the learner observes a stream of unlabeled data and makes spot decisions about whether or not to ask for each point's label. If the data is drawn uniformly from the surface of the unit sphere in R^d, and the hidden labels correspond perfectly to a homogeneous (i.e., through the origin) linear separator from this same distribution, then it is possible to achieve generalization error ϵ after seeing $O((d/\epsilon)\log(1/\epsilon))$ points and requesting just $O(d\log(1/\epsilon))$ labels: an exponential improvement over the usual $O(d/\epsilon)$ sample complexity of learning linear separators in a supervised setting. The query-by-committee algorithm involves random sampling from intermediate version spaces; the complexity of the update step scales polynomially with the number of updates performed.

An information-based approach for active data selection is presented in [71]. In [107], a two-stage sampling scheme for reducing both the bias and variance is given, and based on it, two active learning methods are given.

In a framework for batch mode active learning [56], a number of informative examples are selected for manual labeling in each iteration. The key feature is to reduce the redundancy among the selected examples such that each example provides unique information for model updating. The set of unlabeled examples that can efficiently reduce the Fisher information of the classification model is chosen [56].

2.1.4 Other Learning Methods

Ordinal Regression and Ranking

Regression is the process of learning relationships between inputs and continuous outputs from example data. Regression is a form of supervised learning where the output space is continuous.

Categorical data can generally be classified into ordinal data and nominal data. Ordinal and nominal data both have a set of possible states, and the value of a variable will be in one of those possible states. The difference between them is that the states in ordinal data are ordered, but are unordered in nominal data. A nominal variable can only have two matching results, either match or not match. For instance, hair color is a nominal variable that may have four states: *black*, *blond*, *red*, and *brown*. Service quality assessment is an ordinal variable that may have five states: *very good*, *good*, *medium*, *poor*, *very poor*.

Ordinal regression is generally defined as the task where some input sample vectors are ranked on an ordinal scale. Ordinal regression is commonly formulated as a multiclass problem with ordinal constraints. The aim is to predict variables of ordinal scale. Ordinal regression is between classification and regression. The naive idea is to transform the ordinal scales into numerical values and then solve the problem as a standard regression problem.

An ordinal regression classifier can be used as a ranking function by treating the class labels as scores. The labels exhibit a natural order. During the test phase, the objective is to obtain correct labels or as close as possible to the correct ones. Traditional approaches have a 1-D output space, where ordinality is realized by defining a monotone threshold configuration [74].

Ranking problems aim at ordering a finite set of alternatives (items, actions) from the best to the worst, using a relative comparison approach. Ranking needs to recognize whether the data belong to the same category or not, at the same time, it should preserve the ordinal relationship of the data.

Learning to rank is an important supervised learning technique because of its application to search engines and online advertisement. Learning to rank models can be categorized into three types [26]. Pointwise methods (e.g., decision tree models and linear regression) directly learn the relevance score of each instance; pairwise methods like rankSVM [50] learn to classify preference pairs; and listwise methods such as LambdaMART [22] try to optimize the measurement for evaluating the whole ranking list.

Gradient boosting decision trees (GBDT) [40] and its variant, LambdaMART [22], give competitive performance on web search ranking data.

Preference learning is to learn a binary preference relation [38]. In a comparison of two input points, the preference relation is able to evaluate whether the first point ranks before the second one.

Manifold Learning

Representation learning is a set of methods that allow a machine to automatically discover the representations needed for detection or classification from raw data. The algorithms often serve as a preprocessing step before performing classification or predictions. They are typically unsupervised learning algorithms. Examples include PCA and cluster analysis.

Manifold learning attempts to perform representation learning under the constraint that the learned representation is a low-dimensional manifold that preserves the intrinsic geometric information of data points. Manifold learning preserves the local geometric structure of the manifold using a simple linear approximation to the nonlinear mappings. If two objects are close in the intrinsic geometry of data manifold, they should be close to each other after dimension reduction.

Sparse coding performs manifold learning under the constraint that the learned representation is sparse. Multilinear subspace learning algorithms aim to learn low-dimensional representations directly from tensor representations for multidimensional data. Deep learning discovers multiple levels of representation, or a hierarchy of features, with higher level, more abstract features defined in terms of (or generating) lower level features.

A lot of linear dimension-reduction methods are based on manifold learning, such as locally linear embedding [90], Laplacian eigenmap [13], and orthogonal neighborhood preserving projection [63]. Locally linear embedding [90] is capable of learning the global structure of nonlinear manifolds generated by face images. Orthogonal neighborhood preserving projection [63] is the representative linear extension of locally linear embedding. It aims at preserving the local neighborhood geometry structure of the data. Laplacian eigenmap [12] provides a nonlinear dimensionality reduction approach with properties of local preservation by constructing a representation of data sample.

Transfer Learning

In order to learn accurate models for rare cases, it is desirable to use data and knowledge from similar cases; this is known as *transfer learning* (or *meta-analysis* in the statistical literature). Transfer learning describes the procedure of using data recorded in one task to boost the performance in another related task [80]. It exploits the insight that generalization may occur not only within tasks, but also across tasks. Transfer learning is related in spirit to case-based and analogical learning. A theoretical analysis based on an empirical Bayes perspective exhibits that the number of labeled examples required for learning with the transfer is often significantly smaller than that required for learning each target independently [121].

Transfer learning algorithms can be roughly grouped into homogeneous and heterogeneous transfers. The homogeneous transfer refers to samples in target and source domains that are drawn from the same instance space but different distributions, while the heterogeneous transfer refers to samples in target and source domains that are drawn from different, but related instance spaces. Transfer learning can be grouped into instance transfer, feature representation transfer, parameter transfer, and relational knowledge transfer.

Multi-view Learning

We sense the world in a multimodal way: seeing, hearing, touching, smelling, and tasting. Data is often obtained from multiple sources rather than a single source. Different views provide information complementary to one another. The multi-view data should be consistent. Diversity (due to multimodality) is the key to data fusion. Multi-view (or multimodal) learning explicitly fuses the complementary information from different modalities to achieve improved performance.

Multi-view learning methods can be classified into three groups: co-training, multiple kernel learning, and subspace learning. Co-training methods train alternately to maximize the mutual agreement on two distinct views of the data. Canonical correlation analysis (CCA) [58] can be regarded as the multi-view version of PCA. Through maximizing the correlation between two views in the subspace, CCA outputs one optimal projection on each view. Joint analysis of multiple datasets includes multiset CCA [61], and tensor decomposition [113]. To learn a discriminant common space for two views, the class label information is generally incorporated. The PAC generalization bound for co-training shows that the generalization error of a classifier from each view is upper bounded by the disagreement rate of the classifiers from the two views [32].

Multiple kernel learning [66] straightforwardly corresponds to multiple modalities and elegantly combines kernels of different modalities to achieve improved performance. Subspace learning algorithms aim to obtain a latent subspace shared by multiple views by assuming that the input views are generated from this latent subspace.

Multitask learning improves the generalization performance of learners by leveraging the domain-specific information contained in the related tasks [23]. Multiple related tasks are learned simultaneously using a shared representation. In fact, the training signals for extra tasks serve as an inductive bias [23]. Multitask learning is closely related to multimodal learning.

Multilabel Learning

Multilabel learning distinguishes multilabel classification and multilabel ranking. Multilabel classification [123] is to learn a function from a given dataset, where each example has a number of labels. Multilabel ranking is to learn a scoring function that assigns a real number indicating the relevance of an instance to a class label, achieving an order of the class labels.

A multilabel problem can be naturally decomposed into a set of single-label problems by one of the three methods: binary relevance, label powerset, and pairwise methods. Consider a multilabel problem of up to q labels. Binary relevance uses the one-against-all strategy to convert the multilabel problem into q binary classification problems. However, this approach can result in poor performance when strong label correlations exist. The label powerset method is to transform the multilabel problem into a single multiclass problem with 2^q labels using the power set of labels as the set of possible labels. Label powerset-based methods directly take the label correlations into account, but the space of possible label subsets can be very large. Pairwise method learns $q(q-1)/2$ classifiers that cover all pairs of labels. All classifiers are then combined to make predictions by majority voting.

Multiple-Instance Learning

Multiple-instance learning [34] problem is an extension of supervised learning. In multiple-instance learning, the examples are sets (bags) of feature vectors (instances), and the bag label is a function of the labels of its instances. Typically, this function is the Boolean OR. A unified theoretical analysis for multiple-instance learning and a PAC-learning algorithm are introduced in [94].

Multiple-instance learning receives a set of bags that are labeled positive or negative, rather than receiving a set of instances which have positive or negative labels. In addition, instances in bags have no label information. The goal of multiple-instance learning is to learn a classifier from the training bags such that it can correctly predict the labels of unseen bags. Beyond drug activity prediction, multiple instance learning has been recognized as a state-of-the-art approach to image categorization/annotation, particularly for region-based image categorization/annotation.

Formally, let $\mathcal{X}$ denote the instance space and $\mathcal{Y}$ the set of class labels. The task of multi-instance learning is to learn a function $f : 2^{\mathcal{X}} \rightarrow \{-1, +1\}$ from a given dataset $\{(\mathbf{X}_1, y_1), (\mathbf{X}_2, y_2), \ldots, (\mathbf{X}_m, y_m)\}$, where $\mathbf{X}_i \subseteq \mathcal{X}$ is a set of instances $\{\boldsymbol{x}_1^{(i)}, \boldsymbol{x}_2^{(i)}, \ldots, \boldsymbol{x}_{n_i}^{(i)}\}$, $\boldsymbol{x}_j^{(i)} \in \mathcal{X}(j \in \{1, \ldots, n_i\})$, and $y_i \in \{-1, +1\}$ is the known label of $\mathbf{X}_i$.

Parametric, Semiparametric and Nonparametric Classifications

Pattern classification techniques with numerical inputs can be generally classified into parametric, semiparametric and nonparametric groups. The parametric and semiparametric classifiers need a certain amount of a priori information about the structure of the data in the training set. Parametric techniques assume that the form of the pdf is known in advance except for a vector of parameters, which has to be estimated from the sample of realizations. In this case, smaller sample size can yield good performance if the form of the pdf is properly selected. When some insights about the form of the pdf are available, parametric techniques offer the most valid and efficient approach to density estimation. Examples of parametric classifiers are SVM and logistic regression.

Semiparametric techniques consider models having a number of parameters not growing with the sample size, though greater than that involved in parametric techniques.

Nonparametric techniques aim to retrieve the behavior of the pdf without imposing any a priori assumption on it; therefore, they require a sample size significantly higher than the dimension of the domain of the random variable. Nonparametric methods are more flexible than parametric methods, whereas a parametric model, which has a fixed and finite parameterization, limits the range of functions that can be represented. An example of the nonparametric method is k-nearest neighbor (k-NN) classifier.

Density estimation methods using neural networks or SVMs fall into the category of nonparametric techniques. The Parzen's windows approach [81] is a nonparametric method for estimating the pdf of a finite set of patterns; it has a very high computational cost due to the very large number of kernels required for its representation. A decision tree such as C5.0 (http://www.rulequest.com/see5-info.html) is an efficient nonparametric method. A decision tree is a hierarchical data structure implementing the divide-and-conquer strategy. It is a supervised learning method, and can be used for both classification and regression.

Learning from Imbalanced Data

Learning from imbalanced data is a challenging problem. It is the problem of learning a classification rule from data that are skewed in favor of one class. Many real-world datasets are imbalanced and the majority class has much more training patterns than the minority class. The resultant hyperplane will be shifted toward the majority class. However, the minority class is often the most interesting one for the task.

For the imbalanced datasets, a classifier may fail. The remedies can be divided into two categories. The first category processes the data before feeding them into the classifier, such as the oversampling and undersampling techniques, combining oversampling with undersampling, and synthetic minority oversampling technique (SMOTE) [27]. The oversampling technique duplicates the positive data by interpolation, while undersampling technique removes the redundant negative data to reduce the imbalanced ratio. They are classifier-independent approaches. The second category belongs to the algorithm-based approach such as different error cost algorithms [67], and class-boundary-alignment algorithm [117]. The different cost algorithms suggest that by assigning a heavier penalty to the smaller class, the skew of the optimal separating hyperplane can be corrected.

The one-class classification problem is known as outlier/novelty/anomaly detection. The core of the problem lies in modeling and recognizing patterns that belong only to a target class. All other patterns are termed nontarget.

2.2 Learning and Generalization

From an approximation viewpoint, learning is a hypersurface reconstruction based on existing examples, while generalization means estimating the value on the hypersurface where there is no example. Mathematically, the learning process is a nonlinear curve-fitting process, while generalization is the interpolation and extrapolation of the input data.

The goal of training neural networks is not to learn an exact representation of the training data itself, but rather to build a statistical model of the process which generates the data. The problem of reconstructing the mapping is said to be *well-posed* if an input always generates a unique output, and the mapping is continuous. Learning is an ill-posed inverse problem. Given examples of an input–output mapping, an approximate solution is required to be found for the mapping. The input data may be noisy or imprecise, and also may be insufficient to uniquely construct the mapping. The regularization technique can transform an ill-posed problem into a well-posed one so as to stabilize the solution by adding some auxiliary nonnegative functional for constraints [84, 112].

When a network is overtrained with too many examples, parameters or epochs, it may produce good results for the training data, but has a poor generalization capability. This is the overfitting phenomenon, and is illustrated in Fig. 2.3. In statistics, overfitting applies to the situation wherein a model possesses too many parameters, and fits the noise in the data rather than the underlying function. A simple network with smooth input–output mapping usually has a better generalization capability. Generally, the generalization capability of a network is jointly determined by the size of the training pattern set, the complexity of the problem, and the architecture of the network.

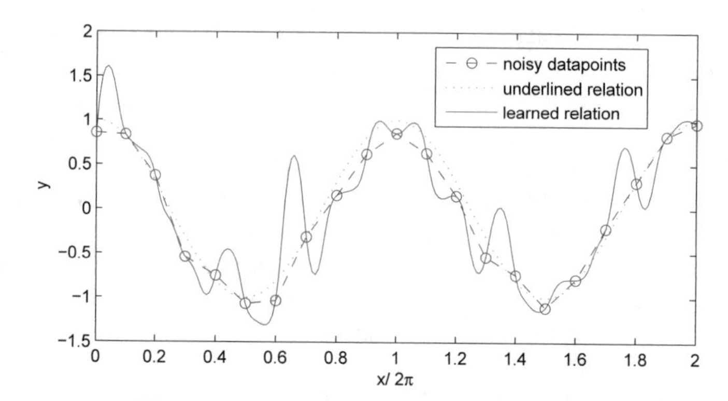

Fig. 2.3 Proper fitting and overfitting. Dashed line corresponds to proper fitting, and solid line corresponds to overfitting

Example 2.1 To approximate a noisy cosine function, with 20 random samples, we employ a 1-30-1 feedforward network. The result is plotted in Fig. 2.3. The noisy samples is represented by the "∘" symbols, and the true network response is given by the solid line. Clearly, the learned network is overfitted, and it does not generalize well. Notice that if the number of parameters in the network is much smaller than the total number of points in the training set, then there is little or no worry of overfitting.

For a given network topology, we can estimate the minimal size of the training set for successfully training the network. For conventional curve-fitting techniques, the required number of examples usually grows with the dimensionality of the input space, namely, the *curse of dimensionality*. Feature extraction can reduce input dimensionality and thus improve the generalization capability of the network.

The training set should be sufficiently large and diverse so that it could represent the problem well. For good generalization, the size of the training set, N, should be at least several times larger than the network's capacity, i.e., $N \gg \frac{N_w}{N_y}$, where N_w the total number of weights or free parameters, and N_y the number of output components [116].

2.2.1 Generalization Error

The generalization error of a trained network can be decomposed into two parts, namely, an *approximation error* that is due to a finite number of parameters of the approximation scheme used and an unknown level of noise in the training data, and an *estimation error* that is due to a finite number of data available [78]. For a feedforward network with J_1 input nodes and a single output node, a bound on the generalization error is associated with the order of hypothesis parameters N_P and the number of examples N [78]

$$O\left(\frac{1}{N_P}\right) + O\left(\left[\frac{N_P J_1 \ln(N_P N) - \ln\delta}{N}\right]^{1/2}\right), \text{ with probability } p > 1 - \delta, \quad (2.2)$$

where $\delta \in (0, 1)$ is the confidence parameter, and N_P is proportional to the number of parameters, such as N_P centers in an RBF network, or N_P sigmoidal hidden units in an MLP. The first term corresponds to the bound on the approximation error, and the second to that on the estimation error.

As N_P increases, the approximation error decreases since a larger model is used; however, the estimation error increases due to overfitting (or alternatively, more data). Thus, one cannot reduce the upper bounds on both the error components simultaneously. Given the amount of data available, the optimal size of the model for the trade-off between the approximation and estimation errors is selected as $N_P \propto N^{\frac{1}{3}}$ [78]. After suitably selecting N_P and N, the generalization error for feedforward networks should be $O\left(\frac{1}{N_P}\right)$. This result is similar to that for an MLP with sigmoidal functions [8].

2.2.2 *Generalization by Stopping Criterion*

Generalization can be controlled during training. Overtraining can be avoided by stopping the training before the absolute minimum is reached. Neural networks trained with iterative gradient-based methods tend to learn a mapping in the hierarchical order of its increasing components of frequency. When training is stopped at an appropriate point, the network will not learn the high-frequency noise. While the training error will always decrease, the generalization error will decrease to a minimum and then begins to rise again as the network is being overtrained. Training should stop at the optimum stopping point. The generalization error is defined in the same form as the learning error, but on a separate validation set of data. Early stopping is the default method for improving generalization.

Example 2.2 In order to use early stopping technique, the available data is divided into three subsets. The first subset is the training set. The second subset is the validation set. The error on the validation set is monitored during the training process. The validation error normally decreases during the initial phase of training, as does the error on the training set. However, when the network begins to overfit the data, the error on the validation set typically begins to rise. When the validation error increases for a specified number of iterations, training is stopped, and the weights and biases at the minimum of the validation error are returned. The error on the test set is not used during training, but it is used to compare different models. It is also useful to plot the error on the test set during the training process. If the error on the test set reaches a minimum at a significantly different iteration number than the error on the validation set, this might indicate a poor division of the dataset. From Example 2.1, the 40 data samples are divided by 60, 20 and 20% of samples as the training, validation, and test sets. The relation is illustrated in Fig. 2.4.

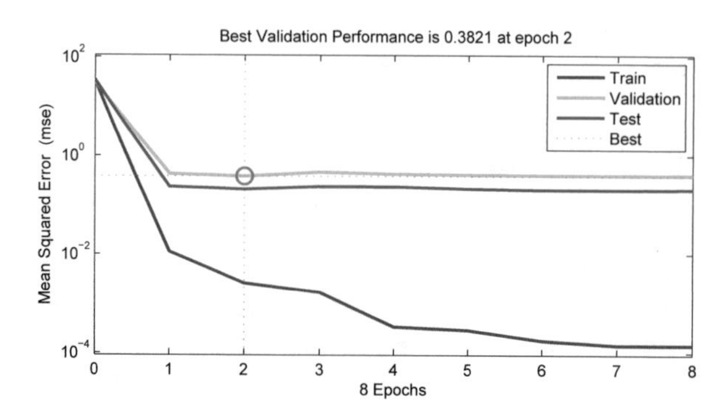

Fig. 2.4 Learning and generalization. When the network is overtrained, its generalization performance degrades

Early stopping is implemented with cross-validation to decide when to stop. Three early stopping criteria are defined and empirically compared in [85]. Slower stopping criteria, which stop later than others, on average lead to small improvements in generalization, but result in a much longer training time [85].

Statistical analysis for the three-layer MLP has been performed in [4]. As far as the generalization performance is concerned, exhaustive learning is satisfactory when $N > 30N_w$. When $N < N_w$, early stopping can really prevent overtraining. When $N < 30N_w$, overtraining may also occur. In the latter two cases, cross-validation can be used to stop training.

An optimized approximation algorithm [69] avoids overfitting in function approximation applications. The algorithm utilizes a quantitative stopping criterion based on the estimation of the signal-to-noise ratio figure (SNRF). Using SNRF, overfitting can be automatically detected from the training error only.

2.2.3 *Generalization by Regularization*

Regularization is a reliable method for improving generalization. The target function is assumed to be smooth, and small changes in the input do not cause large changes in the output. A constraint term E_c, which penalizes poor generalization, is added to the standard training cost function E

$$E_T = E + \lambda_c E_c, \tag{2.3}$$

where λ_c is a positive value that balances the trade-off between error minimization and smoothing. In contrast to the early stopping criterion method, the regularization method is applicable to both the iterative gradient-based techniques and the one-step linear optimization such as the singular value decomposition (SVD) technique.

Network-pruning techniques such as the weight-decay technique also help to improve generalization [18, 86]. At the end of training, there are some weights significantly different from zero, while some other weights are close to zero. Those connections with small weights can be removed from the network. Biases should be excluded from the penalty term so that the network yields an unbiased estimate of the true target mean.

Early stopping has a behavior similar to that of a simple weight-decay technique in the case of the MSE function [17]. The quantity $\frac{1}{\eta t}$, where η is the learning rate and t is the iteration index, plays the role of λ_c. The effective number of weights, that is, the number of weights whose values differ significantly from zero, grows as training proceeds.

Training with a small amount of jitter in the input while keeping the same output can improve generalization. With jitter, the learning problem is equivalent to a smoothing regularization with the noise variance playing the role of the regularization parameter [17, 86]. Training with jitter thus allows regularization within the conventional layered feedforward network architecture. Although large networks are generally trained rapidly, they tend to generalize poorly due to insufficient constraints. Training with jitter helps to prevent overfitting.

In [57], noise is added to the available training set to generate an unlimited source of training samples. This is interpreted as a kernel estimate of the probability density that describes the training vector distribution. It helps to enhance the generalization performance, speed up the BP, and reduce the possibility of local minima entrapment.

In [54], each weight is encoded with a short bit-length to decrease the complexity of the model. The amount of information in a weight can be controlled by adding Gaussian noise and the noise level can be adapted during learning to optimize the trade-off between the expected squared error of the network and the amount of information in the weights.

Weight sharing is to control several weights by a single parameter [92]. This reduces the number of free parameters in a network, and thus improves generalization. A soft weight-sharing method is implemented in [79] by adding a regularization term to the error function, where the learning algorithm decides which of the weights should be tied together.

Regularization decreases the representation capability of the network, but increases the bias (bias–variance dilemma [43]). The principle of regularization is to choose a well-defined regularizer to decrease the variance by affecting the bias as little as possible [17].

2.2.4 Dropout

Overfitting is a serious problem in deep networks. Dropout technique [104] randomly drops units (along with their connections) from the neural network during training. Standard BP learning builds up brittle co-adaptations that work for the training data but do not generalize to unseen data. Random dropout breaks up these co-adaptations by making the presence of any particular hidden unit unreliable. This prevents units from co-adapting too much. During training, dropout samples from an exponential number of different thinned networks. At test time, it is easy to approximate the effect of averaging the predictions of all these thinned networks by simply using a single unthinned network that has smaller weights. In the simplest case, each unit is retained with a fixed probability p independent of the other units. Dropout can be interpreted as a way of regularizing a neural network by adding noise to its hidden units. Dropping out 20% of the input units and 50% of the hidden units were often found to be optimal.

Dropout can be seen as a stochastic regularization technique. Its deterministic counterpart can be obtained by marginalizing out the noise. For linear regression, this regularizer is a modified form of L_2 regularization [104].

Dropconnect [115] goes a step further by randomly setting the elements of the weight matrix $\mathbf{W}$ to zero, instead of randomly setting the outputs of neurons to zero.

A number of possible explanations have been suggested for dropout's success. In [53], dropout is interpreted from a similarity to the theory of the role of sex in evolution. Sexual reproduction involves taking half the genes of each parent and combining them to produce offspring. This explains that dropout training works the best when $p = 0.5$. Dropout is argued to control network complexity by restricting the ability to coadapt weights and it appears to learn simpler functions at the second layer [52]. In [7], dropout is viewed as an ensemble method combining the different network topologies resulting from the random deletion of nodes or connections, with arbitrary probability values. Dropout performs stochastic gradient descent on a regularized error function. In [42], training in deep dropout neural networks is cast as approximate Bayesian inference in deep Gaussian processes. This provides tools to model uncertainty with dropout neural networks. The dropout objective minimizes the Kullback–Leibler divergence between an approximate distribution and the posterior of a deep Gaussian process [42].

Deep dropout networks with rectified linear unit (ReLU) function and the quadratic loss show differences between the behavior of dropout and weight decay [49]. Dropout training can produce negative weights, which supports the suggestion that dropout discourages co-adaptation of weights. The dropout penalty can grow exponentially in the depth of the network, while the weight-decay penalty remains essentially linear. Dropout is insensitive to various re-scalings of the input features, outputs, and network weights, implying that there are no isolated local minima of the dropout training criterion.

Spectral dropout [62] is a regularization method that prevents overfitting by eliminating weak and noisy Fourier coefficients of the neural network activations, leading to remarkably better results than the current regularization methods. Frequency transform can be cast as a convolutional layer using a decorrelation transform with fixed basis functions.

2.2.5 Fault Tolerance and Generalization

Fault tolerance is one property of the human brain. Fault tolerance is strongly associated with generalization. A trained neural network should be able to tolerate weight or neuron failure. However, without a proper training procedure, a trained neural network has a very poor fault-tolerant ability.

Input noise during training improves generalization ability [18], and synaptic noise during training improves fault tolerance [76]. When fault tolerance is improved, the generalization ability is usually better [36], and vice versa [29]. The lower the weight magnitude, the higher the fault tolerance [16, 29] and the generalization ability [16, 65]. Based on the Vapnik–Chervonenkis (VC) dimension, it is qualitatively explained in [82] why adding redundancy can improve fault tolerance and generalization.

Fault tolerance is related to a uniform distribution of the learning among the different neurons, but BP algorithm does not guarantee this good distribution [36]. Just as input noise is introduced to enhance the generalization ability, the perturbation of weights during training also increases the fault tolerance of MLPs [36, 76]. Saliency, used as a measurement of fault tolerance to weight deviations [76], is computed from the diagonal elements of the Hessian matrix of the error with respect to the weight values. A low value of saliency implies a higher fault tolerance.

An analysis of the influence of weight and input perturbations in an MLP is made in [16]. The measurements introduced are explicitly related to the MSE degradation in the presence of perturbations, thus constituting a selection criterion between different alternatives of weight configurations. Quantitative measurements of fault tolerance, noise immunity, and generalization ability are provided, from which several previous conjectures are deduced.

When training the MLP with online node fault injection, hidden nodes randomly output zeros during training. The trained MLP is able to tolerate random node fault. The convergence of the algorithm is proved in [109]. The corresponding objective functions consist of an MSE term, a regularizer term, and a weight-decay term.

Six common fault/noise-injection-based online learning algorithms, namely, injecting additive input noise, injecting additive/multiplicative weight noise, injecting multiplicative node noise, injecting multiweight fault, injecting multinode fault during training, and weight decay with injecting multinode fault, are investigated in [55] for RBF networks. The convergence of the six online algorithms is shown to be almost sure, and their true objective functions being minimized are derived. For injecting additive input noise during training, the objective function is identical to that of the Tikhonov regularizer approach. For injecting additive/multiplicative weight noise during training, the objective function is the simple mean square training error; thus, injecting additive/multiplicative weight noise during training cannot improve the fault tolerance of an RBF network. Similar to injective additive input noise, the objective functions of other fault/noise-injection-based online algorithms contain an MSE term and a specialized regularization term.

Stuck-at fault is a popular fault model to describe node failure, where the output of a faulty node is tied at a value zero, one, or any value between 0 and 1. A well-trained faulty network can tolerate weight or neuron failure [29, 118].

2.2.6 *Sparsity Versus Stability*

Stability establishes the generalization performance of an algorithm [20]. Sparsity and stability are the two desired properties of learning algorithms. Both properties lead to good generalization ability. These two properties are fundamentally at odds with each other and this no-free-lunch theorem is proved in [119, 120]: A sparse algorithm cannot be stable and vice versa. A sparse algorithm can have nonunique optimal solutions and is therefore ill-posed. If an algorithm is sparse, then its uniform stability is lower bounded by a nonzero constant. This also shows that any algorithmically stable algorithm cannot be sparse. Thus, one has to trade-off sparsity and stability in designing a learning algorithm.

In [119, 120], L_1-regularized regression (LASSO) is shown to be not stable, while L_2-regularized regression is known to have strong stability properties and is therefore not sparse. Sparsity promoting algorithms include LASSO, L_1-norm SVM, deep belief network, and sparse PCA.

2.3 Model Selection

Occam's razor was formulated by William of Occam in the late Middle Ages. Occam's razor principle states: *No more things should be presumed to exist than are absolutely necessary*. That is, if two models of different complexity fit the data approximately equally well, the simpler one usually is a better predictive model. From models approximating the noisy data, the ones that have minimal complexity should be chosen.

The objective of model selection is to find a model that is as simple as possible that fits a given dataset with sufficient accuracy, and has a good generalization capability to unseen data. The generalization performance of a network gives a measure of the quality of the chosen model. Model selection approaches can be generally grouped into four categories: cross-validation, complexity criteria, regularization, and network pruning/growing.

The generalization error of a learning method can be estimated via either cross-validation or bootstrap. In cross-validation methods, many networks of different complexity are trained and then tested on an independent validation set. The procedure is computationally demanding and/or requires additional data withheld from the total pattern set. In complexity criterion-based methods, training of many networks is required and hence, computationally demanding, though a validation set is not required. Regularization methods are more efficient than cross-validation techniques, but the results may be suboptimal since the penalty terms damage the representation capability of the network. Pruning/growing methods can be under the framework of regularization, which often makes restrictive assumptions, resulting in networks that are suboptimal.

Occam's Razor

Occam's razor is a philosophical principle of induction. A widely accepted interpretation of Occam's razor is: "Given two classifiers with the same training error, the simpler classifier is more likely to generalize better". Domingos [35] rejects this interpretation and proposes that model complexity is only a confounding factor usually correlated with the number of models from which the learner selects. It is thus hypothesized that the risk of overfitting (poor generalization) follows only from the number of model tests rather than the complexity of the selected model. The confusion between the two factors arises from the fact that a learning algorithm usually conducts a greater amount of testing to fit a more complex model.

Experimental results on real-life datasets confirm Domingos' hypothesis [122]. In particular, the experiments test the following assertions. (i) Models selected from a larger set of tested candidate models overfit more than those selected from a smaller set (assuming constant model complexity). (ii) More complex models overfit more than simpler models (assuming a constant number of candidate models tested). According to Domingos' hypothesis, the first assertion should be true and the second should be false.

2.3.1 Cross-Validation

Cross-validation is a standard model selection method in statistics [60]. The total pattern set is randomly partitioned into a training set and a validation (test) set. The major part of the total pattern set is included in the training set, which is used to train the network. The remaining, typically, 10–20%, is included in the validation set and is used for validation. When only one sample is used for validation, the method is called *leave-one-out* cross-validation. Methods on conducting cross-validation are given in [85]. This kind of hold-out estimate of performance lacks computational efficiency due to the repeated training, but with lower variance of the estimate.

Let $\mathcal{D}_i$ and $\overline{\mathcal{D}}_i$, $i = 1, \ldots, m$, be the data subsets of the total pattern set arising from the ith partitioning, which are, respectively, used for training and testing. The cross-validation process trains the algorithm m times, and is actually to find a suitable model by minimizing the log-likelihood function

$$E_{\mathrm{cv}} = -\frac{1}{m}\sum_{i=1}^{m} \ln\left(L\left(\widehat{\mathbf{W}}\left(\overline{\mathcal{D}}_i\right)\middle|\, \mathcal{D}_i\right)\right), \tag{2.4}$$

where $\widehat{\mathbf{W}}(\overline{\mathcal{D}}_i)$ denotes the maximum likelihood (ML) parameter estimates on $\overline{\mathcal{D}}_i$, and $L(\widehat{\mathbf{W}}(\overline{\mathcal{D}}_i)|\mathcal{D}_i)$ is the likelihood evaluated on the dataset $\mathcal{D}_i$.

Validation uses data different from the training set, thus the validation set is independent from the estimated model. This helps to select the best one among the different model parameters. Since this dataset is independent from the estimated model, the generalization error obtained is a fair estimate. Sometimes it is not optimal if we train the network to perfection on a given pattern set due to the ill-posedness of the finite training pattern set. Cross-validation helps to generate good generalization of the network, when N, the size of the training set, is too large. Cross-validation is effective for finding a large network with a good generalization performance.

The popular K-fold cross-validation [106] employs a nonoverlapping test set selection scheme. The data universe $\mathcal{D}$ is divided into K nonoverlapping data subsets of the same size. Each data subset is then used as a test set, with the remaining $K - 1$ folds acting as a training set, and an error value is calculated by testing the classifier in the remaining fold. Finally, the K-fold cross-validation estimation of the error is the average value of the errors committed in each fold. Thus, the K-fold cross-validation error estimator depends on two factors: the training set and the partitioning into folds. Estimating the variance of K-fold cross-validation can be done from independent realizations or from dependent realizations whose correlation is known. K-fold cross-validation produces dependent test errors. Consequently, there is no universal unbiased estimator of the variance of K-fold cross-validation that is valid under all distributions [15].

The variance estimators of the K-fold cross-validation estimator of the generalization error presented in [72] are almost unbiased in the cases of smooth loss functions and the absolute error loss. The problem of variance estimation is approached as a problem in approximating the moments of a statistic. The estimators depend on the distribution of the errors and on the knowledge of the learning algorithm. Overall, a test set that use 25% of the available data seems to be a reasonable compromise in selecting among the various forms of K-fold cross-validation [72].

In least squares regression, K-fold cross-validation is suboptimal for model selection if K stays bounded, because K-fold cross-validation is biased. A non-asymptotic oracle inequality for K-fold cross-validation and its bias-corrected version (K-fold penalization) are proved in [5]. K-fold penalization is asymptotically optimal in the nonparametric case. The variance of K-fold cross-validation performance increases much from $K = 2$ to $K = 5$ or 10, and then is almost constant. This can explain the common advice to take $K = 5$ [21]. An oracle inequality and exact formulas for the variance are also proved for Monte Carlo cross-validation, also known as repeated cross-validation, where the parameter K is replaced by the number B of random splits of the data [5].

The leave-many-out variants of cross-validation perform better than the leave-one-out versions [83]. Empirically, both types of cross-validation can exhibit high variance in small samples, but this may be alleviated by increasing the level of resampling. Used appropriately, leave-many-out cross-validation is, in general, more robust than leave-one-out cross-validation [83]. Cross-validation is shown to be inconsistent for model selection [98].

The moment approximation estimator [72] performs better in terms of both the variance and the bias than the Nadeau–Bengio estimator [77]. The latter is computationally simpler than the former for general loss functions, as it does not require the computation of the derivatives of the loss function; but it is not an appropriate one to be used for nonrandom test set selection.

Cross-validation and bootstrapping are both resampling methods. Resampling varies the training set numerous times based on one set of available data. One fundamental difference between cross-validation and bootstrapping is that bootstrapping resamples the available data at random with replacement, whereas cross-validation resamples the available data at random without replacement. Cross-validation methods never evaluate the trained networks over examples that appear in the training set, whereas bootstrapping methods typically do that. Cross-validation methods split the data such that a sample does not appear in more than one validation set. Cross-validation is commonly used for estimating generalization error, whereas bootstrapping finds widespread use in estimating error bars and confidence intervals. Cross-validation is commonly believed to be more accurate (less biased) than bootstrapping, but to have a higher variance than bootstrapping does in small samples [83].

2.3.2 Complexity Criteria

An efficient approach for improving the generalization performance is to construct a small network using a parsimonious principle. Statistical model selection with information criteria such as Akaike's final prediction error criterion [1], Akaike information criterion (AIC) [3], Schwartz's Bayesian information criterion (BIC) [97], and Rissanen's minimum description length (MDL) principle [88] are popular and have been widely used for model selection of neural networks. Although the motivations and approaches for these criteria may be very different from one another, most of them can be expressed as a function with two components, one for measuring the training error and the other for penalizing the complexity. These criteria penalize large-size models.

A possible approach to model order selection consists of minimizing the Kullback–Leibler discrepancy between the true pdf of the data and the pdf (or likelihood) of the model, or equivalently maximizing the relative Kullback–Leibler information, which is sometimes called the relative Kullback–Leibler information. Maximizing the asymptotic approximation of the relative Kullback–Leibler information with n, the number of variables, is equivalent to minimizing the AIC function of n. AIC is derived by maximizing an asymptotically unbiased estimate of the relative Kullback–Leibler information I. The BIC rule can be derived from an asymptotically unbiased estimate of the relative Kullback–Leibler information [105]. BIC is the penalized ML method.

AIC minimizes certain loss for prediction purpose, while BIC selects the best model for inference purpose. The AIC and BIC criteria can be, respectively, represented by

$$E_{\text{AIC}} = -\frac{1}{N} \ln \left(L_N \left(\widehat{\mathbf{W}}_N \right) \right) + \frac{N_P}{N}, \tag{2.5}$$

$$E_{\text{BIC}} = -\frac{1}{N} \ln \left(L_N \left(\widehat{\mathbf{W}}_N \right) \right) + \frac{N_P}{2N} \ln N, \tag{2.6}$$

where $L_N(\widehat{\mathbf{W}}_N)$ is the likelihood estimated for a training set of size N and model parameters $\widehat{\mathbf{W}}_N$, and N_P is the number of parameters in the model. More specifically, the two criteria can be expressed by [105]

$$AIC(N_P) = R_{\text{emp}}(N_P) + \frac{2N_P}{N} \hat{\sigma}^2, \tag{2.7}$$

$$BIC(N_P) = R_{\text{emp}}(N_P) + \frac{N_P}{N} \hat{\sigma}^2 \ln N, \tag{2.8}$$

where $\hat{\sigma}^2$ denotes an estimate of noise variance, and the empirical risk is given by

$$R_{\text{emp}}(N_P) = \frac{1}{N} \sum_{i=1}^{N} (y_i - f(\boldsymbol{x}_i, N_P))^2, \tag{2.9}$$

and the noise variance can be estimated, for a linear estimator with N_P parameters, as

$$\hat{\sigma}^2 = \frac{N}{N - N_P} \frac{1}{N} \sum_{i=1}^{N} (y_i - \hat{y}_i)^2. \tag{2.10}$$

This leads to the following form of AIC known as final prediction error [2]:

$$\text{FPE}(N_P) = \frac{1 + \frac{N_P}{N}}{1 - \frac{N_P}{N}} R_{\text{emp}}(N_P). \tag{2.11}$$

MDL principle gives a formal justification to Occam's razor. MDL principle stems from coding theory to find as short a description as possible of a database with as few symbols as possible [88, 89]. The description length of the model characterizes the information needed for simultaneously encoding a description of the model and a description of the prediction errors of the model. The best model is the one with the minimum description length. The total description length E_{MDL} has three terms: code cost for coding the input vectors, model cost for defining the reconstruction method, and reconstruction error due to reconstruction of the input vector from its code. The description length is described by the number of bits. Existing unsupervised learning algorithms such as the competitive learning and PCA can be explained using the MDL principle [54]. Good generalization can be achieved by encoding the weights with short bit-lengths by penalizing the amount of information they contain using the MDL principle [54]. The MDL measure can be regarded as an approximation of the Bayesian measure, and thus has a Bayesian interpretation. BIC rule has also been obtained by an approach based on coding arguments and the MDL principle.

Generalization error Err is characterized by the sum of the training (approximation) error err and the degree of optimism OP inherent in a particular estimate [44], that is, $Err = err + OP$. Complexity criteria such as BIC can be used for estimating OP.

2.4 Bias and Variance

The generalization error can be represented by the sum of the *bias* squared plus the *variance* [43]. Most existing supervised learning algorithms suffer from the bias–variance dilemma [43]. That is, the requirements for small bias and small variance are conflicting and a trade-off must be made.

Let $f(\boldsymbol{x}; \hat{\boldsymbol{w}})$ be the best model in model space. Thus, $\hat{\boldsymbol{w}}$ does not depend on the training data. The bias and variance can be defined by [17]

$$\text{bias} = E_{\mathcal{S}}(f(\boldsymbol{x})) - f(\boldsymbol{x}; \hat{\boldsymbol{w}}), \tag{2.12}$$

$$\text{var} = E_{\mathcal{S}}\left((f(\boldsymbol{x}) - E_{\mathcal{S}}(f(\boldsymbol{x})))^2\right), \tag{2.13}$$

where $f(\boldsymbol{x})$ is the function to be estimated, and $E_{\mathcal{S}}$ denotes the expectation operation over all possible training sets. Bias is caused by an inappropriate choice of the size of a class of models when the number of training samples is assumed infinite, while the variance is the error caused by the finite number of training samples.

Example 2.3 An illustration of the concepts of bias and variance in the two-dimensional space is shown in Fig. 2.5. $f(x; \hat{\boldsymbol{w}})$ is the underlying function; $f_1(x)$ and $f_2(x)$ are used to approximate $f(x; \hat{\boldsymbol{w}})$: $f_1(x)$ is an exact interpolation of the data points, while $f_2(x)$ is a fixed function independent of the data points. For $f_1(x)$, the bias is zero at the data points and is small in the neighborhood of the data points, while the variance is the variance of the noise on the data, which could be significant; for $f_2(x)$, the bias is high while the variance is zero.

The generalized error can be decomposed into a sum of the bias and variance

$$\begin{aligned}
& E_{\mathcal{S}}\left(\left[f(\boldsymbol{x}) - f(\boldsymbol{x}, \hat{\boldsymbol{w}})\right]^2\right) \\
&= E_{\mathcal{S}}\left(\left\{[f(\boldsymbol{x}) - E_{\mathcal{S}}(f(\boldsymbol{x}))] + \left[E_{\mathcal{S}}(f(\boldsymbol{x})) - f(\boldsymbol{x}, \hat{\boldsymbol{w}})\right]\right\}^2\right) \\
&= E_{\mathcal{S}}\left([f(\boldsymbol{x}) - E_{\mathcal{S}}(f(\boldsymbol{x}))]^2\right) + E_{\mathcal{S}}\left(\left[E_{\mathcal{S}}(f(\boldsymbol{x})) - f(\boldsymbol{x}, \hat{\boldsymbol{w}})\right]^2\right) \\
&\quad + 2E_{\mathcal{S}}\left([f(\boldsymbol{x}) - E_{\mathcal{S}}(f(\boldsymbol{x}))]\left[E_{\mathcal{S}}(f(\boldsymbol{x})) - f(\boldsymbol{x}, \hat{\boldsymbol{w}})\right]\right) \\
&= (\text{Bias})^2 + \text{Var}.
\end{aligned} \tag{2.14}$$

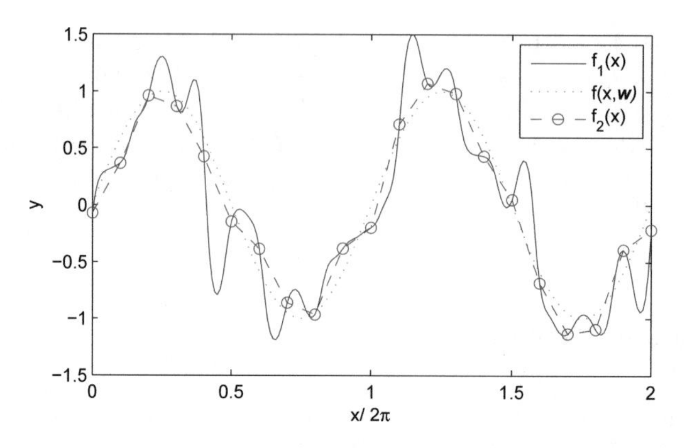

Fig. 2.5 Bias and variance. Circles denote examples from a training set

A network with a small number of adjustable parameters gives poor generalization on new data, since the model has very little flexibility and thus yields underfitting with a high bias and low variance. In contrast, a network with too many adjustable parameters also gives a poor generalization performance, since it is too flexible and fits too much of the noise on the training data, thus yielding overfitting with a low bias but high variance. The best generalization performance is achieved by balancing bias and variance, which optimizes the complexity of the model through either finding a model with an optimal size or by adding a regularization term in an objective function. For nonparametric methods, most complexity criteria-based techniques operate on the variance term in order to get a good compromise between the contributions made by the bias and variance to the error. When the number of hidden cells is increased, the bias term is likely to be reduced, whereas the variance would increase.

For three-layer feedforward networks with N_P hidden sigmoidal units, the bias and variance are upper bounded explicitly [8] by $O\left(\frac{1}{N_P}\right)$ and $O\left(\frac{N_P J_1 \ln N}{N}\right)$, respectively, where N is the size of the training set and J_1 is the dimensionality of the feature vectors. Thus when N_P is large, the bias is small. However, when N is finite, a network with an excessively large space complexity will overfit the training set. The average performance can decrease as N_P gets larger. As a result, a trade-off needs to be made between the bias and variance.

While unbiasedness is a beneficial quality of a model selection criterion, a low variance is at least as important, as a nonnegligible variance introduces the potential for overfitting in model selection as well as in training the model [25]. The effects of this form of overfitting are often comparable to differences in performance between learning algorithms [25]. This could be ameliorated by regularization of the model selection criterion [24].

2.5 Criterion Functions

MSE is by far the most popular measure of error. This error measure ensures that a large error receives much greater attention than a small error. The MSE criterion is optimal and results in an ML estimation of the weights if the distributions of the feature vectors are Gaussian [93]. This is desired for most applications. In some situations, other error measures such as the mean absolute error, maximum absolute error, and median squared error, may be preferred.

The logarithmic error function, which takes the form of the instantaneous relative entropy or Kullback–Leibler divergence criterion, has some merits over the MSE function [11]

$$E_p(\mathbf{W}) = \frac{1}{2}\sum_{i=1}^{J_M}\left[\left(1 + y_{p,i}\right)\ln\left(\frac{1 + y_{p,i}}{1 + \hat{y}_{p,i}}\right) + \left(1 - y_{p,i}\right)\ln\left(\frac{1 - y_{p,i}}{1 - \hat{y}_{p,i}}\right)\right] \quad (2.15)$$

for the tanh activation function, where $y_{p,i} \in (-1, 1)$. For the logistic activation function, the criterion can be written as [73]

$$E_p(\mathbf{W}) = \frac{1}{2}\sum_{i=1}^{J_M}\left[y_{p,i}\ln\left(\frac{y_{p,i}}{\hat{y}_{p,i}}\right) + \left(1 - y_{p,i}\right)\ln\left(\frac{1 - y_{p,i}}{1 - \hat{y}_{p,i}}\right)\right], \quad (2.16)$$

where $y_{p,i} \in (0, 1)$. In the latter case, $y_{p,i}$, $\hat{y}_{p,i}$, $1 - y_{p,i}$, and $1 - \hat{y}_{p,i}$ are regarded as probabilities. These criteria take zero only when $y_{p,i} = \hat{y}_{p,i}$, $i = 1, \ldots, J_M$, and are strictly positive otherwise. Another criterion function obtained by simplifying (2.16) via omitting the constant terms related to the patterns is [103]

$$E_p(\mathbf{W}) = -\frac{1}{2}\sum_{i=1}^{J_M}\left[y_{p,i}\ln\hat{y}_{p,i} + \left(1 - y_{p,i}\right)\ln\left(1 - \hat{y}_{p,i}\right)\right]. \quad (2.17)$$

The problem of loading a set of training examples onto a neural network is nondeterministic polynomial (NP)-complete [19, 101]. As a consequence, existing algorithms cannot be guaranteed to learn the optimal solution in polynomial time. In the case of one neuron, the logistic function paired with the MSE function can lead to $\left(\frac{N}{J_1}\right)^{J_1}$ local minima, for N training patterns and an input dimension of J_1 [6], while with the entropic error function, the error function is convex and thus has only one minimum [11, 103]. The use of the entropic error function considerably reduces the total number of local minima.

The BP algorithm derived from the entropy criteria can partially solve the flat-spot problem. These criteria do not add computation load to calculate the error function. They, however, remarkably reduce the training time, and alleviate the problem of getting stuck at local minima by reducing the density of local minima [73]. Besides, the entropy-based BP is well suited to probabilistic training data, since it can be viewed as learning the correct probabilities of a set of hypotheses represented by the outputs of the neurons.

Traditionally, classification problems are learned through error backpropagation by providing a vector of hard 0/1 target values to represent the class label of a particular pattern. Minimizing an error function with hard target values tends to a saturation of weights, leading to overfitting. The magnitude of the weights plays a more important role in generalization than the number of hidden nodes [10]. Overfitting might be reduced by keeping the weights smaller.

The cross-entropy cost function can also be derived from the ML principle

$$E_{CE} = -\sum_{i=1}^{N}\sum_{k=1}^{C} t_{k,i} \ln(y_{k,i}) \tag{2.18}$$

for training set $\{\boldsymbol{x}_i, \boldsymbol{t}_i\}$, C classes and N samples, and $t_{k,i} \in \{0, 1\}$.

Marked reductions on convergence rates and density of local minima are observed due to the characteristic steepness of the cross-entropy function [73, 103]. As a function of the absolute errors, MSE tends to produce large relative errors for small output values. As a function of the relative errors, cross-entropy is expected to estimate more accurately small probabilities [45, 51, 103]. When a neural network is trained using MSE or cross-entropy minimization, its outputs approximate the posterior probabilities of class membership. Thus, in the presence of large datasets, it tends to produce optimal solutions in the Bayes sense. However, minimization of the error function does not necessarily imply misclassification minimization in practice. Suboptimal solutions may occur due to flat regions in weight space. Thus, minimization of these error functions does not imply misclassification minimization.

MSE function can be obtained by the ML principle assuming the independence and Gaussianity of the target data. However, the Gaussianity assumption of the target data in classification is not valid, due to its discrete nature of class labels. Thus, the MSE function is not the most appropriate one for data classification problems. Nevertheless, when using a 1-out-of-C coding scheme for the targets, with large N and a number of samples in each class, the MSE trained outputs of the network approximate the posterior probabilities of the class membership [45]. The cross-entropy error function and other entropy-based functions are suitable for training neural network classifiers, because when interpreting the outputs as probabilities this is the optimal solution.

Classification-based (CB) error functions [87] heuristically seek to directly minimize classification error by backpropagating network error only on misclassified patterns. In so doing, they perform relatively minimal updates to network parameters in order to discourage premature weight saturation and overfitting.

MSE criterion can be generalized into the Minkowski-r metric [46]

$$E_p = \frac{1}{r}\sum_{i=1}^{J_M} \left|\hat{y}_{p,i} - y_{p,i}\right|^r . \tag{2.19}$$

When $r = 1$, the metric is called the *city block* metric. The Minkowski-r metric corresponds to the MSE criterion for $r = 2$. A small value of r ($r < 2$) reduces the influence of large deviations, thus it can be used in the case of outliers. In contrast, a large r weights large deviations, and generates a better generation surface when the noise is absent in the data or when the data clusters in the training set are compact.

A generalized error function embodying complementary features of other functions, which can emulate the behavior of other error functions by adjustment of a single real-valued parameter, is proposed in [100]. Many other criterion functions can be used for deriving learning algorithms, including correntropy [68] for robust adaptation in case of nonGaussian noise, and those based on robust statistics [59] or regularization [84].

2.6 Robust Learning

When the training data is corrupted by large noise, such as outliers, conventional learning algorithms may not yield acceptable performance since a small number of outliers have a large impact on the MSE. An outlier is an observation that deviates significantly from the other observations; this may be due to erroneous measurements or noisy data from the tail of the noise distribution functions. When noise becomes large or outliers exist, the networks may try to fit those improper data and thus, the learned systems are corrupted. The Student-t distribution has heavier tails than the Gaussian distribution and is, therefore, less sensitive to any departure of the empirical distribution from Gaussianity. For nonlinear regression, the techniques of robust statistics [59] can be applied to deal with the outliers. The M-estimator is derived from the ML estimator to deal with situations, where the exact probability model is unknown. The M-estimator replaces the conventional squared error term by the so-called *loss functions*. The loss function is used to degrade the effects of those outliers in learning. A difficulty is the selection of the scale estimator of the loss function in the M-estimator.

The cost function of a robust learning algorithm is defined by

$$E_r = \sum_{i=1}^{N} \sigma\left(\epsilon_i; \beta\right), \tag{2.20}$$

where $\sigma(\cdot)$ is the loss function, which is a symmetric function with a unique minimum at zero, $\beta > 0$ is the scale estimator, known as the *cutoff parameter*, ϵ_i is the estimated error for the ith training pattern, and N is the size of the training set. The loss function can be typically selected as one of the following functions:

- The logistic function [59]

$$\sigma(\epsilon_i; \beta) = \frac{\beta}{2} \ln\left(1 + \frac{\epsilon_i^2}{\beta}\right). \tag{2.21}$$

- Huber's function [59]

$$\sigma(\epsilon_i;\beta)=\begin{cases}\frac{1}{2}\epsilon_i^2, & |\epsilon_i|\le\beta\\ \beta\,|\epsilon_i|-\frac{1}{2}\beta^2, & |\epsilon_i|>\beta\end{cases}. \tag{2.22}$$

- Talwar's function [30]

$$\sigma(\epsilon_i;\beta)=\begin{cases}\frac{1}{2}\epsilon_i^2, & |\epsilon_i|\le\beta\\ \frac{1}{2}\beta^2, & |\epsilon_i|>\beta\end{cases}. \tag{2.23}$$

- Hampel's tanh estimator [28]

$$\sigma(\epsilon_i;\beta_1,\beta_2)=\begin{cases}\frac{1}{2}\epsilon_i^2, & |\epsilon_i|\le\beta_1\\ \frac{1}{2}\beta_1^2-\frac{2c_1}{c_2}\ln\frac{1+e^{c_2(\beta_2-|\epsilon_i|)}}{1+e^{c_2(\beta_2-\beta_1)}}-c_1\left(|\epsilon_i|-\beta_1\right), & \beta_1<|\epsilon_i|\le\beta_2\\ \frac{1}{2}\beta_1^2-\frac{2c_1}{c_2}\ln\frac{2}{1+e^{c_2(\beta_2-\beta_1)}}-c_1\left(\beta_2-\beta_1\right), & |\epsilon_i|>\beta_2\end{cases}. \tag{2.24}$$

In the tanh estimator, β_1 and β_2 are the two cutoff points, and constants c_1 and c_2 adjust the shape of the influence function (to be defined in (2.26)). When $c_1=\frac{\beta_1}{\tan(c_2(\beta_2-\beta_1))}$, the influence function is continuous. In the interval of the two cutoff points, the influence function can be represented by a hyperbolic tangent relation.

Using the gradient-descent method, the weights are updated by

$$\Delta w_{jk}=-\eta\frac{\partial E_r}{\partial w_{jk}}=-\eta\sum_{i=1}^{N}\varphi(\epsilon_i;\beta)\frac{\partial\epsilon_i}{\partial w_{jk}}, \tag{2.25}$$

where η is a learning rate or step size, and $\varphi(\cdot)$, called the *influence function*, is given by

$$\varphi(\epsilon_i;\beta)=\frac{\partial\sigma(\epsilon_i;\beta)}{\partial\epsilon_i}. \tag{2.26}$$

The conventional MSE function corresponds to $\sigma(\epsilon_i)=\frac{1}{2}\epsilon_i^2$ and $\varphi(\epsilon_i;\beta)=\epsilon_i$. To suppresses the effect of large errors, loss functions used for robust learning are defined such that $\varphi(\epsilon_i;\beta)$ is sublinear.

Example 2.4 The loss functions given above and their respective influence functions are illustrated in Fig. 2.6.

τ-estimator [110] can be viewed as an M-estimator with an adaptive bounded influence function $\varphi(\cdot)$ given by the weighted average of two functions $\varphi_1(\cdot)$ and $\varphi_2(\cdot)$, with $\varphi_1(\cdot)$ corresponding to a very robust estimate and $\varphi_2(\cdot)$ to a highly efficient estimate. τ-estimator simultaneously has a high breakdown point and a high efficiency under Gaussian errors.

When the initial weights are not properly selected, the loss functions may not be able to correctly discriminate against the outliers. The selection of β is also a

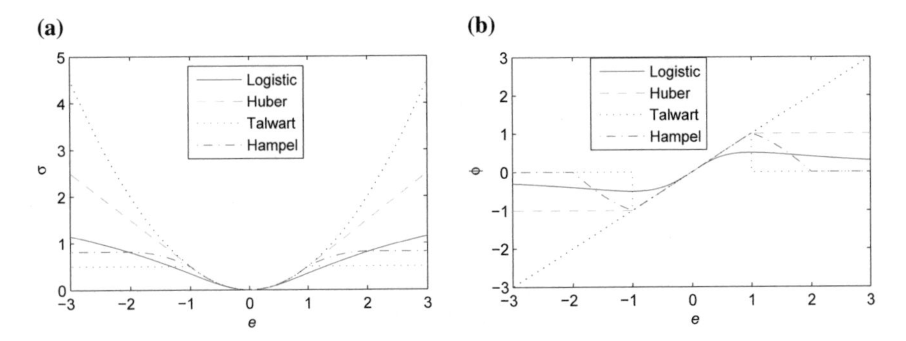

Fig. 2.6 Loss functions and their respective influence functions. For logistic, Huber's, and Talwart's functions, $\beta = 1$. For Hampel's tanh estimator, $\beta_1 = 1$, $\beta_2 = 2$, $c_2 = 1$, and $c_1 = 1.313$. **a** Loss functions σ. **b** Influence functions φ

problem, and one approach is to select β as the median of the absolute deviation (MAD)

$$\beta = c \times \text{median}\left(|\epsilon_i - \text{median}\left(\epsilon_i\right)|\right) \tag{2.27}$$

with c chosen as 1.4826 [59]. Some other methods for selecting β are based on using the median of all errors [59], or counting out a fixed percentage of points as outliers [28].

The correntropy-induced loss (C-loss) function [102] is a bounded, smooth, differentiable, and nonconvex loss that is robust to outliers. C-loss is Bayes consistent [102]. It behaves like L_2-norm for small errors, L_1-norm for medium errors, and L_0-norm for large errors.

2.7 Neural Networks as Universal Machines

The power of neural networks stems from their representation capability. On the one hand, feedforward networks are proved to offer the capability of universal function approximation. On the other hand, recurrent networks using the sigmoidal activation function are Turing equivalent [99] and simulates a universal Turing machine; Thus, recurrent networks can compute whatever function any digital computer can compute.

2.7.1 Boolean Function Approximation

Feedforward networks with binary neurons can be used to represent logic or Boolean functions. In binary neural networks, the input and output values for each neuron are

Boolean variables, denoted by binary (0 or 1) or bipolar (-1 or $+1$) representation. For J_1 independent Boolean variables, there are 2^{J_1} combinations of these variables. This leads to a total of $2^{2^{J_1}}$ different Boolean functions of J_1 variables. An LTG can discriminate between two classes.

The function counting theorem [31, 47] gives the number of linearly separable dichotomies of m points in general position in $\mathcal{R}^n$. It essentially estimates the separating capability of an LTG.

Theorem 2.1 (Function Counting Theorem) *The number of linearly separable dichotomies of m points in general position in R^n is*

$$C(m, n) = \begin{cases} 2\sum_{i=0}^{n} \binom{m-1}{i}, & m > n+1 \\ 2^m, & m \leq n+1 \end{cases}. \tag{2.28}$$

A set of m points in R^n is said to be in *general position* if every subset of m or fewer points is linearly independent.

The total number of possible dichotomies of m points is 2^m. Under the assumption of 2^m equiprobable dichotomies, the probability of a single LTG with n inputs to separate m points in general position is given by

$$P(m, n) = \frac{C(m, n)}{2^m} = \begin{cases} \frac{2}{2^m}\sum_{i=0}^{n} \binom{m-1}{i}, & m > n+1 \\ 1, & m \leq n+1 \end{cases}. \tag{2.29}$$

The fraction $P(m, n)$ is the probability of linear dichotomy. Thus, if $\frac{m}{n+1} \leq 1$, $P = 1$; if $1 < \frac{m}{n+1} < 2$ and $n \to \infty$, $P \to 1$. At $\frac{m}{n+1} = 2$, $P = \frac{1}{2}$. Usually, $m = 2(n+1)$ is used to characterize the statistical capability of a single LTG. Equation (2.29) is plotted in Fig. 2.7.

A three-layer (J_1-2^{J_1}-1) feedforward LTG network can represent any Boolean function with J_1 arguments [33, 75]. To realize an arbitrary function $f : R^{J_1} \to \{0, 1\}$

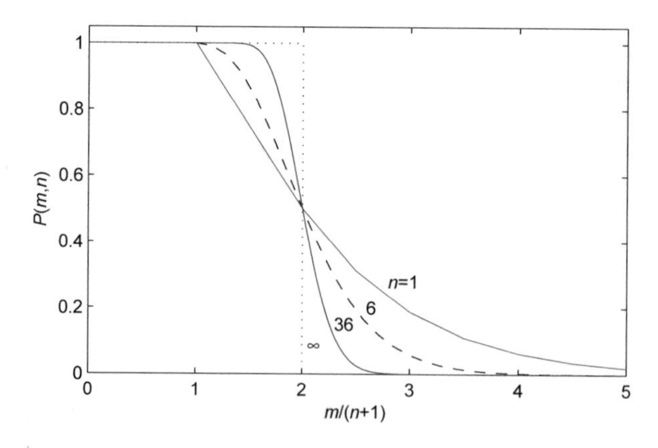

Fig. 2.7 The probability of linear dichotomy of m points in n dimensions

defined on N arbitrary points in R^{J_1}, the lower bound for the number of hidden nodes is derived as $O\left(\frac{N}{J_1 \log_2 \frac{N}{J_1}}\right)$ for $N \geq 3J_1$ and $J_1 \to \infty$ [11, 47]; for N points in general position, the lower bound is $\frac{N}{2J_1}$ when $J_1 \to \infty$ [47]. Networks with two or more hidden layers are found to be potentially more size efficient than networks with a single hidden layer [47].

Binary Radial Basis Function

For three-layer feedforward networks, if the activation function of the hidden neurons is selected as the binary RBF or generalized binary RBF and the output neurons are selected as LTGs, one obtains binary or generalized binary RBF networks. Binary or generalized binary RBF network can be used for the mapping of Boolean functions.

The parameters of the generalized binary RBF neuron are the center $\boldsymbol{c} \in R^n$ and the radius $r \geq 0$. The activation function $\phi : R^n \to \{0, 1\}$ is defined by

$$\phi(\boldsymbol{x}) = \begin{cases} 1, & \|\boldsymbol{x} - \boldsymbol{c}\|_{\mathbf{A}} \leq r \\ 0, & \text{otherwise} \end{cases}, \tag{2.30}$$

where $\mathbf{A}$ is any real, symmetric and positive-definite matrix, and $\|\cdot\|_{\mathbf{A}}$ is the weighted Euclidean norm. When $\mathbf{A}$ is the identity matrix $\mathbf{I}$, the neuron becomes a binary RBF neuron.

Every Boolean function computed by the LTG can also be computed by any generalized binary RBF neuron, and generalized binary RBF neurons are more powerful than LTGs [41]. As an immediate consequence, in any neural network, any LTG that receives only binary inputs can be replaced by a generalized binary RBF neuron having any norm, without any loss of the computational power of the neural network.

2.7.2 Linear Separability and Nonlinear Separability

Definition 2.1 (*Linearly Separable*) Assume that there is a set $\mathcal{X}$ of N patterns $\boldsymbol{x}_i$ of J_1 dimensions, each belonging to one of two classes $\mathcal{C}_1$ and $\mathcal{C}_2$. If there is a hyperplane that separates all the samples of $\mathcal{C}_1$ from $\mathcal{C}_2$, then such a classification problem is said to be *linearly separable*.

A single LTG can realize linearly separable dichotomy function, characterized by a linear separating surface (hyperplane)

$$\boldsymbol{w}^T\boldsymbol{x} + w_0 = 0, \tag{2.31}$$

where $\boldsymbol{w}$ is a J_1-dimensional vector and w_0 is a bias toward the origin. For a pattern, if $\boldsymbol{w}^T\boldsymbol{x} + w_0 > 0$, it belongs to $\mathcal{C}_1$; if $\boldsymbol{w}^T\boldsymbol{x} + w_0 < 0$, it belongs to $\mathcal{C}_2$.

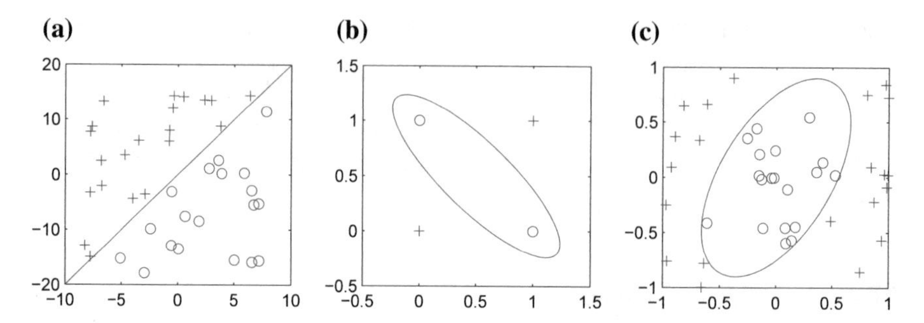

Fig. 2.8 Linearly separable, linearly inseparable, and nonlinearly separable classification in two-dimensional space. Dots and circles denote patterns of different classes

Definition 2.2 (*φ-Separable*) A dichotomy $\{\mathcal{C}_1, \mathcal{C}_2\}$ of set $\mathcal{X}$ is said to be *φ-separable* if there exists a mapping $\varphi : R^{J_1} \rightarrow R^{J_2}$ that satisfies a separating surface [31]

$$\boldsymbol{w}^T \varphi(\boldsymbol{x}) = 0 \tag{2.32}$$

such that $\boldsymbol{w}^T \varphi(\boldsymbol{x}) > 0$ if $\boldsymbol{x} \in \mathcal{C}_1$ and $\boldsymbol{w}^T \varphi(\boldsymbol{x}) < 0$ if $\boldsymbol{x} \in \mathcal{C}_2$. Here $\boldsymbol{w}$ is a J_2-dimensional vector.

A linearly inseparable dichotomy can become nonlinearly separable. As shown in Fig. 2.8, the two linearly inseparable dichotomies become φ-separable.

The nonlinearly separable problem can be realized by using a polynomial threshold gate, which changes the linear term in the LTG into high-order polynomials. The function counting theorem is applicable to polynomial threshold gates; and it still holds true if the set of m points is in general position in φ-space, that is, the set of m points is in φ-general position.

Example 2.5 Some examples of linearly separable classes and linearly inseparable classes in two-dimensional space are illustrated in Fig. 2.8. (a) Two linearly separable classes with $x_1 - x_2 = 0$ as the delimiter. (b) and (c) are linearly inseparable classes, where (b) is the exclusive-or problem. Note that the linearly inseparable classification problems in cases (b) and (c) become nonlinearly separable, when the separating surfaces are ellipses.

Higher order neurons (or Σ-Π units) are simple but powerful extensions of linear neuron models. They introduce the concept of nonlinearity by incorporating monomials, that is, products of input variables, as a hidden layer.

2.7.3 Continuous Function Approximation

A three-layer feedforward network with a sufficient number of hidden units can approximate any continuous function to any degree of accuracy. This is guaranteed by Kolmogorov's theorem [48, 64].

Theorem 2.2 (Kolmogorov) *Any continuous real-valued function* $f(x_1, \ldots, x_n)$ *defined on* $[0, 1]^n$, $n \geq 2$, *can be represented in the form*

$$f(x_1, \ldots, x_n) = \sum_{j=1}^{2n+1} h_j \left(\sum_{i=1}^{n} \psi_{ij}(x_i) \right), \tag{2.33}$$

where h_j *and* ψ_{ij} *are continuous functions of one variable, and* ψ_{ij} *are monotonically increasing functions independent of* f.

Kolmogorov's theorem is the famous solution to Hilbert's 13th problem. According to Kolmogorov's theorem, a continuous multivariate function on a compact set can be expressed using superpositions and compositions of a finite number of single-variable functions. Based on Kolmogorov's theorem, Hecht-Nielsen provided a theorem that is directly related to neural networks [48].

Theorem 2.3 (Hecht-Nielsen) *Any continuous real-valued mapping* $f : [0, 1]^n \rightarrow R^m$ *can be approximated to any degree of accuracy by a feedforward network with* n *input nodes,* $2n + 1$ *hidden units, and* m *output units.*

The Weierstrass theorem asserts that any continuous real-valued multivariate function can be approximated to any accuracy using a polynomial. The Stone–Weierstrass theorem [91] is a generalization of the Weierstrass theorem, and is usually used for verifying a model's approximation capability to dynamic systems.

Theorem 2.4 (Stone–Weierstrass) *Let* $\mathcal{F}$ *be a set of real continuous functions on a compact domain* $\mathcal{U}$ *of* n *dimensions. Let* $\mathcal{F}$ *satisfy the following criteria*

1. Algebraic closure: $\mathcal{F}$ *is closed under addition, multiplication, and scalar multiplication. That is, for any two* $f_1, f_2 \in \mathcal{F}$, *we have* $f_1 f_2 \in \mathcal{F}$ *and* $a_1 f_1 + a_2 f_2 \in \mathcal{F}$, *where* a_1 *and* a_2 *are any real numbers.*
2. Separability on $\mathcal{U}$: *for any two different points* $\boldsymbol{x}_1, \boldsymbol{x}_2 \in \mathcal{U}$, $\boldsymbol{x}_1 \neq \boldsymbol{x}_2$, *there exists* $f \in \mathcal{F}$ *such that* $f(\boldsymbol{x}_1) \neq f(\boldsymbol{x}_2)$;
3. Not constantly zero on $\mathcal{U}$: *for each* $\boldsymbol{x} \in \mathcal{U}$, *there exists* $f \in \mathcal{F}$ *such that* $f(\boldsymbol{x}) \neq 0$.

Then $\mathcal{F}$ *is a dense subset of* $C(\mathcal{U})$, *the set of all continuous real-valued functions on* $\mathcal{U}$. *In other words, for any* $\varepsilon > 0$ *and any function* $g \in C(\mathcal{U})$, *there exists* $f \in \mathcal{F}$ *such that* $|g(\boldsymbol{x}) - f(\boldsymbol{x})| < \varepsilon$ *for any* $\boldsymbol{x} \in \mathcal{U}$.

To date, numerous attempts have been made in searching for suitable forms of activation functions and proving the corresponding network's universal approximation capabilities. Universal approximation to a given nonlinear functional under certain conditions can be realized by using the classical Volterra series or the Wiener series.

2.7.4 Winner-Takes-All

The winner-takes-all (WTA) competition is widely observed in both inanimate and biological media and society. Theoretical analysis [70] shows that WTA is a powerful computational module in comparison with threshold gates and sigmoidal gates (i.e., McCulloch–Pitts neurons). An optimal quadratic lower bound is given in [70] for computing WTA in any feedforward circuit consisting of threshold gates. Arbitrary continuous functions can be approximated by circuits employing a single soft WTA gate as their only nonlinear operation [70].

Theorem 2.5 (Maass, 1 [70]) *Assume that WTA with $n \geq 3$ inputs is computed by some arbitrary feedforward circuit C consisting of threshold gates with arbitrary weights. Then C consists of at least $\binom{n}{2} + n$ threshold gates.*

Theorem 2.6 (Maass, 2 [70]) *Any two-layer feedforward circuit C (with m analog or binary input variables and one binary output variable) consisting of threshold gates can be simulated by a circuit consisting of a single k-winner-takes-all gate (k-WTA) applied to n weighted sums of the input variables with positive weights, except for some set $S \subseteq R^m$ of inputs that has measure 0.*

Any Boolean function $f : \{0, 1\}^m \rightarrow \{0, 1\}$ can be computed by a single k-WTA gate applied to weighted sums of the input bits. If C has polynomial size and integer weights and the size is bounded by a polynomial in m, then n can be bounded by a polynomial in m, and all weights in the simulating circuit are natural numbers and the circuit size is bounded by a polynomial in m.

For real-valued input $(x_1, \ldots, x_n)$, a soft-WTA has its output $(r_1, \ldots, r_n)$, analog numbers r_i reflecting the relative position of x_i within the ordering of x_i. Soft-WTA is plausible as computational function of cortical circuits with lateral inhibition. Single gates from a fairly large class of soft-WTA gates can serve as the only nonlinearity in universal approximators for arbitrary continuous functions.

Theorem 2.7 (Maass, 3 [70]) *Assume that $h : D \rightarrow [0, 1]$ is an arbitrary continuous function with a bounded and closed domain $D \subseteq R^m$. Then for any $\epsilon > 0$ and for any function g satisfying above conditions there exist natural numbers k, n, biases $\alpha_0^j \in R$, and coefficients $\alpha_i^j \geq 0$ for $i = 1, \ldots, m$, $j = 1, \ldots, n$, so that the circuit consisting of the soft-WTA gate soft-WTA$_{n,k}^g$ applied to the n sums $\sum_{i=1}^m \alpha_i^j z_i + \alpha_0^j$ for $j = 1, \ldots, n$ computes a function $f : D \rightarrow [0, 1]$ so that $|f(z) - h(z)| < \epsilon$ for all $z \in D$. Thus, circuits consisting of a single soft-WTA gate applied to positive weighted sums of the input variables are universal approximators for continuous functions.*

Problems

2.1 A distance measure, or metric, between two points, must satisfy three conditions:

- Positivity: $d(\boldsymbol{x}, \boldsymbol{y}) \geq 0$ and $d(\boldsymbol{x}, \boldsymbol{y}) = 0$ if and only if $\boldsymbol{x} = \boldsymbol{y}$;
- Symmetry: $d(\boldsymbol{x}, \boldsymbol{y}) = d(\boldsymbol{y}, \boldsymbol{x})$;
- Triangle inequality: $d(\boldsymbol{x}, \boldsymbol{y}) + d(\boldsymbol{y}, \boldsymbol{z}) \geq d(\boldsymbol{x}, \boldsymbol{z})$.

(a) Show that the Euclidean distance, the city block distance, and the maximum value distance are metrics.
(b) Show that the squared Euclidean distance is not a metric.
(c) How about the Hamming distance?

2.2 Is it possible to use a single neuron to approximate the function $f(x) = x^2$?

2.3 Are the following set of points linearly separable?
Class 1: (0,0,0,0), (1,0,0,1), (0,1,0,1); class 2: (1,1,1,1), (1,1,0,0), (1,0,1,0).

2.4 For a K-class problem, the target $\boldsymbol{t}_k$, $k = 1, \ldots, K$, is a vector of all zeros but for a one in the kth position, show that classifying a pattern to the largest element of $\hat{\boldsymbol{y}}$, if $\hat{\boldsymbol{y}}$ is normalized, is equivalent to choosing the closest target, $\min_k \|\boldsymbol{t}_k - \hat{\boldsymbol{y}}\|$.

2.5 Given a set of N samples $(\boldsymbol{x}_k, t_k), k = 1, \ldots, N$, derive the optimal least squares parameter $\hat{\boldsymbol{w}}$ for the total training loss

$$E = \sum_{n=1}^{N} (t_n - \boldsymbol{w}^T \boldsymbol{x}_n)^2.$$

Compare the expression with that derived from the average loss.

2.6 The C-loss is given by [102]

$$L_C(\epsilon) = \frac{1 - e^{-\frac{(1-\epsilon)^2}{2\sigma^2}}}{1 - e^{-\frac{1}{2\sigma^2}}},$$

where σ is window width.
(a) Plot the C-loss function and its influence function for different σ values.
(b) Find the property of this loss function.

2.7 Describe the difference between K-fold cross-validation and random subsampling.

2.8 Plot the loss functions and their influence functions.
(a) Hypersurface loss $L_h(\epsilon) = \sqrt{1 + \epsilon^2} - 1$.
(b) Cauchy loss $L_c(\epsilon, \gamma) = \ln\left(1 + (\epsilon/\gamma)^2\right)$, γ being a scale parameter.

2.9 For an input vector $\boldsymbol{x}$ with p components and a target y, the projection pursuit regression model has the form

$$f(\boldsymbol{x}) = \sum_{m=1}^{M} g_m(\boldsymbol{w}_m^T \boldsymbol{x}),$$

where $\boldsymbol{w}_m$, $m = 1, 2, \ldots, M$, are unit p-vectors of unknown parameters. The functions g_m are estimated along with the direction $\boldsymbol{w}_m$ using some flexible smoothing method. Neural networks are just nonlinear statistical models. Show how neural networks resemble the projection pursuit regression model.

2.10 The XOR operation is a linearly inseparable problem. Show that a quadratic threshold gate

$$y = \begin{cases} 1, \text{ if } g(x) = \sum_{i=1}^{n} w_i x_i + \sum_{i=1}^{n} \sum_{j=1}^{n} w_{ij} x_i x_j \geq T \\ 0, \text{ otherwise} \end{cases}$$

can be used to separate them. Give an example for $g(x)$, and plot the separating surface.

References

1. Akaike, H. (1969). Fitting autoregressive models for prediction. *Annals of the Institute of Statistical Mathematics, 21*, 425–439.
2. Akaike, H. (1970). Statistical prediction information. *Annals of the Institute of Statistical Mathematics, 22*, 203–217.
3. Akaike, H. (1974). A new look at the statistical model identification. *IEEE Transactions on Automatic Control, 19*, 716–723.
4. Amari, S., Murata, N., Muller, K. R., Finke, M., & Yang, H. (1996). Statistical theory of overtraining: Is cross-validation asymptotically effective? In D. S. Touretzky, M. C. Mozer, & M. E. Hasselmo (Eds.), *Advances in neural information processing systems* (Vol. 8, pp. 176–182). Cambridge, MA: MIT Press.
5. Arlot, S., & Lerasle, M. (2016). Choice of V for V-fold cross-validation in least-squares density estimation. *Journal of Machine Learning Research, 17*, 1–50.
6. Auer, P., Herbster, M., & Warmuth, M. K. (1996). Exponentially many local minima for single neurons. In D. S. Touretzky, M. C. Mozer, & M. E. Hasselmo (Eds.), *Advances in neural information processing systems* (Vol. 8, pp. 316–322). Cambridge, MA: MIT Press.
7. Baldi, P., & Sadowski, P. (2013). Understanding dropout. In *Advances in neural information processing systems* (Vol. 27, pp. 2814–2822).
8. Barron, A. R. (1993). Universal approximation bounds for superpositions of a sigmoidal function. *IEEE Transactions on Information Theory, 39*(3), 930–945.
9. Barto, A. G., Sutton, R. S., & Anderson, C. W. (1983). Neuronlike adaptive elements that can solve difficult learning control problems. *IEEE Transactions on Systems, Man, and Cybernetics, 13*, 834–846.
10. Bartlett, P. L. (1998). The sample complexity of pattern classification with neural networks: The size of the weights is more important than the size of the network. *IEEE Transactions on Information Theory, 44*(2), 525–536.
11. Baum, E. B., & Wilczek, F. (1988). Supervised learning of probability distributions by neural networks. In D. Z. Anderson (Ed.), *Neural information processing systems* (pp. 52–61). New York: American Institute Physics.

12. Belkin, M., & Niyogi, P. (2001). Laplacian eigenmaps and spectral techniques for embedding and clustering. In *Advances neural information processing systems* (Vol. 14, pp. 585–591). Cambridge, MA: MIT Press.
13. Belkin, M., & Niyogi, P. (2003). Laplacian eigenmaps for dimensionality reduction and data representation. *Neural Computation, 15*(6), 1373–1396.
14. Belkin, M., Niyogi, P., & Sindhwani, V. (2006). Manifold regularization: A geometric framework for learning from labeled and unlabeled examples. *Journal of Machine Learning Research, 7*, 2399–2434.
15. Bengio, Y., & Grandvalet, Y. (2004). No unbiased estimator of the variance of K-fold cross-validation. *Journal of Machine Learning Research, 5*, 1089–1105.
16. Bernier, J. L., Ortega, J., Ros, E., Rojas, I., & Prieto, A. (2000). A quantitative study of fault tolerance, noise immunity, and generalization ability of MLPs. *Neural Computation, 12*, 2941–2964.
17. Bishop, C. M. (1995). *Neural networks for pattern recognition*. New York: Oxford Press.
18. Bishop, C. M. (1995). Training with noise is equivalent to Tikhonov regularization. *Neural Computation, 7*(1), 108–116.
19. Blum, A. L., & Rivest, R. L. (1992). Training a 3-node neural network is NP-complete. *Neural Networks, 5*(1), 117–127.
20. Bousquet, O., & Elisseeff, A. (2002). Stability and Generalization. *Journal of Machine Learning Research, 2*, 499–526.
21. Breiman, L., & Spector, P. (1992). Submodel selection and evaluation in regression: The X-random case. *International Statistical Review, 60*(3), 291–319.
22. Burges, C. J. C. (2010). *From RankNet to LambdaRank to LambdaMART: An overview*. Technical Report MSR-TR-2010-82, Microsoft Research.
23. Caruana, R. (1997). Multitask learning. *Machine Learning, 28*, 41–75.
24. Cawley, G. C., & Talbot, N. L. C. (2007). Preventing over-fitting during model selection via Bayesian regularisation of the hyper-parameters. *Journal of Machine Learning Research, 8*, 841–861.
25. Cawley, G. C., & Talbot, N. L. C. (2010). On over-fitting in model selection and subsequent selection bias in performance evaluation. *Journal of Machine Learning Research, 11*, 2079–2107.
26. Chapelle, O., & Chang, Y. (2011). Yahoo! learning to rank challenge overview. In *JMLR workshop and conference proceedings: Workshop on Yahoo! learning to rank challenge* (Vol. 14, pp. 1–24).
27. Chawla, N., Bowyer, K., & Kegelmeyer, W. (2002). SMOTE: Synthetic minority over-sampling technique. *Journal of Artificial Intelligence Research, 16*, 321–357.
28. Chen, D. S., & Jain, R. C. (1994). A robust backpropagation learning algorithm for function approximation. *IEEE Transactions on Neural Networks, 5*(3), 467–479.
29. Chiu, C., Mehrotra, K., Mohan, C. K., & Ranka, S. (1994). Modifying training algorithms for improved fault tolerance. In *Proceedings of IEEE International Conference on Neural Networks*, Orlando, FL, USA (Vol. 4, pp. 333–338).
30. Cichocki, A., & Unbehauen, R. (1992). *Neural networks for optimization and signal processing*. New York: Wiley.
31. Cover, T. M. (1965). Geometrical and statistical properties of systems of linear inequalities with applications in pattern recognition. *IEEE Transactions on Electronic Computers, 14*, 326–334.
32. Dasgupta, S., Littman, M., & McAllester, D. (2002). PAC generalization bounds for co-training. In: *Advances in neural information processing systems* (Vol. 14, pp. 375–382).
33. Denker, J. S., Schwartz, D., Wittner, B., Solla, S. A., Howard, R., Jackel, L., et al. (1987). Large automatic learning, rule extraction, and generalization. *Complex Systems, 1*, 877–922.
34. Dietterich, T. G., Lathrop, R. H., & Lozano-Perez, T. (1997). Solving the multiple instance problem with axis-parallel rectangles. *Artificial Intelligence, 89*, 31–71.
35. Domingos, P. (1999). The role of Occam's razor in knowledge discovery. *Data Mining and Knowledge Discovery, 3*, 409–425.

36. Edwards, P. J., & Murray, A. F. (1998). Towards optimally distributed computation. *Neural Computation*, *10*, 997–1015.
37. Fedorov, V. V. (1972). *Theory of optimal experiments*. San Diego: Academic Press.
38. Freund, Y., Iyer, R., Schapire, R. E., & Singer, Y. (2003). An efficient boosting algorithm for combining preferences. *Journal of Machine Learning Research*, *4*, 933–969.
39. Freund, Y., Seung, H. S., Shamir, E., & Tishby, N. (1997). Selective sampling using the query by committee algorithm. *Machine Learning*, *28*, 133–168.
40. Friedman, J. H. (2001). Greedy function approximation: A gradient boosting machine. *Annals of Statistics*, *29*(5), 1189–1232.
41. Friedrichs, F., & Schmitt, M. (2005). On the power of Boolean computations in generalized RBF neural networks. *Neurocomputing*, *63*, 483–498.
42. Gal, Y., & Ghahramani, Z. (2016). Dropout as a Bayesian approximation: Representing model uncertainty in deep learning. In *Proceedings of the 33rd International Conference on Machine Learning* (Vol. 48, pp. 1050–1059).
43. Geman, S., Bienenstock, E., & Doursat, R. (1992). Neural networks and the bias/variance dilemma. *Neural Computation*, *4*(1), 1–58.
44. Ghodsi, A., & Schuurmans, D. (2003). Automatic basis selection techniques for RBF networks. *Neural Networks*, *16*, 809–816.
45. Gish, H. (1990). A probabilistic approach to the understanding and training of neural network classifiers. In *Proceedings of IEEE International Conference on Acoustics, Speech, and Signal Processing (ICASSP)* (pp. 1361–1364).
46. Hanson, S. J., & Burr, D. J. (1988). Minkowski back-propagation: Learning in connectionist models with non-Euclidean error signals. In D. Z. Anderson (Ed.), *Neural information processing systems* (pp. 348–357). New York: American Institute Physics.
47. Hassoun, M. H. (1995). *Fundamentals of artificial neural networks*. Cambridge, MA: MIT Press.
48. Hecht-Nielsen, R. (1987). Kolmogorov's mapping neural network existence theorem. In *Proceedings of the 1st IEEE International Conference on Neural Networks* (Vol. 3, pp. 11–14). San Diego, CA.
49. Helmbold, D. P., & Long, P. M. (2018). Surprising properties of dropout in deep networks. *Journal of Machine Learning Research*, *18*, 1–28.
50. Herbrich, R., Graepel, T., & Obermayer, K. (2000). Large margin rank boundaries for ordinal regression. In P. J. Bartlett, B. Scholkopf, D. Schuurmans, & A. J. Smola (Eds.), *Advances in large margin classifiers* (pp. 115–132). Cambridge, MA: MIT Press.
51. Hinton, G. E. (1989). Connectionist learning procedure. *Artificial Intelligence*, *40*, 185–234.
52. Hinton, G. E. (2012). Dropout: A simple and effective way to improve neural networks. videolectures.net.
53. Hinton, G. E., Srivastava, N., Krizhevsky, A., Sutskever, I., & Salakhutdinov, R. R. (2012). Improving neural networks by preventing co-adaptation of feature detectors. *The Computing Research Repository (CoRR)*, abs/1207.0580.
54. Hinton, G. E., & van Camp, D. (1993). Keeping neural networks simple by minimizing the description length of the weights. In *Proceedings of the 6th Annual ACM Conference on Computational Learning Theory* (pp. 5–13). Santa Cruz, CA.
55. Ho, K. I.-J., Leung, C.-S., & Sum, J. (2010). Convergence and objective functions of some fault/noise-injection-based online learning algorithms for RBF networks. *IEEE Transactions on Neural Networks*, *21*(6), 938–947.
56. Hoi, S. C. H., Jin, R., & Lyu, M. R. (2009). Batch mode active learning with applications to text categorization and image retrieval. *IEEE Transactions on Knowledge and Data Engineering*, *21*(9), 1233–1248.
57. Holmstrom, L., & Koistinen, P. (1992). Using additive noise in back-propagation training. *IEEE Transactions on Neural Networks*, *3*(1), 24–38.
58. Hotelling, H. (1936). Relations between two sets of variates. *Biometrika*, *28*, 321–377.
59. Huber, P. J. (1981). *Robust statistics*. New York: Wiley.

60. Janssen, P., Stoica, P., Soderstrom, T., & Eykhoff, P. (1988). Model structure selection for multivariable systems by cross-validation. *International Journal of Control*, *47*, 1737–1758.
61. Kettenring, J. (1971). Canonical analysis of several sets of variables. *Biometrika*, *58*(3), 433–451.
62. Khan, S. H., Hayat, M., & Porikli, F. (2019). Regularization of deep neural networks with spectral dropout. *Neural Networks*, *110*, 82–90.
63. Kokiopoulou, E., & Saad, Y. (2007). Orthogonal neighborhood preserving projections: A projection-based dimensionality reduction technique. *IEEE Transactions on Pattern Analysis and Machine Intelligence*, *29*(12), 2143–2156.
64. Kolmogorov, A. N. (1957). On the representation of continuous functions of several variables by superposition of continuous functions of one variable and addition. *Doklady Akademii Nauk USSR*, *114*(5), 953–956.
65. Krogh, A., & Hertz, J. A. (1992). A simple weight decay improves generalization. In *Proceedings of Neural Information and Processing Systems (NIPS) Conference* (pp. 950–957). San Mateo, CA: Morgan Kaufmann.
66. Lanckriet, G. R. G., Cristianini, N., Bartlett, P., El Ghaoui, L., & Jordan, M. I. (2004). Learning the kernel matrix with semidefinite programming. *Journal of Machine Learning Research*, *5*, 27–72.
67. Lin, Y., Lee, Y., & Wahba, G. (2002). Support vector machines for classification in nonstandard situations. *Machine Learning*, *46*, 191–202.
68. Liu, W., Pokharel, P. P., & Principe, J. C. (2007). Correntropy: Properties and applications in non-Gaussian signal processing. *IEEE Transactions on Signal Processing*, *55*(11), 5286–5298.
69. Liu, Y., Starzyk, J. A., & Zhu, Z. (2008). Optimized approximation algorithm in neural networks without overfitting. *IEEE Transactions on Neural Networks*, *19*(6), 983–995.
70. Maass, W. (2000). On the computational power of winner-take-all. *Neural Computation*, *12*, 2519–2535.
71. MacKay, D. (1992). Information-based objective functions for active data selection. *Neural Computation*, *4*(4), 590–604.
72. Markatou, M., Tian, H., Biswas, S., & Hripcsak, G. (2005). Analysis of variance of cross-validation estimators of the generalization error. *Journal of Machine Learning Research*, *6*, 1127–1168.
73. Matsuoka, K., & Yi, J. (1991). Backpropagation based on the logarithmic error function and elimination of local minima. In *Proceedings of the International Joint Conference on Neural Networks* (pp. 1117–1122). Seattle, WA.
74. McCullagh, P. (1980). Regression models for ordinal data. *Journal of the Royal Statistical Society: Series B*, *42*(2), 109–142.
75. Muller, B., Reinhardt, J., & Strickland, M. (1995). *Neural networks: An introduction* (2nd ed.). Berlin: Springer.
76. Murray, A. F., & Edwards, P. J. (1994). Synaptic weight noise euring MLP training: Enhanced MLP performance and fault tolerance resulting from synaptic weight noise during training. *IEEE Transactions on Neural Networks*, *5*(5), 792–802.
77. Nadeau, C., & Bengio, Y. (2003). Inference for the generalization error. *Machine Learning*, *52*, 239–281.
78. Niyogi, P., & Girosi, F. (1999). Generalization bounds for function approximation from scattered noisy data. *Advances in Computational Mathematics*, *10*, 51–80.
79. Nowlan, S. J., & Hinton, G. E. (1992). Simplifying neural networks by soft weight-sharing. *Neural Computation*, *4*(4), 473–493.
80. Pan, S. J., & Yang, Q. (2010). A survey on transfer learning. *IEEE Transactions on Knowledge and Data Engineering*, *22*(10), 1345–1359.
81. Parzen, E. (1962). On estimation of a probability density function and mode. *The Annals of Mathematical Statistics*, *33*(1), 1065–1076.
82. Phatak, D. S. (1999). Relationship between fault tolerance, generalization and the Vapnik-Cervonenkis (VC) dimension of feedforward ANNs. *Proceedings of International Joint Conference on Neural Networks*, *1*, 705–709.

83. Plutowski, M. E. P. (1996). *Survey: Cross-validation in theory and in practice*. Research Report. Princeton, NJ: Department of Computational Science Research, David Sarnoff Research Center.
84. Poggio, T., & Girosi, F. (1990). Networks for approximation and learning. *Proceedings of the IEEE, 78*(9), 1481–1497.
85. Prechelt, L. (1998). Automatic early stopping using cross validation: Quantifying the criteria. *Neural Networks, 11*, 761–767.
86. Reed, R., Marks, R. J, I. I., & Oh, S. (1995). Similarities of error regularization, sigmoid gain scaling, target smoothing, and training with jitter. *IEEE Transactions on Neural Networks, 6*(3), 529–538.
87. Rimer, M., & Martinez, T. (2006). Classification-based objective functions. *Machine Learning, 63*(2), 183–205.
88. Rissanen, J. (1978). Modeling by shortest data description. *Automatica, 14*(5), 465–477.
89. Rissanen, J. (1999). Hypothesis selection and testing by the MDL principle. *Computer Journal, 42*(4), 260–269.
90. Roweis, S. T., & Saul, L. K. (2000). Nonlinear dimensionality reduction by locally linear embedding. *Science, 290*(5500), 2323–2326.
91. Royden, H. L. (1968). *Real analysis* (2nd ed.). New York: Macmillan.
92. Rumelhart, D. E., Hinton, G. E., & Williams, R. J. (1986). Learning internal representations by error propagation. In D. E. Rumelhart & J. L. McClelland (Eds.), *Parallel distributed processing: Explorations in the microstructure of cognition* (Vol. 1, pp. 318–362). Cambridge, MA: MIT Press.
93. Rumelhart, D. E., Durbin, R., Golden, R., & Chauvin, Y. (1995). Backpropagation: the basic theory. In Y. Chauvin & D. E. Rumelhart (Eds.), *Backpropagation: Theory, architecture, and applications* (pp. 1–34). Hillsdale, NJ: Lawrence Erlbaum.
94. Sabato, S., & Tishby, N. (2012). Multi-instance learning with any hypothesis class. *Journal of Machine Learning Research, 13*, 2999–3039.
95. Sarbo, J. J., & Cozijn, R. (2019). Belief in reasoning. *Cognitive Systems Research, 55*, 245–256.
96. Schultz, W. (1998). Predictive reward signal of dopamine neurons. *Journal of Neurophysiology, 80*(1), 1–27.
97. Schwarz, G. (1978). Estimating the dimension of a model. *Annals of Statistics, 6*, 461–464.
98. Shao, J. (1993). Linear model selection by cross-validation. *Journal of the American Statistical Association, 88*, 486–494.
99. Siegelmann, H. T., & Sontag, E. D. (1995). On the computational power of neural nets. *Journal of Computer and System Sciences, 50*(1), 132–150.
100. Silva, L. M., de Sa, J. M., & Alexandre, L. A. (2008). Data classification with multilayer perceptrons using a generalized error function. *Neural Networks, 21*, 1302–1310.
101. Sima, J. (1996). Back-propagation is not efficient. *Neural Networks, 9*(6), 1017–1023.
102. Singh, A., Pokharel, R., & Principe, J. C. (2014). The C-loss function for pattern classification. *Pattern Recognition, 47*(1), 441–453.
103. Solla, S. A., Levin, E., & Fleisher, M. (1988). Accelerated learning in layered neural networks. *Complex Systems, 2*, 625–640.
104. Srivastava, N., Hinton, G., Krizhevsky, A., Sutskever, I., & Salakhutdinov, R. (2014). Dropout: A simple way to prevent neural networks from overfitting. *Journal of Machine Learning Research, 15*, 1929–1958.
105. Stoica, P., & Selen, Y. (2004). A review of information criterion rules. *EEE Signal Processing Magazine, 21*(4), 36–47.
106. Stone, M. (1974). Cross-validatory choice and assessment of statistical predictions. *Journal of the Royal Statistical Society Series B, 36*, 111–147.
107. Sugiyama, M., & Ogawa, H. (2000). Incremental active learning for optimal generalization. *Neural Computation, 12*, 2909–2940.
108. Sugiyama, M., & Nakajima, S. (2009). Pool-based active learning in approximate linear regression. *Machine Learning, 75*, 249–274.

109. Sum, J. P.-F., Leung, C.-S., & Ho, K. I.-J. (2012). On-line node fault injection training algorithm for MLP networks: Objective function and convergence analysis. *IEEE Transactions on Neural Networks and Learning Systems*, *23*(2), 211–222.
110. Tabatabai, M. A., & Argyros, I. K. (1993). Robust estimation and testing for general nonlinear regression models. *Applied Mathematics and Computation*, *58*, 85–101.
111. Tecuci, G., Kaiser, L., Marcu, D., Uttamsingh, C., & Boicu, M. (2018). Evidence-based reasoning in intelligence analysis: Structured methodology and system. *Computing in Science & Engineering*, *20*(6), 9–21.
112. Tikhonov, A. N. (1963). On solving incorrectly posed problems and method of regularization. *Doklady Akademii Nauk USSR*, *151*, 501–504.
113. Tucker, L. R. (1964). The extension of factor analysis to three-dimensional matrices. *Contributions to mathematical psychology* (pp. 109–127). Holt, Rinehardt & Winston: New York, NY.
114. Vapnik, V. N. (1998). *Statistical learning theory*. New York: Wiley.
115. Wan, L., Zeiler, M., Zhang, S., LeCun, Y., Fergus, R. (2013). Regularization of neural networks using dropconnect. In *Proceedings of International Conference on Machine Learning* (pp. 1058–1066).
116. Widrow, B., & Lehr, M. A. (1990). 30 years of adaptive neural networks: Perceptron, Madaline, and backpropagation. *Proceedings of the IEEE*, *78*(9), 1415–1442.
117. Wu, G., & Cheng, E. (2003). Class-boundary alignment for imbalanced dataset learning. In *Proceedings of ICML 2003 Workshop on Learning Imbalanced Data Sets II* (pp. 49–56). Washington, DC.
118. Xiao, Y., Feng, R.-B., Leung, C.-S., & Sum, J. (2016). Objective function and learning algorithm for the general node fault situation. *IEEE Transactions on Neural Networks and Learning Systems*, *27*(4), 863–874.
119. Xu, H., Caramanis, C., & Mannor, S. (2010). Robust regression and Lasso. *IEEE Transactions on Information Theory*, *56*(7), 3561–3574.
120. Xu, H., Caramanis, C., & Mannor, S. (2012). Sparse algorithms are not stable: A no-free-lunch theorem. *IEEE Transactions on Pattern Analysis and Machine Intelligence*, *34*(1), 187–193.
121. Yang, L., Hanneke, S., & Carbonell, J. (2013). A theory of transfer learning with applications to active learning. *Machine Learning*, *90*(2), 161–189.
122. Zahalka, J., & Zelezny, F. (2011). An experimental test of Occam's razor in classification. *Machine Learning*, *82*, 475–481.
123. Zhang, M.-L., & Zhou, Z.-H. (2007). ML-KNN: A lazy learning approach to multi-label learning. *Pattern Recognition*, *40*(7), 2038–2048.

Chapter 3
Elements of Computational Learning Theory

3.1 Introduction

Machine learning makes predictions about the unknown underlying model based on a training set drawn from hypotheses. Due to the finite training set, learning theory cannot provide absolute guarantees of performance of the algorithms. The performance of learning algorithms is commonly bounded by probabilistic terms. Computational learning theory is a statistical tool for the analysis of machine learning algorithms, that is, for characterizing learning and generalization. Computational learning theory addresses the problem of optimal generalization capability for supervised learning. Two popular formalisms of approaches to computational learning theory are the *VC-theory* [43] and the probably approximately correct (PAC) learning [39]. Both approaches are nonparametric and distribution-free learning models.

VC-theory [43], known as *statistical learning theory*, is a dependency-estimation method with finite data. Necessary and sufficient conditions for consistency and fast convergence are obtained based on the empirical risk-minimization (ERM) principle. Uniform convergence for a given class of approximating functions is associated with the capacity of the function class considered [43]. The VC-dimension of a function class quantifies its classification capabilities [43]. It indicates the cardinality of the largest set for which all possible binary-valued classifications can be obtained using functions from the class. The capacity and complexity of the function class are measured in terms of the VC-dimension. ERM principle has been practically applied in SVM [41]. VC-theory provides a general measure of complexity, and gives associated bounds on the optimism.

PAC learning [39] aims to find a hypothesis that is a good approximation to an unknown target concept with a high probability. The PAC learning paradigm is intimately associated with the ERM principle. A hypothesis that minimizes the empirical error, based on a sufficiently large sample, will approximate the target concept with a high probability. The generalization ability of network training can be established estimating the VC-dimension of neural architectures. Boosting [31] is a PAC learning-inspired method for supervised learning.

K.-L. Du and M. N. S. Swamy, *Neural Networks and Statistical Learning*,
https://doi.org/10.1007/978-1-4471-7452-3_3

Kolmogorov complexity is a general notion of complexity that, unfortunately, is not computable. MDL is an approximation of the Kolmogorov complexity. Vapnik and Chervonenkis developed several measures of complexity, such as the VC-entropy, the growth function, and the VC-dimension [42]. The fat-shattering dimension [6] is a generalization of the VC-dimension to real-valued functions. A loose connection between the fat-shattering dimension and the Rademacher complexity is given in [27].

Minimax label complexity is defined as the smallest worst-case number of label requests sufficient for the active learning algorithm to produce a classifier of a specified error rate, in the context of various noise models. Distribution-free upper and lower bounds on the minimax label complexity of active learning with general hypothesis classes, under various noise models, are established in [18].

3.2 Probably Approximately Correct (PAC) Learning

PAC learning paradigm is concerned with learning from examples of a target function called *concept*, by choosing from a set of functions known as the *hypothesis space*, a function that is meant to be a good approximation to the target. In PAC learning framework, a hypothesis space of functions maps the inputs onto $\{0, 1\}$. PAC learning is a classical criterion for supervised learning. It aims to produce a classifier that, with confidence at least $1 - \delta$, has error rate at most ϵ. Such a classifier is said *probably approximately correct*. One of the central questions in PAC learning is determining the sample complexity.

Definition 3.1 (*PAC Learnability*) Let $\mathcal{C}_n$ and $\mathcal{H}_n$, $n \geq 1$, respectively, be a set of target concepts and a set of hypotheses over the instance space $\{0, 1\}^n$, where $\mathcal{C}_n \subseteq \mathcal{H}_n$ for $n \geq 1$. When there exists a polynomial-time learning algorithm that achieves low error with high confidence in approximating all concepts in a class $\mathcal{C} = \{\mathcal{C}_n\}$ by the hypothesis space $\mathcal{H} = \{\mathcal{H}_n\}$ if enough training data are available, the class of concepts $\mathcal{C}$ is said to be *PAC learnable* by $\mathcal{H}$ or simply PAC learnable.

All of the hypotheses form a hypothesis space of a finite size $|\mathcal{H}|$, and the hypothesis space consists of $|\mathcal{H}|$ hypothesis classifiers. Assume the probability that a hypothetical classifier will correctly label a randomly selected example is less than $1 - \epsilon$. Thus, the probability that this classifier will label correctly m random examples is bounded by $P \leq (1 - \epsilon)^m$. This bound is too small for reasonable values of ϵ and m.

Assume that an error rate greater than ϵ is unacceptable. Evaluate all the classifiers on the m training examples, and keep only the $k \leq |\mathcal{H}|$ classifiers that have never made any mistake. By eliminating all classifiers whose error rate exceeds ϵ, the upper bound on the probability that the k classifiers will correctly label the m examples is given by

$$P \leq k(1 - \epsilon)^m \leq |\mathcal{H}|(1 - \epsilon)^m < |\mathcal{H}|e^{-m\epsilon}, \tag{3.1}$$

where $1 - \epsilon < e^{-\epsilon}$.

Assume this probability to be lower than the chance of failure δ, that is, $|\mathcal{H}|e^{-m\epsilon} \leq \delta$. From this, we obtain [20]

$$m > \frac{1}{\epsilon}\left(\ln|\mathcal{H}| + \ln\frac{1}{\delta}\right). \tag{3.2}$$

From (3.2), m grows linearly in $1/\epsilon$, but is less sensitive to changes in δ. However, the derivation is a worst-case analysis since it is too pessimistic to allow the possibility of $k = |\mathcal{H}|$.

A class is said *not PAC learnable* if the number of examples needed to satisfy the given (ϵ, δ)-requirements is too high to be practical.

Example 3.1 **A Boolean function in its general form is not PAC learnable**. For n Boolean attributes, the instant space of 2^n different examples can be created. For such an instant space, we have 2^{2^n} subsets. Thus, the size of the hypothesis space is $|\mathcal{H}| = 2^{2^n}$. From (3.2), we have $m > \frac{1}{\epsilon}(2^n \ln 2 + \ln\frac{1}{\delta})$. Thus, the lower bound grows exponentially in n. Thus, a classifier in general form is not PAC learnable.

3.2.1 Sample Complexity

Definition 3.2 (*Sample Complexity*) The *sample complexity* of a learning algorithm, $m_{\mathcal{H}}$, is defined as the smallest number of samples required for learning $\mathcal{C}$ by $\mathcal{H}$ that achieves a given approximation accuracy ϵ with a probability $1 - \delta$.

Any consistent algorithm that learns $\mathcal{C}$ by $\mathcal{H}$ has a sample complexity with the upper bound [2, 20]

$$m_{\mathcal{H}}(\epsilon, \delta) \leq \frac{1}{\epsilon\left(1 - \sqrt{\epsilon}\right)}\left(2d\ln\frac{6}{\epsilon} + \ln\frac{2}{\delta}\right), \quad \forall 0 < \delta < 1, \tag{3.3}$$

where d is the VC-dimension of the hypothesis class $\mathcal{H}$. In other words, with probability of at least $1 - \delta$, the algorithm returns a hypothesis $h \in \mathcal{H}$ with an error less than ϵ. Note that the number of examples necessary for PAC learning grows linearly in the VC-dimension.

According to (3.2), in terms of $|\mathcal{H}|$, the sample complexity is upper bounded by [20]

$$m_{\mathcal{H}}(\epsilon, \delta) \leq \frac{1}{\epsilon}\left(\ln|\mathcal{H}| + \ln\frac{1}{\delta}\right). \tag{3.4}$$

For most hypothesis spaces on Boolean domains, the second bound gives a better bound. On the other hand, most hypothesis spaces on real-valued attributes are infinite, so only the first bound is applicable. PAC learning is particularly useful for obtaining upper bounds on sufficient training sample size. Linear threshold concepts (perceptrons) are PAC learnable on both Boolean and real-valued instance spaces [20].

Sufficient sample sizes are, respectively, estimated by using the PAC paradigm and the VC-dimension for feedforward networks with sigmoidal neurons [35] and feedforward networks with LTGs [8]. These bounds on sample sizes are dependent on the error rate of hypothesis ϵ and the probability of failure δ. A practical size of the training set for good generalization is $N = O\left(\frac{N_w}{\varepsilon}\right)$ [21].

In [17], sample complexity estimates are provided to uniformly control the empirical average deviation from the expected cost function. This provides a unified perspective on the sample complexity of several popular matrix factorization schemes such as PCA, sparse dictionary learning, NMF, or C-means clustering, for which sample complexity bounds are provided. The derived generalization bounds behave proportional to $(\log(N)/N)^{1/2}$ with respect to the number of samples N for the considered matrix factorization techniques.

A classification algorithm based on a majority vote among classifiers trained on independent datasets is proved to achieve a sample complexity that reduces the logarithmic factor in the upper bound down to a very slowly growing function [38].

3.3 Vapnik–Chervonenkis Dimension

VC-dimension is a combinatorial characterization of the diversity of functions that can be computed by a given neural architecture. It can be viewed as a generalization of the concept of capacity first introduced by Cover [12]. VC-dimension can be regarded as a measure of the capacity or expressive power of a network. It is the measure of model complexity (capacity) used in VC-theory. VC-dimension can be used to estimate the number of training examples for a good generalization capability.

Definition 3.3 (*VC-Dimension*) A subset $\mathcal{S}$ of the domain $\mathcal{X}$ is shattered by a class of functions or neural network $\mathcal{N}$ if every function $f : \mathcal{S} \to \{0, 1\}$ can be computed on $\mathcal{N}$. The VC-dimension of $\mathcal{N}$ is defined as the maximal size of a set $\mathcal{S} \subseteq \mathcal{X}$ that is shattered by $\mathcal{N}$

$$\dim_{\mathrm{VC}}(\mathcal{N}) = \max\left\{|\mathcal{S}| \middle| \mathcal{S} \subseteq \mathcal{X} \text{ is shattered by } \mathcal{N}\right\}, \tag{3.5}$$

where $|\mathcal{S}|$ denotes the cardinality of $\mathcal{S}$.

VC-dimension reflects a combinatorial property of the given class of classifiers. It denotes the maximal sample size that can be shattered by the class. For example, for a neural network with the relation $f(\boldsymbol{x}, \boldsymbol{w}, \theta) = \operatorname{sgn}\left(\boldsymbol{w}^T \boldsymbol{x} + \theta\right)$, it can shatter at most any three points in $\mathcal{X}$, thus its VC-dimension is 3. This is shown in Fig. 3.1. The points are in general position, that is, they are linearly independent.

VC-dimension is a property of a set of functions $\{f(\boldsymbol{\alpha})\}$, and can be defined for various classes of function f. The VC-dimension for the set of functions $\{f(\boldsymbol{\alpha})\}$ is

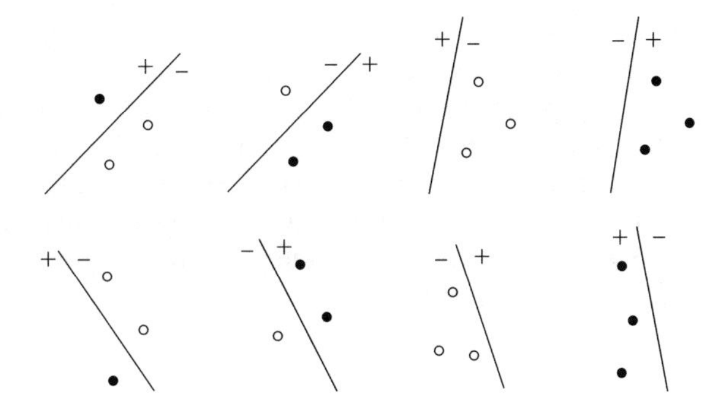

Fig. 3.1 Shatter any three points in $\mathcal{X}$ into two classes

defined as the maximum number of training points that can be shattered by $\{f(\boldsymbol{\alpha})\}$. If the VC-dimension is d, then there exists at least one set of d points that can be shattered, but in general it will not be true that every set of d points can be shattered. For linear estimators, the VC-dimension is equivalent to the number of model parameters.

The Boolean VC-dimension of a neural network $\mathcal{N}$, written $\dim_{\mathrm{BVC}}(\mathcal{N})$, is defined as the VC-dimension of the class of Boolean functions that is computed by $\mathcal{N}$.

In order to show $\dim_{\mathrm{VC}}(\mathcal{H}) = d$, one needs to show that there exists a set C of size d that is shattered by $\mathcal{H}$, and then to show that every set C of size $d + 1$ is not shattered by $\mathcal{H}$.

A hard-limiter function with threshold is typically used as the activation function for binary neurons. The basic function of the McCulloch–Pitts neuron has a linear relation applied by a threshold operation, hence called a *linear threshold gate (LTG)*. A neural network with LTG has a VC-dimension of $O(N_w \log N_w)$ [4], where N_w is the number of weights in a network. VC-dimension has been generalized for neural networks with real-valued output [4]. Feedforward networks with threshold and logistical activation functions have VC-dimensions of $O(N_w \ln N_w)$ [8] and $O\left(N_w^2\right)$ [22], respectively. Higher order neurons with k monomials in n variables have a VC-dimension of at least $nk + 1$ [32].

Given a J_1-J_2-1 feedforward network, whose output neuron is an LTG, we denote the network as $\mathcal{N}_1$, $\mathcal{N}_2$, and $\mathcal{N}_3$, when the J_2 hidden neurons are, respectively, selected as LTGs, binary RBF neurons, and generalized binary RBF neurons. The VC-dimensions of the three networks have the relation [14]

$$\dim_{\mathrm{BVC}}(\mathcal{N}_1) = \dim_{\mathrm{BVC}}(\mathcal{N}_2) \le \dim_{\mathrm{BVC}}(\mathcal{N}_3). \tag{3.6}$$

When $J_1 \ge 3$ and $J_2 \le \frac{2^{J_1+1}}{J_1^2+J_1+2}$, the lower bound for the three neural networks is given by [3, 14]

$$\dim_{\mathrm{BVC}}(\mathcal{N}_1) = J_1 J_2 + 1. \tag{3.7}$$

It is not possible to obtain the analytic estimates of VC-dimension in most cases. Hence, a proposal is to measure the VC-dimension of an estimator experimentally by fitting the theoretical formula to a set of experimental measurements of the frequency of errors on artificially generated datasets of varying sizes [44]. However, with this approach it is difficult to obtain an accurate estimate of VC-dimension due to the variability of random samples. In [34], this problem is addressed by an improved design procedure for specifying the measurement points. This leads to a nonuniform design structure. A more accurate estimation of VC-dimension leads to improved complexity control using analytic VC-generalization bounds and, hence, better prediction accuracy.

3.3.1 Teaching Dimension

Teaching dimension is a learning-theoretic quantity that specifies the minimum training set size to teach a target model to a learner [15, 37]. Consider a teacher who knows both a target model and the learning algorithm used by a machine learner. The teacher wants to teach the target model to the learner by constructing a training set. The training set does not need to contain independent and identically distributed items drawn from some distribution. Furthermore, the teacher can construct any item in the input space.

For a finite hypothesis space $\mathcal{H}$, there is the relation [15]

$$\dim_{VC}(\mathcal{H}) = \log(|\mathcal{H}|) \leq \dim_{TD}(\mathcal{H}) \leq \dim_{VC}(\mathcal{H}) + |\mathcal{H}| - 2^{\dim_{VC}(\mathcal{H})}. \quad (3.8)$$

Originally, teaching dimension focuses on version-space learners which maintain all hypotheses consistent with the training data, and cannot be applied to machine learners which select a specific hypothesis via optimization. The teaching dimension for ridge regression, SVMs, and logistic regression is presented in [25]. The approach generalizes to other linear learners. Optimal training sets that match these teaching dimensions are also exhibited in [25].

3.4 Rademacher Complexity

VC-dimension is a global one and is data independent. It leads to very pessimistic generalization bounds. Rademacher complexity [5, 23] and data-dependent versions of the VC-complexities [36] are data-dependent complexity measures, which allow us to take into account the actual distribution of the data and produce tighter estimates of the complexity of the class.

Rademacher complexity derives a tighter generalization error bounds than those derived using the VC-dimension and cover number.

Definition 3.4 (*Rademacher Complexity* [23]) Let $\mathcal{F}$ be a class of functions on a probability space $(\mathcal{Z}, \mathcal{P})$ and let $z_1 = (x_1, y_1), \ldots, z_N = (x_N, y_N)$ be N training examples independently drawn from $\mathcal{P}$ over $\mathcal{X} \times \mathcal{Y}$. Assume N independent Rademacher random variables $\sigma_1, \ldots, \sigma_N$, i.e., $\Pr\{\sigma_i = 1\} = \Pr\{\sigma_i = -1\} = 1/2$. Introduce the notation

$$R_N \mathcal{F} = \sup_{f \in \mathcal{F}} \frac{1}{N} \sum_{i=1}^{N} \sigma_i f(z_i). \tag{3.9}$$

The Rademacher complexity $E[R_N \mathcal{F}]$ is the expectation of $R_N \mathcal{F}$, and the empirical Rademacher complexity

$$E_\sigma [R_N \mathcal{F}] = E[R_N \mathcal{F} | z_1, \ldots, z_N] \tag{3.10}$$

is defined as the conditional expectation of $R_N \mathcal{F}$.

The above global Rademacher complexity is a global estimation of the complexity of the function class. Based on the notion of global Radmacher complexity, the supervised learning problem has a standard generalization error bound.

Theorem 3.1 (Generalization Error Bound Derived by Rademacher Complexity [5]) *Given $\delta > 0$. Suppose the function $\hat{f}$ is learned over N training points. Then, with probability at least $1 - \delta$, we have*

$$E\left[\hat{f}\right] \leq \inf_{f \in \mathcal{F}} E[f] + 4 R_N \mathcal{F} + \sqrt{\frac{2 \log(2/\delta)}{N}}. \tag{3.11}$$

The generalization error bound in Theorem 3.1 converges at $O(\sqrt{1/N})$.

Local Rademacher complexity characterizes a small subset of the class, i.e., only those functions of the hypothesis class that will be most likely chosen by the learning procedure. For the intesection of the class with a ball centered on the function of interest,

$$E_\sigma \left[R_N \left\{f \in \mathcal{F} : E\left[f^2\right] \leq r\right\}\right] \text{ or } E\left[R_N \left\{f \in \mathcal{F} : E\left[f^2\right] \leq r\right\}\right]. \tag{3.12}$$

Local Rademacher complexities can be viewed as functions of r, and they filter out those functions with large variances.

This local Rademacher complexity is always smaller than the corresponding global Rademacher complexity. It always yields considerably improved estimation error bounds when the variance–expectation relation holds [7].

Theorem 3.2 (Generalization Error Bound Derived by Local Rademacher Complexity [7]) *Given $\delta > 0$, suppose the function $\hat{f}$ is learned over N training points. Assume that there is some $r > 0$ such that for every $f \in \mathcal{F}$, $E\left[f^2\right] \leq r$. Then, with probability at least $1 - \delta$, we have*

$$E[\hat{f}] \leq \inf_{f \in \mathcal{F}} E[f] + 8 L R_N(\mathcal{F}, r) + \sqrt{\frac{8 r \log(2/\delta)}{N}} + \frac{3 \log(2/\delta)}{N}. \tag{3.13}$$

By choosing a much smaller class $\mathcal{F}' \subseteq \mathcal{F}$ with as small a variance as possible while requiring that $\hat{f}$ still lies in $\mathcal{F}'$, the generalization error bound in Theorem 3.2 has a faster convergence rate than that of Theorem 3.1 of up to $O(\log N/N)$.

Once the local Rademacher complexity is known, $E(\hat{f}) - \inf_{f \in \mathcal{F}} E[f]$ can be bounded in terms of the fixed point of the local Redemacher complexity of $\mathcal{F}$.

Direct computation of (local) Rademacher complexity is extremely difficult. Dudley's entropy integral [13] captures the relationship between covering numbers and Rademacher complexities. This classical result is extended to the local Rademacher complexity setting [28].

The Rademacher complexity of ERM-based multi-label learning algorithms can be bounded by the trace norm of the multi-label predictors, which provides a theoretical explanation for the effectiveness of using the trace norm for regularization in multi-label learning [46]. The generalization performance of RBF network is derived by using local Rademacher complexities with the L_1-metric capacity, and substantially improved estimation error bounds are obtained [24].

Annealed VC-entropy [40] allows to bound the generalization error given the empirical one (and vice versa). VC-entropy and Rademacher complexity are related to [1]. This connection allows exploiting Rademacher complexity in Vapnik's general bound, whose convergence rate varies between $O(N^{-1/2})$ and $O(N^{-1})$. This is faster than the ones obtained by exploiting Rademacher complexity, whose convergence rate is $O(N^{-1/2})$.

Local VC-entropy [29] is a localized version of a VC-complexity, and, building on this complexity, a generalization bound for binary classifiers is derived in [29]. It is possible to relate local VC-entropy to local Rademacher complexity by finding an admissible range for one given the other [29]. Local VC-entropy allows one to reduce the computational requirements that arise when dealing with the local Rademacher complexity in binary classification problems. Local VC-entropy can be related to the local Rademacher complexity through an extension of the geometrical framework presented in [1].

Local VC-entropy counts the number of functions in $\mathcal{F}$, given $\mathcal{D}_N$, with perfectly classifying at least one of the possible configurations of $\boldsymbol{\sigma} \in \mathcal{S}$ having Hamming distance from $\boldsymbol{y}$ lower than Nr. The local VC-entropy-based bound improves on the original Vapnik's results because it is able to discard those functions that, most likely, will not be selected during the learning phase.

3.5 Empirical Risk-Minimization Principle

Assume that a set of N samples, $\{(\boldsymbol{x}_i, y_i)\}$, is independently drawn and identically distributed (iid) samples from some unknown probability distribution $p(\boldsymbol{x}, y)$. Assume a machine defined by a set of possible mappings $\boldsymbol{x} \rightarrow f(\boldsymbol{x}, \boldsymbol{\alpha})$, where $\boldsymbol{\alpha}$ contains adjustable parameters. When $\boldsymbol{\alpha}$ is selected, the machine is called a *trained machine*.

The expected risk is the expectation of the generalization error for a trained machine, and is given by

$$R(\boldsymbol{\alpha}) = \int L\left(y, f(\boldsymbol{x}, \boldsymbol{\alpha})\right) \mathrm{d}p(\boldsymbol{x}, y), \tag{3.14}$$

where $L\left(y, f(\boldsymbol{x}, \boldsymbol{\alpha})\right)$ is the loss function, measuring the discrepancy between the output pattern y and the output of the learning machine $f(\boldsymbol{x}, \boldsymbol{\alpha})$. The loss function can be defined in different forms for different purposes:

$$L(y, f(\boldsymbol{x}, \boldsymbol{\alpha})) = \begin{cases} 0, & y = f(\boldsymbol{x}, \boldsymbol{\alpha}) \\ 1, & y \neq f(\boldsymbol{x}, \boldsymbol{\alpha}) \end{cases} \quad \text{(for classification)}, \tag{3.15}$$

$$L(y, f(\boldsymbol{x}, \boldsymbol{\alpha})) = (y - f(\boldsymbol{x}, \boldsymbol{\alpha}))^2 \quad \text{(for regression)}, \tag{3.16}$$

$$L(p(\boldsymbol{x}, \boldsymbol{\alpha})) = -\ln p(\boldsymbol{x}, \boldsymbol{\alpha}) \quad \text{(for density estimation)}. \tag{3.17}$$

The empirical risk $R_{\text{emp}}(\boldsymbol{\alpha})$ is defined to be the measured mean error on a given training set

$$R_{\text{emp}}(\boldsymbol{\alpha}) = \frac{1}{N} \sum_{i=1}^{N} L\left(y_i, f\left(\boldsymbol{x}_i, \boldsymbol{\alpha}\right)\right). \tag{3.18}$$

The ERM principle aims to approximate the loss function by minimizing the empirical risk (3.18) instead of the risk (3.14), with respect to model parameters. ERM is one of the most powerful tools in applied statistics and machine learning. It is recognized as a special form in standard convex optimization. SVMs, linear regression, and logistics regression can be cast as ERM problems.

When the loss function takes the value 0 or 1, with probability $1 - \delta$, there is the upper bound called the *VC-bound* [41]:

$$R(\boldsymbol{\alpha}) \le R_{\text{emp}}(\boldsymbol{\alpha}) + \sqrt{\frac{d\left(\ln \frac{2N}{d} + 1\right) - \ln \frac{\delta}{4}}{N}}, \tag{3.19}$$

where d is the VC-dimension of the machine. The second term on the right-hand side is called the *VC-confidence*, which monotonically increases with increasing d. Reducing d leads to a better upper bound on the actual error. The VC-confidence depends on the class of functions, whereas the empirical risk and actual risk depend on the particular function obtained by the training procedure.

For regression problems, a practical form of the VC-bound is used [42]:

$$R(d) \le R_{\text{emp}}(d) \left(1 - \sqrt{p - p \ln p + \frac{\ln n}{2n}}\right)^{-1}, \tag{3.20}$$

where $p = \frac{d}{n}$ and d is the VC-dimension. The VC-bound (3.20) is a special case of the general analytical bound [41] with appropriately chosen practical values of theoretical constants.

Structural risk-minimization (SRM) principle [41] minimizes the risk functional with respect to both the empirical risk and the VC-dimension of the set of functions; thus, it aims to find the subset of functions that minimizes the bound on the actual risk. SRM principle is crucial to obtain good generalization performances for a variety of learning machines, including SVMs. It finds the function that achieves the minimum of the guaranteed risk for the fixed amount of data. To find the guaranteed risk, one has to use bounds, e.g., VC-bound, on the actual risk. Empirical comparisons between AIC, BIC, and SRM are presented for regression problems [10], based on VC-theory. VC-based model selection consistently outperforms AIC for all the datasets, whereas the SRM and BIC methods show similar predictive performance.

3.5.1 *Function Approximation, Regularization, and Risk Minimization*

Classical statistics and function approximation/regularization rely on the true model that underlies generated data. In contrast, VC-learning theory is based on the concept of risk minimization, and does not use the notion of a true model. The distinction between the three learning paradigms becomes blurred when they are used to motivate practical learning algorithms. Least squares (LS) minimization for function estimation can be derived using the parametric estimation approach via ML arguments under Gaussian noise assumptions, and it can alternatively be introduced under the risk-minimization approach. SVM methodology was originally developed in VC-theory, and later reintroduced in the function approximation/regularization setting [19]. An important conceptual contribution of the VC-approach states that generalization (learning) with finite samples may be possible even if accurate function approximation is not [11]. The regularization program does not yield good generalization for finite sample estimation problems.

In the function approximation theory, the goal is to estimate an unknown true target function in regression problems, or posterior probability $P(y|\boldsymbol{x})$ in classification problems. In VC-theory, it is to find the target function that minimizes prediction risk or achieves good generalization. That is, the result of VC-learning depends on (unknown) input distribution while that of function approximation does not. The important concept of margin was originally introduced under the VC-approach, and later explained and interpreted as a form of regularization. However, the notion of margin is specific to SVM, and it does not exist under the regularization framework. Any of the methodologies (including SRM and SVM) can be regarded as a special case of regularization.

3.6 Fundamental Theorem of Learning Theory

The fundamental theorem of learning theory characterizes PAC learnability of classes of binary classifiers using VC-dimension. A hypothesis class is PAC learnable if and only if its VC-dimension is finite, and the VC-dimension specifies the sample complexity required for PAC learning. Uniform convergence of the empirical error of a function toward the real error on all possible inputs guarantees that all training algorithms that yield a small training error are PAC learnable. If a problem is learnable, then uniform convergence holds and therefore the problem is learnable using the ERM rule.

Theorem 3.3 (Fundamental Theorem of Learning Theory [33]) *Let $\mathcal{H}$ be a hypothesis class of functions from a domain $\mathcal{X}$ to $\{0, 1\}$ and let the loss function be the 0–1 loss. Then, the following are equivalent:*
1. $\mathcal{H}$ has the uniform convergence property.
2. Any ERM rule is a successful (agnostic) PAC learner for $\mathcal{H}$.
3. $\mathcal{H}$ is (agnostic) PAC learnable.
4. $\mathcal{H}$ has a finite VC-dimension.

Agnostic PAC learnability extends the definition of PAC learnability to the more realistic, nonrealizable, learning setting. If the realizability assumption holds, agnostic PAC learning provides the same guarantee as PAC learning. When the realizability assumption does not hold, no learner can guarantee an arbitrarily small error. An agnostic PAC learner can still declare success if its error is not much larger than the best error achievable by a predictor from $\mathcal{H}$.

Let $\mathcal{H}$ be a class of infinite VC-dimension. Then, $\mathcal{H}$ is not PAC learnable. A finite VC-dimension guarantees learnability. Hence, VC-dimension characterizes PAC learnability. The concept of VC-dimension makes it possible to deal with learnability in continuous domains. VC-dimension also determines the sample complexity, as given in (3.3). The sample complexity is proportional to VC-dimension. Some sample complexity results are given in [33].

SRM paradigm is to find a hypothesis that minimizes a certain upper bound on the true risk. Specifying SRM paradigm for countable hypothesis classes yields MDL paradigm. In SRM paradigm, prior knowledge is expressed by specifying preferences over hypotheses within $\mathcal{H} = \bigcup_{n \in N} \mathcal{H}_n$, and assigning a weight to each hypothesis class $\mathcal{H}_n$.

Natarajan dimension is a generalization of VC-dimension to classes of multiclass predictors. The multiclass fundamental theorem is also derived based on Natarajan dimension [33].

3.7 No-Free-Lunch Theorem

Before the no-free-lunch theorem [45] was proposed, people intuitively believed that there exist some universally beneficial algorithms for search, and many people actually made efforts to design some algorithms. The no-free-lunch theorem asserts that there is no universally beneficial algorithm.

The no-free-lunch theorem states that no search algorithm is better than another in locating an extremum of a cost function when averaged over the set of all possible discrete functions. That is, all search algorithms achieve the same performance as random enumeration, when evaluated over the set of all functions.

Theorem 3.4 (No-Free-Lunch Theorem) *Given the set of all functions $\mathcal{F}$ and a set of benchmark functions $\mathcal{F}_1$, if algorithm A_1 is better on average than algorithm A_2 on $\mathcal{F}_1$, then algorithm A_2 must be better than algorithm A_1 on $\mathcal{F} - \mathcal{F}_1$.*

The performance of any algorithm is determined by the knowledge concerning the cost function. Thus, it is meaningless to evaluate the performance of an algorithm without specifying the prior knowledge. Practical problems always contain priors such as smoothness, symmetry, and i.i.d. samples. Prior knowledge can be expressed by restricting the hypothesis class. For example, although neural networks are usually deemed a powerful approach for classification, they cannot solve all classification problems. For some arbitrary classification problems, other methods may be efficient.

The no-free-lunch theorem was later extended to coding methods, early stopping [9], avoidance of overfitting, and noise prediction [26]. Again, it has been asserted that no one method is better than the others for all problems.

Following the no-free-lunch theorem, the inefficiency of leave-one-out cross-validation was demonstrated on a simple problem in [47]. In response to [47], in [16], the strict leave-one-out cross-validation was shown to yield the expected results on this simple problem, and thus leave-one-out cross-validation is not subject to the no-free-lunch criticism [16]. Nonetheless, it is concluded in [30] that the statistical tests are preferable to cross-validation for linear as well as for nonlinear model selection.

Problems

3.1 Show that the sample complexity $m_{\mathcal{H}}$, given by (3.4), is monotonially nonincreasing in terms of ϵ and δ.

3.2 Plot four points that are in general position. Show how they are separated by separating lines.

3.3 Assume that we have a class of functions $\{f(\boldsymbol{x}, \boldsymbol{\alpha})\}$ indexed by a parameter vector $\boldsymbol{\alpha}$, with $\boldsymbol{x} \in R^p$, f being an indicator function, taking value 0 or 1. If $\boldsymbol{\alpha} = (\alpha_0, \alpha_1)$ and f is the linear indicator function $I(\alpha_0 + \alpha_1 x > 0)$, then the complexity of the class f is the number of parameters $p + 1$.

The indicator function $I(\sin(\alpha x) > 0)$ can shatter (separate) an arbitrarily large number of points by choosing an appropriately high frequency α. Show that the set of functions $\{I(\sin(\alpha x) > 0)\}$ can shatter the following points on the line: $x_1 = 2^{-1}, \ldots, x_M = 2^{-M}$, $\forall M$. Hence, the VC-dimension of the class $\{I(\sin(\alpha x) > 0)\}$ is infinite.

3.4 Show the monotonicity of VC-dimension: For any two hypothesis classes, if $\mathcal{H}' \subseteq \mathcal{H}$ then $\dim_{VC}(\mathcal{H}') \leq \dim_{VC}(\mathcal{H})$.

3.5 Describe the implications of the no-free-lunch theorem.

References

1. Anguita, D., Ghio, A., Oneto, L., & Ridella, S. (2014). A deep connection between the Vapnik-Chervonenkis entropy and the Rademacher complexity. *IEEE Transactions on Neural Networks and Learning Systems*, *25*(12), 2202–2211.
2. Anthony, M., & Biggs, N. (1992). *Computational learning theory*. Cambridge, UK: Cambridge University Press.
3. Bartlett, P. L. (1993). Lower bounds on the Vapnik-Chervonenkis dimension of multi-layer threshold networks. In *Proceedings of the 6th Annual ACM Conference on Computational Learning Theory* (pp. 144–150). New York: ACM Press.
4. Bartlett, P. L., & Maass, W. (2003). Vapnik-Chervonenkis dimension of neural nets. In M. A. Arbib (Ed.), *The handbook of brain theory and neural networks* (2nd ed., pp. 1188–1192). Cambridge: MIT Press.
5. Bartlett, P. L., & Mendelson, S. (2003). Rademacher and Gaussian complexities: Risk bounds and structural results. *Journal of Machine Learning Research*, *3*, 463–482.
6. Bartlett, P. L., Long, P. M., & Williamson, R. C. (1994). Fat-shattering and the learnability of real-valued functions. In *Proceedings of the 7th Annual ACM Conference on Computational Learning Theory* (pp. 299–310). New Brunswick, NJ.
7. Bartlett, P. L., Bousquet, O., & Mendelson, S. (2005). Local Rademacher complexities. *Annals of Statistics*, *33*(4), 1497–1537.
8. Baum, E. B., & Haussler, D. (1989). What size net gives valid generalization? *Neural Computation*, *1*, 151–160.
9. Cataltepe, Z., Abu-Mostafa, Y. S., & Magdon-Ismail, M. (1999). No free lunch for early stropping. *Neural Computation*, *11*, 995–1009.
10. Cherkassky, V., & Ma, Y. (2003). Comparison of model selection for regression. *Neural Computation*, *15*, 1691–1714.
11. Cherkassky, V., & Ma, Y. (2009). Another look at statistical learning theory and regularization. *Neural Networks*, *22*, 958–969.
12. Cover, T. M. (1965). Geometrical and statistical properties of systems of linear inequalities with applications in pattern recognition. *IEEE Transactions on Electronic Computers*, *14*, 326–334.
13. Dudley, R. (1967). The sizes of compact subsets of Hilbert space and continuity of Gaussian processes. *Journal of Functional Analysis*, *1*(3), 290–330.
14. Friedrichs, F., & Schmitt, M. (2005). On the power of Boolean computations in generalized RBF neural networks. *Neurocomputing*, *63*, 483–498.
15. Goldman, S., & Kearns, M. (1995). On the complexity of teaching. *Journal of Computer and Systems Sciences*, *50*(1), 20–31.
16. Goutte, C. (1997). Note on free lunches and cross-validation. *Neural Computation*, *9*(6), 1245–1249.

17. Gribonval, R., Jenatton, R., Bach, F., Kleinsteuber, M., & Seibert, M. (2015). Sample complexity of dictionary learning and other matrix factorizations. *IEEE Transactions on Information Theory*, *61*(6), 3469–3486.
18. Hanneke, S., & Yang, L. (2015). Minimax analysis of active learning. *Journal of Machine Learning Research*, *16*, 3487–3602.
19. Hastie, T., Tibshirani, R., & Friedman, J. (2005). *The elements of statistical learning: Data mining, inference, and prediction* (2nd ed.). Berlin: Springer.
20. Haussler, D. (1990). Probably approximately correct learning. In *Proceedings of the 8th National Conference on Artificial Intelligence* (Vol. 2, pp. 1101–1108). Boston, MA.
21. Haykin, S. (1999). *Neural networks: A comprehensive foundation* (2nd ed.). Upper Saddle River, NJ: Prentice Hall.
22. Koiran, P., & Sontag, E. D. (1996). Neural networks with quadratic VC dimension. In D. S. Touretzky, M. C. Mozer, & M. E. Hasselmo (Eds.), *Advances in neural information processing systems* (Vol. 8, pp. 197–203). Cambridge, MA: MIT Press.
23. Koltchinskii, V. (2001). Rademacher penalties and structural risk minimization. *IEEE Transactions on Information Theory*, *47*(5), 1902–1914.
24. Lei, Y., Ding, L., & Zhang, W. (2015). Generalization performance of radial basis function networks. *IEEE Transactions on Neural Networks and Learning Systems*, *26*(3), 551–564.
25. Liu, J., & Zhu, X. (2016). The teaching dimension of linear learners. *Journal of Machine Learning Research*, *17*, 1–25.
26. Magdon-Ismail, M. (2000). No free lunch for noise prediction. *Neural Computation*, *12*, 547–564.
27. Mendelson, S. (2002). Rademacher averages and phase transitions in Glivenko-Cantelli classes. *IEEE Transactions on Information Theory*, *48*(1), 251–263.
28. Mendelson, S. (2003). A few notes on statistical learning theory. In S. Mendelson & A. Smola (Eds.), *Advanced lectures on machine learning (Lecture notes computer science)* (Vol. 2600, pp. 1–40). Berlin: Springer-Verlag.
29. Oneto, L., Anguita, D., & Ridella, S. (2016). A local Vapnik-Chervonenkis complexity. *Neural Networks*, *82*, 62–75.
30. Rivals, I., & Personnaz, L. (1999). On cross-validation for model selection. *Neural Computation*, *11*(4), 863–870.
31. Schapire, R. E. (1990). The strength of weak learnability. *Machine Learning*, *5*, 197–227.
32. Schmitt, M. (2005). On the capabilities of higher-order neurons: A radial basis function approach. *Neural Computation*, *17*, 715–729.
33. Shalev-Shwartz, S., & Ben-David, S. (2014). *Understanding machine learning: From theory to algorithms*. New York, NY: Cambridge University Press.
34. Shao, X., Cherkassky, V., & Li, W. (2000). Measuring the VC-dimension using optimized experimental design. *Neural Computation*, *12*, 1969–1986.
35. Shawe-Taylor, J. (1995). Sample sizes for sigmoidal neural networks. In *Proceedings of the 8th Annual Conference on Computational Learning Theory* (pp. 258–264). Santa Cruz, CA.
36. Shawe-Taylor, J., Bartlett, P. L., Williamson, R. C., & Anthony, M. (1998). Structural risk minimization over data-dependent hierarchies. *IEEE Transactions on Information Theory*, *44*, 1926–1940.
37. Shinohara, A., & Miyano, S. (1991). Teachability in computational learning. *New Generation Computing*, *8*(4), 337–348.
38. Simon, H. U. (2015). An almost optimal PAC algorithm. In *Proceedings of the 28th Conference on Learning Theory* (pp. 1–12). Paris.
39. Valiant, P. (1984). A theory of the learnable. *Communications of the ACM*, *27*(11), 1134–1142.
40. Vapnik, V. N. (1982). *Estimation of dependences based on empirical data*. New York: Springer-Verlag.
41. Vapnik, V. N. (1995). *The nature of statistical learning theory*. New York: Springer.
42. Vapnik, V. N. (1998). *Statistical learning theory*. New York: Wiley.
43. Vapnik, V. N., & Chervonenkis, A. J. (1971). On the uniform convergence of relative frequencies of events to their probabilities. *Theory of Probability & its Applications*, *16*, 264–280.

44. Vapnik, V., Levin, E., & Le Cun, Y. (1994). Measuring the VC-dimension of a learning machine. *Neural Computation*, *6*, 851–876.
45. Wolpert, D. H., & Macready, W. G. (1995). *No free lunch theorems for search*, SFI-TR-95-02-010, Santa Fe Institute.
46. Yu, H.-F., Jain, P., & Dhillon, I. S. (2014). Large-scale multi-label learning with missing labels. In *Proceedings of the 21st International Conference on Machine Learning* (pp. 1–9).
47. Zhu, H. (1996). No free lunch for cross validation. *Neural Computation*, *8*(7), 1421–1426.

Chapter 4
Perceptrons

4.1 One-Neuron Perceptron

The perceptron [38], also referred to as a McCulloch–Pitts neuron or linear threshold gate, is the earliest and simplest neural network model. Rosenblatt used a single-layer perceptron for the classification of linearly separable patterns.

For a one-neuron perceptron, the network topology is shown in Fig. 1.2, and the net input to the neuron is given by

$$net = \sum_{i=1}^{J_1} w_i x_i - \theta = \boldsymbol{w}^T \boldsymbol{x} - \theta, \tag{4.1}$$

where all the symbols are explained in Sect. 1.2. The one-neuron perceptron using the hard-limiter activation function is useful for classification of vector $\boldsymbol{x}$ into two classes. The two decision regions are separated by a hyperplane

$$\boldsymbol{w}^T \boldsymbol{x} - \theta = 0, \tag{4.2}$$

where the threshold θ is a parameter used to shift the decision boundary away from the origin.

The three popular activation functions are the hard-limiter (threshold) function,

$$\phi(x) = \begin{cases} 1, & x \geq 0 \\ -1 (\text{or } 0), & x < 0 \end{cases}, \tag{4.3}$$

the logistic function

$$\phi(x) = \frac{1}{1 + e^{-\beta x}}, \tag{4.4}$$

and the hyperbolic tangent function

K.-L. Du and M. N. S. Swamy, *Neural Networks and Statistical Learning*,
https://doi.org/10.1007/978-1-4471-7452-3_4

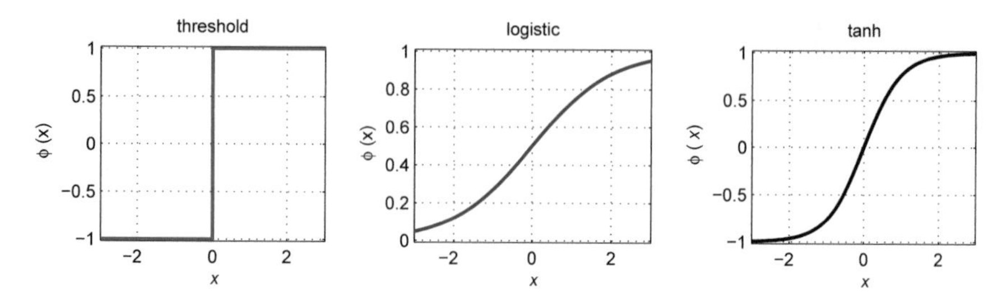

Fig. 4.1 Sigmoidal activation functions

$$\phi(x) = \tanh(\beta x). \tag{4.5}$$

In these functions, β is a gain, typically selected as unity, and is used to control the steepness of the activation function. These activation functions are illustrated in Fig. 4.1.

All the above functions are monotonically increasing with the domain of output $(-1, 1)$ or $(0, 1)$. Sigmoidal functions are usually defined as those monotonically increasing functions satisfying $\lim_{x\to+\infty} \phi(x) = 1$, $\lim_{x\to-\infty} \phi(x) = 0$. Many functions satisfy this definition if stretched out, and they can be treated as sigmoidal functions. Many other sigmoidal activation functions are introduced in [12].

A biologically more plausible perceptron is presented in [40] based on the integrate-and-fire model, with the derived learning rule which enables training of the neuron on nonlinear tasks. The model encodes the mean interspike interval, refractory period, and voltage threshold. It is possible to train such a neuron model by seeking to minimize the output error, and derive a learning rule from the mean interspike interval of the neuron's output.

4.2 Single-Layer Perceptron

When more neurons with the hard-limiter activation function are used, we have a single-layer perceptron, as shown in Fig. 4.2. The single-layer perceptron can be used to classify input vector data $\boldsymbol{x}$ into more classes. For a J_1-J_2 perceptron, the system state is updated by

$$\boldsymbol{net} = \mathbf{W}^T \boldsymbol{x} - \boldsymbol{\theta}, \tag{4.6}$$

$$\hat{\boldsymbol{y}} = \boldsymbol{\phi}(\boldsymbol{net}), \tag{4.7}$$

where the net input vector $\boldsymbol{net} = \left(net_1, \ldots, net_{J_2}\right)^T$, the output vector $\hat{\boldsymbol{y}} = \left(\hat{y}_1, \ldots, \hat{y}_{J_2}\right)^T$, $\boldsymbol{\theta} = \left(\theta_1, \ldots, \theta_{J_m}\right)^T$ corresponds to all the biases in the second layer, and $\boldsymbol{\phi}(\boldsymbol{net}) = \left(\phi_1\left(net_1\right)^T, \ldots, \phi_{J_2}\left(net_{J_2}\right)\right)$ corresponds to all the activation functions of the neurons.

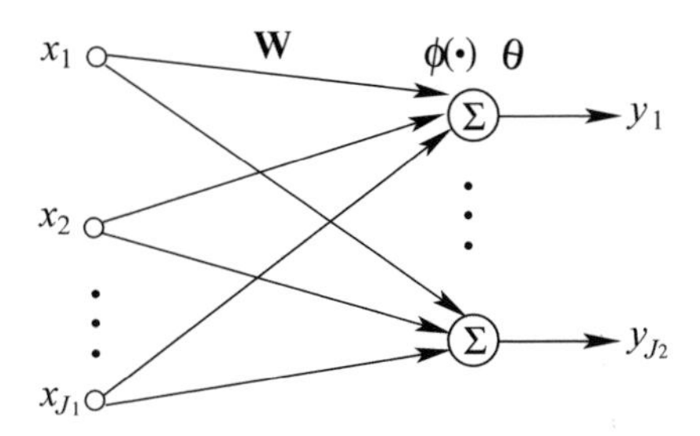

Fig. 4.2 Architecture of the single-layer perceptron

The problem of finding the weights of a single sigmoidal neuron that minimize the quadratic training error proves to be NP-hard [42]. The adaptation of $\mathbf{W}$ is error driven, which can be according to Rosenblatt's perceptron learning algorithm [38, 39] or according to the LMS algorithm based on the adaline model [45].

4.3 Perceptron Learning Algorithm

Rosenblatt proved the perceptron convergence theorem for classification problems [39].

Theorem 4.1 (Perceptron convergence) *Given a one-neuron perceptron and input patterns* $\boldsymbol{x} \in \mathcal{X}$ *from two linearly separable classes. Let the patterns be presented in an arbitrary sequence in each epoch. Then, starting from an arbitrary initial state, the perceptron learning procedure always converges and yields a decision hyperplane between the two classes in finite time.*

From the perceptron convergence theorem, the weights of the perceptron will converge to a fixed point within a finite number of updates for a set of linearly separable input patterns. The perceptron convergence theorem has been extended for the MLP, stating that the pattern mode BP algorithm converges to an optimal solution for linearly separable patterns with no upper bound on the learning rate [21].

The perceptron convergence theorem can be proved by minimizing the following perceptron criterion function using the gradient-descent method:

$$E(\boldsymbol{w}) = \sum_{\boldsymbol{x} \in \overline{\mathcal{X}}} \left(-\boldsymbol{w}^T \boldsymbol{x}\right), \tag{4.8}$$

where $\overline{\mathcal{X}}$ is the set of samples misclassified by $\boldsymbol{w}$. Thus, the weights are modified in such a manner so as to reduce the number of misclassifications. The perceptron convergence theorem can be easily extended to the single-layer perceptron by extending the perceptron learning algorithm from one neuron to multiple neurons.

The perceptron learning algorithm is given as

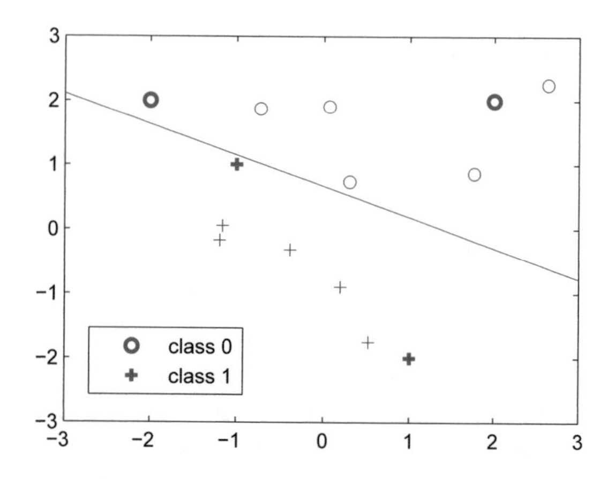

Fig. 4.3 Use perceptron learning for classification

$$net_{t,j} = \sum_{i=1}^{J_1} x_{t,i} w_{ij}(t) - \theta_j = \boldsymbol{w}_j^T \boldsymbol{x}_t - \theta_j, \tag{4.9}$$

$$\hat{y}_{t,j} = \begin{cases} 1, & net_{t,j} > 0 \\ 0, & \text{otherwise} \end{cases}, \tag{4.10}$$

$$e_{t,j} = y_{t,j} - \hat{y}_{t,j}, \tag{4.11}$$

$$w_{ij}(t+1) = w_{ij}(t) + \eta x_{t,i} e_{t,j}, \tag{4.12}$$

for $i = 1, \ldots, J_1$, $j = 1, \ldots, J_2$, where $net_{t,j}$ is the net input of the jth neuron for the tth example, $\boldsymbol{w}_j = \left(w_{1j}, w_{2j}, \ldots, w_{J_1 j}\right)^T$ is the vector collecting all weights terminated at the jth neuron, θ_j is the threshold for the jth neuron, $x_{t,i}$ is the ith input of the tth example, $\hat{y}_{t,j}$ and $y_{t,j}$ are, respectively, the network output and the desired output of the jth neuron for the tth example, with value 0 or 1 representing class membership, and η is the learning rate. All the weights w_{ij} are randomly initialized. The selection of η does not affect the stability of perceptron learning, and affects the convergence speed only for nonzero initial weight vector. η is typically selected as 0.5. The learning process stops when the errors are sufficiently small.

Example 4.1 For a classification problem, the inputs $(2, 2)$, $(-2, 2)$ are in class 0, $(1, -2)$, $(-1, 1)$ are in class 1. Select the initial weights and bias as random numbers between 0 and 1. After training for one epoch, the algorithm converges. The result is illustrated in Fig. 4.3. In the figure, the learning class boundary is $\boldsymbol{w}^T \boldsymbol{x} - \theta = 0.9294 x_1 + 0.7757 x_2 + 0.4868 = 0$. Randomly generate 10 points, and the learned perceptron can correctly classify them. Training can be implemented in adaptive learning mode.

When used for classification, perceptron learning can operate only for linearly separable patterns, and does not terminate for linearly inseparable patterns. The failure of Rosenblatt's and similar methods to converge for linearly inseparable problems is caused by the inability of the methods to detect the minimum of the error function [14].

For a set of nonlinearly separable input patterns, the obtained weights of a perceptron may exhibit a limit cycle behavior. A perceptron exhibiting the limit cycle behavior is actually a neural network with time periodically varying coefficients. The minimum number of updates for the weights of the perceptron to reach the limit cycle depends on the initial weights. The boundedness condition of the perceptron weights is independent of the initial weights [24]. Also, a necessary and sufficient condition for the weights of the perceptron exhibiting a limit cycle behavior is derived, and the range of the number of updates for the weights of the perceptron required to reach the limit cycle is estimated in [24]. In [25], an invariant set of the weights of the perceptron trained by the perceptron training algorithm is defined and characterized. The dynamic range of the steady-state values of the weights can be evaluated by finding the dynamic range of the weights inside the largest invariant set.

The pocket algorithm [19] improves on perceptron learning by adding a checking amendment to stop the algorithm; it optimally dichotomizes the given patterns in the sense of minimizing the erroneous classification rate. It can be applied for the classification of linearly inseparable patterns. The weight vector with the longest unchanged run is identified as the best solution so far and is stored in the *pocket*. The content of the pocket is replaced by any new weight vector with a longer successful run. The pocket convergence theorem guarantees the optimal convergence of the pocket algorithm, if the inputs in the training set are integers or rational [19, 34]. The pocket algorithm with ratchet [19] evaluates the hypotheses on the entire training set and picks the best; it is asserted to find an optimal weight vector with probability one within a finite number of iterations, independently of the given training set [34].

Thermal perceptron learning [17] is obtained by multiplying the second term of (4.12) by a temperature annealing factor $e^{-\frac{|net_j|}{T}}$, where T is an annealing temperature. It finds stable weights for inseparable problems as well as for separable ones. It can be applied for the classification of linearly inseparable patterns.

4.4 Least Mean Squares (LMS) Algorithm

LMS algorithm [45] achieves a robust separation between the patterns of different classes by minimizing the MSE rather than the number of misclassified patterns through the gradient-descent method. Like perceptron learning, it can only be used for the classification of linearly separable patterns. In the LMS algorithm, the activation function is linear, and the error is defined by

$$e_{t,j} = y_{t,j} - net_{t,j}, \tag{4.13}$$

where $net_{t,j}$ is defined by (4.9). The weight update rule is the same as (4.12), and is reproduced here

$$w_{ij}(t+1) = w_{ij}(t) + \eta x_{t,i} e_{t,j}. \tag{4.14}$$

For classification problems, a threshold activation function is further applied to the linear output so as to render the final output to $\{0, 1\}$ or $\{+1, -1\}$

$$\hat{y}_{t,j} = \begin{cases} 1, & net_{t,j} > 0 \\ 0, & \text{otherwise} \end{cases}. \tag{4.15}$$

The whole unit including a linear combiner and the following threshold operation is called an *adaptive linear element (adaline)* [45]. The above LMS rule is also called the *μ-LMS rule*. For practical purposes, η can be selected as $0 < \eta < \frac{2}{\max_t \|\boldsymbol{x}_t\|^2}$ to ensure its convergence.

Widrow–Hoff delta rule, known as the *α-LMS*, is a modification to the LMS rule obtained by normalizing the input vector so that the weights change independently of the magnitude of the input vector [46]

$$w_{ij}(t+1) = w_{ij}(t) + \eta \frac{x_{t,i} e_{t,j}}{\|\boldsymbol{x}_t\|^2}. \tag{4.16}$$

For the convergence of the α-LMS rule, η should be selected as $0 < \eta < 2$, and a practical range for η is $0.1 < \eta < 1.0$ [46]. Unlike perceptron learning, the LMS method can also be used for function approximation. In this case, the threshold operation in the adaline is dropped, and the behavior of the adaline is identical to that of linear regression.

There are also madaline models using layered multiple adalines [46]. Madaline still cannot solve linearly inseparable problems since the adaline network is a linear neural network and consecutive layers can be simplified to a single layer by multiplying the respective weight matrices. The Widrow–Hoff delta rule has become the foundation of modern adaptive signal processing. A complex LMS is given in [6].

Example 4.2 For a classification problem, the inputs $(1, 2)$, $(-2, 1)$ are in class 0, $(1, -1)$, $(-1, 0)$ are in class 1. Use the initial weights and bias as random numbers between 0 and 1. After training for one epoch, the algorithm converges. The result is illustrated in Fig. 4.4. In the figure, the learning class boundary is $\boldsymbol{w}^T\boldsymbol{x} - \theta = 0.0447x_1 - 0.3950x_2 + 0.7080 = 0$. Randomly generate 50 points, and the learned linear model can correctly classify them. Training is implemented in adaptive learning mode. In the model, a threshold function is applied for classification. It classifies a pattern into class 0 if the output is less than 0.5, or class 1 otherwise. Notice that the learned linear model optimizes the MSE but not the classification accuracy.

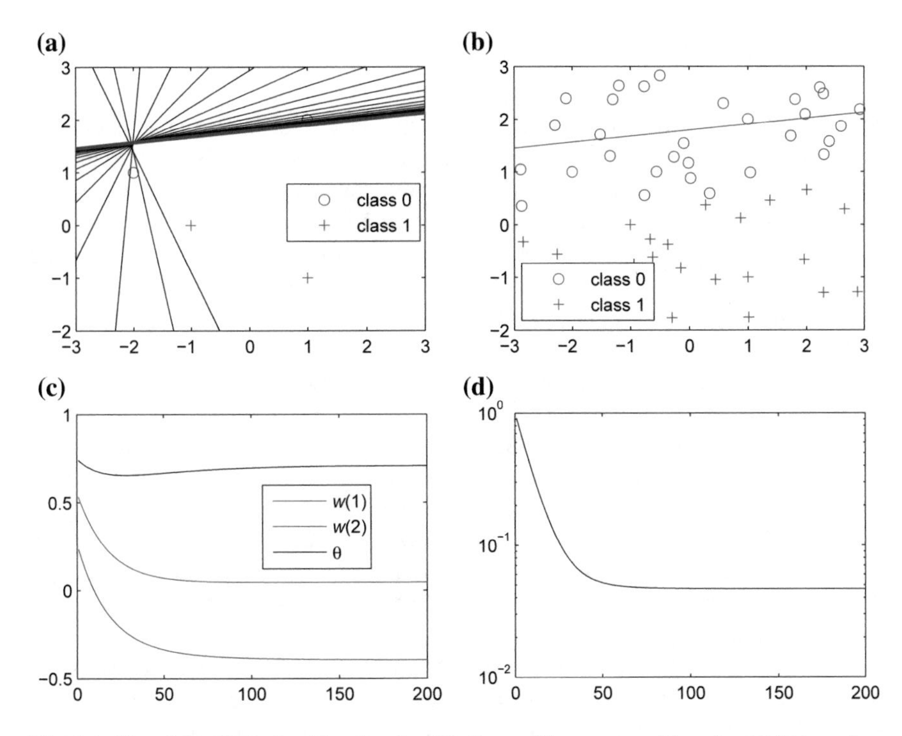

Fig. 4.4 Use of the LMS algorithm for classification. **a** The process of learning LMS boundary. **b** Classification result. **c** The change of the weights and bias. **d** The evolution of the MSE

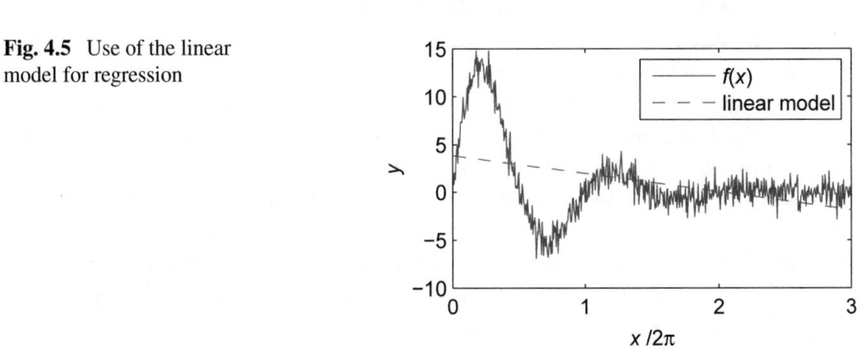

Fig. 4.5 Use of the linear model for regression

Example 4.3 We use the linear networks to approximate $f(x) = 20e^{-0.3x} \sin x + N(0, 1)$, where $N(0, 1)$ is Gaussian noise with zero mean and variance 1. The result is illustrated in Fig. 4.5. In the figure, the learning class boundary is $y = \boldsymbol{w}^T \boldsymbol{x} - \theta = -0.2923x + 3.7344$. The linear model achieves optimum approximation in terms of MSE.

4.5 P-Delta Rule

The perceptron algorithm using margins [28] attempts to establish guarantees on the separation margin during the training process. In most cases, similar performance is obtained by the voted perceptron, which has the advantage that it does not require parameter selection. Techniques using soft margin ideas are runtime intensive and do not give additional performance benefits [26]. In terms of run time, the voted perceptron does not require parameter selection and can therefore be faster to train. Both the voted perceptron and the margin variant reduce the deviation in accuracy in addition to improving the accuracy [26].

The voted perceptron [18] assigns each vector a vote based on the number of sequential correct classifications by that weight vector. Whenever an example is misclassified, the voted perceptron records the number of correct classifications made since the previous misclassification, assigns this number to the current weight vector's vote, saves the current weight vector, and then updates as normal. After training, all the saved vectors are used to classify future examples and their classifications are combined using their votes. When the data are linearly separable and given enough iterations, both these variants will converge to a hypothesis that is very close to the simple perceptron algorithm.

The single-layer perceptron can compute any Boolean function if their majority vote can be viewed as a binary output of the circuit, and they are universal approximators for arbitrary continuous functions with values in [0, 1] if one can apply a simple squashing function to the percentage of votes with value 1 [4]. The parallel perceptron has just binary values as outputs of gates on the hidden layer, implementing a soft-winner-take-all gate. These extremely simple neural networks are also known as *committee machines*.

The parallel delta (p-delta) rule is a simple learning algorithm for parallel perceptrons [4]. It has to tune a single layer of weights, and it does not require the computation and communication of analog values with high precision. These features make the p-delta rule attractive as a biologically more realistic alternative to BP. The p-delta rule also implements gradient descent with regard to a suitable error measure, although it does not require to compute derivatives. The p-delta rule follows a powerful principle from machine learning for committee machines: maximization of the margin of individual perceptrons.

Let $(\boldsymbol{x}, y) \in R^d \times [-1, +1]$ be the current training example and let $\boldsymbol{w}_1, \ldots, \boldsymbol{w}_n \in R^d$ be the current weight vectors of the n individual perceptrons in the parallel perceptron. Thus, the current output of the parallel perceptron is calculated as

$$\hat{y} = s_\rho(p), \quad p = \mathrm{card}\{i : \boldsymbol{w}_i \cdot \boldsymbol{x} \geq 0\} - \mathrm{card}\{i : \boldsymbol{w}_i \cdot \boldsymbol{x} < 0\}, \tag{4.17}$$

where $s_\rho(p)$ is a squashing function analogous to the sigmoidal function, and card denotes the size of the set.

For regression problems, the squashing function could be piecewise linear,

$$s_\rho(p) = \begin{cases} -1, & \text{if } p \le -\rho \\ p/\rho, & \text{if } -\rho < p < \rho \\ +1, & \text{if } p \ge \rho \end{cases}, \tag{4.18}$$

where $1 \le \rho \le n$ denotes the resolution of the squashing function.

The p-delta rule is given by [4]. For all $i = 1, \ldots, n$ and accuracy ϵ:

$$\boldsymbol{w}_i \leftarrow \boldsymbol{w}_i + \eta \Delta_i, \tag{4.19}$$

$$\Delta_i = \begin{cases} -\boldsymbol{x}, & \text{if } \hat{y} > y + \epsilon \text{ and } \boldsymbol{w}_i \cdot \boldsymbol{x} \ge 0 \\ +\boldsymbol{x}, & \text{if } \hat{y} < y - \epsilon \text{ and } \boldsymbol{w}_i \cdot \boldsymbol{x} < 0 \\ +\mu\boldsymbol{x}, & \text{if } \hat{y} \le y + \epsilon \text{ and } 0 \le \boldsymbol{w}_i \cdot \boldsymbol{x} < \gamma \\ -\mu\boldsymbol{x}, & \text{if } \hat{y} \ge y - \epsilon \text{ and } -\gamma < \boldsymbol{w}_i \cdot \boldsymbol{x} < 0 \\ 0, & \text{otherwise} \end{cases}, \tag{4.20}$$

$$\boldsymbol{w}_i \leftarrow \boldsymbol{w}_i / \|\boldsymbol{w}_i\|, \tag{4.21}$$

where $\gamma > 0$ is a margin, and μ, typically selected as 1, measures the importance of a clear margin.

Theorem 4.2 (Universal approximation, [4]) *Parallel perceptrons are universal approximators: Every continuous function $g : R^d \to [-1, 1]$ can be approximated by a parallel perceptron within any given error bound ϵ on any closed and bounded subset of R^d.*

Since any Boolean function from $\{0, 1\}^d$ into $\{0, 1\}$ can be interpolated by a continuous function, any Boolean function can be computed by rounding the output of a parallel perceptron.

Parallel perceptrons trained with the p-delta rule provide results comparable to that of MLP, madaline, decision tree (C4.5), and SVM, despite its simplicity [4]. The p-delta rule can also be applied to biologically realistic integrate-and-fire neuron models. It has already been applied successfully to the training of a pool of spiking neurons [32].

Direct parallel perceptrons [15] use an analytical closed-form expression to directly calculate the weights of parallel perceptrons that globally minimize an error function measuring simultaneously the classification margin and the training error. They have no tunable parameters. They have a computational complexity linear in the number of patterns and in the input dimension. They are tenfold faster than p-delta and two orders of magnitude faster than SVM. They also allow online learning.

4.6 Other Learning Algorithms

There are many other valid learning rules such as Mays' rule [33, 46], the Ho–Kashyap rule [13, 23], and adaptive Ho–Kashyap rules [22]. The Ho–Kashyap algorithm uses the pseudoinverse of the pattern matrix in order to determine the solution vector. The adaptive Ho–Kashyap algorithm does not calculate the pseudoinverse of the pattern matrix, but the empirical choice of the learning parameters is critical for the convergence of the algorithm. Like the perceptron learning algorithm, these algorithms converge only in the case of linearly separable datasets. The one-shot Hebbian learning [43] and nonlinear Hebbian learning [5] have also been used for perceptron learning.

A single-layer complex-valued neural network [1] solves real-valued classification problems by a gradient-descent learning rule: It maps real input values to complex values, and after processing in the complex-valued domain, and the activation function then maps complex values to real values.

Some single-layer perceptron learning algorithms are suitable for both linearly separable and linearly inseparable classification problems. Examples are the convex analysis and nonsmooth optimization-based method [14], the linear programming (LP) method [13, 31], the constrained steepest descent algorithm [37], fuzzy perceptron [9], and the conjugate-gradient (CG) method [35].

The problem of training a single-layer perceptron is to find a solution to a set of linear inequalities, and thus it is known as an LP problem. LP techniques have been applied to single-layer perceptron learning [13, 31]. They can solve linearly inseparable problems. When the training vectors are from $\{-1, +1\}^{J_1}$, the method requires $O\left(J_1^3 \log_2 J_1\right)$ learning cycles in the worst case while the perceptron convergence procedure may require $O\left(2^{J_1}\right)$ learning cycles [31].

The constrained steepest descent algorithm [37] has no free learning parameters. Learning proceeds by iteratively lowering the perceptron cost function following the direction of steepest descent, under the constraint that patterns already correctly classified are not to be affected. A decrease in the error is achieved at each iteration by employing the projection search direction when needed. The training task is decomposed into a succession of small-scale quadratic programming (QP) problems, whose solutions determine the appropriately constrained direction of steepest descent. For linearly separable problems, it always finds a hyperplane that completely separates the patterns belonging to different categories in a finite number of steps. In the case of linearly inseparable problems, the algorithm detects the inseparability in a finite number of steps and terminates, having usually found a good separation hyperplane.

The CG algorithm [35] is also used for perceptron learning, where heuristic techniques based on reinitialization of the CG method is used. A control-inspired approach [11] is applied to the design of iterative steepest descent and CG algorithms for perceptron training in batch mode, by regarding certain parameters of the training/algorithm as controls and then using a control Lyapunov technique to choose appropriate values of these parameters.

The shifting perceptron algorithm [8] is a budget algorithm for shifting hyperplanes. Shifting bounds for online classification algorithms ensure good performance on any sequence of examples that is well predicted by a sequence of changing classifiers.

Classical perceptron algorithm with margin is a member of a broader family of large margin classifiers collectively called the margitron [36]. The margitron, sharing the same update rule with the perceptron, is shown in an incremental setting to converge in a finite number of updates to solutions possessing any desirable fraction of the maximum margin.

Aggressive ROMMA [30] explicitly maximizes the margin on the new example, relative to an approximation of the constraints from previous examples. NORMA [27] performs gradient descent on the soft margin risk resulting in an algorithm that rescales the old weight vector before the additive update. The passive-aggressive algorithm [10] adapts η on each example to guarantee that it is immediately separable with margin. A second-order perceptron called Ballseptron [41] establishes a normalized margin and replaces margin updates with updates on hypothetical examples on which a mistake would be made by using spectral properties of the data in the updates. ALMA [20] renormalizes the weight vector so as to establish a normalized margin. It tunes its parameters automatically during the online session. ALMA has p-norm variants that can lead to other trade-offs improving the performance, for example, when the target is sparse.

The use of nonlinear activation functions causes local minima in the objective functions based on the MSE criterion. The number of such minima can grow exponentially with the input dimension [3]. When using an objective function that measures the errors before the neuron's nonlinear activation function instead of after them, for single-layer neural networks, the new convex objective function does not contain local minima and the global solution is obtained using a system of linear equations [7]. A theoretical analysis of this solution is given in [16], and a new set of linear equations, to obtain the optimal weights for the problem, is derived.

Sign-Constrained Perceptron

The perceptron learning rule and most existing learning algorithms for linear neurons or perceptrons are not true to physiological reality. In these algorithms, weights can take values of any sign. However, biological synapses are either excitatory or inhibitory and usually do not switch between excitation and inhibition. This fact is commonly referred to as *Dale's law*. In fact, many neurophysiologists prefer the assumption that only excitatory synapses are directly used for learning, whereas inhibitory synapses are tuned for other tasks. In the latter case, one arrives at a perceptron with nonnegative weights as a more realistic model. A variation of the perceptron convergence theorem for sign-constrained weights was proven in [2]. It tells us that if a sign-constrained perceptron can implement a given dichotomy, then it can learn it.

An analysis of the classification capability of a sign-constrained perceptron is given in [29]. In particular, the VC-dimension of sign-constrained perceptrons is determined, and a necessary and sufficient criterion is provided that tells us when all 2^m dichotomies over a given set of m patterns can be learned by a sign-constrained perceptron. Uniformity of L_1 norms of input patterns is a sufficient condition for full representation power in the case where all weights are required to be nonnegative. Sparse input patterns improve the classification capability of sign-constrained perceptrons. The VC-dimension is $n + 1$ for an unconstrained perceptron with input dimension n, while that of sign-constrained perceptrons is n [29].

Problems

4.1 Design a McCulloch–Pitts neuron to recognize the letter "X" digitalized in an 8×8 array of pixels.

4.2 Show that the hyperbolic tangent function (4.5) is only a biased and scaled logistic function (4.4).

4.3 Verify that the following functions can be used as sigmoidal functions. Plot these functions.

(a) $\phi(x) = \begin{cases} 1, & x > a \\ \frac{x}{2a} + \frac{1}{2}, & -a \le x \le a \\ 0, & x < -a \end{cases}.$

(b) $\phi(x) = \frac{2}{\pi} \arctan(\beta x)$.

(c) $\phi(x) = \frac{1}{2} + \frac{1}{\pi} \arctan(\beta x)$.

(d) $\phi(x) = e^{-e^{-x}}$ (Gompertz function).

4.4 A Taylor-series approximation of the logistic function is given by [44]

$$\phi(x) = \begin{cases} \frac{1}{b-x+0.5x^2}, & x < 0 \\ 1 - \frac{1}{b+x+0.5x^2}, & x \ge 0 \end{cases},$$

where $b \ge 2$ is a constant. When $b = 2$, $\phi(x)$ is a continuous function. Plot this function.

4.5 Show that the single-layer perceptron is a linear classifier. The perceptron can be used to implement the binary logic functions AND, OR, and COMPLEMENT, but not EXCLUSIVE OR (XOR). Show how it can or cannot implement these logic functions.

4.6 Build perceptrons that construct logical NOT, NAND, and NOR of their inputs.

4.7 The parity problem returns 1 if the number of inputs that are 1 is even, and 0 otherwise.

(a) Try to use a perceptron to learn the parity problem of three inputs.

(b) Show that the parity function of $n > 2$ binary input $x_1, x_2, \ldots, x_n$ cannot be simulated by a perceptron.

4.8 Is it possible to train a perceptron using a perceptron algorithm in which the bias is left unchanged and only the other weights are modified?

4.9 Generate 20 vectors in two dimensions, each belonging to one of two classes. Write a program to implement the perceptron learning algorithm. Plot the decision boundary after each iteration and investigate the behavior of the algorithm. The dataset can be linearly separable or linearly inseparable.

4.10 For a multilayer forward network, if all neurons operate in their linear regions, show that such a network can reduce to a single-layer feedforward network.

4.11 The α-LMS rule is given by

$$\boldsymbol{w}_{k+1} = \boldsymbol{w}_k + \alpha(d_k - y_k)\frac{\boldsymbol{x}_k}{\|\boldsymbol{x}_k\|^2},$$

where $d_k \in R$ is the desired output, $\boldsymbol{x}_k$ is the input vector, and $\alpha > 0$.
(a) Show that the α-LMS rule can be derived from an incremental gradient descent on

$$J(\boldsymbol{w}) = \frac{1}{2}\sum_i \frac{(d_i - y_i)^2}{\|\boldsymbol{x}_i\|}.$$

(b) Show that the Widrow–Hoff rule is stable when $0 < \alpha < 2$, unstable when $\alpha > 2$, and is oscillatory when $\alpha = 2$.

4.12 Given the two-class problem with class 1: $(3, 4)$, $(3, 1)$; class 2: $(-2, -1)$, $(-3, -4)$.
(a) With $\boldsymbol{w}_0 = (1, 0, 0)$, find the separating weight vector.
(b) Plot the decision surface.

References

1. Amin, M. F., & Murase, K. (2009). Single-layered complex-valued neural network for real-valued classification problems. *Neurocomputing*, *72*, 945–955.
2. Amit, D. J., Wong, K. Y. M., & Campbell, C. (1989). Perceptron learning with sign-constrained weights. *Journal of Physics A: Mathematical and General*, *22*, 2039–2045.
3. Auer, P., Hebster, M., & Warmuth, M. K. (1996). Exponentially many local minima for single neurons. In D. S. Touretzky, M. C. Mozer, & M. E. Hasselmo (Eds.), *Advances in neural information processing systems* (Vol. 8, pp. 316–322). Cambridge, MA: MIT Press.
4. Auer, P., Burgsteiner, H., & Maass, W. (2008). A learning rule for very simple universal approximators consisting of a single layer of perceptrons. *Neural Networks*, *21*, 786–795.
5. Bolle, D., & Shim, G. M. (1995). Nonlinear Hebbian training of the perceptron. *Network*, *6*, 619–633.
6. Bouboulis, P., & Theodoridis, S. (2011). Extension of Wirtinger's calculus to reproducing kernel Hilbert spaces and the complex kernel LMS. *IEEE Transactions on Signal Processing*, *59*(3), 964–978.
7. Castillo, E., Fontenla-Romero, O., Alonso-Betanzos, A., & Guijarro-Berdinas, B. (2002). A global optimum approach for one-layer neural networks. *Neural Computation*, *14*(6), 1429–1449.
8. Cavallanti, G., Cesa-Bianchi, N., & Gentile, C. (2007). Tracking the best hyperplane with a simple budget perceptron. *Machine Learning*, *69*, 143–167.
9. Chen, J. L., & Chang, J. Y. (2000). Fuzzy perceptron neural networks for classifiers with numerical data and linguistic rules as inputs. *IEEE Transactions on Fuzzy Systems*, *8*(6), 730–745.
10. Crammer, K., Dekel, O., Shalev-Shwartz, S., & Singer, Y. (2005). Online passive aggressive algorithms. *Journal of Machine Learning Research*, *7*, 551–585.

11. Diene, O., & Bhaya, A. (2009). Perceptron training algorithms designed using discrete-time control Liapunov functions. *Neurocomputing*, *72*, 3131–3137.
12. Duch, W. (2005). Uncertainty of data, fuzzy membership functions, and multilayer perceptrons. *IEEE Transactions on Neural Networks*, *16*(1), 10–23.
13. Duda, R. O., & Hart, P. E. (1973). *Pattern classification and scene analysis*. New York: Wiley.
14. Eitzinger, C., & Plach, H. (2003). A new approach to perceptron training. *IEEE Transactions on Neural Networks*, *14*(1), 216–221.
15. Fernandez-Delgado, M., Ribeiro, J., Cernadas, E., & Ameneiro, S. B. (2011). Direct parallel perceptrons (DPPs): Fast analytical calculation of the parallel perceptrons weights with margin control for classification tasks. *IEEE Transactions on Neural Networks*, *22*(11), 1837–1848.
16. Fontenla-Romero, O., Guijarro-Berdinas, B., Perez-Sanchez, B., & Alonso-Betanzos, A. (2010). A new convex objective function for the supervised learning of single-layer neural networks. *Pattern Recognition*, *43*(5), 1984–1992.
17. Frean, M. (1992). A thermal perceptron learning rule. *Neural Computation*, *4*(6), 946–957.
18. Freund, Y., & Schapire, R. (1999). Large margin classification using the perceptron algorithm. *Machine Learning*, *37*, 277–296.
19. Gallant, S. I. (1990). Perceptron-based learning algorithms. *IEEE Transactions on Neural Networks*, *1*(2), 179–191.
20. Gentile, C. (2001). A new approximate maximal margin classification algorithm. *Journal of Machine Learning Research*, *2*, 213–242.
21. Gori, M., & Maggini, M. (1996). Optimal convergence of on-line backpropagation. *IEEE Transactions on Neural Networks*, *7*(1), 251–254.
22. Hassoun, M. H., & Song, J. (1992). Adaptive Ho-Kashyap rules for perceptron training. *IEEE Transactions on Neural Networks*, *3*(1), 51–61.
23. Ho, Y. C., & Kashyap, R. L. (1965). An algorithm for linear inequalities and its applications. *IEEE Transactions of Electronic Computers*, *14*, 683–688.
24. Ho, C. Y.-F., Ling, B. W.-K., Lam, H.-K., & Nasir, M. H. U. (2008). Global convergence and limit cycle behavior of weights of perceptron. *IEEE Transactions on Neural Networks*, *19*(6), 938–947.
25. Ho, C. Y.-F., Ling, B. W.-K., & Iu, H. H.-C. (2010). Invariant set of weight of perceptron trained by perceptron training algorithm. *IEEE Transactions on Systems, Man, and Cybernetics Part B*, *40*(6), 1521–1530.
26. Khardon, R., & Wachman, G. (2007). Noise tolerant variants of the perceptron algorithm. *Journal of Machine Learning Research*, *8*, 227–248.
27. Kivinen, J., Smola, A. J., & Williamson, R. C. (2004). Online learning with kernels. *IEEE Transactions on Signal Processing*, *52*(8), 2165–2176.
28. Krauth, W., & Mezard, M. (1987). Learning algorithms with optimal stability in neural networks. *Journal of Physics A*, *20*(11), 745–752.
29. Legenstein, R., & Maass, W. (2008). On the classification capability of sign-constrained perceptrons. *Neural Computation*, *20*, 288–309.
30. Li, Y., & Long, P. (2002). The relaxed online maximum margin algorithm. *Machine Learning*, *46*, 361–387.
31. Mansfield, A. J. (1991). *Training perceptrons by linear programming*. NPL Report DITC 181/91, National Physical Laboratory, Teddington, Middlesex, UK.
32. Maass, W., Natschlaeger, T., & Markram, H. (2002). Real-time computing without stable states: A new framework for neural computation based on perturbations. *Neural Computation*, *14*(11), 2531–2560.
33. Mays, C. H. (1963). *Adaptive threshold logic*. PhD thesis, Stanford University.
34. Muselli, M. (1997). On convergence properties of pocket algorithm. *IEEE Transactions on Neural Networks*, *8*(3), 623–629.
35. Nagaraja, G., & Bose, R. P. J. C. (2006). Adaptive conjugate gradient algorithm for perceptron training. *Neurocomputing*, *69*, 368–386.
36. Panagiotakopoulos, C., & Tsampouka, P. (2011). The Margitron: A generalized perceptron with margin. *IEEE Transactions on Neural Networks*, *22*(3), 395–407.

37. Perantonis, S. J., & Virvilis, V. (2000). Efficient perceptron learning using constrained steepest descent. *Neural Networks, 13*(3), 351–364.
38. Rosenblatt, R. (1958). The Perceptron: A probabilistic model for information storage and organization in the brain. *Psychological Review, 65*, 386–408.
39. Rosenblatt, R. (1962). *Principles of neurodynamics*. New York: Spartan Books.
40. Rowcliffe, P., Feng, J., & Buxton, H. (2006). Spiking Perceptrons. *IEEE Transactions on Neural Networks, 17*(3), 803–807.
41. Shalev-Shwartz, S. & Singer, Y. (2005). A new perspective on an old perceptron algorithm. In: *Proceedings of the 16th Annual Conference on Computational Learning Theory* (pp. 264–278).
42. Sima, J. (2002). Training a single sigmoidal neuron Is hard. *Neural Computation, 14*, 2709–2728.
43. Vallet, F. (1989). The Hebb rule for learning linearly separable Boolean functions: learning and generalisation. *Europhysics Letters, 8*(8), 747–751.
44. Werbos, P. J. (1990). Backpropagation through time: What it does and how to do it. *Proceedings of the IEEE, 78*(10), 1550–1560.
45. Widrow, B. & Hoff, M. E. (1960). Adaptive switching circuits. In *Record of IRE Eastern Electronic Show & Convention (WESCON)* (Vol. 4, pp. 96–104).
46. Widrow, B., & Lehr, M. A. (1990). 30 years of adaptive neural networks: Perceptron, Madaline, and backpropagation. *Proceedings of the IEEE, 78*(9), 1415–1442.

Chapter 5
Multilayer Perceptrons: Architecture and Error Backpropagation

5.1 Introduction

MLPs are feedforward networks with one or more layers of units between the input and output layers. The output units represent a hyperplane in the space of the input patterns. The architecture of MLP is illustrated in Fig. 5.1. Assume that there are M layers, each having $J_m, m = 1, \ldots, M$, nodes. The weights from the $(m-1)$th layer to the mth layer are denoted by $\mathbf{W}^{(m-1)}$; the bias, output, and activation function of the ith neuron in the mth layer are, respectively, denoted as $\theta_i^{(m)}$, $o_i^{(m)}$, and $\phi_i^{(m)}(\cdot)$. An MLP trained with the BP algorithm is also called a *BP network*. MLP can be used for classification of linearly inseparable patterns and for function approximation.

From Fig. 5.1, we have the following relations. Notice that a plus sign precedes the bias vector for easy presentation. For $m = 2, \ldots, M$ and the pth example:

$$\hat{\boldsymbol{y}}_p = \boldsymbol{o}_p^{(M)}, \quad \boldsymbol{o}_p^{(1)} = \boldsymbol{x}_p, \tag{5.1}$$

$$\boldsymbol{net}_p^{(m)} = \left[\mathbf{W}^{(m-1)}\right]^T \boldsymbol{o}_p^{(m-1)} + \boldsymbol{\theta}^{(m)}, \tag{5.2}$$

$$\boldsymbol{o}_p^{(m)} = \boldsymbol{\phi}^{(m)}\left(\boldsymbol{net}_p^{(m)}\right), \tag{5.3}$$

where $\boldsymbol{net}_p^{(m)} = \left(net_{p,1}^{(m)}, \ldots, net_{p,J_m}^{(m)}\right)^T$, $\mathbf{W}^{(m-1)}$ is a J_{m-1}-by-J_m matrix, $\boldsymbol{o}_p^{(m-1)} = \left(o_{p,1}^{(m-1)}, \ldots, o_{p,J_{m-1}}^{(m-1)}\right)^T$, $\boldsymbol{\theta}^{(m)} = \left(\theta_1^{(m)}, \ldots, \theta_{J_m}^{(m)}\right)^T$ is the bias vector, and $\boldsymbol{\phi}^{(m)}(\cdot)$ applies $\phi_i^{(m)}(\cdot)$ to the ith component of the vector within.

All $\phi_i^{(m)}(\cdot)$ are typically selected to be the same sigmoidal function; one can also select all $\phi_i^{(m)}(\cdot)$ in the first $M-1$ layers as the same sigmoidal function, and all $\phi_i^{(m)}(\cdot)$ in the Mth layer as another continuous yet differentiable function.

K.-L. Du and M. N. S. Swamy, *Neural Networks and Statistical Learning*,
https://doi.org/10.1007/978-1-4471-7452-3_5

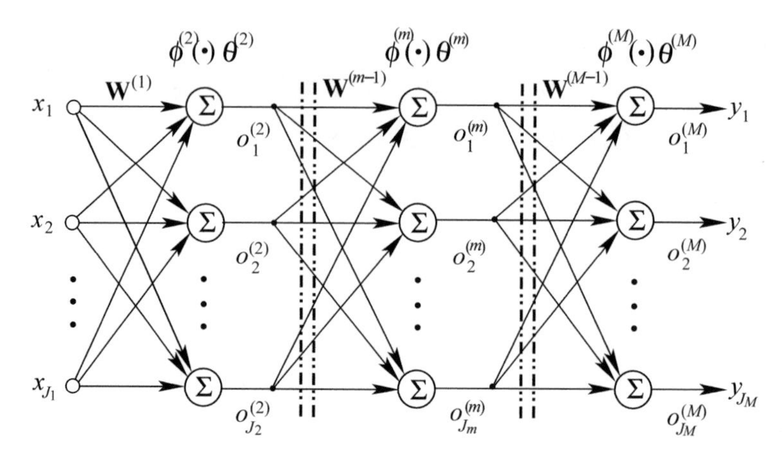

Fig. 5.1 Architecture of MLP

5.2 Universal Approximation

MLP is a universal approximator. Its universal approximation capability stems from the nonlinearities used in the nodes. The universal approximation capability of four-layer MLPs has been addressed in [46, 121]. It has been mathematically proved that a three-layer MLP using sigmoidal activation function can approximate any continuous multivariate function to any accuracy [24, 35, 45, 146]. Usually, a four-layer network can approximate the target with fewer connection weights, but this may, however, introduce extra local minima [18, 121, 146]. Xiang et al. provided a geometrical interpretation of MLP on the basis of the special geometrical shape of the activation function. For the target function with a flat surface located in the domain, a small four-layer MLP can generate better results [146].

MLP is very efficient for function approximation in high-dimensional spaces. The error convergence rate of MLP is independent of the input dimensionality, while conventional linear regression methods suffer from the curse of dimensionality, which results in a decrease of the convergence rate with an increase of the input dimensionality [6]. The necessary number of MLP neurons for approximating a target function depends only upon the basic geometrical shape of the target function, and not on the dimensionality of the input space. Based on a geometrical interpretation of MLP, the minimal number of line segments or hyperplanes that can construct the basic geometrical shape of the target function is suggested as the first trial for the number of hidden neurons of a three-layer MLP [146]. A similar result is given in [160], where the optimal network size can be selected according to the number of extrema and the number of hidden nodes should be selected as the number of extrema.

Approximation of piecewise continuous functions using smooth activation functions requires many hidden nodes and many training iterations but still does not yield very good results due to the Gibbs phenomenon. In [111], a neural network structure is given for approximation of piecewise continuous functions. It consists of neurons having standard sigmoidal functions, plus some additional neurons having a special class of nonsmooth activation functions termed *jump approximation basis function*. This structure can approximate any piecewise continuous function with discontinuities at a finite number of known points. A constructive proof that a real, piecewise continuous function can be almost uniformly approximated by three-layer feedforward networks is given in [73]. The constructive procedure avoids the Gibbs phenomenon.

The approximation of sufficiently smooth multivariable functions with an MLP is considered in [130]. For a given approximation order, explicit formulas for the necessary number of hidden units and its distributions to the hidden layers of MLP are derived. It turns out that more than two hidden layers are not needed for minimizing the number of necessary hidden units. Depending on the number of inputs and the desired approximation order, one or two hidden layers should be used. For high approximation orders (≥ 12), two hidden layers should be used instead of one hidden layer. The same is true for smaller approximation orders and a sufficiently high number of inputs, as long as the approximation order is at least three. A sufficient condition is given for the activation function for which a high approximation order implies a high approximation accuracy.

The Σ-Π network [108] is a generalization of MLP. Unlike MLP, it uses product units as well as summation units to build higher order terms. The BP learning rule can be applied to the learning of the network. The Σ-Π network is known to provide inherently more powerful mapping capabilities than first-order models such as MLP. It is a universal approximator [45]. However, it has a combinatorial increase in the number of product terms and weights.

5.3 Backpropagation Learning Algorithm

Though it is biologically implausible, BP learning is the most popular learning rule for performing supervised learning tasks [108, 141]. It is not only used to train feedforward networks such as MLP but also adapted to RNNs. BP algorithm is a generalization of the delta rule called *LMS algorithm*. Thus, it is also called the *generalized delta rule*. It uses a gradient-search technique to minimize a cost function equivalent to the MSE between the desired and actual network outputs. Due to the BP algorithm, MLP can be extended to many layers.

BP algorithm propagates backward the error between the desired signal and the network output through the network. After providing an input pattern, the output of the network is then compared with a given target pattern and the error of each output unit calculated. This error signal is propagated backward, and a closed-loop control system is thus established. The weights can be adjusted by a gradient-descent based algorithm.

In order to implement the BP algorithm, a continuous, nonlinear, monotonically increasing, differentiable activation function is needed. The logistic function and the hyperbolic tangent function are usually used. In the following, we derive the BP algorithm for MLP. BP algorithms for other neural network models can be derived in a similar manner.

The objective function for optimization is defined as the MSE between the actual network output $\hat{\boldsymbol{y}}_p$ and the desired output $\boldsymbol{y}_p$ for all the training pattern pairs $(\boldsymbol{x}_p, \boldsymbol{y}_p) \in \mathcal{S}$

$$E = \frac{1}{N} \sum_{p \in \mathcal{S}} E_p = \frac{1}{2N} \sum_{p \in \mathcal{S}} \left\| \hat{\boldsymbol{y}}_p - \boldsymbol{y}_p \right\|^2, \tag{5.4}$$

where N is the size of the sample set, and

$$E_p = \frac{1}{2} \left\| \hat{\boldsymbol{y}}_p - \boldsymbol{y}_p \right\|^2 = \frac{1}{2} \boldsymbol{e}_p^T \boldsymbol{e}_p, \tag{5.5}$$

$$\boldsymbol{e}_p = \hat{\boldsymbol{y}}_p - \boldsymbol{y}_p, \tag{5.6}$$

where the ith element of $\boldsymbol{e}_p$ is $e_{p,i} = \hat{y}_{p,i} - y_{p,i}$. Notice that a factor $\frac{1}{2}$ is used in E_p for the convenience of derivation.

All the network parameters $\mathbf{W}^{(m-1)}$ and $\boldsymbol{\theta}^{(m)}, m = 2, \ldots, M$, can be combined and represented by the matrix $\mathbf{W} = \left[w_{ij} \right]$. The error function E or E_p can be minimized by applying the gradient-descent procedure. When minimizing E_p, we have

$$\Delta_p \mathbf{W} = -\eta \frac{\partial E_p}{\partial \mathbf{W}}, \tag{5.7}$$

where η is the learning rate or step size, provided that it is a sufficiently small positive number. Note that the gradient term $\frac{\partial E_p}{\partial \mathbf{W}}$ is a matrix whose (i, j)th entry is $\frac{\partial E_p}{\partial w_{ij}}$.

Applying the chain rule, the derivative in (5.7) can be expressed as

$$\frac{\partial E_p}{\partial w_{uv}^{(m)}} = \frac{\partial E_p}{\partial net_{p,v}^{(m+1)}} \frac{\partial net_{p,v}^{(m+1)}}{\partial w_{uv}^{(m)}}. \tag{5.8}$$

The second factor of (5.8) is derived from (5.2)

$$\frac{\partial net_{p,v}^{(m+1)}}{\partial w_{uv}^{(m)}} = \frac{\partial}{\partial w_{uv}^{(m)}} \left(\sum_{\omega=1}^{J_m} w_{\omega v}^{(m)} o_{p,\omega}^{(m)} + \theta_v^{(m+1)} \right) = o_{p,u}^{(m)}. \tag{5.9}$$

The first factor of (5.8) can again be derived using the chain rule

$$\frac{\partial E_p}{\partial net_{p,v}^{(m+1)}} = \frac{\partial E_p}{\partial o_{p,v}^{(m+1)}} \frac{\partial o_{p,v}^{(m+1)}}{\partial net_{p,v}^{(m+1)}} = \frac{\partial E_p}{\partial o_{p,v}^{(m+1)}} \dot{\phi}_v^{(m+1)} \left(net_{p,v}^{(m+1)} \right), \tag{5.10}$$

where (5.3) is used. To solve the first factor of (5.10), we need to consider two situations for the output units ($m = M - 1$) and for the hidden units ($m = 1, \ldots, M - 2$):

$$\frac{\partial E_p}{\partial o_{p,v}^{(m+1)}} = e_{p,v}, \quad m = M - 1, \tag{5.11}$$

$$\begin{aligned}\frac{\partial E_p}{\partial o_{p,v}^{(m+1)}} &= \sum_{\omega=1}^{J_{m+2}} \left(\frac{\partial E_p}{\partial net_{p,\omega}^{(m+2)}} \frac{\partial net_{p,\omega}^{(m+2)}}{\partial o_{p,v}^{(m+1)}} \right) \\ &= \sum_{\omega=1}^{J_{m+2}} \left[\frac{\partial E_p}{\partial net_{p,\omega}^{(m+2)}} \frac{\partial}{\partial o_{p,v}^{(m+1)}} \left(\sum_{u=1}^{J_{m+1}} w_{u\omega}^{(m+1)} o_{p,u}^{(m+1)} + \theta_{\omega}^{(m+2)} \right) \right] \\ &= \sum_{\omega=1}^{J_{m+2}} \frac{\partial E_p}{\partial net_{p,\omega}^{(m+2)}} w_{v\omega}^{(m+1)}, \quad m = 1, \ldots, M - 2. \end{aligned} \tag{5.12}$$

Define the delta function by

$$\delta_{p,v}^{(m)} = -\frac{\partial E_p}{\partial net_{p,v}^{(m)}}, \quad m = 2, \ldots, M. \tag{5.13}$$

By substituting (5.8), (5.12), and (5.13) into (5.10), we finally obtain for the output units ($m = M - 1$) and for the hidden units ($m = 1, \ldots, M - 2$):

$$\delta_{p,v}^{(M)} = -e_{p,v} \dot{\phi}_v^{(M)} \left(net_{p,v}^{(M)} \right), \quad m = M - 1, \tag{5.14}$$

$$\delta_{p,v}^{(m+1)} = \dot{\phi}_v^{(m+1)} \left(net_{p,v}^{(m+1)} \right) \sum_{\omega=1}^{J_{m+2}} \delta_{p,\omega}^{(m+2)} w_{v\omega}^{(m+1)}, \quad m = 1, \ldots, M - 2. \tag{5.15}$$

Equations (5.14) and (5.15) provide a recursive method to solve $\delta_{p,v}^{(m+1)}$ for the whole network. Thus, $\mathbf{W}$ can be adjusted by

$$\frac{\partial E_p}{\partial w_{uv}^{(m)}} = -\delta_{p,v}^{(m+1)} o_{p,u}^{(m)}. \tag{5.16}$$

For the activation functions, we have the following relations:

$$\dot{\phi}(net) = \beta \phi(net) \left[1 - \phi(net) \right], \quad \text{for logistic function}, \tag{5.17}$$

$$\dot{\phi}(net) = \beta \left[1 - \phi^2(net) \right], \quad \text{for tanh function}. \tag{5.18}$$

The update for the biases can be in two ways. The biases in the $(m + 1)$th layer $\boldsymbol{\theta}^{(m+1)}$ can be expressed as the expansion of the weight $\mathbf{W}^{(m)}$, that is, $\boldsymbol{\theta}^{(m+1)} = \left(w_{0,1}^{(m)}, \ldots, w_{0,J_{m+1}}^{(m)} \right)^T$. Accordingly, the output $\boldsymbol{o}^{(m)}$ is expanded into

$\boldsymbol{o}^{(m)} = \left(1, o_1^{(m)}, \ldots, o_{J_m}^{(m)}\right)^T$. Another way is to use a gradient-descent method with regard to $\boldsymbol{\theta}^{(m)}$, by following the above procedure. Since the biases can be treated as special weights, these are usually omitted in practical applications.

The BP algorithm is defined by (5.7), and is rewritten here as follows:

$$\Delta_p \mathbf{W}(t) = -\eta \frac{\partial E_p}{\partial \mathbf{W}}. \tag{5.19}$$

The algorithm is convergent in the mean if $0 < \eta < \frac{2}{\lambda_{\max}}$, where $\lambda_{\max}$ is the largest eigenvalue of the autocorrelation of the vector $\boldsymbol{x}$, denoted by $\mathbf{R}$ [144]. When η is too small, the possibility of getting stuck at a local minimum of the error function is increased. In contrast, the possibility of falling into oscillatory traps is high when η is too large. By statistically preprocessing the input patterns, namely, decorrelating the input patterns, the excessively large eigenvalues of $\mathbf{R}$ can be avoided and thus, increasing η can effectively speedup the convergence. PCA preconditioning speeds up the BP in most cases, except when the pattern set consists of sparse vectors. In practice, η is usually chosen to be $0 < \eta < 1$ so that successive weight changes do not overshoot the minimum of the error surface. The flowchart of the BP for a three-layer MLP is shown in Algorithm 5.1.

Algorithm 5.1 (BP for a three-layer MLP).

All units have the same activation function $\phi(\cdot)$, and all biases are absorbed into weight matrices.

1. *Initialize $\mathbf{W}^{(1)}$ and $\mathbf{W}^{(2)}$.*
2. *Calculate E using (5.4).*
3. ***for** each epoch:*
   - *Calculate E using (5.4).*
   - ***if** E is less than a threshold ϵ, **return**.*
   - ***for** each $\boldsymbol{x}_p$, $p = 1, \ldots, N$:*
     a. *Forward pass*
        i. *Compute $\boldsymbol{net}_p^{(2)}$ by (5.2) and $\boldsymbol{o}_p^{(2)}$ by (5.3).*
        ii. *Compute $\boldsymbol{net}_p^{(3)}$ by (5.2) and $\hat{\boldsymbol{y}}_p = \boldsymbol{o}_p^{(3)}$ by (5.3).*
        iii *Compute $\boldsymbol{e}_p$ by (5.6).*
     b. *Backward pass, **for** all neurons*
        i. *Compute $\delta_{p,v}^{(3)} = -e_{p,v}\dot{\phi}\left(net_{p,v}^{(3)}\right)$*
        ii. *Update $\mathbf{W}^{(2)}$ by $\Delta w_{uv}^{(2)} = \eta \delta_{p,v}^{(3)} o_{p,u}^{(2)}$*
        iii *Compute $\delta_{p,v}^{(2)} = \left(\sum_{\omega=1}^{J_3} \delta_{p,\omega}^{(3)} w_{v\omega}^{(2)}\right) \dot{\phi}\left(net_{p,v}^{(2)}\right)$*
        iv. *Update $\mathbf{W}^{(1)}$ by $\Delta w_{uv}^{(1)} = \eta \delta_{p,v}^{(2)} o_{p,u}^{(1)}$*
4. ***end***

Fig. 5.2 Descent in weight space. **a** For small learning rate; **b** for large learning rate; **c** for large learning rate with momentum term added

The BP algorithm can be improved by adding a momentum term [108]

$$\Delta_p \mathbf{W}(t) = -\eta \frac{\partial E_p(t)}{\partial \mathbf{W}(t)} + \alpha \Delta \mathbf{W}(t-1), \tag{5.20}$$

where α is the momentum factor, usually $0 < \alpha \leq 1$. The typical value for α is 0.9. This method is usually called the *BP with momentum*. The momentum term can effectively magnify the descent in almost-flat steady downhill regions of the error surface by $\frac{1}{1-\alpha}$. In regions with high fluctuations (due to high learning rates), the momentum has a stabilizing effect. The momentum term actually inserts second-order information in the training process that performs like the CG method. The momentum term effectively smoothens the oscillations and accelerates the convergence. The role of the momentum term is shown in Fig. 5.2. BP with momentum is analyzed and the conditions for convergence are given in [137].

In addition to the gradient and momentum terms, a third term, namely, a proportional term, can be added to the BP update equation. The algorithm can be applied for both batch and incremental learning. For each example, the learning rule can be written as [163]

$$\Delta_p \mathbf{W}(t) = -\eta \frac{\partial E_p(t)}{\partial \mathbf{W}(t)} + \alpha \Delta_p \mathbf{W}(t-1) + \gamma E_p(\mathbf{W}(t))\,\mathbf{1}, \tag{5.21}$$

where the matrix $\mathbf{1}$ has the same size as $\mathbf{W}$ but with all the entries being unity, and γ is a proportional factor. This three-term BP algorithm is analogous to the common PID control algorithm used in feedback control. Three-term BP, having a complexity similar to the BP, significantly outperforms the BP in terms of the convergence speed, and the ability to escape from local minima. It is more robust to the choice of the initial weights, especially when relatively high values for the learning parameters are selected.

The emotional BP modifies the BP with additional emotional weights that are updated using two additional emotional parameters: the anxiety coefficient and the confidence coefficient [55].

In the above, the optimization objective is E_p and the weights are updated after the presentation of each pattern. Thus, the learning is termed as incremental learning, online learning, or pattern learning. When optimizing the average error E, we get the batch learning algorithm, where weights are updated only after all the training patterns are presented.

The essential storage requirement for the BP algorithm consists of all the N_w weights of the network. The computational complexity per iteration of the BP is around N_w multiplications for the forward pass, around $2N_w$ multiplications for the backward pass, and N_w multiplications for multiplying the gradient with η. Thus, four multiplications are required per iteration per weight [52].

Since BP is a gradient-descent technique, it is prone to local minima in the cost function. The performance can be improved and the occurrence of local minima can be reduced by allowing extra hidden units, lowering the gain term, and by training with different initial random weights. The process of presenting all the examples in the pattern set, with each example being presented once, is called an *epoch*. Neural networks are trained by presenting all the examples cyclically by epoch, until the convergence criteria is reached. The training examples should be presented to the network in a random order during each epoch.

Influence of Algorithm Parameters

The step size plays the role of a regularization parameter, whose choice controls the bias and variance properties of the solution [123]. In [71], the learning properties of the stochastic gradient method is analyzed, when multiple passes over the data and mini-batches are allowed. Regularization properties are controlled by the step size, the number of passes, and the mini-batch size. Given the square loss, for a universal step size choice, the number of passes acts as a regularization parameter, and optimal finite sample bounds can be achieved by early stopping. Larger step sizes are allowed when considering mini-batches. Optimal convergence results are derived for batch gradient methods.

The convergence rate and MSE performance of BP with momentum in the constant step size and slow adaptation regime is examined in [157]. Momentum methods are equivalent to standard stochastic gradient method with a rescaled (larger) step size value. The benefits of momentum constructions for deterministic optimization problems do not necessarily carry over to the adaptive online setting when small constant step sizes are used in the presence of persistent gradient noise. By employing a decaying momentum parameter, adaptation is retained without the often-observed degradation in mean-square deviation performance.

Based on the stability analysis for two steepest descent algorithms with momentum for quadratic functions, both the optimal learning rates and the optimal momentum factors are obtained simultaneously which make for the fastest convergence [159].

5.4 Incremental Learning Versus Batch Learning

Incremental learning and batch learning are two methods for BP learning. For incremental learning, the training patterns are presented to the network sequentially. It is a stochastic optimization method. For each training example, the weights are updated by the gradient-descent method

$$\Delta_p w_{ij}^{(m)} = -\eta_{\text{inc}} \frac{\partial E_p}{\partial w_{ij}^{(m)}}. \tag{5.22}$$

The learning algorithm has been proved to minimize the global error E when η_{inc} is sufficiently small [108].

In batch learning, the optimization objective is E, and the weight update is performed at the end of an epoch [108]. It is a deterministic optimization method. The weight incrementals for each example are accumulated over all the training examples before the weights are actually adapted

$$\Delta w_{ij}^{(m)} = -\eta_{\text{batch}} \frac{\partial E}{\partial w_{ij}^{(m)}} = \sum_p \Delta_p w_{ij}^{(m)}. \tag{5.23}$$

For sufficiently small learning rates, incremental learning approaches batch learning, and the two methods produce the same results [33].

Incremental learning can be used when the complete training set is not available, and it is especially effective when the training set is very large, which necessitates large additional storage in the case of batch learning. For small constant learning rates, the randomness introduced provides incremental learning with a quasi-annealing property, and allows for a wider exploration of the search space, which often helps in escaping from local minima [22]. However, incremental learning is hard to parallelize.

Gradient-descent algorithms are only truly gradient descent when their learning rates approach zero; thus, both the batch and incremental learning are using approximations of the true gradient as they move through the weight space. When η_{batch} is sufficiently small, batch learning follows incremental learning quite closely.

Incremental learning tends to be orders of magnitude faster than batch learning, and is at least as accurate as batch learning, especially for large training sets [145]. Online training is able to follow curves in the error surface throughout each cycle, which allows it to safely use a larger learning rate and thus converge with fewer iterations through the training data. For large training sets, batch learning is often completely impractical due to the minuscule η_{batch} required. Incremental training can safely use a larger η_{inc}, and can thus train more quickly.

Example 5.1 As explained in [145], for a training set with 20,000 examples, if η is selected as 0.1 and the average gradient is of the order of ± 0.1 for each weight per example, then the total accumulated weight change for batch learning will be of the order of $\pm 0.1 \times 0.1 \times 20000 = \pm 200$. A change in weight is unreasonably big and will result in wild oscillations across the weight space. When using incremental learning, each weight change will be of the order of $\pm 0.1 \times 0.1 = \pm 0.01$. Thus, for a converging batch learning with η_{batch}, the corresponding incremental learning algorithm can take $\eta_{\text{inc}} = N\eta_{\text{batch}}$, where N is the size of the training set.

It is recommended in [145] that $\eta_{\text{inc}} = \sqrt{N}\eta_{\text{batch}}$. As long as η is small enough to avoid drastic overshooting of curves and local minima, there is a linear relationship between η and the number of epochs required for learning.

Although incremental training has advantages over batch training with respect to the absolute value of the expected difference, it does not, in general, converge to the optimal weight with respect to the expected squared difference [42]. Almost-cyclic learning is a better alternative for batch mode learning than cyclic learning [42]. In [90], the convergence properties of the two schemes applied to quadratic loss functions is analyzed and the rate of convergence for each scheme is given.

Analysis shows that with any analytic sigmoidal function incremental BP training is always convergent under some mild conditions [148]. Incremental training converges to the optimal weight with respect to the expected squared difference, if the variance of the random per-instance gradient decays exponentially with the number of epochs processed during training. With proper η and the decay rate of the variance, incremental training converges to the optimal weight faster than batch training does. If the training set size is sufficiently large, then with regard to the absolute value of the expected difference, batch training converges faster to the globally optimal weight than incremental training does if $\eta < 1.2785$. With respect to the expected squared difference, batch training converges to the globally optimal weight as long as $\eta < 2$. The rate of convergence with respect to the absolute value of the expected difference improves monotonically as η increases up to N for incremental training, whereas batch training fails to converge if $\eta \geq 2$. Based on the estimate of the minimum error, a dynamic learning rate for incremental BP training of three-layer feedforward networks [161] ensures the error sequence to converge to the global minimum error.

Example 5.2 Approximate the function:

$$f(x_1, x_2) = 4x_1 \sin(10x_1)\cos(10x_2) + x_1x_2e^{x_1x_2}\cos(20x_1x_2),$$

where $0 \leq x_1 \leq 1, 0 \leq x_2 \leq 1$.

We use a three-layer network with 30 hidden nodes to approximate the function. The training algorithm is BP in batch mode. The learning rate is selected as 0.004, 441 data points are uniformly generated for training. We also implement BP with momentum in batch mode, and the additional momentum constant is selected as 0.9. Figure 5.3a plots the function. Figure 5.3b plots the MSE evolution for 100,000 epochs. The approximation error for the BP case is shown in Fig. 5.3c, and the result for BP with momentum is even worse. The convergence of BP with momentum is of the same order as that of BP in our simulation.

Example 5.3 We retrain the function shown in Example 5.2. This time we implement BP and BP with momentum in online mode. The simulation setting is the same as that in Example 5.2. The difference is that the learning rate is selected as 0.5, and the momentum constant is selected as 0.9. Figure 5.4a plots the MSE evolution for 100,000 epochs. For a random run, the approximation error for the BP case is shown in Fig. 5.4b, and the result for the BP with momentum is worse. In our experiment, we found the learning rate for online algorithms can be set very large.

Example 5.4 In the iris dataset, shown in Fig. 5.5, 150 patterns are classified into 3 classes. Each pattern has four numeric properties. We use a 4-4-1 MLP to learn

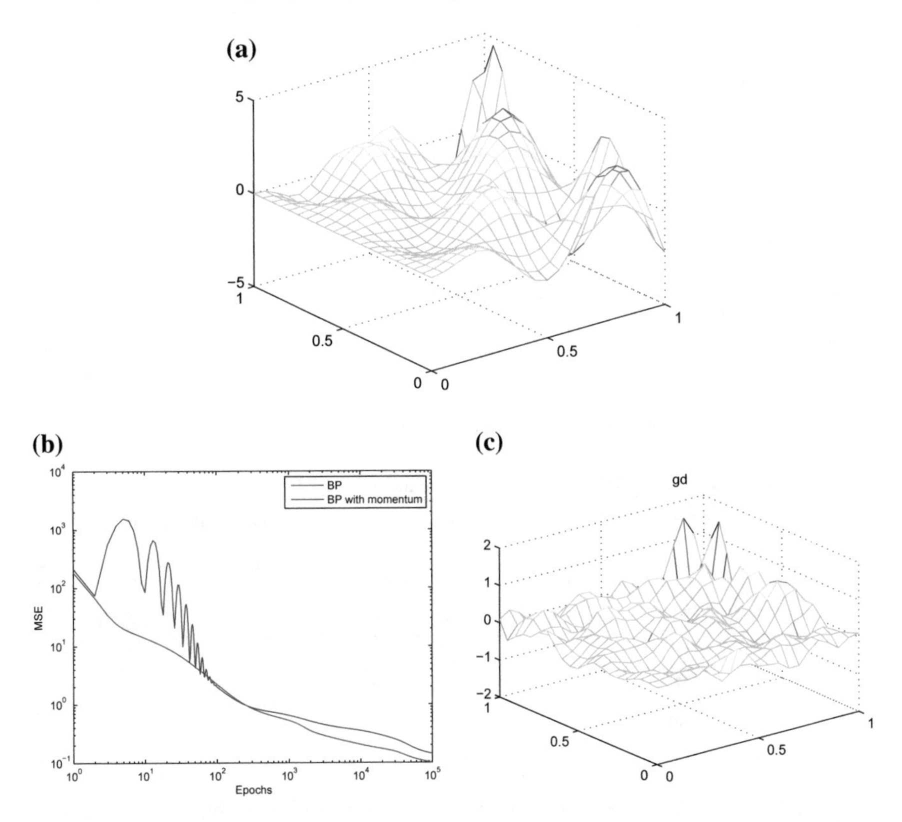

Fig. 5.3 Approximation of a function using MLP. The training algorithms are BP and BP with momentum in batch mode

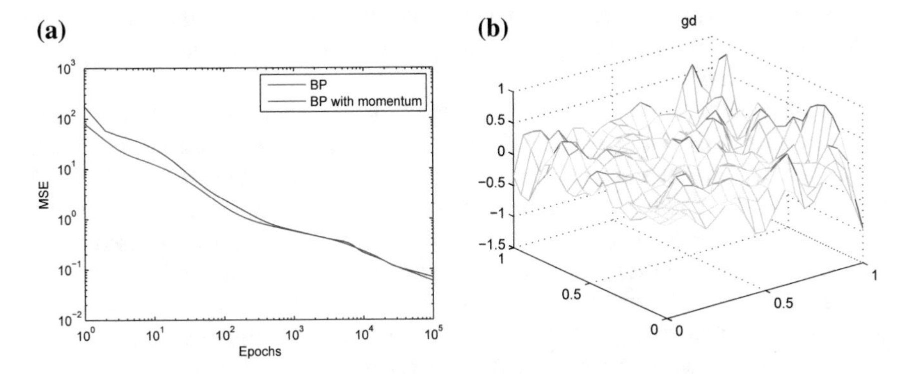

Fig. 5.4 Approximation of a function using MLP. The training algorithms are BP and BP with momentum in online mode

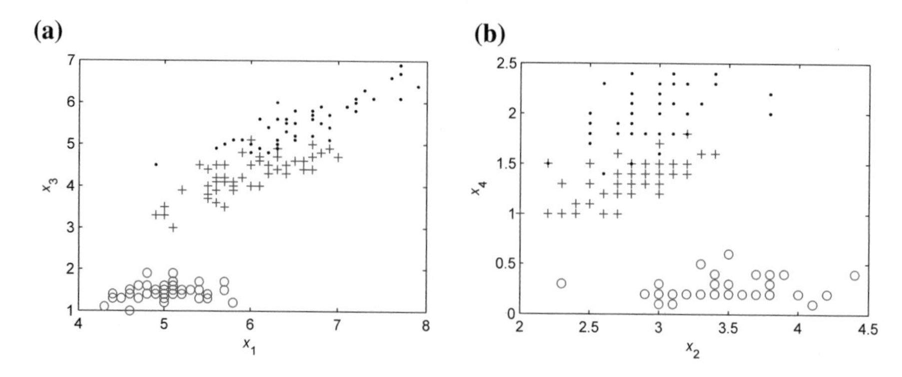

Fig. 5.5 Plot of the iris dataset: **a** x_1 versus x_3. **b** x_2 versus x_4

Table 5.1 Performance comparison of a 4-4-1 MLP trained with BP and BP with momentum

Algorithm	Training MSE	Classification accuracy(%)	std	Mean training time (s)	std (s)
BP, batch	0.1060	92.60	0.1053	9.7086	0.2111
BPM, batch	0.0294	95.60	0.0365	9.7615	0.3349
BP, online	0.1112	86.33	0.1161	11.1486	0.2356
BPM, online	0.1001	86.67	0.0735	11.2803	0.3662

BPM—BP with momentum

this problem with three discrete values representing different classes. The logistic sigmoidal function is selected for the hidden neurons and linear function is used for the output neurons. Two learning schemes are applied. Eighty percent of the dataset is used as training data, and the remaining 20% as testing data. We set the performance goal as 0.001, and the maximum number of epochs as 1000. We simulate and compare BP and BP with momentum.

During generalization, if the network output for an input pattern is closest to one of the attribute values, the pattern is identified as belonging to that class. For batch BP and BP with momentum, η and α both are randomly distributed between 0.3 and 0.9. For online algorithms, η is randomly distributed between 0.5 and 10.5, and α is randomly distributed between 0.3 and 0.9. Table 5.1 lists the results based on an average of 50 random runs. The traces of the training error are plotted in Fig. 5.6 for a random run. For classification, if the distance between the neural network output and the desired output is greater than 0.5, we count in a classification error.

From the simulation, we can see that the performance of BP as well as that of BP with momentum is highly dependent on the learning parameters selected, which are difficult to find for practical problems. There is no clear evidence that BP with momentum is superior to BP or that the algorithms in online mode are superior to their counterparts in batch mode.

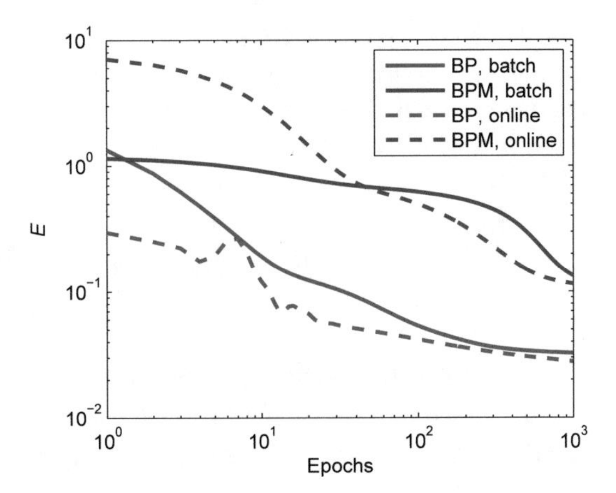

Fig. 5.6 Iris classification trained using 4-4-1 MLP with BP and BP with momentum: the traces of the training error for a random run

5.5 Activation Functions for the Output Layer

Usually, all neurons in MLP use the same sigmoidal activation function. This restricts the outputs of the network to the range of (0, 1) or (−1, 1). For classification problems, this representation is suitable. However, for function approximation problems the output may be far from the desired output, and the training algorithm is actually invalid. A common solution is to apply preprocessing and postprocessing.

The preprocessing and postprocessing procedures are not necessary if the activation function for the neurons in the output layer is selected as a linear function $\phi(x) = x$ to increase the dynamical range of the network output. A three-layer MLP of such networks with J_2 hidden units has a lower bound for the degree of approximation [79]. By suitably selecting an analytic, strictly monotonic, sigmoidal activation function, this lower bound is essentially attainable.

When the activation function of the output layer is selected as $\phi(x) = x$, the network can thus be trained in two steps. With the linearity property of the output units, there is the relation

$$\left[\mathbf{W}^{(M-1)}\right]^T \boldsymbol{o}_p^{(M-1)} = \boldsymbol{y}_p, \tag{5.24}$$

where $\mathbf{W}^{(M-1)}$, a J_{M-1}-by-J_M matrix, can be optimized by the LS method such as the SVD, RLS, or CG method [81]. The CG method converges to the exact solution in J_{M-1} or J_M steps, whichever is larger. BP is then used to update the remaining weights.

The generalized sigmoidal function is introduced in [91, 109] for neurons in the output layer of an MLP used for 1-of-n classification. This unit is sometimes called the *softmax or Potts unit* [109]. The generalized sigmoidal function introduces a behavior that resembles in some respects the behavior of WTA networks. The sum of the outputs of the neurons in the output layer is always equal to unity. The use of the generalized sigmoidal function introduces additional flexibility into the MLP model. Since the response of each output neuron is tempered by the responses of all the output neurons, the competition actually fosters cooperation among the output neurons. A single-layer perceptron using the generalized sigmoidal function can solve linearly inseparable classification problems [91]. The output of the ith output neuron is defined by

$$o_i^{(M)} = \phi\left(net_i^{(M)}\right) = \frac{e^{net_i^{(M)}}}{\sum_{j=1}^{J_M} e^{net_j^{(M)}}}, \tag{5.25}$$

where the summation in the denominator is over all the neurons in the output layer. The derivative of the generalized sigmoidal function is $o_i^{(M)}\left(1 - o_i^{(M)}\right)$, which is identical to the derivative for the logistic sigmoidal function.

5.6 Optimizing Network Structure

Smaller networks that use fewer parameters usually have better generalization capability. When training an MLP, the optimal number of neurons in the hidden layers is unknown and is estimated usually by trial and error. Network pruning and network growing are used to determine the size of the hidden layers.

Network pruning strategy first selects a network with a large number of hidden units, then removes the redundant units during the learning process. Pruning approaches usually fall into two broad groups. In sensitivity-based methods, one estimates the sensitivity of the error function E to the removal of a weight or unit, and removes the least important element. In penalty-based methods, additional terms are added to the error function E so that the new objective function rewards the network for choosing efficient solutions. The BP algorithm derived from this objective function drives unnecessary weights to zero and removes them during training. The two groups overlap if the objective function includes sensitivity terms.

5.6.1 Network Pruning Using Sensitivity Analysis

Network pruning can be performed based on the relevance or sensitivity analysis of the error function E with respect to a weight w. The relevance or sensitivity measure is usually used to quantify the contribution that individual weights or nodes make in solving the network task. The less relevant weights or units can be removed. Mathematically, the normalized sensitivity is defined by

$$S_w^E = \lim_{\Delta w \to 0} \frac{\frac{\Delta E}{E}}{\frac{\Delta w}{w}} = \frac{\partial \ln E}{\partial \ln w} = \frac{w}{E}\frac{\partial E}{\partial w}. \tag{5.26}$$

In the skeletonization technique [89], the sensitivity of E with respect to w is defined as $S_w^E = -w\frac{\partial E}{\partial w}$. This definition of sensitivity has been applied in [54]. In Karnin's method [54], during the training process, the sensitivity for each connection is calculated by making use of the available terms. Upon completion of the training process, those connections that have low sensitivities are pruned, and no retraining procedure is necessary. This method has been further improved in [37] by devising some pruning rules to prevent an input being removed from the network or a particular hidden layer being totally removed. A fast training algorithm is also included to retrain the network after a weight is removed. In [102], Karnin's method has been extended by introducing the local relative sensitivity index within each subgroup or layer of the network. This enables parallel pruning of weights that are relatively redundant in different layers of a feedforward network.

Sensitivity analysis of large-dimensional overtrained networks is conducted in order to assess the relative importance of each hidden unit on the network output by computing the contribution of each hidden unit to the network output. A sensitivity-based method utilizing retraining is described in [114]. The output of each hidden unit is monitored and analyzed for all the training set after the network converges. If the output of a hidden unit is approximately constant for all the training set, this unit actually functions as a bias to all the neurons it feeds, and hence can be removed. Similarly, if two hidden units produce the same or proportional outputs for all the training set, one of the units can be removed. Small weights are assumed to be irrelevant and are pruned. After some units are removed, the network is retrained. This technique leads to a prohibitively long training process for large networks.

A sensitivity-based method that uses linear models for hidden units is developed in [50]. If a hidden unit can be well approximated as a linear model of its net input, then it can be eliminated and replaced by adding biases in subsequent layers and by changing weights that bypass the unit. Thus, such units can be pruned. No retraining of the network is necessary. In [16], an effective hidden unit-pruning algorithm called linear-dependence pruning utilizing sets of linear equations is presented; it improves upon the linear models [50] and includes network retraining. Redundant hidden units are well modeled as linear combinations of the outputs of the other units. Hidden units are modeled as linear combinations of nonlinear units in the same layer and in the earlier layers. The hidden unit that is predicted to increase the training error the least when replaced by its model is identified, and the pruning algorithm replaces it with its model and retrains the weights connecting to the output layer by one iteration of training. A pruning procedure described in [13] iteratively removes hidden units and then adjusts the remaining weights in such a way as to preserve the overall network behavior. The pruning problem is formulated as solving a set of linear equations by a CG algorithm in the LS sense.

In [53], orthogonal transforms such as SVD and QR with column pivoting (QR-cp) are used for pruning neural networks. QR-cp coupled with SVD is used for subset selection and elimination of the redundant set. Based on the transforms on the training set, one can select the optimal sizes of the input and hidden nodes. The reduced-size network is then reinitialized and retrained to the desired convergence. In [124], the significance of increasing the number of neurons in the hidden layer of a feedforward network is evaluated using SVD. A pruning/growing technique based on the singular values of a trained network is then used to estimate the necessary number of neurons in the hidden layer. Practical measures of sensitivities to inputs are developed, and utilized toward deletion of redundant inputs in [162]. When one or more dimensions of the input vectors have relatively small sensitivity in comparison to others, that dimension of the input vectors can be removed, and a smaller size neural network can be successfully retrained in most cases.

A two-phase approach for pruning both the input and hidden units of MLPs based on mutual information is proposed in [147]. All features of the input vectors are first ranked according to their relevance to target outputs through a forward strategy. The salient input units of an MLP are thus determined according to the ranking and their contributions to the network performance, and the irrelevant features of the input vectors can be identified and eliminated. The redundant hidden units are then removed from the trained MLP one after another according to a relevance measure.

Optimal Brain Damage and Optimal Brain Surgeon

The optimal brain damage (OBD) [60] and optimal brain surgeon (OBS) [41] procedures are two network pruning methods based on the perturbation analysis of the second-order Taylor expansion of the error function.

In the following, we use $\vec{\boldsymbol{w}}$ to represent the vector generated by concatenating all entries of $\mathbf{W}$. When the training process converges, the gradient is close to zero, and thus the increase in E due to a change in $\vec{\boldsymbol{w}}$ is given by

$$\Delta E \simeq \frac{1}{2} \Delta \vec{\boldsymbol{w}}^T \mathbf{H} \Delta \vec{\boldsymbol{w}}, \tag{5.27}$$

where $\mathbf{H}$ is the Hessian matrix, $\mathbf{H} = \frac{\partial^2 E}{\partial \vec{\boldsymbol{w}}^2}$.

Removing a weight w_i amounts to equating this weight to zero. Thus, removing a subset of weights, $\mathcal{S}_{\text{prune}}$, results in a change in E by setting $\Delta w_i = w_i$, if $i \in \mathcal{S}_{\text{prune}}$, otherwise $\Delta w_i = 0$. Based on the saliency (5.27), OBD is a special case of OBS, where the Hessian $\mathbf{H}$ is assumed to be a diagonal matrix; in this case, each weight has a saliency

$$(\Delta E)_i \simeq \frac{1}{2} w_i^2 H_{ii}. \tag{5.28}$$

In the procedure, a weight with the smallest saliency is selected for deletion. The calculation of the Hessian $\mathbf{H}$ is fundamental to the OBS procedure.

Optimal cell damage [21] extends OBD to remove irrelevant input and hidden units. The unit-OBS [119] improves OBS by removing one whole unit in each step. The unit-OBS can also conduct feature extraction on the input data by removing unimportant input units. As an intermediate between OBD and OBS, the principal components pruning [69] is based on a block-diagonal approximation of the Hessian; it is based on PCA of the node activations of successive layers of trained feedforward networks for a validation set. The node activation correlation matrix at each layer is required, while the calculation of the full Hessian of the error function is avoided. This method prunes the least salient eigen nodes, and network retraining is not necessary.

In the case of early stopping, OBD and OBS are not suitable since the network is not in a local minimum and the first-order term in the Taylor-series expansion is not zero. Early brain damage [131] is an extension to OBD and OBS in connection with early stopping; furthermore, it allows the revival of the already pruned weights.

A pruning procedure similar to OBD is constructed using the error covariance matrix $\mathbf{P}$ obtained during RLS training [68]. As $\mathbf{P}$ is obtained along with the RLS algorithm, pruning becomes much easier. The RLS-based pruning has a computational complexity of $O\left(N_w^3\right)$, which is much smaller than that of OBD, namely, $O\left(N_w^2 N\right)$, while its performance is very close to that of OBD in terms of the number of pruning weights and generalization ability. In addition, the RLS-based pruning is also suitable for the online situation. Another network pruning technique, based on the training results from the extended Kalman filtering (EKF) technique, is given in [120]. The method prunes a neural network based solely on the obtained error covariance matrix $\mathbf{P}$ and the state (weight) vector.

The variance nullity pruning [28] is based on the sensitivity analysis of the output, rather than that of the error function. If the gradient search and the MSE function are used, then OBD and the output sensitivity analysis are conceptually the same under the assumptions that the Hessain $\mathbf{H}$ is diagonal. Parameter relevance is measured as the variance in sensitivity over the training set, and those hidden or input nodes that are irrelevant are removed. The pruned network is then retrained.

5.6.2 Network Pruning Using Regularization

For the regularization technique, the optimization objective is defined as

$$E_T = E + \lambda_c E_c, \tag{5.29}$$

where E is the error function, E_c is a penalty for the complexity of the structure, and $\lambda_c > 0$ is a regularization parameter, which needs to be appropriately determined for a particular problem. Extra local minima are introduced to the optimization process by the penalty term.

In the weight-decay technique [44, 48, 139], E_c is defined as a function of the weights. In [44], E_c is defined as the sum of the squares of all the weights

$$E_c = \sum_{i,j} w_{ij}^2. \tag{5.30}$$

As a result, the change of each weight is proportional to its value. In [48], E_c is defined as the sum of the absolute values of the weights

$$E_c = \sum_{i,j} \left| w_{ij} \right| . \tag{5.31}$$

Thus, all the weights are decaying at a constant step to zero.

The BP algorithm derived from E_T using a weight-decay term is a structural learning algorithm

$$\Delta w_{ij}^{(m)} = -\eta \frac{\partial E_T}{\partial w_{ij}^{(m)}} = \Delta w_{ij,\mathrm{BP}}^{(m)} - \varepsilon \frac{\partial E_c}{\partial w_{ij}^{(m)}}, \tag{5.32}$$

where $\Delta w_{ij,\mathrm{BP}}^{(m)} = -\eta \frac{\partial E}{\partial w_{ij}^{(m)}}$ is the weight change corresponding to BP learning, and $\varepsilon = \eta \lambda_c$ is the decaying coefficient at each weight change. The amplitudes of the weights decrease continuously toward zero, unless they are reinforced by the BP rule. At the end of training, only the essential weights deviate significantly from zero. By pruning the weights that are close to zero, a skeleton network is obtained. This effectively increases generalization and reduces the danger of overtraining as well. For example, in the modified BP with forgetting [48, 57], the weight-decay term (5.31) is used, and $\frac{\partial E_c}{\partial w_{ij}^{(m)}} = \mathrm{sign}\left(w_{ij}^{(m)} \right)$, where $\mathrm{sign}(\cdot)$ is the signum function. Neural networks trained by weight-decay algorithms are not sensitive to the initial choice of the network.

The weight-decay technique given in [38] is an implementation of a robust network that is insensitive to noise. It decays the weights toward zero by weakening the small weights more rapidly. Because small weights can be used by the network to code noisy patterns, this weight-decay mechanism is especially important in the case of noisy data. The weight-decay technique converges as fast as BP, if not faster, and shows some significant improvement over BP in noisy situations [38].

When selecting weights at a group level using the smoothed approximation of group Lasso penalty, which is the L_1-/L_2-norm of weight vectors, the BP variants outperform weight decay, weight elimination, and approximate smoother, on both generalization and pruning efficiency [138].

The conventional RLS algorithm is essentially a weight-decay algorithm [68], since its objective function is similar to that for the weight-decay technique using (5.30). The error covariance matrix $\mathbf{P}$ obtained during the RLS training possesses properties similar to the Hessian matrix $\mathbf{H}$ of the error function. The initial value of $\mathbf{P}$, namely, $\mathbf{P}(0)$ can be used to control the generalization ability.

The weight-smoothing regularization introduces the constraint of Jacobian profile smoothness during the learning step [2]. Other regularization methods include neural Jacobians like the input perturbation [10], or generalized regular network [101] that minimize the neural Jacobian amplitude to smooth the neural network behavior.

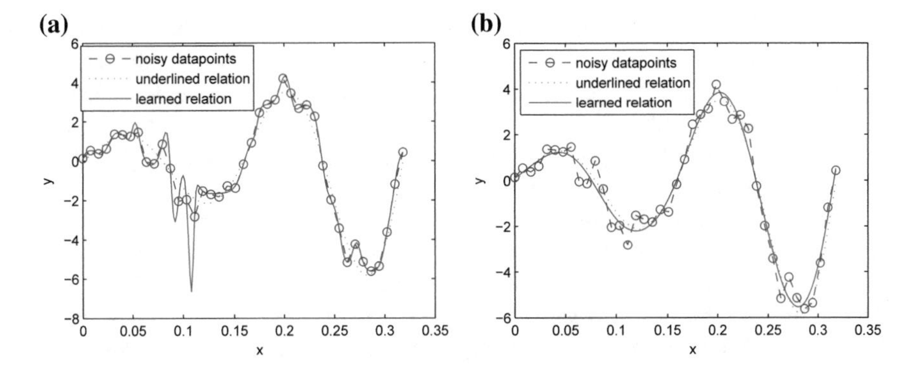

Fig. 5.7 Pruning an MLP by regularization: **a** A 1-20-1 network introduces overfitting. **b** The pruned network has better generalization

Bayesian regularization [75] can determine the optimal regularization parameters in an automated fashion. This eliminates the need to guess the optimum network size. In this framework, the weights and biases of the network are assumed to be random variables with specified distributions. The regularization parameters are related to the unknown variances associated with these distributions. These parameters are then estimated using statistical techniques. For small datasets, Bayesian regularization provides better generalization performance than early stopping does, because Bayesian regularization does not require a validation dataset and thus uses all the data.

Compressing the weights of a layer of an MLP is equivalent to compressing the input of the layer [30]. Thus, some ideas from compressed sensing can be transferred to the training of MLP.

Example 5.5 We generate 41 noisy datapoints from a function $f(x) = \sin(2\pi x)e^x$. We train a 1-20-1 network to approximate the functions, and found the learned function is overfitted, as shown in Fig. 5.7a.

We use E_c defined in (5.30). But E_c is taken as the mean values among the number of all weights, $\lambda_c = 1$. Bayesian regularization is implemented; it provides a measure of how many network parameters (weights and biases) are being effectively used by the network. The final trained network uses approximately 11 parameters out of the 61 total weights and biases. We can see from Fig. 5.7b that the network response is very close to the underlying function (dotted line), and, therefore, the network will generalize well to new inputs.

5.6.3 Network Growing

Another approach to training MLP is the constructive approach, which starts with a small network and then gradually adds hidden units until a given performance is achieved. This helps us in finding a minimal network. Constructive algorithms are computationally more economical than pruning algorithms. The constructive approach also helps in escaping a local minimum by adding a new hidden unit. When the error E does not decrease or decreases too slowly, the network may be trapped in a local minimum, and a new hidden unit is added to change the shape of the error function and thus to escape from the local minimum. The weights of the newly added neurons can be set randomly.

Cascade-correlation learning is a well-known constructive learning approach. It is an efficient technique both computationally and in terms of modeling performance [32]. In the cascaded architecture, each newly recruited unit is connected both to the input nodes and to every preexisting unit. For each newly added hidden unit k, all the weights connected to the previously trained units are frozen, all the weights connected to the newly added unit and the output units are updated. Network construction is based on the one-by-one training and addition of hidden units. Training starts with no hidden unit. If the minimal network cannot solve the problem after a certain number of training cycles, a set of candidate hidden units with random initial weights are generated, from which an additional hidden unit is selected and added to the network. The constructed network has direct connections between the input and output units. Moreover, the depth or the propagation delay through the network is directly proportional to the number of hidden units and can be excessive. Many ideas from the cascade-correlation learning are employed in the constructive algorithms [59, 67].

The dependence identification algorithm constructs and trains an MLP by transforming the training problem into a set of quadratic optimization problems, which are then solved by a succession of sets of linear equations [88]. It is a batch learning process. The method uses the concept of linear dependence to group patterns. The overall convergence speed is orders of magnitude faster than that of BP, although the resulting network is usually large [88]. The algorithm is a faster and more systematic method for developing initial network architectures than the trial-and-error or gradient-based pruning techniques.

A constructive learning algorithm for MLP using an incremental learning procedure has been proposed in [72], which may be useful for real-time learning. Training patterns are learned one-by-one. The algorithm starts with a single training pattern and a single hidden neuron. During training, when the algorithm gets stuck at a local minimum, the weight-scaling technique is applied to help the algorithm to escape from the local minimum. If the algorithm fails in escaping from a local minimum after several consecutive attempts, the network is allowed to grow by adding a hidden neuron. Initial weights for the newly added neuron are selected using an optimization procedure based on the QP and LP techniques.

A constructive training algorithm for three-layer MLPs for classification problems is given in [104]. The Ho–Kashyap algorithm is central to training both the hidden layer nodes and the output layer nodes. A pruning procedure that removes the least important hidden node, one at a time, can be included to increase the generalization ability of the method. When constructing a three-layer MLP using a quasi-Newton method [112], the quasi-Newton method is used to minimize the sequence of error functions associated with the growing network.

Examples of early constructive methods for training feedforward networks with LTG neurons are the tower algorithm [36], the tiling algorithm [86], and the upstart algorithm [34]. These algorithms are based on the pocket algorithm [36], and are used for classification.

5.7 Speeding Up Learning Process

BP is a gradient-descent method and has a slow convergence speed. Numerous measures have been reported in order to speedup the convergence of the BP algorithm.

Preprocessing of a training pattern set relieves the curse of dimensionality, and also improves the generalization ability of the network. Preprocessing is efficient when the training set is very large. A feature selection method particularly suited for feedforward networks has been proposed in [107]. An L_1-norm saliency metric describing the sensitivity of the outputs of the trained network with respect to the jth input is used. In [97], a feature extraction method that exhibits some similarity to PCA has been proposed, where the L_2-norm is used in the saliency metric.

5.7.1 Eliminating Premature Saturation

One major reason for slow convergence is the occurrence of premature saturation of the output of the sigmoidal functions. This can be seen from Fig. 5.8, where the sigmoidal functions and their derivatives are plotted. When the absolute value of net is large, $\dot{\phi}(net)$ is so small that the weight change approaches zero and learning takes an excessively long time. This is the *flat-spot problem*.

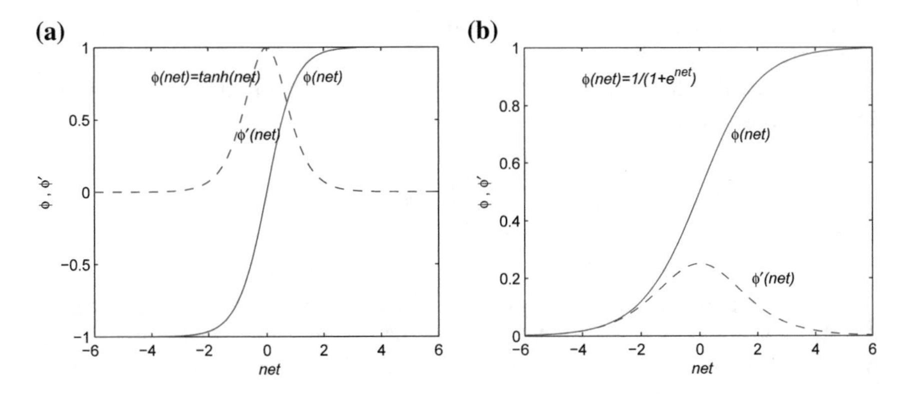

Fig. 5.8 Sigmoidal functions and their derivatives, $\beta = 1$. **a** Hyperbolic tangent function and its derivative. **b** Logistic function and its derivative

Once trapped at saturation, the outputs of saturated units preclude any significant improvement in the training weights directly connected to the units. The premature saturation leads to an increase in the number of training cycles required to release the trapped weights. In order to combat premature saturation, one can modify the slope of the sigmoidal function, or modify the error function E so that when the output unit is saturated, i.e., the slope of the sigmoidal function approaches zero, the backpropagated error is finite.

In [63], premature saturation of network output units has been analyzed as a static phenomenon that occurs at the beginning of the training stage as a consequence of random initialization of weights. The probability of premature saturation at the first training cycle is derived as a function of the initial weights, the number of nodes at each layer, and the slope of the sigmoidal function.

A dynamic mechanism for premature saturation is analyzed in [133]. The momentum term is identified as a leading role in the occurrence of premature saturation. The entire premature saturation process is partitioned into three distinct stages, namely, the beginning of the premature saturation, saturation plateau, and the complete recovery from saturation. For the onset of premature saturation to occur, a set of four necessary conditions must be simultaneously satisfied and usually remain satisfied for a number of consecutive iterations. A method for preventing premature saturation is to temporarily modify the momentum factor α, once the four conditions are satisfied at iteration t. If more than one output unit satisfies the four conditions, α is calculated for each of these units and the smallest α is used to update $\Delta\mathbf{W}(t+1)$. The original α is used again after the $(t+1)$th iteration. The algorithm works like BP unless the four conditions are satisfied simultaneously.

In [64], the BP update equation is revised by adding a term embodying the degree of saturation to prevent premature saturation. This turns out to be adding an additional term embodying the degree of saturation in the energy function. Similarly, in [92], the partial derivatives of the logistic activation function are generalized to $\left[o_{p,i}^m(1-o_{p,i}^m)\right]^{\frac{1}{\rho_0}}$ with $\rho_0 \geq 1$ so that error signals are significantly enlarged when $o_{p,i}^m$ approaches saturation. Other authors also avoid premature saturation using modified energy functions [94, 136].

A modified BP algorithm is derived in [1] based on a criterion with an additional linear quadratic error term

$$E_p = \frac{1}{2}\left\|\hat{\boldsymbol{y}}_p - \boldsymbol{y}_p\right\|^2 + \frac{1}{2}\lambda_c\left\|\boldsymbol{net}^{(M)} - \phi^{-1}(\boldsymbol{y}_p)\right\|^2, \tag{5.33}$$

where $\phi^{-1}(\cdot)$, the inverse of $\phi(\cdot)$, applies to each component of the vector within, and λ_c is a small positive number, usually $0 \leq \lambda_c \leq 1$. For each pattern, the modified BP is slightly more complex than BP, while it always has a significantly faster convergence than BP has in the number of training iterations and in the computation time for a suitably selected λ_c, which can be decreasing from one to zero during the learning process.

The modified BP algorithm for the three-layer MLP [136] significantly improves BP in both the accuracy and the convergence speed. Besides, η can be selected as a large value without the worry of saturation. A new term embodying the saturation degree is added to the conventional criterion function to prevent premature saturation [136]

$$E_p = \frac{1}{2} \left\| \hat{\boldsymbol{y}}_p - \boldsymbol{y}_p \right\|^2 + \frac{1}{2} \left\| \hat{\boldsymbol{y}}_p - \boldsymbol{y}_p \right\|^2 \cdot \sum_{j=1}^{J_H} \left(o_{p,j} - 0.5 \right)^2, \tag{5.34}$$

where J_H is the number of hidden units, and 0.5 is the average value of a sigmoidal activation function. $\sum_{j=1}^{J_H} \left(o_{p,j} - 0.5 \right)^2$ is defined as the saturation degree for all the hidden neurons for pattern p.

While many attempts have been made to change the shape of the flat regions or to avoid them during MLP learning, a constructive approach proposed in [110] exploits the flat regions to stably and successively find excellent solutions.

5.7.2 *Adapting Learning Parameters*

The performances of the BP and BP-with-momentum algorithms are highly dependent upon a suitable selection for η and α. Some heuristics are needed for optimally adjusting η and α to speedup the convergence of the algorithms. Learning parameters are typically adapted once for each epoch.

Globally Adapted Learning Parameters

All the weights in the network are typically updated using the global learning parameters η and α. The optimal η is the inverse of the largest eigenvalue, $\lambda_{\max}$, of the Hessian matrix $\mathbf{H}$ of the error function [62]. The online algorithm for estimating $\lambda_{\max}$ proposed in [62] does not even require a calculation of the Hessian.

A simple and popular method for accelerating the learning is to use the search-then-converge schedule, which starts with a large η and gradually decreases it as the learning proceeds. According to [108], the process of adapting η is similar to that in simulated annealing. The algorithm escapes from a shallow local minimum in early training and converges into a deeper, possibly global minimum. Typically, η is selected as

$$\eta(t) = \frac{\eta_0}{1 + \frac{t}{T_s}}, \tag{5.35}$$

where T_s is the search time.

The bold-driver technique is a heuristic for optimal network performance [7, 134], where at the $(t+1)$th epoch η is updated by

$$\eta(t+1) = \begin{cases} \rho^+ \eta(t), & \Delta E(t) < 0 \\ \rho^- \eta(t), & \Delta E(t) > 0 \end{cases}, \tag{5.36}$$

where ρ^+ is chosen to be slightly larger than unity, typically 1.1, ρ^- is chosen significantly less than unity, typically 0.5, and $\Delta E(t) = E(t) - E(t-1)$. If the error decreases ($\Delta E < 0$), the training is approaching the minimum, and we can increase η to speed up the search process. If, however, the error increases ($\Delta E > 0$), the algorithm must have overshot the minimum and thus η is too large. In this case, the weights are not updated. This process is repeated until a decrease in the error is found. α can be selected as a fixed value. However, at each occurrence of an error increase, the next weight update is needed to be along the negative gradient direction to speed up the convergence and α is set to 0 temporarily. Variants of this heuristic method include those with a fixed increment of η.

The gradient-descent rule can be reformulated as

$$\vec{\boldsymbol{w}}(t+1) = \vec{\boldsymbol{w}}(t) - \eta(t)\nabla E\left(\vec{\boldsymbol{w}}(t)\right), \tag{5.37}$$

where $\vec{\boldsymbol{w}}$ is a vector formed by concatenating all the columns of $\mathbf{W}$. In [95], the learning rate of batch BP is adapted according to the instantaneous value of $E(t)$:

$$\eta(t) = \rho_0 \frac{\rho(E(t))}{\left\|\nabla E\left(\vec{\boldsymbol{w}}(t)\right)\right\|}, \tag{5.38}$$

where ρ_0 is a positive constant, $\rho(E)$ is a function of E, typically $\rho(E) = E$, and

$$\nabla E\left(\vec{\boldsymbol{w}}(t)\right) = \left.\frac{\partial E(\vec{\boldsymbol{w}})}{\partial \vec{\boldsymbol{w}}}\right|_t. \tag{5.39}$$

This adaptation leads to fast convergence. However, $\eta(t)$ is a very large number in the neighborhood of a local or global minimum, leading to jumpy behavior of the weights. The method converges faster than the Quickprop algorithm does [31].

In [77], η is updated according to the local approximation of the Lipschitz constant $L(t)$ based on Armijo's condition for line search. The algorithm is robust against oscillations due to large η and avoids the phenomenon of the nearly constant E value, by ensuring that E is decreased with every weight update. The algorithm results in an improvement in the performance when compared with that of the BP, delta-bar-delta [49], and bold-driver [134] techniques.

The fuzzy inference system is also used to adapt the learning parameters for an MLP with BP [19]. The fuzzy system incorporates Jacobs' heuristics [49] about the unknown learning parameters using fuzzy IF-THEN rules. The heuristics are driven by the behavior of $E(t)$. Change in $E(t)$, denoted by $\Delta E(t)$, is an approximation to the gradient of E, and change in $\Delta E(t)$ is an approximation to the second-order derivatives of E. Fuzzy inference systems are constructed for adjusting η and α, respectively. This fuzzy BP learning is much faster than BP, with a significantly smaller MSE [19].

Locally Adapted Learning Parameters

Each weight $w_{ij}^{(m)}$ can have its own learning rate $\eta_{ij}^{(m)}(t)$ so that

$$\Delta w_{ij}^{(m)}(t) = -\eta_{ij}^{(m)}(t) g_{ij}^{(m)}(t), \tag{5.40}$$

where the gradient

$$g_{ij}^{(m)}(t) = \nabla E\left(w_{ij}^{(m)}\right)\Big|_t = \left.\frac{\partial E}{\partial w_{ij}^{(m)}}\right|_t . \tag{5.41}$$

There are many locally adaptive learning algorithms using weight-specific learning rates such as the heuristics proposed in [115, 125], SuperSAB [128], delta-bar-delta [49], Quickprop [31], equalized error BP [82], and a globally convergent strategy [78].

Due to the nature of the sigmoidal function, a large input may result in saturation that will slow down the adaptation process. In [125], the learning rates for all input weights to a neuron is selected to be inversely proportional to the fan-in of the neuron, namely,

$$\eta_{ij}^{(m-1)}(t) = \frac{\kappa_0}{net_j^{(m)}(t)}, \tag{5.42}$$

where κ_0 is a small positive number. This can maintain a balance among the learning speed of units with different fan-in. The increase in the convergence speed is theoretically justified by studying the eigenvalue distribution of $\mathbf{H}$ [61].

The heuristic proposed in [115] and in SuperSAB [128] is to adapt $\eta_{ij}^{(m)}$ by

$$\eta_{ij}^{(m)}(t+1) = \begin{cases} \eta_0^+ \eta_{ij}^{(m)}(t), & g_{ij}^{(m)}(t) \cdot g_{ij}^{(m)}(t-1) > 0 \\ \eta_0^- \eta_{ij}^{(m)}(t), & g_{ij}^{(m)}(t) \cdot g_{ij}^{(m)}(t-1) < 0 \end{cases}, \tag{5.43}$$

where $\eta_0^+ > 1, 0 < \eta_0^- < 1$. In SuperSAB, $\eta_0^+ \simeq \frac{1}{\eta_0^-}$. Since $\eta_{ij}^{(m)}$ grows and decreases exponentially, too many successive acceleration steps may generate too large or too small $\eta_{ij}^{(m)}$ and thus slow down the learning process. To avoid this, a momentum term is included in SuperSAB.

The delta-bar-delta [49] is similar to the heuristic (5.43), but eliminates its problems by making linear acceleration and exponential deceleration of the learning rates. Individual $\eta_{ij}^{(m)}(t)$ are updated based on a local optimization method, and the change $\Delta\eta_{ij}^{(m)}(t)$ is given by

$$\Delta\eta_{ij}^{(m)}(t) = \begin{cases} \kappa_0, & \overline{g}_{ij}^{(m)}(t-1)g_{ij}^{(m)}(t) > 0 \\ -\beta\eta_{ij}^{(m)}(t), & \overline{g}_{ij}^{(m)}(t-1)g_{ij}^{(m)}(t) < 0 \\ 0, & \text{otherwise} \end{cases}, \tag{5.44}$$

where

$$\overline{g}_{ij}^{(m)}(t) = (1-\varepsilon)g_{ij}^{(m)}(t) + \varepsilon\overline{g}_{ij}^{(m)}(t-1), \tag{5.45}$$

and ε, κ_0, β are positive constants specified by the user. All $\eta_{ij}^{(m)}$s are initialized with small values. Basically, $g_{ij}^{(m)}(t)$ is an exponentially decaying trace of gradient values. Inclusion of a momentum term sometimes causes the delta-bar-delta to diverge, and as such an adaptively changing momentum has been introduced to improve the delta-bar-delta [87]. However, the delta-bar-delta requires a careful selection of the parameters.

In Quickprop method [31, 32], $\alpha_{ij}^{(m)}(t)$ are heuristically adapted. Quickprop is given by

$$\Delta w_{ij}^{(m)}(t) = \begin{cases} \alpha_{ij}^{(m)}(t)\Delta w_{ij}^{(m)}(t-1), & \Delta w_{ij}^{(m)}(t-1) \neq 0 \\ \eta_0 g_{ij}^{(m)}(t), & \Delta w_{ij}^{(m)}(t-1) = 0 \end{cases}, \tag{5.46}$$

where

$$\alpha_{ij}^{(m)}(t) = \min\left\{ \frac{g_{ij}^{(m)}(t)}{g_{ij}^{(m)}(t-1) - g_{ij}^{(m)}(t)}, \alpha_{\max} \right\}, \tag{5.47}$$

$\alpha_{\max}$ is typically 1.75 and $0.01 \leq \eta_0 \leq 0.6$; η_0 is only used at the start or restart of the training. To avoid the flat-spot problems, Quickprop can be improved by adding 0.1 to the derivative of the sigmoidal function [31]. The use of error gradient at two consecutive time steps is a discrete approximation to second-order derivatives, and the method is actually a quasi-Newton method that uses the so-called secant steps. $\alpha_{\max}$ is used to avoid very large Quickprop updates. Quickprop typically performs very reliably and converges very fast [99]. However, the simplification of the Hessian to a diagonal matrix used in Quickprop has not been theoretically justified and convergence problems may occur for certain tasks.

In [154, 155], $\eta_{ij}^{(m)}$ and $\alpha_{ij}^{(m)}$ are optimally tuned using three methods, namely, the second-order-based, first-order-based, and CG-based methods. These methods make use of the derivatives of E with respect to $\eta_{ij}^{(m)}$ and $\alpha_{ij}^{(m)}$, and the information gathered from the forward and backward procedures, but do not need explicit computation of the first- and second-order derivatives in the weight space. The computational and storage burdens are at most triple that of BP, with an order of magnitude faster speed.

A general theoretical result has been derived for developing first-order batch learning algorithms with local learning rates based on Wolfe's conditions for linear search and the Lipschitz condition [78]. This result provides conditions under which global convergence is guaranteed. This globally convergent strategy can be equipped with algorithms of this class to adapt the overall search direction to a descent one at each training iteration. When Quickprop [31] and the algorithms given in [115] are equipped with this strategy, they exhibit a significantly better percentage of success in reaching local minima than their original versions.

5.7.3 *Initializing Weights*

The initial weights of a network play a significant role in the convergence of a training method. Poor initial weight values may result in slow convergence or lead the network stuck at a local minimum. The objective of weight initialization is to find weights that are as close as possible to a global minimum before training, and to increase the convergence speed. By weight initialization, the outputs of the hidden neurons can be assigned in the unsaturation region. Without a priori knowledge of the final weights, it is common practice to initialize all the weights with random small absolute values, or with small zero-mean random numbers [108]. Randomness also helps to break the symmetry of the system, which gradient-based learning algorithms are unable to break, and thus prevents redundancy in the network. Starting from large weights may prematurely saturate the units and slows down the learning process. Theoretically, the probability of prematurely saturated neurons in MLP increases with the maximal value of the weights [63]. By statistical analysis, the maximum amplitude for the initial weights is derived in [26]. For the three-layer MLP, a weight range of $[-0.77, 0.77]$ empirically gives the best mean performance over many existing random weight initialization techniques [126].

There are many heuristics for weight initialization. In [140], the initial weights of the ith unit at the jth layer are selected based on the order of $\frac{1}{\sqrt{n_i^{(j)}}}$, where $n_i^{(j)}$ is the number of weights to the ith unit at the jth layer. When the weights to a unit are uniformly distributed in $\left[-\frac{3}{\sqrt{n_i^{(j)}}}, \frac{3}{\sqrt{n_i^{(j)}}}\right]$, the total input to that unit, $net_i^{(j)}$, is a random variable with zero mean and a standard deviation of unity. This is an empirical optimal initialization of the weights [126]. In [85], the weights are first randomly initialized to the range $[-a_0, a_0]$, $a_0 > 0$, and are then individually scaled to ensure that each neuron is active over its full dynamic range. The scaling factor for the weights connected to the ith neuron at the jth layer is given by $\rho_i^{(j)} = \frac{D_i^{(j)}}{a_0 n_i^{(j)}}$, where $D_i^{(j)}$ is the dynamic range of the activation function. The optimal magnitudes of the initial weights and biases can be determined based on multidimensional geometry [151]. This method ensures that the outputs of the hidden and output layers are well within the active region, while the dynamic range of the activation function is fully utilized. The hidden-layer weights can be initialized in such a way that each hidden node is assigned to approximate a portion of the range of the desired function based on a piecewise-linear approximation of a sigmoidal function at the start of network training [93].

In addition to heuristics, there are many methods for weight initialization using parametric estimation. The sensitivity of BP to the initial weights is discovered to be a complex fractal-like structure for convergence as a function of the initial weights [56]. There are various weight-estimation techniques, where a nonlinear mapping between pattern and target is introduced [25, 26, 65].

Clustering is useful for weight initialization of three-layer MLPs. A three-layer MLP with prototypes is initialized in [25] based on supervised clustering. In [117], the clustering and nearest-neighbor methods are utilized to initialize hidden layer weights, and the output layer weights are then initialized by solving a set of linear equations using SVD. In the initialization method given in [142], the clustering and nearest-neighbor classification technique are used for a number of cluster sets, each representing the training examples with a different degree of accuracy. The orthogonal least squares (OLS) method is used as a practical weight initialization algorithm for MLP in [65]. The maximum covariance initialization method [66] uses a procedure similar to that of the cascade-correlation algorithm [32]. An optimal weight initialization algorithm for the three-layer MLP [152] initializes the hidden-layer weights that extract the salient feature components from the input data based on ICA, whereas the initial output-layer weights are evaluated to keep the output neurons inside the active region.

The optimal initial weights can be evaluated using the LS and linear algebraic method [149, 150]. In [150], the optimal initial weights between layers are evaluated using the LS method by assigning the outputs of hidden neurons with random numbers in the range between 0.1 and 0.9. The actual outputs of the hidden neurons are obtained by propagating the input patterns through the network. The optimal weights between the hidden and output layers can then be evaluated by using the LS method. In [149], the weights connected to the hidden layers are determined by the Cauchy's inequality and the weights connected to the output layer are determined by the LS method.

MLP learning is to estimate a nonlinear mapping between the input and the output of the examples, Φ, by superposition of the sigmoidal functions. By using a Taylor-series development of Φ and the nonlinearity of the sigmoidal function, two weight initialization strategies for the three-layer MLP are obtained based on the first- and second-order identification of Φ [23]. These techniques effectively avoid local minima, significantly speed up the convergence, obtain a better generalization, and estimate the size of the network.

5.7.4 Adapting Activation Function

During training, if a unit has a large net input, net, the output of this unit is close to a saturation region of its sigmoidal function. Thus, if the target value is substantially different from that of the saturated one, the unit has entered a flat spot. Since the first-order derivative of the sigmoidal function $\dot{\phi}(net)$ is very small when net is large in magnitude, the weight update is very slow. Fahlman developed a simple solution by adding a bias, typically 0.1, to $\dot{\phi}(net)$ [31]. Hinton [43] suggested the design of an error function that goes to infinity at points where $\dot{\phi}(net) \to 0$. This leads to a finite nonzero error update.

One way to solve the flat-spot problem is to define an activation function such that [153]

$$\phi_\mu(net) = \mu net + (1 - \mu)\phi(net), \quad \mu \in [0, 1]. \tag{5.48}$$

In the beginning, $\mu = 1$ and all the nodes have linear activation, and BP is used to obtain a local minimum in E. Then μ is decreased gradually and BP is applied until $\mu = 0$. The flat-spot problem does not occur since $\dot{\phi}_\mu(net) > \mu$ and $\mu > 0$ for most of the training time. When $\mu = 1$, $E(\mathbf{W}, \mu)$ is a polynomial of $\mathbf{W}$ and thus has few local minima. This process can be viewed as an annealing process, which helps us in finding a global or good minimum.

For sigmoidal functions, such as the logistic and hyperbolic tangent functions, the gain β represents the steepness (slope) of the activation function. In BP, β is fixed and typically $\beta = 1$. A modified BP with an adaptive β significantly increases the learning speed and improves the generalization [58, 118]. Each neuron has its own variable gain $\beta_i^{(m)}$, which is adapted by gradient descent

$$\Delta\beta_i^{(m)} = -\eta_\beta \frac{\partial E}{\partial \beta_i^{(m)}}, \tag{5.49}$$

where η_β is a small positive learning rate.

A large gain β yields results similar to those with a high learning rate η. Changing β is equivalent to changing η, the weights, and the biases. This is asserted by Theorem 5.1 [29].

Theorem 5.1 (Eom, Jung, and Sirisena [29]) *An MLP with the logistic activation function* $\phi(\cdot)$, *gain* β, *learning rates* $\boldsymbol{\eta}$, *weights* $\mathbf{W}$, *and biases* $\boldsymbol{\theta}$ *is equivalent to a network of identical topology with the activation function* $\phi(\cdot)$, *gain* 1, *learning rates* $\beta^2\boldsymbol{\eta}$, *weights* $\beta\mathbf{W}$ *and biases* $\beta\boldsymbol{\theta}$, *in the sense of BP learning.*

A fuzzy system for automatically tuning the gain β has been proposed in [29] to improve the performance of BP. The inputs of the fuzzy system are the sensitivities of the error with respect to the output and hidden layers, and the output is the appropriate gain of the activation function.

An adaptation rule for the gain β is derived using the gradient-descent method based on a sigmoidal function such as [15]

$$\phi(x) = \left(\frac{1}{1 + e^{-x}}\right)^\beta, \tag{5.50}$$

where $\beta \in (0, \infty)$. For $\beta \neq 1$, the derivative $\dot{\phi}(x)$ is skewed and its maxima shift from the point corresponding to $x = 0$ for $\beta = 1$ and the envelope of the derivatives is also sigmoidal. The method is an order of magnitude faster than standard BP.

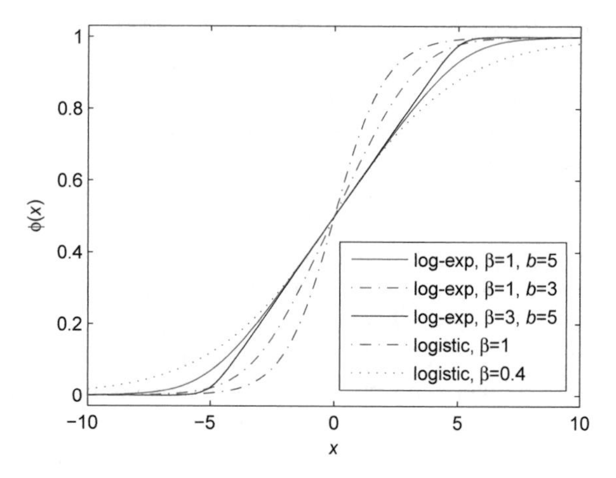

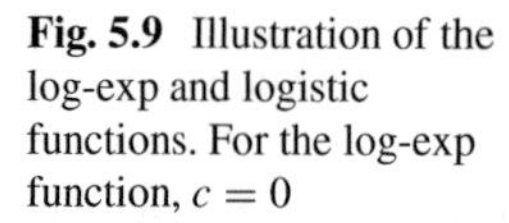
Fig. 5.9 Illustration of the log-exp and logistic functions. For the log-exp function, $c = 0$

A sigmoidal activation function with a wide linear part is derived in [27] by integrating an input distribution of the soft trapezoidal shape to generate a probability of fulfillment

$$\phi(x) = \frac{1}{2\beta b} \ln \left(\frac{1 + e^{\beta[(x-c)+b]}}{1 + e^{\beta[(x-c)-b]}} \right), \tag{5.51}$$

where β is the gain parameter, c decides the center of the shape, and $b > 0$ decides the slope at $x = 0$. A larger b leads to a smaller slope, while a larger β generates a longer linear part. When $b \to 0$, the log-exp function approaches the logistic function $\phi(x) = \frac{1}{1+e^{-\beta x}}$. For the same β, the log-exp function always has a longer linear part and a smaller slope at $x = 0$.

An illustration of the log-exp function as well as the logistic function is shown in Fig. 5.9. For the same β, the log-exp function always has a larger unsaturation zone and a wider linear part than the logistic function. For the log-exp function, the slope at $x = 0$ is decided by b and the width of the linear part is determined by β. The logistic function can extend its unsaturation zone by decreasing β, but the width of its linear part is still limited. The extended linear central part of the log-exp function prevents premature saturation and thus makes training of MLPs quickly.

Consider an algorithm whose time to convergence is unknown. Consider the following strategy. Run the algorithm for a specific time T. If it has not converged for T, rerun it from the start. This restart mechanism [31] is advantageous in problems that are prone to local minima or when there is a large variability in convergence time from run to run, and may lead to a speedup in such cases. It can reduce the overall average and standard deviation of the training time. The restart mechanism has also been applied in many optimization applications. It is theoretically analyzed in [76], where conditions on the probability density of the convergence time for which restart will improve the expected convergence time are obtained and the optimal restart time is derived.

5.8 Some Improved BP Algorithms

BP can be accelerated by extrapolation of each individual weight [52]. This extrapolation procedure is easy to implement and is activated only a few times in between iterations of BP. It leads to significant savings in computation time of BP and the solution is always located in close proximity to the one obtained by the BP procedure. BP by weight extrapolation reduces the required number of iterations at the expense of only a few extrapolation steps embedded in BP and some additional storage for computed weights and gradient vectors.

As an alternative to BP, a gradient-descent method without error backpropagation has been proposed for MLP learning [12]. Unlike BP, the method feeds gradients forward rather than feeding errors backward. The gradients of the final output are determined by feeding the gradients of the intermediate outputs forward at the same time that the outputs of the intermediate layers are fed forward. This method turns out to be equivalent to BP for a three-layer MLP, but is much more readily extended to an arbitrary number of layers without modification. This method has a great potential for concurrency.

Lyapunov stability theory has been applied for weight update [9, 80, 156]. Lyapunov's theory guarantees convergence under certain sufficient conditions. In [156], a generalization of BP is developed for training feedforward networks. The BP, Gauss–Newton, and Levenberg–Marquardt (LM) algorithms are special cases of this general algorithm. The general algorithm has the ability to handle time-varying inputs. The LF I and II algorithms [9], as two adaptive versions of BP, converge much faster than the BP and EKF algorithms to attain the same accuracy. In addition, sliding mode control-based adaptive learning algorithms have been used to train adalines [116] and multilayer networks [96] with good convergence and robustness.

The successive approximation BP algorithm [70] can effectively avoid local minima. Given a set of N pattern pairs $\left\{\left(\boldsymbol{x}_p, \boldsymbol{y}_p\right)\right\}$, all the training patterns are normalized so that $\left|x_{p,j}\right| \leq 1$, $\left|y_{p,k}\right| \leq 1$, for $p = 1, \ldots, N$, $j = 1, \ldots, J_1$, $k = 1, \ldots, J_M$. The training is composed of N_{phase} successive BP training phases, each being terminated when a predefined accuracy δ_i, $i = 1, \ldots, N_{\text{phase}}$, is achieved. At the first phase, the network is trained using BP on the training set. After accuracy δ_1 is achieved, the output of the network for the N input $\left\{\boldsymbol{x}_p\right\}$ are $\left\{\hat{\boldsymbol{y}}_p(1)\right\}$ and the weights are $\mathbf{W}(1)$. Calculate output errors $\delta \boldsymbol{y}_p(1) = \boldsymbol{y}_p - \hat{\boldsymbol{y}}_p(1)$ and normalize each $\delta \boldsymbol{y}_p(1)$ so that $\left|\delta y_{p,k}(1)\right| \leq 1$. In the second phase, the N training patterns are $\left\{\left(\boldsymbol{x}_p, \delta \boldsymbol{y}_p(1)\right)\right\}$. The training terminates at accuracy δ_2, with weights $\mathbf{W}(2)$, and output $\left\{\hat{\boldsymbol{y}}_p(2)\right\}$. Calculate $\delta \boldsymbol{y}_p(2) = \delta \boldsymbol{y}_p(1) - \hat{\boldsymbol{y}}_p(2)$ and normalized $\delta \boldsymbol{y}_p(2)$. This process continues up to phase N_{phase} with accuracy $\delta_{N_{\text{phase}}}$ and weights $\mathbf{W}\left(N_{\text{phase}}\right)$. The final training error is given by

$$E < 2^{N_{\text{phase}}} \prod_{i=1}^{N_{\text{phase}}} \delta_i. \tag{5.52}$$

If all $\delta_i < \frac{1}{2}$, as $N_{\text{phase}} \to \infty$, $E \to 0$. Successive approximation BP empirically, outperforms BP significantly in terms of convergence speed and generalization performance.

5.8.1 BP with Global Descent

The gradient-descent method is a stochastic dynamical system whose stable points only locally minimize the energy (error) function. The global-descent method, which is based on a global optimization technique called *terminal repeller unconstrained subenergy tunneling (TRUST)* [5, 14], is a deterministic dynamic system consisting of a single-vector differential equation. TRUST was introduced for general optimization problems, and it formulates optimization in terms of the flow of a special deterministic dynamical system.

Global descent is a gradient-descent method using a special criterion function. The derived update automatically switches between two phases: the tunneling phase and the local-search phase. At the tunneling phase, the terminal repeller term dominates and the local minimum becomes a repelling unstable equilibrium point, the solution will be repelled from the neighborhood of a local minimum until it reaches a lower basin of attraction. At the local-search phase, the repeller term is identically zero; this phase implements gradient descent and finds a local minimum in a new region. The two phases alternate until a stopping criterion is achieved. BP with tunneling for training MLP is similar to the global descent [14] and can find the global minimum from arbitrary initial choice in the weight space in polynomial time [106].

Another two-phase learning model has a BP phase and a gradient-ascent phase [122]. The BP phase performs steepest descent on an error measure. When BP gets stuck at local minima, the gradient-ascent phase attempts to fill-up the valley by modifying gain parameters in a gradient-ascent direction of the error measure. The two phases are repeated until the network gets out of local minima.

Deterministic global-descent methods usually use a tracing strategy decided by trajectory functions. These can be hybrid global/local minimization methods [5] or based on the concept of the terminal attractor [158].

Nonlinear dynamic systems satisfying the Lipschitz condition have a unique solution for each initial condition, and the trajectory of the state approaches the solution asympototically, but never reaches it. The concept of a terminal attractor was first introduced by Zak [158]. Terminal attractors are fixed points in a dynamic system violating the Lipschitz condition. As a result, a terminal attractor is a singular solution that envelopes the family of regular solutions, while each regular solution approaches such an attractor in finite time. The terminal attractor-based BP algorithm [51, 135] applies the concept of the terminal attractor to enable a finite time convergence to the global minimum. In contrast to BP, η in the terminal attractor-based BP is adapted by

$$\eta = \gamma \frac{h(E)}{\|g\|^2}, \tag{5.53}$$

where $\gamma > 0$, $g = \nabla_{\vec{w}} E$, and $h(E)$ is a nonnegative continuous function of E. This leads to an error function, which evolves by

$$\frac{\mathrm{d}E}{\mathrm{d}t} = -\gamma h(E). \tag{5.54}$$

When $h(E)$ is selected as E^{μ}, with $\frac{1}{2} < \mu < 1$, E will stably reach zero in time [51]

$$T = \frac{E^{1-\mu}(0)}{\gamma(1-\mu)}. \tag{5.55}$$

According to (5.53), at local minima, $\eta \to \infty$, and the algorithm can escape from local minima. By selecting γ and μ, one can tune the time to exactly reach $E = 0$. Terminal attractor-based BP can be three orders of magnitude faster than BP [135]. When $\|g\|$ is sufficiently large so that $\eta < \gamma$, one can, as a heuristic, force $\eta = \gamma$ temporarily to speed up the convergence, that is, switch to BP temporarily.

5.8.2 Robust BP Algorithms

Since BP is a special case of stochastic approximation, the techniques of robust statistics can be applied to BP [143]. In the presence of outliers, M-estimator-based robust learning can be applied. The rate of convergence is improved since the influence of the outliers is suppressed. Robust BP algorithms using M-estimator-based criterion functions are a typical class of robust algorithms, such as the robust BP using Hampel's tanh estimator with time-varying error cutoff points β_1 and β_2 [17], and the annealing robust BP algorithm [20].

Annealing robust BP [20] adopts the annealing concept into robust learning. A deterministic annealing process is applied to the scale estimator. The cost function of annealing robust BP has the same form as (2.20), with $\beta = \beta(t)$ as a deterministic annealing scale estimator. As $\beta(t) \to \infty$, annealing robust BP becomes BP. The basic idea of using an annealing schedule is to use a larger scale estimator in the early training stage and then to use a smaller scale estimator in the later training stage. When $\beta(t) \to 0+$ for $t \to \infty$, the M-estimator is equivalent to the linear L_1-norm estimator. Since the L_1-norm estimator is robust against outliers, the M-estimator equipped with such an annealing schedule is equivalent to the robust mixed-norm learning algorithm [22], where the L_2-norm is used at the beginning, then gradually tending to the L_1-norm according to the total error. An annealing schedule $\beta(t) = \frac{\gamma}{t}$ achieves good performance, where γ is a positive constant [20].

M-estimator-based robust methods have difficulties in the selection of the scale estimator β. Tao-robust BP algorithm [98] overcomes this problem by using a τ-estimator. Tao-robust BP also achieves two important properties: robustness with a high breakdown point and a high efficiency for normally distributed data.

5.9 Resilient Propagation (Rprop)

Rprop [105] is a batch learning algorithm which eliminates the influence of the magnitude of the partial derivative on the step size of the weight update. The update of each weight is according to the sequence of signs of the partial derivatives in each dimension of the weight space.

The update for each weight or bias $w_{ij}^{(m)}$ is given according to the following procedure [105]:

$$C = g_{ij}^{(m)}(t-1) \cdot g_{ij}^{(m)}(t), \tag{5.56}$$

$$\Delta_{ij}^{(m)}(t) = \begin{cases} \min\left\{\eta_0^+ \Delta_{ij}^{(m)}(t-1), \Delta_{\max}\right\}, & C > 0 \\ \max\left\{\eta_0^- \Delta_{ij}^{(m)}(t-1), \Delta_{\min}\right\}, & C < 0 \\ \Delta_{ij}^{(m)}(t-1), & C = 0 \end{cases}, \tag{5.57}$$

$$\Delta w_{ij}^{(m)}(t) = \begin{cases} -\text{sign}\left(g_{ij}^{(m)}(t)\right) \cdot \Delta_{ij}^{(m)}(t), & C \geq 0 \\ -\Delta w_{ij}^{(m)}(t-1), & C < 0 \end{cases}, \tag{5.58}$$

$$g_{ij}^{(m)}(t) = 0, \qquad C < 0, \tag{5.59}$$

$$w_{ij}^{(m)}(t+1) = w_{ij}^{(m)}(t) + \Delta w_{ij}^{(m)}(t), \tag{5.60}$$

where $0 < \eta_0^- < 1 < \eta_0^+$, and typically $\eta_0^+ = 1.2$ and $\eta_0^- = 0.5$. The value of $\Delta_{ij}^{(m)}(0)$ is not critical to the algorithm, and is selected as a positive constant Δ_0. The upper and lower bounds, denoted by $\Delta_{\max}$ and $\Delta_{\min}$, respectively, are used to restrict overflow/underflow problems of floating-point variables. For example, one can select $\Delta_{\max} = 50.00$ and $\Delta_{\min} = 10^{-6}$ [105]. A smaller value of $\Delta_{\max}$ such as 1.0 may result in a smoothened behavior of the decrease in error.

Rprop is robust against the choice of its initial parameters. In comparison with BP, Quickprop [31], and SuperSAB [128], the number of learning steps is significantly reduced and computational complexity of Rprop at each step is considerably smaller [40, 105]. Rprop has a performance comparable to that of the CG method [40]. It is one of the best performing first-order learning methods for neural networks. It is suitable for hardware implementation and is not susceptible to numerical problems. Rprop has also been used for training RBF networks [8] and recurrent fuzzy neural networks [83].

Example 5.6 From the housing dataset, there are a total of 506 example homes with 13 items of geographical and real estate information and their associated market values. We design a network that can predict the value of a house (in $1000s), given 13 inputs. We simulate for Rprop and batch BP for 1000 epochs. The learning rate for BP is 0.00008. The MSE for training is shown in Fig. 5.10. It is shown that Rprop is two orders of magnitude faster than BP.

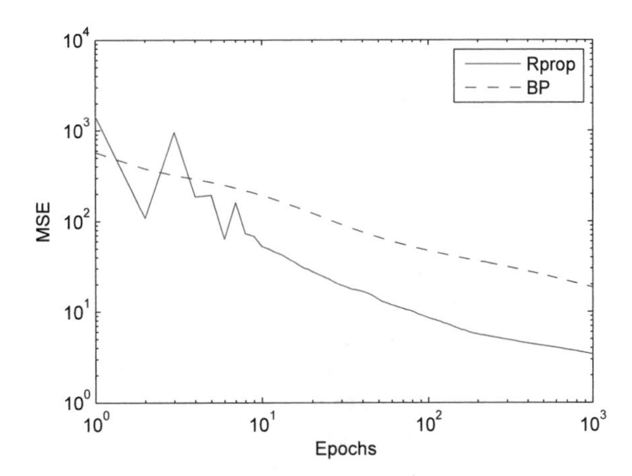

Fig. 5.10 Illustration of the Rprop algorithm

SASS [39] uses the same update rule as Rprop, but the update of $\Delta w_{ij}^{(m)}(t)$ is based on the bisection method for minimization in one dimension and uses two previous signs

$$\Delta_{ij}^{(m)}(t) = \begin{cases} 2.0\Delta_{ij}^{(m)}(t-1), \ C \geq 0 \text{ and } g_{ij}^{(m)}(t-2)\, g_{ij}^{(m)}(t) \geq 0 \\ 0.5\Delta_{ij}^{(m)}(t-1), \ \text{otherwise} \end{cases}. \tag{5.61}$$

SASS provides a performance comparable to that of Rprop [40].

Some variants aiming at improving Rprop are available. QRprop and diagonal estimation Rprop (DERprop) [100] are two similar hybrids of Rprop and second-order search steps. They adaptively switch between the two methods by using the strategy of Rprop and switching to second-order approximation only when the search is in the vicinity of a local minimum. QRprop makes use of local one-dimensional secant steps, which are used in Quickprop. DERprop directly computes the diagonal elements of the Hessian. The addition of simulated annealing in the form of noise and weight decay to Rprop yields the SARprop algorithm [129].

In Rprop, if the partial derivatives in consecutive steps possess the same sign, then weight is updated and, if consecutive partial derivatives possess the opposite sign, that is, $C < 0$, then previous weight update is reverted. The change in sign in successive steps is considered as a jump over the minima. This weight backtracking is counterproductive when the overall error has decreased during the change in sign of the partial derivatives. In improved Rprop [47], the previous weight update is reverted only when it has caused $C < 0$ in case of an overall error increase, that is, $E(t) > E(t-1)$. Improved Rprop outperforms Rprop and CG, and has a performance comparable to that of the BFGS method. The complex Rprop algorithm with error-dependent weight backtracking step [132] extends the improved Rprop algorithm [47] to complex domain.

GRprop [3] is a globally convergent modification of Rprop. It is built on a mathematical framework for convergence analysis [78], which ensures that the adaptive local learning rates of Rprop's schedule generate a descent-search direction at each iteration. GRprop exhibits a better convergence speed and stability than Rprop or improved Rprop does.

ARCprop [4] builds on Rprop, but guarantees convergence by shortening steps as necessary to achieve a sufficient reduction in global error. ARCprop backtracks along all dimensions if a particular step does not achieve a sufficient decrease in error, and offers a proof of the convergence. The basic idea is to take a series of downhill steps from the initial set of weights, until a vanishing error gradient indicates arrival at a local minimum. ARCprop achieves performance similar to Rprop in terms of training speed, training accuracy, and generalization.

RMSprop [127] is a mini-batch version of Rprop, which combines the quality of Rprop's adaptive learning rule for each weight with the efficiency of stochastic gradient descent. A root mean of the squared gradients for each weight over the recent mini-batches is used to normalize the gradient.

5.10 Spiking Neural Network Learning

Spiking neural networks theoretically have Turing-equivalent computing power [74]. The lower bound of computational capacity of spiking neural networks is at least that of perceptron [74]. Supervised learning algorithms for spiking neural networks include SpikeProp [11], ReSuMe [103], QuickProp [84], and Rprop [84].

SpikeProp [11] is an adaptation of BP to train a spiking neural model. The network architecture is a standard three-layered fully connected feedforward network with multiple delayed synaptic connections. Each neuron is assumed to fire exactly once during the time period under consideration. In SpikeProp, the problem of discontinuous membrane potential is tackled immediately after the neuron firing by linearization of membrane potential around the neuron's firing time. ReSuMe [103] adapts the Widrow–Hoff (delta) rule to spiking neural networks.

A spiking neural network training using SpikeProp and its variants is usually affected by a sudden rise in the network training error called *surges*. These surges can change the course of learning and even cause failure in learning as well. SpikePropAd is an adaptive learning rate extension to SpikeProp that assures convergence of the learning process [113]. It achieves good convergence compared to SpikeProp and Rprop.

Problems

5.1 Construct a 2-1-1 network that computes the XOR function of two inputs.

(a) Plot the network structure, and put the learned weights and biases on the graph.
(b) Give the learning steps.

(c) Express the decision regions in a two-dimensional plot.
(d) Construct a truth table for the network operation.

5.2 Parity is cyclic shift invariant. A shift of any bits generates the same parity. Use MLP to

(a) Learn the parity function.
(b) Model 4-bit parity checker (logic problem).

5.3 The correlation matrix of the input vector $\boldsymbol{x}_t$ is given by $\mathbf{R} = \begin{bmatrix} 1 & 0.4 \\ 0.4 & 1 \end{bmatrix}$. For the LMS algorithm, what is the suitable range of the learning rate η for the algorithm to be convergent?

5.4 In Example 5.4, we use different values of a single node to represent different classes. A more common architecture of a neural network classifier is usually defined according to the convention from a competitive learning based classifier. To classify a pattern set with J_1-dimensional inputs into K classes, the neural network is selected usually as having J_1 input nodes, K output nodes with each corresponding to one class, and zero or multiple hidden layers. During training, if the target class is k, the target value of the kth output node is set to 1 and those of all the other output nodes are set to 0. A pattern is considered to be correctly classified if the target output node has the highest output among all the output nodes.
Simulate the iris dataset using a 4-4-3 MLP by BP and BP-with-momentum, and compare the result with that given in Example 5.4. All the learning parameters are selected to be the same as those for Example 5.4.

5.5 Approximate the following functions by using MLP:

(a) $f(x_1, x_2) = \max\{e^{-x_1^2}, e^{-2x_2^2}, 2e^{-0.5(x_1^2+x_2^2)}\}$.
(b) $f(x, y) = 0.5 + 0.1x^2 \cos(y + 3) + 0.4xye^{1-y^2}$.
(c) $f(\boldsymbol{x}) = \sin\left(\sqrt{x_1^2 + x_2^2}\right) / \sqrt{x_1^2 + x_2^2}, \quad \boldsymbol{x} \in [-5, 5]^2$.
(d) $f(x) = \sqrt{2}\sin x + \sqrt{2}\cos x - \sqrt{2}\sin 3x + \sqrt{2}\cos 3x$.

5.6 Why is the MSE not a good measure of performance for classification problems?

5.7 In [139], E_c is defined as

$$E_c = \sum_{i,j} \frac{w_{ij}^2}{w_0^2 + w_{ij}^2},$$

where w_0 is a free parameter. For small w_0, the network prefers large weights. When w_0 is taken as unity, this penalty term decays the small weights more rapidly than the large weights [38]. For large w_0, the network prefers small weights. Derive $\frac{\partial E_c}{\partial w_{ij}^{(m)}}$.

5.8 Derive the gradient descent rule when the error function is defined by $E = \frac{1}{2}\sum_k (t_k - y_k)^2 + \sum_{i,j} w_{ji}^2$, where y_k is the network output for the kth input.

5.9 Derive the forward and backward propagation equations for each of the loss functions:

(a) Kullback–Leibler divergence criterion (2.15), (2.16), (2.17).
(b) The cross-entropy cost function (2.18).
(c) The Minkowski-r metric (2.19).

5.10 Write a program implementing the three-layer MLP with the BP rule, using each of the criteria listed in Problem **1.9**. Test the program using the different criteria on the iris classification problem.

5.11 Write a program implementing the three-layer MLP with the BP rule, incorporating the weight-decaying function.

(a) Generate 200 points from $y = \phi(x_1 + 2x_2) + 0.5(x_1 - x_2)^2 + 0.5N$, where $\phi(\cdot)$ is the logistic sigmoidal function, and N is a number drawn from the standard normal distribution. Apply the program on the samples.
(b) Test the program using 1000 randomly generated samples.
(c) Plot the training and testing errors versus the number of training epochs for differing weight-decay parameters.
(d) Describe the overfitting phenomenon observed.

5.12 Show that the derivative of the softmax function, $y_i = \frac{e^{x_i}}{\sum_j e^{x_j}}$, is $\frac{\partial y_i}{\partial x_j} = y_i(\delta_{ij} - y_j)$, where $\delta_{ij} = 0$ for $i = j$ and 0 otherwise.

5.13 This exercise is excerpted from [10]. Show that adding noise to the input vectors has the same effect as a weight-decaying regularization for a linear network model:

$$y_k = \sum_i w_{ki}x_i + w_{k0}$$

and a sum-of-squares error function

$$E = \frac{1}{2N}\sum_{n=1}^{N}\sum_k \{y_k(\boldsymbol{x}_n) - t_{n,k}\}^2,$$

where N is the training set size, and $t_{n,k}$ is the target values. Assume that random noise components ε_i are Gaussian $\varepsilon_i = \mathcal{N}(0, \nu)$, and $E(\varepsilon_i\varepsilon_j) = \delta_{ij}\nu$.

5.14 Show that the OBD algorithm for network pruning is a special case of the OBS algorithm.

5.15 Show how to find the inverse of an invertible matrix by using the BP algorithm.

References

1. Abid, S., Fnaiech, F., & Najim, M. (2001). A fast feedforward training algorithm using a modified form of the standard backpropagation algorithm. *IEEE Transactions on Neural Networks*, *12*(2), 424–430.
2. Aires, F., Schmitt, M., Chedin, A., & Scott, N. (1999). The "weight smoothing" regularization of MLP for Jacobian stabilization. *IEEE Transactions on Neural Networks*, *10*(6), 1502–1510.
3. Anastasiadis, A. D., Magoulas, G. D., & Vrahatis, M. N. (2005). New globally convergent training scheme based on the resilient propagation algorithm. *Neurocomputing*, *64*, 253–270.
4. Bailey, T. M. (2015). Convergence of Rprop and variants. *Neurocomputing*, *159*, 90–95.
5. Barhen, J., Protopopescu, V., & Reister, D. (1997). TRUST: A deterministic algorithm for global optimization. *Science*, *276*, 1094–1097.
6. Barron, A. R. (1993). Universal approximation bounds for superpositions of a sigmoidal function. *IEEE Transactions on Information Theory*, *39*(3), 930–945.
7. Battiti, R. (1989). Accelerated backpropagation learning: Two optimization methods. *Complex Systems*, *3*, 331–342.
8. Baykal N., & Erkmen, A. M. (2000). Resilient backpropagation for RBF networks. In *Proceedings of 4th International Conference on Knowledge-Based Intelligent Engineering Systems & Allied Technologies* (pp. 624–627). Brighton, UK.
9. Behera, L., Kumar, S., & Patnaik, A. (2006). On adaptive learning rate that guarantees convergence in feedforward networks. *IEEE Transactions on Neural Networks*, *17*(5), 1116–1125.
10. Bishop, C. M. (1995). Training with noise is equivalent to Tikhonov regularization. *Neural Computation*, *7*(1), 108–116.
11. Bohte, S. M., Kok, J. N., & La Poutre, H. (2002). Error-backpropagation in temporally encoded networks of spiking neurons. *Neurocomputing*, *48*, 17–37.
12. Brouwer, R. K. (1997). Training a feed-forward network by feeding gradients forward rather than by back-propagation of errors. *Neurocomputing*, *16*, 117–126.
13. Castellano, G., Fanelli, A. M., & Pelillo, M. (1997). An iterative pruning algorithm for feedforward neural networks. *IEEE Transactions on Neural Networks*, *8*(3), 519–531.
14. Cetin, B. C., Burdick, J. W., & Barhen, J. (1993). Global descent replaces gradient descent to avoid local minima problem in learning with artificial neural networks. In *Proceedings of IEEE International Conference on Neural Networks* (pp. 836–842). San Francisco, CA.
15. Chandra, P., & Singh, Y. (2004). An activation function adapting training algorithm for sigmoidal feedforward networks. *Neurocomputing*, *61*, 429–437.
16. Chandrasekaran, H., Chen, H. H., & Manry, M. T. (2000). Pruning of basis functions in nonlinear approximators. *Neurocomputing*, *34*, 29–53.
17. Chen, D. S., & Jain, R. C. (1994). A robust backpropagation learning algorithm for function approximation. *IEEE Transactions on Neural Networks*, *5*(3), 467–479.
18. Chester, D. L. (1990). Why two hidden layers are better than one. In *Proceedings of International Joint Conference on Neural Networks* (pp. 265–268). Washington, DC.
19. Choi, J. J., Arabshahi, P., Marks II, R. J., & Caudell, T. P. (1992). Fuzzy parameter adaptation in neural systems. In *Proceedings of International Joint Conference on Neural Networks* (Vol. 1, pp. 232–238). Baltimore, MD.
20. Chuang, C. C., Su, S. F., & Hsiao, C. C. (2000). The annealing robust backpropagation (ARBP) learning algorithm. *IEEE Transactions on Neural Networks*, *11*(5), 1067–1077.
21. Cibas, T., Soulie, F. F., Gallinari, P., & Raudys, S. (1996). Variable selection with neural networks. *Neurocomputing*, *12*, 223–248.
22. Cichocki, A., & Unbehauen, R. (1992). *Neural networks for optimization and signal processing*. New York: Wiley.
23. Costa, P., & Larzabal, P. (1999). Initialization of supervised training for parametric estimation. *Neural Processing Letters*, *9*, 53–61.
24. Cybenko, G. (1989). Approximation by superposition of a sigmoid function. *Mathematics of Control, Signals, and Systems*, *2*, 303–314.

25. Denoeux, T., & Lengelle, R. (1993). Initializing backpropagation networks with prototypes. *Neural Networks*, *6*(3), 351–363.
26. Drago, G., & Ridella, S. (1992). Statistically controlled activation weight initialization (SCAWI). *IEEE Transactions on Neural Networks*, *3*(4), 627–631.
27. Duch, W. (2005). Uncertainty of data, fuzzy membership functions, and multilayer perceptrons. *IEEE Transactions on Neural Networks*, *16*(1), 10–23.
28. Engelbrecht, A. P. (2001). A new pruning heuristic based on variance analysis of sensitivity information. *IEEE Transactions on Neural Networks*, *12*(6), 1386–1399.
29. Eom, K., Jung, K., & Sirisena, H. (2003). Performance improvement of backpropagation algorithm by automatic activation function gain tuning using fuzzy logic. *Neurocomputing*, *50*, 439–460.
30. Fabisch, A., Kassahun, Y., Wohrle, H., & Kirchner, F. (2013). Learning in compressed space. *Neural Networks*, *42*, 83–93.
31. Fahlman, S. E. (1988). Fast learning variations on back-propation: An empirical study. In D. S. Touretzky, G. E. Hinton, & T. Sejnowski (Eds.), *Proceedings of 1988 Connectionist Models Summer School* (pp. 38–51) (San Mateo, CA: Morgan Kaufmann). Pittsburgh.
32. Fahlman, S. E., & Lebiere, C. (1990). The cascade-correlation learning architecture. In D. S. Touretzky (Ed.), *Advances in neural information processing systems* (Vol. 2, pp. 524–532). San Mateo, CA: Morgan Kaufmann.
33. Finnoff, W. (1994). Diffusion approximations for the constant learning rate backpropagation algorithm and resistance to local minima. *Neural Computation*, *6*(2), 285–295.
34. Frean, M. (1990). The upstart algorithm: A method for constructing and training feedforward neural networks. *Neural Computation*, *2*(2), 198–209.
35. Funahashi, K. (1989). On the approximate realization of continuous mappings by neural networks. *Neural Networks*, *2*(3), 183–192.
36. Gallant, S. I. (1990). Perceptron-based learning algorithms. *IEEE Transactions on Neural Networks*, *1*(2), 179–191.
37. Goh, Y. S., & Tan, E. C. (1994). Pruning neural networks during training by backpropagation. In *Proceedings of IEEE Region 10's Ninth Annual International Conference (TENCON'94)* (pp. 805–808). Singapore.
38. Gupta, A., & Lam, S. M. (1998). Weight decay backpropagation for noisy data. *Neural Networks*, *11*, 1127–1137.
39. Hannan, J. M., & Bishop, J. M. (1996). *A class of fast artificial NN training algorithms*. Tech. Report JMH-JMB 01/96, Department of Cybernetics, University of Reading, UK.
40. Hannan, J. M., & Bishop, J. M. (1997). A comparison of fast training algorithms over two real problems. In *Proceedings of IEE Conference on Artificial Neural Networks* (pp. 1–6). Cambridge, UK.
41. Hassibi, B., Stork, D. G., & Wolff, G. J. (1992). Optimal brain surgeon and general network pruning. In *Proceedings of IEEE International Conference on Neural Networks* (pp. 293–299). San Francisco, CA.
42. Heskes, T., & Wiegerinck, W. (1996). A theoretical comparison of batch-mode, online, cyclic, and almost-cyclic learning. *IEEE Transactions on Neural Networks*, *7*, 919–925.
43. Hinton, G. E. (1987). *Connectionist Learning Procedures*. Technical Report CMU-CS-87-115, Carnegie-Mellon University, Computer Science Department, Pittsburgh, PA.
44. Hinton, G. E. (1989). Connectionist learning procedure. *Artificial Intelligence*, *40*, 185–234.
45. Hornik, K. M., Stinchcombe, M., & White, H. (1989). Multilayer feedforward networks are universal approximators. *Neural Networks*, *2*, 359–366.
46. Huang, G. B. (2003). Learning capability and storage capacity of two-hidden-layer feedforward networks. *IEEE Transactions on Neural Networks*, *14*(2), 274–281.
47. Igel, C., & Husken, M. (2003). Empirical evaluation of the improved Rprop learning algorithms. *Neurocomputing*, *50*, 105–123.
48. Ishikawa, M. (1995). Learning of modular structured networks. *Artificial Intelligence*, *75*, 51–62.

49. Jacobs, R. A. (1988). Increased rates of convergence through learning rate adaptation. *Neural Networks*, *1*, 295–307.
50. Jiang, X., Chen, M., Manry, M. T., Dawson, M. S., & Fung, A. K. (1994). Analysis and optimization of neural networks for remote sensing. *Remote Sensing Reviews*, *9*, 97–144.
51. Jiang, M., & Yu, X. (2001). Terminal attractor based back propagation learning for feedforward neural networks. In *Proceedings of IEEE International Symposium on Circuits and Systems (ISCAS)* (Vol. 2, pp. 711–714). Sydney, Australia.
52. Kamarthi, S. V., & Pittner, S. (1999). Accelerating neural network training using weight extrapolations. *Neural Networks*, *12*, 1285–1299.
53. Kanjilal, P. P., & Banerjee, D. N. (1995). On the application of orthogonal transformation for the design and analysis of feedforward networks. *IEEE Transactions on Neural Networks*, *6*(5), 1061–1070.
54. Karnin, E. D. (1990). A simple procedure for pruning back-propagation trained neural networks. *IEEE Transactions on Neural Networks*, *1*(2), 239–242.
55. Khashman, A. (2008). A modified backpropagation learning algorithm with added emotional coefficients. *IEEE Transactions on Neural Networks*, *19*(11), 1896–1909.
56. Kolen, J. F., & Pollack, J. B. (1990). Backpropagation is sensitive to initial conditions. *Complex Systems*, *4*(3), 269–280.
57. Kozma, R., Sakuma, M., Yokoyama, Y., & Kitamura, M. (1996). On the accuracy of mapping by neural networks trained by backpropagation with forgetting. *Neurocomputing*, *13*, 295–311.
58. Kruschke, J. K., & Movellan, J. R. (1991). Benefits of gain: Speeded learning and minimal layers in back-propagation networks. *IEEE Transactions on Systems, Man, and Cybernetics*, *21*(1), 273–280.
59. Kwok, T. Y., & Yeung, D. Y. (1997). Objective functions for training new hidden units in constructive neural networks. *IEEE Transactions on Neural Networks*, *8*(5), 1131–1148.
60. Le Cun, Y., Denker, J. S., & Solla, S. A. (1990). Optimal brain damage. In D. S. Touretzky (Ed.), *Advances in neural information processing systems* (Vol. 2, pp. 598–605). San Mateo, CA: Morgan Kaufmann.
61. Le Cun, Y., Kanter, I., & Solla, S. A. (1991). Second order properties of error surfaces: learning time and generalization. In R. P. Lippmann, J. E. Moody, & D. S. Touretzky (Eds.), *Advances in neural information processing systems* (Vol. 3, pp. 918–924). San Mateo, CA: Morgan Kaufmann.
62. Le Cun, Y., Simard, P. Y., & Pearlmutter, B. (1993). Automatic learning rate maximization by on-line estimation of the Hessian's eigenvectors. In S. J. Hanson, J. D. Cowan, & C. L. Giles (Eds.), *Advances in neural information processing systems* (Vol. 5, pp. 156–163). San Mateo, CA: Morgan Kaufmann.
63. Lee, Y., Oh, S. H., & Kim, M. W. (1991). The effect of initial weights on premature saturation in back-propagation training. In *Proceedings of IEEE International Joint Conference on Neural Networks* (Vol. 1, pp. 765–770). Seattle, WA.
64. Lee, H. M., Chen, C. M., & Huang, T. C. (2001). Learning efficiency improvement of back-propagation algorithm by error saturation prevention method. *Neurocomputing*, *41*, 125–143.
65. Lehtokangas, M., Saarinen, J., Huuhtanen, P., & Kaski, K. (1995). Initializing weights of a multilayer perceptron network by using the orthogonal least squares algorithm. *Neural Computation*, *7*, 982–999.
66. Lehtokangas, M., Korpisaari, P., & Kaski, K. (1996). Maximum covariance method for weight initialization of multilayer perceptron networks. In *Proceedings of European Symposium on Artificial Neural Networks (ESANN'96)* (pp. 243–248). Bruges, Belgium.
67. Lehtokangas, M. (1999). Modelling with constructive backpropagation. *Neural Networks*, *12*, 707–716.
68. Leung, C. S., Wong, K. W., Sum, P. F., & Chan, L. W. (2001). A pruning method for the recursive least squared algorithm. *Neural Networks*, *14*, 147–174.
69. Levin, A. U., Leen, T. K., & Moody, J. E. (1994). Fast pruning using principal components. In J. D. Cowan, G. Tesauro, & J. Alspector (Eds.), *Advances in neural information processing systems* (Vol. 6, pp. 35–42). San Francisco, CA: Morgan Kaufman.

70. Liang, Y. C., Feng, D. P., Lee, H. P., Lim, S. P., & Lee, K. H. (2002). Successive approximation training algorithm for feedforward neural networks. *Neurocomputing*, *42*, 311–322.
71. Lin, J., & Rosasco, L. (2017). Optimal rates for multi-pass stochastic gradient methods. *Journal of Machine Learning Research*, *18*, 1–47.
72. Liu, D., Chang, T. S., & Zhang, Y. (2002). A constructive algorithm for feedforward neural networks with incremental training. *IEEE Transactions on Circuits and Systems I*, *49*(12), 1876–1879.
73. Llanas, B., Lantaron, S., & Sainz, F. J. (2008). Constructive approximation of discontinuous functions by neural networks. *Neural Processing Letters*, *27*, 209–226.
74. Maass, W. (1996). Lower bounds for the computational power of networks of spiking neurons. *Neural Computation*, *8*(1), 1–40.
75. MacKay, D. J. C. (1992). Bayesian interpolation. *Neural Computation*, *4*(3), 415–447.
76. Magdon-Ismail, M., & Atiya, A. F. (2000). The early restart algorithm. *Neural Computation*, *12*, 1303–1312.
77. Magoulas, G. D., Vrahatis, M. N., & Androulakis, G. S. (1997). Effective backpropagation training with variable stepsize. *Neural Networks*, *10*(1), 69–82.
78. Magoulas, G. D., Plagianakos, V. P., & Vrahatis, M. N. (2002). Globally convergent algorithms with local learning rates. *IEEE Transactions on Neural Networks*, *13*(3), 774–779.
79. Maiorov, V., & Pinkus, A. (1999). Lower bounds for approximation by MLP neural networks. *Neurocomputing*, *25*, 81–91.
80. Man, Z., Wu, H. R., Liu, S., & Yu, X. (2006). A new adaptive backpropagation algorithm based on Lyapunov stability theory for neural networks. *IEEE Transactions on Neural Networks*, *17*(6), 1580–1591.
81. Manry, M. T., Apollo, S. J., Allen, L. S., Lyle, W. D., Gong, W., Dawson, M. S., et al. (1994). Fast training of neural networks for remote sensing. *Remote Sensing Reviews*, *9*, 77–96.
82. Martens, J. P., & Weymaere, N. (2002). An equalized error backpropagation algorithm for the on-line training of multilayer perceptrons. *IEEE Transactions on Neural Networks*, *13*(3), 532–541.
83. Mastorocostas, P. A. (2004). Resilient back propagation learning algorithm for recurrent fuzzy neural networks. *Electronics Letters*, *40*(1), 57–58.
84. McKennoch, S., Liu, D., & Bushnell, L. G. (2006). Fast modifications of the SpikeProp algorithm. In *Proceedings of the IEEE International Joint Conference on Neural Networks (IJCNN'06)* (pp. 3970–3977).
85. McLoone, S., Brown, M. D., Irwin, G., & Lightbody, G. (1998). A hybrid linear/nonlinear training algorithm for feedforward neural networks. *IEEE Transactions on Neural Networks*, *9*(4), 669–684.
86. Mezard, M., & Nadal, J. P. (1989). Learning in feedforward layered networks: The tiling algorithm. *Journal of Physics A*, *22*, 2191–2203.
87. Minai, A. A., & Williams, R. D. (1990). Backpropagation heuristics: A study of the extended delta-bar-delta algorithm. In *Proceedings of IEEE International Conference on Neural Networks* (Vol. 1, pp. 595–600). San Diego, CA.
88. Moody, J. O., & Antsaklis, P. J. (1996). The dependence identification neural network construction algorithm. *IEEE Transactions on Neural Networks*, *7*(1), 3–13.
89. Mozer, M. C., & Smolensky, P. (1989). Using relevance to reduce network size automatically. *Connection Science*, *1*(1), 3–16.
90. Nakama, T. (2009). Theoretical analysis of batch and on-line training for gradient descent learning in neural networks. *Neurocomputing*, *73*, 151–159.
91. Narayan, S. (1997). The generalized sigmoid activation function: Competitive supervised learning. *Information Sciences*, *99*, 69–82.
92. Ng, S. C., Leung, S. H., & Luk, A. (1999). Fast convergent generalized back-propagation algorithm with constant learning rate. *Neural Processing Letters*, *9*, 13–23.
93. Nguyen, D., & Widrow, B. (1990). Improving the learning speed of 2-layer neural networks by choosing initial values of the adaptive weights. In *Proceedings of International Joint Conference on Neural Networks* (Vol. 3, pp. 21–26). San Diego, CA.

94. Oh, S. H. (1997). Improving the error back-propagation algorithm with a modified error function. *IEEE Transactions on Neural Networks*, *8*(3), 799–803.
95. Parlos, A. G., Femandez, B., Atiya, A. F., Muthusami, J., & Tsai, W. K. (1994). An accelerated learning algorithm for multilayer perceptron networks. *IEEE Transactions on Neural Networks*, *5*(3), 493–497.
96. Parma, G. G., Menezes, B. R., & Braga, A. P. (1998). Sliding mode algorithm for training multilayer artificial neural networks. *Electronics Letters*, *34*(1), 97–98.
97. Perantonis, S. J., & Virvilis, V. (1999). Input feature extraction for multilayered perceptrons using supervised principal component analysis. *Neural Processing Letters*, *10*, 243–252.
98. Pernia-Espinoza, A. V., Ordieres-Mere, J. B., Martinez-de-Pison, F. J., & Gonzalez-Marcos, A. (2005). TAO-robust backpropagation learning algorithm. *Neural Networks*, *18*, 191–204.
99. Pfister, M., & Rojas, R. (1993). Speeding-up backpropagation—A comparison of orthogonal techniques. In *Proceedings of International Joint Conference on Neural Networks* (Vol. 1, pp. 517–523). Nagoya, Japan.
100. Pfister, M., & Rojas, R. (1994). QRprop–a hybrid learning algorithm which adaptively includes second order information. In *Proceedings of the 4th Dortmund Fuzzy Days* (pp. 55–62).
101. Poggio, T., & Girosi, F. (1990). Networks for approximation and learning. *Proceedings of the IEEE*, *78*(9), 1481–1497.
102. Ponnapalli, P. V. S., Ho, K. C., & Thomson, M. (1999). A formal selection and pruning algorithm for feedforward artificial neural network optimization. *IEEE Transactions on Neural Networks*, *10*(4), 964–968.
103. Ponulak, F., & Kasinski, A. (2010). Supervised learning in spiking neural networks with ReSuMe: sequence learning, classification, and spike shifting. *Neural Computation*, *22*(2), 467–510.
104. Rathbun, T. F., Rogers, S. K., DeSimio, M. P., & Oxley, M. E. (1997). MLP iterative construction algorithm. *Neurocomputing*, *17*, 195–216.
105. Riedmiller, M., & Braun, H. (1993). A direct adaptive method for faster backpropagation learning: The RPROP algorithm. In *Proceedings of IEEE International Conference on Neural Networks* (pp. 586–591). San Francisco, CA.
106. RoyChowdhury, P., Singh, Y. P., & Chansarkar, R. A. (1999). Dynamic tunneling technique for efficient training of multilayer perceptrons. *IEEE Transactions on Neural Networks*, *10*(1), 48–55.
107. Ruck, D. W., Rogers, S. K., & Kabrisky, M. (1990). Feature selection using a multilayer perceptron. *Neural Network Computing*, *2*(2), 40–48.
108. Rumelhart, D. E., Hinton, G. E., & Williams, R. J. (1986). Learning internal representations by error propagation. In D. E. Rumelhart & J. L. McClelland (Eds.), *Parallel distributed processing: Explorations in the microstructure of cognition, 1: Foundation* (pp. 318–362). Cambridge, MA: MIT Press.
109. Rumelhart, D. E., Durbin, R., Golden, R., & Chauvin, Y. (1995). Backpropagation: the basic theory. In Y. Chauvin & D. E. Rumelhart (Eds.), *Backpropagation: Theory, architecture, and applications* (pp. 1–34). Hillsdale, NJ: Lawrence Erlbaum.
110. Satoh, S., & Nakano, R. (2013). Fast and stable learning utilizing singular regions of multilayer perceptron. *Neural Processing Letters*, *38*(2), 99–115.
111. Selmic, R. R., & Lewis, F. L. (2002). Neural network approximation of piecewise continuous functions: Application to friction compensation. *IEEE Transactions on Neural Networks*, *13*(3), 745–751.
112. Setiono, R., & Hui, L. C. K. (1995). Use of quasi-Newton method in a feed-forward neural network construction algorithm. *IEEE Transactions on Neural Networks*, *6*(1), 273–277.
113. Shrestha, S. B., & Song, Q. (2015). Adaptive learning rate of SpikeProp based on weight convergence analysis. *Neural Networks*, *63*, 185–198.
114. Sietsma, J., & Dow, R. J. F. (1991). Creating artificial neural networks that generalize. *Neural Networks*, *4*, 67–79.
115. Silva, F. M., & Almeida, L. B. (1990). Speeding-up backpropagation. In R. Eckmiller (Ed.), *Advanced neural computers* (pp. 151–156). Amsterdam: North-Holland.

116. Sira-Ramirez, H., & Colina-Morles, E. (1995). A sliding mode strategy for adaptive learning in adalines. *IEEE Transactions on Circuits and Systems I*, *42*(12), 1001–1012.
117. Smyth, S. G. (1992). Designing multilayer perceptrons from nearest neighbor systems. *IEEE Transactions on Neural Networks*, *3*(2), 329–333.
118. Sperduti, A., & Starita, A. (1993). Speed up learning and networks optimization with extended back propagation. *Neural Networks*, *6*(3), 365–383.
119. Stahlberger, A., & Riedmiller, M. (1997). Fast network pruning and feature extraction using the unit-OBS algorithm. In M. C. Mozer, M. I. Jordan, & T. Petsche (Eds.), *Advances in neural information processing systems* (Vol. 9, pp. 655–661). Cambridge, MA: MIT Press.
120. Sum, J., Leung, C. S., Young, G. H., & Kan, W. K. (1999). On the Kalman filtering method in neural network training and pruning. *IEEE Transactions on Neural Networks*, *10*, 161–166.
121. Tamura, S., & Tateishi, M. (1997). Capabilities of a four-layered feedforward neural network: Four layers versus three. *IEEE Transactions on Neural Networks*, *8*(2), 251–255.
122. Tang, Z., Wang, X., Tamura, H., & Ishii, M. (2003). An algorithm of supervised learning for multilayer neural networks. *Neural Computation*, *15*, 1125–1142.
123. Tarres, P., & Yao, Y. (2014). Online learning as stochastic approximation of regularization paths: Optimality and almost-sure convergence. *IEEE Transactions on Information Theory*, *60*(9), 5716–5735.
124. Teoh, E. J., Tan, K. C., & Xiang, C. (2006). Estimating the number of hidden neurons in a feedforward network using the singular value decomposition. *IEEE Transactions on Neural Networks*, *17*(6), 1623–1629.
125. Tesauro, G., & Janssens, B. (1988). Scaling relationships in back-propagation learning. *Complex Systems*, *2*, 39–44.
126. Thimm, G., & Fiesler, E. (1997). High-order and multilayer perceptron initialization. *IEEE Transactions on Neural Networks*, *8*(2), 349–359.
127. Tieleman, T., & Hinton, G. (2012). Lecture 6.5 – rmsprop: Divide the gradient by a running average of its recent magnitude. In *COURSERA: Neural networks for machine learning*.
128. Tollenaere, T. (1990). SuperSAB: Fast adaptive backpropation with good scaling properties. *Neural Networks*, *3*(5), 561–573.
129. Treadgold, N. K., & Gedeon, T. D. (1998). Simulated annealing and weight decay in adaptive learning: The SARPROP algorithm. *IEEE Transactions on Neural Networks*, *9*(4), 662–668.
130. Trenn, S. (2008). Multilayer perceptrons: Approximation order and necessary number of hidden units. *IEEE Transactions on Neural Networks*, *19*(5), 836–844.
131. Tresp, V., Neuneier, R., & Zimmermann, H. G. (1997). Early brain damage. In M. Mozer, M. I. Jordan, & P. Petsche (Eds.), *Advances in neural information processing systems* (Vol. 9, pp. 669–675). Cambridge, MA: MIT Press.
132. Tripathi, B. K., & Kalra, P. K. (2011). On efficient learning machine with root-power mean neuron in complex domain. *IEEE Transactions on Neural Networks*, *22*(5), 727–738.
133. Vitela, J. E., & Reifman, J. (1997). Premature saturation in backpropagation networks: mechanism and necessary condition. *Neural Networks*, *10*(4), 721–735.
134. Vogl, T. P., Mangis, J. K., Rigler, A. K., Zink, W. T., & Alkon, D. L. (1988). Accelerating the convergence of the backpropagation method. *Biological Cybernetics*, *59*, 257–263.
135. Wang, S. D., & Hsu, C. H. (1991). Terminal attractor learning algorithms for back propagation neural networks. In *Proceedings of International Joint Conference on Neural Networks* (pp. 183–189). Seattle, WA.
136. Wang, X. G., Tang, Z., Tamura, H., & Ishii, M. (2004). A modified error function for the backpropagation algorithm. *Neurocomputing*, *57*, 477–484.
137. Wang, J., Yang, J., & Wu, W. (2011). Convergence of cyclic and almost-cyclic learning with momentum for feedforward neural networks. *IEEE Transactions on Neural Networks*, *22*(8), 1297–1306.
138. Wang, J., Xu, C., Yang, X., & Zurada, J. M. (2018). A novel pruning algorithm for smoothing feedforward neural networks based on group lasso method. *IEEE Transactions on Neural Networks and Learning Systems*, *29*(5), 2012–2024.

139. Weigend, A. S., Rumelhart, D. E., & Huberman, B. A. (1991). Generalization by weight-elimination with application to forecasting. In R. P. Lippmann, J. E. Moody, & D. S. Touretzky (Eds.), *Advances in neural information processing systems* (Vol. 3, pp. 875–882). San Mateo, CA: Morgan Kaufmann.
140. Wessels, L. F. A., & Barnard, E. (1992). Avoiding false local minima by proper initialization of connections. *IEEE Transactions on Neural Networks*, *3*(6), 899–905.
141. Werbos, P. J. (1974). *Beyond regressions: New tools for prediction and analysis in the behavioral sciences*. Cambridge, MA: Harvard University. PhD Thesis.
142. Weymaere, N., & Martens, J. P. (1994). On the initializing and optimization of multilayer perceptrons. *IEEE Transactions on Neural Networks*, *5*, 738–751.
143. White, H. (1989). Learning in artificial neural networks: A statistical perspective. *Neural Computation*, *1*(4), 425–469.
144. Widrow, B., & Stearns, S. D. (1985). *Adaptive signal processing*. Englewood Cliffs, NJ: Prentice-Hall.
145. Wilson, D. R., & Martinez, T. R. (2003). The general inefficiency of batch training for gradient descent learning. *Neural Networks*, *16*, 1429–1451.
146. Xiang, C., Ding, S. Q., & Lee, T. H. (2005). Geometrical interpretation and architecture selection of MLP. *IEEE Transactions on Neural Networks*, *16*(1), 84–96.
147. Xing, H.-J., & Hu, B.-G. (2009). Two-phase construction of multilayer perceptrons using information theory. *IEEE Transactions on Neural Networks*, *20*(4), 715–721.
148. Xu, Z.-B., Zhang, R., & Jing, W.-F. (2009). When does online BP training converge? *IEEE Transactions on Neural Networks*, *20*(10), 1529–1539.
149. Yam, J. Y. F., & Chow, T. W. S. (2000). A weight initialization method for improving training speed in feedforward neural network. *Neurocomputing*, *30*, 219–232.
150. Yam, Y. F., Chow, T. W. S., & Leung, C. T. (1997). A new method in determining the initial weights of feedforward neural networks. *Neurocomputing*, *16*, 23–32.
151. Yam, J. Y. F., & Chow, T. W. S. (2001). Feedforward networks training speed enhancement by optimal initialization of the synaptic coefficients. *IEEE Transactions on Neural Networks*, *12*(2), 430–434.
152. Yam, Y. F., Leung, C. T., Tam, P. K. S., & Siu, W. C. (2002). An independent component analysis based weight initialization method for multilayer perceptrons. *Neurocomputing*, *48*, 807–818.
153. Yang, L., & Yu, W. (1993). Backpropagation with homotopy. *Neural Computation*, *5*(3), 363–366.
154. Yu, X. H., & Chen, G. A. (1997). Efficient backpropagation learning using optimal learning rate and momentum. *Neural Networks*, *10*(3), 517–527.
155. Yu, X. H., Chen, G. A., & Cheng, S. X. (1995). Dynamic learning rate optimization of the backpropagation algorithm. *IEEE Transactions on Neural Networks*, *6*(3), 669–677.
156. Yu, X., Efe, M. O., & Kaynak, O. (2002). A general backpropagation algorithm for feedforward neural networks learning. *IEEE Transactions on Neural Networks*, *13*(1), 251–254.
157. Yuan, K., Ying, B., & Sayed, A. H. (2016). On the influence of momentum acceleration on online learning. *Journal of Machine Learning Research*, *17*, 1–66.
158. Zak, M. (1989). Terminal attractors in neural networks. *Neural Networks*, *2*, 259–274.
159. Zhang, N. (2015). A study on the optimal double parameters for steepest descent with momentum. *Neural Computation*, *27*, 982–1004.
160. Zhang, X. M., Chen, Y. Q., Ansari, N., & Shi, Y. Q. (2004). Mini-max initialization for function approximation. *Neurocomputing*, *57*, 389–409.
161. Zhang, R., Xu, Z.-B., Huang, G.-B., & Wang, D. (2012). Global convergence of online BP training with dynamic learning rate. *IEEE Transactions on Neural Networks and Learning Systems*, *23*(2), 330–341.
162. Zurada, J. M., Malinowski, A., & Usui, S. (1997). Perturbation method for deleting redundant inputs of perceptron networks. *Neurocomputing*, *14*, 177–193.
163. Zweiri, Y. H., Whidborne, J. F., & Seneviratne, L. D. (2003). A three-term backpropagation algorithm. *Neurocomputing*, *50*, 305–318.

Chapter 6
Multilayer Perceptrons: Other Learing Techniques

6.1 Introduction to Second-Order Learning Methods

Training of feedforward networks can be viewed as an unconstrained optimization problem. BP is slow to converge when the error surface is flat along a weight dimension. Second-order optimization techniques have a strong theoretical basis and provide significantly faster convergence. Second-order methods make use of the Hessian matrix $\mathbf{H}$, that is, the second-order derivative of the error E with respect to the N_w-dimensional weight vector $\vec{\boldsymbol{w}}$, which is a vector obtained by concatenating all the weights and biases of a network:

$$\mathbf{H}(t) = \left. \frac{\partial^2 E}{\partial \vec{\boldsymbol{w}}^2} \right|_t . \tag{6.1}$$

It is an $N_w \times N_w$ matrix. This matrix contains information as to how the gradient changes in different directions of the weight space. The calculation of $\mathbf{H}$ can be implemented into the BP algorithm [14]. For feedforward networks, $\mathbf{H}$ is ill-conditioned [74].

Second-order algorithms can either be of matrix or vector type. Matrix-type algorithms require the storage for the Hessian and its inverse. The Broyden–Fletcher–Goldfarb–Shanno (BFGS) method [25] and a class of Newton's methods are matrix-type algorithms. Matrix-type algorithms are typically two orders of magnitude faster than BP. The computational complexity is at least $O\left(N_w^2\right)$ floating-point operations, when used for supervised learning of MLP.

Vector-type algorithms, on the other hand, require the storage of a few vectors. Examples of such algorithms include the limited memory BFGS [7], one-step secant [8, 9], scaled CG [55], and CG methods [39, 85]. They are typically one order of magnitude faster than BP. Vector-type algorithms require iterative computation of the Hessian or implicitly exploit the structure of the Hessian. They are based on line-search or trust-region search methods.

K.-L. Du and M. N. S. Swamy, *Neural Networks and Statistical Learning*,
https://doi.org/10.1007/978-1-4471-7452-3_6

In BP, the selection of the learning parameters by trial-and-error is a daunting task for a large training set. In second-order methods, learning parameters can be automatically adapted. However, second-order methods are required to be used in batch mode due to the numerical sensitivity of the computation of second-order gradients.

6.2 Newton's Methods

Newton's methods [6, 8] require explicit computation and storage of the Hessian. They are variants of the classical Newton's method. These include the Gauss–Newton and Levenberg–Marquardt (LM) methods. Newton's methods achieve the quadratic convergence. They are less sensitive to the learning constant, and a proper learning constant is easily selected.

At step $t+1$, we expand $E(\vec{\boldsymbol{w}})$ into a Taylor series

$$E\left(\vec{\boldsymbol{w}}\right)|_{t+1} = E\left(\vec{\boldsymbol{w}}\right)|_{t} + \left[\vec{\boldsymbol{w}}(t+1) - \vec{\boldsymbol{w}}(t)\right]^T \boldsymbol{g}(t)$$
$$+\frac{1}{2}\left[\vec{\boldsymbol{w}}(t+1) - \vec{\boldsymbol{w}}(t)\right]^T \mathbf{H}(t)\left[\vec{\boldsymbol{w}}(t+1) - \vec{\boldsymbol{w}}(t)\right] + \ldots, \quad (6.2)$$

where the gradient vector is given by

$$\boldsymbol{g}(t) = \nabla E\left(\vec{\boldsymbol{w}}(t)\right) = \nabla E(\vec{\boldsymbol{w}})|_t. \quad (6.3)$$

Equating $\boldsymbol{g}(t+1)$ to zero:

$$\boldsymbol{g}(t+1) = \boldsymbol{g}(t) + \mathbf{H}(t)\left(\vec{\boldsymbol{w}}(t+1) - \vec{\boldsymbol{w}}(t)\right) + \cdots = 0. \quad (6.4)$$

By ignoring the third- and higher order terms, the classical Newton's method is obtained:

$$\vec{\boldsymbol{w}}(t+1) = \vec{\boldsymbol{w}}(t) + \boldsymbol{d}(t), \quad (6.5)$$

$$\boldsymbol{d}(t) = -\mathbf{H}^{-1}(t)\boldsymbol{g}(t). \quad (6.6)$$

For MLP, the Hessian is a singular matrix [87], and thus (6.6) cannot be used. Nevertheless, we can make use of (6.4) and solve the following set of linear equations for the step $\boldsymbol{d}(t)$

$$\boldsymbol{g}(t) = -\mathbf{H}(t)\boldsymbol{d}(t). \quad (6.7)$$

This set of linear equations can be solved by using SVD or QR decomposition.

From second-order conditions, $\mathbf{H}(t)$ must be positive for searching a minimum. At each iteration, E is approximated locally by a second-order Taylor polynomial, which is minimized subsequently. This minimization is computationally prohibitive, since computation of $\mathbf{H}(t)$ needs global information and solution of a set of linear equations is also required [15]. In the classical Newton's method, $O\left(N_w^3\right)$ floating-point operations are needed for computing the search direction; however, it is not suitable when $\vec{\boldsymbol{w}}(t)$ is remote from the solution, since $\mathbf{H}(t)$ may not be positive-definite.

6.2.1 Gauss–Newton Method

Denote $E(\vec{\boldsymbol{w}})$ as

$$E(\vec{\boldsymbol{w}}) = \frac{1}{2}\sum_{i=1}^{N} \epsilon_i^2(\vec{\boldsymbol{w}}) = \frac{1}{2}\boldsymbol{\epsilon}^T\boldsymbol{\epsilon}, \tag{6.8}$$

where $\boldsymbol{\epsilon}\left(\vec{\boldsymbol{w}}\right) = \left(\epsilon_1\left(\vec{\boldsymbol{w}}\right), \epsilon_2\left(\vec{\boldsymbol{w}}\right), \ldots, \epsilon_N\left(\vec{\boldsymbol{w}}\right)\right)^T$, $\epsilon_i = \|\boldsymbol{e}_i\|$, and $\boldsymbol{e}_i = \hat{\boldsymbol{y}}_i - \boldsymbol{y}_i$. Thus, $\epsilon_i^2 = \boldsymbol{e}_i^T\boldsymbol{e}_i$.

The gradient vector is obtained by

$$g(\vec{\boldsymbol{w}}) = \frac{\partial E(\vec{\boldsymbol{w}})}{\partial \vec{\boldsymbol{w}}} = \mathbf{J}^T\left(\vec{\boldsymbol{w}}\right)\boldsymbol{\epsilon}\left(\vec{\boldsymbol{w}}\right), \tag{6.9}$$

where $\mathbf{J}\left(\vec{\boldsymbol{w}}\right)$, an $N \times N_w$ Jacobian matrix, is defined by

$$\mathbf{J}\left(\vec{\boldsymbol{w}}\right) = \frac{\partial \boldsymbol{\epsilon}\left(\vec{\boldsymbol{w}}\right)}{\partial \vec{\boldsymbol{w}}} = \left[J_{ij}\right] = \left[\frac{\partial \epsilon_i}{\partial w_j}\right]. \tag{6.10}$$

Further, the Hessian is obtained by

$$\mathbf{H} = \frac{\partial g(\vec{\boldsymbol{w}})}{\partial \vec{\boldsymbol{w}}} = \mathbf{J}^T(\vec{\boldsymbol{w}})\mathbf{J}\left(\vec{\boldsymbol{w}}\right) + \mathbf{S}\left(\vec{\boldsymbol{w}}\right), \tag{6.11}$$

where

$$\mathbf{S}\left(\vec{\boldsymbol{w}}\right) = \sum_{i=1}^{N} \epsilon_i\left(\vec{\boldsymbol{w}}\right)\nabla^2\epsilon_i\left(\vec{\boldsymbol{w}}\right). \tag{6.12}$$

Assuming that $\mathbf{S}\left(\vec{\boldsymbol{w}}\right)$ is small, we approximate the Hessian using

$$\mathbf{H}_{\mathrm{GN}}(t) = \mathbf{J}^T(t)\mathbf{J}(t), \tag{6.13}$$

where $\mathbf{J}(t)$ denotes $\mathbf{J}\left(\vec{\boldsymbol{w}}(t)\right)$.

In view of (6.5), (6.6) and (6.9), we obtain

$$\vec{\boldsymbol{w}}(t+1) = \vec{\boldsymbol{w}}(t) + \boldsymbol{d}(t), \tag{6.14}$$

$$\boldsymbol{d}(t) = -\mathbf{H}_{\mathrm{GN}}^{-1}(t)\mathbf{J}^{T}(t)\boldsymbol{\epsilon}(t), \tag{6.15}$$

where $\boldsymbol{\epsilon}(t)$ denotes $\boldsymbol{\epsilon}\left(\overrightarrow{\boldsymbol{w}}(t)\right)$. The above procedure is the Gauss–Newton method.

The Gauss–Newton method approximates the Hessian using information from first-order derivatives only. However, far away from the solution, the term $\mathbf{S}$ is not negligible and thus the approximation to the Hessian $\mathbf{H}$ is poor, resulting in slow convergence. The Gauss–Newton method may have an ill-conditioned Jacobian matrix and $\mathbf{H}$ may be noninvertible. In this case, like in the classical Newton's method, one can instead solve $\mathbf{H}_{\mathrm{GN}}(t)\boldsymbol{d}(t) = -\mathbf{J}^{T}(t)\boldsymbol{\epsilon}(t)$ for $\boldsymbol{d}(t)$. For every pattern, the BP algorithm requires only one backpropagation process, while in second-order algorithms the backpropagation process is repeated for every output separately in order to obtain consecutive rows of the Jacobian.

An iterative Gauss–Newton method based on the generalized secant method using Broyden's approach is given as [25]

$$\boldsymbol{q}(t) = \boldsymbol{\epsilon}(t+1) - \boldsymbol{\epsilon}(t), \tag{6.16}$$

$$\mathbf{J}(t+1) = \mathbf{J}(t) + \frac{[\boldsymbol{q}(t) - \mathbf{J}(t)\boldsymbol{d}(t)]\boldsymbol{d}^{T}(t)}{\boldsymbol{d}^{T}(t)\boldsymbol{d}(t)}. \tag{6.17}$$

The method uses the same update given by (6.14) and (6.15).

6.2.2 Levenberg–Marquardt Method

The LM method [56] eliminates the possible singularity of $\mathbf{H}$ by adding a small identity matrix to it. This method is derived by minimizing the quadratic approximation to $E\left(\overrightarrow{\boldsymbol{w}}\right)$ subject to the constraint that the step length $\|\boldsymbol{d}(t)\|$ is within a trust-region at step t. At given $\overrightarrow{\boldsymbol{w}}(t)$, the second-order Taylor approximation of $E\left(\overrightarrow{\boldsymbol{w}}\right)$ is given by

$$\hat{E}\left(\overrightarrow{\boldsymbol{w}}(t) + \boldsymbol{d}(t)\right) = E\left(\overrightarrow{\boldsymbol{w}}(t)\right) + \boldsymbol{g}(t)^{T}\boldsymbol{d}(t) + \frac{1}{2}\boldsymbol{d}^{T}(t)\mathbf{H}(t)\boldsymbol{d}(t). \tag{6.18}$$

The search step $\boldsymbol{d}(t)$ is computed by solving the trust-region subproblem

$$\min_{\boldsymbol{d}(t)} \hat{E}\left(\overrightarrow{\boldsymbol{w}}(t) + \boldsymbol{d}(t)\right) \quad \text{subject to} \quad \|\boldsymbol{d}(t)\| \le \delta_t, \tag{6.19}$$

where δ_t is a positive scalar and $\left\{\boldsymbol{d}(t) \middle| \|\boldsymbol{d}(t)\| \le \delta_t\right\}$ is the trust-region around $\overrightarrow{\boldsymbol{w}}(t)$.

This inequality constrained optimization problem can be solved by using the Karush–Kuhn–Tucker (KKT) theorem [25], which leads to

$$\mathbf{H}_{\mathrm{LM}}(t) = \mathbf{H}(t) + \sigma(t)\mathbf{I}, \tag{6.20}$$

where $\sigma(t)$ is a small positive value, which indirectly controls the size of the trust-region.

The LM modification to the Gauss–Newton method is given as [31]

$$\mathbf{H}_{\text{LM}}(t) = \mathbf{H}_{\text{GN}}(t) + \sigma(t)\mathbf{I}. \tag{6.21}$$

Thus, $\mathbf{H}_{\text{LM}}$ is always invertible. The LM method given in (6.21) can be treated as a trust-region modification to the Gauss–Newton method [8].

The LM method is based on the assumption that such an approximation to the Hessian is valid only inside a trust-region of small radius, controlled by σ. If the eigenvalues of $\mathbf{H}$ are $\lambda_1 \geq \lambda_2 \geq \cdots \geq \lambda_{N_w}$, then the eigenvalues of $\mathbf{H}_{\text{LM}}$ are $\lambda_i + \sigma$, $i = 1, \ldots, N_w$, with the same corresponding eigenvectors. σ is selected so that $\mathbf{H}_{\text{LM}}$ is positive-definite, that is, $\lambda_{N_w} + \sigma > 0$. As a result, the LM method eliminates the singularity of $\mathbf{H}$ for MLP.

The LM method is therefore given by

$$\vec{\boldsymbol{w}}(t+1) = \vec{\boldsymbol{w}}(t) + \boldsymbol{d}(t), \tag{6.22}$$

$$\boldsymbol{d}(t) = -\mathbf{H}_{\text{LM}}^{-1}(t)\mathbf{J}^T(t)\boldsymbol{\epsilon}(t). \tag{6.23}$$

For large σ, the algorithm reduces to BP with $\eta = \frac{1}{\sigma}$. However, when σ is small, the algorithm reduces to the Gauss–Newton method. Thus, there is a trade-off between the fast learning speed of the classical Newton's method and the guaranteed convergence of the gradient descent. σ can be adapted by [31]

$$\sigma(t) = \begin{cases} \sigma(t-1)\gamma, & \text{if} \quad E(t) \geq E(t-1) \\ \frac{\sigma(t-1)}{\gamma}, & \text{if} \quad E(t) < E(t-1) \end{cases}, \tag{6.24}$$

where $\gamma > 1$ is a constant. Typically, $\sigma(0) = 0.01$ and $\gamma = 10$ [31]. The computation of the Jacobian is based on a simple modification to the BP algorithm [31]. Other methods for selecting $\sigma(t)$ inlude the hook step [56], Powell's dogleg method [25], and some rules of thumb [25].

Newton's methods for MLP lack iterative implementation of $\mathbf{H}$, and the computation of $\mathbf{H}^{-1}$ is also expensive. They also suffer from the ill-representability of the diagonal terms of $\mathbf{H}$ and the requirement of a good initial estimate of the weights. The LM method is a trust-region method with a hyperspherical trust-region. It is an efficient algorithm for medium-sized neural networks [31]. The LM method demands large memory space to store the Jacobian, the approximated Hessian, and the inversion of a matrix at each iteration. In [24], backpropagation is used for the matrix–matrix multiplication in the Gauss–Newton matrix; this reduces the running time of the LM method by a factor of $O(J_M)$, where J_M is the number of output nodes.

There are some variants of the LM method. A modified LM method [88] is obtained by modifying the error function and using the slope between the desired and actual outputs in the activation function to replace the standard derivative at the point of the actual output. This method gives a better convergence rate with less computational complexity and reduces the memory requirement from N_w^2 to J_M^2 allocations. The trust-region-based error aggregated training (TREAT) algorithm [20] is similar to the LM method, but uses a different Hessian matrix approximation based on the Jacobian matrix derived from aggregated error vectors. The new Jacobian is significantly smaller. The size of the matrix to be inverted at each iteration is also reduced by using the matrix inversion lemma. A recursive LM algorithm for online training of neural networks is given in [59] for nonlinear system identification.

The disadvantages of the LM method as well as of Newton's methods can be alleviated by the block Hessian based Newton's method [87], where a block Hessian matrix $\mathbf{H}_b$ is defined to approximate and simplify $\mathbf{H}$. Each $\mathbf{W}^{(m)}$, or its vector form $\overrightarrow{\boldsymbol{w}}^{(m)}$, corresponds to a diagonal partition matrix $\mathbf{H}_b^{(m)}$, and

$$\mathbf{H}_b = \text{blockdiag}\left(\mathbf{H}_b^{(1)}, \mathbf{H}_b^{(2)}, \ldots, \mathbf{H}_b^{(M-1)}\right), \tag{6.25}$$

$\mathbf{H}_b$ is proved to be a singular matrix [87]. In LM implementation, the inverse of $\mathbf{H}_b + \sigma\mathbf{I}$ can be decomposed into the inverse of each diagonal block $\mathbf{H}_b^{(m)} + \sigma\mathbf{I}$, and the problem is decomposed into $M-1$ subproblems

$$\Delta\overrightarrow{\boldsymbol{w}}^{(m)} = -\left(\mathbf{H}_b^{(m)} + \sigma\mathbf{I}\right)^{-1} \boldsymbol{g}^{(m)}, \quad m = 1, \ldots, M-1, \tag{6.26}$$

where the gradient partition $\boldsymbol{g}^{(m)} = \frac{\partial E}{\partial \overrightarrow{\boldsymbol{w}}^{(m)}}$. The inverse in each subproblem can be computed recursively according to the matrix inversion lemma.

LM with adaptive momentum (LMAM) and optimized LMAM [3] combine the merits of both the LM and CG techniques, and help the LM method to escape from the local minima. LMAM is derived by optimizing the mutually conjugate property of the two steps subject to a constraint on the error change as well as a different trust-region condition $\boldsymbol{d}^T(t)\mathbf{H}(t)\boldsymbol{d}(t) \leq \delta_t$. This leads to two parameters to be tuned. Optimized LMAM is adaptive, requiring minimal input from the end user. LMAM is globally convergent. Their implementations require minimal additional computations when compared to the LM iteration, and this is, however, compensated by their excellent convergence properties. Both the methods generate better results than LM, BFGS, and Polak–Ribiere CG with restarts [39].

LM training is restricted by the memory requirement for large pattern sets. Its implementations require calculation of the Jacobian matrix, whose size is proportional to the number of training patterns N. In an improved LM algorithm [90], quasi-Hessian matrix and gradient vector are computed directly, without Jacobian matrix multiplication and storage. Memory requirement for quasi-Hessian matrix and gradient vector computation is decreased by $N \times J_M$ times, where J_M is the number of outputs. Exploiting the symmetry of quasi-Hessian matrix, only elements in its upper/lower triangular array are calculated. Therefore, memory requirement and training speed are improved significantly.

A forward-only LM method [91] uses the forward-only computation instead of the traditional forward and backward computation for calculation of the elements of the Jacobian. Information needed for the gradient vector and Jacobian or Hessian matrix is obtained during forward computation. The forward-only method gives an identical number of training iterations and success rates as LM does, since the same Jacobian matrices are obtained. The LM algorithm has been adapted for arbitrarily connected neural networks [89], which can handle a problem of same complexity with a much smaller number of neurons. The forward-only LM method allows for efficiently training arbitrarily connected networks. For networks with multiple outputs, the forward-only LM method (http://www.eng.auburn.edu/users/wilambm/nnt/) has a lower computational complexity than the traditional forward and backward computations do [31, 89].

6.3 Quasi-Newton Methods

Quasi-Newton methods approximate Newton's direction without evaluating second-order derivatives of the cost function. The approximation of the Hessian or its inverse is computed in an iterative process. They are a class of gradient-based methods whose descent direction vector $\boldsymbol{d}(t)$ approximates Newton's direction. Notice that in this subsection, $\boldsymbol{d}(t)$ denotes the descent direction, and $\boldsymbol{s}(t)$ the step size; in Newton's methods, the two vectors are equivalent and are represented by $\boldsymbol{d}(t)$:

$$\boldsymbol{d}(t) = -\mathbf{H}^{-1}(t)\boldsymbol{g}(t). \tag{6.27}$$

Thus, $\boldsymbol{d}(t)$ can be obtained by solving a set of linear equations:

$$\mathbf{H}(t)\boldsymbol{d}(t) = -\boldsymbol{g}(t). \tag{6.28}$$

The Hessian is always symmetric and is often positive-definite. Quasi-Newton methods with positive-definite Hessian are called *variable-metric methods*. Secant methods are a class of variable-metric methods that use differences to obtain an approximation to the Hessian. The memory requirement for quasi-Newton methods is $\frac{1}{2}N_w^2 + O(N_w)$, which is the same as that for Newton's methods. These methods approximate the classical Newton's method, thus convergence is very fast.

The line-search and trust-region methods are two globally convergent strategies. The line-search method tries to limit the step size along Newton's direction until it is unacceptably large, whereas in the trust-region method the quadratic approximation of the cost function can be trusted only within a small region in the vicinity of the current point. Both methods retain the rapid-convergence property of Newton's methods and are generally applicable [25].

In quasi-Newton methods, a line search is applied such that

$$\lambda(t) = \arg\min_{\lambda \geq 0} E\left(\vec{\boldsymbol{w}}(t+1)\right) = \arg\min_{\lambda \geq 0} E\left(\vec{\boldsymbol{w}}(t) + \lambda \boldsymbol{d}(t)\right). \tag{6.29}$$

Line search is used to guarantee that at each iteration the objective function decays, which is dictated by the convergence requirement. The optimal $\lambda(t)$ can be theoretically derived from

$$\frac{\partial}{\partial \lambda} E\left(\vec{\boldsymbol{w}}(t) + \lambda \boldsymbol{d}(t)\right) = 0, \tag{6.30}$$

and this yields a representation using the Hessian. The second-order derivatives are approximated by the difference of the first-order derivatives at two neighboring points, and thus λ is calculated by

$$\lambda(t) = \frac{-\tau \boldsymbol{g}(t)^T \boldsymbol{d}(t)}{\boldsymbol{d}(t)^T \left[\boldsymbol{g}_\tau(t) - \boldsymbol{g}(t)\right] \boldsymbol{d}(t)}, \tag{6.31}$$

where $\boldsymbol{g}_\tau(t) = \nabla_{\vec{\boldsymbol{w}}} E\left(\vec{\boldsymbol{w}}(t) + \tau \boldsymbol{d}(t)\right)$, and the size of neighborhood τ is carefully selected. Some inexact line-search and line-search-free optimization methods are applied to quasi-Newton methods, which are further used for training feedforward networks [10].

There are many secant methods of rank one or rank two. The Broyden family is a family of rank one and rank two methods generated by taking [25]

$$\mathbf{H}(t) = (1 - \vartheta)\mathbf{H}_{\text{DFP}}(t) + \vartheta \mathbf{H}_{\text{BFGS}}(t), \tag{6.32}$$

where $\mathbf{H}_{\text{DFP}}$ and $\mathbf{H}_{\text{BFGS}}$ are, respectively, the Hessain obtained by the Davidon–Fletcher–Powell (DFP) and BFGS methods, and ϑ is a positive constant between 0 and 1. By giving different values for ϑ, one can get DFP ($\vartheta = 0$), BFGS ($\vartheta = 1$), or other rank one or rank two formulae. DFP and BFGS are two dual rank two secant methods, and BFGS emerges as a leading variable-metric contender in theory and practice [57]. Many of the properties of DFP and BFGS are common to the whole family.

6.3.1 BFGS Method

The BFGS method [7, 25, 57] is implemented as follows. Inexact line search can be applied to BFGS, and this significantly reduces the number of evaluations of the error function. The Hessian $\mathbf{H}$ or its inverse is updated by

$$\mathbf{H}(t+1) = \mathbf{H}(t) - \frac{\mathbf{H}(t)\boldsymbol{s}(t)\boldsymbol{s}^T(t)\mathbf{H}(t)}{\boldsymbol{s}^T(t)\mathbf{H}(t)\boldsymbol{s}(t)} + \frac{\boldsymbol{z}(t)\boldsymbol{z}^T(t)}{\boldsymbol{s}^T(t)\boldsymbol{z}(t)}, \tag{6.33}$$

$$\mathbf{H}^{-1}(t+1) = \mathbf{H}^{-1}(t) + \left(1 + \frac{\boldsymbol{z}^T(t)\mathbf{H}^{-1}(t)\boldsymbol{z}(t)}{\boldsymbol{s}^T(t)\boldsymbol{z}(t)}\right)\frac{\boldsymbol{s}(t)\boldsymbol{s}^T(t)}{\boldsymbol{s}^T(t)\boldsymbol{z}(t)} - \left(\frac{\boldsymbol{s}(t)\boldsymbol{z}^T(t)\mathbf{H}^{-1}(t) + \mathbf{H}^{-1}(t)\boldsymbol{z}(t)\boldsymbol{s}^T(t)}{\boldsymbol{s}^T(t)\boldsymbol{z}(t)}\right), \tag{6.34}$$

where

$$\boldsymbol{z}(t) = \boldsymbol{g}(t+1) - \boldsymbol{g}(t), \tag{6.35}$$

$$\boldsymbol{s}(t) = \vec{\boldsymbol{w}}(t+1) - \vec{\boldsymbol{w}}(t). \tag{6.36}$$

For BFGS implementation, $\vec{\boldsymbol{w}}(0)$, $\boldsymbol{g}(0)$, and $\mathbf{H}^{-1}(0)$ are needed to be specified. $\mathbf{H}^{-1}(0)$ is typically selected as the identity matrix. The computational complexity is $O\left(N_w^2\right)$ floating-point operations. The method requires storage of the matrix $\mathbf{H}^{-1}$. By interchanging $\mathbf{H} \leftrightarrow \mathbf{H}^{-1}$, $\boldsymbol{s} \leftrightarrow \boldsymbol{z}$ in (6.33) and (6.34), one can obtain the DFP method [25, 57].

All the secant methods including the BFGS method are derived to satisfy the so-called *quasi-Newton condition* or *secant relation* [25, 57]

$$\mathbf{H}^{-1}(t+1)\boldsymbol{z}(t) = \boldsymbol{s}(t). \tag{6.37}$$

In [75], a small-memory efficient second-order learning algorithm has been proposed for three-layer neural networks. Descent direction is calculated on the basis of a partial BFGS update with $2Nt$ memory space ($t \ll N$), and a reasonably accurate step length is efficiently calculated as the minimal point of a second-order approximation to the objective function with respect to the step length. The search directions are exactly equivalent to those of the original BFGS update during the first $t+1$ iterations.

Limited-memory BFGS methods implement parts of the Hessian approximation by using second-order information from the most recent iterations [7]. A number of limited-memory BFGS algorithms, which have a memory complexity of $O\left(N_w\right)$ and do not require accurate line searches, are listed in [78]. In the trust-region implementation of BFGS [63], Powell's dogleg trust-region method is used to solve the constrained optimization subproblems. Other variants of quasi-Newton methods are the variable-memory BFGS [52] and memory-optimal BFGS methods [54]. A class of limited-memory quasi-Newton methods is given in [16]. These methods utilize an iterative scheme of a generalized BFGS-type method, and suitably approximate the whole Hessian matrix with a rank two formula determined by a fast unitary transform such as the Fourier, Hartley, Jacobi type, or trigonometric transform. It has a computational complexity of $O\left(N_w \log\left(N_w\right)\right)$ and requires $O\left(N_w\right)$ memory allocations. The close relationship between the BFGS and CG methods is important for formulating algorithms with variable storage or limited memory [57]. Memoryless or limited-memory quasi-Newton algorithms can be viewed as a trade-off between the CG and quasi-Newton algorithms, and are closely related to CG.

6.3.2 One-Step Secant Method

One-step secant method [8, 9] is a memoryless BFGS method, and is obtained by resetting $\mathbf{H}^{-1}(t)$ as the identity matrix in the BFGS update Eq. (6.34) at the $(t+1)$th iteration, and multiplying both sides of the update by $-\boldsymbol{g}(t+1)$ to obtain the search direction

$$\boldsymbol{d}(t+1) = -\boldsymbol{g}(t+1) + B(t)\boldsymbol{z}(t) + C(t)\boldsymbol{s}(t), \tag{6.38}$$

where

$$B(t) = \frac{\boldsymbol{s}^T(t)\boldsymbol{g}(t+1)}{\boldsymbol{s}^T(t)\boldsymbol{z}(t)}, \tag{6.39}$$

$$C(t) = -\left(1 + \frac{\boldsymbol{z}^T(t)\boldsymbol{z}(t)}{\boldsymbol{s}^T(t)\boldsymbol{z}(t)}\right) B(t) + \frac{\boldsymbol{z}^T(t)\boldsymbol{g}(t+1)}{\boldsymbol{s}^T(t)\boldsymbol{z}(t)}. \tag{6.40}$$

The one-step secant method does not store the Hessian, and the new search direction can be calculated without computing a matrix inverse. It reduces the computational complexity to $O(N_w)$. However, this results in a considerable reduction of second-order information, and thus yields a slow convergence compared to the BFGS. When exact line search is applied, the one-step secant method generates conjugate directions. Both the BFGS and one-step secant methods are two efficient methods for MLP training. Parallel implementations of the two algorithms are discussed in [53]. A parallel secant method of Broyden's family with parallel inexact searches is developed and applied for the training of feedforward networks [65].

6.4 Conjugate Gradient Methods

The CG method [8, 18, 35, 55] is a popular alternative to BP. It has many tried and tested, linear and nonlinear variants, each using a different search direction and line-search method. Mathematically, the CG method is closely related to the quasi-Newton method. The CG method conducts a series of line searches along noninterfering directions that are constructed to exploit the Hessian structure without explicitly storing it. The storage requirement is $O(N_w)$, and is about four times that for BP [40]. The computation time per weight update cycle is significantly increased due to the line search for an appropriate step size, involving several evaluations of either the error E or its derivative, which requires the presentation of the complete training set.

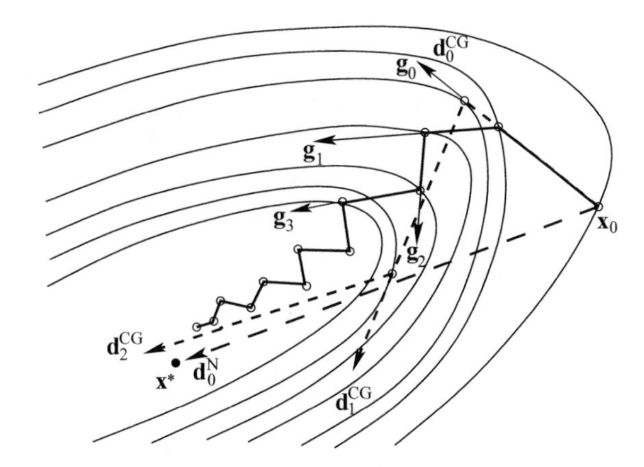

Fig. 6.1 The search directions of the gradient-descent, Newton's, and CG methods

The CG method conducts a special kind of gradient descent. It constructs a set of N_w linearly independent, nonzero search directions $\boldsymbol{d}(t)$, $\boldsymbol{d}(t+1)$, ..., $\boldsymbol{d}(t+N_w-1)$, $t = 0, 1, \ldots$ These search directions are derived from (6.30), and, at the minimum of the line search, satisfy

$$\boldsymbol{g}^T(t+1)\boldsymbol{d}(t) = 0. \tag{6.41}$$

Based on this, one can construct a sequence of N_w successive search directions that satisfy the so-called **H**-conjugate property

$$\boldsymbol{d}^T(t+i)\mathbf{H}\boldsymbol{d}(t+j) = 0, \qquad i \neq j, \forall 0 < |i-j| \leq N_w - 1. \tag{6.42}$$

The CG method is updated by

$$\vec{\boldsymbol{w}}(t+1) = \vec{\boldsymbol{w}}(t) + \lambda(t)\boldsymbol{d}(t), \tag{6.43}$$

$$\boldsymbol{d}(t+1) = -\boldsymbol{g}(t+1) + \beta(t)\boldsymbol{d}(t), \tag{6.44}$$

where $\lambda(t)$ is the exact step to the minimum of $E\left(\vec{\boldsymbol{w}}(t+1)\right)$ along the direction of $\boldsymbol{d}(t)$ and is found by a linear search as given in (6.29), and $\beta(t)$ is a step size to decide $\boldsymbol{d}(t+1)$. A comparison of the search directions of the gradient descent, Newton's methods, and the CG are illustrated in Fig. 6.1, where $\boldsymbol{x}_0$ and $\boldsymbol{x}^*$ are, respectively, the starting point and the local minimum; $\boldsymbol{g}_0, \boldsymbol{g}_1, \ldots$ are negative gradient directions, and are the search directions for the gradient-descent method; $\boldsymbol{d}_0^{\mathrm{N}}$ is the Newton direction, and it generally points toward the local minimum; $\boldsymbol{d}_0^{\mathrm{CG}}, \boldsymbol{d}_1^{\mathrm{CG}}, \ldots$ are the conjugate directions. Notice that $\boldsymbol{d}_0^{\mathrm{CG}}$ and $\boldsymbol{g}_0$ are the same. The contours denote constant values of E.

A practical implementation of the line-search method is to increase λ until $E\left(\vec{\boldsymbol{w}}(t) + \lambda\boldsymbol{d}(t)\right)$ stops being strictly monotonically decreasing and begins to increase. Thus, a minimum is bracketed. The search in the interval between the last two values of λ is then repeated several times until $E\left(\vec{w}(t) + \lambda d(t)\right)$ is sufficiently close to a minimum. Exact line search is computationally expensive, and the speed of the CG method depends critically on the line-search efficiency. Faster CG algorithms with inexact line search [22, 78] are used to train MLP [30]. The scaled CG algorithm [55] avoids line search by introducing a scalar to regulate the positive-definiteness of the Hessian $\mathbf{H}$ as used in the LM method. This is achieved by automatically scaling the weight update vector magnitude using Powell's dogleg trust-region method [25].

The selection of $\beta(t)$ can be one of the following:

$$\beta(t) = \frac{g^T(t+1)z(t)}{d^T(t)z(t)} \quad \text{(Hestenes–Stiefel [35])}, \tag{6.45}$$

$$\beta(t) = \frac{g^T(t+1)g(t+1)}{g^T(t)g(t)} \quad \text{(Fletcher–Reeves [26])}, \tag{6.46}$$

$$\beta(t) = \frac{g^T(t+1)z(t)}{g^T(t)g(t)} \quad \text{(Polak–Ribiere [66])}, \tag{6.47}$$

$$\beta(t) = \frac{g^T(t+1)g(t+1)}{d^T(t)z(t)} \quad \text{(Dai–Yuan [21])}, \tag{6.48}$$

$$\beta(t) = -\frac{g^T(t+1)g(t+1)}{g^T(t)s(t)} \quad \text{(Conjugate descent [25])}, \tag{6.49}$$

where $z(t)$ is defined as in (6.35). In the implementation, $\vec{w}(0)$ is set as a random vector, and we set $d_0 = -g(0)$. When $\|g(t)\|$ is small enough, we terminate the process. The computational complexity of the CG method is $O(N_w)$.

When the objective function is strict convex quadratic and an exact line search is applied, $\beta(t)$ is identical for all the five choices, and termination occurs at most in N_w steps [25, 57]. With periodic restarting, all the above nonlinear CG algorithms are well known to be globally convergent [57]. A globally convergent algorithm is an iterative algorithm that converges to a local minimum from almost any starting point. Polak–Ribiere CG with Powell's restart strategy [67] is considered to be one of the most efficient methods [57, 85]. Polak–Ribiere CG with restarts forces $\beta = 0$ whenever $\beta < 0$. This is equivalent to forgetting the last search direction and restarting it from the direction of steepest descent.

Powell proposed a popular restarting procedure [67]. It tests if there is very little orthogonality between the current gradient and the previous one. If $|g(t)^T g(t-1)| \geq 0.2\|g(t)\|^2$ is satisfied, the CG search direction is restarted by the steepest descent direction $-g(t)$.

In [44], Perry's CG method [64] is used for MLP training. In addition, self-scaled CG is derived from the principles of the Hestenes–Stiefel, Fletcher–Reeves, Polak–Ribiere and Perry's methods. This class is based on the spectral scaling parameter. An efficient line-search technique is incorporated into the CG algorithms based on the Wolfe conditions and on safeguarded cubic interpolation. Finally, an efficient restarting procedure is employed in order to further improve the effectiveness of the CG algorithms.

Empirically, the local minimum achieved with BP will, in general, be a solution that is good enough for most purposes. In contrast, the CG method is easy to be trapped at a bad local minimum, since the CG method moves toward the bottom of whatever valley it reaches. Escaping a local minimum requires an increase in E, and this is excluded by the line-search procedure. Consequently, the convergence condition can never be reached [40]. The CG method is usually applied several times with different random $\vec{\boldsymbol{w}}(0)$ for the minimum error [40].

The CG method can be regarded as an extension of BP with momentum by automatically selecting appropriate learning rate $\eta(t)$ and momentum factor $\alpha(t)$ in each epoch [12, 18, 85, 97]. The CG method can be considered as BP with momentum, which has adjustable $\eta(t)$ and $\alpha(t)$, $\eta(t) = \lambda(t)$, $\alpha(t) = \lambda(t)\beta(t)$ [18]. By an adaptive selection of both $\eta(t)$ and $\alpha(t)$ for a quadratic error function, referred to as *optimally tuned*, BP with momentum is proved to be exactly equivalent to the CG method [12].

The CG method has a regularizing effect with iteration number playing the role of regularization parameter. It has a convergence performance that is comparable to the RLS method, but with a lower complexity.

In [50], MLP is first decomposed into a set of adalines, each having its own local MSE function. The desired local output at each adaline is estimated based on error backpropagation. Each local MSE function has a unique optimum, which can be found within finite steps by using the CG method. By using a modified CG that avoids the line search, the local training method achieves a significant reduction in the number of iterations and the computation time. Given the approximation accuracy, the local method requires a computation time that is typically an order of magnitude less than that of the CG-based global method. The local method is particularly suited to parallel implementation.

Example 6.1 **Iris classification** is revisited. Eighty per cent of the dataset is used as training data, and the remaining 20% as testing data. We set the performance goal as 0.001, and the maximum number of epochs as 1000. We simulate and compare eight popular MLP learning algorithms, namely, Rprop, BFGS, one-step secant, LM, scaled CG, CG with Powell–Beale restarts, Fletcher–Powell CG, and Polak–Ribiere CG algorithms.

We select a 4-4-3 MLP network. At the training stage, for class i, the ith output node has an output 1, and the other two output nodes have value -1. We use the logistic sigmoidal function in the hidden layer and the linear function in the output layer. At the generalization stage, only the output node with the largest output is treated as 1 and outputs at the other nodes are treated as -1. The training results for 50 independent runs are listed in Table 6.1. The learning curves for a random run of these algorithms are shown in Fig. 6.2. For this example, we see that BFGS and LM usually generate better MSE performance, in less time, but more memory. All the algorithms generate good MSE and classification performance.

Example 6.2 **Character recognition** is a classical problem in pattern recognition. A network is to be designed and trained to recognize the 26 capital letters of the

Table 6.1 Performance comparison of a 4-4-3 MLP trained with ten learning algorithms

Algorithm	Mean Epochs	Training MSE	Classification Accuracy (%)	Std	Mean training Time (s)	Std. (s)
RP	985.92	0.0245	96.27	0.0408	9.0727	0.5607
BFGS	146.46	0.0156	96.67	0	2.8061	1.6105
OSS	999.16	0.0267	96.27	0.0354	16.1422	0.4315
LM	328.80	0.0059	93.33	0	5.7625	6.9511
SCG	912.44	0.0144	96.13	0.0371	13.4404	2.9071
CGB	463.54	0.0244	95.07	0.0468	8.4726	5.9076
CGF	574.26	0.0202	95.80	0.0361	10.5986	5.5025
CGP	520.34	0.0216	96.40	0.0349	9.5469	6.0092

RP—Rprop, OSS—one-step secant, SCG—scaled CG, CGB—CG with Powell–Beale restarts, CGF—Fletcher–Powell CG, and CGP—Polak–Ribiere CG

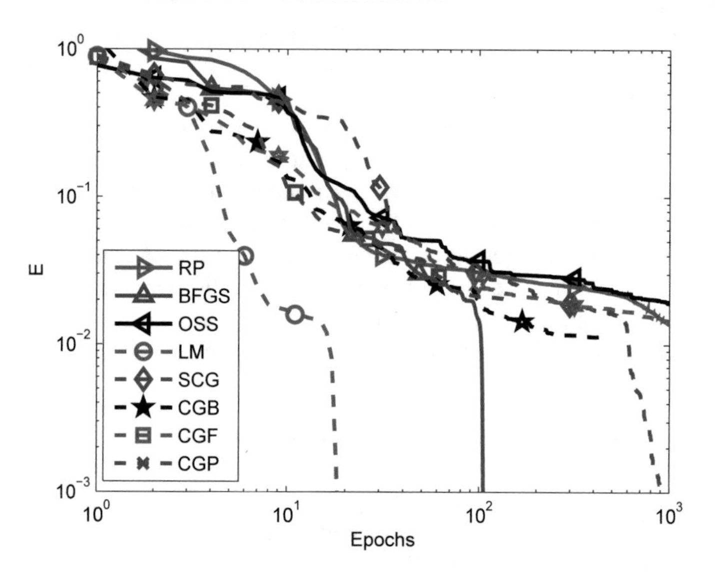

Fig. 6.2 Iris classification using a 4-4-3 MLP trained with ten learning methods: the traces of the training error for a random run. t corresponds to the number of epochs

alphabet. An imaging system that digitizes each letter centered on the system's field of vision is available. The result is that each letter is represented as a 5×7 grid of Boolean values, or 35-element input vectors. The input consists of 26 5×7 arrays of black or white pixels. They are vectorized as a linear array of zeros and ones. For example, for character "A" (shown in Fig. 6.3), we represent it as 00100 01010 01010 10001 11111 10001 10001, or the value of the ASCII code 65_{10}. This is a classification problem of 26 output classes. Each target vector is a 26-element vector with a 1 in the position of the letter it represents, and 0 everywhere else. For example, the letter "A" is to be represented by a 1 in the first element (as "A" is the first letter of the alphabet), and 0 in the remaining 25 elements.

Fig. 6.3 The 5×7 dot matrix image of all the 26 characters

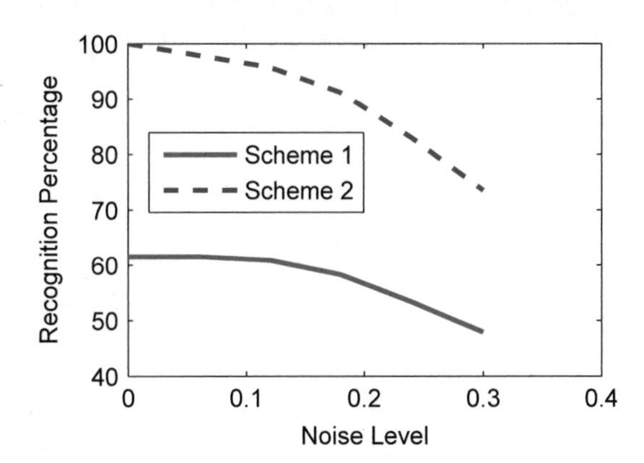

Fig. 6.4 Samples of normalized digits from the testing set

We train the network using BFGS. The goal of training is 0.01 or the maximum number of epochs is 100. We use 40 hidden nodes for training. The logistic sigmoidal function is selected for hidden neurons and linear function is used for the output neurons. We implement two schemes. In the first scheme, we train the network using the desired examples, and we then generalize it using noisy samples. In the second scheme, the network is first trained by using desired examples, and then trained by using noisy examples of 4 different levels with noise of mean 0 and standard deviation of 0.3 or less, and finally trained by the desired samples again. The classification errors for the two schemes are shown in Fig. 6.4. From the result, the network trained with noisy samples have a much better generalization performance.

6.5 Extended Kalman Filtering Methods

EKF method belongs to second-order methods. EKF is an optimum filtering method for a linear system resulting from the linearization of a nonlinear system. It attempts to estimate the state of a system that can be modeled as a linear system driven by an additive white Gaussian noise. It always estimates the optimum step size, namely, the Kalman gain, for weight updating and thus the rate of convergence is significantly increased when compared to that of the gradient-descent method. The EKF approach is an incremental training method. It is a general method for training any feedforward network. The RLS method is a reduced form of the EKF method, and is more widely used in adaptation.

When using EKF, training of an MLP is viewed as a parametric identification problem of a nonlinear system, where the weights are unknown and have to be estimated for the given set of training patterns. The weight vector $\vec{\boldsymbol{w}}$ is now a state vector, and an operator $\boldsymbol{f}(\cdot)$ is defined to perform the function of an MLP that maps the state vector and the input onto the output. The training problem can be posed as a state estimation problem with the following dynamic and observation equations

$$\vec{\boldsymbol{w}}(t+1) = \vec{\boldsymbol{w}}(t) + \boldsymbol{\upsilon}(t), \tag{6.50}$$

$$\boldsymbol{y}_t = \boldsymbol{f}\left(\vec{\boldsymbol{w}}(t), \boldsymbol{x}_t\right) + \boldsymbol{\epsilon}(t), \tag{6.51}$$

where $\boldsymbol{x}_t$ is the input to the network at time t, $\boldsymbol{y}_t$ is the observed (or desired) output of the network, and $\boldsymbol{\upsilon}$ and $\boldsymbol{\epsilon}$ are the observation and measurement noise, assumed white and Gaussian with zero mean and covariance matrices $\mathbf{Q}(t)$ and $\mathbf{R}(t)$, respectively. The state estimation problem is then to determine $\widehat{\vec{\boldsymbol{w}}}$, the estimated weight vector, that minimizes the sum of squared prediction errors of all prior observations.

EKF is a minimum variance estimator based on the Taylor series expansion of $\boldsymbol{f}\left(\vec{\boldsymbol{w}}\right)$ in the vicinity of the previous estimate. The EKF method for estimating $\vec{\boldsymbol{w}}(t)$ is given by [38, 73, 83]

$$\widehat{\vec{\boldsymbol{w}}}(t+1) = \widehat{\vec{\boldsymbol{w}}}(t) + \mathbf{K}(t+1)\left[\boldsymbol{y}_{t+1} - \widehat{\boldsymbol{y}}_{t+1}\right], \tag{6.52}$$

$$\boldsymbol{\Xi}(t+1) = \mathbf{F}(t+1)\mathbf{P}(t)\mathbf{F}^T(t+1) + \mathbf{R}(t+1), \tag{6.53}$$

$$\mathbf{K}(t+1) = \mathbf{P}(t)\mathbf{F}^T(t+1)\boldsymbol{\Xi}^{-1}(t+1), \tag{6.54}$$

$$\mathbf{P}(t+1) = \mathbf{P}(t) - \mathbf{K}(t+1)\mathbf{F}(t+1)\mathbf{P}(t) + \mathbf{Q}(t), \tag{6.55}$$

where $\mathbf{K}$ is the $N_w \times N_y$ Kalman gain, N_y is the dimension of $\boldsymbol{y}$, $\mathbf{P}$ is the $N_w \times N_w$ conditional error covariance matrix, $\widehat{\boldsymbol{y}}$ is estimated output, and

$$\mathbf{F}(t+1) = \left.\frac{\partial \boldsymbol{f}}{\partial \vec{\boldsymbol{w}}}\right|_{\vec{\boldsymbol{w}}=\widehat{\vec{\boldsymbol{w}}}(t)} = \left.\frac{\partial \widehat{\boldsymbol{y}}}{\partial \vec{\boldsymbol{w}}}\right|_{\vec{\boldsymbol{w}}=\widehat{\vec{\boldsymbol{w}}}(t)} \tag{6.56}$$

is an $N_y \times N_w$ matrix.

In the absence of a priori information, the initial state vector can be set randomly and $\mathbf{P}$ can be set as a diagonal matrix: $\mathbf{P}(0) = \frac{1}{\varepsilon}\mathbf{I}$, $\vec{\boldsymbol{w}}(0) = \mathcal{N}(0, \mathbf{P}(0))$, where ε is a small positive number and $\mathcal{N}(0, \mathbf{P}(0))$ denotes a zero-mean Gaussian distribution with covariance matrix $\mathbf{P}(0)$. A relation between the Hessian $\mathbf{H}$ and $\mathbf{P}$ is established in [83].

The method given above is the global EKF method for MLP training [38, 81]. Compared to BP, it needs far fewer training cycles to reach convergence and the quality of the solution is better, at the expense of a much higher computational complexity at each cycle. BP is proved to be a degenerate form of EKF [73]. The fading-memory EKF and a UD factorization based fading-memory EKF, which use an adaptive forgetting factor, are two fast algorithms for learning feedforward networks [99]. The EKF variant reported in [71] performs as efficiently as the LM method.

In order to reduce the complexity of the EKF method, one can partition the global problem into many small-scaled, separate, localized identification subproblems for each neuron in the network so as to solve the individual subproblems. Examples of the localized EKF methods are the multiple extended Kalman algorithm [79] and the decoupled EKF algorithm [68]. The localized algorithms set to zero the off-diagonal terms of the covariance matrix in the global EKF method; this, however, ignores the natural coupling of the weights.

The Kalman filtering method is based on the assumption of the noise being Gaussian, and is thus sensitive to noises of other distributions. According to the H_∞ theory, the maximum energy gain that the Kalman filtering algorithm contributes to the estimation error due to disturbances has no upper bound [60]. The extended H_∞ filtering method, in the form of global and local algorithms, can be treated as an extension to EKF for enhancing the robustness to disturbances. The computational complexity of the extended H_∞ filtering method is typically twice that of EKF.

6.6 Recursive Least Squares

When $\mathbf{R}(t+1)$ in (6.54) and $\mathbf{Q}(t)$ in (6.55), respectively, reduce to the identity matrix $\mathbf{I}$ and zero matrix $\mathbf{O}$ of the same size, the EKF method is reduced to the RLS method. The RLS method is applied for learning of layered feedforward networks in [4, 13, 47, 81]. It is typically an order of magnitude faster than LMS, which is equivalent to BP in one-layer networks. For a given accuracy, RLS is shown to require tenfold fewer epochs than BP [13].

The RLS method is derived from the optimization of the energy function [47]

$$E\left(\vec{\boldsymbol{w}}(t)\right) = \sum_{p=1}^{t} \left[\boldsymbol{y}_p - \boldsymbol{f}\left(\vec{\boldsymbol{w}}(t), \boldsymbol{x}_p\right)\right]^T \left[\boldsymbol{y}_p - \boldsymbol{f}\left(\vec{\boldsymbol{w}}(t), \boldsymbol{x}_p\right)\right] + \left[\vec{\boldsymbol{w}}(t) - \vec{\boldsymbol{w}}(0)\right]^T \mathbf{P}(0) \left[\vec{\boldsymbol{w}}(t) - \vec{\boldsymbol{w}}(0)\right]. \tag{6.57}$$

When $\mathbf{P}(0) = \frac{1}{\varepsilon}\mathbf{I}$ and $\vec{\boldsymbol{w}}(0)$ is a small vector, the second term in (6.57) reduces to $\varepsilon \vec{\boldsymbol{w}}(t)^T \vec{\boldsymbol{w}}(t)$. Thus, the RLS method is implicitly a weight-decay technique whose weight-decay effect is governed by $\mathbf{P}(0)$. Smaller ε usually leads to better training accuracy, while larger ε results in better generalization [47]. At iteration t, the Hessian for the above error function is related to the error covariance matrix $\mathbf{P}(t)$ by [47]

$$\mathbf{H}(t) \approx 2\mathbf{P}^{-1}(t) - 2\mathbf{P}^{-1}(0). \tag{6.58}$$

The RLS method has a computational complexity of $O\left(N_w^2\right)$. It may suffer from numerical instability.

A complex training problem can also be decomposed into separate, localized identification subproblems, each being solved by the RLS method [62, 82]. In the local linearized LS method [82], each subproblem has the objective function as the sum of the squares of the linearized backpropagated error signals for each neuron. In the block RLS algorithm [62], at a step of the algorithm, an M-layer feedforward network is divided into $M - 1$ subproblems, each being an overdetermined system of linear equations for each layer of the network. This is a considerable saving with respect to a global method.

There is no explicit decay in the energy function in the RLS algorithm and the decay effect diminishes linearly with the number of training epochs [47]. In order to speed up the learning process as well as to improve the generalization ability of the trained network, a true weight-decay RLS algorithm [48] combines a regularization term for quadratic weight decay. The generalized RLS model [93] includes a general decay term in the energy function; it can yield a significantly improved generalization ability of the trained networks and a more compact network, with the same computational complexity as that of the RLS algorithm.

6.7 Natural-Gradient-Descent Method

When the parametric space is not Euclidean but has a Riemannian metric, the ordinary gradient does not give the steepest direction of the cost function, while the natural gradient does [1]. Natural gradient descent exploits the *natural* Riemannian metric that the Fisher information matrix defines in the MLP weight space.

Natural gradient descent replaces standard gradient descent by

$$\mathbf{W}_{t+1} = \mathbf{W}_t - \eta_t [\mathbf{G}(\mathbf{W}_t)]^{-1} \nabla e(\boldsymbol{x}_t, y_t; \mathbf{W}_t), \tag{6.59}$$

where $e(\boldsymbol{x}, y; \mathbf{W}) = \frac{1}{2}(f(\boldsymbol{x}; \mathbf{W}) - y)^2$ is the local square error and $\mathbf{G}(\mathbf{W})$ is the metric tensor when the MLP weight space is viewed as an appropriate Riemannian manifold. $\mathbf{G}$ coincides with Levenberg's approximation to the Hessian of a square error function [34]. The most natural way to arrive at $\mathbf{G}$ is to recast MLP training as a log-likelihood maximization problem [1].

The online natural-gradient learning gives the Fisher efficient estimator [1], implying that it is asymptotically equivalent to the optimal batch procedure. This suggests that the flat-spot problem that appears in BP disappears when natural gradient is used [1]. According to the Cramer–Rao bounds, Fisher efficiency is the best asymptotic performance that any unbiased learning algorithm can achieve. However, calculation of the Fisher information matrix and its inverse is practically very difficult. An adaptive method of directly obtaining the inverse of the Fisher information matrix is proposed in [2]. It generalizes the adaptive Gauss–Newton algorithms, and the natural gradient method is equivalent to the Newton method at around the optimal point. In batch natural descent method, the Fisher matrix essentially coincides with the Gauss–Newton approximation of the Hessian of the MLP MSE function and the natural gradient method is closely related to the LM method [34, 37]. Natural gradient descent should have a linear convergence in a Riemannian weight space compared to the superlinear one of the LM method in the Euclidean weight space. A natural conjugate gradient method for MLP training is discussed in [29].

In [34], natural gradients are derived in a slightly different manner and batch mode learning and pruning are linked to existing algorithms such as LM optimization and OBS. The Rprop algorithm with the natural gradient [37] converges significantly faster than Rprop. It shows at least similar performance as the LM and appears to be slightly more robust. Compared to Rprop, in LM optimization and Rprop with the natural gradient, a weight update requires cubic time and quadratic memory, and both methods have additional hyperparameters that are difficult to adjust [37].

6.8 Other Learning Algorithms

The LP method is also used for training feedforward networks [80]. The LP method can be effective for training a small network and can converge rapidly and reliably to a better solution than BP. However, it may take too long a time for each iteration for very large networks. Some measures are considered in [80] so as to extend the method for efficient implementations in large networks.

6.8.1 Layerwise Linear Learning

Feedforward networks can be trained by iterative layerwise learning methods [5, 23, 72, 77]. Weight updating is performed layer by layer, and weight optimization at each layer is reduced to solving a set of linear equations, $\mathbf{A}\vec{\boldsymbol{w}}^{(m)} = \boldsymbol{b}$, where $\vec{\boldsymbol{w}}^{(m)}$ is a weight vector associated with the layer, and $\mathbf{A}$ and $\boldsymbol{b}$ are a matrix and a vector of suitable dimensions, respectively. These algorithms are typically one to two orders of magnitude faster in computational time than BP for a given accuracy.

BP can be combined with a linear algebra method [5] or Kalman filtering method [77]. In [5], sets of linear equations are formed based on the computation of target node values using inverse activation functions. The updated weights need to be transformed to ensure that target values are in the range of the activation functions. An efficient method that combines the layerwise approach and the BP strategy is given in [72]. This layerwise BP algorithm is more accurate and faster than the CG with Powell restarts and the Quickprop.

In [19, 98], a fast algorithm for three-layer MLP, called *OWO-HWO*, a combination of hidden weight optimization (HWO) and output weight optimization (OWO) has been described. HWO is a batch version of the Kalman filtering method given in [77], restricted to hidden units. OWO solves a set of linear equations to optimize the output weights. OWO-HWO is equivalent to a combination of linearly transforming the training data and performing OWO-BP [51], which uses OWO to update the output weights and BP to update the hidden weights. OWO-HWO is superior to OWO-BP in terms of convergence, and converges to the same training error as the LM method does in an order of magnitude less time [19, 98].

The parameterwise algorithm for MLP training [49] is on the basis of the idea of layerwise algorithm. It does not need to calculate the gradient of the error function. In each iteration, the weights or thresholds can be optimized directly one by one with the other variables fixed. The error function is simplified greatly by means of only calculating the changed part of the error function in the training process. In comparisons with BP-with-momentum and the layerwise algorithms, the parameterwise algorithm achieves more than an order of magnitude faster convergence.

6.9 Escaping Local Minima

Conventional first-order and second-order gradient-based methods cannot avoid local minima. The error surface of an MLP has a stair-step appearance with many very flat and very steep regions [36]. For the case of a small number of training examples, there is often a one-to-one correspondence between the individual training examples and the steps on the surface. The surface becomes smoother as the number of training examples is increased. In all directions, there are flat regions extending to infinity, which makes line-search-related learning algorithms useless.

Many strategies have been explored to reduce the chances of getting trapped at a local minimum. One simple and effective technique to avoid local minima in incremental learning is to present examples to the network in a random order from the training set during each epoch. Another way is to run the learning algorithms using initial values in different regions of the weight space, and then to find the best solution. This is especially useful for fast convergent algorithms such as the CG algorithm.

The injection of noise into the learning process is an effective means for escaping from local minima. This also leads to a better generalization capability. Various annealing schemes actually use this strategy. Random noise can be added to the input, to the desired output, or to the weights. The level of the added noise should be decreased as learning progresses. The three methods have the same effect, namely, the inclusion of an extra stochastic term in the weight vector adaptation. A random step size strategy implemented in [86] employs an annealing average step size. The large steps enable the algorithm to jump over local maxima/minima, while the small ones ensure convergence in a local area. An effective way to escape from local minima is realized by incorporating an annealing noise term into the gradient-descent algorithm [17]. This heuristic has also been used in SARprop.

Weight scaling [27, 70] is a technique used for escaping local minima and accelerating convergence. Using the weight scaling process, the weight vector to each neuron $\boldsymbol{w}_j^{(m)}$ is scaled by a factor $\beta_j^{(m)}$, where $\beta_j^{(m)} \in (0, 1)$ is decided by a relation of the degree of saturation at each node $D_j^{(m)} = \left| o_j^{(m)} - 0.5 \right|$ (with $o_j^{(m)}$ being the output of the node), the learning rate, and the maximum error at the output nodes. Weight scaling effectively reduces the degree of saturation of the activation function and thus maintains a relatively large derivative of the activation function. This enables relatively large weight updates, which may eventually lead the training algorithm out of a local minimum.

The natural way to implement Newton's methods is to confine a quadratic approximation of the objective function $E\left(\overrightarrow{w}\right)$ to a trust-region. The trust-region subproblem is then solved to obtain the next iteration. The attractor-based trust-region method is an alternating two-phase algorithm for MLP learning [45]. The first phase is a trust-region based local search for fast training of the network and global convergence, while the second phase is an attractor-based global search for escaping local minima utilizing a quotient gradient system. The trust-region subproblem is solved by applying Powell's dogleg trust-region method [25]. The algorithm outperforms BP with momentum, BP with tunneling, and LM algorithms [45].

Stochastic learning algorithms typically have low convergence when compared to BP, but they can generalize better and effectively avoid local minima. Besides, they are flexible in network topology, error function, and activation function. Gradient information is not required. There are also many heuristic global optimization techniques such as evolutionary algorithm for MLP learning.

6.10 Complex-Valued MLPs and Their Learning

In the real domain, common nonlinear transfer functions are the hyperbolic tangent and logistic functions, which are bounded and analytic everywhere. According to Liouville's theorem, a complex transfer function, which is both bounded and analytic everywhere, has to be a constant. As a result, designing a neural network for processing complex-valued signals is a challenging task, since a complex nonlinear activation function cannot be both analytic and bounded everywhere in the complex plane $\mathcal{C}$. The Cauchy–Riemann equations are necessary and sufficient conditions for a complex function to be analytic at a point $z \in \mathcal{C}$. Complex-valued neural networks are useful for processing complex-valued data, such as equalization and modeling of nonlinear channels. Digital channel equalization can be treated as a classification problem.

The error function E for training complex MLPs is defined by

$$E = \frac{1}{2} \sum_{p=1}^{N} \boldsymbol{e}_p^H \boldsymbol{e}_p, \tag{6.60}$$

where $\boldsymbol{e}_p$ is defined by (5.6), but it is a complex-valued vector. E is not analytic since it is a real-valued function.

6.10.1 Split Complex BP

The conventional approach for learning complex-valued MLPs selects split complex activation functions. Each split complex function consists of a pair of real sigmoidal functions marginally processing the inphase and quadrature components. Based on this split strategy, the complex version of BP is derived for complex-valued MLPs [11, 46, 61, 84]. This approach can avoid the unboundedness of fully complex activation functions. However, the split complex activation function cannot be analytic.

The split complex BP uses the split derivatives of the real and imaginary components instead of relying on well-defined fully complex derivatives. The derivatives cannot fully exploit the correlation between the real and imaginary components of the weighted sum of the input vectors. In the split approach, the activation function is split by

$$\phi(z) = \phi_R\left(\Re(z)\right) + \mathrm{j}\phi_I\left(\Im(z)\right), \tag{6.61}$$

where z is the net input to a neuron, $z = \boldsymbol{w}^T\boldsymbol{x}, \boldsymbol{x}, \boldsymbol{w} \in C^J$ are the J-dimensional complex input and weight vectors, respectively. Typically, $\phi_R(\cdot)$ and $\phi_I(\cdot)$ are selected to be the same sigmoidal function, one can select $\phi_R(x) = \phi_I(x) = x + a_0 \sin(\pi x)$, where constant $a_0 \in (0, 1/\pi)$ [96]. This split complex function satisfies most properties of complex activation functions. This method can reduce the information redundancy among hidden neurons of a complex MLP, and results in a guaranteed weight update when the estimation error is not zero.

The sensitivity of a split complex MLP due to the errors of the inputs and the connection weights between neurons is statistically analyzed in [94]. The sensitivity is affected by the number of the layers and the number of the neurons adopted in each layer, and an efficient algorithm to estimate the sensitivity is developed. When an MLP is trained with split complex BP, it has a relatively strong dependence of the performance on the initial values. For the effective adjustments of the weights and biases in split complex BP, the range of the initial values should be greater than that of the adjustment quantities [95]. This criterion can reduce the misadjustment of the weights and biases. The estimated range of the initial values gives significantly improved performance.

The convergence of split complex BP [92, 100] and split complex BP with momentum and penalty [101] for training complex-valued neural networks have been proved.

6.10.2 Fully Complex BP

Fully complex BP is derived based on a suitably selected complex activation function [28, 42]. In [28], an activation function is defined

$$\phi(z) = \frac{z}{c_0 + \frac{1}{r_0}|z|}, \tag{6.62}$$

where c_0 and r_0 are real positive constants. The function $\phi(\cdot)$ maps a point z on the complex plane to a unique point $\phi(z)$ on the open disc $\{z : |z| < r_0\}$, with the same phase angle, and c_0 controls the steepness of $|\phi(z)|$. This complex function satisfies most of the properties for activation function, and a circuit for such a complex neuron is designed in [28].

In [42], fully complex BP [28] is simplified by using the Cauchy–Riemann equations. It is shown that fully complex BP is the complex conjugate form of BP and that split complex BP is a special case of fully complex BP. This generalization is possible by employing elementary transcendental functions (ETFs) that are almost everywhere bounded and analytic in $\mathcal{C}$. Complex ETFs provide well-defined derivatives for optimization of the fully complex BP algorithm. A list of complex ETFs, including $\sin z$, $\tan z$, $\sinh z$ and $\tanh z$, are suitable nonlinear activation functions in [42]. These ETFs provide a parsimonious structure for processing data in the complex domain. Fully complex MLPs with these ETFs as activation functions are proved to be universal approximators in the complex domain [43].

Fully complex normalized BP [32] is an improvement on the complex BP [28] obtained by including an adaptive normalized learning rate. This is achieved by performing a minimization of the complex-valued instantaneous output error that has been expanded via a Taylor series expansion. The method is valid for any complex activation function discussed in [42].

The minimization criteria used in the complex-valued BP learning algorithms do not approximate the phase of complex-valued output well in function approximation problems. The phase of a complex-valued output is critical in telecommunications, and reconstruction and source localization problems in medical imaging applications. In [76], the convergence of complex-valued neural networks are investigated using a systematic sensitivity study, and the performance of different types of split complex-valued neural networks is compared. A complex-valued BP algorithm with logarithmic performance index is proposed with exponential activation function $f(z) = \exp(z)$; it directly minimizes both the magnitude and phase errors and also provides better convergence characteristics. The exponential function is entire since $f(z) = f'(z) = \exp(z)$ in C. It has an essential singularity at $+\infty$. By restricting the weights of the network to a small ball of radius and the number of hidden neurons to a finite value, the bounded behavior in fully complex-valued MLP can be achieved.

Many other algorithms for training complex-valued MLPs are typically complex versions of some algorithms used for training real-valued MLPs. Split complex EKF [69] has faster convergence than split complex BP [46]. Split complex Rprop [41] outperforms split complex BP [96], and fully complex BP [42] in terms of the computational complexity, convergence speed, and accuracy.

Example 6.3 The Mackey–Glass differential delay equation describes a time series system with chaotic behavior:

$$\frac{dx(t)}{dt} = -0.1x(t) + \frac{0.2x(t-\tau)}{1 + x(t-\tau)^{10}}.$$

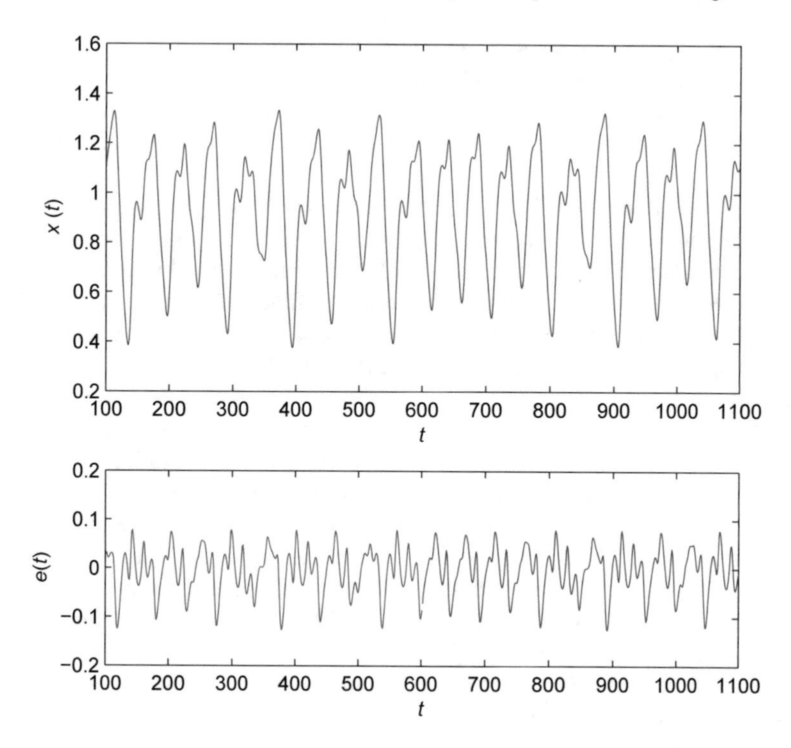

Fig. 6.5 Prediction of the Mackey–Glass time series

The training data are collected by integrating the equation using Euler's method with $\tau = 17$ and a step size of 1. Assume $x(18) = 1.2$, $\tau = 17$, and $x(t) = 0$ for $t < 18$. The samples are shown in Fig. 6.5a.

The task is to predict a future point based on currently available time series points by using MLP with BFGS. Three time series points $x(t)$, $x(t-1)$, and $x(t-2)$ are considered as network input while the output is the sample $x(t+1)$. We use 3 nodes in the hidden layer. We use the first 500 samples for training, and the remaining samples for testing. The training reaches an accuracy of 0.01 with 22 iterations or 2 s. The prediction error is shown in Fig. 6.5. It is shown that after training with the first samples, the prediction is sufficient accuracy for the remaining samples.

Problems

6.1 Show that inclusion of a momentum term in the weight update in the BP algorithm can be considered as an approximation to the CG method.

6.2 Learn the function $f(x) = x \sin x, 0 \leq 0 \leq \pi$ using MLP with the BP and BFGS algorithms. Generate random datasets for training and for testing.

6.3 Matrix inversion lemma given below is usually used for the derivation of the iterative algorithm:

$$(\mathbf{A}+\mathbf{BC})^{-1}=\mathbf{A}^{-1}-\mathbf{A}^{-1}\mathbf{B}(\mathbf{I}+\mathbf{CA}^{-1}\mathbf{B})^{-1}\mathbf{CA}^{-1},$$

where $\mathbf{I}$ is the identity matrix, and $\mathbf{A}$, $\mathbf{B}$, $\mathbf{C}$ are matrices of proper size.
(a) Prove matrix inversion lemma. [Hint: premultiply or postmultiply both sides by $\mathbf{A}+\mathbf{BC}$.]
(b) For MLP, the Hessian matrix for a dataset of N patterns is given by a diagonal approximation [33]

$$\mathbf{H}_N=\sum_{n=1}^{N}\boldsymbol{g}_n\boldsymbol{g}_n^T,$$

where $\boldsymbol{g}_n=\nabla_{\boldsymbol{w}}E_n$, E_n being the sum of squared error for all n patterns. This expression can be written as

$$\mathbf{H}_{n+1}=\mathbf{H}_n+\boldsymbol{g}_{n+1}\boldsymbol{g}_{n+1}^T.$$

Using matrix inverse lemma, show that the inverse Hessian is given by the following interative expression:

$$\mathbf{H}_{n+1}^{-1}=\mathbf{H}_n^{-1}-\frac{\mathbf{H}_n^{-1}\boldsymbol{g}_{n+1}\boldsymbol{g}_{n+1}^T\mathbf{H}_n^{-1}}{1+\boldsymbol{g}_{n+1}^T\mathbf{H}_n^{-1}\boldsymbol{g}_{n+1}}.$$

6.4 Train MLP to solve the character recognition problem. The ASCII characters are printed in 5×7 dot matrix.
(a) Design an MLP network to map the dot matrices of characters "A"–"Z" and "a"–"z" to their corresponding ASCII codes.
(b) Design an MLP network to map the dot matrices of 10 digits to their corresponding ASCII codes.
(c) Describe the network's tolerance to noisy inputs after training is complete.

6.5 For two classes and a single network output, the cross-entropy error function is given by

$$E=-\sum_n\left[t^{(n)}\ln y^{(n)}+(1-t^{(n)})\ln(1-y^{(n)})\right].$$

Derive the Hessian matrix. As training proceeds, the network output approximates the conditional averages of the target data, and some terms in the matrix components vanish.

6.6 The two spirals dataset is a standard benchmark for classification algorithms. Two classes of points in a two-dimensional surface are arranged as interlocked spirals. Separate the two spirals using the MLP network.

References

1. Amari, S. I. (1998). Natural gradient works efficiently in learning. *Neural Computation, 10*, 251–276.
2. Amari, A., Park, H., & Fukumizu, K. (2000). Adaptive method of realizing natiral gradient learning for multilayer perceptrons. *Neural Computation, 12*, 1399–1409.
3. Ampazis, N., & Perantonis, S. J. (2002). Two highly efficient second-order algorithms for training feedforward networks. *IEEE Transactions on Neural Networks, 13*(5), 1064–1074.
4. Azimi-Sadjadi, R., & Liou, R. J. (1992). Fast learning process of multilayer neural networks using recursive least squares method. *IEEE Transactions on Signal Processing, 40*(2), 446–450.
5. Baermann, F., & Biegler-Koenig, F. (1992). On a class of efficient learning algorithms for neural networks. *Neural Networks, 5*(1), 139–144.
6. Barnard, E. (1992). Optimization for training neural nets. *IEEE Transactions on Neural Networks, 3*(2), 232–240.
7. Battiti, R., & Masulli, F. (1990). BFGS optimization for faster automated supervised learning. In *Proceedings of International Neural Network Conference* (Vol. 2, pp. 757–760). Dordrecht, Netherland: Kluwer. Paris, France.
8. Battiti, R. (1992). First- and second-order methods for learning: Between steepest descent and Newton methods. *Neural Computation, 4*(2), 141–166.
9. Battiti, R., Masulli, G., & Tecchiolli, G. (1994). Learning with first, second, and no derivatives: A case study in high energy physics. *Neurocomputing, 6*(2), 181–206.
10. Beigi, H. S. M. (1993). Neural network learning through optimally conditioned quadratically convergent methods requiring no line search. In *Proceedings of IEEE the 36th Midwest Symposium on Circuits and Systems* (Vol. 1, pp. 109–112). Detroit, MI.
11. Benvenuto, N., & Piazza, F. (1992). On the complex backpropagation algorithm. *IEEE Transactions on Signal Processing, 40*(4), 967–969.
12. Bhaya, A., & Kaszkurewicz, E. (2004). Steepest descent with momentum for quadratic functions is a version of the conjugate gradient method. *Neural Networks, 17*, 65–71.
13. Bilski, J., & Rutkowski, L. (1998). A fast training algorithm for neural networks. *IEEE Transactions on Circuits and Systems II, 45*(6), 749–753.
14. Bishop, C. M. (1992). Exact calculation of the Hessian matrix for the multilayer perceptron. *Neural Computation, 4*(4), 494–501.
15. Bishop, C. M. (1995). *Neural networks for pattern recogonition*. New York: Oxford Press.
16. Bortoletti, A., Di Fiore, C., Fanelli, S., & Zellini, P. (2003). A new class of quasi-Newtonian methods for optimal learning in MLP-networks. *IEEE Transactions on Neural Networks, 14*(2), 263–273.
17. Burton, R. M., & Mpitsos, G. J. (1992). Event dependent control of noise enhances learning in neural networks. *Neural Networks, 5*, 627–637.
18. Charalambous, C. (1992). Conjugate gradient algorithm for efficient training of artificial neural networks. *IEE Proceedings—G, 139*(3), 301–310.
19. Chen, H. H., Manry, M. T., & Chandrasekaran, H. (1999). A neural network training algorithm utilizing multiple sets of linear equations. *Neurocomputing, 25*, 55–72.
20. Chen, Y. X., & Wilamowski, B. M. (2002). TREAT: A trust-region-based error-aggregated training algorithm for neural networks. In *Proceedings of International Joint Conference on Neural Networks* (Vol. 2, pp. 1463–1468).
21. Dai, Y. H., & Yuan, Y. (1999). A nonlinear conjugate gradient method with a strong global convergence property. *SIAM Journal on Optimization, 10*, 177–182.
22. Dixon, L. C. W. (1975). Conjugate gradient algorithms: Quadratic termination properties without linear searches. *IMA Journal of Applied Mathematics, 15*, 9–18.
23. Ergezinger, S., & Thomsen, E. (1995). An accelerated learning algorithm for multilayer perceptrons: Optimization layer by layer. *IEEE Transactions on Neural Networks, 6*(1), 31–42.

24. Fairbank, M., Alonso, E., & Schraudolph, N. (2012). Efficient calculation of the Gauss-Newton approximation of the Hessian matrix in neural networks. *Neural Computation*, *24*(3), 607–610.
25. Fletcher, R. (1991). *Practical methods of optimization*. New York: Wiley.
26. Fletcher, R., & Reeves, C. W. (1964). Function minimization by conjugate gradients. *Computer Journal*, *7*, 148–154.
27. Fukuoka, Y., Matsuki, H., Minamitani, H., & Ishida, A. (1998). A modified back-propagation method to avoid false local minima. *Neural Networks*, *11*, 1059–1072.
28. Georgiou, G., & Koutsougeras, C. (1992). Complex domain backpropagation. *IEEE Transactions on Circuits and Systems II*, *39*(5), 330–334.
29. Gonzalez, A., & Dorronsoro, J. R. (2008). Natural conjugate gradient training of multilayer perceptrons. *Neurocomputing*, *71*, 2499–2506.
30. Goryn, D., & Kaveh, M. (1989). Conjugate gradient learning algorithms for multilayer perceptrons. In *Proceedings of the 32nd Midwest Symposium on Circuits and Systems* (pp. 736–739). Champaign, IL.
31. Hagan, M. T., & Menhaj, M. B. (1994). Training feedforward networks with the Marquardt algorithm. *IEEE Transactions on Neural Networks*, *5*(6), 989–993.
32. Hanna, A. I. & Mandic, D. P. (2002). A normalised complex backpropagation algorithm. In *Proceedings of IEEE International Conference on Acoustics, Speech, and Signal Processing (ICASSP)* (pp. 977–980). Orlando, FL.
33. Hassibi, B., Stork, D. G., & Wolff, G. J. (1992). Optimal brain surgeon and general network pruning. In *Proceedings of IEEE International Conference on Neural Networks* (pp. 293–299). San Francisco, CA.
34. Heskes, T. (2000). On "natural" learning and pruning in multilayered perceptrons. *Neural Computation*, *12*, 881–901.
35. Hestenes, M. R., & Stiefel, E. (1952). Methods of conjugate gradients for solving linear systems. *Journal of Research of National Bureau of Standards B*, *49*, 409–436.
36. Hush, D. R., Horne, B., & Salas, J. M. (1992). Error surfaces for multilayer perceptrons. *IEEE Transactions on Systems, Man, and Cybernetics*, *22*(5), 1152–1161.
37. Igel, C., Toussaint, M., & Weishui, W. (2005). Rprop using the natural gradient. In M. G. de Bruin, D. H. Mache, & J. Szabados (Eds.), *Trends and applications in constructive approximation, International series of numerical mathematics* (Vol. 151, pp. 259–272). Basel, Switzerland: Birkhauser.
38. Ilguni, Y., Sakai, H., & Tokumaru, H., (1992). A real-time learning algorithm for a multilayered neural network based on the extended Kalman filter. *IEEE Transactions on Signal Processing*, *40*(4), 959–967.
39. Johansson, E. M., Dowla, F. U., & Goodman, D. M. (1991). Backpropagation learning for multilayer feedforward neural networks using the conjugate gradient method. *International Journal of Neural Systems*, *2*(4), 291–301.
40. Kamarthi, S. V., & Pittner, S. (1999). Accelerating neural network training using weight extrapolations. *Neural Networks*, *12*, 1285–1299.
41. Kantsila, A., Lehtokangas, M., & Saarinen, J. (2004). Complex RPROP-algorithm for neural network equalization of GSM data bursts. *Neurocomputing*, *61*, 339–360.
42. Kim, T., & Adali, T. (2002). Fully complex multi-layer perceptron network for nonlinear signal processing. *Journal of VLSI Signal Processing*, *32*(1), 29–43.
43. Kim, T., & Adali, T. (2003). Approximation by fully complex multilayer perceptrons. *Neural Computation*, *15*, 1641–1666.
44. Kostopoulos, A. E., & Grapsa, T. N. (2009). Self-scaled conjugate gradient training algorithms. *Neurocomputing*, *72*, 3000–3019.
45. Lee, J. (2003). Attractor-based trust-region algorithm for efficient training of multilayer perceptrons. *Electronics Letters*, *39*(9), 727–728.
46. Leung, H., & Haykin, S. (1991). The complex backpropagation algorithm. *IEEE Transactions on Signal Processing*, *3*(9), 2101–2104.

47. Leung, C. S., Wong, K. W., Sum, P. F., & Chan, L. W. (2001). A pruning method for the recursive least squared algorithm. *Neural Networks*, *14*, 147–174.
48. Leung, C. S., Tsoi, A. C., & Chan, L. W. (2001). Two regularizers for recursive least squared algorithms in feedforward multilayered neural networks. *IEEE Transactions on Neural Networks*, *12*, 1314–1332.
49. Li, Y., Zhang, D., & Wang, K. (2006). Parameter by parameter algorithm for multilayer perceptrons. *Neural Processing Letters*, *23*, 229–242.
50. Liu, C. S., & Tseng, C. H. (1999). Quadratic optimization method for multilayer neural networks with local error-backpropagation. *International Journal on Systems Science*, *30*(8), 889–898.
51. Manry, M. T., Apollo, S. J., Allen, L. S., Lyle, W. D., Gong, W., Dawson, M. S., et al. (1994). Fast training of neural networks for remote sensing. *Remote Sensing Reviews*, *9*, 77–96.
52. McLoone, S., & Irwin, G. (1999). A variable memory quasi-Newton training algorithm. *Neural Processing Letters*, *9*, 77–89.
53. McLoone, S. F., & Irwin, G. W. (1997). Fast parallel off-line training of multilayer perceptrons. *IEEE Transactions on Neural Networks*, *8*(3), 646–653.
54. McLoone, S. F., Asirvadam, V. S., & Irwin, G. W. (2002). A memory optimal BFGS neural network training algorithm. In *Proceedings of International Joint Conference on Neural Networks* (Vol. 1, pp. 513–518). Honolulu, HI.
55. Moller, M. F. (1993). A scaled conjugate gradient algorithm for fast supervised learning. *Neural Networks*, *6*(4), 525–533.
56. More, J. J. (1977). The Levenberg-Marquardt algorithm: Implementation and theory. In G. A. Watson (Ed.), *Numerical analysis* (Vol. 630, pp. 105–116)., Lecture notes in mathematics Berlin: Springer-Verlag.
57. Nazareth, J. L. (2003). *Differentiable optimization and equation solving*. New York: Springer.
58. Ng, S. C., Leung, S. H., & Luk, A. (1999). Fast convergent generalized back-propagation algorithm with constant learning rate. *Neural Processing Letters*, *9*, 13–23.
59. Ngia, L. S. H., & Sjoberg, J. (2000). Efficient training of neural nets for nonlinear adaptive filtering using a recursive Levenberg-Marquardt algorithm. *IEEE Transactions on Signal Processing*, *48*(7), 1915–1927.
60. Nishiyama, K., & Suzuki, K. (2001). H_∞-learning of layered neural networks. *IEEE Transactions on Neural Networks*, *12*(6), 1265–1277.
61. Nitta, T. (1997). An extension to the back-propagation algorithm to complex numbers. *Neural Networks*, *10*(8), 1391–1415.
62. Parisi, R., Di Claudio, E. D., Orlandim, G., & Rao, B. D. (1996). A generalized learning paradigm exploiting the structure of feedforward neural networks. *IEEE Transactions on Neural Networks*, *7*(6), 1450–1460.
63. Perantonis, S. J., Ampazis, N., & Spirou, S. (2000). Training feedforward neural networks with the dogleg method and BFGS Hessian updates. In *Proceedings of International Joint Conference on Neural Networks* (pp. 138–143). Como, Italy.
64. Perry, A. (1978). A modified conjugate gradient algorithm. *Operations Research*, *26*, 26–43.
65. Phua, P. K. H., & Ming, D. (2003). Parallel nonlinear optimization techniques for training neural networks. *IEEE Transactions on Neural Networks*, *14*(6), 1460–1468.
66. Polak, E. (1971). *Computational methods in optimization: A unified approach*. New York: Academic Press.
67. Powell, M. J. D. (1977). Restart procedures for the conjugate gradient method. *Mathematical Programming*, *12*, 241–254.
68. Puskorius, G. V., & Feldkamp, L. A. (1991). Decoupled extended Kalman filter training of feedforward layered networks. In *Proceedings of International Joint Conference on Neural Networks* (Vol. 1, pp. 771–777). Seattle, WA.
69. Rao, K. D., Swamy, M. N. S., & Plotkin, E. I. (2000). Complex EKF neural network for adaptive equalization. In *Proceedings of IEEE International Symposium on Circuits and Systems* (pp. 349–352). Geneva, Switzerland.

70. Rigler, A. K., Irvine, J. M., & Vogl, T. P. (1991). Rescaling of variables in back propagation learning. *Neural Networks*, *4*(2), 225–229.
71. Rivals, I., & Personnaz, L. (1998). A recursive algorithm based on the extended Kalman filter for the training of feedforward neural models. *Neurocomputing*, *20*, 279–294.
72. Rubanov, N. S. (2000). The layer-wise method and the backpropagation hybrid approach to learning a feedforward neural network. *IEEE Transactions on Neural Networks*, *11*(2), 295–305.
73. Ruck, D. W., Rogers, S. K., Kabrisky, M., Maybeck, P. S., & Oxley, M. E. (1992). Comparative analysis of backpropagation and the extended Kalman filter for training multilayer perceptrons. *IEEE Transactions on Pattern Analysis and Machine Intelligence*, *14*(6), 686–691.
74. Saarinen, S., Bramley, R., & Cybenko, G. (1993). Ill conditioning in neural network training problems. *SIAM Journal on Scientific Computing*, *14*(3), 693–714.
75. Saito, K., & Nakano, R. (1997). Partial BFGS update and efficient step-length calculation for three-layer neural networks. *Neural Computation*, *9*, 123–141.
76. Savitha, R., Suresh, S., Sundararajan, N., & Saratchandran, P. (2009). A new learning algorithm with logarithmic performance index for complex-valued neural networks. *Neurocomputing*, *72*, 3771–3781.
77. Scalero, R. S., & Tepedelenlioglu, N. (1992). A fast new algorithm for training feedforward neural networks. *IEEE Transactions on Signal Processing*, *40*(1), 202–210.
78. Shanno, D. (1978). Conjugate gradient methods with inexact searches. *Mathematics of Operations Research*, *3*, 244–256.
79. Shah, S., & Palmieri, F. (1990). MEKA–A fast, local algorithm for training feedforward neural networks. In *Proceedings of International Joint Conference on Neural Networks (IJCNN)* (Vol. 3, pp. 41–46). San Diego, CA.
80. Shawe-Taylor, J. S., & Cohen, D. A. (1990). Linear programming algorithm for neural networks. *Neural Networks*, *3*(5), 575–582.
81. Singhal, S., & Wu, L. (1989). Training feedforward networks with the extended Kalman algorithm. In *Proceedings of IEEE International Conference on Acoustics, Speech, and Signal Processing (ICASSP)* (Vol. 2, 1187–1190). Glasgow, UK.
82. Stan, O., & Kamen, E. (2000). A local linearized least squares algorithm for training feedforward neural networks. *IEEE Transactions on Neural Networks*, *11*(2), 487–495.
83. Sum, J., Leung, C. S., Young, G. H., & Kan, W. K. (1999). On the Kalman filtering method in neural network training and pruning. *IEEE Transactions on Neural Networks*, *10*, 161–166.
84. Uncini, A., Vecci, L., Campolucci, P., & Piazza, F. (1999). Complex-valued neural networks with adaptive spline activation functions. *IEEE Transactions on Signal Processing*, *47*(2), 505–514.
85. van der Smagt, P. (1994). Minimisation methods for training feed-forward neural networks. *Neural Networks*, *7*(1), 1–11.
86. Verikas, A., & Gelzinis, A. (2000). Training neural networks by stochastic optimisation. *Neurocomputing*, *30*, 153–172.
87. Wang, Y. J., & Lin, C. T. (1998). A second-order learning algorithm for multilayer networks based on block Hessian matrix. *Neural Networks*, *11*, 1607–1622.
88. Wilamowski, B. M., Iplikci, S., Kaynak, O., & Efe, M.O.(2001). An algorithm for fast convergence in training neural networks. In *Proceedings of International Joint Conference on Neural Networks* (Vol 3, pp. 1778–1782). Washington, DC.
89. Wilamowski, B. M., Cotton, N. J., Kaynak, O., & Dundar, G. (2008). Computing gradient vector and Jacobian matrix in arbitrarily connected neural networks. *IEEE Transactions on Industrial Electronics*, *55*(10), 3784–3790.
90. Wilamowski, B. M., & Yu, H. (2010). Improved computation for Levenberg-Marquardt training. *IEEE Transactions on Neural Networks*, *21*(6), 930–937.
91. Wilamowski, B. M., & Yu, H. (2010). Neural network learning without backpropagation. *IEEE Transactions on Neural Networks*, *21*(11), 1793–1803.
92. Xu, D., Zhang, H., & Liu, L. (2010). Convergence analysis of three classes of split-complex gradient algorithms for complex-valued recurrent neural networks. *Neural Computation*, *22*(10), 2655–2677.

93. Xu, Y., Wong, K.-W., & Leung, C.-S. (2006). Generalized RLS approach to the training of neural networks. *IEEE Trans Neural Netw*, *17*(1), 19–34.
94. Yang, S.-S., Ho, C.-L., & Siu, S. (2007). Sensitivity analysis of the split-complex valued multilayer perceptron due to the errors of the i.i.d. inputs and weights. *IEEE Transactions on Neural Networks*, *18*(5), 1280–1293.
95. Yang, S.-S., Siu, S., & Ho, C.-L. (2008). Analysis of the initial values in split-complex backpropagation algorithm. *IEEE Transactions on Neural Networks*, *19*(9), 1564–1573.
96. You, C., & Hong, D. (1998). Nonlinear blind equalization schemes using complex-valued multilayer feedforward neural networks. *IEEE Transactions on Neural Networks*, *9*(6), 1442–1455.
97. Yu, X. H., Chen, G. A., & Cheng, S. X. (1995). Dynamic learning rate optimization of the backpropagation algorithm. *IEEE Transactions on Neural Networks*, *6*(3), 669–677.
98. Yu, C., Manry, M. T., Li, J., & Narasimha, P. L. (2006). An efficient hidden layer training method for the multilayer perceptron. *Neurocomputing*, *70*, 525–535.
99. Zhang, Y., & Li, X. (1999). A fast U-D factorization-based learning algorithm with applications to nonlinear system modeling and identification. *IEEE Transactions on Neural Networks*, *10*, 930–938.
100. Zhang, H., Zhang, C., & Wu, W. (2009). Convergence of batch split-complex backpropagation algorithm for complex-valued neural networks. *Discrete Dynamics in Nature and Society*, *2009*, 1–16.
101. Zhang, H., Xu, D., & Zhang, Y. (2014). Boundedness and convergence of split-complex backpropagation algorithm with momentum and penalty. *Neural Processing Letters*, *39*, 297–307.

Chapter 7
Hopfield Networks, Simulated Annealing, and Chaotic Neural Networks

7.1 Hopfield Model

Hopfield model [27, 28] is biologically plausible since it functions like the human retina [36]. It is a fully interconnected recurrent network with J McCulloch–Pitts neurons. The Hopfield model is usually represented by using a J-J layered architecture, as illustrated in Fig. 7.1. The input layer only collects and distributes feedback signals from the output layer. The network has a symmetric architecture with a symmetric zero-diagonal real weight matrix, that is, $w_{ij} = w_{ji}$ and $w_{ii} = 0$. Each neuron in the second layer sums the weighted inputs from all the other neurons to calculate its current net activation net_i, then applies an activation function to net_i and broadcasts the result along the connections to all the other neurons. In the figure, $w_{ii} = 0$ is represented by a dashed line; $\phi(\cdot)$ and $\boldsymbol{\theta}$ are, respectively, a vector comprising the activation functions for all the neurons and a vector comprising the biases for all the neurons.

The Hopfield model operates in an unsupervised manner. The dynamics of the network are described by a system of nonlinear ordinary differential equations. The discrete form of the dynamics is defined by

$$net_i(t+1) = \sum_{j=1}^{J} w_{ji} x_j(t) + \theta_i, \tag{7.1}$$

$$x_i(t+1) = \phi(net_i(t+1)), \tag{7.2}$$

where net_i is the weighted net input of the ith neuron, $x_i(t)$ is the output of the ith neuron, θ_i is a bias to the neuron, and $\phi(\cdot)$ is the sigmoidal function. The discrete-time variable t in (7.1) and (7.2) takes values $0, 1, 2, \ldots$

K.-L. Du and M. N. S. Swamy, *Neural Networks and Statistical Learning*,
https://doi.org/10.1007/978-1-4471-7452-3_7

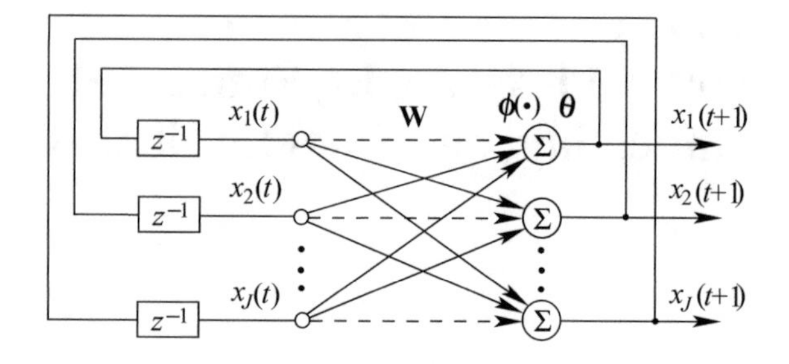

Fig. 7.1 Architecture of the Hopfield network

Correspondingly, the continuous-time Hopfield model is given by

$$\frac{\mathrm{d}net_i(t)}{\mathrm{d}t} = \sum_{j=1}^{J} w_{ji} x_j(t) + \theta_i, \tag{7.3}$$

$$x_i(t) = \phi(net_i(t)), \tag{7.4}$$

where t denotes the continuous-time variable.

In order to characterize the performance of the network, the concept of energy is introduced and the following energy function is defined [27]:

$$\begin{aligned} E &= -\frac{1}{2} \sum_{i=1}^{J} \sum_{j=1}^{J} w_{ij} x_i x_j - \sum_{i=1}^{J} \theta_i x_i \\ &= -\frac{1}{2} \boldsymbol{x}^T \mathbf{W} \boldsymbol{x} - \boldsymbol{x}^T \boldsymbol{\theta}, \end{aligned} \tag{7.5}$$

where $\boldsymbol{x} = (x_1, x_2, \ldots, x_J)^T$ is the input and state vector, and $\boldsymbol{\theta} = (\theta_1, \theta_2, \ldots, \theta_J)^T$ is the bias vector.

Theorem 7.1 (Stability) *The continuous-time Hopfield network always converges to a local minimum.*

The proof is sketched here. From (7.5),

$$\frac{\mathrm{d}E}{\mathrm{d}t} = \sum_{i=1}^{J} \frac{\mathrm{d}E}{\mathrm{d}x_i} \frac{\mathrm{d}x_i}{\mathrm{d}t} = \sum_{i=1}^{J} \frac{\mathrm{d}E}{\mathrm{d}x_i} \frac{\mathrm{d}x_i}{\mathrm{d}net_i} \frac{\mathrm{d}net_i}{\mathrm{d}t}.$$

where $\frac{\mathrm{d}E}{\mathrm{d}x_i} = -\left(\sum_{j=1}^{J} w_{ji} x_j + \theta_i \right) = -\frac{\mathrm{d}net_i}{\mathrm{d}t}$. Thus,

$$\frac{\mathrm{d}E}{\mathrm{d}t} = -\sum_{i=1}^{J} \left(\frac{\mathrm{d}net_i}{\mathrm{d}t} \right)^2 \frac{\mathrm{d}x_i}{\mathrm{d}net_i}.$$

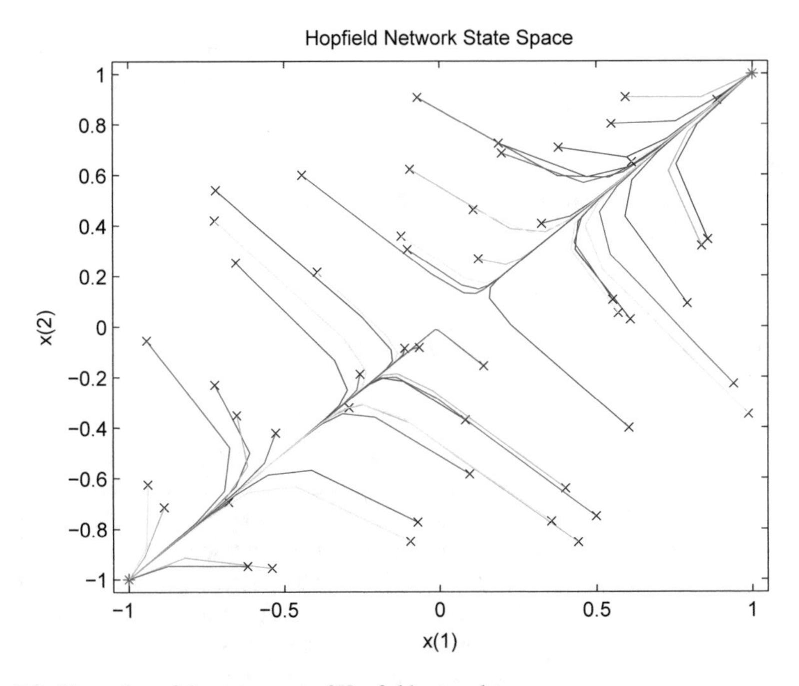

Fig. 7.2 Illustration of the state space of Hopfield network

Since the sigmoidal function $\phi(net_i)$ is monotonically increasing, $\frac{\mathrm{d}x_i}{\mathrm{d}net_i}$ is always positive. Thus, we have $\frac{\mathrm{d}E}{\mathrm{d}t} \leq 0$. According to Lyapunov's theorem, the Hopfield network always converges to a local minimum, and thus is stable. The dynamic equations of the Hopfield network actually implement a gradient-descent algorithm based on the cost function E [28].

The updating of all neurons of the Hopfield model can be carried out synchronously (Little dynamics) at each time step or asynchronously (Glauber dynamics), updating them one at a time. The Hopfield model is asymptotically stable when running in asynchronous or serial update mode. In asynchronous mode, only one neuron at a time updates itself in the output layer, and the energy function either decreases or stays the same after each iteration. However, if the Hopfield memory is running in the synchronous or parallel update mode, that is, all neurons update themselves at the same time, it may not converge to a fixed point, but may instead become oscillatory between two states [7, 12, 41].

The Hopfield network with the signum activation has a smaller degree of freedom compared to that using the sigmoidal activation, since it is constrained to changing the states along the edges of a J-dimensional hypercube $\mathcal{O} = \{-1, 1\}^J$. The use of sigmoidal functions helps in smoothing out some of the local minima.

Due to recurrence, the dynamics of the network are described by a system of ordinary differential equations and by an associated energy function to be minimized. The Hopfield model is a dynamic model that is suitable for hardware implementation and can converge to the result in the same order of time as the circuit time constant. The Hopfield network can be used for converting analog signals into digital format, for associative memory and for solving COPs.

Example 7.1 We store two stable points $(-1, -1)$, $(1, 1)$ as the two fixed points. For random states, they will finally converge to one of the two states, and this process is shown in Fig. 7.2.

7.2 Continuous-Time Hopfield Network

The high interconnection in physical topology makes the Hopfield network especially suitable for analog VLSI implementation. The convergence time of the network dynamics is decided by a circuit time constant, which is of the order of a few nanoseconds. The Hopfield network can be implemented by interconnecting an array of resistors, nonlinear operational amplifiers with symmetrical outputs, capacitors, and external bias current sources. Each neuron can be implemented by a capacitor, a resistor, and a nonlinear amplifier. A current source is necessary for representing the bias. The circuit structure of the neuron is shown in Fig. 7.3. v_i, $i = 1, \ldots, J$, is the output voltage of neuron i, I_i is the external bias current source for neuron i, u_i is the voltage at the interconnection point, C_i is a capacitor, and R_{ik}, $k = 0, 1, \ldots, J$, are resistors. The sigmoidal function $\phi(\cdot)$ is used as the transfer function of the amplifiers. A drawback of the Hopfield network is the necessity to update the complete set of network coefficients caused by the signal change, and this causes difficulties in its circuit implementation.

By applying Ohm's law and Kirchhoff's current law to the ith neuron, we obtain

$$C_i \frac{\mathrm{d}u_i}{\mathrm{d}t} = -\frac{u_i}{R_i} + \sum_{j=1}^{J} \frac{v_j}{R_{ij}} + I_i, \tag{7.6}$$

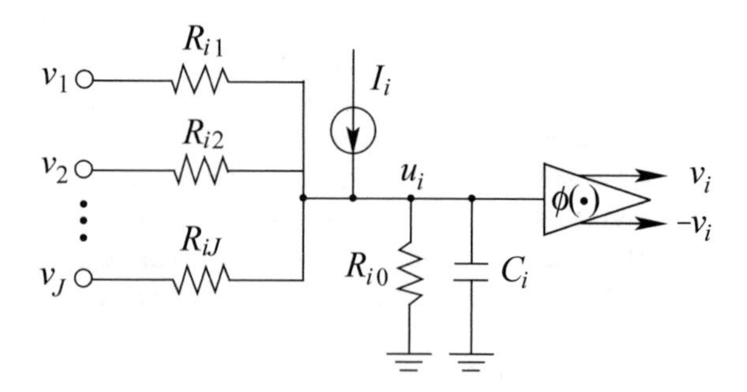

Fig. 7.3 A circuit for neuron i in the Hopfield model

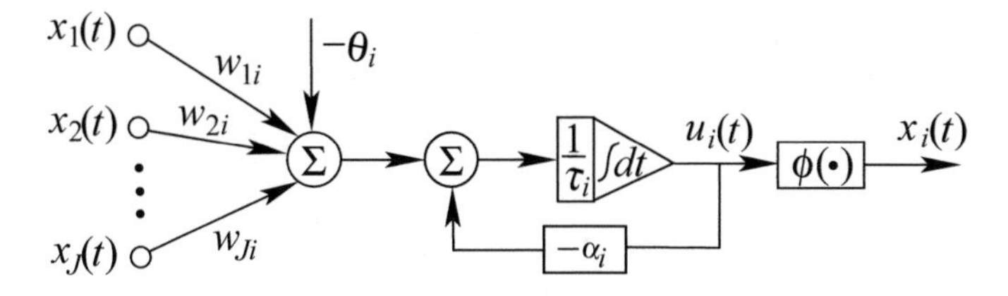

Fig. 7.4 Continuous-time Hopfield structure

where $v_i = \phi(u_i)$, $\phi(u_i)$ is the sigmoidal function, and

$$\frac{1}{R_i} = \frac{1}{R_{i0}} + \sum_{j=1}^{J} \frac{1}{R_{ij}} = G_{i0} + \sum_{j=1}^{J} G_{ij}, \tag{7.7}$$

G_{ij} being the conductance. In the circuits shown in Fig. 7.3, the inverting output of each neuron is used to generate negative weights since the conductance G_{ij} is always positive.

Equation (7.6) can be written as

$$\begin{aligned} \tau_i \frac{\mathrm{d}u_i}{\mathrm{d}t} &= -\alpha_i u_i + \left(\sum_{j=1}^{J} w_{ji} x_j + \theta_i \right) \\ &= -\alpha_i u_i + \boldsymbol{w}_i^T \boldsymbol{x} + \theta_i, \end{aligned} \tag{7.8}$$

where $x_i = v_i = \phi(u_i)$ is the input signal, $\tau_i = r_i C_i$ is the circuit time constant, r_i is a scaling resistance, $\alpha_i = \frac{r_i}{R_i}$ is a damping coefficient of the integrator, which forces the internal signal u_i to zero for a zero input, $w_{ji} = \frac{r_i}{R_{ij}} = r_i G_{ij}$ is the synaptic weight, $\theta_i = r_i I_i$ is the external bias signal, and $\boldsymbol{w}_i = (w_{1i}, w_{1i}, \ldots, w_{Ji})^T$ is the weight vector feeding neuron i. Equation (7.8) is known as the continuous-time Hopfield network, and its circuit is shown in Fig. 7.4.

The dynamics of the whole network can be written as

$$\boldsymbol{\tau} \frac{\mathrm{d}\boldsymbol{u}}{\mathrm{d}t} = -\boldsymbol{\alpha u} + \mathbf{W}^T \boldsymbol{x} + \boldsymbol{\theta}, \tag{7.9}$$

where the circuit time constant matrix $\boldsymbol{\tau} = \mathrm{diag}(\tau_1, \tau_2, \ldots, \tau_J)$, the interconnect-point voltage vector $\boldsymbol{u} = (u_1, u_2, \ldots, u_J)^T$, the damping coefficient matrix $\boldsymbol{\alpha} = \mathrm{diag}(\alpha_1, \alpha_2, \ldots, \alpha_J)$, the input and state vector $\boldsymbol{x} = (x_1, x_2, \ldots, x_J)^T$, the bias vector $\boldsymbol{\theta} = (\theta_1, \theta_2, \ldots, \theta_J)^T$, and the $J \times J$ weight matrix $\mathbf{W} = [\boldsymbol{w}_1 \boldsymbol{w}_2 \ldots \boldsymbol{w}_J]$.

At the equilibrium of the system, $\frac{\mathrm{d}\boldsymbol{u}}{\mathrm{d}t} = \mathbf{0}$, thus

$$\boldsymbol{\alpha u} = \mathbf{W}^T \boldsymbol{x} + \boldsymbol{\theta}. \tag{7.10}$$

The dynamics of the network are controlled by C_i and R_{ij}. A sufficient condition for the Hopfield network to be stable is that $\mathbf{W}$ is a symmetric matrix with diagonal elements being zero [28]. The stable states correspond to the local minima of the Lyapunov function [28]

$$E(\boldsymbol{x}) = -\frac{1}{2}\boldsymbol{x}^T \mathbf{W} \boldsymbol{x} - \boldsymbol{x}^T \boldsymbol{\theta} + \sum_{i=1}^{J} \alpha_i \int_0^{x_i} \phi^{-1}(\xi) \mathrm{d}\xi, \tag{7.11}$$

where $\phi^{-1}(\cdot)$ is the inverse of $\phi(\cdot)$. Equation (7.11) is a special case of the Cohen–Grossberg model [16].

From (7.8) and (7.3), it is seen that α is zero in the basic Hopfield model. This term corresponds to an integral related to $\phi^{-1}(\cdot)$ in the energy function. When the gain of the sigmoidal function $\beta \to \infty$, that is, when the sigmoidal function is selected as the hard-limiter function and the nonlinear amplifiers function as switches, the integral terms are insignificant and $E(\boldsymbol{x})$ in (7.11) approaches (7.5). In this case, the circuit model is exact for the basic Hopfield model. The stable states of the basic Hopfield network are the corners of the hypercube, namely, the local minima of (7.5) are in $\{-1, +1\}^J$ [28]. For large but finite gains, a sigmoidal function leads to a large positive contribution near the hypercube boundaries, but to a negligible contribution far from the boundaries. This leads to an energy surface that still has its maxima at the corners, but the minima slightly move inward from the corners of the hypercube. As β decreases, each minimum moves further inward and disappears one at a time. When β gets small enough, the energy minima start to disappear.

Discrete-time symmetric Hopfield networks are essentially as powerful computationally as general asymmetric networks, despite their Lyapunov function-constrained dynamics. In the binary-state case, symmetric networks can simulate asymmetric ones with only a linear increase in the network size and in the analog-state case, finite symmetric networks are also Turing universal, provided they are supplied with a simple external clock to prevent them from converging. The energy minimization problem remains NP-hard for analog networks also [54]. Continuous-time symmetric Hopfield networks are capable of general computation. Such networks have very constrained Lyapunov function-controlled dynamics. They are universal and efficient computational devices: any convergent synchronous fully parallel computation by a recurrent network of n discrete-time binary neurons, within general asymmetric coupling-weights, can be simulated by a symmetric continuous-time Hopfield network containing only $18n + 7$ units employing the saturated-linear activation function [55]. In terms of standard discrete computation models, any polynomially space-bounded Turing machine can be simulated by a family of polynomial-size continuous-time symmetric Hopfield nets [55].

Linear System in a Saturated Mode

For the realization given in Fig. 7.3, it is not possible to independently adjust the network parameters, since the coefficient α_i is nonlinearly related to all the weights w_{ij}.

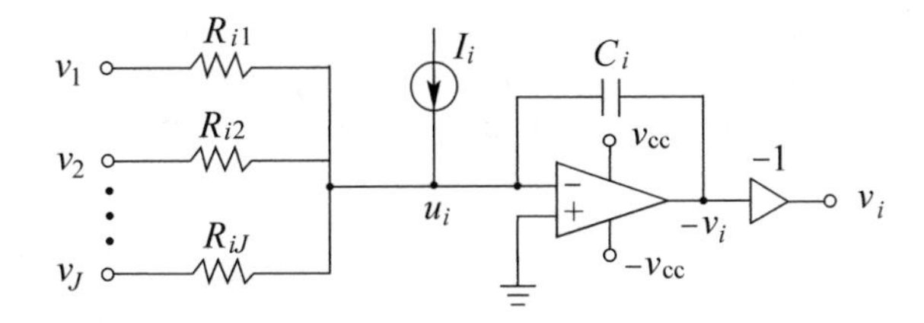

Fig. 7.5 A modified circuit for neuron i in the Hopfield model

In another circuit implementation of the Hopfield model [40], α_i are removed by replacing the integrators and nonlinear amplifiers in the previous model by ideal integrators with saturation. The circuit of such a neuron is illustrated in Fig. 7.5. Notice that the integrator and the nonlinear amplifier in Fig. 7.3 are replaced by an ideal integrator with saturation; v_{cc} is the power voltage. The dynamic equation of this neuron can be described by

$$\frac{\mathrm{d}v_i}{\mathrm{d}t} = \frac{1}{C_i}\left(\sum_{i=1}^{J} \frac{1}{R_{ij}} v_i + I_i\right), \tag{7.12}$$

where $|v_i| \leq 1$. Comparing (7.12) with (7.3), we have $w_{ji} = \frac{1}{C_i R_{ij}}$ and $\theta_i = \frac{I_i}{C_i}$. This model is referred to as a *linear system in a saturated mode*, which retains the basic structure of the Hopfield model and is easier to analyze, synthesize, and implement than the Hopfield model. The energy function of the model is exactly the same as (7.5).

7.3 Simulated Annealing

Annealing is referred to as tempering certain alloys of metal by heating and then gradually cooling them. The simulation of this process is known as simulated annealing. A metal is first heated above its melting point and then cooled slowly until it solidifies into a perfect crystalline structure. The defect-free crystal state corresponds to the global minimum energy configuration. The Metropolis algorithm is a simple method for simulating the evolution to the thermal equilibrium of a solid for a given temperature [45]. Simulated annealing [34] is a variant of the Metropolis algorithm, where the temperature is changing from high to low. It is a descent algorithm modified by random ascent moves in order to escape local minima which are not global minima. The annealing algorithm simulates a nonstationary finite state Markov chain whose state space is the domain of the cost function to be minimized. The idea of annealing is a general optimization principle.

Simulated annealing is a general, serial algorithm for finding a global minimum. The solutions by this technique are close to the global minimum within a polynomial upper bound for the computational time and are independent of the initial conditions. Simulated annealing is a popular Monte Carlo algorithm for combinatorial optimization. Some parallel algorithms for simulated annealing have been proposed aiming to improve the accuracy of the solutions by applying parallelism [18].

According to statistical thermodynamics, P_α, the probability of a physical system being in state α with energy E_α at temperature T satisfies the Boltzmann distribution (also known as the Boltzmann–Gibbs distribution):

$$P_\alpha = \frac{1}{Z} \mathrm{e}^{\frac{-E_\alpha}{k_B T}}, \tag{7.13}$$

where k_B is Boltzmann's constant, T is the absolute temperature, and Z is the partition function, defined by

$$Z = \sum_\beta \mathrm{e}^{-\frac{E_\beta}{k_B T}}, \tag{7.14}$$

the summation being taken over all states β with energy E_β at temperature T. At high T, the Boltzmann distribution exhibits uniform preference for all the states, regardless of the energy. When T approaches zero, only the states with minimum energy have nonzero probability of occurrence.

In simulated annealing, we omit the constant k_B. T is a control parameter called the *computational temperature*, which controls the magnitude of the perturbations of the energy function $E(\boldsymbol{x})$. At high T, the system ignores small changes in the energy and approaches thermal equilibrium rapidly, that is, it performs a coarse search of the space of global states and finds a good minimum. As T is lowered, the system responds to small changes in the energy, and performs a fine search in the neighborhood of the already determined minimum and finds a better minimum. At $T = 0$, any change in the system states does not lead to an increase in the energy, and thus, the system must reach equilibrium if $T = 0$.

When performing simulated annealing, theoretically a global minimum is guaranteed to be reached with a high probability. The artificial thermal noise is gradually decreased in time. The probability of a state change is determined by the Boltzmann distributions of the energy difference of the two states

$$P = \mathrm{e}^{-\frac{\Delta E}{T}}. \tag{7.15}$$

The probability of uphill moves in the energy function ($\Delta E > 0$) is large at high T, and is low at low T. Simulated annealing allows uphill moves in a controlled fashion: It attempts to improve on greedy local search by occasionally taking a risk and accepting a worse solution. It can be performed by Algorithm 7.1 [34].

Algorithm 7.1 (Simulated annealing).

1. *Initialize the system configuration. Randomize* $\boldsymbol{x}(0)$.
2. *Initialize* T *with a large value.*
3. *Repeat until* T *is small enough:*

   – *Repeat* `until` *the number of accepted transitions is below a threshold:*
   a. *Apply random perturbations* $\boldsymbol{x} = \boldsymbol{x} + \Delta \boldsymbol{x}$.
   b. *Evaluate* $\Delta E(\boldsymbol{x}) = E(\boldsymbol{x} + \boldsymbol{x}) - E(\boldsymbol{x})$*:*

  * `if` $\Delta E(\boldsymbol{x}) < 0$, *keep the new state;*
  * `otherwise`, *accept the new state with probability* $P = \mathrm{e}^{-\frac{\Delta E}{T}}$.
- *Set* $T = T - \Delta T$.

Classical simulated annealing is known as *Boltzmann annealing*. The cooling schedule for T is critical to the efficiency of the algorithm. If T is reduced too rapidly, a premature convergence to a local minimum may occur. In contrast, if it is reduced too slowly, the algorithm is very slow to converge. Based on a Markov-chain analysis on the simulated annealing process, a simple necessary and sufficient condition on the cooling schedule for the algorithm state to converge in probability to the set of globally minimum cost states is that T must be decreased by [24]

$$T(t) \geq \frac{T_0}{\ln(1+t)}, \qquad t = 1, 2, \ldots \tag{7.16}$$

to ensure convergence to the global minimum with probability one, where T_0 is an sufficiently large initial temperature. In [25], T_0 is proved to be greater than or equal to the depth, suitably defined, of the deepest local minimum which is not a global minimum state. In other words, in order to guarantee the Boltzmann annealing to converge to the global minimum with probability one, $T(t)$ is needed to decrease logarithmically with time. This is practically too slow. In practice, one usually applies a fast schedule $T(t) = \alpha T(t-1)$ with $0.85 \leq \alpha \leq 0.96$, to achieve a suboptimal solution.

Classical simulated annealing is a slow stochastic search method. The search has been accelerated in Cauchy annealing [57], simulated reannealing [30], generalized simulated annealing [60], and simulated annealing with known global value [43]. Some VLSI designs of simulated annealing are also available [36].

In Cauchy annealing [57], the Cauchy distribution, also known as the *Cauchy–Lorentz distribution*, is used to replace the Boltzmann distribution. The infinite variance provides a better ability to escape from local minima and allows for the use of faster schedules, such as T decreasing by $T(t) = \frac{T_0}{t}$. A stochastic neural network trained with Cauchy annealing is also called a *Cauchy machine*. Generalized simulated annealing [60] generalizes Cauchy annealing and Boltzmann annealing within a unified framework inspired by the generalized thermostatistics. In simulated reannealing [30], T decreases exponentially with t.

In the fuzzy annealing scheme [49], fuzzification is performed by adding an entropy term. The fuzziness at the beginning of the entire procedure is used to prevent the optimization process from getting stuck at an inferior local optimum. Fuzziness is reduced step by step. Fuzzy annealing results in an increase in computation speed by a factor of one hundred or more compared to simulated annealing [49].

Deterministic annealing [50, 51] is a method where randomness is incorporated into the energy or cost function, which is then deterministically optimized at a sequence of decreasing temperature. The approach is derived within the framework of information theory and probability theory. The annealing process is equivalent to the computation of Shannon's rate-distortion function, and the annealing temperature is inversely proportional to the slope of the curve. The application-specific cost is minimized subject to a constraint on the randomness (Shannon entropy) of the solution, which is gradually lowered. The iterative procedure is monotonely nonincreasing in the cost function. Unlike simulated annealing, it is a deterministic method that replaces stochastic simulations by the use of expectation. It has been used for nonconvex optimization problems such as clustering, MLP training, and RBF network training [50, 51]. The reduced-complexity deterministic annealing algorithms [20] use simple low-complexity distributions to mimic the Gibbs distribution used in standard deterministic annealing, yielding an acceleration of over 100 times with negligible performance difference for vector quantizer design.

Parallel simulated annealing takes advantage of parallel processing. In [6], each of a fixed set of samplers operates at different temperature. A solution that costs less is propagated from the higher temperature sampler to the neighboring sampler operating at a lower temperature. Therefore, the best solution at a given time is propagated to all samplers operating at a lower temperature. Sample-Sort [59] has a fixed set of samplers each operating at different static temperatures. The set of samplers uses a biased generator to sample the same distribution of a serial simulated annealing algorithm to maintain the same convergence property. It propagates a less-cost solution to other samplers, but does it probabilistically by permitting the samplers to exchange solutions with neighboring samplers. The samplers are lined up in a row and exchange solutions with samplers that are one or more hops away. It adjusts the probability of accepting a higher cost solution dependent on the temperature of the neighboring sampler.

Multiobjective simulated annealing uses the domination concept and the annealing scheme for efficient search [47]. In [56], the proposed multiobjective simulated annealing maps the optimization of multiple objectives to a single-objective optimization using the true trade-off surface, maintaining the convergence properties of simulated annealing, while encouraging exploration of the full trade-off surface.

7.4 Hopfield Networks for Optimization

The Hopfield network is usually used for solving optimization problems. In general, the continuous model is superior to the discrete one in terms of the local minimum problem, because of its smoother energy surface. Hence, the continuous Hopfield network has dominated the techniques for optimization problems, especially for combinatorial problems [29].

From the computational aspect, the operation of Hopfield network for an optimization problem manages a dynamic system characterized by an energy function, which is a combination of the objective function and constraints of the original problem. Three common techniques, namely penalty functions, Lagrange multipliers, and primal and dual methods, are utilized to construct an energy function. These techniques are suitable for solving various optimization problems such as LP, nonlinear programming and mixed-integer LP.

For the penalty method, there is always a compromise between good-quality solution and convergence. For a feasible solution, the weighting factors for the penalty terms should be sufficiently large, which however causes the constraints on the original problem to become relatively weaker, resulting in a deterioration of the quality of the solution. A trial-and-error process for choosing some of the penalty parameters is inevitable in order to obtain feasible solutions. Moreover, the gradient-descent method often leads to a local minimum of the energy landscape.

The local minima of (7.5) correspond to the attractors in the phase space, which are nominal memories of the network. A large class of COPs can be expressed in this form of QP optimization problems, and thus can be solved using the Hopfield network. The Hopfield network can be used as an effective interface between analog and digital devices, where the input signals to the network are analog and the output signals are discrete values. The neural interface has the capability of learning. The neural-based analog-to-digital (A/D) converter adapts to compensate for initial device mismatches or long-term drifts [58].

Many neural network models for linear, quadratic programming, least squares, and many matrix algebraic, constrained optimization, discrete, and combinatorial optimization problems are described in [15].

7.4.1 Combinatorial Optimization Problems

Any problem that has a large set of discrete solutions and a cost function for rating those solutions relative to one another is a COP. COPs are known to be NP-complete, namely, nondeterministic polynomial-time complete. In COPs, the number of solutions grows exponentially with n, the size of the problem, at $O(n!)$ or $O(e^n)$ such that no algorithm can find the global minimum solution in polynomial computational time. The goal for COPs is to find an optimal solution or sometimes a nearly optimal solution. Exhaustive search of all the possible solutions for the optimum is impractical.

The Hopfield network can be effectively used to deal with COPs with the objective functions of the linear or quadratic form, linear equalities and/or inequalities as the constraints, and binary variable values so that the constructed energy function can be of quadratic form. It can be used to solve the two well-known COPs: traveling salesman problem (TSP) and the location–allocation problem.

Traveling Salesman Problem (TSP)

TSP is the most notorious NP-complete problem. The definition is simple: Find the shortest closed-path through all points. Given a set of points, either nodes on a graph or cities on a map, find the shortest possible tour that visits every point exactly once and then returns to its starting point. There are $(n-1)!/2$ possible tours for an n-city TSP. For symmetric TSPs, the distances between nodes are independent of the direction, i.e., $d_{ij} = d_{ji}$ for every pair of nodes. In the asymmetric TSP, at least one pair of nodes satisfies $d_{ij} \neq d_{ji}$. The Hopfield network was the first neural network used for TSP, and it achieves a near-optimum solution [29]. TSP arises in numerous optimization problems, from routing of wires on a printed circuit board, to VLSI circuit design, to fast food delivery, to parts placement in electronic circuits, to routing in communication network systems, and to resource planning.

For TSP, any individual city is indicated by the output states of a set of n neurons, and can be in any one of the n positions in the tour list. For n cities, a total of n independent sets of n neurons are needed to describe a complete tour. Hence, this is a total of $N = n^2$ neurons which are displayed as an $n \times n$ square array for the network. Since the representation of neural outputs of the network in terms of n rows of n neurons, the N symbols of outputs will be represented by double indices x_{Xj}, denoting the Xth city in the jth position in a tour. To permit the N neurons in the network to compute a solution to the problem, the lowest value of the energy function corresponds to the best path. The space over which the energy function is minimized in this limit is the 2^n corners of the N-dimensional hypercube. A benchmark set for the TSP community is TSPLIB, a growing collection of sample instances.

The problem can be described as:

$$\min \sum_X \sum_{Y \neq X} \sum_i d_{XY} x_{Xi} (x_{Y,i+1} + x_{Y,i-1}) \tag{7.17}$$

subject to

$$\sum_X \sum_i \sum_{j \neq i} x_{Xi} x_{Xj} = 0, \tag{7.18}$$

$$\sum_i \sum_X \sum_{X \neq Y} x_{Xi} x_{Yi} = 0, \tag{7.19}$$

$$\left(\sum_X \sum_i x_{Xi} - n \right)^2 = 0. \tag{7.20}$$

where all indices X, Y, i, j run from 1 to n. The objective is to find the shortest tour. The first constraint is satisfied if and only if each city row X contains no more than one 1, i.e., the rest of the entries should be zero. The second constraint is satisfied if and only if each "position in tour" column contains no more than one 1, i.e., the rest of the entries are zero. The third constraint is satisfied if and only if there are n entries of one in the entire matrix.

Consider those corners of this space which are the local minima of the energy function [29]

$$\begin{aligned} E_{\text{TSP}} = {} & \lambda_1 \sum_X \sum_i \sum_{j \neq i} x_{Xi} x_{Xj} + \lambda_1 \sum_i \sum_X \sum_{X \neq Y} x_{Xi} x_{Yi} + \lambda_3 \left(\sum_X \sum_i x_{Xi} - n \right)^2 \\ & + \lambda_4 \sum_X \sum_{Y \neq X} \sum_i d_{XY} x_{Xi} (x_{Y,i+1} + x_{Y,i-1}), \end{aligned} \tag{7.21}$$

where $\lambda_1, \lambda_2, \lambda_3, \lambda_4$ are positive parameters, chosen by trial and error for a particular problem. The first three terms describe the feasibility requirements and result in a valid tour by a value of zero [29]. The last term represents the objective function of TSP.

The multiple TSP is a generalization of TSP: given a set of intermediate cities and a depot, m salesmen must visit all intermediate cities according to the constraints that the route formed by each salesman must start and end at the depot; each intermediate city must be visited once and by a single salesman; and the cost of the routes must be minimum. Multiple TSP is NP-complete as it includes TSP as a special case. It can be applied to vehicle routing and job scheduling.

A Lagrange multiplier and Hopfield-type barrier function method is proposed in [19] for approximating a solution of TSP. The method is more effective and efficient than the soft-assign algorithm. The introduced Hopfield-type barrier term is given by [19, 28]

$$d(x_{ij}) = x_{ij} \ln x_{ij} - (1 - x_{ij}) \ln(1 - x_{ij}) \tag{7.22}$$

to incorporate $0 \le x_{ij} \le 1$ into the objective function.

Location–Allocation Problem

The location–allocation problem can be stated as follows. Given a set of facilities, each of which serves a certain number of nodes on a graph, the objective is to place the facilities on the graph so that the average distance between each node and its serving facility is minimized.

A class of COPs including the location–allocation problem can be formulated as [44]

$$\min \sum_{i=1}^{m} \sum_{j=1}^{n} c_{ij} x_{ij} \tag{7.23}$$

subject to

$$\sum_{j=1}^{n} a_{ij} x_{ij} = b_i, \quad i = 1, 2, \ldots, m, \tag{7.24}$$

$$\sum_{i=1}^{m} x_{ij} = 1, \quad j = 1, 2, \ldots, n, \tag{7.25}$$

$$x_{ij} \in \{0, 1\}, \quad i = 1, \ldots, m, j = 1, \ldots, n, \tag{7.26}$$

where $\mathbf{X} = \left[x_{ij}\right] \in \{0, 1\}^{m \times n}$ is a variable matrix, $c_{ij} \ge 1$, $a_{ij} \ge 1$, and $b_i \ge 1$, are constant integers.

To make use of the Hopfield network, one needs first to convert a COP into a constrained optimization problem and solve the latter using the penalty method. The COP defined by (7.23) through (7.26) can be transformed into the minimization of the following total cost

$$E = \frac{\lambda_1}{2} \sum_{i=1}^{m} \left(\sum_{j=1}^{n} a_{ij} x_{ij} - b_i \right)^2 + \frac{\lambda_2}{2} \sum_{j=1}^{n} \left(\sum_{i=1}^{m} x_{ij} - 1 \right)^2 + \frac{\lambda_3}{2} \sum_{i=1}^{m} \sum_{j=1}^{n} x_{ij} \left(1 - x_{ij}\right) + \frac{\lambda_4}{2} \sum_{i=1}^{m} \sum_{j=1}^{n} c_{ij} x_{ij}. \quad (7.27)$$

where λ_1, λ_2, λ_3 and λ_4 are weights of individual constraints, which can be tuned for an optimal or good solution. When the first three terms are all zeros, the solution is a feasible one. The cost E has the same form as that of the energy function of the Hopfield network, and thus can be solved by using the Hopfield network.

By minimizing the square of (7.23), the network distinguishes optimal solutions more sharply than with (7.27) and this greatly overcomes many of the weaknesses of the network with (7.27) [44].

Combinatorial Optimization Problems with Equality and Inequality Constraints

Hopfield network can be used to solve COPs under equality as well as inequality constraints, as long as the constructed energy function is of the form of (7.5). Constraints are treated by introducing in the objective function some additional energy terms that penalize any infeasible state. Some extensions to the Hopfield model are necessary in order to handle both equality and inequality constraints [2, 58].

Assume that we have both linear equality and inequality constraints

$$\boldsymbol{r}_i^T \boldsymbol{x} = s_i, \quad i = 1, \ldots, l, \quad (7.28)$$

$$\boldsymbol{q}_j^T \boldsymbol{x} \le h_j, \quad j = 1, \ldots, k, \quad (7.29)$$

where $\boldsymbol{r}_i = \left(r_{i,1}, \ldots, r_{i,J}\right)^T$, $\boldsymbol{q}_j = \left(q_{j,1}, \ldots, q_{j,J}\right)^T$, s_i is a constant, and $h_j > 0$.

In the extended Hopfield model [2], each inequality constraint is converted to an equality constraint by introducing an additional variable managed by a new neuron, known as the *slack neuron*. Each slack neuron is connected to the initial neurons, where their corresponding variables occur in a linear combination. The extended Hopfield model has the drawback of being frequently stabilized in neuron states far from the suitable ones, *i.e.,* zero and one. To deal with this drawback, a new penalty energy term is derived to significantly reduce the number of neurons with unsuitable states [38].

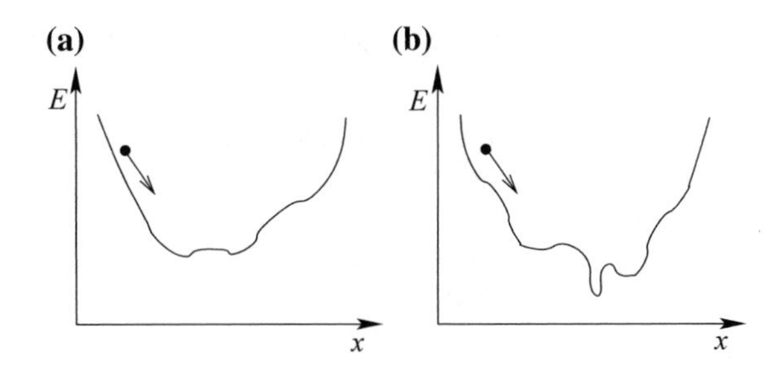

Fig. 7.6 Schematics of the landscapes of the energy function with one-dimensional variable x and different values of gain β. **a** Low β smoothes the surface. **b** High β reveals more details in the surface

7.4.2 *Escaping Local Minima*

Simulated annealing is a popular method for any optimization problem including COPs. However, due to its Monte Carlo nature, it would require even more iterations than complete enumeration, for some problems, in order to guarantee convergence to an exact solution. For example, for an n-city TSP, simulated annealing using the logarithmic cooling schedule needs a computational complexity of $O(n^{n^{2n-1}})$, which is far more than $O((n-1)!)$ for complete enumeration and $O(n^2 2^n)$ for dynamic programming [1, 9]. Thus, one has to apply heuristic fast cooling schedules to improve the convergence speed.

The Hopfield network is desirable for solving COPs that can be formulated into quadratic functions. The Hopfield network converges very fast, and it can also be easily implemented using RC circuits. However, due to its gradient-descent nature, it always gets trapped at the nearest local minimum of the initial random state.

To help the Hopfield network escape from the local minima, a popular strategy is to change the sigmoidal gain β, by starting from a low gain and gradually increasing it. When β is low, the energy landscape is smooth, and the algorithm can easily find a good local minimum. As β increases, more details of the energy landscape are revealed, and the algorithm can find a better solution. This is illustrated in Fig. 7.6. This process is usually called *gain annealing*, as it is analogous to the cooling process of simulated annealing. In the limit, when $\beta \to \infty$, the hypobolic tangent function becomes the signum function.

In order to use the Hopfield network for solving optimization problems, one needs to construct an energy function using the Lagrange multiplier method. By adaptively adjusting the balance between the constraint and objective terms, the network can avoid falling into a local minimum and continue to update in a gradient-descent direction of energy [64]. In the learning strategy given in [64], the minimum found is always a global or a near-global one. The method is capable of finding an optimal or near-optimal solution in a short time for TSP.

7.4.3 Solving Other Optimization Problems

Tank and Hopfield first proposed the Hopfield network structure to solve the linear programming (LP) problem [58]. Their work is extended in [33] to solve a nonlinear programming problem. A class of Lagrange neural networks appropriate for general nonlinear programming, i.e., problems including both equality and inequality constraints, is analyzed in [68].

Matrix inversion can be performed using the Hopfield network [31]. Given a nonsingular $n \times n$ matrix $\mathbf{A}$, the energy function can be defined by $\|\mathbf{AV} - \mathbf{I}\|_F^2$, where $\mathbf{V}$ denotes the inverse of $\mathbf{A}$ and the subscript F denotes the Frobenius norm. This energy function can be decomposed into n energy functions, and n similar networks are required, each optimizing an energy function. This method can be used to solve a system of n linear equations with n variables, $\mathbf{A}\boldsymbol{x} = \boldsymbol{b}$, where $\mathbf{A} \in R^{n \times n}$ and $\boldsymbol{x}$, $\boldsymbol{b} \in R^n$, if $\mathbf{A}$ is nonsingular. In [8], this set of linear equations is solved by using a continuous Hopfield network with n nodes. The Hopfield network is designed to minimize the energy function $E = \frac{1}{2}\|\mathbf{A}\boldsymbol{x} - \boldsymbol{b}\|^2$, and the activation function is selected as a linear transfer function. This method is also applicable when there exists infinitely many solutions and $\mathbf{A}$ is singular. Another neural LS estimator that uses continuous Hopfield network and a nonlinear activation function have been proposed in [22].

A Hopfield network with linear transfer functions augmented by an additional feedforward layer can be used to solve a set of linear equations [61] and to compute the pseudoinverse of a matrix [39]. The resultant augmented linear Hopfield network can be used to solve constrained LS optimization problems.

LP network [58] is designed based on the Hopfield model for solving LP problems

$$\min \boldsymbol{a}^T \boldsymbol{x} \tag{7.30}$$

subject to

$$\boldsymbol{d}_j^T \boldsymbol{x} \geq h_j, \quad j = 1, \ldots, M, \tag{7.31}$$

where $\boldsymbol{d}_j = (d_{j,1}, d_{j,2}, \ldots, d_{j,J})^T$, $\mathbf{D} = [d_{j,i}]$ is an $M \times J$ matrix, and h_j is a constant. Each inequality constraint is modeled by a slack neuron. The network contains a signal plane with J neurons and a constraint plane with M neurons. The energy function decreases until the net reaches a state where all time derivatives are zero.

With some modifications, the LP network [58] can be used to solve least squares error problems [66]. In [17], a circuit based on a modification of the LP network [58] is designed for computing the discrete Hartley transform. A circuit for computing the discrete Fourier transform (DFT) is obtained by simply adding a few adders to the discrete Hartley transform circuit.

Based on the inherent properties of convex quadratic minimax problems, a neural network model for a class of convex quadratic minimax problems is presented in [23]. The model is stable in the sense of Lyapunov and will converge to an exact saddle point in finite time by defining a proper convex energy function. Furthermore, global exponential stability of the model is shown under mild conditions.

7.5 Chaos and Chaotic Neural Networks

7.5.1 Chaos, Bifurcation, and Fractals

If a system seems to behave randomly, it is said to be *chaotic* or to demonstrate *chaos*. Chaos is a complicated behavior of a nonlinear dynamical system. It is also a self-organized process subject to some underlying rules. Chaos is an omnipresent phenomenon in biological behavior and evolution. There is evidence that parts of the brain, as well as individual neurons, exhibit chaotic behavior. Chaos, together with the theory of relativity and quantum mechanics, was considered one of the three monumental discoveries of the twentieth century. In fact, chaos theory is closely related to Heisenberg's uncertainty principle.

Chaotic systems can be either deterministic chaos or nondeterministic chaos. For deterministic chaos, the system's behavior can be approximately or exactly represented by a mathematically or heuristically expressed function. For nondeterministic chaos, the system's behavior is not expressible by a deterministic function and therefore is not at all predictable.

A dynamic system is a system whose state varies over time. It can be represented by state equations. A stable system usually has fixed-point equilibriums, called attractors. Strange attractors are a class of attractors that exhibit a chaotic behavior. A chaotic process can be classified according to its fractal dimensions and Lyapunov exponent. In a typical chaotic system, there exists bifurcation points that lead to chaos, and self-similarity and fractals. Time delays can be the source of instabilities and bifurcations in dynamical systems and are frequently observed in biological systems such as neural networks.

Bifurcation is a common phenomenon found in chaotic systems which indicates sudden, qualitative changes in the dynamics of the system either from one kind of periodic case (with limit cycle(s) and fixed point(s)) to another kind of periodic situation, or from a periodic stage to a chaotic stage.

Example 7.2 A well-known chaotic function is the one used to model fish population growth: $x(n+1) = x(n)(1 - x(n))$. We can represent this logistic function in a slightly modified form as: $x(n+1) = rx(n)(1 - x(n))$, for the bifurcation parameter $r \in [0, 4]$; iteration starts with a randomly selected $x(0)$. Figure 7.7 depicts the bifurcation diagram (x versus r).

The behavior of this first-order differential equation changes dramatically as r is altered. For $r < 1$ the output x goes to zero. For $1 < r < 3$ the output converges to single nonzero value. These are stable regions. When r goes beyond 3, the process begins alternating between two different points without converging to either; the output begins to oscillate between two values initially, then between four values, then between eight values, and so on. After $r = 3.44948\ldots$ the two curves split further into four whereas the iteration oscillate between four different points. The horizontal distance between the split points grows shorter and shorter, until the bifurcation becomes so fast at the point $r = 3.568$ that iterates race all over a segment instead of

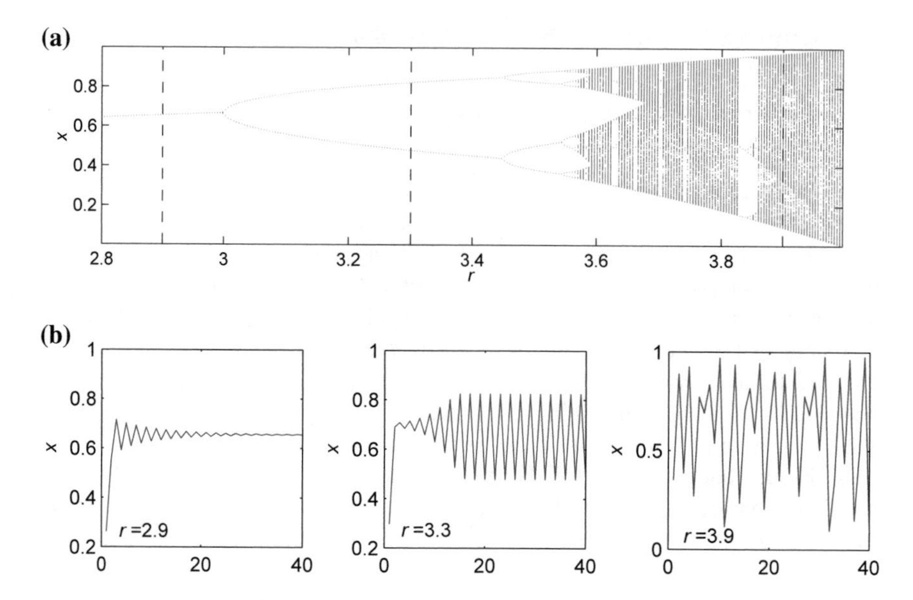

Fig. 7.7 Bifurcation diagram of the logistic equation

alternating between a few fixed points. For $r > 3.568$, the output becomes chaotic. The behavior is chaotic in the sense that it's absolutely impossible to predict where the next iterate will appear. Figure 7.7a shows the local minima and maxima of the asymptotic time series against the parameter r. Figure 7.7b shows representative time series with fixed point at $r = 2.9$, limit cycle at $r = 3.3$ and chaotic attractors at $r = 3.9$.

The phase space of a chaotic process is the feature space where the process is traced over time. A chaotic process goes around areas or points of its phase space, but without repeating the same trajectory; such areas, or points, from the phase space are called chaotic attractors. From a geometrical point of view, all chaotic attractors are fractals, while all fractals have their bifurcations and chaotic features. Fractals are a graphical presentation of chaos while chaos is the physical dynamics of fractals. Fractal dimension indicates the complexity of a chaotic system, for example, a low-dimensional attractor (3–4) would suggest that the problem is solvable. A chaotic process has a fractal dimension, which is defined by its attractors.

7.5.2 *Chaotic Neural Networks*

Chaotic neural networks with the sigmoidal activation functions were successfully implemented for solving various practical problems. Chaotic simulated annealing was utilized to solve COPs, such as TSP [9, 11]. It has a much better searching ability in solving TSP, in comparison with the Hopfield network, the Boltzmann machine, and the Gaussian machine [48].

A recurrent network such as the Hopfield network, when introduced with chaotic dynamics, is sometimes called a *chaotic neural network*. The chaotic dynamics are temporarily generated for searching and self-organizing, and eventually vanish with the autonomous decrease of a bifurcation parameter corresponding to the temperature in the simulated annealing process. Thus, the chaotic neural network gradually approaches to a dynamical structure of the recurrent network.

Since a chaotic neural network operates in a manner similar to that of simulated annealing, but in a deterministically chaotic way, the operation is known as *chaotic simulated annealing*. More specifically, the transiently chaotic dynamics are used for searching a basin containing the global optimum, followed by a stable and convergent phase when the chaotic noise decreases to zero. As a result, the chaotic neural network has a high ability for searching globally optimal or near-optimal solutions [9].

Simulated annealing, employing the Monte Carlo scheme, searches all the possible states by temporally changing the probability distributions. In contrast, chaotic simulated annealing searches a possible fractal subspace with continuous states by temporally changing invariant measures that are determined by its dynamics. Thus, the search region in chaotic simulated annealing is very small compared with the state space, and chaotic simulated annealing can perform an efficient search.

A small amount of chaotic noise can be injected into the output of the neurons and/or to the weights during the operation of the Hopfield network. In [26], a chaotic neural network is obtained by adding chaotic noise to each neuron of the discrete-time continuous-output Hopfield network and gradually reducing the noise so that it is initially chaotic, but eventually convergent.

A chaotic neural network based on a modified Nagumo–Sato neuron model was proposed in [4] in order to explain complex dynamics observed in a biological neural system. The chaotic neural network introduced in [4, 9] is obtained by adding a negative self-coupling to the Hopfield network. By gradually removing the self-coupling, the transient chaos is used for searching and self-organizing. The updating rule for the chaotic neural network is given by [9]

$$net_i(t+1) = \left(1 - \frac{\alpha_i}{\tau_i}\right) net_i(t) + \frac{1}{\tau_i}\left(\boldsymbol{w}_i^T \boldsymbol{x} + \theta_i\right) - c(t)\left(x_i - b\right), \tag{7.32}$$

$$x_i(t) = \phi(net_i(t)), \tag{7.33}$$

where $c(t+1) = \beta c(t)$, $\beta \in [0, 1]$, the bias $b > 0$, and other parameters are the same as for (7.8). A large initial value of $c(t)$ is used so that self-coupling is strong enough to generate chaotic dynamics for searching the global minima. The damping of $c(t)$ produces successive bifurcations so that the neurodynamics eventually converge from strange attractors to a stable equilibrium point.

It is shown in [48] that the Euler approximation of the continuous-time Hopfield network with a negative neuronal self-coupling exhibits chaotic dynamics and that this model is equivalent to a special case of a chaotic network proposed in [4] after a variable transformation.

The chaotic simulated annealing approach is derived by varying the time step Δt of an Euler discretization of the Hopfield network described by (7.8) [62]

$$net_i(t + \Delta t) = net_i(t)\left(1 - \frac{\alpha_i \Delta t}{\tau_i}\right) + \frac{\Delta t}{\tau_i}\left(\boldsymbol{w}_i^T \boldsymbol{x} + \theta_i\right). \tag{7.34}$$

The time step is analogous to the temperature parameter in simulated annealing, and the method starts with large Δt, where the dynamics are chaotic, and gradually decreases it. When $\Delta t \to 0$, the system approaches the Hopfield model (7.8) and minimizes its energy function. When $\Delta t = 1$, the Euler-discretized Hopfield network is identical to the chaotic neural network given in [9]. The simulation results for COPs are comparable to that of the method proposed in [9].

Many chaotic approaches [9, 26, 62] can be unified and compared under the framework of adding an extra energy term E_{CSA} into the original computational energy (7.11) of the Hopfield model [35]. The extra energy term modifies the original Hopfield energy landscape to accommodate transient chaos. For example, for the method proposed in [9], E_{CSA} can be selected as

$$E_{\mathrm{CSA}} = \frac{c(t)}{2} \sum_i x_i\left(x_i - 1\right). \tag{7.35}$$

This results in many logistic maps being added to the Hopfield energy function. E_{CSA} is convex, and hence drives $\boldsymbol{x}$ toward the interior of the hypercube. This driving force is diminished as $E_{\mathrm{CSA}} \to 0$ when $\lambda(t) \to 0$.

A theoretical explanation for the global searching ability of the chaotic neural network is given in [11]: its attracting set contains all global and local optima of the optimization problem under certain conditions, and since the chaotic attracting set has a fractal structure and covers only a very small fraction of the entire state space, chaotic simulated annealing is more efficient in searching for good solutions for optimization problems compared to simulated annealing.

However, a number of network parameters must be subtly adjusted so as to guarantee the convergence of the chaotic network. Chaotic simulated annealing is not guaranteed to settle down at a global optimum no matter how slowly annealing is carried out, because the chaotic dynamics are completely deterministic. Stochastic chaotic simulated annealing [63] is a combination of simulated annealing and chaotic simulated annealing by using a noisy chaotic network, which is obtained by adding decaying stochastic noise into the chaotic network proposed in [9]. Stochastic chaotic simulated annealing restricts the random search to a subspace of chaotic attracting sets, and this subspace is much smaller than the entire state space searched by simulated annealing.

The Wang-oscillator, or Wang's chaotic neural oscillator [65], is a neural oscillatory model consisting of two neurons, one excitatory and one inhibitory. It encodes the input stimulus and gives the responses by altering the behavior of the neural dynamics. Inspired by the Wang-oscillator, the Lee-oscillator [37] provides a transient chaotic progressive growth in its neural dynamics, which solves the fundamental shortcoming of the Wang-oscillator. The main purpose of Aihara's chaotic network is to realize the dynamical pattern association, so it does not converge to any particular pattern, but rather oscillating against different stored patterns in a chaotic manner. Using Lee-oscillators upon pattern association, a chaotic auto-associative network, namely Lee-associator, is constructed. Lee-oscillator provides gradual and progressive changes of neural dynamics in the transition region while Aihara's chaotic network provides an abrupt change in the neural dynamics in this region. Compared with the chaotic auto-associative networks developed in [3, 65], Lee-associator produces a robust progressive memory recalling scheme. Lee-associator provides a remarkable progressive memory association scheme during the chaotic memory association. This is completely consistent with the latest research in psychiatry and perception psychology on dynamic memory recalling schemes, as well as the implications and analogues to human perception as illustrated by the remarkable Rubin-vase experiment on visual psychology.

7.6 Multistate Hopfield Networks

The multilevel Hopfield network [21, 67, 69, 72] and the complex-valued multistate Hopfield network [32, 46] are two direct generalizations of the Hopfield network. The multilevel Hopfield network uses neurons with an increasing multistep function as the activation function [21], while the complex-valued multistate Hopfield network uses a multivalued complex-signum function as the activation function.

The multilevel sigmoidal function has typically been used as the activation function in the multilevel Hopfield network. In [67], a multilevel Hopfield-like network is obtained by using a new neuron with self-feedback and the multilevel sigmoidal activation function. The multilevel model has been applied for A/D conversion, and a circuit implementation for the neural A/D converter has been fabricated [67]. For an activation function of N levels, sufficient conditions are given in [42] ensuring that the networks have $2N + 1$ equilibriums with $N + 1$ of them as locally stable points, as well as some criteria guaranteeing global and complete stability of the Hopfield networks with multilevel activation functions by using the continuity property of neuron state variables and Lyapunov functional.

The use of multistate neurons leads to a network architecture that is significantly smaller than that of the Hopfield network, and hence, a simple hardware implementation. The reduction in the network size is highly desirable in large-scale applications such as image restoration and TSP. In addition, the complex-valued multistate Hopfield network is also more efficient and convenient than the Hopfield network in the manipulation of complex-valued signals. Storage capacity can be improved by using complex-valued multistate Hopfield networks [32, 46], or replacing bilevel activation functions with multilevel ones [42, 69].

The complex-valued multistate Hopfield network [32, 46] employs the multivalued complex-signum activation function that is defined as an L-stage phase quantizer for complex numbers

$$\mathrm{csign}(u) \widehat{=} \begin{cases} z^0, & \arg(u) \in [0, \varphi_0) \\ z^1, & \arg(u) \in [\varphi_0, 2\varphi_0) \\ \vdots & \\ z^{L-1}, & \arg(u) \in [(L-1)\varphi_0, L\varphi_0) \end{cases}, \tag{7.36}$$

where $z = \mathrm{e}^{j\varphi_0}$ with $\varphi_0 = \frac{2\pi}{L}$ is the Lth root of unity. Each state takes one of the equally spaced L points on the unit circle of the complex plane.

Similar to the Hopfield network, the network dynamics are defined by

$$net_i(t) = \sum_{k=1}^{J} w_{ki} x_k(t), \quad i = 1, \ldots, J, \tag{7.37}$$

$$x_i(t+1) = \mathrm{csign}\left(net_i(t) \cdot z^{\frac{1}{2}}\right), \quad i = 1, \ldots, J, \tag{7.38}$$

where J is the number of neurons, and the factor $z^{\frac{1}{2}} = \mathrm{e}^{j\frac{\varphi_0}{2}}$ places the resulting states in the angular centers of each sector. A sufficient condition for the stability of the dynamics is that the weight matrix is Hermitian with nonnegative diagonal entries, that is, $\mathbf{W} = \mathbf{W}^H$, $w_{ii} \geq 0$ [32]. The energy can be defined as

$$E(\boldsymbol{x}) = -\frac{1}{2}\boldsymbol{x}^H \mathbf{W} \boldsymbol{x}. \tag{7.39}$$

7.7 Cellular Neural Networks

A cellular network is a two- or higher dimensional array of cells, with only local interactions that can be programmed by a template matrix [13]. It is made of a massive aggregate of regularly spaced circuit clones, called *cells*, which communicate with one another directly only through its nearest neighbors. Each cell is made of a linear capacitor, a nonlinear voltage-controlled current source, and a few resistive linear circuit elements. Thus, each cell has its own dynamics whose evolution is dependent on its circuit time constant $\tau = RC$.

The cellular network is a generalization of the Hopfield network, and can be used to solve a more generalized optimization problem. It overcomes the massive interconnection problem of parallel distributed processing. The key features are asynchronous parallel processing, continuous-time dynamics, local interactions among the network elements, and VLSI implementation.

Due to its local interconnectivity property, cellular network chips can have high-density cells, and some physical implementations, such as analog CMOS, emulated digital CMOS and optical implementations, are available. The cellular network universal machine [52] is the analog cellular computer for processing analog array signals. It has a computational power of tera (10^{12}) or peta (10^{15}) analog operations per second on a single CMOS chip [14].

The cellular network with a two-dimensional array architecture is a natural candidate for image processing or simulation of partial differential equations. Any input–output function to be realized by the cellular network can also be visualized as an image processing task, where the external input, the initial condition, and the output, arranged as two-dimensional arrays, are, respectively, the external input, initial and output images. The external input image together with the initial image constitutes the input images of the cellular network. Using different cloning templates, namely, the representation of the local interconnection patterns, different operations can be conducted on an image. The cellular network has become an important method for image processing, and has been applied to many signal processing applications such as image compression, filtering, learning, and pattern recognition.

The image processing tasks using cellular networks were mainly developed for black and white output images, since a cell in the cellular network has a stable equilibrium point at the two saturation regions of the piecewise linear output function after the transient has decayed toward equilibrium. Due to the cellular network characteristics, each pixel of an image corresponds to a cell of a cellular network. In contrast, the nonlinear dynamics of a cellular network with a two-level output function converts the input images into bilevel pulse digital sequences. Cellular network can convert the multi-bit image into an optimal binary halftone image. This significant characteristic of a cellular network suggests the possibility of a spatial domain sigma-delta modulation [5]. The proposed system can be treated as a very large-scale and super-parallel sigma-delta modulator.

Problems

7.1 Form the energy function of the Hopfield network, and show that if $\boldsymbol{x}^*$ is a local minimum of E, then $-\boldsymbol{x}^*$ is also a local minimum of E. This explains the reason for negative fundamental memories.

7.2 Lyapunov's second theorem is usually used to verify the stability of a linear system $\dot{\boldsymbol{x}} = \mathbf{A}\boldsymbol{x}$. The Lyapunov function can be selected as $E(\boldsymbol{x}) = \boldsymbol{x}^T\mathbf{Q}\boldsymbol{x}$, where $\mathbf{Q}$ can be obtained from $\mathbf{A}^T\mathbf{Q} + \mathbf{Q}\mathbf{A} = -\mathbf{I}$, $\mathbf{I}$ being the identity matrix. Verify the stability of the linear system with $\mathbf{A} = \begin{bmatrix} 0 & -1 \\ -2 & +1 \end{bmatrix}$.

7.3 For the assignment problem given by (7.23)–(7.26), assume $m = n$, $a_{ij} = 1$, $b_i = 1$. The energy function can be constructed as

$$E(t, \mathbf{x}(t)) = \frac{1}{2}\left\{\sum_{i=1}^{n}\left(\sum_{j=1}^{n} x_{ij}(t) - 1\right)^2 + \sum_{j=1}^{n}\left(\sum_{i=1}^{n} x_{ij}(t) - 1\right)^2\right\}$$
$$+ \sum_{i=1}^{n}\sum_{j=1}^{n} x_{ij}\left(1 - x_{ij}\right) + \alpha e^{-t/\tau}\sum_{i=1}^{n}\sum_{j=1}^{n} c_{ij}x_{ij}(t),$$

where α and τ are positive constants. $\alpha e^{-t/\tau}$ is an annealing factor to gradually balance the cost and constraints.

(a) Design a Hopfield network to solve this problem.

(b) What is the optimal solution?

7.4 Write a program to solve a TSP of N cities using a Hopfield network. The objective is to minimize the total length of a tour, L_P, where P is a permutation of the N cities.

7.5 Use simulated annealing to solve TSP of N cities.

7.6 A typical chaotic function is the Mackey–Glass chaotic time series is generated from the following delay differential equation:

$$dx(t)/dt = [0.2x(t - D)]/[1 + x^{10}(t - D)] - 0.1x(t),$$

where D is a delay.

(a) Plot the time series for different D values.

(b) Plot the bifurcation diagram.

(c) Verify that for $D > 17$ the function shows a chaotic behavior.

7.7 For Aihara's chaotic neural network, the neural dynamics of a single neuron is given by [10]:

$$x(t+1) = \frac{1}{1 + \exp\left(-\frac{1}{\epsilon}\left[k\epsilon \ln \frac{x(t)}{1-x(t)} + \omega x(t) - \omega a_0 + I\right]\right)},$$

where $x(t)$ is output of the neural oscillator, ϵ is the steepness parameter, k is the damping factor and I is the external stimulus. By using $\epsilon = 1/250$, $k = 0.9$, $a_0 = 0.5$ and $\omega = -0.07$ [10], plot the bifurcation diagram (x versus I) of Aihara's cellular network (Hint: Refer to [37]).

7.8 Solve TSP using:

(a) The Hopfield network.

(b) Chen and Aihara's method [9] (Hint: Refer to [63]).

7.9 A 3-bit A/D converter can convert a continuous analog signal $y(t) \in [0, 7]$ into 3-bit representation (x_2, x_1, x_0), where x_0 is the least significant bit. The quadratic objective function can be given by

$$J(x_0, x_1, x_2) = \frac{1}{2}\left(y - \sum_{i=0}^{2} 2^i x_i\right)^2$$

subject to $x_i \in \{0, 1\}$, or $x_i(1 - x_i) = 0$, $i = 0, 1, 2$.
We need to define a Lagrangian that matches the energy function of the Hopfield network. This can be achieved by setting the value of Lagrange multipliers λ_i, $i = 0, 1, 2$, and omitting constants that are not associated with x_is.
(a) Solve for the weights and inputs of a Hopfield network.
(b) Give the network dynamics for solving this problem.

7.10 Generalize the design in Problem 1.8 to an n-bit A/D converter. Show that the weights and inputs can be set by $w_{ij} = -2^{i+j}$ when $i \neq j$, or 0 when $i = j$, and $I_i = -2^{2i-1} + 2^i x$, $i, j = 0, 1, \ldots, n-1$. According to Tank and Hopfield [58], operation of this Hopfield A/D converter for every new analog input x requires reintialization of all states to zero.

7.11 A saddle node is an illustrative bifurcation. For $\frac{dz}{dt} = a + z^2$, investigate the fixed points for $a < 0$, $a = 0$, and $a > 0$. Plot z versus a, using solid lines to denotes fixed-point attractors and dashed lines for fixed-point repellors.

References

1. Aarts, E., & Korst, J. (1989). *Simulated annealing and Boltzmann machines*. Chichester: John Wiley.
2. Abe, S., Kawakami, J., & Hirasawa, K. (1992). Solving inequality constrained combinatorial optimization problems by the Hopfield neural networks. *Neural Networks*, *5*, 663–670.
3. Adachi, M., & Aihara, K. (1997). Associative dynamics in a chaotic neural network. *Neural Networks*, *10*(1), 83–98.
4. Aihara, K., Takabe, T., & Toyoda, M. (1990). Chaotic neural networks. *Physics Letters A*, *144*, 333–340.
5. Aomori, H., Otake, T., Takahashi, N., & Tanaka, M. (2008). Sigma-delta cellular neural network for 2D modulation. *Neural Networks*, *21*, 349–357.
6. Azencott, R. (1992). *Simulated annealing: Parallelization techniques*. New York: Wiley.
7. Bruck, J. (1990). On the convergence properties of the Hopfield model. *Proceedings of the IEEE*, *78*(10), 1579–1585.
8. Chakraborty, K., Mehrotra, K. G., Mohan, C. K., & Ranka, S. (1992). An optimization network for solving a set of simultaneous linear equations. In *Proceedings of the International Joint Conference on Neural Networks* (Vol. 2, pp. 516–521). Baltimore, MD.
9. Chen, L., & Aihara, K. (1995). Chaotic simulated annealing by a neural-network model with transient chaos. *Neural Networks*, *8*(6), 915–930.
10. Chen, L., & Aihara, K. (1997). Chaos and asymptotical stability in discrete-time neural networks. *Physica D*, *104*, 286–325.
11. Chen, L., & Aihara, K. (1999). Global searching ability of chaotic neural networks. *IEEE Transactions on Circuits and Systems I*, *48*(8), 974–993.
12. Chiueh, T. D., & Goodman, R. M. (1991). Recurrent correlation associative memories. *IEEE Transactions on Neural Networks*, *2*(2), 275–284.
13. Chua, L. O., & Yang, L. (1988). Cellular neural network—Part I: Theory; Part II: Applications. *IEEE Transactions on Circuits and Systems*, *35*, 1257–1290.

14. Chua, L. O., & Roska, T. (2002). *Cellular neural network and visual computing—Foundation and applications*. Cambridge, UK: Cambridge University Press.
15. Cichocki, A., & Unbehauen, R. (1993). *Neural networks for optimization and signal processing*. New York: Wiley.
16. Cohen, M. A., & Grossberg, S. (1983). Absolute stability of global pattern formation and parallel memory storage by competitive neural networks. *IEEE Transactions on Systems, Man, and Cybernetics*, *13*, 815–826.
17. Culhane, A. D., Peckerar, M. C., & Marrian, C. R. K. (1989). A neural net approach to discrete Hartley and Fourier transforms. *IEEE Transactions on Circuits and Systems*, *36*(5), 695–702.
18. Czech, Z. J. (2001). Three parallel algorithms for simulated annealing. In: R. Wyrzykowski, J. Dongarra, M. Paprzycki, & J. Waniewski (Eds.), *Proceedings of the 4th International Conference on Parallel Processing and Applied Mathematics, LNCS* (Vol. 2328, pp. 210–217). Naczow, Poland; London: Springer.
19. Dang, C., & Xu, L. (2001). A Lagrange multiplier and Hopfield-type barrier function method for the traveling salesman problem. *Neural Computation*, *14*, 303–324.
20. Demirciler, K., & Ortega, A. (2005). Reduced-complexity deterministic annealing for vector quantizer design. *EURASIP Journal on Applied Signal Processing*, *2005*(12), 1807–1820.
21. Fleisher, M. (1988). The Hopfield model with multi-level neurons. In D. Z. Anderson (Ed.), *Neural information processing systems* (pp. 278–289). New York: American Institute Physics.
22. Gao, K., Ahmad, M. O., & Swamy, M. N. S. (1990). A neural network least-square estimator. In *Proceedings of the International Joint Conference on Neural Networks* (Vol. 3, pp. 805–810). Washington, DC.
23. Gao, X.-B., & Liao, L.-Z. (2006). A novel neural network for a class of convex quadratic minimax problems. *Neural Computation*, *18*, 1818–1846.
24. Geman, S., & Geman, D. (1984). Stochastic relaxation, Gibbs distributions, and the Bayesian restoration of images. *IEEE Transactions on Pattern Analysis and Machine Intelligence*, *6*, 721–741.
25. Hajek, B. (1988). Cooling schedules for optimal annealing. *Mathematical Operations Research*, *13*(2), 311–329.
26. He, Y. (2002). Chaotic simulated annealing with decaying chaotic noise. *IEEE Transactions on Neural Networks*, *13*(6), 1526–1531.
27. Hopfield, J. J. (1982). Neural networks and physical systems with emergent collective computational abilities. *Proceedings of the National Academy of Sciences of the United States of America*, *79*, 2554–2558.
28. Hopfield, J. J. (1984). Neurons with graded response have collective computational properties like those of two-state neurons. *Proceedings of the National Academy of Sciences of the United States of America*, *81*, 3088–3092.
29. Hopfield, J. J., & Tank, D. W. (1985). Neural computation of decisions in optimization problems. *Biological Cybernetics*, *52*, 141–152.
30. Ingber, L. (1989). Very fast simulated re-annealing. *Mathematical and Computer Modelling*, *12*(8), 967–973.
31. Jang, J. S., Lee, S. Y., & Shin, S. Y. (1988). An optimization network for matrix inversion. In D. Z. Anderson (Ed.), *Neural information processing systems* (pp. 397–401). New York: American Institute Physics.
32. Jankowski, S., Lozowski, A., & Zurada, J. M. (1996). Complex-valued multi-state neural associative memory. *IEEE Transactions on Neural Networks*, *7*(6), 1491–1496.
33. Kennedy, M. P., & Chua, L. O. (1988). Neural networks for nonlinear programming. *IEEE Transactions on Circuits and Systems*, *35*, 554–562.
34. Kirkpatrick, S., Gelatt, C. D, Jr., & Vecchi, M. P. (1983). Optimization by simulated annealing. *Science*, *220*, 671–680.
35. Kwok, T., & Smith, K. A. (1999). A unified framework for chaotic neural-network approaches to combinatorial optimization. *IEEE Transactions on Neural Networks*, *10*(4), 978–981.
36. Lee, B. W., & Shen, B. J. (1992). Design and analysis of analog VLSI neural networks. In B. Kosko (Ed.), *Neural networks for signal processing* (pp. 229–284). Englewood Cliffs, NJ: Prentice-Hall.

37. Lee, R. S. T. (2006). Lee-Associator: A chaotic auto-associative network for progressive memory recalling. *Neural Networks*, *19*, 644–666.
38. Le Gall, A., & Zissimopoulos, V. (1999). Extended Hopfield models for combinatorial optimization. *IEEE Transactions on Neural Networks*, *10*(1), 72–80.
39. Lendaris, G. G., Mathia, K., & Saeks, R. (1999). Linear Hopfield networks and constrained optimization. *IEEE Transactions on Systems, Man, and Cybernetics, Part B*, *29*(1), 114–118.
40. Li, J. H., Michel, A. N., & Parod, W. (1989). Analysis and synthesis of a class of neural networks: Linear systems operating on a closed hypercube. *IEEE Transactions on Circuits and Systems*, *36*(11), 1405–1422.
41. Little, W. A. (1974). The existence of persistent states in the brain. *Mathematical Biosciences*, *19*, 101–120.
42. Liu, Y., & You, Z. (2008). Stability analysis for the generalized Hopfield neural networks with multi-level activation functions. *Neurocomputing*, *71*, 3595–3601.
43. Locatelli, M. (2001). Convergence and first hitting time of simulated annealing algorithms for continuous global optimization. *Mathematical Methods of Operations Research*, *54*, 171–199.
44. Matsuda, S. (1998). "Optimal" Hopfield network for combinatorial optimization with linear cost function. *IEEE Transactions on Neural Networks*, *9*(6), 1319–1330.
45. Metropolis, N., Rosenbluth, A., Rosenbluth, M., Teller, A., & Teller, E. (1953). Equations of state calculations by fast computing machines. *Journal of Chemical Physics*, *21*(6), 1087–1092.
46. Muezzinoglu, M. K., Guzelis, C., & Zurada, J. M. (2003). A new design method for the complex-valued multistate Hopfield associative memory. *IEEE Transactions on Neural Networks*, *14*(4), 891–899.
47. Nam, D. K., & Park, C. H. (2000). Multiobjective simulated annealing: A comparative study to evolutionary algorithms. *International Journal of Fuzzy Systems*, *2*(2), 87–97.
48. Nozawa, H. (1992). A neural network model as a globally coupled map and applications based on chaos. *Chaos*, *2*(3), 377–386.
49. Richardt, J., Karl, F., & Muller, C. (1998). Connections between fuzzy theory, simulated annealing, and convex duality. *Fuzzy Sets and Systems*, *96*, 307–334.
50. Rose, K., Gurewitz, E., & Fox, G. C. (1990). A deterministic annealing approach to clustering. *Pattern Recognition Letters*, *11*(9), 589–594.
51. Rose, K. (1998). Deterministic annealing for clustering, compression, classification, regression, and related optimization problems. *Proceedings of the IEEE*, *86*(11), 2210–2239.
52. Roska, T., & Chua, L. O. (1993). The CNN universal machine: An analogic array computer. *IEEE Transactions on Circuits and Systems II*, *40*(3), 163–173.
53. Si, J., & Michel, A. N. (1995). Analysis and synthesis of a class of discrete-time neural networks with multilevel threshold neurons. *IEEE Transactions on Neural Networks*, *6*(1), 105–116.
54. Sima, J., Orponen, P., & Antti-Poika, T. (2000). On the computational complexity of binary and analog symmetric Hopfield nets. *Neural Computation*, *12*, 2965–2989.
55. Sima, J., & Orponen, P. (2003). Continuous-time symmetric Hopfield nets are computationally universal. *Neural Computation*, *15*, 693–733.
56. Smith, K. I., Everson, R. M., Fieldsend, J. E., Murphy, C., & Misra, R. (2008). Dominance-based multiobjective simulated annealing. *IEEE Transactions on Evolutionary Computation*, *12*(3), 323–342.
57. Szu, H. H., & Hartley, R. L. (1987). Nonconvex optimization by fast simulated annealing. *Proceedings of the IEEE*, *75*, 1538–1540.
58. Tank, D. W., & Hopfield, J. J. (1986). Simple "neural" optimization networks: An A/D converter, signal decision circuit, and a linear programming circuit. *IEEE Transactions on Circuits and Systems*, *33*, 533–541.
59. Thompson, D. R., & Bilbro, G. L. (2005). Sample-sort simulated annealing. *IEEE Transactions on Systems, Man, and Cybernetics, Part B*, *35*(3), 625–632.
60. Tsallis, C., & Stariolo, D. A. (1996). Generalized simulated annealing. *Physica A*, *233*, 395–406.
61. Wang, J., & Li, H. (1994). Solving simultaneous linear equations using recurrent neural networks. *Information Sciences*, *76*, 255–277.

62. Wang, L., & Smith, K. (1998). On chaotic simulated annealing. *IEEE Transactions on Neural Networks*, *9*, 716–718.
63. Wang, L., Li, S., Tian, F., & Fu, X. (2004). A noisy chaotic neural network for solving combinatorial optimization problems: Stochastic chaotic simulated annealing. *IEEE Transactions on Systems, Man, and Cybernetics, Part B*, *34*(5), 2119–2125.
64. Wang, R. L., Tang, Z., & Cao, Q. P. (2002). A learning method in Hopfield neural network for combinatorial optimization problem. *Neurocomputing*, *48*, 1021–1024.
65. Wang, X. (1991). Period-doublings to chaos in a simple neural network: An analytic proof. *Complex Systems*, *5*, 425–441.
66. Yan, H. (1991). Stability and relaxation time of Tank and Hopfield's neural network for solving LSE problems. *IEEE Transactions on Circuits and Systems*, *38*(9), 1108–1110.
67. Yuh, J. D., & Newcomb, R. W. (1993). A multilevel neural network for A/D conversion. *IEEE Transactions on Neural Networks*, *4*(3), 470–483.
68. Zhang, S., & Constantinides, A. G. (1992). Lagrange programming neural networks. *IEEE Transactions on Circuits and Systems II*, *39*(7), 441–452.
69. Zurada, J. M., Cloete, I., & van der Poel, E. (1996). Generalized Hopfield networks for associative memories with multi-valued stable states. *Neurocomputing*, *13*, 135–149.

Chapter 8
Associative Memory Networks

8.1 Introduction

Mammals have lifelong learning ability. Brain regions such as hippocampus and neocortex are thought to operate as associative memories. Complementary learning systems theory [46, 57] holds that the hippocampal system exhibits short-term adaptation for rapid learning of novel information, which will be played back over time to the neocortical system for long-term retention. This allows for effectively generalizing across experiences while retaining specific memories in a lifelong manner. The interplay of an episodic memory (specific experience) and a semantic memory (general structured knowledge) enlightens the mechanisms of knowledge consolidation in the absence of sensory input.

The human brain stores the information in synapses or in reverberating loops of electrical activity. Most of the existing associative memory models store information in synapses. However, loop-based algorithms can learn complex control tasks faster, with exponentially fewer neurons, and avoid the problem of weight transport, but with long feedback delays [28]. They explain aspects of consolidation, the role of attention, and the relapses.

Association is a salient feature of human memory. The brain recalls by association, that is, the brain associates the recalled item with a piece of information or with another item. Associative memory models, known as *content-addressable memories*, are well analyzed. A memory is a system with three functions or stages: recording—storing the information, preservation—keeping the information safely, and recalling–retrieving the information. A pattern can be stored in memory through a learning process. For an imperfect input pattern, associative memory has the capability to recall the stored pattern correctly by performing a collective relaxation search. Associative memories can be either heteroassociative or autoassociative. For heteroassociation, the input and output vectors range over different vector spaces, while for autoassociation, both the input and output vectors range over the same vector space. Neural associative memories have applications in different fields, such as image processing, pattern recognition, and optimization.

K.-L. Du and M. N. S. Swamy, *Neural Networks and Statistical Learning*,
https://doi.org/10.1007/978-1-4471-7452-3_8

Episodic memory allows one to remember his own experiences in an explicit and conscious manner. Episodic memory is crucial in supporting many cognitive capabilities, including concept formation, representation of events in spatiotemporal dimension, and record of progress in goal processing [20]. Two basic elements of episodic memory are events and episodes. An event can be described as a snapshot of experience. Usually, a remembered event can be used to answer critical questions such as what, where, and when. An episode can be considered as a temporal sequence of events. Three major tasks in episodic memory retrieval are event detection, episode recognition, and episode recall. Forgetting should exist in memory to avoid information overflow.

Recognition memory is involved with two types of retrieval processes: familiarity and recollection. When presented with an item, one might have a sense of recognition but cannot recall the detail of the stimulus encountered before. This is called familiarity memory. Familiarity distinguishes whether the stimulus was previously encountered. The medial temporal lobe and the prefrontal cortex play a critical role in familiarity memory [23]. Recollection retrieves detailed information about an experienced event. Familiarity capacity is typically proportional to the number of synapses within the network [11], whereas the capacity for recollection is typically proportional to the square root of the number of synapses, that is, the number of neurons in a fully connected network [6]. Mean-field analysis indicates that the capacity of the familiarity discriminators that are based on a neural network model is bigger than that of recollection capacity [23].

Research on neural associative memories originated in the 1950s with matrix associative memories [77]. In 1972, the linear associative memory was introduced, independently, by several authors, where correlation or Hebbian learning is used to synthesize the synaptic weight matrix [3, 8, 42]. The brain-state-in-a-box (BSB) network is a discrete-time nonlinear dynamical system as a memory model based on neurophysiological considerations. The Hopfield model [35] is a continuous-time, continuous-state dynamic associative memory model. The binary Hopfield network is a well-known model for nonlinear associative memories. It can retrieve a pattern stored in memory in response to the presentation of a corrupted version of the pattern. This is done by mapping a fundamental memory $\boldsymbol{x}$ onto a stable point of a dynamical system. Kosko [44] extended the Hopfield associative memory to bidirectional associative memory (BAM) by incorporating an additional layer to perform recurrent autoassociation or heteroassociation.

Linear associative memories, BSB [9] and BAM [10, 44], can be used as both autoassociative and heteroassociative memories, while the Hopfield model, the Hamming network [51], and the Boltzmann machine can only be used as autoassociative models. Perfect recall can be guaranteed by imposing an orthogonality condition on the stored patterns. The optimal linear associative memory, which employs the projection recording recipe, is not subject to this constraint [43]. The optimal linear associative memory, though exhibits a better storage capacity than the linear associative memory model, has low noise tolerance.

An autoassociator is a brain-like distributed network that learns from the samples in a category to reproduce each sample at the output with a mapping. An autoassociator learns normal examples; when an unknown pattern is presented, the reconstruction error by the autoassociator will be compared with a threshold to signal whether it is a novel pattern (with larger error) or a normal pattern (with smaller error). This classification methodology has been applied to various detection problems such as face detection, network security, and natural language grammar learning.

Memory is important for transforming a static network into a dynamic one. Memories can be long-term or short-term. A long-term memory is used to store stable system information, while a short-term memory is useful for simulating a dynamic system with a temporal dimension. For a Hopfield network, the states of the neurons can be considered as short-term memories while the synaptic weights can be treated as long-term memories. Feedforward networks can become dynamic by embedding memory into the network using time delay. Recurrent network models such as the Hopfield model and the Boltzmann machine are popular associative memory models.

Other associative memories' models with unlimited storage capacity include the morphological associative memory [68, 69]. Morphological associative memories employ the min and max operations that are used in mathematical morphology. The morphological models are very efficient to recall patterns corrupted either with additive noise or subtractive noise.

8.2 Hopfield Model: Storage and Retrieval

Operation of the Hopfield network as an associative memory includes two phases: storage and retrieval. Bipolar coding is often used for associative memory in that bipolar vectors have a greater probability of being orthogonal than binary vectors. We use bipolar coding in this chapter.

We now store in the network a set of N bipolar patterns, $\{\boldsymbol{x}_p\}$, where $\boldsymbol{x}_p = (x_{p,1}, x_{p,2}, \ldots, x_{p,J})^T$, $x_{p,i} = \pm 1$. These patterns are called *fundamental memories*. Storage is implemented by using a learning algorithm, while retrieval is based on the dynamics of the network.

8.2.1 Generalized Hebbian Rule

Conventional algorithms for associative storage are typically local algorithms based on the Hebbian rule. Hebbian rule is known as the *outer product rule of storage* in connection with associative learning. Using this method, $\boldsymbol{\theta}$ is chosen as the zero vector. A generalized Hebbian rule for training the Hopfield network is defined by [35]

$$w_{ij} = \frac{1}{J} \sum_{p=1}^{N} x_{p,i} x_{p,j}, \quad \text{for all } i \neq j, \tag{8.1}$$

and $w_{ii} = 0$. In matrix form

$$\mathbf{W} = \frac{1}{J}\left(\sum_{p=1}^{N} \boldsymbol{x}_p \boldsymbol{x}_p^T - N\mathbf{I}_J\right), \tag{8.2}$$

where $\mathbf{I}_J$ denotes the $J \times J$ identity matrix.

The generalized Hebbian rule can be written in an incremental form

$$w_{ij}(t) = w_{ij}(t-1) + \eta x_{t,i} x_{t,j}, \quad \text{for all } i \neq j, \tag{8.3}$$

where the step size $\eta = \frac{1}{J}$, $t = 1, \ldots, N$, and $w_{ij}(0) = 0$. As such, learning is completed after each pattern $\boldsymbol{x}_t$ in the pattern set is presented exactly once.

The generalized Hebbian rule is both local and incremental. It has an absolute storage capability of $N_{\max} = \frac{J}{2 \ln J}$ [58]. The storage capability of an associative memory network is defined by the maximum number of fundamental memories, $N_{\max}$, that can be stored and retrieved reliably. For reliable retrieval, $N_{\max}$ is dropped to approximately $\frac{J}{4 \ln J}$ [58]. Generalized Hebbian rule, however, suffers severe degradation and $N_{\max}$ decreases significantly, if the training patterns are correlated. For example, time series usually include significant correlations in the measurements of adjacent samples. Some variants of the Hebbian rule, such as the weighted Hebbian rule [3] and the Hebbian rule with decay [43], can increase the storage capability.

When training associative memory networks using classical Hebbian learning, an additional term called *crosstalk* may arise. When crosstalk becomes too large, spurious states other than the negative stored patterns appear [70]. The number of negative stored patterns is always equivalent to the number of stored patterns. Hebbian learning produces good results when the stored patterns are nearly orthogonal. This is the case when N bipolar vectors are randomly selected from R^J and $N \ll J$. In practice, patterns are usually correlated and the incurred crosstalk may reduce the capacity of the network. The storage capability of the network is expected to decrease if the Hamming distance between the fundamental memories becomes smaller.

An improved Hebbian rule is given by local and incremental learning rule [73, 74]

$$w_{ij}(t) = w_{ij}(t-1) + \eta \left[x_{t,i} x_{t,j} - h_{ji}(t) x_{t,i} - h_{ij}(t) x_{t,j}\right], \tag{8.4}$$

$$h_{ij}(t) = \sum_{u=1, u \neq i, j}^{J} w_{iu}(t-1) x_{t,u}, \tag{8.5}$$

where $\eta = \frac{1}{J}$, $t = 1, 2, \ldots, N$, $w_{ij}(0) = 0$ for all i and j, and h_{ij} is a form of local field at neuron i. This rule has an absolute capacity of $\frac{J}{\sqrt{2 \ln J}}$ for uncorrelated patterns. It also performs better than the generalized Hebbian rule for correlated patterns [73]. It does not suffer significant capacity loss when patterns with medium correlation are stored.

8.2.2 *Pseudoinverse Rule*

The pseudoinverse solution targets at minimizing the crosstalk between the stored patterns. The pseudoinverse rule uses the pseudoinverse of the pattern matrix, while classical Hebbian learning uses the correlation matrix of the patterns [40, 67].

Denoting $\mathbf{X} = [\boldsymbol{x}_1, \boldsymbol{x}_2, \ldots, \boldsymbol{x}_N]$, the autoassociative memory is defined as

$$\mathbf{X}^T\mathbf{W} = \mathbf{X}^T. \tag{8.6}$$

Using pseudoinverse, we actually minimize $E = \left\|\mathbf{X}^T\mathbf{W} - \mathbf{X}^T\right\|_F$, thus minimizing the crosstalk in the associative network. The pseudoinverse solution for the weight matrix is given by

$$\mathbf{W} = \left(\mathbf{X}^T\right)^{\dagger}\mathbf{X}^T. \tag{8.7}$$

The pseudoinverse rule, also called the *projection learning rule*, is neither incremental nor local. It involves inverting an $N \times N$ matrix, thus training is very slow and impractical.

The pseudoinverse solution performs better than Hebbian learning when the patterns are correlated. Both the Hebbian and pseudoinverse rules are general-purpose methods for training associative memory networks that can be represented as $\mathbf{X}^T\mathbf{W} = \overline{\mathbf{X}}^T$, where $\mathbf{X}$ and $\overline{\mathbf{X}}$ are, respectively, the stored and associated pattern matrices. For an autoassociated pattern $\boldsymbol{x}_i$, the weights generated from Hebbian learning project the whole input space into the linear subspace spanned by $\boldsymbol{x}_i$. The projection, however, is not orthogonal. Instead, the pseudoinverse solution provides orthogonal projection to the linear subspace spanned by the stored patterns [70]. Theoretically, for $N < J$ and uncorrelated patterns, the pseudoinverse solution has a zero error, and the storage capability in this case is $N_{\max} = J - 1$ [40, 70]. It is shown in [74] that the Hebbian rule is the zeroth-order expansion of the pseudoinverse rule, and the improved Hebbian rule given by (8.4) and (8.5) is one form of the first-order expansion of the pseudoinverse rule.

The pseudoinverse rule is also adapted to sorting sequences of prototypes, where an input $\boldsymbol{x}_i$ leads to an output $\boldsymbol{x}_{i+1}$. The MLP with BP can be used to compute the pseudoinverse solution when the dimension J is large, since direct methods to solve the pseudoinverse will use up the memory and the convergence time is intolerably large [40].

8.2.3 *Perceptron-Type Learning Rule*

The rules addressed above are *one-shot* methods, in which the network training is completed in a single epoch. A learning problem in a Hopfield network with J units can be transformed into a learning problem for a perceptron of dimension

$\frac{J(J+1)}{2}$ [70]. This equivalence between Hopfield networks and perceptrons leads to the conclusion that every learning algorithm for perceptrons can be transformed into a learning algorithm for Hopfield networks.

Perceptron learning algorithms for storing bipolar patterns in Hopfield networks have been discussed in [40]. They are simple, online, local algorithms. Unlike Hebbian rule-based algorithms, perceptron learning-based algorithms work over multiple epochs and often reduce the error nonmonotonically over the epochs. The perceptron-type learning rule is given by [38, 40]

$$w_{ij}(t) = w_{ij}(t-1) + \eta \left[x_{t,i} x_{t,j} - \frac{1}{2} \left(y_{t,i} x_{t,j} + y_{t,j} x_{t,i} \right) \right], \tag{8.8}$$

where $\boldsymbol{y}_t = \text{sgn}\,(\mathbf{W}\boldsymbol{x}_t), t = 1, \ldots, N$, the learning rate $\eta > 0$, and $w_{ji}(0)$s are small random numbers or zero. In this chapter, the signum function $\text{sgn}(x)$ is defined as 1 for $x \geq 0$ and -1 for $x < 0$. If selecting $w_{ji}(0) = 0$ for all i, j, η can be selected as any positive number; otherwise, η can be selected as a number of the same order of magnitude or larger than the weights. This accelerates the convergence process. Notice that $w_{ji}(t) = w_{ij}(t)$.

However, when the signum vector is not realizable, the perceptron-type rule does not converge but oscillates indefinitely. The perceptron-type rule can be viewed as a supervised extension of the Hebbian rule by incorporating a term for correcting unstable bits. For a recurrent network, the storage capability of the perceptron-type algorithm can reach the upper bound $N_{\max} = J$, for uncorrelated patterns.

An extensive experimental comparison between a perceptron-type learning rule [40] and the generalized Hebbian rule has been made in [38] on a wide range of conditions on the library patterns: the number of patterns N, the pattern density p, and the amount of correlation of the bits in a pattern, decided by block size B. In terms of stability of the library patterns and error correction ability during the recall phase, the perceptron-type rule is found to be perfect in ensuring stability of the stored library patterns under all the evaluated conditions, while the generalized Hebbian rule degrades rapidly as N is increased, or p is decreased, or B is increased. In many cases, the perceptron-type rule works much better than the generalized Hebbian rule in correcting pattern errors.

8.2.4 Retrieval Stage

After the bipolar words have been stored, the network can be used for information retrieval. When a J-dimensional vector (bipolar word) $\boldsymbol{x}$, representing a corrupted or incomplete memory of the network, is presented to the network as its state, information retrieval is performed automatically according to the network dynamics given by (7.1) and (7.2), or (7.3) and (7.4). For hard-limiting activation function, the discrete form of the network dynamics can be written as

$$x_i(t+1) = \text{sgn}\left(\sum_{j=1}^{J} w_{ji} x_j(t) + \theta_i\right), \quad i = 1, 2, \ldots, J, \tag{8.9}$$

or in matrix form

$$\boldsymbol{x}(t+1) = \text{sgn}(\mathbf{W}\boldsymbol{x}(t) + \boldsymbol{\theta}), \tag{8.10}$$

where $\boldsymbol{x}(0)$ is the input corrupted memory, and $\boldsymbol{x}(t)$ represents the retrieved memory at time t. The retrieval process continues until the state vector $\boldsymbol{x}$ remains unchanged. The convergent $\boldsymbol{x}$ is a fixed point or the retrieved memory.

Models such as the complex-valued Hopfield network are not as tolerant with respect to incomplete patterns and salt/pepper noise. However, they perform better in the presence of Gaussian noise. An essential feature of the noise acting on a pattern is its local nature. If a pattern is split into enough sub-patterns, a few of them will be less or more affected by noise, and others will remain intact. A simple but effective methodology exploits this fact for efficient restoration of a pattern [24]. A pattern is restored if enough of its sub-patterns are restored. Since several patterns can share the same sub-patterns, the final decision is accomplished by means of a voting mechanism. Before deciding if a sub-pattern belongs to a pattern, sub-pattern restoration in the presence of noise is done by an associative memory.

8.3 Storage Capability of Hopfield Model

In Sect. 8.2, the storage capability for each of the four storage algorithms is given. In practice, there are some upper bounds on the storage capability of general recurrent networks. An upper bound on the storage capability of a class of recurrent networks with zero-diagonal weight matrix is derived deterministically in [1].

Theorem 8.1 (Upper bound [1]) *For any subset of N binary J-vectors, in order to find a corresponding zero-diagonal weight matrix* $\mathbf{W}$ *and a bias vector* $\boldsymbol{\theta}$ *such that these vectors are fixed points of the network*

$$\boldsymbol{x}_i = \text{sgn}(\mathbf{W}\boldsymbol{x}_i + \boldsymbol{\theta}), \quad i = 1, 2, \ldots, N, \tag{8.11}$$

one needs to have $N \le J$.

Thus, the upper bound on the storage capability is $N_{\max} = J$. This bound is valid for any learning algorithm for recurrent networks with a zero-diagonal weight matrix. The Hopfield network, having a symmetric zero-diagonal weight matrix, is one such network, and as a result, the Hopfield network can at most stably store J patterns.

The upper bound introduced in Theorem 8.1 is too tight, since it requires that all the N-tuple subsets of bipolar J-vectors are retrievable. It is also noted in [79] that any two patterns differing in precisely one component cannot be jointly stored as stable states in the Hopfield network.

An in-depth analysis of the Hopfield model's storage capacity has been done by [5] by relying on a mean-field approach and on replica methods originally developed for spin-glass models. Hopfield networks, when coupled with this learning rule, are unlikely to store more than $0.14N$ uncorrelated random patterns.

A better way of storing patterns is given by an iterative version of the Hebbian rule [29]. At each learning iteration, the stability of every nominal pattern ψ^{μ} is tested. Whenever one pattern has not yet reached stability, the responsible neuron i reinforces its connectivity by adding a Hebbian term to all the synaptic connections impinging on it,

$$w_{ij}(t+1) = w_{ij}(t) + \eta \psi_i^{\mu} \psi_j^{\mu}. \tag{8.12}$$

All patterns to be learned are repeatedly tested for stability, and once all are stable, learning is complete. This learning algorithm is incremental since learning of new information can be done by preserving all information that has already been learned. By using this procedure, the capacity can be increased up to $2N$ uncorrelated random patterns [29].

When permitting a small fraction ϵ of a set of N bipolar J-vectors irretrievable, the upper bound approximates $2J$ when $J \to \infty$. This is given by a theorem derived from the function counting theorem (Theorem 2.1) [30, 40, 79].

Theorem 8.2 (Asymptotical upper bound) *For N prototype vectors in general position, the storage capacity $N_{\max}$ can approach $2J$, in the sense that, for any $\epsilon > 0$ the probability of retrieving a fraction $(1-\epsilon)$ of any set of $2J$ vectors tends to unity when $J \to \infty$.*

N prototype vectors in *general position* means that any subset of up to N vectors is linearly independent. Theorem 8.2 is more general than Theorem 8.1, since there is no constraint on $\mathbf{W}$. This recurrent network is sometimes referred to as the *generalized Hopfield network*. Both theorems hold true irrespective of the updating mode, be it synchronous or asynchronous.

The generalized Hopfield network with a general, zero-diagonal weight matrix has stable states in randomly asynchronous mode [55]. The asymptotic storage capacity of such a network using the perceptron learning scheme has been analyzed in [56]. The perceptron learning rule with zero bias is used to compute the columns of $\mathbf{W}$ for each neuron independently, and as such the entire $\mathbf{W}$ is constructed. A lower and an upper bound of the asymptotic storage capacity are obtained as $J-1$ and $2J$, respectively.

In a special case of the generalized Hopfield network with zero bias vector, some spectral strategies are used for constructing $\mathbf{W}$ [79]. All the spectral storage algorithms have a storage capacity of J for uncorrelated patterns [79]. A recursive implementation of the pseudoinverse spectral storage algorithm has also been given.

Example 8.1 This example is designed to check the storage capabilities of the Hopfield network trained with three local algorithms, namely, the generalized Hebbian, improved Hebbian, and perceptron-type learning rules. After the Hopfield network is trained with a pattern set, we present the same pattern set and examine the average retrieval bit error rates and the average storage error rates for a number of random runs.

A set of bipolar patterns each having a bit-length of $J = 50$ is given. The pattern set $\{\boldsymbol{x}_i\}$ is generated randomly. Theoretically, the storage capacities for the generalized Hebbian, improved Hebbian, and perceptron-type learning rules are $\frac{J}{2 \ln J} = 6.39$, $\frac{J}{\sqrt{2 \ln J}} = 17.88$, and $J = 50$, respectively. These capacities have been verified during our experiments. After the Hopfield network is trained, the maximum number of iterations at the performing stage is set as 30. The bit error rates and storage error rates are calculated based on 50 random runs.

Simulation is conducted for the case of N uncorrelated patterns as well as N slightly correlated patterns. In the case of N uncorrelated patterns, the matrix composed of the randomly generated patterns is of full rank, that is, having a rank of N. In the case of N slightly correlated patterns, $N - 1$ patterns are randomly generated and are uncorrelated; the remaining one pattern is generated by linearly combining any three of the $N - 1$ patterns and then applying the signum function, until the corresponding matrix has a rank of $N - 1$.

For perceptron-type learning, we can select $\mathbf{W}(0)$ as a symmetrical, random, zero-diagonal matrix with each entry in the range of $(-0.1, 0.1)$ or as the zero matrix. Our empirical results show that the latter scheme can generate better results, and it is used here. η is selected as 0.2. The maximum number of epochs is set as 50. Training terminates when the relative energy change between two epochs is below 10^{-4}.

We store a set of $N = 20$ patterns. The training and performing results are shown in Table 8.1, and the evolution of the system energy during the training process for a random run is illustrated in Fig. 8.1.

During the retrieval stage, if the fundamental memories are presented to the network, the desired patterns can usually be produced by the network after one iteration if N is less than the capacity of the network. The network trained by the general-

Table 8.1 Comparison of three associative memory algorithms ($J = 50$, $N = 20$)

Algorithm	Uncorrelated			
	Training epochs	Performing iterations	Bit error rate	Storage error rate
GH	1	29.64	0.2226	0.9590
IH	1	5.66	0.0038	0.0920
PT	4.70	1	0	0
Algorithm	**Correlated**			
	Training epochs	Performing iterations	Bit error rate	Storage error rate
GH	1	23.28	0.2932	0.8660
IH	1	3.44	0.0026	0.0680
PT	4.50	1	0	0

GH—generalized Hebbian rule; IH—improved Hebbian rule; PT—perceptron-type rule

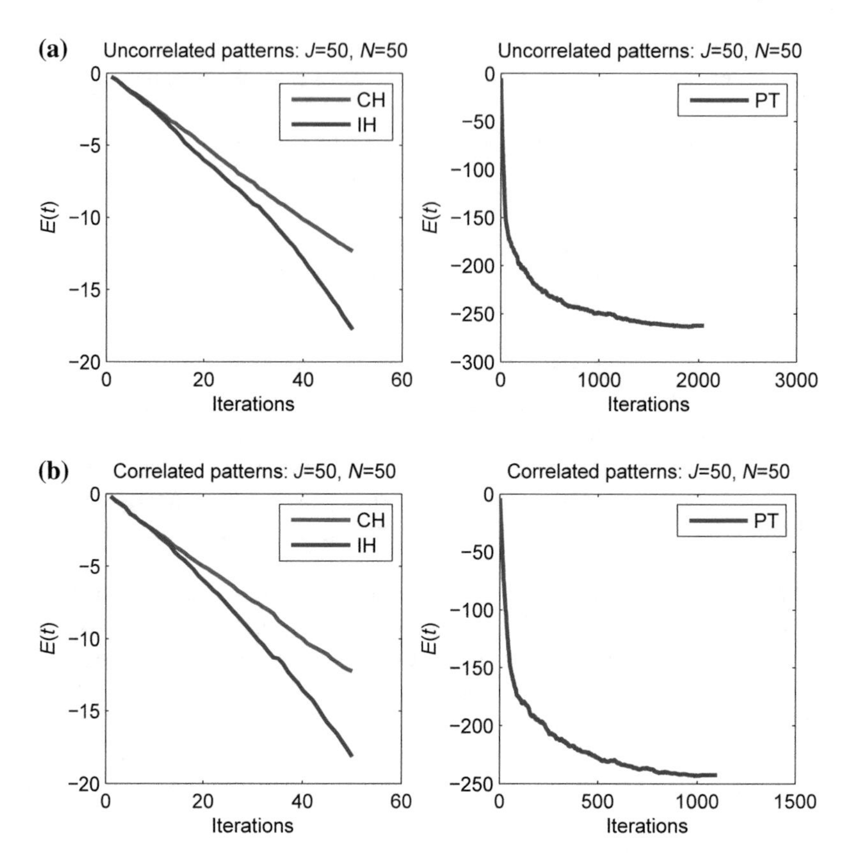

Fig. 8.1 Energy evolutions: $J = 50$, $N = 20$. **a** Uncorrelated patterns. **b** Slightly correlated patterns. t is the number of iterations. Notice that one epoch corresponds to 20 iterations

ized Hebbian rule cannot correctly retrieve any of the patterns, since the number of patterns is much greater than its storage capacity. The network trained with the improved Hebbian rule can, on average, correctly retrieve 18 patterns, which is close to its theoretical capacity. The perceptron-type rule can correctly retrieve all the patterns. It is noted that the results for the uncorrelated and slightly correlated cases are very close to each other for all these algorithms.

Example 8.2 Now, let us increase N to 50; the corresponding results are listed in Table 8.2 and shown in Fig. 8.2. The average capacity of perceptron-like learning is 38 for training 50 epochs. By increasing the number of epochs to 100, the storage capability of perceptron-like learning can be further improved, the average iteration for the retrieval stage is 1, and the storage error rate is close to 0. Thus, the storage capability can reach 50 for both the uncorrelated and slightly correlated cases. Perceptron-type learning can retrieve all the 50 patterns with a small storage error rate, while the improved Hebbian and generalized Hebbian rules actually fail. The

Table 8.2 Comparison of three associative memory algorithms ($J = 50$, $N = 50$)

Algorithm	Uncorrelated			
	Training epochs	Performing iterations	Bit error rate	Storage error rate
GH	1	30	0.2942	1.00
IH	1	30	0.3104	0.9720
PT	35.44	17.90	0.0280	0.0688
Algorithm	Correlated			
	Training epochs	Performing iterations	Bit error rate	Storage error rate
GH	1	30	0.3072	0.9848
IH	1	30	0.3061	0.9688
PT	35.98	14.92	0.0235	0.0568

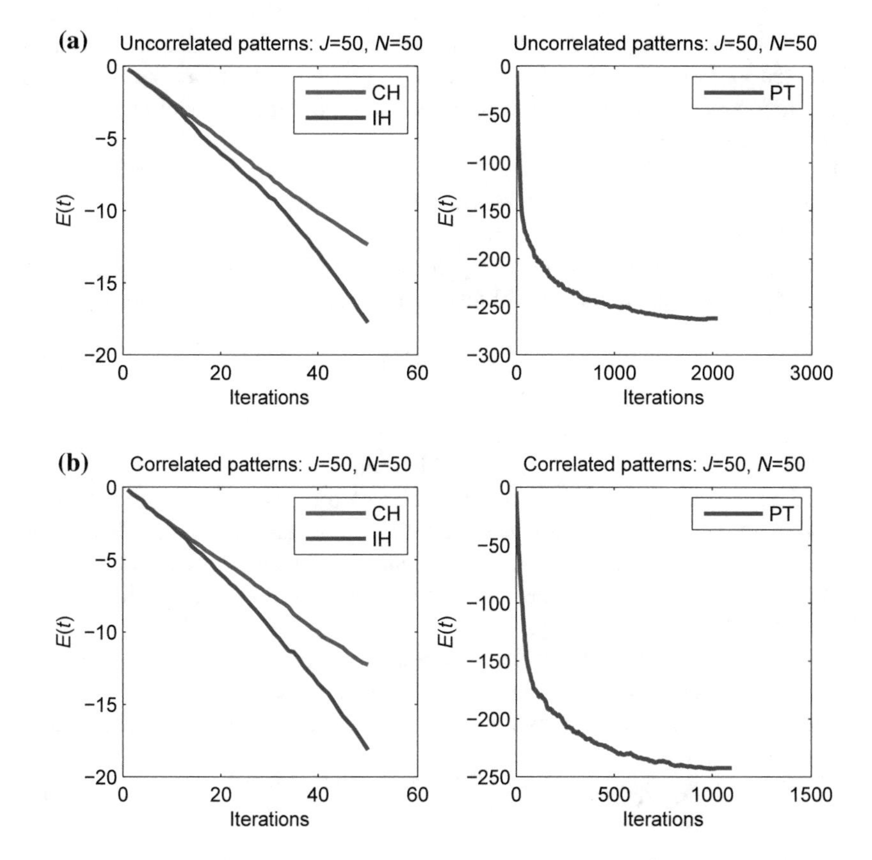

Fig. 8.2 Energy evolutions: $J = 50$, $N = 50$. **a** Uncorrelated patterns. **b** Slightly correlated patterns. t is the number of iterations. Notice that one epoch corresponds to 50 iterations

perceptron-type rule can almost correctly retrieve all the patterns; this accuracy can be further improved by training with more epochs.

8.4 Increasing Storage Capacity

When the Hopfield network is used as associative memory, there are cases where the fundamental memories are not stable. In addition, spurious states, which are other stable states different from the fundamental memories and their negative counterparts, may arise [2]. The Hopfield network trained with the generalized Hebbian rule can have a large number of spurious states, depending exponentially on N, the number of fundamental memories, even in the case when these vectors are orthogonal [12]. These spurious states are the corners of the unit hypercube that lie on or near the subspace spanned by the N fundamental memories. The presence of spurious states and limited storage capacity are the two major restrictions for the Hopfield network being used as associative memory. It has been proved that as long as $N \ll J$, J being the number of neurons in the network, the fundamental memories are stable in a probabilistic sense [2].

The Gardner conditions [30] are often used as a measure of the stability of the patterns. Associative learning can be designed to enhance the basin of attraction for every pattern to be stored by optimizing these conditions. The Gardner algorithm [30] combines maximal storage with a predefined level of stability for the patterns. Based on the Gardner conditions, the inverse Hebbian rule [21, 22] is given by

$$w_{ij} = -\left(\mathbf{R}^{-1}\right)_{ij}, \tag{8.13}$$

where the correlation matrix $\mathbf{R} = \frac{1}{N}\sum_{p=1}^{N} \boldsymbol{x}_p \boldsymbol{x}_p^T$.

Unlike the generalized Hebbian rule, which can only store unbiased patterns, the inverse Hebbian rule is capable of storing N patterns, biased or unbiased, in a Hopfield network of N neurons. The patterns have zero basins of attraction, and $\mathbf{R}$ must be nonsingular. Matrix inversion can be implemented using a local learning algorithm. The inverse Hebbian rule provides ideal initial conditions for any algorithm capable of increasing the pattern stability.

Nonmonotonic activation functions are much more beneficial to the enlargement of the memory capacity of a neural network model. In [41], by using the Gardner algorithm for training the weights and using a nonmonotonic activation function

$$\phi(net) = \begin{cases} +1, & net \in (-\infty, -b_1) \cup (0, b_1) \\ -1, & net \in (b_1, \infty) \cup (-b_1, 0) \end{cases}, \tag{8.14}$$

the storage capacity of the network can be made to be always larger than $2J$ and reach its maximum value of $10.5J$ when $b_1 = 1.22$. In [60, 83], a continuous nonmonotonic activation function is used to improve the performance of the Hopfield network. The exact form of the nonmonotonic activation and its parameters are not very critical. The storage capacity of the Hopfield network can be improved to approximately $0.4J$, and spurious states can be totally eliminated. When it fails to recall a memory, a chaotic behavior will occur. In the application to realizing the autocorrelation associative memory, the chaotic neural network model with sinusoidal activation functions possesses a large memory capacity as well as a remarkable ability of retrieving the stored patterns, better than the chaotic model with only monotonic activation functions such as sigmoidal functions [62]. It is shown in [50] that any finite-dimensional network model with periodic activation functions and properly selected parameters has much more abundant chaotic dynamics that truly determine the model's memory capacity and pattern-retrieval ability.

The eigenstructure learning rule [49] is developed for continuous-time Hopfield models in linear saturated mode. The design method allows linear combinations of the prototype vectors to be stored as asymptotically stable equilibrium points as well. The storage capacity is better than those of the pseudoinverse solution and the generalized Hebbian rule. All the desired patterns are guaranteed to be stored as asymptotically stable equilibrium points. The method has been extended to discrete-time neural networks in [59].

A quantum learning algorithm, which is a combination of quantum computation with the Hopfield network, has been developed in [80]. The quantum associative memory has a capacity that is exponential in the number of neurons, namely, offering a storage capacity of $O\left(2^J\right)$. It employs simple spin-$1/2$ (two-state) quantum systems and represents patterns as quantum operators.

Complex-valued associative memories such as complex-valued Hopfield networks are used for storing complex-valued patterns. In [63], a complex perceptron learning algorithm has also been studied for associative memory by using complex weights and a decision circle in the complex plane for the output function.

Other Associative Memories

Morphological associative memories involve a very low computational effort in synthesizing the weight matrix by the use of minimax algebra. An exact characterization of the fixed points and the basins of attraction of grayscale autoassociative morphological memories is made in terms of the eigenvectors of the weight matrix in [75]. The set of fixed points consists exactly of all linear combinations of the fundamental eigenvectors. If J-dimensional patterns are binary, then 2^J patterns can be stored.

For the generalized BSB model [36], the synthesis for optimal performance is performed in [65], given a set of desired binary patterns to be stored as asymptotically stable equilibrium points. The synthesis problem is formulated as a constrained optimization problem, which can be converted into a quasi-convex optimization problem (generalized eigenvalue problem) in the form of an LMI (linear matrix inequality)-based optimization problem. In [13], the design of recurrent associative memories based on the generalized BSB model is formulated as a set of independent classification tasks which can be efficiently solved by using a pool of SVMs.

The storage capacity of recurrent attractor neural networks with sign-constrained weights was investigated in [7].

8.5 Multistate Hopfield Networks as Associative Memories

In Sect. 7.6, we have described multistate Hopfield networks. In [27], a multilevel Hopfield network modifies the generalized Hebbian rule. The storage capability of the multilevel Hopfield network is proved to be $O(J^3)$ bits for a network of J neurons, which is of the same order as that of the Hopfield network [1]. Given a network of J neurons, the number of patterns that the multilevel network can reliably store and retrieve may be considerably less than that for the Hopfield network, since each codeword in the multilevel Hopfield network typically contains more bits. In [72], a storage procedure for the multilevel Hopfield network in the synchronous mode is derived based on the LS solution, and also examined by using an image restoration example.

The complex-valued multistate Hopfield network [39, 61] employs the multivalued complex-signum activation function that is defined as an L-stage phase quantizer for complex numbers. In order to store a set of N patterns, $\{\boldsymbol{x}_i\} \subset \{0, 1, \ldots, L-1\}^J$, $\boldsymbol{x}_i$ is first encoded to its complex memory state $\boldsymbol{\epsilon}_i = \left(\epsilon_{i,1}, \ldots, \epsilon_{i,J}\right)^T$ with

$$\epsilon_{i,j} = z^{x_{i,j}}. \tag{8.15}$$

The decoding of a memory state to a pattern is the inverse of (8.15). The complex-valued pattern set $\{\boldsymbol{\epsilon}_i\}$ can be stored in weights by the generalized Hebbian rule [35]

$$w_{ji} = \frac{1}{J} \sum_{p=1}^{N} \epsilon_{p,i} \epsilon^*_{p,j}, \quad i, j = 1, 2, \ldots, J, \tag{8.16}$$

where superscript $*$ denotes conjugate operation. Thus, $\mathbf{W}$ is Hermitian.

The storage capability of the memory, $N_{\max}$, is dependent upon the resolution L for an acceptable level of the error probability $P_{\max}$. As L increases, $N_{\max}$ decreases, but each pattern contains more information. In [39], the capacity of a complex-valued multistate neural associative memory is estimated to be less than $0.15J$. With a prohibitively complicated learning rule, the capacity is increased to J [61].

Due to the use of generalized Hebbian rule, the storage capacity of the network is very low and the problem of spurious memories is very pronounced. In [47], a gradient-descent learning rule has been proposed to enhance the storage capacity and also reduce the number of spurious memories. In [61], an LP method has been proposed for storing into the network each pattern in an integer set $M \subset \{0, 1, 2, \ldots, L-1\}^J$ as a fixed point. A set of inequalities are employed to render each memory pattern as a strict local minimum of a quadratic energy landscape, and the LP method is employed to obtain the weight matrix and the threshold vector. The LP method significantly reduces the number of spurious memories.

Since grayscale images can be represented by integer vectors, reconstruction of such images from their distorted versions constitutes a straightforward application of multistate associative memory. The complex-valued Hopfield network is particularly suitable for interpreting images transformed by two-dimensional Fourier transform and two-dimensional autocorrelation functions [39].

8.6 Multilayer Perceptrons as Associative Memories

Most recurrent network-based associative memories have low storage capacity as well as poor retrieval ability. Recurrent networks exhibit asymptotic behavior and as such are difficult to analyze. MLP-based autoassociative memories with equal numbers of input and output nodes have been introduced to overcome these limitations [17, 81].

The recurrent correlation associative memory uses a J-N-J MLP-based recurrent architecture [17], as shown in Fig. 8.3. Notice that the number of hidden units is taken as N, the number of stored patterns. At each time instant, the hidden layer computes an intermediate mapping, while the output layer completes an association of the input pattern to an approximate prototype pattern. The approximated pattern is fed back to the network and the process continues until convergence to a prototype is achieved. The activation function for the ith neuron in the hidden layer is $\phi_i(\cdot)$, and the activation function at the output layer is the signum function.

The matrix $\mathbf{W}^{(1)}$, a $J \times N$ matrix, is made up of the N J-bit bipolar memory patterns $\boldsymbol{x}_i$, $i = 1, 2, \ldots, N$, that is, $\mathbf{W}^{(1)} = [\boldsymbol{x}_1, \boldsymbol{x}_2, \ldots, \boldsymbol{x}_N]$. And $\mathbf{W}^{(2)} = \left[\mathbf{W}^{(1)}\right]^T$. When presenting pattern $\boldsymbol{x}$, the net input to neuron j in the hidden layer is $net_j^{(2)} = \boldsymbol{x}_j^T \boldsymbol{x}$. We have

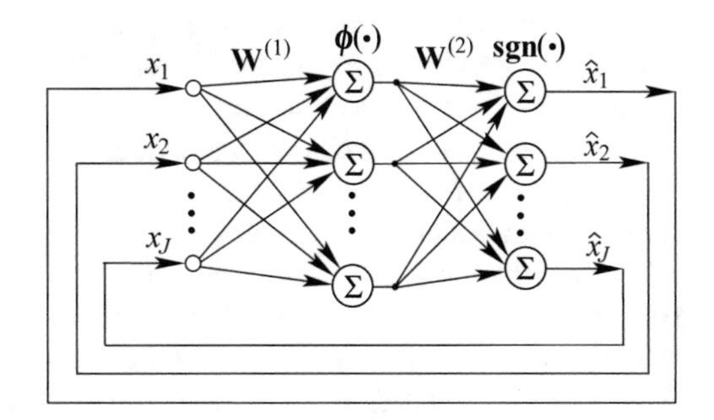

Fig. 8.3 Architecture of the J-N-J MLP-based recurrent correlation associative memory

$$\boldsymbol{x}(t+1) = \operatorname{sgn}\left(\sum_{j=1}^{N} \phi_j\left(\boldsymbol{x}_j^T \boldsymbol{x}(t)\right) \cdot \boldsymbol{x}_j\right). \tag{8.17}$$

The correlation of two patterns, $\boldsymbol{x}_1^T \boldsymbol{x}_2 = J - 2d_H(\boldsymbol{x}_1, \boldsymbol{x}_2)$, where $d_H(\cdot)$ is the Hamming distance between two binary vectors within, which is the number of bits in the two vectors that do not match each other.

In the case when all $\phi_i(net) = \phi(net)$, $\phi(net)$ being any continuous, monotonic nondecreasing weighting function over $[-J, J]$, (8.17) is proved to be asymptotically stable in both the synchronous and asynchronous update modes [17]. This property is especially suitable for hardware implementation, since there are faults in the manufacture of any physical device.

When all $\phi_i(net) = net$, the recurrent correlation associative memory model is equivalent to the correlation matrix associative memory [8, 42], that is, the connection corresponding to the case of the Hopfield network can be written as $\mathbf{W} = \sum_{p=1}^{N} \boldsymbol{x}_p \boldsymbol{x}_p^T$. By suitably selecting $\phi_i(\cdot)$, the model is reduced to some existing associative memories, which have a storage capacity that grows polynomially or exponentially with J [17].

In particular, when all $\phi_i(net) = a^{net}$ with radix $a > 1$, an exponential correlation associative memory [17] is obtained. The exponential activation function stretches the ratios among the weights and makes the largest weight more overwhelming. This significantly increases the storage capacity. The exponential correlation associative memory exhibits an asymptotic storage capacity that scales exponentially with J. Under noise-free condition, this storage capacity is 2^J patterns [17]. A VLSI chip for this memory has been fabricated and tested [17]. The multivalued recurrent correlation associative memory [18] can increase the error correction capability with large storage capability and less interconnection complexity.

The local identical index model [81] is an autoassociative memory model that uses the J-N-J MLP architecture. The weight matrices $\mathbf{W}^{(1)}$ and $\mathbf{W}^{(2)}$ are the same as those defined in the recurrent correlation associative memory model. It utilizes the signum activation function and biases in both the hidden and output layers. The local identical index model utilizes the local characteristics of the fundamental memories through two metrics, namely, the global identical index and the local identical index. Using the minimum Hamming distance as the underlying association principle, the scheme can be viewed as an approximate Hamming decoder. The local identical index model exhibits low structural as well as operational complexity. It is a one-shot associative memory and can accommodate up to 2^J prototype patterns. This model outperforms the *linear system in a saturated mode* [49] and its discrete version [59] in recognition accuracy at the presentation of the corrupted patterns, controlled by using the Hamming distance. It can successfully associate input patterns that are even loosely correlated with the corresponding prototype pattern.

For a J-J_2-J MLP-based autoassociative memory, the hidden layer is a bottleneck layer with fewer nodes, $J_2 < J$. This bottleneck layer is used to discover a limited set of unique prototypes that cluster the training set. The neurons at the bottleneck layer use the sigmoidal function.

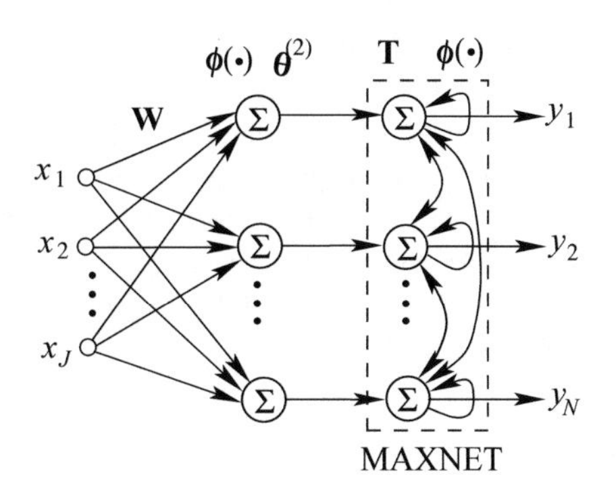

Fig. 8.4 Architecture of the J-N-N Hamming network

In a three-layer MLP with a feedback connection that links input and output layers through delay elements [4], the MLP is initially trained to store the real-valued patterns as an autoassociative memory. A sigmoidal function is chosen for the hidden-layer nodes, while at the output layer there is a linear summation. The model has compact size to store numerous stable fixed points, but it is also able to learn asymmetric arrangement of fixed points, whereas the self-feedback neural network model is limited to orthogonal arrangements.

The bipolar recurrent correlation associative memory [17] has been generalized to the complex-valued case in [78]. The complex-valued recurrent correlation networks are implemented by the fully connected two-layer recurrent networks. The model always converges to a stationary state. It outperforms other complex-valued models.

8.7 Hamming Network

The Hamming network [51] is a straightforward associative memory. It calculates the Hamming distance between the input pattern and each memory pattern and selects the memory with the smallest Hamming distance. The network output is the index of a prototype pattern and thus the network can be used as a pattern classifier. The Hamming network is used as the classical Hamming decoder or Hamming associative memory. It provides the minimum-Hamming-distance solution.

The Hamming network has a J-N-N layered architecture, as illustrated in Fig. 8.4. The activation function at each of the units in each layer, denoted by vector ϕ, is the signum function, and $\boldsymbol{\theta}^{(2)}$ is a vector comprising the biases for all neurons at the hidden layer. The weights in the third layer $\mathbf{T} = [t_{kl}]$, $k, l = 1, \ldots, N$.

The third layer is called the *memory layer*, each of whose neurons corresponds to a prototype pattern. The input and hidden layers are feedforward, fully connected, while each hidden node has a feedforward connection to its corresponding node in the memory layer. Neurons in the memory layer are fully interconnected and form a competitive subnetwork known as the MAXNET. The MAXNET responds to an input pattern by generating a winner neuron through iterative competitions. The

Hamming network is implicitly recurrent due to the interconnections in the memory layer.

The second layer generates matching scores that are equal to J minus the Hamming distances to the stored patterns, that is, $J - d_H(\boldsymbol{x}, \boldsymbol{x}_i)$, $i = 1, \ldots, N$, for pattern $\boldsymbol{x}$. These matching scores range from 0 to J. The unit with the highest matching score corresponds to the stored pattern that best matches the input. The weights between the input and hidden layers and the biases of the hidden layer are, respectively, set as

$$w_{ij} = \frac{x_{j,i}}{2}, \quad \theta_j^{(2)} = \frac{J}{2}, \quad j = 1, \ldots, N, i = 1, \ldots, J. \tag{8.18}$$

All the thresholds and the weights t_{kl} in the MAXNET are fixed. The thresholds are set as zero. The weights from each node to itself are set as unity, and weights between nodes are inhibitory, that is,

$$t_{kl} = \begin{cases} 1, & k = l \\ -\varepsilon, & k \neq l \end{cases}, \tag{8.19}$$

where $\varepsilon < \frac{1}{N}$.

When a binary pattern is presented to the network, the network first generates an initial input for the MAXNET

$$y_j(0) = \phi\left(\sum_{i=1}^{N} w_{ij} x_i - \theta_j^{(2)}\right), \quad j = 1, \ldots, N, \tag{8.20}$$

where $\phi(\cdot)$ is the threshold-logic nonlinear function.

The input pattern is then removed and the MAXNET continues the iteration

$$y_j(t+1) = \phi\left(\sum_{k=1}^{N} t_{kj} y_k(t)\right) = \phi\left(y_j(t) - \varepsilon \sum_{k=1, k\neq j}^{N} y_k(t)\right), \quad j = 1, \ldots, N, \tag{8.21}$$

until the output of only one node is positive. This node corresponds to the selected class.

The Hamming network implements the minimum error classifier, when the bit errors are random and independent. For a J-N-N Hamming network, there are $J \times N + N^2$ connections, while for the Hopfield network the number of connections is J^2. When $J \gg N$, the number of connections in the Hamming network is significantly less than that in the Hopfield network. In addition, the Hamming network offers a storage capacity that is exponential in the input dimension [33], and it does not have any spurious state that corresponds to a no-match result. Under the noise-free condition, the Hamming network has a storage capacity of 2^J patterns [33]. For a sufficiently large but finite radix a, the exponential correlation associative memory operates as a Hamming associative memory [33].

The Hamming network suffers from difficulties in hardware implementation and low retrieval speed. Based on the correspondence between the Hamming network and the exponential correlation associative memory [33], the exponential correlation associative memory can be used to compute the minimum Hamming distance, in a distributed fashion by analog exponentiation and thresholding devices. The two-level Hamming network [37] generalizes the Hamming memory by providing for local Hamming distance computations in the first level and a voting mechanism in the second level. It allows for a much more practical hardware implementation and a faster retrieval.

8.8 Bidirectional Associative Memories

BAM is created by adapting the nonlinear feedback of the Hopfield model to a heteroassociative memory [45]. It is used to store N bipolar pairs $(\boldsymbol{x}_p, \boldsymbol{y}_p)$, $p = 1, \ldots, N$, where $\boldsymbol{x}_p = (x_{1p}, \ldots, x_{J_1 p})^T$, $\boldsymbol{y}_p = (y_{1p}, \ldots, y_{J_2 p})^T$, and $x_{ip}, y_{jp} \in \{+1, -1\}$. The first layer has J_1 neurons and the second has J_2 neurons.

BAM learning is accomplished with a simple Hebbian rule. The weight matrix is

$$\mathbf{W} = \mathbf{Y}\mathbf{X}^T = \sum_{p=1}^{N} \boldsymbol{y}_p \boldsymbol{x}_p^T, \tag{8.22}$$

where $\mathbf{X}$ and $\mathbf{Y}$ are matrices that represent the sets of bipolar vector pairs to be associated. BAM effectively uses a signum function to recall noisy patterns, despite the fact that the weight matrix developed is not optimal. This one-shot learning rule leads to poor memory storage capacity, is sensitive to noise, and is subject to spurious steady states during recall.

The retrieval process is an iterative feedback process that starts with $\mathbf{X}^{(0)}$ in layer 1:

$$\mathbf{Y}^{(v+1)} = \mathrm{sgn}\left(\mathbf{W}\mathbf{X}^{(v)}\right), \quad \mathbf{X}^{(v+1)} = \mathrm{sgn}\left(\mathbf{W}\mathbf{Y}^{(v+1)}\right). \tag{8.23}$$

For any real connection matrix, one of fixed points $(\boldsymbol{x}_f, \boldsymbol{y}_f)$ can be obtained from this iterative process. A fixed point has the properties:

$$\boldsymbol{x}_f = \mathrm{sgn}\left(\mathbf{W}^T \boldsymbol{y}_f\right), \quad \boldsymbol{y}_f = \mathrm{sgn}\left(\mathbf{W}^T \boldsymbol{x}_f\right). \tag{8.24}$$

With the iterative process, BAM can achieve both heteroassociative and autoassociative data recollections. The final state in layer 1 represents the autoassociative recall, and the final state in layer 2 represents the heteroassociative recall.

Kosko's encoding method has the ability of incremental learning: encoding new pattern pairs to the model is based on only the current connection matrix. This method can correctly store up to $\frac{\min(J_1, J_2)}{2 \log \min(J_1, J_2)}$ pattern pairs [48]. When the total number exceeds this, all pattern pairs may not be stored as fixed points. To avoid this, one can forget past pattern pairs such that BAM can correctly store as many as possible of the most recent learning pattern pairs. Forgetting learning is an incremental learning rule in associative memories. The storage behavior of BAM under the forgetting learning is analyzed in [48].

In [15], a BAM model is introduced that uses a simple self-convergent iterative learning rule and a nonlinear output function. In addition, the output function enables it to learn and recall gray-level patterns in a bidirectional way. The heteroassociative neural network can learn bipolar and real-valued patterns. The network is able to stabilize its weights to fixed-point attractors in a local fashion. The model is immune to overlearning and develops fewer spurious attractors compared to other BAMs.

The storage capacity of Kosko's BAM is $0.15J_1$, while that of asymmetrical BAM [82], generalized asymmetrical BAM [26], and the model proposed in [15] is J_1, and that for general BAM [71] is greater than J_1. Global exponential stability criteria are established for BAM networks with time delays in [54].

Most BAM networks use a symmetrical output function for dual fixed-point behavior. In [16], by introducing an asymmetry parameter into a chaotic BAM output function, prior knowledge can be used to momentarily disable desired attractors from memory, hence biasing the search space to improve recall performance. This property allows control of chaotic wandering, favoring given subspaces over others. In addition, reinforcement learning can then enable a dual BAM architecture to store and recall nonlinearly separable patterns. The same BAM framework is allowed to model three different types of learning: supervised, reinforcement, and unsupervised.

8.9 Cohen–Grossberg Model

A general class of neural networks defined in [19] is globally stable. The general form of the model is

$$\frac{d}{dt}x_i = a_i(x_i)\left(b_i(x_i) - \sum_{j=1}^{J} c_{ij} d_j(x_j)\right), \tag{8.25}$$

where J is the number of units in the network, $\boldsymbol{x} = \{x_1, x_2, \ldots, x_J\}$ is the state of the network, $\mathbf{C} = [c_{ij}]$ is a coefficient matrix, $a_i(\cdot)$ is the amplification function, $b_i(\cdot)$ is the self-signal function, and $d_j(\cdot)$ is the other signal function [32].

Theorem 8.3 (Cohen–Grossberg theorem) *If a network can be written in the general form (8.25) and also obeys the three conditions, (C1) symmetry:* $c_{ij} = c_{ji}$, *(C2)* *positivity:* $a_i(x_i) \geq 0$, *(C3) monotonicity: the derivative* $d'_j(x_j) \geq 0$, *then its Lyapunov function can be defined by*

$$L(x) = -\sum_{i=1}^{N} \int_0^{x_i} b_i(\xi_i) d'_i(\xi_i) d\xi_i + \frac{1}{2} \sum_{j=1}^{N} \sum_{k=1}^{N} c_{jk} d_j(x_j) d_k(x_k). \tag{8.26}$$

Many neural networks fall in this class of models [32], e.g., BSB [9] and the Boltzmann machine. The global stability of the continuous-time, continuous-state BSB dynamical systems with real symmetric weight matrices is proved in [19]. Nonlinear dynamic recurrent associative memory [14, 15] is an instance of the Cohen–Grossberg model when $\delta \leq 1/2$ [34], and analysis of the energy function shows that the transmission is stable in the entire domain of the model [34]; it is a nonlinear synchronous attractor neural network, converging to a set of real-valued attractors in single-layer neural networks [14] and bidirectional associative memories [15].

Applications of networks with time delays often require that the network has a unique equilibrium point which is globally exponentially stable, if the network is to be suitable for solving problems in real time. In hardware implementation of neural networks, time delays even time-varying delays in neuron signal transmission or processing are often inevitable. It is more realistic to design neural networks which are robust on delays.

8.10 Cellular Networks as Associative Memories

Cellular networks are suitable for hardware implementation and, consequently, for their employment in applications such as in real-time image processing and in construction of efficient associative memories. Adjustment of cellular network parameters is a complex problem involved in the configuration of cellular network for associative memories.

A cellular network is a high-dimensional array of cells that are locally interconnected. The neighborhood of a cell $c(i, j)$ is denoted by

$$V_r(i, j) = \{c(k, l) : \max(|k - i|, |l - j|) \leq r\}, \quad k = 1, \ldots, M, l = 1, \ldots, N, \tag{8.27}$$

where the subscript r indicates the neighborhood radius around the cell and $M \times N$ is the total number of cells in the array.

The dynamics of a cellular network are given by

$$\dot{\boldsymbol{x}} = -\boldsymbol{x} + \mathbf{T}\boldsymbol{y} + \boldsymbol{I}, \tag{8.28}$$

$$\boldsymbol{y} = \mathrm{sat}(\boldsymbol{x}), \tag{8.29}$$

where $\boldsymbol{x} = (x_1, x_2, \ldots, x_n)^T \in R^n$ denotes the cell states, $n = M \times N$ represents the number of network cells, $\boldsymbol{y} = (y_1, y_2, \ldots, y_n)^T \in R^n$, $y_i \in [-1, 1]$, $i = 1, \ldots, n$, is the cell outputs (bipolar), $\mathbf{T} \in R^{n \times n}$ represents the interconnection matrix (sparse), $\boldsymbol{I} \in R^n$ is a vector of bias, and $\text{sat}(\boldsymbol{x}) = (\text{sat}(x_1), \ldots, \text{sat}(x_n))^T$ is a saturation function,

$$\text{sat}(x_i) = \begin{cases} -1, & \text{if } x_i < -1 \\ x_i, & \text{if } -1 \le x_i \le 1 \\ 1, & \text{if } x_i > 1 \end{cases}, \quad i = 1, \ldots, n. \tag{8.30}$$

Cellular Networks for Image Processing

For image processing, cellular networks are generally used for movement detection, contour extraction, image smoothing, and detection of directional stimuli. An operation is in general applied to each image pixel (cell). Noise and contours can produce crisp variations. Image smoothing must eliminate noise and preserve contours, and this can be achieved using Laplacian operator.

For image smoothing, the image is provided as initial condition $\boldsymbol{x}(0)$ of the network and $\boldsymbol{I}$ must be zero. Output $\boldsymbol{y}(t)$ represents the processed image, when $t \to \infty$.

Example 8.3 In order to illustrate the elimination of noise by a cellular network, an original image (Fig. 8.5a) is corrupted by Gaussian noise of variance 0.01 generating the image shown in Fig. 8.5b.

The range of $x_{ij}(0)$ is $[-1.0, 1.0]$ to codify pixel intensities in a grayscale, with -1.0 corresponding to white and 1.0 to black. The neighborhood can be stored by a 3×3 matrix, which corresponds to a neighborhood of $r = 1$. Equations (8.28) and (8.29) can be numerically integrated using the fourth-order Runge–Kutta algorithm for the time period $[0, 0.1]$ and for 10 iterations, and the bias is set to 0.

In a cellular network, for each pixel of an image ψ, $\mathbf{T}$ can be determined from a mask $\mathbf{A}$ with an neighborhood of radius 1:

$$\mathbf{T} = \begin{pmatrix} A_{-1,1} & A_{-1,0} & A_{-1,1} \\ A_{0,-1} & A_{0,0} & A_{0,1} \\ A_{1,-1} & A_{1,0} & A_{1,1} \end{pmatrix} = \begin{pmatrix} 0 & 1 & 0 \\ 1 & -4 & 1 \\ 0 & 1 & 0 \end{pmatrix}.$$

The central position ($A_{0,0}$) denotes the connection from a cell (i, j) to itself. The representation of $\mathbf{T}$ using a mask highlights the relationship of a cell and its neighbors. Moreover, the mask synthesizes how a cellular network processes signals.

Figure 8.5c presents the image obtained by the application of the Laplacian operator. Figure 8.5d presents the image obtained by using Laplacian operator directly on the noisy image. It is seen that image smoothing using the cellular network is better than conventional way of applying image operator.

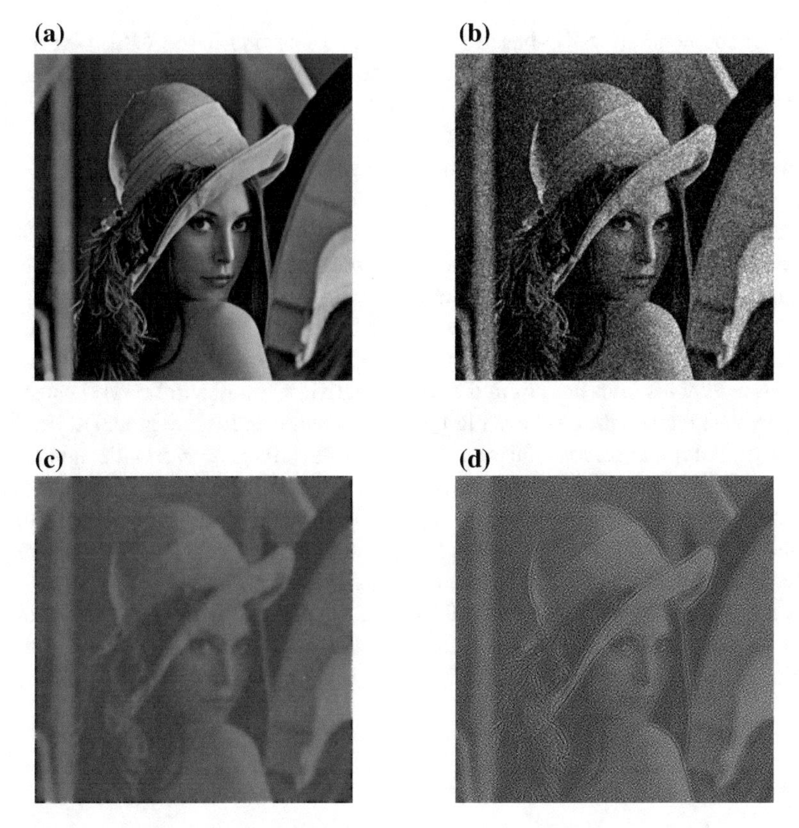

Fig. 8.5 Image smoothing using cellular network. **a** Original image. **b** Image corrupted by Gaussian noise; **c** Image restored using Laplacian operator implemented in a cellular network. **d** Image obtained by applying Laplacian operator directly on the noisy image

Cellular Networks for Associative Memories

To use cellular networks as associative memories, parameters $\mathbf{T}$ and $\boldsymbol{I}$ need to be properly adjusted. The equilibrium points of (8.28) correspond to the different patterns of the input to be stored. In the pseudoinverse method [31], pseudoinverse matrices are utilized for solving the equilibrium equations (8.28): For all patterns $p = 1, \ldots, N$,

$$\boldsymbol{x}^p = \mathbf{T}\boldsymbol{y}^p + \boldsymbol{I}, \tag{8.31}$$

$$\boldsymbol{y}^p = \text{sat}(\boldsymbol{x}^p), \tag{8.32}$$

where $\boldsymbol{y}^p \in \{-1, 1\}^n$ represents the pth pattern to be stored and $\boldsymbol{x}^p$ are the respective equilibrium points.

Other methods are SVD-based [52], Hebbian learning-based [76], perceptron-based [53], and LMI-based [66] methods. The five algorithms and their comparison are given in [25]. LMI and Hebbian methods show superior performance for tests involving binary noise. For tests with Gaussian noise, perceptron, SVD, and Hebbian approaches present similar performance with almost 100% of pattern retrieval. The LMI method is not adequate for cellular networks with small neighbor radius. The pseudoinverse method presents poor performance for cellular networks with larger neighbor ratio. In general, these approaches present better performance for patterns corrupted with Gaussian than binary noise.

Associative memories can be synthesized based on discrete-time recurrent networks [84] or continuous-time cellular networks with time delays [85]. The design procedure enables both hetero- and autoassociative memories to be synthesized. In [84], the designed memories have high storage capacity and assure global asymptotic stability. As typical representatives, discrete-time cellular networks designed with space-invariant cloning templates are examined in detail. The storage capacity of the designed associative memories is as high as $2J$ bipolar patterns [84]. In [85], the synthesizing procedure solves a set of linear inequalities with few design parameters and retrieval probes feeding from external inputs instead of initial states. The designed associative memories are robust in terms of design parameter selection. In addition, the hosting cellular networks are guaranteed to be globally exponentially stable.

Problems

8.1 For the network trained in Example 8.1, randomly flip up to 10% of the J bits, and then retrieve using the learned network. Calculate the bit error rate and storage error rate for the case of the three algorithms.

8.2 Find the weights and thresholds for a Hopfield network that stores the patterns 0101, 1010, 0011, 0110.

8.3 Store three fundamental memories into a five-neuron Hopfield network:

$$\xi_1 = (+1, +1, +1, +1, -1), \xi_2 = (-1, -1, -1, +1, -1), \xi_3 = (+1, -1, -1, +1, +1).$$

The Hamming distance between these memories are 2 or 3.

(a) Solve for the 5×5 weight matrix.
(b) Verify that the three fundamental memories can be correctly retrieved using asynchronous updating. What about synchorous updating?
(c) When presenting a noisy version of ξ_2, with one element's polarity being reversed, verify the retrieval performance.
(d) Show that $-\xi_1, -\xi_2, -\xi_3$ are also fundamental memories.
(e) If the second element of ξ_2 is unknown at retrieval stage, find out the retrieval result.
(f) Calculate the energy of the network.

8.4 Consider the Hopfield network.

(a) Plot storage capacity versus the number of neurons N for N up to 20. Consider the cases for different storage algorithms and for different bounds.
(b) If we want to store 12 patterns, how many neurons are needed?

8.5 Write a program to implement the Hamming network. Use the program to retrieve ten 5×7 numeric digits.

8.6 The BSB model can be characterized by

$$\boldsymbol{y}(n) = \boldsymbol{x}(n) + \beta \mathbf{W} \boldsymbol{x}(n),$$

$$\boldsymbol{x}(n+1) = \varphi(\boldsymbol{y}(n)),$$

where β is a small positive constant, $\mathbf{W}$ is a symmetric weight matrix whose largest eigenvalues have positive real components, the activation function $\varphi(x)$ is a piecewise-linear function which operates on the components of a vector, and it is $+1$ if $x > +1$, -1 if $x < -1$, and is linear inbetween. The Lyapunov function for the BSB model is $E = -\frac{\beta}{2}\boldsymbol{x}^T \mathbf{W} \boldsymbol{x}$. Prove the stability of the BSB model by using the Cohen–Grossberg theorem.

8.7 Store five 100×100 images into a Hopfield network. For example, you can store the pictures of five different shapes, or different objects.

(a) Add noise to one of the pictures and retrieve the correct object.
(b) Erase some parts of the picture and then retrieve the correct object.

8.8 Consider the BAM model.

(a) Train it using samples (0010010010, 01), (0100111101, 10).
(b) Test the trained network, with a corrupted sample.

8.9 This problem is adapted from [15]. Convert the 26 English characters from lower case to upper case. The network associates 26 correlated patterns consisting of 7×7 binary images. Figure 8.6 illustrates the stimuli used for the simulation.

(a) Check whether Kosko's BAM can handle the storage. Notice that the images were converted into vectors of 49 dimensions and this corresponds to a memory load of 53% (26/49) of the space capacity.
(b) Select another BAM algorithm with higher storage capacity.
(c) Test the network performance on a noisy recall task. The task was to recall the correct associated stimulus from a noisy input obtained by randomly flipping from 0 to 10 pixels in the input pattern, corresponding to a noise proportion of 0 to 20%.

8.10 The Laplacian of a function $\psi(x, y)$ in the plane is given by $\Delta^2 \psi = \frac{\partial^2 \psi}{\partial x^2} + \frac{\partial^2 \psi}{\partial y^2}$. For an image ψ to be processed by a cellular network, verify that the mask for

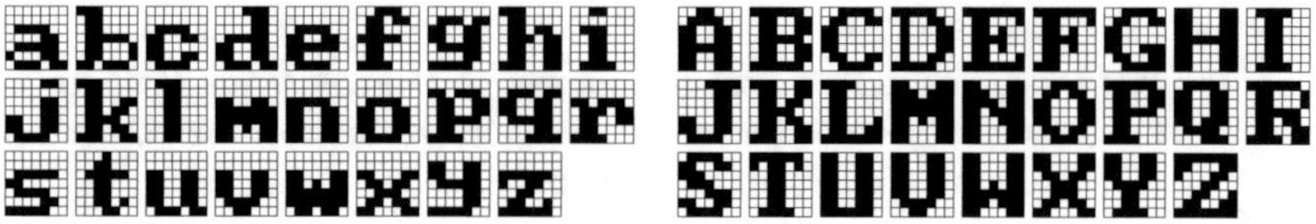

Fig. 8.6 Pattern pairs to be associated

each cell (pixel) is $\mathbf{A} = \begin{pmatrix} 0 & 1 & 0 \\ 1 & -4 & 1 \\ 0 & 1 & 0 \end{pmatrix}$. [Hint: Apply an approximation of central differences by Taylor series expansion.]

References

1. Abu-Mostafa, Y., & St Jacques, J. (1985). Information capability of the Hopfield network. *IEEE Transactions on Information Theory*, *31*(4), 461–464.
2. Aiyer, S. V. B., Niranjan, N., & Fallside, F. (1990). A theoretical investigation into the performance of the Hopfield model. *IEEE Transactions on Neural Networks*, *1*(2), 204–215.
3. Amari, S. I. (1972). Learning patterns and pattern sequences by self-organizing nets of threshold elements. *IEEE Transactions on Computers*, *21*, 1197–1206.
4. Amiri, M., Saeb, S., Yazdanpanah, M. J., & Seyyedsalehi, S. A. (2008). Analysis of the dynamical behavior of a feedback auto-associative memory. *Neurocomputing*, *71*, 486–494.
5. Amit, D. J., Gutfreund, G., & Sompolinsky, H. (1987). Statistical mechanics of neural networks near saturation. *Annals of Physics*, *173*, 30–67.
6. Amit, D. J. (1989). *Modeling brain function: The world of attractor neural networks*. Cambridge: Cambridge University Press.
7. Amit, D. J., Campbell, C., & Wong, K. Y. M. (1989). The interaction space of neural networks with sign-constrained synapses. *Journal of Physics A: General Physics*, *22*, 4687–4693.
8. Anderson, J. A. (1972). A simple neural network generating interactive memory. *Mathematical Biosciences*, *14*, 197–220.
9. Anderson, J. A., Silverstein, J. W., Ritz, S. A., & Jones, R. S. (1977). Distinctive features, categorical perception, and probability learning: Some applications of a neural model. *Psychological Review*, *84*, 413–451.
10. Baird, B. (1990). Associative memory in a simple model of oscillating cortex. In D. S. Touretzky (Ed.), *Advances in neural information processing systems* (Vol. 2, pp. 68–75). San Mateo: Morgan Kaufmann.
11. Bogacz, R., & Brown, M. W. (2003). Comparison of computational models of familiarity discrimination in the perirhinal cortex. *Hippocampus*, *13*, 494–524.
12. Bruck, J., & Roychowdhury, W. P. (1990). On the number of spurious memories in the Hopfield model. *IEEE Transactions on Information Theory*, *36*(2), 393–397.
13. Casali, D., Costantini, G., Perfetti, R., & Ricci, E. (2006). Associative memory design using support vector machines. *IEEE Transactions on Neural Networks*, *17*(5), 1165–1174.
14. Chartier, S., & Proulx, R. (2005). NDRAM: Nonlinear dynamic recurrent associative memory for learning bipolar and non-bipolar correlated patterns. *IEEE Transactions on Neural Networks*, *16*, 1393–1400.
15. Chartier, S., & Boukadoum, M. (2006). A bidirectional heteroassociative memory for binary and grey-level patterns. *IEEE Transactions on Neural Networks*, *17*(2), 385–396.

16. Chartier, S., Boukadoum, M., & Amiri, M. (2009). BAM learning of nonlinearly separable tasks by using an asymmetrical output function and reinforcement learning. *IEEE Transactions on Neural Networks*, *20*(8), 1281–1292.
17. Chiueh, T. D., & Goodman, R. M. (1991). Recurrent correlation associative memories. *IEEE Transactions on Neural Networks*, *2*(2), 275–284.
18. Chiueh, T. D., & Tsai, H. K. (1993). Multivalued associative memories based on recurrent networks. *IEEE Transactions on Neural Networks*, *4*(2), 364–366.
19. Cohen, M. A., & Grossberg, S. (1983). Absolute stability of global pattern formation and parallel memory storage by competitive neural networks. *IEEE Transactions on Systems, Man, and Cybernetics*, *13*, 815–826.
20. Conway, M. A. (2008). Exploring episodic memory. *Handbook of behavioral neuroscience* (Vol. 18, pp. 19–29). Amsterdam: Elsevier.
21. Coombes, S., & Taylor, J. G. (1994). Using generalized principal component analysis to achieve associative memory in a Hopfield net. *Network*, *5*, 75–88.
22. Coombes, S., & Campbell, C. (1996). *Efficient learning beyond saturation by single-layered neural networks*. Technical report 96.6, Bristol Center for Applied Nonlinear Mathematics, University of Bristol, UK.
23. Cortes, J. M., Greve, A., Barrett, A. B., & van Rossum, M. C. W. (2010). Dynamics and robustness of familiarity memory. *Neural Computation*, *22*, 448–466.
24. Cruz, B., Sossa, H., & Barron, R. (2007). A new two-level associative memory for efficient pattern restoration. *Neural Processing Letters*, *25*, 1–16.
25. Delbem, A. C. B., Correa, L. G., & Zhao, L. (2009). Design of associative memories using cellular neural networks. *Neurocomputing*, *72*, 2180–2188.
26. Eom, T., Choi, C., & Lee, J. (2002). Generalized asymmetrical bidirectional associative memory for multiple association. *Applied Mathematics and Computation*, *127*, 221–233.
27. Fleisher, M. (1988). The Hopfield model with multi-level neurons. In D. Z. Anderson (Ed.), *Neural information processing systems* (pp. 278–289). New York: American Institute Physics.
28. Fortney, K., Tweed, D. B., & Sejnowski, T. (2012). Computational advantages of reverberating loops for sensorimotor learning. *Neural Computation*, *24*(3), 611–634.
29. Gardner, E. (1987). Maximum storage capacity in neural networks. *Europhysics Letters*, *4*, 481–485.
30. Gardner, E. (1988). The space of the interactions in neural network models. *Journal of Physics A*, *21*, 257–270.
31. Grassi, G. (2001). On discrete-time cellular neural networks for associative memories. *IEEE Transactions on Circuits and Systems*, *48*(1), 107–111.
32. Grossberg, S. (1988). Nonlinear neural networks: Principles, mechanisms, and architectures. *Neural Networks*, *1*, 17–61.
33. Hassoun, M. H., & Watta, P. B. (1996). The Hamming associative memory and its relation to the exponential capacity DAM. In *Proceedings of IEEE International Conference on Neural Networks* (Vol. 1, pp. 583–587). Washington, DC.
34. Helie, S. (2008). Energy minimization in the nonlinear dynamic recurrent associative memory. *Neural Networks*, *21*, 1041–1044.
35. Hopfield, J. J. (1982). Neural networks and physical systems with emergent collective computational abilities. *Proceedings of the National Academy of Sciences of the USA*, *79*, 2554–2558.
36. Hui, S., & Zak, S. H. (1992). Dynamic analysis of the brain-state-in-a-box (BSB) neural models. *IEEE Transactions on Neural Networks*, *3*, 86–100.
37. Ikeda, N., Watta, P., Artiklar, M., & Hassoun, M. H. (2001). A two-level Hamming network for high performance associative memory. *Neural Networks*, *14*, 1189–1200.
38. Jagota, A., & Mandziuk, J. (1998). Experimental study of Perceptron-type local learning rule for Hopfield associative memory. *Information Sciences*, *111*, 65–81.
39. Jankowski, S., Lozowski, A., & Zurada, J. M. (1996). Complex-valued multi-state neural associative memory. *IEEE Transactions on Neural Networks*, *7*(6), 1491–1496.
40. Kamp, Y., & Hasler, M. (1990). *Recursive neural networks for associative memory*. New York: Wiley.

41. Kobayashi, K. (1991). On the capacity of a neuron with a non-monotone output function. *Network*, *2*, 237–243.
42. Kohonen, T. (1972). Correlation matrix memories. *IEEE Transactions on Computers*, *21*(4), 353–359.
43. Kohonen, T. (1989). *Self-organization and associative memory*. Berlin: Springer.
44. Kosko, B. (1987). Adaptive bidirectional associative memories. *Applied Optics*, *26*, 4947–4960.
45. Kosko, B. (1988). Bidirectional associative memories. *IEEE Transactions on Systems, Man, and Cybernetics*, *18*(1), 49–60.
46. Kumaran, D., Hassabis, D., & McClelland, J. L. (2016). What learning systems do intelligent agents need? Complementary learning systems theory updated. *Trends in Cognitive Sciences*, *20*(7), 512–534.
47. Lee, D. L. (2001). Improving the capacity of complex-valued neural networks with a modified gradient descent learning rule. *IEEE Transactions on Neural Networks*, *12*(2), 439–443.
48. Leung, C. S., & Chan, L. W. (1997). The behavior of forgetting learning in bidirectional associative memory. *Neural Computation*, *9*, 385–401.
49. Li, J. H., Michel, A. N., & Parod, W. (1989). Analysis and synthesis of a class of neural networks: Linear systems operating on a closed hypercube. *IEEE Transactions on Circuits and Systems*, *36*(11), 1405–1422.
50. Lin, W., & Chen, G. (2009). Large memory capacity in chaotic artificial neural networks: A view of the anti-integrable limit. *IEEE Transactions on Neural Networks*, *20*(8), 1340–1351.
51. Lippman, R. P. (1987). An introduction to computing with neural nets. *IEEE ASSP Magazine*, *4*(2), 4–22.
52. Liu, D., & Michel, A. N. (1994). Sparsely interconnected neural networks for associative memories with applications to cellular neural networks. *IEEE Transactions on Circuits and Systems*, *41*, 295–307.
53. Liu, D., & Lu, Z. (1997). A new synthesis approach for feedback neural networks based on the perceptron training algorithm. *IEEE Transactions on Neural Networks*, *8*(6), 1468–1482.
54. Liu, X.-G., Martin, R. R., Wu, M., & Tang, M.-L. (2008). Global exponential stability of bidirectional associative memory neural networks with time delays. *IEEE Transactions on Neural Networks*, *19*(3), 397–407.
55. Ma, J. (1997). The stability of the generalized Hopfield networks in randomly asynchronous mode. *Neural Networks*, *10*, 1109–1116.
56. Ma, J. (1999). The asymptotic memory capacity of the generalized Hopfield network. *Neural Networks*, *12*, 1207–1212.
57. McClelland, J. L., McNaughton, B. L., & O'Reilly, R. C. (1995). Why there are complementary learning systems in the hippocampus and neocortex: Insights from the successes and failures of connectionist models of learning and memory. *Psychological Review*, *102*, 419–457.
58. McEliece, R. J., Posner, E. C., Rodemich, E. R., & Venkatesh, S. S. (1987). The capacity of the Hopfield associative memory. *IEEE Transactions on Information Theory*, *33*(4), 461–482.
59. Michel, A. N., Si, J., & Yen, G. (1991). Analysis and synthesis of a class of discrete-time neural networks described on hypercubes. *IEEE Transactions on Neural Networks*, *2*(1), 32–46.
60. Morita, M. (1993). Associative memory with nonmonotonicity dynamics. *Neural Networks*, *6*, 115–126.
61. Muezzinoglu, M. K., Guzelis, C., & Zurada, J. M. (2003). A new design method for the complex-valued multistate Hopfield associative memory. *IEEE Transactions on Neural Networks*, *14*(4), 891–899.
62. Nakagawa, M. (1996). A parameter controlled chaos neural network. *Journal of the Physical Society of Japan*, *65*, 1859–1867.
63. Nemoto, I., & Kubono, M. (1996). Complex associative memory. *Neural Networks*, *9*(2), 253–261.
64. Oja, E. (1982). A simplified neuron model as a principal component analyzer. *Journal of Mathematical Biology*, *15*, 267–273.
65. Park, J., & Park, Y. (2000). An optimization approach to design of generalized BSB neural associative memories. *Neural Computation*, *12*, 1449–1462.

66. Park, J., Kim, H. Y., & Lee, S. W. (2001). A synthesis procedure for associative memories based on space-varying cellular neural networks. *Neural Networks*, *14*, 107–113.
67. Personnaz, L., Guyon, I., & Dreyfus, G. (1986). Collective computational properties of neural networks: New learning mechanism. *Physical Review A*, *34*(5), 4217–4228.
68. Ritter, G. X., Sussner, P., & de Leon, J. L. D. (1998). Morphological associative memories. *IEEE Transactions on Neural Networks*, *9*(2), 281–293.
69. Ritter, G. X., de Leon, J. L. D., & Sussner, P. (1999). Morphological bidirectional associative memories. *Neural Networks*, *6*(12), 851–867.
70. Rojas, R. (1996). *Neural networks: A systematic introduction*. Berlin: Springer.
71. Shi, H., Zhao, Y., & Zhuang, X. (1998). A general model for bidirectional associative memories. *IEEE Transactions on Systems, Man, and Cybernetics Part B*, *28*(4), 511–519.
72. Si, J., & Michel, A. N. (1995). Analysis and synthesis of a class of discrete-time neural networks with multilevel threshold neurons. *IEEE Transactions on Neural Networks*, *6*(1), 105–116.
73. Storkey, A. J. (1997). Increasing the capacity of the Hopfield network without sacrificing functionality. In W. Gerstner, A. Germond, M. Hastler, & J. Nicoud (Eds.), *Proceedings of International Conference on Artificial Neural Networks (ICANN)*, LNCS (Vol. 1327, pp. 451–456). Berlin: Springer.
74. Storkey, A. J., & Valabregue, R. (1997). Hopfield learning rule with high capacity storage of time-correlated patterns. *Electronics Letters*, *33*(21), 1803–1804.
75. Sussner, P., & Valle, M. E. (2006). Gray-scale morphological associative memories. *IEEE Transactions on Neural Networks*, *17*(3), 559–570.
76. Szolgay, P., Szatmari, I., & Laszlo, K. (1997). A fast fixed-point learning method to implement associative memoryon CNNs. *IEEE Transactions on Circuits and Systems*, *44*(4), 362–366.
77. Taylor, W. (1956). Eletrical simulation of some nervous system functional activities. *Information Theory*, *3*, 314–328.
78. Valle, M. E. (2014). Complex-valued recurrent correlation neural networks. *IEEE Transactions on Neural Networks and Learning Systems*, *25*(9), 1600–1612.
79. Venkatesh, S. S., & Psaltis, D. (1989). Linear and logarithmic capacities in associative memory. *IEEE Transactions on Information Theory*, *35*, 558–568.
80. Ventura, D., & Martinez, T. (2000). Quantum associative memory. *Information Sciences*, *124*, 273–296.
81. Wu, Y., & Batalama, S. N. (2000). An efficient learning algorithm for associative memories. *IEEE Transactions on Neural Networks*, *11*(5), 1058–1066.
82. Xu, Z. B., Leung, Y., & He, X. W. (1994). Asymmetric bidirectional associative memories. *IEEE Transactions on Systems, Man, and Cybernetics*, *24*(10), 1558–1564.
83. Yoshizawa, S., Morita, M., & Amari, S. I. (1993). Capacity of associative memory using a nonmonotonic neuron model. *Neural Networks*, *6*, 167–176.
84. Zeng, Z., & Wang, J. (2008). Design and analysis of high-capacity associative memories based on a class of discrete-time recurrent neural networks. *IEEE Transactions on Systems, Man, and Cybernetics Part B*, *38*(6), 1525–1536.
85. Zeng, Z., & Wang, J. (2009). Associative memories based on continuous-time cellular neural networks designed using space-invariant cloning templates. *Neural Networks*, *22*, 651–657.

Chapter 9
Clustering I: Basic Clustering Models and Algorithms

9.1 Vector Quantization

Vector quantization is a classical method that produces an approximation to a continuous pdf $p(\boldsymbol{x})$ of the vector variable $\boldsymbol{x} \in \mathcal{R}^n$ using a finite number of prototypes. That is, vector quantization represents a set of feature vectors $\boldsymbol{x}$ by a finite set of prototypes $\{\boldsymbol{c}_1, \ldots, \boldsymbol{c}_K\} \subset \mathcal{R}^n$. The finite set of prototypes is referred to as the *codebook*. Codebook design can be performed by using clustering algorithms. Once the codebook is specified, the approximation of $\boldsymbol{x}$ involves finding the reference vector $\boldsymbol{c}$ closest to $\boldsymbol{x}$ such that

$$\|\boldsymbol{x} - \boldsymbol{c}\| = \min_i \|\boldsymbol{x} - \boldsymbol{c}_i\| . \tag{9.1}$$

This is the nearest neighbor paradigm, and the procedure is actually the simple competitive learning.

The codebook can be designed by minimizing the expected squared quantization error

$$E = \int \|\boldsymbol{x} - \boldsymbol{c}\|^2 p(\boldsymbol{x}) \mathrm{d}\boldsymbol{x}, \tag{9.2}$$

where $\boldsymbol{c}$ satisfies (9.1), that is, $\boldsymbol{c}$ is a function of $\boldsymbol{x}$ and $\boldsymbol{c}_i$.

An iterative approximation scheme for finding the codebook is derived from criterion (9.2) [72]

$$\boldsymbol{c}_i(t+1) = \boldsymbol{c}_i(t) + \eta(t)\delta_{wi} \left[\boldsymbol{x}(t) - \boldsymbol{c}_i(t)\right], \tag{9.3}$$

where subscript w denotes the index of the prototype closest to $\boldsymbol{x}(t)$, termed the *winning prototype*, δ_{wi} is the Kronecker delta ($\delta_{wi} = 1$ for $w = i$, and 0 otherwise), and $\eta > 0$ is a small learning rate, satisfying the classical Robbins–Monro conditions

$$\sum \eta(t) = \infty \quad \text{and} \quad \sum \eta^2(t) < \infty. \tag{9.4}$$

K.-L. Du and M. N. S. Swamy, *Neural Networks and Statistical Learning*,
https://doi.org/10.1007/978-1-4471-7452-3_9

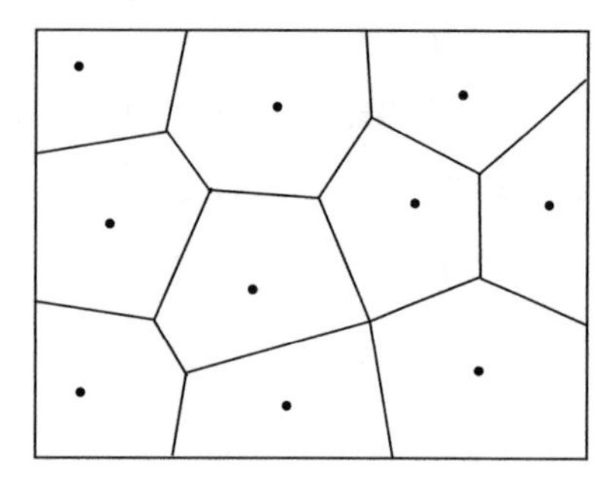

Fig. 9.1 Voronoi tessellation in two-dimensional space. Codebook vectors are denoted by black points

Typically, η is selected to be decreasing monotonically in time. For instance, one can select $\eta(t) = \eta_0 \left(1 - \frac{t}{T}\right)$, where $\eta_0 \in (0, 1]$ and T is the maximum number of iterations.

Voronoi tessellation, also called a *Voronoi diagram*, is useful for the illustration of vector quantization results. The space is partitioned into a finite number of regions bordered by hyperplanes. Each region is represented by a codebook vector, which is the nearest neighbor to any point within the same region. An illustration of Voronoi tessellation in the two-dimensional space is shown in Fig. 9.1. All vectors in one of the regions constitute a *Voronoi set*. For a smooth underlying probability density $p(\boldsymbol{x})$ and large K, all regions in an optimal Voronoi partition have the same within-region variance σ^2 [45].

9.2 Competitive Learning

Competitive learning can be implemented using a J-K neural network. The output layer is called the *competition layer* whose neurons are fully connected to the input nodes. In the competition layer, lateral connections are used to perform lateral inhibition. The architecture of the competitive learning network is shown in Fig. 9.2. For input $\boldsymbol{x}$, the network selects one of the K prototypes (weights) $\boldsymbol{c}_i$ by setting $y_i = 1$ and $y_j = 0$, $j \neq i$.

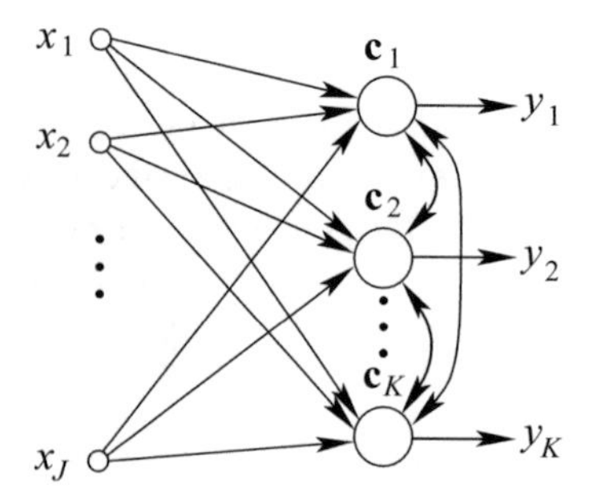

Fig. 9.2 Architecture of the J-K competitive learning network

The basic principle underlying competitive learning is the mathematical statistics problem called *cluster analysis*. Competitive learning is usually based on the minimization of a functional such as

$$E = \frac{1}{N} \sum_{p=1}^{N} \sum_{k=1}^{K} \mu_{kp} \left\| \boldsymbol{x}_p - \boldsymbol{c}_k \right\|^2, \tag{9.5}$$

where N is the size of the pattern set, and μ_{kp} is the connection weight assigned to prototype $\boldsymbol{c}_k$ with respect to $\boldsymbol{x}_p$, denoting the membership of pattern p in cluster k.

Minimization of (9.5) can lead to batch algorithms, but it is difficult to apply the gradient-descent method, since the winning prototypes must be determined with respect to each pattern $\boldsymbol{x}_p$. By using the functional

$$E_p = \sum_{k=1}^{K} \mu_{kp} \left\| \boldsymbol{x}_p - \boldsymbol{c}_k \right\|^2, \tag{9.6}$$

the gradient-descent method leads to sequential updating of the prototypes with respect to pattern $\boldsymbol{x}_p$. When $\boldsymbol{c}_k$ is the winning prototype of $\boldsymbol{x}_p$ in terms of the Euclidean metric, $\mu_{kp} = 1$; otherwise, $\mu_{kp} < 1$.

Simple competitive learning is derived by minimizing (9.5) under the assumption that the weights are obtained according to the nearest prototype condition

$$\mu_{kp} = \begin{cases} 1, & k = \arg_k \min \left\| \boldsymbol{x}_p - \boldsymbol{c}_k \right\| \\ 0, & \text{otherwise} \end{cases}. \tag{9.7}$$

Thus (9.5) becomes

$$E = \frac{1}{N} \sum_{p=1}^{N} \left\{ \min_{1 \le k \le K} \left\| \boldsymbol{x}_p - \boldsymbol{c}_k \right\|^2 \right\}. \tag{9.8}$$

This is the average of the squared Euclidean distances between the inputs $\boldsymbol{x}_p$ and their closest prototypes $\boldsymbol{c}_k$. The minimization of (9.8) implies that each input attracts only its winning prototype and has no effect on its nonwinning prototypes.

Based on the squared error criterion (9.6) and the gradient-descent method, assuming $\boldsymbol{c}_w(t)$ to be the winning prototype of $\boldsymbol{x}_t$, we get the simple competitive learning as

$$\boldsymbol{c}_w(t+1) = \boldsymbol{c}_w(t) + \eta(t) \left[\boldsymbol{x}_t - \boldsymbol{c}_w(t) \right], \tag{9.9}$$

$$\boldsymbol{c}_i(t+1) = \boldsymbol{c}_i(t), \quad i \neq w, \tag{9.10}$$

where $\eta(t)$ can be selected according to (9.4). The process is known as *winner-takes-all (WTA)*. In the WTA process, agents in a group compete with each other and only the one with the highest input stays active while all the others are deactivated. This phenomenon widely exists in nature and society. The WTA mechanism plays an important role in the design of unsupervised learning neural networks. If each cluster

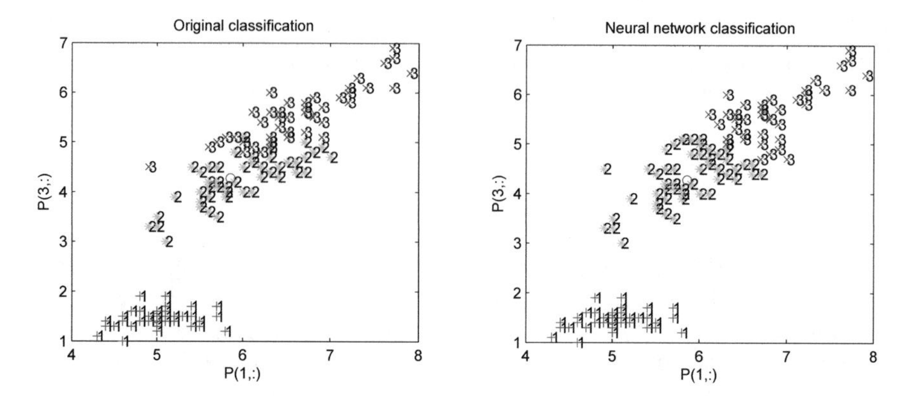

Fig. 9.3 Classification result for the iris dataset

has its own learning rate as $\eta_i = \frac{1}{N_i}$, where N_i is the number of samples assigned to the ith cluster, the algorithm achieves the minimum output variance [117]. k-winner-takes-all (k-WTA) is a process of selecting k winners from the codebook for a training sample $\boldsymbol{x}_t$.

Winner-kill-loser is another rule for competitive learning and has been applied to the neocognitron [43]. Every time when a training sample is presented, nonsilent cells compete with each other. The winner not only takes all but also kills losers. In other words, the winner learns the training sample, and losers are removed from the network. If all cells are silent, a new cell is generated and it learns the training sample.

Example 9.1 For the iris dataset, we use simple competitive learning model to classify the dataset. We set the number of training epochs to 100, and the learning rate to 0.3. The original classification and the output of the neural network classification are shown in Fig. 9.3. The rate of correct classification is 89.33%.

9.3 Self-Organizing Maps

The different regions of the cerebral cortex respond to different sensory inputs (e.g., visual-, auditory-, motor-, or somatosensory), and topographically ordered mappings are widely observed in the cortex. A cytoarchitectural map of the cerebral cortex is presented in Fig. 9.4 [14]. The primary sensory regions of the cortical maps are established genetically in a predetermined manner, and more detailed associative areas between the primary sensory areas are gradually developed through topographical self-organization during life [68]. There exist two main types of brain maps [74]: pointwise-ordered projections from a receptive surface onto a cortical area (e.g., the somatotopic and visual maps), and abstract or computational maps, which are ordered along with some sensory feature value or a computed entity (e.g., the color map in area 4 of the visual cortex and the target-range map in the mustache bat auditory cortex).

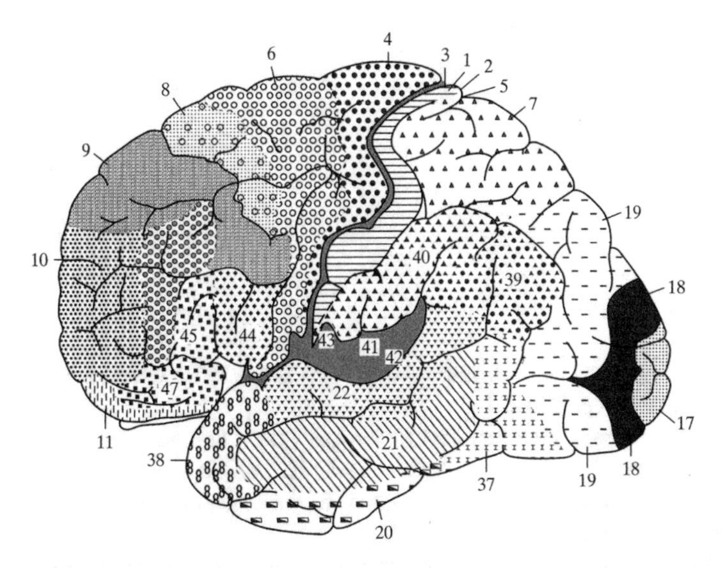

Fig. 9.4 Cytoarchitectural map of the cerebral cortex. The different areas of the cerebral cortex have different layer thicknesses and types of cells. Some of the sensory areas are motor cortex (area 4), premotor area (area 6), frontal eye fields (area 8), somatosensory cortex (areas 1, 2, 3), visual cortex (areas 17, 18, 19), and auditory cortex (areas 41, 42). ©A. Brodal, 1981, Oxford University Press [14]

Von der Malsburg's line-detector model [112] and Kohonen's SOM [66] are two well-known topology-preserving competitive learning models. They are of abstract or computational maps. The line-detector model is based on fixed excitatory and inhibitory lateral connections and the Hebbian rule of synaptic plasticity of the afferent connections; however, the natural signal patterns are usually more complex. SOM models the sensory-to-cortex mapping and is an unsupervised, associative memory mechanism. In [74], a pointwise-ordered projection from the input layer to the output layer is created in a self-organized fashion relating to SOM. If the input layer consists of feature detectors, the output layer forms a feature map of the inputs.

SOM is well known for its ability to perform clustering while preserving topology. It compresses information while preserving the most important topological and metric relationships of the primary data elements. SOM can be regarded as competitive learning with a topological constraint. It is useful for vector quantization, clustering, feature extraction, and data visualization. The Kohonen learning rule is a major development of competitive learning.

9.3.1 Kohonen Network

The Kohonen network is a J-K feedforward structure with fully interconnected processing units that compete for signals. The output layer is called the *Kohonen layer*. Input nodes are fully connected to output neurons with their associated weights. Lateral connections between neurons are used as a form of feedback whose magnitude

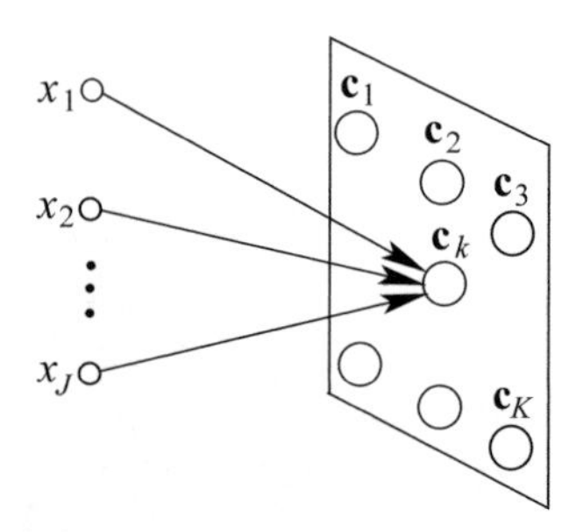

Fig. 9.5 Architecture of the two-dimensional J-K Kohonen network

is dependent on the lateral distance from a specific neuron, which is characterized by a neighborhood parameter.

The Kohonen network defined on R^n is a one-, two-, or higher dimensional grid $\mathcal{A}$ of neurons characterized by prototypes $\boldsymbol{c}_k \in R^n$ [67, 68]. $\boldsymbol{c}_k$ can also be viewed as the weight vector to neuron k. The architecture of the network in the two-dimensional space is illustrated in Fig. 9.5.

The Kohonen network uses competitive learning. Patterns are presented sequentially in time through the input layer, without specifying the desired output. The Kohonen network is extended to SOM when the lateral feedback is more sophisticated than the WTA rule. For example, the lateral feedback used in SOM can be selected as the so-called *Mexican-hat function*, which is observed in the visual cortex.

9.3.2 Basic Self-Organizing Maps

SOM employs the Kohonen network topology. An SOM not only categorizes the input data but also recognizes which input patterns are nearby one another in stimulus space. For each neuron k, compute the Euclidean distance to input pattern $\boldsymbol{x}$, and find the neuron whose prototype is closest to $\boldsymbol{x}$:

$$\|\boldsymbol{x}_t - \boldsymbol{c}_w\| = \min_{k \in \mathcal{A}} \|\boldsymbol{x}_t - \boldsymbol{c}_k\|, \tag{9.11}$$

where subscript w denotes the winning neuron, called the *excitation center*, which becomes the center of a group of input vectors that lie closest to $\boldsymbol{c}_w$.

For all the input vectors closest to $\boldsymbol{c}_w$, update all the prototype vectors by

$$\boldsymbol{c}_k(t+1) = \boldsymbol{c}_k(t) + \eta(t) h_{kw}(t) \left[\boldsymbol{x}_t - \boldsymbol{c}_k(t)\right], \quad k = 1, \ldots, K, \tag{9.12}$$

where $\eta(t)$ is selected according to (9.4), and $h_{kw}(t)$ is the so-called *excitation response* or *neighborhood function*, which defines the response of neuron k when $\boldsymbol{c}_w$ is the excitation center. Equation (9.12) is known as the *Kohonen learning rule* [68].

If $h_{kw}(t) = 1$ for $k = w$ and 0 otherwise, (9.12) reduces to simple competitive learning. $h_{kw}(t)$ can be selected as a function that decreases with an increasing

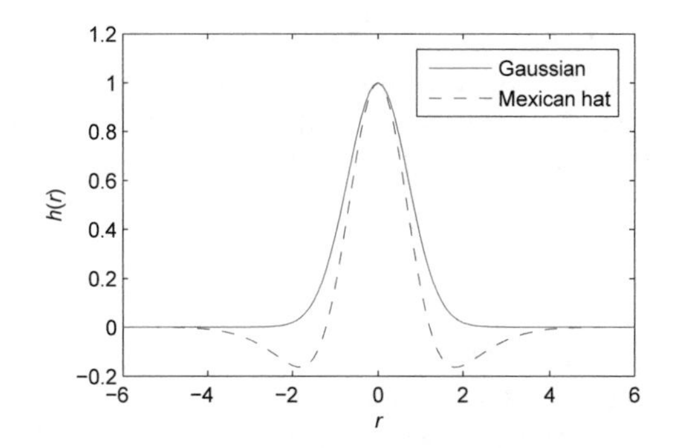

Fig. 9.6 The Gaussian and Mexican-hat functions: $h(r) = \mathrm{e}^{-r^2}$ and $h(r) = \frac{1}{2}\left(3\mathrm{e}^{-r^2} - \mathrm{e}^{-\frac{r^2}{4}}\right)$

distance between $\boldsymbol{c}_k$ and $\boldsymbol{c}_w$, and is typically selected as the Gaussian function

$$h_{kw}(t) = h_0 \mathrm{e}^{-\frac{\|\boldsymbol{c}_k - \boldsymbol{c}_w\|^2}{\sigma^2(t)}}, \tag{9.13}$$

where $h_0 > 0$ is a constant. In SOM, the topological neighborhood shrinks with time, thus $\sigma(t)$ is a decreasing function of t, and a popular choice is the exponential decay with time [95]

$$\sigma(t) = \sigma_0 \mathrm{e}^{-\frac{t}{\tau}}, \tag{9.14}$$

where σ_0 is a positive constant and τ is a time constant.

Another popular neighborhood function is the Mexican-hat function. A Mexican-hat function is well suited to bipolar stimuli and is described by

$$h_{kw}(t) = \frac{1}{2} h_0 \left(3\mathrm{e}^{-\frac{\|\boldsymbol{c}_k - \boldsymbol{c}_w\|^2}{\sigma^2}} - \mathrm{e}^{-\frac{\|\boldsymbol{c}_k - \boldsymbol{c}_w\|^2}{4\sigma^2}}\right), \tag{9.15}$$

where $\sigma(t)$ is defined by (9.14). The Gaussian and Mexican-hat functions are plotted in Fig. 9.6.

The Gaussian topological neighborhood is biologically more reasonable than a rectangular one. SOM using the Gaussian neighborhood converges more quickly than SOM using a rectangular one [86]. SOM is given by Algorithm 9.1.

Algorithm 9.1 (SOM)

1. *Set* $t = 0$.
2. *Initialize all* $\boldsymbol{c}_k(0)$ *and learning parameters* $\eta(0)$, h_0, σ_0, *and* τ.
3. **Repeat until** *a criterion is satisfied:*

   a. *Present pattern* $\boldsymbol{x}_t$ *at time* t.
   b. *Select the winning neuron for* $\boldsymbol{x}_t$ *by (9.11).*
   c. *Update the prototypes for all neurons by (9.12).*
   d. *Set* $t = t + 1$.

The algorithm can be stopped when the map achieves an equilibrium with a given accuracy or when a specified number of iterations is reached. In the convergence phase, h_{wk} can be selected as time-invariant, and each prototype is recommended to be updated by using an individual learning rate η_k [72]

$$\eta_k(t+1) = \frac{\eta_k(t)}{1 + h_{wk}\eta_k(t)}. \tag{9.16}$$

Normalization of $\boldsymbol{x}$ is suggested since the resulting reference vectors tend to have the same dynamic range. This may improve the numerical accuracy [68].

After learning is completed, the network is ready for generalization. When a new pattern $\boldsymbol{x}$ is presented to the map, the corresponding output $\boldsymbol{c}$ is determined according to the mapping: $\boldsymbol{x} \to \boldsymbol{c}$ such that $\|\boldsymbol{x} - \boldsymbol{c}\| = \min_{r \in \mathcal{A}} \|\boldsymbol{x} - \boldsymbol{c}_r\|$. The mapping performs vector quantization of the input space into the map $\mathcal{A}$.

Compared with the symmetric neighborhood function, an asymmetric neighborhood function for SOM accelerates the ordering process of SOM [4], though this asymmetry tends to distort the generated ordered map. The number of learning steps required for perfect ordering in the case of the one-dimensional SOM is numerically shown to be reduced from $O(N^3)$ to $O(N^2)$ with an asymmetric neighborhood function, even when the improved algorithm is used to get the final map without distortion.

SOM is deemed to converge to an organized configuration in one- or higher dimensional SOM with probability one. In the literature, there are some proofs for the convergence of one-dimensional SOM based on Markov chain analysis [40]. However, no general proof of convergence for multidimensional SOM is available. SOM suffers from several major problems such as forced termination, unguaranteed convergence, nonoptimized procedures, and the output being often dependent on the sequence of data. SOM is not derived from any known objective function, and its termination is not based on optimizing any model of the process or its data. It is closely related to C-means clustering [85]. SOM is shown to be an asymptotically optimal vector quantization [120]. With neighborhood learning, it is an error-tolerant vector quantization [87] and a Bayesian vector quantization [88].

SOM with dynamic learning [25] improves SOM training on signals with sparse events which allows for more representative prototype vectors to be found, and consequently better signal reconstruction. The training rule is given by [25]

$$\boldsymbol{c}_k(t+1) = \boldsymbol{c}_k(t) + \eta(t)h_{kw}(t)\text{sgn}(\boldsymbol{x}_t - \boldsymbol{c}_k(t))\|\boldsymbol{x}_t - \boldsymbol{c}_k(t)\|^2, \quad k = 1, \ldots, K. \tag{9.17}$$

Parameterless SOM [10] calculates the learning rate and neighborhood size based on the local quadratic fitting error of the map to the input space. This allows the map to make large adjustments in response to unfamiliar inputs, while making small changes in response to inputs it is already well adjusted to. It markedly decreases the number of iterations required to get a stable and ordered map. Parameterless SOM is measurably less ordered than a properly tuned SOM, and edge shrinking is also more

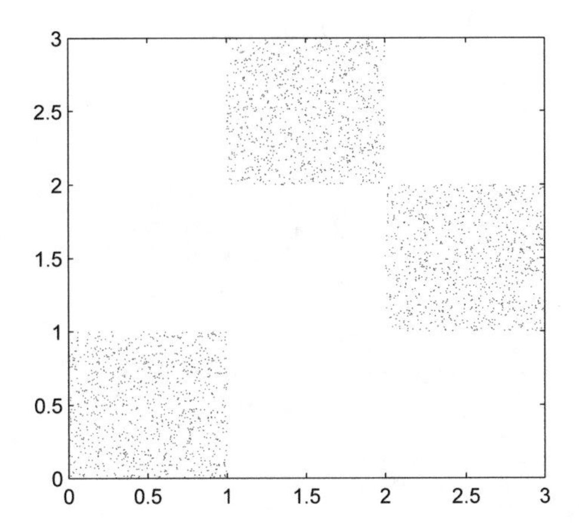

Fig. 9.7 Random data points in two-dimensional space. In each of the two quarters, there are 1000 uniformly random points

marked in parameterless SOM. It is able to handle input probability distributions that lead to failure of SOM. It is guaranteed to achieve ordering under certain conditions.

Like classical vector quantization method, SOM was originally intended to approximate input signals or their pdfs by quantified codebook vectors that are localized in the input space to minimize a quantization error functional [68]. SOM is related to adaptive C-means, but performs a topological feature map that is more complex than just cluster analysis. The topology-preservation property makes SOM a popular choice in data analysis. However, SOM is not a good choice in terms of clustering performance compared to other popular clustering algorithms such as C-means [89], neural gas [91], and ART 2A [57]. Besides, for large output dimensions, the number of nodes in the adaptive SOM grid increases exponentially with the number of function parameters. The prespecified standard grid topology may not be able to match the structure of the distribution and can thus lead to poor topological mappings.

When M^2 is the size of a feature map, the number of compared weight vectors for one input vector to search a winner vector by exhaustive search is equivalent to M^2. In [77], the proposed SOM algorithm with $O(\log_2 M)$ complexity is composed of a subdividing method and a binary search method. Only winner vectors are trained. The algorithm subdivides the map repeatedly, and new nodes of weight vectors emerge in every step.

Complex-valued SOM performs adaptive clustering of the feature vector in the complex-amplitude space [53].

Example 9.2 We implement vector quantization using SOM with a grid of cells. The dataset is composed of 1500 random data points in the two-dimensional space: 500 uniformly random points in each of the three unit squares, as shown in Fig. 9.7.

The link distance[1] is employed. All prototypes of the cells are initialized at the center of the range of the dataset, namely (1.5, 1.5). The ordering phase starts from

[1]The link distance between two points A and B inside a polygon P is defined to be the minimum number of edges required to connect A and B inside P.

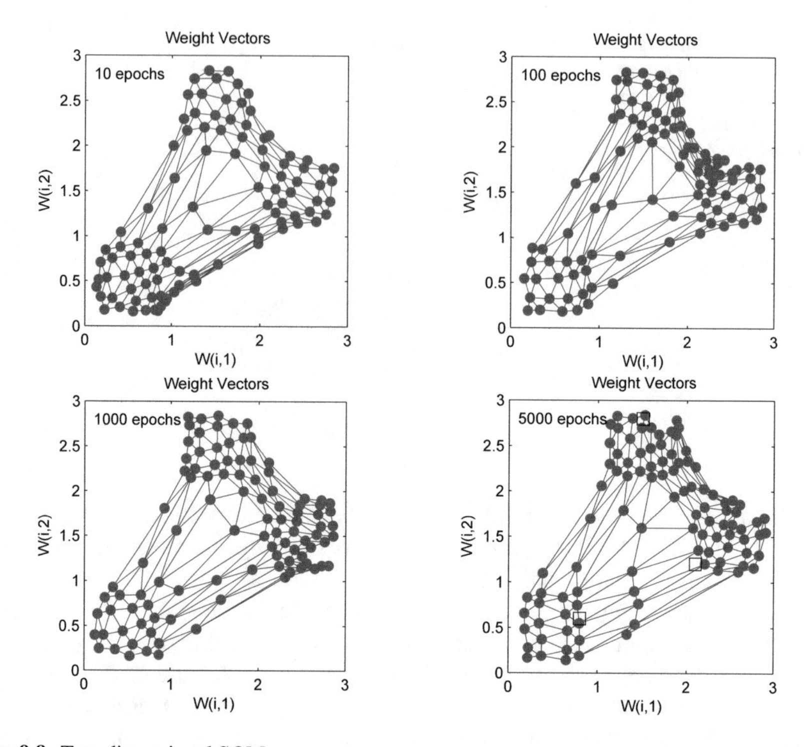

Fig. 9.8 Two-dimensional SOM

a learning rate of 0.9 and decreases to the tuning-phase learning rate 0.02 in 1000 epochs, and then the tuning phase lasts much longer time with a slowly decreasing learning rate. In the tuning phase, the neighborhood distance is set as 1. When training is completed, two points, $\boldsymbol{p}_1 = (0.8, 0.6)$, $\boldsymbol{p}_2 = (1.5, 2.8)$ and $\boldsymbol{p}_3 = (2.1, 1.2)$, are used as test points.

In the first group of simulations, the output cells are arranged in a 10×10 grid. The hexagonal neighborhood topology is employed. The training results for 10, 100, 1000, and 5000 epochs are shown in Fig. 9.8. At 5000 epochs, we tested $\boldsymbol{p}_1$, $\boldsymbol{p}_2$, and $\boldsymbol{p}_3$, and found that they, respectively, belong to the 25th, 93rd, and 27th clusters.

In the second group of simulations, the output cells are arranged in a one-dimensional grid of 100 nodes. The corresponding results are shown in Fig. 9.9. In this case, $\boldsymbol{p}_1$, $\boldsymbol{p}_2$, and $\boldsymbol{p}_3$, respectively, belong to the 68th, 61st, and 48th clusters.

Example 9.3 SOM can be applied to solve the TSP [41]. The process results in a neural encoding that gradually relaxes toward a valid tour. Assume that 40 cities are randomly located in a unit square. The objective is to find the shortest route that passes through all the cities, each city being visited exactly once. No constraint is applied on the Kohonen network since the topology of the solution is contained in the network topology. A one-dimensional grid of 80 units is used by the SOM. The desired solution is that all the cities are covered by nodes, and all the additional

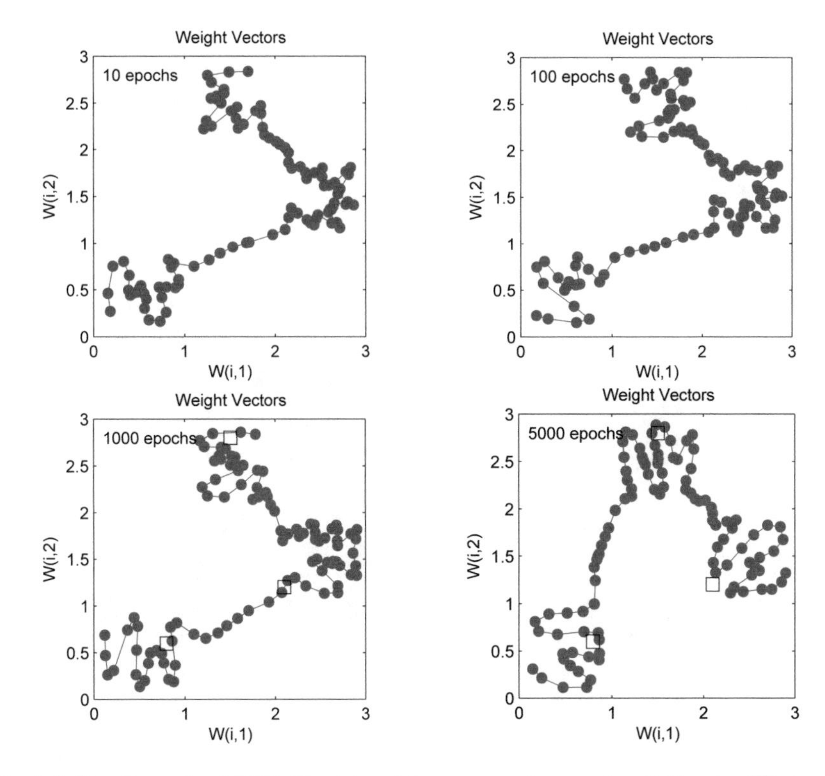

Fig. 9.9 One-dimensional SOM. The data points $\boldsymbol{p}_1$, $\boldsymbol{p}_2$, and $\boldsymbol{p}_3$ are denoted by plus (□) signs

nodes are along the lines between cities. The Euclidean distance is employed. Other parameters are the same as those for Example 9.2. The search results at the 10th, 100th, 1,000th, and 10,000th epochs are illustrated in Fig. 9.10, and the total map length at the 10,000th epoch is 5.2464.

It is seen that the results from SOM are not satisfactory, and the routes are not feasible solutions since some cities are not covered by nodes. Nevertheless, SOM can be used to find a preliminary search for a suboptimal route, which can be modified manually to obtain a feasible solution. The SOM solution can be used as an initialization of other TSP solvers. In this case, we do not need to run SOM for many epochs.

Classical SOM is not efficient for searching suboptimal solution for the TSP. Many practical TSP solvers have been developed using self-organizing neural network models based on SOM, among which some solvers can find a suboptimal solution for a TSP of hundred cities within dozens of epochs. Most of them are based on the concept of *elastic ring* [60]. A large-scale TSP can be rapidly solved by a divide-and-conquer technique, where clustering methods are first used to group the cities and a local optimization algorithm is used to find the minimum in each group [94]. This speedup is offset by a slight loss in tour quality, but the structure is suitable for parallel implementation.

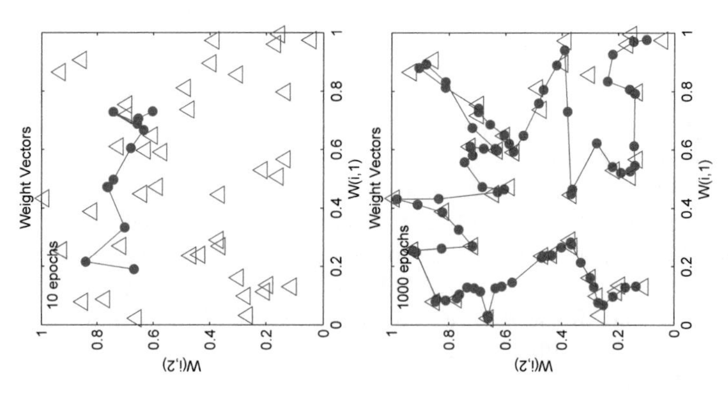

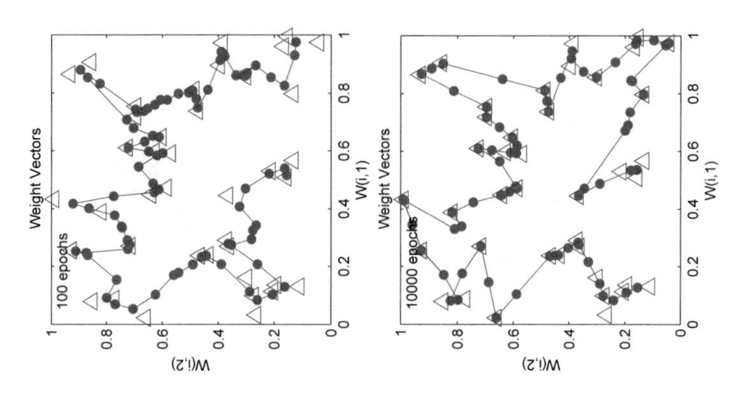

Fig. 9.10 The TSP using one-dimensional SOM. Circles denote positions of the cities

Batch-Mode SOM

In the batch SOM algorithm [70], the prototypes are updated once for each epoch:

$$\boldsymbol{c}_k(t_f) = \frac{\sum_{t=t_0}^{t_f} h_{wk}(t)\boldsymbol{x}_t}{\sum_{t=t_0}^{t_f} h_{wk}(t)}, \tag{9.18}$$

where t_0 and t_f denote the start and finish of the present epoch, respectively, and $\boldsymbol{c}_k(t_f)$ are the prototype vectors computed at the end of the present epoch. The winning node at each presentation is computed using

$$d_w(t) = \min_k \|\boldsymbol{x}_t - \boldsymbol{c}_k(t_0)\|^2, \tag{9.19}$$

where $\boldsymbol{c}_k(t_0)$ are the prototype vectors computed at the end of the previous epoch. The neighborhood functions $h_{wk}(t)$ are computed from (9.13), but with the winning nodes determined from (9.19).

Compared with the conventional online SOM method, batch SOM offers no dependence upon the order in which the input records are presented. In addition to facilitating the development of data-partitioned parallel methods, this also eliminates concerns that input records encountered later in the training sequence may overly influence the final results. The learning rate coefficient $\alpha(t)$ does not appear in batch SOM. Batch C-means and batch SOM optimize the same cost functions as their online variants [21]. The batch training algorithm is generally much faster than the incremental algorithm.

Adaptive-Subspace SOM (ASSOM)

Adaptive-subspace SOM (ASSOM) [71, 72] is a modular neural network model comprising an array of topologically ordered SOM submodels. ASSOM creates a set of local subspace representations by competitive selection and cooperative learning. Each submodel is responsible for describing a specific region of the input space by its local principal subspace and represents a manifold such as a linear subspace with small dimensionality, whose basis vectors are determined adaptively. ASSOM not only inherits the topological representation property of SOM but provides learning results that reasonably describe the kernels of various transformation groups like PCA. ASSOM is used to learn a number of invariant features, usually pieces of elementary one- or two-dimensional waveforms with different frequencies called *wavelets*, independent of their phases. Two fast implementations of ASSOM are proposed in [124] based on the basis rotation operator of ASSOM.

9.4 Learning Vector Quantization

LVQ [67, 68] is a widely used approach to classification. LVQ employs exactly the same network architecture as the Kohonen network with the exception that each output neuron is specified with a class membership and no assumption is made concerning the topological structure. The LVQ network is associated with the two-layer competitive learning network shown in Fig. 9.2.

LVQ is based on the known classification of feature vectors and can be treated as a supervised version of SOM. It is used for vector quantization and classification, as well as for fine-tuning of SOM. LVQ algorithms define near-optimal decision borders between classes, even in the sense of classical Bayesian decision theory.

LVQ minimizes the functional (9.5), where $\mu_{kp} = 1$ if neuron k is the winner and zero otherwise, when pattern pair p is presented. LVQ works on a set of N pattern pairs $(\boldsymbol{x}_p, \boldsymbol{y}_p)$, where $\boldsymbol{x}_p \in R^J$ is the input vector and $\boldsymbol{y}_p \in R^K$ is the binary target vector that codes the class membership, that is, only one entry of $\boldsymbol{y}_p$ takes value unity, while all its other entries are zero. Kohonen proposed a family of LVQ algorithms including LVQ1, LVQ2, and LVQ3 [68]. Assuming that pattern p is presented at time t, LVQ1 is given as

$$\begin{aligned} \boldsymbol{c}_w(t+1) &= \boldsymbol{c}_w(t) + \eta(k)\left[\boldsymbol{x}_t - \boldsymbol{c}_w(t)\right], & y_{p,w} &= 1, \\ \boldsymbol{c}_w(t+1) &= \boldsymbol{c}_w(t) - \eta(t)\left[\boldsymbol{x}_t - \boldsymbol{c}_w(k)\right], & y_{p,w} &= 0, \\ \boldsymbol{c}_i(t+1) &= \boldsymbol{c}_i(t), & i &\neq w, \end{aligned} \tag{9.20}$$

where w is the index of the winning neuron, $\boldsymbol{x}_t = \boldsymbol{x}_p$, $y_{p,w} = 1$ and 0 represent the cases of correct and incorrect classifications of $\boldsymbol{x}_p$, respectively, and $\eta(t)$ is defined as in earlier formulations. When it is used to fine-tune SOM, one can start with small $\eta(0)$, usually less than 0.1. LVQ1 tends to reduce the point density of $\boldsymbol{c}_i$ around the Bayesian decision surfaces.

LVQ1 can be considered a modified version of online C-means in which class labels affect the way that the clustering process is performed and online gradient descent is used over a cost function [11]. LVQ1 is given by Algorithm 9.2.

Algorithm 9.2 (LVQ1)

1. *Set* $t = 0$.
2. *Initialize all* $\mathbf{c}_k(0)$ *and* $\eta(0)$.
3. **Repeat until** *a criterion is satisfied:*

   a. *Present pattern* $\boldsymbol{x}_t$.
   b. *Select the winning neuron for* $\boldsymbol{x}_t$ *by (9.1).*
   c. *Update the prototypes for all neurons by (9.20).*
   d. *Decrease* $\alpha(t)$.
   e. *Set* $t = t + 1$.

OLVQ1 is an optimized version of LVQ1. In OLVQ1, each codebook vector $\boldsymbol{c}_i$ is assigned an individual adaptive learning rate [73]

$$\eta_i(t) = \frac{\eta_i(t-1)}{1 + s(t)\eta_i(t-1)}, \tag{9.21}$$

where $s(t) = +1$ for correct classification and $s(t) = -1$ for wrong classification. Since $\eta_i(t)$ may increase, it should be limited to be less than 1. One can restrict $\eta_i(t) < \eta_i(0)$, and set $\eta_i(0) = 0.3$. The convergence of OLVQ1 may be up to one order of magnitude faster than that of LVQ1.

LVQ2 and LVQ3 comply better with the Bayesian decision surface. In LVQ1, only one codebook vector $\boldsymbol{c}_i$ is updated at each step, while LVQ2 and LVQ3 change two codebook vectors simultaneously. Different LVQ algorithms can be combined in the clustering process. However, both LVQ2 and LVQ3 have the problem of reference vector divergence [105]. In a generalization of LVQ2 [105], this problem is eliminated by applying gradient descent on a nonlinear cost function.

Addition of training counters to individual neurons of LVQ can effectively record the training statistics of LVQ [96]. This allows for dynamic self-allocation of the neurons to classes. At the generalization stage, these counters provide an estimate of the reliability of classification of the individual neurons. The method turns out to be especially valuable in handling strongly overlapping class distributions in pattern space.

Example 9.4 We generate 50 data points of two classes that are nonlinear separable. An LVQ network can solve this problem with no difficulty. LVQ1 is used in this example. We set the number of training epochs to 200 and the learning rate to 0.02. The original classification and the output of the LVQ network are shown in Fig. 9.11. For this trial, a training MSE of 0 is achieved, and the classification for 10 test points generates reasonable results.

LVQ and its variants are purely heuristically motivated local learners for adaptive nearest prototype classification. They suffer from the problem of instabilities for overlapping classes. They are sensitive to the initialization of prototypes and are restricted to classification scenarios in Euclidean space. Generalized relevance LVQ copes with these problems by integrating neighborhood cooperation to deal with local optima [52]. It shows very robust behavior. It obeys gradient dynamics, and the chosen objective is related to margin optimization.

Using concepts from statistical physics and online learning, a mathematical framework is presented in [13, 46] to analyze the performance of different LVQ algorithms including LVQ1, LVQ2.1 [69], and learning-from-mistakes in terms of their dynamics, sensitivity to initial conditions, and generalization ability. LVQ1 shows near-optimal asymptotic generalization error for all choices of the prior distribution in the equal class variance case, independent of initialization. Learning-from-mistakes is a crisp version of robust soft LVQ [107]. A global cost function is lacking for LVQ1, whereas a cost function is available for a soft version like for LVQ2.1 and learning-from-mistakes. Soft LVQ algorithms [107] are derived from an objective function

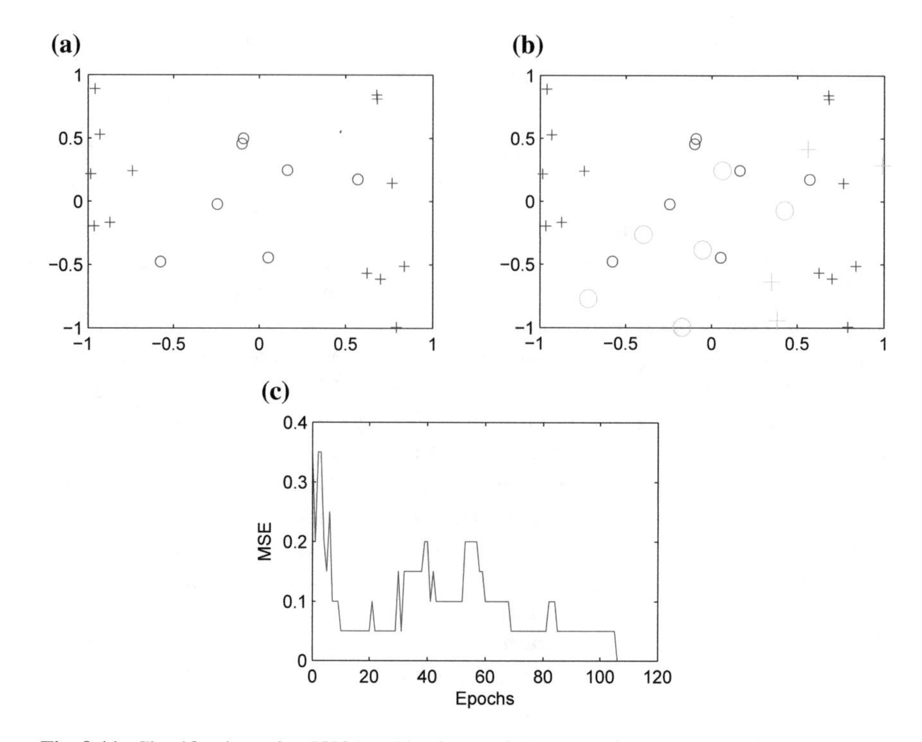

Fig. 9.11 Classification using LVQ1. **a** The dataset. **b** The classification result. **c** MSE

based on a likelihood ratio using gradient descent, leading to better classification performance than LVQ2.1. The behavior of LVQ2.1 is unstable.

9.5 Nearest Neighbor Algorithms

The nearest neighbor classifier is the earliest and also a widely used classifier [39]. It is a local learning system: it fits the training data only in a region around the location of an input pattern. The k-nearest neighbor (k-NN) approach is statistically inspired in the estimation of the posterior probability $p(H_i|\boldsymbol{x})$ of the hypothesis H_i, conditioned on an observation point $\boldsymbol{x}$. It is among the simplest, nonparametric, and most successful classification methods.

A k-NN classifier classifies an input by identifying the k examples with the closest inputs and assigning the class label from the majority of those examples. The algorithm is simple to implement, and it works fast for small training sets. The Parzen window approach has the drawback that the data are very sensitive to the choice of cell size, while k-NN solves this by letting the cell volume be a function of the data. The classification performance of k-NN varies significantly with k. Therefore, the optimal value of k can be found by using a trial-and-error procedure. In k-NN, all neighbors receive equal importance.

k-NN is also used for outlier detection. All training patterns are used as prototypes, and an input pattern is assigned to the class with the closest prototype. It generalizes well for large training sets, and the training set can be extended at any time.

k-NN converges to Bayes' classifier as the number of neighbors k and the number of prototypes M tend to infinity at an appropriate rate for all distributions. The theoretical asymptotic classification error is upper bounded by twice Bayes' error. Under mild regularity assumptions, the error rate of the 1-NN classifier is asymptotically bounded by twice the Bayesian error rate, when the sample size tends to infinity [28].

However, it requires the storage of the whole training set which may be excessive for large datasets and has a computational complexity of $O(N^2)$ for a set of N patterns. It also takes a long time for recall. Thus, k-NN is impractical for large training sets. To classify pattern $\boldsymbol{x}$, k-NN is given by Algorithm 9.3.

Algorithm 9.3 **(k-NN)**

1. *Find the k nearest patterns to* $\boldsymbol{x}$ *in the set of prototypes* $\mathcal{P} = \{(\boldsymbol{m}_j, cl(\boldsymbol{m}_j)), j = 0, \ldots, M-1\}$, *where* $\boldsymbol{m}_j$ *is a prototype that belongs to one of the M classes and* $cl(\boldsymbol{m}_j)$ *is the class indicator variable.*
2. *Classify by a majority vote amongst these k patterns.*

By sorting the data with respect to each attribute as well as using complex data structures such as kd-trees, significant gain in performance can be obtained. By replacing the sort operation with the calculation of the order statistics, the k-NN method can further be improved in speed and its stability with respect to the order of presentation of the data [8].

Nearest neighbor classification is a lazy learning method because training data are not preprocessed in any way. The class assigned to a pattern is the class of the nearest pattern known to the system, measured in terms of a distance defined on the feature space. On this space, each pattern defines its Voronoi region. For the Euclidean distance, Voronoi regions are delimited by linear borders. In practice, $k = 1$ is a common choice, since the Euclidean 1-NN classifier forms class boundaries with piecewise-linear hyperplanes and any border can be approximated by a series of locally defined hyperplanes. Due to using only the training point closest to the query point, the bias of the 1-NN estimate is often low, but the variance is high.

Considering a larger number of codebook vectors close to an input sample may lead to lower error rates than using the nearest prototype only, thus the k-NN rule usually outperforms the 1-NN rule. PAC error bounds for k-NN classifiers are $O(N^{-2/5})$ for a training set of size N [7].

Nearest neighbor classifier has infinite VC-dimension [108], implying that it tends to overfit the data. This problem can be mitigated by taking the majority vote among $k > 1$ nearest neighbors [108], or by deleting some sample points so as to attain a larger margin [47].

The set of prototypes $\mathcal{P}$ is computed from training data. A simple method is to select the whole training set as $\mathcal{P}$, but this results in large memory and execution

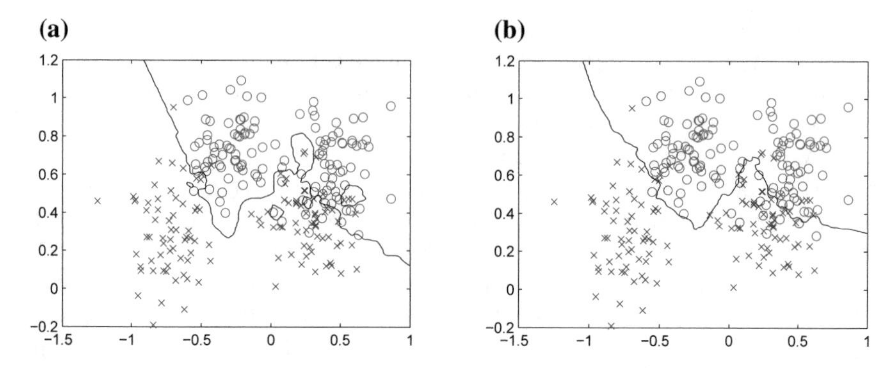

Fig. 9.12 The decision boundary of the k-NN classifier: **a** $k = 4$, **b** $k = 16$

requirements for large databases. Therefore, in practice, a small set of prototypes of size M is mandatory.

Example 9.5 By using STPRtool (http://cmp.felk.cvut.cz/cmp/software/stprtool/), we create the k-NN classification rule using the Riply's data for training and testing. The training data and the decision boundary are plotted in Fig. 9.12. The testing error is 0.15 for $k = 1$, 0.127 for $k = 4$, and 0.082 for $k = 16$.

Improving k-NN

A set of unlabeled prototypes can be first obtained from training data by clustering. These prototypes can then be used to divide the input space into k-NN cells. Finally, labels are assigned to prototypes according to a majority vote by the training data in each cell. However, a one-step learning strategy such as LVQ is more efficient to compute labeled centroids.

LVQ1 does not minimize the classification error. In [11], LVQ1 is generalized for nearest neighbor classifiers. It is based on a regularizing parameter that monotonically decreases the upper bound of the training classification error toward a minimum. LVQ1 places prototypes around Bayes' borders and consequently the resulting nearest neighbor classifier estimates Bayes' classifier. The regularized LVQ1 algorithm improves the classification rate of LVQ1 [11].

There are two procedures for reducing the number of prototypes. *Editing* processes the training set to increase generalization capabilities by removing prototypes that contribute to the misclassification rate, for example, removing outlier patterns or removing patterns that are surrounded mostly by others of different classes [113]. *Condensing* is to obtain a small template that is a subset of the training set without changing the nearest neighbor decision boundary substantially. This can be established by reducing the number of prototypes that are centered in dense areas of the same class [54]. The condensing algorithm for k-NN, namely, the template reduction for k-NN [38], drops patterns that are far away from the boundary. A chain is defined as a sequence of nearest neighbors from alternating classes. Patterns further down the chain are close to the classification boundary.

Learning k-NN rules [78] is suggested as a refinement of LVQ1. LVQ1 performs better than any of the learning k-NN rules if the codebook size is rather small. However, those extensions to k neighbors achieve lower error rates than LVQ1 when the codebook size is substantially increased. These results agree with the asymptotical theoretical analysis for the k-NN rule, in the sense that this can reach the optimal classification accuracy only when a sufficiently large number of codebook vectors are available.

A method proposed in [5] enhances the estimate of the posterior probabilities for pattern classification. It is based on observing the k nearest neighbors and weighting their contribution to the posterior probabilities differently. The weights are estimated using the ML procedure.

9.6 Neural Gas

Neural gas [91] is a vector quantization model that minimizes a known cost function and converges to the C-means quantization error via a soft-to-hard competitive model transition. The soft-to-hard annealing process helps the algorithm to escape from local minima. Neural gas is a topology-preserving network. It is particularly useful for three-dimensional (3-D) reconstruction. It can be treated as an extension to C-means. It has a fixed number K of processing units with no lateral connection.

The goal of neural gas is to find prototypes $\boldsymbol{c}_i \in R^m$, $i = 1, \ldots, K$, such that these prototypes represent the underlying distribution P as accurately as possible, minimizing the cost function [91]:

$$E_{\mathrm{NG}}(\boldsymbol{c}_i) = \frac{1}{2C(\lambda)} \sum_{i=1}^{K} \int h(r_i(\boldsymbol{x}, \boldsymbol{c}_i)) d(\boldsymbol{x}, \boldsymbol{c}_i) P(d\boldsymbol{x}), \tag{9.22}$$

where $d(\cdot, \cdot)$ denotes the Euclidean distance, $r_i(\boldsymbol{x}, \boldsymbol{c}_i) = |\{\boldsymbol{c}_j | d(\boldsymbol{x}, \boldsymbol{c}_j) < d(\boldsymbol{x}, \boldsymbol{c}_i)\}|$ is the rank of the prototypes sorted according to the distances, $h(t) = e^{-t/\lambda}$ with $\lambda > 0$ decreasing with time, and $C(\lambda) = \sum_{i=1}^{K} h(r_i)$. The learning rule is derived from gradient descent.

A data optimal topological ordering is achieved by using neighborhood ranking within the input space at each training step. To find its neighborhood rank, each neuron compares its distance to the input vector with the distances of all the other neurons to the input vector. Neighborhood ranking provides a training strategy with mechanisms related to robust statistics, and neural gas does not suffer from the prototype underutilization problem (see Sect. 10.1). At each step t, the Euclidean distances between an input vector $\boldsymbol{x}_t$ and all the prototype vectors $\boldsymbol{c}_k(t)$, $k = 1, \ldots, K$, are calculated by

$$d_k(\boldsymbol{x}_t) = \|\boldsymbol{x}_t - \boldsymbol{c}_k(t)\| \tag{9.23}$$

and $\boldsymbol{d}(t) = (d_1(\boldsymbol{x}_t), \ldots, d_K(\boldsymbol{x}_t))^T$. Each prototype $\boldsymbol{c}_k(t)$ is assigned a rank $r_k(t)$, which takes an integer value from 0 to $K-1$, with 0 for the smallest and $K-1$ for the largest $d_k(\boldsymbol{x}_t)$.

The prototypes are updated by

$$\boldsymbol{c}_k(t+1) = \boldsymbol{c}_k(t) + \eta h(r_k(t))(\boldsymbol{x}_t - \boldsymbol{c}_k(t)), \tag{9.24}$$

where $h(r) = \mathrm{e}^{-\frac{r}{\rho(t)}}$ realizes soft-competition, and $\rho(t)$ is the neighborhood width. When $\rho(t) \to 0$, (9.24) reduces to the C-means update rule (9.34). During the iterations, both $\rho(t)$ and $\eta(t)$ decrease exponentially from their initial positive values

$$\eta(t) = \eta_0 \left(\frac{\eta_f}{\eta_0}\right)^{\frac{t}{T_f}}, \qquad \rho(t) = \rho_0 \left(\frac{\rho_f}{\rho_0}\right)^{\frac{t}{T_f}}, \tag{9.25}$$

where η_0 and ρ_0 are the initial decay parameters, η_f and ρ_f are the final decay parameters, and T_f is the maximum number of iterations.

The prototypes $\boldsymbol{c}_k$ are initialized by randomly assigning vectors from the training set. Neural gas is given by Algorithm 9.4.

Algorithm 9.4 (Neural gas)

1. *Initialize* K, $\boldsymbol{c}_k$, $k = 1, \ldots, K$, ρ_0, η_0, ρ_f, η_f *and* T_f.
2. *Set* $t = 1$.
3. ***Repeat until*** *a stopping criterion is satisfied:*
   a. *Calculate distances* $d_k(\boldsymbol{x}_t)$, $k = 1, \ldots, K$, *by (9.23).*
   b. *Sort the components of* $\boldsymbol{d}(t)$ *and assign each prototype* $\boldsymbol{c}_k$ *with a rank* $r_k(t)$, *which is a unique value from* 0 *to* $K-1$.
   c. *Calculate* $\eta(t)$, $\rho(t)$ *by (9.25).*
   d. *Update* $\boldsymbol{c}_k$, $k = 1, \ldots, K$, *by (9.24).*
   e. *Set* $t = t + 1$.

Unlike SOM, neural gas determines a dynamical neighborhood relation as learning proceeds. Neural gas can be derived from a gradient-descent procedure on a potential function associated with the framework of fuzzy clustering. It is not sensitive to neuron initialization. Neural gas automatically determines a data optimum lattice, such that a small quantization error can be achieved.

Neural gas converges faster to a smaller error E than C-means, maximum-entropy clustering [104], and SOM. This advantage is achieved at the price of a higher computational effort. In a serial implementation, the complexity for neural gas is $O(K \log K)$ while the other three methods all have a complexity of $O(K)$, where K is the number of prototypes. Nevertheless, in parallel implementation, all the four algorithms have the same complexity, $O(\log K)$ [91]. In a fast implementation of sequential neural gas [26], a truncated exponential function is used as the neighborhood function and neighborhood ranking is implemented without evaluating and sorting all the distances. Given the same quality of the resulting codebook, this fast realization gains a speedup of five times over the original neural gas for codebook design in image vector quantization.

An exact mathematical analysis of vector quantization dynamics is presented in [114]. In case of no suboptimal local minima of the quantization error, WTA always converges to the best quantization error, but the search speed is sensitive to prototype initialization. Neural gas can improve convergence speed and achieve robustness to initial conditions. However, depending on the structure of the data, neural gas does not always obtain the best asymptotic quantization error.

Lack of an output space has limited the application of neural gas to data projection and visualization. Curvilinear component analysis [31] first performs vector quantization of the data manifold in input space using SOM, and then makes a nonlinear projection of the quantizing vectors by minimizing a cost function based on the inter-point distances. The computational complexity is $O(N)$. The output is not a fixed grid but a continuous space that is able to take the shape of the data manifold. Online visualization neural gas [37] concurrently adjusts the codebook vectors in input space and the codebook positions in a continuous output space. The method has a complexity of $O(N \log N)$. It outperforms SOM-based and neural gas-based curvilinear component analysis methods, in both their batch and online versions for neighborhood sizes smaller than 20 or 30. In general, neural gas-based curvilinear component analysis exhibits much better performance than its SOM-based counterpart.

Single-pass extensions of neural gas and SOM [1] are based on a simple patch decomposition of the dataset and fast batch optimization schemes of the underlying cost function. The algorithms require fixed memory space and maintain the benefits of the original ones including easy implementation and interpretation as well as large flexibility and adaptability.

Based on the cost function of neural gas, a batch variant of neural gas [27] shows much faster convergence and can be interpreted as optimization of the cost function by the Newton method. Based on the notion of the generalized median in analogy to median SOM, a variant for non-vectorial proximity data can be introduced. Convergence of batch and median versions of neural gas, SOM and C-means are proved in a unified formulation in [27].

Competitive Hebbian Learning

In a Voronoi tessellation, when the prototype of each Voronoi region is connected to all the prototypes of its bordering Voronoi regions, a Delaunay triangulation is obtained. Competitive Hebbian learning [90, 92] is a method that generates a subgraph of the Delaunay triangulation of the prototypes, called an *induced Delaunay triangulation*, by masking the Delaunay triangulation with a data distribution $P(\boldsymbol{x})$. Induced Delaunay triangulation has been proved to be optimally topology-preserving in a general sense [90].

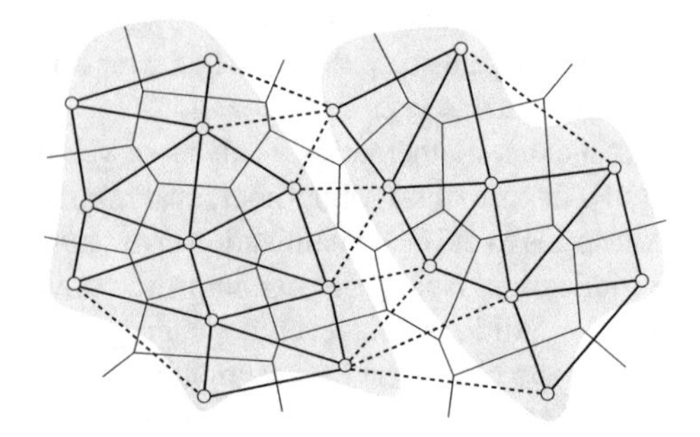

Fig. 9.13 Illustration of the Delaunay triangulation and an induced Delaunay triangulation

Given a number of prototypes in R^J, competitive Hebbian learning successively adds connections among them by evaluating input data drawn from a distribution $P(\boldsymbol{x})$. The method does not change the prototypes but only generates topology according to them. For each input vector $\boldsymbol{x}$, the two closest prototypes are connected by an edge. This leads to an induced Delaunay triangulation, which is limited to those regions of the input space R^J, where $P(\boldsymbol{x}) > 0$. The Delaunay triangulation and the induced Delaunay triangulation are illustrated in Fig. 9.13. The Delaunay triangulation is represented by a mix of thick and thick-dashed lines, the induced Delaunay triangulation by thick lines, Voronoi tessellation by thin lines, prototypes by circles, and a data distribution $P(\boldsymbol{x})$ by shaded regions. To generate an induced Delaunay triangulation, two prototypes are connected only if at least a part of the common border of their Voronoi polygons lies in a region where $P(\boldsymbol{x}) > 0$.

The topology-representing network [92] is obtained by alternating the learning steps of neural gas and competitive Hebbian learning, where neural gas is used to distribute a certain number of prototypes and competitive Hebbian learning is then used to generate a topology. An edge aging scheme is used to remove obsolete edges. Competitive Hebbian learning avoids the topological defects observed for SOM.

9.7 ART Networks

Adaptive resonance theory (ART) [48] is biologically motivated and is a major development of the competitive learning paradigm. The theory leads to an evolving series of real-time unsupervised network models for clustering, pattern recognition, and associative memory [16–18, 20]. These models are capable of stable category recognition in response to arbitrary input sequences with either fast or slow learning. ART models are characterized by systems of differential equations, which formulate stable self-organizing learning methods. Instar and outstar learning rules are the two learning rules used. ART has the ability to adapt, yet not forget the past training, and this is referred to as the *stability–plasticity dilemma* [17, 48].

At the training stage, the stored prototype of a category is adapted when an input pattern is sufficiently similar to the prototype. When novelty is detected, ART adaptively and autonomously creates a new category with the input pattern as the prototype. The meaning of being sufficiently similar is dependent on a vigilance parameter $\rho \in (0, 1]$. If ρ is large, the similarity condition becomes stringent and many finely divided categories are formed. In contrast, smaller ρ gives coarser categorization, resulting in fewer categories.

The stability and plasticity properties as well as the ability to efficiently process dynamic data make ART attractive for clustering large, rapidly changing sequences of input patterns, such as in the case of data mining. However, the ART approach does not correspond to C-means and vector quantization in a global optimization sense [85]. The ART model family is sensitive to the order of presentation of the input patterns. ART models tend to build clusters of the same size, independently of the distribution of the data.

9.7.1 ART Models

The ART model family includes a series of unsupervised learning models. ART networks employ a J-K recurrent architecture. The input layer F1, called a *comparing layer*, has J neurons, while the output layer F2, called a *recognizing layer*, has K neurons. Layers F1 and F2 are fully interconnected in both the directions. Layer F2 acts as a WTA network. The feedforward weights connecting to F2 neuron j are represented by vector $\boldsymbol{w}_j$, while the feedback weights from the same neuron are represented by vector $\boldsymbol{c}_j$. The vector $\boldsymbol{c}_j$ stores the prototype of cluster j. J is the number of features used to represent a pattern and the number of clusters K varies with the size of the problem. The architecture of the ART model is shown in Fig. 9.14. The feedforward weights connecting to F2 neuron j are represented by $\boldsymbol{w}_j = \left(w_{1j}, \ldots, w_{Jj}\right)^T$, while the feedback weights from the same neuron are represented by $\boldsymbol{c}_j = \left(c_{j1}, c_{j2}, \ldots, c_{jJ}\right)^T$. The output selects one of the K prototypes, $\boldsymbol{c}_i$, by setting $y_i = 1$ and $y_j = 0$, $j \neq i$.

The ART models are characterized by a set of short-term memory and long-term memory time-domain nonlinear differential equations. The short-term memory equations describe the evolution of the neurons and the interactions between them, while the long-term memory equations describe the change of the interconnection

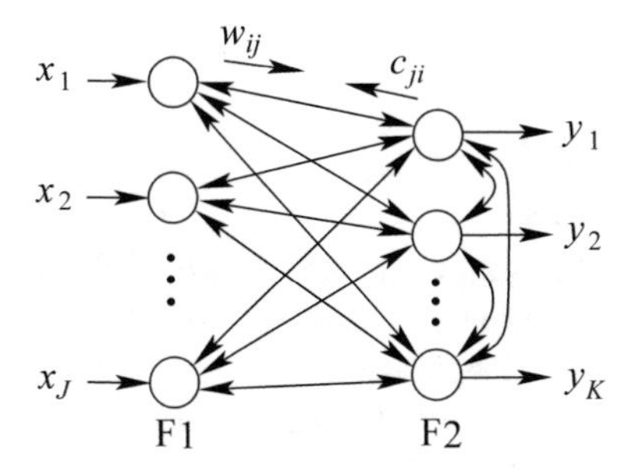

Fig. 9.14 Architecture of the ART model

weights with time as a function of the system state. Layer F1 stores the short-term memory for the current input pattern, while F2 stores the prototypes of clusters as the long-term memory.

Three types of ART implementations can be distinguished, namely, full-mode, short-term memory steady-state mode, and fast-learning mode [18, 106]. In full-mode implementation, both the short-term memory and long-term memory differential equations are realized. The short-term memory steady-state mode only implements the long-term memory differential equations, while the short-term memory behavior is governed by nonlinear algebraic equations. In fast-learning mode, both the short-term memory and the long-term memory are implemented by their steady-state nonlinear algebraic equations, and thus proper sequencing of short-term memory and long-term memory events is required. The fast-learning mode is inexpensive and is most popular.

The simplest and most popular ART model is ART 1 [17] for learning to categorize arbitrarily many, complex binary input patterns presented in an arbitrary order. ART 2 [18] is designed to categorize analog or binary random input sequences. ART 2 has a more complex F1 field that allows it to stably categorize sequences of analog inputs that can be arbitrarily close to one another. By characterizing the clustering behavior of ART 2, Burke [15] has found similarity between ART-based clustering and C-means clustering. In ART 2A [16], only feedforward connection between F1 and F2 is used in ART 2A learning. An implementation of ART 2A is given in [57]. ART-C 2A [57] applies a constraint-reset mechanism on ART 2A to allow a direct control on the number of output clusters generated during the self-organizing process. ART 2A and ART-C 2A have clustering quality comparable to that of C-means and SOM, but with an advantage in computation time [57].

The ARTMAP model family is a class of supervised learning methods. ARTMAP, also termed *predictive ART*, autonomously learns to classify arbitrarily many, arbitrarily ordered vectors into recognition categories based on predictive success [19]. ARTMAP is self-organizing, self-stabilizing, match learning, and real-time. It learns orders of magnitude more quickly and also is more accurate than BP. These are achieved by using an internal controller that jointly maximizes predictive generalization and minimizes predictive error by linking predictive success to category size on a trial-by-trial basis, using only local operations. However, ARTMAP is very sensitive to the order of presentation of the training patterns. Fuzzy ARTMAP [20] is shown to be a universal approximator [111]. Many popular ART and ARTMAP models and algorithms are reviewed in [34, 35].

9.7.2 ART 1

The main elements of basic ART 1 model are shown in Fig. 9.15. The two fields of neurons, F1 and F2, are linked both bottom-up and top-down by adaptive filters. The unsupervised two-layer feedforward (bottom-up) pattern recognition network is termed an *attentional subsystem*. There is also an auxiliary subsystem, called the *orienting subsystem*, that becomes active during search.

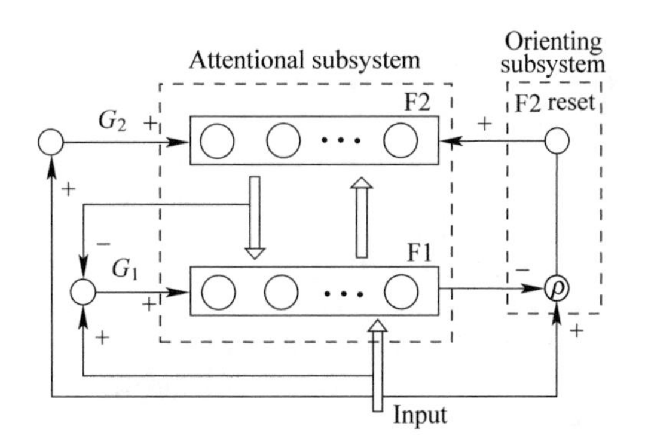

Fig. 9.15 Architecture of ART 1 with supplemental units. G_1 and G_2 are outputs of gain control units. The F2 reset unit controls vigilance matching

Since ART 1 in fast-learning mode is most widely used, we only discuss this mode here. To begin with, all cluster categories are set as *uncommitted*. When a new pattern is presented at time t, the net input to F2 neuron j is given by

$$net_j(t) = \boldsymbol{x}_t^T \boldsymbol{w}_j(t), \quad j = 1, \ldots, K, \tag{9.26}$$

where $\boldsymbol{w}_j = \left(w_{1j}, \ldots, w_{Jj}\right)^T$.

Competition between F2 neurons is performed to select the winning neuron w such that

$$net_w(t) = \max_{j=1,\ldots,K} net_j(t). \tag{9.27}$$

Neuron w then undergoes a vigilance test so as to determine whether it is close enough to $\boldsymbol{x}_t$:

$$\frac{\|\boldsymbol{x}_t \wedge \boldsymbol{c}_w(t-1)\|}{\|\boldsymbol{x}_t\|} \geq \rho, \tag{9.28}$$

where $\wedge$ denotes logical and operation. For binary values of x_i, the Euclidean norm $\|\boldsymbol{x}\| = \sum_i x_i$.

If neuron w passes the vigilance test, the system enters the resonance mode, and the weights for the winning neuron are updated by

$$\boldsymbol{c}_w(t) = \boldsymbol{c}_w(t-1) \wedge \boldsymbol{x}_t, \tag{9.29}$$

$$\boldsymbol{w}_w(t) = \frac{L\left[\boldsymbol{c}_w(t-1) \wedge \boldsymbol{x}_t\right]}{L - 1 + \|\boldsymbol{c}_w(t-1) \wedge \boldsymbol{x}_t\|}, \tag{9.30}$$

where $L > 1$ is a constant parameter.

Otherwise, the F2 neuron reset mechanism is applied to remove neuron w from the current search by setting $net_w = -1$ and the system enters the search mode. If all the stored categories cannot pass the vigilance test, one of the uncommitted categories of the K categories is assigned to this pattern.

A popular fast-learning implementation is given in [93, 106]. For initialization, select $0 < \rho \leq 1$ and $L > 1$, $\boldsymbol{w}_j(0) = \frac{1}{1+J}\mathbf{1}$ and $\boldsymbol{c}_j(0) = \mathbf{1}$, $j = 1, \ldots, K$, where $\mathbf{1}$ denotes a J-dimensional vector whose entries are all unity. In [106], $\boldsymbol{w}_j(0) = \frac{L}{L-1+J}\mathbf{1}$. In the algorithm, ρ determines the level of abstraction at which ART discovers clusters. The minimal number of clusters present in the data can be determined by $\rho_{\min} < \frac{1}{J}$. Initial bottom-up weights are usually selected as $0 < w_{jk}(0) \leq \frac{L}{L-1+J}$. Larger values of $w_{jk}(0)$ favor the creation of new nodes, while smaller values attempt to put a pattern into an existing cluster. The order of the training patterns may influence the final prototypes and clusters. Unlike many alternative methods such as SOM and the Hopfield network, ART 1 can deal with an arbitrary combination of binary input patterns. In addition, ART 1 has no restriction on memory capacity since its memory matrices are not square.

ART models are typically governed by differential equations, which result in a high computational complexity for numerical implementations. Implementations using analog or optical hardware are more desirable. A modified ART 1 algorithm in fast-learning mode is used for easy hardware implementation [106]. The method has also been extended for full-mode and short-term memory steady-state mode. A number of hardware implementations of ART 1 in different modes are also surveyed in [106].

9.8 *C*-Means Clustering

The most well-known data-clustering technique is the statistical C-means (also known as k-means) algorithm [89]. The C-means algorithm approximates the ML solution for determining the locations of the means of a mixture density of component densities. It is closely related to simple competitive learning and is a special case of SOM. The algorithm partitions a set of N input patterns, $\mathcal{X}$, into K separated subsets $\mathcal{C}_k$, each containing N_k input patterns by minimizing the MSE function

$$E\left(\boldsymbol{c}_1, \ldots, \boldsymbol{c}_K\right) = \frac{1}{N} \sum_{k=1}^{K} \sum_{\boldsymbol{x}_n \in \mathcal{C}_k} \|\boldsymbol{x}_n - \boldsymbol{c}_k\|^2, \tag{9.31}$$

where $\boldsymbol{c}_k$ is the prototype or *center* of cluster $\mathcal{C}_k$. To improve the similarity of samples in each cluster, one can minimize E with respect to $\boldsymbol{c}_k$ by setting $\frac{\partial E}{\partial \boldsymbol{c}_k} = 0$; thus, the optimal location of $\boldsymbol{c}_k$ is the mean of the samples in the cluster

$$\boldsymbol{c}_k = \frac{1}{N_k} \sum_{\boldsymbol{x}_i \in \mathcal{C}_k} \boldsymbol{x}_i. \tag{9.32}$$

C-means clustering can be implemented in either batch mode [84] or incremental mode [89]. Batch C-means [84], frequently called the *Linde–Buzo–Gray*, *LBG* or *generalized Lloyd algorithm*, is applied when the whole training set is available. When the training set is obtained online, incremental C-means is commonly applied.

In batch C-means, the initial partition is arbitrarily defined by placing each input pattern into a randomly selected cluster. The prototypes are defined to be the average of the patterns in the individual clusters. When C-means is performed, at each step the patterns keep changing from one cluster to the closest cluster $\boldsymbol{c}_k$ according to the simple competitive learning rule

$$\|\boldsymbol{x}_i - \boldsymbol{c}_k\| = \min_j \left\|\boldsymbol{x}_i - \boldsymbol{c}_j\right\| \tag{9.33}$$

and the prototypes are then recalculated according to (9.32).

In incremental C-means, each cluster is initialized with a random pattern as its prototype. C-means continues to update the prototypes upon the presentation of each new pattern. If at time t the kth prototype is $\boldsymbol{c}_k(t)$ and the input pattern is $\boldsymbol{x}_t$, then at time $t+1$ incremental C-means gives the new prototype as

$$\boldsymbol{c}_k(t+1) = \begin{cases} \boldsymbol{c}_k(t) + \eta(t)\left(\boldsymbol{x}_t - \boldsymbol{c}_k(t)\right), & k = \arg_j \min \left\|\boldsymbol{x} - \boldsymbol{c}_j\right\| \\ \boldsymbol{c}_k(t), & \text{otherwise} \end{cases}, \tag{9.34}$$

where $\eta(t)$ should slowly decrease to zero, and typically $\eta(0) < 1$.

Neighborhood cooperation such as for SOM and neural gas offers one biologically plausible solution. Unlike SOM and neural gas, C-means is very sensitive to initialization of the prototypes since it adapts the prototypes only locally according to their nearest data points. The general procedure for C-means clustering is given by Algorithm 9.5.

Algorithm 9.5 (C-means)

1. *Set K.*
2. *Arbitrarily select an initial cluster partition.*
3. ***Repeat until** the change in all $\boldsymbol{c}_k$ is sufficiently small:*
   a. *Decide K cluster prototypes $\boldsymbol{c}_k$.*
   b. *Redistribute patterns among the clusters using criterion (9.31).*

After the algorithm converges, we can calculate the variance vector, $\boldsymbol{\sigma}_k = \left(\sigma_{k,1}, \ldots, \sigma_{k,J}\right)^T$, for each cluster

$$\sigma_{k,i} = \sqrt{\frac{\sum_{\boldsymbol{x}_j \in C_k} \left(x_{j,i} - c_{k,i}\right)^2}{N_k - 1}}, \quad k = 1, \ldots, K, i = 1, \ldots, J. \tag{9.35}$$

The relation between PCA and C-means is established in [33]. Principal components have been proved to be the continuous solutions to the discrete cluster membership indicators for C-means clustering, with a clear simplex cluster structure [33].

Lower bounds for the C-means objective function (9.31) are derived as the total variance minus the eigenvalues of the data covariance matrix [33].

In [123], the classic alternating loop in C-means has been simplified to a pure stochastic optimization procedure. The procedure of C-means becomes simpler and converges to considerably better local optima. Neither the costly initial assignment nor the search of the closest centroid for each sample in the iteration is necessary. This leads to higher speed and considerably lower clustering distortion.

Improvements on C-Means

In [22], incremental C-means is improved by adding two mechanisms, one for biasing the clustering toward an optimal Voronoi partition by using a cluster variance-weighted MSE as the objective function and the other for adjusting the learning rate dynamically according to the current variances in all partitions. The method always converges to an optimal or near-optimum configuration.

Enhanced LBG [99] is derived directly from LBG with a negligible overhead. The concept of utility of a codeword is a powerful instrument to overcome the problem of bad local minima arising from a bad choice of the initial codebook. The utility allows the identification of those badly positioned codewords and guides their movement from the proximity of a local minimum in the error function. Enhanced LBG outperforms LBG with utility [42] both in terms of accuracy and number of required iterations.

To deal with the initialization problem, the global C-means algorithm [82] obtains near-optimal solutions in terms of clustering error by employing C-means as a local search procedure. It is an incremental-deterministic algorithm. It incrementally solves the M-clustering problem by solving all intermediate problems with $1, \ldots, M$ clusters using C-means. Global C-means is better than C-means with multiple restarts.

In an efficient implementation of LBG [62], the data points are stored by a k-d tree. The algorithm is typically one order of magnitude faster than LBG. A fast C-means clustering algorithm [79] uses the cluster center displacements between two successive partition processes to reject unlikely candidates for a data point. The computing time increases linearly with the data dimension d, whereas the computational complexity of k-d tree-based algorithms increases exponentially with d.

SYNCLUS [32] is a method for variable weighting in C-means clustering. Starting from an initial set of weights, it first uses C-means to partition data into K clusters. It then estimates a new set of optimal weights by optimizing a weighted mean-squares, stress-like cost function. The two stages alternate until they converge to an optimal set of weights. W-C-means [58] can automatically weight variables based on the importance of the variables in clustering. It adds a step to C-means to update the variable weights based on the current partition of data. W-C-means outperforms C-means in recovering clusters in data. The computational complexity of the algorithm is $O(tmNK)$ for t iterations, K clusters, m attributes, and N objects.

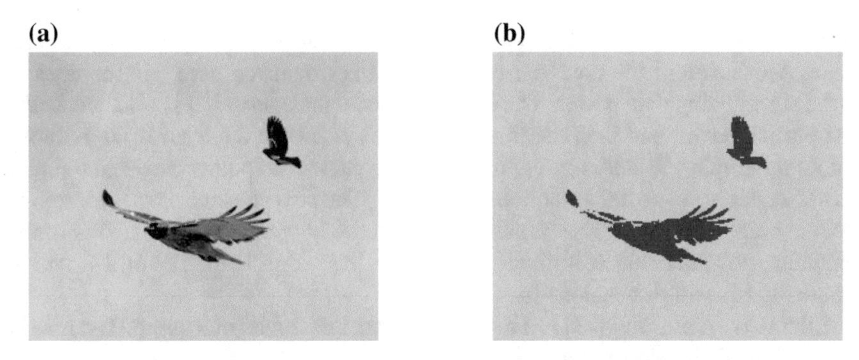

Fig. 9.16 Image segmentation using clustering. **a** Original image. **b** Segmented image

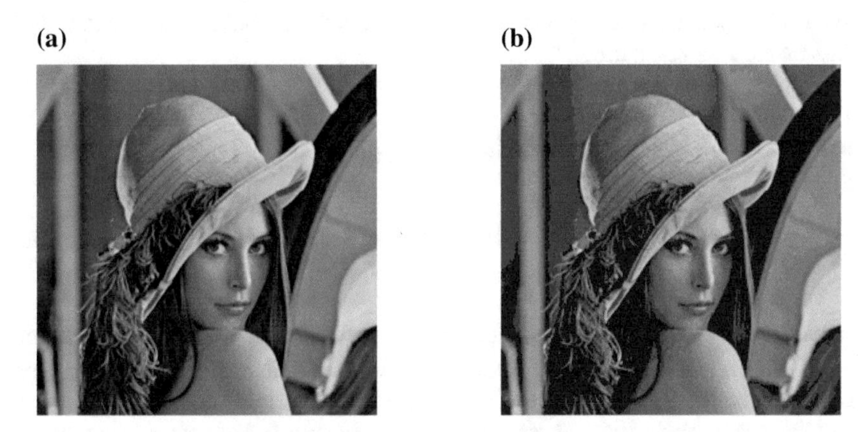

Fig. 9.17 Image segmentation using clustering. **a** Original image. **b** Quantized image

Example 9.6 Clustering can be used for image segmentation. We apply C-means with $K = 2$, and the result is shown in Fig. 9.16. It indicates that the two birds are clearly segmented.

Example 9.7 Figure 9.17 shows the result of using C-means for image quantization. We select $K = 16$. It is shown that by quantizing a grayscale image from 256 levels to 16 levels the image quality is still acceptable. The picture SNR for quantization is 36.8569 dB.

9.9 Subtractive Clustering

Mountain clustering [116] is a simple and effective method for estimating the count of clusters and the initial locations of the cluster centers, which are the difficulties faced by most conventional methods. The method grids the data space and computes a potential value for each grid point based on its distance to the actual data points.

Each grid point is treated as a potential cluster center depending on its potential value. A measure of the potential for each grid is calculated based on the density of the surrounding data points. The grid with the highest potential is selected as the first cluster center, and then the potential values of all the other grids are reduced according to their distances to the first cluster center. Grid points closer to the first cluster center have greater reduction in potential. The next cluster center is located at the grid point with the highest remaining potential. This process is repeated until the remaining potential values of all the grids fall below a threshold. The grid structure causes the curse of dimensionality.

Subtractive clustering [23] is a modified form of mountain clustering. The idea is to use all the data points to replace all the grid points as potential cluster centers. By this means, the effective number of grid points is reduced to the size of the pattern set, which is independent of the dimensionality of the problem. Subtractive clustering is a fast method for estimating clusters in the data.

Subtractive clustering assumes each of the N data points in the pattern set, $\boldsymbol{x}_i$, to be a potential cluster center, and the potential measure is defined as a function of the Euclidean distances to all the other input data points

$$P(i) = \sum_{j=1}^{N} \mathrm{e}^{-\alpha \|\boldsymbol{x}_i - \boldsymbol{x}_j\|^2}, \quad i = 1, \ldots, N, \tag{9.36}$$

where $\alpha = \frac{4}{r_a^2}$, the constant $r_a > 0$ being effectively a normalized radius defining the neighborhood. Data points outside this radius have insignificant influence on the potentials. A data point surrounded by many neighboring data points has a high potential value. Thus, the mountain and subtractive clustering techniques are less sensitive to noise than other clustering algorithms, such as C-means and FCM [12].

After the data point with the highest potential, $\boldsymbol{x}_u$ with $u = \arg_i \max P(i)$, is selected as the kth cluster center, that is, $\boldsymbol{c}_k = \boldsymbol{x}_u$ with $\overline{P}(k) = P(u)$ as its potential value, the potential of each data point $\boldsymbol{x}_i$ is revised by subtracting a term associated with $\boldsymbol{c}_k$

$$P(i) = P(i) - \overline{P}(k)\mathrm{e}^{-\beta \|\boldsymbol{x}_i - \boldsymbol{c}_k\|^2}, \tag{9.37}$$

where $\beta = \frac{4}{r_b^2}$, and the constant $r_b > 0$ is a normalized radius defining the neighborhood. In order to avoid closely located cluster centers, select $r_b > r_a$, typically $r_b = 1.25 r_a$.

The algorithm continues until the remaining potential of all the data points is below some fraction of the potential of the first cluster center, that is,

$$\overline{P}(k) = \max_i P(i) < \varepsilon \overline{P}(1), \tag{9.38}$$

where $\varepsilon \in (0, 1)$. When ε is close to 0, a large number of hidden nodes will be generated. On the contrary, a value of ε close to 1 will lead to a small network structure. Typically, ε is selected as 0.15.

Subtractive clustering is described by Algorithm 9.6 [23, 24].

Algorithm 9.6 (Subtractive clustering)

1. *Set* r_a, r_b *and* ε.
2. *Calculate the potential values* $P(i)$, $i = 1, \ldots, N$.
3. *Set* $k = 1$.
4. ***Repeat until*** $\overline{P}(k) < \varepsilon \overline{P}(1)$:
   a. *Find data point* $\boldsymbol{x}_u$ *with* $u = \arg_i \max P(i)$.
   b. *Set the kth cluster center as* $\boldsymbol{x}_u$, *that is,* $\boldsymbol{c}_k = \boldsymbol{x}_u$ *and* $\overline{P}(k) = P(u)$.
   c. *Revise the potential of each data point* $\boldsymbol{x}_i$ *by (9.37).*
   d. *Set* $k = k + 1$.

The training data $\boldsymbol{x}_i$ can be scaled before applying the method. This helps in selecting proper values for α and β. Since it is difficult to select suitable ε, additional criteria for accepting/rejecting cluster centers can be used. One method is to select two thresholds [23, 24], namely, $\overline{\varepsilon}$ and $\underline{\varepsilon}$. Above $\overline{\varepsilon}$, $\boldsymbol{c}_k$ is definitely accepted as a cluster center, while below $\underline{\varepsilon}$ it is definitely rejected. If $\overline{P}(k)$ falls between the two thresholds, a trade-off between a reasonable potential and its distance to the existing cluster centers must be examined

$$R = \frac{d_{\min}}{r_a} + \frac{\overline{P}(k)}{\overline{P}(1)}, \tag{9.39}$$

where $d_{\min}$ is the shortest of the distances between $\boldsymbol{c}_k$ and $\boldsymbol{c}_i$, $i = 1, \ldots, k-1$. If $R \geq 1$, accept $\boldsymbol{c}_k$ and continue the algorithm. If $R < 1$, reject $\boldsymbol{c}_k$ and set $\overline{P}(k) = P(u) = 0$, and select the data point with the next highest potential as $\boldsymbol{c}_k$ and retest.

Unlike C-means and FCM, which require iterations of many epochs, subtractive clustering requires only one pass of the training data. Besides, the number of clusters does not need to be specified *a priori*. Subtractive clustering is a deterministic method: For the same network structure, the same network parameters are always obtained.

Both C-means and FCM require $O(KNT)$ computations, where T is the total number of epochs and each computation requires the calculation of the distance and the memberships. The computational load for subtractive clustering is $O\left(N^2 + KN\right)$, each computation involving calculation of an exponential function. Thus, for small- or medium-size training sets, subtractive clustering is relatively fast. However, when $N \gg KT$, subtractive clustering requires more training time [29].

Subtractive clustering provides only rough estimates of the cluster centers, since the cluster centers obtained are situated at some data points. Moreover, since α and β are not determined from the dataset and no cluster validity is used, the clusters produced may not appropriately represent the clusters. For small datasets, one can try a number of values for α, β, and ε and select a proper network structure. The results by subtractive clustering can be used to determine the number of clusters and their initial values for initializing iterative clustering algorithms such as C-means and FCM.

Subtractive clustering can be improved by performing a search over α and β, which makes it essentially equivalent to the least-biased fuzzy clustering algorithm [9]. The least-biased fuzzy clustering, based on deterministic annealing [103, 104], tries to minimize the clustering entropy of each cluster, namely, the entropy of the centroid with respect to the clustering membership distribution of data points, under the assumption of unbiased centroids. Subtractive clustering can be realized by replacing the Gaussian potential function with a Cauchy-type function of first order [3].

In [98], mountain clustering and subtractive clustering are improved by tuning the prototypes obtained using the gradient-descent method so as to maximize the potential function. By modifying the potential function, mountain clustering can also be used to detect other types of clusters like circular shells [98].

Another density-based clustering method, namely, density peaks clustering [102], computes two metrics for every point p: ρ, the local density, which is the number of points within a specified distance from p; and δ, the minimum distance from p to other points with higher densities. (ρ, δ) provides a two-dimensional representation of the input point data. The density peaks are distinguished from other points as they have the highest local density ρ and a large δ. Cluster centers can be determined from the density peaks. The method supports arbitrarily shaped clusters. It does not require a priori knowledge about the point distribution. It is deterministic and is robust against the initial choice of algorithm parameters. The method has a computational cost of $O(N^2)$.

In belief-peaks evidential clustering [109], all data objects in the neighborhood of each sample provide pieces of evidence that induce belief on the possibility of such sample being a cluster center. A sample having a local maximal belief and located far away from the other local maxima is characterized as a cluster center. Finally, a credal partition is created by minimizing an objective function with the fixed cluster centers. An adaptive distance metric is used to fit for unknown shapes of data structures.

9.10 Fuzzy Clustering

Fuzzy clustering is an important class of clustering algorithms. It helps to find natural vague boundaries in data. We introduce some fuzzy clustering algorithms in this section. Preliminaries of fuzzy sets and fuzzy logic are given in Chap. 26.

9.10.1 Fuzzy C-Means Clustering

The discreteness of each cluster endows the C-means algorithm with analytical and algorithmic intractabilities. Partitioning the dataset in a fuzzy manner helps to circumvent such difficulties. FCM clustering [12], also known as the *fuzzy ISODATA* [36], considers each cluster as a fuzzy set, and each feature vector may be assigned to multiple clusters with some degree of certainty measured by the membership function taking values in [0, 1].

FCM optimizes the objective function

$$E = \sum_{j=1}^{K} \sum_{i=1}^{N} \mu_{ji}^{m} \left\| \boldsymbol{x}_i - \boldsymbol{c}_j \right\|^2, \tag{9.40}$$

where N is the size of the input pattern set, $\mathbf{U} = \left\{ \mu_{ji} \right\}$ denotes the membership matrix whose element μ_{ji} denotes the membership of $\boldsymbol{x}_i$ into cluster j and $\mu_{ji} \in [0, 1]$. The parameter $m \in (1, \infty)$ is a weighting factor called a *fuzzifier*. For better interpretation, the following condition must be satisfied:

$$\sum_{j=1}^{K} \mu_{ji} = 1, \quad i = 1, \ldots, N. \tag{9.41}$$

By minimizing (9.40) subject to (9.41), the optimal membership function μ_{ji} and cluster centers are derived as

$$\mu_{ji} = \frac{\left(\frac{1}{\|\boldsymbol{x}_i - \boldsymbol{c}_j\|^2} \right)^{\frac{1}{m-1}}}{\sum_{l=1}^{K} \left(\frac{1}{\|\boldsymbol{x}_i - \boldsymbol{c}_l\|^2} \right)^{\frac{1}{m-1}}}, \quad i = 1, \ldots, N, \; j = 1, \ldots, K, \tag{9.42}$$

$$\boldsymbol{c}_j = \frac{\sum_{i=1}^{N} \left(\mu_{ji} \right)^m \boldsymbol{x}_i}{\sum_{i=1}^{N} \left(\mu_{ji} \right)^m}, \quad j = 1, \ldots, K. \tag{9.43}$$

Equation (9.42) corresponds to a softmax rule and (9.43) is similar to the mean of the data points in a cluster. The two equations are dependent on each other. The iterative optimization procedure is known as *alternating optimization*.

The iteration process terminates when the change in the prototypes

$$e(t) = \sum_{j=1}^{K} \left\| \boldsymbol{c}_j(t) - \boldsymbol{c}_j(t-1) \right\|^2 \tag{9.44}$$

is sufficiently small. FCM is summarized in Algorithm 9.7.

Algorithm 9.7 (Fuzzy C-means)

1. *Set* $t = 0$.
2. *Initialize* K, ε *and* m.
3. *Randomize and normalize* $\mathbf{U}(0)$ *according to (9.41), and then calculate* $\boldsymbol{c}_j(0)$, $j = 1, \ldots, K$, *by (9.43).*
   Or alternatively, set $\boldsymbol{c}_j(0)$, $j = 1, \ldots, K$, *and then calculate* $\mathbf{U}(0)$ *by (9.42).*
4. ***Repeat until*** $e(t) \leq \varepsilon$:
   a. *Set* $t = t + 1$.
   b. *Calculate* $\mu_{ji}(t)$ *and* $\boldsymbol{c}_j(t)$ *according to (9.42) and (9.43).*
   c. *Calculate* $e(t)$ *by (9.44).*

In FCM, the fuzzifier m determines the fuzziness of the partition produced and reduces the influence of small membership values. If $m \to 1+$, the resulting partition asymptotically approaches a hard or crisp partition. On the other hand, the partition becomes a maximally fuzzy partition if $m \to \infty$. FCM with a high degree of fuzziness diminishes the probability of getting stuck at local minima [12]. A typical value for m is 1.5 or 2.0. Interval type-2 FCM accepts an interval-valued fuzzifier $[m_L, m_R]$ [61], and general type-2 FCM [83] extends interval type-2 FCM via the α-planes representation theorem.

FCM needs to store the membership matrix $\mathbf{U}$ and all the prototypes $\boldsymbol{c}_i$. The alternating estimation of $\mathbf{U}$ and $\boldsymbol{c}_i$'s causes a computational and storage burden for large-scale datasets. Computation can be accelerated by combining their updates [75], and consequently storage of $\mathbf{U}$ is avoided. The single iteration timing of the accelerated method grows linearly with K, while that of FCM grows quadratically with K since the norm calculation introduces another nested summation [75].

C-means is a special case of FCM, when u_{ji} is unity for only one class and zero for all the other classes. Like C-means, FCM may find a local optimum for a specified number of centers. The result is dependent on the initial membership matrix $\mathbf{U}(0)$ or cluster centers $\boldsymbol{c}_j(0)$, $j = 1, \ldots, K$.

Single-pass FCM and online FCM [51] facilitate scaling to very large numbers of examples while providing partitions that very closely approximate those one would obtain using FCM.

FCM has been generalized by introducing the generalized Boltzmann distribution to escape local minima [101]. In [121], the relation between the stability of the fixed points of FCM, $(\mathbf{U}^*, \overline{\boldsymbol{x}})$, and the dataset is given. This relation provides a theoretical basis for selecting the weighting exponent in FCM.

Penalized FCM [118] is a convergent generalized FCM obtained by adding a penalty term associated with μ_{ji}. A weighted FCM [110] is used for fuzzy modeling toward developing a Takagi–Sugeno–Kang (TSK) fuzzy model of optimal structure. All these and many other generalizations of FCM can be analyzed in a unified framework, termed the *generalized FCM* [122], by using the Lagrange multiplier method from an objective function that comprises a generalization of the FCM criterion and a regularization term representing the constraints.

In an agglomerative FCM algorithm [81], a penalty entropy term is introduced to the objective function of FCM to make the clustering process not sensitive to the initial cluster centers. The initial number of clusters is set to be larger than the true number of clusters in a dataset. With the entropy cost function, each initial cluster centers will move to the dense centers of the clusters in a dataset. These initial cluster centers are merged in the same location, and the number of the determined clusters is just the number of the merged clusters in the output of the algorithm.

ε-insensitive FCM (εFCM) is an extension to FCM that is obtained by introducing the robust statistics using Vapnik's ε-insensitive estimator as the loss function to reduce the effect of outliers [80]. It is based on L_1-norm clustering [65]. Other robust extensions to FCM include L_p-norm clustering ($0 < p < 1$) [56] and L_1-norm clustering [65].

The concept of α-cut implementation can be used to form cluster cores such that the data points inside the cluster core will have a membership value of 1. FCMα [119] can achieve robustness for suitably large m values with the same computational complexity as FCM. The cluster cores generated by FCMα are suitable for nonspherical shape clusters. FCMα is equivalent to FCM when $\alpha = 1$. When the weighting exponent m becomes larger, FCMα clustering trims most noisy points.

Neutrosophic C-means and neutrosophic evidential C-means [49] are based on neutrosophic logic. Neutrosophic C-means alleviates the limitations of FCM by introducing an objective function that contains two types of rejection for both noise and outlier rejections. Ambiguity rejection concerns patterns lying near the cluster boundaries, while distance rejection deals with patterns that are far away from the clusters.

When the dataset is a blend of unlabeled and labeled patterns, FCM with partial supervision [100] can be applied. The classification information is added to the objective function used in FCM, and FCM with partial supervision is derived following the same procedure as that of FCM. Conditional FCM and deterministic annealing clustering [103] consider various contributions of different samples and take account of sample weighting. Locality-weighted C-means and locality-weighted FCM are two locality-sensitive algorithms [59], where the neighborhood structure information between objects are transformed into weights of objects. The weight between a point and a center is in form of a Gaussian function. In addition, two semi-supervised extensions of locality-weighted FCM are proposed to better use some given partial supervision information in data objects.

9.10.2 Other Fuzzy Clustering Algorithms

There are numerous other clustering algorithms based on the concept of fuzzy membership. Two early fuzzy clustering algorithms are Gustafson–Kessel clustering [50] and adaptive fuzzy clustering [2]. Gustafson–Kessel clustering extends FCM by using the Mahalanobis distance and is suited for hyperellipsoidal clusters of equal volume. This algorithm takes typically fivefold the time for FCM to complete cluster formation [63]. Adaptive fuzzy clustering [2] also employs the Mahalanobis distance and is suitable for ellipsoidal or linear clusters. Gath–Geva clustering [44] is derived from a combination of FCM and fuzzy ML estimation. The method incorporates the hypervolume and density criteria as cluster validity measures and performs well in situations of large variability of cluster shapes, densities, and number of data points in each cluster.

C-means and FCM are based on the minimization of the trace of the (fuzzy) within-cluster scatter matrix. The minimum scatter volume and minimum cluster volume algorithms are two iterative clustering algorithms based on determinant (volume) criteria [76]. The minimum scatter volume algorithm minimizes the determinant of the sum of the scatter matrices of the clusters, while the minimum cluster volume algorithm minimizes the sum of the volumes of the individual clusters. The behavior of the minimum scatter volume algorithm is similar to that of C-means, whereas the minimum cluster volume algorithm is more versatile. The minimum cluster volume algorithm in general gives better results than the C-means, minimum scatter volume, and Gustafson–Kessel algorithms do, and is less sensitive to initialization than the EM algorithm.

A cluster represented by a volume prototype implies that all the data points close to a cluster center belong fully to that cluster. In [64], Gustafson–Kessel clustering and FCM have been extended by using the volume prototypes and similarity-driven merging of clusters.

Soft-competitive learning in clustering algorithms has the same function as fuzzy clustering [6]. The soft-competition scheme [117] is another soft version of LVQ. Soft-competition scheme asympototically evolves into the Kohonen learning algorithm. It is a sequential, deterministic vector quantization algorithm, which is realized by modifying the neighborhood mechanism of the Kohonen learning algorithm and incorporating the stochastic relaxation principles. Soft-competition scheme consistently provides better codebooks than incremental C-means, even for the same computation time. The learning rates of the soft-competition scheme are partially based on posterior probabilities.

Generalized LVQ [97] introduces soft-competition into LVQ by updating every prototype for each input vector. If there is a perfect match between the incoming input and the winner node, then generalized LVQ reduces to LVQ. On the other hand, the greater the mismatch to the winner, the larger the impact of an input vector on the update of the nonwinner nodes. Generalized LVQ is very sensitive to simple scaling of the input data, since its learning rates are reciprocally dependent on the sum of the squares of the distances from an input vector to the node weight vectors.

FCM-DFCV [30] is a method based on an adaptive quadratic distance for each class defined by a diagonal matrix and is a special case of Gustafson–Kessel clustering [50] based on a quadratic adaptive distance of each cluster defined by a fuzzy covariance matrix. The methods based on adaptive distances outperform FCM. There are also many fuzzy ART and ARTMAP models, and fuzzy clustering algorithms

Table 9.1 Comparison of C-means, FCM, and subtractive clustering for an artificial dataset. $\overline{d}_{\text{wc}}$ stands for the mean within-cluster distance, and $\overline{d}_{\text{bc}}$ for the mean between-cluster distance. A smaller value of $\overline{d}_{\text{wc}}/\overline{d}_{\text{bc}}$ corresponds to better performance

	C-means	FCM	Subtractive
$\overline{d}_{\text{wc}}$	1.2274	1.2275	1.2542
$\overline{d}_{\text{bc}}$	4.6077	4.6480	4.5963
$\overline{d}_{\text{wc}}/\overline{d}_{\text{bc}}$	0.2664	0.2641	0.2729
Time	0.06384	0.06802	0.4978

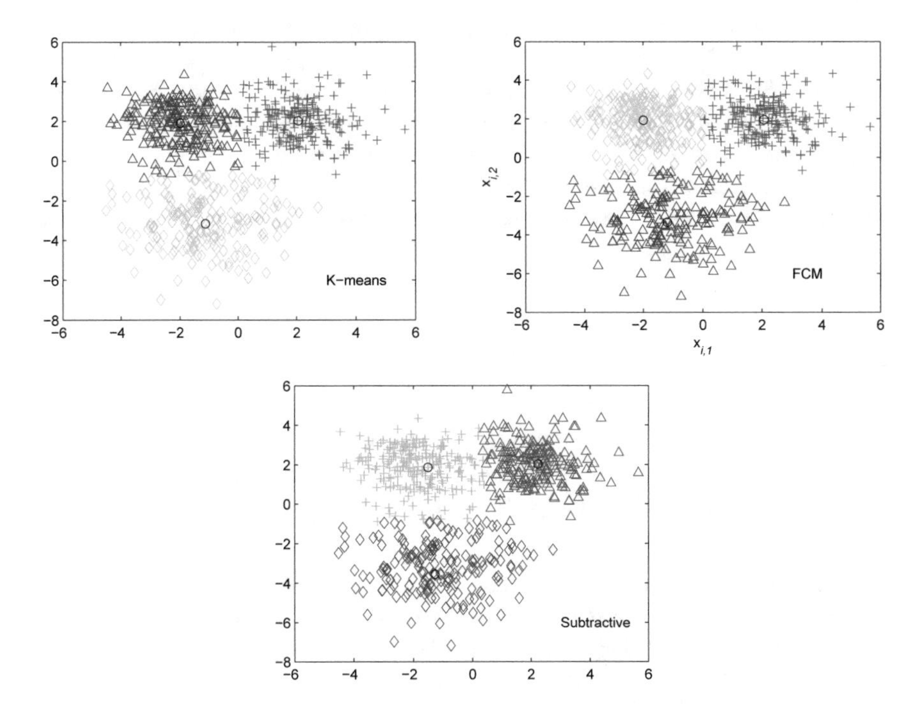

Fig. 9.18 Clustering using three methods: C-means, FCM, and subtractive clustering. The stars denote the cluster centers

based on the Kohonen Network and LVQ, ART networks, or the Hopfield network. These algorithms are reviewed in [34, 35].

Fuzzy C-regressions [55] embeds FCM into switching regression. The method always depends heavily on the initial values. Mountain C-regressions [115] solves the initial-value problem. Using a modified mountain clustering to extract C cluster centers in the transformed data space, which correspond to C regression models in the original dataset, mountain C-regressions can form well-estimated C regression models for switching regression datasets. According to the properties of transformation, mountain C-regressions is also robust to noise and outliers.

Example 9.8 In this example, we illustrate three popular clustering algorithms: C-means, FCM, and subtractive clustering. The input data represent three clusters centered at $(2, 2)$, $(-2, 2)$, and $(-1, -3)$, each having 200 data with a Gaussian distribution $\mathcal{N}(0, 1)$ in both x- and y-directions. The initial cluster centers for C-means are randomly sampled from the dataset. The fuzzifier of FCM is selected as $m = 2$. For C-means and FCM, the termination criterion is that the error in the objective function for two adjacent iterations is less than 10^{-5} and the maximum number of epochs is 100. We specify the number of clusters as 3. For subtractive clustering, the parameter ε is selected as 0.4.

Simulations are performed based on averaging of 1000 random runs. The simulation results are listed in Table 9.1. As far as this artificial dataset is concerned, C-means and FCM have almost the same performance, which is considerably superior to that of subtractive clustering. However, subtractive clustering is a deterministic method, and it can automatically detect a suitable number of clusters for a wide range of ε. The clustering results for a random run are illustrated in Fig. 9.18. There are minor differences in the cluster boundaries and the cluster centers between the algorithms.

Problems

9.1 A two-dimensional grid of neurons is trained with three-dimensional input data randomly distributed in a volume defined by $0 < x_1, x_2, x_3 < 1$. Use SOM to find the network topology of 8×8 neurons after 10, 1,000, and 10,000 iterations.

9.2 Failure of topological ordering may usually occur during SOM iterations. This may cause a rapid decay of the neighborhood size of the winning node. Verify this statement by using computer simulation.

9.3 For SOM, if a node fails, what is its effect:

(a) during the learning process;
(b) after learning is completed?

9.4 Define an energy function to be minimized for clustering or vector quantization problems. Is it possible to solve it using the Hopfield network?

9.5 For k-NN, k should take an odd number in a two-class setting. Give the reason.

9.6 Write a MATLAB program to implement the k-NN algorithm. Classify the Iris problem using the program.

9.7 Both SOM and neural gas are well-known topology-preserving clustering algorithms. Implement neural gas on the dataset for Problem 9.1. Try to draw conclusion from the simulation result.

9.8 Clustering is used to partition image pixels into K clusters. Although the intensity-based FCM algorithm functions well on segmenting most noise-free images, it fails to segment images corrupted by noise, outliers, and other imaging artifacts. Show this on a magnetic resonance imaging (MRI) image by computer experiments.

9.9 Show that the computational complexity of C-means is $O(NKdT)$.

9.10 Consider the quantization of the Lena image of 512×512 pixels with 256 gray levels. LBG is applied to the quantization of the 4×4 subimages extracted from the original pictures.

(a) Calculate the MSE when 32 codewords are used.
(b) Draw the original and reconstructed images.
(c) Calculate the picture SNR.

References

1. Alex, N., Hasenfuss, A., & Hammer, B. (2009). Patch clustering for massive data sets. *Neurocomputing*, *72*, 1455–1469.
2. Anderson, I. A., Bezdek, J. C., & Dave, R. (1982). Polygonal shape description of plane boundaries. In L. Troncale (Ed.), *Systems science and science* (Vol. 1, pp. 295–301). Louisville, KY: SGSR.
3. Angelov, P. P., & Filev, D. P. (2004). An approach to online identification of Takagi–Sugeno fuzzy models. *IEEE Transactions on Systems, Man, and Cybernetics, Part B*, *34*(1), 484–498.
4. Aoki, T., & Aoyagi, T. (2007). Self-organizing maps with asymmetric neighborhood function. *Neural Computation*, *19*, 2515–2535.
5. Atiya, A. F. (2005). Estimating the posterior probabilities using the K-nearest neighbor rule. *Neural Computation*, *17*, 731–740.
6. Baraldi, A., & Blonda, P. (1999). A survey of fuzzy clustering algorithms for pattern recognition—Part II. *IEEE Transactions on Systems, Man, and Cybernetics, Part B*, *29*(6), 786–801.
7. Bax, E. (2012). Validation of k-nearest neighbor classifiers. *IEEE Transactions on Information Theory*, *58*(5), 3225–3234.
8. Beliakov, G., & Li, G. (2012). Improving the speed and stability of the k-nearest neighbors method. *Pattern Recognition Letters*, *33*, 1296–1301.
9. Beni, G., & Liu, X. (1994). A least biased fuzzy clustering method. *IEEE Transactions on Pattern Analysis and Machine Intelligence*, *16*(9), 954–960.
10. Berglund, E., & Sitte, J. (2006). The parameterless self-organizing map algorithm. *IEEE Transactions on Neural Networks*, *17*(2), 305–316.
11. Bermejo, S. (2006). The regularized LVQ1 algorithm. *Neurocomputing*, *70*, 475–488.
12. Bezdek, J. (1981). *Pattern recognition with fuzzy objective function algorithms*. New York: Plenum Press.
13. Biehl, M., Ghosh, A., & Hammer, B. (2007). Dynamics and generalization ability of LVQ algorithms. *Journal of Machine Learning Research*, *8*, 323–360.
14. Brodal, A. (1981). *Neurological anatomy in relation to clinical medicine* (3rd ed.). New York: Oxford University Press.
15. Burke, L. I. (1991). Clustering characterization of adaptive resonance. *Neural Networks*, *4*(4), 485–491.

16. Carpenter, G., Grossberg, S., & Rosen, D. B. (1991). ART 2-A: An adaptive resonance algorithm for rapid category learning and recognition. *Neural Networks*, *4*, 493–504.
17. Carpenter, G. A., & Grossberg, S. (1987). A massively parallel architecture for a self-organizing neural pattern recognition machine. *Computer Vision, Graphics, and Image Processing*, *37*, 54–115.
18. Carpenter, G. A., & Grossberg, S. (1987). ART 2: Self-organization of stable category recognition codes for analog input patterns. *Applied Optics*, *26*, 4919–4930.
19. Carpenter, G. A., Grossberg, S., & Reynolds, J. H. (1991). ARTMAP: Supervised real-time learning and classification of nonstationary data by a self-organizing neural network. *Neural Networks*, *4*(5), 565–588.
20. Carpenter, G. A., Grossberg, S., Markuzon, N., Reynolds, J. H., & Rosen, D. B. (1992). Fuzzy ARTMAP: A neural network architecture for incremental supervised learning of analog multidimensional maps. *IEEE Transactions on Neural Networks*, *3*, 698–713.
21. Cheng, Y. (1997). Convergence and ordering of Kohonen's batch map. *Neural Computation*, *9*, 1667–1676.
22. Chinrunrueng, C., & Sequin, C. H. (1995). Optimal adaptive k-means algorithm with dynamic adjustment of learning rate. *IEEE Transactions on Neural Networks*, *6*(1), 157–169.
23. Chiu, S. (1994). Fuzzy model identification based on cluster estimation. *Journal of Intelligent and Fuzzy Systems*, *2*(3), 267–278.
24. Chiu, S. L. (1994). A cluster estimation method with extension to fuzzy model identification. In *Proceedings of the IEEE International Conference on Fuzzy Systems* (Vol. 2, pp. 1240–1245). Orlando, FL.
25. Cho, J., Paiva, A. R. C., Kim, S.-P., Sanchez, J. C., & Principe, J. C. (2007). Self-organizing maps with dynamic learning for signal reconstruction. *Neural Networks*, *20*, 274–284.
26. Choy, C. S. T., & Siu, W. C. (1998). Fast sequential implementation of "neural-gas" network for vector quantization. *IEEE Transactions on Communications*, *46*(3), 301–304.
27. Cottrell, M., Hammer, B., Hasenfuss, A., & Villmann, T. (2006). Batch and median neural gas. *Neural Networks*, *19*, 762–771.
28. Cover, T. M., & Hart, P. E. (1967). Nearest neighbor pattern classification. *IEEE Transactions on Information Theory*, *13*, 21–27.
29. Dave, R. N., & Krishnapuram, R. (1997). Robust clustering methods: A unified view. *IEEE Transactions on Fuzzy Systems*, *5*(2), 270–293.
30. de Carvalho, F. A. T., Tenorio, C. P., & Cavalcanti, N. L, Jr. (2006). Partitional fuzzy clustering methods based on adaptive quadratic distances. *Fuzzy Sets and Systems*, *157*, 2833–2857.
31. Demartines, P., & Herault, J. (1997). Curvilinear component analysis: A self-organizing neural network for nonlinear mapping of data sets. *IEEE Transactions on Neural Networks*, *8*(1), 148–154.
32. Desarbo, W. S., Carroll, J. D., Clark, L. A., & Green, P. E. (1984). Synthesized clustering: A method for amalgamating clustering bases with differential weighting variables. *Psychometrika*, *49*, 57–78.
33. Ding, C., & He, X. (2004). Cluster structure of k-means clustering via principal component analysis. In *Proceedings of the 8th Pacific-Asia Conference on Advances in Knowledge Discovery and Data Mining (PAKDD), LNCS* (Vol. 3056, pp. 414–418). Sydney, Australia; Berlin: Springer.
34. Du, K.-L., & Swamy, M. N. S. (2006). *Neural networks in a softcomputing framework*. London: Springer.
35. Du, K.-L. (2010). Clustering: A neural network approach. *Neural Networks*, *23*(1), 89–107.
36. Dunn, J. C. (1974). Some recent investigations of a new fuzzy partitioning algorithm and its applicatiopn to pattern classification problems. *Journal of Cybernetics*, *4*, 1–15.
37. Estevez, P. A., & Figueroa, C. J. (2006). Online data visualization using the neural gas network. *Neural Networks*, *19*, 923–934.
38. Fayed, H. A., & Atiya, A. F. (2009). A novel template reduction approach for the K-nearest neighbor method. *IEEE Transactions on Neural Networks*, *20*(5), 890–896.

39. Fix, E., & Hodges, J. L., Jr. (1951). *Discriminatory analysis—Nonparametric discrimination: Consistency properties*. Project No. 2-49-004, Report No. 4. Randolph Field, TX: USAF School of Aviation (Reprinted in *International Statistical Review*, *57*(3), 238–247 (1989)).
40. Flanagan, J. A. (1996). Self-organization in Kohonen's SOM. *Neural Networks*, *9*(7), 1185–1197.
41. Fort, J. C. (1988). Solving a combinatorial problem via self-organizing process: An application of Kohonen-type neural networks to the travelling salesman problem. *Biology in Cybernetics*, *59*, 33–40.
42. Fritzke, B. (1997). The LBG-U method for vector quantization—An improvement over LBG inspired from neural networks. *Neural Processing Letters*, *5*(1), 35–45.
43. Fukushima, K. (2010). Neocognitron trained with winner-kill-loser rule. *Neural Networks*, *23*, 926–938.
44. Gath, I., & Geva, A. B. (1989). Unsupervised optimal fuzzy clustering. *IEEE Transactions on Pattern Analysis and Machine Intelligence*, *11*(7), 773–781.
45. Gersho, A. (1979). Asymptotically optimal block quantization. *IEEE Transactions on Information Theory*, *25*(4), 373–380.
46. Ghosh, A., Biehl, M., & Hammer, B. (2006). Performance analysis of LVQ algorithms: A statistical physics approach. *Neural Networks*, *19*, 817–829.
47. Gottlieb, L.-A., Kontorovich, A., & Krauthgamer, R. (2014). Efficient classification for metric data. *IEEE Transactions on Information Theory*, *60*(9), 5750–5759.
48. Grossberg, S. (1976). Adaptive pattern classification and universal recording: I. Parallel development and coding of neural feature detectors; II. Feedback, expectation, olfaction, and illusions. *Biological Cybernetics*, *23*, 121–34, 187–202.
49. Guo, Y., & Sengur, A. (2015). NCM: Neutrosophic c-means clustering algorithm. *Pattern Recognition*, *48*(8), 2710–2724.
50. Gustafson, D. E., & Kessel, W. (1979). Fuzzy clustering with a fuzzy covariance matrix. In *Proceedings of the IEEE Conference on Decision and Control* (pp. 761–766). San Diego, CA.
51. Hall, L. O., & Goldgof, D. B. (2011). Convergence of the single-pass and online fuzzy C-means algorithms. *IEEE Transactions on Fuzzy Systems*, *19*(4), 792–794.
52. Hammer, B., Strickert, M., & Villmann, T. (2005). Supervised neural gas with general similarity measure. *Neural Processing Letters*, *21*(1), 21–44.
53. Hara, T., & Hirose, A. (2004). Plastic mine detecting radar system using complex-valued self-organizing map that deals with multiple-frequency interferometric images. *Neural Networks*, *17*, 1201–1210.
54. Hart, P. E. (1968). The condensed nearest neighbor rule. *IEEE Transactions on Information Theory*, *14*(3), 515–516.
55. Hathaway, R. J., & Bezdek, J. C. (1993). Switching regression models and fuzzy clustering. *IEEE Transactions on Fuzzy Systems*, *1*, 195–204.
56. Hathaway, R. J., & Bezdek, J. C. (2000). Generalized fuzzy c-means clustering strategies using L_p norm distances. *IEEE Transactions on Fuzzy Systems*, *8*(5), 576–582.
57. He, J., Tan, A. H., & Tan, C. L. (2004). Modified ART 2A growing network capable of generating a fixed number of nodes. *IEEE Transactions on Neural Networks*, *15*(3), 728–737.
58. Huang, J. Z., Ng, M. K., Rong, H., & Li, Z. (2005). Automated variable weighting in k-means type clustering. *IEEE Transactions on Pattern Analysis and Machine Intelligence*, *27*(5), 657–668.
59. Huang, P., & Zhang, D. (2010). Locality sensitive C-means clustering algorithms. *Neurocomputing*, *73*, 2935–2943.
60. Hueter, G. J. (1988). Solution of the traveling salesman problem with an adaptive ring. In *Proceedings of International Conference on Neural Networks* (pp. 85–92). San Diego, CA.
61. Hwang, C., & Rhee, F. (2007). Uncertain fuzzy clustering: Interval type-2 fuzzy approach to C-means. *IEEE Transactions on Fuzzy Systems*, *15*(1), 107–120.
62. Kanungo, T., Mount, D. M., Netanyahu, N. S., Piatko, C. D., Silverman, R., & Wu, A. Y. (2002). An efficient k-means clustering algorithm: Analysis and implementation. *IEEE Transactions on Pattern Analysis and Machine Intelligence*, *24*(7), 881–892.

63. Karayiannis, N. B., & Randolph-Gips, M. M. (2003). Soft learning vector quantization and clustering algorithms based on non-Euclidean norms: Multinorm algorithms. *IEEE Transactions on Neural Networks*, *14*(1), 89–102.
64. Kaymak, U., & Setnes, M. (2002). Fuzzy clustering with volume prototypes and adaptive cluster merging. *IEEE Transactions on Fuzzy Systems*, *10*(6), 705–712.
65. Kersten, P. R. (1999). Fuzzy order statistics and their application to fuzzy clustering. *IEEE Transactions on Fuzzy Systems*, *7*(6), 708–712.
66. Kohonen, T. (1982). Self-organized formation of topologically correct feature maps. *Biological Cybernetics*, *43*, 59–69.
67. Kohonen, T. (1989). *Self-organization and associative memory*. Berlin: Springer.
68. Kohonen, T. (1990). The self-organizing map. *Proceedings of the IEEE*, *78*, 1464–1480.
69. Kohonen, T. (1990). Improved versions of learning vector quantization. In *Proceedings of the International Joint Conference on Neural Networks (IJCNN)* (Vol. 1, pp. 545–550). San Diego, CA.
70. Kohonen, T. (1990). Derivation of a class of training algorithms. *IEEE Transactions on Neural Networks*, *1*, 229–232.
71. Kohonen, T. (1996). Emergence of invariant-feature detectors in the adaptive-subspace self-organizing map. *Biological Cybernetics*, *75*, 281–291.
72. Kohonen, T. (2001). *Self-organizing maps* (3rd ed.). Berlin: Springer.
73. Kohonen, T., Kangas, J., Laaksonen, J., & Torkkola, K. (1992). LVQPAK: A program package for the correct application of learning vector quantization algorithms. In *Proceedings of the International Joint Conference on Neural Networks (IJCNN)* (Vol. 1, pp. 725–730). Baltimore, MD.
74. Kohonen, T. (2006). Self-organizing neural projections. *Neural Networks*, *19*, 723–733.
75. Kolen, J., & Hutcheson, T. (2002). Reducing the time complexity of the fuzzy C-means algorithm. *IEEE Transactions on Fuzzy Systems*, *10*(2), 263–267.
76. Krishnapuram, R., & Kim, J. (2000). Clustering algorithms based on volume criteria. *IEEE Transactions on Fuzzy Systems*, *8*(2), 228–236.
77. Kusumoto, H., & Takefuji, Y. (2006). $O(\log_2 M)$ self-organizing map algorithm without learning of neighborhood vectors. *IEEE Transactions on Neural Networks*, *17*(6), 1656–1661.
78. Laaksonen, J., & Oja, E. (1996). Classification with learning k-nearest neighbors. In *Proceedings of the International Conference on Neural Networks* (pp. 1480–1483). Washington, DC.
79. Lai, J. Z. C., Huang, T.-J., & Liaw, Y.-C. (2009). A fast k-means clustering algorithm using cluster center displacement. *Pattern Recognition*, *42*, 2551–2556.
80. Leski, J. (2003). Towards a robust fuzzy clustering. *Fuzzy Sets and Systems*, *137*, 215–233.
81. Li, M. J., Ng, M. K., Cheung, Y.-M., & Huang, J. Z. (2008). Agglomerative fuzzy K-means clustering algorithm with selection of number of clusters. *IEEE Transactions on Knowledge and Data Engineering*, *20*(11), 1519–1534.
82. Likas, A., Vlassis, N., & Verbeek, J. J. (2003). The global k-means clustering algorithm. *Pattern Recognition*, *36*(2), 451–461.
83. Linda, O., & Manic, M. (2012). General type-2 fuzzy C-means algorithm for uncertain fuzzy clustering. *IEEE Transactions on Fuzzy Systems*, *20*(5), 883–897.
84. Linde, Y., Buzo, A., & Gray, R. M. (1980). An algorithm for vector quantizer design. *IEEE Transactions on Communications*, *28*, 84–95.
85. Lippman, R. P. (1987). An introduction to computing with neural nets. *IEEE ASSP Magazine*, *4*(2), 4–22.
86. Lo, Z. P., & Bavarian, B. (1991). On the rate of convergence in topology preserving neural networks. *Biological Cybernetics*, *65*, 55–63.
87. Luttrell, S. P. (1990). Derivation of a class of training algorithms. *IEEE Transactions on Neural Networks*, *1*, 229–232.
88. Luttrell, S. P. (1994). A Bayesian analysis of self-organizing maps. *Neural Computation*, *6*, 767–794.

89. MacQueen, J. B. (1967). Some methods for classification and analysis of multivariate observations. In *Proceedings of the 5th Berkeley Symposium on Mathematical Statistics and Probability* (pp. 281–297). Berkeley, CA: University of California Press.
90. Martinetz, T. M. (1993). Competitive Hebbian learning rule forms perfectly topology preserving maps. In *Proceedings of the International Conference on Artificial Neural Networks (ICANN)* (pp. 427–434). Amsterdam, The Netherlands.
91. Martinetz, T. M., Berkovich, S. G., & Schulten, K. J. (1993). Neural-gas network for vector quantization and its application to time-series predictions. *IEEE Transactions on Neural Networks*, *4*(4), 558–569.
92. Martinetz, T. M., & Schulten, K. J. (1994). Topology representing networks. *Neural Networks*, *7*, 507–522.
93. Moore, B. (1988). ART and pattern clustering. In D. Touretzky, G. Hinton & T. Sejnowski (Eds.), *Proceedings of the 1988 Connectionist Model Summer School* (pp. 174–183). San Mateo, CA: Morgan Kaufmann.
94. Mulder, S. A., & Wunsch, D. C, I. I. (2003). Million city traveling salesman problem solution by divide and conquer clustering with adaptive resonance neural networks. *Neural Networks*, *16*, 827–832.
95. Obermayer, K., Ritter, H., & Schulten, K. (1991). Development and spatial structure of cortical feature maps: A model study. In R. P. Lippmann, J. E. Moody, & D. S. Touretzky (Eds.), *Advances in neural information processing systems* (Vol. 3, pp. 11–17). San Mateo, CA: Morgan Kaufmann.
96. Odorico, R. (1997). Learning vector quantization with training count (LVQTC). *Neural Networks*, *10*(6), 1083–1088.
97. Pal, N. R., Bezdek, J. C., & Tsao, E. C. K. (1993). Generalized clustering networks and Kohonen's self-organizing scheme. *IEEE Transactions on Neural Networks*, *4*(2), 549–557.
98. Pal, N. R., & Chakraborty, D. (2000). Mountain and subtractive clustering method: Improvements and generalizations. *International Journal of Intelligent Systems*, *15*, 329–341.
99. Patane, G., & Russo, M. (2001). The enhanced LBG algorithm. *Neural Networks*, *14*(9), 1219–1237.
100. Pedrycz, W., & Waletzky, J. (1997). Fuzzy clustering with partial supervision. *IEEE Transactions on Systems, Man, and Cybernetics, Part B*, *27*(5), 787–795.
101. Richardt, J., Karl, F., & Muller, C. (1998). Connections between fuzzy theory, simulated annealing, and convex duality. *Fuzzy Sets and Systems*, *96*, 307–334.
102. Rodriguez, A., & Laio, A. (2014). Clustering by fast search and find of density peaks. *Science*, *344*(6191), 1492–1496.
103. Rose, K. (1998). Deterministic annealing for clustering, compression, classification, regression, and related optimization problems. *Proceedings of the IEEE*, *86*(11), 2210–2239.
104. Rose, K., Gurewitz, E., & Fox, G. C. (1990). A deterministic annealing approach to clustering. *Pattern Recognition Letters*, *11*(9), 589–594.
105. Sato, A., & Yamada, K. (1995). Generalized learning vector quantization. In G. Tesauro, D. Touretzky, & T. Leen (Eds.), *Advances in neural information processing systems* (Vol. 7, pp. 423–429). Cambridge, MA: MIT Press.
106. Serrano-Gotarredona, T., & Linares-Barranco, B. (1996). A modified ART 1 algorithm more suitable for VLSI implementations. *Neural Networks*, *9*(6), 1025–1043.
107. Seo, S., & Obermayer, K. (2003). Soft learning vector quantization. *Neural Computation*, *15*, 1589–1604.
108. Shalev-Shwartz, S., & Ben-David, S. (2014). *Understanding machine learning: From theory to algorithms*. Cambridge, UK: Cambridge University Press.
109. Su, Z., & Denoeux, T. (2019). BPEC: Belief-peaks evidential clustering. *IEEE Transactions on Fuzzy Systems*, *27*(1), 111–123.
110. Tsekouras, G., Sarimveis, H., Kavakli, E., & Bafas, G. (2004). A hierarchical fuzzy-clustering approach to fuzzy modeling. *Fuzzy Sets and Systems*, *150*(2), 245–266.
111. Verzi, S. J., Heileman, G. L., Georgiopoulos, M., & Anagnostopoulos, G. C. (2003). Universal approximation with fuzzy ART and fuzzy ARTMAP. In *Proceedings of the International Joint Conference on Neural Networks (IJCNN)* (Vol. 3, pp. 1987–1892). Portland, OR.

112. von der Malsburg, C. (1973). Self-organizing of orientation sensitive cells in the striata cortex. *Kybernetik*, *14*, 85–100.
113. Wilson, D. L. (1972). Asymptotic properties of nearest neighbor rules using edited data. *IEEE Transactions on Systems, Man, and Cybernetics*, *2*(3), 408–420.
114. Witoelar, A., Biehl, M., Ghosh, A., & Hammer, B. (2008). Learning dynamics and robustness of vector quantization and neural gas. *Neurocomputing*, *71*, 1210–1219.
115. Wu, K.-L., Yang, M.-S., & Hsieh, J.-N. (2010). Mountain c-regressions method. *Pattern Recognition*, *43*, 86–98.
116. Yager, R. R., & Filev, D. (1994). Approximate clustering via the mountain method. *IEEE Transactions on Systems, Man, and Cybernetics*, *24*(8), 1279–1284.
117. Yair, E., Zeger, K., & Gersho, A. (1992). Competitive learning and soft competition for vector quantizer design. *IEEE Transactions on Signal Processing*, *40*(2), 294–309.
118. Yang, M. S. (1993). On a class of fuzzy classification maximum likelihood procedures. *Fuzzy Sets and Systems*, *57*, 365–375.
119. Yang, M.-S., Wu, K.-L., Hsieh, J.-N., & Yu, J. (2008). Alpha-cut implemented fuzzy clustering algorithms and switching regressions. *IEEE Transactions on Systems, Man, and Cybernetics, Part B*, *38*(3), 588–603.
120. Yin, H., & Allinson, N. M. (1995). On the distribution and convergence of the feature space in self-organizing maps. *Neural Computation*, *7*(6), 1178–1187.
121. Yu, J., Cheng, Q., & Huang, H. (2004). Analysis of the weighting exponent in the FCM. *IEEE Transactions on Systems, Man, and Cybernetics, Part B*, *34*(1), 634–639.
122. Yu, J., & Yang, M. S. (2005). Optimality test for generalized FCM and its application to parameter selection. *IEEE Transactions on Fuzzy Systems*, *13*(1), 164–176.
123. Zhao, W.-L., Deng, C.-H., & Ngo, C.-W. (2018). k-means: A revisit. *Neurocomputing*, *291*, 195–206.
124. Zheng, H., Lefebvre, G., & Laurent, C. (2008). Fast-learning adaptive-subspace self-organizing map: An application to saliency-based invariant image feature construction. *IEEE Transactions on Neural Networks*, *19*(5), 746–757.

Chapter 10
Clustering II: Topics in Clustering

10.1 Underutilization Problem

Conventional competitive learning-based clustering algorithms like C-means and LVQ are plagued by a severe initialization problem [58, 109]. If the initial values of the prototypes are not in the convex hull formed by the input data, clustering may not produce meaningful results. This is the so-called *prototype underutilization* or *dead-unit problem* since some prototypes, called *dead units*, may never win the competition. The underutilization problem is caused by the fact that the algorithm updates only the winning prototype for every input.

The underutilization problem is illustrated in Fig. 10.1. There are three clusters in the dataset. If the three prototypes $\boldsymbol{c}_1$, $\boldsymbol{c}_2$ and $\boldsymbol{c}_3$ are initialized at A, B, and C, respectively, they will correctly move to the centers of the three clusters. However, if they are initialized at A, B, and C′, respectively, C′ will never become a winner and thus becomes a dead unit. In the latter case, the system divides the three data clusters into two clusters, and the prototypes $\boldsymbol{c}_1$ and $\boldsymbol{c}_2$ will, respectively, move to the centroids of the two clusters.

In order to alleviate the sensitivity of competitive learning to the initialization of the clustering centers, many efforts have been made to solve the underutilization problem. Initializing the prototypes with random input vectors can reduce the probability of the underutilization problem but does not eliminate it. In the leaky learning strategy [58, 109], all the prototypes are updated. The winning prototype is updated by employing a high learning rate, while all the losing prototypes move toward the input vector with a much smaller learning rate.

10.1.1 Competitive Learning with Conscience

To avoid the underutilization problem, one can assign each processing unit with a threshold, and then increase the threshold if a unit wins, or decrease it otherwise [109].

K.-L. Du and M. N. S. Swamy, *Neural Networks and Statistical Learning*,
https://doi.org/10.1007/978-1-4471-7452-3_10

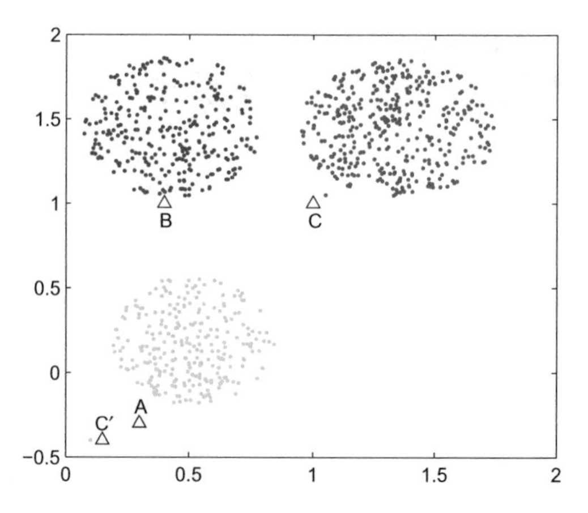

Fig. 10.1 Illustration of the underutilization problem for competitive learning-based clustering

A similar idea is embodied in the conscience strategy, which reduces the winning rate of the frequent winners [38]. The frequent winner receives a bad conscience by adding a penalty term to its distance from the input signal. This leads to an entropy maximization, that is, each unit wins at an approximately equal probability. Thus, the probability of underutilized neurons being selected as winners is increased.

Frequency-sensitive competitive learning (FSCL) [6] employs this conscience strategy. It reduces the underutilization problem by introducing a distortion measure that ensures all codewords in the codebook to be updated with a similar probability. The codebooks obtained by FSCL have sufficient entropy so that Huffman coding of the vector quantization indices would not provide significant additional compression.

In FSCL, each prototype incorporates a count of the times it has been the winner, $u_i, i = 1, \ldots, K$. The distance measure is modified to give prototypes with a lower count value a chance to win a competition. The algorithm is similar to the vector quantization algorithm, with the only difference being that the winning neuron is found by [6]

$$\boldsymbol{c}_w(t) = \arg_{\boldsymbol{c}_i} \min_{i=1,\ldots,K} \left\{ u_i(t-1) \left\| \boldsymbol{x}_t - \boldsymbol{c}_i(t-1) \right\| \right\}, \tag{10.1}$$

where w is the index of the winning node, u_i is updated by

$$u_i(t) = \begin{cases} u_i(t-1) + 1, & i = w \\ u_i(t-1), & \text{otherwise} \end{cases} \tag{10.2}$$

and $u_i(0) = 0$, $i = 1, \ldots, K$. In (10.1), $u_i \left\| \boldsymbol{x}_t - \boldsymbol{c}_i \right\|$ can be generalized as $F(u_i) \left\| \boldsymbol{x}_t - \boldsymbol{c}_j \right\|$. When selecting the fairness function as $F(u_i) = u_i^{\beta_0 e^{-t/T_0}}$, with constants β_0 and T_0, FSCL emphasizes the winning uniformity of codewords initially and gradually turns into competitive learning as training proceeds to minimize the MSE function.

In multiplicatively biased competitive learning [29], the competition among the neurons is biased by a multiplicative term. The method avoids neuron underutilization with probability one, as time goes to infinity. Only one weight vector is updated per learning step. FSCL is a member of this family. Fuzzy FSCL [30] combines the frequency sensitivity with fuzzy competitive learning. Since both FSCL and fuzzy FSCL use a non-Euclidean distance to determine the winner, they may lead to the problem of shared clusters in the sense that a number of prototypes may be updated into the same cluster during the learning process. SOM may yield grid units that are never active, and a number of topographic map formation models [11] add a conscience to the weight update process so that every grid unit is used equally.

The habituation mechanism consists in a reversible decrement of the neural response to a repetitive stimulus. The response recovers only after a period in which there is no activity, and the longer the presynaptic neuron is active, the slower it recovers. In [97], habituation has been used to build a growing neural network: New units are added to the network when the same neural unit answers to many input patterns. In [106], habituation is implemented in an SOM network and its effects on learning speed and vector quantization are analyzed. The conscience learning mechanism follows roughly the same principle but is less sophisticated. The proposed habituation mechanism is simple to implement and more flexible than conscience learning because it can be used to manage the learning process in a fine-grained way, also allowing multilayer self-organizing structures to be built. Moreover, while conscience learning modifies the comparison between the input pattern and the neuron weights, habituation only affects the activation of the neuron and it adds other variables that allow the learning process to be fine-tuned.

10.1.2 *Rival Penalized Competitive Learning*

The problem of shared clusters for FSCL and fuzzy FSCL has been considered in the rival penalized competitive learning (RPCL) algorithm [132]. RPCL adds a new mechanism to FSCL by creating a rival penalizing force. For each input, not only is the winning unit modified to adapt to the input but also the second-place winner called a *rival* is updated by a smaller learning rate along the opposite direction, all the other prototypes being unchanged:

$$\boldsymbol{c}_i(t+1) = \begin{cases} \boldsymbol{c}_i(t) + \eta_w \left(\boldsymbol{x}_t - \boldsymbol{c}_i(t)\right), & i = w \\ \boldsymbol{c}_i(t) - \eta_r \left(\boldsymbol{x}_t - \boldsymbol{c}_i(t)\right), & i = r \\ \boldsymbol{c}_i(t), & \text{otherwise} \end{cases}, \tag{10.3}$$

where w and r are the indices of the winning and rival prototypes, which are decided by (10.1), and η_w and η_r are their respective learning rates. In practice, $\eta_w(t) \gg \eta_r$. η_r is also called the *delearning rate* for the rival. The principle of RPCL is shown in Fig. 10.2.

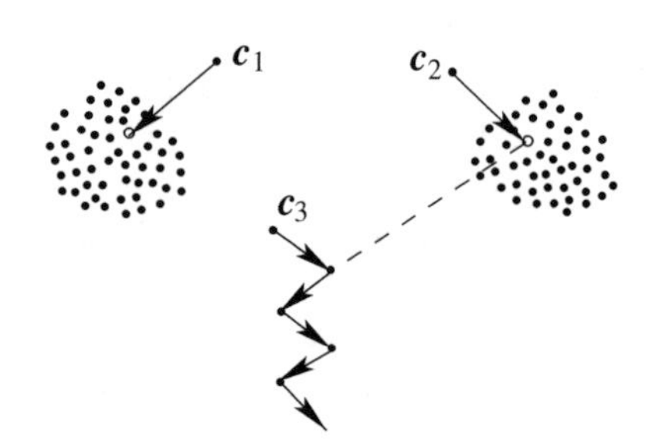

Fig. 10.2 Illustration of RPCL. The rival $\boldsymbol{c}_3$ of both $\boldsymbol{c}_1$ and $\boldsymbol{c}_2$ is driven out to infinity along a zigzag path

This actually pushes the rival away from the sample pattern so as to prevent its interference in the competition. RPCL automatically allocates an appropriate number of prototypes for an input dataset, and all the extra candidate prototypes will finally be pushed to infinity. It provides a better performance than FSCL. RPCL can be regarded as an unsupervised extension of supervised LVQ2. It simultaneously modifies the weight vectors of both the winner and its rival, when the winner is in a wrong class but the rival is in a correct class for an input vector [132]. Lotto-type competitive learning [91] can be treated as a generalization of RPCL, where instead of just penalizing the nearest rival, all the losers are penalized equally. Generalized lotto-type competitive learning [92] modifies lotto-type competitive learning by allowing more than one winner, which are divided into tiers, with each tier being rewarded differently. However, RPCL may encounter the overpenalization and underpenalization problems due to an inappropriate delearning rate [140]. Generalized RPCL [131] distributes learning and penalization to groups of agents proportionally to their activation strengths, that is, RPCL is extended to a soft-competitive mechanism comprising multiple winners and rivals.

Stepwise automatic rival penalized (STAR) C-means [25] consists of two separate phases. The first phase implements FSCL, which assigns each cluster with at least a prototype. The second phase is derived from a Kullback–Leibler divergence-based criterion and adjusts the units adaptively by a learning rule that automatically penalizes the winning chance of all rival prototypes in subsequent competitions while tuning the winning one to adapt to the input. STAR C-means has a mechanism similar to RPCL but penalizes the rivals in an implicit way, whereby circumventing the determination of the rival delearning rate of RPCL. STAR C-means is applicable to ellipse-shaped data clusters as well as sphere-shaped ones without the underutilization problem and without having to predetermine the correct cluster number.

The convergence problem of RPCL is investigated via a cost-function approach in [93]. A general form of RPCL, called distance-sensitive RPCL, is proved to be associated with the minimization of a cost function on the weight vectors of a competitive learning network. As a distance-sensitive RPCL process decreases the cost to a local minimum, a number of weight vectors eventually fall into a hypersphere surrounding the sample data, while the other weight vectors diverge to infinity. If the cost reduces into the global minimum, a correct number of weight vectors are automatically selected and located around the centers of the actual clusters.

The performance of RPCL is very sensitive to the selection of the delearning rate. Rival penalized controlled competitive learning (RPCCL) [26] generalizes RPCL; it dynamically controls the rival penalizing forces. RPCCL always sets the delearning rate as the learning rate. A rival seed point is always penalized with the delearning rule. The RPCCL mechanism fully penalizes the rival if the winner suffers from severe competition from the rival; otherwise, the penalization strength should be proportional to the degree of competition level. RPCCL can be implemented in a stochastic version, called stochastic RPCL. Stochastic RPCL penalizes the rivals by using the same rule as RPCL, but the penalization is performed stochastically.

Competitive repetition-suppression (CoRe) learning [7] is inspired by a cortical memory mechanism called *repetition suppression*. CoRe learning produces sharp feature detectors and compact neural representations of the input stimuli through a mechanism of neural suppression and strengthening that is dependent on the frequency of the stimuli. CoRe clustering can automatically estimate the unknown cluster number from the data without a priori information of the input distribution. It is a robust extension of RPCL, by combining the biological inspiration with the robustness properties of the M-estimators. CoRe clustering generalizes RPCL by allowing winner and rival competitors to refer to the set of units (instead of single neuron) and implementing a soft-competitive rival penalization scheme. Each CoRe neuron acts as a cluster detector and the repetition-suppression mechanism is used to selectively suppress irrelevant neurons, consequently determining the unknown cluster number. The repetition-suppression mechanism offers a mean for adaptively controlling rival penalization.

Rival-model penalized SOM [27], for each input, adaptively chooses several rivals of the best-matching unit and penalizes their associated prototypes a little far away from the input. Each node on the grid is associated with a prototype. Rival-model penalized SOM utilizes a constant learning rate but still reaches a robust result. The rival is not selected as the second nearest neuron, because the one-neighborhood neurons of the best-matching unit are usually topologically close to the best-matching unit, and penalizing the second nearest neuron violates the intrinsic characteristic of SOM.

10.1.3 Soft-Competitive Learning

By relaxing the WTA criterion, clustering methods can treat more than a single neuron as winners to a certain degree and update their prototypes accordingly, resulting in the winner-takes-most paradigm, namely, soft-competitive learning. Examples are soft-competition scheme [133], SOM, neural gas, maximum-entropy clustering [108], generalized LVQ, FCM, and fuzzy competitive learning [30]. The winner-takes-most criterion, however, detracts some prototypes from their corresponding clusters, since all the prototypes are attracted to each input pattern; consequently, it becomes biased toward the global mean of the clusters [90].

Fuzzy competitive learning [30] is a class of sequential algorithms obtained by fuzzifying competitive learning algorithms, such as simple competitive learning, unsupervised LVQ, and FSCL. The concept of winning is formulated as a fuzzy set, and the network outputs become the winning memberships of the competing neurons. Enhanced sequential fuzzy clustering [144] is a modification to fuzzy competitive learning, obtained by introducing a nonunity weighting on the winning centroid and an excitation–inhibition mechanism so as to better overcome the underutilization problem.

SOM is a learning process, which takes the winner-takes-most strategy at the early stages and becomes a WTA approach, while its neighborhood size reduces to unity as a function of time in a predetermined manner. Due to the soft-competitive strategy, SOM, growing neural gas (GNG) [51] and fuzzy clustering algorithms are less likely to be trapped at local minima and to generate dead units than hard competitive alternatives [10].

Maximum-entropy clustering [108] avoids the underutilization problem and local minima in the error function by using soft-competitive learning and deterministic annealing. The prototypes are updated by

$$\boldsymbol{c}_i(t+1) = \boldsymbol{c}_i(t) + \eta(t) \left[\frac{\mathrm{e}^{-\beta \|\boldsymbol{x}_t - \boldsymbol{c}_i(t)\|^2}}{\sum_{j=1}^{K} \mathrm{e}^{-\beta \|\boldsymbol{x}_t - \boldsymbol{c}_j(t)\|^2}} \right] (\boldsymbol{x}_t - \boldsymbol{c}_i(t)), \quad i = 1, \ldots, K, \tag{10.4}$$

where η is the learning rate and the parameter $\frac{1}{\beta}$ anneals from a large parameter to zero, and the term within the brackets turns out to be the Boltzmann distribution. Soft-competition scheme [133] employs a similar soft-competitive strategy, but $\beta = 1$.

10.2 Robust Clustering

Datasets usually contain noisy points or outliers. The robust statistics approach has been integrated into robust clustering methods [19, 35, 49, 65, 81]. C-median clustering [19] is derived by solving a bilinear programming problem that utilizes the L_1-norm distance. Fuzzy C-median clustering [81] is a robust FCM method that uses L_1-norm with the exemplar estimation based on fuzzy median. An approach to fuzzy clustering proposed in [23] combines the benefits of local dimensionality reduction within each cluster using factor analysis and the robustness to outliers of Student-t distributions.

The clustering of a vectorial dataset with missing entries belongs to the category of robust clustering. In [66], four strategies, namely, the whole data, partial distance, optimal completion, and nearest prototype strategies, have been discussed for implementing FCM clustering for incomplete data.

Robust competitive agglomeration [49] combines the advantages of hierarchical and partitional clustering techniques. The objective function also contains a constraint term given in the form of (10.11). An optimum number of clusters is determined via a process of competitive agglomeration, while the knowledge of the global shape of the clusters is incorporated by using prototypes. Robust statistics like the M-estimator is incorporated to handle outliers. Overlapping clusters are handled by the use of fuzzy memberships.

Robust clustering algorithms can be derived by optimizing a specially designed objective function, which usually has two terms:

$$E_T = E + E_c, \tag{10.5}$$

where E is the cost for the conventional algorithms such as (9.40), and the constraint term E_c characterizes the outliers.

The noise-clustering algorithm can be treated as a robustified FCM. In the noise-clustering approach [33], noise and outliers are collected into a separate, amorphous noise cluster, whose prototype has the same distance, δ, from all the data points. The other points are collected into K clusters. The threshold δ is a relatively high value compared to the distances of the *good* points to the cluster prototypes. If a noisy point is far away from all the K clusters, it is attracted to the noise cluster. When all the K clusters have about the same size, noise clustering is very effective. However, a single threshold is too restrictive if the cluster size varies widely in the dataset.

In noise clustering, the second term is given by

$$E_c = \sum_{i=1}^{N} \delta^2 \left(1 - \sum_{j=1}^{K} \mu_{ji} \right)^m . \tag{10.6}$$

Following the procedure for the derivation of FCM, we have

$$\mu_{ji} = \frac{\left(\frac{1}{\|\boldsymbol{x}_i - \boldsymbol{c}_j\|^2} \right)^{\frac{1}{m-1}}}{\sum_{k=1}^{K} \left(\frac{1}{\|\boldsymbol{x}_i - \boldsymbol{c}_k\|^2} \right)^{\frac{1}{m-1}} + \left(\frac{1}{\delta^2} \right)^{\frac{1}{m-1}}}. \tag{10.7}$$

The second term in the denominator, which is due to outliers, leads to small μ_{ji}. The formula for the prototypes is the same as that for FCM, given by (9.43).

In a fuzzified PCA-guided robust C-means method [71], a robust C-means partition is derived by using a noise-rejection mechanism based on the noise-clustering approach. The responsibility weight of each sample for the C-means process is estimated by considering the noise degree of the sample, and cluster indicators are calculated in a fuzzy PCA-guided manner, where fuzzy PCA-guided robust C-means is performed by considering responsibility weights of samples.

Relational data can be clustered by using non-Euclidean relational FCM [64]. A number of fuzzy clustering algorithms for relational data have been reviewed in [36]. The introduction of the concept of noise clustering into these relational clustering techniques leads to their robust versions [36]. Weighted non-Euclidean relational FCM [67] reduces the original dataset to a smaller one, assigns each selected datum a weight reflecting the number of nearby data, clusters the weighted reduced dataset using a weighted version of the feature or relational data FCM, and if desired, extends the reduced data results back to the original dataset.

The three important issues associated with competitive learning clustering are initialization, adaptation to clusters of different size and sparsity, and elimination of the disturbance caused by outliers. Energy-based competitive learning [127] simultaneously tackles these problems. Initialization is achieved by extracting samples of high energy to form a core point set, whereby connected components are obtained as initial clusters. To adapt to clusters of different sizes and sparsities, size–sparsity balance of clusters is developed to select a winning prototype, and a prototype energy-weighted squared distance objective function is defined. For eliminating the disturbance caused by outliers, adaptive learning rate based on samples' energy is proposed to update the winner.

10.2.1 Possibilistic C-Means

Possibilistic C-means (PCM) [84], as opposed to FCM, does not require that the sum of the memberships of a data point across the clusters be unity. This allows the membership functions to represent a possibility of belonging rather than a relative degree of membership between clusters. As a result, the derived degree of membership does not decrease as the number of clusters increases. Due to the elimination of this constraint, the modified objective function is decomposed into many individual objective functions, one for each cluster, which can be optimized separately.

The constraint term for PCM is given by a sum associated with the fuzzy complements of all the K clusters

$$E_c = \sum_{j=1}^{K} \beta_j \sum_{i=1}^{N} \left(1 - \mu_{ji}\right)^m , \tag{10.8}$$

where β_j are suitable positive numbers. The individual objective functions are given as

$$E_T^j = \sum_{i=1}^{N} \mu_{ji}^m \left\|\boldsymbol{x}_i - \boldsymbol{c}_j\right\|^2 + \beta_j \sum_{i=1}^{N} \left(1 - \mu_{ji}\right)^m , \quad j = 1, \ldots, K. \tag{10.9}$$

Differentiating (10.9) with respect to μ_{ji} and setting it to zero leads to the solution

$$\mu_{ji} = \frac{1}{1 + \left(\frac{\left\|\boldsymbol{x}_i - \boldsymbol{c}_j\right\|^2}{\beta_j}\right)^{\frac{1}{m-1}}}, \tag{10.10}$$

where the second term in the denominator is large for outliers, leading to small μ_{ji}. Some heuristics for selecting β_j have also been given in [84].

Given a number of clusters K, FCM will arbitrarily split or merge real clusters in the dataset to produce exactly the specified number of clusters. PCM, in contrast to FCM, can find those natural clusters in the dataset. When K is smaller than the number of actual clusters, only K good clusters are found, and the other data points are treated as outliers. When K is larger than the number of actual clusters, all the actual clusters can be found and some clusters will coincide. Thus K can be specified somewhat arbitrarily.

In the noise-clustering algorithm, there is only one noise cluster, while in PCM there are K noise clusters. PCM functions as a collection of K independent noise-clustering algorithms, each looking for a single cluster. The performance of PCM, however, relies heavily on good initialization of cluster prototypes and estimation of β_j, and PCM tends to converge to coincidental clusters [35, 84]. There have been many efforts to improve the stability of possibilistic clustering [141]. Possibilistic FCM [102] provides a hybrid model of FCM and PCM; it performs effectively for low-dimensional data clustering.

Most fuzzy clustering methods can only process the spatial data instead of the nonspatial data. In [123], similarity-based PCM is proposed to cluster nonspatial data without requesting users to specify the cluster number. It extends PCM for similarity-based clustering applications by integration with the mountain method. Rough-fuzzy possibilistic C-means [94] comprises a judicious integration of the principles of rough and fuzzy sets. It incorporates both probabilistic and possibilistic memberships simultaneously to avoid the noise sensitivity of FCM and the coincident clusters of PCM.

10.2.2 A Unified Framework for Robust Clustering

By extending the idea of treating outliers as the fuzzy complement, a family of robust clustering algorithms has been obtained [134]. Assume that a noise cluster exists outside each data cluster. The fuzzy complement of μ_{ji}, denoted as $f\left(\mu_{ji}\right)$, may be interpreted as the degree to which $\boldsymbol{x}_i$ does not belong to the ith data cluster. Thus, the fuzzy complement can be viewed as the membership of $\boldsymbol{x}_i$ in the noise cluster with a distance β_j. Based on this, one can propose many different implementations of the probabilistic approach [35, 134]. For robust fuzzy clustering, a general form of E_c is given as a generalization of that for PCM [134]

$$E_c = \sum_{i=1}^{N} \sum_{j=1}^{K} \beta_j \left[f(\mu_{ji}) \right]^m . \tag{10.11}$$

Notice that PCM uses the standard fuzzy complement $f\left(\mu_{ji}\right) = 1 - \mu_{ji}$.

By setting to zero the derivatives of E_T with respect to the variables, a fuzzy clustering algorithm is obtained. For example, by setting $m = 1$ and $f\left(\mu_{ji}\right) = \mu_{ji} \ln\left(\mu_{ji}\right) - \mu_{ji}$ [35] or $f\left(\mu_{ji}\right) = 1 + \mu_{ji} \ln\left(\mu_{ji}\right) - \mu_{ji}$ [134], we can obtain

$$\mu_{ji} = \mathrm{e}^{-\frac{\left\|x_i - c_j\right\|^2}{\beta_j}}, \tag{10.12}$$

and $\boldsymbol{c}_j$ has the same form as that for FCM. The alternating cluster estimation method [110] is a simple extension of the general method given in [35, 134]. β_j can be adjusted by [110]

$$\beta_j = \min_k \left\|\boldsymbol{c}_k - \boldsymbol{c}_j\right\|^2, \quad k \neq j. \tag{10.13}$$

The fuzzy robust C-spherical shells algorithm [134] searches the clusters belonging to the spherical shells by combining the concept of fuzzy complement and the fuzzy C-spherical shells algorithm [83]. The hard robust clustering algorithm [134] is obtained by setting $\beta_j = \infty$ if neuron j is the winner and setting β_j by (10.13) if it is a loser. In these robust algorithms, the initial values and adjustment of β_j are very important.

10.3 Supervised Clustering

Conventional clustering methods are unsupervised clustering, where unlabeled patterns are involved. When output patterns are used in clustering, this yields supervised clustering methods. The locations of the cluster centers are influenced not only by the input pattern spread but also by the output pattern deviations.

For classification problems, the classmembership of each pattern in the training set is available and can be used for clustering. Supervised clustering develops clusters preserving the homogeneity of the clustered patterns with regard to their similarity in the input space, as well as their respective values assumed in the output space. Examples of supervised clustering methods include LVQ family, ARTMAP family, and conditional FCM [103]. For classification problems, supervised clustering significantly improves the decision accuracy.

In the case of supervised learning using kernel-based neural networks such as the RBF network, the structure (kernels) is usually determined by using unsupervised clustering. This method, however, is not effective for finding a parsimonious network. Supervised clustering can be implemented by augmenting the input pattern with its output pattern, $\widetilde{\boldsymbol{x}}_i = \left[\boldsymbol{x}_i^T, \beta \boldsymbol{y}_i^T\right]^T$, so as to obtain an improved distribution of the cluster centers by unsupervised clustering [103, 110]. A scaling factor β balances between the underlying similarity in the input space and the similarity in the output space. The resulting objective function in the case of FCM is given by

$$E=\sum_{j=1}^{K}\sum_{i=1}^{N}\mu_{ji}^{m}\left\|\boldsymbol{x}_i-\boldsymbol{c}_{x,j}\right\|^2+\sum_{j=1}^{K}\sum_{i=1}^{N}\mu_{ji}^{m}\left\|\beta\boldsymbol{y}_i-\boldsymbol{c}_{y,j}\right\|^2, \tag{10.14}$$

where the new cluster center $\boldsymbol{c}_j=\left[\boldsymbol{c}_{x,j}^T,\boldsymbol{c}_{y,j}^T\right]^T$. The first term corresponds to FCM, and the second term applies to supervised learning. The resulting cluster codebook vectors are rescaled and projected onto the input space to obtain the centers.

Conditional FCM [103] is based on FCM but requires the output variable of a cluster to satisfy a particular condition. This condition can be treated as a fuzzy set, defined via the corresponding membership. A family of generalized weighted conditional FCM clustering algorithms has been derived in [88]. Semi-supervised enhancement of FCM is given in [17], where a kernel-based distance is applied.

Based on enhanced LBG, clustering for function approximation [57] is specially designed for function approximation problems. The method increases the density of the prototypes in the input areas where the target function presents a more variable response, rather than just in the zones with more input examples [57]. In [117], a prototype regression function is built as a linear combination of local linear regression models, one for each cluster, and is then inserted into the FCM functional. In this way, the prototypes can be adjusted according to both the input distribution and the regression function in the output space.

10.4 Clustering Using Non-Euclidean Distance Measures

Conventional clustering methods are based on the Euclidean distance, which favors hyperspherically shaped clusters of equal size. The Euclidean distance measure results in the undesirable property of splitting big and elongated clusters. Other distance measures can be defined to search for clusters of specific shapes in the feature space.

The Mahalanobis distance can be used to look for hyperellipsoid-shaped clusters. It is used in Gustafson–Kessel clustering [60] and adaptive fuzzy clustering. However, C-means with the Mahalanobis distance tends to produce unusually large or unusually small clusters. The hyperellipsoidal clustering network [96] integrates PCA and clustering into one network and can adaptively estimate the hyperellipsoidal shape of each cluster. Hyperellipsoidal clustering implements clustering using a regularized Mahalanobis distance, which is a linear combination of the Mahalanobis distance and the Euclidean distance, to prevent from producing unusually large or unusually small clusters.

Symmetry-based C-means [118] can effectively find clusters with symmetric shapes, such as the human face. The method employs the point-symmetry distance as the dissimilarity measure. The point-symmetry distance is defined by

$$d_{j,i}=d\left(\boldsymbol{x}_j,\boldsymbol{c}_i\right)=\min_{p=1,\ldots,N,\ p\neq j}\frac{\left\|\left(\boldsymbol{x}_j-\boldsymbol{c}_i\right)+\left(\boldsymbol{x}_p-\boldsymbol{c}_i\right)\right\|}{\left\|\boldsymbol{x}_j-\boldsymbol{c}_i\right\|+\left\|\boldsymbol{x}_p-\boldsymbol{c}_i\right\|}, \tag{10.15}$$

where $\boldsymbol{c}_i$ is a prototype vector, and the pattern set $\{\boldsymbol{x}_i\}$ is of size N. Notice that $d_{j,i} = 0$ only when $\boldsymbol{x}_p = 2\boldsymbol{c}_i - \boldsymbol{x}_j$. Symmetry-based C-means uses C-means as a coarse search for the K cluster centroids. A fine-tuning procedure is then performed based on the point-symmetry distance using the nearest neighbor paradigm.

To deal efficiently with pattern recognition and image segmentation in which we could encounter various geometrical shapes of the clusters, a number of shell clustering algorithms for detecting circles and hyperspherical shells have been proposed as extensions of C-means and FCM.

Fuzzy C-varieties method [15] can be regarded as a simultaneous algorithm of fuzzy clustering and PCA, in which the prototypes are multidimensional linear varieties represented by some local principal component vectors. Fuzzy C-shells method is successful in clustering spherical shells, and it has been further generalized to adaptive fuzzy C-shells for the case of elliptical shells [32, 34]. Fuzzy C-spherical shells method [83] reduces the computational cost of fuzzy C-shells by introducing an algebraic distance measure. For two-dimensional cases, fuzzy C-rings method [95] is used for clustering ring data, while fuzzy C-ellipses method [55] is for elliptical data. Fuzzy C-quadric shells method [85] detects quadrics-like circles, ellipses, hyperbolas, or lines. The clustering algorithms for detecting rectangular shells include norm-induced shell prototypes [16] and fuzzy C-rectangular shells [70].

The above approaches are listed in Table 10.1. Like FCM, they suffer from three problems: lack of robustness against noisy points, sensitivity to prototype initialization, and requiring a priori knowledge of the optimal cluster number. Based on fuzzy C-spherical shells, information fuzzy C-spherical shells [116] are for robust fuzzy clustering of spherical shells of outlier detection, prototype initialization, and cluster validity in a unified framework of information clustering.

These algorithms for shell clustering are based on iterative optimization of objective functions similar to that for FCM but defines the distance from a prototype $\boldsymbol{\lambda}_i = (\boldsymbol{c}_i, r_i)$ to the point $\boldsymbol{x}_j$ as

$$d_{j,i}^2 = d^2\left(\boldsymbol{x}_j, \boldsymbol{\lambda}_i\right) = \left(\left\|\boldsymbol{x}_j - \boldsymbol{c}_i\right\| - r_i\right)^2, \tag{10.16}$$

where $\boldsymbol{c}_i$ is the center of the hypersphere and r_i is the radius. They can effectively estimate the optimal number of substructures in the dataset by using some validity criteria such as spherical shell thickness [83], fuzzy hypervolume, and fuzzy density [54, 95]. By using different distance measures, many clustering algorithms can be derived for detecting clusters of various shapes such as lines and planes [80], and ellipsoids [80].

Directional clustering deals with a clustering problem where feature vectors are clustered depending on the angle between feature vectors. This directional distance measure arises in document classification and human brain imaging. Spherical C-means [72, 145] deals with directional clustering, using the similarity measure which scales features $\boldsymbol{x}_j$ and cluster centers $\boldsymbol{c}_i$ to unit length

Table 10.1 A family of shell clustering algorithms

Algorithm	Cluster shape
Fuzzy C-varieties [15]	Line segments and lines
Fuzzy C-shells [32, 34]	Circles and ellipses
Hard C-spherical shells [83]	Circles and spheres
Unsupervised C-spherical shells [83]	Circles and spheres
Fuzzy C-spherical shells [83]	Circles and spheres
Possibilistic C-spherical shells [84]	Circles and spheres
Fuzzy C-rings [95]	Circles
Fuzzy C-ellipses [55]	Ellipses
Fuzzy C-means	Spheres
Gustafson–Kessel [60]	Ellipsoids and possibly lines
Gath–Geva [54]	Ellipsoids and possibly lines
Fuzzy C-elliptotype [111]	Ellipsoids and possibly lines
Fuzzy C-quadric shells [85]	Linear and quadric shell
Norm-induced shell prototypes [16]	Rectangular shells
Fuzzy C-rectangular shells [70]	Rectangular shells, rectangles/polygons (approximation of circle, lines, ellipses)

$$d_{j,i} = d\left(\boldsymbol{x}_j, \boldsymbol{c}_i\right) = 2 - 2\frac{< \boldsymbol{x}_j, \boldsymbol{c}_i >}{\|\boldsymbol{x}_j\| \|\boldsymbol{c}_i\|} = \left\| \frac{\boldsymbol{x}_j}{\|\boldsymbol{x}_j\|} - \frac{\boldsymbol{c}_i}{\|\boldsymbol{c}_i\|} \right\|^2. \tag{10.17}$$

10.5 Partitional, Hierarchical, and Density-Based Clustering

Existing clustering algorithms are broadly classified into partitional, hierarchical, and density-based clustering. Clustering methods discussed thus far primarily belong to partitional clustering, and density-based clustering has also been introduced in Sect. 9.9.

Partitional clustering can be either hard clustering or fuzzy clustering. Fuzzy clustering can deal with overlapping cluster boundaries. Partitional clustering is dynamic, where points can move from one cluster to another. Knowledge of the shape or size of the clusters can be incorporated by using appropriate prototypes and distance measures. Due to the optimization of a certain criterion function, partitional clustering is sensitive to initialization and susceptible to local minima. It has difficulty in determining the suitable number of clusters K. In addition, it is also sensitive to noise and outliers. Typical partitional clustering algorithms have a computational complexity of $O(N)$, for a training set of size N.

Hierarchical clustering consists of a sequence of partitions in a hierarchical structure, which can be represented graphically as a clustering tree, called a *dendrogram*. It can be either an agglomerative or a divisive technique. Hierarchical clustering usually takes the form of agglomerative clustering. New clusters are formed by reallocating the membership degree of one point at a time, based on some measure of similarity or distance. It is suitable for data with dendritic substructure. Divisive clustering performs in a way opposite to that of agglomerative clustering but is computationally more costly.

Hierarchical clustering has a number of advantages over partitional clustering. In hierarchical clustering, outliers can be easily identified, since they merge with other points less often due to their larger distances from the other points. Consequently, the number of points in a collection of outliers is typically much less than the number in a cluster. In addition, the number of clusters K does not need to be specified, and the local minimum problem arising from initialization is no longer a problem anymore. However, prior knowledge of the shape or size of the clusters cannot be incorporated, and consequently overlapping clusters cannot always be separated. Moreover, hierarchical clustering is static, and points committed to a given cluster in the early stages cannot move to a different cluster. It typically has a computational complexity of at least $O\left(N^2\right)$, which makes it impractical for larger datasets.

Density-based clustering groups neighboring objects of a dataset into clusters based on density conditions. Clusters are dense regions of objects in the data space and are separated by regions of low density. Density-based clustering is robust against outliers since an outlier affects clustering only in the neighborhood of this data point. It can handle outliers and discover clusters of arbitrary shape. The computational complexity of density-based clustering is of the same order of magnitude as that of hierarchical algorithms.

10.6 Hierarchical Clustering

10.6.1 Distance Measures, Cluster Representations, and Dendrograms

The two simplest and well-known agglomerative clustering algorithms are the single linkage [115] and complete linkage [82] algorithms. The single-linkage algorithm, also called the *nearest neighbor paradigm*, is a bottom-up approach that generates clusters by sequentially merging pairs of similar clusters. The technique calculates the intercluster distance using the two closest data points in different clusters:

$$d\left(\mathcal{C}_1, \mathcal{C}_2\right) = \min_{\boldsymbol{x} \in \mathcal{C}_1, \boldsymbol{y} \in \mathcal{C}_2} d(\boldsymbol{x}, \boldsymbol{y}), \tag{10.18}$$

where $d\left(\mathcal{C}_1, \mathcal{C}_2\right)$ denotes the distance between clusters $\mathcal{C}_1$ and $\mathcal{C}_2$, and $d(\boldsymbol{x}, \boldsymbol{y})$ the distance between data points $\boldsymbol{x}$ and $\boldsymbol{y}$. The single-linkage technique is more suitable for finding well-separated stringy clusters.

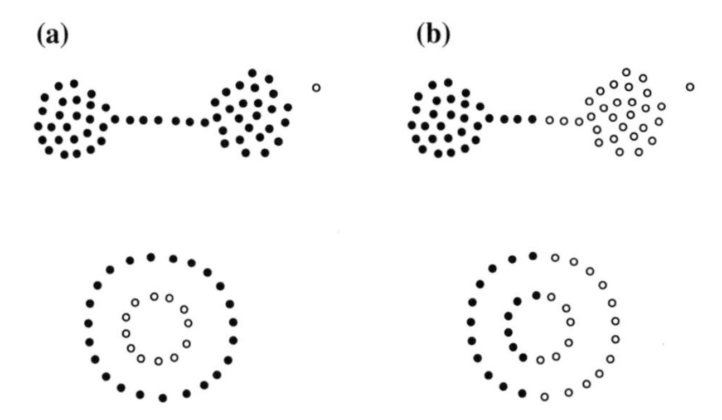

Fig. 10.3 Agglomerative clustering results. **a** The single-link algorithm. **b** The complete-link algorithm

In contrast, the complete linkage method uses the farthest distance between any two data points in different clusters to define the intercluster distance. Both single-linkage and complete linkage algorithms require time complexity of $O(N^2 \log N)$, for N points. The clustering results for the two methods are illustrated in Fig. 10.3. Other more complicated methods are group average linkage, median linkage, and centroid linkage methods.

The representation of clusters is also necessary for hierarchical clustering. Agglomerative clustering can be based on the centroid [142], all-points [138], and scatter-points [59] representations. The shape and extent of a cluster are conventionally represented by its centroid or prototype. This is desirable only for spherically shaped clusters but causes cluster splitting for a large or arbitrarily shaped cluster, since the centroids of its subclusters can be far apart. At the other extreme, all data points in a cluster are used as its representatives, and this makes the clustering algorithm extremely sensitive to noisy data points and outliers. This all-points representation can cluster arbitrary shapes. The scatter-points representation, as a trade-off between the two extremes, represents each cluster by a certain fixed number of points that are generated by selecting well-scattered points from the cluster and then shrinking them toward the center of the cluster by a specified fraction. This reduces the adverse effects of the outliers since the outliers are typically farther away from the mean and are thus shifted by a larger distance due to shrinking. The scatter-points representation achieves robustness to outliers and identifies clusters having nonspherical shape and wide variations in size. For large datasets, storage or multiple input/output scans of the data points are bottlenecks for existing clustering algorithms.

Agglomerative clustering starts from N clusters, each containing exactly one data point. A series of nested merging is performed until finally all the data points are grouped into one cluster. Agglomerative clustering processes a set of N^2 numerical relationships between the N data points, and agglomerates according to their similarity, usually measured by a distance. It is based on a local connectivity criterion. The runtime is $O\left(N^2\right)$. The process of agglomerative clustering can be easily illustrated by using a dendrogram, as shown in Fig. 10.4. The process of successive merging

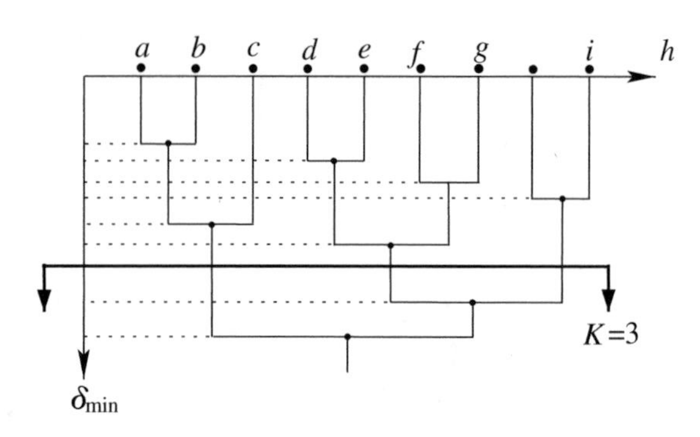

Fig. 10.4 A single-linkage dendrogram

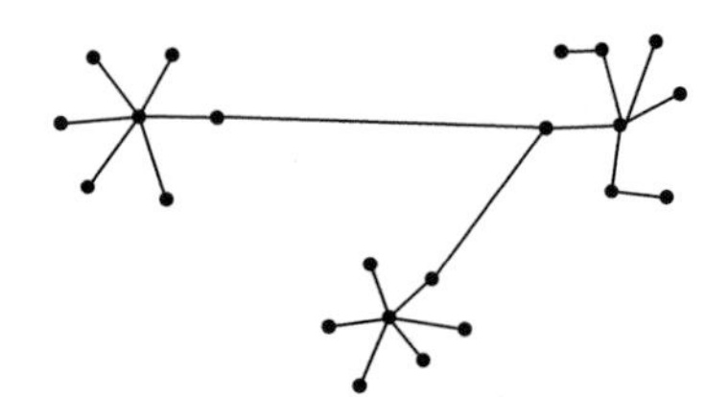

Fig. 10.5 Illustration of an MST. By removing the two longest edges, three clusters are obtained

of the clusters is guided by the set distance $\delta_{\min}$. At a cross section with $\delta_{\min}$, the number of clusters can be decided. At the cross section shown in Fig. 10.4, there are three clusters: $\{a, b, c\}$, $\{d, e, f, g\}$, and $\{h, i\}$.

10.6.2 Minimum Spanning Tree (MST) Clustering

MST-based clustering [138] is a conventional agglomerative clustering technique. The MST method is a graph-theoretical technique [120]. It uses the all-points representation. MST clustering uses single linkage and first finds an MST for the input data. Then, by removing the longest $K - 1$ edges, K clusters are obtained, as shown in Fig. 10.5. Initially, each point is a separate cluster. An agglomerative algorithm starts with the disjoint set of clusters. Pairs of clusters with minimum distance are then successively merged until a criterion is satisfied. MST clustering is good at clustering arbitrary shapes, and it has the ability to detect clusters with irregular boundaries. It has a complexity of $O(N^2)$.

In an MST graph, two points or vertices can be connected either by a direct edge, or by a sequence of edges called a path. The length of a path is the number of edges on it. The degree of link of a vertex is the number of edges that link to this vertex. A loop in a graph is a closed path. A connected graph has one or more paths between every pair of points. A tree is a connected graph with no closed loop. A spanning tree is a tree that contains every point in the dataset. When a value is assigned to each edge in a tree, we get a weighted tree. The weight for each edge can be the distance between its two end points. The weight of a tree is the total sum of the edge weights in

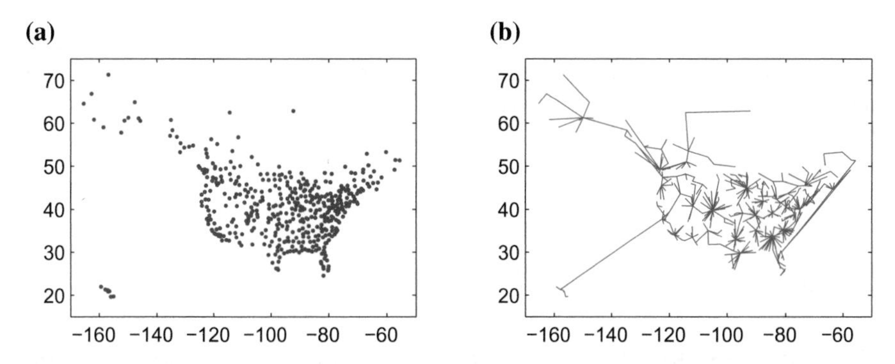

Fig. 10.6 The MST for the data. **a** The locations of the airports. **b** The generated MST

the tree. An MST is a spanning tree that has the minimal total weight. Two properties used to identify edges provably in an MST are the cut and cycle properties. The cut property states that the edge with the smallest weight crossing any two partitions of the vertex set belongs to the MST. The cycle property states that the edge with the largest weight in any cycle in a graph cannot be in the MST. For MST clustering, the weight associated with each edge denotes a distance between the two end points.

An MST can be constructed using either Prim's algorithm [104] or Kruskal's algorithm [86]. Both algorithms grow the tree by adding one edge at a time. The cost of constructing an MST is $O(m \log n)$ for n vertices and m edges [86, 104]. The reverse-delete algorithm is the reverse of Kruskal's algorithm; it starts with the full graph and delete edges in order of nonincreasing weights based on the cycle property as long as doing so does not disconnect the graph. In MST clustering, the inputs are a set of N data points and a distance measure defined upon them, and the time complexity of Kruskal's algorithm, Prim's algorithm, and the reverse-delete algorithm is $O(N^2)$ [128]. *k-d* tree and Delaunay triangulation have been employed in the construction of an MST to reduce the time complexity to near $O(N \log N)$, but they work well only for dimensions no more than 5 [13].

Example 10.1 Consider a dataset of 456 airports in U.S. We have the mean travel time between those airports, along with their latitude and longitude. The locations of the airports and the corresponding MST in terms of travel time are shown in Fig. 10.6. The MST is generated using Prim's algorithm, and the result is based on the Gaimc package at MATLAB Central, provided by David Gleich.

A fast MST-inspired clustering algorithm [128] tries to identify the relatively small number of inconsistent edges and remove them to form clusters before the complete MST is constructed. It can have a much better performance than $O(N^2)$ by using an efficient implementation of the cut and the cycle properties of the MSTs. A more efficient method that can quickly identify the longest edges in an MST is also presented in [128].

MST-based clustering is very sensitive to the outliers, and it may merge two clusters due to the existence of a chain of outliers connecting them. With an MST being constructed, the next step is to define an edge inconsistency measure so as to partition the tree into clusters. In real-world tasks, outliers often exist, and this makes the longest edges an unreliable indication of cluster separations. In these cases, all the edges that satisfy the inconsistency measure are removed and the data points in the smallest clusters are regarded as outliers.

10.6.3 BIRCH, CURE, CHAMELEON, and DBSCAN

The BIRCH (balanced iterative reducing and clustering using hierarchies) method [142, 143] first performs an incremental and approximate preclustering phase in which dense regions of points are represented by compact summaries, and then a centroid-based hierarchical algorithm is used to cluster the set of summaries. During preclustering, the entire database is scanned, and cluster summaries are stored in an in-memory data structure called a *clustering feature tree*. BIRCH uses cluster features to represent a subcluster. Given the cluster features of a subcluster, one can obtain the centroid, radius, and diameter of that subcluster easily (in constant time). Furthermore, the cluster feature vector of a new cluster formed by merging two subclusters can be directly derived from the cluster feature vectors of the two subclusters by algebraic operations. On several large datasets, BIRCH is significantly superior to CLARANS [98] and C-means in terms of quality, speed, stability, and scalability overall on large datasets. Robustness is achieved by eliminating outliers from the summaries via the identification of sparsely distributed data points in feature space. The clustering feature tree grows by aggregation with only one pass over the data, thus having a complexity of $O(N)$. Only one scan is needed to obtain good clustering results. One or more additional passes can be used to further improve the clustering qualities. BIRCH is not sensitive to the input order of the data. However, BIRCH fails to identify clusters with nonspherical shapes or wide variation in size by splitting larger clusters and merging smaller clusters.

CURE (clustering using representation) [59] is a robust clustering algorithm based on the scatter-points representation. It is an improvement of the single-linkage algorithm. CURE selects several scattered data points carefully as the representatives for each cluster and shrinks these representatives toward the centroid in order to eliminate the effects of outliers and avoid the chaining effect. The distance between two clusters in CURE is defined as the minimal distance between the two representatives of each cluster. In each iteration, it merges the two closest clusters. CURE clusters data of any shape. To handle large databases, CURE employs a combination of random sampling and partitioning to reduce the computational complexity. Random samples drawn from the dataset are first partitioned and each partition is partially clustered. The partial clusters are then clustered in a second pass to yield the desired clusters. CURE uses the k-d tree to search the nearest representatives and heap data structures. However, the k-d tree searching structure does not work well in a high-dimensional dataset [13]. CURE has a computational complexity of $O\left(N^2\right)$ for low-dimensional data, which is no worse than that of the centroid-based hierar-

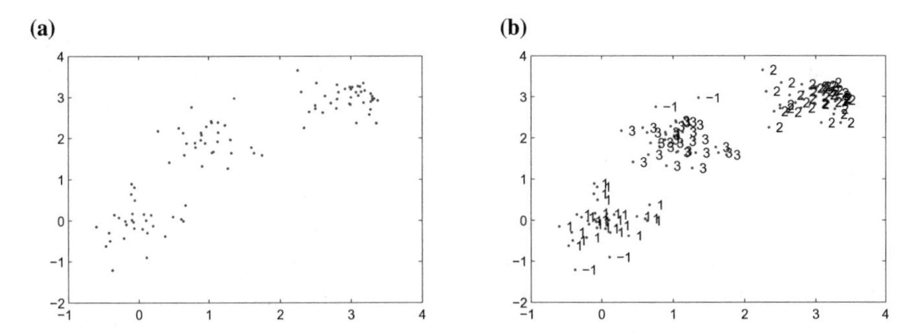

Fig. 10.7 Clustering using DBSCAN. **a** The dataset. **b** The clustering result

chical algorithm. It provides better performance with less execution time compared to BIRCH. CURE can discover clusters with interesting shapes and is less sensitive than MST to the outliers.

CHAMELEON [78] is a hybrid clustering algorithm. It first creates a graph, where each node represents a pattern and the edges between each node and all the other nodes exist according to the k-NN paradigm. A graph-partitioning algorithm is used to recursively partition the graph into many small unconnected subgraphs, each partitioning yielding two subgraphs of roughly equal size. Agglomerative clustering is applied, each subcluster being used as an initial subcluster. CHAMELEON merges two subclusters only when the interconnectivity (the number of links between two clusters) as well as the closeness (the length of those links) of the individual clusters is very similar. It automatically adapts to the characteristics of the clusters being merged. CHAMELEON is more effective than CURE in discovering clusters of arbitrary shapes and varying densities. It has a computational complexity of $O\left(N^2\right)$ [78].

DBSCAN (density-based spatial clustering of applications with noise) [43] is a well-known density-based clustering algorithm. DBSCAN is designed to discover clusters of arbitrary shape as well as to distinguish noise. In DBSCAN, a region is defined as the set of points that lie in the ϵ-neighborhood of some point p. Cluster label propagation from p to the other points in a region $\mathcal{R}$ happens if $|\mathcal{R}|$, the cardinality of $\mathcal{R}$, exceeds a given threshold for the minimal number of points. Generalized DBSCAN [112] can cluster point objects as well as spatially extended objects according to both their spatial and nonspatial attributes. DBSCAN defines two specified parameters for a single density, and thus it does not perform well to datasets with varying densities. DDBSCAN (double-density-based SCAN) [24] needs two different densities as input parameters.

Example 10.2 We apply DBSCAN on a dataset of three clusters. The threshold for the minimal number of points is 3, and the neighborhood radius for DBSCAN ϵ is obtained by analysis method. The clustering result is shown in Fig. 10.7. Those points with -1 are identified as outliers.

Affinity propagation [47] is a distance-based algorithm for identifying exemplars in a dataset. The method executes a message-passing process as well as an iterative process to find out the final exemplars in a dataset. It selects some existing points in the dataset as exemplars. It does not need to input any parameter in advance. However, affinity propagation usually breaks the shapes of the clusters and partitions them into patches. APSCAN (affinity propagation spatial clustering of applications with noise) [24] is a parameter-free clustering algorithm that combines affinity propagation and DDBSCAN. APSCAN does not need to predefine the two parameters as required in DBSCAN; it not only can cluster datasets with varying densities but also preserve the nonlinear data structure for such datasets. APSCAN can take the shape of clusters into account and preserve the irregular structure of clusters.

DBSCAN and BIRCH give promising results for low-dimensional data. Online divisive-agglomerative clustering [107] is an incremental system for clustering streaming time series. The system is designed to process thousands of data streams that flow at high rate. The main features of the system include update time and memory consumption that do not depend on the number of examples in the stream. Moreover, the time and memory required to process an example decrease whenever the cluster structure expands.

A fast agglomerative clustering method proposed in [46] uses an approximate nearest neighbor graph for reducing the number of distance calculations. The computational complexity is improved from $O(\tau N^2)$ to $O(\tau N \log N)$ at the cost of a slight increase in distortion, where τ denotes the number of nearest neighbor updates required at each iteration.

Partitioning around medoids (PAM) [79] is based on finding k representative objects (medoids) that minimize the sum of the within-cluster dissimilarities. PAM works satisfactorily only for small datasets. Clustering large applications (CLARA) [79] is a modified PAM that can handle very large datasets. CLARA draws a sample of the dataset, applies PAM on the sample, and finds the medoids of the sample; it draws multiple samples and gives the best clustering as the output. CLARANS (clustering large applications based on randomized search) [98] is a variant of CLARA that makes the search for the k-medoids more efficient. The runtime of a single call of CLARANS is close to quadratic. For small datasets, CLARANS [99] is a few times faster than PAM; the performance gap for larger datasets is even larger. Compared with CLARA, CLARANS has a search space that is not localized to a specific subgraph chosen a priori and can produce clustering results with much better quality. CLARANS can handle not only point objects but also polygon objects efficiently.

Example 10.3 We apply vector quantization for image compression. We use the Lena image of 512×512 pixels as example. The training vectors are obtained by dividing the image into 4×4 blocks (arranged as a vector of 16 dimensions), and the desired codebook size is set to 256. By applying LBG, the peak signal-to-noise ratio (PSNR) is 30.4641 dB. The compression ratio is $1/16$, and each vector is coded by 8 bits. When applying entropy coding, entropy is defined as the average number of bits needed for encoding a training vector, and the entropy for Lena is 7.2904.

Fig. 10.8 Vector quantization of an image. **a** The original image. **b** The quantized image

Applying the same procedure on the baboon image of 512×512 pixels, we obtain an entropy of 7.5258. Figure 10.8 shows the results for the original and compressed images.

10.6.4 Hybrid Hierarchical/Partitional Clustering

Most partitional clustering algorithms run in linear time. However, the clustering quality of a partitional algorithm is not as good as that of hierarchical algorithms. Some methods exploit the advantages of both the hierarchical and partitional clustering techniques [49, 56, 124, 125].

The VQ-clustering and VQ-agglomeration methods [125] involve a vector quantization process followed, respectively, by a clustering algorithm and an agglomerative clustering algorithm that treat codewords as initial prototypes. The agglomeration algorithm requires that each codeword be moved directly to the centroid of its neighboring codewords. A similar two-stage clustering procedure that uses SOM for vector quantization and an agglomerative clustering or C-means for further clustering is given in [124]. These two-stage methods have performance comparable to those of direct methods, with significantly reduced computational complexity [124, 125].

A two-stage clustering algorithm given in [119] clusters data with arbitrary shapes without knowing the number of clusters in advance. An ART-like algorithm is first used to partition data into a set of small multidimensional hyperellipsoids, and a dendrogram is then built to sequentially merge those hyperellipsoids. Hierarchical unsupervised fuzzy clustering [56] has the advantages of both hierarchical clustering and fuzzy clustering. Robust competitive agglomeration [49] employs competitive agglomeration to find the optimum number of clusters, uses prototypes to represent the global shape of the clusters, and integrates robust statistics to achieve noise immunity.

Cohesion-based self-merging [89] runs in time linear to the size of input dataset. Cohesion is a similarity measure to measure the intercluster distances. The method partitions the input dataset into several small subclusters, and then continuously merges the subclusters based on cohesion in a hierarchical manner. The method is very robust and possesses excellent tolerance to outliers in various workloads. It is able to cluster the datasets of arbitrary shapes very efficiently.

10.7 Constructive Clustering Techniques

Conventional partitional clustering algorithms assume a network with a fixed number of clusters (nodes) K, which needs to be prespecified. However, selecting an appropriate value of K is a difficult task without prior knowledge of the input data. This difficulty can be resolved by using constructive clustering. A simple strategy for determining the optimal K is to perform clustering for a range of K, and select the value of K that minimizes a cluster validity measure.

The leader algorithm [63] is the fastest clustering algorithm. It requires one pass through the data to put each input pattern in a particular cluster or group of patterns. Associated with each cluster is a leader, which is a pattern against which new patterns will be compared to determine whether the new pattern belongs to this particular cluster. The leader algorithm starts with zero prototypes and adds the current input pattern as a prototype called *leader* whenever none of the existing prototypes is close enough to it. The cosine of the angle between an input vector and each prototype is used as a similarity measure. The clusters that are created first will tend to be very large. To determine the cluster that a new pattern will be mapped to, the modified leader algorithm searches for the cluster leader closest to the pattern. If that closest cluster leader is close enough to the new pattern, then the new pattern belongs to that cluster; otherwise, the new pattern becomes the leader for a new cluster. In this way, each cluster has an equal chance of having the new pattern fit into it rather than clusters that are created earlier having an undue advantage. The choice of this threshold is critical. The training time is of the same order but lower than that of C-means. The drawbacks of the modified leader algorithm are the same as that of C-means, and additionally the first pattern presented will always be a cluster leader.

ISODATA [8] is a popular early statistical method for data clustering. It can be treated as a variant of incremental C-means by incorporating some heuristics for merging and splitting clusters, and for handling outliers. Thus, ISODATA has a variable number of clusters K.

The self-creating mechanism in the competitive learning process can adaptively determine the natural number of clusters. Each node is associated with a local statistical variable, which is used to control the growing and pruning of the network architecture. The self-creating and organizing neural network [28] employs adaptively modified node thresholds to control its self-growth. At the presentation of a new input, if the winning node is active, the winning node is updated; otherwise, a new node is recruited from the winning node. The method avoids the underutilization problem and has vector quantization accuracy and speed advantage over SOM and batch C-means. Self-splitting competitive learning [140] can find the natural number of clusters based on the one-prototype-take-one-cluster paradigm and a validity measure for self-splitting. The paradigm enables each prototype to be situated at the centroid of one natural cluster when the number of clusters is greater than that of the prototypes. Self-splitting competitive learning starts with a single prototype and splits adaptively during the learning process until all the clusters are found.

The growing cell structures (GCS) network [50] can be regarded as a modification of SOM by integrating the node-recruiting and pruning functions. GCS assigns each node with a local accumulated statistical variable called a *signal counter* u_i. At each pattern presentation, only the winning node increases its signal counter u_w by 1, and then all the signal counters u_i decay with a forgetting factor. After a fixed number of learning iterations, the node with the largest signal counter gets the right to insert a new node between itself and its farthest neighbor. The network occasionally prunes a node whose signal counter is less than a specified threshold during a complete epoch. The growing grid network [52] is strongly related to GCS but has a strictly rectangular topology. By inserting complete rows or columns of units, the grid may adapt its height/width ratio to a given pattern distribution. The semi-supervised learning method for growing SOMs [73] allows fast visualization of data class structure on the two-dimensional feature map, based on Fritzke's supervised learning architecture used on GCS.

The GNG model [51, 53] is based on GCS and neural gas. It is an SOM without a fixed global network dimensionality, i.e., GNG is able to adapt to the local dimensionality as well as to the local density of the input data. In addition, the topology of the neurons reflects the topology of the input distribution. GNG is capable of generating and removing neurons and lateral connections dynamically. In GNG, lateral connections are generated according to the competitive Hebbian learning rule. GNG achieves robustness against noise and performs perfect topology-preserving mapping. By integrating an online criterion so as to identify and delete useless neurons,

GNG with utility criterion [53] is also able to track nonstationary data input. GNG-T [48], extended from GNG, performs vector quantization continuously over a distribution that changes over time. It deals with both sudden changes and continuous ones, and is thus suited to the video tracking framework.

The dynamic cell structures (DCS) model [20] uses a modified Kohonen learning rule to adjust the prototypes and the competitive Hebbian rule to establish a dynamic lateral connection structure. Applying the principle of DCS to GCS yields the DCS-GCS algorithm, which has a behavior similar to that of GNG. The lifelong learning cell structures (LLCS) algorithm [61] is an online clustering and topology-representation method. It employs a strategy similar to that of ART. A similarity-based unit pruning and an aging-based edge pruning procedures are incorporated.

In adaptive incremental LBG [113], new codewords are inserted in regions of the input vector space where the distortion error is highest until the desired number of codewords (or a distortion error threshold) is achieved. The adaptive distance function is adopted to improve the quantization process. During the incremental process, a removal–insertion technique is used to fine-tune the codebook to make the proposed method independent of the initial conditions. The method works better than enhanced LBG. It can also be used for such tasks: with fixed distortion error, to minimize the number of codewords and find a suitable codebook.

Growing hierarchical tree SOM [44] has an SOM-like self-organizing process that allows the network to adapt the topology of each layer of the hierarchy to the characteristics of the training set. The network grows as a tree, it starts with a triangle SOM and every neuron grows adding one new triangle SOM. Moreover, the training process considers the possibility of deleting nodes.

By employing a lattice with a hyperbolic grid topology, hyperbolic SOM combines the virtues of the SOM and hyperbolic spaces for adaptive data visualization. However, due to the exponential growth of its hyperbolic lattice, it also exacerbates the need for addressing the scaling problem of SOMs comprising very large numbers of nodes. Hierarchically growing hyperbolic SOM [101] combines the virtues of hierarchical data organization, adaptive growing to a required granularity, good scaling behavior and smooth, map-based browsing.

10.8 Cluster Validity

The number of clusters is application-specific and is usually specified by a user. An optimal number of clusters or a good clustering algorithm is only in the sense of a certain cluster validity criterion. Many cluster validity measures are defined for this purpose.

10.8.1 Measures Based on Compactness and Separation of Clusters

A good clustering algorithm should generate clusters with small intracluster deviations and large intercluster separations. Cluster compactness and cluster separation are two measures for describing the performance of clustering. Given the clustered result of a dataset $\mathcal{X}$, the cluster compactness, cluster separation, and overall cluster quality measures are, respectively, defined by [68]

$$E_{\text{CMP}} = \frac{1}{K} \frac{\sum_{i=1}^{K} \sigma(\boldsymbol{c}_i)}{\sigma(\mathcal{X})}, \tag{10.19}$$

$$E_{\text{SEP}} = \frac{1}{K(K-1)} \sum_{i=1}^{K} \sum_{j=1, j \neq i}^{K} \mathrm{e}^{-\frac{d^2(\boldsymbol{c}_i, \boldsymbol{c}_j)}{2\sigma_0^2}}, \tag{10.20}$$

$$E_{\text{OCQ}}(\gamma) = \gamma E_{\text{CMP}} + (1-\gamma) E_{\text{SEP}}, \tag{10.21}$$

where K is the number of clusters, $\boldsymbol{c}_i$ and $\boldsymbol{c}_j$ are, respectively, the centers of clusters i and j, $\sigma(\boldsymbol{c}_i)$ denotes the standard deviation of cluster i, $\sigma(\mathcal{X})$ is the standard deviation of dataset $\mathcal{X}$, σ_0 is the deviation of a Gaussian distribution, $d\left(\boldsymbol{c}_i, \boldsymbol{c}_j\right)$ is the distance between $\boldsymbol{c}_i$ and $\boldsymbol{c}_j$, and $\gamma \in [0, 1]$. Small E_{CMP} means that all the clusters have small deviations, and smaller E_{SEP} corresponds to better separation performance.

Another popular cluster validity criterion is defined as a function of the ratio of the sum of the within-cluster scatters to the between-cluster separation [37]

$$E_{\text{WBR}} = \frac{1}{K} \sum_{k=1}^{K} \max_{l \neq k} \left\{ \frac{d_{\text{WCS}}(\boldsymbol{c}_k) + d_{\text{WCS}}(\boldsymbol{c}_l)}{d_{\text{BCS}}(\boldsymbol{c}_k, \boldsymbol{c}_l)} \right\}, \tag{10.22}$$

where the within-cluster scatter for cluster k, denoted as $d_{\text{WCS}}(\boldsymbol{c}_k)$, and the between-cluster separation for cluster k and cluster l, denoted as $d_{\text{BCS}}(\boldsymbol{c}_k, \boldsymbol{c}_l)$, are, respectively, calculated by

$$d_{\text{WCS}}(\boldsymbol{c}_k) = \frac{\sum_i \|\boldsymbol{x}_i - \boldsymbol{c}_k\|}{N_k}, \tag{10.23}$$

$$d_{\text{BCS}}(\boldsymbol{c}_k, \boldsymbol{c}_l) = \|\boldsymbol{c}_k - \boldsymbol{c}_l\|, \tag{10.24}$$

N_k being the number of data points in cluster k. The best clustering minimizes E_{WBR}. This measure indicates good clustering results for spherical clusters [124].

A cluster validity criterion has been defined in [130] for evaluating fuzzy clustering; it minimizes the ratio of compactness E_{CMP1} and separation E_{SEP1}, $E_{\text{XB}} = \frac{E_{\text{CMP1}}}{E_{\text{SEP1}}}$, defined by

$$E_{\text{CMP1}} = \frac{1}{N} \sum_{i=1}^{K} \sum_{p=1}^{N} \mu_{ip}^m \left\|\boldsymbol{x}_p - \boldsymbol{c}_i\right\|_{\mathbf{A}}^2, \tag{10.25}$$

$$E_{\text{SEP1}} = \min_{i \neq j} \left\| \boldsymbol{c}_i - \boldsymbol{c}_j \right\|_{\mathbf{A}}^2, \tag{10.26}$$

where $\| \cdot \|_{\mathbf{A}}$ denotes a weighted norm, and $\mathbf{A}$ is a positive-definite symmetric matrix. Notice that E_{CMP1} is equal to the criterion function for FCM given by (9.40) when $m = 2$ and $\mathbf{A}$ is the identity matrix.

In [139], a cluster validity criterion is defined based on the ratio and summation of compactness and overlap measures. Both measures are calculated from membership degrees. The maximal value of the criterion denotes the optimal fuzzy partition that is expected to have high compactness and a low degree of overlap among clusters. The criterion is reliable and effective, especially when evaluating partitions with clusters that widely differ in size or density.

10.8.2 Measures Based on Hypervolume and Density of Clusters

A good partitioning of the data usually leads to a small total hypervolume and a large average density of the clusters. Cluster validity criteria can thus be selected as the hypervolume and average density of the clusters. The fuzzy hypervolume criterion is defined by [54, 83]

$$E_{\text{FHV}} = \sum_{i=1}^{K} V_i, \tag{10.27}$$

where V_i is the volume of the ith cluster

$$V_i = \left[\det\left(\mathbf{F}_i\right)\right]^{\frac{1}{2}}, \tag{10.28}$$

and $\mathbf{F}_i$, the fuzzy covariance matrix of the ith cluster, is defined by [60]

$$\mathbf{F}_i = \frac{1}{\sum_{j=1}^{N} \mu_{ij}^m} \sum_{j=1}^{N} \mu_{ij}^m \left(\boldsymbol{x}_j - \boldsymbol{c}_i\right) \left(\boldsymbol{x}_j - \boldsymbol{c}_i\right)^T. \tag{10.29}$$

The average fuzzy density criterion is defined by

$$E_{\text{AFD}} = \frac{1}{K} \sum_{i=1}^{K} \frac{S_i}{V_i}, \tag{10.30}$$

where S_i sums the membership degrees of only the members within the hyperellipsoid

$$S_i = \sum_{j=1}^{N} \mu_{ij}, \quad \forall \boldsymbol{x}_j \in \left\{ \boldsymbol{x}_j \middle| \left(\boldsymbol{x}_j - \boldsymbol{c}_i\right)^T \mathbf{F}_i^{-1} \left(\boldsymbol{x}_j - \boldsymbol{c}_i\right) < 1 \right\}. \tag{10.31}$$

The fuzzy hypervolume criterion typically has a clear extremum; the average fuzzy density criterion is not desirable when there is substantial cluster overlapping and large variability in the compactness of the clusters [54]. The average fuzzy density criterion averages the fuzzy densities of individual clusters, and a partitioning that results in both dense and loose clusters may lead to a large average fuzzy density.

Measures for Shell Clustering

For shell clustering, the hypervolume and average density measures are still applicable. However, the distance vector between a pattern and a prototype needs to be redefined. In the case of spherical shell clustering, the distance vector between pattern $\boldsymbol{x}_j$ and prototype $\boldsymbol{\lambda}_i = (\boldsymbol{c}_i, r_i)$ is defined by

$$\boldsymbol{d}_{j,i} = \left(\boldsymbol{x}_j - \boldsymbol{c}_i\right) - r_i \frac{\boldsymbol{x}_j - \boldsymbol{c}_i}{\left\|\boldsymbol{x}_j - \boldsymbol{c}_i\right\|}, \tag{10.32}$$

where r_i is the radius of the shell. The fuzzy hypervolume and average fuzzy density measures for spherical shell clustering are obtained by replacing $\boldsymbol{x}_j - \boldsymbol{c}_i$ in (10.29) and (10.31) by $\boldsymbol{d}_{j,i}$.

For shell clustering, the shell thickness measure can be used to describe the compactness of a shell. In the case of fuzzy spherical shell clustering, the fuzzy shell thickness of a cluster can be defined by [83]

$$T_j = \frac{\sum_{i=1}^{N} \left(\mu_{ji}\right)^m \left(\left\|\boldsymbol{x}_i - \boldsymbol{c}_j\right\| - r_j\right)^2}{r_j \sum_{i=1}^{N} \left(\mu_{ji}\right)^m}. \tag{10.33}$$

The average shell thickness of all clusters can be used as a cluster validity criterion for shell clustering

$$E_{\text{THK}} = \frac{1}{K} \sum_{j=1}^{K} T_j. \tag{10.34}$$

10.8.3 Crisp Silhouette and Fuzzy Silhouette

The average silhouette width criterion, or crisp silhouette [79], is defined as

$$E_{\text{CS}} = \frac{1}{N} \sum_{j=1}^{N} s_j, \tag{10.35}$$

where s_j is the silhouette of pattern j according to

$$s_j = \frac{b_{pj} - a_{pj}}{\max\{a_{pj}, b_{pj}\}}, \tag{10.36}$$

a_{pj} is the average distance of pattern j to all other patterns belonging to cluster p, d_{qj} is the average distance of pattern j to all patterns in cluster q, $q \neq p$, and $b_{pj} = \min_{q=1,\ldots,K, q\neq p} d_{qj}$, which represents the dissimilarity of pattern j to its closest neighboring cluster.

Fuzzy silhouette [22] improves crisp silhouette in detecting regions with higher data density when the dataset involves overlapping clusters, besides being more appealing in the context of fuzzy cluster analysis. It is defined as

$$E_{\text{FS}} = \frac{\sum_{j=1}^{N} (\mu_{mj} - \mu_{nj})^{\alpha} s_j}{\sum_{j=1}^{N} (\mu_{mj} - \mu_{nj})^{\alpha}}, \tag{10.37}$$

where μ_{mj} and μ_{nj} are the first and second largest elements of the jth column of the fuzzy partition matrix, respectively, and $\alpha \geq 0$ is a weighting coefficient.

Fuzzy silhouette is computationally much less intensive than the fuzzy hyper-volume and average partition density criteria, especially when the dataset involves many attributes. It is straightforwardly applicable as an objective function for global optimization methods, designed for automatically finding the right number of clusters in a dataset.

Other Measures

Several robust-type validity measures are proposed in [129] by analyzing the robustness of a validity measure using the ϕ-function of M-estimate. Median-type validity measures [129] are robust to noise and outliers, and work better than the mean-type validity measures.

A review of fuzzy cluster validity measures is given in [126]. Moreover, extensive comparisons of many measures in conjunction with FCM are conducted on a number of widely used datasets. It is concluded that none of the measures correctly recognizes optimal cluster numbers K for all test datasets.

Nearest neighbor clustering [21] is a baseline algorithm to minimize arbitrary clustering objective functions. It is statistically consistent for all commonly used clustering objective functions. An empirical risk approximation approach for unsupervised learning is proposed along the line of empirical risk minimization for the supervised case. The clustering quality is an expectation with respect to the true underlying probability distribution, and the empirical quality is the corresponding empirical expectation. Then, generalization bounds can be derived using VC-dimensions.

Bregman divergences include a large number of useful distortion functions such as squared loss, Kullback–Leibler divergence, logistic loss, Mahalanobis distance, Itakura–Saito distance, and I-divergence [9]. C-means, LBG for clustering speech data, and information-theoretic clustering for clustering probability distributions [39] are special cases of Bregman hard clustering for squared Euclidean distance, Itakura–Saito distance, and Kullback–Leibler divergence, respectively. This is achieved by first posing the hard clustering problem in terms of minimizing the loss in Bregman information, a quantity motivated by rate-distortion theory, and then deriving an iterative algorithm that monotonically decreases this loss.

10.9 Projected Clustering

Sparsity is a phenomenon of high-dimensional data. In text data, documents related to a particular topic are categorized by one subset of terms. Such a situation also occurs in supplier categorization. Subspace clustering seeks to group objects into clusters on subsets of dimensions or attributes of a dataset. Clustering high-dimensional data has been a major challenge due to the inherent sparsity of the points. The similarity between different members of a cluster can only be recognized in the specific subspace.

CLIQUE [5] identifies dense clusters in subspaces of maximum dimensionality. It works in a levelwise manner, exploring k-dimensional projected clusters after clusters of dimensionality $k-1$ have been discovered. CLIQUE automatically finds subspaces with high-density clusters. It produces identical results irrespective of the order of the input presentation. CLIQUE scales $O(N)$ for N samples and scales exponentially regarding the cluster dimensionality.

The partitional approach PROCLUS [3] is similar to iterative clustering techniques such as C-means or k-medoids [79]. It is a medoid-based projected clustering algorithm that improves the scalability of CLIQUE by selecting a number of good candidate medoids and exploring the clusters around them. Some patterns are initially chosen as the medoids. But, before assigning every pattern in the dataset to the nearest medoid, each medoid is first assigned a set of neighboring patterns that are close to it in the input space to form a tentative cluster. The technique iteratively groups the patterns into clusters and eliminates the least relevant dimensions from each of the clusters. Since PROCLUS optimizes a criterion similar to that of C-means, it can find only spherically shaped clusters. Both the number of clusters and the average number of dimensions per cluster are user-defined. ORCLUS [2] improves PROCLUS by adding a merging process of clusters and selecting for each cluster principal components instead of attributes. ORCLUS can discover arbitrarily oriented clusters. It still relies on user-supplied values in deciding the number of dimensions to select for each cluster.

Halite is a fast, deterministic subspace clustering method [31]. It analyzes the point distribution in the full space by performing a multiresolution, recursive partition of that space so as to find clusters covering regions with varying sizes, shapes, density, correlated axes, and number of points. Halite uses MDL to automatically tune a density threshold with regard to the data distribution. It is robust to noise. Halite is linear or quasi-linear in time and space in terms of the data size and dimensionality.

HARP [136] does not depend on user inputs in determining the relevant dimensions of clusters. It utilizes the relevance index, histogram-based validation, and dynamic threshold loosening to adaptively adjust the merging requirements according to the clustering status. HARP has high accuracy and usability, even when handling noisy data. It exploits the clustering status to adjust the internal thresholds dynamically without the assistance of user parameters.

Based on the analogy between mining frequent itemsets and discovering dense projected clusters around random points, DOC [137] performs iterative greedy projected clustering. Several techniques that employ the branch-and-bound paradigm are proposed to efficiently discover the projected clusters [137]. DOC can automatically discover the number of clusters K, and it can discover a set of clusters with large size variations. A density-based projective clustering algorithm (DOC/FastDOC) [105] requires to set the maximum distance between attribute values and pursues an optimality criterion defined in terms of density of each cluster in its corresponding subspace. In practice, it may be difficult to set the parameters of DOC, as each relevant attribute can have different local variances.

Soft subspace clustering [74] is to cluster data objects in the entire data space but assign different weighting values to different dimensions of clusters in the clustering process. EWKM [77] extends C-means to calculate a weight for each dimension in each cluster and uses the weight values to identify the subsets of important dimensions that categorize different clusters. This is achieved by including the weight entropy in the objective function that is minimized in C-means. An additional step is added to automatically compute the weights of all dimensions in each cluster. EWKM outperforms PROCLUS and HARP.

High-dimensional projected stream (HPStream) clustering [4] incorporates a fading cluster structure and projection-based clustering methodology. It is incrementally updatable and is highly scalable on both the number of dimensions and the size of the data streams, and it achieves better clustering quality than previous stream clustering methods does.

Projected clustering based on the k-means algorithm (PCKA) [18] is a robust partitional distance-based projected clustering algorithm. Interactive projected clustering (IPCLUS) [1] performs high-dimensional clustering by cooperation between the human and the computer.

10.10 Spectral Clustering

Spectral clustering arises from concepts in spectral graph theory. The basic idea is to construct a weighted graph from the initial dataset where each node represents a pattern and each weighted edge accounts for the similarity between two patterns; the clustering problem is configured as a graph-cut problem, where an appropriate objective function has to be optimized. Spectral clustering is considered superior to C-means in terms of having a deterministic polynomial time solution and the ability to model arbitrary shaped clusters. It has a computational complexity of $O(N^3)$ and a memory complexity of $O(N^2)$ for N samples. It is an approach for finding non-convex clusters.

Spectral clustering algorithm amounts to embedding the data into a feature space by using the eigenvectors of the Laplacian matrix in such a way that the clusters may be separated by hyperplanes. The Laplacian or similarity matrix $\mathbf{L}$ of the weighted graph is obtained from data:

$$\mathbf{L} = \mathbf{D} - \mathbf{W}, \tag{10.38}$$

where $\mathbf{W} \in R^{N \times N}$ is a sparse affinity matrix between data points, with a diagonal degree matrix $\mathbf{D} \in R^{N \times N}$,

$$D_{ii} = \sum_j W_{ij}. \tag{10.39}$$

A popular implementation of spectral clustering is to compute the smallest m eigenvectors, $\mathbf{Q}_{N \times m} = \left[\boldsymbol{q}_1, \boldsymbol{q}_2, \ldots, \boldsymbol{q}_m\right]$, of the normalized Laplacian matrix [114]

$$\mathbf{L} = \mathbf{D}^{-1/2}(\mathbf{D} - \mathbf{W})\mathbf{D}^{-1/2}. \tag{10.40}$$

Each row of $\mathbf{Q}$ is a data point, and apply a standard method like C-means to get the clusters.

A direct connection between kernel PCA and spectral methods has been shown in [14]. A unifying view of kernel C-means and spectral clustering methods has been pointed out in [41]. A general weighted kernel C-means objective is mathematically equivalent to a weighted graph clustering objective. Based on this equivalence, a fast multilevel algorithm is developed that directly optimizes various weighted graph clustering objectives [41]. This eliminates the need for eigenvector computation for graph clustering problems. The multilevel algorithm removes the restriction of equal-sized clusters by using kernel C-means to optimize weighted graph cuts.

Incremental spectral clustering handles not only insertion/deletion of data points but also similarity changes between existing points. In [100], spectral clustering is extended to evolving data, by introducing the incidence vector/matrix to represent two kinds of dynamics in the same framework and by incrementally updating the eigensystem.

A Markov random walks view of spectral clustering is given in [69]. This interpretation shows that many properties of spectral clustering methods can be expressed in terms of a stochastic transition matrix $\mathbf{P}$ obtained by normalizing the affinity matrix such that its rows sum to 1.

10.11 Coclustering

Coclustering, or biclustering, simultaneously clusters patterns and their features. Among the advantages of coclustering are its good performance in high dimension and its ability to provide more interpretable clusters than its clustering counterpart. Coclustering can perform well in high-dimensional space because its feature clustering process can be seen as a dynamic dimensionality reduction for the pattern space and vice versa. A bicluster is a subset of rows that exhibit similar behavior across a subset of columns, and vice versa.

Hartigan [62] pioneered this type of analysis using two-way analysis of variance to locate constant-valued submatrices within datasets. The method refers to the simultaneous clustering of both rows and columns of a data matrix.

Soft model dual fuzzy possibilistic coclustering [122] is inspired by possibilistic FCM. The model targets robustness to outliers and richer representations of coclusters. It preserves the desired properties of possibilistic FCM and has the same time complexity as that of possibilistic FCM and FCM, for the number of (co-)clusters.

An information-theoretic coclustering algorithm [40] views a nonnegative matrix as the estimate of a (scaled) empirical joint probability distribution of two discrete random variables and poses the coclustering problem as an optimization problem in information theory, where the optimal coclustering maximizes the mutual information between the clustered random variables subject to constraints on the number of row and column clusters.

10.12 Handling Qualitative Data

A majority of the real-world data is described by a combination of numeric and qualitative (nominal, ordinal) features such as categorical data, which is the case in survey data. There are a number of challenges in clustering categorical data. First, lack of an inherent order on the domains of the individual attributes prevents the definition of a notion of similarity, which catches resemblance between categorical data objects. A typical approach to processing categorical values is to resort to a preprocess such as binary encoding, which transforms each categorical attribute into a set of binary attributes in such a way that each distinct categorical value is associated with one of the binary attributes. Consequently, after the transformation, all categorical attributes become binary attributes, which can thus be treated as numeric attributes with the domain of $\{0, 1\}$.

Some algorithms for clustering categorical data are k-modes [75], fuzzy centroid, and fuzzy k-partitions [135]. k-modes is an extension of C-means to categorical domains and domains with mixed numeric and categorical values [75]. It uses a simple matching dissimilarity measure to deal with categorical patterns, replaces the means of clusters with modes, and uses a frequency-based method to update the modes in the clustering process to minimize the cost function. The k-prototypes algorithm, through the definition of a combined dissimilarity measure, further integrates C-means and k-modes to allow for clustering patterns described by mixed numeric and categorical attributes. A fuzzy k-partitions model [135] is based on the likelihood function of multivariate multinomial distributions. FCM has also been extended for clustering symbolic data [42]. All these algorithms have linear complexity.

KAMILA (KAymeans for MIxed LArge data) clustering method [45] combines the best features of C-means algorithm and Gaussian-multinomial mixture models [76]. Like C-means, KAMILA does not make strong parametric assumptions about the continuous variables. Like Gaussian-multinomial mixture models, KAMILA can balance the contribution of continuous and categorical variables without specifying weights and transforming categorical variables. KAMILA is based on an appropriate density estimator computed from the data, effectively relaxing the Gaussian assumption.

10.13 Bibliographical Notes

Hebbian learning-based data clustering using spiking neurons [87] is capable of distinguishing between clusters and noisy background data and finds an arbitrary number of clusters of arbitrary shape. The clustering ability is more powerful than C-means and linkage clustering, and the time complexity of the method is also more modest than that of its generally used strongest competitor. The robust, locally mediated, self-organizing design makes it a genuinely nonparametric approach. The algorithm does not require any information about the number or shape of clusters.

Constrained clustering algorithms incorporate known information about the desired data partitions into the clustering process [12]. The must-link (the two data points must be in the same cluster) and cannot-link (the two data points cannot be in the same cluster) constraints are two common types of constraints about pairs of objects [12].

Information bottleneck [121] is an information-theoretic principle. The information bottleneck principle can be motivated from Shannon's rate-distortion theory, which provides lower bounds on the number of classes we can divide a source given a distortion constraint. Among all the possible clusterings of a given pattern set into a fixed number of clusters, information bottleneck clustering minimizes the loss of mutual information between the patterns and the features extracted from them.

Problems

10.1 For competitive clustering, name and describe a few heuristics to avoid the dead-unit problem.

10.2 Implement the RPCL algorithm and apply it for image segmentation on an image.

10.3 Consider the grayscale Lena image of size 512×512. Apply image quantization for the following two cases:

(a) The image is divided into 4×4 blocks and the resulting 16, 384 16-dimensional vectors are the input vector data.
(b) The image is divided into 8×8 blocks and the resulting 4, 096 16-dimensional vectors are the input vector data.
Use PSNR to evaluate the reconstructed images after encoding and decoding.

10.4 Clustering Algorithms' Referee Package (CARP, http://www.mloss.org) is an open-source C package for evaluating clustering algorithms. CARP generates datasets of different clustering complexities and assesses the performance of the concerned algorithm in terms of its ability to classify each dataset relative to the true grouping. Download CARP and use it to evaluate the performance of different clustering algorithms.

10.5 Randomly generate 20 points in the square $x_1, x_2 \in [2, 8]$.
(a) Create an MST of the weighted graph based on the Euclidean distance.
(b) Remove the inconsistent edges and example from the formed clusters.
(c) Write a program to complete the clustering.

10.6 In Example 10.3, both the Lena and baboon images are compressed by using LBG. Now, consider to get the codebook by applying LBG on the Lena image. Then apply the codebook on the baboon image and get the quantized image. Calculate the PSNR of the compressed baboon image. Calculate the entropy of the codebook vector.

10.7 Evaluate three different cluster validity measures on the iris dataset by searching the number of clusters $2 \leq C < \sqrt{N}$.
(a) Using FCM.
(b) Using GK clustering.

References

1. Aggarwal, C. C. (2004). A human-computer interactive method for projected clustering. *IEEE Transactions on Knowledge and Data Engineering*, *16*(4), 448–460.
2. Aggarwal, C. C., &. Yu, P. S. (2000). Finding generalized projected clusters in high dimensional spaces. In *Proceedings of the ACM SIGMOD International Conference on Management of Data* (pp. 70–81).
3. Aggarwal, C. C., Procopiuc, C., Wolf, J. L., Yu, P. S., & Park, J. S. (1999). Fast algorithms for projected clustering. In *Proceedings of the ACM SIGMOD International Conference on Management of Data* (pp. 61–72).
4. Aggarwal, C. C., Han, J., Wang, J., & Yu, P. S. (2005). On high dimensional projected clustering of data streams. *Data Mining and Knowledge Discovery*, *10*, 251–273.
5. Agrawal, R., Gehrke, J., Gunopulos, D., & Raghavan, P. (1998). Automatic subspace clustering of high dimensional data for data mining applications. In *Proceedings of the ACM SIGMOD International Conference on Management of Data* (pp. 94–105).
6. Ahalt, S. C., Krishnamurty, A. K., Chen, P., & Melton, D. E. (1990). Competitive learning algorithms for vector quantization. *Neural Networks*, *3*(3), 277–290.

7. Bacciu, D., & Starita, A. (2008). Competitive repetition suppression (CoRe) clustering: A biologically inspired learning model with application to robust clustering. *IEEE Transactions on Neural Networks*, *19*(11), 1922–1941.
8. Ball, G. H., & Hall, D. J. (1967). A clustering technique for summarizing multivariate data. *Behavioral Sciences*, *12*, 153–155.
9. Banerjee, A., Merugu, S., Dhillon, I. S., & Ghosh, J. (2005). Clustering with Bregman divergences. *Journal of Machine Learning Research*, *6*, 1705–1749.
10. Baraldi, A., & Blonda, P. (1999). A survey of fuzzy clustering algorithms for pattern recognition-Part II. *IEEE Transactions on Systems, Man, and Cybernetics Part B*, *29*(6), 786–801.
11. Bauer, H.-U., Der, R., & Herrmann, M. (1996). Controlling the magnification factor of self-organizing feature maps. *Neural Computation*, *8*, 757–771.
12. Basu, S., Davidson, I., & Wagstaff, K. L. (2008). *Constrained clustering: Advances in algorithms, theory, and applications*. New York: Chapman & Hall/CRC.
13. Bay, S. D., & Schwabacher, M. (2003). Mining distance-based outliers in near linear time with randomization and a simple pruning rule. In *Proceedings of the 9th ACM SIGKDD International Conference on Knowledge Discovery and Data Mining* (pp. 29–38).
14. Bengio, Y., Delalleau, O., Le Roux, N., Paiement, J. F., Vincent, P., & Ouimet, M. (2004). Learning eigenfunctions links spectral embedding and kernel PCA. *Neural Computation*, *16*(10), 2197–2219.
15. Bezdek, J. C., Coray, C., Gunderson, R., & Watson, J. (1981). Detection and characterization of cluster substructure: Fuzzy c-varieties and convex combinations thereof. *SIAM Journal of Applied Mathematics*, *40*(2), 358–372.
16. Bezdek, J. C., Hathaway, R. J., & Pal, N. R. (1995). Norm-induced shell-prototypes (NIPS) clustering. *Neural, Parallel, and Scientific Computations*, *3*, 431–450.
17. Bouchachia, A., & Pedrycz, W. (2006). Enhancement of fuzzy clustering by mechanisms of partial supervision. *Fuzzy Sets and Systems*, *157*, 1733–1759.
18. Bouguessa, M., & Wang, S. (2009). Mining projected clusters in high-dimensional spaces. *IEEE Transactions on Knowledge and Data Engineering*, *21*(4), 507–522.
19. Bradley, P. S., Mangasarian, O. L., & Steet, W. N. (1996). Clustering via concave minimization. In D. S. Touretzky, M. C. Mozer, & M. E. Hasselmo (Eds.), *Advances in neural information processing systems* (Vol. 8, pp. 368–374). Cambridge: MIT Press.
20. Bruske, J., & Sommer, G. (1995). Dynamic cell structure. In G. Tesauro, D. S. Touretzky, & T. K. Leen (Eds.), *Advances in neural information processing systems* (Vol. 7, pp. 497–504). Cambridge: MIT Press.
21. Bubeck, S., & von Luxburg, U. (2009). Nearest neighbor clustering: A baseline method for consistent clustering with arbitrary objective functions. *Journal of Machine Learning Research*, *10*, 657–698.
22. Campello, R. J. G. B., & Hruschka, E. R. (2006). A fuzzy extension of the silhouette width criterion for cluster analysis. *Fuzzy Sets and Systems*, *157*, 2858–2875.
23. Chatzis, S., & Varvarigou, T. (2009). Factor analysis latent subspace modeling and robust fuzzy clustering using t-distributions. *IEEE Transactions on Fuzzy Systems*, *17*(3), 505–517.
24. Chen, X., Liu, W., Qiu, H., & Lai, J. (2011). APSCAN: A parameter free algorithm for clustering. *Pattern Recognition Letters*, *32*(7), 973–986.
25. Cheung, Y. M. (2003). $k*$-Means: A new generalized k-means clustering algorithm. *Pattern Recognition Letters*, *24*, 2883–2893.
26. Cheung, Y. M. (2005). On rival penalization controlled competitive learning for clustering with automatic cluster number selection. *IEEE Transactions on Knowledge and Data Engineering*, *17*(11), 1583–1588.
27. Cheung, Y. M., & Law, L. T. (2007). Rival-model penalized self-organizing map. *IEEE Transactions on Neural Networks*, *18*(1), 289–295.
28. Choi, D. I., & Park, S. H. (1994). Self-creating and organizing neural network. *IEEE Transactions on Neural Networks*, *5*(4), 561–575.

29. Choy, C. S. T., & Siu, W. C. (1998). A class of competitive learning models which avoids neuron underutilization problem. *IEEE Transactions on Neural Networks, 9*(6), 1258–1269.
30. Chung, F. L., & Lee, T. (1994). Fuzzy competitive learning. *Neural Networks, 7*(3), 539–551.
31. Cordeiro, R. L. F., Traina, A. J. M., Faloutsos, C., & Traina, C, Jr. (2013). Halite: Fast and scalable multiresolution local-correlation clustering. *IEEE Transactions on Knowledge and Data Engineering, 25*(2), 387–401.
32. Dave, R. N. (1990). Fuzzy shell-clustering and applications to circle detection in digital images. *International Journal of General Systems, 16*(4), 343–355.
33. Dave, R. N. (1991). Characterization and detection of noise in clustering. *Pattern Recognition Letters, 12*, 657–664.
34. Dave, R. N., & Bhaswan, K. (1992). Adaptive fuzzy C-shells clustering and detection of ellipse. *IEEE Transactions on Neural Networks, 3*(5), 643–662.
35. Dave, R. N., & Krishnapuram, R. (1997). Robust clustering methods: A unified view. *IEEE Transactions on Fuzzy Systems, 5*(2), 270–293.
36. Dave, R. N., & Sen, S. (2002). Robust fuzzy clustering of relational data. *IEEE Transactions on Fuzzy Systems, 10*(6), 713–727.
37. Davies, D. L., & Bouldin, D. W. (1979). A cluster separation measure. *IEEE Transactions on Pattern Analysis and Machine Intelligence, 1*(4), 224–227.
38. Desieno, D. (1988). Adding a conscience to competitive learning. *Proceedings of IEEE International Conference on Neural Networks, 1*, 117–124.
39. Dhillon, I., Mallela, S., & Kumar, R. (2003). A divisive information-theoretic feature clustering algorithm for text classification. *Journal of Machine Learning Research, 3*(4), 1265–1287.
40. Dhillon, I. S., Mallela, S., & Modha, D. S. (2003). Information-theoretic co-clustering. In *Proceedings of the 9th ACM SIGKDD International Conference on Knowledge Discovery and Data Mining* (pp. 89–98).
41. Dhillon, I. S., Guan, Y., & Kulis, B. (2007). Weighted graph cuts without eigenvectors: A multilevel approach. *IEEE Transactions on Pattern Analysis and Machine Intelligence, 29*(11), 1944–1957.
42. El-Sonbaty, Y., & Ismail, M. (1998). Fuzzy clustering for symbolic data. *IEEE Transactions on Fuzzy Systems, 6*(2), 195–204.
43. Ester, M., Kriegel, H. P., Sander, J., & Xu, X. (1996). A density-based algorithm for discovering clusters in large spatial databases with noise. In *Proceedings of the 2nd International Conference on Knowledge Discovery and Data Mining (KDD)* (pp. 226–231). Portland, OR.
44. Forti, A., & Foresti, G. L. (2006). Growing hierarchical tree SOM: An unsupervised neural network with dynamic topology. *Neural Networks, 19*, 1568–1580.
45. Foss, A., Markatou, M., Ray, B., & Heching, A. (2016). A semiparametric method for clustering mixed data. *Machine Learning, 105*, 419–458.
46. Franti, P., Virmajoki, O., & Hautamaki, V. (2006). Fast agglomerative clustering using a k-nearest neighbor graph. *IEEE Transactions on Pattern Analysis and Machine Intelligence, 28*(11), 1875–1881.
47. Frey, B. J., & Dueck, D. (2007). Clustering by passing message between data points. *Science, 315*, 972–976.
48. Frezza-Buet, H. (2008). Following non-stationary distributions by controlling the vector quantization accuracy of a growing neural gas network. *Neurocomputing, 71*, 1191–1202.
49. Frigui, H., & Krishnapuram, R. (1999). A robust competitive clustering algorithm with applications in computer vision. *IEEE Transactions on Pattern Analysis and Machine Intelligence, 21*(5), 450–465.
50. Fritzke, B. (1994). Growing cell structures - A self-organizing neural networks for unsupervised and supvised learning. *Neural Networks, 7*(9), 1441–1460.
51. Fritzke, B. (1995). A growing neural gas network learns topologies. In G. Tesauro, D. S. Touretzky, & T. K. Leen (Eds.), *Advances in neural information processing systems* (Vol. 7, pp. 625–632). Cambridge: MIT Press.
52. Fritzke, B. (1995). Growing grid-A self-organizing network with constant neighborhood range and adaptation strength. *Neural Processing Letters, 2*(5), 9–13.

53. Fritzke, B. (1997). A self-organizing network that can follow nonstationary distributions. In W. Gerstner, A. Germond, M. Hasler, & J. D. Nicoud (Eds.), *Proceedings of International Conference on Artificial Neural Networks, LNCS* (Vol. 1327, pp. 613–618). Lausanne, Switzerland. Berlin: Springer.
54. Gath, I., & Geva, A. B. (1989). Unsupervised optimal fuzzy clustering. *IEEE Transactions on Pattern Analysis and Machine Intelligence, 11*(7), 773–781.
55. Gath, I., & Hoory, D. (1995). Fuzzy clustering of elliptic ring-shaped clusters. *Pattern Recognition Letters, 16*, 727–741.
56. Geva, A. B. (1999). Hierarchical unsupervised fuzzy clustering. *IEEE Transactions on Fuzzy Systems, 7*(6), 723–733.
57. Gonzalez, J., Rojas, I., Pomares, H., Ortega, J., & Prieto, A. (2002). A new clustering technique for function approximation. *IEEE Transactions on Neural Networks, 13*(1), 132–142.
58. Grossberg, S. (1987). Competitive learning: From iterative activation to adaptive resonance. *Cognitive Science, 11*, 23–63.
59. Guha, S., Rastogi, R., & Shim, K. (2001). CURE: An efficient clustering algorithm for large databases. *Information Systems, 26*(1), 35–58.
60. Gustafson, D. E., & Kessel, W. (1979). Fuzzy clustering with a fuzzy covariance matrix. In *Proceedings of the IEEE Conference on Decision and Control* (pp. 761–766). San Diego, CA.
61. Hamker, F. H. (2001). Life-long learning cell structures–Continuously learning without catastrophic interference. *Neural Networks, 14*, 551–573.
62. Hartigan, J. A. (1972). Direct clustering of a data matrix. *Journal of the American Statistical Association, 67*(337), 123–129.
63. Hartigan, J. A. (1975). *Clustering algorithms*. New York: Wiley.
64. Hathaway, R. J., & Bezdek, J. C. (1994). NERF *c*-means: Non-euclidean relational fuzzy clustering. *Pattern Recognition, 27*, 429–437.
65. Hathaway, R. J., & Bezdek, J. C. (2000). Generalized fuzzy *c*-means clustering strategies using L_p norm distances. *IEEE Transactions on Fuzzy Systems, 8*(5), 576–582.
66. Hathaway, R. J., & Bezdek, J. C. (2001). Fuzzy *c*-means clustering of incomplete data. *IEEE Transactions on Systems, Man, and Cybernetics Part B, 31*(5), 735–744.
67. Hathaway, R. J., & Hu, Y. (2009). Density-weighted fuzzy *c*-means clustering. *IEEE Transactions on Fuzzy Systems, 17*(1), 243–252.
68. He, J., Tan, A. H., & Tan, C. L. (2004). Modified ART 2A growing network capable of generating a fixed number of nodes. *IEEE Transactions on Neural Networks, 15*(3), 728–737.
69. Hein, M., Audibert, J.-Y., & von Luxburg, U. (2007). Graph Laplacians and their convergence on random neighborhood graphs. *Journal of Machine Learning Research, 8*, 1325–1370.
70. Hoeppner, F. (1997). Fuzzy shell clustering algorithms in image processing: Fuzzy *C*-rectangular and 2-rectangular shells. *IEEE Transactions on Fuzzy Systems, 5*(4), 599–613.
71. Honda, K., Notsu, A., & Ichihashi, H. (2010). Fuzzy PCA-guided robust *k*-means clustering. *IEEE Transactions on Fuzzy Systems, 18*(1), 67–79.
72. Hornik, K., Feinerer, I., Kober, M., & Buchta, C. (2012). Spherical *k*-means clustering. *Journal of Statistical Software, 50*(10), 1–22.
73. Hsu, A., & Halgamuge, S. K. (2008). Class structure visualization with semi-supervised growing self-organizing maps. *Neurocomputing, 71*, 3124–3130.
74. Huang, J. Z., Ng, M. K., Rong, H., & Li, Z. (2005). Automated variable weighting in *k*-means type clustering. *IEEE Transactions on Pattern Analysis and Machine Intelligence, 27*(5), 1–12.
75. Huang, Z. (1998). Extensions to the *k*-means algorithm for clustering large data sets with categorical values. *Data Mining and Knowledge Discovery, 2*, 283–304.
76. Hunt, L., & Jorgensen, M. (2011). Clustering mixed data. *WIREs Data Mining and Knowledge Discovery, 1*, 352–361.
77. Jing, L., Ng, M. K., & Huang, J. Z. (2007). An entropy weighting *k*-means algorithm for subspace clustering of high-dimensional sparse data. *IEEE Transactions on Knowledge and Data Engineering, 19*(8), 1026–1041.

78. Karypis, G., Han, E. H., & Kumar, V. (1999). Chameleon: Hierarchical clustering using dynamic modeling cover feature. *Computer, 12*, 68–75.
79. Kaufman, L., & Rousseeuw, P. J. (1990). *Finding groups in data: An introduction to cluster analysis*. New York: Wiley.
80. Kaymak, U., & Setnes, M. (2002). Fuzzy clustering with volume prototypes and adaptive cluster merging. *IEEE Transactions on Fuzzy Systems, 10*(6), 705–712.
81. Kersten, P. R. (1999). Fuzzy order statistics and their application to fuzzy clustering. *IEEE Transactions on Fuzzy Systems, 7*(6), 708–712.
82. King, B. (1967). Step-wise clustering procedures. *Journal of the American Statistical Association, 69*, 86–101.
83. Krishnapuram, R., Nasraoui, O., & Frigui, H. (1992). The fuzzy c-spherical shells algorithm: A new approach. *IEEE Transactions on Neural Networks, 3*(5), 663–671.
84. Krishnapuram, R., & Keller, J. M. (1993). A possibilistic approach to clustering. *IEEE Transactions on Fuzzy Systems, 1*(2), 98–110.
85. Krishnapuram, R., Frigui, H., & Nasraoui, O. (1995). Fuzzy and possibilistic shell clustering algorithms and their application to boundary detection and surface approximation: Part 1 & 2. *IEEE Transactions on Fuzzy Systems, 3*(1), 44–60.
86. Kruskal, J. (1956). On the shortest spanning subtree and the traveling salesman problem. In *Proceedings of the American Mathematical Society* (pp. 48–50).
87. Landis, F., Ott, T., & Stoop, R. (2010). Hebbian self-organizing integrate-and-fire networks for data clustering. *Neural Computation, 22*, 273–288.
88. Leski, J. M. (2003). Generalized weighted conditional fuzzy clustering. *IEEE Transactions on Fuzzy Systems, 11*(6), 709–715.
89. Lin, C.-R., & Chen, M.-S. (2005). Combining partitional and hierarchical algorithms for robust and efficient data clustering with cohesion self-merging. *IEEE Transactions on Knowledge and Data Engineering, 17*(2), 145–159.
90. Liu, Z. Q., Glickman, M., & Zhang, Y. J. (2000). Soft-competitive learning paradigms. In Z. Q. Liu & S. Miyamoto (Eds.), *Soft Computing and Human-Centered Machines* (pp. 131–161). New York: Springer.
91. Luk, A., & Lien, S. (1998). Learning with lotto-type competition. In *Proceedings of International Joint Conference on Neural Networks* (Vol. 2, pp. 1143–1146). Anchorage, AK.
92. Luk, A., & Lien, S. (1999). Lotto-type competitive learning and its stability. In *Proceedings of International Joint Conference on Neural Networks* (Vol. 2, pp. 1425–1428). Washington, DC.
93. Ma, J., & Wang, T. (2006). A cost-function approach to rival penalized competitive learning (RPCL). *IEEE Transactions on Systems, Man, Cybernetics Part B, 36*(4), 722–737.
94. Maji, P., & Pal, S. K. (2007). Rough set based generalized fuzzy C-means algorithm and quantitative indices. *IEEE Transactions on Systems, Man, Cybernetics Part B, 37*(6), 1529–1540.
95. Man, Y., & Gath, I. (1994). Detection and separation of ring-shaped clusters using fuzzy clustering. *IEEE Transactions on Pattern Analysis and Machine Intelligence, 16*(8), 855–861.
96. Mao, J., & Jain, A. K. (1996). A self-organizing network for hyperellipsoidal clustering (HEC). *IEEE Transactions on Neural Networks, 7*(1), 16–29.
97. Marshland, S., Shapiro, J., & Nehmzow, U. (2002). A self-organizing network that grows when required. *Neural Networks, 15*, 1041–1058.
98. Ng, R. T., & Han, J. (1994). Efficient and effective clustering methods for spatial data mining. In *Proceedings of the 20th International Conference on Very Large Data Bases* (pp. 144–155). Santiago, Chile.
99. Ng, R. T., & Han, J. (2002). CLARANS: A method for clustering objects for spatial data mining. *IEEE Transactions on Knowledge and Data Engineering, 14*(5), 1003–1016.
100. Ning, H., Xu, W., Chi, Y., Gong, Y., & Huang, T. S. (2010). Incremental spectral clustering by efficiently updating the eigen-system. *Pattern Recognition, 43*(1), 113–127.

101. Ontrup, J., & Ritter, H. (2006). Large-scale data exploration with the hierarchically growing hyperbolic SOM. *Neural Networks, 19*, 751–761.
102. Pal, N. R., Pal, K., Keller, J. M., & Bezdek, J. C. (2005). A possibilistic fuzzy c-means clustering algorithm. *IEEE Transactions on Fuzzy Systems, 13*(4), 517–530.
103. Pedrycz, W. (1998). Conditional fuzzy clustering in the design of radial basis function neural networks. *IEEE Transactions on Neural Networks, 9*(4), 601–612.
104. Prim, R. (1957). Shortest connection networks and some generalization. *Bell System Technical Journal, 36*, 1389–1401.
105. Procopiuc, C. M., Jones, M., Agarwal, P.K., & Murali, T. M. (2002). A Monte Carlo algorithm for fast projective clustering. In *Proceedings of the ACM SIGMOD International Conference on Management of Data* (pp. 418–427).
106. Rizzo, R., & Chella, A. (2006). A comparison between habituation and conscience mechanism in self-organizing maps. *IEEE Transactions on Neural Networks, 17*(3), 807–810.
107. Rodrigues, P. P., Gama, J., & Pedroso, J. P. (2008). Hierarchical clustering of time series data streams. *IEEE Transactions on Knowledge and Data Engineering, 20*(5), 615–627.
108. Rose, K., Gurewitz, E., & Fox, G. C. (1990). A deterministic annealing approach to clustering. *Pattern Recognition Letters, 11*(9), 589–594.
109. Rumelhart, D. E., & Zipser, D. (1985). Feature discovery by competititve learning. *Cognitive Sciences, 9*, 75–112.
110. Runkler, T. A., & Bezdek, J. C. (1999). Alternating cluster estimation: A new tool for clustering and function approximation. *IEEE Transactions on Fuzzy Systems, 7*(4), 377–393.
111. Runkler, T. A., & Palm, R. W. (1996). Identification of nonlinear systems using regular fuzzy *c*-elliptotype clustering. In *Proceedings of the 5th IEEE Conference on Fuzzy Systems* (pp. 1026–1030).
112. Sander, J., Ester, M., Kriegel, H.-P., & Xu, X. (1998). Density-based clustering in spatial databases: The algorithm GDBSCAN and its applications. *Data Mining and Knowledge Discovery, 2*, 169–194.
113. Shen, F., & Hasegawa, O. (2006). An adaptive incremental LBG for vector quantization. *Neural Networks, 19*, 694–704.
114. Shi, J., & Malik, J. (2000). Normalized cuts and image segmentation. *IEEE Transactions on Pattern Analysis and Machine Intelligence, 22*(8), 888–905.
115. Sneath, P. H. A., & Sokal, R. R. (1973). *Numerical taxonomy*. London: Freeman.
116. Song, Q., Yang, X., Soh, Y. C., & Wang, Z. M. (2010). An information-theoretic fuzzy *C*-spherical shells clustering algorithm. *Fuzzy Sets and Systems, 161*, 1755–1773.
117. Staiano, A., Tagliaferri, R., & Pedrycz, W. (2006). Improving RBF networks performance in regression tasks by means of a supervised fuzzy clustering. *Neurocomputing, 69*, 1570–1581.
118. Su, M. C., & Chou, C. H. (2001). A modified version of the *K*-means algorithm with a distance based on cluster symmetry. *IEEE Transactions on Pattern Analysis and Machine Intelligence, 23*(6), 674–680.
119. Su, M. C., & Liu, Y. C. (2005). A new approach to clustering data with arbitrary shapes. *Pattern Recognition, 38*, 1887–1901.
120. Thulasiraman, K., & Swamy, M. N. S. (1992). *Graphs: Theory and algorithms*. New York: Wiley.
121. Tishby, N., Pereira, F., & Bialek, W. (1999). The information bottleneck method. In *Proceedings of the 37th Annual Allerton Conference on Communication, Control, and Computing* (pp. 368–377).
122. Tjhi, W.-C., & Chen, L. (2009). Dual fuzzy-possibilistic coclustering for categorization of documents. *IEEE Transactions on Fuzzy Systems, 17*(3), 532–543.
123. Tseng, V. S., & Kao, C.-P. (2007). A novel similarity-based fuzzy clustering algorithm by integrating PCM and mountain method. *IEEE Transactions on Fuzzy Systems, 15*(6), 1188–1196.
124. Vesanto, J., & Alhoniemi, E. (2000). Clustering of the self-organizing map. *IEEE Transactions on Neural Networks, 11*(3), 586–600.

125. Wang, J. H., & Rau, J. D. (2001). VQ-agglomeration: A novel approach to clustering. *IEE Proceedings - Vision, Image and Signal Processing*, *148*(1), 36–44.
126. Wang, W., & Zhang, Y. (2007). On fuzzy cluster validity indices. *Fuzzy Sets and Systems*, *158*, 2095–2117.
127. Wang, C.-D., & Lai, J.-H. (2011). Energy based competitive learning. *Neurocomputing*, *74*, 2265–2275.
128. Wang, X., Wang, X., & Wilkes, D. M. (2009). A divide-and-conquer approach for minimum spanning tree-based clustering. *IEEE Transactions on Knowledge and Data Engineering*, *21*(7), 945–958.
129. Wu, K.-L., Yang, M.-S., & Hsieh, J.-N. (2009). Robust cluster validity indexes. *Pattern Recognition*, *42*, 2541–2550.
130. Xie, X. L., & Beni, G. (1991). A validity measure for fuzzy clustering. *IEEE Transactions on Pattern Analysis and Machine Intelligence*, *13*(8), 841–847.
131. Xu, L. (2007). A unified perspective and new results on RHT computing, mixture based learning and multi-learner based problem solving. *Pattern Recognition*, *40*, 2129–2153.
132. Xu, L., Krzyzak, A., & Oja, E. (1993). Rival penalized competitive learning for clustering analysis, RBF net, and curve detection. *IEEE Transactions on Neural Networks*, *4*(4), 636–649.
133. Yair, E., Zeger, K., & Gersho, A. (1992). Competitive learning and soft competition for vector quantizer design. *IEEE Transactions on Signal Processing*, *40*(2), 294–309.
134. Yang, T. N., & Wang, S. D. (2004). Competitive algorithms for the clustering of noisy data. *Fuzzy Sets and Systems*, *141*, 281–299.
135. Yang, M.-S., Chiang, Y.-H., Chen, C.-C., & Lai, C.-Y. (2008). A fuzzy k-partitions model for categorical data and its comparison to the GoM model. *Fuzzy Sets and Systems*, *159*, 390–405.
136. Yip, K. Y., Cheung, D. W., & Ng, M. K. (2004). HARP: A practical projected clustering algorithm. *IEEE Transactions on Knowledge and Data Engineering*, *16*(11), 1387–1397.
137. Yiu, M. L., & Mamoulis, N. (2005). Iterative projected clustering by subspace mining. *IEEE Transactions on Knowledge and Data Engineering*, *17*(2), 176–189.
138. Zahn, C. T. (1971). Graph-theoretical methods for detecting and describing gestalt clusters. *IEEE Transactions on Computers*, *20*(1), 68–86.
139. Zalik, K. R. (2010). Cluster validity index for estimation of fuzzy clusters of different sizes and densities. *Pattern Recognition*, *43*(10), 3374–3390.
140. Zhang, Y. J., & Liu, Z. Q. (2002). Self-splitting competitive learning: A new on-line clustering paradigm. *IEEE Transactions on Neural Networks*, *13*(2), 369–380.
141. Zhang, J.-S., & Leung, Y.-W. (2004). Improved possibilistic c-means clustering algorithms. *IEEE Transactions on Fuzzy Systems*, *12*(2), 209–217.
142. Zhang, T., Ramakrishnan, R., & Livny, M. (1996). BIRCH: An efficient data clustering method for very large databases. In *Proceedings of ACM SIGMOD Conference on Management of Data* (pp. 103–114). Montreal, Canada.
143. Zhang, T., Ramakrishnan, R., & Livny, M. (1997). BIRCH: A new data clustering algorithm and its applications. *Data Mining and Knowledge Discovery*, *1*, 141–182.
144. Zheng, G. L., & Billings, S. A. (1999). An enhanced sequential fuzzy clustering algorithm. *International Journal of Systems Science*, *30*(3), 295–307.
145. Zhong, S. (2005). Efficient online spherical k-means clustering. In *Proceedings of International Joint Conference on Neural Networks (IJCNN)* (pp. 3180–3185). Montreal, Canada.

Chapter 11
Radial Basis Function Networks

11.1 Introduction

Learning is an approximation problem, which is closely related to the conventional approximation techniques, such as generalized splines and regularization techniques. The RBF network has its origin in performing exact interpolation of a set of data points in a multidimensional space [78]. The RBF network is a universal approximator, and it is a popular alternative to the MLP, since it has a simpler structure and a much faster training process. Both models are widely used for classification and function approximation.

The RBF network has a network architecture similar to that of the classical regularization network [77], where the basis functions are Green's functions of the Gram operator associated with the stabilizer. If the stabilizer exhibits radial symmetry, the basis functions are radially symmetric as well and hence, an RBF network is obtained. From the viewpoint of approximation theory, the regularization network has three desirable properties [30, 77]: It can approximate any multivariate continuous function on a compact domain to an arbitrary accuracy, given a sufficient number of units; it has the best-approximation property since the unknown coefficients are linear; and the solution is optimal in the sense that it minimizes a functional that measures how much it oscillates.

The RBF network with a localized RBF is a receptive-field or localized network. The localized approximation method provides the strongest output when the input is near the prototype of a node. For a suitably trained localized RBF network, similar input vectors always generate similar outputs, while distant input vectors produce nearly independent outputs. This is the intrinsic local generalization property. A receptive-field network is an associative network in that only a small subspace is determined by the input to the network. The domain of receptive-field functions is practically a finite real interval defined by the parameters of the function. This property is particularly attractive since the receptive-field function produces a local effect. Thus, receptive-field networks can be conveniently constructed by adjusting the parameters of the receptive-field functions and/or adding or removing neurons.

K.-L. Du and M. N. S. Swamy, *Neural Networks and Statistical Learning*,
https://doi.org/10.1007/978-1-4471-7452-3_11

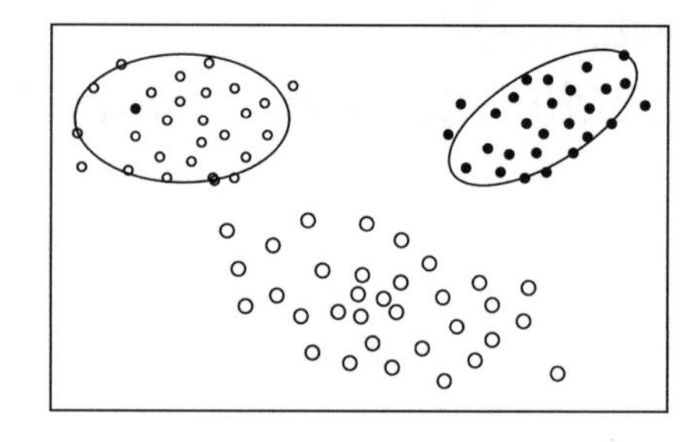

Fig. 11.1 The RBF network for classification: each class is fitted by a kernel function

RBF Networks and Classification

In a classification problem, we model the posterior probabilities $p(\mathcal{C}_k|\boldsymbol{x})$ for each of the classes. We have Bayes' theorem

$$P(\mathcal{C}_k|\boldsymbol{x}) = \frac{p(\boldsymbol{x}|\mathcal{C}_k)P(\mathcal{C}_k)}{p(\boldsymbol{x})} = \frac{p(\boldsymbol{x}|\mathcal{C}_k)P(\mathcal{C}_k)}{\sum_j p(\boldsymbol{x}|\mathcal{C}_j)P(\mathcal{C}_j)}, \tag{11.1}$$

where $P(\mathcal{C}_k)$ is the prior probability. If we model $p(\boldsymbol{x}|\mathcal{C}_k)$ as an RBF kernel, we can define a normalized RBF kernel to model Bayes' theorem

$$\phi_k(\boldsymbol{x}) = \frac{p(\boldsymbol{x}|\mathcal{C}_k)}{\sum_j p(\boldsymbol{x}|\mathcal{C}_j)P(\mathcal{C}_j)}. \tag{11.2}$$

Therefore, the RBF network has a natural resemblance with Bayes' theorem. This is illustrated in Fig. 11.1.

In practice, each class-conditional distribution $p(\boldsymbol{x}|\mathcal{C}_k)$ can be represented by a mixture of models, as a linear combination of kernel functions.

Related Neural Network Models

The Gaussian RBF network is a popular receptive-field network. Another well-known receptive-field network is the cerebellar model articulation controller (CMAC) [2, 67] associative memory network inspired by the neurophysiological properties of the cerebellum. CMAC is a distributed look-up-table system and is also suitable for VLSI realization. It can approximate slow-varying functions and is orders of magnitude faster than BP. However, CMAC may fail in approximating highly nonlinear or rapidly oscillating functions [10]. Pseudo-self-evolving CMAC [92], inspired by the cerebellar experience-driven synaptic plasticity phenomenon observed in the cerebellum, nonuniformly allocates its computing cells to overcome the architectural deficiencies encountered by the CMAC network. This is where significantly higher densities of synaptic connections are located in the frequently accessed regions.

The generalized single-layer network [34], also known as the *generalized linear discriminant*, has a three-layer architecture similar to the RBF network. Each node in the hidden layer has a nonlinear activation function $\phi_i(\cdot)$, and the output nodes implement linear combinations of the nonlinear kernel functions of the inputs. The RBF network is a type of generalized single-layer network. Like OLS method, orthogonal methods in conjunction with some information criteria are usually used for self-structuring the generalized single-layer network to generate a parsimonious, yet accurate, network [1]. The generalization ability of the generalized single-layer network is analyzed in [34] by using PAC learning theory and the concept of VC-dimension. Necessary and sufficient conditions on the number of training examples are derived to guarantee a particular generalization performance of the generalized single-layer network [34].

The wavelet neural network [108, 109] has the same structure as the RBF network but uses wavelet functions as the activation function for the hidden units. Due to the localized properties in both the time and frequency domains of wavelet functions, wavelets are locally receptive-field functions that approximate discontinuous or rapidly changing functions. The wavelet neural network has become a popular tool for function approximation. Wavelets with coarse resolution can capture the global or low-frequency feature easily, while wavelets with fine resolution can capture the local or high-frequency feature of the function accurately. This distinguished characteristic leads the wavelet neural network to fast convergence, easy training, and high accuracy.

11.2 RBF Network Architecture

The RBF network, shown in Fig. 11.2, is a J_1-J_2-J_3 feedforward network. Each node in the hidden layer uses an RBF $\phi(r)$ as its nonlinear activation function. $\phi_0(\boldsymbol{x}) = 1$ corresponds to the bias in the output layer, while $\phi_i(\boldsymbol{x}) = \phi(\boldsymbol{x} - \boldsymbol{c}_i)$, where $\boldsymbol{c}_i$ is the center of the ith node and $\phi(\boldsymbol{x})$is an RBF. The hidden layer performs a nonlinear transform of the input, and the output layer is a linear combiner mapping the nonlinearity into a new space. The biases of the output layer neurons can be modeled by an additional neuron in the hidden layer, which has a constant activation function $\phi_0(r) = 1$. The RBF network can achieve a global optimal solution to the adjustable weights in the minimum MSE sense by using the linear optimization method.

For input pattern $\boldsymbol{x}$, the output of the network is given by

$$y_i(\boldsymbol{x}) = \sum_{k=1}^{J_2} w_{ki}\phi\left(\|\boldsymbol{x} - \boldsymbol{c}_k\|\right), \quad i = 1, \ldots, J_3, \tag{11.3}$$

where $y_i(\boldsymbol{x})$ is the ith output of the RBF network, w_{ki} is the connection weight from the kth hidden unit to the ith output unit, $\boldsymbol{c}_k$ is the prototype or center of the kth

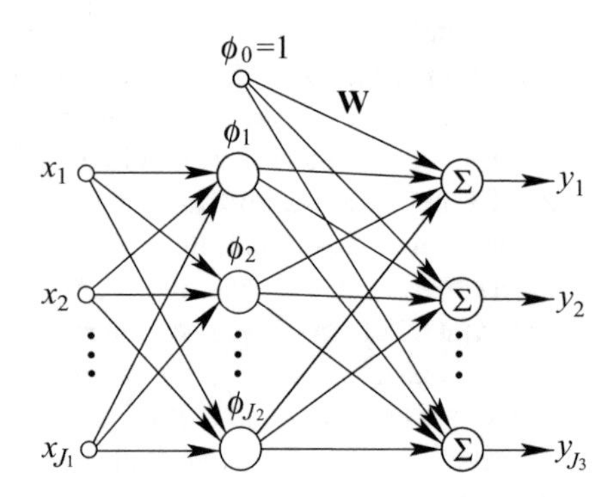

Fig. 11.2 Architecture of the RBF network

hidden unit, and $\|\cdot\|$ denotes the Euclidean norm. $\phi(\cdot)$ is typically selected as the Gaussian function.

For a set of N pattern pairs $\left\{\left(\boldsymbol{x}_p, \boldsymbol{y}_p\right)\right\}$, (11.3) can be expressed in matrix form

$$\mathbf{Y} = \mathbf{W}^T \mathbf{\Phi}, \tag{11.4}$$

where $\mathbf{W} = \left[\boldsymbol{w}_1, \ldots, \boldsymbol{w}_{J_3}\right]$ is a $J_2 \times J_3$ weight matrix, $\boldsymbol{w}_i = \left(w_{1i}, \ldots, w_{J_2 i}\right)^T$, $\mathbf{\Phi} = \left[\boldsymbol{\phi}_1, \ldots, \boldsymbol{\phi}_N\right]$ is a $J_2 \times N$ matrix, $\boldsymbol{\phi}_p = \left(\phi_{p,1}, \ldots, \phi_{p,J_2}\right)^T$ is the output of the hidden layer for the pth sample, $\phi_{p,k} = \phi\left(\left\|\boldsymbol{x}_p - \boldsymbol{c}_k\right\|\right)$, $\mathbf{Y} = \left[\boldsymbol{y}_1, \boldsymbol{y}_2, \ldots, \boldsymbol{y}_N\right]$ is a $J_3 \times N$ matrix, and $\boldsymbol{y}_p = \left(y_{p,1}, \ldots, y_{p,J_3}\right)^T$.

11.3 Universal Approximation of RBF Networks

The RBF network has universal approximation and regularization capabilities. Theoretically, the RBF network can approximate any continuous function arbitrarily well, if the RBF is suitably chosen [60, 74, 77].

Micchelli considered the solution of the interpolation problem $s(\boldsymbol{x}_k) = y_k$, $k = 1, \ldots, J_2$, by functions of the form $s(\boldsymbol{x}) = \sum_{k=1}^{J_2} w_k \phi\left(\|\boldsymbol{x} - \boldsymbol{x}_k\|^2\right)$, and proposed Micchelli's interpolation theorem [66]. $\phi(\cdot)$ is required to be completely monotonic on $(0, \infty)$, that is, it is continuous on $(0, \infty)$ and its lth-order derivative $\phi^{(l)}(x)$ satisfies $(-1)^l \phi^{(l)}(x) \geq 0$, $\forall x \in (0, \infty)$ and $l = 0, 1, 2, \ldots$. A less restrictive condition has been given in [47], where $\phi(\cdot)$ is continuous on $(0, \infty)$ and its derivatives satisfy $(-1)^l \phi^{(l)}(x) > 0$, $\forall x \in (0, \infty)$ and $l = 0, 1, 2$.

RBFs possess excellent mathematical properties. In the context of the exact interpolation problem, many properties of the interpolating function are relatively insensitive to the precise form of the nonlinear function $\phi(\cdot)$ [78]. The choice of RBF is not crucial to the performance of the RBF network [13].

The Gaussian RBF network can approximate, to any degree of accuracy, any continuous function by a sufficient number of centers $\boldsymbol{c}_i$, $i = 1, \ldots, J_2$, and a common standard deviation $\sigma > 0$ in L_p-norm, $p \in [1, \infty]$ [74]. A class of RBF networks can achieve universal approximation when the RBF is continuous and integrable [74].

The requirement of the integrability of the RBF is relaxed in [60]. For an RBF that is continuous almost everywhere, locally essentially bounded, and not a polynomial, the RBF network can approximate any continuous function with respect to the uniform norm [60]. From this result, such RBFs as $\phi(r) = e^{-\frac{r}{\sigma^2}}$ and $\phi(r) = e^{\frac{r}{\sigma^2}}$ also lead to universal approximation capability [60].

In [39], it is proved in an incremental constructive method that three-layer feedforward networks with randomly generated hidden nodes are universal approximators, and only the weights linking the hidden and output layers need to be adjusted. The proof itself gives an efficient incremental construction of the network. Theoretically, the network with any bounded nonlinear piecewise continuous activation can function as a universal approximator.

A bound on the generalization error for feedforward networks is given by (2.2) [70]. This bound has been considerably improved to $O\left(\left(\frac{\ln N}{N}\right)^{\frac{1}{2}}\right)$ in [50] for RBF network regression with the MSE function.

A decaying RBF $\phi(x)$ is not zero at $x = 0$ but approaches zero as $x \to \infty$. It is clear that the Gaussian function and the wavelet functions such as Mexican-hat wavelet $\phi(x) = \frac{2}{\sqrt{3}}\pi^{-1/4}(1 - x^2)e^{-x^2/2}$ are decaying RBFs. A constructive proof is given in [36] for the fact that a decaying RBF network with $n + 1$ hidden neurons can interpolate $n + 1$ multivariate samples with zero error. The given decaying RBFs can uniformly approximate any continuous multivariate function with arbitrary precision without training [36], thus giving faster convergence and better generalization performance than conventional RBF algorithm, BP, extreme learning machine, and SVMs.

11.4 Formulation for RBF Network Learning

Like MLP learning, the learning of the RBF network is formulated as the minimization of the MSE function:

$$E = \frac{1}{N}\sum_{i=1}^{N}\left\|\boldsymbol{y}_p - \mathbf{W}^T\boldsymbol{\phi}_p\right\|^2 = \frac{1}{N}\left\|\mathbf{Y} - \mathbf{W}^T\boldsymbol{\Phi}\right\|_F^2, \tag{11.5}$$

where $\mathbf{Y} = \left[\boldsymbol{y}_1, \boldsymbol{y}_2, \ldots, \boldsymbol{y}_N\right]$, $\boldsymbol{y}_i$ is the target output for the ith sample in the training set, and $\|\cdot\|_F$ is the Frobenius norm defined as $\|\mathbf{A}\|_F^2 = \mathrm{tr}\left(\mathbf{A}^T\mathbf{A}\right)$.

RBF network learning requires the determination of the RBF centers and weights. The selection of the RBF centers is most critical to a successful RBF network implementation. The centers can be placed on a random subset or all of the training examples or determined by clustering or via a learning procedure. For some RBFs such as the Gaussian, it is also necessary to determine the smoothness parameter σ. The RBF network using the Gaussian RBF is usually termed the *Gaussian RBF network*. Existing learning algorithms are mainly developed for the Gaussian RBF network and can be modified accordingly when other RBFs are used.

The Gaussian RBF network can be regarded as an improved alternative to the four-layer probabilistic neural network [89], which is based on the Parzen classifier. In a probabilistic neural network, a Gaussian RBF node is placed at the position of each training pattern so that the unknown density can be well interpolated and approximated. This technique yields optimal decision surfaces in a Bayesian sense. Training is to associate each node with its target class. This approach, however, severely suffers from the curse of dimensionality and results in poor generalization.

11.5 Radial Basis Functions

A number of functions can be used as the RBF [60, 66, 77]

$$\phi(r) = \mathrm{e}^{-\frac{r^2}{2\sigma^2}}, \qquad \text{Gaussian,} \tag{11.6}$$

$$\phi(r) = r^2 \ln(r), \qquad \text{thin-plate spline,} \tag{11.7}$$

$$\phi(r) = \frac{1}{1 + \mathrm{e}^{\frac{r}{\sigma^2} - \theta}}, \qquad \text{logistic function,} \tag{11.8}$$

where $r > 0$ denotes the distance from data point $\boldsymbol{x}$ to center $\boldsymbol{c}$, σ is used to control the smoothness of the interpolating function, and θ in (11.8) is an adjustable bias. These RBFs are illustrated in Fig. 11.3.

The Gaussian (11.6) and the logistic function (11.8) are localized RBFs with the property that $\phi(r) \to 0$ as $r \to \infty$. Physiologically, there exist Gaussian-like receptive fields in cortical cells [77]. As a result, $\phi(r)$ is typically selected as the Gaussian or other localized RBFs.

The RBF network conventionally uses the Gaussian function (11.6) as the RBF. The Gaussian is compact and positive. It is motivated from the point of view of kernel regression and kernel density estimation. In fitting data in which there is normally distributed noise with the inputs, the Gaussian is the optimal basis function in the LS

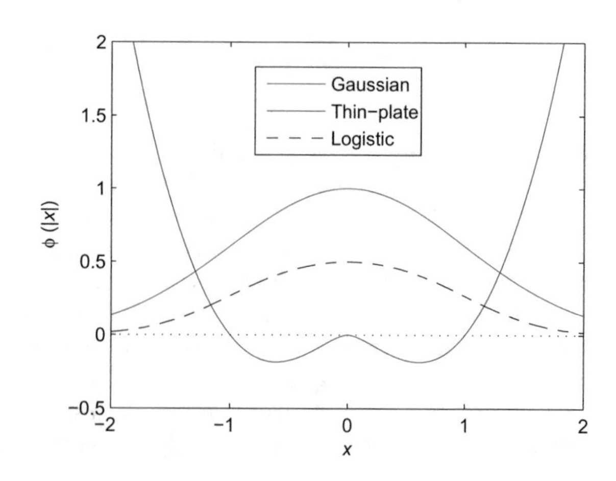

Fig. 11.3 Illustration of RBFs. $\sigma = 1, \theta = 0$

sense [97]. The Gaussian is the only factorizable RBF, and this property is desirable for hardware implementation of the RBF network.

The thin-plate spline function (11.7) is another popular RBF for universal approximation. The use of the thin-plate spline is motivated from a curve-fitting perspective [61]. It diverges at infinity and is negative over the region of $r \in (0, 1)$. However, for training purposes, the approximated function needs to be defined only over a specified range. There is some empirical evidence to suggest that the thin-plate spline better fits the data in high-dimensional settings [61].

A pseudo-Gaussian function in one-dimensional space is introduced by selecting the standard deviation σ in the Gaussian (11.6) as two different positive values, namely, σ_- for $x < 0$ and σ_+ for $x > 0$ [80]. In n-dimensional space, the pseudo-Gaussian function can be defined by

$$\phi(\boldsymbol{x}) = \prod_{i=1}^{n} \varphi_i (x_i), \tag{11.9}$$

$$\varphi_i (x_i) = \begin{cases} \mathrm{e}^{-\frac{(x_i - c_i)^2}{\sigma_-^2}}, & x_i < c_i \\ \mathrm{e}^{-\frac{(x_i - c_i)^2}{\sigma_+^2}}, & x_i \geq c_i \end{cases}, \tag{11.10}$$

where $\boldsymbol{c} = (c_1, \ldots, c_n)^T$ is the center vector, and index i runs over the dimension of the input space n. The pseudo-Gaussian function is not strictly an RBF due to its radial asymmetry, and this, however, eliminates the symmetry restriction and provides the hidden units with greater flexibility with respect to function approximation.

When utilized to approximate the functional behavior with sharp noncircular features, many circular-shaped Gaussian basis functions may be required. In order to reduce the size of the RBF network, direction-dependent scaling, shaping, and rotation of Gaussian RBFs are introduced in [88] for maximal trend sensing with minimal parameter representations for function approximation. Shaping and rotation of the RBFs help in reducing the total number of function units required to approximate any given input–output data, while improving accuracy.

Radial Basis Functions for Approximating Constant Values

Approximating functions with constant-valued segments using localized RBFs is most difficult. If a function has nearly constant values in some intervals, the Gaussian RBF network is inefficient in approximating these values unless its variance is very large approaching infinity. The sigmoidal RBF, as a composite of a set of sigmoidal functions, can be used to deal with this problem [52]

$$\phi(x) = \frac{1}{1 + \mathrm{e}^{-\beta[(x-c)+\theta]}} - \frac{1}{1 + \mathrm{e}^{-\beta[(x-c)-\theta]}}, \tag{11.11}$$

where the bias $\theta > 0$ and the gain $\beta > 0$. $\phi(x)$ is radially symmetric with the maximum at the center c. β controls the steepness and θ controls the width of the function. The shape of $\phi(x)$ is approximately rectangular if $\beta\theta$ is large. For large β and θ, it has a soft trapezoidal shape, while for small β and θ it is bell-shaped. $\phi(x)$ can be extended to an n-dimensional approximation

$$\phi(\boldsymbol{x}) = \prod_{i=1}^{n} \varphi_i(x_i), \tag{11.12}$$

$$\varphi_i(x_i) = \frac{1}{1 + \mathrm{e}^{-\beta_i[(x_i - c_i) + \theta_i]}} - \frac{1}{1 + \mathrm{e}^{-\beta_i[(x_i - c_i) - \theta_i]}}, \tag{11.13}$$

where $\boldsymbol{x} = (x_1, \ldots, x_n)^T$, $\boldsymbol{c} = (c_1, \ldots, c_n)^T$, $\boldsymbol{\theta} = (\theta_1, \ldots, \theta_n)^T$, and $\boldsymbol{\beta} = (\beta_1, \ldots, \beta_n)^T$.

When β_i and θ_i are small, the sigmoidal RBF $\phi(\boldsymbol{x})$ will be close to zero and the corresponding node will have little contribution to the approximation task regardless of the tuning of the other parameters thereafter. To accommodate constant values of the desired output and to avoid diminishing the kernel functions, $\phi(\boldsymbol{x})$ can be modified by adding an additional term to the product term $\varphi_i(x_i)$ [53]

$$\phi(\boldsymbol{x}) = \prod_{i=1}^{n} \left(\varphi_i(x_i) + \widetilde{\varphi}_i(x_i)\right), \tag{11.14}$$

where

$$\widetilde{\varphi}_i(x_i) = [1 - \varphi_i(x_i)]\,\mathrm{e}^{-a_i(x_i - c_i)^2} \tag{11.15}$$

with $a_i \geq 0$, and $\widetilde{\varphi}_i(x_i)$ being used as a compensating function to keep the product term from decreasing to zero when $\varphi_i(x_i)$ is small. β_i and a_i are, respectively, associated with the steepness and sharpness of the product term and θ_i controls the width of the product term. The parameters are adjusted by the gradient-descent method.

An alternative approach is to use the raised-cosine RBF [85]

$$\phi(x) = \begin{cases} \cos^2\left(\frac{\pi x}{2}\right) & |x| \leq 1 \\ 0 & |x| > 1 \end{cases}, \tag{11.16}$$

where $\phi(x)$ is a zero-centered function with compact support since $\phi(0) = 1$ and $\phi(x) = 0$ for $|x| \geq 1$. The raised-cosine RBF can represent a constant function exactly using two terms. This RBF can be generalized to n dimensions [85]

$$\phi(\boldsymbol{x}) = \prod_{i=1}^{n} \phi(x_i - c_i). \tag{11.17}$$

Notice that $\phi(\boldsymbol{x})$ is nonzero only when $\boldsymbol{x}$ is in the $(-1, 1)^n$ vicinity of $\boldsymbol{c}$.

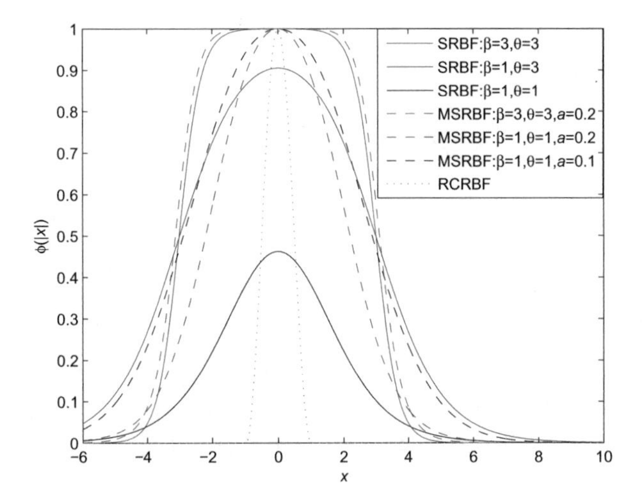

Fig. 11.4 RBFs for approximating constant-valued segments in the one-dimensional space: SRBF—sigmoidal RBF (11.11), MSRBF—modified sigmoidal RBF (11.14), and RCRBF—raised-cosine RBF (11.16). The center is selected as $c_i = c = 0$

Figure 11.4 illustrates the sigmoidal, modified sigmoidal, and raised-cosine RBFs with different selections of β and θ. In Chap. 26, we will introduce some popular fuzzy membership functions, which can be used as RBFs by suitably constraining some parameters.

11.6 Learning RBF Centers

RBF network learning is usually implemented using a two-phase strategy. The first phase specifies and fixes suitable centers $\boldsymbol{c}_i$ and their respective RBF parameters σ_i. For the Gaussian RBF, σ_i's denote standard deviations, also known as *widths* or *radii*. The second phase adjusts the network weights $\mathbf{W}$. In this section, we describe the first phase.

Selecting RBF Centers Randomly from Training Sets

A simple method to specify the RBF centers is to randomly select a subset of the input patterns from the training set as the RBF centers. If the training set is representative of the learning problem, this method is appropriate. This method is relatively insensitive to the use of pseudoinverse, and hence it may be a regularization method.

The Gaussian RBF network using the same σ for all RBF centers has universal approximation capability [74]. This global width can be selected as the average of all the Euclidian distances between the ith RBF center and its nearest neighbor

$$\sigma = \left\langle \left\| \boldsymbol{c}_i - \boldsymbol{c}_j \right\| \right\rangle, \tag{11.18}$$

where

$$\left\| \boldsymbol{c}_i - \boldsymbol{c}_j \right\| = \min_{k=1,\ldots,J_2; k \neq i} \left\| \boldsymbol{c}_i - \boldsymbol{c}_k \right\|. \tag{11.19}$$

One can also select σ by $\sigma = d_{\max}/\sqrt{2J_2}$, where $d_{\max}$ is the maximum distance between the selected centers, $d_{\max} = \max_{i,k=1,\ldots,J_2;k>i} \left\| \boldsymbol{c}_i - \boldsymbol{c}_k \right\|$ [9]. This choice makes the Gaussian RBF neither too steep nor too flat.

In practice, the width of each RBF $\sigma_i, i = 1, \ldots, J_2$, can be determined according to the data distribution in the region of the corresponding RBF center. A heuristic for selecting σ_i is to average the distances between the ith RBF center and its L nearest neighbors. Alternatively, σ_i is selected according to the distance of unit i to its nearest neighbor unit j, $\sigma_i = a \left\| \boldsymbol{c}_i - \boldsymbol{c}_j \right\|$, $a \in [1.0, 1.5]$.

Selecting RBF Centers by Clustering Training Sets

Clustering is usually used for determining the RBF centers. The training set is grouped into appropriate clusters, whose prototypes are then used as RBF centers. The number of clusters can be specified or determined automatically depending on the clustering algorithm.

Unsupervised clustering such as C-means is popular for clustering RBF centers [68]. RBF centers determined by supervised clustering are usually more efficient for RBF network learning than those determined by unsupervised clustering [16], since the distribution of the output patterns in the training set is also considered. When the RBF network is trained for classification, LVQ1 is a popular method for clustering the RBF centers.

The relationship between the augmented unsupervised clustering process and the MSE of RBF network learning has been investigated in [95]. In the case of the Gaussian RBF and any Lipschitz continuous RBF, a weighted MSE for supervised quantization yields an upper bound on the MSE of RBF network learning. This upper bound and consequently the output error can be made arbitrarily small by decreasing the quantization error, which can be accomplished by increasing the number of hidden units.

After the RBF centers are determined, the covariance matrices of the RBFs are set to the covariances of the input patterns in each cluster. In this case, the Gaussian RBF network is extended to the generalized RBF network using the Mahalanobis distance, defined by the weighted norm [77]

$$\phi \left(\left\| \boldsymbol{x} - \boldsymbol{c}_k \right\|_{\mathbf{A}} \right) = \mathrm{e}^{-\frac{1}{2}(\boldsymbol{x}-\boldsymbol{c}_k)^T \boldsymbol{\Sigma}^{-1} (\boldsymbol{x}-\boldsymbol{c}_k)}, \tag{11.20}$$

where the squared weighted norm $\|\boldsymbol{x}\|_{\mathbf{A}}^2 = (\mathbf{A}\boldsymbol{x})^T(\mathbf{A}\boldsymbol{x}) = \boldsymbol{x}^T\mathbf{A}^T\mathbf{A}\boldsymbol{x}$ and $\boldsymbol{\Sigma}^{-1} = 2\mathbf{A}^T\mathbf{A}$. When the Euclidean distance is employed, one can also select the width of the Gaussian RBF network using the heuristics for selecting RBF centers randomly from training sets.

11.7 Learning the Weights

After RBF centers and their widths or covariance matrices are determined, learning of the weights **W** is reduced to a linear optimization problem, which can be solved using the LS method or the gradient-descent method.

11.7.1 Least Squares Methods for Weights Learning

After the parameters related to the RBF centers are determined, the weight matrix **W** is then trained to minimize the MSE (11.5). This LS problem requires a computational complexity of $O\left(NJ_2^2\right)$ for $N > J_2$ when popular orthogonalization techniques such as SVD and QR decomposition are applied. A simple representation of the solution of the batch LS method is given explicitly by [9]

$$\mathbf{W} = \left(\boldsymbol{\Phi}^T\right)^{\dagger} \mathbf{Y}^T = \left(\Phi\Phi^T\right)^{-1} \boldsymbol{\Phi}\mathbf{Y}^T, \tag{11.21}$$

where $[\cdot]^{\dagger}$ is the pseudoinverse of the matrix within. The over- or underdetermined linear LS system is an ill-conditioned problem. SVD is an efficient and numerically robust technique for dealing with such an ill-conditioned problem and is preferred.

According to (11.21), if $\boldsymbol{\Phi}^T\Phi = \mathbf{I}$, inversion operation is unnecessary. The optimum weight can be computed by

$$\boldsymbol{w}_k = \boldsymbol{\Phi}\overline{\boldsymbol{y}}_k, \quad k = 1, \ldots, J_3, \tag{11.22}$$

where $\overline{\boldsymbol{y}}_k = \left(y_{1,k}, \ldots, y_{N,k}\right)^T$ corresponds to the kth row of $\mathbf{Y}$. Based on this observation, an efficient, noniterative weight learning technique has been introduced by applying GSO on RBFs [45]. The RBFs are first transformed into a set of orthonormal RBFs for which the optimum weights are computed. These weights are then recomputed in such a way that their values can be fitted back into the original RBF network structure, i.e., with kernel functions unchanged. In addition, the method has low storage requirements, and the computation procedure can be organized in a parallel manner. Incorporation of new hidden nodes aimed at improving the network performance does not require recomputation of the network weights already calculated. The contribution of each RBF to the overall network output can be evaluated.

When the full dataset is not available and samples are obtained online, the RLS method can be used to train the weights online

$$\boldsymbol{w}_i(t) = \boldsymbol{w}_i(t-1) + \mathbf{K}(t)e_i(t), \quad i = 1, \ldots, J_3, \tag{11.23}$$

$$\mathbf{K}(t) = \frac{\mathbf{P}(t-1)\boldsymbol{\phi}_t}{\boldsymbol{\phi}_t^T\mathbf{P}(t-1)\boldsymbol{\phi}_t + \mu}, \tag{11.24}$$

$$e_i(t) = y_{t,i} - \boldsymbol{\phi}_t^T \boldsymbol{w}_i(t-1), \quad i = 1, \ldots, J_3, \tag{11.25}$$

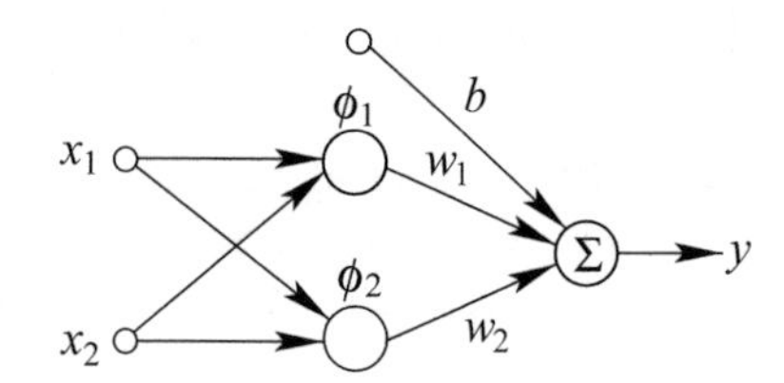

Fig. 11.5 Solve the XOR problem using a 2-2-1 RBF network

$$\mathbf{P}(t) = \frac{1}{\mu}\left[\mathbf{P}(t-1) - \mathbf{K}(t)\phi_t^T\mathbf{P}(t-1)\right], \tag{11.26}$$

where $\mu \in (0, 1]$ is the forgetting factor. Typically, $\mathbf{P}(0) = a_0\mathbf{I}_{J_2}$, where a_0 is a sufficiently large number and $\mathbf{I}_{J_2}$ is the $J_2 \times J_2$ identity matrix, and $\boldsymbol{w}_i(0)$ is selected as a small random matrix.

Example 11.1 We solve the XOR problem using the RBF network. The objective is to classify the input patterns $(1, 1)$ and $(0, 0)$ as class 0, and classify $(1, 0)$ and $(0, 1)$ as class 1.

We employ a 2-2-1 RBF network, as shown in Fig. 11.5. We set the bias $b = 0.4$. We select two of the points $(0, 1)$ and $(1, 0)$ as the RBF centers $\boldsymbol{c}_i$, $i = 1, 2$, and select the Gaussian RBF $\phi_i(\boldsymbol{x}) = \mathrm{e}^{-\frac{\|\boldsymbol{x}-\boldsymbol{c}_i\|^2}{2\sigma^2}}$, and $\sigma = 0.5$. Given the input $\boldsymbol{x}_1 = (0, 0)$, $\boldsymbol{x}_2 = (0, 1)$, $\boldsymbol{x}_3 = (1, 0)$, $\boldsymbol{x}_4 = (1, 1)$, we get their mappings in the feature space. The input points and their mappings in the feature space are shown in Fig. 11.6. It is seen that the input patterns are linearly inseparable in the input space, whereas they are linearly separable in the feature space.

By using (11.21), we can solve the weights as $w_1 = w_2 = 0.3367$. We thus get the decision boundary in the feature space as $0.4528\phi_1 + 0.4528\phi_2 - 0.4 = 0$ or $0.4528\,\mathrm{e}^{-2(x_1^2+(x_2-1)^2)} + 0.4528\,\mathrm{e}^{-2((x_1-1)^2+x_2^2)} - 0.4 = 0$ in the input space.

Extreme Learning Machines

Extreme learning machine (ELM) [40] is a framework for estimating the parameters of single-hidden-layer feedforward networks. The hidden-layer node parameters are selected randomly and the output weights are determined analytically. It tends to provide good generalization performance at fast learning speed. ELM can be used to train single-hidden-layer feedforward networks with nondifferentiable activation functions such as threshold functions. As it is based on the pseudoinverse of a matrix, it can learn thousands of times faster than gradient-based learning algorithms and SVM, and tends to achieve similar or better generalization performance. The low computational burden of ELM is due to the fact that only a few neurons are utilized to synthesize the estimator. However, a sparse model with a satisfactory generalization performance may not be obtainable. Similar ideas of using random features have been implemented in reservoir computing (in Chap. 12) and in deep learning (in Chap. 24).

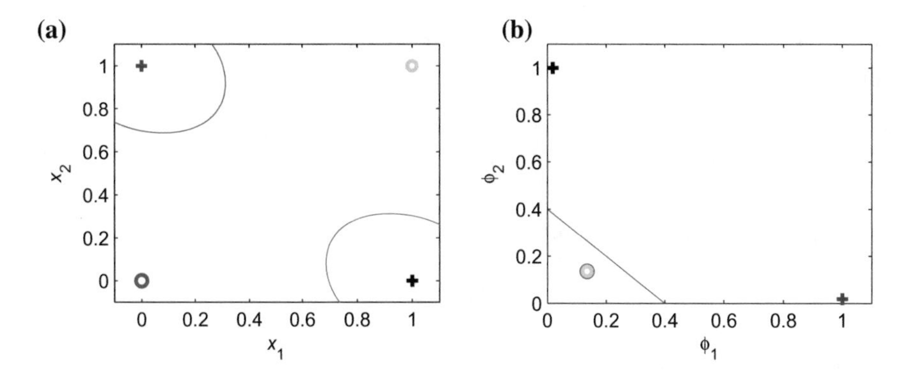

Fig. 11.6 Solve the XOR problem using the RBF network. **a** The input (x_1-x_2) patterns. **b** The patterns in the feature (φ_1-φ_2) space

11.8 RBF Network Learning Using Orthogonal Least Squares

Optimal subset selection techniques are computationally prohibitive. OLS method [13–15] is an efficient way for subset model selection. The approach chooses and adds RBF centers one by one until an adequate network is constructed. All the training examples are considered as candidates for the centers, and the one that reduces the MSE the most is selected as a new hidden unit. GSO is first used to construct a set of orthogonal vectors in the space spanned by the vectors of the hidden unit activation ϕ_p, and a new RBF center is then selected by minimizing the residual MSE. Model selection criteria are used to determine the size of the network.

Batch OLS method can not only determine the weights but also choose the number and the positions of the RBF centers. Batch OLS can employ the forward [14, 15] and backward [35] center selection approaches.

When the RBF centers are distinct, $\mathbf{\Phi}^T$ is of full rank. The orthogonal decomposition of $\mathbf{\Phi}^T$ is performed using QR decomposition

$$\mathbf{\Phi}^T = \mathbf{Q}\begin{bmatrix} \mathbf{R} \\ \mathbf{O} \end{bmatrix}, \tag{11.27}$$

where $\mathbf{Q} = \left[\boldsymbol{q}_1 \ldots \boldsymbol{q}_N\right]$ is an $N \times N$ orthogonal matrix, $\mathbf{R}$ is a $J_2 \times J_2$ upper triangular matrix, and $\mathbf{O}$ is a zero matrix of an appropriate size. By minimizing the MSE given by (11.5), one can make use of the invariant property of the Frobenius norm

$$E = \frac{1}{N}\left\|\mathbf{YQ} - \mathbf{W}^T\mathbf{\Phi}\mathbf{Q}\right\|_F^2 = \frac{1}{N}\left\|\mathbf{Q}^T\mathbf{Y}^T - \mathbf{Q}^T\mathbf{\Phi}^T\mathbf{W}\right\|_F^2. \tag{11.28}$$

Let

$$\mathbf{Q}^T\mathbf{Y}^T = \begin{bmatrix} \widetilde{\mathbf{B}} \\ \overline{\mathbf{B}} \end{bmatrix}, \tag{11.29}$$

where $\widetilde{\mathbf{B}} = \left[\tilde{b}_{ij}\right]$ and $\overline{\mathbf{B}} = \left[\overline{b}_{ij}\right]$ are, respectively, a $J_2 \times J_3$ and an $(N - J_2) \times J_3$ matrices. We then have

$$E = \frac{1}{N} \left\| \begin{bmatrix} \widetilde{\mathbf{B}} \\ \overline{\mathbf{B}} \end{bmatrix} - \begin{bmatrix} \mathbf{R} \\ \mathbf{O} \end{bmatrix} \mathbf{W} \right\|_F^2 = \frac{1}{N} \left\| \begin{bmatrix} \widetilde{\mathbf{B}} - \mathbf{RW} \\ \overline{\mathbf{B}} \end{bmatrix} \right\|_F^2 . \tag{11.30}$$

Thus, the optimal $\mathbf{W}$ that minimizes E is derived from

$$\mathbf{RW} = \widetilde{\mathbf{B}}. \tag{11.31}$$

In this case, the residual

$$E = \frac{1}{N} \left\| \overline{\mathbf{B}} \right\|_F^2 . \tag{11.32}$$

This is batch OLS.

Due to the orthogonalization procedure, it is very convenient to implement the forward and backward center selection approaches. The forward selection approach is to build up a network by adding, one at a time, centers at the data points that result in the largest decrease in the network output error at each stage. The backward selection algorithm is an alternative approach that sequentially removes from the network, one at a time, those centers that cause the smallest increase in the residual.

The error reduction ratio due to the kth RBF neuron is defined by [15]

$$ERR_k = \frac{\left(\sum_{i=1}^{J_3} \tilde{b}_{ki}^2\right) \boldsymbol{q}_k^T \boldsymbol{q}_k}{\mathrm{tr}\left(\mathbf{Y}\mathbf{Y}^T\right)}, \quad k = 1, \ldots, N. \tag{11.33}$$

Error reduction ratio is a performance-oriented criterion. RBF network training can be in a constructive way and the centers with the largest error reduction ratio values are recruited until

$$1 - \sum_{k=1}^{J_2} ERR_k < \rho, \tag{11.34}$$

where $\rho \in (0, 1)$ is a tolerance.

An alternative terminating criterion can be based on AIC [15], which balances between the performance and the complexity. The weights are determined at the same time. The criterion used to stop center selection is a simple threshold on the error reduction ratio. To improve generalization, regularized forward OLS methods can be implemented by penalizing large weights [72].

The computational complexity of the orthogonal decomposition of an information matrix $\mathbf{\Phi}^T$ is $O\left(N J_2^2\right)$. When the size of a training dataset N is large, batch OLS is computationally demanding and also needs a large amount of computer memory.

The RBF center clustering method based on the Fisher ratio class separability measure [63] is similar to the forward selection OLS algorithm [14, 15].

Recursive OLS algorithms have been proposed for updating the weights of single-input single-output [6] and multi-input multi-output systems [31, 106]. Recursive OLS can be used to sequentially select the centers to minimize the network output error. Forward and backward center selection methods are developed from this information, and Akaike's final prediction error criterion is used in model selection [31].

11.9 Supervised Learning of All Parameters

The preceding methods for selecting the network parameters are practical, but by no means optimal. The gradient-descent method is the simplest method for finding the minimum value of E. In this section, we apply the gradient-descent-based supervised methods to RBF network learning.

11.9.1 Supervised Learning for General RBF Networks

To derive the supervised learning algorithm for the general RBF network, we rewrite the error function (11.5) as

$$E = \frac{1}{N} \sum_{n=1}^{N} \sum_{i=1}^{J_3} \left(e_{n,i}\right)^2, \tag{11.35}$$

where $e_{n,i}$ is the approximation error at the ith output node for the nth example

$$e_{n,i} = y_{n,i} - \sum_{m=1}^{J_2} w_{mi} \phi\left(\|\boldsymbol{x}_n - \boldsymbol{c}_m\|\right) = y_{n,i} - \boldsymbol{w}_i^T \boldsymbol{\phi}_n. \tag{11.36}$$

Taking the derivative of E with respect to w_{mi} and $\boldsymbol{c}_m$, respectively, we have

$$\frac{\partial E}{\partial w_{mi}} = -\frac{2}{N} \sum_{n=1}^{N} e_{n,i} \phi\left(\|\boldsymbol{x}_n - \boldsymbol{c}_m\|\right), \quad m = 1, \ldots, J_2, i = 1, \ldots, J_3, \tag{11.37}$$

$$\frac{\partial E}{\partial \boldsymbol{c}_m} = \frac{2}{N} \sum_{i=1}^{J_3} w_{mi} \sum_{n=1}^{N} e_{n,i} \dot{\phi}\left(\|\boldsymbol{x}_n - \boldsymbol{c}_m\|\right) \frac{\boldsymbol{x}_n - \boldsymbol{c}_m}{\|\boldsymbol{x}_n - \boldsymbol{c}_m\|}, \quad m = 1, \ldots, J_2, \tag{11.38}$$

where $\dot{\phi}(\cdot)$ is the first derivative of $\phi(\cdot)$.

The gradient-descent method is defined by the update equations

$$\Delta w_{mi} = -\eta_1 \frac{\partial E}{\partial w_{mi}}, \tag{11.39}$$

$$\Delta \boldsymbol{c}_m = -\eta_2 \frac{\partial E}{\partial \boldsymbol{c}_m}, \tag{11.40}$$

where η_1 and η_2 are learning rates.

To prevent the situation that two or more centers are too close or coincide with one another during the learning process, one can add a term such as $\sum_{\alpha \neq \beta} \psi \left(\left\| \boldsymbol{c}_\alpha - \boldsymbol{c}_\beta \right\| \right)$ to E, where $\psi(\cdot)$ is an appropriate repulsive potential. The gradient-descent method given by (11.39) and (11.40) is modified accordingly.

A simple strategy for initialization is to select the RBF centers based on a random subset of the examples and the weights $\mathbf{W}$ as a matrix with small random components. To accelerate the search process, one can use clustering to find the initial RBF centers and LS to find the initial weights, and then apply the gradient-descent procedure to refine the learning result.

Setting the gradients to zero, the optimal solutions to the weights and centers can be derived. The gradient-descent procedure is the iterative approximation to the optimal solutions. For each sample n, if we set $e_{n,i} = 0$, then the right-hand side of (11.37) is zero, we then achieve the global optimum and get

$$\boldsymbol{y}_n = \mathbf{W}^T \boldsymbol{\phi}_n. \tag{11.41}$$

For all samples,

$$\mathbf{Y} = \mathbf{W}^T \boldsymbol{\Phi}. \tag{11.42}$$

This is exactly the same set of linear equations as (11.4). The optimum solution to weights is given by (11.21). Equating (11.38) to zero leads to

$$\boldsymbol{c}_m = \frac{\sum_{i=1}^{J_3} w_{mi} \sum_{n=1}^{N} \frac{e_{n,i} \dot{\phi}_{mn}}{\|\boldsymbol{x}_n - \boldsymbol{c}_m\|} \boldsymbol{x}_n}{\sum_{i=1}^{J_3} w_{mi} \sum_{n=1}^{N} \frac{e_{n,i} \dot{\phi}_{mn}}{\|\boldsymbol{x}_n - \boldsymbol{c}_m\|}}, \tag{11.43}$$

where $\dot{\phi}_{mn} = \dot{\phi} \left(\|\boldsymbol{x}_n - \boldsymbol{c}_m\| \right)$. Thus, the optimal centers are weighted sums of the data points, corresponding to a task-dependent clustering problem.

11.9.2 Supervised Learning for Gaussian RBF Networks

For the Gaussian RBF network, the RBF at each center can be assigned a different width σ_i

$$\phi_i(\boldsymbol{x}) = \mathrm{e}^{-\frac{\|\boldsymbol{x} - \boldsymbol{c}_i\|^2}{2\sigma_i^2}}. \tag{11.44}$$

The RBFs can be further generalized:

$$\phi_i(\boldsymbol{x}) = \mathrm{e}^{-\frac{1}{2}(\boldsymbol{x} - \boldsymbol{c}_i)^T \Sigma_i^{-1} (\boldsymbol{x} - \boldsymbol{c}_i)}, \tag{11.45}$$

where $\Sigma_i \in R^{J_1 \times J_1}$ is a positive-definite, symmetric covariance matrix. When Σ_i^{-1} is in general form, the shape and orientation of the axes of the hyperellipsoid are arbitrary in the feature space.

If Σ_i^{-1} is a diagonal matrix with nonconstant diagonal elements, Σ_i^{-1} is completely defined by a vector $\boldsymbol{\sigma}_i \in R^{J_1}$, and each ϕ_i is a hyperellipsoid whose axes are along the axes of the feature space

$$\Sigma_i^{-1} = \text{diag}\left(\frac{1}{\sigma_{i,1}^2}, \ldots, \frac{1}{\sigma_{i,J_1}^2}\right). \tag{11.46}$$

For the J_1-dimensional input space, each RBF of the form (11.45) has a total of $\frac{J_1(J_1+3)}{2}$ independent adjustable parameters, while each RBF of the form (11.44) and each RBF of the form (11.46) have $J_1 + 1$ and $2J_1$ independent parameters, respectively. There is a trade-off between using a small network with many adjustable parameters and using a large network with fewer adjustable parameters.

The gradients $\frac{\partial E}{\partial c_m}$ and $\frac{\partial E}{\partial \sigma_m}$ for the RBF (11.44), and the gradients $\frac{\partial E}{\partial c_{m,j}}$ and $\frac{\partial E}{\partial \sigma_{m,j}}$ for the RBF (11.46) can be derived accordingly. The results are given in [26, 27]. Adaptations for $\boldsymbol{c}_i$ and Σ_i are along the negative gradient directions.

The weights $\mathbf{W}$ are updated by (11.37) and (11.39). To prevent unreasonable radii, the updating algorithms can also be derived by adding to E a constraint term that penalizes small radii, $E_c = \sum_i \frac{1}{\sigma_i}$ or $E_c = \sum_{i,j} \frac{1}{\sigma_{i,j}}$.

In [103], the improved LM algorithm [102] is applied for training RBF networks to adjust all the parameters. The proposed improved second-order algorithm can normally reach smaller training/testing error with much less number of RBF units. During the computation process, quasi-Hessian matrix and gradient vector are accumulated as the sum of related submatrices and vectors, respectively. Only one Jacobian row is stored and used for multiplication, instead of the entire Jacobian matrix storage and multiplication.

11.9.3 Discussion on Supervised Learning

The gradient-descent algorithms introduced thus far are batch learning algorithms. As discussed in Chap. 5, by dropping $\frac{1}{N}\sum_{p=1}^{N}$ in the error function E and accordingly in the algorithms, one can update the parameters at each example $\left(\boldsymbol{x}_p, \boldsymbol{y}_p\right)$. This yields incremental learning algorithms, which are typically much faster than their batch counterparts for suitably selected learning parameters.

Although the RBF network trained by the gradient-descent method is capable of providing equivalent or better performance compared to that of the MLP trained with BP, the training time for the two methods are comparable [100]. The gradient-descent method is slow in convergence since it cannot efficiently use the locally tuned representation of the hidden-layer units. When the hidden unit receptive fields, controlled by the widths σ_i, are narrow, for a given input only a few of the total number of hidden units will be activated and hence only these units need to be updated. However, in the gradient-descent method, there is no limitation on σ_i, and thus there is no guarantee that the RBF network remains localized after supervised learning [68]. As a result, the computational advantage of locality is not utilized.

The gradient-descent method is prone to finding local minima of the error function. For reasonably well-localized RBF, an input will generate a significant activation in a small region, and the opportunity of getting stuck at a local minimum is small. Unsupervised methods can be used to determine σ_i. Unsupervised learning is used to initialize the network parameters, and supervised learning is usually used for the fine-tuning of the network parameters. The ultimate RBF network learning is typically a blend of unsupervised and supervised algorithms.

11.10 Various Learning Methods

All general-purpose unconstrained optimization methods, including those introduced for the MLP, are applicable for RBF network learning, since the RBF network is a special feedforward network and RBF network learning is an unconstrained optimization problem. These include popular second-order approaches like LM, CG, BFGS, and EKF, and heuristic global optimization methods.

The LM method is used for RBF network learning [32, 64, 75]. In [64, 75], the LM method is used for estimating nonlinear parameters, and the SVD-based LS method is used for linear weight estimation at each iteration. In [75], at each iteration, the weights are updated many times during the process of looking for the search direction to update the nonlinear parameters.

EKF can be used for RBF network learning [87]. After the number of centers is chosen, EKF simultaneously solves for the prototype vectors and the weight matrix. A decoupled EKF further decreases the computational complexity of the training algorithm [87]. In [22], a pair of parallel running extended Kalman filters are used to sequentially update both the output weights and the RBF centers.

In [47], the RBF network is reformulated by using RBFs formed in terms of admissible generator functions and provides a fully supervised gradient-descent training method. LP models with polynomial time complexity are also employed to train the RBF network [82]. In [33], a multiplication-free Gaussian RBF network with a gradient-based nonlinear learning algorithm is described for adaptive function approximation.

The learning algorithm for training cosine RBF networks given in [48] trains reformulated RBF networks by updating selected adjustable parameters to minimize the class-conditional variances at the outputs of their RBFs. Cosine RBF networks trained by such a learning algorithm are capable of identifying uncertainty in data classification. The classification accuracy of cosine RBF networks is also improved by rejecting ambiguous feature vectors based on their responses.

The RBF network using regression weights can significantly reduce the number of hidden units and is effectively used for approximating nonlinear dynamic systems

[51, 80, 85]. For a J_1-J_2-1 RBF network, the linear regression weights are defined by [51]

$$w_i = \boldsymbol{a}_i^T \tilde{\boldsymbol{x}} + \xi_i, \quad i = 1, \ldots, J_2, \tag{11.47}$$

where w_i is the weight from the ith hidden unit to the output unit, $\boldsymbol{a}_i = \left(a_{i,0}, a_{i,1}, \ldots, a_{i,J_1}\right)^T$ is the regression parameter vector, $\tilde{\boldsymbol{x}} = \left(1, \boldsymbol{x}^T\right)^T$ is the augmented input vector, and ξ_i is a zero-mean Gaussian white-noise process. For the Gaussian RBF network, the RBF centers $\boldsymbol{c}_i$ and their widths σ_i can be selected by C-means and the nearest neighbor heuristic [68], while the parameters of the regression weights are estimated by the EM method [51]. In [85], a simple but fast computational procedure is achieved by using a high-dimensional raised-cosine RBF. Storage space is also reduced by allowing the RBF centers to be situated at a nonuniform grid of points.

Some regularization techniques for improving the generalization capability of the MLP and the RBF network have been discussed in [65]. As in the MLP, the favored weight quadratic penalty term $\sum w_{ij}^2$ is also appropriate for the RBF network. The widths of the RBFs are widely known to be a major source of ill-conditioning in RBF network training, and large width parameters are desirable for better generalization. Some suitable penalty terms for widths are given in [65].

Hyper basis function (HyperBF) networks [77] are generalized RBF networks with a radial function of a Mahalanobis-like distance. HyperBF networks can be constructed with three learning phases [86], where regular two-phase methods are used to initialize the network and in the third phase, means, scaling factors, and weights are estimated simultaneously by gradient-descent and backpropagation using a single variable learning factor for all parameters that are estimated adaptively. In a regularization method that performs soft local dimension reduction in addition to weight decay [62], hierarchical clustering is used to initialize neurons followed by a multiple-step-size gradient optimization using a scaled version of Rprop with a localized partial backtracking step. The training provides faster and smoother convergence than regular Rprop.

The probabilistic RBF network [93] constitutes a probabilistic version of the RBF network for classification that extends the typical mixture model approach to classification by allowing the sharing of mixture components among all classes. It is an alternative approach for class-conditional density estimation. A typical learning method employs the EM algorithm and depends strongly on the initial parameter values. In [23], a technique for incremental training of the probabilistic RBF network for classification is proposed, based on criteria for detecting a region that is crucial for the classification task. After the addition of all components, the algorithm splits every component of the network into subcomponents, each corresponding to a different class.

When a training set contains outliers, robust statistics can be applied for robust learning of the RBF network. Robust learning algorithms are usually derived from the M-estimator method [21, 84]. The robust RBF network learning algorithm [84] is based on Hampel's tanh-estimator function. The network architecture is initialized by using the conventional SVD-based learning method. The robust part of the learning method is implemented iteratively using the CG method. The annealing robust RBF network [21] improves the robustness of the RBF network against outliers for function approximation by using the M-estimator and the annealing robust learning algorithm [20]. The median RBF algorithm [8] is based on robust parameter estimation of the RBF centers and employs the Mahalanobis distance.

11.11 Normalized RBF Networks

The normalized RBF network is defined by normalizing the vector composing the responses of all the RBF units [68]

$$y_i(\boldsymbol{x}) = \sum_{k=1}^{J_2} w_{ki}\hat{\phi}_k(\boldsymbol{x}), \quad i = 1, \ldots, J_3, \tag{11.48}$$

where

$$\hat{\phi}_k(\boldsymbol{x}) = \frac{\phi\left(\boldsymbol{x} - \boldsymbol{c}_k\right)}{\sum_{j=1}^{J_2} \phi\left(\boldsymbol{x} - \boldsymbol{c}_j\right)}. \tag{11.49}$$

The normalization operation is nonlocal, since each hidden node is required to know about the outputs of other hidden nodes. Hence, the convergence process is computationally costly.

The normalized RBF network given by (11.48) can be presented in another form [11, 51]. The network output is defined by

$$y_i(\boldsymbol{x}) = \frac{\sum_{j=1}^{J_2} w_{ji}\phi\left(\boldsymbol{x} - \boldsymbol{c}_j\right)}{\sum_{j=1}^{J_2} \phi\left(\boldsymbol{x} - \boldsymbol{c}_j\right)}. \tag{11.50}$$

Now, normalization is performed in the output layer. As it already receives information from all the hidden units, the locality of the computational processes is preserved. The two forms of the normalized RBF network, (11.48) and (11.50), are equivalent.

In the normalized RBF network of the form (11.50), the traditional roles of the weights and activities in the hidden layer are exchanged. In the RBF network, the weights determine as to how much each hidden node contributes to the output, while in the normalized RBF network, the activities of the hidden nodes determine which weights contribute most to the output. The normalized RBF network provides better smoothness than the RBF network does. Due to the localized property of the receptive fields, for most data points, there is usually only one hidden node that contributes significantly to (11.50). The normalized RBF network (11.50) can be trained using a procedure similar to that for the RBF network. The normalized Gaussian RBF network exhibits superiority in supervised classification due to its soft modification rule. It is also a universal approximator in the space of continuous functions with compact support in the space $L^p\left(R^p, \mathrm{d}\boldsymbol{x}\right)$ [4].

The normalized RBF network is an RBF network with a quasi-linear activation function with a squashing coefficient decided by the activations of all the hidden units. The normalized RBF network loses the localized characteristics of the localized RBF network and exhibits excellent generalization properties. Thus, it softens the curse of dimensionality associated with localized RBF networks [11]. The normalized Gaussian RBF network outperforms the Gaussian RBF network in terms of training and generalization errors and exhibits a more uniform error over the training domain. In addition, the normalized Gaussian RBF network is not sensitive to the RBF widths.

11.12 Optimizing Network Structure

The optimum structure of an RBF network is to determine the number and locations of the RBF centers automatically by using constructive and pruning methods.

11.12.1 Constructive Methods

The constructive approach gradually increases the number of RBF centers until a criterion is satisfied. In [46], a new prototype is created in a region of the input space by splitting an existing prototype $\boldsymbol{c}_j$ selected by a splitting criterion, and splitting is performed by adding the perturbation vectors $\pm\epsilon_j$ to $\boldsymbol{c}_j$. The resulting vectors $\boldsymbol{c}_j \pm \epsilon_j$ together with the existing centers form the initial set of centers for the next growing cycle. Existing algorithms for updating the centers $\boldsymbol{c}_j$, widths σ_j, and weights can be used. The process continues until a stopping criterion is satisfied. In a heuristic incremental algorithm [28], the training phase adds a hidden node $\boldsymbol{c}_t$ at each epoch t by an error-driven rule.

The incremental RBF network architecture using hierarchical gridding of the input space [7] allows for a uniform approximation without wasting resources. Additional layers of Gaussians at lower scales are added where the residual error is higher. The method shows a high accuracy in the reconstruction, and it can deal with nonevenly spaced data points and is fully parallelizable.

Hierarchical RBF network [29] is a multiscale version of the RBF network. It is constituted by hierarchical layers, each containing a Gaussian grid at a decreasing scale. The grids are not completely filled, but units are inserted only where the local error is over a threshold. The constructive approach is based only on the local operations, which do not require any iteration on the data. It allows for an effective network to be built in a very short time. The coarse-to-fine approach enables the hierarchical RBF network to grow until the reconstructed surface meets the required quality.

The forward OLS algorithm [13] is a well-known constructive algorithm. Based on the OLS algorithm, a constructive algorithm for the generalized Gaussian RBF network is given in [96]. RBF network learning based on a modification to the cascade-correlation algorithm works in a way similar to OLS method but with a significantly faster convergence [56].

The dynamic decay adjustment algorithm is a fast constructive training method for the RBF network when used for classification [5]. It has independent adjustment for the decay factor or width σ_i of each prototype. The method is faster and also achieves higher classification accuracy than the RBF network does.

Error correction algorithm [107] is an offline algorithm for incrementally constructing and training RBF networks. In each iteration, one RBF unit is added to fit and then eliminate the highest peak (or lowest valley) in the error surface. Each new RBF unit is trained using an improved LM method to compensate for residual errors. This process is repeated until a desired error level is reached. The method generates very compact networks. It is possible to reach an acceptable solution just with one try.

Constructive Methods with Pruning

The normalized RBF network can be trained by using a constructive method with pruning strategy based on the novelty of the data and the overall behavior of the network [80]. The network starts from one neuron and adds a new neuron if an example passes two novelty criteria, until a specified maximum number of neurons is reached. The first criterion is the same as (11.52), and the second one deals with the activation of the nonlinear neurons $\max_i \phi_i(\boldsymbol{x}_t) < \zeta$, where ζ is a threshold. A sequential learning algorithm is derived from the gradient-descent method. After the whole pattern set is presented at an epoch, the algorithm starts to remove those neurons that meet any of the three cases, namely, neurons with a very small mean activation for the whole pattern set, neurons with a very small activation region, or neurons having an activation very similar to that of other neurons.

The dynamic decay adjustment [5] may result in too many neurons. Dynamic decay adjustment with temporary neurons [73] introduces online pruning of neurons after each dynamic decay adjustment training epoch. After each training epoch, if the individual neurons cover a sufficient number of samples, they are marked as *permanent*; otherwise, they are deleted. In dynamic decay adjustment with selective pruning and model selection [71], only a portion of the neurons that cover only one training example is pruned and pruning is carried out only after the last epoch of the dynamic decay adjustment training. The method improves the generalization performance of dynamic decay adjustment and dynamic decay adjustment with temporary neurons but yields a larger network size than dynamic decay adjustment with temporary neurons.

The resource-allocating network (RAN) [76] and RAN algorithms with pruning strategy are well-known RBF network construction methods, which are described in Sect. 11.12.2.

11.12.2 Resource-Allocating Networks

RAN [76] is a sequential learning method for the localized RBF network such as the Gaussian RBF network, which is suitable for online modeling of nonstationary processes. The network begins with no hidden units. As pattern pairs are received during training, a new hidden unit may be added according to the novelty in the data. The novelty in the data is defined by two conditions

$$\|\boldsymbol{x}_t - \boldsymbol{c}_i\| > \varepsilon(t), \tag{11.51}$$

$$\|\boldsymbol{e}(t)\| = \left\|\boldsymbol{y}_t - \boldsymbol{f}(\boldsymbol{x}_t)\right\| > e_{\min}, \tag{11.52}$$

where $\boldsymbol{c}_i$ is the center nearest to $\boldsymbol{x}_t$, the prediction error $\boldsymbol{e} = \left(e_1, \ldots, e_{J_3}\right)^T$, and $\varepsilon(t)$ and $e_{\min}$ are thresholds to be selected appropriately. The algorithm starts with $\varepsilon(t) = \varepsilon_{\max}$, where $\varepsilon_{\max}$ is chosen as the largest scale in the input space, typically the entire input space of nonzero probability. The distance $\varepsilon(t)$ shrinks exponentially as $\varepsilon(t) = \max\left\{\varepsilon_{\max} e^{-\frac{t}{\tau}}, \varepsilon_{\min}\right\}$, where τ is a decay constant. $\varepsilon(t)$ is decayed until it reaches $\varepsilon_{\min}$. Assuming that there are k nodes at time $t-1$, for the Gaussian RBF network, the newly added hidden unit at time t can be initialized as

$$\boldsymbol{c}_{k+1} = \boldsymbol{x}_t, \tag{11.53}$$

$$w_{(k+1)j} = e_j(t), \quad j = 1, \ldots, J_3, \tag{11.54}$$

$$\sigma_{k+1} = \alpha \|\boldsymbol{x}_t - \boldsymbol{c}_i\|, \tag{11.55}$$

where σ_{k+1} is selected based on the nearest neighbor heuristic and α is a parameter defining the size of neighborhood. If pattern pair $\left(\boldsymbol{x}_t, \boldsymbol{y}_t\right)$ does not pass the novelty criteria, no hidden unit is added and the existing network parameters are adapted using the LMS method. The RAN method performs much better than RBF network learning using random centers and that using the centers clustered by C-means [68] do in terms of network size and MSE. It achieves roughly the same performance as the MLP trained with BP does, but with much less computation.

EKF-based RAN [43] replaces the LMS method by the EKF method for the network parameter adaptation so as to generate a more parsimonious network. Two geometric criteria, namely, the prediction error criterion, which is the same as (11.52), and the angle criterion, are also obtained from a geometric viewpoint. The angle criterion attempts to assign RBFs that are nearly orthogonal to all the other existing RBFs. These criteria are proved equivalent to Platt's criteria [76]. In [44], the statistical novelty criterion is defined. By using the EKF method and using the statistical novelty criterion to replace the criteria (11.51) and (11.52), for a given task, more compact networks and smaller MSEs are achieved than RAN and EKF-based RAN.

Resource-Allocating Networks with Pruning

The RAN method can be improved by integrating node-pruning procedure [37, 38, 81, 83, 94, 104, 105]. Minimal RAN [104, 105] is based on EKF-based RAN and achieves a more compact network with equivalent or better accuracy by incorporating a pruning strategy to remove inactive nodes and augmenting the basic growth criterion of RAN. The output of each RBF unit is scaled as

$$\hat{o}_i(\boldsymbol{x}) = \frac{|o_i(\boldsymbol{x})|}{\max_{1 \le j \le J_2} \left\{ \left| o_j(\boldsymbol{x}) \right| \right\}}, \quad i = 1, \ldots, J_2. \tag{11.56}$$

If $\hat{o}_i(\boldsymbol{x})$ is below a predefined threshold δ for a given number of iterations, this node is idle and can be removed. For a given accuracy, the minimal RAN achieves a smaller complexity than the MLP trained with Rprop does, and achieves a more compact network and requiring less training time than the MLP constructed by dependence identification does.

In [81], RAN is improved by using the Givens QR decomposition-based RLS for the adaptation of the weights and integrating a node-pruning strategy. The error reduction ratio criterion in [14] is used to select the most important regressors. In [83], RAN is improved by using in each iteration the combination of the SVD and QR-cp methods for determining the structure as well as for pruning the network. SVD and QR-cp determine a subset of RBFs that is relevant to the linear output combination. In the early phase of learning, the addition of RBFs is in small groups, and this leads to an increased rate of convergence.

The growing and pruning algorithm for RBF (GAP-RBF) [37] and the generalized GAP-RBF [38] are RAN-based sequential learning algorithms for realizing parsimonious RBF networks. These algorithms make use of the notion of *significance* of a hidden neuron, which is defined as a neuron's statistical contribution over all the inputs seen so far to the overall performance of the network. In addition to the two growing criteria of RAN, a new neuron is added only when its significance is also above a chosen learning accuracy. If during training the significance of a neuron becomes less than the learning accuracy, that neuron will be pruned. For each new pattern, only its nearest neuron is checked for growing, pruning, or updating using EKF. Generalized GAP-RBF enhances the significance criterion such that it is applicable for training examples with arbitrary sampling density. GAP-RBF and generalized GAP-RBF outperform RAN, EKF-based RAN, and minimal RAN in terms of learning speed, network size, and generalization performance.

In [94], EKF-based RAN with statistical novelty criterion [44] is extended by incorporating an online pruning procedure, which is derived using the parameters and innovation statistics estimated from EKF. The online pruning method is analogous to saliency-based OBS and OBD. IncNet and IncNet Pro [41] are RAN-EKF networks with statistically controlled growth criterion. The pruning method is similar to OBS, but based on the result of the EKF algorithm.

11.12.3 Pruning Methods

Well-known pruning methods are OBD and OBS, which are described in Chap. 5. Pruning algorithms based on the regularization technique are also popular, since additional terms that penalize the complexity of the network are incorporated into the MSE criterion.

The pruning method proposed in [57] starts from a big RBF network and achieves a compact network through an iterative procedure of training and selection. The training procedure adaptively changes the centers and the width of the RBFs and trains the linear weights. The selection procedure performs the elimination of the redundant RBFs using an objective function based on the MDL principle. In [69], all the data vectors are initially selected as centers. Redundant centers are eliminated by merging two centers at each adaptation cycle by using an iterative clustering method. The technique is superior to the traditional RBF network algorithms, particularly in terms of processing speed and the solvability of nonlinear patterns. In [49], two methods are described for reducing the size of the probabilistic neural network while preserving the classification performance as good as possible.

In classical training methods for node-open fault, one needs to consider many potential faulty networks. In [58], the Kullback–Leibler divergence is used to define an objective function for improving the fault tolerance of RBF networks. Compared with some conventional approaches, including weight-decay-based regularizers, this approach has better fault-tolerant ability. In [90], an objective function is presented for training a functional-link network to tolerate multiplicative weight noise. Under some mild conditions, the derived regularizer is essentially the same as a weight-decay regularizer. This explains why applying weight decay can also improve the fault-tolerant ability of an RBF with multiplicative weight noise.

Sparse RBF networks for classification problems [79] apply a two-phase construction algorithm by using L_1 regularization. An improved maximum data coverage algorithm is first implemented for the initialization of RBF centers and widths. Then a specialized orthant-wise limited-memory quasi-Newton method is employed to perform simultaneous network pruning and parameter optimization. Better generalization performance is guaranteed with higher model sparsity.

11.13 Complex RBF Networks

In a complex RBF network, the input and output of the network are complex values, whereas the activation function of the hidden nodes is the same as that for the RBF network. The Euclidean distance in the complex domain is defined by [17]

$$d\left(\boldsymbol{x}_t, \boldsymbol{c}_i\right) = \left[\left(\boldsymbol{x}_t - \boldsymbol{c}_i\right)^H \left(\boldsymbol{x}_t - \boldsymbol{c}_i\right)\right]^{\frac{1}{2}}, \tag{11.57}$$

where $\boldsymbol{c}_i$ is a J_1-dimensional complex center vector. The output weights are complex-valued. Most of the existing RBF network learning algorithms can be easily extended for training various versions of the complex RBF network [12, 17, 42, 54]. When using clustering techniques to determine the RBF centers, the similarity measure can be based on the distance defined by (11.57). The Gaussian RBF is usually used in the complex RBF network.

The Mahalanobis distance (11.45) defined for the Gaussian RBF can be extended to the complex domain [54]

$$d\left(\boldsymbol{x}_t, \boldsymbol{c}_i\right) = \left(\left[\boldsymbol{x}_t - \boldsymbol{c}_i(t-1)\right]^H \boldsymbol{\Sigma}_i^{-1}(t-1)\left[\boldsymbol{x}_t - \boldsymbol{c}_i(t-1)\right]\right)^{\frac{1}{2}}, \quad i = 1, \ldots, J_2. \tag{11.58}$$

Notice that transpose T in (11.45) is changed into Hermitian transpose H.

Learning of the complex Gaussian RBF network can be performed in two phases, where the RBF centers are first selected by using incremental C-means and the weights are then solved by fixing the RBF parameters [54]. At each iteration t, C-means first finds the winning node $\boldsymbol{c}_w$ by using the nearest neighbor paradigm, and then updates both the center and the variance of the winning node by

$$\boldsymbol{c}_w(t) = \boldsymbol{c}_w(t-1) + \eta\left[\boldsymbol{x}_t - \boldsymbol{c}_w(t-1)\right], \tag{11.59}$$

$$\Sigma_w(t) = \Sigma_w(t-1) + \eta\left[\boldsymbol{x}_t - \boldsymbol{c}_w(t-1)\right]\left[\boldsymbol{x}_t - \boldsymbol{c}_w(t-1)\right]^H, \tag{11.60}$$

where η is the learning rate. C-means is repeated until the changes in all $\boldsymbol{c}_i(t)$ and $\Sigma_i(t)$ are within specified accuracy, that is,

$$\|\boldsymbol{c}_i(t) - \boldsymbol{c}_i(t-1)\| \leq \varepsilon_0, \tag{11.61}$$

$$\|\Sigma_i(t) - \Sigma_i(t-1)\|_F \leq \varepsilon_1, \tag{11.62}$$

where ε_0 and ε_1 are predefined small positive numbers. After complex RBF centers are determined, the weight matrix $\mathbf{W}$ is determined using the LS or the RLS algorithm.

In the complex-valued RBF network [18], each RBF node has a real-valued response that can be interpreted as a conditional pdf. Because the RBF node's response is real-valued, this complex-valued RBF network is essentially two separate real-valued RBF networks.

A fully complex-valued RBF network [19] has a complex-valued response at each RBF node. For regression problems, the locally regularized OLS algorithm aided with the D-optimality experimental design is extended to the fully complex-valued RBF network. A complex-valued orthogonal forward selection algorithm based on the multiclass Fisher ratio of class separability measure is derived for constructing sparse complex-valued RBF classifiers that generalize well.

The complex RBF network [12] adopts the stochastic gradient learning algorithm to adjust the parameters. A complex-valued minimal RAN equalizer is developed in [42]. Applying the growing and pruning criteria, it realizes a more compact structure and obtains better performance than complex RBF and many other equalizers do. Although the inputs and centers of complex RBF and complex-valued minimal RAN are complex-valued, the basis functions still remain real-valued.

Complex-valued self-regulating RAN [91] is an incremental learning algorithm for a complex-valued RAN with a self-regulating scheme to select the appropriate number of hidden neurons. It uses a *sech* activation function in the hidden layer. The network is updated using a complex-valued EKF algorithm.

The fully complex ELM [59] uses any ETF as activation function. The fully complex ELM-based channel equalizer significantly outperforms other equalizers based on complex-valued minimal RAN, complex RBF network [12], and complex BP in terms of symbol error rate and learning speed.

11.14 A Comparison of RBF Networks and MLPs

Both the MLP and the RBF network are used for supervised learning. In the RBF network, the activation of an RBF unit is determined by the distance between the input and prototype vectors. For classification problems, RBF units map input patterns from a nonlinearly separable space to a linearly separable space, and the responses of the RBF units form new feature vectors. Each RBF prototype is a cluster serving mainly a certain class. When the MLP with a linear output layer is applied to classification problems, minimizing the error at the output of the network is equivalent to maximizing the so-called *network discriminant function* at the output of the hidden units [98]. A comparison between the MLP and the localized RBF network is as follows.

Global Method Versus Local Method

The use of the sigmoidal activation function makes the MLP a global method. For an input pattern, many hidden units will contribute to the network output. On the other hand, in the localized RBF network, each localized RBF covers a very small local zone. The local method satisfies the minimal disturbance principle [101], that is, the adaptation not only reduces the output error for the current example but also minimizes disturbance to those already learned.

Local Minima

Due to the sigmoidal function, the crosscoupling between hidden units of the MLP or recurrent networks results in high nonlinearity in the error surface, resulting in the problem of local minima or nearly flat regions. This problem gets worse as the network size increases. In contrast, the RBF network has a simple architecture with linear weights, and therefore has a unique solution to the weights.

Approximation and Generalization

Due to the global activation function, the MLP has greater generalization for each training example, and thus the MLP is a good candidate for extrapolation. On the contrary, the extension of a localized RBF to its neighborhood is determined by its variance. This localized property prevents the RBF network from extrapolation beyond the training data.

Network Resources and Curse of Dimensionality

The localized RBF network, like most kernel-type approximation methods, suffers from the problem of curse of dimensionality. It typically requires much more data and more hidden units to achieve an accuracy similar to that of the MLP. In order to approximate a wide class of smooth functions, the number of hidden units required for the three-layer MLP is polynomial with respect to the input dimensions, while that for the localized RBF network is exponential [3]. The curse of dimensionality can be alleviated by using smaller networks with more adaptive parameters [77] or by progressive learning [25]. This requires a high number of training data and often leads to a poor ability to generalize.

Hyperplanes Versus Hyperellipsoids

For the MLP, the response of a hidden unit is constant on a surface that consists of parallel $(J_1 - 1)$-dimensional hyperplanes. As a result, the MLP is preferable for linearly separable problems. On the other hand, in the RBF network, the activation of the hidden units is constant on concentric $(J_1 - 1)$-dimensional hyperspheres or hyperellipsoids. Thus, the RBF network may be more efficient for linearly inseparable classification problems.

Training Speed and Performing Speed

The error surface of the MLP has many local minima or large flat regions called *plateaus*, which lead to slow convergence of the training process. The MLP frequently gets trapped at local minima. For the localized RBF network, only a few hidden units have significant activations for a given input, and thus the network modifies the weights only in the vicinity of the sample point and retains constant weights in the other regions. The RBF network requires orders of magnitude less training time for convergence than the MLP trained with BP to achieve comparable performance [9, 68]. For equivalent generalization performance, the MLP requires far fewer hidden units than the localized RBF network, and thus the trained MLP is much faster in performing.

Generally speaking, the MLP is a better choice if the training data are expensive. However, when the training data are cheap and plentiful or online training is required, the RBF network is desirable.

Some properties of the MLP and the RBF network are combined for modeling purposes. In the centroid-based MLP [55], a centroid layer is inserted into the MLP as the second layer. The conic-sectional function network [24] generalizes the activation function to include both the bounded (hypersphere) and unbounded (hyperplane) decision regions in one network. It can make automatic decisions with respect to the two decision regions. It combines the speed of the RBF network and the error minimization of the MLP. In [99], a hybrid RBF sigmoid neural network with a three-step training algorithm that utilizes both global search and gradient-descent training

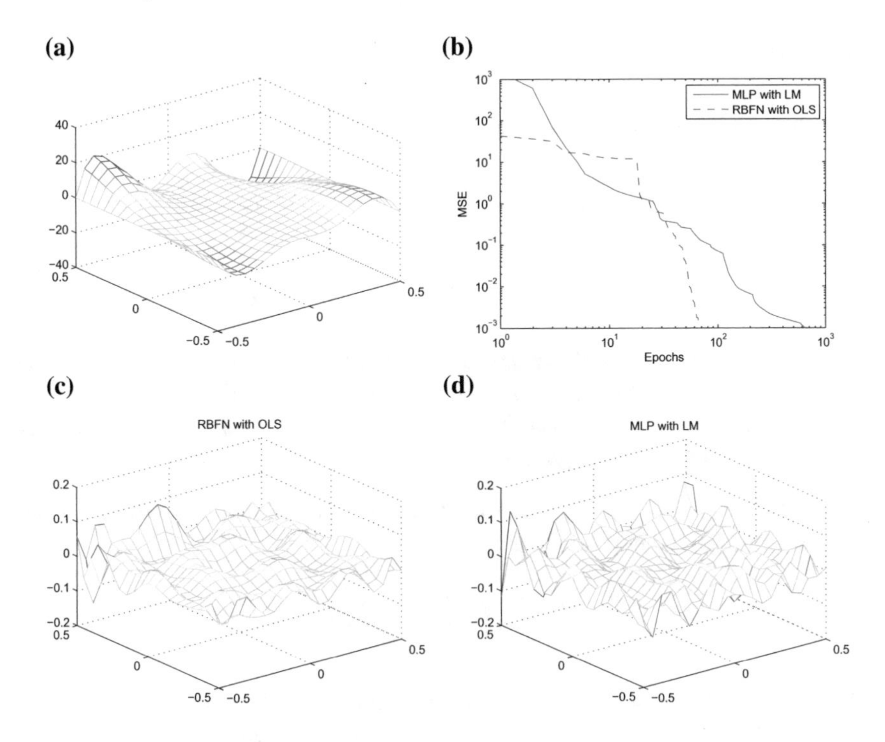

Fig. 11.7 A comparison between the MLP and the RBF network. **a** The function to be approximated. **b** The MSE obtained by using the MLP and the RBF network. **c** The approximation error by the MLP. **d** The approximation error by the RBF network

is proposed. The algorithm identifies global features of an input–output relationship before adding local detail to the approximating function.

Example 11.2 Approximate the following function using the MLP and the RBF network:

$$f(x_1, x_2) = 4(0.1 + x_2\, \mathrm{e}^{x_2}(0.05 + x_1^4 - 10x_1^2x_2^2 + 5x_2^4))\cos(2\pi x_1),$$

$$x_1, x_2 \in [-0.5, 0.5].$$

The number of epochs is selected as 1000, and the goal MSE is selected to be 10^{-3}. A grid of 21×21 samples is generated as the training set. For the MLP, we select 25 nodes and LM as the training algorithm. For the RBF network, we select 70 nodes, $\sigma = 0.5$, and OLS as the training algorithm. The simulation result is demonstrated in Fig. 11.7. The training and testing times for the MLP are 18.4208 and 0.0511 s, respectively, while those for the RBFN are 2.0622 s and 0.0178 s.

Problems

11.1 Plot the following two RBFs: [66, 77]

$$\phi(r) = \frac{1}{\left(\sigma^2 + r^2\right)^{\alpha}}, \qquad \alpha > 0,$$

$$\phi(r) = \left(\sigma^2 + r^2\right)^{\beta}, \qquad 0 < \beta < 1,$$

where $r > 0$ denotes the distance from data point $\boldsymbol{x}$ to center $\boldsymbol{c}$, σ is used to control the smoothness of the interpolating function. When $\beta = \frac{1}{2}$, the RBF becomes Hardy's multiquadric function, which is extensively used in surface interpolation with very good results [77].

11.2 Consider a 2-4-1 RBF network for XOR problem. Compute the linear weight $\boldsymbol{w}$, if the Gaussian RBF is used:

$$\phi_i(\boldsymbol{x}) = \exp\left(-\frac{\|\boldsymbol{x} - \boldsymbol{x}_i\|^2}{2\sigma^2}\right), \quad \sigma = 1.$$

11.3 Use the RBF network to approximate the following functions:
(a) $h(x) = (1 - x + 2x^2)\exp(-x^2/2), \quad x \in [-10, 10]$.
(b) $f(x) = 2\sin\left(\frac{x_1}{4}\right)\cos\left(\frac{x_2}{2}\right), \quad x \in [0, 10]$.

11.4 For the XOR problem, investigate the samples in the transformed space when using
(a) The logistic function $\varphi(r) = \frac{1}{1+e^{\frac{r}{\sigma^2}}}$.
(b) The thin-plate spline $\phi(r) = r^2 \ln(r)$.
(c) The multiquadratic function $\varphi(x) = (x^2 + 1)^{1/2}$.
(d) The inverse multiquadratic function $\varphi(x) = \frac{1}{(x^2+1)^{1/2}}$.
(e) Are the four points linearly separable in the transformed space?

11.5 For the Gaussian RBF network, derive the Jacobian matrix $\mathbf{J} = \left[\frac{\partial y_j}{\partial x_i}\right]$ and the Hessian matrix $\mathbf{H} = \left[\frac{\partial^2 E}{\partial w_i \partial w_j}\right]$, where w_i, w_j denote any two parameters such as network weights, RBF centers, and width.

References

1. Adeney, K. M., & Korenberg, M. J. (2000). Iterative fast orthogonal search algorithm for MDL-based training of generalized single-layer networks. *Neural Networks*, *13*, 787–799.
2. Albus, J. S. (1975). A new approach to manipulator control: Cerebellar model articulation control (CMAC). *Transactions of the ASME: Journal of Dynamic System, Measurement and Control*, *97*, 220–227.
3. Barron, A. R. (1993). Universal approximation bounds for superpositions of a sigmoidal function. *IEEE Transactions on Information Theory*, *39*(3), 930–945.
4. Benaim, M. (1994). On functional approximation with normalized Gaussian units. *Neural Computation*, *6*(2), 319–333.
5. Berthold, M. R., & Diamond, J. (1995). Boosting the performance of RBF networks with dynamic decay adjustment. In G. Tesauro, D. S. Touretzky, & T. Leen (Eds.), *Advances in Neural Information Processing Systems* (Vol. 7, pp. 521–528). Cambridge: MIT Press.
6. Bobrow, J. E., & Murray, W. (1993). An algorithm for RLS identification of parameters that vary quickly with time. *IEEE Transactions on Automatic Control*, *38*(2), 351–354.
7. Borghese, N. A., & Ferrari, S. (1998). Hierarchical RBF networks and local parameter estimate. *Neurocomputing*, *19*, 259–283.
8. Bors, G. A., & Pitas, I. (1996). Median radial basis function neural network. *IEEE Transactions on Neural Networks*, *7*(6), 1351–1364.
9. Broomhead, D. S., & Lowe, D. (1988). Multivariable functional interpolation and adaptive networks. *Complex System*, *2*, 321–355.
10. Brown, M., Harris, C. J., & Parks, P. (1993). The interpolation capabilities of the binary CMAC. *Neural Networks*, *6*(3), 429–440.
11. Bugmann, G. (1998). Normalized Gaussian radial basis function networks. *Neurocomputing*, *20*, 97–110.
12. Cha, I., & Kassam, S. A. (1995). Channel equalization using adaptive complex radial basis function networks. *IEEE Journal on Selected Areas in Communications*, *13*(1), 122–131.
13. Chen, S., Billings, S. A., Cowan, C. F. N., & Grant, P. M. (1990). Practical identification of NARMAX models using radial basis functions. *International Journal of Control*, *52*(6), 1327–1350.
14. Chen, S., Cowan, C., & Grant, P. (1991). Orthogonal least squares learning algorithm for radial basis function networks. *IEEE Transactions on Neural Networks*, *2*(2), 302–309.
15. Chen, S., Grant, P. M., & Cowan, C. F. N. (1992). Orthogonal least squares learning algorithm for training multioutput radial basis function networks. *IEE Proceedings - F*, *139*(6), 378–384.
16. Chen, C. L., Chen, W. C., & Chang, F. Y. (1993). Hybrid learning algorithm for Gaussian potential function networks. *IEE Proceedings - D*, *140*(6), 442–448.
17. Chen, S., Grant, P. M., McLaughlin, S., & Mulgrew, B. (1993). Complex-valued radial basis function networks. In *Proceedings of the 3rd IEE International Conference on Artificial Neural Networks* (pp. 148–152). Brighton, UK.
18. Chen, S., McLaughlin, S., & Mulgrew, B. (1994). Complex-valued radial basis function network, part I: network architecture and learning algorithms. *Signal Processing*, *35*, 19–31.
19. Chen, S., Hong, X., Harris, C. J., & Hanzo, L. (2008). Fully complex-valued radial basis function networks: Orthogonal least squares regression and classification. *Neurocomputing*, *71*, 3421–3433.
20. Chuang, C. C., Su, S. F., & Hsiao, C. C. (2000). The annealing robust backpropagation (ARBP) learning algorithm. *IEEE Transactions on Neural Networks*, *11*(5), 1067–1077.
21. Chuang, C. C., Jeng, J. T., & Lin, P. T. (2004). Annealing robust radial basis function networks for function approximation with outliers. *Neurocomputing*, *56*, 123–139.
22. Ciocoiu, I. B. (2002). RBF networks training using a dual extended Kalman filter. *Neurocomputing*, *48*, 609–622.
23. Constantinopoulos, C., & Likas, A. (2006). An incremental training method for the probabilistic RBF network. *IEEE Transactions on Neural Networks*, *17*(4), 966–974.

24. Dorffner, G. (1994). Unified framework for MLPs and RBFNs: Introducing conic section function networks. *Cybernetics and Systems*, *25*, 511–554.
25. Du, K.-L., Huang, X., Wang, M., Zhang, B., & Hu, J. (2000). Robot impedance learning of the peg-in-hole dynamic assembly process. *International Journal of Robotics and Automation*, *15*(3), 107–118.
26. Du, K.-L., & Swamy, M. N. S. (2006). *Neural networks in a softcomputing framework*. London: Springer.
27. Du, K.-L., & Swamy, M. N. S. (2014). *Neural networks and statistical learning*. London: Springer.
28. Esposito, A., Marinaro, M., Oricchio, D., & Scarpetta, S. (2000). Approximation of continuous and discontinuous mappings by a growing neural RBF-based algorithm. *Neural Networks*, *13*, 651–665.
29. Ferrari, S., Maggioni, M., & Borghese, N. A. (2004). Multiscale approximation with hierarchical radial basis functions networks. *IEEE Transactions on Neural Networks*, *15*(1), 178–188.
30. Girosi, F., & Poggio, T. (1990). Networks and the best approximation property. *Biological Cybernetics*, *63*, 169–176.
31. Gomm, J. B., & Yu, D. L. (2000). Selecting radial basis function network centers with recursive orthogonal least squares training. *IEEE Transactions on Neural Networks*, *11*(2), 306–314.
32. Gorinevsky, D. (1997). An approach to parametric nonlinear least square optimization and application to task-level learning control. *IEEE Transactions on Automatic Control*, *42*(7), 912–927.
33. Heiss, M., & Kampl, S. (1996). Multiplication-free radial basis function network. *IEEE Transactions on Neural Networks*, *7*(6), 1461–1464.
34. Holden, S. B., & Rayner, P. J. W. (1995). Generalization and PAC learning: some new results for the class of generalized single-layer networks. *IEEE Transactions on Neural Networks*, *6*(2), 368–380.
35. Hong, X., & Billings, S. A. (1997). Givens rotation based fast backward elimination algorithm for RBF neural network pruning. *IEE Proceedings - Control Theory and Applications*, *144*(5), 381–384.
36. Hou, M., & Han, X. (2010). Constructive approximation to multivariate function by decay RBF neural network. *IEEE Transactions on Neural Networks*, *21*(9), 1517–1523.
37. Huang, G. B., Saratchandran, P., & Sundararajan, N. (2004). An efficient sequential learning algorithm for growing and pruning RBF (GAP-RBF) Networks. *IEEE Transactions on Systems, Man, and Cybernetics Part B*, *34*(6), 2284–2292.
38. Huang, G. B., Saratchandran, P., & Sundararajan, N. (2005). A generalized growing and pruning RBF (GGAP-RBF) neural network for function approximation. *IEEE Transactions on Neural Networks*, *16*(1), 57–67.
39. Huang, G.-B., Chen, L., & Siew, C.-K. (2006). Universal approximation using incremental constructive feedforward networks with random hidden nodes. *IEEE Transactions on Neural Networks*, *17*(4), 879–892.
40. Huang, G.-B., Zhu, Q.-Y., & Siew, C.-K. (2006). Extreme learning machine: Theory and applications. *Neurocomputing*, *70*, 489–501.
41. Jankowski, N., & Kadirkamanathan, V. (1997). Statistical control of growing and pruning in RBF-like neural networks. In *Proceedings of the 3rd Conference on Neural Networks and Their Applications* (pp. 663–670). Kule, Poland.
42. Jianping, D., Sundararajan, N., & Saratchandran, P. (2002). Communication channel equalization using complex-valued minimal radial basis function neural networks. *IEEE Transactions on Neural Networks*, *13*(3), 687–696.
43. Kadirkamanathan, V., & Niranjan, M. (1993). A function estimation approach to sequential learning with neural network. *Neural Computation*, *5*(6), 954–975.
44. Kadirkamanathan, V. (1994). A statistical inference based growth criterion for the RBF network. In *Proceedings of the IEEE Workshop on Neural Networks for Signal Processing* (pp. 12–21). Ermioni, Greece.

45. Kaminski, W., & Strumillo, P. (1997). Kernel orthonormalization in radial basis function neural networks. *IEEE Transactions on Neural Networks*, *8*(5), 1177–1183.
46. Karayiannis, N. B., & Mi, G. W. (1997). Growing radial basis neural networks: Merging supervised and unsupervised learning with network growth techniques. *IEEE Transactions on Neural Networks*, *8*(6), 1492–1506.
47. Karayiannis, N. B. (1999). Reformulated radial basis neural networks trained by gradient descent. *IEEE Transactions on Neural Networks*, *10*(3), 657–671.
48. Karayiannis, N. B., & Xiong, Y. (2006). Training reformulated radial basis function neural networks capable of identifying uncertainty in data classification. *IEEE Transactions on Neural Networks*, *17*(5), 1222–1234.
49. Kraaijveld, M. A., & Duin, R. P. W. (1991). Generalization capabilities of minimal kernel-based networks. In *Proceedings of the International Joint Conference on Neural Networks* (Vol. 1, pp. 843–848). Seattle, WA.
50. Krzyzak, A., & Linder, T. (1998). Radial basis function networks and complexity regularization in function learning. *IEEE Transactions on Neural Networks*, *9*(2), 247–256.
51. Langari, R., Wang, L., & Yen, J. (1997). Radial basis function networks, regression weights, and the expectation-maximization algorithm. *IEEE Transactions on Systems, Man, and Cybernetics Part A*, *27*(5), 613–623.
52. Lee, C. C., Chung, P. C., Tsai, J. R., & Chang, C. I. (1999). Robust radial basis function neural networks. *IEEE Transactions on Systems, Man, and Cybernetics Part B*, *29*(6), 674–685.
53. Lee, S. J., & Hou, C. L. (2002). An ART-based construction of RBF networks. *IEEE Transactions on Neural Networks*, *13*(6), 1308–1321.
54. Lee, K. Y., & Jung, S. (1999). Extended complex RBF and its application to M-QAM in presence of co-channel interference. *Electronics Letters*, *35*(1), 17–19.
55. Lehtokangas, M., & Saarinen, J. (1998). Centroid based multilayer perceptron networks. *Neural Processing Letters*, *7*, 101–106.
56. Lehtokangas, M., Saarinen, J., & Kaski, K. (1995). Accelerating training of radial basis function networks with cascade-correlation algorithm. *Neurocomputing*, *9*, 207–213.
57. Leonardis, A., & Bischof, H. (1998). An efficient MDL-based construction of RBF networks. *Neural Networks*, *11*, 963–973.
58. Leung, C.-S., & Sum, J. P.-F. (2008). A fault-tolerant regularizer for RBF networks. *IEEE Transactions on Neural Networks*, *19*(3), 493–507.
59. Li, M.-B., Huang, G.-B., Saratchandran, P., & Sundararajan, N. (2005). Fully complex extreme learning machine. *Neurocomputing*, *68*, 306–314.
60. Liao, Y., Fang, S. C., & Nuttle, H. L. W. (2003). Relaxed conditions for radial-basis function networks to be universal approximators. *Neural Networks*, *16*, 1019–1028.
61. Lowe, D. (1995). On the use of nonlocal and non-positive definite basis functions in radial basis function networks. In *Proceedings of the IEE International Conference on Artificial Neural Networks* (pp. 206–211). Cambridge, UK.
62. Mahdi, R. N., & Rouchka, E. C. (2011). Reduced hyperBF networks: Regularization by explicit complexity reduction and scaled Rprop-based training. *IEEE Transactions on Neural Networks*, *22*(5), 673–686.
63. Mao, K. Z. (2002). RBF neural network center selection based on Fisher ratio class separability measure. *IEEE Transactions on Neural Networks*, *13*(5), 1211–1217.
64. McLoone, S., Brown, M. D., Irwin, G., & Lightbody, G. (1998). A hybrid linear/nonlinear training algorithm for feedforward neural networks. *IEEE Transactions on Neural Networks*, *9*(4), 669–684.
65. McLoone, S., & Irwin, G. (2001). Improving neural network training solutions using regularisation. *Neurocomputing*, *37*, 71–90.
66. Micchelli, C. A. (1986). Interpolation of scattered data: Distance matrices and conditionally positive definite functions. *Constructive Approximation*, *2*, 11–22.
67. Miller, W. T., Glanz, F. H., & Kraft, L. G. (1990). CMAC: An associative neural network alternative to backpropagation. *Proceedings of the IEEE*, *78*(10), 1561–1567.

68. Moody, J., & Darken, C. J. (1989). Fast learning in networks of locally-tuned processing units. *Neural Computation*, *1*(2), 281–294.
69. Musavi, M. T., Ahmed, W., Chan, K. H., Faris, K. B., & Hummels, D. M. (1992). On the training of radial basis function classifiers. *Neural Networks*, *5*(4), 595–603.
70. Niyogi, P., & Girosi, F. (1999). Generalization bounds for function approximation from scattered noisy data. *Advances in Computational Mathematics*, *10*, 51–80.
71. Oliveira, A. L. I., Melo, B. J. M., & Meira, S. R. L. (2005). Improving constructive training of RBF networks through selective pruning and model selection. *Neurocomputing*, *64*, 537–541.
72. Orr, M. J. L. (1995). Regularization in the selection of radial basis function centers. *Neural Computation*, *7*(3), 606–623.
73. Paetz, J. (2004). Reducing the number of neurons in radial basis function networks with dynamic decay adjustment. *Neurocomputing*, *62*, 79–91.
74. Park, J., & Sanberg, I. W. (1991). Universal approximation using radial-basis-function networks. *Neural Computation*, *3*, 246–257.
75. Peng, H., Ozaki, T., Haggan-Ozaki, V., & Toyoda, Y. (2003). A parameter optimization method for radial basis function type models. *IEEE Transactions on Neural Networks*, *14*(2), 432–438.
76. Platt, J. (1991). A resource allocating network for function interpolation. *Neural Computation*, *3*(2), 213–225.
77. Poggio, T., & Girosi, F. (1990). Networks for approximation and learning. *Proceedings of the IEEE*, *78*(9), 1481–1497.
78. Powell, M. J. D. (1987). Radial basis functions for multivariable interpolation: A review. In J. C. Mason & M. G. Cox (Eds.), *Algorithms for Approximation* (pp. 143–167). Oxford: Clarendon Press.
79. Qian, X., Huang, H., Chen, X., & Huang, T. (2017). Efficient construction of sparse radial basis function neural networks using L_1-regularization. *Neural Networks*, *94*, 239–254.
80. Rojas, I., Pomares, H., Bernier, J. L., Ortega, J., Pino, B., Pelayo, F. J., et al. (2002). Time series analysis using normalized PG-RBF network with regression weights. *Neurocomputing*, *42*, 267–285.
81. Rosipal, R., Koska, M., & Farkas, I. (1998). Prediction of chaotic time-series with a resource-allocating RBF network. *Neural Processing Letters*, *7*, 185–197.
82. Roy, A., Govil, S., & Miranda, R. (1995). An algorithm to generate radial basis functions (RBF)-like nets for classification problems. *Neural Networks*, *8*(2), 179–201.
83. Salmeron, M., Ortega, J., Puntonet, C. G., & Prieto, A. (2001). Improved RAN sequential prediction using orthogonal techniques. *Neurocomputing*, *41*, 153–172.
84. Sanchez, A., & V. D., (1995). Robustization of a learning method for RBF networks. *Neurocomputing*, *9*, 85–94.
85. Schilling, R. J., Carroll, J. J, Jr., & Al-Ajlouni, A. F. (2001). Approximation of nonlinear systems with radial basis function neural networks. *IEEE Transactions on Neural Networks*, *12*(1), 1–15.
86. Schwenker, F., Kestler, H. A., & Palm, G. (2001). Three learning phases for radial-basis-function networks. *Neural Networks*, *14*, 439–458.
87. Simon, D. (2002). Training radial basis neural networks with the extended Kalman filter. *Neurocomputing*, *48*, 455–475.
88. Singla, P., Subbarao, K., & Junkins, J. L. (2007). Direction-dependent learning approach for radial basis function networks. *IEEE Transactions on Neural Networks*, *18*(1), 203–222.
89. Specht, D. F. (1990). Probabilistic neural networks. *Neural Networks*, *3*, 109–118.
90. Sum, J. P.-F., Leung, C.-S., & Ho, K. I.-J. (2009). On objective function, regularizer, and prediction error of a learning algorithm for dealing with multiplicative weight noise. *IEEE Transactions on Neural Networks*, *20*(1), 124–138.
91. Suresh, S., Savitha, R., & Sundararajan, N. (2011). A sequential learning algorithm for complex-valued self-regulating resource allocation network - CSRAN. *IEEE Transactions on Neural Networks*, *22*(7), 1061–1072.
92. Teddy, S. D., Quek, C., & Lai, E. M.-K. (2008). PSECMAC: A novel self-organizing multiresolution associative memory architecture. *IEEE Transactions on Neural Networks*, *19*(4), 689–712.

93. Titsias, M. K., & Likas, A. (2001). Shared kernel models for class conditional density estimation. *IEEE Transactions on Neural Networks, 12*(5), 987–997.
94. Todorovic, B., & Stankovic, M. (2001). Sequential growing and pruning of radial basis function network. In *Proceedings of the International Joint Conference on Neural Networks (IJCNN)* (pp. 1954–1959). Washington, DC.
95. Uykan, Z., Guzelis, C., Celebi, M. E., & Koivo, H. N. (2000). Analysis of input-output clustering for determining centers of RBFN. *IEEE Transactions on Neural Networks, 11*(4), 851–858.
96. Wang, X. X., Chen, S., & Brown, D. J. (2004). An approach for constructing parsimonious generalized Gaussian kernel regression models. *Neurocomputing, 62*, 441–457.
97. Webb, A. R. (1994). Functional approximation in feed-forward networks: A least-squares approach to generalization. *IEEE Transactions on Neural Networks, 5*, 363–371.
98. Webb, A. R., & Lowe, D. (1990). The optimized internal representation of multilayer classifier networks performs nonlinear discriminant analysis. *Neural Networks, 3*, 367–375.
99. Wedge, D., Ingram, D., McLean, D., Mingham, C., & Bandar, Z. (2006). On global-local artificial neural networks for function approximation. *IEEE Transactions on Neural Networks, 17*(4), 942–952.
100. Wettschereck, D., & Dietterich, T. (1992). Improving the performance of radial basis function networks by learning center locations. In J. E. Moody, S. J. Hanson, & R. P. Lippmann (Eds.), *Advances in neural information processing systems* (Vol. 4, pp. 1133–1140). San Mateo: Morgan Kaufmann.
101. Widrow, B., & Lehr, M. A. (1990). 30 years of adaptive neural networks: Perceptron, madaline, and backpropagation. *Proceedings of the IEEE, 78*(9), 1415–1442.
102. Wilamowski, B. M., & Yu, H. (2010). Improved computation for Levenberg-Marquardt training. *IEEE Transactions on Neural Networks, 21*(6), 930–937.
103. Xie, T., Yu, H., Hewlett, J., Rozycki, P., & Wilamowski, B. (2012). Fast and efficient second-order method for training radial basis function networks. *IEEE Transactions on Neural Networks and Learning Systems, 23*(4), 609–619.
104. Yingwei, L., Sundararajan, N., & Saratchandran, P. (1997). A sequential learning scheme for function approximation by using minimal radial basis function neural networks. *Neural Computation, 9*(2), 461–478.
105. Yingwei, L., Sundararajan, N., & Saratchandran, P. (1998). Performance evaluation of a sequential minimal radial basis function (RBF) neural network learning algorithm. *IEEE Transactions on Neural Networks, 9*(2), 308–318.
106. Yu, D. L., Gomm, J. B., & Williams, D. (1997). A recursive orthogonal least squares algorithm for training RBF networks. *Neural Processing Letters, 5*, 167–176.
107. Yu, H., Reiner, P. D., Xie, T., Bartczak, T., & Wilamowski, B. M. (2014). An incremental design of radial basis function networks. *IEEE Transactions on Neural Networks and Learning Systems, 25*(10), 1793–1803.
108. Zhang, Q. (1997). Using wavelet networks in nonparametric estimation. *IEEE Transactions on Neural Networks, 8*(2), 227–236.
109. Zhang, Q., & Benveniste, A. (1992). Wavelet networks. *IEEE Transactions on Neural Networks, 3*(6), 899–905.

Chapter 12
Recurrent Neural Networks

12.1 Introduction

The brain is a strongly recurrent structure. This massive recurrence suggests a major role of self-feeding dynamics in the processes of perceiving, acting, and learning, and in maintaining the organism alive. Recurrent networks harness the power of brain-like computing. There is at least one feedback connection in recurrent networks. When recurrent networks are used, the network size is significantly compact compared with feedforward networks for the same approximation accuracy of dynamic systems. MLP is fundamentally limited in its ability to solve topological relation problems. Recurrent networks can also be used as associative memories to build attractors $\boldsymbol{y}_p$ from input–output association $\{\boldsymbol{x}_p, \boldsymbol{y}_p\}$.

MLP is purely static and is incapable of processing time information. One can add a time window over the data to act as a memory for the past. In the applications of dynamical systems, we need to forecast an input at time $t+1$ from the network state at time t. The resulting network model for modeling a dynamical process is referred to as a *temporal association network*. Temporal association networks must have a recurrent architecture so as to handle the time-dependent nature of the association.

To generate a dynamic neural network, memory must be introduced. The simplest memory element is the unit time delay, which has the transfer function $H(z) = z^{-1}$. The simplest memory architecture is the tapped delay line consisting of a series of unit time delays. Tapped delay lines are the basis of traditional linear dynamical models such as finite impulse response (FIR) or infinite impulse response (IIR) models. An MLP may be made dynamic by introducing time-delay loops to the input, hidden, and/or output layers. The memory elements can be either fully or sparsely interconnected. A network architecture incorporating time delays is the time-delay neural network [61].

Since the recurrent networks are modeled by systems of ordinary differential equations, they are suitable for digital implementation using standard software for the integration of ordinary differential equations. The Hopfield model and the Cohen–Grossberg model are the two common recurrent network models. The Hopfield model

K.-L. Du and M. N. S. Swamy, *Neural Networks and Statistical Learning*,
https://doi.org/10.1007/978-1-4471-7452-3_12

can store information in a dynamically stable structure. The Boltzmann machine is a generalization of Hopfield model.

Recurrent networks can generally be classified into globally recurrent networks, in which feedback connections between every neuron are allowed, and locally recurrent, globally feedforward networks [59] with the dynamics realized inside neuron models. Both classes of models can be universal approximators for dynamical systems [27, 28]. In general, globally recurrent networks suffer from stability problems during training and require complicated and time-consuming training algorithms. In contrast, locally recurrent networks are designed with dynamic neuron models which contain inner feedbacks, but interconnections between neurons are strict feedforward ones just as in the case of MLP. They have a less complicated structure and yield simpler training. They allow for easy checking of the stability by examining poles of their internal filters. Explicit incorporation of past information into an architecture can be easily implemented. Analytical results show that a locally recurrent network with two hidden layers is able to approximate a state-space trajectory produced by any Lipschitz continuous function with arbitrary accuracy [39].

Two discrete-time formulations of recurrent networks are the time-delayed recurrent network and the simultaneous recurrent network. The time-delayed recurrent network is trained so as to minimize the error in prediction. By contrast, the simultaneous recurrent network is not intended to provide better forecasting over time or to provide memory of past or future but rather uses recurrence to provide general function approximation capability, based on concepts in Turing theory and complexity theory. The simultaneous recurrent network is a powerful function approximator [67]. It has been shown experimentally that an arbitrary function generated by an MLP can always be learned by a simultaneous recurrent network. However, the opposite is not true.

The cellular structure-based simultaneous recurrent network has some interesting similarity to the hippocampus [67]. It is a function approximator that is more powerful than the MLP [21]; it can realize a desired mapping with much lower complexity than the MLP can. A generic cellular simultaneous recurrent network is implemented by training the network with EKF [21]. The cell is a generalized MLP. Each cell has the same weights, and this allows for arbitrarily large networks without increasing the number of weight parameters.

Genetic regulatory networks can be described by nonlinear differential equations with time delays. Delay-independent stability of two genetic regulatory networks, namely, a real-life repressilatory network with three genes and three proteins, and a synthetic gene regulatory network with five genes and seven proteins, is analyzed in [70].

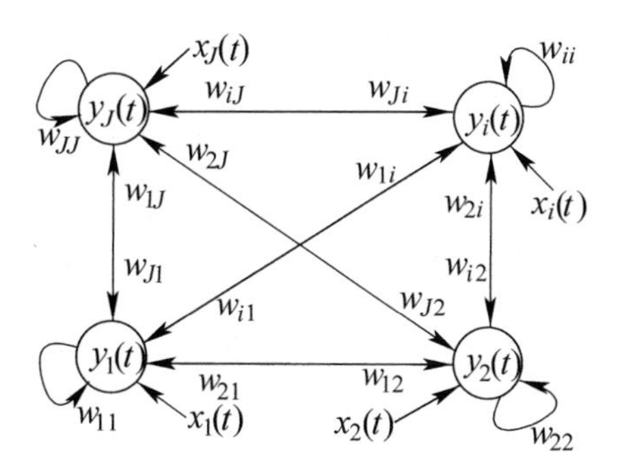

Fig. 12.1 Architecture of a fully connected recurrent network of J neurons

12.2 Fully Connected Recurrent Networks

For a recurrent network of J units, we denote the input of unit i by x_i and the output or state of unit i as y_i. The architecture of a fully connected recurrent network is illustrated in Fig. 12.1.

The dynamics of unit i with sound biological and electronic motivation are given by [20]

$$net_i(t) = \sum_{j=1}^{J} w_{ji} y_j(t) + x_i(t), \quad i = 1, \ldots, J, \tag{12.1}$$

$$\tau_i \frac{\mathrm{d}y_i(t)}{\mathrm{d}t} = -y_i(t) + \phi(net_i) + x_i(t), \quad i = 1, \ldots, J, \tag{12.2}$$

where τ_i is a time constant, net_i is the net input to unit i, $\phi(\cdot)$ is a sigmoidal function, and input $x_i(t)$ and output $y_i(t)$ are continuous functions of time. In (12.2), $-y_i(t)$ denotes natural signal decay.

Any continuous state-space trajectory can be approximated to any desired degree of accuracy by the output of a sufficiently large continuous-time recurrent network described by (12.1) and (12.2) [13]. In other words, the recurrent network is a universal approximator of dynamical systems. The universal approximation capability of recurrent networks has been investigated in [27, 29, 47]. A fully connected discrete-time recurrent network with the sigmoidal activation function is a universal approximator of discrete- or continuous-time trajectories on compact time intervals [27]. A continuous-time recurrent network with the sigmoidal activation function and external input can approximate any finite-time trajectory of a dynamical time-variant system [29]. Recurrent networks are proved to be universal approximators to represent any open dynamical system arbitrarily well [47].

Turing Capability

The neural Moore machine is the most general recurrent network architecture. It is the neural network version of the Moore machine, which is a type of finite-state machine. Elman's simple recurrent net [10] is a widely used neural Moore machine. All general digital computers have some common features. The programs executable on them form the class of recursive functions, and the model describing them is called a Turing machine. In a Turing machine, a finite automaton is used as a control or main computing unit, but this unit has access to potentially infinite storage space.

In fact, while hidden Markov models (HMMs) and traditional discrete symbolic grammar learning devices are limited to discrete state spaces, recurrent networks are in principle suited to all sequence learning tasks due to their Turing capabilities. Recurrent networks are Turing equivalent [51] and can therefore compute whatever function any digital computer can compute. The simple recurrent network is proved to have a computational power equivalent to that of any finite-state machine [51].

Theorem 12.1 (Siegelmann and Sontag, 1995 *[51]*) *All Turing machines can be simulated by fully connected recurrent networks built on neurons with sigmoidal activation functions.*

12.3 Time-Delay Neural Networks

The time-delay neural network [61] maps a finite-time sequence $\{x(t), x(t-1), \ldots, x(t-m)\}$ into a single output $y(t)$. It is a feedforward network equipped with time-delayed versions of a signal $x(t)$ as input. BP can be used to train the network. The architecture of a time-delay neural network using a three-layer MLP is illustrated in Fig. 12.2. The input to the network is a vector composing $m+1$ continuous samples. If the input to the network is $\boldsymbol{x}_t = (x(t), x(t-1), \ldots, x(t-m))^T$ at time t, then it is $\boldsymbol{x}_{t+i} = (x(t+i), x(t+i-1), \ldots, x(t+i-m))^T$ at time $t+i$. The model has been successfully applied to speech recognition [61] and time series prediction. The architecture can be generalized when the input and output are vectors. This network practically functions as an FIR filter. A time-delay neural network is not a recurrent network since there is no feedback and it preserves its dynamic properties by unfolding the input sequence over time.

In Fig. 12.2, if the single output $y(t+1)$ is applied to a tapped-delayed-line memory of p units and the p delayed replicas of $y(t+1)$ are fed back to the input of the network, the input to this new recurrent network is then $\left(\boldsymbol{x}^T(t); \boldsymbol{y}^T(t)\right)^T = (x(t), x(t-1), \ldots, x(t-m); y(t), y(t-1), \ldots, y(t-p))^T$ and the output of the network is $y(t+1)$. Vector $\boldsymbol{x}(t)$ is an exogenous input originating from outside the network, and $\boldsymbol{y}(t)$ is regression of the model output $y(t+1)$. The model is called a *nonlinear autoregressive with exogenous inputs (NARX)* model. It is a delay recurrent network with feedback.

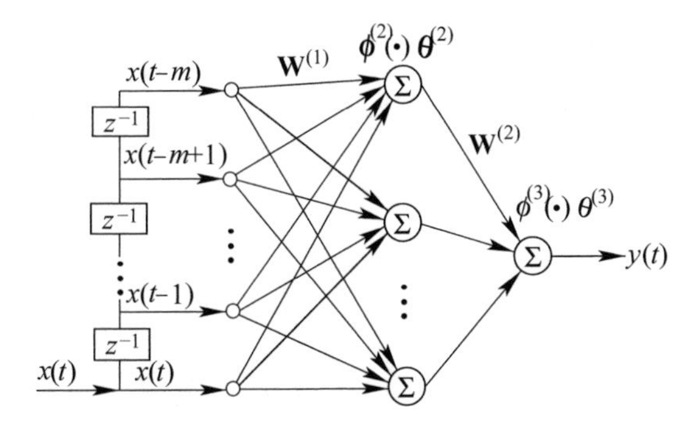

Fig. 12.2 Architecture of the time-delay neural network

As opposed to other recurrent networks, NARX networks have a limited feedback which comes only from the output neuron rather than from the hidden states. The NARX networks with a finite number of parameters are constructively proved to be computationally as strong as fully connected recurrent networks, and thus the Turing machines [52]. The computational power of NARX network is at least as great as that of Turing machines [52].

Theorem 12.2 (Siegelmann, Horne and Giles, 1997 *[52]*) *NARX networks with one layer of hidden neurons with bounded, one-sided saturated activation functions and a linear output neuron can simulate fully connected recurrent networks with bounded, one-sided saturated activation functions, except for a linear slowdown.*

A linear slowdown means that if the fully connected recurrent network with N neurons computes a task of interest in time T, then the total time taken by the equivalent NARX network is $(N+1)T$. By a minor modification, the logistic function can be made a bounded, one-sided saturated function. From Theorems 12.1 and 12.2, NARX networks with one hidden layer of neurons with bounded, one-sided saturated activation functions and a linear output neuron are Turing equivalent.

By replacing each synapse weight with a linear, time-invariant filter, the MLP can be used for temporal processing [49]. When the filter is an FIR filter, we get an FIR neural network. The FIR MLP can be implemented as a resistance–capacitance model [49]. One can use a temporal extension of BP to train the FIR MLP [63, 64]. Once the network is trained, all the weights are fixed and the network can be used as an MLP. The time-delay neural network is functionally equivalent to the FIR network [64]. It can be easily related to a multilayer network by replacing the static synaptic weights with FIR filters [64].

The time-delay neural network can be an MLP-based [61] or an RBF network-based [3] temporal neural network for nonlinear dynamics and time series learning. Both approaches use the same spatial representation of time.

Example 12.1 One of the most well-known applications of the MLP is the speech synthesis system NETtalk [48]. NETtalk is a three-layer classification network, as illustrated in Fig. 12.3, that translates English letters into phonemes. A string of

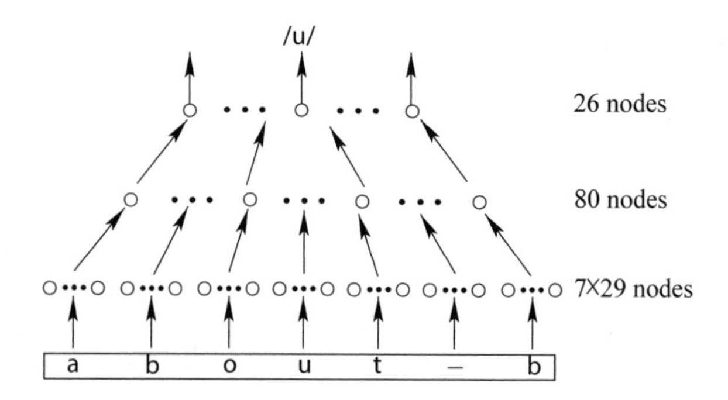

Fig. 12.3 Schematic drawing of the NETtalk architecture

characters forming English text is then converted into a string of phonemes, which can be further sent to an electronic speech synthesizer to produce speech.

The MLP-based time-delay neural network is applied. The network has a total of 309 nodes and 18, 629 weights. Both the BP and Boltzmann learning algorithms are applied to train the network. The hidden units play the same role as a rule extractor. The training set consists of a corpus from a dictionary, and phonetic transcriptions from informal, continuous speech of a child. NETtalk is suitable for fast implementation without any domain knowledge, while the development of conventional rule-based expert systems such as DECtalk needs years of group work.

Example 12.2 The adaline network is a widely used neural network found in practical applications. Adaptive filtering is one of its major application areas. We use the adaline network with the tapped delay line. The input signal enters from the left and passes through $N-1$ delays. The output of the tapped delay line is an N-dimensional vector, made up of the input signal at the current and previous instances. The network is just an adaline neuron. In digital signal processing, this neuron is referred to as an FIR filter.

We use an adaptive filter to predict the next value of a stationary random process, $p(t)$. Given the target function $p(t)$, we use a time-delay adaline network to train the network. The input to the network is the previous five data samples $(p(t-1), p(t-2), \ldots, p(t-5))^T$, and the output approximates the target $p(t)$. The learning rate is set as 0.05. The approximation result for 10 adaptation passes is shown in Fig. 12.4a.

We solve the same problem using MLP-based time-delay neural network with 1 hidden neuron. BP with a learning rate 0.05 is used for training, and training is performed for 10 adaptation passes. The result is shown in Fig. 12.4b. The performance of the MLP-based method is poor but can be improved by using more hidden nodes, a better training method, and more epochs.

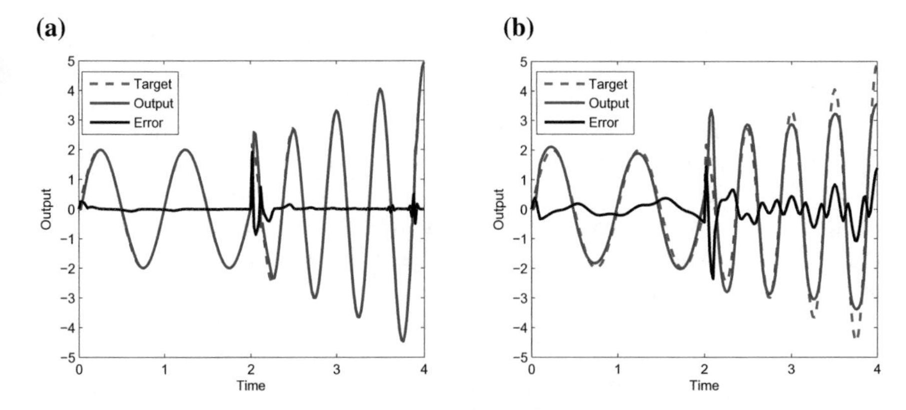

Fig. 12.4 Prediction using a time-delay network. **a** Result for 10 adaption passes using a time-delay adaline network. **b** Result for 10 adaptation passes using an MLP-based time-delay neural network

12.4 Backpropagation for Temporal Learning

When recurrent feature is integrated into the MLP architecture, the new model is capable of learning dynamic systems. The BP algorithm is required to be modified accordingly.

Temporal behavior can be modeled by a linear, time-invariant filter. The temporal behavior of synapse i of neuron j may be described by an impulse response $h_{ji}(t)$ that is a function of continuous time t, which corresponds to the weight w_{ji} [49]. These filters can be implemented by using RC circuits. When the hidden and output neurons of an MLP all use the FIR model, such a neural network is referred to as an FIR MLP.

An FIR MLP can be unfolded in time. This removes all the time delays in the network by expanding it into an equivalent but larger network so that standard BP algorithm can be applied to compute the instantaneous error gradients. One can implement forward or backward unfolding in time. The forward choice results in a network size that is linear with the total number of time delays and the total number of free parameters in the network, while the backward case yields a network size that grows geometrically with the number of time delays and layers. The forward choice is thus preferred. Temporal backpropagation learning [63] overcomes the drawbacks of the unfolded method.

Time-lagged recurrent networks subsume many conventional signal processing structures, e.g., tapped delay lines, FIR, and IIR filters. At the same time, they can represent dynamic systems with strongly hidden states and possess universal approximation capability for dynamic systems. Recurrent MLPs are a special class of time-lagged recurrent networks; they have a layered connectivity pattern. The recurrent MLP architecture, in which the output of the MLP is fed back to the input of the network, is a universal approximator for any dynamical system [28]. A BP-like learning algorithm with robust modification is used to train the recurrent MLP.

Recurrent BP is used to train fully connected recurrent networks, where the units are assumed to have continuous states [1, 43]. After training on a set of N patterns $\{(\boldsymbol{x}_k, \boldsymbol{y}_k)\}$, the presentation of $\boldsymbol{x}_k$ will drive the network output to a fixed attractor state $\boldsymbol{y}_k$. Thus, the algorithm learns static mappings and may be used as associative memories. The computational complexity is $O(J^2)$ for a recurrent network of J units. When the initial weights are selected as small values, the network almost always converges to a stable point.

The time-dependent recurrent learning (TDRL) algorithm is an extension of recurrent BP to dynamic sequences that produce time-dependent trajectories [40]. It is a gradient-descent method that searches for the weights of a continuous recurrent network to minimize the error function of the temporal trajectory of the states.

Real-Time Recurrent Learning

The real-time recurrent learning (RTRL) algorithm [68] is used for training fully connected recurrent networks with discrete-time states. It is a modified BP algorithm and is an online algorithm without the need for allocating memory proportional to the maximum sequence length. RTRL decomposes the error function in time and evaluates instantaneous error derivatives with respect to the weights at each time step. The performance criterion for RTRL is the minimization of the total error over the entire temporal interval. In RTRL, calculation of the derivatives of node outputs with respect to the network weights must be carried out during the forward propagation of signals in a network. RTRL is suitable for tasks that require retention of information over fixed or indefinite time length, and it is best suitable for real-time applications. The normalized RTRL [37] normalizes the learning rate of RTRL at each step so that one has the optimal adaptive learning rate for every discrete-time instant. The algorithm has a posteriori learning in the recurrent networks. Normalized RTRL is faster and more stable than RTRL.

For a time-lagged recurrent network, the computational complexity of RTRL scales by $O(J^4)$ for J nodes (in the worst case), with the storage of all variables scaling by $O(J^3)$ [69]. Furthermore, RTRL requires that the dynamic derivatives be computed at every time step for which the time-lagged recurrent network is executed. Such coupling of forward propagation and derivative calculation is due to the fact that in RTRL both the derivatives and the time-lagged recurrent network node outputs evolve recursively.

The delayed recurrent network is a continuous-time recurrent network having time-delayed feedbacks. Due to the presence of time delay in the differential equation, the system has an infinite number of degrees of freedom. The TDRL and RTRL algorithms are introduced into the delayed recurrent network [58]. A comparative study of the recurrent network and the time-delay neural network has been made in terms of the learning algorithms, learning capability, and robustness against noise in [58]. The existence of time delays usually causes divergence, oscillation, or even instability of neural networks. Therefore, it is necessary to perform stability analysis of neural networks with time delays.

In [25], RTRL is extended to its complex-valued form where the inputs, outputs, weights, and activation functions are complex-valued. Complex-valued RTRL algorithms using split complex activation functions [9, 25] do not follow the generic form of their real RTRL counterparts. A complex-valued RTRL algorithm using a general fully complex activation function [14] represents a natural extension of the real-valued RTRL. Based on augmented complex statistics, augmented complex-valued RTRL [15] significantly outperforms complex-valued RTRL [14] for noncircular (or improper) signals. An augmented complex-valued EKF algorithm is also given in [16]. Kurtosis-based complex-valued RTRL and kurtosis-based augmented complex-valued RTRL [38] provide a faster convergence rate as well as a lower steady-state error.

BP Through Time

BP through time (BPTT) is the most popular method for performing supervised learning of recurrent networks [46, 66]. It is an adapted version of BP for recurrent networks. BPTT is a method for unfolding a recurrent network in time to make an equivalent feedforward network each time a sequence is processed so that the derivatives can be computed via standard BP. The main limitation of BPTT is the static interval unfolded in time, red which is unable to accommodate the processing of newly arrived information. Normally, BPTT truncates the continuous input–output sequence by a length n, which defines the number of time intervals to unfold and is the size of buffer memory to train the unfolded layers. It means that BPTT cannot take care of the sequences before n time steps and that the main memory is in external buffer memory except for the weights.

The use of normal recurrent network and BPTT in incremental learning destroys the memory of past sequences. For a long sequence, the unfolded network may be very large, and BPTT will be inefficient. Truncated BPTT alleviates this problem by ignoring all the past contributions to the gradient beyond certain time into the past [69]. For BPTT(h), the computational complexity scales by $O(hJ^2)$ and the required memory is $O(hJ)$, for truncation length h. BPTT(h) leads to more stable computation of dynamic derivatives than forward methods do because it utilizes only the most recent information in a trajectory. The use of BPTT(h) permits training to be carried out asynchronously with execution of the time-lagged recurrent network.

Simultaneous recurrent networks trained by truncated BPTT with EKF are used for training weights in WTA networks with a smooth, nonlinear activation function [8]. BPTT is used for obtaining temporal derivatives, whereas EKF is the weight update method utilizing these derivatives.

For an overview of various gradient-based learning algorithms for recurrent networks, see [41]. A comprehensive analysis and comparison of BPTT, recurrent BP, and RTRL is given in [69].

12.5 RBF Networks for Modeling Dynamic Systems

The sequential RBF network learning algorithms, such as the RAN family, are capable of modifying both the network structure and the output weights online; thus, these algorithms are particularly suitable for modeling dynamical time-varying systems, where not only the dynamics but also the operating region changes with time.

In order to model complex nonlinear dynamical systems, the state-dependent autoregressive (AR) model with functional coefficients is often used. The RBF network can be used as a nonlinear AR time series model for forecasting application [50]. It can also be used to approximate the coefficients of a state-dependent AR model, thus yielding the RBF-AR model [60]. The RBF-ARX model is an RBF-AR model with an exogenous variable [42]. The RBF-ARX model usually uses far fewer RBF centers when compared to the RBF network.

For time series applications, the input to the network is $\boldsymbol{x}(t) = (y(t-1), \ldots, y(t-n_y))^T$, and the network output is $y(t)$. The dual-orthogonal RBF network algorithm is specially designed for nonlinear time series prediction [5].

For online adaptation of nonlinear systems, a constant exponential forgetting factor is commonly applied to all the past data uniformly. This is incorrect for nonlinear systems whose dynamics are different in different operating regions. In [71], online adaptation of the Gaussian RBF network is implemented using a localized forgetting method, which sets different forgetting factors in different regions according to the response of the local prototypes to the current input vector. The method is applied in conjunction with recursive OLS and the computation is very efficient.

The spatial representation of time in the time-delay neural network model is inconvenient and also the use of temporal window imposes a limit on the sequence length. Recurrent RBF networks, which combine features from the recurrent network and the RBF network, are suitable for the modeling of nonlinear dynamic systems [12]. The recurrent RBF network introduced in [12] has a four-layer architecture, with an input layer, an RBF layer, a state layer, and a single-neuron output layer. The state and output layers use the sigmoidal activation function.

Real-time approximators for continuous-time dynamical systems with many inputs are presented in [30]. These approximators employ a self-organizing RBF network, whose structure varies dynamically to keep a specified approximation accuracy by adding or pruning online. The performance of this variable structure RBF network approximator with both the Gaussian RBF and the raised-cosine RBF is analyzed. The compact support of the raised-cosine RBF enables faster training and easier output evaluation of the network, compared to the case with the Gaussian RBF.

12.6 Some Recurrent Models

In [26], the complex-valued recurrent network is used as an associative memory of temporal sequences. It has a much superior ability to deal with temporal sequences than the real-valued counterpart does. One of the examples is memorization of melodies. The network can memorize plural melodies and recall them correctly from any part.

A special kind of recurrent network for online matrix inversion is designed based on a matrix-valued error function instead of a scalar-valued error function in [72]. Its discrete-time model is investigated in [73]. When the linear activation function and a unit step size are used, the discrete-time model reduces exactly to Newton iteration for matrix inversion.

The backpropagation-decorrelation rule [55] combines three principles: one-step backpropagation of errors, the use of the temporal memory in the network dynamics, and the utilization of a reservoir of inner neurons. The algorithm adapts only the output weights of a possibly large network and therefore can learn with a complexity of $O(N)$. A stability analysis of the algorithm is provided based on nonlinear feedback theory in [56]. Backpropagation-decorrelation learning is further enhanced with an efficient online rescaling algorithm to stabilize the network while adapting.

The stability of dynamic BP training is studied by the Lyapunov method in [31]. A robust adaptive gradient-descent algorithm [54] for the recurrent network is similar, in some ways, to the RTRL algorithm in terms of using a specifically designed derivative based on the extended recurrent gradient to approximate the true gradient for real-time learning. It switches the training patterns between standard online BP and RTRL according to the derived convergence and stability conditions so as to optimize the convergence speed of robust adaptive gradient descent and to make an optimal trade-off between the online BP and RTRL training strategies to maximize the learning speed. The optimized adaptive learning maximizes the training speed of the recurrent network for each weight update without violating the stability and convergence criteria. Robust adaptive gradient descent provides improved training speed over RTRL with less discrete-time steps of transit and smaller steady-state error. The method uses three adaptive parameters to adjust the effective adaptive learning rate and to provide guaranteed weight convergence and system stability for training.

The simplex and interior-point algorithms are two effective methods for the LP problem. A one-layer recurrent network with a discontinuous activation function is proposed for LP in [32]. The number of neurons is equal to that of the decision variables. The neural network with a sufficiently high gain is proven to be globally convergent to the optimal solution.

Elman Networks

A simultaneous recurrent network in its simplest form is a three-layer feedforward network with the self-recurrent hidden layer [10]. The Elman recurrent network topology consists of feedback connections from every hidden neuron output to every hidden neuron input via a context layer. It is a special case of a general recurrent network and could thus be trained with full BPTT and its variants. Instead of regarding the hidden layer as self-recurrent, the activities of the hidden neurons are stored into a context layer in each time step and the context layer acts as an additional input to the hidden layer in the next time step. Elman BP is a variant of BPTT(n), with $n = 1$. An Elman recurrent network can simulate any given deterministic finite-state automaton. The attention-gated reinforcement learning scheme [45] affects an amalgamation of BP and reinforcement learning for feedforward networks in classification tasks. These ideas are recast to simultaneous recurrent networks in prediction tasks, resulting in the reimplementation of Elman BP as a reinforcement scheme [19].

Elman networks are not as reliable as some other kinds of networks, because both training and adaptation happen using an approximation of the error gradient due to the delays in Elman networks. For Elman networks, we do not recommend algorithms that take large step sizes, such as trainlm and trainrp. An Elman network needs more hidden neurons in its hidden layer than are actually required for a solution by an other method. While a solution might be available with fewer neurons, the Elman network is less able to find the most appropriate weights for hidden neurons due to the approximated error gradient.

A general recursive Bayesian LM algorithm is derived to sequentially update the weights and the covariance (Hessian) matrix of the Elman network for improved time series modeling in [36]. The approach employs a principled handling of the regularization hyperparameters. The recursive Bayesian LM algorithm outperforms standard RTRL and EKF algorithms for training recurrent networks on time series modeling.

Example 12.3 A problem where temporal patterns are recognized and classified with a spatial pattern is amplitude detection. Amplitude detection requires that a waveform be presented to a network through time, and that the network output the amplitude of the waveform. It demonstrates the Elman network design process.

We create an Elman network with one hidden layer of 10 nodes, and the maximum number of epochs is 1000. The training algorithm is gradient descent. The performance for training and generalization is shown in Fig. 12.5. It is shown that the trained Elman network has good generalization performance for an input signal of varying amplitude.

12.7 Reservoir Computing

Reservoir computing is a general paradigm for state-dependent computation in cortical networks [6]. It is a technique for efficient training of recurrent networks. It refers to a class of state-space models with a fixed state transition structure (the reservoir) and an adaptable readout form the state space. The reservoir is supposed to be sufficiently complex so as to capture a large number of features of the input stream that can be exploited by the reservoir-to-output readout mapping.

Reservoir computing models are dynamical models for processing time series that make a conceptual separation of the temporal data processing into two parts: representation of temporal structure in the input stream through a nonadaptable dynamic

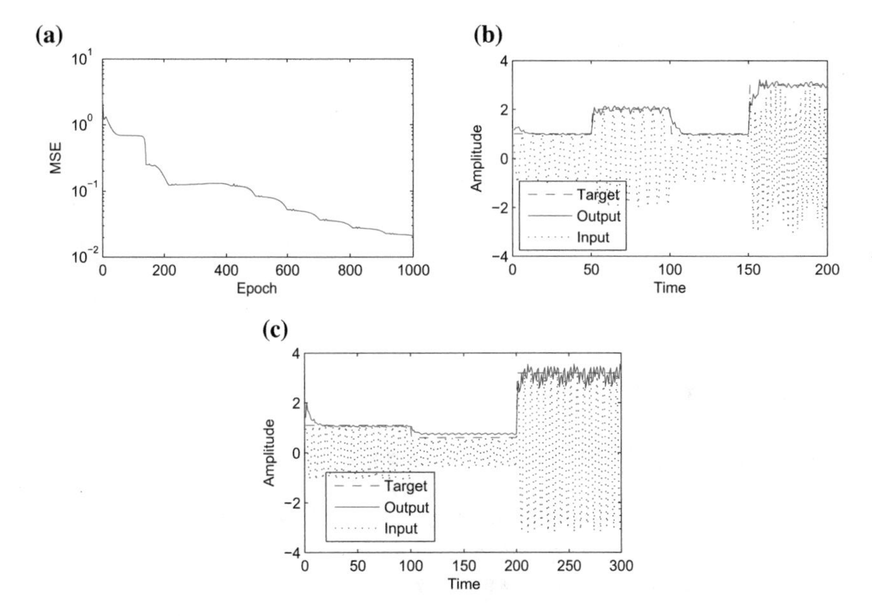

Fig. 12.5 Training the Elman network for amplitude detection. **a** The training MSE. **b** The network output after training. **c** The generalization of the trained network for different input amplitudes

reservoir, and a memoryless easy-to-adapt readout from the reservoir. Reservoir computing subsumes the idea of using general dynamical systems, the so-called reservoirs, in conjunction with trained memoryless readout functions as computational devices. Linear reservoir systems with either polynomial or neural network readout maps are universal approximators [17].

Two notable issues plaguing recurrent networks trained using direct gradient-descent methods are the vanishing and exploding gradient problems [2]. For sigmoid and tanh activation functions, the error flow tends to vanish for learning long-term dependence. As an alternative method, reservoir computing [22, 34] is free from the problems associated with gradient-based training, such as slow convergence, local optima, and computational complexity.

The idea of using a randomly connected recurrent network for online computation on an input sequence was introduced in the echo state network [22] and liquid state machine [34]. A large number of randomly connected neurons form a reservoir providing memory for different aspects of the signals. Readout neurons extract from this reservoir stable information in real time. Echo state networks use analog sigmoid neurons, and liquid state machines use spiking neurons.

An input/output system is said to have fading memory when the outputs associated to inputs that are close in the recent past are close. Echo state networks are universal uniform approximants in the context of discrete-time fading memory filters with uniformly bounded inputs defined on negative infinite times [18].

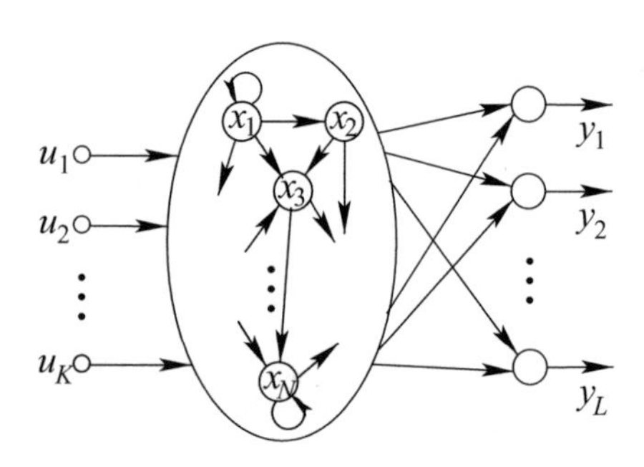

Fig. 12.6 Echo state network architecture

Echo State Networks

For efficient training of large recurrent networks, instead of training all the weights, the weights are initialized randomly and the desired function is implemented by a full instantaneous linear mapping of the neuron states. An echo state network is a recurrent discrete-time network with a nontrainable sparse recurrent part (reservoir) and a simple linear readout, as shown in Fig. 12.6. It uses analog sigmoidal neurons as network units.

Assume that the echo state network has K input units, N internal (reservoir) units and L output units, where $\boldsymbol{u} = (u_1, u_2, \ldots, u_K)^T$ is the input vector, $\boldsymbol{x} = (x_1, x_2, \ldots, x_N)^T$ is the internal state vector, and $\boldsymbol{y} = (y_1, y_2, \ldots, y_L)^T$ is the output vector. The discrete form of the dynamics is defined by

$$\boldsymbol{x}(t) = \sigma\big(\mathbf{A}\boldsymbol{x}(t-1) + \mathbf{C}\boldsymbol{u}(t) + \boldsymbol{b}\big) \tag{12.3}$$

$$\boldsymbol{y}(t) = \mathbf{W}\boldsymbol{x}(t), \tag{12.4}$$

where $\mathbf{C} \in R^{N \times K}$ is the input mask, $\boldsymbol{b} \in R^N$ is the input shift, and $\mathbf{A} \in R^{N \times N}$ is called *reservoir matrix*, and σ is the activation function with componentwise application of a sigmoid function. The readout matrix $\mathbf{W} \in R^{L \times N}$.

The performance is largely independent of the sparsity of the network or the exact network topology. Connection weights in the reservoir and the input weights are randomly generated. The reservoir weights are scaled so as to ensure the echo state property: the reservoir state is an echo of the entire input history. The echo state property is a condition of asymptotic state convergence of the reservoir network, influenced by driving input. A memoryless readout device is then trained in order to approximate from this echo a given time-invariant target operator with fading memory, whereas the network itself remains untrained.

The reservoir of a quantized echo state network is defined as a network of discrete-valued units, where the number of admissible states of a single unit is controlled by a parameter called state resolution, which is measured in bits. Liquid state machines and echo state networks can thus be interpreted as the two limiting cases of quantized echo state networks for low and high state resolution, respectively.

Reservoir construction is largely driven by a series of randomized model-building stages, which rely on a series of trials and errors. Typical model construction decision of an echo state network involves setting the reservoir size, the sparsity of the reservoir and input connections, the ranges for random input and reservoir weights, and the reservoir matrix scaling parameter α. A simple, deterministically constructed cycle reservoir is comparable to the standard echo state network methodology [44]. The (short-term) memory capacity of linear cyclic reservoirs can be made arbitrarily close to the proved optimal value.

The echo state property [22] asserts that initial conditions will be asymptotically washed out. It guarantees stability of the dynamic system and prevents chaotic behavior. For hyperbolic tangent activation units, the echo state property is empirically observed to hold if the spectral radius (the largest eigenvalue of $\mathbf{W}$) is less than one, $\rho < 1$. The closer ρ is to 1, the longer the temporal correlations between the reservoir states and hence, the memory of the system.

Online echo state Gaussian process (OESGP) and online infinite echo state Gaussian process (OIESGP) [53] are two reservoir inspired methods. Both algorithms are iterative fixed-budget methods that learn from noisy time series. OESGP combines the echo state network with Bayesian online learning for Gaussian processes. Extending this to infinite reservoirs yields OIESGP, which uses a recursive kernel with automatic relevance determination that enables spatial and temporal feature weighting. OIESGP obviates the need to create and maintain an explicit reservoir. Using natural-gradient descent to adapt the hyperparameters, OIESGP learns the impact of not only the spatial features but also the relevancy of the past.

Small-world architectures are common in biological neuronal networks such as cortical neural connectivity and many real-world networks. A small-world network [65] is a type of graph in which the nodes are highly clustered compared to a random graph, and a short path length exists between any two nodes in the network. An echo state network having the small-world topology as a reservoir provides a stable echo state property over a broad range of the weight matrix and efficiently propagates input signals to the output nodes [24]. Small-world topology is essential for maintaining the echo state property, which is the appropriate neural dynamics between input and output brain regions.

Liquid State Machines

Liquid state machine has been introduced as a spiking neural network model, inspired by the structural and functional organization of the mammalian neocortex [34]. It uses spiking neurons connected by dynamic synapses to project inputs into a high-dimensional feature space, allowing classification of inputs by linear separation, similar to the SVM approach. The probability that two neurons are connected depends on the distance between their positions. Such a reservoir is called a *liquid*, as it exhibits ripples in response to stimulation inputs. Any time-invariant filter with the fading memory property can be approximated by liquid state machines to any accuracy [33, 34].

The key idea is to use a large but fixed recurrent part as a reservoir of dynamic features and to train only the output layer to extract the desired information. The training thus consists of a linear regression problem. Liquid state machines exploit the power of recurrent spiking neural networks without training the spiking neural network. They can yield competitive results; however, the process can require numerous time-consuming epochs.

Short-Term Memory Capacity of Reservoir Computing

Reservoir computing has been used to quantify the short-term memory capacity in several network architectures, such as spiking networks [34, 62], discrete-time networks [23], and continuous-time networks [7]. These analyses show that even under optimal conditions, the short-term memory capacity (i.e., the length of the stimulus the network is able to recover) scales linearly with the number of nodes in the reservoir.

For echo state networks, short-term memory capacity is defined to measure the network ability to reconstruct the past information fed to the network from the reservoir on the network output by computing correlations [23]:

$$MC = \sum_{k=1}^{k_{\max}} MC_k = \sum_{k=1}^{k_{\max}} \frac{\mathrm{cov}^2(u(t-k), y_k(t))}{\mathbf{var}(u(t)) \cdot \mathbf{var}(y_k(t))}, \tag{12.5}$$

where cov denotes the covariance of the two time series, var denotes variance, $k_{\max} = \infty$, $u(t-k)$ is the input presented k steps before the current input, and $y_k(t) = \boldsymbol{w}_k^{out}\boldsymbol{x}(t) = \tilde{u}(t-k)$ is its reconstruction at the network output (using linear readout), with $\boldsymbol{w}_k^{out}$ being the weight vector of the kth output unit. Computation of MC is approximated using $k_{\max} = L$ (the number of the output neurons).

A reservoir with $L+1$ units with the last neuron serving as a mere delay of the input can mimic the behavior of a reservoir which allows direct input–output connections. It is proved in [23] that the memory capacity of recalling an independent, identically distributed (i.i.d.) input by an L-unit echo state network with identity activation function is bounded by L.

In discrete echo state networks, the uniqueness of trajectories is guaranteed by the echo state property [23, 35] which, however, does not ensure robustness. The short-term memory capacity still scales linearly with the network size [57]. In case of randomly connected spiking networks, short-term memory is much longer in the neutrally stable regime than in the converging or diverging regimes [4, 7]. The short-term memory capacity is affected by the reservoir connectivity. However, these studies assume low connectivity.

In case of higher reservoir connectivity, the short-term memory capacity is reduced [62]. In contrast, linear echo state networks with high connectivities (appropriately normalized) [7] can have relatively large short-term memory capacities, reaching the number of nodes [57].

Networks are able to exhibit longest memories if they operated near the edge of chaos, at the critical border between a stable and an unstable (chaotic) dynamics regime [4, 7].

In [11], the memory capacity is systematically analyzed from the input and reservoir weights scaling, reservoir size, and its sparsity. Two gradient-descent-based orthogonalization procedures for recurrent weights matrix are derived to considerably increase the memory capacity, approaching the upper bound, which is equal to the reservoir size, as proved for linear reservoirs [11].

Problems

12.1 Consider the Machey–Glass chaotic time series, generated by

$$\frac{dx(t)}{dt} = \frac{ax(t-\tau)}{1+x^{10}(t-\tau)} - bx(t)$$

with the parameter $a = 0.2$, $b = 0.1$, and $\tau = 21$.

(a) Design a time-delay neural network to predict 5 samples into the future.
(b) Train the time-delay neural network with 500 points of the time series, and then use the trained network to predict on the next 200 points. Plot the prediction results, and give the prediction performance.

12.2 In order to use BPTT, construct a multilayer feedforward network by unfolding the recurrent network shown in Fig. 12.7.

12.3 Consider the following problem in discrete time:

$$y(k+1) = 0.5y(k) + 0.1y(k)\left[\sum_{i=0}^{7} y(k-i)\right] + 1.2u(k-9)u(k) + 0.2,$$

where random input $u(k)$ is uniformly drawn from [0, 1] as input. Predict the next output $y(k+1)$.

12.4 An autoregressive AR(4) complex process is given by

$$r(k) = 1.8r(k-1) - 2.0r(k-2) + 1.2r(k-3) - 0.4r(k-4) + n(k),$$

where the input $n(k)$ is colored noise. Predict the output using the RBF-AR model.

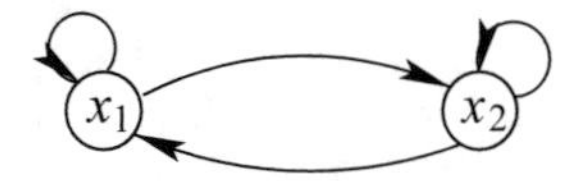

Fig. 12.7 Figure for Problem 12.2

References

1. Almeida, L. B. (1987). A learning rule for asynchronous perceptrons with feedback in combinatorial environment. In *Proceedings of the IEEE 1st International Conference on Neural Networks* (pp. 609–618). San Diego, CA.
2. Bengio, Y., Simard, P., & Frasconi, P. (1994). Learning long-term dependencies with gradient descent is difficult. *IEEE Transactions on Neural Networks*, *5*(2), 157–166.
3. Berthold, M. R. (1994). A time delay radial basis function network for phoneme recognition. In *Proceedings of IEEE International Conference on Neural Networks* (Vol. 7, pp. 4470–4473). Orlando, FL.
4. Bertschinger, N., & Natschlager, T. (2004). Real-time computation at the edge of chaos in recurrent neural networks. *Neural Computation*, *16*, 1413–1436.
5. Billings, S. A., & Hong, X. (1998). Dual-orthogonal radial basis function networks for nonlinear time series prediction. *Neural Networks*, *11*, 479–493.
6. Buonomano, D. V., & Maass, W. (2009). State-dependent computations: spatiotemporal processing in cortical networks. *Nature Reviews Neuroscience*, *10*, 113–125.
7. Busing, L., Schrauwen, B., & Legenstein, R. (2010). Connectivity, dynamics, and memory in reservoir computing with binary and analog neurons. *Neural Computation*, *22*, 1272–1311.
8. Cai, X., Prokhorov, D. V., & Wunsch, D. C, I. I. (2007). Training winner-take-all simultaneous recurrent neural networks. *IEEE Transactions on Neural Networks*, *18*(3), 674–684.
9. Coelho, P. H. G. (2001). A complex EKF-RTRL neural network. In *Proceedings of the International Joint Conference on Neural Networks (IJCNN)* (Vol. 1, pp. 120–125).
10. Elman, J. L. (1990). Finding structure in time. *Cognitive Science*, *14*, 179–211.
11. Farkas, I., Bosak, R., & Gergel, P. (2016). Computational analysis of memory capacity in echo state networks. *Neural Networks*, *83*, 109–120.
12. Frascon, P., Cori, M., Maggini, M., & Soda, G. (1996). Representation of finite state automata in recurrent radial basis function networks. *Machine Learning*, *23*, 5–32.
13. Funahashi, K. I., & Nakamura, Y. (1993). Approximation of dynamical systems by continuous time recurrent neural networks. *Neural Networks*, *6*(6), 801–806.
14. Goh, S. L., & Mandic, D. P. (2004). A complex-valued RTRL algorithm for recurrent neural networks. *Neural Computation*, *16*, 2699–2713.
15. Goh, S. L., & Mandic, D. P. (2007). An augmented CRTRL for complex-valued recurrent neural networks. *Neural Networks*, *20*(10), 1061–1066.
16. Goh, S. L., & Mandic, D. P. (2007). An augmented extended Kalman filter algorithm for complex-valued recurrent neural networks. *Neural Computation*, *19*, 1039–1055.
17. Gonon, L., & Ortega, J.-P. (2019). Reservoir computing universality with stochastic inputs. *IEEE Transactions on Neural Networks and Learning Systems*. https://doi.org/10.1109/TNNLS.2019.2899649.
18. Grigoryeva, L., & Ortega, J.-P. (2018). Echo state networks are universal. *Neural Networks*, *108*, 495–508.
19. Gruning, A. (2007). Elman backpropagation as reinforcement for simple recurrent networks. *Neural Computation*, *19*, 3108–3131.
20. Hassoun, M. H. (1995). *Fundamentals of artificial neural networks*. Cambridge: MIT Press.
21. Ilin, R., Kozma, R., & Werbos, P. J. (2008). Beyond feedforward models trained by backpropagation: A practical training tool for a more efficient universal approximator. *IEEE Transactions on Neural Networks*, *19*(6), 929–937.
22. Jaeger, H. (2001). *The "echo state" approach to analyzing and training recurrent neural networks*. GMD Technical Report 148. Sankt Augustin, Germany: German National Research Center for Information Technology.
23. Jaeger, H. (2001). *Short term memory in echo state networks*. GMD Technical Report 152. Sankt Augustin, Germany: German National Research Center for Information Technology.
24. Kawai, Y., Park, J., & Asada, M. (2019). A small-world topology enhances the echo state property and signal propagation in reservoir computing. *Neural Networks*, *112*, 15–23.

25. Kechriotis, G., & Manolakos, E. (1994). Training fully recurrent neural networks with complex weights. *IEEE Transactions on Circuits and Systems II*, *41*(3), 235–238.
26. Kinouchi, M., & Hagiwara, M. (1996). Memorization of melodies by complex-valued recurrent neural network. In *Proceedings of International Conference on Neural Networks* (Vol. 3, pp. 1324–1328). Washington, DC.
27. Li, L. K. (1992). Approximation theory and recurrent networks. In *Proceedings of the International Joint Conference on Neural Networks (IJCNN)* (pp. 266–271). Baltimore, MD.
28. Li, X., & Yu, W. (2002). Dynamic system identification via recurrent multilayer perceptrons. *Information Sciences*, *147*, 45–63.
29. Li, X. D., Ho, J. K. L., & Chow, T. W. S. (2005). Approximation of dynamical time-variant systems by continuous-time recurrent neural networks. *IEEE Transactions on Circuits and Systems*, *52*(10), 656–660.
30. Lian, J., Lee, Y., Sudhoff, S. D., & Zak, S. H. (2008). Self-organizing radial basis function network for real-time approximation of continuous-time dynamical systems. *IEEE Transactions on Neural Networks*, *19*(3), 460–474.
31. Liang, J., & Gupta, M. M. (1999). Stable dynamic backpropagation learning in recurrent neural networks. *IEEE Transactions on Neural Networks*, *10*(6), 1321–1334.
32. Liu, Q., & Wang, J. (2008). A one-layer recurrent neural network with a discontinuous activation function for linear programming. *Neural Computation*, *20*, 1366–1383.
33. Maass, W., & Markram, H. (2004). On the computational power of circuits of spiking neurons. *Journal of Computer and System Sciences*, *69*(4), 593–616.
34. Maass, W., Natschlager, T., & Markram, H. (2002). Real-time computing without stable states: A new framework for neural computation based on perturbations. *Neural Computation*, *14*(11), 2531–2560.
35. Manjunath, G., & Jaeger, H. (2013). Echo state property linked to an input: Exploring a fundamental characteristic of recurrent neural networks. *Neural Computation*, *25*, 671–696.
36. Mirikitani, D. T., & Nikolaev, N. (2010). Recursive Bayesian recurrent neural networks for time-series modeling. *IEEE Transactions on Neural Networks*, *21*(2), 262–274.
37. Mandic, D. P., & Chambers, J. A. (2000). A normalised real time recurrent learning algorithm. *Signal Processing*, *80*, 1909–1916.
38. Menguc, E. C., & Acir, N. (2018). Kurtosis-based CRTRL algorithms for fully connected recurrent neural networks. *IEEE Transactions on Neural Networks and Learning Systems*, *29*(12), 6123–6131.
39. Patan, K. (2008). Approximation of state-space trajectories by locally recurrent globally feedforward neural networks. *Neural Networks*, *21*, 59–64.
40. Pearlmutter, B. A. (1989). Learning state space trajectories in recurrent neural networks. In *Proceedings of the IEEE International Joint Conference on Neural Networks (IJCNN)* (pp. 365–372). Washington, DC.
41. Pearlmutter, B. A. (1995). Gradient calculations for dynamic recurrent neural networks: A survey. *IEEE Transactions on Neural Networks*, *6*(5), 1212–1228.
42. Peng, H., Ozaki, T., Haggan-Ozaki, V., & Toyoda, Y. (2003). A parameter optimization method for radial basis function type models. *IEEE Transactions on Neural Networks*, *14*(2), 432–438.
43. Pineda, F. J. (1987). Generalization of back-propagation to recurrent neural networks. *Physical Review Letters*, *59*, 2229–2232.
44. Rodan, A., & Tino, P. (2011). Minimum complexity echo state network. *IEEE Transactions on Neural Networks*, *22*(1), 131–144.
45. Roelfsema, P. R., & van Ooyen, A. (2005). Attention-gated reinforcement learning of internal representations for classification. *Neural Computation*, *17*, 1–39.
46. Rumelhart, D. E., Hinton, G. E., & Williams, R. J. (1986). Learning internal representations by error propagation. In D. E. Rumelhart & J. L. McClelland (Eds.), *Parallel distributed processing: Explorations in the microstructure of cognition, 1: Foundation* (pp. 318–362). Cambridge: MIT Press.
47. Schafer, A. M., & Zimmermann, H. G. (2006). Recurrent neural networks are universal approximators. In *Proceedings of the 16th International Conference on Artificial Neural Networks (ICANN), LNCS* (Vol. 4131, pp. 632–640). Berlin: Springer.

48. Sejnowski, T., & Rosenberg, C. (1986). *NETtalk: A parallel network that learns to read aloud*. Technical Report JHU/EECS-86/01, Johns Hopkins University.
49. Shamma, S. A. (1989). Spatial and temporal processing in cellular auditory network. In C. Koch & I. Segev (Eds.), *Methods in neural modeling* (pp. 247–289). Cambridge: MIT Press.
50. Sheta, A. F., & De Jong, K. (2001). Time series forecasting using GA-tuned radial basis functions. *Information Sciences*, *133*, 221–228.
51. Siegelmann, H. T., & Sontag, E. D. (1995). On the computational power of neural nets. *Journal of Computer and System Sciences*, *50*(1), 132–150.
52. Siegelmann, H. T., Horne, B. G., & Giles, C. L. (1997). Computational capabilities of recurrent NARX neural networks. *IEEE Transactions on Systems, Man, and Cybernetics Part B*, *27*(2), 208–215.
53. Soh, H., & Demiris, Y. (2015). Spatio-temporal learning with the online finite and infinite echo-state Gaussian processes. *IEEE Transactions on Neural Networks and Learning Systems*, *26*(3), 522–536.
54. Song, Q., Wu, Y., & Soh, Y. C. (2008). Robust adaptive gradient-descent training algorithm for recurrent neural networks in discrete time domain. *IEEE Transactions on Neural Networks*, *19*(11), 1841–1853.
55. Steil, J. J. (2004). Backpropagation-decorrelation: Recurrent learning with $O(N)$ complexity. In *Proceedings of the IEEE International Joint Conference on Neural Networks (IJCNN)* (Vol. 1, pp. 843–848).
56. Steil, J. J. (2006). Online stability of backpropagation-decorrelation recurrent learning. *Neurocomputing*, *69*, 642–650.
57. Strauss, T., Wustlich, W., & Labahn, R. (2012). Design strategies for weight matrices of echo state networks. *Neural Computation*, *24*, 3246–3276.
58. Tokuda, I., Tokunaga, R., & Aihara, K. (2003). Back-propagation learning of infinite-dimensional dynamical systems. *Neural Networks*, *16*, 1179–1193.
59. Tsoi, A. C., & Back, A. D. (1994). Locally recurrent globally feedforward networks: A critical review of architectures. *IEEE Transactions on Neural Networks*, *5*(2), 229–239.
60. Vesin, J. (1993). An amplitude-dependent autoregressive signal model based on a radial basis function expansion. In *Proceedings of International Conference on Acoustics, Speech, and Signal Processing (ICASSP)* (Vol. 3, pp. 129–132). Minneapolis, MN.
61. Waibel, A., Hanazawa, T., Hinton, G., Shikano, K., & Lang, K. J. (1989). Phoneme recognition using time-delay neural networks. *IEEE Transactions on Acoustics, Speech and Signal Processing*, *37*(3), 328–339.
62. Wallace, E., Hamid, R., & Latham, P. (2013). Randomly connected networks have short temporal memory. *Neural Computation*, *25*, 1408–1439.
63. Wan, E. A. (1990). Temporal backpropagation for FIR neural networks. In *Proceedings of the IEEE International Joint Conference on Neural Networks (IJCNN)* (pp. 575–580). San Diego, CA.
64. Wan, E. A. (1994). Time series prediction by using a connectionist network with internal delay lines. In A. S. Weigend & N. A. Gershenfeld (Eds.), *Time series prediction: Forcasting the future and understanding the past* (pp. 195–217). Reading: Addison-Wesley.
65. Watts, D. J., & Strogatz, S. H. (1998). Collective dynamics of small-world networks. *Nature*, *393*, 440–442.
66. Werbos, P. J. (1990). Backpropagation through time: What it does and how to do it. *Proceedings of the IEEE*, *78*(10), 1550–1560.
67. Werbos, P. J., & Pang, X. (1996). Generalized maze navigation: SRN critics solve what feedforward or Hebbian cannot. In *Proceedings of the IEEE International Conference on Systems, Man, and Cybernetics* (Vol. 3, pp. 1764–1769).
68. Williams, R. J., & Zipser, D. (1989). A learning algorithm for continually running fully recurrent neural networks. *Neural Computation*, *1*(2), 270–280.
69. Williams, R. J., & Zipser, D. (1995). Gradient-based learning algorithms for recurrent networks and their computational complexity. In Y. Chauvin & D. E. Rumelhart (Eds.), *Backpropagation: Theory, architecture, and applications* (pp. 433–486). Hillsdale: Lawrence Erlbaum.

70. Wu, F.-X. (2011). Delay-independent stability of genetic regulatory networks. *IEEE Transactions on Neural Networks*, *22*(11), 1685–1692.
71. Yu, D. L. (2004). A localized forgetting method for Gaussian RBFN model adaptation. *Neural Processing Letters*, *20*, 125–135.
72. Zhang, Y., Jiang, D., & Wang, J. (2002). A recurrent neural network for solving Sylvester equation with time-varying coefficients. *IEEE Transactions on Neural Networks*, *13*(5), 1053–1063.
73. Zhang, Y., Ma, W., & Cai, B. (2009). From Zhang neural network to Newton iteration for matrix inversion. *IEEE Transactions on Circuits and Systems Part I*, *56*(7), 1405–1415.

Chapter 13
Principal Component Analysis

13.1 Introduction

Most signal-processing problems can be reduced to some form of eigenvalue or singular value problems. EVD and SVD are usually used for solving these problems. PCA is a classical statistical method for data analysis [50]. It is related to EVD and SVD. Minor component analysis (MCA), as a variant of PCA, is most useful for solving the total least squares (TLS) problem. There are many neural network models and algorithms for PCA, MCA, and SVD. These algorithms are typically based on unsupervised learning. They significantly reduce the cost for adaptive signal, speech, image and video processing, pattern recognition, data compression and coding, high-resolution spectrum analysis, and array signal processing [26].

Stochastic approximation theory [81], first introduced by Robbins and Monro in [115], is now an important tool for analyzing stochastic discrete-time systems including the classical gradient-descent method.

Given a stochastic discrete-time system of the form

$$\Delta \boldsymbol{z}(t) = \boldsymbol{z}(t+1) - \boldsymbol{z}(t) = \eta(t) \left(\boldsymbol{f}(\boldsymbol{z}, t) + \boldsymbol{n}(t) \right), \tag{13.1}$$

where $\boldsymbol{z}$ is the state vector, $\boldsymbol{f}(\boldsymbol{z}, t)$ is a finite nonzero vector with functions as entries, and $\boldsymbol{n}(t)$ is an unbiased noisy term at a particular instant. The continuous-time representation is very useful for analyzing the asymptotic behavior of the algorithm. According to stochastic approximation theory, assuming that $\{\eta(t)\}$ is a sequence of positive numbers satisfying the Robbins–Monro conditions [81]

$$\sum_{t=1}^{\infty} \eta(t) = \infty, \qquad \sum_{t=1}^{\infty} \eta^2(t) < \infty, \tag{13.2}$$

the analysis of the stochastic system (13.1) can be transformed into the analysis of a deterministic differential equation

K.-L. Du and M. N. S. Swamy, *Neural Networks and Statistical Learning*,
https://doi.org/10.1007/978-1-4471-7452-3_13

$$\frac{dz}{dt} = f(z, t). \tag{13.3}$$

If all the trajectories of (13.3) converge to a fixed-point z^*, the discrete-time system $z(t) \to z^*$ with probability one as $t \to \infty$. By (13.2), it is required that $\eta(t) \to 0$ as $t \to \infty$. $\eta(t)$ is typically selected as $\eta(t) = \frac{1}{\alpha+t}$ with a constant $\alpha \geq 0$, or as $\eta(t) = \frac{1}{t^\beta}$, $\frac{1}{2} \leq \beta \leq 1$ [99, 120].

13.1.1 Hebbian Learning Rule

The classical Hebbian synaptic modification rule [47] states that biological synaptic weights change in proportion to the correlation between the pre- and postsynaptic signals. For a single neuron, the Hebbian rule can be written as

$$\boldsymbol{w}(t+1) = \boldsymbol{w}(t) + \eta y(t)\boldsymbol{x}_t, \tag{13.4}$$

where the learning rate $\eta > 0$, $\boldsymbol{w} \in R^n$ is the weight vector, $\boldsymbol{x}_t \in R^n$ is an input vector at time t, and $y(t)$ is the neuron output defined by

$$y(t) = \boldsymbol{w}^T(t)\boldsymbol{x}_t. \tag{13.5}$$

For stochastic input vector $\boldsymbol{x}$, assuming that $\boldsymbol{x}$ and $\boldsymbol{w}$ are uncorrelated, the expected weight change is given by

$$\mathrm{E}[\Delta \boldsymbol{w}] = \eta \mathrm{E}[y\boldsymbol{x}] = \eta \mathrm{E}\left[\boldsymbol{x}\boldsymbol{x}^T\boldsymbol{w}\right] = \eta \mathbf{C}\mathrm{E}[\boldsymbol{w}], \tag{13.6}$$

where $\mathrm{E}[\cdot]$ is the expectation operator and $\mathbf{C} = \mathrm{E}\left[\boldsymbol{x}\boldsymbol{x}^T\right]$ is the autocorrelation matrix of $\boldsymbol{x}$.

At equilibrium, $\mathrm{E}[\Delta \boldsymbol{w}] = \mathbf{0}$; hence, we have a deterministic equation $\mathbf{C}\boldsymbol{w} = \mathbf{0}$. Due to the effect of noise terms, $\mathbf{C}$ is a full-rank positive-definite Hermitian matrix with positive eigenvalues λ_i, $i = 1, 2, \ldots, n$, and the corresponding orthogonal eigenvectors $\boldsymbol{c}_i$, where $n = \mathrm{rank}(\mathbf{C})$. Thus, $\boldsymbol{w} = \mathbf{0}$ is the only equilibrium state.

Equation (13.4) can be represented in continuous-time form

$$\dot{\boldsymbol{w}} = y\boldsymbol{x}. \tag{13.7}$$

Taking statistical averaging, we have

$$\mathrm{E}[\dot{\boldsymbol{w}}] = \mathrm{E}[y\boldsymbol{x}] = \mathbf{C}\mathrm{E}[\boldsymbol{w}]. \tag{13.8}$$

This can be derived by minimizing the average instantaneous criterion [44]

$$\mathrm{E}\left[E_{\text{Hebb}}\right] = -\frac{1}{2}\mathrm{E}\left[y^2\right] = -\frac{1}{2}\mathrm{E}\left[\boldsymbol{w}^T\right]\mathbf{C}\mathrm{E}[\boldsymbol{w}], \tag{13.9}$$

where E_{Hebb} is the instantaneous criterion. At equilibrium, $\mathrm{E}\left[\frac{\partial E_{\text{Hebb}}}{\partial \boldsymbol{w}}\right] = -\mathbf{C}\mathrm{E}[\boldsymbol{w}] = \mathbf{0}$, thus $\boldsymbol{w} = \mathbf{0}$. Since the Hessian $\mathrm{E}[\mathbf{H}(\boldsymbol{w})] = \mathrm{E}\left[\frac{\partial E_{\text{Hebb}}^2}{\partial^2 \boldsymbol{w}}\right] = -\mathbf{C}$ is nonpositive, the solution $\boldsymbol{w} = \mathbf{0}$ is unstable, which drives $\boldsymbol{w}$ to infinite magnitude with a direction parallel to that of the eigenvector of $\mathbf{C}$ corresponding to the largest eigenvalue [44].

To prevent the divergence of the Hebbian rule, one can normalize $\|\boldsymbol{w}\|$ to unity after each iteration [117], and this leads to the normalized Hebbian rule. Other methods such as Oja's rule [97], Yuille's rule [152], Linsker's rule [78, 79], and Hassoun's rule [44] add a weight-decay term to the Hebbian rule to stabilize the algorithm.

13.1.2 Oja's Learning Rule

Oja's rule introduces a weight-decay term into the Hebbian rule and is given by [97]

$$\boldsymbol{w}(t+1) = \boldsymbol{w}(t) + \eta y(t)\boldsymbol{x}_t - \eta y^2(t)\boldsymbol{w}(t). \tag{13.10}$$

Oja's rule converges to a state that minimizes (13.9) subject to $\|\boldsymbol{w}\| = 1$. The solution is the principal eigenvector of $\mathbf{C}$. For small η, Oja's rule is equivalent to the normalized Hebbian rule [97].

The continuous-time version of Oja's rule is given by a nonlinear stochastic differential equation

$$\dot{\boldsymbol{w}} = \eta\left(y\boldsymbol{x} - y^2\boldsymbol{w}\right). \tag{13.11}$$

The corresponding deterministic equation based on statistical averaging is thus derived as

$$\dot{\boldsymbol{w}} = \eta\left[\mathbf{C}\boldsymbol{w} - \left(\boldsymbol{w}^T\mathbf{C}\boldsymbol{w}\right)\boldsymbol{w}\right]. \tag{13.12}$$

At equilibrium,

$$\mathbf{C}\boldsymbol{w} = \left(\boldsymbol{w}^T\mathbf{C}\boldsymbol{w}\right)\boldsymbol{w}. \tag{13.13}$$

It is easily seen that the solutions are $\boldsymbol{w} = \pm\boldsymbol{c}_i$, $i = 1, 2, \ldots, n$ with the corresponding eigenvalues λ_i arranged in a descending order $\lambda_1 \geq \lambda_2 \geq \ldots \lambda_n \geq 0$.

Notice that the average Hessian

$$\begin{aligned}\mathbf{H}(\boldsymbol{w}) &= \frac{\partial}{\partial \boldsymbol{w}}\left[-\mathbf{C}\boldsymbol{w} + \left(\boldsymbol{w}^T\mathbf{C}\boldsymbol{w}\right)\boldsymbol{w}\right] \\ &= -\mathbf{C} + \boldsymbol{w}^T\mathbf{C}\boldsymbol{w}\mathbf{I} + 2\boldsymbol{w}\boldsymbol{w}^T\mathbf{C}\end{aligned} \tag{13.14}$$

is positive definite only at $\boldsymbol{w} = \pm\boldsymbol{c}_1$, where $\mathbf{I}$ is the $n \times n$ identity matrix, if $\lambda_1 \neq \lambda_2$ [44]. Thus, Oja's rule always converges to the principal component of $\mathbf{C}$.

The convergence analysis of the stochastic discrete-time algorithms such as the gradient-descent method is conventionally based on stochastic approximation theory [81]. A stochastic discrete-time algorithm is first converted into deterministic continuous-time ordinary differential equations, and then analyzed by using Lyapunov's second theorem. This conversion is based on the Robbins–Monro conditions, which require the learning rate to gradually approach zero as $t \to \infty$. This limitation is not practical for implementation, especially for learning nonstationary data. The stochastic discrete-time algorithms can be converted into their deterministic discrete-time formulations that characterize their average evolution from a conditional expectation perspective [158]. This method has been applied to Oja's rule. Analysis based on this method guarantees the convergence of Oja's rule by selecting some constant learning rate for fast convergence. Oja's rule is proved to almost always converge exponentially to the unit eigenvector associated with the largest eigenvalue of $\mathbf{C}$, starting from points in an invariant set [151]. The initial vectors have been suggested to be selected from the domain of a unit hypersphere to guarantee convergence. It is suggested that $\eta = \frac{0.618}{\lambda_1}$ [151].

13.2 PCA: Conception and Model

PCA is based on the spectral analysis of the second-order moment matrix, called the *correlation matrix*, that statistically characterizes a random vector. In the zero-mean case, the correlation matrix becomes the covariance matrix. For image coding, PCA is known as *Karhunen–Loeve transform* [82], which is an optimal scheme for data compression based on the exploitation of correlation between neighboring pixels or groups of pixels. PCA is directly related to SVD, and the most common way to perform PCA is via SVD of the data matrix.

PCA allows the removal of the second-order correlation among given random processes. By calculating the eigenvectors of the covariance matrix of the input vector, PCA linearly transforms a high-dimensional input vector into a low-dimensional one whose components are uncorrelated. PCA is often based on optimization of some information criterion, such as maximization of the variance of the projected data or minimization of the reconstruction error. The objective of PCA is to extract m orthonormal directions $\overline{\boldsymbol{w}}_i \in R^n, i = 1, 2, \ldots, m$, in the input space that account for as much of the data's variance as possible. Subsequently, an input vector $\boldsymbol{x} \in R^n$ may be transformed into a lower m-dimensional space without losing essential intrinsic information. The vector $\boldsymbol{x}$ can be represented by being projected onto the m-dimensional subspace spanned by $\overline{\boldsymbol{w}}_i$ using the inner products $\boldsymbol{x}^T\overline{\boldsymbol{w}}_i$, yielding dimension reduction.

PCA finds those unit directions $\overline{\boldsymbol{w}} \in R^n$ along which the projections of the input vectors, known as the *principal components*, $y = \boldsymbol{x}^T\overline{\boldsymbol{w}}$, have the largest variance

$$E_{\mathrm{PCA}}(\boldsymbol{w}) = \mathrm{E}\left[y^2\right] = \overline{\boldsymbol{w}}^T\mathbf{C}\overline{\boldsymbol{w}} = \frac{\boldsymbol{w}^T\mathbf{C}\boldsymbol{w}}{\|\boldsymbol{w}\|^2}, \tag{13.15}$$

where $\overline{\boldsymbol{w}} = \frac{\boldsymbol{w}}{\|\boldsymbol{w}\|}$. $E_{\text{PCA}}(\boldsymbol{w})$ is a positive-semidefinite function. Setting $\frac{\partial E_{\text{PCA}}}{\partial \boldsymbol{w}} = \mathbf{0}$, we get

$$\mathbf{C}\boldsymbol{w} = \frac{(\boldsymbol{w}^T \mathbf{C}\boldsymbol{w})}{\|\boldsymbol{w}\|^2}\boldsymbol{w}. \tag{13.16}$$

The solutions to (13.16) are $\boldsymbol{w} = \alpha \boldsymbol{c}_i$, $i = 1, 2, \ldots, n$, where $\alpha \in R$. When $\alpha = 1$, $\boldsymbol{w}$ becomes a unit vector. In PCA, principal components are sometimes called *factors* or *latent variables* of the data.

We now examine the positive definiteness of the Hessian of $E_{\text{PCA}}(\boldsymbol{w})$ at $\boldsymbol{w} = \boldsymbol{c}_i$. Multiplying the Hessian by $\boldsymbol{c}_j$ leads to [44]

$$\mathbf{H}(\boldsymbol{c}_i)\boldsymbol{c}_j = \begin{cases} \mathbf{0} & i = j \\ \left(\lambda_i - \lambda_j\right)\boldsymbol{c}_j & i \neq j \end{cases}. \tag{13.17}$$

Thus, $\mathbf{H}(\boldsymbol{w})$ has the same eigenvectors as $\mathbf{C}$ but with different eigenvalues. $\mathbf{H}(\boldsymbol{w})$ is positive semidefinite only when $\boldsymbol{w} = \boldsymbol{c}_1$. As a result, $\boldsymbol{w}$ will eventually point in the direction of $\boldsymbol{c}_1$ and $E_{\text{PCA}}(\boldsymbol{w})$ takes its maximum value.

By repeating maximization of $E_{\text{PCA}}(\boldsymbol{w})$ but forcing $\boldsymbol{w}$ orthogonal to $\boldsymbol{c}_1$, the maximum of $E_{\text{PCA}}(\boldsymbol{w})$ is equal to λ_2 at $\boldsymbol{w} = \alpha \boldsymbol{c}_2$. Following this deflation procedure, all the m principal directions $\overline{\boldsymbol{w}}_i$ can be derived [44]. The projections $y_i = \boldsymbol{x}^T \overline{\boldsymbol{w}}_i$, $i = 1, 2, \ldots, m$, are the principal components of $\boldsymbol{x}$. The result for two-dimensional input data is illustrated in Fig. 13.1. Each data point is accurately characterized by its projections on the two principal directions $\overline{\boldsymbol{w}}_1 = \frac{\boldsymbol{w}_1}{\|\boldsymbol{w}_1\|}$ and $\overline{\boldsymbol{w}}_2 = \frac{\boldsymbol{w}_2}{\|\boldsymbol{w}_2\|}$. If the data is compressed to one-dimensional space, each data point is then represented by its projection on eigenvector $\overline{\boldsymbol{w}}_1$.

A linear LS estimate $\hat{\boldsymbol{x}}$ can be constructed for the original input $\boldsymbol{x}$

$$\hat{\boldsymbol{x}} = \sum_{i=1}^{m} y_i \overline{\boldsymbol{w}}_i. \tag{13.18}$$

The reconstruction error $\boldsymbol{e}$ is defined by

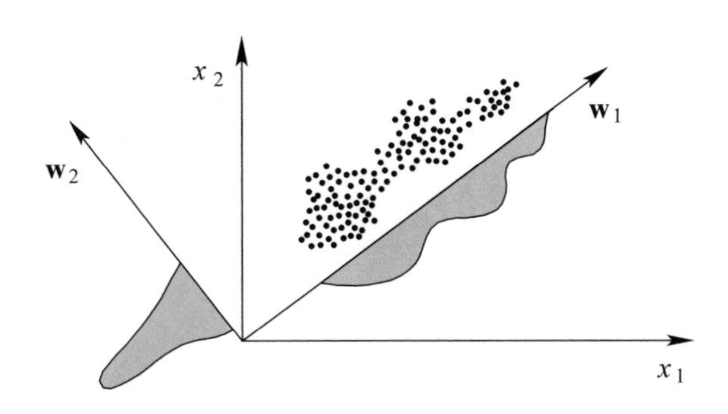

Fig. 13.1 Illustration of PCA in two dimensions

$$e = x - \hat{x} = \sum_{i=m+1}^{n} y_i \overline{w}_i. \tag{13.19}$$

Naturally, e is orthogonal to $\hat{x}$. Each principal component y_i is a Gaussian with zero mean and variance $\sigma_i^2 = \lambda_i$. The variances of x, $\hat{x}$, and e can be, respectively, expressed as

$$\mathrm{E}\left[\|x\|^2\right] = \sum_{i=1}^{n} \sigma_i^2 = \sum_{i=1}^{n} \lambda_i, \tag{13.20}$$

$$\mathrm{E}\left[\|\hat{x}\|^2\right] = \sum_{i=1}^{m} \sigma_i^2 = \sum_{i=1}^{m} \lambda_i, \tag{13.21}$$

$$\mathrm{E}\left[\|e\|^2\right] = \sum_{i=m+1}^{n} \sigma_i^2 = \sum_{i=m+1}^{n} \lambda_i. \tag{13.22}$$

When we use only the first m_1 among the extracted m principal components to represent the raw data, we need to evaluate the error by replacing m by m_1.

Neural PCA originates from the seminal work by Oja [97]. Oja's single-neuron PCA model is illustrated in Fig. 13.2. The output of the neuron is updated by

$$y = w^T x, \tag{13.23}$$

where $w = \left(w_1, \ldots, w_{J_1}\right)^T$. Notice that the activation function is the linear function $\phi(x) = x$.

The PCA network model was first proposed by Oja [99], where a J_1-J_2 feedforward network, as shown in Fig. 13.3, is used to extract the first J_2 principal components. The architecture of the PCA network is a simple expansion of the single-neuron PCA model. The output of the network is given by

$$y = \mathbf{W}^T x, \tag{13.24}$$

where $y = \left(y_1, y_2, \ldots, y_{J_2}\right)^T$, $x = \left(x_1, x_2, \ldots, x_{J_1}\right)^T$, $\mathbf{W} = \left[w_1, w_2, \ldots, w_{J_2}\right]$, and $w_i = \left(w_{1i}, w_{2i}, \ldots, w_{J_1 i}\right)^T$.

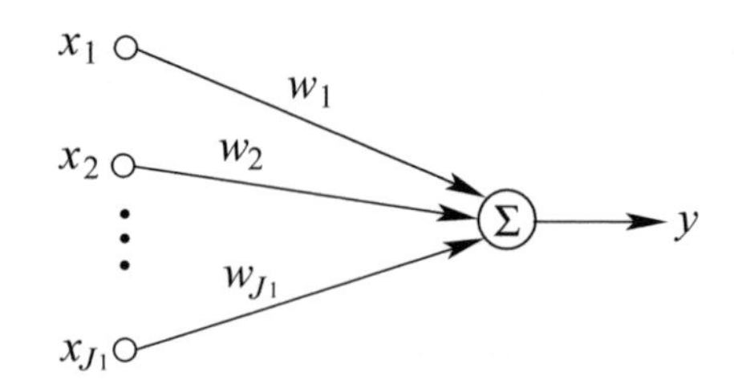

Fig. 13.2 The single-neuron PCA model extracts the first principal component of **C**

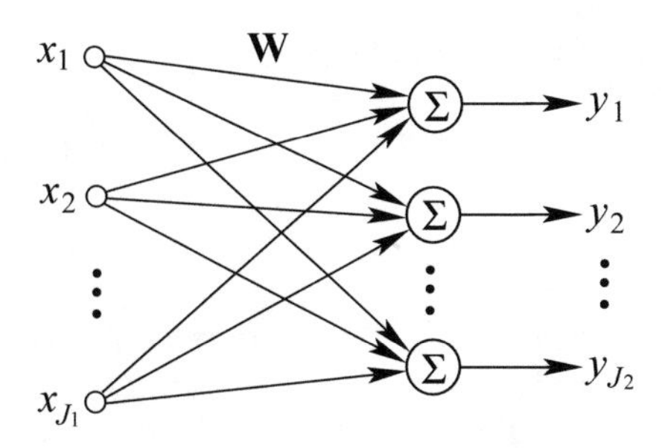

Fig. 13.3 Architecture of the PCA network

13.3 Hebbian Rule-Based PCA

PCA maximizes the output variances $\mathrm{E}\left[y_i^2\right] = \mathrm{E}\left[\left(\boldsymbol{w}_i^T \boldsymbol{x}\right)^2\right] = \boldsymbol{w}_i^T \mathbf{C} \boldsymbol{w}_i$ of the linear network under orthonormality constraints. In the hierarchical case, the constraints take the form $\boldsymbol{w}_i^T \boldsymbol{w}_j = \delta_{ij}$, $j \leq i$, δ_{ij} being the Kronecker delta. In the symmetric case, symmetric orthonormality constraints $\boldsymbol{w}_i^T \boldsymbol{w}_j = \delta_{ij}$ are applied. The subspace learning algorithm (SLA) and generalized Hebbian algorithm (GHA) are corresponding to the symmetric and hierarchical network structures, respectively.

13.3.1 Subspace Learning Algorithms

By using Oja's rule (13.10), $\boldsymbol{w}$ will converge to a unit eigenvector of the correlation matrix $\mathbf{C}$, and the variance of the output y is maximized. For zero-mean input data, this extracts the first principal component [97]. We rewrite (13.10) for convenience of presentation

$$\boldsymbol{w}(t+1) = \boldsymbol{w}(t) + \eta\left[y(t)\boldsymbol{x}_t - y^2(t)\boldsymbol{w}(t)\right], \tag{13.25}$$

where $y(t)\boldsymbol{x}_t$ is the Hebbian term, and $-y^2(t)\boldsymbol{w}(t)$ is a decaying term, which is used to prevent instability. In order to keep the algorithm convergent, $0 < \eta(t) < \frac{1}{1.2\lambda_1}$ is required [99], where λ_1 is the largest eigenvalue of $\mathbf{C}$. If $\eta(t) \geq \frac{1}{\lambda_1}$, $\boldsymbol{w}$ will not converge to $\pm\boldsymbol{c}_1$ [12]. One can select $\eta(t) = 0.5\left[\boldsymbol{x}_t^T \boldsymbol{x}_t\right]$ at the beginning and gradually decrease it [99].

Symmetrical SLA for the PCA network [99] can be derived by maximizing

$$E_{\mathrm{SLA}} = \frac{1}{2}\mathrm{tr}\left(\mathbf{W}^T \mathbf{C} \mathbf{W}\right) \tag{13.26}$$

subject to

$$\mathbf{W}^T \mathbf{W} = \mathbf{I}, \tag{13.27}$$

where $\mathbf{I}$ is the $J_2 \times J_2$ identity matrix.

SLA is given by [99]

$$\boldsymbol{w}_i(t+1) = \boldsymbol{w}_i(t) + \eta(t) y_i(t) \left[\boldsymbol{x}_t - \hat{\boldsymbol{x}}_t \right], \tag{13.28}$$

$$\hat{\boldsymbol{x}}_t = \mathbf{W} \boldsymbol{y}. \tag{13.29}$$

After the algorithm converges, $\mathbf{W}$ is roughly orthonormal and the columns of $\mathbf{W}$, namely, $\boldsymbol{w}_i, i = 1, \ldots, J_2$, converge to some linear combination of the first J_2 principal eigenvectors of $\mathbf{C}$ [99], which is a rotated basis of the dominant eigenvector subspace. This analysis is called the *principal subspace analysis (PSA)*. The value of $\boldsymbol{w}_i$ is dependent on the initial conditions and training samples.

The corresponding eigenvalues $\lambda_i, i = 1, \ldots, J_2$, approximate $E\left[y_i^2\right]$, which can be adaptively estimated by

$$\hat{\lambda}_i(t+1) = \left(1 - \frac{1}{t+1}\right) \hat{\lambda}_i(t) + \frac{1}{t+1} y_i^2(t+1). \tag{13.30}$$

Weighted SLA can be derived by maximizing the same criterion (13.26), but the constraint (13.27) can be modified as [100]

$$\mathbf{W}^T \mathbf{W} = \boldsymbol{\alpha}, \tag{13.31}$$

where $\boldsymbol{\alpha} = \operatorname{diag}\left(\alpha_1, \alpha_2, \ldots, \alpha_{J_2}\right)$ is an arbitrary diagonal matrix with $\alpha_1 > \alpha_2 > \cdots > \alpha_{J_2} > 0$.

Weighted SLA is given by [98, 100]

$$\boldsymbol{w}_i(t+1) = \boldsymbol{w}_i(t) + \eta(t) y_i(t) \left[\boldsymbol{x}_t - \gamma_i \hat{\boldsymbol{x}}_t \right], \quad i = 1, \ldots, J_2, \tag{13.32}$$

$$\hat{\boldsymbol{x}} = \mathbf{W} \boldsymbol{y}, \tag{13.33}$$

where γ_i, $i = 1, \ldots, J_2$, are any coefficients that satisfy $0 < \gamma_1 < \gamma_2 < \ldots < \gamma_{J_2}$. Due to the asymmetry introduced by γ_i, $\boldsymbol{w}_i$ almost surely converges to the eigenvectors of $\mathbf{C}$. Weighted SLA can perform PCA; however, norms of the weight vectors are not equal to unity.

SLA and weighted SLA are nonlocal algorithms that rely on the calculation of the errors and the backward propagation of the values between the layers.

By adding one more term to the PSA algorithm, a PCA algorithm can be obtained [55]. This additional term rotates the basis vectors in the principal subspace toward the principal eigenvectors. PCA derived from SLA is given as [55]

$$\boldsymbol{w}_i(t+1) = \boldsymbol{w}_i(t) + \eta(t) y_i(t) \left[\boldsymbol{x}_t - \hat{\boldsymbol{x}}_t \right] \tag{13.34}$$

$$+ \eta(t) \rho_i \left(y_i(t) \boldsymbol{x}_t - \boldsymbol{w}_i(t) y_i^2(t) \right), \tag{13.35}$$

where $1 > |\rho_1| > |\rho_2| > \ldots > |\rho_{J_2}|$. This PCA algorithm generates weight vectors of unit length.

The adaptive learning algorithm [12] is a PCA algorithm based on SLA. In this method, each neuron adaptively updates its learning rate by

$$\eta_i(t) = \frac{\beta_i(t)}{\hat{\lambda}_i(t)}, \tag{13.36}$$

where $\hat{\lambda}_i(t)$ is the estimated eigenvalue, which can be estimated using (13.30), $\beta_i(t)$ is set to be smaller than $2(\sqrt{2}-1)$ and decreases to zero as $t \to \infty$. If $\beta_i(t)$ is the same for all i, $\boldsymbol{w}_i(t)$ will quickly converge, at nearly the same rate, to $\boldsymbol{c}_i$ for all i in an order of descending eigenvalues. The adaptive learning algorithm [12] converges to the desired target both in the large eigenvalue case as well as in the small eigenvalue case, with performance better than that of GHA [120].

The modulated Hebbian rule [54] is a biologically inspired PSA. In order to achieve orthonormality, the modulated Hebb–Oja rule [54] performs Oja's rule on separate weight vectors with only one difference: learning factor η is specifically programmed by the network. In this case, η decreases as time approaches infinity. Unlike some other recursive PCA/PSA methods that use local feedback connections in order to maintain stability, the modulated Hebb–Oja rule uses global feedback connection. Number of global calculation circuits is two in the modulated Hebb–Oja algorithm and J_1 in SLA. Number of calculations required for SLA is lower than, but very close to, the number of calculations required for modulated Hebb–Oja. For the modulated Hebb–Oja algorithm, we have [56]

$$\Delta w_{kl} = \hat{\eta}(x_k y_l - w_{kl} y_l^2), \tag{13.37}$$

$$\hat{\eta} = \eta(\boldsymbol{x}^T\boldsymbol{x} - \boldsymbol{y}^T\boldsymbol{y}), \tag{13.38}$$

$$\boldsymbol{y} = \mathbf{W}^T\boldsymbol{x}. \tag{13.39}$$

For J_1 inputs and J_2 outputs, $J_2 < J_1$. The modulated Hebb–Oja algorithm is given by

$$\mathbf{W}(t+1) = \mathbf{W}(t) + \eta(t)(\boldsymbol{x}_t^T\boldsymbol{x}_t - \boldsymbol{y}^T(t)\boldsymbol{y}(t)) \cdot (\boldsymbol{x}_t\boldsymbol{y}^T(t) - \mathbf{W}(t)\mathrm{diag}(\boldsymbol{y}(t)\boldsymbol{y}^T(t))). \tag{13.40}$$

Example 13.1 The concept of subspace is involved in many signal-processing problems. This requires the EVD of the autocorrelation matrix of a dataset or the SVD of the cross-correlation matrix of two datasets. This example illustrates the use of weighted SLA for extracting the multiple principal components.

Given a dataset of 2000 vectors $\{\boldsymbol{x}_p \in R^3\}$, where $\boldsymbol{x}_p = (x_{p,1}, x_{p,2}, x_{p,3})^T$. We take $x_{p,1} = 1 + 0.5N(0,1)$, $x_{p,2} = (-1 + 2N(0,1))x_{p,1}$, and $x_{p,3} = (1 + 2N(0,1))x_{p,2}$, where $N(0,1)$ denotes a Gaussian noise term with zero-mean and unit variance. The autocorrelation matrix is calculated as $\mathbf{C} = \frac{1}{2000}\sum_{p=1}^{2000}\boldsymbol{x}_p\boldsymbol{x}_p^T$.

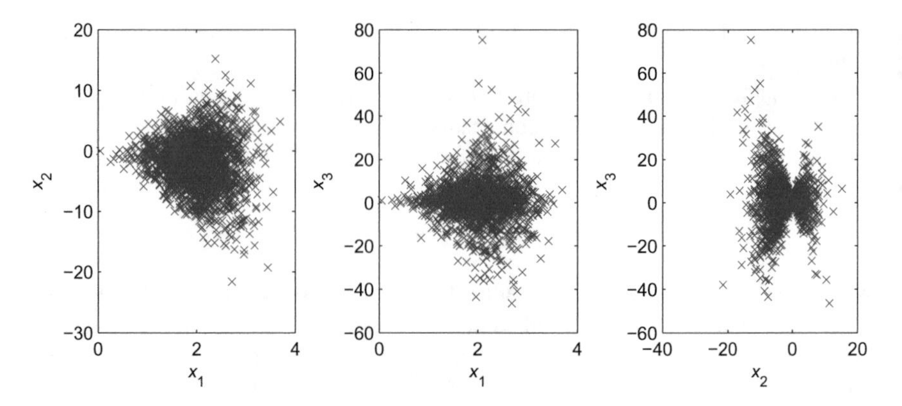

Fig. 13.4 The benchmark dataset

The dataset is shown in Fig. 13.4. Applying SVD, we get three eigenvalues in a descending order:

$$\boldsymbol{w}_1 = (0.0277, -0.0432, 0.9987)^T, \quad \|\boldsymbol{w}_1\| = 1.0000, \quad \lambda_1 = 86.3399,$$

$$\boldsymbol{w}_2 = (-0.2180, 0.9748, 0.0482)^T, \quad \|\boldsymbol{w}_2\| = 1.0000, \quad \lambda_2 = 22.9053,$$

$$\boldsymbol{w}_3 = (0.9756, 0.2191, -0.0176)^T, \quad \|\boldsymbol{w}_3\| = 1.0000, \quad \lambda_3 = 3.2122.$$

In this example, the simulation results slightly deviate from these values since we use only 2000 samples.

For weighted SLA, we select $\boldsymbol{\gamma} = (1, 2, 3)^T$. We select the learning rate $\eta_i = \frac{1}{2000+t}$, where each time t corresponds to the presentation of a new sample. Training is performed for 10 epochs, and the training samples are provided in a fixed deterministic sequence. We calculate the adaptations for $\|\boldsymbol{w}_i\|$, λ_i, and the cosine of the angle θ_i between $\boldsymbol{w}_i$ and $\boldsymbol{c}_i$:

$$\lambda_1 = 84.2226, \quad \lambda_2 = 20.5520, \quad \lambda_3 = 3.8760.$$

$$\cos\theta_1 = 1.0000, \quad \cos\theta_2 = -1.0000, \quad \cos\theta_3 = -1.0000.$$

The adaptations for a random run are shown in Fig. 13.5.

We selected $\gamma_i = 0.1i$, $i = 1, 2, 3$. The weight vectors converge to the directions or directions opposite to those of the principal eigenvectors, that is, $\cos\theta_i \to \pm 1$ as $t \to \infty$. The converging λ_i and $\|\boldsymbol{w}_i\|$ do not converge to their respective theoretical values.

From Fig. 13.5, we see that the convergence to smaller eigenvalues is slow. $\|\boldsymbol{w}_i\|$s do not converge to unity, and accordingly λ_i's do not converge to their theoretical

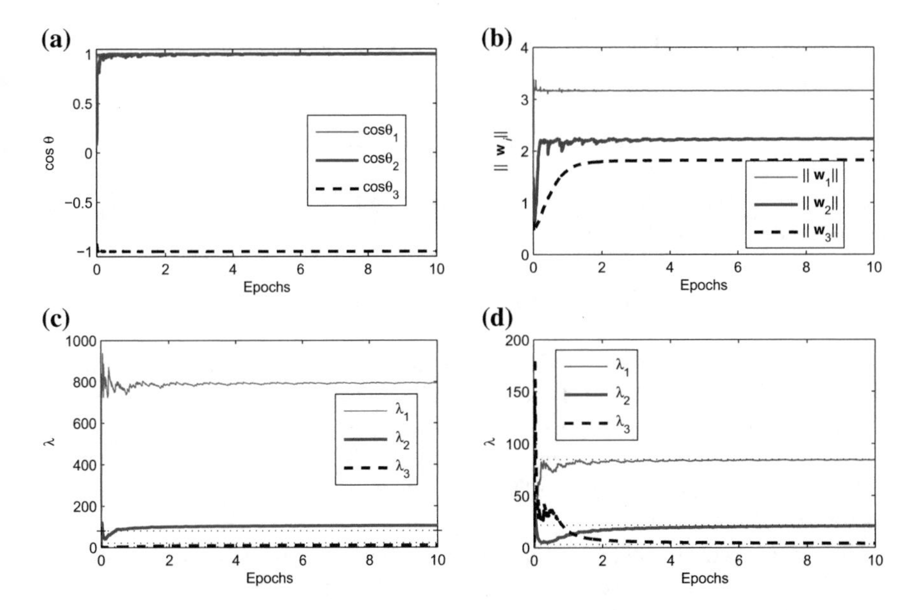

Fig. 13.5 Adaptation of weighted SLA. **a** $\cos\theta_i$. **b** $\|\boldsymbol{w}_i\|$. **c** λ_i. **d** λ_i obtained after normalization. The dotted lines correspond to the theoretical eigenvalues

values. By applying a normalization step at each iteration, we normalize $\|\boldsymbol{w}_i\|$; the resulting $\lambda_i' = \frac{\lambda_i}{\|\boldsymbol{w}_i\|^2}$ is plotted in Fig. 13.5d.

For weighted SLA, we also test for different γ_i. When all γ_i are selected as unity, weighted SLA reduces to SLA. $\cos\theta_i$ could be any value between $[-1, +1]$. In case of SLA, $\|\boldsymbol{w}_i\|$s converge to unity very rapidly, but λ_is converge to some values different from their theoretical eigenvalues and also not in a descending order.

13.3.2 Generalized Hebbian Algorithm

By combining Oja's rule and the GSO procedure, Sanger proposed GHA for extracting the first J_2 principal components [120]. GHA can extract the first J_2 eigenvectors in an order of decreasing eigenvalues.

GHA is given by [120]

$$\boldsymbol{w}_i(t+1) = \boldsymbol{w}_i(t) + \eta_i(t) y_i(t) \left[\boldsymbol{x}_t - \hat{\boldsymbol{x}}_i(t)\right], \quad i = 1, 2, \ldots, J_2, \tag{13.41}$$

$$\hat{\boldsymbol{x}}_i(t) = \sum_{j=1}^{i} \boldsymbol{w}_j(t) y_j(t), \quad i = 1, 2, \ldots, J_2. \tag{13.42}$$

GHA becomes a local algorithm by solving the summation term in (13.42) in a recursive form

$$\hat{\boldsymbol{x}}_i(t) = \hat{\boldsymbol{x}}_{i-1}(t) + \boldsymbol{w}_i(t) y_i(t), \quad i = 1, 2, \ldots, J_2, \tag{13.43}$$

where $\hat{\boldsymbol{x}}_0(t) = 0$. Usually, one selects $\eta_i = \eta$ for all neurons i, and accordingly the algorithm can be written in matrix form

$$\mathbf{W}(t+1) = \mathbf{W}(t) - \eta \mathbf{W}(t) \mathrm{LT}\left[\boldsymbol{y}(t)\boldsymbol{y}^T(t)\right] + \eta \boldsymbol{x}_t \boldsymbol{y}^T(t), \tag{13.44}$$

where operator LT[·] selects the lower triangle of the matrix contained within. In GHA, the mth neuron converges to the mth principal component, and all the neurons tend to converge together. $\boldsymbol{w}_i \to \boldsymbol{c}_i$ and $\mathrm{E}\left[y_i^2\right] \to \lambda_i$, as $t \to \infty$.

Both SLA and GHA employ implicit or explicit GSO to decorrelate the connection weights from one another. Weighted SLA performs well for extracting less-dominant components [98].

Traditionally, the learning rates of GHA are required to satisfy the Robbins–Monro conditions so that its convergence can be analyzed by studying the corresponding deterministic continuous-time equations. Based on analyzing the corresponding deterministic discrete-time equations, the global convergence of GHA is guaranteed by using the adaptive learning rates [87]

$$\eta_i(k) = \frac{\zeta}{y_i^2(k)}, \quad k > 0, \tag{13.45}$$

where the constant $0 < \zeta < 1$. These learning rates converge to some positive constants, which speed up the algorithm considerably and also enable the convergence speed in all eigendirections to be approximately equal.

In addition to popular SLA, weighted SLA, and GHA, there are also some other Hebbian rule-based PCA algorithms such as LEAP [11]. LEAP [11] is a local PCA algorithm for extracting all the J_2 principal components and their corresponding eigenvectors. It performs GSO among all weights at each iteration. Unlike SLA and GHA, whose stability analyses are based on the stochastic approximation theory, the stability analysis of LEAP is based on Lyapunov's first theorem, and η can be selected as a small positive constant. LEAP is capable of tracking nonstationary processes, and can satisfactorily extract principal components even for ill-conditioned autocorrelation matrices [11].

Example 13.2 Using the same dataset given in Example 13.1, we conduct simulation for GHA. We select the same learning rate $\eta_i = \frac{1}{2000+t}$, where each time t corresponds to the presentation of a new sample. The dataset is repeated 10 times, and the training samples are provided in a fixed deterministic sequence. We calculate the adaptations for $\|\boldsymbol{w}_i\|$, λ_i, and the cosine of the angle θ_i between $\boldsymbol{w}_i$ and $\boldsymbol{c}_i$.

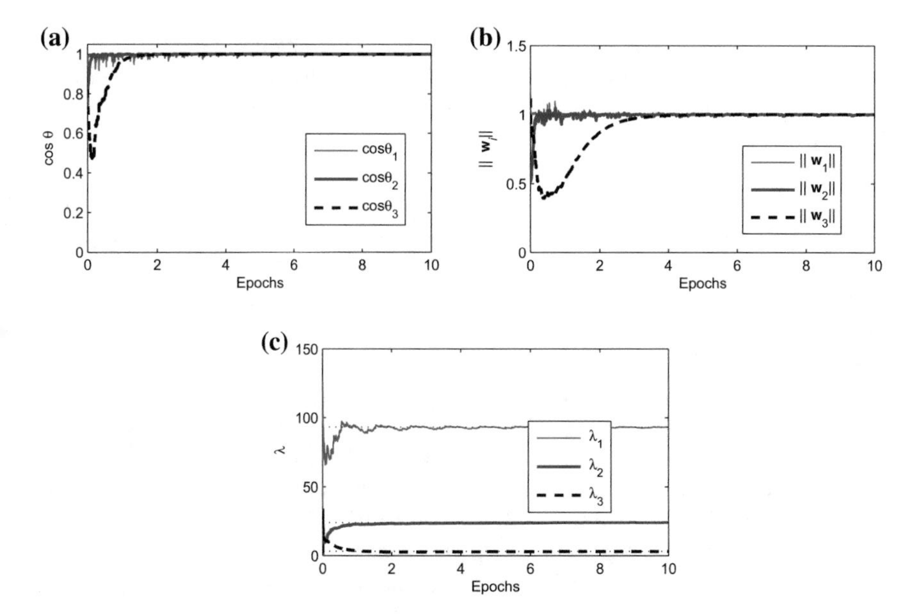

Fig. 13.6 Adaptation of GHA. **a** $\cos\theta_i$. **b** $\|\boldsymbol{w}_i\|$. **c** λ_i. The dotted lines in (**c**) correspond to the theoretical eigenvalues

The performances of the algorithms are evaluated by averaging 50 random runs. The adaptations for a random run are shown in Fig. 13.6. Our empirical results show that for GHA a larger starting η can be used.

From Fig. 13.6, we see that the convergence to smaller eigenvalues is slow. The strategy used in the adaptive learning algorithm [12] can be applied to make the algorithms converge to all the eigenvalues at the same speed. $\boldsymbol{w}_i$ converges to the directions of $\boldsymbol{c}_i$. Unlike weighted SLA, all $\|\boldsymbol{w}_i\|$s converge to unity, and accordingly λ_is converge to their theoretical values.

13.4 Least Mean Squared Error-Based PCA

Existing PCA algorithms including Hebbian rule-based algorithms can be derived by optimizing an objective function using the gradient-descent method. The least mean squared error (LMSE)-based methods are derived from the modified MSE function

$$E(\mathbf{W}) = \sum_{t_1=1}^{t} \mu^{t-t_1} \left\|\boldsymbol{x}_{t_1} - \mathbf{W}\mathbf{W}^T\boldsymbol{x}_{t_1}\right\|^2, \tag{13.46}$$

where $0 < \mu \leq 1$ is a forgetting factor used for nonstationary observation sequences, and t is the current instant. Many adaptive PCA algorithms actually optimize (13.46) by using the gradient-descent method [138, 142] and the RLS method [3, 91, 102, 104, 142].

The gradient-descent or Hebbian rule-based algorithms are highly sensitive to parameters such as η. It is difficult to choose proper parameters guaranteeing both small misadjustment and fast convergence. To overcome these drawbacks, applying RLS to minimization of (13.46) yields RLS-based algorithms such as adaptive principal components extraction (APEX) [72], Kalman-type RLS [3], projection approximation subspace tracking (PAST) [142], PAST with deflation (PASTd) [142], and robust RLS algorithm [102].

All RLS-based PCA algorithms exhibit fast convergence and high tracking accuracy, and are suitable for slowly varying nonstationary vector stochastic processes. All these algorithms correspond to a three-layer J_1-J_2-J_1 linear autoassociative network model, and they can extract all the J_2 principal components in a descending order of the eigenvalues, where a GSO-like orthonormalization procedure is used.

In [76], a regularization term $\mu^t \overrightarrow{\boldsymbol{w}}^T \mathbf{P}_0^{-1} \overrightarrow{\boldsymbol{w}}$ is added to (13.46), where $\overrightarrow{\boldsymbol{w}}$ is a stack vector of $\mathbf{W}$ and $\mathbf{P}_0$ is a diagonal $J_1 J_2 \times J_1 J_2$ matrix. As t is sufficiently large, this term is negligible. This term ensures that the entries of $\mathbf{W}$ do not become too large. Without this term, some matrices in the recursive updating equations may become indefinite. Gauss–Seidel recursive PCA and Jacobi recursive PCA are derived in [76].

The least mean squared error reconstruction (LMSER) algorithm [138] is derived from the MSE criterion using the gradient-descent method. LMSER reduces to Oja's algorithm when $\mathbf{W}(t)$ is orthonormal, namely, $\mathbf{W}^T(t)\mathbf{W}(t) = \mathbf{I}$. In this sense, Oja's algorithm can be treated as an approximate stochastic gradient rule to minimize the MSE. LMSER has been compared with weighted SLA and GHA in [8]. The learning rates for all the algorithms are selected as $\eta(t) = \frac{\delta}{t^\alpha}$, where $\delta > 0$ and $\frac{1}{2} < \alpha \leq 1$. A trade-off is obtained: Increasing the values of γ and δ results in a larger asymptotic MSE but faster convergence and *vice versa*, namely, the stability–speed problem. LMSER uses nearly twice as much computation as weighted SLA and GHA, for each update of the weight. However, it leads to a smaller asymptotic MSE and faster convergence for the minor eigenvectors [8].

PASTd Algorithm

PASTd [142] is a well-known subspace tracking algorithm updating the signal eigenvectors and eigenvalues. PASTd is based on PAST. Both PAST and PASTd are derived for complex-valued signals, which are common in signal processing. At iteration t, PASTd is given as follows [142]: for $i = 1, \ldots, J_2$,

$$y_i(t) = \boldsymbol{w}_i^H(t-1)\boldsymbol{x}_i(t), \tag{13.47}$$

$$\delta_i(t) = \mu\delta_i(t-1) + |y_i(t)|^2, \tag{13.48}$$

$$\hat{\boldsymbol{x}}_i(t) = \boldsymbol{w}_i(t-1)y_i(t), \tag{13.49}$$

$$\boldsymbol{w}_i(t) = \boldsymbol{w}_i(t-1) + \left[\boldsymbol{x}_i(t) - \hat{\boldsymbol{x}}_i(t)\right]\frac{y_i^*(t)}{\delta_i(t)}, \tag{13.50}$$

$$\boldsymbol{x}_{i+1}(t) = \boldsymbol{x}_i(t) - \boldsymbol{w}_i(t)y_i(t), \tag{13.51}$$

where $\boldsymbol{x}_1(t) = \boldsymbol{x}_t$, and superscript $*$ denotes the conjugate operator.

$\boldsymbol{w}_i(0)$ and $\delta_i(0)$ should be suitably selected. $\mathbf{W}(0)$ should contain J_2 orthonormal vectors, which can be calculated from an initial block of data or from arbitrary initial data. A simple way is to set $\mathbf{W}(0)$ to the J_2 leading unit vectors of the $J_1 \times J_1$ identity matrix. $\delta_i(0)$ can be set as unity. The choice of these initial values affects the transient behavior, but not the steady-state performance of the algorithm. $\boldsymbol{w}_i(t)$ provides an estimate of the ith eigenvector, and $\delta_i(t)$ is an exponentially weighted estimate of the corresponding eigenvalue.

Both PAST and PASTd have linear computational complexity, that is, $O(J_1 J_2)$ operations every update, as in the cases of SLA, GHA, LMSER, and the novel information criterion (NIC) algorithm [91]. PAST computes an arbitrary basis of the signal subspace, while PASTd is able to update the signal eigenvectors and eigenvalues. Both the algorithms produce nearly orthonormal, but not exactly orthonormal, subspace basis, or eigenvector estimates. If perfectly orthonormal eigenvector estimates are required, an orthonormalization procedure is necessary.

Kalman-type RLS [3] combines RLS with the GSO procedure in a manner similar to that of GHA. Kalman-type RLS and PASTd are exactly identical if the inverse of the covariance of the ith neuron's output, $\mathbf{P}_i(t)$, in Kalman-type RLSA is set as $\frac{1}{\delta_i(t)}$ in PASTd. In the one-unit case, both PAST and PASTd are identical to Oja's rule except that PAST and PASTd have a self-tuning learning rate $\frac{1}{\delta_1(t)}$. Both PAST and PASTd provide much more robust estimates than EVD, and converge much faster than SLA. PASTd has been extended for tracking both the rank and the subspace by using information-theoretic criteria such as AIC and MDL [143].

The constrained PAST algorithm [131] is for tracking the signal subspace recursively. Based on an interpretation of the signal subspace as the solution of a constrained minimization task, it guarantees the orthonormality of the estimated signal subspace basis at each update, hence avoiding the orthonormalization process. To reduce the computational complexity, fast constrained PAST that has a complexity of $O(J_1 J_2)$ is introduced. For tracking the signal sources with abrupt change in their parameters, an alternative implementation with truncated window is proposed. Furthermore, a signal subspace rank estimator is employed to track the number of sources.

A perturbation-based fixed-point algorithm for subspace tracking [48] has fast tracking capability due to the recursive nature of the complete eigenvector matrix updates. They avoid deflation and the optimization of a cost function using gradients.

It recursively updates the eigenvector and eigenvalue matrices simultaneously with every new sample.

Robust RLS Algorithm

The robust RLS algorithm [102] is more robust than PASTd. It can be implemented in a sequential or parallel form. Given the ith neuron, $i = 1, \ldots, J_2$, the sequential algorithm is given for all the patterns as [102]

$$\overline{\boldsymbol{w}}_i(t-1) = \frac{\boldsymbol{w}_i(t-1)}{\|\boldsymbol{w}_i(t-1)\|}, \tag{13.52}$$

$$y_i(t) = \overline{\boldsymbol{w}}_i^T(t-1)\boldsymbol{x}_t, \tag{13.53}$$

$$\hat{\boldsymbol{x}}_i(t) = \sum_{j=1}^{i-1} y_j(t)\overline{\boldsymbol{w}}_j(t-1), \tag{13.54}$$

$$\boldsymbol{w}_i(t) = \mu\boldsymbol{w}_i(t-1) + \left[\boldsymbol{x}_t - \hat{\boldsymbol{x}}_i(t)\right] y_i(t), \tag{13.55}$$

$$\hat{\lambda}_i(t) = \frac{\|\boldsymbol{w}_i(t)\|}{t}, \tag{13.56}$$

where y_i is the output of the ith hidden unit, and $\boldsymbol{w}_i(0)$ is initialized as a small random value. By changing (13.54) into a recursive form, the robust RLS algorithm becomes a local algorithm.

Robust RLS has the same flexibility as Kalman-type RLS [3], PASTd, and APEX, in that increasing the number of neurons does not affect the previously extracted principal components. It naturally selects the inverse of the output energy to be the adaptive learning rate for the Hebbian rule. The Hebbian and Oja rules are closely related to the robust RLS algorithm by a suitable selection of the learning rates [102]. Robust RLS can also be derived from the adaptive learning algorithm [12] by using the first-order Taylor approximation [93].

Robust RLS is also robust to the error accumulation from the previous components, which exists in the sequential PCA algorithms like Kalman-type RLS and PASTd. Robust RLS converges rapidly, even if the eigenvalues extend over several orders of magnitude. According to the empirical results [102], robust RLS provides the best performance in terms of convergence speed as well as steady-state error, whereas Kalman-type RLS and PASTd have similar performance, which is inferior to that of robust RLS, and the adaptive learning algorithm [12] exhibits the poorest performance.

13.4.1 Other Optimization-Based PCA

PCA can be derived from many optimization methods based on a properly defined objective function. This leads to many other algorithms, including gradient-descent-based algorithms [9, 78, 79, 108, 152], the CG method [37], and the quasi-Newton method [62, 105]. The gradient-descent method usually converges to a local minimum. Second-order algorithms such as the CG and quasi-Newton methods typically converge much faster than first-order methods, but have a computational complexity of $O\left(J_1^2 J_2\right)$ per iteration.

The infomax principle [78, 79] derives the principal subspace by maximizing the mutual information criterion. Other examples of information criterion-based algorithms are the NIC algorithm [91] and coupled PCA [93].

The NIC algorithm [91] is obtained by applying the gradient-descent method to maximize the NIC, which is a cost function very similar to the mutual information criterion, but integrates a soft constraint on weight orthogonalization. The NIC has a steep landscape along the trajectory from a small weight matrix to the optimum one; it has a single global maximum, and all the other stationary points are unstable saddle points. At the global maximum, $\mathbf{W}$ yields an arbitrary orthonormal basis of the principal subspace, and thus the NIC algorithm is a PSA method. It can extract the principal eigenvectors when the deflation technique is incorporated. The NIC algorithm converges much faster than SLA and LMSER, and is able to globally converge to the PSA solution from almost any weight initialization. Reorthormalization can be applied so as to perform true PCA [91, 142]. The NIC algorithm has a computational complexity of $O\left(J_1^2 J_2\right)$ for each iteration. By selecting a well-defined adaptive learning rate, the NIC algorithm also generalizes some well-known PSA/PCA algorithms such as PAST. For online implementation, an RLS version of the NIC algorithm has also been given in [91]. Weighted information criterion (WINC) [104] is obtained by adding a weight to the NIC to break its symmetry. The gradient-ascent-based WINC algorithm can be viewed as an extended weighted SLA with an adaptive step size, leading to a much faster convergence speed. The RLS-based WINC algorithm provides fast convergence, high accuracy as well as low computational complexity.

In PCA algorithms, the eigenmotion depends on the principal eigenvalue of the covariance matrix, while in MCA algorithms it depends on all the eigenvalues [93]. Coupled learning rules can be derived by applying the Newton method to a common information criterion. In coupled PCA/MCA algorithms, both the eigenvalues and the eigenvectors are simultaneously adapted. The Newton method yields averaged systems with identical speed of convergence in all eigendirections. The derived Newton-descent-based PCA and MCA algorithms are, respectively, called *nPCA* and *nMCA*. The robust PCA algorithm [93], derived from nPCA, is shown to be closely related to robust RLS algorithm by applying the first-order Taylor approximation on the robust PCA. In order to extract multiple principal components, one has to apply an orthonormalization procedure, which is GSO, or its first-order approximation as in SLA, or deflation as in GHA. In the coupled learning rules, multiple principal components are simultaneously estimated by a coupled system of equations. In the

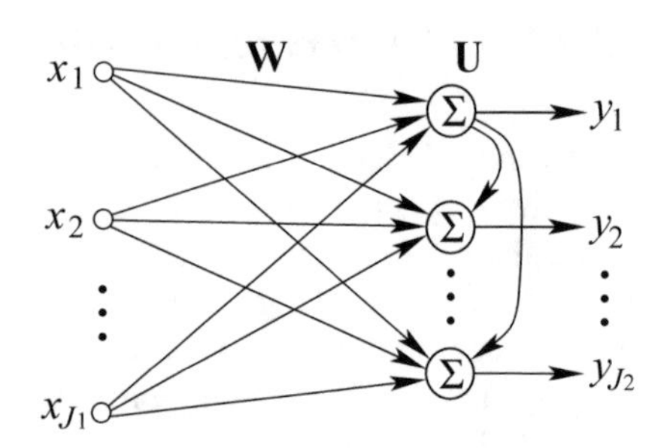

Fig. 13.7 The PCA network with hierarchical lateral connections

coupled learning rules, a first-order approximation of GSO is superior to the standard deflation procedure in terms of the orthonormality error and the quality of the eigenvectors and eigenvalues generated [94]. An additional normalization step that enforces unit length of the eigenvectors further improves the orthonormality of the weight vectors [94].

13.5 Anti-Hebbian Rule-Based PCA

When the update of a synaptic weight is proportional to the correlation of the pre- and postsynaptic activities, but the direction of the change is opposite to that in the Hebbian rule, we get an *anti-Hebbian learning rule*. The anti-Hebbian rule can be used to remove correlations between units receiving correlated inputs [35, 116, 117]; it is inherently stable [116, 117].

Anti-Hebbian rule-based PCA algorithms can be derived by using a J_1-J_2 feedforward network with lateral connections among the output units [35, 116, 117]. The lateral connections can be in a symmetrical or hierarchical topology. A hierarchical lateral connection topology is illustrated in Fig. 13.7, based on which the Rubner–Tavan PCA algorithm [116, 117] and APEX [71] are proposed. The lateral weight matrix $\mathbf{U}$ is an upper triangular matrix with the diagonal elements being zero. In [35], the local PCA algorithm is based on a symmetrical lateral connection topology. The feedforward weight matrix $\mathbf{W}$ is described in the preceding sections, and the lateral weight matrix $\mathbf{U} = \left[\boldsymbol{u}_1, \ldots, \boldsymbol{u}_{J_2}\right]$ is a $J_2 \times J_2$ matrix, where $\boldsymbol{u}_i = \left(u_{1i}, u_{2i}, \ldots, u_{J_2 i}\right)^T$ includes all the lateral weights connected to neuron i and u_{ji} denotes the lateral weight from neuron j to neuron i.

The Rubner–Tavan PCA algorithm is based on the PCA network with hierarchical lateral connection topology. The algorithm extracts the first J_2 principal components in an order of decreasing eigenvalues. The output of the network is given by [116, 117]

$$y_i = \boldsymbol{w}_i^T \boldsymbol{x} + \boldsymbol{u}_i^T \boldsymbol{y}, \quad i = 1, \ldots, J_2. \tag{13.57}$$

Notice that $u_{ji} = 0$ for $j \geq i$ and $\mathbf{U}$ is a $J_2 \times J_2$ upper triangular matrix.

The weights $\boldsymbol{w}_i$ are trained by Oja's rule, while the lateral weights $\boldsymbol{u}_i$ are updated by the anti-Hebbian rule

$$\boldsymbol{w}_i(t+1) = \boldsymbol{w}_i(t) + \eta_1(t) y_i(t) \left[\boldsymbol{x}_t - \hat{\boldsymbol{x}}(t) \right], \tag{13.58}$$

$$\hat{\boldsymbol{x}} = \mathbf{W}^T \boldsymbol{y}, \tag{13.59}$$

$$\boldsymbol{u}_i(t+1) = \boldsymbol{u}_i(t) - \eta_2 y_i(t) \boldsymbol{y}(t). \tag{13.60}$$

This is a nonlocal algorithm. Typically, $\eta_1 = \eta_2 > 0$ is selected as a small number or by the Robbins–Monro conditions. During the training process, the outputs of the neurons are gradually uncorrelated and the lateral weights approach zero. The network should be trained until the lateral weights $\boldsymbol{u}_i$ are below a specified level. The PCA algorithm proposed in [35] has the same form as the Rubner–Tavan PCA, but $\mathbf{U}$ is a full matrix.

13.5.1 APEX Algorithm

The APEX algorithm is used to recursively and adaptively extract the principal components [71]. Given $i-1$ principal components, it can produce the ith principal component iteratively. The hierarchical structure of lateral connections among the output units serves the purpose of weight orthogonalization. This structure also allows the network to grow or shrink without retraining the old units. The convergence analysis of APEX is based on stochastic approximation theory, and APEX is proved to have the property of exponential convergence.

Assuming that the correlation matrix $\mathbf{C}$ has distinct eigenvalues arranged in decreasing order as $\lambda_1 > \lambda_2 > \cdots > \lambda_{J_2}$ with the corresponding eigenvectors $\boldsymbol{w}_1, \ldots, \boldsymbol{w}_{J_2}$, the algorithm is given as [71, 72]

$$\boldsymbol{y} = \mathbf{W}^T \boldsymbol{x}, \tag{13.61}$$

$$y_i = \boldsymbol{w}_i^T \boldsymbol{x} + \boldsymbol{u}^T \boldsymbol{y}, \tag{13.62}$$

where $\boldsymbol{y} = (y_1, \ldots, y_{i-1})^T$ is the output vector, $\boldsymbol{u} = \left(u_{1i}, u_{2i}, \ldots, u_{(i-1)i} \right)^T$, and $\mathbf{W} = \left[\boldsymbol{w}_1, \ldots, \boldsymbol{w}_{i-1} \right]$ is the weight matrix of the first $i-1$ neurons. These definitions are for the first i neurons, which are different from their respective definitions given in the preceding sections. The iteration is given by [71, 72]

$$\boldsymbol{w}_i(t+1) = \boldsymbol{w}_i(t) + \eta_i(t) \left[y_i(t) \boldsymbol{x}_t - y_i^2(t) \boldsymbol{w}_i(t) \right], \tag{13.63}$$

$$\boldsymbol{u}(t+1) = \boldsymbol{u}(t) - \eta_i(k) \left[y_i(t) \boldsymbol{y}(t) + y_i^2(t) \boldsymbol{u}(t) \right]. \tag{13.64}$$

Equations (13.63) and (13.64) are, respectively, the Hebbian and anti-Hebbian parts of the algorithm. y_i tends to be orthogonal to all the previous components due to the anti-Hebbian rule, also called the *orthogonalization rule*.

APEX can also be derived from the RLS method using the MSE criterion. Based on the RLS method, the optimum learning rate in terms of convergence speed is given by [72]

$$\eta_i(t) = \frac{1}{\sum_{l=0}^{t} \mu^{t-l} y_i^2(l)} = \frac{\eta_i(t-1)}{\mu - y_i^2(t)\eta_i(t-1)}, \tag{13.65}$$

where $0 < \mu \leq 1$ is a forgetting factor, which induces an effective time window of size $M = \frac{1}{1-\mu}$. The optimal learning rate can also be written as [71]

$$\eta_i(t) = \frac{1}{M\sigma_i^2}, \tag{13.66}$$

where $\sigma_i^2 = E\left[y_i^2(t)\right]$ is the average output power or variance of neuron i. According to [71], $\sigma_i^2(t) \to \lambda_i$, as $t \to \infty$. A practical value of η_i is selected by

$$\eta_i(t) = \frac{1}{M\lambda_{i-1}}, \tag{13.67}$$

since $\lambda_{i-1} > \lambda_i$ and λ_i is not easy to get.

Both sequential and parallel APEX algorithms are given in [72]. In parallel APEX, all the J_2 output neurons work simultaneously. In sequential APEX, the output neurons are added one by one. Sequential APEX is more attractive in practical applications since one can decide a desirable number of neurons during the learning process. APEX is especially useful when the number of required principal components is not known *a priori*. When the environment changes over time, a new principal component can be added to compensate for the change without affecting previously computed principal components. Thus, the network structure can be expanded if necessary.

The stopping criterion can be that for each i the changes in $\boldsymbol{w}_i$ and $\boldsymbol{u}$ are below a threshold. At this time, $\boldsymbol{w}_i$ converges to the eigenvector of the correlation matrix $\mathbf{C}$ corresponding to the ith largest eigenvalue, and $\boldsymbol{u}$ converges to zero. The stopping criterion can also be the change of the average output variance $\sigma_i^2(t)$ being sufficiently small.

Most existing linear complexity methods including GHA, SLA, and PCA with the lateral connections require a computational complexity of $O(J_1 J_2)$ per iteration. For recursive computation of each additional principal component, APEX requires $O(J_1)$ operations per iteration, while GHA utilizes $O(J_1 J_2)$ per iteration.

In contrast to the heuristic derivation of APEX, a class of learning algorithms called the *ψ-APEX*, is presented based on criterion optimization [34]. ψ can be selected as any function that guarantees the stability of the network. Some members in the class have better numerical performance and require less computational effort compared to that of both GHA and APEX.

Fig. 13.8 Sample images for face recognition

Example 13.3 **Face recognition using PCA**. PCA is a classical approach to face recognition. Each sample is a face image, normalized to the same size. In this example, we randomly select 60 samples from 30 persons, 2 samples for each person, for training. The testing set includes 30 samples, one for each person. The samples are of size 200×180 pixels, and they are reshaped in vector form. These samples are excerpted from Spacek's Faces94 collection (http://cswww.essex.ac.uk/mv/allfaces/faces94.html). Some of the face samples are shown in Fig. 13.8.

After applying PCA on the training set, the corresponding weights called eigenfaces are obtained. When a new sample is presented, the projection on the weights is derived. The projection of the presented sample is compared with those of all the training samples, and the training sample that has the minimum difference from the test sample is classified as the correct class. We test the trained PCA method using all the 30 testing samples, and the classification rate for this example is 100%.

Image Compression Using PCA

Image compression is performed to remove the redundancy in an image for storage and/or transmission purposes. Image compression is usually implemented by partitioning an image into many nonoverlaping 8×8 pixel blocks and then compressing them one by one. For example, if we compress each of the 64-pixel patch into 8 values, we achieve a compression ratio of $1 : 8$. This work can be performed by using a PCA network. Based on the statistics of all the regions, one can use PCA to compress the image. Each region is concatenated into a vector, and all the vectors constitute a training set. PCA is then applied to extract those prominent principal components, as such the image is compressed. Sanger used GHA for image compression [120].

Similar results using a three-layer autoassociative network with BP learning has been reported in [44].

PCA as well as LDA achieves the same results for an original dataset and its orthonormally transformed version [16]. Thus, PCA and LDA can be directly implemented in DCT domain and the results are exactly the same as that obtained from spatial domain. For images compressed using DCT, such as in the JPEG or MPEG standard, PCA and LDA can be directly implemented in DCT domain so that inverse DCT transform can be avoided and computation conducted on a reduced data dimension.

Example 13.4 We use the Lena image of 512×512 pixels with 256 gray levels for training the PCA network. A linear 64–8 PCA network is used to learn the image. By 8×8 partitioning, we get $64 \times 64 = 4,096$ samples. Each of the output nodes is connected by 64 weights, denoted by an 8×8 mask. The training results for the eight codewords are illustrated in Fig. 13.9, where the positive weights are shown as white, the negative weights as black, and zero weights as gray.

After the learning algorithm converges, the PCA network can be used to code the image. An 8×8 block is multiplied by each of the eight weight masks, and this yields eight coefficients. The reconstruction of the image from the coefficients can be conducted by multiplying the weights by those coefficients, and combining the

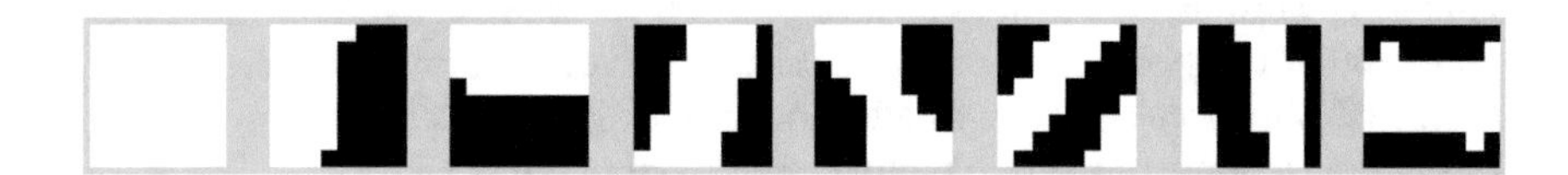

Fig. 13.9 Network weights after training with the Lina image

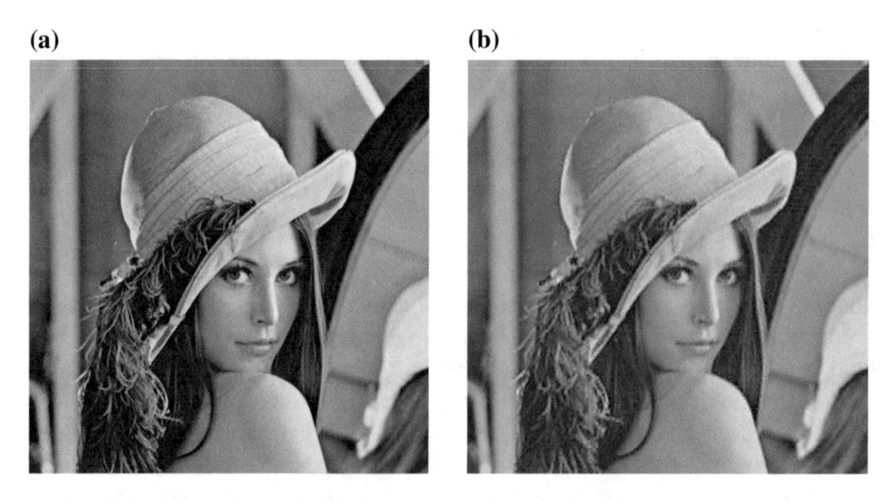

Fig. 13.10 The Lena picture (**a**) and its restored version (**b**)

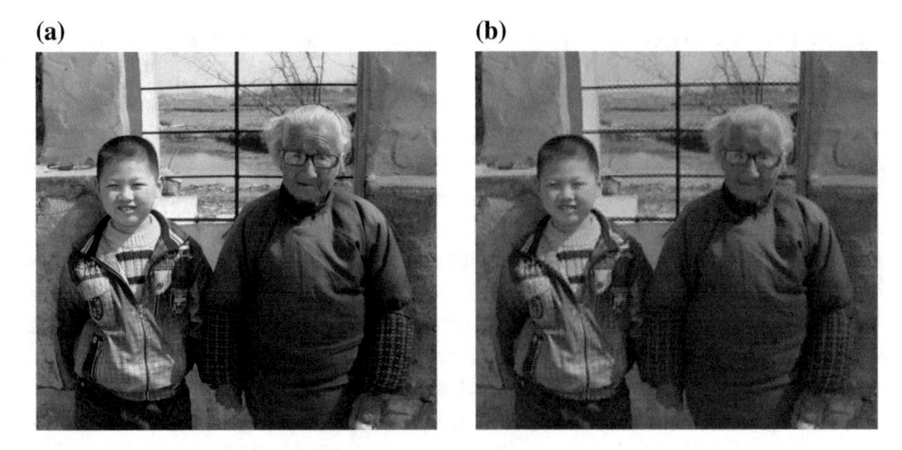

Fig. 13.11 The original picture (**a**) and its reconstructed version (**a**)

reconstructed blocks into an image. The reconstructed image, illustrated in Fig. 13.10, is as good to the human eye as the network without any quantization.

We now use the trained network to encode the family picture of 512×560, and the result is shown in Fig. 13.11. The reconstructed image is of good quality to the human eye.

In the above two examples, an 8×8 block is encoded by only eight coefficients. In consideration of the 64×8 codebook for the entire image, the coarse compression rate is close to $8:1$ if the coefficients are quantized into 8 bits. By using entropy, each of the eight coefficients can be further uniformly quantized using less bits, which are proportional to the logarithm of the variance of that coefficient over the whole image. For example, in [120] for an image the first two coefficients require 5 bits each, the third coefficient requires 3 bits, and the remaining five coefficients require 2 bits each, and a total of 23 bits are used to code each 8×8 block, that is, a bit rate of 0.36 bits per pixel. This achieves a compression ratio of $\frac{64\times 8}{23} = 22.26$ to 1.

13.6 Nonlinear PCA

For non-Gaussian data distributions, PCA is not able to capture complex nonlinear correlations, and nonlinear processing of the data is usually more efficient. Nonlinearities introduce higher order statistics into the computation in an implicit way. Higher order statistics, defined by cumulants or higher-than-second moments, are needed for good characterization of non-Gaussian data. For non-Gaussian input data, nonlinear PCA permits extraction of higher order components and provides a sufficient representation. Nonlinear PCA networks and learning algorithms can be classified into symmetric and hierarchical ones similar to those for PCA networks. After training, the lateral connections between output units are not needed, and the network becomes purely feedforward.

Several popular PCA algorithms have been generalized into robust versions by applying a statistical-physics approach [141], where the defined objective function can be regarded as a soft generalization of the M-estimator. Robust PCA algorithms are defined so that the optimization criterion grows less than quadratically and the constraint conditions are the same as for PCA algorithms [64]. To derive robust PCA algorithms, the variance maximization criterion is generalized as $\mathrm{E}\left[\sigma(\boldsymbol{w}_i^T\boldsymbol{x})\right]$ for the ith neuron, subject to hierarchical or symmetric orthonormality constraints, where $\sigma(x)$ is the M-estimator assumed to be a valid differentiable cost function that grows less than quadratically, at least for large values of x. Examples of such functions are $\sigma(x) = \ln\cosh(x)$ and $\sigma(x) = |x|$. Robust/nonlinear PCA can be obtained by minimizing the MSE that introduces nonlinearity using the gradient-descent procedure [64, 138]. Robust/nonlinear PCA algorithms have better stability properties than the corresponding PCA algorithms if the (odd) nonlinearity $\varphi(x)$ grows less than linearly, namely, $|\varphi(x)| < |x|$ [64].

In SOM, lateral inhibitory connections for output neurons are usually used to induce WTA competition among all the output neurons. It is capable of performing dimension reduction on the input. SOM is inherently nonlinear, and is viewed as a nonlinear PCA [114]. ASSOM can be treated as a hybrid of vector quantization and PCA.

Principal curves [45] are nonlinear generalizations of the notion of the first principal component of PCA. A principal curve is a parameterized curve passing through the "middle" of a data cloud.

13.6.1 Autoassociative Network-Based Nonlinear PCA

The MLP can be used to perform nonlinear dimension reduction and hence nonlinear PCA. Both the input and output layers of the MLP have J_1 units, and one of its hidden layers, known as the *bottleneck* or *representation layer*, has J_2 units, $J_2 < J_1$. The network is trained to reproduce its input vectors. This kind of network is called the *autoassociative MLP*. After the network is trained, it performs a projection onto the J_2-dimensional subspace spanned by the first J_2 principal components of the data. The vectors of weights leading to the hidden units form a basis set that spans the principal subspace, and data compression therefore occurs in the bottleneck layer. Many applications of the MLP in autoassociative mode for PCA are available in the literature [6, 68].

The three-layer autoassociative J_1-J_2-J_1 feedforward network or MLP network can also be used to extract the first J_2 principal components of J_1-dimensional data. If nonlinear activation functions are applied in the hidden layer, the network performs as a nonlinear PCA network. In the case of nonlinear units, local minima certainly appear. However, if linear units are used in the output layer, nonlinearity in the hidden layer is theoretically meaningless [6]. This is due to the fact that the network tries to approximate a linear mapping.

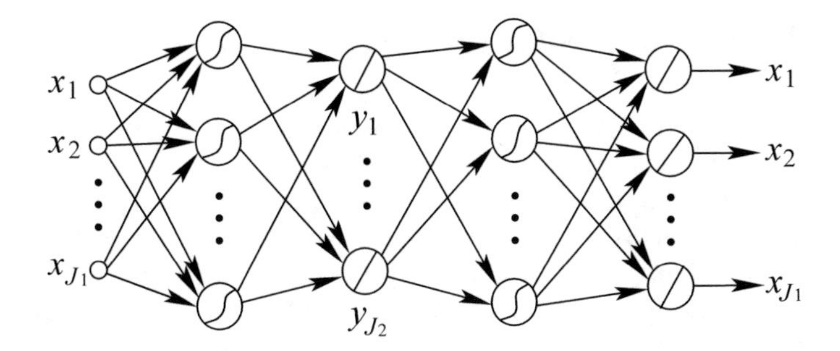

Fig. 13.12 Architecture of Kramer's nonlinear PCA network

Kramer's nonlinear PCA network [68] is a five-layer autoassociative MLP, whose architecture is illustrated in Fig. 13.12. It has J_1 input and J_1 output nodes. The third layer has J_2 nodes. y_i, $i = 1, \ldots, J_2$, is the ith output of the bottleneck layer. Nonlinear activation functions such as the sigmoidal functions are used in the second and fourth layers, while the nodes in the bottleneck and output layers usually have linear activation functions though they can be nonlinear. The network is trained by BP. Kramer's nonlinear PCA fits a lower dimensional surface through the training data.

A three-layer MLP can approximate arbitrarily well any continuous function. The input, second, and bottleneck layers constitute a three-layer MLP, which projects the training data onto the surface giving principal components. Likewise, the combination of the bottleneck, fourth, and output layers defines the surface that inversely maps the principal components into the training data. The outputs of the network are trained to approximate the inputs. After the network is trained, the nodes in the bottleneck layer give a lower dimensional representation of the inputs. Usually, data compression achieved in the bottleneck layer in such networks is somewhat better than that provided by the respective PCA solution [61]. This is actually a nonlinear PCA network. However, BP is prone to local minima and often requires excessive time for convergence.

With very noisy data, having plentiful samples eliminates overfitting in nonlinear regression, but not in nonlinear PCA. To overcome this problem in Kramer's nonlinear PCA, an information criterion is proposed for selecting the best model among multiple models with different complexities and regularizations [53].

A hierarchical nonlinear PCA network composed of a number of independent subnetworks can extract ordered nonlinear principal components [119]. Each subnetwork extracts one principal component, and can be selected as Kramer's nonlinear PCA network. The subnetworks are hierarchically arranged and trained.

In contrast to autoassociative networks, the output pattern in heteroassociative networks is not the same as the input pattern for each training pair. Heteroassociative networks develop arbitrary internal representations in the hidden layers to associate inputs to class identifiers, usually in the context of pattern classification. The rate of dimension reduction is not as high as that of autoassociative networks.

13.7 Minor Component Analysis

In contrast to PCA, MCA is to find the smallest eigenvalues and their corresponding eigenvectors of the autocorrelation matrix $\mathbf{C}$ of the signals. MCA is closely associated with the curve and surface fitting under the TLS criterion [139]. MCA provides an alternative solution to the TLS problem [39]. The TLS technique achieves a better optimal objective than the LS technique [40]. Both the solutions to the TLS and LS problems can be obtained by SVD. However, the TLS technique is computationally much more expensive than the LS technique. MCA is useful in many fields including spectrum estimation, optimization, TLS parameter estimation in adaptive signal processing, and eigen-based bearing estimation.

Minor components can be extracted in ways similar to that for principal components. A simple idea is to reverse the sign of the PCA algorithms, since in many algorithms principal and minor components correspond to the maximum and minimum of a cost function, respectively. However, this idea does not work in general [98].

13.7.1 Extracting the First Minor Component

The anti-Hebbian learning rule and its normalized version can be used for MCA [140]. The anti-Hebbian algorithm tends rapidly to infinite magnitudes of the weights. The normalized anti-Hebbian algorithm leads to better convergence, but it may also lead to infinite magnitudes of weights before the algorithm converges. To avoid this, one can renormalize the weight vector at each iteration. The constrained anti-Hebbian learning algorithm [38, 39] has a simple structure, and requires a low computational complexity per update. It can be used to solve the TLS parameter estimation [39], and has been extended for complex-valued TLS problem [38]. However, as in the anti-Hebbian algorithm, the convergence of the magnitudes of the weights cannot be guaranteed unless the initial weights take special values.

The total least mean squares (TLMS) algorithm [28] is a random adaptive algorithm for extracting the minor component, which has an equilibrium point under persistent excitation conditions. The TLMS algorithm requires about $4J_1$ multiplications per iteration, which is twice the complexity of the LMS algorithm. An adaptive step-size learning algorithm [101] is derived for extracting the minor component by introducing information criterion. The algorithm globally converges asymptotically to a stable equilibrium point, which corresponds to the minor component and its corresponding eigenvector. The algorithm outperforms TLMS in terms of both convergence speed and estimation accuracy.

In [109, 110], learning algorithms for estimating minor component from input signals are proposed, and their dynamics are analyzed via a deterministic discrete-time method. Some sufficient conditions are obtained to guarantee convergence of the algorithm. A convergent algorithm is given by [110]

$$\boldsymbol{w}(t+1) = \boldsymbol{w}(t) - \eta \left[\boldsymbol{w}^T(t)\boldsymbol{w}(t)y(t)\boldsymbol{x}_t - \frac{y^2(t)\boldsymbol{w}(t)}{\boldsymbol{w}^T(t)\boldsymbol{w}(t)} \right], \tag{13.68}$$

where $\eta < 0.5/\lambda_1$, $\lambda_1 \geq \lambda_2 \geq \ldots \geq \lambda_n > 0$ are all the eigenvalues of $\mathbf{C}$, $\|\boldsymbol{w}(0)\|^2 \leq 0.5/\eta$, and $\boldsymbol{w}^T(0)\boldsymbol{v}_n \neq 0$, $\boldsymbol{v}_n$ is the eigenvector associated with the smallest eigenvalue of $\mathbf{C}$. $\boldsymbol{w}(t)$ will converge to the minor component of the input data.

A class of self-stabilizing MCA algorithms, which is proved to be a globally asymptotically convergent using Lyapunov's theorem, is given by [149]

$$\boldsymbol{w}(t+1) = \boldsymbol{w}(t) - \eta(t)y(t) \left[\boldsymbol{x}_t - \frac{y(t)}{\|\boldsymbol{w}(t)\|^{\alpha+2}} \boldsymbol{w}(t) \right] \tag{13.69}$$

for integer $\alpha \geq 0$. It reduces to normalized Oja for $\alpha = 0$ [98, 140], to the one given in [148] for $\alpha = 1$, and to a modified Oja's algorithm [31] for $\alpha = 2$. To improve the convergence speed, select large α if $\|\boldsymbol{w}(t)\| \geq 1$, and set $\alpha = 1$ if $\|\boldsymbol{w}(t)\| < 1$.

13.7.2 Self-Stabilizing Minor Component Analysis

A general algorithm that can extract in parallel, principal, and minor eigenvectors of arbitrary dimensions is derived based on the natural-gradient method in [14]. The difference between PCA and MCA lies in the sign of the learning rate. The MCA algorithm can be written as [14]

$$\mathbf{W}(t+1) = \mathbf{W}(t) - \eta \left[\boldsymbol{x}_t \boldsymbol{y}^T(t)\mathbf{W}^T(t)\mathbf{W}(t) - \mathbf{W}(t)\boldsymbol{y}(t)\boldsymbol{y}^T(t) \right]. \tag{13.70}$$

At initialization, $\mathbf{W}^T(0)\mathbf{W}(0)$ is required to be diagonal. It suffers from a marginal instability, and thus it requires intermittent normalization such that $\|\boldsymbol{w}_i\| = 1$ [25].

A self-stabilizing MCA algorithm is given in [25] as

$$\mathbf{W}(t+1) = \mathbf{W}(t) - \eta \left[\boldsymbol{x}_t \boldsymbol{y}^T(t)\mathbf{W}^T(t)\mathbf{W}(t)\mathbf{W}^T(t)\mathbf{W}(t) - \mathbf{W}(t)\boldsymbol{y}(t)\boldsymbol{y}^T(t) \right]. \tag{13.71}$$

Algorithm (13.71) is self-stabilizing, such that none of $\|\boldsymbol{w}_i(t)\|$ deviates significantly from unity. It diverges for PCA when $-\eta$ is changed to $+\eta$.

A class of self-stabilizing MCA algorithms is obtained, and its convergence and stability are analyzed via a deterministic discrete-time method [67]. Some sufficient conditions are obtained to guarantee the convergence of these learning algorithms. These self-stabilizing algorithms can efficiently extract the minor component, and they outperform several self-stability MCA algorithms, a modified Oja algorithm [31], and algorithm (13.68) [110].

13.7.3 Oja-Based MCA

Oja's minor subspace analysis (MSA) algorithm can be formulated by reversing the sign of the learning rate of SLA for PSA [99]. However, Oja's MSA algorithm is known to diverge [2, 25, 98].

The orthogonal Oja algorithm consists of Oja's MSA plus an orthogonalization of $\mathbf{W}(t)$ at each iteration [1]

$$\mathbf{W}^T(t)\mathbf{W}(t) = \mathbf{I}. \tag{13.72}$$

In this case, Oja's MSA [99], (13.70) [14], and (13.71) [25] are equivalent. A Householder transform-based implementation of the MCA algorithm is given in [1]. Orthogonal Oja is numerically very stable. By reversing the sign of η, we extract J_2 principal components.

Normalized Oja [2] is derived by optimizing the MSE subject to an approximation to the orthonormal constraint (13.72). This leads to the optimal learning rate. Normalized orthogonal Oja is an orthogonal version of normalized Oja such that (13.72) is perfectly satisfied [2]. Both algorithms offer, as compared to SLA, faster convergence, orthogonality, and better numerical stability with a slight increase in computational complexity. By switching the sign of η in given learning algorithms, both normalized Oja and normalized orthogonal Oja can be used for the estimation of minor and principal subspaces of a vector sequence.

Oja's MSA, (13.70), (13.71), orthogonal Oja, normalized Oja, and normalized orthogonal Oja all have a complexity of $O(J_1 J_2)$ [1, 25]. Orthogonal Oja, normalized Oja, and normalized orthogonal Oja require less computation load than algorithms (13.70) and (13.71) [1, 2].

13.7.4 Other Algorithms

In [15], the proposed MCA algorithm for extracting multiple minor components utilizes the idea of sequential addition, and a conversion method between MCA and PCA is also discussed.

Based on a generalized differential equation for the generalized eigenvalue problem, a class of algorithms can be obtained for extracting the first principal or minor component by selecting different parameters and functions [155]. Many PCA algorithms [97, 133, 152] and MCA algorithms [133] are special cases of this class. All the algorithms of this class have the same order of convergence speed and are robust to implementation error.

A rapidly convergent quasi-Newton method is applied to extract multiple minor components in [90]. The algorithm has a complexity of $O\left(J_2 J_1^2\right)$ but with quadratic convergence. The algorithm makes use of the implicit orthogonalization procedure that is built into it through an inflation technique.

The adaptive minor component extraction (AMEX) algorithm [103] extends the work in [101] to extract multiple minor components corresponding to distinct eigenvalues. AMEX comes from the TLMS algorithm [28]. Unlike TLMS, AMEX is developed by unconstrained minimization of an information criterion using the gradient search approach. The criterion has a unique global minimum at the minor subspace and all other equilibrium points are saddle points. The algorithm automatically performs multiple minor component extraction in parallel without an inflation procedure. AMEX has the merit that increasing the number of the desired minor components does not affect the previously extracted minor components.

Several minor component algorithms, including (13.69), normalized Oja [98, 140], and an algorithm given in [103], are extended to those for tracking multiple minor components or the minor subspace in [31].

In [80, 150], simple neural network models, described by differential equations, calculate the largest and smallest eigenvalues as well as their corresponding eigenvectors of any real symmetric matrix.

13.8 Constrained PCA

When certain subspaces are less preferred than others, this yields the constrained PCA [70]. The optimality criterion for constrained PCA is variance maximization, as in PCA, but with an external subspace orthogonality constraint such that extracted principal components are orthogonal to some undesired subspace.

Given a J_1-dimensional stationary stochastic input vector $\boldsymbol{x}_t$ and an l-dimensional $(l < J_1)$ constraint vector $\boldsymbol{q}(t)$, such that

$$\boldsymbol{q}(t) = \mathbf{Q}\boldsymbol{x}_t, \tag{13.73}$$

where $\mathbf{Q}$ is an orthonormal constraint matrix, spanning an undesirable subspace $\mathcal{L}$. The task is to find, in the principal component sense, the most representative J_2-dimensional subspace $\mathcal{L}^{J_2}$ that is constrained to be orthogonal to $\mathcal{L}$, where $l + J_2 \leq J_1$. That is, we are required to find the optimal linear transform

$$\boldsymbol{y}(t) = \mathbf{W}^T \boldsymbol{x}_t, \tag{13.74}$$

where $\mathbf{W}$ is orthonormal, such that

$$E_{\text{CPCA}} = \text{E}\left[\left\|\boldsymbol{x} - \hat{\boldsymbol{x}}\right\|^2\right] = \text{E}\left[\|\boldsymbol{x} - \mathbf{W}\boldsymbol{y}\|^2\right] \tag{13.75}$$

is minimized subject to

$$\mathbf{QW} = \mathbf{0}. \tag{13.76}$$

The optimal solution to the constrained PCA problem is given by [70, 72]

$$\mathbf{W}^* = \left[\tilde{\boldsymbol{c}}_1 \ldots \tilde{\boldsymbol{c}}_{J_2}\right], \tag{13.77}$$

where $\tilde{\boldsymbol{c}}_i, i = 1, \ldots, J_2$, are the principal eigenvectors of the skewed autocorrelation matrix

$$\mathbf{C}_s = \left(\mathbf{I} - \mathbf{Q}\mathbf{Q}^T\right)\mathbf{C}. \tag{13.78}$$

At the optimum, E_{CPCA} takes its minimum

$$E_{\text{CPCA}}^* = \sum_{i=J_2+1}^{J_1} \tilde{\lambda}_i, \tag{13.79}$$

where $\tilde{\lambda}_i, i = 1, \ldots, J_1$, are the eigenvalues of $\mathbf{C}_s$ in descending order. Like PCA, the components now maximize the output variance, but under the additional constraint (13.76).

PCA usually obtains the best fixed-rank approximation to the data in the LS sense. On the other hand, constrained PCA allows specifying metric matrices that modulate the effects of rows and columns of a data matrix. This actually is weighted LS estimation. Constrained PCA first decomposes the data matrix by projecting the data matrix onto the spaces spanned by matrices of external information and then applies PCA to decomposed matrices, which involves generalized SVD. APEX can be applied to recursively solve the constrained PCA problem [72].

Given a sample covariance matrix, we examine the problem of maximizing the variance accounted for by a linear combination of the input variables while constraining the number of nonzero coefficients in this combination. This is known as *sparse PCA*. The problem is to find sparse factors that account for a maximum amount of variance. A semidefinite relaxation to this problem is formulated in [20] and a greedy algorithm that computes a full set of good solutions for all target numbers of nonzero coefficients is derived.

13.8.1 Sparse PCA

One would like to express as much variability in the data as possible, using components constructed from as few variables as possible. There are two kinds of sparse PCA: sparse loading PCA and sparse variable PCA.

Sparse variable PCA removes some measured variables completely by simultaneously zeroing out all their loadings. In [57], a sparse variable PCA method that is based on selecting a subset of measured variables with largest sample variances and then performing a PCA on the selected subset is given. Sparse variable PCA is capable of huge additional dimension reduction beyond PCA.

Sparse loading PCA focuses on zeroing out individual PCA loadings but keeps all the variables. One can simply set to zero the PCA loadings which are in absolute value smaller than some threshold constants [7]. SCoTLASS [58] directly puts

L_1 constraints on the PCA loadings. A greedy sparse loading PCA algorithm is developed in [20].

In an L_1 penalized likelihood approach to sparse variable PCA [129], smooth approximation of the L_1 penalty is used, and the optimization by geodesic steepest descent on a Stiefel manifold is carried out. A vector L_0 penalized likelihood approach to sparse variable PCA is considered; the proposed penalized EM algorithm, in an L_0 setting, leads to a closed-form M-step, and a convergence analysis is provided in [130]. Thus, one does not need to approximate the vector L_0 penalty and does not need to use Stiefel optimization.

In an approach to sparse PCA [59], two single-unit and two block optimization formulations of the sparse PCA problem are proposed, aimed at extracting a single sparse dominant principal component of a data matrix or more components at once, respectively. The dimension of the search space is decreased enormously if the data matrix has many more columns (variables) than rows.

Sparse solutions to a generalized EVD problem is obtained by solving the generalized EVD problem while constraining the cardinality of the solution. Instead of relaxing the cardinality constraint using an L_1-norm approximation, a tighter approximation that is related to the negative log-likelihood of a Student-t distribution is considered [123]. The problem is solved as a sequence of convex programs by invoking the majorization–minimization method. The resulting algorithm is proved to exhibit global convergence behavior. Three specific examples of sparse generalized EVD problems are sparse PCA, sparse CCA, and sparse LDA. The majorization–minimization method can be thought of as a generalization of the EM algorithm.

Compressive-projection PCA [36] is driven by projections at the sensor onto lower dimensional subspaces chosen at random, while the decoder, given only these random projections, recovers not only the coefficients associated with PCA but also an approximation to PCA basis itself. This makes possible an excellent dimension-reduction performance in a light-encoder/heavy-decoder system architecture, particularly in satellite-borne remote-sensing applications.

13.9 Localized PCA, Incremental PCA, and Supervised PCA

Localized PCA

The nonlinear PCA problem can be overcome using localized PCA. The data space is partitioned into a number of disjunctive regions, followed by the estimation of the principal subspace within each partition by linear PCA. The distribution is collectively modeled by a collection or a mixture of linear PCA models, each characterizing a partition. Localized PCA is different from local PCA. In local PCA, the update at each node makes use of only local information. Localized PCA provides an efficient means to decompose high-dimensional data-compression problems into low-dimensional ones.

VQ-PCA [61] is a locally linear model that uses vector quantization to define the Voronoi regions for localized PCA. The algorithm builds a piecewise linear model of the data. It performs better than the global models implemented by the linear PCA model and Kramer's nonlinear PCA, and is significantly faster than Kramer's nonlinear PCA. The localized PCA method is commonly used in image compression. An image is often transformation-coded by PCA, followed by coefficient quantization.

An online localized PCA algorithm [92] is developed by extending the neural gas method. Instead of the Euclidean distance measure, a combination of a normalized Mahalanobis distance and the squared reconstruction error guides the competition between the units. The unit centers are updated as in neural gas, while subspace learning is based on the robust RLS algorithm. ASSOM is another localized PCA for unsupervised extraction of invariant local features from the input data. It associates a subspace instead of a single weight vector to each node of SOM.

Incremental PCA

Incremental PCA algorithm [42] can update eigenvectors and eigenvalues incrementally. It is applied to a single training sample at a time, and the intermediate eigenproblem must be solved repeatedly for every training sample. Chunk incremental PCA [106] processes a chunk of training samples at a time. It can reduce the training time effectively as compared with incremental PCA unless the number of input attributes is too large. It can obtain major eigenvectors with fairly good approximation. In chunk incremental PCA, the update of an eigenspace is completed by performing single eigenvalue decomposition. The SVD updating-based incremental PCA algorithm [156] gives a close approximation to the batch-mode PCA method, and the approximation error is proved to be bounded.

Candid covariance-free IPCA [135] is a fast incremental PCA algorithm used to compute the principal components of a sequence of samples incrementally without estimating the covariance matrix. It is motivated by the concept of statistical efficiency (the estimate has the smallest variance given the observed data). Some links between incremental PCA and the development of the cerebral cortex are discussed in [135].

In a probabilistic online algorithm for PCA [132], in each trial the current instance is centered and projected onto a probabilistically chosen low-dimensional subspace. The total expected quadratic compression loss of the online algorithm minus the total quadratic compression loss of the batch algorithm is bounded by a term whose dependence on the dimension of the instances is only logarithmic. The running time is $O(n^2)$ per trial, where n is the dimension of the instances.

Other PCA Methods

Like supervised clustering, supervised PCA [13] is achieved by augmenting the input of PCA with the class label of the dataset. Class-augmented PCA [107] is a supervised feature extraction method, it is composed of processes for encoding the class information, augmenting the encoded information to data, and extracting features from class-augmented data by applying PCA.

PCA can be generalized to distributions of the exponential family [19]. This generalization is based on a generalized linear model and criterion functions using the Bregman distance. This approach permits hybrid dimension reduction in which different distributions are used for different attributes of the data.

13.10 Complex-Valued PCA

Complex PCA is a generalization of PCA in complex-valued datasets [49]. It has been widely applied to complex-valued data and two-dimensional vector fields. Complex PCA employs the same neural network architecture as that for PCA, but with complex weights. The objective functions for PCA can also be adapted to complex PCA by changing the transpose into the Hermitian transpose. For example, for complex PCA, one can minimize the MSE function

$$E = \frac{1}{N} \sum_{i=1}^{N} \left\| \boldsymbol{z}_i - \mathbf{W}\mathbf{W}^H \boldsymbol{z}_i \right\|^2, \tag{13.80}$$

where $\boldsymbol{z}_i$, $i = 1, \ldots, N$, are the input complex vectors. By minimizing (13.80), the first complex principal component is extracted.

Complex-domain GHA [153] extends GHA for complex principal component extraction. Complex-domain GHA is very similar to GHA except that complex notations are introduced. The updating rule for $\boldsymbol{w}_j$ is [153]:

$$\boldsymbol{w}_j(n+1) = \boldsymbol{w}_j(n) + \mu(n)\text{conj}[y_j(n)][\boldsymbol{x}(n) - y_j(n)\boldsymbol{w}_j(n) - \sum_{i<j} y_i(n)\boldsymbol{w}_i(n), \tag{13.81}$$

$$y_j(n) = \boldsymbol{w}_j^H(n)\boldsymbol{x}(n), \tag{13.82}$$

where H denotes the Hermitian transpose. With any initial $\boldsymbol{w}_j$, it is proved to converge to the jth normalized eigenvector of $\mathbf{C} = \mathrm{E}[\boldsymbol{x}\boldsymbol{x}^H]$.

In [112], a complex-valued neural network model is developed for nonlinear complex PCA. Nonlinear complex PCA has the ability to extract nonlinear features missed by PCA. It uses the architecture of Kramer's nonlinear PCA network, but with complex weights and biases. For a similar number of model parameters, it captures more variance of a dataset than the alternative real approach, where each complex variable is replaced by two real variables and applied to Kramer's nonlinear PCA. The complex hyperbolic tangent $\tanh(z)$ with $|z| < \frac{\pi}{2}$ is selected as the transfer function. Complex-valued BP or quasi-Newton method can be used for training.

Both PAST and PASTd are, respectively, the PSA and PCA algorithms derived for complex-valued signals [142]. Complex-valued APEX [17] actually allows extracting a number of principal components from a complex-valued signal. The robust complex PCA algorithms have also been derived in [18] for hierarchically extracting principal components of complex-valued signals based on a robust statistics-based loss function.

As far as complex MCA is concerned, the constrained anti-Hebbian learning algorithm [38, 39] has been extended for the complex-valued TLS problem [38].

Unlike L_1-PCA of real-valued data, which has an optimal polynomial-cost algorithm, complex L_1-PCA is formally NP-hard [128]. Two suboptimal algorithms are presented in [128].

13.11 Two-Dimensional PCA

Because of the small-sample-size problem for image representation, PCA is prone to be overfitted to the training set. In PCA, an $m \times n$ image $\mathbf{X}$ should be mapped into a high-dimensional $mn \times 1$ vector $\boldsymbol{x}$ in advance. Two-dimensional PCA can address these problems. In two-dimensional PCA, an image covariance matrix is constructed directly using the original image matrices instead of the transformed vectors, and its eigenvectors are derived for image feature extraction.

For $m \times n$ images, the size of the image covariance (scatter) matrix using 2DPCA [144] is $n \times n$, whereas for PCA the size is $mn \times mn$. 2DPCA evaluates the covariance matrix more accurately than PCA does. 2DPCA is a row-based PCA, and it only reflects the information between rows. It treats an image as m row vectors of dimension $1 \times n$ and performs PCA on all row vectors in the training set. In 2DPCA, the actual vector dimension is n and the actual sample size is mN, where $n \ll mN$. Thus, the small-sample-size problem is resolved. Despite its advantages, 2DPCA still suffers from the high feature dimension problem.

Diagonal PCA [154] improves 2DPCA by defining the image scatter matrix as the covariances between the variations of the rows and those of the columns of the images, and is more accurate than PCA and 2DPCA. $(\text{PC})^2$A [136] adopts image preprocessing plus PCA.

In modular PCA [41], an image is divided into n_1 subimages and PCA is performed on all these subimages. Since modular PCA divides an image into a number of subimages, the actual vector dimension in modular PCA will be much lower than in PCA. The number of training vectors used in modular PCA is much higher than the number used in PCA. Thus, modular PCA can be used to solve the overfitting problem. The feature dimension increases as the number of subimages is increased.

2DPCA and modular PCA both solve the overfitting problems by reducing the dimension and by increasing the training vectors yet introduce the high feature dimension problem.

Bidirectional PCA [159] reduces the dimension in both column and row directions for image feature extraction. The feature dimension of BD-PCA is much less than that of 2DPCA. Bidirectional PCA is a straightforward image projection technique where a $k_{\mathrm{col}} \times k_{\mathrm{row}}$ feature matrix $\mathbf{Y}$ of an $m \times n$ image $\mathbf{X}(k_{\mathrm{col}} \ll m, k_{\mathrm{row}} \ll n)$ can be obtained by

$$\mathbf{Y} = \mathbf{W}_{\mathrm{col}}^T \mathbf{X} \mathbf{W}_{\mathrm{row}}, \tag{13.83}$$

where $\mathbf{W}_{\mathrm{col}}$ is the column projector and $\mathbf{W}_{\mathrm{row}}$ is the row projector. 2DPCA can be regarded as a special bidirectional PCA with $\mathbf{W}_{\mathrm{col}}$ being an $m \times m$ identity matrix. Bidirectional PCA has to be performed in batch mode. The scatter matrices of bidirectional PCA are formulated as the sum of K (sample size) image covariance matrices, making incremental learning directly on the scatters impossible. With the concepts of tensor, k-mode unfolding and matricization, an SVD-revision-based incremental learning method of bidirectional PCA [113] gives a close approximation to bidirectional PCA, but using less time.

PCA-L1 [74] is a fast and robust L_1-norm-based PCA method. L_1-norm-based two-dimensional PCA (2DPCA-L1) [77] is a two-dimensional generalization of PCA-L1 [74]. It avoids computation of the eigendecomposition process and its iteration step is easy to perform. The generalized low-rank approximation of matrices (GLRAM) [147] is another two-dimensional PCA method.

The uncorrelated multilinear PCA algorithm [84] is used for unsupervised subspace learning of tensorial data. It is a multilinear extension of PCA. Through successive variance maximization, uncorrelated multilinear PCA seeks a tensor-to-vector projection that captures most of the variation in the original tensorial input while producing uncorrelated features. This work offers a way to systematically determine the maximum number of uncorrelated multilinear features that can be extracted by the method. The method not only obtains features that maximize the variance captured but also enforces a zero-correlation constraint, thus extracting uncorrelated features. It is the only multilinear extension of PCA that can produce uncorrelated features in a fashion similar to that of PCA, in contrast to other multilinear PCA extensions, such as 2DPCA [144] and multilinear PCA (MPCA) [83].

13.12 Generalized Eigenvalue Decomposition

Generalized EVD is a statistical tool that is extremely useful in feature extraction, pattern recognition as well as signal estimation and detection. The generalized EVD problem is to find a pair $(\lambda, \boldsymbol{x})$ such that

$$\mathbf{R}_1 \boldsymbol{x} = \lambda \mathbf{R}_2 \boldsymbol{x}, \tag{13.84}$$

where $\mathbf{R}_1 \in R^{n \times n}$, $\mathbf{R}_2 \in R^{n \times n}$, and $\lambda \in R$. Generalized EVD aims to find multiple principal or minor generalized eigenvectors of a positive-definite symmetric matrix pencil $(\mathbf{R}_1, \mathbf{R}_2)$. PCA, CCA, and LDA are specific instances of generalized EVD problems.

The generalized EVD problem involves the matrix equation

$$\mathbf{R}_1 \boldsymbol{w}_i = \lambda_i \mathbf{R}_2 \boldsymbol{w}_i, \tag{13.85}$$

where $\mathbf{R}_1, \mathbf{R}_2 \in R^{J_1 \times J_1}$, and λ_i, $\boldsymbol{w}_i$, $i = 1, \ldots, J_2$, are, respectively, the ith generalized eigenvalue and its corresponding generalized eigenvector of the matrix pencil $(\mathbf{R}_1, \mathbf{R}_2)$.

For real symmetric and positive-definite matrices, all the generalized eigenvectors are real and the corresponding generalized eigenvalues are positive.

Generalized EVD achieves simultaneous diagonalization of $\mathbf{R}_1$ and $\mathbf{R}_2$

$$\mathbf{W}^T \mathbf{R}_1 \mathbf{W} = \mathbf{\Lambda}, \quad \mathbf{W}^T \mathbf{R}_2 \mathbf{W} = \mathbf{I}, \tag{13.86}$$

where $\mathbf{W} = \left[\boldsymbol{w}_1, \ldots, \boldsymbol{w}_{J_2}\right]$ and $\mathbf{\Lambda} = \mathrm{diag}\left(\lambda_1, \ldots, \lambda_{J_2}\right)$. Typically, $\mathbf{R}_1$ and $\mathbf{R}_2$ are, respectively, the full covariance matrices of zero-mean stationary random signals $\boldsymbol{x}_1, \boldsymbol{x}_2 \in R^{J_1}$. In this case, iterative generalized EVD algorithms can be obtained by using two PCA steps. Alternatively, generalized EVD is also referred to as oriented PCA [24]. When $\mathbf{R}_2$ becomes an identity matrix, generalized EVD reduces to PCA.

Any generalized eigenvector $\boldsymbol{w}_i$ is a stationary point of the criterion function

$$E_{\mathrm{GEVD}}(\boldsymbol{w}) = \frac{\boldsymbol{w}^T \mathbf{R}_1 \boldsymbol{w}}{\boldsymbol{w}^T \mathbf{R}_2 \boldsymbol{w}}. \tag{13.87}$$

The LDA problem is a typical generalized EVD problem. The three-layer LDA network [88] is obtained by the concatenation of two Rubner–Tavan PCA subnetworks. Each subnetwork is trained by the Rubner–Tavan PCA algorithm [116, 117]. Based on the Rubner–Tavan PCA network architecture, online local learning algorithms for LDA and generalized EVD are given in [21].

Generalized EVD methods for extracting multiple principal generalized eigenvectors from two sequences of sample vectors are typically adaptive ones for online implementation. These include the LDA-based gradient-descent algorithm [10, 137], a gradient-based adaptive algorithm for estimating the largest principal generalized eigenvector [95], quasi-Newton type generalized EVD algorithm [89, 145, 146], RLS-like fixed-point generalized EVD algorithm [111], error-correction learning [21], Hebbian learning [21], and a self-stabilizing algorithm for extracting multiple generalized eigenpairs of a matrix pencil of two vector sequences adaptively [33]. Fixed-point algorithms do not require any external step-size parameters like the gradient-based methods. These algorithms may be sequential algorithms or parallel ones. As in case of PCA algorithms, sequential algorithms also use a deflation procedure. This causes error propagation, leading to slow convergence of minor generalized eigenvectors.

Implementation of generalized EVD algorithms can employ a neural network architecture, such as a two-layer linear heteroassociative network [10] or a lateral inhibition network [137]. Multiple generalized eigenvectors can be extracted by the deflation technique. A recurrent network with invariant B-norm proposed in [126] computes the largest or smallest generalized eigenvalue and the corresponding eigenvector of any symmetric positive pair, which can be simply extended to compute the second largest or smallest generalized eigenvalue and the corresponding eigenvector.

Most noncoupled algorithms suffer from the stability–speed problem. Two adaptive coupled algorithms for extracting the principal and minor generalized eigenpairs, respectively, are derived from two deterministic discrete-time systems [96], which characterize a generalized eigenpair as a stationary point of a certain function. The algorithms are natural combinations of the normalization and quasi-Newton steps for finding the solution. The two coupled algorithms were extended to multiple generalized eigenpairs extraction [32].

13.13 Singular Value Decomposition

SVD is among the most important tools in numerical analysis for solving a wide scope of approximation problems in signal processing, model reduction, and data compression. The cross-correlation asymmetric PCA/MCA networks can be used to extract the singular values of the cross-correlation matrix of two stochastic signal vectors, or to implement SVD of a general matrix.

The SVD updating algorithm [75] provides an efficient way to carry out SVD of a larger matrix. By exploiting the orthonormal properties and block structure, the SVD computation of $[\mathbf{A}, \mathbf{B}]$ can be efficiently carried out by using the smaller matrices and SVD of the smaller matrix.

13.13.1 Cross-Correlation Asymmetric PCA Networks

Given two sets of random vectors with zero mean, $\boldsymbol{x}_t \in R^{n_1}$ and $\boldsymbol{y}_t \in R^{n_2}$, the cross-correlation matrix is defined by

$$\mathbf{C}_{xy} = \mathrm{E}\left[\boldsymbol{x}_t \boldsymbol{y}_t^T\right] = \sum_{i=1}^{n} \sigma_i \boldsymbol{v}_i^x \left(\boldsymbol{v}_i^y\right)^T, \tag{13.88}$$

where $\sigma_i > 0$ is the ith singular value, $\boldsymbol{v}_i^x$ and $\boldsymbol{v}_i^y$ are its corresponding left and right singular vectors, and $n = \min\{n_1, n_2\}$. The cross-correlation asymmetric PCA network is a method for extracting multiple principal singular components of $\mathbf{C}_{xy}$.

The cross-correlation asymmetric PCA network consists of two sets of neurons that are laterally hierarchically connected [22]. The asymmetric PCA network, shown in Fig. 13.13, is composed of two hierarchical PCA networks. $\boldsymbol{x} \in R^{n_1}$ and $\boldsymbol{y} \in R^{n_2}$ are input vectors, $\boldsymbol{a}, \boldsymbol{b} \in R^m$ are the output vectors of the hidden layers. The $n_1 \times m$ matrix $\underline{\mathbf{W}} = \left[\underline{\boldsymbol{w}}_1 \ldots \underline{\boldsymbol{w}}_m\right]$ and the $n_2 \times m$ matrix $\overline{\mathbf{W}} = [\overline{\boldsymbol{w}}_1 \ldots \overline{\boldsymbol{w}}_m]$ are the feedforward weights, while $\underline{\mathbf{U}} = \left[\underline{\boldsymbol{u}}_1 \ldots \underline{\boldsymbol{u}}_m\right]$ and $\overline{\mathbf{U}} = [\overline{\boldsymbol{u}}_1 \ldots \overline{\boldsymbol{u}}_m]$ are the $n_2 \times m$ matrices of lateral connection weights, where $\underline{\boldsymbol{u}}_i = \left(\underline{u}_{1i}, \ldots, \underline{u}_{mi}\right)^T$, $\overline{\boldsymbol{u}}_i = (\overline{u}_{1i}, \ldots, \overline{u}_{mi})^T$, and $m \le \min\{n_1, n_2\}$. This model performs SVD of $\mathbf{C}_{xy}$.

The network has the following relations:

$$\boldsymbol{a} = \underline{\mathbf{W}}^T \boldsymbol{x}, \quad \boldsymbol{b} = \overline{\mathbf{W}}^T \boldsymbol{y}, \tag{13.89}$$

where $\boldsymbol{a} = (a_1, \ldots, a_m)^T$ and $\boldsymbol{b} = (b_1, \ldots, b_m)^T$.

The objective function for extracting the first principal singular value of the covariance matrix is given by

$$E_{\text{APCA}}\left(\underline{\boldsymbol{w}}, \overline{\boldsymbol{w}}\right) = \frac{\mathrm{E}\left[a_1(t) b_1(t)\right]}{\left\|\underline{\boldsymbol{w}}\right\| \left\|\overline{\boldsymbol{w}}\right\|} = \frac{\underline{\boldsymbol{w}}^T \mathbf{C}_{xy} \overline{\boldsymbol{w}}}{\left\|\underline{\boldsymbol{w}}\right\| \left\|\overline{\boldsymbol{w}}\right\|}. \tag{13.90}$$

It is an indefinite function. When $\boldsymbol{y} = \boldsymbol{x}$, it reduces to PCA. After the principal singular component is extracted, a deflation transformation is introduced to nullify the principal singular value so as to make the next singular value principal. Thus, $\mathbf{C}_{xy}$ in the criterion (13.90) can be replaced by one of the following three transformed forms so as to extract the $(i+1)$th principal singular component

$$\mathbf{C}_{xy}^{(i+1)} = \mathbf{C}_{xy}^{(i)} \left(\mathbf{I} - \boldsymbol{v}_i^y \left(\boldsymbol{v}_i^y\right)^T\right), \tag{13.91}$$

$$\mathbf{C}_{xy}^{(i+1)} = \left(\mathbf{I} - \boldsymbol{v}_i^x \left(\boldsymbol{v}_i^x\right)^T\right) \mathbf{C}_{xy}^{(i)}, \tag{13.92}$$

$$\mathbf{C}_{xy}^{(i+1)} = \left(\mathbf{I} - \boldsymbol{v}_i^x \left(\boldsymbol{v}_i^x\right)^T\right) \mathbf{C}_{xy}^{(i)} \left(\mathbf{I} - \boldsymbol{v}_i^y \left(\boldsymbol{v}_i^y\right)^T\right) \tag{13.93}$$

for $i = 1, \ldots, m-1$, where $\mathbf{C}_{xy}^{(1)} = \mathbf{C}_{xy}$. These are, respectively, obtained by the transforms on the data:

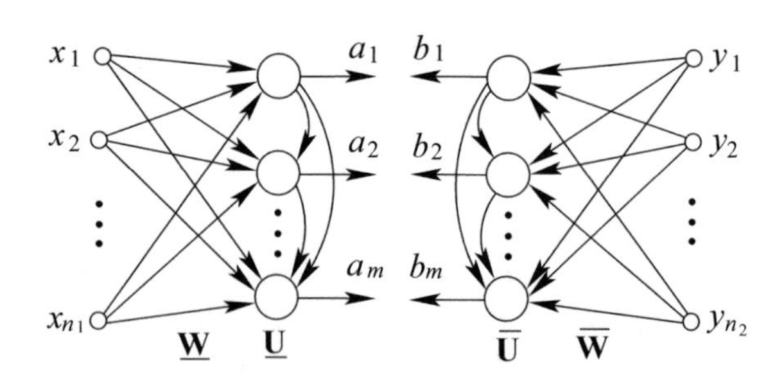

Fig. 13.13 Architecture of the cross-correlation asymmetric PCA network

$$\boldsymbol{x} \leftarrow \boldsymbol{x}, \qquad \boldsymbol{y} \leftarrow \boldsymbol{y} - \boldsymbol{v}_i^x \left(\boldsymbol{v}_i^x\right)^T \boldsymbol{y}, \tag{13.94}$$

$$\boldsymbol{x} \leftarrow \boldsymbol{x} - \boldsymbol{v}_i^y \left(\boldsymbol{v}_i^y\right)^T \boldsymbol{x}, \qquad \boldsymbol{y} \leftarrow \boldsymbol{y}, \tag{13.95}$$

$$\boldsymbol{x} \leftarrow \boldsymbol{x} - \boldsymbol{v}_i^y \left(\boldsymbol{v}_i^y\right)^T \boldsymbol{x}, \qquad \boldsymbol{y} \leftarrow \boldsymbol{y} - \boldsymbol{v}_i^x \left(\boldsymbol{v}_i^x\right)^T \boldsymbol{y}. \tag{13.96}$$

Using a deflation transformation, the two sets of neurons are trained with the cross-coupled Hebbian learning rules, which, for $j = 1, \ldots, m$, are given by [22]

$$\underline{\boldsymbol{w}}_j(t+1) = \underline{\boldsymbol{w}}_j(t) + \eta \left[\boldsymbol{x}(t) - \underline{\boldsymbol{w}}_j(t) a_j(t)\right] b_j'(t), \tag{13.97}$$

$$\overline{\boldsymbol{w}}_j(t+1) = \overline{\boldsymbol{w}}_j(t) + \eta \left[\boldsymbol{y}(t) - \overline{\boldsymbol{w}}_j(t) b_j(t)\right] a_j'(t), \tag{13.98}$$

where the learning rate $\eta > 0$ is selected as a small constant or according to the Robbins–Monro conditions,

$$a_j' = a_j - \sum_{i=1}^{j-1} \underline{u}_{ij} a_i, \quad b_j' = b_j - \sum_{i=1}^{j-1} \overline{u}_{ij} b_i, \tag{13.99}$$

$$a_i = \underline{\boldsymbol{w}}_i^T \boldsymbol{x}, \quad b_i = \overline{\boldsymbol{w}}_i^T \boldsymbol{y}, \quad i = 1, \ldots, j \tag{13.100}$$

and the lateral weights should be equal to

$$\underline{u}_{ij} = \underline{\boldsymbol{w}}_i^T \underline{\boldsymbol{w}}_j, \quad \overline{u}_{ij} = \overline{\boldsymbol{w}}_i^T \overline{\boldsymbol{w}}_j, \quad i = 1, \ldots, j-1. \tag{13.101}$$

A set of lateral connections among the units is called a *lateral orthogonaliztion network*. Hence, $\overline{\mathbf{U}}$ and $\underline{\mathbf{U}}$ are upper triangular matrices. The stability of the algorithm is proved based on Lyapunov's second theorem.

A local algorithm for calculating $\underline{u}_{ij}$ and $\overline{u}_{ij}$ is derived by premultiplying (13.97) with $\underline{\boldsymbol{w}}_i^T$, premultiplying (13.98) with $\overline{\boldsymbol{w}}_i^T$, and then employing (13.101)

$$\underline{u}_{ij}(t+1) = \underline{u}_{ij}(t) + \eta \left[a_i(t) - \overline{u}_{ij}(t) a_j(t)\right] b_j'(t), \tag{13.102}$$

$$\overline{u}_{ij}(t+1) = \overline{u}_{ij}(t) + \eta \left[b_i(t) - \overline{u}_{ij}(t) b_j(t)\right] a_j'(t). \tag{13.103}$$

We can select $\underline{u}_{ij}(0) = \underline{\boldsymbol{w}}_i^T(0) \underline{\boldsymbol{w}}_j(0)$ and $\overline{u}_{ij}(0) = \overline{\boldsymbol{w}}_i^T(0) \overline{\boldsymbol{w}}_j(0)$. However, this initial condition is not critical to the convergence of the algorithm.

$\underline{\boldsymbol{w}}_i$ and $\overline{\boldsymbol{w}}_i$ approximate the ith left and right principal singular vectors of $\mathbf{C}_{xy}$, respectively, and σ_i approximates its corresponding criterion E_{APCA}, as $t \to \infty$; that is, the algorithm extracts the first m principal singular values in descending order and their corresponding left and right singular vectors. Like APEX, asymmetric PCA incrementally adds nodes without retraining the learned nodes. Exponential convergence has been observed by simulation [22].

When m in the asymmetric PCA network is selected as unity, the principal singular component of $\mathbf{C}_{xy}$ can be extracted by modifying the cross-coupled Hebbian rule [27]

$$\underline{\boldsymbol{w}}_1(t+1) = \underline{\boldsymbol{w}}_1(t) + \eta \left[b_1(t)\boldsymbol{x}(t) - \left\| \underline{\boldsymbol{w}}_1(t) \right\|^2 \underline{\boldsymbol{w}}_1(t) \right], \tag{13.104}$$

$$\overline{\boldsymbol{w}}_1(t+1) = \overline{\boldsymbol{w}}_1(t) + \eta \left[a_1(t)\boldsymbol{y}(t) - \left\| \overline{\boldsymbol{w}}_1(t) \right\|^2 \overline{\boldsymbol{w}}_1(t) \right]. \tag{13.105}$$

This algorithm can efficiently extract the principal singular component, and is proved to have global asymptotic convergence. When $\boldsymbol{y}(t) = \boldsymbol{x}(t)$, it reduces to Oja's PCA algorithm [97].

13.13.2 Extracting Principal Singular Components for Nonsquare Matrices

When the cross-correlation matrix is replaced by a general nonsquare matrix, (13.104) and (13.105) can be directly transformed into the algorithm for extracting the principal singular component of a general matrix $\mathbf{A} \in R^{n_1 \times n_2}$ [27]

$$\underline{\boldsymbol{w}}_1(t+1) = \underline{\boldsymbol{w}}_1(t) + \eta \left[\mathbf{A}\overline{\boldsymbol{w}}_1(t) - \left\| \underline{\boldsymbol{w}}_1(t) \right\|^2 \underline{\boldsymbol{w}}_1(t) \right], \tag{13.106}$$

$$\overline{\boldsymbol{w}}_1(t+1) = \overline{\boldsymbol{w}}_1(t) + \eta \left[\mathbf{A}^T \underline{\boldsymbol{w}}_1(t) - \left\| \overline{\boldsymbol{w}}_1(t) \right\|^2 \overline{\boldsymbol{w}}_1(t) \right]. \tag{13.107}$$

Using (13.106) and (13.107) and a deflation transformation, one can extract multiple principal singular components for the nonsquare matrix $\mathbf{A}$ [27]

$$\underline{\boldsymbol{w}}_i(t+1) = \underline{\boldsymbol{w}}_i(t) + \eta_i \left[\mathbf{A}_i\overline{\boldsymbol{w}}_i(t) - \left\| \underline{\boldsymbol{w}}_i(t) \right\|^2 \underline{\boldsymbol{w}}_i(t) \right], \quad i = 1, \ldots, m, \tag{13.108}$$

$$\overline{\boldsymbol{w}}_i(t+1) = \overline{\boldsymbol{w}}_i(t) + \eta_i \left[\mathbf{A}_i^T \underline{\boldsymbol{w}}_i(t) - \left\| \overline{\boldsymbol{w}}_i(t) \right\|^2 \overline{\boldsymbol{w}}_i(t) \right], \quad i = 1, \ldots, m, \tag{13.109}$$

$$\mathbf{A}_{i+1} = \mathbf{A}_i - \underline{\boldsymbol{w}}_i \overline{\boldsymbol{w}}_i^T, \quad i = 1, \ldots, m-1, \tag{13.110}$$

where $\mathbf{A}_1 = \mathbf{A}$, and the learning rates are suggested to be $\eta_i < \frac{1}{4\sigma_i}$. As $t \to \infty$, $\underline{\boldsymbol{w}}_i(t)$ and $\overline{\boldsymbol{w}}_i(t)$ represent the left and right singular vectors corresponding to the ith singular value, arranged in descending order $\sigma_1 \geq \sigma_2 \geq \ldots \geq \sigma_m > 0$.

The measures of stopping iteration can be given as

$$e_i(t) = \left\| \mathbf{A}_i\overline{\boldsymbol{w}}_i(t) - \left\| \underline{\boldsymbol{w}}_i(t) \right\|^2 \underline{\boldsymbol{w}}_i(k) \right\|^2 + \left\| \mathbf{A}_i^T \underline{\boldsymbol{w}}_i(t) - \left\| \overline{\boldsymbol{w}}_i(t) \right\|^2 \overline{\boldsymbol{w}}_i(t) \right\|^2 < \varepsilon, \quad i = 1, \ldots, m, \tag{13.111}$$

where ε is a small number such as 10^{-20}. The algorithm is proved convergent.

The algorithm can efficiently perform SVD of an ill-posed matrix. It can be used to solve the smallest singular component of the general matrix **A**. Although the method is indirect for computing the smallest singular component of a nonsquare matrix, it is efficient and robust. The algorithm is particularly useful for TLS problems.

13.13.3 Extracting Multiple Principal Singular Components

The double generalized Hebbian algorithm [121] is derived from a twofold optimization problem: the left singular vector estimate is adapted by GHA, whereas the right singular vector estimate is adapted by the Widrow–Hoff rule. The linear approximation asymmetric PCA network [23] or orthogonal asymmetric encoder [121] is a two-layer feedforward linear network. Training is performed by BP. A stochastic online algorithm has been suggested [23, 121]. The network has a bottleneck topology. The cross-associative network for single component learning is given in [29], where fixed points are the first principal left and right singular vectors of **A**. The cross-associative neural network [30] is derived from a non-quadratic objective function which incorporates a matrix logarithm for extracting multiple principal singular components.

Coupled online learning rules [60] are derived for SVD of a cross-covariance matrix of two correlated data streams. The coupled SVD rule is derived by applying Newton's method to an objective function which is neither subject to minimization nor to maximization. Newton's method guarantees nearly equal convergence speed in all directions, independent of the singular structure of **A**, and turns the saddle point into an attractor [93]. The online learning rules resemble PCA rules [93] and the cross-coupled Hebbian rule [22].

A first-order approximation of GSO is used as a decorrelation method for estimation of multiple singular vectors and singular values. By inserting the first-order approximation or deflation, we can obtain the corresponding Hebbain SVD algorithms and the coupled SVD algorithms.

Coupled learning rules for SVD produce better results than Hebbian learning rules. Combined with first-order approximation of GSO, precise estimates of singular vectors and singular values with only small deviations from orthonormality are produced. Double deflation is clearly superior to standard deflation but inferior to first-order approximation of GSO, both with respect to orthonormality and diagonalization errors. Coupled learning rules converge faster than Hebbian learning rules, and the first-order approximation of GSO produces more precise estimates and better orthonormality than standard deflation [60]. Many SVD algorithms are reviewed in [60].

13.14 Factor Analysis

Factor analysis [122] is a powerful multivariate analysis technique that identifies the common characteristics among a set of variables. It has been widely used in many disciplines such as botany, biology, psychology, social sciences, economics, and engineering.

Factor analysis is a linear latent variable scheme which is also used to capture local substructures. It models correlations in multidimensional data by expressing the correlations in a lower dimensional subspace, and hence reducing the attribute space from a larger number of observable variables to a smaller number of latent variables called factors.

Statistical factor analysis is given by a common latent variable model [127]

$$\boldsymbol{x} = \mathbf{A}\boldsymbol{y} + \boldsymbol{\mu} + \boldsymbol{n}, \tag{13.112}$$

where the observed vector $\boldsymbol{x}$ is a J_1-dimensional vector, the parameter matrix $\mathbf{A} = [a_{ij}]$ is a $J_1 \times J_2$ matrix (with $J_2 < J_1$) containing the factor loadings a_{ij}, the latent variable vector $\boldsymbol{y}$ is J_2-dimensional vector whose entries are termed *common factors*, *factors* or *latent variables*, $\boldsymbol{\mu}$ is a mean vector, and $\boldsymbol{n}$ is a noise term. The entries in $\boldsymbol{\mu}$ and $\boldsymbol{n}$ are known as *specific factors*. When $\boldsymbol{y}$ and $\boldsymbol{n}$ are Gaussian, $\boldsymbol{x}$ also has a normal distribution.

Factor analysis is related to PCA. It can be seen that PCA has exactly the same probabilistic linear model for (13.112). However, the factors estimated are not defined uniquely, but only up to a rotation. The ML method for fitting factor analysis is very popular and EM has slow convergence.

Group factor analysis [66] extends the formulation into linear factors that explain relationships between groups of variables, where each group represents either a set of related variables or a dataset. The model naturally extends CCA to more than two sets. Group factor analysis implements a Gibbs sampling scheme for inferring the posterior. The solution is formulated as a variational inference of a latent variable model with structural sparsity, and it consists of two hierarchical levels: the higher level models the relationships between the groups and the lower models the observed variables given the higher level. A structured Bayesian group factor analysis model [157] extends the factor model to multiple coupled observation matrices; in the case of two observations, this reduces to a Bayesian model of CCA.

Boolean factor analysis implies that all the components of the data, factor loadings, and factor scores are binary variables. Boolean factor analysis may be based on Boolean matrix factorization [5, 85] or on dataset likelihood maximization according to some generative model of data [86].

13.15 Canonical Correlation Analysis

CCA [51], proposed by Hotelling in 1936, is a multivariate statistical technique. It makes use of two views of the same set of objects and projects them onto a lower dimensional space in which they are maximally correlated. CCA seeks prominently correlated projections between two views of data and it has been long known to be equivalent to LDA when the data features are used in one view and the class labels are used in the other view [4, 46]. In other words, LDA is a special case of CCA. CCA is equivalent to LDA for binary-class problems [46], and it can be formulated as an LS problem for binary-class problems.

CCA leads to a generalized EVD problem. Thus, we can employ a kernelized version of CCA to compute a flexible contrast function for ICA. Generalized CCA consists of a generalization of CCA to more than two sets of variables [65].

Given two centered random multivariables $\boldsymbol{x} \in R^{n_x}$ and $\boldsymbol{y} \in R^{n_y}$, the goal of CCA is to find a pair of directions called *canonical vectors* $\boldsymbol{\omega}_x$ and $\boldsymbol{\omega}_y$ such that the correlation $\rho(\boldsymbol{x}, \boldsymbol{y})$ between the two projections $\boldsymbol{\omega}_x^T \boldsymbol{x}$ and $\boldsymbol{\omega}_y^T \boldsymbol{y}$ is maximized.

Suppose that we are given a sample of instances $S = \{(\boldsymbol{x}_1, \boldsymbol{y}_1), \ldots, (\boldsymbol{x}_n, \boldsymbol{y}_n)\}$ of $(\boldsymbol{x}, \boldsymbol{y})$. Let S_x denote $(\boldsymbol{x}_1, \ldots, \boldsymbol{x}_n)$ and similarly S_y denote $(\boldsymbol{y}_1, \ldots, \boldsymbol{y}_n)$. We can consider defining a new coordinate for $\boldsymbol{x}$ by choosing direction $\boldsymbol{w}_x$ and projecting $\boldsymbol{x}$ onto that direction, $\boldsymbol{x} \to \boldsymbol{w}_x^T \boldsymbol{x}$. If we do the same for $\boldsymbol{y}$ by choosing direction $\boldsymbol{w}_y$, we obtain a sample of the new mapping for $\boldsymbol{y}$. Let

$$S_{x,\boldsymbol{w}_x} = (\boldsymbol{w}_x^T \boldsymbol{x}_1, \ldots, \boldsymbol{w}_x^T \boldsymbol{x}_n), \tag{13.113}$$

with the corresponding values of the mapping for $\boldsymbol{y}$ being

$$S_{y,\boldsymbol{w}_y} = (\boldsymbol{w}_y^T \boldsymbol{y}_1, \ldots, \boldsymbol{w}_y^T \boldsymbol{y}_n). \tag{13.114}$$

The first stage of canonical correlation is to choose $\boldsymbol{w}_x$ and $\boldsymbol{w}_y$ to maximize the correlation between the two vectors

$$\rho = \max_{\boldsymbol{w}_x, \boldsymbol{w}_y} \operatorname{corr}\left(S_{x\boldsymbol{w}_x}, S_{y\boldsymbol{w}_y}\right) = \max_{\boldsymbol{w}_x, \boldsymbol{w}_y} \frac{\langle S_{x\boldsymbol{w}_x}, S_{y\boldsymbol{w}_y} \rangle}{\left\| S_{x\boldsymbol{w}_x} \right\| \left\| S_{y\boldsymbol{w}_y} \right\|}. \tag{13.115}$$

After manipulation, we have

$$\rho = \max_{\boldsymbol{w}_x, \boldsymbol{w}_y} \frac{\boldsymbol{w}_x^T \mathrm{E}\left[\boldsymbol{x}\boldsymbol{y}^T\right] \boldsymbol{w}_y}{\sqrt{\boldsymbol{w}_x^T \mathrm{E}\left[\boldsymbol{x}\boldsymbol{x}^T\right] \boldsymbol{w}_x \boldsymbol{w}_y^T \mathrm{E}\left[\boldsymbol{y}\boldsymbol{y}^T\right] \boldsymbol{w}_y}} = \frac{\boldsymbol{w}_x^T \mathbf{C}_{xy} \boldsymbol{w}_y}{\sqrt{\boldsymbol{w}_x^T \mathbf{C}_{xx} \boldsymbol{w}_x \boldsymbol{w}_y^T \mathbf{C}_{yy} \boldsymbol{w}_y}}, \tag{13.116}$$

where the covariance matrix of $(\boldsymbol{x}, \boldsymbol{y})$ is defined by

$$\boldsymbol{C}(x, y) = \mathrm{E}\left[\begin{pmatrix} \boldsymbol{x} \\ \boldsymbol{y} \end{pmatrix} \begin{pmatrix} \boldsymbol{x} \\ \boldsymbol{y} \end{pmatrix}^T\right] = \begin{bmatrix} \mathbf{C}_{xx} & \mathbf{C}_{xy} \\ \mathbf{C}_{yx} & \mathbf{C}_{yy} \end{bmatrix} = \mathbf{C}. \tag{13.117}$$

The problem can be transformed into

$$\max_{\boldsymbol{w}_x, \boldsymbol{w}_y} \boldsymbol{w}_x^T \mathbf{C}_{xy} \boldsymbol{w}_y \tag{13.118}$$

subject to

$$\boldsymbol{w}_x^T \mathbf{C}_{xx} \boldsymbol{w}_x = 1, \quad \boldsymbol{w}_y^T \mathbf{C}_{yy} \boldsymbol{w}_y = 1. \tag{13.119}$$

This optimization problem can be solved by a generalized eigenvalue problem:

$$\mathbf{C}_{xy} \boldsymbol{w}_y = \mu \mathbf{C}_{xx} \boldsymbol{w}_x, \quad \mathbf{C}_{yx} \boldsymbol{w}_x = \nu \mathbf{C}_{yy} \boldsymbol{w}_y, \tag{13.120}$$

where μ and ν are Lagrange multipliers. It can be derived that $\boldsymbol{w}_x$ and $\boldsymbol{w}_y$ are the eigenvectors of $\mathbf{C}_x = \mathbf{C}_{\boldsymbol{xx}}^{-1} \mathbf{C}_{xy} \mathbf{C}_{yy}^{-1} \mathbf{C}_{xy}^T$ and $\mathbf{C}_y = \mathbf{C}_{yy}^{-1} \mathbf{C}_{xy}^T \mathbf{C}_{xx}^{-1} \mathbf{C}_{xy}$ corresponding to their largest eigenvalues, respectively.

Under a mild condition which tends to hold for high-dimensional data, CCA in the multilabel case can be formulated as an LS problem [125]. Based on this, efficient algorithms for solving LS problems can be applied to scale CCA to very large datasets. In addition, several CCA extensions, including the sparse CCA formulation based on L_1-norm regularization, are proposed in [125]. The LS formulation of CCA and its extensions can be solved efficiently. The LS formulation is extended to orthonormalized partial least squares by establishing the equivalence relationship between CCA and orthonormalized partial least squares [125]. The CCA projection for one set of variables is independent of the regularization on the other set of multidimensional variables.

In [73], a strategy for reducing LDA to CCA is proposed. Within-class coupling CCA (WCCCA) is to apply CCA to pairs of data samples that are most likely to belong to the same class. Each one of the samples of a class, serving as the first view, is paired with every other sample of that class serving as the second view. The equivalence between LDA and such an application of CCA is proved.

Two-dimensional CCA seeks linear correlation based on images directly. Motivated by locality-preserving CCA [124] and spectral clustering, a manifold learning method called local two-dimensional CCA [134] identifies the local correlation by weighting images differently according to their closeness. That is, the correlation is measured locally, which makes local two-dimensional CCA more accurate in finding correlative information. Local two-dimensional CCA is formulated as solving generalized eigenvalue equations tuned by Laplacian matrices.

CCArc [69] is a two-dimensional CCA that is based on representing the image as the sets of its rows and columns and implementation of CCA using these sets. CCArc does not require preliminary downsampling procedure, it is not iterative and it is applied along the rows and columns of input image. Size of covariance matrices in CCArc is equal to $\max\{M, N\}$. Small-sample-size problem in CCArc does not occur, because we actually use N images of size $M \times 1$ and M images of size $N \times 1$; this always meets the condition $\max\{M, N\} < (M + N)$.

A method for solving CCA in a sparse convex framework [43] is proposed using an LS approach. Sparse CCA minimizes the number of features used in both the primal and dual projections while maximizing the correlation between the two views. When the number of the original features is large, sparse CCA outperforms kernel CCA, learning the common semantic space from a sparse set of features. Least squares canonical dependency analysis [63] is an extension of CCA that can effectively capture complicated nonlinear correlations through maximization of the statistical dependency between two projected variables.

Multi-view CCA is a generalization of CCA to multi-view scenario [118]. The goal is to find a set of linear transforms $\{\boldsymbol{w}_1, \ldots, \boldsymbol{w}_v\}$ to respectively project the samples of v views $\{\mathbf{X}_1, \ldots, \mathbf{X}_v\}$ to one common space, i.e., $\{\boldsymbol{w}_1^T \mathbf{X}_1, \ldots, \boldsymbol{w}_v^T \mathbf{X}_v\}$. In multi-view CCA, v view-specific transforms, one for each view, are obtained by maximizing the total of the correlations in the common space between any two views:

$$\max_{\boldsymbol{w}_1, \boldsymbol{w}_2, \ldots, \boldsymbol{w}_v} \sum_{i<j} \boldsymbol{w}_i^T \mathbf{X}_i \mathbf{X}_j^T \boldsymbol{w}_j, \tag{13.121}$$

$$\text{s.t.} \boldsymbol{w}_i^T \mathbf{X}_i \mathbf{X}_i^T \boldsymbol{w}_i = 1, i = 1, \ldots, v, \tag{13.122}$$

where $\mathbf{X}_i \in R^{p_i \times n}$ is the data matrix of ith view with n samples of p_i dimensions. Like CCA, the number of samples for each view is the same.

Problems

13.1 Show that the Oja's algorithm is asymptotically stable.

13.2 Show that the discrete-time Oja rule is a good approximation of the normalized Hebbian rule.

13.3 Show that the average Hessian $\mathbf{H}(\boldsymbol{w})$ in (13.14) is positive definite only at $\boldsymbol{w} = \pm \boldsymbol{c}_1$.

13.4 For the data generated by the augoregressive process

$$x_k = 0.8 x_{k-1} + e_k,$$

where e_k is a zero-mean uncorrelated Gaussian driving sequence with unit variance. The data points are arranged in blocks of size $N = 6$. Extract the first two minor components.

13.5 The grayscale Lenna picture of the 512×512 pixels is split into nonoverlapping 8×8 blocks. Each block constructs a 64-dimensional vector. The vectors are selected randomly form an input sequence $\boldsymbol{x}(k) \in \mathcal{R}^{64}$. Compute six principal directions and their direction cosines by using GHA with learning rates given by (13.45). Plot the reconstructed image and the picture SNR.

13.6 Redo Example 13.1 by using APEX.

13.7 Explain the function of the bottleneck layer in the five-layer autoassociative neural network.

13.8 Generate 400 observations of three variates X_1, X_2, X_3 according to $X_1 = Z_1, X_2 = X_1 + 0.1Z_2, X_3 = 10Z_3$, where Z_1, Z_2, Z_3 are independent standard normal variates. Compute and plot the leading principal component and factor analysis directions.

13.9 In Example 13.4, PCA is used for image coding. Complete the example by quantizing each coefficient using a suitable number of bits. Calculate the compression ratio of the image.

13.10 Given a dataset generated by

$$\boldsymbol{x}_i = \begin{pmatrix} \sin y_i \\ \sin 2y_i \end{pmatrix}, \quad y_i = \frac{\pi i}{2n}, \quad i = 1, \ldots, 100.$$

(a) For the samples, plot x_2 against x_1.
(b) Plot the discovered coordinates of $\boldsymbol{x}_i$ obtained by CCA against y_i.

References

1. Abed-Meraim, K., Attallah, S., Chkeif, A., & Hua, Y. (2000). Orthogonal Oja algorithm. *IEEE Signal Processing Letters*, *7*(5), 116–119.
2. Attallah, S., & Abed-Meraim, K. (2001). Fast algorithms for subspace tracking. *IEEE Signal Processing Letters*, *8*(7), 203–206.
3. Bannour, S., & Azimi-Sadjadi, M. R. (1995). Principal component extraction using recursive least squares learning. *IEEE Transactions on Neural Networks*, *6*(2), 457–469.
4. Bartlett, M. S. (1938). Further aspects of the theory of multiple regression. *Proceedings of the Cambridge Philosophical Society*, *34*, 33–40.
5. Belohlavek, R., & Vychodil, V. (2010). Discovery of optimal factors in binary data via a novel method of matrix decomposition. *Journal of Computer and System Sciences*, *76*(1), 3–20.
6. Bourlard, H., & Kamp, Y. (1988). Auto-association by multilayer perceptrons and singular value decomposition. *Biological Cybernetics*, *59*, 291–294.
7. Cadima, J., & Jolliffe, I. (1995). Loadings and correlations in the interpretation of principal component analysis. *Journal of Applied Statistics*, *22*(2), 203–214.
8. Chatterjee, C., Roychowdhury, V. P., & Chong, E. K. P. (1998). On relative convergence properties of principal component analysis algorithms. *IEEE Transactions on Neural Networks*, *9*(2), 319–329.
9. Chauvin, Y. (1989). Principal component analysis by gradient descent on a constrained linear Hebbian cell. In *Proceedings of the International Joint Conference on Neural Networks* (pp. 373–380). Wanshington, DC.
10. Chatterjee, C., Roychowdhury, V. P., Ramos, J., & Zoltowski, M. D. (1997). Self-organizing algorithms for generalized eigen-decomposition. *IEEE Transactions on Neural Networks*, *8*(6), 1518–1530.
11. Chen, H., & Liu, R. W. (1994). An on-line unsupervised learning machine for adaptive feature extraction. *IEEE Transactions on Circuits and Systems II*, *41*(2), 87–98.

12. Chen, L. H., & Chang, S. (1995). An adaptive learning algorithm for principal component analysis. *IEEE Transactions on Neural Networks*, *6*(5), 1255–1263.
13. Chen, S., & Sun, T. (2005). Class-information-incorporated principal component analysis. *Neurocomputing*, *69*, 216–223.
14. Chen, T., Amari, S. I., & Lin, Q. (1998). A unified algorithm for principal and minor components extraction. *Neural Networks*, *11*, 385–390.
15. Chen, T., Amari, S. I., & Murata, N. (2001). Sequential extraction of minor components. *Neural Processing Letters*, *13*, 195–201.
16. Chen, W., Er, M. J., & Wu, S. (2005). PCA and LDA in DCT domain. *Pattern Recognition Letters*, *26*, 2474–2482.
17. Chen, Y., & Hou, C. (1992). High resolution adaptive bearing estimation using a complex-weighted neural network. In *Proceedings of IEEE International Conference on Acoustics, Speech, and Signal Processing (ICASSP)* (Vol. 2, pp. 317–320). San Francisco, CA.
18. Cichocki, A., Swiniarski, R. W., & Bogner, R. E. (1996). Hierarchical neural network for robust PCA computation of complex valued signals. In *Proceedings of the World Congress Neural Networks* (pp. 818–821). San Diego, CA.
19. Collins, M., Dasgupta, S., & Schapire, R. E. (2002). A generalization of principal component analysis to the exponential family. In T. D. Dietterich, S. Becker, & Z. Ghahramani (Eds.), *Advances in neural information processing systems* (Vol. 14, pp. 617–624). Cambridge, MA: MIT Press.
20. d'Aspremont, A., Bach, F., & El Ghaoui, L. (2008). Optimal solutions for sparse principal component analysis. *Journal of Machine Learning Research*, *9*, 1269–1294.
21. Demir, G. K., & Ozmehmet, K. (2005). Online local learning algorithms for linear discriminant analysis. *Pattern Recognition Letters*, *26*, 421–431.
22. Diamantaras, K. I., & Kung, S. Y. (1994). Cross-correlation neural network models. *IEEE Transactions on Signal Processing*, *42*(11), 3218–3323.
23. Diamantaras, K. I., & Kung, S.-Y. (1994). Multilayer neural networks for reduced-rank approximation. *IEEE Transactions on Neural Networks*, *5*(5), 684–697.
24. Diamantaras, K. I., & Kung, S. Y. (1996). *Principal component neural networks: Theory and applications*. New York: Wiley.
25. Douglas, S. C., Kung, S., & Amari, S. (1998). A self-stabilized minor subspace rule. *IEEE Signal Processing Letters*, *5*(12), 328–330.
26. Du, K.-L., & Swamy, M. N. S. (2004). Simple and practical cyclostationary beamforming algorithms. *IEE Proceedings - Vision, Image and Signal Processing*, *151*(3), 175–179.
27. Feng, D.-Z., Bao, Z., & Shi, W.-X. (1998). Cross-correlation neural network model for the smallest singular component of general matrix. *Signal Processing*, *64*, 333–346.
28. Feng, D.-Z., Bao, Z., & Jiao, L.-C. (1998). Total least mean squares algorithm. *IEEE Transactions on Signal Processing*, *46*(8), 2122–2130.
29. Feng, D.-Z., Bao, Z., & Zhang, X.-D. (2001). A crossassociative neural network for SVD of nonsquared data matrix in signal processing. *IEEE Transactions on Neural Networks*, *12*(5), 1215–1221.
30. Feng, D.-Z., Zhang, X.-D., & Bao, Z. (2004). A neural network learning for adaptively extracting crosscorrelation features between two high-dimensional data streams. *IEEE Transactions on Neural Networks*, *15*(6), 1541–1554.
31. Feng, D.-Z., Zheng, W.-X., & Jia, Y. (2005). Neural network learning algorithms for tracking minor subspace in high-dimensional data stream. *IEEE Transactions on Neural Networks*, *16*(3), 513–521.
32. Feng, X., Kong, X., Duan, Z., & Ma, H. (2016). Adaptive generalized eigenpairs extraction algorithms and their convergence analysis. *IEEE Transactions on Signal Processing*, *64*(11), 2976–2989.
33. Feng, X., Kong, X., Ma, H., & Si, X. (2017). A novel unified and self-stabilizing algorithm for generalized eigenpairs extraction. *IEEE Transactions on Neural Networks and Learning Systems*, *28*(12), 3032–3044.

34. Fiori, S., & Piazza, F. (1998). A general class of ψ-APEX PCA neural algorithms. *IEEE Transactions on Circuits and Systems I*, *47*(9), 1394–1397.
35. Foldiak, P. (1989). Adaptive network for optimal linear feature extraction. In *Proceedings of the International Joint Conference on Neural Networks (IJCNN)* (Vol. 1, pp. 401–405). Washington, DC.
36. Fowler, J. E. (2009). Compressive-projection principal component analysis. *IEEE Transactions on Image Processing*, *18*(10), 2230–2242.
37. Fu, Z., & Dowling, E. M. (1995). Conjugate gradient eigenstructure tracking for adaptive spectral estimation. *IEEE Transactions on Signal Processing*, *43*(5), 1151–1160.
38. Gao, K., Ahmad, M. O., & Swamy, M. N. S. (1992). A modified Hebbian rule for total least-squares estimation with complex valued arguments. In *Proceedings of IEEE International Symposium on Circuits and Systems* (pp. 1231–1234). San Diego, CA.
39. Gao, K., Ahmad, M. O., & Swamy, M. N. S. (1994). A constrained anti-Hebbian learning algorithm for total least-square estimation with applications to adaptive FIR and IIR filtering. *IEEE Transactions on Circuits and Systems II*, *41*(11), 718–729.
40. Golub, G. H., & van Loan, C. F. (1989). *Matrix computation* (2nd ed.). Baltimore, MD: John Hopkins University Press.
41. Gottumukkal, R., & Asari, V. K. (2004). An improved face recognition technique based on modular PCA approach. *Pattern Recognition Letters*, *25*(4), 429–436.
42. Hall, P., & Martin, R. (1998). Incremental eigenanalysis for classification. In *Proceedings of British Machine Vision Conference*, (Vol. 1, pp. 286–295).
43. Hardoon, D. R., & Shawe-Taylor, J. (2011). Sparse canonical correlation analysis. *Machine Learning*, *83*, 331–353.
44. Hassoun, M. H. (1995). *Fundamentals of artificial neural networks*. Cambridge, MA: MIT Press.
45. Hastie, T., & Stuetzle, W. (1989). Principal curves. *Journal of the American Statistical Association*, *84*, 502–516.
46. Hastie, T., Buja, A., & Tibshirani, R. (1995). Penalized discriminant analysis. *Annals of Statistics*, *23*(1), 73–102.
47. Hebb, D. O. (1949). *The organization of behavior*. New York: Wiley.
48. Hegde, A., Principe, J. C., Erdogmus, D., & Ozertem, U. (2006). Perturbation-based eigenvector updates for on-line principal components analysis and canonical correlation analysis. *Journal of VLSI Signal Processing*, *45*, 85–95.
49. Horel, J. D. (1984). Complex principal component analysis: Theory and examples. *Journal of Applied Meteorology and Climatology*, *23*, 1660–1673.
50. Hotelling, H. (1933). Analysis of a complex of statistical variables into principal components. *Journal of Educational Psychology*, *24*, 417–441.
51. Hotelling, H. (1936). Relations between two sets of variates. *Biometrika*, *28*, 321–377.
52. Hoyle, D. C. (2008). Automatic PCA dimension selection for high dimensional data and small sample sizes. *Journal of Machine Learning Research*, *9*, 2733–2759.
53. Hsieh, W. W. (2007). Nonlinear principal component analysis of noisy data. *Neural Networks*, *20*, 434–443.
54. Jankovic, M., & Ogawa, H. (2003). A new modulated hebb learning rule-Biologically plausible method for local computation of principal subspace. *International Journal of Neural Systems*, *13*(4), 215–224.
55. Jankovic, M., & Ogawa, H. (2004). Time-oriented hierarchical method for computation of principal components using subspace learning algorithm. *International Journal of Neural Systems*, *14*(5), 313–323.
56. Jankovic, M. V., & Ogawa, H. (2006). Modulated Hebb-Oja learning rule-A method for principal subspace analysis. *IEEE Transactions on Neural Networks*, *17*(2), 345–356.
57. Johnstone, I. M., & Lu, A. (2009). On consistency and sparsity for principal components analysis in high dimensions. *Journal of the American Statistical Association*, *104*(486), 682–693.

58. Jolliffe, I., & Uddin, M. (2003). A modified principal component technique based on the lasso. *Journal of Computational and Graphical Statistics*, *12*(3), 531–547.
59. Journee, M., Nesterov, Y., Richtarik, P., & Sepulchre, R. (2010). Generalized power method for sparse principal component analysis. *Journal of Machine Learning Research*, *11*, 517–553.
60. Kaiser, A., Schenck, W., & Moller, R. (2010). Coupled singular value decomposition of a cross-covariance matrix. *International Journal of Neural Systems*, *20*(4), 293–318.
61. Kambhatla, N., & Leen, T. K. (1993). Fast non-linear dimension reduction. In *Proceedings of IEEE International Conference on Neural Networks* (Vol. 3, pp. 1213–1218). San Francisco, CA.
62. Kang, Z., Chatterjee, C., & Roychowdhury, V. P. (2000). An adaptive quasi-Newton algorithm for eigensubspace estimation. *IEEE Transactions on Signal Processing*, *48*(12), 3328–3333.
63. Karasuyama, M., & Sugiyama, M. (2012). Canonical dependency analysis based on squared-loss mutual information. *Neural Networks*, *34*, 46–55.
64. Karhunen, J., & Joutsensalo, J. (1995). Generalizations of principal component analysis, optimization problems, and neural networks. *Neural Networks*, *8*(4), 549–562.
65. Kettenring, J. R. (1971). Canonical analysis of several sets of variables. *Biometrika*, *58*(3), 433–451.
66. Klami, A., Virtanen, S., Leppaaho, E., & Kaski, S. (2015). Group factor analysis. *IEEE Transactions on Neural Networks and Learning Systems*, *26*(9), 2136–2147.
67. Kong, X., Hu, C., & Han, C. (2010). On the discrete-time dynamics of a class of self-stabilizing MCA extraction algorithms. *IEEE Transactions on Neural Networks*, *21*(1), 175–181.
68. Kramer, M. A. (1991). Nonlinear principal component analysis using autoassociative neural networks. *AIChE Journal*, *37*(2), 233–243.
69. Kukharev, G., & Kamenskaya, E. (2010). Application of two-dimensional canonical correlation analysis for face image processing and recognition. *Pattern Recognition and Image Analysis*, *20*(2), 210–219.
70. Kung, S. Y. (1990). Constrained principal component analysis via an orthogonal learning network. In *Proceedings of the IEEE International Symposium on Circuits and Systems* (Vol. 1, pp. 719–722). New Orleans, LA.
71. Kung, S. Y., & Diamantaras, K. I. (1990). A neural network learning algorithm for adaptive principal components extraction (APEX). In *Proceedings of International Conference on Acoustics, Speech, and Signal Processing (ICASSP)* (pp. 861–864). Albuquerque, NM.
72. Kung, S. Y., Diamantaras, K. I., & Taur, J. S. (1994). Adaptive principal components extraction (APEX) and applications. *IEEE Transactions on Signal Processing*, *42*(5), 1202–1217.
73. Kursun, O., Alpaydin, E., & Favorov, O. V. (2011). Canonical correlation analysis using within-class coupling. *Pattern Recognition Letters*, *32*, 134–144.
74. Kwak, N. (2008). Principal component analysis based on L1-norm maximization. *IEEE Transactions on Pattern Analysis and Machine Intelligence*, *30*(9), 1672–1680.
75. Kwok, J. T., & Zhao, H. (2003). Incremental eigendecomposition. In *Proceedings of International Conference on Artificial Neural Networks (ICANN)* (pp. 270–273). Istanbul, Turkey.
76. Leung, A. C. S., Wong, K. W., & Tsoi, A. C. (1997). Recursive algorithms for principal component extraction. *Network*, *8*, 323–334.
77. Li, X., Pang, Y., & Yuan, Y. (2010). L1-norm-based 2DPCA. *IEEE Transactions on Systems, Man, and Cybernetics Part B*, *40*(4), 1170–1175.
78. Linsker, R. (1986). From basic network principles to neural architecture. *Proceedings of the National Academy of Sciences of the USA, 83*, 7508–7512, 8390–8394, 9779–8783
79. Linsker, R. (1988). Self-organization in a perceptual network. *IEEE Computer*, *21*(3), 105–117.
80. Liu, Y., You, Z., & Cao, L. (2005). A simple functional neural network for computing the largest and smallest eigenvalues and corresponding eigenvectors of a real symmetric matrix. *Neurocomputing*, *67*, 369–383.
81. Ljung, L. (1977). Analysis of recursive stochastic algorithm. *IEEE Transactions on Automatic Control*, *22*, 551–575.
82. Loeve, M. (1963). *Probability theory* (3rd ed.). New York: Van Nostrand.

83. Lu, H., Plataniotis, K. N., & Venetsanopoulos, A. N. (2008). MPCA: Multilinear principal component analysis of tensor objects. *IEEE Transactions on Neural Networks*, *19*(1), 18–39.
84. Lu, H., Plataniotis, K. N. K., & Venetsanopoulos, A. N. (2009). Uncorrelated multilinear principal component analysis for unsupervised multilinear subspace learning. *IEEE Transactions on Neural Networks*, *20*(11), 1820–1836.
85. Lucchese, C., Orlando, S., & Perego, R. (2014). A unifying framework for mining approximate top-k binary patterns. *IEEE Transactions on Knowledge and Data Engineering*, *26*(12), 2900–2913.
86. Lucke, J., & Sahani, M. (2008). Maximal causes for non-linear component extraction. *Journal of Machine Learning Research*, *9*, 1227–1267.
87. Lv, J. C., Yi, Z., & Tan, K. K. (2007). Global convergence of GHA learning algorithm with nonzero-approaching adaptive learning rates. *IEEE Transactions on Neural Networks*, *18*(6), 1557–1571.
88. Mao, J., & Jain, A. K. (1995). Artificial neural networks for feature extraction and multivariate data projection. *IEEE Transactions on Neural Networks*, *6*(2), 296–317.
89. Mathew, G., & Reddy, V. U. (1996). A quasi-Newton adaptive algorithm for generalized symmetric eigenvalue problem. *IEEE Transactions on Signal Processing*, *44*(10), 2413–2422.
90. Mathew, G., Reddy, V. U., & Dasgupta, S. (1995). Adaptive estimation of eigensubspace. *IEEE Transactions on Signal Processing*, *43*(2), 401–411.
91. Miao, Y., & Hua, Y. (1998). Fast subspace tracking and neural network learning by a novel information criterion. *IEEE Transactions on Signal Processing*, *46*(7), 1967–1979.
92. Moller, R., & Hoffmann, H. (2004). An extension of neural gas to local PCA. *Neurocomputing*, *62*, 305–326.
93. Moller, R., & Konies, A. (2004). Coupled principal component analysis. *IEEE Transactions on Neural Networks*, *15*(1), 214–222.
94. Moller, R. (2006). First-order approximation of Gram-Schmidt orthonormalization beats deflation in coupled PCA learning rules. *Neurocomputing*, *69*, 1582–1590.
95. Morgan, D. R. (2004). Adaptive algorithms for solving generalized eigenvalue signal enhancement problems. *Signal Processing*, *84*(6), 957–968.
96. Nguyen, T. D., & Yamada, I. (2013). Adaptive normalized quasi-Newton algorithms for extraction of generalized eigen-pairs and their convergence analysis. *IEEE Transactions on Signal Processing*, *61*(6), 1404–1418.
97. Oja, E. (1982). A simplified neuron model as a principal component analyzer. *Journal of Mathematical Biology*, *15*, 267–273.
98. Oja, E. (1992). Principal components, minor components, and linear neural networks. *Neural Networks*, *5*, 929–935.
99. Oja, E., & Karhunen, J. (1985). On stochastic approximation of the eigenvectors and eigenvalues of the expectation of a random matrix. *Journal of Mathematical Analysis and Applications*, *104*, 69–84.
100. Oja, E., Ogawa, H., & Wangviwattana, J. (1992). Principal component analysis by homogeneous neural networks. *IEICE Transactions on Information and Systems, E75-D*, 366–382.
101. Ouyang, S., Bao, Z., & Liao, G. (1999). Adaptive step-size minor component extraction algorithm. *Electronics Letters*, *35*(6), 443–444.
102. Ouyang, S., Bao, Z., & Liao, G. (2000). Robust recursive least squares learning algorithm for principal component analysis. *IEEE Transactions on Neural Networks*, *11*(1), 215–221.
103. Ouyang, S., Bao, Z., Liao, G. S., & Ching, P. C. (2001). Adaptive minor component extraction with modular structure. *IEEE Transactions on Signal Processing*, *49*(9), 2127–2137.
104. Ouyang, S., & Bao, Z. (2002). Fast principal component extraction by a weighted information criterion. *IEEE Transactions on Signal Processing*, *50*(8), 1994–2002.
105. Ouyang, S., Ching, P. C., & Lee, T. (2003). Robust adaptive quasi-Newton algorithms for eigensubspace estimation. *IEE Proceedings—Vision, Image and Signal Processing*, *150*(5), 321–330.
106. Ozawa, S., Pang, S., & Kasabov, N. (2008). Incremental learning of chunk data for online pattern classification systems. *IEEE Transactions on Neural Networks*, *19*(6), 1061–1074.

107. Park, M. S., & Choi, J. Y. (2009). Theoretical analysis on feature extraction capability of class-augmented PCA. *Pattern Recognition*, *42*, 2353–2362.
108. Pearlmutter, B. A., & Hinton, G. E. (1986). G-maximization: An unsupervised learning procedure for discovering regularities. In J. S. Denker (Ed.), *AIP Conference Proceedings on Neural Networks for Computing* (Vol. 151, pp. 333–338). Snowbird, UT: American Institute of Physics.
109. Peng, D., Yi, Z., & Luo, W. (2007). Convergence analysis of a simple minor component analysis algorithm. *Neural Networks*, *20*, 842–850.
110. Peng, D., Yi, Z., Lv, J. C., & Xiang, Y. (2008). A neural networks learning algorithm for minor component analysis and its convergence analysis. *Neurocomputing*, *71*, 1748–1752.
111. Rao, Y. N., Principe, J. C., & Wong, T. F. (2004). Fast RLS-like algorithm for generalized eigendecomposition and its applications. *Journal of VLSI Signal Processing*, *37*, 333–344.
112. Rattan, S. S. P., & Hsieh, W. W. (2005). Complex-valued neural networks for nonlinear complex principal component analysis. *Neural Networks*, *18*, 61–69.
113. Ren, C.-X., & Dai, D.-Q. (2010). Incremental learning of bidirectional principal components for face recognition. *Pattern Recognition*, *43*, 318–330.
114. Ritter, H. (1995). Self-organizing feature maps: Kohonen maps. In M. A. Arbib (Ed.), *The handbook of brain theory and neural networks* (pp. 846–851). Cambridge, MA: MIT Press.
115. Robbins, H., & Monro, S. (1951). A stochastic approximation method. *Annals of Mathematical Statistics*, *22*(3), 400–407.
116. Rubner, J., & Schulten, K. (1990). Development of feature detectors by self-organization. *Biological Cybernetics*, *62*, 193–199.
117. Rubner, J., & Tavan, P. (1989). A self-organizing network for principal-component analysis. *Europhysics Letters*, *10*, 693–698.
118. Rupnik, J., & Shawe-Taylor, J. (2010). Multi-view canonical correlation analysis. In *Proceedings of Slovenian KDD Conference on Data Mining and Data Warehouses (SiKDD)* (pp. 1–4).
119. Saegusa, R., Sakano, H., & Hashimoto, S. (2004). Nonlinear principal component analysis to preserve the order of principal components. *Neurocomputing*, *61*, 57–70.
120. Sanger, T. D. (1989). Optimal unsupervised learning in a single-layer linear feedforward neural network. *Neural Networks*, *2*, 459–473.
121. Sanger, T. D. (1994). Two iterative algorithms for computing the singular value decomposition from input/output samples. In J. D. Cowan, G. Tesauro, & J. Alspector (Eds.), *Advances in neural information processing systems* (Vol. 6, pp. 144–151). San Francisco, CA: Morgan Kaufmann.
122. Spearman, C. (1904). General intelligence, objectively determined and measured. *American Journal of Psychology*, *15*, 201–293.
123. Sriperumbudur, B. K., Torres, D. A., & Lanckriet, G. R. G. (2011). A majorization-minimization approach to the sparse generalized eigenvalue problem. *Machine Learning*, *85*, 3–39.
124. Sun, T., & Chen, S. (2007). Locality preserving CCA with applications to data visualization and pose estimation. *Image and Vision Computing*, *25*, 531–543.
125. Sun, L., Ji, S., & Ye, J. (2011). Canonical correlation analysis for multilabel classification: A least-squares formulation, extensions, and analysis. *IEEE Transactions on Pattern Analysis and Machine Intelligence*, *33*(1), 194–200.
126. Tang, Y., & Li, J. (2010). Notes on "Recurrent neural network model for computing largest and smallest generalized eigenvalue". *Neurocomputing*, *73*, 1006–1012.
127. Tipping, M. E., & Bishop, C. M. (1999). Mixtures of probabilistic principal component analyzers. *Neural Computation*, *11*, 443–482.
128. Tsagkarakis, N., Markopoulos, P. P., Sklivanitis, G., & Pados, D. A. (2018). L_1-norm principal-component analysis of complex data. *IEEE Transactions on Signal Processing*, *66*(12), 3256–3267.
129. Ulfarsson, M. O., & Solo, V. (2008). Sparse variable PCA using geodesic steepest descent. *IEEE Transactions on Signal Processing*, *56*(12), 5823–5832.

130. Ulfarsson, M. O., & Solo, V. (2011). Vector l_0 sparse variable PCA. *IEEE Transactions on Signal Processing*, *59*(5), 1949–1958.
131. Valizadeh, A., & Karimi, M. (2009). Fast subspace tracking algorithm based on the constrained projection approximation. *EURASIP Journal on Advances in Signal Processing, 2009*, Article ID 576972, 16 pages.
132. Warmuth, M. K., & Kuzmin, D. (2008). Randomized online PCA algorithms with regret bounds that are logarithmic in the dimension. *Journal of Machine Learning Research*, *9*, 2287–2320.
133. Wang, L., & Karhunen, J. (1996). A simplified neural bigradient algorithm for robust PCA and MCA. *International Journal of Neural Systems*, *7*(1), 53–67.
134. Wang, H. (2010). Local two-dimensional canonical correlation analysis. *IEEE Signal Processing Letters*, *17*(11), 921–924.
135. Weng, J., Zhang, Y., & Hwang, W.-S. (2003). Candid covariance-free incremental principal component analysis. *IEEE Transactions on Pattern Analysis Machine Intelligence*, *25*(8), 1034–1040.
136. Wu, J., & Zhou, Z. H. (2002). Face recognition with one training image per person. *Pattern Recognition Letters*, *23*(14), 1711–1719.
137. Xu, D., Principe, J. C., & Wu, H. C. (1998). Generalized eigendecomposition with an on-line local algorithm. *IEEE Signal Processing Letters*, *5*(11), 298–301.
138. Xu, L. (1993). Least mean square error reconstruction principle for self-organizing neural-nets. *Neural Networks*, *6*, 627–648.
139. Xu, L., Krzyzak, A., & Oja, E. (1993). Rival penalized competitive learning for clustering analysis, RBF net, and curve detection. *IEEE Transactions on Neural Networks*, *4*(4), 636–649.
140. Xu, L., Oja, E., & Suen, C. Y. (1992). Modified Hebbian learning for curve and surface fitting. *Neural Networks*, *5*, 441–457.
141. Xu, L., & Yuille, A. L. (1995). Robust principal component analysis by self-organizing rules based on statistical physics approach. *IEEE Transactions on Neural Networks*, *6*(1), 131–143.
142. Yang, B. (1995). Projection approximation subspace tracking. *IEEE Transactions on Signal Processing*, *43*(1), 95–107.
143. Yang, B. (1995). An extension of the PASTd algorithm to both rank and subspace tracking. *IEEE Signal Processing Letters*, *2*(9), 179–182.
144. Yang, J., Zhang, D., Frangi, A. F., & Yang, J. Y. (2004). Two-dimensional PCA: A new approach to appearance-based face representation and recognition. *IEEE Transactions on Pattern Analysis Machine Intelligence*, *26*(1), 131–137.
145. Yang, J., Zhao, Y., & Xi, H. (2011). Weighted rule based adaptive algorithm for simultaneously extracting generalized eigenvectors. *IEEE Transactions on Neural Networks*, *22*(5), 800–806.
146. Yang, J., Chen, X., & Xi, H. (2013). Fast adaptive extraction algorithm for multiple principal generalized eigenvectors. *International Journal of Intelligent Systems*, *28*, 289–306.
147. Ye, J. (2005). Generalized low rank approximations of matrices. *Machine Learning*, *61*, 167–191.
148. Ye, M. (2005). Global convergence analysis of a self-stabilizing MCA learning algorithm. *Neurocomputing*, *67*, 321–327.
149. Ye, M., Fan, X.-Q., & Li, X. (2006). A class of self-stabilizing MCA learning algorithms. *IEEE Transactions on Neural Networks*, *17*(6), 1634–1638.
150. Yi, Z., Fu, Y., & Tang, H. J. (2004). Neural networks based approach for computing eigenvectors and eigenvalues of symmetric matrix. *Computers & Mathematics with Applications*, *47*, 1155–1164.
151. Yi, Z., Ye, M., Lv, J. C., & Tan, K. K. (2005). Convergence analysis of a deterministic discrete time system of Oja's PCA learning algorithm. *IEEE Transactions on Neural Networks*, *16*(6), 1318–1328.
152. Yuille, A. L., Kammen, D. M., & Cohen, D. S. (1989). Quadrature and development of orientation selective cortical cells by Hebb rules. *Biological Cybernetics*, *61*, 183–194.

153. Zhang, Y., & Ma, Y. (1997). CGHA for principal component extraction in the complex domain. *IEEE Transactions on Neural Networks*, *8*(5), 1031–1036.
154. Zhang, D., Zhou, Z. H., & Chen, S. (2006). Diagonal principal component analysis for face recognition. *Pattern Recognition*, *39*, 140–142.
155. Zhang, Q., & Leung, Y. W. (2000). A class of learning algorithms for principal component analysis and minor component analysis. *IEEE Transactions on Neural Networks*, *11*(1), 200–204.
156. Zhao, H., Yuen, P. C., & Kwok, J. T. (2006). A novel incremental principal component analysis and its application for face recognition. *IEEE Transactions on Systems, Man, and Cybernetics*, *36*(4), 873–886.
157. Zhao, S., Gao, C., Mukherjee, S., & Engelhardt, B. E. (2016). Bayesian group factor analysis with structured sparsity. *Journal of Machine Learning Research*, *17*, 1–47.
158. Zufiria, P. J. (2002). On the discrete-time dynamics of the basic Hebbian neural-network node. *IEEE Transactions on Neural Networks*, *13*(6), 1342–1352.
159. Zuo, W., Zhang, D., & Wang, K. (2006). Bidirectional PCA with assembled matrix distance metric for image recognition. *IEEE Transactions on Systems, Man, and Cybernetics Part B*, *36*(4), 863–872.

Chapter 14
Nonnegative Matrix Factorization

14.1 Introduction

SVD is a classical method for matrix factorization, which gives the optimal low-rank approximation to a real-valued matrix in terms of the squared error. Many application areas, including information retrieval, pattern recognition, and data mining, require processing of binary rather than real data.

Many real-life data or physical signals, such as pixel intensities, amplitude spectra, text corpora, gene expressions, air quality, information, and occurrence counts, are naturally represented by nonnegative numbers. In the analysis of mixtures of such data, nonnegativity of the individual components is a reasonable constraint. A variety of techniques are available for analysis of such data, such as nonnegative PCA, nonnegative ICA, and nonnegative matrix factorization (NMF) [41]. The goal of all of these techniques is to express the given nonnegative data as a guaranteed nonnegative linear combination of a set of nonnegative bases.

NMF [41], also known as nonnegative matrix approximation or positive matrix factorization [53], is an unsupervised learning method for factorizing a matrix as a product of two matrices, in which all the elements are nonnegative. In NMF, the nonnegative constraint prevents mutual cancellation between basis functions and yields parts-based representations. NMF has become an established method for performing tasks such as BSS of images and nonnegative signals [2], spectra recovery [56], feature extraction [41], dimension reduction, segmentation and clustering [13], language modeling, text mining, neurobiology (gene separation), and gene expression profiles.

The NMF problem is described as follows. Given a nonnegative matrix $\mathbf{X}$, find nonnegative matrix factors $\mathbf{A}$ and $\mathbf{S}$ such that the difference measure between $\mathbf{X}$ and $\mathbf{AS}$ is the minimum according to some cost function:

$$\mathbf{X} = \mathbf{AS}, \tag{14.1}$$

K.-L. Du and M. N. S. Swamy, *Neural Networks and Statistical Learning*,
https://doi.org/10.1007/978-1-4471-7452-3_14

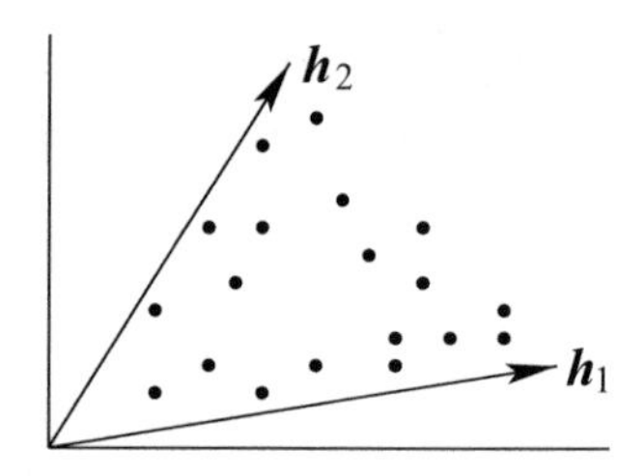

Fig. 14.1 Nonuniqueness of NMF

where $\mathbf{X} \in R^{m \times n}$, the coefficient matrix $\mathbf{A} \in R^{m \times k}$, the matrix of sources $\mathbf{S} \in R^{k \times n}$ ($k \ll n, m$), elements in both $\mathbf{A}$ and $\mathbf{S}$ are nonnegative, and the rows in $\mathbf{S}$ may be statistically dependent to some extent. In other words, $\mathbf{A}$ can be seen as a basis that is optimized for linear approximation of the data in $\mathbf{X}$ by minimizing $E = \|\mathbf{X} - \mathbf{AS}\|_F^2$, where $\| \cdot \|_F^2$ denotes the squared Frobenius norm (the sum of squared elements). It is usually selected that $km + kn \ll mn$ for data reduction.

For a set of nonnegative multivariate data points $\mathbf{X} = \{\boldsymbol{x}_1, \ldots, \boldsymbol{x}_n\}$, each data point $\boldsymbol{x}_i$ is a m-dimensional vector. Arranging the data points as columns, $\mathbf{X} \in R^{m \times n}$. NMF finds two matrices $\mathbf{A} \in R^{d \times k}$, $\mathbf{S} \in R^{k \times n}$ such that (14.1) is satisfied.

It can be written as $\boldsymbol{x}_i \approx \mathbf{A}\boldsymbol{s}_i$, where $\boldsymbol{s}_i$ is the ith column of $\mathbf{S}$. $\boldsymbol{x}_i$ can be seen as a linear combination of each column of $\mathbf{A}$, parameterized by the corresponding elements of $\boldsymbol{s}_i$. Thus, $\mathbf{A}$ is a basic vector matrix, and $\mathbf{S}$ is the coefficient matrix. Usually, $k \ll m, n$. Thus, NMF results in a compressed representation of the original matrix. If k is the number of clusters, then the jth column of $\mathbf{A}$ is a representation of the jth cluster, and $\boldsymbol{x}_i$ should be assigned to the cluster having the largest weight.

NMF may not directly yield a unique decomposition [16]. It is characterized by a scale and permutation indeterminacies. Although the uniqueness may be improved by imposing some constraints to the factors, it is still a challenging problem to uniquely identify the sources in general cases [40]. Figure 14.1 illustrates the nonuniqueness problem in two dimensions. There is open space between the data points and the coordinate axes. We can choose the basis vectors $\boldsymbol{h}_1$ and $\boldsymbol{h}_2$ anywhere in this open space between the coordinate axes and data, and represent each data point exactly with a nonnegative linear combination of these vectors. Some well-posed NMF problems are obtained and the solutions are optimal and sparse under the separability assumption [23].

NMF and nonnegative tensor factorization decompose a nonnegative data matrix into a product of lower rank nonnegative matrices or tensors. Boolean matrix factorization or Boolean factor analysis is the factorization of data sets in binary alphabet based on Boolean algebra [34]. Although both NMF and sparse coding learn sparse representation, they are different because NMF learns low-rank representation while sparse coding usually learns the full-rank representation.

NMF is NP-hard in general [57], and highly ill-posed. NMF has been shown to be tractable under the separability assumption, under which all the columns of the input data matrix M belong to the convex cone generated by only a few of these columns. Separability means that there exists a rank-r NMF $(W, H) \geq 0$ of M where each column of W is equal to some column of M [24].

14.2 Algorithms for NMF

NMF optimization problems are usually nonconvex. NMF is usually performed with an alternating gradient-descent technique that is applied to the squared Euclidean distance or Kullback–Leibler divergence. The two measures are unified by using the parameterized cost functions such as β-divergence [37] or a broader class called Bregman divergence [12]. Bregman divergences include Frobenius norm and KL-divergence. This approach belongs to a class of multiplicative iterative algorithms [41]. In spite of low complexity, it converges slowly, gives only a strictly positive solution, and can easily fall into local minima of a nonconvex cost function. Another popular algorithm is alternating nonnegative least squares [53].

An algorithm for NMF can be applied to BSS by adding two suitable regularization terms in the original objective function of NMF to increase sparseness and/or smoothness of the estimated components [10]. In a pattern-expression NMF approach [75] to BSS, two regularization terms are added to the original loss function of standard NMF for effective expression of patterns with basis vectors in the pattern-expression NMF. Nonsmooth NMF finds localized, parts-based representations of nonnegative multivariate data items [54].

Projected gradient methods are highly efficient in solving large-scale convex minimization problems subject to linear constraints. The NMF projected gradient algorithms [9, 74] are quite efficient for solving large-scale minimization problems subject to nonnegativity and sparsity constraints.

Smooth component analysis [70] imposes smoothness constraints not on the raw data, but rather on the hidden components, i.e., vectors of the factor matrix, and/or on basis vectors of the mixing matrix.

14.2.1 Multiplicative Update Algorithm and Alternating Nonnegative Least Squares

The NMF problem can be formulated as

$$\min_{\mathbf{A},\mathbf{S}} \|\mathbf{X} - \mathbf{A}\mathbf{S}\|_F, \tag{14.2}$$

subject to the constraints that all elements of $\mathbf{A}$ and $\mathbf{S}$ are nonnegative.

The multiplicative update rule for NMF is given by [41]:

$$\mathbf{S} \leftarrow \mathbf{S} \otimes \frac{\mathbf{A}^T\mathbf{X}}{\mathbf{A}^T\mathbf{A}\mathbf{S}}, \tag{14.3}$$

$$\mathbf{A} \leftarrow \mathbf{A} \otimes \frac{\mathbf{X}\mathbf{S}^T}{\mathbf{A}\mathbf{S}\mathbf{S}^T}, \tag{14.4}$$

where $\otimes$ and / denote element-wise multiplication and division, respectively. The matrices **A** and **S** are initialized with positive random values. These equations iterate, guaranteeing monotonical convergence to a local maximum of the objective function [41]:

$$F = \sum_{i=1}^{m} \sum_{\mu=1}^{n} (X_{i\mu} \ln(\mathbf{AS})_{i\mu} - (\mathbf{AS})_{i\mu}). \tag{14.5}$$

After learning the NMF basis vectors **A**, new data in matrix $\mathbf{X}'$ are mapped to k-dimensional space by fixing **A** and then randomly initializing **S** and iterating until convergence; or by fixing **A** and then solving an LS problem $\mathbf{X}' = \mathbf{AS}'$ for $\mathbf{S}'$ using pseudoinverse. The LS solution can produce negative entries of $\mathbf{S}'$. One can enforce nonnegativity through setting negative values to zero or by using nonnegative LS. Setting negative values to zero is much computationally simpler than solving LS with nonnegativity constraints, but some information is lost after zeroing.

The multiplicative update algorithm may fail to converge to a stationary point [2, 48]. In a modified strategy [48], if a whole column of **A** is zero, then it as well as the corresponding row in **S** are unchanged, the convergence of this strategy to a stationary point is guaranteed. This modified strategy can ensure the modified sequence to be bounded. Both algorithms are very slow, due to their gradient-decent nature. When minimizing the Euclidean distance between the approximate and true values, the derived multiplicative update algorithm [44] is proved to converge to a stationary point, and also at faster speed.

Example 14.1 By using the software package NMFPACK (http://www.cs.helsinki.fi/patrik.hoyer/ [32]), we implement the multiplicative update algorithm with the Euclidean objective for a parts-based representation of the ORL face image database. For the ORL database, $m = 92$, $k = 25$ and $n = 400$. Basis images derived from the ORL face image database are shown as well as the objective function throughout the optimization is shown in Fig. 14.2. It is shown that the multiplicative update algorithm converges very slowly.

Alternating nonnegative LS is a block coordinate descent in bound-constrained optimization [53]:

$$\min_{\mathbf{A},\mathbf{S}} F(\mathbf{A}, \mathbf{S}) = \frac{1}{2} \|\mathbf{X} - \mathbf{AS}\|_F^2, \tag{14.6}$$

subject to the constraints that all elements of **A** and **S** are nonnegative.

The algorithm can be implemented as two alternating convex optimization problems

$$\begin{aligned} \mathbf{A}_{k+1} &= \arg\min_{\mathbf{A}\geq 0} F(\mathbf{A}, \mathbf{S}_k), \\ \mathbf{S}_{k+1} &= \arg\min_{\mathbf{S}\geq 0} F(\mathbf{A}_{k+1}, \mathbf{S}). \end{aligned} \tag{14.7}$$

The method can have fast convergence, work well in practice, but its convergence is not warranted [2]. A modified strategy proposed in [48] is applied to ensure that the

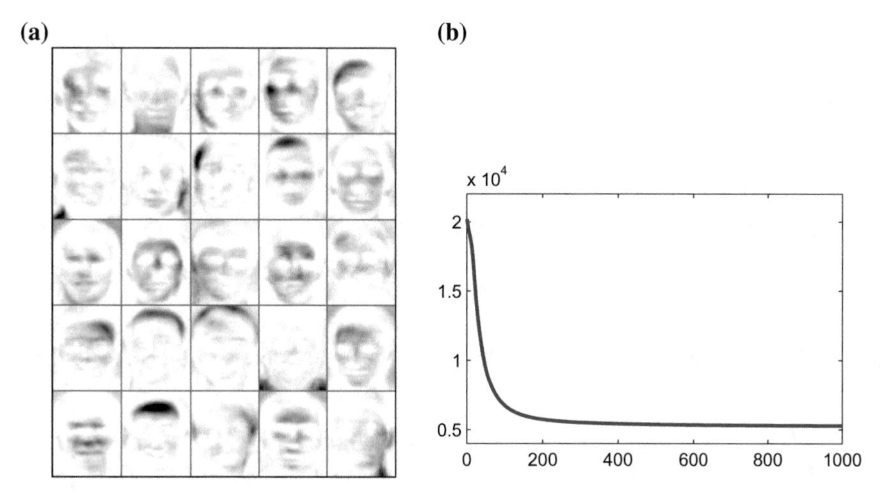

Fig. 14.2 NMF applied to the ORL face image database. **a** The basis images for matrix **A**. **b** The evolution of the objective function

sequence generated has at least one limit point, and this limit point is a stationary point of NMF [50].

Gradient-based algorithms for NMF learning are of slow convergence for large-scale problems. Some efforts have been made to apply projected gradient method [47], projected alternating LS method [2] or projected Newton method [73] to NMF for speeding up its convergence. Projected gradient algorithm [47] solves the NMF problem by solving nonnegative LS problems for **A** and **S** in an alternating manner. Unlike the multiplicative update algorithm which uses a fixed step-size in the gradient descent, projected gradient algorithm uses a flexible step-size.

NMF may not yield unique decomposition. Some modifications of NMF have been reported for the decomposition to be unique by imposing the sparseness constraint on mixing matrix or source matrix, or both [32]. Possible circumstances under which NMF yields unique decomposition can also be found in [40].

An improved algorithmic framework for the LS NMF problem [35] overcomes many deficiencies of gradient descent-based methods including the multiplicative update algorithm and the alternating LS heuristic. This framework readily admits powerful optimization techniques such as the Newton, BFGS and CG methods, and includes regularization and box-constraints, thus overcoming deficiencies without sacrificing convergence guarantees.

Since **A** and/or **S** are usually sparse matrices, a hybrid approach called the gradient projection CG algorithm is adapted for NMF. The α-divergence is used to unify many well-known cost functions. In a projected quasi-Newton method [73], a regularized Hessian with the LM approach is inverted with the Q-less QR decomposition. The method uses the quasi-Newton iterates for updating **A** and the fixed-point regularized LS algorithm for computing **S**. The best result is obtained with the quasi-Newton fixed-point algorithm. The gradient projection CG gives slightly worse results. The algorithms are implemented in the MATLAB toolbox: NMFLAB for Signal and Image Processing [8].

Without the explicit assumption of independence, NMF for BSS discussed in [10] achieves the estimations of the original sources from the mixtures. Using the invariant set method, the convergence area for the NMF based BSS algorithm is obtained [66].

NMF can also be implemented incrementally for data streams; two examples are the online NMF algorithm [26] and the incremental orthogonal projective NMF algorithm [58]. Two online NMF solvers in the presence of outliers [78] are based on projected gradient descent and the alternating direction method of multipliers. The sequence of objective values are proved to converge almost surely, and the sequence of learned dictionaries converges to the set of stationary points of the expected loss function almost surely.

Example 14.2 The approximate error of NMF in Frobenius norm is generally larger than that of SVD if the same approximate rank is taken, due to the nonnegative constraint. An example given in [77] illustrates this phenomenon. For a rank-3 nonnegative matrix

$$\mathbf{X} = \begin{bmatrix} 2 & 0 & 1 & 1? \\ 1 & 1 & 1 & 0 \\ 1 & 1 & 0 & 2 \\ 0 & 2 & 1 & 1 \end{bmatrix}, \tag{14.8}$$

SVD produces the optimal rank-3 factorization of $\mathbf{X}$ without errors. However, the optimal NMF approximation $\mathbf{AS}$ of rank 3 has the error $\|\mathbf{X} - \mathbf{AS}\|_F \approx 0.4823$.

14.3 Other NMF Methods

Sparse NMF can be implemented by constraining or penalizing the L_1-norm of the factor matrices into the NMF cost function, or by constraining the L_0-norm of either of the factor matrices [55]. Doubly sparse NMF arises when both factor matrices have the sparsity constraint. It can benefit the document–word co-clustering task. NMF with minimum-volume-constraint [79] can improve the sparseness of the results of NMF. This sparseness is L_0-norm oriented and can give desirable results even in very weak sparseness situations. The model based on quadratic programming is quite efficient for small-scale problems, and the model using multiplicative updates incorporating natural gradient is more suitable for large-scale problems.

For NMF, the learned basis is theoretically proved to be unnecessarily parts-based [16]. By introducing the manifold regularization and the margin maximization to NMF, the manifold regularized discriminative NMF can produce parts-based basis by using a Newton-based fast gradient descent algorithm.

Projective NMF approximates a data matrix by its nonnegative subspace projection [68, 69]

$$\mathbf{X} \approx \mathbf{P}\mathbf{X}, \tag{14.9}$$

where $\mathbf{P}$ is a positive low-rank matrix. The objective function can be measured by Frobenius matrix norm or a modified Kullback–Leibler divergence. Both measures are minimized by multiplicative update rules, whose convergence is proven in [69]. A nonnegative multiplicative version of Oja's learning rule can be used for computing projective NMF [68].

Compared with NMF, $\mathbf{X} \approx \mathbf{A}\mathbf{S}$, projective NMF replaces $\mathbf{S}$ with $\mathbf{A}^T\mathbf{X}$. This brings projective NMF close to nonnegative PCA. The term projective refers to the fact that $\mathbf{A}\mathbf{A}^T$ is indeed a projection matrix if $\mathbf{A}$ is an orthogonal matrix: $\mathbf{A}^T\mathbf{A} = \mathbf{I}$. In projective NMF learning, $\mathbf{A}$ becomes approximately orthogonal. This has positive consequences in sparseness of the approximation, orthogonality of the factorizing matrix, decreased computational complexity in learning, close equivalence to clustering, generalization of the approximation to new data without heavy recomputations, and easy extension to a nonlinear kernel method with wide applications for optimization problems.

Object representation in the inferior temporal cortex, an area of visual cortex critical for object recognition in the primate, exhibits two prominent properties [31]: objects are represented by the combined activity of columnar clusters of neurons, with each cluster representing component features or parts of objects, and closely related features are continuously represented along the tangential direction of individual columnar clusters. Topographic NMF [31] is a learning model that reflects these properties of parts-based representation and topographic organization in a unified framework. Topographic NMF incorporates neighborhood connections between NMF basis functions arranged on a topographic map. With this extension, the nonnegative constraint leads to an overlapping of basis functions along neighboring structures. Nonnegativity of NMF has been related to the network properties such as firing rate representation and signed synaptic weight [41]. Topographic NMF represents an input by multiple activity peaks to describe diverse information, whereas conventional topographic models such as SOM represent an input by a single activity peak in a topographic map. Topographic NMF reconstructs the neuronal responses better than SOM.

In the case of the data being highly nonlinearly distributed, it is desirable to kernelize NMF. Projected gradient kernel NMF [71] is a nonlinear nonnegative component analysis method using kernels. Arbitrary positive definite kernels can be adopted. Use of projected gradient procedure [47] guarantees that the limit point of the algorithm is a stationary point of the optimization procedure. The proposed method leads to better classification rates when compared with kernel PCA and kernel ICA [71].

Discriminant NMF [72] is a supervised NMF approach to enhancing the classification accuracy by introducing Fisher's discriminative information to NMF. Max-min distance NMF [60] is an iterative algorithm that uses the class labels. It minimizes the maximum distance of the within-class pairs in the new NMF space, and meanwhile maximizes the minimum distance of the between-class pairs, with an alternating strategy.

Graph-regularized NMF [5] takes into account the geometrically-based regularizer to determine the low-dimension manifold structure of the data. Smoothing of the encoding vectors is applied to increase the sparseness of the basis vectors. This problem is solved iteratively by multiplicative rules. The update algorithms will converge to their fixed points under some given conditions [67]. Analysis based on Lyapunov indirect method shows that the NMF learning algorithm has multiple fixed points and that at any of the fixed points, the update algorithms are stable.

A model of low-rank matrix factorization [77] incorporates manifold regularization to the matrix factorization. Superior to the graph-regularized NMF, this regularization model has globally optimal and closed-form solutions. A direct algorithm (for data with small number of points) and an alternate iterative algorithm with inexact inner iteration (for large scale data) are proposed to solve the model. A convergence analysis establishes the global convergence of the alternating iterative algorithm.

Semi-supervised NMF [42] is formulated as a joint factorization of the data matrix and the label matrix, sharing a common factor matrix $\mathbf{S}$ for consistency. Constrained NMF is a semi-supervised matrix decomposition method, which incorporates the label information as additional constraints [49]. Combining label information improves the discriminating power of the resulting matrix decomposition.

To deal with mixed-sign data, semi-NMF, convex-NMF, and cluster-NMF algorithms have been proposed [14]. Semi-NMF is defined by $\mathbf{X} = \mathbf{AS}$, where the elements of $\mathbf{S}$ are nonnegative but $\mathbf{X}$ and $\mathbf{A}$ are not constrained [14]. Convex-NMF is obtained when the basis vectors of $\mathbf{A}$ are convex combinations of the data points [14]. This is used for a kernel extension of NMF. Convex-NMF applies to both nonnegative and mixed-sign data matrices, and both factor matrices tend to be very sparse.

Another matrix factorization method is latent semantic indexing [11]. CX algorithm is a column-based matrix decomposition algorithm, while CUR algorithm is a column-row-based one [20]. In CX algorithm [19, 20], a matrix $\mathbf{A}$ composing sample vectors is decomposed into two matrices $\mathbf{C}$ and $\mathbf{X}$. For term-document data and binary image data, the columns of $\mathbf{A}$ are sparse and nonnegative. The prototype-preserving property of CX algorithm makes the columns of $\mathbf{C}$ sparse and nonnegative too. CX algorithm samples a small number of columns by randomly sampling columns of the data matrix according to a constructed nonuniform probability which is derived from SVD. A deterministic version of CX algorithm [46] selects columns in a deterministic manner, which well-approximates SVD. Each selected column is related to an eigenvector of PCA or incremental PCA.

Hierarchical alternating LS algorithm [7] solves a set of column-wise nonlinear LS problems for each column and updates $\mathbf{A}$ and $\mathbf{S}$ column-wisely. The method converges very fast.

Greedy coordinate descent algorithm for NMF [33] is an element-wise update algorithm. It selects and updates the most contributable variables for minimization. It cannot work with such a constraint that affects all elements of one column at the same time, such as the graph regularized constraint. Greedy coordinate descent is not applicable to orthogonal NMF because the orthogonal condition requires an interaction between different rows.

Scalar block coordinate descent for Bregman divergence NMF (http://www.cc.gatech.edu/grads/l/lli86/sbcd.zip) [43] is a column-wise update algorithm. Bregman divergences used Taylor series expansion to derive the element-wise problem. The complexity is the same as that of column-wise update algorithms.

Truncated CauchyNMF [27] robustly learns the subspace on noisy datasets contaminated by outliers by using truncated Cauchy loss. Robust NMF using $L_{2,1}$-norm loss function is available for data with Laplacian noise [38].

14.3.1 NMF Methods for Clustering

Under certain assumption, NMF is equivalent to clustering [13, 32]. The matrix **A** is considered to be a centroid matrix as every column represents a cluster center, while **S** is the cluster membership matrix.

Orthogonal NMF is NMF with orthogonality constraint on either the factor **A** or **S** [13]. C-means is equivalent to NMF with orthogonality on **S** [15]. Orthogonal NMF is equivalent to C-means clustering, and orthogonal NMF based on the factor **A** or **S** is identical to clustering the rows or columns of an input data matrix, respectively [64].

Bregman divergence orthogonal NMF problem [36] is equivalent to Bregman hard clustering [1]. The constrained matrix **A** can be considered as an indicator matrix in orthogonal NMF. Let **X** be an instance feature matrix factorized by **AS**. Then ith row of **A** can be considered as a membership vector of instance i to k groups (features). Especially, a solution **AS** in orthogonal NMF is expected to have a crisp membership of a single one. We assign the ith instance to kth cluster such as $k = \arg\max_j A_{ij}$.

Cluster-NMF [14] is an idea similar to projective NMF; it is based on Frobenius norm and is a particular case of convex-NMF. Cluster-NMF is close to C-means clustering.

Many clustering methods can be interpreted in terms of a matrix factorization problem. Interpreting non-negative factors of matrices as describing a data clustering was proposed in [41, 53]. For n feature vectors gathered as the m-dimensional columns of a matrix $\mathbf{X} \in R^{m \times n}$, the C-means problem can be written as a matrix factorization problem (14.2), subject to $\mathbf{S} \in \{0, 1\}^{k \times n}$, $\mathbf{S}^T \mathbf{1}_k = \mathbf{1}_n$. The matrix **A** can be interpreted as a matrix whose columns are the k cluster centroids. $S_{ij} = 1$ if sample j belongs in cluster i, and 0 otherwise.

Orthogonal NMF imposes an orthogonal constraint onto NMF, under an orthogonal constraint on the first factor matrix:

$$\min_{\mathbf{A},\mathbf{S}} \|\mathbf{X} - \mathbf{AS}\|_F^2 \tag{14.10}$$

subject to

$$\mathbf{A} \geq 0, \mathbf{S} \geq 0, \mathbf{A}^T\mathbf{A} = \mathbf{I}, \tag{14.11}$$

where $\mathbf{I}$ is the identity matrix. The optimization problem is stated as a Lagrangian formulation.

The minimization problem of Bregman divergence orthogonal NMF is given by

$$\min_{\mathbf{A},\mathbf{S}} \mathbf{D}_{\phi}(\mathbf{X} \| \mathbf{AS}) \tag{14.12}$$

subject to (14.11).

Orthogonal NMF can be solved using multiplicative update algorithm [13]. In [52], the proposed multiplicative update algorithm for orthogonal NMF is convergent. In this algorithm, the zero-lock problem was forcibly avoided by replacing zero values with a small positive value. These matrix-wise alternating block coordinate descent based algorithms require a relatively large number of iterations to converge.

Scalar block coordinate descent as a column-wise update algorithm for Bregman divergence orthogonal NMF is given by incorporating the column-wise orthogonal constraint into their scalar block coordinate descent algorithm [36].

This scalar block coordinate descent for orthogonal NMF algorithm is an extension of the hierarchical alternating LS algorithm for orthogonal NMF, but its convergence is slower because scalar block coordinate descent for orthogonal NMF needs to update the column-wise residual in addition to the updating of each column, while hierarchical alternating LS algorithm for orthogonal NMF does not need to do so for the residual. The proposed algorithms converge faster than the other orthogonal NMF algorithms, due to their smaller numbers of iterations.

Fast hierarchical alternating LS algorithm for orthogonal NMF [36] is based on hierarchical alternating LS algorithm [7]. The columnwise orthogonal constraint can also be applied to scalar block coordinate descent for solving Bregman divergence NMF [43].

NMF and symmetric NMF are effective for clustering linearly separable data and nonlinearly separable data, respectively. Symmetric NMF is closely related to spectral clustering [15, 39].

Symmetric NMF produces a symmetric and nonnegative low-rank approximation to the graph matrix. It can be seen as a graph clustering algorithm. Symmetric NMF [39] is a special case of NMF, $\mathbf{X} \approx \mathbf{PP}^T$, in which $\mathbf{P}$ is a nonnegative factor and $\mathbf{X}$ is completely positive. Weighted symmetric NMF or symmetric nonnegative tri-factorization is defined by $\mathbf{X} = \mathbf{PQP}^T$, where $\mathbf{Q}$ is a symmetric nonnegative matrix. Parallel multiplicative update algorithms [30] minimize the Euclidean distance, with proved convergence under mild conditions. These algorithms are applied to probabilistic clustering.

In the constrained clustering problem, limited domain knowledge in the form of must-link and cannot-link is available. Semi-supervised NMF algorithms, such as an NMF-based constrained clustering algorithm [59] and a symmetric NMF-based constrained clustering algorithm [6], reward the satisfaction of must-link constraints but penalize the violation of cannot-link constraints. The similarity between two points on a must-link is enforced to approximate 1, and that on a cannot-link is enforced to approximate 0. NMF based constrained clustering and symmetric NMF based constrained clustering [76] use multiplicative update rules to, respectively, solve clustering of linearly separable data and nonlinearly separable data for the constrained clustering problem.

A directional clustering approach is derived in [3] using ideas from the field of constrained low-rank matrix factorization and sparse approximation.

14.3.2 Concept Factorization

Concept factorization [65] is an extension of NMF for data clustering. It models each cluster (concept) $\boldsymbol{u}_c$, $c = 1, \ldots, k$, as a nonnegative linear combination of the data points $\boldsymbol{x}_j$, $j = 1, \ldots, n$, and each data point $\boldsymbol{x}_j$ as a nonnegative linear combination of all the cluster centers (concepts) $\boldsymbol{u}_c$. Data clustering is then accomplished by computing the two sets of linear coefficients, which is carried out by finding the nonnegative solution that minimizes the reconstruction error of the data points. The superiority of concept factorization over NMF is shown for document clustering in [65].

In concept factorization, each cluster center (concept) $\boldsymbol{u}_c, c = 1, \ldots, k$, is modeled as a linear combination of the data point $\boldsymbol{x}_j$, $j = 1, 2, \ldots, n$, and each data point is modeled as a linear combination of the cluster centers, that is,

$$\boldsymbol{u}_c = \sum_{j=1}^{n} w_{jc} \boldsymbol{x}_j, \tag{14.13}$$

$$\boldsymbol{x}_j = \sum_{c=1}^{k} v_{jc} \boldsymbol{u}_c, \tag{14.14}$$

where w_{jc} is a non-negative weight to represent the degree of representativeness of $\boldsymbol{x}_j$ in the cth concept, and v_{jc} is a nonnegative number to represent its degree belonging to the cth concept.

Let $\mathbf{W} = \left[w_{jc}\right] \in R^{n \times k}$, $\mathbf{V} = \left[v_{jc}\right] \in R^{n \times k}$. Concept factorization essentially tries to find the approximation

$$\mathbf{X} \approx \mathbf{X}\mathbf{W}\mathbf{V}^T, \tag{14.15}$$

with elements of $\mathbf{W}$ and $\mathbf{V}$ being nonnegative.

Similar to NMF, it aims to minimize

$$E_{\text{CF}} = \frac{1}{2} \left\|\mathbf{X} - \mathbf{X}\mathbf{W}\mathbf{V}^T\right\|^2. \tag{14.16}$$

When fixing $\mathbf{W}$ and $\mathbf{V}$ alternately, multiplicative updating rules are given as [65]

$$w_{jc} \longleftarrow w_{jc} \frac{(\mathbf{KV})_{jc}}{(\mathbf{V}\mathbf{W}^T\mathbf{K}\mathbf{W})_{jc})}, \tag{14.17}$$

$$v_{jc} \longleftarrow v_{jc} \frac{(\mathbf{KW})_{jc}}{(\mathbf{K}\mathbf{W}\mathbf{V}^T\mathbf{V})_{jc})}, \tag{14.18}$$

where $\mathbf{K} = \mathbf{X}^T\mathbf{X}$. With the inner product of $\mathbf{X}$, concept factorization can be easily kernelized.

Pairwise constrained concept factorization [29] is a semi-supervised method that incorporates pairwise constraints into the concept factorization framework. Data points which have pairwise must-link constraints should have the same class label, while data points with pairwise cannot-link constraints should have different class labels.

Locally consistent concept factorization algorithm [4] encodes the geometrical information of the data space by constructing a nearest-neighbor graph to model the local manifold structure. When the label information is provided, it can be directly encoded into the graph structure.

Based on the geometric interpretation of NMF that it seeks a proper simplicial cone to try to contain $\mathbf{X}$ [16], large-cone NMF [51] applies the large-cone penalty for NMF to simultaneously obtain a small empirical reconstruction error, and a good generalization ability. Large-cone NMF algorithms are derived using two large-cone penalties for NMF. Compared with NMF, large-cone NMF will obtain bases comprising a larger simplicial cone. The obtained bases have a low-overlapping property, which enables the bases to be sparse and makes the large-cone NMF algorithms very robust.

14.4 Nystrom Method

Standard approaches for low-rank approximation include rank-revealing QR, and truncated SVD. PCA is a truncated SVD applied on recentered data. Truncated SVD achieves the best approximation in Frobenius norm (or 2-norm, equivalently):

$$\min_{\mathbf{A},\mathbf{S}} \|\mathbf{X} - \mathbf{AS}\|_F^2, \tag{14.19}$$

subject to

$$\mathbf{A}^T\mathbf{A} = \mathbf{I}_r, \tag{14.20}$$

for a given matrix $\mathbf{X}$ and a preindicated rank r of the approximation.

Standard eigenvalue decomposition takes $O(n^3)$ time. Symmetric positive semidefinite matrix approximation methods have been extensively used to speed up large-scale eigenvalue computation and kernel learning methods. Typical methods efficiently form a low-rank decomposition

$$\mathbf{K} \approx \mathbf{CUC}^T, \tag{14.21}$$

where $\mathbf{C} \in R^{n\times c}$ is a sketch of $\mathbf{K}$ (e.g., randomly sampled c columns of $\mathbf{K}$) and $\mathbf{U} \in R^{c\times c}$ can be computed in different ways.

It takes $O(nc^2)$ additional time to approximately compute the rank k ($k \leq c$) eigenvalue decomposition or the matrix inversion. Therefore, if $\mathbf{C}$ and $\mathbf{U}$ are obtained in linear time (w.r.t. n) and c is independent of n, then the eigenvalue decomposition and matrix inversion can be approximately solved in linear time.

A more general alternative is to use Nystrom method [17, 63]. Nystrom method is a sampling-based algorithm for approximating large kernel matrices and their eigensystems. Nystrom method is highly efficient, but can only achieve low accuracy. It originated from solving integral equations and was introduced to the machine learning community [63].

Nystrom method selects a subset of $c \leq n$ columns from $\mathbf{K}$, and then uses the correlations between the sampled columns and the remaining columns to form a low-rank approximation of the full matrix. This makes Nystrom method highly scalable. Nystrom method only has to decompose a much smaller $c \times c$ matrix, which consists of the intersection of the selected columns and their corresponding rows. However, to ensure an accurate approximation, a sufficient number of columns have to be sampled.

Let $\mathbf{P}$ be an $n \times c$ sketching matrix such as uniform sampling [63]. Nystrom method computes

$$\mathbf{C} = \mathbf{KP}, \quad \mathbf{U} = (\mathbf{P}^T\mathbf{C})^{\dagger}. \tag{14.22}$$

where $\dagger$ denotes pseudoinverse. One can sample $c = O(k/\epsilon)$ columns of $\mathbf{K}$ to form $\mathbf{C}$ such that $\min_{\mathbf{U}} \|\mathbf{K} - \mathbf{CUC}^T\|_F^2 \leq (1+\epsilon)\|\mathbf{K} - \mathbf{K}_k\|_F^2$, $\mathbf{K}_k$ being the estimated $\mathbf{K}$ of rank-k.

Standard sketch-based method is randomized SVD for symmetric matrices [28]. It also extends the Monte Carlo algorithms, on which the analysis of Nystrom method is based. Unlike Nystrom method which simply samples a column subset for approximation, it first constructs a low-dimensional subspace that captures the action of the input matrix, and then standard decomposition (e.g., QR and SVD) is performed on the subspace matrix. The algorithm needs at least one pass over the whole input matrix. It achieves much higher accuracy by solving $\mathbf{U}$ from $\min_{\mathbf{U}} \|\mathbf{K} - \mathbf{CUC}^T\|_F^2$, but at a high time complexity of $O(n^2c)$. Unlike Nystrom method, standard sketch based method does not require c to grow with n.

Fast symmetric positive semi-definite matrix approximation model [62] is nearly as efficient as the Nystrom method and as accurate as the standard sketch-based method. It avoids computing the entire kernel matrix $\mathbf{K}$. The time complexity for computing $\mathbf{U}$ is linear in n.

An accurate and scalable Nystrom scheme [45] first samples a large column subset from the input matrix, then only performs an approximate SVD on the inner submatrix using the randomized low-rank matrix approximation algorithms. As accurate as the Nystrom method, its time complexity is only as low as performing a small SVD.

14.5 CUR Decomposition

SVD or QR decomposition for large sparse matrices does not preserve sparsity in general. Also, the basis vectors resulting from SVD have no physical meaning. It is useful to compute a low-rank matrix decomposition which preserves such structural properties of the original matrix. Matrix column selection and CUR matrix decomposition are two techniques that represent a matrix in terms of a small number of actual columns and/or actual rows of the matrix.

Matrix column selection method is also known as $\mathbf{A} \approx \mathbf{CX}$ *decomposition*. Given a large sparse matrix $\mathbf{A}$, its submatrix $\mathbf{C}$ is sparse, but the coefficient matrix $\mathbf{X} \in R^{c\times n}$ is not sparse in general.

CUR matrix decomposition [17, 25] seeks to find a subset of c columns of any matrix $\mathbf{A} \in R^{m\times n}$ to form a matrix $\mathbf{C} \in R^{m\times c}$, a subset of r rows to form a matrix $\mathbf{R} \in R^{r\times n}$, and computes a matrix $\mathbf{U} \in R^{c\times r}$ such that

$$\min \|\mathbf{A} - \mathbf{CUR}\|_{\psi}. \tag{14.23}$$

It costs time $O(mn \min\{c, r\})$ to compute the optimal $\mathbf{U} = \mathbf{C}^{\dagger}\mathbf{A}\mathbf{R}^{\dagger}$.

CUR decomposition usually requires a large number of columns and rows to be chosen. Most CUR algorithms [18, 21] work in a two-stage manner where the first stage is a standard column selection procedure. Subspace sampling algorithm [21] is a two-stage randomized CUR algorithm which has a relative-error bound with high probability. It selects columns/rows according to the statistical leverage scores. The computational cost is at least equal to the cost of the truncated SVD of A. Fast approximation to statistical leverage scores can be used to speed up the subspace sampling algorithm heuristically [22].

Nystrom method approximates a symmetric positive semidefinite matrix in terms of a small number of its columns, while CUR decomposition approximates an arbitrary matrix by a small number of its columns and rows. Thus, CUR decomposition is an extension of the standard sketch based method from symmetric matrices to general matrices. It is troubled by the same computational problem.

The lower error bounds of the Nystrom method and the ensemble Nystrom method are even much worse than the upper bounds of some CUR algorithms. More accurate CUR and Nystrom algorithms with expected relative-error bounds are derived based on a more general error bound for the adaptive column/row sampling algorithm [61]. The algorithms have low time complexity and can avoid maintaining the whole matrix in RAM.

Fast symmetric positive semi-definite matrix approximation model [62] is applied to improve CUR decomposition of the general matrices. The time cost drops to $O(cr\epsilon^{-1} \min\{m, n\} \min\{c, r\})$, while the approximation quality is nearly the same.

Problems

14.1 Characterize the following two cases as NMF with spareness constraints [32].
(a) A doctor analyzing disease patterns might assume that most diseases are rare (hence sparse) but that each disease can cause a large number of symptoms. Assuming that symptoms make up the rows of her matrix and the columns denote different individuals, in this case it is the coefficients which should be sparse and the basis vectors unconstrained.
(b) When trying to learn useful features from a database of images, it might make sense to require both **A** and **S** to be sparse, signifying that any given object is present in few images and affects only a small part of the image.

14.2 Sparseness measures quantify as to how much energy of a vector is contained in only a few of its components. One of the sparseness measures is defined based on the relationship between the L_1-norm and the L_2-norm [32]:

$$sparseness(\boldsymbol{x}) = \frac{\sqrt{n} - \frac{\sum |x_i|}{\sqrt{\sum x_i^2}}}{\sqrt{n} - 1},$$

where n is the dimensionality of $\boldsymbol{x}$. Explain why this definition is reasonable.

14.3 Give features of the ORL face image database using nmfpack (http://www.cs.helsinki.fi/u/phoyer/software.html).

14.4 Show that C-means is equivalent to NMF with orthogonality on **S**.

References

1. Banerjee, A., Merugu, S., Dhillon, I. S., & Ghosh, J. (2005). Clustering with Bregman divergences. *Journal of Machine Learning Research*, *6*, 1705–1749.
2. Berry, M. W., Browne, M., Langville, A. N., Pauca, V. P., & Plemmons, R. J. (2007). Algorithms and applications for approximate nonnegative matrix factorization. *Computational Statistics and Data Analysis*, *52*(1), 155–173.
3. Blumensath, T. (2016). Directional clustering through matrix factorization. *IEEE Transactions on Neural Networks and Learning Systems*, *27*(10), 2095–2107.
4. Cai, D., He, X., & Han, J. (2011). Locally consistent concept factorization for document clustering. *IEEE Transactions on Knowledge and Data Engineering*, *23*(6), 902–913.
5. Cai, D., He, X., Han, J., & Huang, T. S. (2011). Graph regularized nonnegative matrix factorization for data representation. *IEEE Transactions on Pattern Analysis and Machine Intelligence*, *33*(8), 1548–1560.
6. Chen, Y., Rege, M., Dong, M., & Hua, J. (2008). Non-negative matrix factorization for semi-supervised data clustering. *Knowledge and Information Systems*, *17*(3), 355–379.
7. Cichocki, A., & Anh-Huy, P. (2009). Fast local algorithms for large scale nonnegative matrix and tensor factorizations. *IEICE Transactions on Fundamentals of Electronics, Communications and Computer Sciences*, *92*(3), 708–721.
8. Cichocki, A., & Zdunek, R. (2006). *NMFLAB for signal and image processing*. Technical Report, Laboratory for Advanced Brain Signal Processing, BSI RIKEN, Saitama, Japan.

9. Cichocki, A., & Zdunek, R. (2007). Multilayer nonnegative matrix factorization using projected gradient approaches. *International Journal of Neural Systems, 17*(6), 431–446.
10. Cichocki, A., Zdunek, R., & Amari, S. (2006). New algorithms for non-negative matrix factorization in applications to blind source separation. In *Proceedings of IEEE International Conference on Acoustics, Speech, and Signal Processing (ICASSP)* (Vol. 5, pp. 621–624). Toulouse, France.
11. Deerwester, S. C., Dumais, S. T., Landauer, T. K., Furnas, G. W., & Harshman, R. A., (1990). Indexing by latent semantic analysis. *Journal of the American Society for Information Science, 416*, 391–407.
12. Dhillon, I. S., & Sra, S. (2006). Generalized nonnegative matrix approximations with Bregman divergences. In *Advances in neural information processing systems* (Vol. 18, pp. 283–290).
13. Ding, C., Li, T., Peng, W., & Park, H. (2006). Orthogonal nonnegative matrix tri-factorizations for clustering. In *Proceedings of the 12th ACM SIGKDD International Conference on Knowledge Discovery and Data Mining (KDD'06)* (pp. 126–135). Philadelphia, PA.
14. Ding, C., Li, T., & Jordan, M. I. (2010). Convex and semi-nonnegative matrix factorizations. *IEEE Transactions on Pattern Analysis and Machine Intelligence, 32*(1), 45–55.
15. Ding, C. H., He, X., & Simon, H. D. (2005). On the equivalence of nonnegative matrix factorization and spectral clustering. In *Proceedings of the SIAM International Conference on Data Mining* (pp. 606–610). Newport Beach, CA.
16. Donoho, D., & Stodden, V. (2003). When does nonnegative matrix factorization give a correct decomposition into parts? In *Advances in neural information processing systems* (Vol. 16, pp. 1141–1148). Vancouver, Canada; Cambridge, MA: MIT Press.
17. Drineas, P., & Mahoney, M. W. (2005). On the Nystrom method for approximating a gram matrix for improved kernel-based learning. *Journal of Machine Learning Research, 6*, 2153–2175.
18. Drineas, P., Kannan, R., & Mahoney, M. W. (2006). Fast Monte Carlo algorithms for matrices. III: Computing a compressed approximate matrix decomposition. *SIAM Journal on Computing, 36*(1), 184–206.
19. Drineas, P., Mahoney, M. W., & Muthukrishnan, S. (2006). Subspace sampling and relative-error matrix approximation: Column-based methods. In *Proceedings of the 10th Annual International Workshop on Randomization and Computation (RANDOM), LNCS* (Vol. 4110, pp. 316–326). Berlin: Springer.
20. Drineas, P., Mahoney, M. W., & Muthukrishnan, S. (2007). *Relative-error CUR matrix decompositions*. Technical Report, Department of Mathematics, Yale University.
21. Drineas, P., Mahoney, M. W., & Muthukrishnan, S. (2008). Relative-error CUR matrix decompositions. *SIAM Journal on Matrix Analysis and Applications, 30*(2), 844–881.
22. Drineas, P., Magdon-Ismail, M., Mahoney, M. W., & Woodruff, D. P. (2012). Fast approximation of matrix coherence and statistical leverage. *Journal of Machine Learning Research, 13*, 3441–3472.
23. Gillis, N. (2012). Sparse and unique nonnegative matrix factorization through data preprocessing. *Journal of Machine Learning Research, 13*, 3349–3386.
24. Gillis, N., & Luce, R. (2014). Robust near-separable nonnegative matrix factorization using linear optimization. *Journal of Machine Learning Research, 15*, 1249–1280.
25. Goreinov, S. A., Tyrtyshnikov, E. E., & Zamarashkin, N. L. (1997). A theory of pseudoskeleton approximations. *Linear Algebra and Its Applications, 261*, 1–21.
26. Guan, N., Tao, D., Luo, Z., & Yuan, B. (2012). Online nonnegative matrix factorization with robust stochastic approximation. *IEEE Transactions on Neural Networks and Learning Systems, 23*(7), 1087–1099.
27. Guan, N., Liu, T., Zhang, Y., Tao, D., & Davis, L. S. (2019). Truncated Cauchy non-negative matrix factorization. *IEEE Transactions on Pattern Analysis and Machine Intelligence, 41*(1), 246–259.
28. Halko, N., Martinsson, P. G., & Tropp, J. A. (2011). Finding structure with randomness: Probabilistic algorithms for constructing approximate matrix decompositions. *SIAM Review, 53*(2), 217–288.

29. He, Y., Lu, H., Huang, L., & Xie, S. (2014). Pairwise constrained concept factorization for data representation. *Neural Networks*, *52*, 1–17.
30. He, Z., Xie, S., Zdunek, R., Zhou, G., & Cichocki, A. (2011). Symmetric nonnegative matrix factorization: Algorithms and applications to probabilistic clustering. *IEEE Transactions on Neural Networks*, *22*(12), 2117–2131.
31. Hosoda, K., Watanabe, M., Wersing, H., Korner, E., Tsujino, H., Tamura, H., et al. (2009). A model for learning topographically organized parts-based representations of objects in visual cortex: Topographic nonnegative matrix factorization. *Neural Computation*, *21*, 2605–2633.
32. Hoyer, P. O. (2004). Nonnegative matrix factorization with sparseness constraints. *Journal of Machine Learning Research*, *5*, 1457–1469.
33. Hsieh, C.-J., & Dhillon, I. S. (2011). Fast coordinate descent methods with variable selection for non-negative matrix factorization. In *Proceedings of the 17th ACM SIGKDD International Conference on Knowledge Discovery and Data Mining* (pp. 1064–1072).
34. Keprt, A., & Snasel, V. (2005). Binary factor analysis with genetic algorithms. In *Proceedings of the 4th IEEE International Workshop on Soft Computing as Transdisciplinary Science and Technology (WSTST), AINSC* (Vol. 29, pp. 1259–1268). Muroran, Japan; Berlin: Springer.
35. Kim, D., Sra, S., & Dhillon, I. S. (2008). Fast projection-based methods for the least squares nonnegative matrix approximation problem. *Statistical Analysis and Data Mining*, *1*, 38–51.
36. Kimura, K., Kudo, M., & Tanaka, Y. (2016). A column-wise update algorithm for nonnegative matrix factorization in Bregman divergence with an orthogonal constraint. *Machine Learning*, *103*, 285–306.
37. Kompass, R. (2007). A generalized divergence measure for nonnegative matrix factorization. *Neural Computation*, *19*(3), 780–791.
38. Kong, D., Ding, C., & Huang, H. (2011). Robust nonnegative matrix factorization using $l_{2,1}$-norm. In *Proceedings of the 20th ACM International Conference on Information and Knowledge Management* (pp. 673–682). Glasgow, UK.
39. Kuang, D., Ding, C., & Park, H. (2012). Symmetric nonnegative matrix factorization for graph clustering. In *Proceedings of the 12th SIAM International Conference on Data Mining* (pp. 106–117). Anaheim, CA.
40. Laurberg, H., Christensen, M. G., Plumbley, M. D., Hansen, L. K., & Jensen, S. H. (2008). Theorems on positive data: on the uniqueness of NMF. *Computational Intelligence and Neuroscience*, *2008*, Article ID 764206.
41. Lee, D. D., & Seung, H. S. (1999). Learning the parts of objects by nonnegative matrix factorization. *Nature*, *401*(6755), 788–791.
42. Lee, H., Yoo, J., & Choi, S. (2010). Semi-supervised nonnegative matrix factorization. *IEEE Signal Processing Letters*, *17*(1), 4–7.
43. Li, L., Lebanon, G., & Park, H. (2012). Fast Bregman divergence NMF using Taylor expansion and coordinate descent. In *Proceedings of the 18th ACM SIGKDD International Conference on Knowledge Discovery and Data Mining* (pp. 307–315).
44. Li, L.-X., Wu, L., Zhang, H.-S., & Wu, F.-X. (2014). A fast algorithm for nonnegative matrix factorization and its convergence. *IEEE Transactions on Neural Networks and Learning Systems*, *25*(10), 1855–1863.
45. Li, M., Bi, W., Kwok, J. T., & Lu, B.-L. (2015). Large-scale Nystrom kernel matrix approximation using randomized SVD. *IEEE Transactions on Neural Networks and Learning Systems*, *26*(1), 152–164.
46. Li, X., & Pang, Y. (2010). Deterministic column-based matrix decomposition. *IEEE Transactions on Knowledge and Data Engineering*, *22*(1), 145–149.
47. Lin, C.-J. (2007). Projected gradients for non-negative matrix factorization. *Neural Computation*, *19*(10), 2756–2779.
48. Lin, C.-J. (2007). On the convergence of multiplicative update algorithms for non-negative matrix factorization. *IEEE Transactions on Neural Networks*, *18*(6), 1589–1596.
49. Liu, H., Wu, Z., Li, X., Cai, D., & Huang, T. S. (2012). Constrained nonnegative matrix factorization for image representation. *IEEE Transactions on Pattern Analysis and Machine Intelligence*, *34*(7), 1299–1311.

50. Liu, H., Li, X., & Zheng, X. (2013). Solving non-negative matrix factorization by alternating least squares with a modified strategy. *Data Mining and Knowledge Discovery*, *26*(3), 435–451.
51. Liu, T., Gong, M., & Tao, D. (2017). Large-cone nonnegative matrix factorization. *IEEE Transactions on Neural Networks and Learning Systems*, *28*(9), 2129–2142.
52. Mirzal, A. (2014). A convergent algorithm for orthogonal nonnegative matrix factorization. *Journal of Computational and Applied Mathematics*, *260*, 149–166.
53. Paatero, P., & Tapper, U. (1994). Positive matrix factorization: A nonnegative factor model with optimal utilization of error estimates of data values. *Environmetrics*, *5*(2), 111–126.
54. Pascual-Montano, A., Carazo, J. M., Kochi, K., Lehmann, D., & Pascual-Marqui, R. D. (2006). Nonsmooth nonnegative matrix factorization (nsNMF). *EEE Transactions on Pattern Analysis and Machine Intelligence*, *28*(3), 403–415.
55. Peharz, R., & Pernkopf, F. (2012). Sparse nonnegative matrix factorization with l^0-constraints. *Neurocomputing*, *80*, 38–46.
56. Sajda, P., Du, S., Brown, T. R., Stoyanova, R., Shungu, D. C., Mao, X., et al. (2004). Nonnegative matrix factorization for rapid recovery of constituent spectra in magnetic resonance chemical shift imaging of the brain. *IEEE Transactions on Medical Imaging*, *23*(12), 1453–1465.
57. Vavasis, S. A. (2009). On the complexity of nonnegative matrix factorization. *SIAM Journal on Optimization*, *20*(3), 1364–1377.
58. Wang, D., & Lu, H. (2013). On-line learning parts-based representation via incremental orthogonal projective non-negative matrix factorization. *Signal Processing*, *93*, 1608–1623.
59. Wang, F., Li, T., & Zhang, C. (2008). Semi-supervised clustering via matrix factorization. In *Proceedings of the SIAM International Conference on Data Mining* (pp. 1–12). Atlanta, GA.
60. Wang, J. J.-Y., & Gao, X. (2015). Max–min distance nonnegative matrix factorization. *Neural Networks*, *61*, 75–84.
61. Wang, S., & Zhang, Z. (2013). Improving CUR matrix decomposition and the Nystrom approximation via adaptive sampling. *Journal of Machine Learning Research*, *14*, 2729–2769.
62. Wang, S., Zhang, Z., & Zhang, T. (2016). Towards more efficient SPSD matrix approximation and CUR matrix decomposition. *Journal of Machine Learning Research*, *17*, 1–49.
63. Williams, C., & Seeger, M. (2001). Using the Nystrom method to speedup kernel machines. In T. Leen, T. Dietterich, & V. Tresp (Eds.), *Advances in neural information processing systems* (Vol. 13, pp. 682–690). Cambridge, MA, USA: MIT Press.
64. Wu, X., Kumar, V., Quinlan, J. R., Ghosh, J., Yang, Q., Motoda, H., et al. (2008). Top 10 algorithms in data mining. *Knowledge and Information Systems*, *14*(1), 1–37.
65. Xu, W., & Gong, Y. (2004). Document clustering by concept factorization. In *Proceedings of the 27th Annual International ACM SIGIR Conference on Research and Development in Information Retrieval* (pp. 202–209).
66. Yang, S., & Yi, Z. (2010). Convergence analysis of non-negative matrix factorization for BSS algorithm. *Neural Processing Letters*, *31*, 45–64.
67. Yang, S., Yi, Z., Ye, M., & He, X. (2014). Convergence analysis of graph regularized nonnegative matrix factorization. *IEEE Transactions on Knowledge and Data Engineering*, *26*(9), 2151–2165.
68. Yang, Z., & Laaksonen, J. (2007). Multiplicative updates for non-negative projections. *Neurocomputing*, *71*, 363–373.
69. Yang, Z., & Oja, E. (2010). Linear and nonlinear projective nonnegative matrix factorization. *IEEE Transactions on Neural Networks*, *21*(5), 734–749.
70. Yokota, T., Zdunek, R., Cichocki, A., & Yamashita, Y. (2015). Smooth nonnegative matrix and tensor factorizations for robust multi-way data analysis. *Signal Processing*, *113*, 234–249.
71. Zafeiriou, S., & Petrou, M. (2010). Nonlinear non-negative component analysis algorithms. *IEEE Transactions on Image Processing*, *19*(4), 1050–1066.
72. Zafeiriou, S., Tefas, A., Buciu, I., & Pitas, I. (2006). Exploiting discriminant information in nonnegative matrix factorization with application to frontal face verification. *IEEE Transactions on Neural Networks*, *17*(3), 683–695.
73. Zdunek, R., & Cichocki, A. (2007). Nonnegative matrix factorization with constrained second-order optimization. *Signal Processing*, *87*, 1904–1916.

74. Zdunek, R., & Cichocki, A. (2008). Fast nonnegative matrix factorization algorithms using projected gradient approaches for large-scale problems. *Computational Intelligence and Neuroscience, 2008*, Article ID 939567.
75. Zhang, J., Wei, L., Feng, X., Ma, Z., & Wang, Y. (2008). Pattern expression nonnegative matrix factorization: Algorithm and applications to blind source separation. *Computational Intelligence and Neuroscience, 2008*, Article ID 168769.
76. Zhang, X., Zong, L., Liu, X., & Luo, J. (2016). Constrained clustering with nonnegative matrix factorization. *IEEE Transactions on Neural Networks and Learning Systems, 27*(7), 1514–1526.
77. Zhang, Z., & Zhao, K. (2013). Low-rank matrix approximation with manifold regularization. *IEEE Transactions on Pattern Analysis and Machine Intelligence, 35*(7), 1717–1729.
78. Zhao, R., & Tan, V. Y. F. (2017). Online nonnegative matrix factorization with outliers. *IEEE Transactions on Signal Processing, 65*(3), 555–570.
79. Zhou, G., Xie, S., Yang, Z., Yang, J.-M., & He, Z. (2011). Minimum-volume-constrained nonnegative matrix factorization: Enhanced ability of learning parts. *IEEE Transactions on Neural Networks, 22*(10), 1626–1637.

Chapter 15
Independent Component Analysis

15.1 Introduction

Imagine that you are attending a cocktail party, the surrounding is full of chatting and noise, and somebody is talking about you. In this case, your ears are particularly sensitive to this speaker. This is the cocktail-party problem, which can be solved by blind source separation (BSS).

BSS is a very active and commercially driven topic, and has found wide applications in a variety of areas including mobile communications, signal processing, separating audio signals, demixing multispectral images, biomedical systems, and seismic signal processing. In medical systems, BSS is used to identify artifacts and signals of interest from the analysis of functional brain imaging signals, such as electrical recordings of brain activity as given by magnetoencephalography (MEG) or electroencephalography (EEG), and functional magnetic resonance imaging (fMRI). BSS has been applied to extract the fetal electrocardiography (FECG) from the electrocardiography (ECG) recordings measured on the mother's skin.

ICA [34], as a generalization of PCA, is a statistical model. The goal of ICA is to recover the latent components from observations. ICA finds a linear representation of non-Gaussian data so that the components are statistically independent, or as independent as possible. ICA has now been widely used for BSS, feature extraction, and signal detection.

For BSS applications, the ICA model is required to have model identifiability and separability [34]. ICA corresponds to a class of methods with the objective of recovering underlying latent factors present in the data. The observed variables are linear mixtures of the components which are assumed to be mutually independent. Instead of obtaining uncorrelated components as in PCA, ICA attempts to linearly transform the original inputs into features that are statistically mutually independent. Independence is a stronger condition than uncorrelatedness, and is equivalent to uncorrelatedness only in the case of Gaussian distributions. The first neural network model with a heuristic learning algorithm, which is related to ICA, was developed for online BSS of linearly mixed signals in [67].

K.-L. Du and M. N. S. Swamy, *Neural Networks and Statistical Learning*,
https://doi.org/10.1007/978-1-4471-7452-3_15

BSS is not identical to ICA since methods using second-order statistics can be used for BSS. These second-order statistics approaches are not restricted by the Gaussianity of the sources but rather, require that the sources, although independent of one another, are colored in the temporal domain. BSS methods are typically decorrelation based and ICA based. Decorrelation methods minimize the squared crosscorrelation between all pairs of source estimates at two or more lags. They are useful for BSS when the sources possess sufficient spectral diversity even if all the sources are Gaussian distributed. Conversely, ICA methods minimize the statistical dependence of the source estimates at lag 0. A spatiotemporal BSS method would be appropriate for either of the two aforementioned cases since it combines ICA and decorrelation criteria into one algorithm.

Most algorithms for ICA, directly or indirectly, minimize the mutual information between the component estimates, which corresponds to maximization of the negentropy, a measure of non-Gaussianity of the components [61]. The exact maximization of the negentropy is difficult and computationally demanding because a correct estimation of the source densities is required. Most of the existing ICA algorithms can be viewed as approximating negentropy through simple measures, such as high-order cumulants [54, 61]. Most of the ICA algorithms based on unsupervised learning belong to the Hebb-type rule or its generalization with adopting nonlinear functions.

15.2 ICA Model

Let a J_1-vector $\boldsymbol{x}$ denote a linear mixture and a J_2-vector $\boldsymbol{s}$, whose components have zero mean and are statistically mutually independent, denote the original source signals. The ICA model can be defined by

$$\boldsymbol{x} = \mathbf{A}\boldsymbol{s} + \boldsymbol{n}, \tag{15.1}$$

where $\mathbf{A}$ is a constant full-rank $J_1 \times J_2$ mixing matrix whose elements are the unknown coefficients of the mixtures, and $\boldsymbol{n}$ denotes an additive noise term, which is often omitted since it is usually impossible to separate noise from the sources. ICA takes one of three forms, namely, square ICA for $J_1 = J_2$, overcomplete ICA for $J_1 < J_2$, and undercomplete ICA for $J_1 > J_2$. While undercomplete ICA is useful for feature extraction, overcomplete ICA may be applied to signal and image processing methods based on multiscale and redundant basis sets.

The goal of ICA is to estimate $\boldsymbol{s}$ by

$$\boldsymbol{y} = \mathbf{W}^T \boldsymbol{x} \tag{15.2}$$

such that the components of $\boldsymbol{y}$, which is the estimate of $\boldsymbol{s}$, are statistically as independent as possible. $\mathbf{W}$ is a $J_1 \times J_2$ demixing matrix. In the ICA model, two ambiguities hold: one cannot determine the variances (energies) of the independent components and one cannot determine the order of the independent components. ICA can be considered a variant of projection pursuit.

The statistical independence property implies that the joint probability density of the components of $\boldsymbol{s}$ equals the product of the marginal densities of the individual components. Each component of $\boldsymbol{s}$ is a stationary stochastic process and only one of the components is allowed to be Gaussian distributed. The higher order statistics of the original inputs is required for estimating $\boldsymbol{s}$, rather than the second-order moment or covariance of the samples as used in PCA. Notice that the MSE for any estimate of a nonrandom parameter has a lower bound, called the *Cramer–Rao bound*. This lower bound defines the ultimate accuracy of any estimator, and is closely related to the ML estimator. To estimate a vector of parameters $\boldsymbol{\theta}$ from a data vector $\boldsymbol{x}$ that has a probability density, by using some unbiased estimator $\hat{\boldsymbol{\theta}}$, the Cramer–Rao bound, which is the lower bound for the variance of $\hat{\boldsymbol{\theta}}$ on estimating the source signals in ICA is derived in [79], based on the assumption that all independent components have finite variance.

Two distinct characteristics exist between PCA and ICA. The components of the signal extracted by ICA are statistically independent, not merely uncorrelated as in PCA. The demixing matrix $\mathbf{W}$ of ICA is not orthogonal, while in PCA the components of the weights are represented on an orthonormal basis. ICA provides in many cases a more meaningful representation of the data than PCA does. ICA can be realized by adding nonlinearity to linear PCA networks such that they are able to improve the independence of their outputs. In [71], an efficient ICA algorithm is derived by minimizing a nonlinear PCA criterion using the RLS approach. A conceptual comparison of PCA and ICA is illustrated in Fig. 15.1. $\boldsymbol{w}_i^{\text{ICA}}$ and $\boldsymbol{w}_i^{\text{PCA}}$, $i = 1, 2$, are the ith principal and ith independent directions, respectively.

15.3 Approaches to ICA

A well-known two-phase approach to ICA is to preprocess the data by PCA, and then to estimate the necessary rotation matrix. A generic approach to ICA consists of preprocessing the data, defining measures of non-Gaussianity, and optimizing an objective function, known as a *contrast function*. Some measures of non-Gaussianity

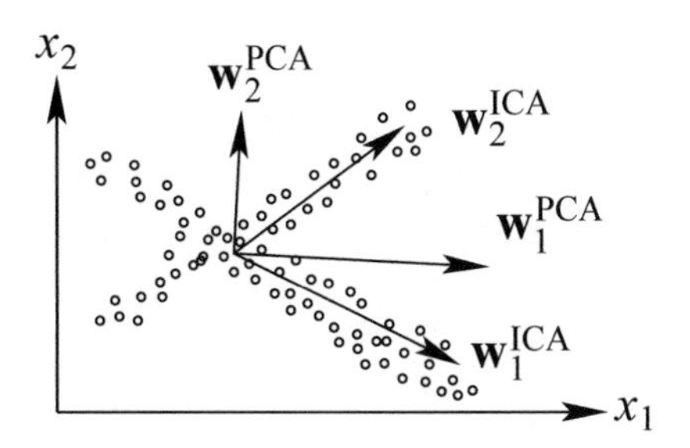

Fig. 15.1 Illustration of PCA and ICA for a two-dimensional non-Gaussian dataset

are kurtosis, differential entropy, negentropy, and mutual information, which can be derived from one another. For example, one approach is to minimize the mutual information between the components of the output vector

$$I(\boldsymbol{y}) = \sum_{i=1}^{J_2} H(y_i) - H(\boldsymbol{y}), \tag{15.3}$$

where $H(\boldsymbol{y}) = -\int p(\boldsymbol{y}) \ln p(\boldsymbol{y}) \mathrm{d}\boldsymbol{y}$ is the joint entropy, $H(y_i) = -\int p_i(y_i) \ln p_i(y_i) \mathrm{d}y_i$ is the marginal entropy of component i, $p(\boldsymbol{y})$ being the joint pdf of all the elements of $\boldsymbol{y}$, and $p_i(y_i)$ is the marginal pdf of y_i. The mutual information $I \geq 0$ and is zero only when the components are mutually independent.

The classical measure of non-Gaussianity is kurtosis. Kurtosis is the degree of peakedness of a distribution, based on the fourth central moment of the distribution. The kurtosis of y_i is classically defined by

$$\text{kurt}(y_i) = \mathrm{E}\left[y_i^4\right] - 3\left(\mathrm{E}\left[y_i^2\right]\right)^2. \tag{15.4}$$

If $\text{kurt}(y_i) < 0$, $y_i(t)$ is a sub-Gaussian source, while for super-Gaussian sources $\text{kurt}(y_i) > 0$. For a Gaussian y_i, the fourth moment equals $3\left(\mathrm{E}\left[y_i^2\right]\right)^2$, and thus the kurtosis of Gaussian sources is zero. Super-Gaussian random variables have typically a spiky pdf with heavy tails, i.e., the pdf is relatively large at zero and at large values of the variable, while being small for intermediate values. A typical example is the Laplace distribution. Sub-Gaussian random variables have typically a flat pdf, which is rather constant near zero, and very small for larger values of the variable. Typically non-Gaussianity is measured by the absolute value of kurtosis. However, kurtosis has to be estimated from a measured sample, and it can be very sensitive to outliers.

Negentropy is a measure of non-Gaussianity that is zero for a Gaussian variable and always nonnegative:

$$J(\boldsymbol{y}) = H(\boldsymbol{y}_{\text{gauss}}) - H(\boldsymbol{y}), \tag{15.5}$$

where $\boldsymbol{y}_{\text{gauss}}$ is a Gaussian random variable of the same covariance matrix as $\boldsymbol{y}$. Negentropy is invariant for invertible linear transformations [34]. Entropy is small for distributions that are clearly concentrated on certain values, i.e., when the variable is clearly clustered or has a pdf that is very spiky. In fact, negentropy is in some sense the optimal estimator of non-Gaussianity, as far as statistical properties are concerned. Computation of negentropy is very difficult. Therefore, simpler approximations of negentropy are very useful.

Two common approaches in ICA algorithms are minimum mutual information and maximum output entropy approaches. A natural measure of independence is mutual information [34], a nonnegative scalar that equals zero when the signals are independent. Mutual information estimation of continuous signals is notoriously difficult but when prewhitening of the observed signal space is performed, mutual information minimization becomes equivalent to finding the orthogonal directions for which the negentropy is maximized [34]; thus much research has focused on developing one-dimensional approximations of negentropy and differential entropy. Minimization of output mutual information is the canonical contrast for BSS [22]. Three best-known methods for ICA, namely, JADE [22], infomax [13], and FastICA [53, 57], use diagonalization of cumulant matrices, maximization of output entropy, and fourth-order cumulants, respectively. Two other popular ICA algorithms are the natural gradient [5] and the equivariant adaptive separation via independence (EASI) [23]. These methods can be easily extended to the complex domain by using Hermitian transpose and complex nonlinear functions.

In the context of BSS, higher order statistics are necessary only for temporally uncorrelated stationary sources. Second-order statistics-based source separation exploits temporally correlated stationary sources [31] and the nonstationarity of the sources [31, 94]. Many natural signals are inherently nonstationary with time-varying variances since the source signals incorporate time delays into the basic BSS model.

The Overlearning Problem

In the presence of insufficient samples, most ICA algorithms produce very similar types of overlearning [58]. These consist of source estimates that have a single spike or bump and are practically zero everywhere else, regardless of the observations $\boldsymbol{x}$. The overlearning problem in ICA algorithms based on marginal distribution information is discussed in [112]. The resulting overlearned components have a single spike or bump and are practically zero everywhere else. This is similar to the classical overlearning in linear regression. The solutions to this spike problem include the acquisition of more samples, or the reduction of dimensions. The reduction of dimensions is a more efficient way to avoid the problem, provided that there are more sensors than sources.

The overlearning problem cannot be solved by acquiring more samples nor by dimension reduction, when the data has strong time dependencies, such as a $1/f$ power spectrum. This overlearning is better characterized by bumps instead of spikes. This spectrum characteristic is typical of MEG as well as many other natural data. Due to its $1/f$ nature, MEG data tends to show a bump-like overlearning, much like the one in the random walk dataset, rather than the spike type observed for the Gaussian i.i.d. dataset. Asymptotically, the kurtoses of the spikes and bumps tend to zero, when the number of samples increases.

15.4 Popular ICA Algorithms

15.4.1 Infomax ICA

The infomax approach [13] aims to maximize the mutual information between the observations and the nonlinearly transformed outputs of a set of linear filters. It is a gradient-based technique implementing entropy maximization in a single-layer feedforward network.

A certain transformation is applied to the inputs,

$$y_i = f_i\left(\boldsymbol{w}_i^T \boldsymbol{x}\right), \quad i = 1, \ldots, J_2, \tag{15.6}$$

where $f_i(\cdot)$ is a monotone squashing function such as a sigmoid. Maximizing the mutual information between outputs $\boldsymbol{y}$ and inputs $\boldsymbol{x}$, which is equivalent to maximizing the entropy of $\boldsymbol{y}$ due to the deterministic relation (15.6), would lead to independent components. This effect can be understood through the decomposition

$$H(\boldsymbol{y}) = \sum_{i=1}^{J_2} H_i\left(y_i\right) - I\left(y_1, \ldots, y_{J_2}\right), \tag{15.7}$$

where $H(\boldsymbol{y})$ is the joint entropy of $\boldsymbol{y}$, $H_i\left(y_i\right)$ is the individual entropy, and I is the mutual information among y_i's. Maximizing the joint entropy thus involves maximizing the individual entropies of y_i's and minimizing the mutual information between y_i's.

For square representations, the infomax approach turns out to be equivalent to the causal generative one if we interpret $f_i(\cdot)$ to be the cumulative distribution function of $p_i(\cdot)$ [24].

The natural gradient solution is obtained as [13] follows:

$$\mathbf{W}^T(t+1) = \mathbf{W}^T(t) + \eta\left[\mathbf{I} - 2g(\boldsymbol{y})\boldsymbol{y}^T\right]\mathbf{W}^T(t), \tag{15.8}$$

where $\eta > 0$ is the learning rate, and $g(\boldsymbol{y}) = \left(g_1(y_1), \ldots, g_n(y_n)\right)^T$, $g(y) = \tanh(y)$ for source signals with positive kurtosis.

Infomax can be seen as an ML one [24] or as a mutual information-based one. It can be interpreted as assuming some given, a priori marginal distributions for y_i. MISEP [2] is an infomax-based ICA technique for linear and nonlinear mixtures, but estimates the marginal distributions in a different way, based on a maximum entropy criterion. MISEP generalizes infomax in two ways: to deal with nonlinear mixtures, and to be able to adapt to the actual statistical distributions of the sources, by dynamically estimating the nonlinearities to be used at the outputs. MISEP optimizes a network with a specialized architecture, with the output entropy as objective function. A number of components of $\boldsymbol{s}$ and $\boldsymbol{y}$ are assumed to be the same.

Infomax is better suited for the estimation of super-Gaussian sources: sharply peaked pdfs with heavy tails. It fails to separate sources that have negative kurtosis [13]. An extension of the infomax algorithm [84] is able to blindly separate mixed signals with sub- and super-Gaussian source distributions, by using a simple type of learning rule by choosing negentropy as a projection pursuit index. Parameterized probability distributions with sub- and super-Gaussian regimes are used to derive a general learning rule. This general learning rule preserves the simple architecture proposed by [13], is optimized using the natural gradient [7], and uses the stability analysis given in [23] to switch between sub- and super-Gaussian regimes.

The BSS algorithm in [105] is a trade-off between gradient infomax and natural gradient infomax. Desired equilibrium points are locally stable by choosing appropriate score functions and step sizes. The algorithm provides better performance than the gradient algorithm, and it is free from approximation error and the small-step-size restriction of the natural gradient algorithm. The proof of local stability of the desired equilibrium points for BSS is given by using monotonically increasing and odd score functions.

15.4.2 EASI, JADE, and Natural Gradient ICA

The EASI rule is given by [23]

$$\mathbf{W}^T(t+1) = \mathbf{W}^T(t) + \eta\left[\boldsymbol{y}g(\boldsymbol{y}^T) - g(\boldsymbol{y})\boldsymbol{y}^T\right]\mathbf{W}^T(t), \tag{15.9}$$

where η is the learning rate, and nonlinearity is usually simple cubic polynomials, $g(y) = y^3$. In [111], the optimal choice of these nonlinearities is addressed. This optimal nonlinearity is the output score function difference. It is a multivariate function which depends on the output distributions. The resulting quasi-optimal EASI can achieve better performance than standard EASI, but requires an accurate estimation of score function difference. However, the method has a great advantage to converge for any source, contrary to standard EASI whose convergence assumes a condition on the source statistics [23].

JADE [22] is an exact algebraic approach to perform ICA. It is based on joint diagonalization of the fourth-order cumulant tensors of prewhitened input data. The bottleneck with JADE when dealing with high-dimensional problems is the algebraic determination of the mixing matrices. JADE is based on the estimation of kurtosis via cumulants. A neural implementation of JADE [134] adaptively determines the mixing matrices to be jointly diagonalized with JADE. The learning rule uses higher order neurons and generalizes Oja's PCA rule.

Natural gradient learning for $J_1 = J_2$ [5, 7] is the true steepest-descent method in the Riemannian parametric space of the nonsingular matrices. It is proved to be Fisher-efficient in general, having the equivariant property. Natural gradient learning is extended to the overcomplete and undercomplete cases in [8]. The observed signals are assumed to be whitened by preprocessing, so that we can use the natural Riemannian gradient in Stiefel manifolds. The objective function is given by

$$E_1 = -\ln|\det \mathbf{W}| - \sum_{i=1}^{n} \log p_i(y_i(t)), \tag{15.10}$$

where $\mathbf{y} = \mathbf{W}\mathbf{x}(t)$ and $p_i(\cdot)$ represents the hypothesized pdf for the latent variable $s_i(t)$ (or its estimate $y_i(t)$).

The natural gradient ICA algorithm, which iteratively finds a minimum of (15.10), has the form [5]

$$\mathbf{W}(t+1) = \mathbf{W}(t) + \eta \left[\mathbf{I} - g(\mathbf{y}(t))\mathbf{y}^T(t) \right] \mathbf{W}(t), \tag{15.11}$$

where $\eta > 0$ and $g(\mathbf{y}) = (g_1(y_1), \ldots, g_n(y_n))^T$, each element of which corresponds to the negative score function, i.e., $g_i(y_i) = -d \log p_i(y_i)/dy_i$. Function $g(\cdot)$ is given by the polynomial $g(y) = \frac{29}{4}y^3 - \frac{47}{4}y^5 - \frac{14}{3}y^7 + \frac{25}{4}y^9 + \frac{3}{4}y^{11}$.

Differential ICA [32] is a variation of natural gradient ICA, where learning relies on the concurrent change of output variables. Differential learning is interpreted as the ML estimation of parameters with latent variables represented by the random walk model. The differential anti-Hebb rule is a modification of the anti-Hebb rule. It updates the synaptic weights in a linear feedback network in such a way that the concurrent change of neurons is minimized.

The relative gradient [23] or the natural gradient [7] is efficient in learning the parameters when the parametric space belongs to the Riemannian manifold. The relative gradient leads to algorithms having the equivariant property which produces the uniform performance, regardless of the condition of the mixing matrix in the task of BSS or ICA. When the mixing matrix is ill-conditioned, the relative gradient ICA algorithms outperform other types of algorithms.

15.4.3 FastICA Algorithm

FastICA is a well-known fixed-point ICA algorithm [53, 57]. It is derived from the optimization of the kurtosis or the negentropy measure by using Newton's method. FastICA achieves reliable and at least quadratic convergence. It can be considered as a fixed-point algorithm for ML estimation of the ICA model. FastICA is parallel, distributed, computationally simple, and requires little memory space.

FastICA estimates multiple independent components one by one using a GSO-like deflation scheme. It first prewhitens the observed data to remove any second-order correlations, and then performs an orthogonal rotation of the whitened data to find the directions of the sources. FastICA is very simple, does not depend on any user-defined parameters, and rapidly converges to the most accurate solution allowed by the data. The algorithm finds, one at a time, all non-Gaussian independent components, regardless of their probability distributions. It is performed in either batch mode or a semiadaptive manner. The convergence of the algorithm is rigorously proved, and the convergence speed is shown to be cubic.

The original FastICA [53] based on kurtosis nonlinearity is nonrobust due to the sensitivity of the sample fourth-order moment to outliers. Consequently, nonlinearities are offered as more robust choices [57]. A rigorous statistical analysis of the deflation-based FastICA estimator is provided in [103]. The derived compact closed-form expression of the influence function reveals the vulnerability of the FastICA estimator to outliers regardless of the nonlinearity used. The influence function allows the derivation of a compact closed-form expression for the asymptotic covariance matrix of the FastICA estimator and subsequently its asymptotic relative efficiencies.

The mixtures $\boldsymbol{x}_t$ are first prewhitened according to Appendix A.2

$$\boldsymbol{v}(t) = \mathbf{V}^T \boldsymbol{x}_t, \tag{15.12}$$

where $\boldsymbol{v}(t)$ is the whitened mixture and $\mathbf{V}$ denotes a $J_1 \times J_2$ whitening matrix. The components of $\boldsymbol{v}(t)$ are mutually uncorrelated with unit variances, namely, $\mathrm{E}\left[\boldsymbol{v}(t)\boldsymbol{v}^T(t)\right] = \mathbf{I}_{J_2}$.

The demixing matrix $\mathbf{W}$ is factorized by

$$\mathbf{W}^T = \mathbf{U}^T \mathbf{V}^T, \tag{15.13}$$

where $\mathbf{U} = \left[\boldsymbol{u}_1, \ldots, \boldsymbol{u}_{J_2}\right]$ is the $J_2 \times J_2$ orthogonal separating matrix, that is, $\mathbf{U}^T\mathbf{U} = \mathbf{I}_{J_2}$. The vectors $\boldsymbol{u}_i$ can be obtained by iteration:

$$\widetilde{\boldsymbol{u}}_i = \mathrm{E}\left[\boldsymbol{v} g\left(\boldsymbol{u}_i^T \boldsymbol{v}\right)\right] - \mathrm{E}\left[\dot{g}\left(\boldsymbol{u}_i^T \boldsymbol{v}\right)\right] \boldsymbol{u}_i, \quad i = 1, \ldots, J_2, \tag{15.14}$$

$$\boldsymbol{u}_i = \frac{\widetilde{\boldsymbol{u}}_i}{\|\widetilde{\boldsymbol{u}}_i\|}, \quad i = 1, \ldots, J_2, \tag{15.15}$$

where $g(\cdot)$ can be selected as $g_1(x) = \tanh(ax)$ and $g_2(x) = x\mathrm{e}^{-\frac{x^2}{2}}$. The independent components can be estimated in a hierarchical fashion, that is, estimated one by one. After the ith independent component is estimated, $\boldsymbol{u}_i$ is orthogonalized by an orthogonalization procedure.

FastICA can also be implemented in a symmetric mode, where all the independent components are extracted and orthogonalized at the same time [57]. A similar fixed-point algorithm based on the nonstationary property of signals is derived in [60]. FastICA is easy to use and there are no step-size parameters to choose, while gradient-descent-based algorithms seem to be preferable only if fast adaptivity in a changing environment is required. FastICA directly finds the independent components of practically any non-Gaussian distribution using any nonlinearity $g(\cdot)$ [57].

For FastICA, Hyvarinen suggested three different functions [57]:

$$g_4(s) = \log\cosh(s), \tag{15.16}$$

$$g_5(s) = -\exp(-s^2/2), \tag{15.17}$$

$$g_6(s) = s^4/4. \tag{15.18}$$

g_4 is a good general-purpose function and g_5 is justified if robustness is very important. For sources of fixed variance, the maximum of g_6 coincides with the maximum of kurtosis. All these contrast functions can be viewed as approximations of negentropy. Cumulant-based approximations, such as g_6, are mainly sensitive to the tails of the distributions, and thus sensitive to outliers as well [54].

In [48], a comprehensive experimental comparison has been conducted on different classes of ICA algorithms including FastICA, infomax, natural gradient, EASI, and an RLS-based nonlinear PCA [70]. The fixed-point FastICA with symmetric orthogonalization and tanh nonlinearity $g_1(x)$ is concluded as the best trade-off for ICA since it provides results similar to that of infomax and natural gradient, which are optimal with respect to minimizing the mutual information, but with a clearly smaller computational load. When $g(x) = g_3(x) = x^3$, the fixed-point FastICA algorithm achieves cubic convergence; however, the algorithm is less accurate than the case when tanh nonlinearity is used [48].

FastICA can approach the Cramer–Rao lower bound in two situations [79], namely, when the distribution of the sources is nearly Gaussian and the algorithm is in symmetric mode using the nonlinear function $g_1(x)$, $g_2(x)$, or $g_3(x)$, and when the distribution of the sources is very different from Gaussian and the nonlinear function equals the score function of each independent component. A closed-form expression for the Cramer–Rao bound on estimating the source signals in the linear ICA problem is derived in [79], assuming that all independent components have finite variance. An asymptotic performance analysis of FastICA in [120] derives the exact expression for this error variance. The accuracy of FastICA is very close, but not equal to the Cramer–Rao bound. The condition for this is that the nonlinearity $g(\cdot)$ in the FastICA contrast function is the integral of the score function $\psi(s)$ of the original signals, or the negative log density

$$g(s) = \int \psi(s)ds = -\int \frac{p_i'(s)}{p_i(s)}ds = -\log p_i(s). \tag{15.19}$$

Efficient FastICA [80] improves FastICA, and it can attain the Cramer–Rao bound. This result is rigorously proven under the assumption that the probability distribution of the independent signal components belongs to the class of generalized Gaussian distributions with parameter α, denoted GG(α) for $\alpha > 2$. The algorithm is about three times faster than that of symmetric FastICA.

FastICA can be implemented in the deflation (or sequential extraction) and the symmetric (or simultaneous extraction) modes. In the deflation mode, the constraint of uncorrelatedness with the previously found sources is required to prevent the algorithm from converging to previously found components. In the symmetric mode, the components are estimated simultaneously. The deflation-based FastICA can estimate a single or a subset of the original independent components one by one, with reduced computational load, but errors can accumulate in successive deflation stages. Symmetric FastICA recovers all source signals simultaneously [53]. It is widely used in practice for BSS, due to its good accuracy and convergence speed. This algorithm shows local quadratic convergence to the correct solution with a generic cost function. For the kurtosis cost function, the convergence is cubic. Thus, the one-unit behavior generalizes to the parallel case as well. The chosen nonlinearity in the cost function has very little effect on the behavior. The score function of the sources, which is optimal for ML and minimum entropy criteria for ICA, seems not to offer any advantages for the speed of convergence. However, the true score function does minimize the residual error in the finite sample case [79].

The local convergence properties of FastICA have been investigated under a general setting in [101]. Unfortunately, for FastICA, the major difficulty of local convergence analysis is due to the well-known sign-flipping phenomenon of FastICA, which causes the discontinuity of the corresponding FastICA map on the unit sphere, and the approach taken is not mathematically rigorous. There have been a few attempts to generalize FastICA to solve the problem of independent subspace analysis [63]. In [116], by using the mathematical concept of principal fiber bundles, FastICA is proven to be locally quadratically convergent to a correct separation. Higher order local convergence properties of FastICA are also investigated in the framework of a scalar shift strategy. As a parallelized version of FastICA, QR FastICA [116], which employs the GSO process instead of the polar decomposition, shares similar local convergence properties with FastICA.

The Huber M-estimator cost function is introduced as a contrast function for use within prewhitened BSS algorithms such as FastICA [39]. Key properties regarding the local stability of the algorithm for general non-Gaussian source distributions are established, and its separating capabilities are shown through analysis to be insensitive to the threshold parameter. The use of the Huber M-estimator cost as a criterion for successful separation of large-scale and ill-conditioned signal mixtures with reduced dataset requirements.

A family of flexible score functions for BSS is based on the family of generalized gamma densities. Flexible FastICA [78] uses FastICA to blindly extract the independent source signals, while an efficient ML-based method is used to adaptively estimate the parameters of such score functions. A FastICA algorithm suitable for the separation of quaternion-valued signals from an observed linear mixture is proposed in [66].

Example 15.1 Assume that we have five independent signals, as shown in Fig. 15.2a. If we have more than five input sources that are obtained by linearly mixing the five sources, by using the ICA procedure, only five independent sources can be obtained. Figure 15.2b illustrates $J_1 = 6$ sources that are obtained by linearly mixing the five sources. We apply FastICA. After the iteration, $\mathbf{WA}$ becomes a 5×5 identity matrix. The separated sources are shown in Fig. 15.2c. The separated $J_2 = 5$ signals are very close to the original independent signals, and there are ambiguities in the order of the separated independent components and in the amplitude of some independent components. In comparison with the original signals, some separated signals are multiplied by -1. When $J_1 > J_2$, ICA can be used for both BSS and feature extraction.

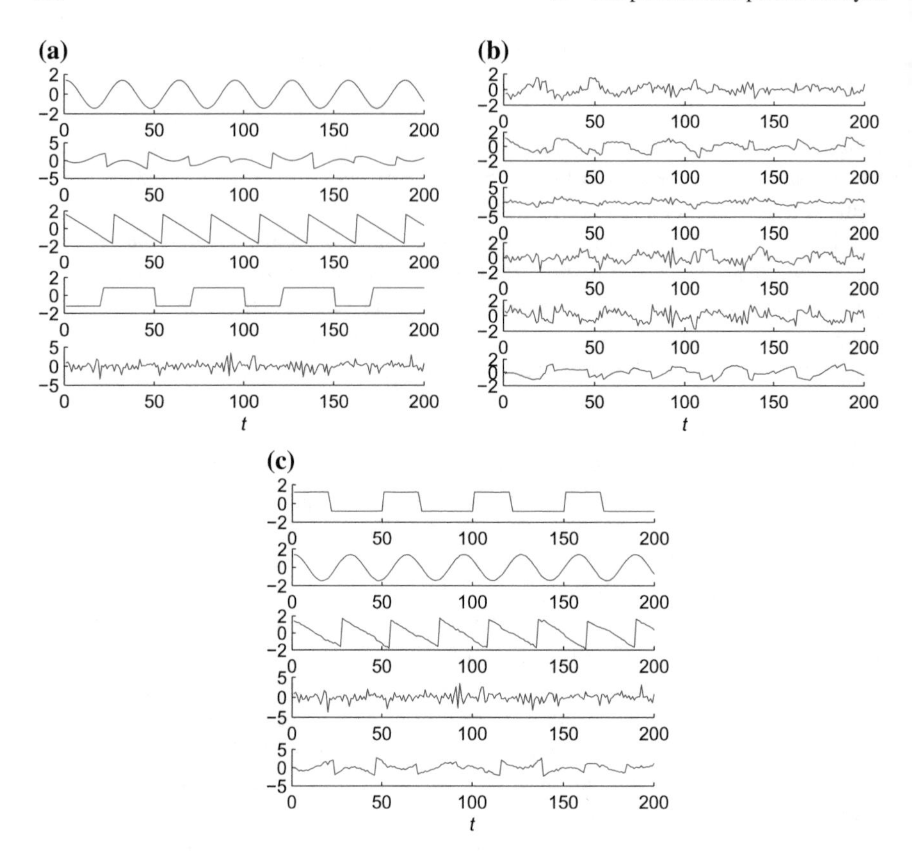

Fig. 15.2 An example of ICA using FastICA. **a** Five independent sources. **b** Mixed sources. **c** Separated sources

Example 15.2 We use the MATLAB package (http://research.ics.aalto.fi/ica/imageica/ [59]) for estimating ICA basis windows from image data. The statistical analysis is performed on 13 natural grayscale images. The model size is set as 128 and we apply FastICA to get the ICA model. We implement ICA estimation for 200 iterations. For each input image patch $\boldsymbol{x}$, $\boldsymbol{x} = \mathbf{A}\boldsymbol{s} = [\boldsymbol{a}_1^T, \boldsymbol{a}_2^T, \ldots, \boldsymbol{a}_{128}^T]^T (s_1, s_2, \ldots, s_{128})^T$. The basis row vector $\boldsymbol{a}_i$ are represented by a basis window of the same size of image patch, are localized both in space and in frequency, resembling the wavelets. The obtained basis windows of the model are shown in Fig. 15.3. The obtained basis windows can be used for image coding.

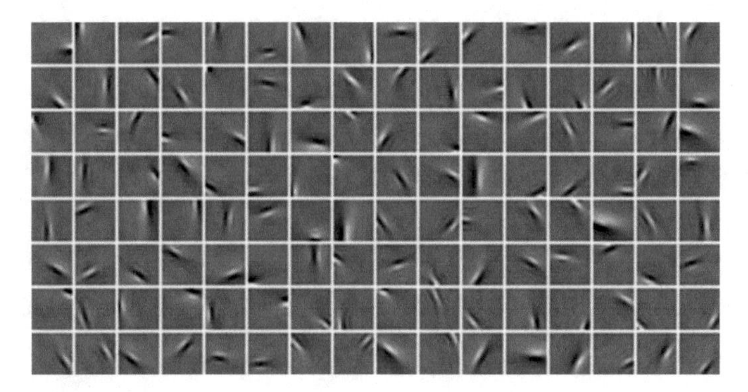

Fig. 15.3 The basis windows of the ICA model

15.5 ICA Networks

ICA does not require a nonlinear network for linear mixtures, but its basis vectors are usually nonorthogonal and the learning algorithm must contain some nonlinearities for higher order statistics. A three-layer J_1-J_2-J_1 linear autoassociative network that is used for PCA can also be used as an ICA network, as long as the outputs of the hidden layer are independent. For the ICA network, the weight matrix between the input and hidden layers corresponds to the $J_1 \times J_2$ demixing matrix $\mathbf{W}$, and the weight matrix from the hidden to the output layer corresponds to the $J_2 \times J_1$ mixing matrix $\mathbf{A}$. In [69], $\mathbf{W}$ is further factorized into two parts according to (15.13) and the network becomes a four-layer J_1-J_2-J_2-J_1 network, as shown in Fig. 15.4. The weight matrices between the layers are, respectively, $\mathbf{V}$, $\mathbf{U}$, and $\mathbf{A}^T$. $y_i, i = 1, \ldots, J_2$, is the ith output of the bottleneck layer, and $\hat{x}_j, j = 1, \ldots, J_1$, is the estimate of x_j.

Each of the three weight matrices performs one of the processing tasks required for ICA, namely, whitening, separation, and estimation of the basis vectors of ICA. It can be used for both BSS and estimation of the basis vectors of ICA, which is useful, for example, in projection pursuit. If the task is merely BSS, the last ICA basis vector estimation layer is not needed.

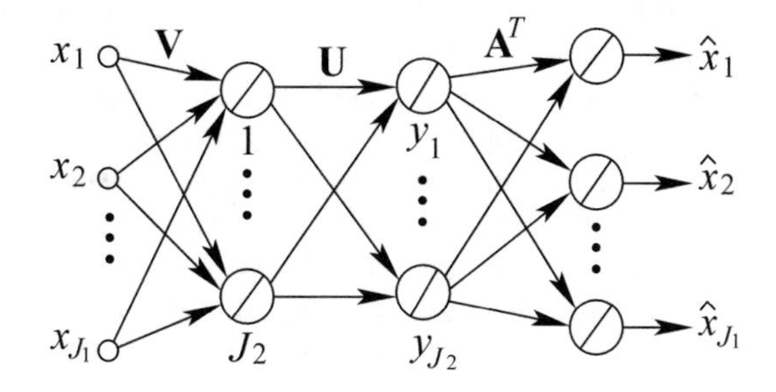

Fig. 15.4 Architecture of the ICA network

The weights between the input and second layers perform prewhitening

$$\boldsymbol{v}(t) = \mathbf{V}^T \boldsymbol{x}_t. \tag{15.20}$$

The weights between the second and third layers perform separation

$$\boldsymbol{y}(t) = \mathbf{U}^T \boldsymbol{v}(t), \tag{15.21}$$

and the weights between the last two layers estimate the basis vectors of ICA

$$\hat{\boldsymbol{x}}(t) = \mathbf{A}\boldsymbol{y}(t). \tag{15.22}$$

When the three-layer linear autoassociative network is used for PCA, we have the relations $\hat{\boldsymbol{x}} = \mathbf{A}\mathbf{W}^T\boldsymbol{x}$ and $\mathbf{A} = \mathbf{W}(\mathbf{W}^T\mathbf{W})^{-1}$ [12]. For PCA, $\mathbf{W}^T\mathbf{W} = \mathbf{I}_{J_2}$, thus $\hat{\boldsymbol{x}} = \mathbf{W}\mathbf{W}^T\boldsymbol{x}$. The ICA solution can be obtained by imposing the additional constraint that the components of the bottleneck layer output vector $\boldsymbol{y} = \mathbf{W}^T\boldsymbol{x}$ must be mutually independent or as independent as possible. For ICA, $\hat{\boldsymbol{x}} = \mathbf{A}\mathbf{W}^T\boldsymbol{x} = \mathbf{A}\left(\mathbf{A}^T\mathbf{A}\right)^{-1}\mathbf{A}^T\boldsymbol{x}$ is the LS approximation [69]. Notice that $\mathbf{A}$ and $\mathbf{W}$ are the pseudoinverses of each other.

Prewhitening

By performing prewhitening, we have

$$\mathbf{V}^T = \Lambda^{\frac{1}{2}}\mathbf{E}^T, \tag{15.23}$$

where $\Lambda = \text{diag}\left(\lambda_1, \ldots, \lambda_{J_2}\right)$, $\mathbf{E} = \left[\boldsymbol{c}_1, \ldots, \boldsymbol{c}_{J_2}\right]$, with λ_i and $\boldsymbol{c}_i$ as the ith largest eigenvalue and the corresponding eigenvector of the covariance matrix $\mathbf{C}$. PCA can be applied to solve for the eigenvalues and eigenvectors.

A simple local algorithm for learning the whitening matrix is given by [23]

$$\mathbf{V}(t+1) = \mathbf{V}(t) - \eta(t)\mathbf{V}(t)\left[\boldsymbol{v}(t)\boldsymbol{v}^T(t) - \mathbf{I}\right]. \tag{15.24}$$

It is used as part of the EASI separation algorithm [23]. This algorithm does not have any optimality properties in data compression, and it sometimes suffers from stability problems. The validity of the algorithm can be justified by observing the whiteness condition $\mathrm{E}\left[\boldsymbol{v}(t)\boldsymbol{v}^T(t)\right] = \mathbf{I}_{J_2}$ after convergence.

Prewhitening usually makes separation algorithms converge faster and often have better stability properties. However, if the mixing matrix $\mathbf{A}$ is ill-conditioned, whitening can make separation of sources more difficult or even impossible [23, 69].

β-prewhitening [95] minimizes the empirical β-divergence over the space of all Gaussian distributions. β-divergence reduces to Kullback–Leibler divergence when $\beta \to 0$. β-prewhitening with $\beta = 0$ is equivalent to standard prewhitening, if the dataset is not seriously corrupted by noises or outliers. For datasets seriously corrupted by noise or outliers, β-prewhitening with $\beta > 0$ is much better than standard prewhitening.

Separating Algorithms

The separating algorithms can be based on robust PCA, nonlinear PCA, or bigradient nonlinear PCA [69]. $\mathbf{W}$ can also be calculated iteratively without prewhitening as in EASI [23] or generalized EASI [69]. The nonlinear PCA algorithm is given as [69]

$$\mathbf{U}(t+1) = \mathbf{U}(t) + \eta(t)\left[\boldsymbol{v}(t) - \mathbf{U}(t)g(\boldsymbol{y}(t))\right] g\left(\boldsymbol{y}^T(t)\right), \tag{15.25}$$

where $g(\cdot)$ is usually selected to be odd for stability and separation reasons, such as $g(t) = t^3$, and $\eta > 0$ slowly reduces to zero or is a small constant.

Estimation of Basis Vectors

The estimation of the basis vectors can be based on the LS solution

$$\hat{\boldsymbol{x}}(t) = \widehat{\mathbf{A}}\boldsymbol{y}(t) = \mathbf{W}\left(\mathbf{W}^T\mathbf{W}\right)^{-1}\boldsymbol{y}(t), \tag{15.26}$$

where $\widehat{\mathbf{A}} = \left[\hat{\boldsymbol{a}}_1, \ldots, \hat{\boldsymbol{a}}_{J_2}\right] = \mathbf{W}\left(\mathbf{W}^T\mathbf{W}\right)^{-1}$ is a $J_1 \times J_2$ matrix. If prewhitening, (15.23) is applied, $\widehat{\mathbf{A}}$ can be simplified as

$$\widehat{\mathbf{A}} = \mathbf{E}\Lambda^{\frac{1}{2}}\mathbf{U}. \tag{15.27}$$

Thus, the unnormalized ith basis vector of ICA is $\hat{\boldsymbol{a}}_i = \mathbf{E}\Lambda^{\frac{1}{2}}\boldsymbol{u}_i$ where $\boldsymbol{u}_i$ is the ith column of $\mathbf{U}$, and its squared norm becomes $\left\|\hat{\boldsymbol{a}}_i\right\|^2 = \boldsymbol{u}_i^T\Lambda\boldsymbol{u}_i$. Local algorithms for estimating the basis vectors can be derived by minimizing the MSE $E[\|\boldsymbol{x} - \mathbf{A}\boldsymbol{y}\|^2]$ using the gradient-descent method [69]

$$\mathbf{A}(t+1) = \mathbf{A}(t) + \eta\boldsymbol{y}(t)\left[\boldsymbol{x}_t^T - \boldsymbol{y}^T(t)\mathbf{A}(t)\right]. \tag{15.28}$$

For any of the last three layers of the ICA network, it is possible to use either a local or a nonlocal learning method.

The quality of separation can be measured in terms of the performance index defined as [5] follows:

$$J = \sum_{k=1}^{p}\left(\sum_{l=1}^{p}\frac{|c_{kl}|}{\max_n |c_{kn}|} - 1\right) + \sum_{l=1}^{p}\left(\sum_{k=1}^{p}\frac{|c_{kl}|}{\max_n |c_{nl}|} - 1\right), \tag{15.29}$$

where $\mathbf{C} = [c_{kl}] = \mathbf{WA}$. The performance index is always nonnegative, and zero value means perfect separation.

15.6 Some BSS Methods

15.6.1 Nonlinear ICA

ICA algorithms discussed so far are linear ICA methods for separating original sources from linear mixtures. Blind separation of the original signals in nonlinear mixtures has many difficulties such as intrinsic indeterminacy, the unknown distribution of the sources as well as the mixing conditions, and the presence of noise. It is impossible to separate the original sources using only the source independence assumption of some unknown nonlinear transformations of the sources [56]. In the nonlinear case, however, ICA has an infinite number of solutions that are not related in any simple way to one another [56]. Nonlinear BSS is an ill-posed problem: further knowledge is applied to the problem through a suitable form of regularization. In spite of these difficulties, some methods are effective for nonlinear BSS. Nonlinear ICA can be modeled by a parameterized neural network whose parameters can be determined under the criterion of independence of its outputs.

The inverse of the nonlinear mixing model can be modeled by using the three-layer MLP [2, 19] or the RBF network [119]. SOM provides a parameter-free method to the nonlinear ICA problem [51], but suffers from the exponential growth of the network complexity with the dimensions of the output lattice.

The minimal nonlinear distortion principle [132] tackles the ill-posedness of nonlinear ICA problems. It prefers the nonlinear ICA solution that is as close as possible to the linear solution, among all possible solutions. It also helps to avoid local optima in the solutions. To achieve minimal nonlinear distortion, a regularization term is exploited to minimize the MSE between the nonlinear mixing mapping and the best-fitting linear one.

15.6.2 Constrained ICA

ICA is an ill-posed problem because of the indeterminacy of scaling and permutation of the solution. The recovered independent components can have an arbitrary permutation of the original sources, and the estimated independent components may also be dilated from the originals [34]. Incorporation of prior knowledge and further requirements converts the ill-posed ICA problem into a well-posed problem. In some BSS applications, knowledge about the sources or the mixing channels may be available: for example, statistical properties of speeches or physical distances between the location of the microphones and the speakers in the cocktail-party problem.

Constrained ICA is a framework that incorporates additional requirements and prior information in the form of constraints into the ICA contrast function [90]. The approach given in [89] sorts independent components according to some statistic and normalizes the demixing matrix or the energies of separated independent components. With some prior knowledge, the algorithm is able to identify and extract the original sources perfectly from their mixtures. Adaptive solutions using Newton-like learning are given in [90].

ICA with reference [91] extracts an interesting subset of independent sources from their linear mixtures when some *a priori* information of the sources is available in the form of rough templates (references), in a single process. A neural algorithm is proposed using a Newton-like approach to obtain an optimal solution to the constrained optimization problem. ICA with reference converges at least quadratically.

Constrained ICA can be formulated in mutual information terms directly [122]. As an estimate of mutual information, a robust version of the Edgeworth expansion is used, on which gradient descent is performed. Another way of constraining ICA has been introduced in [1]. ICA is applied to two datasets separately. The corresponding dependent components between the two sets are then determined using a natural gradient-type learning rule, thus performing an ICA-style generalization of CCA.

Prior information on the sparsity of the mixing matrix can be a constraint [62]. In a biological interpretation, sparseness of mixing matrix means sparse connectivity of the neural network. The criterion functions for ICA can be modified if prior information about the entries of the mixing matrix is available [64].

15.6.3 Nonnegativity ICA

The constraint of nonnegative sources, perhaps with an additional constraint of nonnegativity on the mixing matrix **A**, is often known as NMF. We refer to the combination of nonnegativity and independence assumptions on the sources as nonnegative ICA.

Nonnegative PCA and nonnegative ICA algorithms are given in [107], where the sources $\boldsymbol{s}_i$ are assumed to be nonnegative. The nonnegative ICA algorithms [107] are based on a two-stage process common to many ICA algorithms, prewhitening, and rotation. However, instead of using the usual non-Gaussianity measures such as kurtosis in the rotation stage, the nonnegativity constraint is used.

Some algorithms for nonnegative ICA [100, 107] are based on the assumption that the sources s_i are well grounded except for independence and nonnegativity. We call a source s_i well grounded if $\Pr(s_i < \epsilon) > 0$ for any $\epsilon > 0$, i.e., s_i has nonzero pdf all the way down to zero. However, many real-world nonnegative sources are not well grounded, e.g., images.

In [100], a gradient algorithm is derived from a cost function whose minimum coincides with nonnegativity under the whitening constraint, under which the separating matrix is orthogonal. In the Stiefel manifold of orthogonal matrices, the cost function is a Lyapunov function for the matrix gradient flow, implying global convergence [100]. A nonnegative PCA algorithm has good performance of separation [100], and a discrete-time version of the algorithm developed is shown to be globally convergent under certain conditions [129]. Nonnegative ICA proposed in [133] can work efficiently even when the source signals are not well grounded. This method is insensitive to the particular underlying distribution of the source data.

Stochastic nonnegative ICA method of [10] minimizes mutual information between recovered components by using a nonnegativity-constrained simulated annealing algorithm. Convex analysis of mixtures of nonnegative sources [28] has been theoretically proven to achieve perfect separation by searching for all the extreme points of an observation-constructed polyhedral set. In [124], a joint correlation function of multiple signals confirms that the observations after nonnegative mixing would have higher joint correlation than the original unknown sources. Accordingly, a nonnegative least-correlated component analysis method designs the unmixing matrix by minimizing the joint correlation function among the estimated nonnegative sources. The general algorithm is developed based on an iterative volume maximization principle and LP. The source identifiability and required conditions are discussed and proven [124].

15.6.4 ICA for Convolutive Mixtures

ICA for convolutive mixtures is computationally expensive for long FIR filters because it includes convolution operations. Time-domain BSS applies ICA directly to the convolutive mixture model [6, 73]. The approach achieves good separation once the algorithm converges since ICA correctly evaluates the independence of separated signals.

Frequency-domain BSS applies complex-valued ICA for instantaneous mixtures in each frequency bin [96, 115, 117]. ICA can be performed separately at each frequency. Also, any complex-valued instantaneous ICA algorithm can be employed. However, frequency-domain BSS involves a permutation problem: the permutation ambiguity of ICA in each frequency bin should be aligned so that a separated signal in the time domain contains frequency components of the same source signal. A robust and precise method is presented in [113] for solving the permutation problem, based on two approaches: direction of arrival (DoA) estimation for sources and the interfrequency correlation of signal envelopes.

Independent vector analysis solves frequency-domain BSS effectively without suffering from the permutation problem between the frequencies by utilizing dependencies of frequency bins. It performs successfully under most conditions including the ill-posed condition such as the case where the mixing filters of the sources are very similar [75].

In an ML spatiotemporal BSS algorithm [52], the temporal dependencies are analyzed by assuming that each source is an autoregressive process and the distribution is described using a mixture of Gaussians. Optimization is performed by using the EM method to maximize the likelihood, and the update equations have a simple, analytical form. The method has excellent performance for artificial mixtures of real audio.

15.6.5 Other BSS/ICA Methods

Independent subspace analysis [108] is a generalization of ICA. There are several groups of sources; within each such group (or subspace) the sources may have dependencies, but groups are mutually independent. ICA is a case of independent subspace analysis with only one source per subspace.

The boundedness property is utilized as an additional assumption in ICA framework [106]. The mutual information cost function is reformulated in terms of order statistics. In the bounded case, this formulation leads to the effective minimization of the separator output ranges. Under the boundedness constraint on sources, boundedness can be utilized to replace mutual statistical independence assumption with a weaker assumption [37]. Bounded component analysis [37] enables separation of independent and dependent (even correlated) sources. The bounded component analysis approach proposed in [40] enables separation of both independent and dependent, including correlated, bounded sources from their instantaneous mixtures. As an extension of instantaneous bounded component analysis [40], a family of convolutive bounded component analysis criteria and corresponding algorithms are introduced in [65]. The algorithms are capable of separating not only the independent sources but also the sources that are dependent/correlated in both component (space) and sample (time) dimensions.

NMF is a BSS problem that assumes the sources and the mixing matrix being nonnegative. NMF-based methods can deal with sources that are either independent or spatially correlated. However, these methods have local minima problem that results from using the alternate LS iteration scheme, and thus BSS is guaranteed.

Synchronous source separation is a linear and instantaneous BSS problem assuming that all the sources are fully synchronous. This problem cannot be addressed through ICA methods because synchronous sources are statistically dependent. Independent phase analysis [3] and phase-locked matrix factorization [4] are two-step algorithms for synchronous source separation. Independent phase analysis performs well in the noiseless case and with moderate levels of added Gaussian white noise. In comparison, phase-locked matrix factorization has no singular solutions and can deal with higher amounts of noise and with nonsquare mixing matrices (more measurements than sources).

A direct application of the trust-region method to ICA can be found in [29]. Relative trust-region learning [30] jointly exploits the trust-region method and relative optimization. The method finds a direction and a step size with the help of a quadratic model of the objective function and updates parameters in a multiplicative fashion. The resulting relative trust-region ICA algorithm achieves a faster convergence than the relative gradient and even Newton-type algorithms do. FastICA seemed a little bit faster than relative trust-region ICA. However, in the case of small datasets, FastICA does not work well, whereas relative trust-region ICA works fine. Moreover, in an ill-conditioned high-dimensional dataset, the relative trust-region converges much faster than FastICA does.

When a linear mixture of independent sources is contaminated by multiplicative noise, the problems of BSS and feature extraction are highly complex. This is called the *multiplicative ICA model*. This noise commonly exists in coherent images as ultrasound, synthetic aperture radar or laser images. The approach followed by ICA does not produce proper results, because the output of a linear transformation of the noisy data cannot be independent. However, the statistic of this output possesses a special structure that can be used to obtain the original mixture. In [15], this statistical structure is studied and a general approach to solving the problem is stated. The FMICA method [16] obtains the unmixing matrix from a linear mixture of independent sources in the presence of multiplicative noise, without any limitation in the nature of the sources or the noise. The statistical structure of a linear transformation of the noisy data is studied up to the fourth order, and then this structure is used to find the inverse of the mixing matrix through the minimization of a cost function.

A recurrent network is described for performing robust BSS in [33].

RADICAL (Robust, Accurate, Direct ICA aLgorithm) is an ICA algorithm based on an efficient entropy estimator [83]. It directly minimizes the measure of departure from independence according to the estimated Kullback–Leibler divergence between the joint distribution and the product of the marginal distributions. In particular, the entropy estimator used is consistent and exhibits rapid convergence. The estimator's relative insensitivity to outliers translates into superior performance by RADICAL on outlier tests.

RobustICA [130] is a simple deflationary ICA method. It consists of performing exact line search optimization of the kurtosis contrast function. RobustICA can avoid prewhitening and deals with real- and complex-valued mixtures of possibly noncircular sources alike. The algorithm targets sub-Gaussian or super-Gaussian sources in the order specified by the user. RobustICA proves faster and more efficient than FastICA with asymptotic cubic global convergence. In the real-valued two-signal case, the algorithm converges in a single iteration.

The minimax mutual information ICA algorithm [41] is an efficient and robust ICA algorithm motivated by the maximum entropy principle. The optimality criterion is the minimum output mutual information, where the estimated pdfs are from the exponential family and are approximate solutions to a constrained entropy maximization problem. This approach yields an upper bound for the actual mutual information of the output signals. One approach that is commonly taken in designing information-theoretic ICA algorithms is to use some form of polynomial expansion to approximate the pdf of the signals [5, 34, 54]. A sequential, fixed-point subspace ICA algorithm given in [121] is based on the gradient of a robust version of the Edgeworth expansion of mutual information and a constrained ICA version is also based on goal programming of mutual information objectives.

In [26], the connections between mutual information, entropy, and non-Gaussianity in a larger framework are explored without resorting to a somewhat arbitrary decorrelation constraint. A key result is that the mutual information can be decomposed, under linear transforms, as the sum of two terms: one term expressing the decorrelation of the components and the other expressing their non-Gaussianity.

L_1-norm PCA can perform ICA under the whitening assumption [92]. ICA can actually be performed by L_1-PCA after L2-norm whitening. However, when the source probability distributions fulfill certain conditions, the L_1-norm criterion needs to be minimized rather than maximized, which can be accomplished by simple modifications on existing optimal algorithms for L_1-PCA. If the sources have symmetric distributions, that L_1-PCA is linked to kurtosis optimization [92]. ICA can be performed using optimal algorithms for L_1-PCA with guaranteed global convergence while inheriting the increased robustness to outliers of the L_1-norm criterion.

Second-order methods for BSS cannot be easily implemented by neural models. Results on the nonlinear PCA criterion in BSS [71] clarify how the nonlinearity should be chosen optimally. The connections of the nonlinear PCA learning rule with the infomax algorithm and the adaptive EASI algorithm are also discussed in [71]. A nonlinear PCA criterion can be minimized using LS approaches, leading to computationally efficient and fast converging algorithms.

BSS can be achieved by using the sparsity of the sources in the time domain [88], frequency domain [18], or time–frequency domain [74]. Some smooth dense signals often can be very sparse in a transform domain. For example, EEG signals are much more sparse in the frequency domain than in the time domain.

Sparse component analysis [93] allows latent source signals to be recovered by exploiting the sparsity of the source signals and/or the mixing matrix. ICA of sparse signals (sparse ICA) may be done by a combination of a clustering algorithm and PCA [11]. The final algorithm is easy to implement for any number of sources. This, however, requires an exponential growing of the sample number as the number of sources increases.

Localized ICA is used to characterize nonlinear ICA [72]. Clustering is first used for an overall coarse nonlinear representation of the underlying data and linear ICA is then applied in each cluster so as to describe local features of the data. The data are grouped in several clusters based on the similarities between the observed data ahead of the preprocessing of linear ICA using some clustering algorithms. This leads to a better representation of the data than in linear ICA in a computationally feasible manner.

In practice, the estimated independent components are often not at all independent. This residual dependence structure could be used to define a topographic order for the components [61]. A distance between two components could be defined using their higher order correlations, and this distance could be used to create a topographic representation. Topographic ICA obtains a linear decomposition into approximately independent components. It can be considered a generalization of independent subspace analysis [59].

A linear combination of the separator output fourth-order marginal cumulants (kurtoses) is a valid contrast function for eliminating the permutation ambiguity of ICA with prewhitening. The analysis confirms that the method presented in [34], despite arising from the mutual information principle, presents ML-optimality features [131].

The BSS problem can be solved in a single step by the CCA approach [114]. With CCA, the objective is to find a transformation matrix, which when applied to the mixtures, maximizes the autocorrelation of each of the recovered signals.

Separation of independent sources using ICA requires prior knowledge of the number of independent sources. In the overcomplete situation, performing ICA can result in incorrect separation and poor quality. Undercomplete situation is often encountered for applications such as sensor networks, where the numbers of sensors may often exceed the number of components such as in sensor networks for environmental or defense monitoring, or when the components are not independent. Normalized determinant of the global matrix $|\mathbf{G}| = |\mathbf{WA}|$ is a measure of the number of independent sources in a given mixture, N, in a mixture of M recordings [97].

In [104], a contrast for BSS of natural signals is proposed, which measures the algorithmic complexity of the sources and also the complexity of the mixing mapping. The approach can be seen as an application of the MDL principle. The complexity is then taken as the length of the compressed signal in bits. No assumption about underlying pdfs of the sources is necessary. Instead, it is required that the independent source signals have low complexity, which is generally true for natural signals. Minimum mutual information coincides with minimizing complexity in a special case. The complexity minimization method gives clearly more accurate results for separating correlated signals than the reference method utilizing ICA does. It can be applied to nonlinear BSS and nonlinear exploratory projection pursuit.

15.7 Complex-Valued ICA

ICA for separating complex-valued sources is needed for convolutive source separation in the frequency domain, or for performing source separation on complex-valued data, such as fMRI or radar data. Split complex infomax [117] uses nonanalytic nonlinearity since the real and imaginary values are split into separate channels. Fully complex infomax [20] simply uses an analytic (and hence unbounded) complex nonlinearity for infomax for processing complex-valued sources. When compared to split complex approaches, the shape of the performance surface is improved resulting in better convergence characteristics.

In the complex ICA model, all sources s_j are zero mean and have unit variance with uncorrelated real and imaginary parts of equal variance. That is, $\mathrm{E}[\boldsymbol{s}\boldsymbol{s}^H] = \mathbf{I}$ and $\mathrm{E}[\boldsymbol{s}\boldsymbol{s}^T] = \mathbf{O}$. For algorithms such as JADE, the extension to the complex case is straightforward due to the algorithm's use of fourth-order cumulants. An Edgeworth expansion is used in [34] to approximate negentropy based on third- and fourth-order cumulants, and hence again, it can be relatively easily applied to the complex case. However, using higher order cumulants typically results in an estimate sensitive to outliers.

The theory of complex-weighted network learning has led to effective ICA algorithms for complex-valued statistically independent signals [45, 46]. The ψ-APEX algorithms and GHA are, respectively, extended to the complex-valued case [45,

46]. Based on a suitably selected nonlinear function, these algorithms can be used for BSS of complex-valued circular source signals.

FastICA has been extended to complex-valued sources, leading to c-FastICA [14]. c-FastICA is shown to keep the cubic global convergence property of its real counterpart [110]. c-FastICA, however, is only valid for second-order circular sources. Stability analysis shows that practically any nonquadratic even function can be used to construct a cost function for ICA through non-Gaussianity maximization [55]. This observation is extended to complex sources in c-FastICA by using the cost function [14]

$$J(\boldsymbol{w}) = E\left[g\left(\left|\boldsymbol{w}^H \boldsymbol{z}\right|^2\right)\right], \tag{15.30}$$

where $g(\cdot)$ is a smooth even function, e.g., $g(y) = y^2$.

Recent efforts extend the usefulness of the algorithm to noncircular sources [38, 85, 98, 99]. Complex ICA can be performed by maximization of the complex kurtosis cost function using gradient update, fixed-point update, or Newton update [85]. FastICA is also derived in [38] for the blind separation of complex-valued mixtures of independent, noncircularly symmetric, and non-Gaussian source signals on a kurtosis-based contrast. In [99], the whitened observation pseudocovariance matrix is incorporated into the FastICA update rule to guarantee local stability at the separating solutions even in the presence of noncircular sources. For kurtosis-based nonlinearity, the resulting algorithm bears close resemblance to that derived in [38] through an approach sparing differentiation. Similar algorithms are proposed in [98] through a negentropy-based family of cost functions preserving phase information and thus adapted to noncircular sources. Both a gradient-descent and a quasi-Newton algorithm are derived by using the full second-order statistics, providing superior performance with circular and noncircular sources.

The kurtosis or fourth-order cumulant of a zero-mean complex random variable is defined as a real number [25]

$$\text{kurt}(y) = \text{cum}\left(y, y^*, y, y^*\right) = E\left[|y|^4\right] - 2\left(E\left[|y|^2\right]\right)^2 - \left|E\left[y^2\right]\right|^2, \tag{15.31}$$

and can be shown to be zero for any complex Gaussian variable, circular or noncircular. This result also implies that any source with zero kurtosis will not be separated well under this criterion as well as noncircular Gaussian sources, which can be separated using ML [27] or the strongly uncorrelating transform algorithm [42].

A linear transformation called strong-uncorrelating transform [42] uses second-order statistics information through the covariance and pseudocovariance matrices and performs ICA by joint diagonalization of these matrices. Although efficient, the algorithm restricts the sources to be noncircular with distinct spectra of the pseudo-covariance matrix. It can be viewed as an extension of the conventional whitening transform for complex random vectors. The strong-uncorrelating transform is just the ordinary whitening transform for a second-order circular complex random vector ($\mathrm{E}[\boldsymbol{s}\boldsymbol{s}^T] = \mathbf{O}$). The method can be used as a fast ICA method for complex signals. It is able to separate almost all mixtures, if the sources belong to a class of complex noncircular random variables. Strong-uncorrelating transform is used as a prewhitening step in some ICA algorithms, e.g., in [38].

Extending the theorems proved for the real-valued instantaneous ICA model [34], theorems given in [43] states the conditions for identifiability, separability, and uniqueness of complex-valued linear ICA models. Both circular (proper) and noncircular complex random vectors are covered by the theorems. The conditions for identifiability and uniqueness are sufficient and the separability condition is necessary.

In [87], natural gradient complex ML ICA update rule and its variant with a unitary constraint on demixing matrix, as well as a Newton algorithm are derived. The conditions for local stability are derived using a generalized Gaussian density source model.

Complex ICA by entropy-bound minimization [86] uses an entropy estimator for complex random variables by approximating the entropy estimate using a numerically computed maximum entropy bound and a line search optimization procedure. It has superior separation performance and computational efficiency in separation of complex sources that come from a wide range of bivariate distributions.

Generalized uncorrelating transform [102] is a generalization of the strong-uncorrelating transform [43] based on generalized estimators of the scatter matrix and spatial pseudoscatter matrix. It is a separating matrix estimator for complex-valued ICA when at most one source random variable possess circularly symmetric distribution and sources do not have identical distribution.

NMF, as a BSS approach, has been extended to the complex domain, and is used in the separation of audio signals [68, 76].

15.8 Source Separation for Time Series

Stationary Subspace Analysis

In many settings, the observed signals are a mixture of underlying stationary and nonstationary sources. Stationary subspace analysis decomposes a multivariate time series into its stationary and nonstationary parts [123]. The observed time series $\boldsymbol{x}(t)$ is generated as a linear mixture of stationary source $\boldsymbol{s}^s(t)$ and nonstationary source $\boldsymbol{s}^n(t)$ with a time-constant mixing matrix $\mathbf{A}$,

$$\boldsymbol{x}(t) = \mathbf{A}\boldsymbol{s} = \mathbf{A}\begin{bmatrix} \boldsymbol{s}^s(t) \\ \boldsymbol{s}^n(t) \end{bmatrix}, \tag{15.32}$$

and the objective is to recover these two groups of underlying sources given only samples from $\boldsymbol{x}(t)$. Stationary subspace analysis can be used for change-point detection in high-dimensional time series [17]. The dimensionality of the data can be reduced to the most nonstationary directions, which are most informative for detecting state changes in the time series.

Analytic stationary subspace analysis [50] solves a generalized eigenvalue problem. The solution is guaranteed to be optimal under the assumption that the covariance between stationary and nonstationary sources is time constant. Analytic stationary subspace analysis finds a sequence of projections, ordered by their degree of stationarity. It is more than 100 times faster than the Kullback–Leibler divergence-based method [123].

Slow Feature Analysis

Slow feature analysis [127] was originally developed to model aspects of the visual cortex. It has been successfully applied to different problems such as age estimation.

Slow features encode spatiotemporal regularities, which are information-rich explanatory factors underlying the high-dimensional input streams. Slow feature analysis [126, 127] aims to find a set of scalar functions that generate output signals that vary as slowly as possible. To ensure that the output signals carry significant information about the input, they are required to be uncorrelated and have zero mean and unit variance.

Let $\{\boldsymbol{x}_t\}_{t=1}^n \subset \mathcal{X}$ be a sequence of n observations. Slow feature analysis aims to find a set of mappings $\phi_i : \mathcal{X} \to R$, $i = 1, \ldots, p$, such that $\phi_i(\boldsymbol{x}_t)$ changes slowly over time. The updating complexity is cubic with respect to the input dimensionality.

For a set of mappings $\boldsymbol{\phi} : \mathcal{X} \to R^p$ such that $\phi_i(\boldsymbol{x}(t))$ changes slowly over an observed Markov chain $\{\boldsymbol{x}(t)\}_{t=1}^n \subset \mathcal{X}$. The objective called *slowness* S is defined as the expectation of the squared discrete temporal derivative:

$$\inf_{\phi \in \mathcal{F}^p} \sum_{i=1}^{p} S(\phi_i) = \sum_{i=1}^{p} \tilde{E}_t[\dot{\phi}_i^2(\boldsymbol{x}(t))]. \tag{15.33}$$

To ensure each slow feature encodes unique information and can be calculated in an iterative fashion, the following constraints must hold: $\forall i = 1, \ldots, p$,

$$\tilde{E}_t[\phi_i(\boldsymbol{x}(t))] = 0 \quad \text{(zero mean)}, \tag{15.34}$$

$$\tilde{E}_t[\phi_i^2(\boldsymbol{x}(t))] = 1 \quad \text{(Unit variance)}, \tag{15.35}$$

$$\forall j \neq i : \tilde{E}_t[\phi_i(\boldsymbol{x}(t))\phi_j(\boldsymbol{x}(t))] = 0 \quad \text{(decorrelation)}, \tag{15.36}$$

$$\forall j > i : S(\phi_i) \leq S(\phi_j) \quad \text{(order)}. \tag{15.37}$$

The principle of slowness has been used for a long time in the context of neural networks [47].

Slow feature analysis can be implemented as PCA on the derivative data. That is, assuming derivatives $\dot{\boldsymbol{x}} = (\dot{x}_1, \ldots, \dot{x}_n)$, slow feature analysis minimizes the trace of the covariance of the projection $tr(\mathbf{M}^T \dot{\boldsymbol{x}} \dot{\boldsymbol{x}}^T \mathbf{M})$.

It is possible to reformulate the slowness principle implemented by slow feature analysis in terms of graph embedding, for instance, to incorporate label information into the optimization problem [44].

Slow features can be learned incrementally. Incremental slow feature analysis [81] combines candid covariance-free incremental PCA and covariance-free incremental MCA. It has simple Hebbian and anti-Hebbian updates with a linear complexity in terms of the input dimensionality.

Following Schmidhuber's theory of artificial curiosity, an agent should always concentrate on the area where it can learn the easiest-to-learn set of features that it has not already learned. Using curiosity-driven modular incremental slow feature analysis [82], the agent will learn slow feature representations in order of increasing learning difficulty, under certain mild conditions.

Slow feature analysis is extended as an approach to nonlinear BSS [118]. The algorithm relies on temporal correlations and iteratively reconstructs a set of statistically independent sources from arbitrary nonlinear instantaneous mixtures.

Predictable Feature Analysis

Predictable feature analysis [109, 125] is an unsupervised learning approach that builds up the desired model through autoregressive processes. Inspired by slow feature analysis, the method aims to find the projection of high-dimensional time series into a subspace to extract subsignals that behave as predictable as possible. These predictable features are highly relevant for modeling. High predictability implies low variance in the distribution of the next data point given the previous ones. In slow feature analysis, slowly varying signals can be seen as a special case of predictable features [36].

Forecastable Component Analysis

Forecastable component analysis [49] is based on the idea that predictable signals can be recognized by their low entropy in the power spectrum while white noise would result in a power spectrum with maximal entropy. Assume $\{\boldsymbol{X}_t\}$ to be a stationary second-order process. The method aims to find an extraction vector $\boldsymbol{a}$ such that the projected signals $Y_t = \boldsymbol{a}^T \boldsymbol{X}_t$ are as forecastable as possible, that is, having a low entropy in their power spectrum. Like slow feature analysis, forecastable component analysis has the advantage of being completely model- and parameter-free.

15.9 EEG, MEG, and fMRI

The human brain exhibits relevant dynamics on all spatial scales, ranging from a single neuron to the entire cortex. Extending Hodgkin–Huxley neuron model from a patch of cell membrane to whole neurons and to populations of neurons in order to predict macroscopic signals such as EEG is a dominant focus in this field. Brain–computer interfaces translate brain activities into control signals for devices like computers, robots, and so forth. They have a huge potential in medical and industrial applications for both disabled and normal people, where the learning burden has shifted from a subject to a computer. Experimental and theoretical studies of functional connectivity in humans require noninvasive techniques such as EEG, MEG,

ECG, and fMRI. High-density EEG and/or MEG data model the event-related dynamics of many cortical areas that contribute distinctive information to the recorded signals.

The electrical potentials measured at the scalp constitute EEG. When EEG is time-locked to stimulation—such as the presentation of a word—and averaged over many such presentations, the event-related potential (ERP) is obtained. Several classifiers are applied to ERPs representing the response of individuals to a stream of text designed to be idiosyncratically familiar to different individuals [9]. There are robustly identifiable features of the ERP that enable labeling of ERPs as belonging to individuals with accuracy reliably above chance. Further, these features are stable over time.

EEG and MEG provide the most direct measure of cortical activity with high temporal resolution (1 ms), but with spatial resolution (1–10 cm) limited by the locations of sensors on the scalp. In contrast, fMRI has low temporal resolution (1–10 s), but high spatial resolution (1–10 mm). To the extent that functional activity among brain regions in the cortex may be conceptualized as a large-scale brain network with diffuse nodes, fMRI may delineate the anatomy of these networks, perhaps most effectively in identifying major network hubs. These technologies provide a complete view of dynamical brain activity both spatially and temporally.

EEG data consists of recordings of electrical potentials in many different locations on the scalp. The amplitude of EEG potential is in the range from 10 to 200 μV. In general, the informative features lie within the frequency range of 0.5–45 Hz, composed of delta (0.5–4 Hz), theta (4–8 Hz), alpha (8–12 Hz), beta (12–30 Hz), and gamma (>30 Hz) bands [21]. These potentials are presumably generated by mixing some underlying components of brain activity. Automatic detection of seizures in the intracranial EEG recordings is implemented in [128].

ECG recordings contain contributions from several bioelectric phenomena which include maternal and fetal heart activity and various kinds of noise. fMRI determines the spatial distribution of brain activities evoked by a given stimuli in a noninvasive manner, for the study of cognitive function of the brain. fMRI provides only an indirect view of neural activity via the blood oxygen level dependent (BOLD) functional imaging in primary visual cortex.

ICA has become an important tool to untangle the components of signals in multichannel EEG data. Subjects wear a cap embedded with a lattice of EEG electrodes, which record brain activity at different locations on the scalp. Stochastically spiking neurons with refractoriness could in principle learn in an unsupervised manner to carry out both information bottleneck optimization and the extraction of independent components [77]. Suitable learning rules are derived, which simultaneously keep the firing rate of the neuron within a biologically realistic range.

In [35], extended infomax, FastICA, JADE in a MATLAB-based toolbox, and group ICA of fMRI toolbox (GIFT) (http://icatb.sourceforge.net) are compared in terms of fMRI analysis, incorporating the implementations from ICALAB toolbox (http://www.bsp.brain.riken.jp/ICALAB). fMRI is a technique that produces complex-valued data.

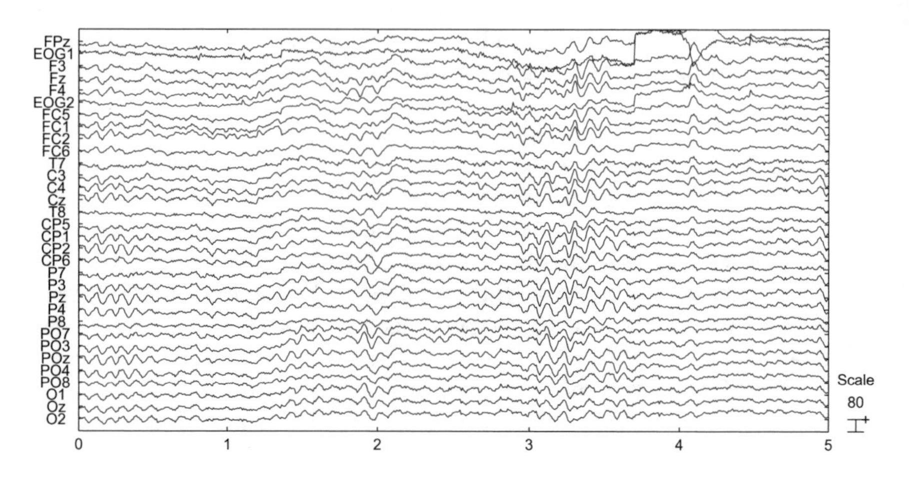

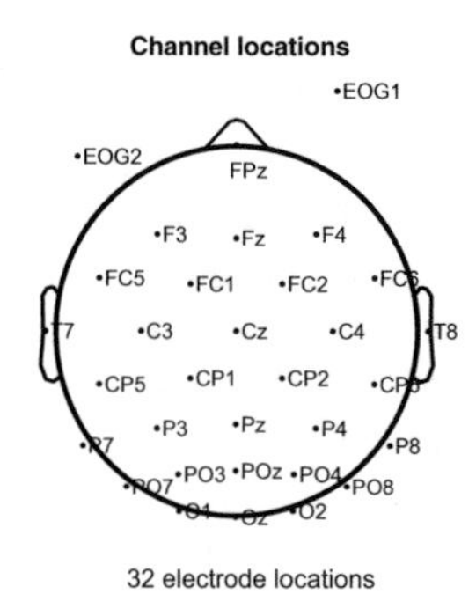

Fig. 15.5 An EEG dataset using 32 channels and the channel locations on the human scalp. Two of the channel locations are in the front of the head

Example 15.3 EEGLAB is MATLAB-based software for processing continuous or event-related EEG or other physiological data. We illustrate EEG principle by using a dataset provided by EEGLAB package. There are 32 channels for EEG measurements, and the sampling rate is 128 Hz. The dataset as well as their measurement locations is shown in Fig. 15.5. ICA can be used to separate out several important types of nonbrain artifacts from EEG data—only those associated with fixed scalp-amp projections, including eye movements and eye blinks, temporal muscle activity, and line noise.

Problems

15.1 Make a comparison between PCA and ICA.

15.2 BSS can be performed by ICA or decorrelation. Explain the two approaches.

15.3 In textile industry, one needs to monitor the uniformness of the thread radius. A solution is to let the thread pass through a capacitor and measure the capacitance variation.
(a) Derive the sensitivity of the capacitance change on the thread radius change.
(b) The noise is much higher than the capacitance variation caused by the nonuniformness of the thread. Consider how to reduce its influence.

15.4 Consider three independent sources:

$$s_1(n) = 0.5\sin(20n)\cos(10n),$$

$$s_2(n) = 0.01\exp(\sin(10n)\cos(2n) + \cos(16n)),$$

and $s_3(n)$ is a random noise, uniformly drawn from $[-1, 1]$. The mixing matrix

$$\mathbf{A} = \begin{bmatrix} 0.30 & 3.1 & -1.60 \\ -0.65 & 0.49 & 3.54 \\ 0.25 & 0.42 & -0.68 \end{bmatrix}.$$

Select an ICA algorithm to solve for the demixing matrix $\mathbf{W}$. Plot the waveforms produced and compare them with the source signals.

15.5 Two image patches of size 320×320 are selected from a set of images of natural scenes [59], and downsampled by a factor of 4 in both directions to yield 80×80 images. The third image is an artificial one containing only noisy signals. Each of the images is treated as one source with 6400 pixel samples. The three sources are then mixed using a randomly chosen mixing matrix

$$\mathbf{A} = \begin{bmatrix} 0.8762 & 0.2513 & 0.3564 \\ 0.2864 & -0.5615 & 0.3241 \\ -0.3523 & 0.7614 & 0.5234 \end{bmatrix}.$$

Recover the pictures.

15.6 Four digitized, grayscale facial images are used as the source signals. The images are linearly mixed using a randomly chosen square mixing matrix

$$\mathbf{A} = \begin{bmatrix} 0.6829 & -0.4246 & 1.8724 & 0.8260 \\ 1.2634 & 1.4520 & -0.5582 & 0.7451 \\ -0.7754 & 0.3251 & 0.5721 & 1.3774 \\ -0.7193 & 1.2051 & 0.2823 & 0.6821 \end{bmatrix}.$$

Separate these images using an ICA algorithm.

15.7 Consider the application of ICA to wireless sensor networks.

15.8 Demo of real-room blind separation/deconvolution of two speech sources are available at http://cnl.salk.edu/~tewon/Blind/blind_audio.html. Synthetic benchmarks for speech signals are available at http://sound.media.mit.edu/ica-bench/. Select two speech signals from these websites. Use the instantaneous linear mixtures of the two speech signals. The mixing matrix $\mathbf{A} = \begin{bmatrix} 0.25 & 0.85 \\ 0.72 & 0.18 \end{bmatrix}$. Reconstruct the original speech signals from the mixed signals.

15.9 The ECG recordings of a pregnant woman can be downloaded from the SISTA Identification Database (http://homes.esat.kuleuven.be/~tokka/daisydata.html). The data contains eight channels of recordings, where the first five channels record the abdominal measure and the last three channels serve as the thoracic measure. Separate the signals of independent sources from the measurements.

15.10 Investigate the use of ICA for edge detection of an image.

References

1. Akaho, S., Kiuchi, Y., & Umeyama, S. (1999). MICA: Multidimensional independent component analysis. In *Proceedings of the International Joint Conference on Neural Networks (IJCNN)* (pp. 927–932). Washington, DC.
2. Almeida, L. B. (2003). MISEP—Linear and nonlinear ICA based on mutual information. *Journal of Machine Learning Research*, *4*, 1297–1318.
3. Almeida, M., Schleimer, J.-H., Bioucas-Dias, J., & Vigario, R. (2011). Source separation and clustering of phase-locked subspaces. *IEEE Transactions on Neural Networks*, *22*(9), 1419–1434.
4. Almeida, M., Vigario, R., & Bioucas-Dias, J. (2011). Phase locked matrix factorization. In *Proceedings of the 19th European Signal Processing Conference (EUSIPCO)* (pp. 1728–1732). Barcelona, Spain.
5. Amari, S. I., Cichocki, A., & Yang, H. (1996). A new learning algorithm for blind signal separation. In D. S. Touretzky, M. C. Mozer, & M. E. Hasselmo (Eds.), *Advances in neural information processing systems* (Vol. 8, pp. 757–763). Cambridge, MA: MIT Press.
6. Amari, S., Douglas, S. C., Cichocki, A., & Yang, H. H. (1997). Multichannel blind deconvolution and equalization using the natural gradient. In *Proceedings of the IEEE Workshop on Signal Processing Advances in Wireless Communications* (pp. 101–104).
7. Amari, S. (1998). Natural gradient works efficiently in learning. *Neural Computation*, *10*(2), 251–276.
8. Amari, S. (1999). Natural gradient learning for over- and under-complete bases in ICA. *Neural Computation*, *11*, 1875–1883.
9. Armstrong, B. C., Ruiz-Blondet, M. V., Khalifian, N., Kurtz, K. J., Jin, Z., & Laszlo, S. (2015). Brainprint: Assessing the uniqueness, collectability, and permanence of a novel method for ERP biometrics. *Neurocomputing*, *166*, 59–67.
10. Astakhov, S. A., Stogbauer, H., Kraskov, A., & Grassberger, P. (2006). Monte Carlo algorithm for least dependent non-negative mixture decomposition. *Analytical Chemistry*, *78*(5), 1620–1627.
11. Babaie-Zadeh, M., Jutten, C., & Mansour, A. (2006). Sparse ICA via cluster-wise PCA. *Neurocomputing*, *69*, 1458–1466.
12. Baldi, P., & Hornik, K. (1989). Neural networks for principal component analysis: Learning from examples without local minima. *Neural Networks*, *2*, 53–58.

13. Bell, A. J., & Sejnowski, T. J. (1995). An information-maximization approach to blind separation and blind deconvolution. *Neural Computation*, *7*(6), 1129–1159.
14. Bingham, E., & Hyvarinen, A. (2000). ICA of complex valued signals: A fast and robust deflationary algorithm. In *Proceedings of the International Joint Conference on Neural Networks (IJCNN)* (Vol. 3, pp. 357–362). Como, Italy.
15. Blanco, D., Mulgrew, B., McLaughlin, S., Ruiz, D. P., & Carrion, M. C. (2006). The use of ICA in multiplicative noise. *Neurocomputing*, *69*, 1435–1441.
16. Blanco, D., Mulgrew, B., Ruiz, D. P., & Carrion, M. C. (2007). Independent component analysis in signals with multiplicative noise using fourth-order statistics. *Signal Processing*, *87*, 1917–1932.
17. Blythe, D. A. J., von Bunau, P., Meinecke, F. C., & Muller, K.-R. (2012). Feature extraction for change-point detection using stationary subspace analysis. *IEEE Transactions on Neural Networks and Learning Systems*, *23*(4), 631–643.
18. Bofill, P., & Zibulevsky, M. (2001). Underdetermined blind source separation using sparse representations. *Signal Processing*, *81*(11), 2353–2362.
19. Burel, G. (1992). Blind separation of sources: A nonlinear neural algorithm. *Neural Networks*, *5*, 937–947.
20. Calhoun, V., & Adali, T. (2006). Complex infomax: Convergence and approximation of infomax with complex nonlinearities. *Journal of VLSI Signal Processing*, *44*, 173–190.
21. Campisi, P., & La Rocca, D. (2014). Brain waves for automatic biometric based user recognition. *IEEE Transactions on Information Forensics and Security*, *9*(5), 782–800.
22. Cardoso, J.-F., & Souloumiac, A. (1993). Blind beamforming for non-Gaussian signals. *IEE Proceedings—F*, *140*(6), 362–370.
23. Cardoso, J.-F., & Laheld, B. H. (1996). Equivariant adaptive source separation. *IEEE Transactions on Signal Processing*, *44*(12), 3017–3030.
24. Cardoso, J.-F. (1997). Infomax and maximum likelihood for blind source separation. *Signal Processing Letters*, *4*(4), 112–114.
25. Cardoso, J.-F. (1999). High-order contrasts for independent component analysis. *Neural Computation*, *11*, 157–192.
26. Cardoso, J.-F. (2003). Dependence, correlation and Gaussianity in independent component analysis. *Journal of Machine Learning Research*, *4*, 1177–1203.
27. Cardoso, J.-F., & Adali, T. (2006). The maximum likelihood approach to complex ICA. In *Proceedings of IEEE International Conference on Acoustics, Speech, and Signal Processing (ICASSP)* (pp. 673–676). Toulouse, France.
28. Chan, T.-H., Ma, W.-K., Chi, C.-Y., & Wang, Y. (2008). A convex analysis framework for blind separation of non-negative sources. *IEEE Transactions on Signal Processing*, *56*(10), 5120–5134.
29. Choi, H., Kim, S., & Choi, S. (2004). Trust-region learning for ICA. In *Proceedings of the International Joint Conference on Neural Networks (IJCNN)* (pp. 41–46). Budapest, Hungary.
30. Choi, H., & Choi, S. (2007). A relative trust-region algorithm for independent component analysis. *Neurocomputing*, *70*, 1502–1510.
31. Choi, S., Cichocki, A., & Amari, S. (2002). Equivariant nonstationary source separation. *Neural Networks*, *15*, 121–130.
32. Choi, S. (2006). Differential learning algorithms for decorrelation and independent component analysis. *Neural Networks*, *19*, 1558–1567.
33. Cichocki, A., Douglas, S. C., & Amari, S. (1998). Robust techniques for independent component analysis (ICA) with noisy data. *Neurocomputing*, *22*, 113–129.
34. Comon, P. (1994). Independent component analysis—A new concept? *Signal Processing*, *36*(3), 287–314.
35. Correa, N., Adali, T., Li, Y., & Calhoun, V. D. (2005). Comparison of blind source separation algorithms for fMRI using a new matlab toolbox: GIFT. In *Proceedings of IEEE International Conference on Acoustics, Speech, and Signal Processing (ICASSP)* (pp. 401–404). Philadelphia, PA.

36. Creutzig, F., & Sprekeler, H. (2008). Predictive coding and the slowness principle: An information-theoretic approach. *Neural Computation*, *20*(4), 1026–1041.
37. Cruces, S. (2010). Bounded component analysis of linear mixtures: A criterion of minimum convex perimeter. *IEEE Transactions on Signal Processing*, *58*(4), 2141–2154.
38. Douglas, S. C. (2007). Fixed-point algorithms for the blind separation of arbitrary complex-valued non-Gaussian signal mixtures. *EURASIP Journal on Advances in Signal Processing*, *2007*, Article ID 36525, 15 pp.
39. Douglas, S. C., & Chao, J.-C. (2007). Simple, robust, and memory-efficient FastICA algorithms using the Huber M-estimator cost function. *Journal of VLSI Signal Processing*, *48*, 143–159.
40. Erdogan, A. T. (2012). A family of bounded component analysis algorithms. In *Proceedings of IEEE International Conference on Acoustics, Speech, and Signal Processing (ICASSP)* (pp. 1881–1884). Kyoto, Japan.
41. Erdogmus, D., Hild, II, K. E., Rao, Y. N., & Principe, J. C. (2004). Minimax mutual information approach for independent component analysis. *Neural Computation*, *16*, 1235–1252.
42. Eriksson, J., & Koivunen, V. (2004). Identifiability, separability and uniqueness of linear ICA models. *Signal Processing Letters*, *11*(7), 601–604.
43. Eriksson, J., & Koivunen, V. (2006). Complex random vectors and ICA models: Identifiability, uniqueness, and separability. *IEEE Transactions on Information Theory*, *52*(3), 1017–1029.
44. Escalante, B., Alberto, N., & Wiskott, L. (2013). How to solve classification and regression problems on high-dimensional data with a supervised extension of slow feature analysis. *Journal of Machine Learning Research*, *14*(1), 3683–3719.
45. Fiori, S. (2000). Blind separation of circularly-distributed sources by neural extended APEX algorithm. *Neurocomputing*, *34*, 239–252.
46. Fiori, S. (2003). Extended Hebbian learning for blind separation of complex-valued sources sources. *IEEE Transactions on Circuits and Systems II*, *50*(4), 195–202.
47. Foldiak, P. (1991). Learning invariance from transformation sequences. *Neural Computation*, *3*(2), 194–200.
48. Giannakopoulos, X., Karhunen, J., & Oja, E. (1999). An experimental comparison of neural algorithms for independent component analysis and blind separation. *International Journal of Neural Systems*, *9*(2), 99–114.
49. Goerg, G. (2013). Forecastable component analysis. In *Proceedings of the 30th International Conference on Machine Learning (ICML 2013)* (Vol. 28, pp. 64–72).
50. Hara, S., Kawahara, Y., Washio, T., von Bunau, P., Tokunaga, T., & Yumoto, K. (2012). Separation of stationary and non-stationary sources with a generalized eigenvalue problem. *Neural Networks*, *33*, 7–20.
51. Haritopoulos, M., Yin, H., & Allinson, N. M. (2002). Image denoising using self-organizing map-based nonlinear independent component analysis. *Neural Networks*, *15*, 1085–1098.
52. Hild, II, K. E., Attias, H. T., & Nagarajan, S. S. (2008). An expectation-maximization method for spatio-temporal blind source separation using an AR-MOG source model. *IEEE Transactions on Neural Networks*, *19*(3), 508–519.
53. Hyvarinen, A., & Oja, E. (1997). A fast fixed-point algorithm for independent component analysis. *Neural Computation*, *9*(7), 1483–1492.
54. Hyvarinen, A. (1998). New approximations of differential entropy for independent component analysis and projection pursuit. In M. I. Jordan, M. J. Kearns, & S. A. Solla (Eds.), *Advances in neural information processing* (Vol. 10, pp. 273–279). Cambridge, MA: MIT Press.
55. Hyvarinen, A., & Oja, E. (1998). Independent component analysis by general non-linear Hebbian-like learning rules. *Signal Processing*, *64*, 301–313.
56. Hyvarinen, A., & Pajunen, P. (1999). Nonlinear independent component analysis: Existence and uniqueness results. *Neural Networks*, *12*, 429–439.
57. Hyvarinen, A. (1999). Fast and robust fixed-point algorithms for independent component analysis. *IEEE Transactions on Neural Networks*, *10*(3), 626–634.
58. Hyvarinen, A., Sarela, J., & Vigario, R. (1999). Spikes and bumps: Artefacts generated by independent component analysis with insufficient sample size. In *Proceedings of the International Workshop on ICA* (pp. 425–429). Aussois, France.

59. Hyvarinen, A., & Hoyer, P. O. (2000). Emergence of phase and shift invariant features by decomposition of natural images into independent feature subspaces. *Neural Computation, 12*(7), 1705–1720.
60. Hyvarinen, A. (2001). Blind source separation by nonstationarity of variance: A cumulant-based approach. *IEEE Transactions on Neural Networks, 12*(6), 1471–1474.
61. Hyvarinen, A., Hoyer, P. O., & Inki, M. (2001). Topographic independent component analysis. *Neural Computation, 13*, 1527–1558.
62. Hyvarinen, A., & Raju, K. (2002). Imposing sparsity on the mixing matrix in independent component analysis. *Neurocomputing, 49*, 151–162.
63. Hyvarinen, A., & Koster, U. (2006). FastISA: A fast fixed-point algorithm for independent subspace analysis. In *Proceedings of the 14th European Symposium on Artificial Neural Networks* (pp. 371–376). Bruges, Belgium.
64. Igual, J., Vergara, L., Camacho, A., & Miralles, R. (2003). Independent component analysis with prior information about the mixing matrix. *Neurocomputing, 50*, 419–438.
65. Inan, H. A., & Erdogan, A. T. (2015). Convolutive bounded component analysis algorithms for independent and dependent source separation. *IEEE Transactions on Neural Networks and Learning Systems, 26*(4), 697–708.
66. Javidi, S., Took, C. C., & Mandic, D. P. (2011). Fast independent component analysis algorithm for quaternion valued signals. *IEEE Transactions on Neural Networks, 22*(12), 1967–1978.
67. Jutten, C., & Herault, J. (1991). Blind separation of sources, Part I. An adaptive algorithm based on a neuromimetic architecture. *Signal Processing, 24*(1), 1–10.
68. Kameoka, H., Ono, N., Kashino, K., & Sagayama, S. (2009). Complex NMF: A new sparse representation for acoustic signals. In *Proceedings of IEEE International Conference on Acoustics, Speech, and Signal Processing (ICASSP)* (pp. 3437–3440).
69. Karhunen, J., Oja, E., Wang, L., Vigario, R., & Joutsensalo, J. (1997). A class of neural networks for independent component analysis. *IEEE Transactions on Neural Networks, 8*(3), 486–504.
70. Karhunen, J., & Pajunen, P. (1997). Blind source separation and tracking using nonlinear PCA criterion: A least-squares approach. In *Proceedings of the IEEE Internal Conference on Neural Networks* (Vol. 4, pp. 2147–2152). Houston, TX.
71. Karhunen, J., Pajunen, P., & Oja, E. (1998). The nonlinear PCA criterion in blind source separation: Relations with other approaches. *Neurocomputing, 22*(1), 5–20.
72. Karhunen, J., Malaroiu, S., & Ilmoniemi, M. (1999). Local linear independent component analysis based on clustering. *International Journal of Neural Systems, 10*(6), 439–451.
73. Kawamoto, M., Matsuoka, K., & Ohnishi, N. (1998). A method of blind separation for convolved nonstationary signals. *Neurocomputing, 22*, 157–171.
74. Kim, S., & Yoo, C. D. (2009). Underdetermined blind source separation based on subspace representation. *IEEE Transactions on Signal Processing, 57*(7), 2604–2614.
75. Kim, T. (2010). Real-time independent vector analysis for convolutive blind source separation. *IEEE Transactions on Circuits and Systems I, 57*(7), 1431–1438.
76. King, B. J., & Atlas, L. (2011). Single-channel source separation using complex matrix factorization. *IEEE/ACM Transactions on Audio, Speech, and Language Processing, 19*(8), 2591–2597.
77. Klampfl, S., Legenstein, R., & Maass, W. (2009). Spiking neurons can learn to solve information bottleneck problems and extract independent components. *Neural Computation, 21*, 911–959.
78. Kokkinakis, K., & Nandi, A. K. (2007). Generalized gamma density-based score functions for fast and flexible ICA. *Signal Processing, 87*, 1156–1162.
79. Koldovsky, Z., Tichavsky, P., & Oja, E. (2005). Cramer–Rao lower bound for linear independent component analysis. In *Proceedings of International Conference on Acoustics, Speech, and Signal Processing (ICASSP)* (Vol. 3, pp. 581–584). Philadelphia, PA.
80. Koldovsky, Z., Tichavsky, P., & Oja, E. (2006). Efficient variant of algorithm FastICA for independent component analysis attaining the Cramer–Rao lower bound. *IEEE Transactions on Neural Networks, 17*(5), 1265–1277.

81. Kompella, V. R., Luciw, M. D., & Schmidhuber, J. (2012). Incremental slow feature analysis: Adaptive low-complexity slow feature updating from high-dimensional input streams. *Neural Computation*, *24*(11), 2994–3024.
82. Kompella, V. R., Luciw, M., Stollenga, M. F., & Schmidhuber, J. (2016). Optimal curiosity-driven modular incremental slow feature analysis. *Neural Computation*, *28*(8), 1599–1662.
83. Learned-Miller, E. G., & Fisher, III, J. W. (2003). ICA using spacings estimates of entropy. *Journal of Machine Learning Research*, *4*, 1271–1295.
84. Lee, T.-W., Girolami, M., & Sejnowski, T. J. (1999). Indepedent component analysis using an extended infomax algorithm for mixed subgaussian and supergaussian sources. *Neural Computation*, *11*(2), 417–441.
85. Li, H., & Adali, T. (2008). A class of complex ICA algorithms based on the kurtosis cost function. *IEEE Transactions on Neural Networks*, *19*(3), 408–420.
86. Li, X.-L., & Adali, T. (2010). Complex independent component analysis by entropy bound minimization. *IEEE Transactions on Circuits and Systems I*, *57*(7), 1417–1430.
87. Li, H., & Adali, T. (2010). Algorithms for complex ML ICA and their stability analysis using Wirtinger calculus. *IEEE Transactions on Signal Processing*, *58*(12), 6156–6167.
88. Li, Y., Cichocki, A., & Amari, S.-I. (2004). Analysis of sparse representation and blind source separation. *Neural Computation*, *16*(6), 1193–1234.
89. Lu, W., & Rajapakse, J. C. (2003). Eliminating indeterminacy in ICA. *Neurocomputing*, *50*, 271–290.
90. Lu, W., & Rajapakse, J. C. (2005). Approach and applications of constrained ICA. *IEEE Transactions on Neural Networks*, *16*(1), 203–212.
91. Lu, W., & Rajapakse, J. C. (2006). ICA with reference. *Neurocomputing*, *69*, 2244–2257.
92. Martin-Clemente, R., & Zarzoso, V. (2017). On the link between L_1-PCA and ICA. *IEEE Transactions on Pattern Analysis and Machine Intelligence*, *39*(3), 515–528.
93. Marvasti, F., Amini, A., Haddadi, F., Soltanolkotabi, M., Khalaj, B. H., Aldroubi, A., et al. (2012). A unified approach to sparse signal processing. *EURASIP Journal on Advances in Signal Processing*, *2012*(44), 1–45.
94. Matsuoka, K., Ohya, M., & Kawamoto, M. (1995). A neural net for blind separation of nonstationary signals. *Neural Networks*, *8*(3), 411–419.
95. Mollah, M. N. H., Eguchi, S., & Minami, M. (2007). Robust prewhitening for ICA by minimizing β-divergence and its application to FastICA. *Neural Processing Letters*, *25*, 91–110.
96. Murata, N., Ikeda, S., & Ziehe, A. (2001). An approach to blind source separation based on temporal structure of speech signals. *Neurocomputing*, *41*, 1–24.
97. Naik, G. R., & Kumar, D. K. (2009). Determining number of independent sources in undercomplete mixture. *EURASIP Journal on Advances in Signal Processing*, *2009*, Article ID 694850, 5 pp.
98. Novey, M., & Adali, T. (2008). Complex ICA by negentropy maximization. *IEEE Transactions on Neural Networks*, *19*(4), 596–609.
99. Novey, M., & Adali, T. (2008). On extending the complex FastICA algorithm to noncircular sources. *IEEE Transactions on Signal Processing*, *56*(5), 2148–2154.
100. Oja, E., & Plumbley, M. (2004). Blind separation of positive sources by globally convergent gradient search. *Neural Computation*, *16*, 1811–1825.
101. Oja, E., & Yuan, Z. (2006). The FastICA algorithm revisited: Convergence analysis. *IEEE Transactions on Neural Networks*, *17*(6), 1370–1381.
102. Ollila, E., & Koivunen, V. (2009). Complex ICA using generalized uncorrelating transform. *Signal Processing*, *89*, 365–377.
103. Ollila, E. (2010). The deflation-based FastICA estimator: Statistical analysis revisited. *IEEE Transactions on Signal Processing*, *58*(3), 1527–1541.
104. Pajunen, P. (1998). Blind source separation using algorithmic information theory. *Neurocomputing*, *22*, 35–48.
105. Park, H.-M., Oh, S.-H., & Lee, S.-Y. (2006). A modified infomax algorithm for blind signal separation. *Neurocomputing*, *70*, 229–240.

106. Pham, D.-T. (2000). Blind separation of instantenaous mixtrures of sources based on order statistics. *IEEE Transactions on Signal Processing*, *48*(2), 363–375.
107. Plumbley, M. D. (2003). Algorithms for nonnegative independent component analysis. *IEEE Transactions on Neural Networks*, *14*(3), 534–543.
108. Poczos, B., & Lorincz, A. (2005). Independent subspace analysis using geodesic spanning trees. In *Proceedings of the 22nd International Conference on Machine Learning* (pp. 673–680). Bonn, Germany.
109. Richthofer, S., & Wiskott, L. (2015). Predictable feature analysis. *Proceedings of the IEEE 14th International Conference on Machine Learning and Applications*. Miami, FL.
110. Ristaniemi, T., & Joutsensalo, J. (2002). Advanced ICA-based receivers for block fading DS-CDMA channels. *Signal Processing*, *82*(3), 417–431.
111. Samadi, S., Babaie-Zadeh, M., & Jutten, C. (2006). Quasi-optimal EASI algorithm based on the Score Function Difference (SFD). *Neurocomputing*, *69*, 1415–1424.
112. Sarela, J., & Vigario, R. (2003). Overlearning in marginal distribution-based ICA: Analysis and solutions. *Journal of Machine Learning Research*, *4*, 1447–1469.
113. Sawada, H., Mukai, R., Araki, S., & Makino, S. (2004). A robust and precise method for solving the permutation problem of frequency-domain blind source separation. *IEEE Transactions on Speech and Audio Processing*, *12*(5), 530–538.
114. Schell, S. V., & Gardner, W. A. (1995). Programmable canonical correlation analysis: A flexible framework for blind adaptive spatial filtering. *IEEE Transactions on Signal Processing*, *42*(12), 2898–2908.
115. Schobben, L., & Sommen, W. (2002). A frequency domain blind signal separation method based on decorrelation. *IEEE Transactions on Signal Processing*, *50*, 1855–1865.
116. Shen, H., Kleinsteuber, M., & Huper, K. (2008). Local convergence analysis of FastICA and related algorithms. *IEEE Transactions on Neural Networks*, *19*(6), 1022–1032.
117. Smaragdis, P. (1998). Blind separation of convolved mixtures in the frequency domain. *Neurocomputing*, *22*, 21–34.
118. Sprekeler, H., Zito, T., & Wiskott, L. (2014). An extension of slow feature analysis for nonlinear blind source separation. *Journal of Machine Learning Research*, *15*, 921–947.
119. Tan, Y., Wang, J., & Zurada, J. M. (2001). Nonlinear blind source separation using a radial basis function network. *IEEE Transactions on Neural Networks*, *12*(1), 134–144.
120. Tichavsky, P., Koldovsky, Z., & Oja, E. (2005). Asymptotic performance of the FastICA algorithm for independent component analysis and its improvements. In *Proceedings of IEEE Workshop on Statistical Signal Processing* (pp. 1084–1089). Bordeaux, France.
121. Van Hulle, M. M. (2008). Sequential fixed-point ICA based on mutual information minimization. *Neural Computation*, *20*, 1344–1365.
122. Van Hulle, M. M. (2008). Constrained subspace ICA based on mutual information optimization directly. *Neural Computation*, *20*, 964–973.
123. von Bunau, P., Meinecke, F. C., Kiraly, F. C., & Muller, K.-R. (2009). Finding stationary subspaces in multivariate time series. *Physical Review Letters*, *103*(21), 214101.
124. Wang, F.-Y., Chi, C.-Y., Chan, T.-H., & Wang, Y. (2010). Nonnegative least-correlated component analysis for separation of dependent sources by volume maximization. *IEEE Transactions on Pattern Analysis and Machine Intelligence*, *32*(5), 875–888.
125. Weghenkel, B., Fischer, A., & Wiskott, L. (2017). Graph-based predictable feature analysis. *Machine Learning*, *106*, 1359–1380.
126. Wiskott, L. (2003). Slow feature analysis: A theoretical analysis of optimal free responses. *Neural Computation*, *15*(9), 2147–2177.
127. Wiskott, L., & Sejnowski, T. (2002). Slow feature analysis: Unsupervised learning of invariances. *Neural Computation*, *14*(4), 715–770.
128. Yadav, R., Agarwal, R., & Swamy, M. N. S. (2012). Model-based seizure detection for intracranial EEG recordings. *IEEE Transactions on Biomedical Engineering*, *59*, 1419–1428.
129. Ye, M. (2006). Global convergence analysis of a discrete time nonnegative ICA algorithm. *IEEE Transactions on Neural Networks*, *17*(1), 253–256.

130. Zarzoso, V., & Comon, P. (2010). Robust independent component analysis by iterative maximization of the kurtosis contrast with algebraic optimal step size. *IEEE Transactions on Neural Networks*, *21*(2), 248–261.
131. Zarzoso, V., Comon, P., & Phlypo, R. (2010). A contrast function for independent component analysis without permutation ambiguity. *IEEE Transactions on Neural Networks*, *21*(5), 863–868.
132. Zhang, K., & Chan, L. (2008). Minimal nonlinear distortion principle for nonlinear independent component analysis. *Journal of Machine Learning Research*, *9*, 2455–2487.
133. Zheng, C.-H., Huang, D.-S., Sun, Z.-L., Lyu, M. R., & Lok, T.-M. (2006). Nonnegative independent component analysis based on minimizing mutual information technique. *Neurocomputing*, *69*, 878–883.
134. Ziegaus, Ch., & Lang, E. W. (2004). A neural implementation ofthe JADE algorithm (nJADE) using higher-order neurons. *Neurocomputing*, *56*, 79–100.

Chapter 16
Discriminant Analysis

16.1 Linear Discriminant Analysis

LDA is a supervised dimension-reduction technique. It projects the data into an effective low-dimensional linear subspace while finding directions that maximize the ratio of between-class scatter to within-class scatter of the projected data. In the statistics community, LDA is equivalent to a t-test or F-test for significant difference between the mean of discriminants for two sampled classes; in fact, the statistic is designed to have the largest possible value [48]. LDA utilizes EVD to find an orientation which projects high-dimensional feature vectors of different classes to a low-dimensional space in the most discriminative way for classification. C-means can be used to generate cluster labels, which can be further used for LDA to do subspace selection [16].

LDA creates a linear combination of the given independent features that yield the largest mean differences between the desired classes [19]. Given a dataset $\{\boldsymbol{x}_i\}$ of size N, which is composed of J_1-dimensional vectors, for all the samples of all the C classes, the within-class scatter matrix $\mathbf{S}_w$, the between-class scatter matrix $\mathbf{S}_b$, and the mixture or total scatter matrix $\mathbf{S}_t$ are, respectively, defined by

$$\mathbf{S}_w = \frac{1}{N}\sum_{j=1}^{C}\sum_{i=1}^{N_j}\left(\boldsymbol{x}_i^{(j)} - \boldsymbol{\mu}_j\right)\left(\boldsymbol{x}_i^{(j)} - \boldsymbol{\mu}_j\right)^T, \tag{16.1}$$

$$\mathbf{S}_b = \frac{1}{N}\sum_{j=1}^{C} N_j \left(\boldsymbol{\mu}_j - \boldsymbol{\mu}\right)\left(\boldsymbol{\mu}_j - \boldsymbol{\mu}\right)^T, \tag{16.2}$$

$$\mathbf{S}_t = \frac{1}{N}\sum_{j=1}^{N}\left(\boldsymbol{x}_j - \boldsymbol{\mu}\right)\left(\boldsymbol{x}_j - \boldsymbol{\mu}\right)^T, \tag{16.3}$$

K.-L. Du and M. N. S. Swamy, *Neural Networks and Statistical Learning*,
https://doi.org/10.1007/978-1-4471-7452-3_16

where $\boldsymbol{x}_i^{(j)}$ is the ith sample of class j, $\boldsymbol{\mu}_j$ is the mean of class j, N_j is the number of samples in class j, and $\boldsymbol{\mu}$ represents the mean of all classes. Note that $\sum_{j=1}^{C} N_j = N$. All the scatter matrices are of size $J_1 \times J_1$, and are related by

$$\mathbf{S}_t = \mathbf{S}_w + \mathbf{S}_b. \tag{16.4}$$

The objective of LDA is to maximize the between-class measure while minimizing the within-class measure after applying a $J_1 \times J_2$ transform matrix $\mathbf{W}$, J_2 being the number of features, $J_1 > J_2$, which transforms the $J_1 \times J_1$ scatter matrices into $J_2 \times J_2$ matrices $\widetilde{\mathbf{S}}_w$, $\widetilde{\mathbf{S}}_b$, and $\widetilde{\mathbf{S}}_t$,

$$\begin{aligned} \widetilde{\mathbf{S}}_w &= \mathbf{W}^T \mathbf{S}_w \mathbf{W}, \\ \widetilde{\mathbf{S}}_b &= \mathbf{W}^T \mathbf{S}_b \mathbf{W}, \\ \widetilde{\mathbf{S}}_t &= \mathbf{W}^T \mathbf{S}_t \mathbf{W}. \end{aligned} \tag{16.5}$$

tr $(\mathbf{S}_w)$ measures the closeness of the samples within the clusters and tr $(\mathbf{S}_b)$ measures the separation between the clusters, where $\mathrm{tr}(\cdot)$ is trace operator. An optimal $\mathbf{W}$ should preserve a given cluster structure, and simultaneously maximize $\mathrm{tr}(\widetilde{\mathbf{S}}_b)$ and minimize $\mathrm{tr}(\widetilde{\mathbf{S}}_w)$. This is equivalent to maximizing [49]

$$E_{\mathrm{LDA},1}(\mathbf{W}) = \mathrm{tr}\left(\widetilde{\mathbf{S}}_w^{-1} \widetilde{\mathbf{S}}_b\right), \tag{16.6}$$

when $\widetilde{\mathbf{S}}_w$ is a nonsingular matrix.

Assuming that $\mathbf{S}_w$ is a nonsingular matrix, one can maximize the Rayleigh coefficient [52]

$$E_{\mathrm{LDA},2}(\boldsymbol{w}) = \frac{\boldsymbol{w}^T \mathbf{S}_b \boldsymbol{w}}{\boldsymbol{w}^T \mathbf{S}_w \boldsymbol{w}} \tag{16.7}$$

to find the principal projection direction $\boldsymbol{w}_1$. Conventionally, the following Fisher's determinant ratio criterion is maximized for finding the projection directions [14, 32]

$$E_{\mathrm{LDA},3}(\mathbf{W}) = \frac{\det(\widetilde{\mathbf{S}}_b)}{\det(\widetilde{\mathbf{S}}_w)} = \frac{\det\left(\mathbf{W}^T \mathbf{S}_b \mathbf{W}\right)}{\det\left(\mathbf{W}^T \mathbf{S}_w \mathbf{W}\right)}, \tag{16.8}$$

where the column vectors $\boldsymbol{w}_i$, $i = 1, \ldots, J_2$, of the projection matrix $\mathbf{W}$, are the first J_2 principal eigenvectors of $\mathbf{S}_w^{-1} \mathbf{S}_b$.

LDA is equivalent to ML classification assuming normal distribution for each class with a common covariance matrix [29]. When each class has more complex structure, LDA may fail.

LDA has $O(N J_2 m + m^3)$ cubic computational complexity and requires $O(N J_2 + Nt + J_2 m)$ memory, where $m = \min(N, J_2)$. It is infeasible to apply LDA when both N and J_2 are large.

LDA is targeted to find a set of weights $\boldsymbol{w}$ and a threshold θ such that the discriminant function

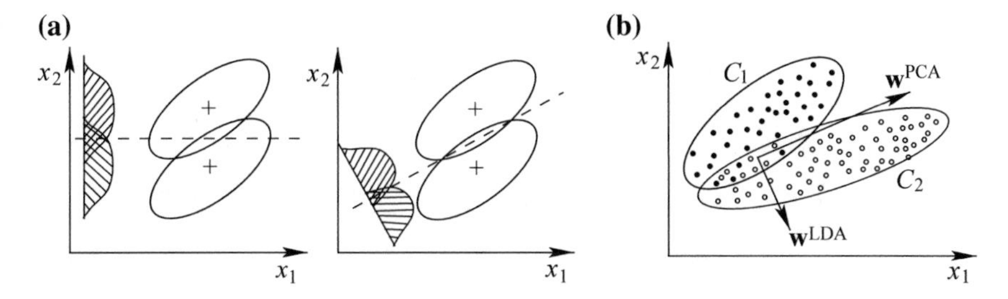

Fig. 16.1 **a** The line joining the centroids defines the direction of greatest centroid spread, but the projected data overlap because of the covariance (left). The discriminant direction minimizes this overlap for Gaussian data (right). **b** The projections of PCA and LDA for a dataset

$$t(\boldsymbol{x}_i) = \boldsymbol{w}^T \boldsymbol{x}_i - \theta \tag{16.9}$$

maximizes a discrimination criterion such that the between-class variance is maximized relative to the within-class variance. The between-class variance is the variance of the class means of $\{t(\boldsymbol{x}_i)\}$, and the within-class variance is the pooled variance about the means. Figure 16.1a shows why this criterion makes sense.

In the two-class problem, a data vector $\boldsymbol{x}_i$ is assigned to one class if $t(\boldsymbol{x}_i) > 0$ and to the other class if $t(\boldsymbol{x}_i) < 0$. Methods for determining $\boldsymbol{w}$ and θ can be the perceptron, LDA, and regression. The simplicity of the LDA model makes it a good candidate for classification in situations where training data are very limited.

An illustration of PCA and LDA for a two-dimensional dataset is shown in Fig. 16.1b. Two Gaussian classes $\mathcal{C}_1$ and $\mathcal{C}_2$ are represented by two ellipses. The principal direction obtained from PCA, namely, $\boldsymbol{w}^{\text{PCA}}$, cannot discriminate the two classes, while $\boldsymbol{w}^{\text{LDA}}$, the principal direction obtained from LDA, can discriminate the two classes. It is clearly seen that PCA is purely descriptive, while LDA is discriminative.

LDA is asymptotically Bayes optimal under the assumption that the class distributions are homoscedastic (identically distributed) Gaussians [52]. The lower bound of the generalization discrimination power is applicable when the dimensionality D and training sample size N are proportionally large, that is, the ratio $\gamma = D/N$ is a constant [3]. In such a case, the generalization ability of LDA (with respect to the Bayes optimum) is independent of the spectral structure of the population covariance, given its nonsingularity and above conditions. The discrimination power bound also leads to an upper bound on the generalization error of binary classification with LDA.

Example 16.1 We compare PCA and LDA by using STPRtool (http://cmp.felk.cvut.cz/cmp/software/stprtool/). LDA and PCA are trained on the synthetical data generated from a Gaussian mixture model. The LDA and PCA directions are shown in Fig. 16.2a. The extracted data using LDA and PCA are shown with the Gaussians fitted by the ML method (see Fig. 16.2b). It is indicated that LDA effectively separates the samples whereas PCA fails.

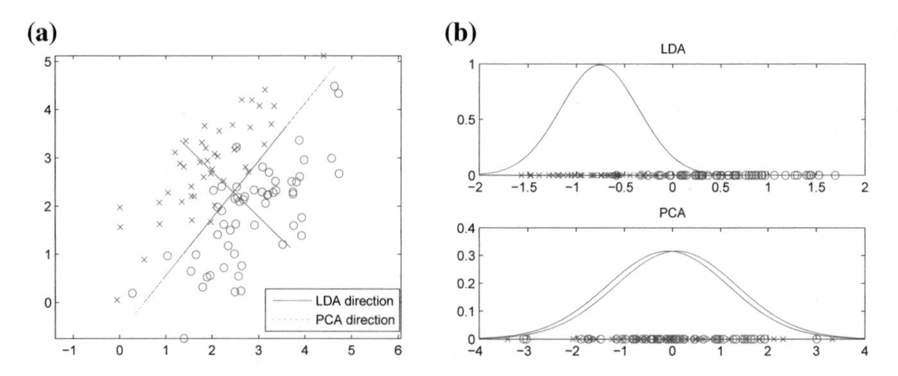

Fig. 16.2 PCA versus LDA. **a** The input data and the found LDA and PCA directions. **b** The class-conditional Gaussians estimated from data projected onto LDA and PCA directions

For the binary classification problem, LDA has been shown to be equivalent to regression [29]. This relation is extended to the multiclass case [4]. LS-LDA [79] puts LDA into the framework of multivariate regression and gives the LS solution to LDA. By using spectral graph analysis, spectral regression discriminant analysis [4] casts discriminant analysis into a multiple linear regression framework that facilitates both efficient computation and the use of regularization techniques. Specifically, the method only needs to solve a set of regularized LS problems. This allows for additional constraints (e.g., sparsity) on the LDA solution. It can be computed with $O(ms)$ time and $O(ms)$ memory, where $s(\leq n)$ is the average number of nonzero features in each sample.

Based on a single-layer linear feedforward network, LDA algorithms are also given in [6, 15]. The $\mathbf{Q}^{-1/2}$ network [7] is another neural network-based LDA. This algorithm adaptively computes $\mathbf{Q}^{-1/2}$, where $\mathbf{Q}$ is the correlation or covariance matrix. LDA is also extended to regression problems [38].

Logistic Regression

Logistic regression is a classical statistical method that measures the relationship between the categorical-dependent variable and one or more independent variables by estimating probabilities using the cumulative logistic distribution. It is a probabilistic statistical classification model that predicts the probability of the occurrence of an event.

Logistic regression is a special case of the generalized linear model and thus analogous to linear regression. It is a low-bias classifier, which tends to have lower bias than naive Bayes [81]. Logistic regression is an alternative to LDA. If the assumptions of LDA hold, the conditioning can be reversed to produce logistic regression. In contrast, logistic regression does not require the multivariate normal assumption of LDA.

16.2 Solving Small Sample Size Problem

Since there are at most $C - 1$ nonzero generalized eigenvalues for the LDA problem, an upper bound on J_2 is $C - 1$. The rank of $\mathbf{S}_w$ is at most $N - C$, and thus at least $N = J_1 + C$ samples are needed to guarantee $\mathbf{S}_w$ to be nonsingular. This requirement on the number of samples may be severe for some problems like image processing. Typically, the number of images from each class is considerably limited in face recognition: only several faces can be acquired from each person. The dimension of the sample space is typically much larger than the number of samples in a training set. For instance, an image of 32-by-32 pixels is represented by a $1,024$-dimensional vector, and consequently $\mathbf{S}_w$ is singular and LDA cannot be applied directly. This problem is known as the *small sample size*, *singularity*, or *undersampled* problem.

Pseudoinverse LDA [63] is based on the pseudoinverse of the scatter matrices. It applies the eigendecomposition to the matrix $\mathbf{S}_b^{\dagger}\mathbf{S}_w$, $\mathbf{S}_w^{\dagger}\mathbf{S}_b$, or $\mathbf{S}_t^{\dagger}\mathbf{S}_b$. The criterion F_1 is an extension of the classical one (16.6), with the inverse of a matrix replaced by the pseudoinverse:

$$\max F_1(\mathbf{W}) = \mathrm{tr}\left(\widetilde{\mathbf{S}}_t^{\dagger}\widetilde{\mathbf{S}}_b\right). \tag{16.10}$$

The pseudoinverse-based methods are competitive compared to regularized LDA [21] and Fisherfaces [67].

16.3 Fisherfaces

PCA is known as the *eigenfaces* method when it is applied to a large set of image depicting different human faces. Eigenfaces are a set of standardized face ingredients, and any human face can be considered to be a combination of these standard faces. This technique is also used for handwriting analysis, lipreading, voice recognition, sign language/hand gestures interpretation, and medical imaging analysis. Therefore, some prefer to call this method as eigenimage.

A common way to deal with the small sample size problem is to apply an intermediate dimension-reduction stage, such as PCA, to reduce the dimension of the original data before LDA is applied. This method is known as PCA+LDA or Fisherfaces [2, 67]. In order to avoid the complication of singular $\mathbf{S}_w$, the Fisherfaces method discards the smallest principal components. The overall performance of the two-stage approach is sensitive to the reduced dimension in the first stage, and the optimal value of the reduced dimension for PCA is difficult to determine. The method discards the discriminant information contained in the null space of the within-class covariance matrix.

When the training dataset is small, the eigenfaces method outperforms the Fisherfaces method [47]. It might be because the Fisherfaces method uses all the principal components, but the components with the small eigenvalues correspond to high-frequency components and usually encode noise [43]. In line with this, two enhanced LDA models improve the generalization capability of LDA by decomposing the LDA procedure into simultaneous diagonalization of $\mathbf{S}_w$ and $\mathbf{S}_b$ [43]. The simultaneous diagonalization is stepwise equivalent to two operations: whitening $\mathbf{S}_w$ and applying PCA on $\mathbf{S}_b$ using the transformed data. As an alternative to the Fisherfaces method, a two-stage LDA [89] avoids computation of the inverse $\mathbf{S}_w$ by decomposing $\mathbf{S}_w^{-1}\mathbf{S}_b$.

Example 16.2 We have implemented the eigenfaces method in Example 14.2. We now implement the Fisherfaces method for solving the same face recognition problem by using the same training and test sets. In this example, we randomly select $N = 60$ samples from $C = 30$ persons, 2 samples for each person, for training. The test set includes 30 samples, one for each person.

At first, a centered image in vector form is mapped onto an $(N - C)$-dimensional linear subspace by PCA weight matrix, and the output is further projected onto a $(C - 1)$-dimensional linear subspace by Fisher weight matrix, so that images of the same class move closer together and images of difference classes move further apart.

After applying the two-step procedure on the training set, the corresponding weights are called Fisherfaces. When a new sample is presented, its projection on the weights is compared with those of all the training samples, and the training sample that has the minimum difference from the test sample is classified as the decision. On the test samples, the classification rate is 86.67%. It is found that for this small training set, the classification accuracy of the eigenfaces method is better than that of the Fisherfaces method.

16.4 Regularized LDA

Another common way is to add some constant value $\mu > 0$ to the diagonal elements of $\mathbf{S}_w$, as $\mathbf{S}_w + \mu\mathbf{I}_d$, where $\mathbf{I}_d$ is an identity matrix [21]. $\mathbf{S}_w + \mu\mathbf{I}_d$ is positive definite, hence nonsingular. This approach is called regularized LDA. Regularization reduces the high variance related to the eigenvalue estimates of $\mathbf{S}_w$, at the expense of potentially increased bias. The optimal value of the regularization parameter μ is difficult to determine, and cross-validation is commonly applied for estimating the optimal μ. By adjusting μ, a set of LDA variants are obtained, such as DLDA [46] for $\mu = 1$. The trade-off between variance and bias, depending on the severity of the small sample size problem, is controlled by the strength of regularization.

In regularized LDA, an appropriate regularization parameter is selected from a given parameter candidate set by using cross-validation for classification. In regularized orthogonal LDA, the regularization parameter is selected by using a mathematical criterion [10].

Quadratic discriminant analysis models the likelihood of each class as a Gaussian distribution, and then uses the posterior distributions to estimate the class for a given test point [29]. The Gaussian parameters can be estimated from training points with ML estimation. Unfortunately, when the number N of training samples is small compared to the dimension d of each training sample, the ML covariance estimation can be ill-posed.

Regularized quadratic discriminant analysis [21] shrinks the covariances of quadratic discriminant analysis toward a common covariance to achieve a compromise between LDA and quadratic discriminant analysis. Regularized quadratic discriminant analysis performs well when the true Gaussian distribution matches one of their regularization covariance models (e.g., diagonal, identity), but can fail when the generating distribution has a full covariance matrix, particularly when features are correlated. The single-parameter regularized quadratic discriminant analysis algorithm [9] reduces the computational complexity from $O(N^3)$ to $O(N)$. Bayesian quadratic discriminant analysis [64] performs similar to ML quadratic discriminant analysis in terms of error rates. Its performance is very sensitive to the choice of the prior.

In penalized discriminant analysis [28], a symmetric and positive-semidefinite penalty matrix Λ is added to $\mathbf{S}_w$. Flexible discriminant analysis [27] extends LDA to the nonlinear and multiclass classification via a penalized regression setting. To do that, it reformulates the discriminant analysis problem as a regression one and, then, uses a nonlinear function to fit the data. It is based on encoding the class labels into response scores and then a nonparametric regression technique, such as neural networks, is used to fit the response scores.

Null-space LDA [8] attempts to solve the small sample size problems directly. The null space of $\mathbf{S}_w$ contains useful discriminant information. The method first projects the data onto the null space of $\mathbf{S}_w$ and it then applies PCA to maximize $\mathbf{S}_b$ in the transformed space. Based on the eigendecomposition of the original scatter matrices, null-space LDA may ignore some useful information by considering the null space of $\mathbf{S}_w$ only. The discriminative common vector method [5] addresses computational difficulties encountered in null-space LDA. In a fast implementation for null-space-based LDA [12], the optimal transformation matrix is obtained by orthogonal transformations by using QR factorization and QR factorization with column pivoting of the data matrix.

Gradient LDA [61] is based on gradient-descent method but the convergence is fast and reliable. It does not discard any null spaces of $\mathbf{S}_w$ and $\mathbf{S}_b$ matrices and thus preserves discriminative information which is useful for classification.

Weighted piecewise LDA [39] first creates subsets of features and applies LDA to each subset. It then combines the resulting piecewise linear discriminants to produce an overall solution. Initially, a set of weighted piecewise discriminant hyperplanes are used in order to provide a more accurate discriminant decision than the one produced by LDA.

Linear discriminants are computed by minimizing the regularization functional [58]. The common regularization technique [21] for resolving the singularity problem is well justified in the framework of statistical learning theory. The resulting discriminants capture both regular and irregular information, where regular discriminants reside in the range space of $\mathbf{S}_w$, while irregular discriminants reside in the null space of $\mathbf{S}_w$. Linear discriminants are computed by regularized LS regression. The method and its nonlinear extension belong to the same framework where SVMs are formulated.

16.5 Uncorrelated LDA and Orthogonal LDA

Uncorrelated LDA [34] computes the optimal discriminant vectors that are $\mathbf{S}_t$-orthogonal. It extracts features that are uncorrelated in the dimension-reduced space. It overcomes the small sample size problem by optimizing a generalized Fisher criterion. If there are a large number of samples in each class, uncorrelated LDA may overfit noise in the data.

The solution to uncorrelated LDA can be found by optimizing [76]:

$$\mathbf{W} = \arg\max_{\mathbf{W}} \{\mathrm{tr}((\mathbf{W}^T \mathbf{S}_t \mathbf{W})^\dagger \mathbf{W}^T \mathbf{S}_b \mathbf{W})\}, \tag{16.11}$$

where superscript $\dagger$ denotes the pseudoinverse. Specifically, suppose that r vectors $\boldsymbol{w}_1, \boldsymbol{w}_2, \ldots, \boldsymbol{w}_r$ are obtained, then the $(r+1)$-th vector $\boldsymbol{w}_{r+1}$ of uncorrelated LDA is the one that maximizes the Fisher criterion function $E_{\mathrm{LDA},2}(\boldsymbol{w})$ given by (16.7) subject to the constraints

$$\boldsymbol{w}_{r+1}^T \mathbf{S}_t \boldsymbol{w}_i = 0, \quad i = 1, \ldots, r. \tag{16.12}$$

The algorithm given in [34] finds $\boldsymbol{w}_i$ successively.

The uncorrelated LDA transformation maps all data points from the same class to a common vector. Both regularized LDA and PCA+LDA are regularized versions of uncorrelated LDA. A unified framework for generalized LDA [33] elucidates the properties of various algorithms and their relationships via a transfer function.

To overcome the rank limitation of LDA, the Foley–Sammon optimal discriminant vector method [20] aims to find an optimal set of orthonormal discriminant vectors that maximize the Fisher discriminant criterion under the orthogonal constraint. The Foley–Sammon method outperforms classical LDA in the sense that it can obtain more discriminant vectors for recognition, but its solution is more complicated than other LDA methods. The multiclass Foley–Sammon method can only extract the linear features of the input patterns, and the algorithm can be based on subspace decomposition [56] or be an analytic method based on Lagrange multipliers [17]. The Foley–Sammon method does not show good performance when having to deal with nonlinear patterns, such as face patterns. Both uncorrelated LDA and Foley–Sammon LDA use the same Fisher criterion function, and the main difference is that the optimal discriminant vectors generated by uncorrelated LDA are $\mathbf{S}_t$-orthogonal to one another, while the optimal discriminant vectors of Foley–Sammon LDA are orthogonal to one another.

An uncorrelated optimal discrimination vector method [34] uses the constraint of statistical uncorrelation. Orthogonal LDA [17, 76] enforces $\mathbf{W}$ in Fisher's criterion (16.8) to be orthogonal: $\mathbf{W}^T\mathbf{W} = \mathbf{I}$. Orthogonal LDA provides a simple and efficient way for computing orthogonal transformations in the framework of LDA. The discriminant vectors of orthogonal LDA are orthogonal to one another, i.e., the transformation matrix of orthogonal LDA is orthogonal. Orthogonal LDA often leads to better performance than uncorrelated LDA in classification. The features in the reduced space of uncorrelated LDA [34] are uncorrelated, while the discriminant vectors of orthogonal LDA [76] are orthogonal to one another. Geometrically, both uncorrelated LDA and orthogonal LDA project the data onto the subspace spanned by the centroids. Uncorrelated LDA may be sensitive to the noise in the data. Regularized orthogonal LDA is proposed in [80].

The approach of common vectors extracts the common properties of classes in the training set by eliminating the differences of the samples in each class [25]. A common vector for each individual class is obtained by removing all the features that are in the direction of the eigenvectors corresponding to the nonzero eigenvalues of the scatter matrix of its own class. The discriminative common vectors are obtained from the common vectors. Every sample in a given class produces the same unique common vector when they are projected onto the null space of $\mathbf{S}_w$. The optimal projection vectors are found by using the common vectors and the discriminative common vectors are determined by projecting any sample from each class onto the span of optimal projection vectors. The discriminative common vector method [5] finds optimal orthonormal projection vectors in the optimal discriminant subspace. It is equivalent to the null space method but omitting the dimension-reduction step; therefore, the method exploits the original high-dimensional space. It combines kernel-based methodologies with the optimal discriminant subspace concept.

Compared with Fisherfaces, exponential discriminant analysis [82] can extract the most discriminant information that is contained in the null space of $\mathbf{S}_w$. Compared with null-space LDA, the discriminant information that is contained in the non-null space of $\mathbf{S}_w$ is not discarded. Exponential discriminant analysis is equivalent to transforming the original data into a new space by distance diffusion mapping, and LDA is then applied in such a new space. Diffusion mapping enlarges the margin between different classes, improving the classification accuracy.

16.6 LDA/GSVD and LDA/QR

A generalization of LDA by using generalized SVD (LDA/GSVD) [32, 73] can be used to solve the problem of singularity of $\mathbf{S}_w$. LDA/GSVD has numerical advantages over the two-stage approach, and is a special case of pseudoinverse LDA, where the pseudoinverse is applied to $\mathbf{S}_t$. It avoids the inversion of $\mathbf{S}_w$ by applying generalized SVD. The nonsingularity of $\mathbf{S}_w$ is not required, and it solves the eigendecomposition of $\mathbf{S}_t^{\dagger}\mathbf{S}_b$ [73]. The solution to LDA/GSVD can be obtained by computing the eigendecomposition on the matrix $\mathbf{S}_t^{\dagger}\mathbf{S}_b$. LDA/GSVD computes the solution exactly without losing any information, but with high computational cost.

The criterion F_0 used in [73] is

$$F_0(\mathbf{W}) = \mathrm{tr}[\mathbf{S}_b^{\dagger}\mathbf{S}_w]. \tag{16.13}$$

LDA/GSVD aims to find the optimal $\mathbf{W}$ that minimizes $F_0(\mathbf{W})$, subject to the constraint that $\mathrm{rank}(\mathbf{W}^T\mathbf{H}_b) = q$, where q is the rank of $\mathbf{S}_b$.

LDA/QR [77] is also a special case of pseudoinverse LDA, where the pseudoinverse is applied to $\mathbf{S}_b$ instead. It is a two-stage LDA extension. The first stage maximizes the separation between different classes by applying QR decomposition to a small-size matrix. The distinct property of this stage is its low time/space complexity. The second stage incorporates both between-class and within-class information by applying LDA to the reduced scatter matrices resulting from the first stage. The computational complexity of LDA/QR is $O(Nd)$ for N training examples of d dimensions. LDA/QR scales to large datasets since it does not require the entire data in main memory. Both LDA/QR and Fisherfaces are approximations of LDA/GSVD but LDA/QR is much more efficient than PCA+LDA.

16.7 Incremental LDA

Examples of incremental LDA (ILDA) algorithms are neural network-based LDA [49], IDR/QR [75], and GSVD-ILDA [86]. Iterative algorithms for neural network-based LDA [7, 49] require $O(d^2)$ time for one-step update, where d is the dimension of the data.

IDR/QR [75] applies QR decomposition at the first stage to maximize the separability between different classes. The second stage incorporates both between-class and within-class information by applying LDA on the reduced scatter matrices resulting from the first stage. IDR/QR does not require that the whole data matrix be in main memory, which allows it to scale to very large datasets. The computational complexity of IDR/QR is $O(NdK)$ for N training examples, K classes, and d dimensions. IDR/QR can be an order of magnitude faster than SVD or generalized SVD-based LDA algorithms.

Based on LDA/GSVD, GSVD-ILDA [86] determines the projection matrix in full space. GSVD-ILDA can incrementally learn an adaptive subspace instead of recomputing LDA/GSVD. It gives the same performance as LDA/GSVD but with much smaller computational complexity. GSVD-ILDA yields a better classification performance than the other incremental LDA algorithms do [86].

In an incremental LDA algorithm [57], $\mathbf{S}_b$ and $\mathbf{S}_w$ are incrementally updated, and then the eigenaxes of a feature space are obtained by solving an eigenproblem. In [87], the proposed algorithm for solving generalized discriminant analysis applies QR decomposition rather than SVD. It incrementally updates the discriminant vectors when new classes are inserted into the training set.

LS-ILDA [44], as an incremental version of LS-LDA [79], performs exact incremental update whenever a new data sample is added. ILDA/QR [11] is the exact incremental version of LDA/QR. It can easily handle the update from one new sample or a chunk of new samples. An incremental implementation of the maximum margin criterion method can be found in [72].

16.8 Other Discriminant Methods

Neighborhood component analysis [23] is a nonparametric learning method that handles the tasks of distance learning and dimension reduction. It maximizes the between-class separability by maximizing a stochastic variant of the leave-one-out k-NN score on the training set. A Mahalanobis distance measure is learned for k-NN classification. Nearest neighbor discriminant analysis [59] is another nonparametric linear feature extraction method proposed from the view of the nearest neighbor classification.

Subclass discriminant analysis [88] uses a single formulation for most distribution types by approximating the underlying distribution of each class with a mixture of Gaussians. The method resolves the problem of multimodally distributed classes. It is easy to use generalized EVD to find those discriminant vectors that best linearly classify the data. The major problem is to determine the optimal number of Gaussians per class, i.e., the number of subclasses. For data with Gaussian homoscedastic subclass structure, it does not guarantee to provide the discriminant subspace that minimizes the Bayes error. Mixture subclass discriminant analysis [22] alleviates this shortcoming by modifying the objective function and utilizes a partitioning procedure to aid discrimination of data with Gaussian homoscedastic subclass structure.

The performance of LDA-based methods degrades when the actual distribution is non-Gaussian. To address this problem, a formulation of scatter matrices extends the two-class nonparametric discriminant analysis to multiclass cases. Multiclass nonparametric subspace analysis [42] has two complementary methods that are based on the principal space and the null space of the intraclass scatter matrix, respectively. Corresponding multiclass nonparametric feature analysis methods are derived as enhanced versions of their nonparametric subspace analysis counterparts. In another extension of LDA to multiclass [45], the approximate pairwise accuracy criteria, which weight the contribution of individual class pairs in terms of Bayes error, replace Fisher's criterion.

A Bayes-optimal LDA algorithm [26] provides the one-dimensional subspace, where the Bayes error is minimized for the C-class problem with homoscedastic Gaussian distributions by using standard convex optimization. The algorithm is then extended to the minimization of Bayes error in the more general case of heteroscedastic distributions by means of an appropriate kernel mapping function.

Recursive LDA [70] determines the discriminant direction for separating different classes by maximizing the generalized Rayleigh quotient, and generates a new sample set by projecting the samples into a subspace that is orthogonal to this discriminant direction. The second step is repeated. The kth discriminating vector extracted can be interpreted as the kth best direction for separation by the nature of the optimization process involved. The recursive process naturally stops when the between-class scatter is zero. The total number of discriminating vectors from recursive LDA is independent of the number of classes C while that of LDA is limited to $C - 1$. All the discriminating vectors found may not form a complete basis of even finite-dimensional feature space.

Linear boundary discriminant analysis [53] increases class separability by reflecting the different significances of nonboundary and boundary patterns. This is achieved by defining two scatter matrices and solving the eigenproblem on the criterion described by these scatter matrices. The possible number of features obtained is larger than that by LDA, and it brings better classification performance. To distinguish the boundary patterns from the nonboundary patterns, relevant patterns election technique can be employed, by selecting boundary patterns according to a proximity measure.

Rotational LDA [62] applies an additional rotational transform prior to dimension-reduction transformation. The rotational transform rotates the feature vectors in the original feature space around their respective class centroids in such a way that the overlap between the classes in the reduced feature space is further minimized.

IDA technique [55] is based on a numerical optimization of an information-theoretic objective function, which can be computed analytically. If the classes conform to the homoscedastic Gaussian conditions, IDA reduces to LDA and is an optimal feature extraction technique in the sense of Bayes. When class-conditional pdfs are highly overlapped, IDA outperforms other second-order techniques. In [83], LDA method is proposed by maximizing a nonparametric estimate of the mutual information between linearly transformed input data and the class labels. The method can produce linear transformations that can significantly boost class separability, especially for nonlinear classification.

Maximum margin criterion (MMC) [40] is applied to dimension reduction. The optimal transformation is computed by maximizing the sum of all interclass distances. MMC does not involve the inversion of scatter matrices, and thus avoids the small sample size problem implicitly. The MMC is defined as [40]

$$J_{\text{MMC}}(\mathbf{W}) = \text{tr}(\mathbf{W}^T(\mathbf{S}_b - \mathbf{S}_w)\mathbf{W}). \tag{16.14}$$

The projection matrix $\mathbf{W}$ can be found as the eigenvectors of $\mathbf{S}_b - \mathbf{S}_w$ corresponding to the largest eigenvalues. MMC is not equivalent to the Fisher criterion. The discriminant vectors using the two criteria are different. The MMC method is an efficient algorithm to compute the projection matrix of MMC under the constraint that $\mathbf{W}^T\mathbf{S}_t\mathbf{W} = \mathbf{I}$. It is found to be the same as uncorrelated LDA [76].

Maximum margin projection [68] aims to project data samples into the most discriminative subspace, where clusters are most well separated. It projects input patterns onto the normal of the maximum margin separating hyperplanes. As a result, the method only depends on the geometry of the optimal decision boundary. The problem is a nonconvex one, which can be decomposed into a series of convex subproblems using the constrained concave–convex procedure (CCCP). The computation time is linear in the size of dataset. Maximum margin projection extracts a subspace more suitable for discrimination than geometry-based methods.

Multi-view discriminant analysis [35] seeks a single discriminant common space for multiple views by jointly learning multiple view-specific linear transforms. It is formulated as the optimization of a generalized Rayleigh quotient, i.e., maximizing the between-class variations and minimizing the within-class variations from both intra-view and inter-view in the common space. An analytical solution is obtained through generalized EVD.

Besides interpretability, sparse LDA may also be motivated by robustness to noise and computational efficiency in prediction. Some algorithms are exact sparse LDA [51], greedy sparse LDA [51], penalized LDA [69], sparse discriminant analysis (http://www2.imm.dtu.dk/pubdb/views/publication_details.php?id=5671) [13], group-lasso optimal scoring solver [50], and sparse uncorrelated LDA [84]. Sparsity is typically introduced by adding the L_1-norm of the transformation matrix to the objective function. Sparse uncorrelated LDA [84] incorporates sparsity into the uncorrelated LDA transformation by seeking the solution with minimum L_1-norm from all minimum dimension solutions of the generalized uncorrelated LDA.

One-class LDA [18] isolates normal data from outliers. The discrimination outlier/dominant is formalized from the properties of the diagonal elements of the hat matrix which are traditionally used as outlier diagnostics in linear regression. Considering a Gaussian model for the data, one-class LDA is derived but is limited to deal with linear target data.

16.9 Nonlinear Discriminant Analysis

The nonlinear discriminant analysis network proposed in [14] uses the MLP architecture and Fisher's determinant ratio criterion. After an MLP-like nonlinear mapping of input vectors, the eigenvector-based linear map of Fisher's analysis is applied to the last hidden-layer outputs. When compared with MLP, the nonlinear discriminant analysis network can provide better results in imbalanced-class problems. For these problems, MLP tends to underemphasize the small class samples, while the target-free training of nonlinear discriminant analysis gives more balanced classifiers. Natural-gradient training for nonlinear discriminant analysis network [24] is comparable with those obtained with CG training, although CG has reduced complexity.

A layered lateral network-based LDA network and an MLP-based nonlinear discriminant analysis network are proposed in [49]. The two-layer LDA network determines the LDA projection matrix, where each layer is a Rubner–Tavan PCA network. This algorithm performs a simultaneous diagonalization of two matrices: $\mathbf{S}_w$ and $\mathbf{S}_t$. The network output gives the eigenvectors of $\mathbf{S}_w^{-1}\mathbf{S}_t$. This method has slow convergence, particularly when the input dimension is high [49].

Generalized discriminant analysis [1] extends LDA from linear domain to a nonlinear domain via the kernel trick. It aims to solve a generalized eigenvalue problem, which is always implemented by SVD. Semi-supervised generalized discriminant analysis [85] utilizes unlabeled data to maximize an optimality criterion of generalized discriminant analysis and formulates the problem as an optimization problem that is solved using CCCP. Kernel subclass discriminant analysis [88] can resolve the problem of nonlinearly separable classes by using the kernel between-subclass scatter matrix. Many kernel-based nonlinear discriminant analysis methods are expounded in Sect. 20.4.

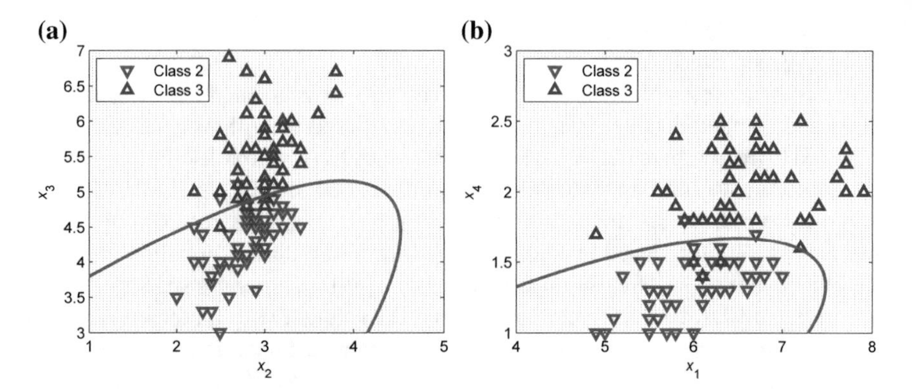

Fig. 16.3 Classification of the iris dataset using discriminant analysis. **a** x_2 vs. x_3. **b** x_1 vs. x_4

Locally linear discriminant analysis [36] is an approach to nonlinear discriminant analysis that involves a set of locally linear transformations. Input vectors are projected into multiple local feature spaces by the corresponding linear transformations to yield classes that maximize the between-class covariance while minimizing the within-class covariance. For nonlinear multiclass discrimination, the method is computationally highly efficient compared to generalized discriminant analysis [1]. The method does not suffer from overfitting due to the linear base structure of the solution.

Semi-supervised local LDA [66] preserves the global structure of unlabeled samples in addition to separating labeled samples in different classes from one another. It has an analytic form of the globally optimal solution and can be computed based on eigendecomposition.

LDA tends to give undesired results if samples in a class form several separate clusters (multimodal) or there are outliers. Locality-preserving projection [30] is an unsupervised dimension-reduction method that works well with multimodal labeled data due to its locality-preserving property. It seeks a transformation matrix such that nearby data pairs in the original space are kept close in the embedding space. The idea of Laplacian score is to evaluate each feature by its locality-preserving power, showing similarity in spirit to locality-preserving projection [31]. Local LDA [65] effectively combines the ideas of LDA and locality-preserving projection, that is, local LDA maximizes between-class separability and preserves within-class local structure, and thus works well even when within-class multimodality or outliers exist. The solution can be easily computed just by solving a generalized eigenvalue problem, thus resulting in high scalability in data visualization and classification tasks.

Example 16.3 Discriminant analysis fits a parametric model to the training data and interpolates to classify new data. Linear discrimination fits a multivariate normal density to each group, with a pooled estimate of covariance. Quadratic discrimination fits multivariable normal densities with covariance estimates stratified by group. Both methods use likelihood ratios to assign observations to groups. By applying quadratic discriminant analysis on Fisher's iris dataset, the classification between class 2 and class 3 is shown in Fig. 16.3.

16.10 Two-Dimensional Discriminant Analysis

Some feature extraction methods have been developed by representing images with matrix directly. Two-dimensional PCA and generalized low-rank approximations of matrices [78] reduce the reconstruction error without considering the classification.

Two-dimensional LDA algorithms [37, 41, 71, 74] provide an efficient approach to image feature extraction and can overcome the small sample size problem. Based on image matrix, two-dimensional LDA aims to minimize the within-class distances and maximize between-class distances, and fails when each class does not belong to single Gaussian distribution or their centers overlap.

Two-dimensional nearest neighbor discriminant analysis [60] extracts features to improve the performance of nearest neighbor classification. The method can be regarded as a two-dimensional extension of nearest neighbor discriminant analysis [59] with matrix-based image representation.

Two-directional two-dimensional LDA [54] is proposed for object/face image representation and recognition to straighten out the problem of massive memory requirements of the two-dimensional LDA method. It has the advantage of higher recognition rate, less memory requirements, and better computing performance than standard PCA, 2D-PCA, and two-dimensional LDA methods.

Problems

16.1 Consider four points in two classes: class 1: $(2, 4)$, $(1, -4)$; class 2: $(-2, 3)$, $(-4, 2)$. Compute the scatter matrices for LDA.

16.2 Show how to transform the generalized eigenvalue problem

$$\max \boldsymbol{x}^T \mathbf{S}_b \boldsymbol{x} \quad \text{subject to } \boldsymbol{x}^T \mathbf{S}_w \boldsymbol{x} = 1$$

into a standard eigenvalue problem.

16.3 Consider a dataset $\{(1, 1, 1), (1, 2, 1), (1.5, 1, 1), (1, 3, 1), (4, 4, 2), (3, 5, 2), (5, 4, 2), (6, 4, 2)\}$, where each pattern consists of a x-coordinate, a y-coordinate, and a class label. Find the projection directions associated with LDA.

References

1. Baudat, G., & Anouar, F. (2000). Generalized discriminant analysis using a kernel approach. *Neural Computation*, *12*, 2385–2404.
2. Belhumeur, P. N., Hespanha, J. P., & Kriegman, D. J. (1997). Eigenfaces vs. fisherface: Recognition using class specific linear projection. *IEEE Transactions on Pattern Analysis and Machine Intelligence*, *19*(7), 711–720.
3. Bian, W., & Tao, D. (2014). Asymptotic generalization bound of Fisher's linear discriminant analysis. *IEEE Transactions on Pattern Analysis and Machine Intelligence*, *36*(12), 2325–2337.
4. Cai, D., He, X., & Han, J. (2008). SRDA: An efficient algorithm for large-scale discriminant analysis. *IEEE Transactions on Knowledge and Data Engineering*, *20*(1), 1–12.
5. Cevikalp, H., Neamtu, M., Wilkes, M., & Barkana, A. (2005). Discriminative common vectors for face recognition. *IEEE Transactions on Pattern Analysis and Machine Intelligence*, *27*(1), 4–13.
6. Chatterjee, C., Roychowdhury, V. P., Ramos, J., & Zoltowski, M. D. (1997). Self-organizing algorithms for generalized eigen-decomposition. *IEEE Transactions on Neural Networks*, *8*(6), 1518–1530.
7. Chatterjee, C., & Roychowdhury, V. P. (1997). On self-organizing algorithms and networks for class-separability features. *IEEE Transactions on Neural Networks*, *8*(3), 663–678.
8. Chen, L.-F., Liao, H.-Y. M., Ko, M.-T., Lin, J.-C., & Yu, G.-J. (2000). A new LDA-based face recognition system which can solve the small sample size problem. *Pattern Recognition*, *33*(10), 1713–1726.
9. Chen, W. S., Yuen, P. C., & Huang, J. (2005). A new regularized linear discriminant analysis methods to solve small sample size problems. *International Journal of Pattern Recognition and Artificial Intelligence*, *19*(7), 917–935.
10. Ching, W.-K., Chu, D., Liao, L.-Z., & Wang, X. (2012). Regularized orthogonal linear discriminant analysis. *Pattern Recognition*, *45*, 2719–2732.
11. Chu, D., Liao, L.-Z., Ng, M. K.-P., & Wang, X. (2015). Incremental linear discriminant analysis: A fast algorithm and comparisons. *IEEE Transactions on Neural Networks and Learning Systems*, *26*(11), 2716–2735.
12. Chu, D., & Thye, G. S. (2010). A new and fast implementation for null space based linear discriminant analysis. *Pattern Recognition*, *43*(4), 1373–1379.
13. Clemmensen, L., Hastie, T., Witten, D., & Ersbøll, B. (2011). Sparse discriminant analysis. *Technometrics*, *53*(4), 406–413.
14. Cruz, C. S., & Dorronsoro, J. R. (1998). A nonlinear discriminant algorithm for feature extraction and data classification. *IEEE Transactions on Neural Networks*, *9*(6), 1370–1376.
15. Demir, G. K., & Ozmehmet, K. (2005). Online local learning algorithms for linear discriminant analysis. *Pattern Recognition Letters*, *26*, 421–431.
16. Ding, C., & Li, T. (2007). Adaptive dimension reduction using discriminant analysis and K-means clustering. In *Proceedings of the 24th International Conference on Machine Learning* (pp. 521–528). Corvallis, OR.
17. Duchene, L., & Leclerq, S. (1988). An optimal transformation for discriminant and principal component analysis. *IEEE Transactions on Pattern Analysis and Machine Intelligence*, *10*(6), 978–983.
18. Dufrenois, F., & Noyer, J. C. (2013). Formulating robust linear regression estimation as a one-class LDA criterion: Discriminative hat matrix. *IEEE Transactions on Neural Networks and Learning Systems*, *24*(2), 262–273.
19. Fisher, R. A. (1936). The use of multiple measurements in taxonomic problems. *Annals of Eugenics*, *7*, 179–188.
20. Foley, D. H., & Sammon, J. W. (1975). An optimal set of discriminant vectors. *IEEE Transactions on Computers*, *24*(3), 281–289.
21. Friedman, J. H. (1989). Regularized discriminant analysis. *Journal of the American Statistical Association*, *84*(405), 165–175.

22. Gkalelis, N., Mezaris, V., & Kompatsiaris, I. (2011). Mixture subclass discriminant analysis. *IEEE Signal Processing Letters, 18*(5), 319–322.
23. Goldberger, J., Roweis, S., Hinton, G., & Salakhutdinov, R. (2005). Neighbourhood components analysis. In L. K. Saul, Y. Weiss, & L. Bottou (Eds.), *Advances in neural information processing systems* (Vol. 17, pp. 513–520). Cambridge: MIT Press.
24. Gonzalez, A., & Dorronsoro, J. R. (2007). Natural learning in NLDA networks. *Neural Networks, 20*, 610–620.
25. Gulmezoglu, M. B., Dzhafarov, V., Keskin, M., & Barkana, A. (1999). A novel approach to isolated word recognition. *IEEE Transactions on Speech and Audio Processing, 7*(6), 620–628.
26. Hamsici, O. C., & Martinez, A. M. (2008). Bayes optimality in linear discriminant analysis. *IEEE Transactions on Pattern Analysis and Machine Intelligence, 30*(4), 647–657.
27. Hastie, T., Tibshirani, R., & Buja, A. (1994). Flexible discriminant analysis by optimal scoring. *Journal of the American Statistical Association, 89*(428), 1255–1270.
28. Hastie, T., & Tibshirani, R. (1995). Penalized discriminant analysis. *Annals of Statistics, 23*(1), 73–102.
29. Hastie, T., Tibshirani, R., & Friedman, J. H. (2001). *The elements of statistical learning: Data mining, inference, and prediction*. New York: Springer.
30. He, X., & Niyogi, P. (2004). Locality preserving projections. In S. Thrun, L. Saul, & B. Scholkopf (Eds.), *Advances in neural information processing systems* (Vol. 16, pp. 153–160). Cambridge: MIT Press.
31. He, X., Yan, S., Hu, Y., Niyogi, P., & Zhang, H.-J. (2005). Face recognition using Laplacianfaces. *IEEE Transactions on Pattern Analysis and Machine Intelligence, 27*(3), 328–340.
32. Howland, P., & Park, H. (2004). Generalizing discriminant analysis using the generalized singular value decomposition. *IEEE Transactions on Pattern Analysis and Machine Intelligence, 26*(8), 995–1006.
33. Ji, S., & Ye, J. (2008). Generalized linear discriminant analysis: A unified framework and efficient model selection. *IEEE Transactions on Neural Networks, 19*(10), 1768–1782.
34. Jin, Z., Yang, J., Hu, Z., & Lou, Z. (2001). Face recognition based on the uncorrelated discrimination transformation. *Pattern Recognition, 34*(7), 1405–1416.
35. Kan, M., Shan, S., Zhang, H., Lao, S., & Chen, X. (2016). Multi-view discriminant analysis. *IEEE Transactions on Pattern Analysis and Machine Intelligence, 38*(1), 188–194.
36. Kim, T.-K., & Kittler, J. (2005). Locally linear discriminant analysis for multimodally distributed classes for face recognition with a single model image. *IEEE Transactions on Pattern Analysis and Machine Intelligence, 27*(3), 318–327.
37. Kong, H., Wang, L., Teoh, E. K., Wang, J. G., & Venkateswarlu, R. (2004). A framework of 2D Fisher discriminant analysis: Application to face recognition with small number of training samples. In *Proceedings of the IEEE Computer Society Conference on Computer Vision and Pattern Recognition (CVPR)* (Vol. 2, pp. 1083–1088). San Diego, CA.
38. Kwak, N., & Lee, J.-W. (2010). Feature extraction based on subspace methods for regression problems. *Neurocomputing, 73*, 1740–1751.
39. Kyperountas, M., Tefas, A., & Pitas, I. (2007). Weighted piecewise LDA for solving the small sample size problem in face verification. *IEEE Transactions on Neural Networks, 18*(2), 506–519.
40. Li, H., Jiang, T., & Zhang, K. (2006). Efficient and robust feature extraction by maximum margin criterion. *IEEE Transactions on Neural Networks, 17*(1), 157–165.
41. Li, M., & Yuan, B. (2005). 2D-LDA: A statistical linear discriminant analysis for image matrix. *Pattern Recognition Letters, 26*, 527–532.
42. Li, Z., Lin, D., & Tang, X. (2009). Nonparametric discriminant analysis for face recognition. *IEEE Transactions on Pattern Analysis and Machine Intelligence, 31*(4), 755–761.
43. Liu, C., & Wechsler, H. (2000). Robust coding scheme for indexing and retrieval from large face databases. *IEEE Transactions on Image Processing, 9*, 132–137.
44. Liu, L.-P., Jiang, Y., & Zhou, Z.-H. (2009). Least square incremental linear discriminant analysis. In *Proceedings of the 9th IEEE International Conference on Data Mining* (pp. 298–306). Miami, FL.

45. Loog, M., Duin, R. P. W., & Haeb-Umbach, R. (2001). Multiclass linear dimension reduction by weighted pairwise Fisher criteria. *IEEE Transactions on Pattern Analysis and Machine Intelligence*, *23*, 762–766.
46. Lu, J., Plataniotis, K. N., & Venetsanopoulos, A. N. (2003). Face recognition using LDA based algorithms. *IEEE Transactions on Neural Networks*, *14*(1), 195–200.
47. Martinez, A. M., & Kak, A. C. (2001). PCA versus LDA. *IEEE Transactions on Pattern Analysis and Machine Intelligence*, *23*, 228–233.
48. Michie, D., Spiegelhalter, D. J., & Taylor, C. C. (Eds.). (1994). *Machine learning, neural and statistical classification*. New York: Ellis Horwood.
49. Mao, J., & Jain, A. K. (1995). Artificial neural networks for feature extraction and multivariate data projection. *IEEE Transactions on Neural Networks*, *6*(2), 296–317.
50. Merchante, L., Grandvalet, Y., & Govaert, G. (2012). An efficient approach to sparse linear discriminant analysis. In *Proceedings of the 29th International Conference on Machine Learning* (pp. 1167-1174). Edinburgh, UK.
51. Moghaddam, B., Weiss, Y., & Avidan, S. (2006). Generalized spectral bounds for sparse LDA. In *Proceedings of the 23th International Conference on Machine Learning* (pp. 641–648). Pittsburgh, PA.
52. Muller, K. R., Mika, S., Ratsch, G., Tsuda, K., & Scholkopf, B. (2001). An introduction to kernel-based learning algorithms. *IEEE Transactions on Neural Networks*, *12*(2), 181–201.
53. Na, J. H., Park, M. S., & Choi, J. Y. (2010). Linear boundary discriminant analysis. *Pattern Recognition*, *43*, 929–936.
54. Nagabhushan, P., Guru, D. S., & Shekar, B. H. (2006). $(2D)^2$ FLD: An efficient approach for appearance based object recognition. *Neurocomputing*, *69*, 934–940.
55. Nenadic, Z. (2007). Information discriminant analysis: Feature extraction with an information-theoretic objective. *IEEE Transactions on Pattern Analysis and Machine Intelligence*, *29*(8), 1394–1407.
56. Okada, T., & Tomita, S. (1985). An optimal orthonormal system for discriminant analysis. *Pattern Recognition*, *18*(2), 139–144.
57. Pang, S., Ozawa, S., & Kasabov, N. (2004). One-pass incremental membership authentication by face classification. In D. Zhang & A. K. Jain (Eds.), *Biometric authentication* (pp. 155–161). New York: Springer.
58. Peng, J., Zhang, P., & Riedel, N. (2008). Discriminant learning analysis. *IEEE Transactions on Systems, Man, and Cybernetics Part B*, *38*(6), 1614–1625.
59. Qiu, X., & Wu, L. (2006). Nearest neighbor discriminant analysis. *International Journal of Pattern Recognition and Artificial Intelligence*, *20*(8), 1245–1259.
60. Qiu, X., & Wu, L. (2007). Two-dimensional nearest neighbor discriminant analysis. *Neurocomputing*, *70*, 2572–2575.
61. Sharma, A., & Paliwal, K. K. (2008). A gradient linear discriminant analysis for small sample sized problem. *Neural Processing Letters*, *27*, 17–24.
62. Sharma, A., & Paliwal, K. K. (2008). Rotational linear discriminant analysis technique for dimensionality reduction. *IEEE Transactions on Knowledge and Data Engineering*, *20*(10), 1336–1347.
63. Skurichina, M., & Duin, R. P. W. (1996). Stabilizing classifiers for very small sample size. In *Proceedings of International Conference on Pattern Recognition* (pp. 891–896).
64. Srivastava, S., & Gupta, M. R. (2006). Distribution-based Bayesian minimum expected risk for discriminant analysis. In *Proceedings of IEEE International Symposium on Information Theory* (pp. 2294–2298).
65. Sugiyama, M. (2007). Dimensionality reduction of multimodal labeled data by local Fisher discriminant analysis. *Journal of Machine Learning Research*, *8*, 1027–1061.
66. Sugiyama, M., Ide, T., Nakajima, S., & Sese, J. (2010). Semi-supervised local Fisher discriminant analysis for dimensionality reduction. *Machine Learning*, *78*, 35–61.
67. Swets, D. L., & Weng, J. (1996). Using discriminant eigenfeatures for image retrieval. *IEEE Transactions on Pattern Analysis and Machine Intelligence*, *18*(8), 831–836.

68. Wang, F., Zhao, B., & Zhang, C. (2011). Unsupervised large margin discriminative projection. *IEEE Transactions on Neural Networks*, *22*(9), 1446–1456.
69. Witten, D. M., & Tibshirani, R. (2011). Penalized classification using Fisher's linear discriminant. *Journal of the Royal Statistical Society Series B*, *73*(5), 753–772.
70. Xiang, C., Fan, X. A., & Lee, T. H. (2006). Face recognition using recursive Fisher linear discriminant. *IEEE Transactions on Image Processing*, *15*(8), 2097–2105.
71. Xiong, H., Swamy, M. N. S., & Ahmad, M. O. (2005). Two-dimension FLD for face recognition. *Pattern Recognition*, *38*(7), 1121–1124.
72. Yan, J., Zhang, B., Yan, S., Yan, S., Yang, Q., Li, H., Chen, Z., Xi, W., Fan, W., Ma, W.-Y., & Cheng, Q. (2004). IMMC: Incremental maximum margin criterion. In *Proceedings of the 10th ACM SIGKDD International Conference on Knowledge Discovery and Data Mining* (pp. 725–730). Seattle, WA.
73. Ye, J., Janardan, R., Park, C. H., & Park, H. (2004). An optimization criterion for generalized discriminant analysis on undersampled problems. *IEEE Transactions on Pattern Analysis and Machine Intelligence*, *26*(8), 982–994.
74. Ye, J., Janardan, R., & Li, Q. (2004). Two-dimensional linear discriminant analysis. *Advances in neural information processing systems* (Vol. 17, pp. 1569–1576). Vancouver, Canada.
75. Ye, J., Li, Q., Xiong, H., Park, H., Janardan, R., & Kumar, V. (2005). IDR/QR: An incremental dimension reduction algorithm via QR decomposition. *IEEE Transactions on Knowledge and Data Engineering*, *17*(9), 1208–1222.
76. Ye, J. (2005). Characterization of a family of algorithms for generalized discriminant analysis on undersampled problems. *Journal of Machine Learning Research*, *6*, 483–502.
77. Ye, J., & Li, Q. (2005). A two-stage linear discriminant analysis via QR-decomposition. *IEEE Transactions on Pattern Analysis and Machine Intelligence*, *27*(6), 929–941.
78. Ye, J. (2005). Generalized low rank approximations of matrices. *Machine Learning*, *61*(1), 167–191.
79. Ye, J. (2007). Least squares linear discriminant analysis. In *Proceedings of the 24th International Conference on Machine Learning* (pp. 1087–1094). Corvallis, OR.
80. Ye, J., & Xiong, T. (2006). Computational and theoretical analysis of null space and orthogonal linear discriminant analysis. *Journal of Machine Learning Research*, *7*, 1183–1204.
81. Zaidi, N. A., Webb, G. I., Carman, M. J., Petitjean, F., & Cerquides, J. (2016). ALR^n: Accelerated higher-order logistic regression. *Machine Learning*, *104*(2), 151–194.
82. Zhang, T., Fang, B., Tang, Y. Y., Shang, Z., & Xu, B. (2010). Generalized discriminant analysis: A matrix exponential approach. *IEEE Transactions on Systems, Man, and Cybernetics Part B*, *40*(1), 186–197.
83. Zhang, H., Guan, C., & Li, Y. (2011). A linear discriminant analysis method based on mutual information maximization. *Pattern Recognition*, *44*(4), 877–885.
84. Zhang, X., Chu, D., & Tan, R. C. E. (2016). Sparse uncorrelated linear discriminant analysis for undersampled problems. *IEEE Transactions on Neural Networks and Learning Systems*, *27*(7), 1469–1485.
85. Zhang, Y., & Yeung, D.-Y. (2011). Semisupervised generalized discriminant analysis. *IEEE Transactions on Neural Networks*, *22*(8), 1207–1217.
86. Zhao, H., & Yuen, P. C. (2008). Incremental linear discriminant analysis for face recognition. *IEEE Transactions on Systems, Man, and Cybernetics Part B*, *38*(1), 210–221.
87. Zheng, W. (2006). Class-incremental generalized discriminant analysis. *Neural Computation*, *18*, 979–1006.
88. Zhu, M., & Martinez, A. M. (2006). Subclass discriminant analysis. *IEEE Transactions on Pattern Analysis and Machine Intelligence*, *28*(8), 1274–1286.
89. Zhu, M., & Martinez, A. M. (2008). Pruning noisy bases in discriminant analysis. *IEEE Transactions on Neural Networks*, *19*(1), 148–157.

Chapter 17
Reinforcement Learning

17.1 Introduction

Reinforcement learning has its origin in the psychology of animal learning. It awards the learner (agent) for correct actions and punishes for wrong actions. In the mammalian brain, learning by reinforcement is a function of brain nuclei known as the basal ganglia. The basal ganglia use this reward-related information to modulate sensory-motor pathways so as to render future behaviors more rewarding [37]. Dopaminergic neurons provide a prediction error akin to the error computed in the temporal-difference (TD) learning models of reinforcement learning [16].

Reinforcement learning is a type of machine learning in which an agent (e.g., a real or simulated robot) seeks an effective policy for solving a sequential decision task. Such a policy dictates how the agent should behave in each state it may encounter in order to maximize total expected reward (or minimize punishment) by trial-and-error interaction with a dynamic environment [3, 45]. There is no need to specify how the task is to be achieved. The computed difference, termed reward-prediction error, has been shown to correlate very well with the phasic activity of dopamine-releasing neurons projecting from the substantia nigra in nonhuman primates [42].

Reinforcement learning can be subdivided into two fundamental problems: learning and planning. Learning is for an agent to improve its policy from interactions with its environment, and planning is for an agent to improve its policy without further interaction with its environment.

The reinforcement learning problem is defined by three features, namely, agent–environment interface, function for evaluative feedback, and Markov property of the learning process. The agent is connected to its environment via sensors. An agent acts in an unknown or partly known environment with the goal of maximizing an external reward signal. This is in keeping with the learning situations an animal encounters in the real world, where there is no supervision but rewards and penalties such as hunger, satiety, pain, and pleasure abound.

K.-L. Du and M. N. S. Swamy, *Neural Networks and Statistical Learning*,
https://doi.org/10.1007/978-1-4471-7452-3_17

Reinforcement learning is a special case of supervised learning, where the exact desired output is unknown. A learner must explicitly explore its environment. The teacher supplies only feedback about success or failure of an answer. This is cognitively more plausible than supervised learning since a fully specified correct answer might not always be available to the learner or even the teacher. It is based only on an evaluative feedback, which can be noisy or sparse: the information as to whether or not the actual output is close to the estimate. Reinforcement learning is a learning procedure that *rewards* the agent for its *good* output result and *punishes* it for the *bad* output result. Explicit computation of derivatives is not required. This, however, presents a slower learning process. For a control system, if the controller still works properly after an input, the output is judged as *good*; otherwise, it is considered as *bad*. The evaluation of the binary output, called *external reinforcement*, is used as the error signal.

When considering the base elements of decision optimization (states, actions, and reinforcements) from a system-theoretic perspective, the reinforcement learning model could be implied together with the interpretation as a decision tree. The objective of reinforcement learning is to find a path through the decision tree which maximizes the sum of rewards. Reinforcement learning is a practical tool for solving sequential decision problems that can be modeled as Markov decision problems. It is among the most general frameworks of learning control to create truly autonomous learning systems. Reinforcement learning is widely used in robot control and artificial intelligence. It has influenced a number of fields, including operations research, cognitive science, optimal control, psychology, neuroscience, and others.

In an offline implementation, the algorithm runs on the simulator before being implemented on the real system. On the other hand, in an online implementation, the algorithm runs on a real-time basis in the real system.

In discrete reinforcement learning, numbers of states and actions are finite and countable, and the values of states (or state–action pairs) are saved in a value table whose elements are adjusted independently [48]. In contrast, continuous reinforcement learning has infinite numbers of states and actions, and function approximators are used to approximate the value function. Changing an approximator parameter may cause changes in the approximate values of the entire space.

A fundamental problem in reinforcement learning is the exploration–exploitation problem. A suitable strategy is to have higher exploration and lower exploitation at the early stage of learning, and then to decrease exploration and increase exploitation gradually. Before exploiting, however, adequate exploration and accurate estimate of action-value function should have been achieved. Although in discrete reinforcement learning, having longer exploration leads to more accurate action-value function, it may not be the case in continuous reinforcement learning. In fact, in continuous reinforcement learning, the approximation accuracy of action-value function depends on the distribution of data.

Multiobjective reinforcement learning problems have two or more objectives to be achieved by the agent, each with its own associated reward signal. It generates multiple policies rather than a single policy.

Reinforcement Learning Versus Dynamic Programming

Reinforcement learning aims to solve the same problem as optimal control. However, in reinforcement learning, the agent needs to learn about the consequences of actions in the environment by trial and error as a model of the state transition dynamics is not available to the agent. In contrast, such a model is available in optimal control.

Dynamic programming is a mathematical method for finding an optimal control and its solution using a value function in a dynamic system. It is guaranteed to give optimal solutions to Markov decision problems (MDPs) and semi-Markov decision problems. The two main algorithms of dynamic programming, value iteration and policy iteration, are based on the Bellman equation, which contains the elements of the value function as the unknowns. In dynamic programming, the transition probability and the transition reward matrices are first generated, and these matrices are then used to generate a solution. All dynamic programming algorithms are model based. The mechanism of model building is to construct the transition probability model in a simulator by straightforward counting. Implementation of dynamic programming is often a difficult and tedious process that involves a theoretical model.

Reinforcement learning is essentially a form of simulation-based dynamic programming and is primarily used to solve Markov and semi-Markov decision problems. As such, it is often called a heuristic dynamic programming technique. The environment can be modeled as a finite Markov decision process where the goal of the agent is to obtain near-optimal discounted return. In reinforcement learning, we do not estimate the transition probability or the reward matrices; instead, we simulate the system using the distributions of the governing random variables. It can learn the system structure by trial and error, and is suitable for online learning. Most reinforcement learning algorithms calculate the value function of dynamic programming. The value function is stored in the form of the so-called Q-values. Most reinforcement learning algorithms are based on the Q-value version of the Bellman equation. An algorithm that does not use transition probabilities in its updating equations is called a *model-free* algorithm. A suitable reinforcement learning algorithm can obtain a near-optimal solution.

Reinforcement learning can also avoid the computational burden encountered by dynamic programming. A major difficulty associated with the Bellman equation-based approaches is the curse of dimensionality. Consider a problem with a million state–action pairs. Using model-free algorithms, one can avoid storing the huge transition probability matrices. Nonetheless, one must still find some way to store the one million Q-values. Function approximation is a strategy for reducing the storage; it can be done by state aggregation, function fitting, and function interpolation.

17.2 Learning Through Awards

Reinforcement learning is illustrated in Fig. 17.1. It learns a mapping from situation to actions by maximizing the scalar reward or reinforcement signal, fed back from the environment or an external evaluator. An agent receives sensory inputs (as the

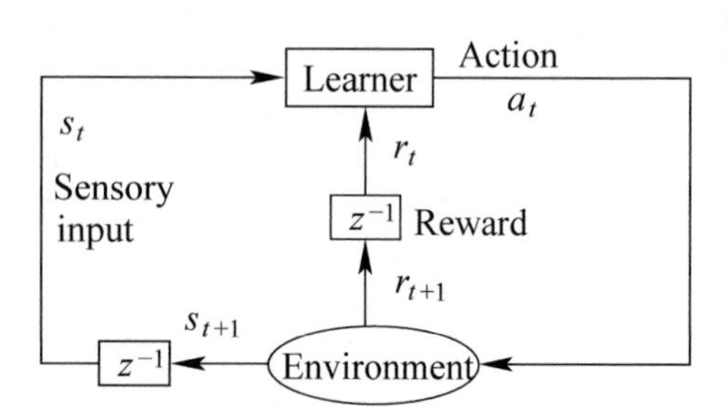

Fig. 17.1 Principle of reinforcement learning

state of the environment) from its environment. Once an action is executed, the agent receives a reinforcement signal or reward. A negative reward punishes the agent for a bad action. This loop can help the algorithm to stabilize. The trial-and-error search helps to find better action while a memory of good actions helps to keep good solutions such that a reward can be assigned. This is an exploration–exploitation trade-off.

A reinforcement learning agent has several components.

- A *policy* π is a decision-making function that specifies the next action from current state. It is a mapping from state to action. In general, the policy is a mapping from states to a probability distribution over actions. An optimal policy is any policy that maximizes the expected return in the environment.
- A *reward function* estimates the instantaneous desirability of the perceived state of the environment for an attempted control action. Design of a reward function is of critical importance.
- A *value function* predicts future reward. The agent finds a policy π that maximizes the value function.
- A *critic* evaluates the performance of the current state in response to an action taken according to current policy. The critic-agent shapes the policy by making continuous and ongoing corrections.
- A *model* is a planning tool that aids in predicting the future course of action by constructing possible future states.

A concept underlying reinforcement learning is the Markov property—only the current state affects the next state. In every step of interaction, the agent receives a feedback about the state of the environment s_{t+1} and the reward r_{t+1} for its latest action a_t. The agent chooses an action a_{t+1} representing the output function, which changes the state s_{t+1} of environment and thus leads to state s_{t+2}. The agent receives new feedback from reinforcement signal r_{t+2}. The reinforcement signal r is often delayed since it is a result of network outputs in the past. This is solved by learning a critic network which represents a cost function J predicting future reinforcement.

Reinforcement learning tries to compute an optimal policy (control strategy) π^* that selects actions a to maximize an agent's total expected future reward (called *value*) from all states.

$$\pi^* = \arg\max_{\pi} E[r|\pi].$$

Optimality for both the state-value function V_π and the state action value function Q_π is governed by the Bellman optimality equation.

The basic idea behind many reinforcement learning algorithms is to estimate the action-value function by using the Bellman equation as an iterative update,

$$Q_{i+1}(s,a) = E_{s'}\left[r + \gamma \max_{a'} Q_i(s',a')|s,a\right]. \tag{17.1}$$

Such value iteration algorithms converge to the optimal action-value function, $Q_i \to Q^*$ as $i \to \infty$.

This basic approach is impractical because the action-value function is estimated separately for each sequence. It is common to use a function approximator to estimate the action-value function, $Q(s,a;\theta) \approx Q^*(s,a)$. This is typically a linear function approximator but sometimes a nonlinear function approximator is used instead such as a neural network. We refer to a neural network function approximator with weights θ as a Q-network. A Q-network can be trained by adjusting the parameters θ_i at iteration i to reduce the mean squared error in the Bellman equation.

Reinforcement learning is solved by methods based on value functions and methods based on policy search. A hybrid actor–critic approach employs both value functions and policy search.

Reward shaping [21] provides an additional reward that does not come from the environment. The additional reward is extra information that is incorporated by the system designer and estimated on the basis of knowledge of the problem.

The scalability of reinforcement learning to high-dimensional continuous state–action systems suffers from the curse of dimensionality and credit assignment problem. Hierarchical reinforcement learning scales to problems with large state spaces by using the task (or action) structure to restrict the space of policies [13]. It deals with the credit assignment problem by using higher level policies to determine useful subgoals and learning to apply them when appropriate [2]. Free energy-based reinforcement learning [39] is capable of handling high-dimensional inputs and actions, and a credit assignment problem.

17.3 Actor–Critic Model

Reinforcement learning is also called *learning with acritic*. The learning agent can be split into two separate entities: the actor (policy) and the critic (value function). As such, reinforcement learning algorithms can be divided into three groups [4]: actor-only, critic-only, and actor–critic methods. Actor-only methods typically work with a parameterized policy over which optimization can be used directly. The optimization methods used suffer from high variance in the estimates of the gradient. Due to its gradient-descent nature, actor-only methods have strong convergence property but slow learning. Compared to critic-only methods, actor-only methods allow the policy to generate actions in the complete continuous action space.

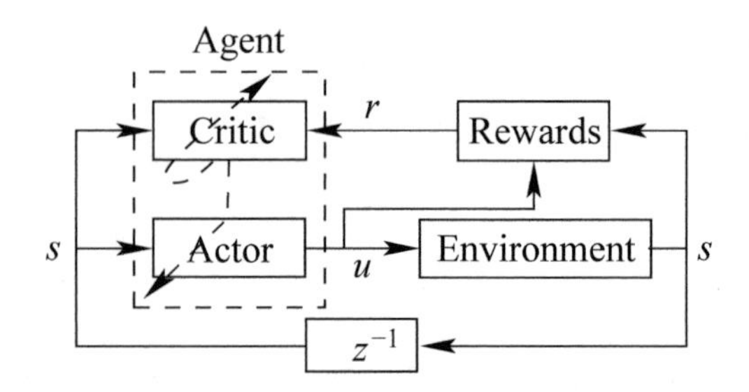

Fig. 17.2 Illustration of an actor–critic network. The actor is responsible for generating a control input u, given the current state s. The critic is responsible for updating the actor and itself

Critic-only methods have a lower variance in the estimates of expected returns [43]. A policy can be derived by selecting greedy actions [45]. Therefore, critic-only methods usually discretize the continuous action space, after which the optimization over the action space becomes a matter of enumeration. Critic-only methods, such as Q-learning [50, 51] and SARSA [45], use a state–action value function and no explicit function for the policy. For continuous state and action spaces, this will be an approximate state–action value function. For Q-learning, the system estimates an action-value function $Q(s, a)$ for all state–action pairs, and selects the optimal control based on $Q(s, a)$.

Actor–critic methods combine the merits of both the actor-only and critic-only methods. In actor–critic methods [3], the critic evaluates the quality of the current policy prescribed by the actor. An explicit representation is used for the value function of the policy [3, 23]. The actor is a separate memory structure to explicitly represent the control policy. The critic approximates and updates the value function using the rewards it receives. The state-value function $V(s)$ is then used to update the actor's policy parameters for the best control actions in the direction of performance improvement. Thus, the critic addresses the problem of prediction, whereas the actor is concerned with control. Policy gradient-based actor–critic algorithms are popular for being able to search for optimal policies using low-variance gradient estimates. The actor typically updates its policy along an estimate of the gradient (or natural gradient) of some measure of performance with respect to the policy parameters.

Figure 17.2 gives the structure of an actor–critic network. The basic building blocks are an actor which uses a stochastic method to determine the correct relation between the input and the output and an adaptive critic which learns from the TD-error to update itself and the actor, based on a prediction of future reward or punishment [3]. The external reinforcement signal r can be generated by a special sensor or be derived from the state vector.

The binary external reinforcement provides very limited information for the learning algorithm. An additional adaptive critic network [52] is usually used to predict the future reinforcement signal called *internal reinforcement*. This assures avoiding *bad* states from happening. In most formulations, reinforcement learning is achieved by using reward-prediction errors, i.e., the difference between an agent's current reward prediction and the actual reward obtained, to update the agent's reward predictions. As the reward predictions are learned, the predictions can also be used by an agent to select its next action.

Actor–critic methods have achieved superhuman skills in complex real tasks. They are prone to instability, partly due to the interaction between the actor and critic during learning. The stability of the critic can be improved by using a slowly changing critic or a low-variance critic. TD-regularized actor–critic method [33] regularizes the learning objective of the actor by penalizing the TD-error of the critic.

Actor–critic methods rely on Monte Carlo techniques in estimating the gradient of the performance measure, resulting in slow convergence. They were largely supplanted in the 1990s by TD(λ) [43] and Sarsa(λ) [38] methods that estimate action-value functions and use them directly to select actions without maintaining an explicit representation of the policy.

17.4 Model-Free and Model-Based Reinforcement Learning

Agents may use value functions or direct policy search, be model based or model free. Model-free approaches obtain an optimal decision-making policy by directly mapping environmental states to actions. They learn to predict the utility of each action in different situations but they do not learn the effects of actions. Model-based methods attempt to construct a model of the environment, typically in the form of a Markov decision process, followed by a selection of optimal actions based on that model. As a consequence, model-based methods often make better use of a limited amount of experience and thus achieve a better policy with fewer environmental interactions. Model-free methods are simpler and require less computational resources.

Policy Iteration Method

Policy iteration and policy search are two popular formulations of model-free reinforcement learning. In the policy iteration approach [3], the value function is first estimated by solving a set of linear equations and then policies are determined and updated based on the learned value function. The value function of a policy is just the expected infinite discounted reward that will be gained, at each state, by executing that policy. The optimal value function can be determined by value iteration.

Policy iteration discovers a deterministic optimal policy by generating a sequence of monotonically improving policies. Each iteration k consists of two phases: policy evaluation in which the action-value function Q^{π_k} of the current policy π_k is computed (the expectation step), and policy improvement in which the new (improved) policy π_{k+1} is generated as the greedy policy w.r.t. Q^{π_k} (the maximization step).

In approximate policy iteration, a function approximation scheme is usually employed in the policy evaluation phase. The most common approach is to find a good approximation of the value function of π_k in a real-valued function space [6]. Least squares policy iteration [24] provides good properties in convergence, stability, and sample complexity.

Slow feature analysis is used on a random walk of observations to generate basis functions from samples for reinforcement learning [25, 29]. It is shown in [5] that slow feature analysis minimizes the average bound on the value approximation error by least squares TD learning [6]. In [5], an importance sampling algorithm is proposed for least squares policy iteration in both discrete and continuous state spaces.

Although policy iteration can naturally deal with continuous states by function approximation, continuous actions are hard to handle due to the difficulty of finding maximizers of value functions with respect to actions. Control policies can vary drastically in each iteration, causing severe instability in a physical system.

Policy Gradient and Policy Search Methods

In the policy search approach, control policies are directly learned so that the return is maximized. Policy gradient methods adapt a parameterized policy by following a performance gradient estimate. They maintain a parameterized action selection policy and update the policy parameters by moving them in the direction of an estimate of the gradient of a performance measure.

Conventional policy gradient methods use Monte Carlo techniques to estimate this gradient, resulting in slow convergence. While stochastic policy gradients integrate over both the state and action spaces, deterministic policy gradients only integrate over the state space, requiring fewer samples in problems with large action spaces.

An effective policy gradient method [53] combines policy gradients with parameter-based exploration, an importance sampling technique, and an optimal baseline. A Gauss–Newton algorithm for policy optimization [10] is closely related to both the EM algorithm and the natural gradient ascent applied to Markov decision processes.

Free energy-based reinforcement learning [39] uses a restricted Boltzmann machine as a function approximator to handle high-dimensional state and action spaces. The action-value function Q is approximated by the negative free energy of the network state. It is able to solve a temporal credit assignment problem in Markov decision processes with large state and action spaces. Two energy-based actor–critic policy gradient algorithms of [15] are more robust and more effective than standard free energy-based reinforcement learning. In [9], value function approximation by restricted Boltzmann machine can improve significantly by approximating the value function by the negative expected energy, and the latter is also able to handle continuous state input.

Actor–critic algorithms are a class of policy gradient algorithms for reducing the variance of policy gradient estimates by using an explicit representation for the value function of the policy. Actor–critic algorithms follow policy iteration in spirit, and differ in many ways from the regular policy iteration algorithm that is based on the Bellman equation. They start with a stochastic policy in the simulator, where actions are selected in a probabilistic manner. Many actor–critic algorithms approximate the policy and the value function using neural networks. In policy iteration, the improved policy is greedy in the value function over the action variables. In contrast, actor–critic methods employ gradient rules to update the policy in a direction that increases the received returns. The gradient estimate is constructed using the value function.

Adaptive heuristic critic algorithm [21] is an adaptive policy iteration method in which the value function is computed by the TD(0) algorithm [43]. The method consists of a critic for learning the value function V_π for a policy and a reinforcement learning component for learning a new policy π' that maximizes the new value function. The work of the two components can be accomplished in a unified manner by Q-learning algorithm. The two components operate in an alternating or a simultaneous manner. It can be hard to select the relative learning rates so that the two components converge together. The alternating implementation is guaranteed to converge to the optimal policy under appropriate conditions [21].

Standard approach to reinforcement learning specifies feedback in the form of real-valued rewards. Preference-based policy search uses a qualitative preference signal as feedback for driving the policy learner toward better policies [11]. Other forms of feedback, most notably external advice, can also be incorporated.

Approximate policy evaluation is a difficult problem because it involves finding an approximate solution to a Bellman equation. An explicit representation of the policy can be avoided by computing improved actions on demand from the current value function. Alternatively, the policy can be represented explicitly by policy approximation.

Model-Based Methods

In model-based methods, an agent uses its experience to learn an internal model as to how the actions affect the agent and its environment. Such a model can be used in conjunction with dynamic programming to perform offline planning, often achieving better performance with fewer environmental samples than model-free methods do. The agent simultaneously uses experience to build a model and to adjust the policy, and uses the model to adjust the policy. Most of the model-building algorithms are based on the idea of computing the value function rather than computing Q-values. However, since the updating within a simulator is asynchronous and step-size based, Q-value versions are perhaps more appropriate. Model-building algorithms require more storage space in comparison to their model-free counterparts.

Dyna architecture [44, 45] is a hybrid model-based and model-free reinforcement learning algorithm, in which interactions with the environment are used both for a direct policy update with a model-free reinforcement learning algorithm, and for an update of an environmental model. Dyna applies temporal-difference learning both to real experience and to simulated experience. Dyna-2 combines temporal-difference learning with temporal-difference search, using long- and short-term memories. The long-term memory is updated from real experience, and the short-term memory is updated from simulated experience, both using the TD(λ) algorithm. The Dyna-style system proposed in [20] utilizes a temporal-difference method for direct learning and relative values for planning between two successive direct learning cycles. A simple predictor of average rewards is introduced to the actor–critic architecture in the simulation (planning) mode. The accumulated difference between the immediate reward and the average reward is used to steer the process in the right direction.

17.5 Learning from Demonstrations

As a wrong action can result in unrecoverable effects, this poses a safety–exploration dilemma, especially for a model-free approach. A common approach to safety consists of assigning negative rewards for undesired transitions, such that the most reliable policy maximizes the minimal sum of reward in the presence of uncertainties and stochasticity, yielding a worst-case or minimax problem. By assigning a negative reward, the variance of the return can be taken into account by adopting risk-sensitivity approaches [31]. There are three approaches to safe exploration [12]:

- providing initial knowledge, directing the learning in its initial stage toward more profitable and safer regions of the state space;
- deriving a policy from demonstrations through learning from demonstration; and
- providing teacher advice through interrupting exploration and providing expert knowledge by a teacher, or by consulting a teacher, when an agent is confronted with unexpected situations.

Learning from demonstrations is a paradigm by which an apprentice agent learns a control policy for a dynamic environment by observing demonstrations delivered by an expert agent. It is usually implemented as either imitation learning or inverse reinforcement learning. Imitation learning and inverse reinforcement learning can be redefined in a way that they are equivalent, in the sense that there exists an explicit bijective operator (namely, the inverse optimal Bellman operator) between their respective spaces of solutions [35].

Inverse reinforcement learning is a paradigm relying on the Markov decision processes, where the goal of the apprentice agent is to find a reward function from the expert demonstrations that could explain the expert behavior. This is a natural way to examine animal and human behaviors. Most of the existing inverse reinforcement learning algorithms assume that the environment is modeled as a Markov decision process. For more realistic partially observable environments that can be modeled as a partially observable Markov decision process, inverse reinforcement learning poses a greater challenge since it is ill-posed and computationally intractable. These obstacles are overcome in [7]. The representation of a given expert's behavior can be the case in which the expert's policy is explicitly given, or the case in which the expert's trajectories are available instead.

Inverse reinforcement learning aims at finding a reward function R that could explain the expert policy π_E from demonstrations. Demonstrations are provided in the form of sampled transitions of the expert policy. It computes a reward R for which all the expert actions and only the expert actions are optimal. The model-based reinforcement learning approach learns a transition model of the environment from data, and then derives the optimal policy using the transition model.

In general, inverse reinforcement learning occurs in a batch setting where the Markov decision process is unknown and no interaction with the Markov decision process is possible while learning. Only transitions sampled from a Markov decision process and no rewards are available to the apprentice.

Imitation learning consists in directly generalizing the expert strategy, observed in the demonstrations, to unvisited states. Imitation learning is similar to supervised learning [41]. It learns a policy which reproduces demonstrated trajectories. Imitation learning designates methods trying to generalize the policy π_E observed in the expert dataset $\mathcal{D}_E$ to any situation. The inverse reinforcement learning and imitation learning batch settings thus share the same input sets. However, the demonstrated trajectories are not always optimal, and thus the policy mimicking the demonstrated trajectories is not necessarily optimal. It is common to further improve the mimicked policy by reinforcement learning.

17.6 Temporal-Difference Learning

When the agent receives a reward (or penalty), a major problem is how to distribute the reinforcement among the decisions that led to it; this is known as the temporal credit assignment problem. Temporal-difference (TD) learning is a particularly effective model-free reinforcement learning method of solving this problem. Policy evaluation has been dominated by TD methods due to their data efficiency.

Temporal-difference methods (Python, http://github.com/chrodan/tdlearn) [43] are a class of incremental learning procedures specialized for prediction problems, that is, for using past experience with an incompletely known system to predict its future behavior. They can be viewed as gradient descent in the space of the parameters by minimizing an overall error measure. The steps in a sequence should be evaluated and adjusted according to their immediate or near-immediate successors, rather than according to the final outcome.

Whereas conventional prediction-learning methods assign credit by means of the difference between predicted and actual outcomes, temporal-difference methods assign credit by means of the difference between temporally successive predictions, and learning occurs whenever there is a change in prediction over time. In temporal-difference learning, reward estimates at successive times are compared. By comparing reward estimates rather than waiting for a reward from the environment, a temporal-difference learning system is effective for solving tasks where the reward is sparse. Temporal-difference methods require low memory and peak computation but produce accurate predictions [43]. The unique feature of temporal-difference learning is its use of bootstrapping, where predictions are used as targets during the course of learning.

With temporal-difference learning it is possible to learn good estimates of the expected return quickly by bootstrapping from other expected-return estimates. While the update target is no longer unbiased, the variance is typically much smaller and learning much faster. Temporal-difference learning uses the Bellman equations as its mathematical foundation.

Actor–critic methods [3] are a special case of temporal-difference methods. Temporal-difference error depends also on the reward signal obtained from the environment as a result of the control action. A spiking neural network model for implementing actor–critic temporal-difference learning that combines local plasticity rules with a global reward signal is given in [36]. The synaptic plasticity underlying the learning process relies on biologically plausible measures of pre- and postsynaptic activity and a global reward signal.

17.6.1 TD(λ)

TD(λ) [43] is a state-value function learning algorithm. It learns an estimate of the state-value function, V_π.

We define (s, a, r, s') to be an experience tuple summarizing a single transition in the environment, where s and s' are the states of the agent before and after the transition, a is its choice of action, and r the instantaneous reward it receives. The value of a policy is learned using the TD(0) algorithm [43]

$$V(s) \leftarrow V(s) + \eta \left[r + V(s') - V(s) \right]. \tag{17.2}$$

Whenever a state s is visited, its estimated value is updated to be closer to $r + V(s')$. If the learning rate η is adjusted properly and the policy is held fixed, TD(0) is guaranteed to converge to the optimal value function.

TD(0) rule is an instance of a class of algorithms called TD(λ), with $\lambda \in [0, 1]$ as the trace-decay rate. TD(0) looks only one step ahead when adjusting value estimates, and the convergence is very slow. At the other extreme, TD(1) updates the value of a state from the final return; it is equivalent to Monte Carlo evaluation. TD(λ) rule is applied to every state u according to its eligibility $e(u)$, rather than just to the immediately previous state s [21]:

$$V(u) \leftarrow V(u) + \eta \left[r + V(s') - V(s) \right] e(u). \tag{17.3}$$

The eligibility trace can be defined by

$$e(u) = \sum_{k=1}^{t} (\lambda\gamma)^{t-k} \delta_{u,s_k}, \tag{17.4}$$

where $\delta_{u,s_k} = 1$ if $u = s_k$ or 0 otherwise, and γ is the discount factor. The eligibility can be updated online by

$$e(u) \leftarrow \begin{cases} \gamma\lambda e(u) + 1, & \text{if } u = \text{current state} \\ \gamma\lambda e(u), & \text{otherwise} \end{cases}. \tag{17.5}$$

The eligibility trace represents the total credit assigned to a state for any subsequent errors in evaluation. If all states are visited infinitely many times, and with appropriate step sizes, TD(λ) converges to the value of the policy V_π for any λ [8]. TD(λ) often converges considerably faster for large λ, but is computationally more expensive.

TD(λ) is simple and has low cost. The estimate at each time step is moved toward an update target known as the λ-return, with λ determining the fundamental trade-off between bias and variance of the update target. TD(λ) closely approximates the forward view only for small η.

kNN-TD(λ) algorithms [30] are a series of general-purpose reinforcement learning algorithms for linear function approximation based on temporal-difference learning and weighted k-NN. These algorithms are able to learn quickly, to generalize properly over continuous state spaces, and also to be robust to a high degree of environmental noise.

17.6.2 Sarsa(λ)

Sarsa(λ) [38] is an action-value function learning algorithm. It learns an estimate of the action-value function, Q_π, while the agent follows policy π. Sarsa(λ) takes the same form as TD(λ).

Parameterizing the value functions $V_t \approx V_\pi$ and $Q_t \approx Q_\pi$ by the parameter vector $\boldsymbol{\theta}_t$, the parameters are updated by gradient descent

$$\boldsymbol{\theta}_{t+1} = \boldsymbol{\theta}_t + \eta \delta_t \boldsymbol{e}_t, \tag{17.6}$$

where TD-error δ_t is

$$\delta_t = r_t + \gamma V_t(s_{t+1}) - V_t(s_t), \quad \text{for TD}(\lambda), \tag{17.7}$$

$$\delta_t = r_t + \gamma Q_t(s_{t+1}, a_{t+1}) - Q_t(s_t, a_t), \quad \text{for Sarsa } (\lambda), \tag{17.8}$$

the eligibility trace vector $\boldsymbol{e}_t$ is given by

$$\boldsymbol{e}_t = \gamma\lambda \boldsymbol{e}_{t-1} + \nabla_{\theta_t} V_t(s_t), \quad \boldsymbol{e}_0 = \boldsymbol{0}, \quad \text{for TD}(\lambda), \tag{17.9}$$

$$\boldsymbol{e}_t = \gamma\lambda \boldsymbol{e}_{t-1} + \nabla_{\theta_t} Q_t(s_t, a_t), \quad \boldsymbol{e}_0 = \boldsymbol{0}, \quad \text{for Sarsa}(\lambda), \tag{17.10}$$

s_t is the state at time t, a_t is the action at time t, r_t is the reward for action a_t in state s_t, and $\nabla_{\theta_t} V_t$ and $\nabla_{\theta_t} Q_t$ are the vectors of partial derivatives of the function approximators with respect to each component of $\boldsymbol{\theta}_t$.

True online TD(λ) and true online Sarsa(λ) [49] maintain an exact equivalence with the forward view at all times, whereas their traditional versions only approximate it for small step sizes. True online TD(λ)/Sarsa(λ) methods dominate the regular TD(λ)/Sarsa(λ) methods. Across all domains/representations, the learning speed of the true online methods are often better, but never worse than that of the regular methods. An additional advantage is that no choice between traces has to be made for the true online methods.

17.7 Q-Learning

Q-learning algorithm [51] is a model-free reinforcement learning algorithm that is guaranteed to find the optimal policy for a given Markov decision process. In Q-learning [50], the approximation to the optimal action-value function takes place independently of the evaluation policy by using only the path with the greatest action value to calculate a one-periodic difference. Q-learning is the most widely used reinforcement learning algorithm for addressing the control problem because of its off-policy update, which makes the convergence control easier.

In Q-learning, the objective is to learn the expected discounted reinforcement values, $Q^*(s, a)$, of taking action a in state s by always choosing actions optimally. Assuming the best action is taken initially, we have the value of s as $V^*(s) = \max_a Q^*(s, a)$ and the optimal policy $\pi^*(s) = \arg\max_a Q^*(s, a)$, which chooses an action just by taking the one with the maximum Q-value for the current state. The Q-values can be estimated online using a method essentially the same as TD(0), and are used to define the policy.

The Q-learning rule is given by

$$Q(s, a) \leftarrow Q(s, a) + \eta \left(r + \gamma \max_{a_{t+1} \in \mathcal{A}} Q(s_{t+1}, a_{t+1}) - Q(s, a) \right), \tag{17.11}$$

where r is the immediate reward, $\mathcal{A}$ is the action space, s is the current state, and s_{t+1} is the future state. The next action is the one with the highest Q-value. The state–action value $Q(s, a)$ is also known as the quality function.

If each action is executed in each state an infinite number of times on an infinite run and η is decayed appropriately, the Q-values will converge to the optimal values with probability 1 to Q^* [50], independent of how the agent behaves while the data are being collected. For these reasons, Q-learning is a popular and effective model-free algorithm for learning from delayed reinforcement. However, it does not address any of the issues involved in generalizing over large state and/or action spaces. In addition, it may converge quite slowly to a good policy.

Q-learning can also be extended to update states that occurred more than one step previously, as in TD(λ) [34]. When the Q-values nearly converge to their optimal values, it is appropriate for the agent to act greedily, taking the action with the highest Q-value in each situation. However, it is difficult to make an exploitation–exploration trade-off during learning.

Dyna-Q learning [44] repeats the Q-learning update using the learned model, which is independent of the underlying policy. It can be orders of magnitudes slower than Q-learning.

The policy-bias problem appears in Q-learning because the value of an action is only updated when the action is executed. Consequently, the effective rate of updating an action value directly depends on the probability of choosing the action for execution. This dependency is not resolved by changing the policy. The policy bias may cause a temporary decrease in the probability of choosing optimal actions with higher expected payoffs [22, 26].

The policy-bias problem of Q-learning can be addressed. Individual Q-learning [26] and frequency adjusted Q-learning [22] are two equivalent modifications of Q-learning by scaling the learning rate inversely proportional to the policy, thereby approximating the simultaneous updating of all actions every time a state is visited. Repeated update Q-learning [1] repeats the Q-learning update rule as inversely proportional to the policy. It maintains the convergence guarantee of Q-learning in stationary environments while relaxing the coupling between the execution policy and the learning dynamics.

Q-Learning Versus Sarsa

Q-learning and Sarsa have some similarity. The value policy is updated by temporal-difference learning [43]

$$Q_{t+1}(s,a) = Q_t(s,a) + \eta\left(r_{t+1} + \gamma x_t - Q_t(s,a)\right), \tag{17.12}$$

where the state–action value $Q(s, a)$ represents the possible reward received in the next step for taking action a in state s, plus the discounted future reward received from the next state–action observation, and

$$x_t = \max_a Q_t(s_{t+1}, a), \quad (Q\text{-learning [52]}), \tag{17.13}$$

$$x_t = Q_t(s_{t+1}, a_{t+1}), \quad (\text{Sarsa [46]}), \tag{17.14}$$

with r_{t+1} being the immediate reward, s and a correspond to the current state and action, and s_{t+1} and a_{t+1} denote the future state and action. Dynamic programming allows an optimal action to be selected for the next state on a decision level, as long as all actions have been evaluated until time t.

When interacting with the environment, a Sarsa agent updates the policy based on actions taken, while Q-learning updates the policy based on the maximum reward of available actions. Q-learning learns Q-values with better exploitation policy compared with Sarsa. Sarsa employs on-policy update, while Q-learning uses off-policy update. Sarsa is also called *approximate policy iteration*, and Q-learning is the off-policy variant of the policy iteration algorithm.

Example 17.1 A Java applet of Q-learning (http://thierry.masson.free.fr/IA/en/qlearning_applet.htm, written by Thierry Masson) allows the user to construct a grid with danger (red), neutral, and target (green) cells, and to modify various learning parameters.

An agent (or robot) has to learn as to how to move on a grid map: it learns to avoid dangerous cells and to reach target cells as quickly as possible. The agent is able to move from cell to cell, using one of the four directions (north, east, south, and west). It is also able to determine the nature of the cell it currently occupies. The grid is a square closed domain—the agent cannot escape but may hit the domain bounds.

As long as it explores the grid, the agent receives a reward for each move it makes. It receives a reward of 0.0 if entering a neutral cell, a penalty of −5000.0 if entering

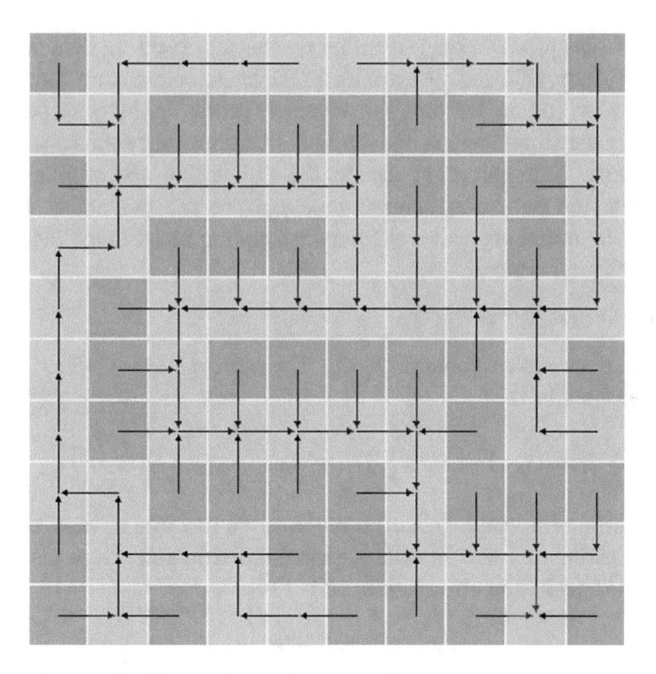

Fig. 17.3 Learning result of Q-learning: the learned policy

a dangerous cell, and a reward of $+1000.0$ if entering a target cell. The maximum iterations is set to 1,000,000, the exploration probability is set to $\epsilon = 0.8$, the learning rate is $\eta = 0.05$, and the discount factor is $\gamma = 0.9$. The exploration strategy chooses a random action with probability ϵ, and leaves the original action unchanged with probability $1 - \epsilon$. The route is set to be digressive. These parameters can be adjusted.

At the beginning of the learning process, or anytime the agent has hit a target cell, a new exploration process begins, starting from a randomly chosen position on the grid. And then, the agent restarts to explore the domain. After running the Java applet on a PC with Core2 Duo CPU at 2.10 GHz and 1 GB memory, the target has been hit 119,934 times, and the approximate learning time is 1.828 s. Figure 17.3 gives the learned policy which is represented as arrows overlaying the grid. The policy denotes which direction the agent should move in from each square of the grid.

17.8 Multiagent Reinforcement Learning

A multiagent system is composed of a group of autonomous agents that sense and interact with an environment and act for defined goals. It is a paradigm for designing AI applications such as robotic teams, manufacturing control, and networks.

Nash equilibrium is an equilibrium concept in competitive games and is a point in the joint policy space where all the agents will be locked if all agents follow the same rational learning algorithm that tries to learn best response to opponents' actions.

Multiagent reinforcement learning includes equilibrium-based multiagent reinforcement learning and learning automata. Due to the curse of dimensionality, it is difficult for many multiagent reinforcement learning algorithms to scale up.

17.8.1 Equilibrium-Based Multiagent Reinforcement Learning

Equilibrium-based multiagent reinforcement learning adopts Markov games as the framework and introduces various equilibrium solution concepts in game theory to learn equilibrium policies.

Minimax-Q [27] uses a minimax rule for action selection and value function update. It focuses on two-agent zero-sum Markov games. The action-value function $Q(s, a, o)$ considers the agent's action a and the opponent's action o. The value of state s is defined as

$$V(s) = \max_{\pi} \min_{o} \sum_{a} Q(s, a, o)\pi(s, a), \tag{17.15}$$

where π is any policy of the agent, $\pi(s, a)$ is the probability of taking action a in state s, and

$$Q(s, a, o) \leftarrow (1 - \eta)Q(s, a, o) + \eta[r + \gamma V(s')], \tag{17.16}$$

with $\eta \in [0, 1]$, r is the reward received by the agent, and s' is the next state.

Based on the concept of Nash equilibrium, Nash Q-learning [17] is derived for general-sum Markov games. The convergence condition of Nash Q-learning requires each state's one-shot game to have a global optimal equilibrium point or a saddle point. Friend-or-foe Q-learning [28] learns two special kinds of Nash equilibria (adversarial equilibrium and coordination equilibrium) and it has less strict convergence condition compared with Nash Q-learning. Based on the concept of correlated equilibrium, correlated Q-learning [14] generalizes Nash Q-learning and allows for dependencies among agents' actions.

During the learning process of equilibrium-based multiagent reinforcement learning, the one-shot games corresponding to each state's successive visits often have the same or similar equilibria [18]. Equilibrium transfer is used to accelerate learning [18]. It reuses previously computed equilibria when each agent has a small incentive to deviate.

Equilibrium-based multiagent reinforcement learning typically requires agents to replicate the other agents' value functions and adopt mixed strategy equilibria. Without the need for agents to share value functions, negotiation-based Q-learning [19] adopts pure strategy equilibrium solution concepts and uses a multi-step negotiation process for finding pure strategy equilibria since value functions are not shared among agents.

17.8.2 Learning Automata

Learning automata, a branch of the theory of adaptive control, were originally described explicitly as finite-state automata [47]. A learning automaton is an agent situated in a random environment that learns the optimal action through repeated interactions with its environment. The actions are chosen according to a specific probability distribution, which is updated based on the environment response the agent obtains by performing a particular action. Learning automata is a reinforcement learning method that directly manipulates the policy.

The linear reinforcement scheme is a simple sample algorithm. Let p_i be the agent's probability of taking action i. When action a_i succeeds, it gets rewarded, and its probability is increased while the probabilities of all other actions are decreased:

$$p_i(t+1) = p_i(t) + \alpha(1 - p_i(t)),$$
$$p_j(t+1) = (1-\alpha)p_j(t), \quad \text{for } j \neq i, \tag{17.17}$$

where $\alpha \in (0, 1)$ is the reward parameter. Similarly, when action a_i fails, the probabilities are defined by

$$p_i(t+1) = (1-\beta)p_i(t),$$
$$p_j(t+1) = \beta/(n_a - 1) + (1-\beta)p_j(t), \quad \text{for } j \neq i, \tag{17.18}$$

where $\beta \in (0, 1)$ is the penalty parameter and n_a is the number of actions.

When $\beta = \alpha$, we get the linear reward-penalty (L_{R-P}) algorithm. When $\beta = 0$, p_i remains unchanged for all i, and we get the linear reward-inaction (L_{R-I}) algorithm [21]. In the L_{R-I} scheme, the action probabilities are updated in the case of a reward response from the environment but no penalties are assessed.

Linear reward-inaction (L_{R-I}) converges with probability 1 to a probability vector containing a single 1 and the rest 0's, that is, choosing a particular action with probability 1. Unfortunately, it does not always converge to the correct action, but the probability that it converges to the wrong one can be made arbitrarily small by making α small [32]. Linear reward-inaction (L_{R-I}) is proven to be ϵ-optimal [32].

A single automaton is generally sufficient for learning the optimal value of one parameter. However, for multidimensional optimization problems, a system consisting of as many automata as the number of parameters is needed. Such a system of automata can be a game of automata.

Let $A_1, \ldots, A_N$ be the automata involved in an N-player game. Each play of the game consists of all automaton players choosing their actions and then getting the payoffs (or reinforcements) from the environment for the actions. Let $\boldsymbol{p}_1(k), \ldots, \boldsymbol{p}_N(k)$ be the action probability distributions of the N automata at the kth play. Each automaton A_i chooses an action $a^i(k)$ independently and at random according to $\boldsymbol{p}_i(k)$, $1 \leq i \leq N$. Essentially, each player wants to maximize its payoff $r^i(k)$. Since there are multiple payoff functions, learning can be targeted to reach a Nash equilibrium. It is known that if each of the automata uses an L_{R-I} algorithm for updating action probabilities, then the game would converge to a Nash equilibrium [40]. Learning automata are thoroughly overviewed in [32, 46].

Problems

17.1 Show how the idea of reinforcement learning is implemented in the LVQ2 algorithm.

17.2 The Rprop algorithm employs the idea of reinforcement learning. Describe how Rprop implement this idea.

17.3 Reinforcement learning is very useful for guiding robots through obstacles. Describe the process.

References

1. Abdallah, S., & Kaisers, M. (2016). Addressing environment non-stationarity by repeating Q-learning updates. *Journal of Machine Learning Research, 17*, 1–31.
2. Bakker, B., & Schmidhuber, J. (2004). Hierarchical reinforcement learning based on subgoal discovery and subpolicy specialization. In *Proceedings of the 8th Conference on Intelligent Autonomous Systems* (pp. 438–445). Amsterdam, The Netherlands.
3. Barto, A. G., Sutton, R. S., & Anderson, C. W. (1983). Neuron-like adaptive elements that can solve difficult learning control problems. *IEEE Transaction on Systems, Man and Cybernetics, 13*(5), 834–846.
4. Barto, A. G. (1992). Reinforcement learning and adaptive critic methods. In D. A. White & D. A. Sofge (Eds.), *Handbook of intelligent control: Neural, fuzzy, and adaptive approaches* (pp. 469–471). New York: Van Nostrand Reinhold.
5. Bohmer, W., Grunewalde, S., Shen, Y., Musial, M., & Obermayer, K. (2013). Construction of approximation spaces for reinforcement learning. *Journal of Machine Learning Research, 14*, 2067–2118.
6. Bradtke, S. J., & Barto, A. G. (1996). Linear least-squares algorithms for temporal difference learning. *Machine Learning, 22*, 33–57.
7. Choi, J., & Kim, K.-E. (2011). Inverse reinforcement learning in partially observable environments. *Journal of Machine Learning Research, 12*, 691–730.
8. Dayan, P., & Sejnowski, T. (1994). TD(λ) converges with probability 1. *Machine Learning, 14*(1), 295–301.
9. Elfwing, S., Uchibe, E., & Doya, K. (2016). From free energy to expected energy: Improving energy-based value function approximation in reinforcement learning. *Neural Networks, 84*, 17–27.
10. Furmston, T., Lever, G., & Barber, D. (2016). Approximate Newton methods for policy search in Markov decision processes. *Journal of Machine Learning Research, 17*, 1–51.
11. Furnkranz, J., Hullermeier, E., Cheng, W., & Park, S.-H. (2012). Preference-based reinforcement learning: A formal framework and a policy iteration algorithm. *Machine Learning, 89*, 123–156.
12. Garcia, J., & Fernandez, F. (2015). A comprehensive survey on safe reinforcement learning. *Journal of Machine Learning Research, 16*(1), 1437–1480.

13. Ghavamzadeh, M., & Mahadevan, S. (2007). Hierarchical average reward reinforcement learning. *Journal of Machine Learning Research*, *8*, 2629–2669.
14. Greenwald, A., Hall, K., & Serrano, R. (2003). Correlated Q-learning. In *Proceedings of the 20th International Conference on Machine Learning* (pp. 242–249). Washington, DC.
15. Heess, N., Silver, D., & Teh, Y. W. (2012). Actor–critic reinforcement learning with energy-based policies. In *JMLR Workshop and Conference Proceedings: 10th European Workshop on Reinforcement Learning (EWRL)* (Vol. 24, pp. 43–57).
16. Houk, J., Adams, J. L., & Barto, A. G. (1995). A model of how the basal ganglia generate and use neural signals that predict reinforcement. In J. C. Houk, J. Davis, & D. Beiser (Eds.), *Models of information processing in the basal ganglia* (pp. 250–268). Cambridge, MA: MIT Press.
17. Hu, J., & Wellman, M. P. (2003). Nash Q-learning for general-sum stochastic games. *Journal of Machine Learning Research*, *4*, 1039–1069.
18. Hu, Y., Gao, Y., & An, B. (2015). Accelerating multiagent reinforcement learning by equilibrium transfer. *IEEE Transactions on Cybernetics*, *45*(7), 1289–1302.
19. Hu, Y., Gao, Y., & An, B. (2015). Multiagent reinforcement learning with unshared value functions. *IEEE Transactions on Cybernetics*, *45*(4), 647–662.
20. Hwang, K.-S., & Lo, C.-Y. (2013). Policy improvement by a model-free Dyna architecture. *IEEE Transactions on Neural Networks and Learning Systems*, *24*(5), 776–788.
21. Kaelbling, L. P., Littman, M. I., & Moore, A. W. (1996). Reinforcement lerning: A survey. *Journal of Artificial Intelligence Research*, *4*, 237–285.
22. Kaisers, M., & Tuyls, K. (2010). Frequency adjusted multi-agent Q-learning. In *Proceedings of International Conference on Autonomous Agents and Multiagent Systems (AAMAS)* (pp. 309–316).
23. Konda, V., & Tsitsiklis, J. (2000). Actor–critic algorithms. In *Advances in neural information processing systems* (Vol. 12, pp. 1008–1014).
24. Lagoudakis, M. G., & Parr, R. (2003). Least-squares policy iteration. *Journal of Machine Learning Research*, *4*, 1107–1149.
25. Legenstein, R., Wilbert, N., & Wiskott, L. (2010). Reinforcement learning on slow features of high-dimensional input streams. *PLoS Computational Biology*, *6*(8), e1000894.
26. Leslie, D. S., & Collins, E. J. (2005). Individual Q-learning in normal form games. *SIAM Journal on Control and Optimization*, *44*(2), 495–514.
27. Littman, M. L. (1994). Markov games as a framework for multi-agent reinforcement learning. In *Proceedings of the 11th International Conference on Machine Learning* (pp. 157–163). New Brunswick, NJ.
28. Littman, M. L. (2001). Friend-or-foe Q-learning in general-sum games. In *Proceedings of the 18th International Conference on Machine Learning* (pp. 322–328). San Francisco, CA: Morgan Kaufmann.
29. Luciw, M., & Schmidhuber, J. (2012). Low complexity proto-value function learning from sensory observations with incremental slow feature analysis. In *Proceedings of International Conference on Artificial Neural Networks, LNCS* (Vol. 7553, pp. 279–287). Berlin: Springer.
30. Martin, H. J. A., de Lope, J., & Maravall, D. (2011). Robust high performance reinforcement learning through weighted k-nearest neighbors. *Neurocomputing*, *74*(8), 1251–1259.
31. Mihatsch, O., & Neuneier, R. (2014). Risk-sensitive reinforcement learning. *Machine Learning*, *49*(2), 267–290.
32. Narendra, K. S., & Thathachar, M. A. L. (1974). Learning automata: A survey. *IEEE Transactions on Systems, Man, and Cybernetics*, *4*(4), 323–334.
33. Parisi, S., Tangkaratt, V., Peters, J., & Khan, M. E. (2019). TD-regularized actor-critic methods. *Machine Learning*. https://doi.org/10.1007/s10994-019-05788-0.
34. Peng, J., & Williams, R. J. (1994). Incremental multi-step Q-learning. In: *Proceedings of the 11th International Conference on Machine Learning* (pp. 226–232). San Francisco, CA: Morgan Kaufmann.
35. Piot, B., Geist, M., & Pietquin, O. (2017). Bridging the gap between imitation learning and inverse reinforcement learning. *IEEE Transactions on Neural Networks and Learning Systems*, *28*(8), 1814–1826.

36. Potjans, W., Morrison, A., & Diesmann, M. (2009). A spiking neural network model of an actor–critic learning agent. *Neural Computation*, *21*, 301–339.
37. Reynolds, J. N., Hyland, B. I., & Wickens, J. R. (2001). A cellular mechanism of reward-related learning. *Nature*, *413*, 67–70.
38. Rummery, G., & Niranjan, M. (1994). *On-line Q-learning using connectionist systems*. Technical Report CUED/F-INFENG/TR 166, Engineering Department, Cambridge University.
39. Sallans, B., & Hinton, G. E. (2004). Reinforcement learning with factored states and actions. *Journal of Machine Learning Research*, *5*, 1063–1088.
40. Sastry, P. S., Phansalkar, V. V., & Thathachar, M. A. L. (1994). Decentralised learning of Nash equilibria in multiperson stochastic games with incomplete information. *IEEE Transactions on Systems, Man, and Cybernetics*, *24*, 769–777.
41. Schaal, S. (1999). Is imitation learning the route to humanoid robots? *Trends in Cognitive Sciences*, *3*(6), 233–242.
42. Schultz, W. (1998). Predictive reward signal of dopamine neurons. *Journal of Neurophysiology*, *80*(1), 1–27.
43. Sutton, R. S. (1988). Learning to predict by the method of temporal difference. *Machine Learning*, *3*(1), 9–44.
44. Sutton, R. S. (1990). Integrated architectures for learning, planning, and reacting based on approximating dynamic programming. In *Proceedings of the 7th International Conference on Machine Learning* (pp. 216–224). Austin, TX.
45. Sutton, R. S., & Barto, A. G. (1998). *Reinforcement learning: An introduction*. Cambridge, MA: MIT Press.
46. Thathachar, M. A. L., & Sastry, P. S. (2002). Varieties of learning automata: An overview. *IEEE Transactions on Systems, Man, and Cybernetics B*, *32*(6), 711–722.
47. Tsetlin, M. L. (1973). *Automata theory and modeling of biological systems*. New York: Academic.
48. Tsitsiklis, J. N., & Van Roy, B. (1997). An analysis of temporal-difference learning with function approximation. *IEEE Transactions on Automatic Control*, *42*(5), 674–690.
49. van Seijen, H., Mahmood, A. R., Pilarski, P. M., Machado, M. C., & Sutton, R. S. (2016). True online temporal-difference learning. *Journal of Machine Learning Research*, *17*, 1–40.
50. Watkins, C. J. H. C. (1989). *Learning from Delayed Rewards*. Ph.D. thesis, Department of Computer Science, King's College, Cambridge University, UK.
51. Watkins, C. J. C. H., & Dayan, P. (1992). Q-learning. *Machine Learning*, *8*(3), 279–292.
52. Werbos, P. J. (1990). Consistency of HDP applied to a simple reinforcement learning problem. *Neural Networks*, *3*, 179–189.
53. Zhao, T., Hachiya, H., Tangkaratt, V., Morimoto, J., & Sugiyama, M. (2013). Efficient sample reuse in policy gradients with parameter-based exploration. *Neural Networks*, *25*(6), 1512–1547.

Chapter 18
Compressed Sensing and Dictionary Learning

18.1 Introduction

Sparsity is a natural property of many real signals, for example, activities of the brain, eye movements, and heartbeat. A rational behind sparse coding is the sparse connectivity between neurons in human brain where only 1–4% of neurons are active at any time [11]. The sparse coding model has been shown to account for many observed response properties in the primary visual cortex (V1), including classic receptive field structure [101] and nonclassical response modulations [143]. In the sparse coding model for the primary visual cortex, a small subset of learned dictionary elements will encode most natural images, and only a small subset of the cortical neurons needs to be active for representing the high-dimensional visual inputs [101].

The goal of sparse coding is to learn an overcomplete basis set that represents each data point as a sparse combination of the basis vectors. In a sparse representation, a small number of coefficients contain a large portion of the energy. Sparse recovery aims to reconstruct sparse signals from a set of underdetermined compressed linear measurements. Many signals like audio, images, and video can be efficiently represented by sparse coding. Sparse representations of signals are of fundamental importance in fields such as blind source separation, compression, sampling, and signal analysis.

Compressed sensing, or compressed sampling, [20, 46] is a recent sampling algorithm stating that if a signal is sparse, it can be faithfully reconstructed from a small number of random measurements. Compressed sensing integrates the signal acquisition and compression steps into a single process. It is an alternative to Shannon/Nyquist sampling for the acquisition of sparse or compressible signals that can be well approximated by just $K \ll N$ elements from an N-dimensional basis. Compressed sensing allows perfect recovery of sparse signals (or signals sparse in some basis) using only a small number of random measurements. In practice, signals tend to be compressible, rather than sparse. Compressible signals are well approximated by sparse signals. Compressed sensing is a framework for signal acquisition and

K.-L. Du and M. N. S. Swamy, *Neural Networks and Statistical Learning*,
https://doi.org/10.1007/978-1-4471-7452-3_18

smart sensor design. It becomes an effective data acquisition technique, when communication bandwidth is limited.

Compressed sensing is not appropriate for non-sparse signals. Compressive covariance sensing [107] can compress a non-sparse signal and recover its second-order statistics from the compressed signal without any sparsity constraint.

18.2 Compressed Sensing

Compressed sensing theory guarantees the exact recovery of sparse signals from a small number of linear projections. It seeks to recover sparse or compressible signals from undersampled linear measurements [25, 46]; it asserts that the number of measurements should be proportional to the information content of the signal, rather than its dimension. According to compressed sensing, we can project a sparse or compressible high-dimensional signal into a low-dimensional space by applying a random observation matrix.

Compressed sensing relies on two fundamental properties: compressibility of the data and acquiring incoherent measurements.

Definition 18.1 A signal $\boldsymbol{x}$ is said to be *compressible* if there exists a dictionary Φ such that $\boldsymbol{\alpha} = \Phi^T \boldsymbol{x}$ are sparsely distributed.

In compressed sensing, a signal $\boldsymbol{x} \in \mathcal{C}^N$ is acquired by collecting data of linear measurements

$$\boldsymbol{y} = \mathbf{A}\boldsymbol{x} + \boldsymbol{n}, \tag{18.1}$$

where the random matrix $\mathbf{A}$ is an $M \times N$ sampling/sensing/measurement matrix, with the number of measurements M in $\boldsymbol{y} \in \mathcal{C}^M$ smaller than the number of samples N in $\boldsymbol{x}$: $M < N$, and $\boldsymbol{n}$ is a noise term. It focuses on underdetermined problems where the forward operator $\mathbf{A} \in \mathcal{C}^{M \times N}$ has unit-norm columns and forms an incomplete basis with $M \ll N$.

In compressed sensing, random distributions for generating $\mathbf{A}$ have to satisfy the so-called *restricted isometry property (RIP)* in order to preserve the information in sparse and compressible signals, and ensure a stable recovery of both sparse and compressible signals $\boldsymbol{x}$ [20]. A large class of random matrices has RIP with high probability.

The matrix $\mathbf{A}$ should satisfy an incoherence property. If $\mathbf{A}$ is perfectly incoherent (e.g., random Gaussian measurements or Fourier coefficients drawn uniformly at random), then only $M = O(k \ln(N))$ measurements are sufficient to perfectly reconstruct the k-sparse vector $\boldsymbol{x}$.

Compressed sensing can achieve stable recovery of compressible, noisy signals through the solution of L_0 regularized inverse problem, or the computationally tractable L_1 regularized inverse problem

$$\min_{\boldsymbol{x}} \|\boldsymbol{x}\|_1 \quad \text{subject to} \quad \|\mathbf{A}\boldsymbol{x} - \boldsymbol{y}\|_2^2 \le \epsilon^2. \tag{18.2}$$

LP is the reconstruction method that achieves the best sparsity-undersampling trade-off, but having a high computational cost for large-scale applications. LASSO [116] and approximate message passing [48] are well-known low-complexity reconstruction procedures.

The popular least absolute selection and shrinkage operator (LASSO) minimizes a weighted sum of the residual norm and a regularization term $\|\boldsymbol{x}\|_1$. LASSO has the ability to reconstruct sparse solutions when sampling occurs far below the Nyquist rate, and also to recover the sparsity pattern exactly with probability one, asymptotically as the number of observations increases. Approximate message-passing algorithm [48] is an iterative thresholding algorithm, leading to the sparsity-undersampling trade-off equivalent to that of the corresponding LP procedure while running dramatically faster.

Standard compressed sensing dictates that robust signal recovery is possible from $O(K \log(N/K))$ measurements. A model-based compressed sensing theory [8] provides concrete guidelines on how to create model-based recovery algorithms with provable performance guarantees. Wavelet trees and block sparsity are integrated into two compressed sensing recovery algorithms (Matlab toolbox, http://dsp.rice.edu/software) and they are proved to offer robust recovery from just $O(K)$ measurements [8].

Model-based compressed sensing refers to compressed sensing with extra structure about the underlying sparse signal known a priori. Weighted L_1-minimization improves the performance in the nonuniform sparse model of signals. When the measurements are obtained using a matrix with independent identically distributed Gaussian entries, weighted L_1-minimization successfully recovers the sparse signal from its measurements with overwhelming probability [92]. A method is also provided to choose these weights for any general signal model from the nonuniform sparse class of signal models [92].

18.2.1 Restricted Isometry Property

A universal sufficient condition for reliable reconstruction of sparse signals is given by the restricted isometry property (RIP) of sampling matrices [46]. Constructing RIP matrices from binary vectors was proposed in [44]. To connect to error-correction codes, both its algorithmic and constructive aspects were pursued [18, 19, 39].

Definition 18.2 (*K-Sparse*) A vector $\boldsymbol{x}$ is said to be *K-sparse* when $\boldsymbol{x}$ has at most K nonzero entries, i.e., when $\|\boldsymbol{x}\|_0 \leq K$, where $\|\cdot\|_0$ is L_0-norm, which returns the number of nonzero elements in its argument.

Definition 18.3 (*K-Restricted Isometry Property (K-RIP),* [20]) The $M \times N$ sensing/sampling matrix $\mathbf{A}$ is said to satisfy the K-restricted isometry property (K-RIP) if there exists a constant $\delta \in (0, 1)$ such that

$$(1-\delta)\|\boldsymbol{x}\|_2^2 \leq \|\mathbf{A}\boldsymbol{x}\|_2^2 \leq (1+\delta)\|\boldsymbol{x}\|_2^2 \tag{18.3}$$

holds for any K-sparse vector $\boldsymbol{x}$ of length N. The minimum of all constants $\delta \in (0, 1)$ that satisfy (18.3) is referred to as the *restricted isometry constant (RIC)* δ_K of the sensing matrix.

In general, a smaller RIC implies that the transformation is closer to an isometry. The RIC satisfies the monotonicity property, i.e., $\delta_K \leq \delta_{2K}$. The maximum sparsity order K guaranteeing recovery of all sparse vectors are found by RIP analysis.

In other words, K-RIP ensures that all submatrices of $\mathbf{A}$ of size $M \times K$ are close to an isometry, and therefore distance (and information) preserving. The goal is to push M as close as possible to K in order to perform as much signal compression during acquisition as possible. RIP measures the orthogonality of column vectors of a dictionary. It is NP-hard to determine whether a measurement matrix possesses RIP for any accuracy parameter [126]. This implies that it is NP-hard to approximate the range of parameters for which a matrix possesses RIP with accuracy better than some constant. Constructing RIP matrices deterministically is a hard problem [6], but there are simple random methods to generate RIP matrices with high probability [7, 66].

Based on RIP analysis, Gaussian measurement matrices have been proven to be information-theoretically optimal in terms of minimizing the required number of measurements for sparse recovery [25, 47].

Definition 18.4 (*Statistical RIP,* [9]) A matrix is said to have a statistical RIP of order K if most submatrices with K columns define a near-isometric map of R^K into R^m.

Statistical RIP is closely related to sparse signal recovery of deterministic sensing matrices of size $m \times N$. Sufficient conditions for the statistical RIP of a matrix in terms of the mutual coherence and mean square coherence have been established in [9]. For many existing deterministic families of sampling matrices, $m = O(k)$ rows suffice for k-statistical RIP [9]. Imposing the statistical RIP condition together with the statistical incoherence condition suffices to prove stable sparse recovery by basis pursuit [9].

18.2.2 Sparse Recovery

Signals can be represented as a linear combination of a small number of atoms in a well-chosen basis. Recover a k-sparse signal $\boldsymbol{x}^* \in R^N$ from the linear measurements

$$\boldsymbol{y} = \mathbf{A}\boldsymbol{x}^* + \boldsymbol{n} \in R^M, \tag{18.4}$$

where the sensing matrix $\mathbf{A} \in R^{M \times N} (M \leq N)$, $\boldsymbol{y} \in R^M$ represents the measurements vector, and $\boldsymbol{n} \in R^M$ is an unknown noise vector and $2k \leq M \leq N$.

A key feature ensuring recovery is the use of nonlinear reconstruction algorithms. The L_0-norm problem can be tackled directly

$$\min_{x \in R^N} \Psi(\boldsymbol{x}) = \frac{1}{2}\|\mathbf{A}\boldsymbol{x} - \boldsymbol{y}\|^2 \quad \text{subject to} \quad \|\boldsymbol{x}\|_0 \leq k, \tag{18.5}$$

where the L_0-norm $\|\boldsymbol{x}\|_0$ counts the nonzero entries in $\boldsymbol{x}$.

The L_0-norm regularized problem in compressed sensing reconstruction is non-convex with NP-hard computational complexity. It has many local minimizers, and in the case of zero noise, with a unique global minimizer at the k-sparse vector $\boldsymbol{x}^*$ [100].

A common approach to the L_0-norm problem (18.5) is to solve by an optimization relaxation. The problem can be reconstructed via the L_1-norm minimization problem [23, 46]:

$$\min_{x \in R^N} \|\boldsymbol{x}\|_1 \quad \text{subject to} \quad \mathbf{A}\boldsymbol{x} = \boldsymbol{y}, \tag{18.6}$$

where L_1-norm is the convex envelope of L_0-norm, $\|\boldsymbol{x}\|_1 = \sum_{i=1}^{N} |x_i|$. This method is known as *basis pursuit* [34]. Under appropriate constraints on the sensing matrix $\mathbf{A}$, basis pursuit yields exact recovery of the sparse signal.

L_p-norm is an approximation to L_0-norm. Alternatively, the lack of sparsity of $\boldsymbol{x}$ can be penalized by means of L_p-norms with $0 < p < 1$.

$$\min_{x \in R^N} \|\boldsymbol{x}\|_p^p \quad \text{subject to} \quad \mathbf{A}\boldsymbol{x} = \boldsymbol{y}, \tag{18.7}$$

where the L_p-norm is non-convex.

Suboptimal fast signal recovery methods available for such problems are greedy pursuit methods, thresholding methods, and convex relaxation algorithms. Greedy pursuit methods, such as matching pursuit [90], orthogonal matching pursuit (OMP) [103], OLS [106], subspace pursuit [38], and compressive sampling matching pursuit (CoSaMP) [97], can be used to tackle the L_0-problem directly. An example of thresholding method is iterative hard thresholding (IHT) for L_0-norm problem [12].

Examples of convex relaxation algorithms that solve the L_1-norm problem are gradient projection method [56], accelerated proximal gradient method [99], iterative reweighted method [26], homotopy method [89], and least angle regression method [50]. There are also L_p-norm $(0 < p < 1)$ relaxation methods [30].

Basis pursuit [33] is a greedy sparse approximation technique for decomposing a signal into a superposition of dictionary elements (basis functions), which has the smallest L_1 norm of coefficients among all such decompositions. It solves the optimization problem for a given set of basis vectors, and then greedily searches for vectors to add or remove. It is implemented as `pdco` and `SolveBP` in the SparseLab toolbox (http://sparselab.stanford.edu).

OMP [103] is a simple and effective greedy algorithm for sparse recovery/approximation. Inspired by the success of OMP, CoSaMP [97] and subspace pursuit [38] made improvements by selecting multiple coordinates followed by a pruning step in each iteration, and the recovery condition was framed under the RIP [25]. The more careful selection strategy of CoSaMP and subspace pursuit leads to an optimal sample complexity. CoSaMP prunes the gradient at the beginning of each iteration, followed by solving an LS program restricted on a small support set. In particular, in the last step, CoSaMP applies hard thresholding to form a k-sparse iterate for future updates. Hard thresholding pursuit [57] is a greedy method adopting a strategy of adding as well as pruning indices from the list.

Examples of algorithms for the L_p-norm problem are iterative reweighted methods [31], approximate operator [91, 134], and projected gradient [32]. They converge to a global minimum for properly chosen initial points. For larger p, it is easier to choose a good initial point so that the method can converge to a global minimum [32]. To guarantee that the unique optimum of the L_0-norm problem is the unique solution to the L_p-norm problem, it requires a smaller number of samples M for the L_p-norm problem than that for the L_1-norm case [59].

Compressed sensing relies on sparse representation of a signal in a finite discrete dictionary, while the true parameters may be actually specified in a continuous dictionary. This basis mismatch results in loss of sparsity due to spectral leakage along the Dirichlet kernel [37]. Spectral compressed sensing aims to recover a spectrally sparse signal from a small random subset of its n time-domain samples. The signal is assumed to be a superposition of r multidimensional complex sinusoids, while the underlying frequencies can assume any continuous values in the normalized frequency domain. When the frequencies of the signal indeed fall on a grid, compressed sensing algorithms based on L_1 minimization [23, 47] assert that it is possible to recover the spectrally sparse signal from $O(r \log n)$ random time-domain samples. These algorithms admit faithful recovery even when the samples are contaminated by bounded noise [22] or arbitrary sparse outliers [83]. Total-variation norm minimization algorithm [21] super resolves a sparse signal from frequency samples at the low end of the spectrum.

Sparse signal reconstruction is considered in [122], when linear random measurements are corrupted by highly impulsive environments. By adopting the family of symmetric alpha-stable distributions, a greedy method is designed to solve a minimum dispersion optimization problem based on fractional lower order moments.

Some efficient algorithms for sparse signal recovery are L_1-LS (http://www.stanford.edu/~boyd/l1_ls/), SpaRSA (http://www.lx.it.pt/~mtf/SpaRSA/), FISTA (http://statweb.stanford.edu/~candes/nesta/) [10], SPGL1 (http://www.math.ucdavis.edu/~mpf/software.html).

18.2.3 Iterative Hard Thresholding

Iterative thresholding [12, 13, 42] is a simple greedy technique that generates feasible steepest descent steps for L_1-norm problem (18.6), obtained by projecting steps along the negative gradient direction of Ψ onto the L_0-norm constraint by means of the hard threshold operator, which simply sets all but the k largest-in-magnitude elements of a vector to zero. IHT performs hard thresholding after gradient update. It uses the proximal-point technique at each iteration. IHT performs gradient projection with constant step size, while normalized IHT [14] employs a self-adaptive step-size strategy to guarantee stability and performance.

IHT converges to a fixed point of IHT/local minimizer of (18.6) provided the spectral norm of the measurement matrix **A** is less than one, a somewhat restrictive condition [12]. Stable recovery is guaranteed provided **A** satisfies RIP [13]. Other RIP-based recovery conditions have been obtained for IHT [57], and for normalized IHT [14].

For an arbitrary measurement matrix **A**, a sufficient condition for the convergence of IHT to a fixed point and a necessary condition for the existence of fixed points were derived in [27]. These conditions allow a sparse signal recovery analysis to be performed in the deterministic noiseless case by implying that the sparse signal is the unique fixed point and limit point of IHT, and in the case of Gaussian measurement matrices and noise by generating a bound on the approximation error of the IHT limit as a multiple of the noise level. The analysis has been extended to normalized IHT.

IHT [12] and iterative soft-thresholding [42] algorithms are for the L_0- and L_1-norm problems, respectively. When $p = 1/2$, half thresholding [134] is given through an analytical thresholding form. The iterative thresholding algorithms are efficient for high-dimensional problems, and it is relatively easy to specify the regularization parameter.

Proximal gradient homotopy method calculates and traces the solutions of the regularized problem along a continuous path without the need for selecting a regularization parameter λ. Homotopy technique is combined with IHT to overcome the difficulty of the IHT method on the choice of the regularization parameter value, by tracing solutions of the regularized problem along a homotopy path [45]. Any accumulation point of the sequences generated by the homotopy technique is proved to be a feasible solution of the problem [45].

Hard thresholding pursuit [57] combines the idea of CoSaMP and IHT, and is superior to both in terms of the RIP condition. It is an iterative greedy selection procedure for finding sparse solutions of underdetermined linear systems. In [136], hard thresholding pursuit is generalized to a generic problem setup of sparsity-constrained convex optimization. The algorithm iterates between a standard gradient-descent step and a hard thresholding step with or without debiasing.

Compared to the convex programs, hard thresholding-based algorithms are always orders-of-magnitude computationally more efficient, and hence more practical for large-scale problems [121]. It is hard for the L_1-based stochastic algorithms to preserve the sparse structure of the solution as the batch solvers do [79].

In [111], a tight bound is established to quantitatively characterize the deviation of the thresholded solution from a given signal. The RIP and the sparsity parameter are bridged for many hard thresholding-based algorithms, and this improves on the RIP condition especially in case of unknown sparsity. The presented stochastic algorithm performs hard thresholding in each iteration, hence ensuring parsimonious solutions. Global linear convergence is proved for a number of prevalent statistical models under mild assumptions, even though the problem is non-convex [111].

18.2.4 Orthogonal Matching Pursuit

OMP [103], also referred to as *forward stepwise regression*, is a simple and fast greedy algorithm, and it has sparse solution. At each iteration, a column of $\mathbf{A}$ that is maximally correlated with the residual is chosen, the index of this column is added to the list, and then the vestige of columns in the list is eliminated from the measurements, generating a new residual for the next iteration. OMP adds one new element of the dictionary and makes one orthogonal projection at each iteration. While OMP may fail for some deterministic sensing matrices, it recovers the true signal with high probability when using random matrices such as Gaussian [119, 120].

OMP is capable of recovering a k-sparse signal $\boldsymbol{x} \in R^N$ based on a set of incomplete measurements $\boldsymbol{y} \in R^M$ obeying the linear model

$$\boldsymbol{y} = \mathbf{A}\boldsymbol{x} + \boldsymbol{n}, \tag{18.8}$$

where the sensing matrix $\mathbf{A} \in R^{M \times N}$ with $N \gg M (\gg k)$, and the measurement noise $\boldsymbol{n} \in R^M$.

The signal reconstruction problem can be formulated as

$$\min \|\hat{\boldsymbol{x}}\|_0 \tag{18.9}$$

subject to

$$\hat{\boldsymbol{y}} = \hat{\mathbf{A}}\hat{\boldsymbol{x}}, \tag{18.10}$$

where $\hat{\boldsymbol{x}}$ is the reconstructed sparse vector, $\hat{\boldsymbol{y}}$ is the reconstructed measurement vector, and $\hat{\mathbf{A}}$ is the reconstructed sparse representation matrix.

Greedy methods that add multiple indices per iteration include stagewise OMP [49], regularized OMP [98], and generalized OMP [124] (also known as orthogonal super greedy algorithm [85]). These algorithms identify candidates at each iteration according to correlations between columns of the sensing matrix and the residual vector. Specifically, stagewise OMP picks indices whose magnitudes of correlation exceed a deliberately designed threshold. Regularized OMP first chooses a set of K indices with strongest correlations and then narrows down the candidates to a subset based on a predefined regularization rule. Generalized OMP finds a fixed number of indices with strongest correlations in each selection. Similarly, orthogonal super greedy algorithm adds multiple new elements of the dictionary and makes one orthogonal projection at each iteration.

Multipath matching pursuit [77] is an extension of OMP that recovers sparse signals with a tree-searching strategy. Unlike OMP which identifies one index per iteration, it enhances accuracy through tracing and extending multiple promising candidate paths for each iteration, and finally chooses one candidate that minimizes the residual power.

Block OMP [52] is a greedy algorithm for recovering block sparse signals whose nonzero entries occur in few blocks.

OMP builds the dictionary in a constructive way, like OLS. Both OMP and OLS identify the support of the underlying sparse signal by adding one index to the list at a time, and estimate the sparse coefficients over the enlarged support. While OLS seeks a candidate which results in the most significant decrease in the residual power, OMP chooses a column that is most strongly correlated with the signal residual. OLS has better convergence property, but is computationally more expensive than OMP [114].

Multiple OLS [125] extends OLS by allowing multiple L indices to be selected per iteration. It often converges in fewer iterations. Stable recovery of sparse signals can be achieved with multiple OLS when the signal-to-noise ratio (SNR) scales linearly with the sparsity level of input signals.

In order to solve the sparse approximation problem, a single sufficient condition is developed in [119] under which both OMP and basis pursuit [33] can recover an exactly sparse signal. For every input signal, OMP can calculate a sparse approximant whose error is only a small factor worse than the optimal error which can be attained with the same number of terms [119]. OMP can reliably recover a signal with K nonzero entries in dimension N given $O(K \ln N)$ random linear measurements of that signal [120].

Reconstruction of a class of structured sparse signals modeled by trigonometric polynomials via OMP is addressed in [105].

18.2.5 Restricted Isometry Property for Signal Recovery Methods

Under the noiseless setting, the study of sufficient conditions for perfect signal recovery using OMP received considerable attention in the area of compressive sensing. In the noiseless case, OMP can exactly identify the support of a K-sparse signal in K iterations if $\mathbf{A}$ satisfies RIP of order $K+1$ with RIC $\delta_{K+1} < 1/(3\sqrt{K})$ [43]. Since a smaller RIC results in a better signal reconstruction performance, the achievable performance of sparse signal recovery with interference-nulling is actually better than that predicted by [28]. A less restricted sufficient condition can be further relaxed to $\delta_{K+1} < 1/(\sqrt{K}+1)$ [69]. Sufficient conditions specified in terms of the RIC bound $\delta_{K+1} < 1/(\sqrt{K}+1)$ along with certain requirements on the minimal signal entry magnitude have been derived to guarantee exact support identification under measurement noise [130].

In the noiseless case, $\delta_{K+1} < 1/(\sqrt{K}+1)$ is sufficient for OMP to recover any K-sparse $\boldsymbol{x}$ in K iterations [94, 124]. In [29], it is shown that, in the presence of noise, a relaxed upper bound $\delta_{K+1} < (\sqrt{4K+1}-1)/(2K)$ together with a relaxed requirement on the minimal signal entry magnitude suffices to achieve perfect support identification using OMP. In the noiseless case, such a relaxed bound can ensure exact support recovery in K iterations.

For any K-sparse signal $\boldsymbol{x}$, if a sensing matrix $\mathbf{A}$ satisfies the RIP with $\delta_{K+1} < 1/\sqrt{K+1}$, then under some constraints on the minimum magnitude of nonzero elements of $\boldsymbol{x}$, OMP exactly recovers any K-sparse signal $\boldsymbol{x}$ from its measurements in K iterations [93, 128]. This sufficient condition is sharp in terms of δ_{K+1}. Also, the constraints on the minimum magnitude of nonzero elements of $\boldsymbol{x}$ are weaker than the existing ones. If the sampling matrix $\mathbf{A}$ satisfies the RIP with $\delta_{K+1} < \frac{1}{\sqrt{K+1}}$, then OLS exactly recovers the support of any K-sparse vector $\boldsymbol{x}$ from its samples in K iterations [127].

RIP-based analysis of an OMP-like algorithm, in which the square-error metric for support identification in each iteration is replaced by a general convex objective function, is considered in [141]. Notably, by allowing more than K iterations in OMP, the RIP-based recovery condition is similar to that in the L_1-minimization algorithm [51].

In [17], it is proved that for a K-sparse $\boldsymbol{x}$, if the RIC of $\mathbf{A}$, $\delta_{tK} \le \sqrt{\frac{t-1}{t}}$, where $t \ge \frac{4}{3}$, then the L_1-norm problem can recover $\boldsymbol{x}$ exactly.

Multiple OLS ($L > 1$) [125] performs exact recovery of K-sparse signals ($K > 1$) in at most K iterations, provided that the sensing matrix $\mathbf{A}$ obeys the RIP with isometry constant $\delta_{LK} < \sqrt{L}/(\sqrt{K} + 2\sqrt{L})$. When $L = 1$, multiple OLS reduces to OLS and exact recovery is guaranteed under $\delta_{K+1} < 1/(\sqrt{K} + 2)$. This condition is nearly optimal with respect to δ_{K+1} in the sense that, even with a small relaxation (e.g., $\delta_{K+1} = 1/\sqrt{K}$), exact recovery with OLS may not be guaranteed.

Multipath matching pursuit exactly recovers all K-sparse signals if the sampling matrix $\mathbf{A}$ satisfies the RIP of order $K + L$ with RIC [77] $\delta_{K+L} < \frac{\sqrt{L}}{\sqrt{K}+2\sqrt{L}}$, where L is the number of child paths per candidate. This bound is improved to an optimal bound as $\delta_{K+L} < \sqrt{\frac{L}{K+L}}$ in [82].

Using block-RIP, some sufficient conditions for exact or stable recovery of block sparse signals with block OMP are investigated in [129].

RIP-based performance guarantees for subspace pursuit, in both noiseless and noisy cases, have been reported in [38, 63].

The L_p-norm problem for sparse recovery has been studied by using RIP [131, 140]. For any $0 < p \le 1$, the L_p-norm problem can recover any K-sparse signal, if $\delta_{2K} \le$ a positive constant decided by p [140].

RIP is not the only framework for analyzing the performance of sparse recovery. Some related notions are restricted orthogonality constant (ROC) [25] and null space property [64]. The L_1- and L_p-norm problems for sparse recovery have been studied by using the null space property [59, 64, 104, 123]. The condition of the null space constant being less than 1 is proved to be sufficient and necessary to guarantee that the L_p-norm problem can recover any K-sparse signal [64]. Based on theoretical analyses by null space property, if L_p-norm problem with $p = p_1$ can recover any K-sparse vector, then L_p-norm problem with $p \le p_1$ can also recover any K-sparse vector [59]. It is proved in [104] that L_p-minimization with sufficiently small p is equivalent to L_0-minimization for sparse recovery. In [123], L_p-norm problem with

$p < p^*$, where p^* is obtained from the derived upper bound on null space constant, can be guaranteed for exact recovery.

Sufficient conditions for exact recovery have also been obtained in terms of the RIC for L_p-minimization [58], and for CoSaMP [97], regularized OMP [98].

18.2.6 Tensor Compressive Sensing

A tensor is a multidimensional array. The order of a tensor is the number of modes. Many data types, such as those in color imaging, video sequences, and multisensor networks, are intrinsically represented by higher order tensors. An image matrix is a tensor of order 2.

Multi-way compressed sensing [112] for sparse and low-rank tensors suggests a two-step recovery process: fitting a low-rank model in compressed domain, followed by per-mode decompression. However, the performance of multi-way compressed sensing relies highly on the estimation of the tensor rank, which is an NP-hard problem.

Generalized tensor compressive sensing [61] is a unified framework for compressed sensing of higher order tensors, which preserves the intrinsic structure of the tensor data with reduced computational complexity at reconstruction. The method offers an efficient means for representation of multidimensional data by providing simultaneous acquisition and compression from all tensor modes. It outperforms multi-way compressive sensing [112] in terms of both reconstruction accuracy and processing speed.

18.3 Sparse Coding and Dictionary Learning

Sparse coding allows a complex and dense signal to be represented using only a few elements of an overcomplete dictionary [101]. Given a set of homogeneous signals, dictionary learning aims to recover the elementary signals (atoms) that efficiently represent the data. Generally, this goal is achieved by imposing some sparseness constraint on the coefficients of the representation. Thus, sparse coding and dictionary learning are both sparse approximation methods with similar objective. Dictionary learning is related to matrix factorization problems.

Dictionary learning is an effective means for finding a sparse, patch-level representation of an image [2]. Sparse dictionary learning has been applied in image de-noising, based on the idea that a clean image patch can be sparsely represented by an image dictionary, but the noise cannot be.

Dictionary learning problems encompass sparse coding [102], nonnegative sparse coding [68, 75], L_p-sparse coding [76], K-SVD [2], hierarchical sparse coding [71], elastic-net-based dictionary learning [88], and fused-LASSO-based dictionary learning [118].

With a formulation similar to that of compressed sensing, sparse approximation has a different objective. Assume that a target signal $\boldsymbol{y} \in R^M$ can be represented exactly (or at least approximated with sufficient accuracy) by a linear combination of exemplars in the overcomplete dictionary $\mathbf{A} = [\boldsymbol{a}_1, \boldsymbol{a}_2, \ldots, \boldsymbol{a}_N]$, with $\boldsymbol{a}_i$'s being words:

$$\boldsymbol{y} = \mathbf{A}\boldsymbol{x}, \tag{18.11}$$

where $\mathbf{A}$ is a real $M \times N$ basis matrix with $N \gg M$ whose columns have unit Euclidean norm: $\|\boldsymbol{a}_j\|_2 = 1$, $j = 1, 2, \ldots, N$, and $\boldsymbol{x} \in R^N$ is a representation of $\boldsymbol{y}$ under the basis matrix. In fact, since the dictionary is overcomplete, any vector can be represented as a linear combination of vectors from the dictionary.

The problem of recovering both $\mathbf{A}$ and $\boldsymbol{x}$ given $\boldsymbol{y}$ is underdetermined. When $\boldsymbol{x}$ is sparse and random, one can recover both $\mathbf{A}$ and $\boldsymbol{x}$ efficiently from $\boldsymbol{y}$ with high probability, provided that p (the number of samples) is sufficiently large [115]. Exact Recovery of Sparsely-Used Dictionaries (ER-SpUD) is a polynomial-time algorithm that is proved to probably recover the dictionary and coefficient matrix when the coefficient matrix is sufficiently sparse. The approach consists of an ER-SpUD step and a greedy step [115]. The method works for $p \geq CN^2 \log^2 N$ and they conjectured that $p \geq CN \log N$ suffices. In [1], a version of Er-SpUD algorithm recovers $\mathbf{A}$ and $\boldsymbol{x}$ exactly with high probability, provided that the number of observations $p \geq CN \log N$.

Sparsity has emerged as a fundamental type of regularization. Sparse approximation [33] seeks an approximate solution to (18.11) while requiring that the number K of nonzero entries of $\boldsymbol{x}$ is only a few relative to its dimension N. Compressive sensing is a specific type of sparse approximation problem.

Although the system of linear equations in (18.11) has no unique solution, if $\boldsymbol{x}$ is sufficiently sparse, $\boldsymbol{x}$ can be uniquely determined by solving [46]

$$\boldsymbol{x} = \arg\min_{\tilde{\boldsymbol{x}} \in R^N} \|\tilde{\boldsymbol{x}}\|_0 \quad \text{subject to} \quad \boldsymbol{y} = \mathbf{A}\tilde{\boldsymbol{x}}. \tag{18.12}$$

The L_0-norm problem (18.12) is NP-hard [96].

Matching pursuit [90] is often used to approximately solve the L_0 problem. OMP [103] is the simplest effective greedy algorithm for sparse approximation.

With weak conditions on $\mathbf{A}$, the solution of the L_0-norm minimization given by (18.12) is equal to the solution of an L_1-norm minimization [47]

$$\boldsymbol{x} = \arg\min_{\tilde{\boldsymbol{x}} \in R^N} \|\tilde{\boldsymbol{x}}\|_1 \quad \text{subject to} \quad \boldsymbol{y} = \mathbf{A}\tilde{\boldsymbol{x}}. \tag{18.13}$$

This convex minimization problem is the same as that given by (18.2). This indicates that the problems of recovering sparse signals from compressed measurements and constructing sparse approximation are the same in nature.

The solution of the problem (18.13) is approximately equal to the solution of the problem (18.12) if the optimal solution is sufficiently sparse [24, 34, 47, 103]. The problem (18.13) can be effectively solved by LP methods.

Compared with L_1 regularization, sparsity is better achieved with L_0 penalties based on prediction and false discovery rate arguments [84]. Under the RIP condition, the L_1 and L_0 solutions are equal [47]. However, L_1 regularization may cause biased estimation for large coefficients since it over-penalizes true large coefficients [55, 139].

The leading sparse coding methods are a wide variety of convex optimization algorithms, e.g., alternating direction method of multipliers (ADMM) that alternatively minimizes over the coefficients and atoms separately, solving (18.12), and greedy algorithms, e.g., matching pursuit [90] and OMP [103]) for approximate solution of (18.13).

The most popular algorithm for solving L_0-norm-based problems is K-SVD method [2], which calls OMP for solving the sparse approximation subproblem. K-SVD method uses an alternating iteration scheme between $\mathbf{A}$ and $\{\boldsymbol{x}_i\}$: with the dictionary fixed, it uses OMP to find sparse coefficients $\{\boldsymbol{x}_i\}$, and then with sparse coefficients $\{\boldsymbol{x}_i\}$ fixed, atoms in the dictionary $\mathbf{A}$ are sequentially updated via SVD. However, the computational burden of OMP is not trivial.

The sequence generated by K-SVD is not always convergent. Proximal alternating method [3] and proximal alternating linearized method [15] solve a class of non-convex optimization problems, with global convergence. Accelerated plain dictionary learning [5] is a multi-block alternating iteration scheme with global convergence property tailored for L_0-norm sparse coding problems. Motivated by multi-block coordinate descent, proximal alternating method [3], and proximal alternating linearized method [15], a multi-block hybrid proximal alternating iteration scheme [5] is further combined with an acceleration technique from K-SVD method.

Motivated by K-SVD algorithm [2], a formulation for dictionary learning over positive definite matrices is proposed through alternating minimization [113]. Tensor Sparse Library (http://www.ece.umn.edu/users/sival001/research.html) [113] comprises C++ binaries for sparse coding, sparse classification, dictionary learning, and discriminative dictionary learning algorithms. Eigenlibrary (http://eigen.tuxfamily.org) with OpenMP is used. The sparse coding algorithms are implemented using a coordinate descent approach which works much faster than interior point methods using generic solvers.

When both signal and noise are sparse but in different domains, the signal recovery problem is non-convex and NP-hard. In [138], the problem is solved by two methods. The first method replaces L_0-norm by L_1-norm and then applies ADMM. The second method replaces L_0-norm by a smoothed L_0-norm and then applies the gradient projection method.

Enforcing the information to be spread uniformly over representation coefficients exhibits relevant properties in various applications such as robust encoding in digital communications. Antisparse regularization can be naturally expressed through an L_∞-norm penalty. A probabilistic formulation of such representations is derived in [53]. A probability distribution called the democratic prior is used as a prior to promote antisparsity in a Gaussian linear model, yielding a fully Bayesian formulation of antisparse coding.

Sparse coding with latents described by discrete instead of continuous prior distributions is studied in [54]. The latents (while being sparse) can take on any value of a finite set of possible values and the prior probability of any value is learned from data. As the prior probabilities are learned, the approach then allows for estimating the prior shape without assuming specific functional forms. Discrete sparse coding algorithms can scale efficiently to work with realistic datasets and provide statistical quantities to describe the structure of the data.

18.4 LASSO

Least absolute shrinkage and selection operator (LASSO) [116, 117] was originally proposed for estimation in linear models. With the L_1-norm constraint, it tends to produce sparse solutions and interpretable models. It is a popular supervised learning technique for recovering sparse signals from high-dimensional measurements. LASSO allows computationally efficient feature selection based on the assumption of linear dependency between input features and output values. Thanks to its ability to perform variable selection and model estimation simultaneously, LASSO (and L_1-regularization in general) is widely used in applications involving a huge number of candidate features or predictors.

In the context of supervised learning, LASSO problem can be reduced to an equivalent SVM formulation [70]. For unsupervised learning, the idea of LASSO regression has been used in [80] for bi-clustering in biological research. LASSO is particularly useful when the number of features is larger than the number of training samples.

LASSO minimizes the sum of squared errors, given a fixed bound on the sum of absolute value of the regression coefficients. LASSO can be formulated as an L_1-regularized LS problem, and large-scale instances must usually be tackled by means of an efficient first-order algorithm. The L_1-regularizer in LASSO tends to produce a sparse solution, leading the regression coefficients for irrelevant features to zero, thus giving interpretable models that are sparse. It is implemented as `SolveLasso` in the SparseLab toolbox [50].

The convex minimization problem given by (18.13) or (18.2) can be cast as an LS problem with L_1 penalty, also referred to as LASSO [116]

$$\boldsymbol{x} = \arg\min_{\tilde{\boldsymbol{x}} \in R^N} \{ \|\mathbf{A}\tilde{\boldsymbol{x}} - \boldsymbol{y}\|_2^2 + \lambda \|\tilde{\boldsymbol{x}}\|_1 \}, \tag{18.14}$$

where $\boldsymbol{x} = (x_1, \ldots, x_d)^T$ is a regression coefficient vector, x_k denotes the regression coefficient of the kth feature, and the regularization parameter $\lambda > 0$.

Stochastic algorithms such as stochastic gradient descent and stochastic coordinate descent have been proposed for the LASSO problem [79, 109]. Public domain software packages exist to solve problem (18.14) efficiently.

Least angle regression [50] is a greedy algorithm for stepwise variable selection. It allows us to generate the entire LASSO regularization path with the same computational cost as standard LS via QR decomposition. At each step, it identifies the variable most correlated with the residuals of the model obtained so far and includes it in the set of active variables, moving the current iterate in a direction equiangular with the rest of the active predictors. Unlike OMP, which permanently maintains a variable once it is selected into the model, least angle regression adjusts the coefficient of the most correlated variable until that variable is no longer the most correlated with the recent residual.

LASSO is part of a powerful family of regularized linear regression methods for high-dimensional data analysis, which also includes ridge regression [65] and elastic-net [144]. Ridge regression is obtained by substituting the L_1-norm by the squared L_2-norm. From a statistical point of view, they can be viewed as methods for trading off the bias and variance of the coefficient estimates in order to find a model with better predictive performance. From a machine learning perspective, they allow us to adaptively control the capacity of the model space in order to prevent overfitting. It is well known that LASSO does not only reduce the variance of coefficient estimates but also is able to perform variable selection by shrinking many of these coefficients to zero. Elastic-net regularization trades L_1- and L_2-norms using a linearly mixed penalty [144].

LASSO cannot capture nonlinear dependency. To handle nonlinearity, instance-wise nonlinear LASSO [108] transforms the instance $\boldsymbol{x}$ by a nonlinear function $\psi(\cdot) : R^d \rightarrow R^d$. It gives a sparse solution in terms of instances but not features. To obtain sparsity in terms of features, featurewise nonlinear LASSO (also known as feature vector machine) [81] applies a nonlinear transformation in a featurewise manner. Both methods use the kernel trick.

Frank–Wolfe method is tailored to solve large-scale LASSO regression problems, based on a randomized iteration. The method outperforms the coordinate descent method [60]. It yields an $O(1/k)$ convergence rate (in terms of expected value). Frank–Wolfe method obtains solutions that are significantly more sparse in terms of the number of features compared with those from various competing methods while retaining the same optimization accuracy.

LASSO implicitly does model selection and shares many connections with forward stagewise regression [50]. Sparsity-inducing algorithms like LASSO are not (uniformly) algorithmically stable [133]. That is, leave-one-out versions of LASSO estimator are not uniformly close to one another. The data must be used twice, for choosing the tuning parameter and for estimating the model, respectively. Under some restrictions on the design matrix, LASSO estimator is still risk consistent with a tuning parameter empirically chosen via cross-validation [67].

The sparsity and consistency of LASSO are shown based on its robustness interpretation [132]. The robust optimization formulation is shown to be related to kernel density estimation. A no-free-lunch theorem proved in [132] states that sparsity and algorithmic stability contradict each other, and hence LASSO is not stable. An asymptotic analysis shows that the asymptotic variances of some of the robust versions of LASSO estimators are stabilized in the presence of large variance noise, compared with the unbounded asymptotic variance of the ordinary LASSO estimator [35].

Group Lasso has become a popular method for variable selection [135]. It deals with the variable selection at group level. Group Lasso penalty can be viewed as an intermediate mode between the L_1-type and L_2-type penalties.

LS ridge regression estimates linear regression coefficients, with the L_2 ridge regularization on coefficients. To better identify important features in the data, LASSO uses the L_1 penalty instead of using the L_2 ridge regularization. LASSO, or L_1 regularized LS, has been explored extensively for its remarkable sparsity properties. For sparse and high-dimensional regression problems, marginal regression, where each dependent variable is regressed separately on each covariate, computes the estimates roughly two orders-of-magnitude faster than LASSO solutions [62].

18.5 Other Sparse Algorithms

Sparse PCA attempts to find a sparse basis to make the result more interpretable. Thus, there is a trade-off between statistical fidelity and interpretability. Interpretability can be gained by post-processing the PCA loadings.

Loading rotation [72], simplified component technique-LASSO (SCoTLASS) [73], and augmented Lagrangian sparse PCA [86] have considered the orthogonality of loadings. SCoTLASS [73] optimizes the classical objective of PCA while imposing a sparsity constraint on each loading. Loading rotation [72] applies various criteria to rotate the PCA loadings so that simple structure appears, e.g., varimax criterion drives the entries to be either small or large, which is close to a sparse structure. Simple thresholding [16] obtains sparse loadings via directly setting entries of PCA loadings below a small threshold to zero. Augmented Lagrangian sparse PCA [86] solves the problem based on an augmented Lagrangian optimization. Augmented Lagrangian sparse PCA simultaneously considers the explained variance, orthogonality, and correlation among principal components.

SCoTLASS, rSVD [110], the greedy methods [40, 95], GPower [74], and TPower [137] belong to the deflation group. rSVD [110] obtains sparse loadings by solving a sequence of rank-1 matrix approximations, with an imposed sparsity penalty. In [95], greedy search and branch-and-bound methods were used to solve small instances of the problem exactly, leading to a total complexity of $O(p^4)$ for a full set of solutions. Another greedy algorithm PathSPCA [40] further approximates the solution process of [95], resulting in complexity of $O(p^3)$ for a full set of solutions. GPower method [74] formulates the problem as maximization of a convex objective function, and the solution is obtained by generalizing the power method used to compute the PCA loadings. TPower [137] and a related power method, iterative thresholding sparse PCA [87], aim at recovering sparse principal subspace.

In [145], sparse PCA formulates the problem as a regression-type optimization, to facilitate the use of LASSO or elastic-net techniques to solve the problem. The method has been extended to multilinear sparse PCA [78].

Direct sparse PCA [41] transforms the problem into a semidefinite convex relaxation problem, so that global optimality of the solution is guaranteed. The computational complexity is $O(p^4(\log p)^{1/2})$ for p variables, which is expensive for most applications. A variable elimination method [142] of complexity $O(p^3)$ was developed to make the application feasible on large-scale problems.

A sparse LMS algorithm [36] incorporates two sparsity constraints into the quadratic cost function of LMS algorithm. Recursive L_1-regularized LS algorithm [4] is developed for the estimation of a sparse tap-weight vector in the adaptive filtering setting. It exploits noisy observations of the tap-weight vector output stream and produces its estimate using an EM-type algorithm. The method converges to a near-optimal estimate in a stationary environment. It has significant improvement over RLS algorithm in terms of MSE but with lower computational requirements.

References

1. Adamczak, R. (2016). A Note on the sample complexity of the Er-SpUD algorithm by Spielman, Wang and Wright for exact recovery of sparsely used dictionaries. *Journal of Machine Learning Research*, *17*, 1–18.
2. Aharon, M., Elad, M., & Bruckstein, A. (2006). K-SVD: An algorithm for designing overcomplete dictionaries for sparse representation. *IEEE Transactions on Signal Processing*, *54*(11), 4311–4322.
3. Attouch, H., Bolte, J., Redont, P., & Soubeyran, A. (2010). Proximal alternating minimization and projection methods for nonconvex problems: An approach based on the Kurdyka-Lojasiewicz inequality. *Mathematics of Operational Research*, *35*(2), 438–457.
4. Babadi, B., Kalouptsidis, N., & Tarokh, V. (2010). SPARLS: The sparse RLS algorithm. *IEEE Transactions on Signal Processing*, *58*(8), 4013–4025.
5. Bao, C., Ji, H., Quan, Y., & Shen, Z. (2016). Dictionary learning for sparse coding: Algorithms and convergence analysis. *IEEE Transactions on Pattern Analysis and Machine Intelligence*, *38*(7), 1356–1369.
6. Bandeira, A. S., Fickus, M., Mixon, D. G., & Wong, P. (2013). The road to deterministic matrices with the restricted isometry property. *Journal of Fourier Analysis and Applications*, *19*(6), 1123–1149.
7. Baraniuk, R., Davenport, M., DeVore, R., & Wakin, M. (2008). A simple proof of the restricted isometry property for random matrices. *Constructive Approximation*, *28*(3), 253–263.
8. Baraniuk, R. G., Cevher, V., Duarte, M. F., & Hegde, C. (2010). Model-based compressive sensing. *IEEE Transactions on Information Theory*, *56*(4), 1982–2001.
9. Barg, A., Mazumdar, A., & Wang, R. (2015). Restricted isometry property of random subdictionaries. *IEEE Transactions on Information Theory*, *61*(8), 4440–4450.
10. Beck, A., & Teboulle, M. (2012). Smoothing and first order methods: a unified framework. *SIAM Journal on Optimization*, *22*(2), 557–580.
11. Bengio, Y. (2009). Learning deep architectures for AI. *Foundations and Trends in Machine Learning*, *2*(1), 1–127.
12. Blumensath, T., & Davies, M. E. (2008). Iterative thresholding for sparse approximations. *Journal of Fourier Analysis and Applications*, *14*, 629–654.
13. Blumensath, T., & Davies, M. E. (2009). Iterative hard thresholding for compressed sensing. *Applied and Computational Harmonic Analysis*, *27*(3), 265–274.
14. Blumensath, T., & Davies, M. E. (2010). Normalized iterative hard thresholding: Guaranteed stability and performance. *IEEE Journal of Selected Topics in Signal Processing*, *4*(2), 298–309.
15. Bolte, J., Sabach, S., & Teboulle, M. (2014). Proximal alternating linearized minimization for nonconvex and nonsmooth problems. *Mathematical Programming*, *146*, 459–494.

16. Cadima, J., & Jolliffe, I. T. (1995). Loading and correlations in the interpretation of principle compenents. *Applied Statistics*, *22*(2), 203–214.
17. Cai, T. T., & Zhang, A. (2014). Sparse representation of a polytope and recovery of sparse signals and low-rank matrices. *IEEE Transactions on Information Theory*, *60*(1), 122–132.
18. Calderbank, R., Howard, S., & Jafarpour, S. (2010). Construction of a large class of deterministic sensing matrices that satisfy a statistical isometry property. *IEEE Journal of Selected Topics in Signal Processing*, *4*(2), 358–374.
19. Calderbank, R., & Jafarpour, S. (2010). Reed Muller sensing matrices and the LASSO. In C. Carlet & A. Pott (Eds.), *Sequences and their applications* (Vol. 6338, pp. 442–463). LNCS. Berlin: Springer.
20. Candes, E. J. (2006). Compressive sampling. In *Proceedings of the International Congress of Mathematicians* (Vol. 3, pp. 1433–1452). Madrid, Spain.
21. Candes, E. J., & Fernandez-Granda, C. (2014). Towards a mathematical theory of super-resolution. *Communications on Pure and Applied Mathematics*, *67*(6), 906–956.
22. Candes, E. J., Romberg, J. K., & Tao, T. (2006). Stable signal recovery from incomplete and inaccurate measurements. *Communications on Pure and Applied Mathematics*, *59*(8), 1207–1223.
23. Candes, E. J., Romberg, J., & Tao, T. (2006). Robust uncertainty principles: Exact signal reconstruction from highly incomplete frequency information. *IEEE Transactions on Information Theory*, *52*(2), 489–509.
24. Candes, E. J., & Tao, T. (2006). Near-optimal signal recovery from random projections: Universal encoding strategies? *IEEE Transactions on Information Theory*, *52*(12), 5406–5425.
25. Candes, E. J., & Tao, T. (2005). Decoding by linear programming. *IEEE Transactions on Information Theory*, *51*(12), 4203–4215.
26. Candes, E. J., Wakin, M. B., & Boyd, S. P. (2008). Enhancing sparsity by reweighted l_1 minimization. *Journal of Fourier Analysis and Applications*, *14*, 877–905.
27. Cartis, C., & Thompson, A. (2015). A new and improved quantitative recovery analysis for iterative hard thresholding algorithms in compressed sensing. *IEEE Transactions on Information Theory*, *61*(4), 2019–2042.
28. Chang, L.-H., & Wu, J.-Y. (2012). *Compressive-domain interference cancellation via orthogonal projection: How small the restricted isometry constant of the effective sensing matrix can be? In Proceedings of IEEE Wireless Communications and Networking Conference (WCNC)* (pp. 256–261). Shanghai: China.
29. Chang, L.-H., & Wu, J.-Y. (2014). An improved RIP-based performance Guarantee for sparse signal recovery via orthogonal matching pursuit. *IEEE Transactions on Information Theory*, *60*(9), 5702–5715.
30. Chartrand, R. (2007). Exact reconstruction of sparse signals via nonconvex minimization. *IEEE Signal Processing Letters*, *14*(10), 707–710.
31. Chartrand, R., & Yin, W. (2008). Iteratively reweighted algorithms for compressive sensing. In *Proceedings of the IEEE International Conference on Acoustics, Speech, and Signal Processing* (pp. 3869–3872). Las Vegas, NV.
32. Chen, L., & Gu, Y. (2014). The convergence guarantees of a non-convex approach for sparse recovery. *IEEE Transactions on Signal Processing*, *62*(15), 3754–3767.
33. Chen, S. S., Donoho, D. L., & Saunders, M. A. (1999). Atomic decomposition by basis pursuit. *SIAM Journal on Scientific Computing*, *20*(1), 33–61.
34. Chen, S. S., Donoho, D. L., & Saunders, M. A. (2001). Atomic decomposition by basis pursuit. *SIAM Review*, *43*(1), 129–159.
35. Chen, X., Wang, Z. J., & McKeown, M. J. (2010). Asymptotic analysis of robust LASSOs in the presence of noise with large variance. *IEEE Transactions on Information Theory*, *56*(10), 5131–5149.
36. Chen, Y., Gu, Y., & Hero III, A. O. (2009). Sparse LMS for system identification. In *Proceedings of IEEE International Conference on Acoustics, Speech and Signal Processing*. Taipei, Taiwan.

37. Chi, Y., Scharf, L. L., Pezeshki, A., & Calderbank, A. R. (2011). Sensitivity to basis mismatch in compressed sensing. *IEEE Transactions on Signal Processing*, *59*(5), 2182–2195.
38. Dai, W., & Milenkovic, O. (2009). Subspace pursuit for compressive sensing signal reconstruction. *IEEE Transactions on Information Theory*, *55*(5), 2230–2249.
39. Dai, W., & Milenkovic, O. (2009). Weighted superimposed codes and constrained integer compressed sensing. *IEEE Transactions on Information Theory*, *55*(5), 2215–2229.
40. d'Aspremont, A., Bach, F., & El Ghaoui, L. (2008). Optimal solutions for sparse principal component analysis. *Journal of Machine Learning Research*, *9*, 1269–1294.
41. d'Aspremont, A., El Ghaoui, L., Jordan, M. I., & Lanckriet, G. R. G. (2007). A direct formulation for sparse PCA using semidefinite programming. *SIAM Review*, *49*(3), 434–448.
42. Daubechies, I., Defrise, M., & De Mol, C. (2004). An iterative thresholding algorithm for linear inverse problems with a sparsity constraint. *Communications on Pure and Applied Mathematics*, *57*(11), 1413–1457.
43. Davenport, M. A., & Wakin, M. B. (2010). Analysis of orthogonal matching pursuit using the restricted isometry property. *IEEE Transactions on Information Theory*, *56*(9), 4395–4401.
44. DeVore, R. A. (2007). Deterministic constructions of compressed sensing matrices. *Journal of Complexity*, *23*, 918–925.
45. Dong, Z., & Zhu, W. (2018). Homotopy methods based on l_0-norm for compressed sensing. *IEEE Transactions on Neural Networks and Learning Systems*, *29*(4), 1132–1146.
46. Donoho, D. L. (2006). Compressed sensing. *IEEE Transactions on Information Theory*, *52*(4), 1289–1306.
47. Donoho, D. L. (2006). For most large underdetermined systems of linear equations the minimal l_1-norm solution is also the sparsest solution. *Communications on Pure and Applied Mathematics*, *59*, 797–829.
48. Donoho, D. L., Maleki, A., & Montanari, A. (2009). Message-passing algorithms for compressed sensing. *Proceedings of the National Academy of Sciences of the USA*, *106*(45), 18914–18919.
49. Donoho, D. L., Tsaig, Y., Drori, I., & Starck, J.-L. (2012). Sparse solution of underdetermined systems of linear equations by stagewise orthogonal matching pursuit. *IEEE Transactions on Information Theory*, *58*(2), 1094–1121.
50. Efron, B., Hastie, T., Johnstone, I., & Tibshirani, R. (2004). Least angle regression. *Annals of Statistics*, *32*(2), 407–499.
51. Eldar, Y. C., & Kutyniok, G. (2012). *Compressed sensing: Theory and applications*. Cambridge: Cambridge University Press.
52. Eldar, Y. C., Kuppinger, P., & Bolcskei, H. (2010). Block-sparse signals: Uncertainty relations and efficient recovery. *IEEE Transactions on Signal Processing*, *58*(6), 3042–3054.
53. Elvira, C., Chainais, P., & Dobigeon, N. (2017). Bayesian antisparse coding. *IEEE Transactions on Signal Processing*, *65*(7), 1660–1672.
54. Exarchakis, G., & Lucke, J. (2017). Discrete sparse coding. *Neural Computation*, *29*(11), 2979–3013.
55. Fan, J., & Li, R. (2001). Variable selection via nonconcave penalized likelihood and its oracle properties. *Journal of the American Statistical Association*, *96*, 1348–1360.
56. Figueiredo, M. A. T., Nowak, R. D., & Wright, S. J. (2007). Gradient projection for sparse reconstruction: Application to compressed sensing and other inverse problems. *IEEE Journal of Selected Topics in Signal Processing*, *1*(4), 586–597.
57. Foucart, S. (2011). Hard thresholding pursuit: An algorithm for compressive sensing. *SIAM Journal on Numerical Analysis*, *49*(6), 2543–2563.
58. Foucart, S., & Lai, M.-J. (2009). Sparsest solutions of underdetermined linear systems via l_q-minimization for $0<q \leq 1$. *Applied and Computational Harmonic Analysis*, *26*(3), 395–407.
59. Foucart, S., & Rauhut, H. (2013). *A mathematical introduction to compressive sensing*. Cambridge: Birkhauser.
60. Frandi, E., Nanculef, R., Lodi, S., Sartori, C., & Suykens, J. A. K. (2016). Fast and scalable Lasso via stochastic Frank-Wolfe methods with a convergence guarantee. *Machine Learning*, *104*(2), 195–221.

61. Friedland, S., Li, Q., & Schonfeld, D. (2014). Compressive sensing of sparse tensors. *IEEE Transactions on Image Processing*, *23*(10), 4438–4447.
62. Genovese, C. R., Jin, J., Wasserman, L., & Yao, Z. (2012). A comparison of the lasso and marginal regression. *Journal of Machine Learning Research*, *13*, 2107–2143.
63. Giryes, R., & Elad, M. (2012). RIP-based near-oracle performance guarantees for SP, CoSaMP, and IHT. *IEEE Transactions on Signal Processing*, *60*(3), 1465–1568.
64. Gribonval, R., & Nielsen, M. (2007). Highly sparse representations from dictionaries are unique and independent of the sparseness measure. *Applied and Computational Harmonic Analysis*, *22*(3), 335–355.
65. Hastie, T., Tibshirani, R., & Friedman, J. (2009). *The elements of statistical learning*. New York: Springer.
66. Haviv, I., & Regev, O. (2016). The restricted isometry property of subsampled Fourier matrices. In *Proceedings of the 27th Annual ACM-SIAM Symposium on Discrete Algorithms* (pp. 288–297). Arlington, TX.
67. Homrighausen, D., & McDonald, D. J. (2014). Leave-one-out cross-validation is risk consistent for lasso. *Machine Learning*, *97*, 65–78.
68. Hoyer, P. (2002). Non-negative sparse coding. In *Proceedings of the 12th IEEE Workshop on Neural Networks for Signal Processing* (pp. 557–565).
69. Huang, S., & Zhu, J. (2011). Recovery of sparse signals using OMP and its variants: Convergence analysis based on RIP. *Inverse Problems*, *27*(3), 035003.
70. Jaggi, M. (2014). An equivalence between the Lasso and support vector machines. In J. A. K. Suykens, M. Signoretto, & A. Argyriou (Eds.), *Regularization, optimization, kernels, and support vector machines* (Chap. 1, pp. 1–26). Boca Raton: Chapman & Hall/CRC.
71. Jenatton, R., Mairal, J., Obozinski, G., & Bach, F. (2011). Proximal methods for hierarchical sparse coding. *Journal of Machine Learning Research*, *12*, 2297–2334.
72. Jolliffe, I. T. (1989). Rotation of ill-defined principal components. *Applied Statistics*, *38*(1), 139–147.
73. Jolliffe, I. T., Trendafilov, N. T., & Uddin, M. (2003). A modified principal component technique based on the LASSO. *Journal of Computational and Graphical Statistics*, *12*(3), 531–547.
74. Journee, M., Nesterov, Y., Richtarik, P., & Sepulchre, R. (2010). Generalized power method for sparse principal component analysis. *Journal of Machine Learning Research*, *11*, 517–553.
75. Kim, H., & Park, H. (2008). Nonnegative matrix factorization based on alternating nonnegativity constrained least squares and active set method. *SIAM Journal on Matrix Analysis and Applications*, *30*(2), 713–730.
76. Kreutz-Delgado, K., Murray, J. F., Rao, B. D., Engan, K., Lee, T.-W., & Sejnowski, T. J. (2003). Dictionary learning algorithms for sparse representation. *Neural Computation*, *15*(2), 349–396.
77. Kwon, S., Wang, J., & Shim, B. (2014). Multipath matching pursuit. *IEEE Transactions on Information Theory*, *60*(5), 2986–3001.
78. Lai, Z., Xu, Y., Chen, Q., Yang, J., & Zhang, D. (2014). Multilinear sparse principal component analysis. *IEEE Transactions on Neural Networks and Learning Systems*, *25*(10), 1942–1950.
79. Langford, J., Li, L., & Zhang, T. (2009). Sparse online learning via truncated gradient. *Journal of Machine Learning Research*, *10*, 777–801.
80. Lee, M., Shen, H., Huang, J. Z., & Marron, J. S. (2010). Biclustering via sparse singular value decomposition. *Biometrics*, *66*(4), 1087–1095.
81. Li, F., Yang, Y., & Xing, E. (2006). FromLasso regression to feature vector machine. In Y. Weiss, B. Scholkopf, & J. Platt (Eds.), *Advances in neural information processing systems* (Vol. 18, pp. 779–786). Cambridge: MIT Press.
82. Li, H., Wang, J., & Yuan, X. (2018). On the fundamental limit of multipath matching pursuit. *IEEE Journal of Selected Topics in Signal Processing*, *12*(5), 916–927.
83. Li, X. (2013). Compressed sensing and matrix completion with constant proportion of corruptions. *Constructive Approximation*, *37*(1), 73–99.

84. Lin, D., Pitler, E., Foster, D. P., & Ungar, L. H. (2008). In defense of l_0. In *Proceedings of ICML/UAI/COLT Workshop on Sparse Optimization and Variable Selection*. Helsinki, Finland.
85. Liu, E., & Temlyakov, V. N. (2012). The orthogonal super greedy algorithm and applications in compressed sensing. *IEEE Transactions on Information Theory*, *58*(4), 2040–2047.
86. Lu, Z., & Zhang, Y. (2012). An augmented Lagrangian approach for sparse principal component analysis. *Mathematical Programming*, *135*, 149–193.
87. Ma, Z. (2013). Sparse principal component analysis and iterative thresholding. *Annals of Statistics*, *41*(2), 772–801.
88. Mairal, J., Bach, F., & Ponce, J. (2012). Task-driven dictionary learning. *IEEE Transactions on Pattern Analysis and Machine Intelligence*, *34*(4), 791–804.
89. Malioutov, D. M., Cetin, M., & Willsky, A. S. (2005). Homotopy continuation for sparse signal representation. In *Proceedings of IEEE International Conference on Acoustics, Speech, and Signal Processing* (pp. 733–736). Philadelphia, PA.
90. Mallat, S. G., & Zhang, Z. (1993). Matching pursuits with timefrequency dictionaries. *IEEE Transactions on Signal Processing*, *41*(12), 3397–3415.
91. Marjanovic, G., & Solo, V. (2012). On l_q optimization and matrix completion. *IEEE Transactions on Signal Processing*, *60*(11), 5714–5724.
92. Misra, S., & Parrilo, P. A. (2015). Weighted l_1-minimization for generalized non-uniform sparse model. *IEEE Transactions on Information Theory*, *61*(8), 4424–4439.
93. Mo, Q. (2015). A sharp restricted isometry constant bound of orthogonal matching pursuit. https://arxiv.org/pdf/1501.01708.pdf.
94. Mo, Q., & Yi, S. (2012). A remark on the restricted isometry property in orthogonal matching pursuit. *IEEE Transactions on Information Theory*, *58*(6), 3654–3656.
95. Moghaddam, B., Weiss, Y., & Avidan, S. (2006). Spectral bounds for sparse PCA: Exact and greedy algorithms. *Advances in neural information processing systems* (Vol. 18, pp. 915-922). Cambridge: MIT Press.
96. Natarajan, B. K. (1995). Sparse approximate solutions to linear systems. *SIAM Journal on Computing*, *24*(2), 227–234.
97. Needell, D., & Tropp, J. A. (2009). CoSaMP: Iterative signal recovery from incomplete and inaccurate samples. *Applied and Computational Harmonic Analysis*, *26*(3), 301–321.
98. Needell, D., & Vershynin, R. (2010). Signal recovery from incomplete and inaccurate measurements via regularized orthogonal matching pursuit. *IEEE Journal of Selected Topics in Signal Processing*, *4*(2), 310–316.
99. Nesterov, Y. (2013). Gradient methods for minimizing composite functions. *Mathematical Programming*, *140*(1), 125–161.
100. Nikolova, M. (2013). Description of the minimizers of least squares regularized with l_0-norm. Uniqueness of the global minimizer. *SIAM Journal on Imaging Sciences*, *6*(2), 904–937.
101. Olshausen, B. A., & Field, D. J. (1996). Emergence of simple-cell receptive field properties by learning a sparse code for natural images. *Nature*, *381*(6583), 607–609.
102. Olshausen, B. A., & Field, D. J. (1997). Sparse coding with an overcomplete basis set: A strategy employed by V1? *Vision Research*, *37*(23), 3311–3325.
103. Pati, Y. C., Rezaiifar, R., & Krishnaprasad, P. S. (1993). Orthogonal matching pursuit: Recursive function approximation with applications to wavelet decomposition. In *Proceedings of the 27th Asilomar Conference on Signals, Systems and Computers* (Vol. 1, pp. 40–44). Los Alamitos, CA.
104. Peng, J., Yue, S., & Li, H. (2015). NP/CMP equivalence: A phenomenon hidden among sparsity models l_0 minimization and l_p minimization for information processing. *IEEE Transactions on Information Theory*, *61*(7), 4028–4033.
105. Rauhut, H. (2008). Stability results for random sampling of sparse trigonometric polynomials. *IEEE Transactions on Information Theory*, *54*(12), 5661–5670.
106. Rebollo-Neira, L., & Lowe, D. (2002). Optimized orthogonal matching pursuit approach. *IEEE Signal Processing Letters*, *9*(4), 137–140.

107. Romero, D., Ariananda, D. D., Tian, Z., & Leus, G.(2016). Compressive covariance sensing: Structure-based compressive sensing beyond sparsity. *IEEE Signal Processing Magazine, 33*(1), 78–93.
108. Roth, V. (2004). The generalized Lasso. *IEEE Transactions on Neural Networks, 15*(1), 16–28.
109. Shalev-Shwartz, S., & Tewari, A. (2011). Stochastic methods for l_1-regularized loss minimization. *Journal of Machine Learning Research, 12*, 1865–1892.
110. Shen, H., & Huang, J. Z. (2008). Sparse principal component analysis via regularized low rank matrix approximation. *Journal of Multivariate Analysis, 99*(6), 1015–1034.
111. Shen, J., & Li, P. (2018). A tight bound of hard thresholding. *Journal of Machine Learning Research, 18*, 1–42.
112. Sidiropoulos, N. D., & Kyrillidis, A. (2012). Multi-way compressed sensing for sparse low-rank tensors. *IEEE Signal Processing Letters, 19*(11), 757–760.
113. Sivalingam, R., Boley, D., Morellas, V., & Papanikolopoulos, N. (2015). Tensor dictionary learning for positive definite matrices. *IEEE Transactions on Image Processing, 24*(11), 4592–4601.
114. Soussen, C., Gribonval, R., Idier, J., & Herzet, C. (2013). Joint k-step analysis of orthogonal matching pursuit and orthogonal least squares. *IEEE Transactions on Information Theory, 59*(5), 3158–3174.
115. Spielman, D., Wang, H., & Wright, J. (2012). Exact recovery of sparsely-used dictionaries. In: *JMLR: Workshop and Conference Proceedings of the 25th Annual Conference on Learning Theory* (Vol. 23, pp. 37.1–37.18).
116. Tibshirani, R. (1996). Regression shrinkage and selection via the lasso. *Journal of the Royal Statistical Society: Series B, 58*(1), 267–288.
117. Tibshirani, R. (2011). Regression shrinkage and selection via the lasso: A retrospective. *Journal of the Royal Statistical Society Series B, 73*(3), 273–282.
118. Tibshirani, R., Saunders, M., Rosset, S., Zhu, J., & Knight, K. (2005). Sparsity and smoothness via the fused lasso. *Journal of the Royal Statistical Society: Series B, 67*(1), 91–108.
119. Tropp, J. A. (2004). Greed is good: Algorithmic results for sparse approximation. *IEEE Transactions on Information Theory, 50*(10), 2231–2242.
120. Tropp, J. A., & Gilbert, A. C. (2007). Signal recovery from random measurements via orthogonal matching pursuit. *IEEE Transactions on Information Theory, 53*(12), 4655–4666.
121. Tropp, J. A., & Wright, S. J. (2010). Computational methods for sparse solution of linear inverse problems. *Proceedings of the IEEE, 98*(6), 948–958.
122. Tzagkarakis, G., Nolan, J. P., & Tsakalides, P. (2019). Compressive sensing using symmetric alpha-stable distributions for robust sparse signal reconstruction. *IEEE Transactions on Signal Processing, 67*(3), 808–820.
123. Wang, C., Yue, S., & Peng, J., (2015). When is P such that l_0-minimization equals to l_p-minimization. *CoRR*, arxiv:abs/1511.07628.
124. Wang, J., Kwon, S., & Shim, B. (2012). Generalized orthogonal matching pursuit. *IEEE Transactions on Signal Processing, 60*(12), 6202–6216.
125. Wang, J., & Li, P. (2017). Recovery of sparse signals using multiple orthogonal least squares. *IEEE Transactions on Signal Processing, 65*(8), 2049–2062.
126. Weed, J. (2018). Approximately certifying the restricted isometry property is hard. *IEEE Transactions on Information Theory, 64*(8), 5488–5497.
127. Wen, J., Wang, J., & Zhang, Q. (2017). Nearly optimal bounds for orthogonal least squares. *IEEE Transactions on Signal Processing, 65*(20), 5347–5356.
128. Wen, J., Zhou, Z., Wang, J., Tang, X., & Mo, Q. (2017). A sharp condition for exact support recovery with orthogonal matching pursuit. *IEEE Transactions on Signal Processing, 65*(6), 1370–1382.
129. Wen, J., Zhou, Z., Liu, Z., Lai, M.-J., & Tang, X. (2019). Sharp sufficient conditions for stable recovery of block sparse signals by block orthogonal matching pursuit. *Applied and Computational Harmonic Analysis, 47*(3), 948–974.
130. Wu, R., & Chen, D.-R. (2013a). The improved bounds of restricted isometry constant for recovery via ℓ_p-minimization. *IEEE Transactions on Information Theory, 59*(9), 6142–6147.

131. Wu, R., Huang, W., & Chen, D.-R. (2013b). The exact support recovery of sparse signals with noise via orthogonal matching pursuit. *IEEE Signal Processing Letters*, *20*(4), 403–406.
132. Xu, H., Caramanis, C., & Mannor, S. (2010). Robust regression and Lasso. *IEEE Transactions on Information Theory*, *56*(7), 3561–3574.
133. Xu, H., Mannor, S., & Caramanis, C. (2008). Sparse algorithms are not stable: A no-free-lunch theorem. In *Proceedings of the IEEE 46th Annual Allerton Conference on Communication, Control, and Computing* (pp. 1299–1303).
134. Xu, Z., Chang, X., Xu, F., & Zhang, H. (2012). $L_{1/2}$ regularization: A thresholding representation theory and a fast solver. *IEEE Transactions on Neural Networks and Learning Systems*, *23*(7), 1013–1027.
135. Yuan, M., & Lin, Y. (2006). Model selection and estimation in regression with grouped variables. *Journal of the Royal Statistical Society Series B*, *68*(1), 49–67.
136. Yuan, X.-T., Li, P., & Zhang, T. (2018). Gradient hard thresholding pursuit. *Journal of Machine Learning Research*, *18*, 1–43.
137. Yuan, X.-T., & Zhang, T. (2013). Truncated power method for sparse eigenvalue problems. *Journal of Machine Learning Research*, *14*(1), 899–925.
138. Zarmehi, N., & Marvasti, F. (2019). Removal of sparse noise from sparse signals. *Signal Processing*, *158*, 91–99.
139. Zhang, C. H. (2010). Nearly unbiased variable selection under minimax concave penalty. *Annals of Statistics*, *38*(2), 894–942.
140. Zhang, R., & Li, S. (2019). Optimal RIP bounds for sparse signals recovery via ℓ_p minimization. *Applied and Computational Harmonic Analysis*, *47*(3), 466–584.
141. Zhang, T. (2011). Sparse recovery with orthogonal matching pursuit under RIP. *IEEE Transactions on Information Theory*, *57*(9), 6215–6221.
142. Zhang, Y., & El Ghaoui, L. (2011). Large-scale sparse principal component analysis with application to text data. In *Advances in neural information processing systems* (Vol. 24, pp. 532–539). Red Hook: Curran & Associates Inc.
143. Zhu, M., & Rozell, C. J. (2013). Visual nonclassical receptive field effects emerge from sparse coding in a dynamical system. *PLOS Computational Biology*, *9*, e1003191.
144. Zou, H., & Hastie, T. (2005). Regularization and variable selection via the elastic net. *Journal of the Royal Statistical Society: Series B*, *67*(2), 301–320.
145. Zou, H., Hastie, T., & Tibshirani, R. (2006). Sparse principal component analysis. *Journal of Computational and Graphical Statistics*, *15*(2), 265–286.

Chapter 19
Matrix Completion

19.1 Introduction

Low-rank representation is a popular method for recovering data from corruptions or outliers. Low-rank captures the global structure of the data. PCA, which projects data to fixed-rank low-dimensional space by minimizing L_2-norm error, is the most popular low-rank method. Robust PCA [12] and GoDec [35, 104] decompose data into low-rank components and corruptions.

Matrix completion aims to recover a matrix from a small subset of its entries. The matrix completion problem is prevalent in many applications, including computer vision, collaborative filtering, sensor network localization, learning and content analytics, rank aggregation, and manifold learning.

In many applications, the entries of the matrix are not real-valued, but discrete or quantized, e.g., binary-valued or multiple-valued. For example, in the Netflix problem, a subset of the users' ratings, which takes integer values between 1 and 5, is observed. The Netflix problem is a low-rank matrix completion problem. Classical matrix completion treats these values as real-valued with good results. However, performance improvement can be achieved when the observations are treated as discrete [23, 57].

In the area of recommender systems, users submit ratings on a subset of entries in a database, and the vendor provides recommendations based on the user's preferences [78, 84]. Users are given the opportunity to rate movies, but users typically rate only very few movies so that there are very few scattered entries in this matrix. In this case, the matrix may be low-rank since only a few factors contribute to an individual's opinion. A collaborative filtering dataset can be interpreted as the incomplete observation of a ratings matrix with columns corresponding to users and rows corresponding to items. The goal is to infer the unobserved entries of this ratings matrix. For M items and N users, this gives us an $M \times N$ user-item matrix $\mathbf{X}$, where $x_{m,n}$ represents the nth user's recommendation for item m.

K.-L. Du and M. N. S. Swamy, *Neural Networks and Statistical Learning*,
https://doi.org/10.1007/978-1-4471-7452-3_19

Completion of an arbitrary matrix is an ill-posed problem. A commonly adopted assumption is that the underlying matrix comes from a restricted class. It is well known that most real-world data such as images have a low rank or an approximately low-rank structure. Recovery of a matrix from a subset of its entries is known as the matrix completion problem. Most existing methods formulate the task as a low-rank matrix approximation problem. The results are connected with compressed sensing, and show that objects other than signals and images can be perfectly reconstructed from very limited information.

19.2 Matrix Completion

Given an incomplete data matrix $\mathbf{M} \in R^{m\times n}$ of low rank, the matrix completion problem can be formulated as

$$\min \operatorname{rank}(\mathbf{X}) \quad \text{subject to} \quad X_{ij} = M_{ij}, (i, j) \in \Omega, \tag{19.1}$$

where $\mathbf{X} \in R^{m\times n}$ is the decision variable, Ω is the set of locations corresponding to the observed entries, and $\operatorname{rank}(\cdot)$ denotes the rank of a matrix.

This problem seeks the simplest explanation fitting the observed data. It is an NP-hard problem due to the discontinuous and nonconvex nature of the rank function [10]. Low-rank matrix completion problem is ill-posed in general.

The missing entries of a low-rank matrix can be exactly recovered with high probability under certain constraints of missing rate, matrix rank, and sampling scheme [6, 10, 21].

Let $\mathbf{X}$ be the observed data corrupted by errors $\mathbf{E}$. The problem is to recover the low-rank matrix $\mathbf{Z} \in R^{m_1\times n_1}$ from the measurement $\mathbf{X} = F(\mathbf{Z}) + \mathbf{E} \in R^{m\times n}$ with the given linear operator $F : R^{m_1\times n_1} \rightarrow R^{m\times n}$:

$$\min_{\mathbf{Z},\mathbf{E}} \operatorname{rank}(\mathbf{Z}) + \lambda\|\mathbf{E}\|_L \tag{19.2}$$

subject to

$$\mathbf{X} = F(\mathbf{Z}) + \mathbf{E}, \tag{19.3}$$

where λ is a a given parameter, and $\|\cdot\|_L$ is the L_0-norm [12] or $L_{2,0}$-norm [62] for promoting sparsity.

Applications of low-rank matrix recovery problems include low-rank and sparse matrix decomposition [12] ($F(.)$ as an identity operator), low-rank representation [62, 65] ($F(\mathbf{X}) = \mathbf{AX}$ and $\mathbf{A}$ as a given data dictionary), and low-rank matrix completion [10] ($F(.)$ as a sampling operator).

The optimization problem (19.3) is NP-hard due to the discrete nature of the rank function and L_0-norm (or $L_{2,0}$-norm). For rank r, there are only r nonzero singular values. A widely used convex relaxation of rank is *nuclear norm* (also known as *trace norm*), the sum of the singular values of matrix [29]. The rank function counts the number of nonvanishing singular values, and the nuclear norm sums their amplitude. The nuclear norm is the tightest convex lower bound of the rank function of matrices [77]. The nuclear norm is a convex function, as such can be optimized efficiently via semidefinite programming. A convex relaxation of L_0-norm (or $L_{2,0}$-norm) is L_1-norm (or $L_{2,1}$-norm). In some sense, the nuclear norm is to the rank functional what the L_1 norm is to the counting L_0 norm in the area of sparse signal recovery [77]. L_∞-norm is used as a convex relaxation for the rank in [30] for matrix completion under uniform sampling distribution.

Two major categories of matrix completion methods are rank minimization-based methods [10, 19, 91] and matrix factorization-based methods [84]. Matrix factorization-based methods factorize an $m \times n$ matrix of rank-r into two smaller matrices of size $m \times r$ and $r \times n$, where $r < \min(m, n)$. Hence the missing entries can be recovered through finding such pairwise matrices [84]. Another approach to low-rank matrix completion is spectral methods [47].

Some methods for solving low-rank matrix problems are singular value thresholding [8], accelerated proximal gradient [61], and augmented Lagrange multiplier method [60].

By assuming a Laplacian noise model instead of a Gaussian noise model, many weighted low-rank matrix approximation methods in the presence of missing data are based on L_1-norm [27, 46, 60]. They are computationally too intensive and it is hard to find a good solution of the L_1-norm-based cost function because of its nonconvexity and nonsmoothness. In [46], convex programming and weighted median approaches are presented based on alternating minimization for the L_1-norm-based cost function. The methods given in [27] find the solution using convex LP. Robust PCA based on the L_1-norm and nuclear norm for a nonfixed rank problem [60] is solved using augmented Lagrange method. It performs SVD at each iteration.

Conventional matrix completion methods are not effective when the data are from nonlinear transformations of lower dimensional latent subspace. Matrices consisting of such nonlinear data are always of high rank or even full rank. Nonlinear matrix completion recovers missing entries of data matrices with nonlinear structures. The method proposed in [28] minimizes the rank (approximated by Schatten p-norm) of a matrix in the feature space given by a nonlinear mapping of the data (input) space, where kernel trick is used to avoid carrying out the unknown nonlinear mapping explicitly.

19.2.1 Minimizing the Nuclear Norm

A matrix with missing values can be recovered exactly via nuclear-norm minimization under some general conditions [9–11, 76]. When the observed entries are noiseless, perfect recovery of a low-rank matrix is possible [10]. This result has been extended to noisy measurements in [9]: with high probability, the recovery is subject to an error bound proportional to the noise level.

The problem is formulated as [8, 9, 29, 77]:

$$\min_{\mathbf{X}} \|\mathbf{X}\|_* = \sum_{k=1}^{\min(m,n)} \sigma_k(\mathbf{X}) \quad \text{subject to} \quad X_{ij} = M_{ij}, (i, j) \in \Omega, \tag{19.4}$$

where the nuclear norm $\| \cdot \|_*$ denotes the convex envelope of the rank function, and $\sigma_k(\mathbf{X})$ is the kth largest singular value of $\mathbf{X}$.

The convex relaxed trace-norm minimization problems (19.4) have to be solved iteratively and involve SVD at each iteration. Such algorithms suffer from high computation cost. Variations of alternating minimization strategies are popular for matrix completion [19, 54, 105].

A proof of global convergence of gradient search for low-rank matrix approximation is given based on the optimization on the Grassmann manifold and Fubiny–Study distance on this space [74].

Singular value thresholding [8], nuclear-norm regularized LS [91], and robust PCA [12, 97] are based on the nuclear norm.

Singular value thresholding algorithm is a gradient-descent method that applies Uzawa method to efficiently solve the optimization problem [8]:

$$\min_{\mathbf{X}} \|\mathbf{X}\|_* + \alpha\|\mathbf{X}\|_F^2 \tag{19.5}$$

subject to

$$P_\Omega(\mathbf{X}) = P_\Omega(\mathbf{M}), \tag{19.6}$$

where $P_\Omega(\cdot)$ is a submatrix extracted from the original matrix, with the set of locations Ω, and α is a a given parameter.

The nuclear-norm regularized LS problem is given by [45, 91]

$$\min_{\mathbf{X}} \frac{1}{2}\|P_\Omega(\mathbf{X}) - P_\Omega(\mathbf{M})\|_F^2 + \mu\|\mathbf{X}\|_*, \tag{19.7}$$

where μ is a given parameter.

An accelerated proximal gradient optimization technique is applied for solving the nuclear-norm regularized LS problem [45, 91]. The primal error of their algorithms is smaller than ϵ after $O(1/\sqrt{\epsilon})$ iterations [45, 91].

Robust PCA introduces the nuclear norm to recover the subspace structure from the data corrupted by noises or occlusions [12]. Principal component pursuit [12] recovers a matrix observed with mostly noiseless entries and otherwise, a small amount of outliers. This is done by modeling the observed matrix as a sum of a low-rank matrix and a sparse matrix. For the matrix completion problem when the observed entries are noisy and contain outliers, the method presented in [96] uses Huber function to downweigh the effects of outliers. The method is fast and monotonically convergent.

However, nuclear norm is not a good approximation to the rank function. The incoherence property of the nuclear norm heuristic is very hard to satisfy in practice [10, 11].

Minimizing the Truncated Nuclear Norm

In nuclear-norm minimization-based methods, all the singular values are simultaneously minimized, and thus the rank may not be well approximated in practice. Truncated nuclear norm, given by a nuclear norm subtracted by the sum of the largest few singular values, achieves a better approximation to the rank of a matrix than a nuclear norm [42]. Truncated nuclear-norm regularization method converges better than its nuclear-norm-based counterpart for accurate matrix completion.

A truncated nuclear norm $\|\mathbf{X}\|_r$ is defined as the sum of $\min(m, n) - r$ minimum singular values, i.e., $\|\mathbf{X}\|_r = \sum_{i=r+1}^{\min(m,n)} \sigma_i(\mathbf{X})$. Thus, truncated nuclear-norm regularization method [42] formulates

$$\min_{\mathbf{X}} \|\mathbf{X}\|_r \quad \text{subject to} \quad P_\Omega(\mathbf{X}) = P_\Omega(\mathbf{M}). \tag{19.8}$$

Since the largest r nonzero singular values will not affect the rank of the matrix, the sum of the smallest $\min(m, n) - r$ singular values is minimized.

Since $\|\mathbf{X}\|_r$ is nonconvex, it is not easy to solve the problem (19.8). A two-step iterative scheme [42] is designed, where an alternating direction method of multipliers (ADMM) is implemented to optimize the convex subproblem in the second step with excellent convergence accuracy. However, the method is not robust to the number of subtracted singular values and requires a large number of iterations to converge. As it is hard to choose a suitable penalty parameter for ADMM, an adaptive penalty is used to achieve a faster convergence rate.

19.2.2 Matrix Factorization-Based Methods

Matrix completion problems can also be solved based on matrix factorization [83]. A maximum-margin factorization method [84] uses a factor model $\mathbf{X} = \mathbf{U}\mathbf{V}^T$ for the original matrix. Consider the problem:

$$\min_{\mathbf{U},\mathbf{V}} \sum_{(i,j)\in\Omega} (X_{ij} - (\mathbf{U}\mathbf{V}^T)_{ij})^2 + \beta(\|\mathbf{U}\|_F^2 + \|\mathbf{V}\|_F^2), \tag{19.9}$$

where β is a scaling parameter.

This formulation and its extensions have been explored in [78, 84, 86]. Biconvex methods get stuck in suboptimal local minimum if the rank r is small [84]. The computational cost is very high for large r and dimensions m, n.

An efficient matrix bi-factorization method [64] approximates the original trace-norm minimization problem and mitigates the computation cost of performing SVDs. Matrix bi-factorization method can be used to address a wide range of low-rank matrix recovery and completion problems, such as low-rank and sparse matrix decomposition, low-rank representation, and low-rank matrix completion. Two linearized proximal alternative optimization algorithms are developed for solving the three problems [64].

An efficient algorithm for large matrix factorization and completion (R package softImpute, http://CRAN.R-project.org/package=softImpute) [37] brings nuclear-norm minimization method and maximum-margin matrix factorization method together, and it outperforms both the methods. It is a stylized variant of block coordinate descent.

Focusing on noisy matrix completion [9] and noisy robust matrix factorization [12, 15], a scalable divide-and-conquer framework for noisy matrix factorization is introduced in [69], with attainable near-linear to superlinear speed-ups. The method randomly divides the matrix factorization task into cheaper subproblems, solves those subproblems in parallel using a base matrix factorization algorithm for nuclear-norm regularized formulations, and combines the solutions to the subproblems using efficient techniques from randomized matrix approximation.

Two low-rank factorization methods are proposed in [51] for L_1-based low-rank matrix approximation problems. They find proper projection and coefficient matrices by using alternating rectified gradient method, at low running time and memory. After finding an update direction, weighted median algorithm is used to find the step size for updating a matrix. Weighted median algorithm is applied to the entire matrix at once to reduce the computational burden, while it is applied columnwise in [46].

19.2.3 Theoretical Guarantees on Exact Matrix Completion

Suppose that we observe m entries selected uniformly at random from a matrix $\mathbf{M}$. One can perfectly recover most low-rank matrices from what appears to be an incomplete set of entries [10]. It is proved that if the number m of sampled entries obeys $m \geq C\, n^{1.25} r \log n$ for some positive numerical constant C, then with very high probability, most $n \times n$ matrices of rank r can be perfectly recovered by solving a simple convex optimization program. This program finds the matrix with minimum nuclear norm that fits the data. Similar results hold for arbitrary rectangular matrices as well.

The first algorithm and theoretical guarantees for exact low-rank matrix completion appeared in [10]; it was shown that nuclear-norm minimization works when the low-rank matrix is incoherent, and sampling is uniformly random and independent of the matrix. Subsequent works have refined provable completion results for incoherent matrices under the uniformly random sampling model, both via nuclear-norm minimization [11, 17, 76], and other methods like SVD followed by local descent

[47] and alternating minimization [44]. The setting with sparse errors and additive noise is considered in [11, 12, 16, 20, 71].

Most of the existing sufficient conditions [10, 11] require that the subset of observed elements should be uniformly randomly chosen, and the low-rank matrix be incoherent or not spiky (i.e., its row and column spaces should be diffuse, having low inner products with the standard basis vectors). Under these conditions, the matrix is provably recoverable via methods based on convex optimization [10], alternating minimization [44], iterative thresholding [8] using as few as $O(nr \log n)$ observed elements for an $n \times n$ matrix of rank r.

In [55] matrix completion is considered when the row space is allowed to be coherent but the column space is required to be incoherent with parameter μ_0. The proposed adaptive sampling algorithm requires $O(\mu_0 r^{3/2} n \log(2r/\delta))$ observed elements with a success probability $1 - \delta$, which is superlinear in r. A corollary of the results guarantees a sample complexity that is linear in r in this row-coherent setting. The sample complexity is improved to $O(\mu_0 rn \log^2(r^2/\delta))$ in [56].

If an $n \times n$ matrix of rank r satisfies certain incoherence properties, then it is possible to exactly reconstruct the matrix with high probability from $nr\text{polylog}(n) \ll n^2$ randomly sampled entries using efficient polynomial-time algorithms [11, 34, 44, 47, 76]. With $\Omega(nr \log^2 n)$ (i.e., bounded below by $nr \log^2 n$ asymptotically) uniformly sampled entries, one can recover a matrix that satisfies standard incoherence condition but is not jointly incoherent (e.g., a positive semidefinite matrix).

All these works require the joint incoherence condition. The work in [12, 20, 59] prove the success of specific algorithms assuming the standard and joint incoherence conditions. For exact matrix completion, the joint incoherence condition is not necessary [17]. The sample complexity of recovering a semidefinite matrix is improved to $O(nr \log^2 n)$, and the highest allowable rank to $\Theta(n/\log^2 n)$ [17].

Compressed sensing paradigms suffer from basis mismatch when imposing a discrete dictionary on the Fourier representation. Enhanced matrix completion [18] addresses this issue based on structured matrix completion that does not require prior knowledge of the model order. The algorithm arranges the data into a low-rank enhanced form exhibiting multifold Hankel structure, and then attempts recovery via nuclear-norm minimization. Under mild incoherence conditions, the method allows perfect recovery as soon as the number of the samples exceeds $O(r \log^4 n)$, and is stable against bounded noise. Even if a constant portion of samples are corrupted with arbitrary magnitude, it still allows exact recovery, provided that the sample complexity exceeds $O(r^2 \log^3 n)$.

Matrix completion with no assumptions on the incoherence of the underlying matrix is studied in [21]. An n-by-n matrix of rank r can be exactly recovered from as few as $O(nr \log^2 n)$ randomly chosen elements, provided this random choice is made according to a specific biased distribution suitably dependent on the coherence structure of the matrix [21]. The presented two-phase sampling algorithm does not require knowledge of underlying structure of the matrix. Exact recovery guarantees are provided for the weighted nuclear-norm minimization approach when the observed entries are given and distributed nonuniformly.

For trace-norm regularization, most nontrivial guarantees assume that the observed entries are sampled uniformly at random [10, 11, 83]. This is an unrealistic assumption. For example, in the Netflix challenge dataset, where the matrix contains the ratings of users (rows) for movies (columns), the number and distribution of ratings differ drastically between users. In practice, standard trace-norm regularization works quite well even for nonuniform data. By adding very mild assumptions, which correspond to matrix completion as performed in practice, it is possible to learn in a distribution-free manner by observing $O(n^{3/2})$ entries from an $m \times n$ matrix (where $m \leq n$, and for a reasonable trace-norm regime) [81]. This bound is tight.

19.2.4 Discrete Matrix Completion

In the Netflix problem, the ratings of movies are quantized as integers from 1 to 5. In the recommender systems, only a single bit of rating standing for a thumbs-up or thumbs-down is recorded at each occurrence.

A L_∞-norm constrained maximum likelihood estimate [7] is introduced for noisy 1-bit matrix completion under a general nonuniform sampling distribution. The minimax upper and lower bounds together yield the optimal rate of convergence for the Frobenius norm loss.

1-bit matrix completion under the uniform sampling model is analyzed in [23]. The trace-norm constrained approach achieves minimax rate-optimal of convergence. It recovers an approximately low-rank matrix M from a set of noise-corrupted sign (1-bit) measurements by using a trace-norm constrained maximum likelihood estimator.

In collaborative filtering, the sampling distribution is nonuniform. As such, standard trace-norm regularized method might fail, specifically when certain rows or columns are sampled with very high probabilities [80]. Weighted trace norm takes the sampling distribution into account [80]. Rigorous recovery guarantees for learning with standard weighted, smoothed weighted and smoothed empirically weighted trace norms are provided in [31]. Theoretical guarantees on approximate low-rank matrix completion using weighted trace norm in general sampling case are provided in [71], assuming that each row/column is sampled with positive probability.

Upper bounds on the error norm for discrete matrix completion is given in [7, 23, 57]. In [13] matrix completion for categorical data is investigated and the results of [23] are extended to multilevel observations.

Recovery of a low-rank real-valued matrix $\mathbf{M}$ given a subset of noisy discrete measurements is considered in [5]. Under a constraint on the L_∞-norm of $\mathbf{M}$ and an exact rank constraint, a globally convergent constrained maximum likelihood estimation algorithm is proposed based on low-rank factorization of $\mathbf{M}$. The likelihood comes from any strictly log-concave distribution, which includes distributions of bounded discrete random variables from the exponential family.

19.3 Low-Rank Representation

PCA recovers the best low-rank representation in terms of L_2 errors. Robust PCA aims to recover the low-rank clean data from given noisy data. Low-rank representation methods have robustness on the noise/corrupted data. For a given set of observed data corrupted with sparse errors, low-rank representation [62, 65, 78] learns a lowest-rank representation of all data jointly.

Low-rank matrix approximation/recovery [12] aims at decomposing a data matrix $\mathbf{X}$ into $\mathbf{Z}+\mathbf{E}$, in which $\mathbf{Z}$ is a low-rank matrix and $\mathbf{E}$ is the associated sparse error. The problem is formulated as

$$\min_{\mathbf{Z},\mathbf{E}} \text{rank}(\mathbf{Z}) + \lambda\|\mathbf{E}\|_0 \quad \text{subject to} \quad \mathbf{X}=\mathbf{Z}+\mathbf{E}. \tag{19.10}$$

Solving its nuclear-norm version with L_1-norm of $\mathbf{E}$ is equivalent to solving (19.10), as long as the rank of $\mathbf{Z}$ is not too large, and $\mathbf{E}$ is sufficiently sparse [12]. Augmented Lagrange multipliers method [60] is applied due to its computational efficiency.

For problem (19.10), we may solve for the clean data $\mathbf{Z}$ directly:

$$\min_{\mathbf{X},\mathbf{E}} \|\mathbf{Z}\|_* + \lambda\|\mathbf{E}\|_1 \quad \text{subject to} \quad \mathbf{X}=\mathbf{Z}+\mathbf{E}, \tag{19.11}$$

where $\|\mathbf{Z}\|_*$ is nuclear norm, and $\|\mathbf{E}\|_1$ is L_1-norm.

This links low-rank representation with robust PCA [10]. Unfortunately, none of the robust PCA yields a polynomial-time algorithm under broad conditions [12]. The resulting problem, which minimizes a combination of the nuclear norm and the L_1-norm, is convex and can be solved in polynomial time [12].

To capture the global structure of the data, low-rank representation considers 2D sparsity (low rankness) for subspace segmentation. It can better capture the global structure of the data, while the local manifold structure is ignored. Low-rank representation can get the block-diagonal solution, which is perfect for subspace segmentation, when the subspaces are independent and data sampling is sufficient [62, 64]. Low-rank representation is a subspace clustering method [26].

Based on the observation that a data point can be represented by the data points from the same subspace, the basic model of low-rank representation can be formulated as

$$\min_{\mathbf{Z}} \text{rank}(\mathbf{Z}) + \lambda\|\mathbf{E}\|_0 \quad \text{subject to} \quad \mathbf{X}=\mathbf{A}\mathbf{Z}+\mathbf{E}, \tag{19.12}$$

where $\mathbf{X}$ is the sample set, $\mathbf{A}$ is the dictionary, $\mathbf{E}$ denotes the error components, and λ is a penalty parameter.

Low-rank representation for subspace segmentation can be formulated as a convex relaxation of (19.12):

$$\min_{\mathbf{Z}} \|\mathbf{Z}\|_* + \lambda\|\mathbf{E}\|_1 \quad \text{subject to} \quad \mathbf{X}=\mathbf{A}\mathbf{Z}+\mathbf{E}, \tag{19.13}$$

In noise-free case, the solution to (19.13) is also the solution to (19.12) [65].

Low-rank representation can accurately recover the row space of the original data and detect outliers under mild conditions [65]. Only the row space information is recovered and the column space information of the input data matrix is not sufficiently exploited for learning subspace structure. Double low-rank representation [100] simultaneously learns the row space and column space information embedded in a given dataset.

Low-rank representation does not consider nonlinear geometric structures within data, thus the locality and similarity information among data may be missing in the learning process. Manifold learning methods, such as the locally linear embedding [79], ISOMAP [90], locality-preserving projection [39], neighborhood-preserving embedding [38] and Laplacian Eigenmap [4], preserve local geometric structures embedded in a high-dimensional space. A nonnegative sparse hyper-Laplacian regularized low-rank representation model [101] introduces a hypergraph Laplacian regularizer into a general Laplacian regularized low-rank representation framework for data representation.

GoDec and GoDec+

GoDec (https://sites.google.com/site/godecomposition/code) [104] is an efficient, robust low-rank matrix decomposition algorithm. It decomposes a matrix $\mathbf{X}$ into

$$\mathbf{X} = \mathbf{L} + \mathbf{S} + \mathbf{G}, \tag{19.14}$$

where $\mathbf{L}$ is a low-rank matrix, $\mathbf{S}$ is a sparse corruption, and $\mathbf{G}$ is a Gaussian noise.

GoDec alternatively assigns a low-rank approximation of $\mathbf{X} - \mathbf{S}$ to $\mathbf{L}$ and a sparse approximation of $\mathbf{X} - \mathbf{L}$ to $\mathbf{S}$. It can be significantly accelerated by bilateral random projections. The objective $\|\mathbf{X} - \mathbf{L} - \mathbf{S}\|_F^2$ converges to a local minimum, while $\mathbf{L}$ and $\mathbf{S}$ linearly converge to local optimums [35].

GoDec+ [35] is a more robust and faster low-rank decomposition algorithm that solves a maximum correntropy criterion using half-quadratic optimization and greedy bilateral paradigm. Correntropy is a robust local similarity measure to describe the corruptions. GoDec+ is efficient and robust to different corruptions on images, including Gaussian noise, Laplacian noise, salt & pepper noise, and occlusion.

19.4 Tensor Factorization and Tensor Completion

A tensor (i.e., multiway array) is a multidimensional array which is a higher order generalization of vectors and matrices. Tensor can provide an efficient and faithful representation of structural and correlation properties for multidimensional data. A tensor is very powerful to represent the interactions controlled by multiple factors. The order, also known as *way* or *mode*, of a tensor is the number of dimensions. The tensor representation of data reflects the relationships between different features. Tensor can also be treated as the multi-view data. An image is a second-order tensor, whereas a video is a third-order tensor. Color images are three-dimensional objects with column, row and color modes.

Multi-linear algebra is the algebra of higher order tensors. It reduces the computational complexity and the requirements in terms of memory. Tensor-based approaches avoid overtraining, especially in small size problems [88].

Higher order tensor decomposition [24] has become an important technique in computer vision and pattern recognition. Common tensor factorization algorithms include parallel factor (Parafac) model [36], Tucker decomposition [92, 93], tensor CCA [68], and multi-linear PCA [66]. Among them, multi-linear PCA is the most popular one, which follows PCA paradigm and determines a multi-linear projection onto a tensor subspace of lower dimension.

The tensor completion problem is to recover a low-n-rank tensor from a subset of its entries. In computer vision and graphics, it is known as image and video inpainting problem [32, 63].

CP-rank and mode-d rank are the two main types of tensor rank.

Definition 19.1 (*CP-Rank*) *CP-rank* is defined as the minimum positive integer R such that for a tensor $\mathcal{X} \in \mathcal{T}$, it can be factorized as a sum of R rank-one tensors. This is CP decomposition.

Finding CP decomposition exactly is NP-hard for higher order tensors [52].

Definition 19.2 (*Mode-d Rank*) *Mode-d rank*, also known as *Tucker rank*, of a tensor $\mathcal{X} \in \mathcal{T}$ is defined as the rank of the mode-d unfolding matrix $\mathbf{X}(d)$. $\mathcal{X}$ is said to be rank-$(R_1, \ldots, R_N)$ if the mode-d rank of $\mathcal{X}$ is $(R_1, \ldots, R_N)$.

For a rank-$(R_1, \ldots, R_N)$ tensor $\mathcal{X}$, Tucker tensor decomposition aims to find the matrices and core tensor by minimizing the Frobenius norm:

$$\min_{\mathbf{U}^i, i=1,2,\ldots,N; \mathcal{G}} \left\| \mathcal{X} = \mathcal{G} \times_1 \mathbf{U}^1 \times_2 \ldots \times_N \mathbf{U}^N \right\|_F^2, \tag{19.15}$$

subject to

$$(\mathbf{U}^i)^T \mathbf{U}^i = \mathbf{I}, \forall i = 1, 2, \ldots, N, \tag{19.16}$$

where $\mathcal{G} \in R^{R_1 \times \cdots \times R_N}$ is called the *core tensor*, an Nth-order tensor that contains the 1-mode, 2-mode, …, and N-mode singular values of $\mathcal{X}$, which are defined as the Frobenius norm of the 1-mode, 2-mode, …, and N-mode slices of tensor $\mathcal{G}$ respectively; and $\mathbf{U}^i \in R^{R_i \times R_i}$ are unitary *factor matrices*. The j-mode product ($j = 1, 2, \ldots, N$) of $\mathcal{G}$ by $\mathbf{U}^j$ is denoted by $\mathcal{G} \times_j \mathbf{U}^j$.

The noncovex Tucker decomposition problem can be effectively solved using higher order SVD [25, 92]. Tucker decomposition decomposes a three-dimensional signal directly using three-dimensional PCA, which is a multi-linear generalization of SVD to multidimensional data. For video frames, this higher order SVD decomposes the dynamic texture as a multidimensional signal (tensor) without unfolding the video frames on column vectors. Higher order SVD requires, on average, five times less parameters than SVD. The analysis part is more expensive, but the synthesis has the same cost as that for the existing algorithms [22].

Concurrent subspaces analysis [99], multi-linear PCA [66] and its uncorrelated variation [67] have been proposed for face and gait recognition tasks. Multi-linear discriminant analysis generalizes LDA to tensor-based LDA. Unfortunately, the ratio-based multi-linear discriminant analysis does not converge and appears to be extremely sensitive to parameter settings [98]. Therefore, general tensor discriminant analysis [89] is proposed for gait recognition using the differential scatter discriminant criterion, and the tensor maximum margin criterion [41] is proposed for object recognition using the maximum margin criterion.

Sparse tensor discriminant analysis [58] extends multi-linear discriminant analysis to a sparse case. By introducing the L_1 and L_2-norms into the objective function, multiple interrelated sparse discriminant subspaces for feature extraction can be obtained. All the projection matrices derived from different modes are sparse. The optimal multi-linear sparse projections are obtained by iterating the elastic-net regression and SVD instead of solving the eigenequations as in multi-linear PCA, multi-linear discriminant analysis, and general tensor discriminant analysis.

19.4.1 Tensor Factorization

Tensor decomposition gives a concise representation of the underlying structure of tensor. For general tensors, tensor decomposition does not deliver best low-rank approximation.

The two popular tensor factorization methods, namely CANDECOMP/Parafac (CP) or canonical polyadic (CP) decomposition [14, 36], and Tucker decomposition [92], can be considered as higher order extensions of SVD. CP decomposition assumes a trilinear structure in the data and is easier to interpret, while the Tucker family defines more generic models for complex interactions. Other well-known tensor decompositions are tensor-train decomposition [72], and tubal rank decomposition [49].

The main solution strategy is based on the extensions of trace norm for the minimization of tensor rank via convex optimization. This strategy bears the computational cost required by SVD. A multi-linear low-n-rank Tucker decomposition model [87] is solved by applying the nonlinear Gauss–Seidel method that only requires solving an LS problem per iteration. It can reliably solve a wide range of problems several times faster than trace-norm minimization algorithm.

Tucker decomposition is a form of higher order PCA. It decomposes a tensor into a core tensor multiplied by a matrix along each mode. The method suffers from the curse of dimensionality and lack of uniqueness, which indicates that the size of the core tensor increases exponentially with the dimension.

Nonnegative Tucker decomposition [50], an extension of NMF, extracts nonnegative parts-based and physically meaningful latent components from high-dimensional tensor data while preserving the natural multi-linear structure of data. Its multiplicative algorithm is based on minimization of squared Euclidean distance and KL divergence. The decomposition is more likely to be unique and provides physically meaningful components. Moreover, the core tensor is often very sparse. Low (multi-linear) rank approximation of tensors is able to significantly simplify the computation of the gradients of the cost function, upon which a family of efficient first-order algorithms are developed [106].

Higher order SVD for tensor decomposition [25] is a multi-linear generalization of SVD. It involves eigenvalue decomposition of very large matrices. Nonnegative multi-linear PCA [73] is a tensor factorization method to find a tensor-to-tensor projection via multi-linear subspace learning for music genre classification. A series of unsupervised and supervised nonnegative tensor factorization algorithms are introduced in [103].

For a nonnegative tensor, a best nonnegative rank-r approximation is almost always unique, its best rank-one approximation may always be chosen to be a best nonnegative rank-one approximation, and the set of nonnegative tensors with nonunique best rank-one approximations forms an algebraic hypersurface [75].

Multi-view tensor factorization [48] performs a joint CP decomposition of multiple tensors to find dependencies between datasets. Tensorfaces [94] is a multi-linear analysis method that decomposes the modes due to identity, pose, and illumination.

19.4.2 Tensor Completion

The goal of tensor recovery is to find a tensor $\mathcal{X} \in \mathcal{T}$ satisfying

$$\mathcal{A}(\mathcal{X}) = \boldsymbol{b}, \tag{19.17}$$

where $\boldsymbol{b} \in R^p$ is a vector, and $\mathcal{A} : \mathcal{T} \to R^p$ with $p \leq \prod_{i=1}^{N} n_i$ is a linear map defined as

$$\mathcal{A}(\cdot) = \left[< \mathcal{A}_1, \cdot >, < \mathcal{A}_2, \cdot >, \ldots, < \mathcal{A}_p, \cdot >\right]^T \tag{19.18}$$

with $\mathcal{A}_i \in \mathcal{T}, i = 1, \ldots, p$.

The low-rank tensor recovery problem reads:

$$\min \operatorname{rank}(\mathcal{X}) \quad \text{subject to} \quad \mathcal{A}(\mathcal{X}) = \boldsymbol{b}. \tag{19.19}$$

Since the mode-d rank is easier to compute than the CP-rank, the problem becomes [32]

$$\min_{\mathcal{X} \in \mathcal{T}} \sum_{i=1}^{N} \operatorname{rank}(\mathbf{X}_{(i)}) \quad \text{subject to} \quad \mathcal{A}(\mathcal{X}) = \boldsymbol{b}. \tag{19.20}$$

A special case of (19.20) is the tensor completion problem, which aims at recovering a low-rank tensor from partial observations

$$\min_{\mathcal{X}\in\mathcal{T}} \sum_{i=1}^{N} \text{rank}(\mathbf{X}_{(i)}) \quad \text{subject to} \quad \mathcal{X}_{i_1\ldots i_N} = \mathcal{B}_{i_1\ldots i_N}, \quad (i_1\ldots i_N)\in\Omega, \tag{19.21}$$

where $\mathcal{B}\in\mathcal{T}$ represents the observed tensor and Ω denotes the set of multi-indices that correspond to the observed entries.

Low-rank tensor completion is a higher order extension of matrix completion. The problem of low-rank tensor completion is generally NP-hard [40]. Most existing tensor completion solutions are based on CP rank [32, 55], and the problem of computing CP rank of a tensor is NP-hard [40].

Based on low tensor tubal rank, tensor completion also implements an orthogonal pursuit on tubal rank-1 tensors [85]. The tensor tubal rank has similar properties like that of matrix rank derived from SVD.

Since multi-linear rank of a tensor is defined as the rank of its mode-n matricizations, it can be optimized by applying nuclear-norm-based framework, yielding an extension to tensor completion [63]. To get a tractable problem, the objective functions in the problems (19.20) and (19.21) are replaced by the sum of nuclear norms of mode matrices, which is also called the *tensor nuclear norm* [32, 63]. The tensor nuclear norm is suboptimal when the dimension of the tensor is greater than 3. The square norm [70] and the tensor-train norm [95] can be used to recover high-dimensional tensors.

Tensor completion algorithms can be local algorithms or global algorithms. Local algorithms [2, 53] assume that the dependence of two points is dependent on their distance and the missing entries mainly depend on their neighbors. The rank is a powerful tool to capture the global information. Based on the extensions of trace norm for the minimization of tensor rank, some global algorithms [32, 63] solve the tensor completion problem via convex optimization.

When the data are partially observed, tensor factorization can be applied for tensor completion. Low-rank CP factorization with missing values employs CP weighted optimization [1] and CP nonlinear LS [82].

Higher order robust PCA [33] is a tensor completion method that can handle both missing data and outliers. It formulates the problem by a convex optimization framework in which nuclear norm and L_1-norm are exploited as regularization terms on the low-rank tensor and residual errors, respectively. However, it essentially optimizes the multi-linear rank and the predictive performance is sensitive to tuning parameters.

Some other works provide a completion of the given rank using alternating minimization, e.g., for low-CP-rank tensor [43] and low-tensor-train-rank tensor [95].

Consider the low-CP-rank tensor completion problem. In [3], the number of sampled entries that guarantees finite completability with high probability is derived by a combinatorial method. The number of samples required for finite or unique completability obtained by the analysis on the CP manifold is orders of magnitude lower than that obtained by the existing analysis on the Grassmannian manifold.

The low-rank assumption is not sufficient for the recovery of visual data, when the ratio of missing data is extremely high. Smooth PARAFAC tensor completion method [102] considers smoothness constraints as well as low-rank approximations. It integrates smooth PARAFAC decomposition for incomplete tensors and the efficient selection of models in order to minimize the tensor rank.

Problem

19.1 Assume a rank-1 matrix is given by

$$\mathbf{X} = \begin{bmatrix} 1 & 3 & 8 & ? \\ 3 & 9 & ? & 18 \\ 4 & ? & 32 & ? \end{bmatrix}$$

where "?" denotes a missing entry. (a) Recover the missing entries by rank minimization.

(b) Verify the result by representing each column by a linear combination of other columns.

References

1. Acar, E., Dunlavy, D. M., Kolda, T. G., & Morup, M. (2011). Scalable tensor factorizations for incomplete data. *Chemometrics and Intelligent Laboratory Systems*, *106*(1), 41–56.
2. Argyriou, A., Evgeniou, T., & Pontil, M. (2007). Multi-task feature learning. *Advances in neural information processing systems* (Vol. 20, pp. 243–272).
3. Ashraphijuo, M., & Wang, X. (2017). Fundamental conditions for low-CP-rank tensor completion. *Journal of Machine Learning Research*, *18*, 1–29.
4. Belkin, M., & Niyogi, P. (2003). Laplacian eigenmaps for dimensionality reduction and data representation. *Neural Computation*, *15*, 1373–1396.
5. Bhaskar, S. A. (2016). Probabilistic low-rank matrix completion from quantized measurements. *Journal of Machine Learning Research*, *17*, 1–34.
6. Bhojanapalli, S., & Jain, P. (2014). Universal matrix completion. In *Proceedings of the 31st International Conference on Machine Learning* (pp. 1881–1889). Beijing, China.
7. Cai, T., & Zhou, W.-X. (2013). A max-norm constrained minimization approach to 1-bit matrix completion. *Journal of Machine Learning Research*, *14*, 3619–3647.
8. Cai, J.-F., Candes, E. J., & Shen, Z. (2010). A singular value thresholding algorithm for matrix completion. *SIAM Journal on Optimization*, *20*(4), 1956–1982.
9. Candes, E. J., & Plan, Y. (2010). Matrix completion with noise. *Proceedings of the IEEE*, *98*(6), 925–936.
10. Candes, E. J., & Recht, B. (2009). Exact matrix completion via convex optimization. *Foundations of Computational Mathematics*, *9*(6), 717–772.
11. Candes, E. J., & Tao, T. (2010). The power of convex relaxation: Near-optimal matrix completion. *IEEE Transactions on Information Theory*, *56*(5), 2053–2080.
12. Candes, E. J., Li, X., Ma, Y., & Wright, J. (2011). Robust principal component analysis? *Journal of the ACM*, *58*(3), 1–37.

13. Cao, Y., & Xie, Y. (2015). Categorical matrix completion. In *Proceedings of IEEE International Workshop on Computational Advances in Multi-Sensor Adaptive Processing (CAMSAP)* (pp. 369–372). Cancun, Mexico.
14. Carroll, J. D., & Chang, J.-J. (1970). Analysis of individual differences in multidimensional scaling via an N-way generalization of Eckart-Young decomposition. *Psychometrika*, *35*(3), 283–319.
15. Chandrasekaran, V., Sanghavi, S., Parrilo, P. A., & Willsky, A. S. (2009). Sparse and low-rank matrix decompositions. In *Proceedings of the 47th Annual Allerton Conference on Communication, Control, and Computing* (pp. 962–967). Monticello, IL.
16. Chandrasekaran, V., Sanghavi, S., Parrilo, P. A., & Willsky, A. S. (2011). Rank-sparsity incoherence for matrix decomposition. *SIAM Journal on Optimization*, *21*(2), 572–596.
17. Chen, Y. (2015). Incoherence-optimal matrix completion. *IEEE Transactions on Information Theory*, *61*(5), 2909–2923.
18. Chen, Y., & Chi, Y. (2014). Robust spectral compressed sensing via structured matrix completion. *IEEE Transactions on Information Theory*, *60*(10), 6576–6601.
19. Chen, C., He, B., & Yuan, X. (2012). Matrix completion via an alternating direction method. *IMA Journal of Numerical Analysis*, *32*(1), 227–245.
20. Chen, Y., Jalali, A., Sanghavi, S., & Caramanis, C. (2013). Low-rank matrix recovery from errors and erasures. *IEEE Transactions on Information Theory*, *59*(7), 4324–4337.
21. Chen, Y., Bhojanapalli, S., Sanghavi, S., & Ward, R. (2015). Completing any low-rank matrix, provably. *Journal of Machine Learning Research*, *16*, 2999–3034.
22. Costantini, R., Sbaiz, L., & Susstrunk, S. (2008). Higher order SVD analysis for dynamic texture synthesis. *IEEE Transactions on Image Processing*, *17*(1), 42–52.
23. Davenport, M. A., Plan, Y., van den Berg, E., & Wootters, M. (2014). 1-bit matrix completion. *Information and Inference*, *3*, 189–223.
24. De Lathauwer, L., De Moor, B., & Vandewalle, J. (2000). On the best rank-1 and rank-(R1,R2,...,RN) approximation of high-order tensors. *SIAM Journal on Matrix Analysis and Applications*, *21*(4), 1324–1342.
25. De Lathauwer, L., De Moor, B., & Vandewalle, J. (2000). A multilinear singular value decomposition. *SIAM Journal on Matrix Analysis and Applications*, *21*(4), 1253–1278.
26. Elhamifar, E., & Vidal, R. (2013). Sparse subspace clustering: Algorithm, theory, and applications. *IEEE Transactions on Pattern Analysis and Machine Intelligence*, *35*(11), 2765–2781.
27. Eriksson, A., & van den Hengel, A. (2012). Efficient computation of robust weighted low-rank matrix approximations using the L_1 norm. *IEEE Transactions on Pattern Analysis and Machine Intelligence*, *34*(9), 1681–1690.
28. Fan, J., & Chow, T. W. S. (2018). Non-linear matrix completion. *Pattern Recognition*, *77*, 378–394.
29. Fazel, M. (2002). *Matrix rank minimization with applications*. Ph.D. thesis, Stanford University.
30. Foygel, R., & Srebro, N. (2011). Concentration-based guarantees for low-rank matrix reconstruction. In *JMLR: Workshop and Conference Proceedings* (Vol. 19, pp. 315–339).
31. Foygel, R., Shamir, O., Srebro, N., & Salakhutdinov, R. (2011). Learning with the weighted trace-norm under arbitrary sampling distributions. *Advances in neural information processing systems* (Vol. 24, pp. 2133–2141).
32. Gandy, S., Recht, B., & Yamada, I. (2011). Tensor completion and low-n-rank tensor recovery via convex optimization. *Inverse Problems*, *27*(2), 1–19.
33. Goldfarb, D., & Qin, Z. (2014). Robust low-rank tensor recovery: Models and algorithms. *SIAM Journal on Matrix Analysis and Applications*, *35*(1), 225–253.
34. Gross, D. (2011). Recovering low-rank matrices from few coefficients in any basis. *IEEE Transactions on Information Theory*, *57*(3), 1548–1566.
35. Guo, K., Liu, L., Xu, X., Xu, D., & Tao, D. (2018). Godec+: Fast and robust low-rank matrix decomposition based on maximum correntropy. *IEEE Transactions on Neural Networks and Learning Systems*, *29*(6), 2323–2336.

36. Harshman, R. A. (1970). Foundations of the PARAFAC procedure: Models and conditions for an "explanatory" multimodal factor analysis. *UCLA Working Papers in Phonetics* (Vol. 16, pp. 1–84).
37. Hastie, T., Mazumder, R., Lee, J. D., & Zadeh, R. (2015). Matrix completion and low-rank SVD via fast alternating least squares. *Journal of Machine Learning Research, 16*, 3367–3402.
38. He, X., Cai, D., Yan, S., & Zhang, H.-J. (2005). Neighborhood preserving embedding. In *Proceedings of the 10th IEEE International Conference on Computer Vision* (pp. 1208–1213). Beijing, China.
39. He, X., Yan, S., Hu, Y., Niyogi, P., & Zhang, H. J. (2005). Face recognition using Laplacianfaces. *IEEE Transactions on Pattern Analysis and Machine Intelligence, 27*(3), 328–340.
40. Hillar, C. J., & Lim, L.-H. (2013). Most tensor problems are NP-hard. *Journal of the ACM, 60*(6), Article No. 45, 39 p.
41. Hu, R.-X., Jia, W., Huang, D.-S., & Lei, Y.-K. (2010). Maximum margin criterion with tensor representation. *Neurocomputing, 73*, 1541–1549.
42. Hu, Y., Zhang, D., Ye, J., Li, X., & He, X. (2013). Fast and accurate matrix completion via truncated nuclear norm regularization. *IEEE Transactions on Pattern Analysis and Machine Intelligence, 35*(9), 2117–2130.
43. Jain, P., & Oh, S. (2014). Provable tensor factorization with missing data. In *Advances in neural information processing systems* (Vol. 27, pp. 1431–1439).
44. Jain, P., Netrapalli, P., & S. Sanghavi, (2013). Low-rank matrix completion using alternating minimization. In *Proceedings of the 45th Annual ACM Symposium on Theory of Computing* (pp. 665–674).
45. Ji, S., & Ye, J. (2009). An accelerated gradient method for trace norm minimization. In *Proceedings of the 26th Annual International Conference on Machine Learning* (pp. 457–464). Montreal, Canada.
46. Ke, Q., & Kanade, T. (2005). Robust L_1 norm factorization in the presence of outliers and missing data by alternative convex programming. In *Proceedings of IEEE Conference on Computer Vision and Pattern Recognition* (pp. 739–746). San Diego, CA.
47. Keshavan, R. H., Montanari, A., & Oh, S. (2010). Matrix completion from a few entries. *IEEE Transactions on Information Theory, 56*(6), 2980–2998.
48. Khan, S. A., & Kaski, S. (2014). Bayesian multi-view tensor factorization. In T. Calders, F. Esposito, E. Hullermeier, & R. Meo (Eds.), *Proceedings of Joint European Conference on Machine Learning and Knowledge Discovery in Databases* (pp. 656-671). Berlin: Springer.
49. Kilmer, M. E., Braman, K., Hao, N., & Hoover, R. C. (2013). Third-order tensors as operators on matrices: A theoretical and computational framework with applications in imaging. *SIAM Journal on Matrix Analysis and Applications, 34*(1), 148–172.
50. Kim, Y.-D., & Choi, S. (2007). Nonnegative Tucker decomposition. In *Proceedings of IEEE Conference on Computer Vision and Pattern Recognition* (pp. 1–8). Minneapolis, MN.
51. Kim, E., Lee, M., Choi, C.-H., Kwak, N., & Oh, S. (2015). Efficient l_1-norm-based low-rank matrix approximations for large-scale problems using alternating rectified gradient method. *IEEE Transactions on Neural Networks and Learning Systems, 26*(2), 237–251.
52. Kolda, T. G., & Bader, B. W. (2009). Tensor decompositions and applications. *SIAM Review, 51*(3), 455–500.
53. Komodakis, N., & Tziritas, G. (2006). Image completion using global optimization. In *Proceedings of IEEE Conference on Computer Vision and Pattern Recognition (CVPR)* (pp. 417–424). New York, NY.
54. Koren, Y., Bell, R., & Volinsky, C. (2009). Matrix factorization techniques for recommender systems. *Computer, 42*(8), 30–37.
55. Krishnamurthy, A., & Singh, A. (2013). Low-rank matrix and tensor completion via adaptive sampling. *Advances in neural information processing systems* (Vol. 26, pp. 836–844).
56. Krishnamurthy, A., & Singh, A. (2014). On the power of adaptivity in matrix completion and approximation. arXiv preprint arXiv:1407.3619.
57. Lafond, J., Klopp, O., Moulines, E., & Salmon, J. (2014). Probabilistic low-rank matrix completion on finite alphabets. *Advances in neural information processing systems* (Vol. 27, pp. 1727–1735). Cambridge: MIT Press.

58. Lai, Z., Xu, Y., Yang, J., Tang, J., & Zhang, D. (2013). Sparse tensor discriminant analysis. *IEEE Transactions on Image Processing*, *22*(10), 3904–3915.
59. Li, X. (2013). Compressed sensing and matrix completion with constant proportion of corruptions. *Constructive Approximation*, *37*(1), 73–99.
60. Lin, Z., Chen, M., Wu, L., & Ma, Y. (2009). *The augmented Lagrange multiplier method for exact recovery of corrupted low-rank matrices*. Technical Report UILU-ENG-09-2215. Champaign, IL: Department of Electrical and Computer Engineering, University of Illinois at Urbana-Champaign.
61. Lin, Z., Ganesh, A., Wright, J., Wu, L., Chen, M., & Ma, Y. (2009). *Fast convex optimization algorithms for exact recovery of a corrupted low-rank matrix*. Technical Report UILU-ENG-09-2214. Champaign, IL: University of Illinois at Urbana-Champaign.
62. Liu, G., Lin, Z., & Yu, Y. (2010). Robust subspace segmentation by low-rank representation. In *Proceedings of the 25th International Conference on Machine Learning* (pp. 663–670). Haifa, Israel.
63. Liu, J., Musialski, P., Wonka, P., & Ye, J. (2013). Tensor completion for estimating missing values in visual data. *IEEE Transactions on Pattern Analysis and Machine Intelligence*, *35*(1), 208–220.
64. Liu, Y., Jiao, L. C., Shang, F., Yin, F., & Liu, F. (2013). An efficient matrix bi-factorization alternative optimization method for low-rank matrix recovery and completion. *Neural Networks*, *48*, 8–18.
65. Liu, G., Lin, Z., Yan, S., Sun, J., Yu, Y., & Ma, Y. (2013c). Robust recovery of subspace structures by low-rank representation. *IEEE Transactions on Pattern Analysis and Machine Intelligence*, *35*(1), 171–184.
66. Lu, H., Plataniotis, K. N., & Venetsanopoulos, A. N. (2008). MPCA: Multilinear principal component analysis of tensor objects. *IEEE Transactions on Neural Networks*, *19*(1), 18–39.
67. Lu, H., Plataniotis, K. N., & Venetsanopoulos, A. N. (2009). Uncorrelated multilinear principal component analysis for unsupervised multilinear subspace learning. *IEEE Transactions on Neural Networks*, *20*(11), 1820–1836.
68. Luo, Y., Tao, D., Ramamohanarao, K., & Xu, C. (2015). Tensor canonical correlation analysis for multi-view dimension reduction. *IEEE Transactions on Knowledge and Data Engineering*, *27*(11), 3111–3124.
69. Mackey, L., Talwalkar, A., & Jordan, M. I. (2015). Distributed matrix completion and robust factorization. *Journal of Machine Learning Research*, *16*, 913–960.
70. Mu, C., Huang, B., Wright, J., & Goldfarb, D. (2014). Square deal: Lower bounds and improved relaxations for tensor recovery. In *JMLR W&CP: Proceedings of the 31st International Conference on Machine Learning* (Vol. 32). Beijing, China.
71. Negahban, S., & Wainwright, M. J. (2012). Restricted strong convexity and weighted matrix completion: Optimal bounds with noise. *Journal of Machine Learning Research*, *13*, 1665–1697.
72. Oseledets, I. V. (2011). Tensor-train decomposition. *SIAM Journal on Scientific Computing*, *33*(5), 2295–2317.
73. Panagakis, Y., Kotropoulos, C., & Arce, G. R. (2010). Non-negative multilinear principal component analysis of auditory temporal modulations for music genre classification. *IEEE/ACM Transactions on Audio, Speech, and Language Processing*, *18*(3), 576–588.
74. Pitaval, R.-A., Dai, W., & Tirkkonen, O. (2015). Convergence of gradient descent for low-rank matrix approximation. *IEEE Transactions on Information Theory*, *61*(8), 4451–4457.
75. Qi, Y., Comon, P., & Lim, L.-H. (2016). Uniqueness of nonnegative tensor approximations. *IEEE Transactions on Information Theory*, *62*(4), 2170–2183.
76. Recht, B. (2011). A simpler approach to matrix completion. *Journal of Machine Learning Research*, *12*, 3413–3430.
77. Recht, B., Fazel, M., & Parrilo, P. A. (2010). Guaranteed minimum-rank solutions of linear matrix equations via nuclear norm minimization. *SIAM Review*, *52*(3), 471–501.
78. Rennie, J. D. M., & Srebro, N. (2005). Fast maximum margin matrix factorization for collaborative prediction. In *Proceedings of the 22nd International Conference of Machine Learning* (pp. 713–719). Bonn, Germany.

79. Roweis, S. T., & Saul, L. K. (2000). Nonlinear dimensionality reduction by locally linear embedding. *Science*, *290*, 2323–2326.
80. Salakhutdinov, R., & Srebro, N. (2010). Collaborative filtering in a non-uniform world: Learning with the weighted trace norm. In J. LaFerty, C. K. I. Williams, J. Shawe-Taylor, R. S. Zemel, & A. Culotta (Eds.), *Advances in neural information processing systems* (Vol. 23, pp. 2056–2064). Cambridge: MIT Press.
81. Shamir, O., & Shalev-Shwartz, S. (2014). Matrix completion with the trace norm: Learning, bounding, and transducing. *Journal of Machine Learning Research*, *15*, 3401–3423.
82. Sorber, L., Van Barel, M., & De Lathauwer, L. (2013). Optimization-based algorithms for tensor decompositions: Canonical polyadic decomposition, decomposition in rank-$(L_r, L_r, 1)$ terms, and a new generalization. *SIAM Journal on Optimization*, *23*(2), 695–720.
83. Srebro, N., & Shraibman, A. (2005). Rank, trace-norm and max-norm. In *Proceedings of the 18th Annual Conference on Learning Theory (COLT)* (pp. 545–560). Berlin: Springer.
84. Srebro, N., Rennie, J. D. M., & Jaakkola, T. S. (2004). Maximum-margin matrix factorization. *Advances in neural information processing systems* (Vol. 17, pp. 1329–1336).
85. Sun, W., Huang, L., So, H. C., & Wang, J. (2019). Orthogonal tubal rank-1 tensor pursuit for tensor completion. *Signal Processing*, *157*, 213–224.
86. Takacs, G., Pilaszy, I., Nemeth, B., & Tikk, D. (2009). Scalable collaborative filtering approaches for large recommender systems. *Journal of Machine Learning Research*, *10*, 623–656.
87. Tan, H., Cheng, B., Wang, W., Zhang, Y.-J., & Ran, B. (2014). Tensor completion via a multi-linear low-n-rank factorization model. *Neurocomputing*, *1*(33), 161–169.
88. Tao, D., Li, X., Wu, X., Hu, W., & Maybank, S. J. (2007). Supervised tensor learning. *Knowledge and Information Systems*, *13*(1), 1–42.
89. Tao, D., Li, X., Wu, X., & Maybank, S. J. (2007). General tensor discriminant analysis and gabor features for gait recognition. *IEEE Transactions on Pattern Analysis and Machine Intelligence*, *29*(10), 1700–17015.
90. Tenenbaum, J. B., de Silva, V., & Langford, J. C. (2000). A global geometric framework for nonlinear dimensionality reduction. *Science*, *290*(5500), 2319–2323.
91. Toh, K.-C., & Yun, S. (2010). An accelerated proximal gradient algorithm for nuclear norm regularized least squares problems. *Pacific Journal of Optimization*, *6*(3), 615–640.
92. Tucker, L. R. (1966). Some mathematical notes on three-mode factor analysis. *Psychometrika*, *31*(3), 279–311.
93. Tucker, L. R., & Harris, C. W. (1963). Implication of factor analysis of three-way matrices for measurement of change. In C. W. Harris (Ed.), *Problems in measuring change* (pp. 122–137). Madison: University Wisconsin Press.
94. Vasilescu, M. A. O., & Terzopoulos, D. (2002). Multilinear analysis of image ensembles: Tensorfaces. In *Proceedigs of European Conference on Computer Vision, LNCS* (Vol. 2350, pp. 447–460). Copenhagen, Denmark. Berlin: Springer.
95. Wang, W., Aggarwal, V., & Aeron, S. (2016). Tensor completion by alternating minimization under the tensor train (TT) model. arXiv:1609.05587.
96. Wong, R. K. W., & Lee, T. C. M. (2017). Matrix completion with noisy entries and outliers. *Journal of Machine Learning Research*, *18*, 1–25.
97. Wright, J., Ganesh, A., Rao, S., Peng, Y., & Ma, Y. (2009). Robust principal component analysis: Exact recovery of corrupted low-rank matrices via convex optimization. *Advances in neural information processing systems* (Vol. 22, pp. 2080–2088). Vancouver, Canada.
98. Xu, D., Yan, S., Tao, D., Zhang, L., Li, X., & Zhang, H. (2006). Human gait recognition with matrix representation. *IEEE Transactions on Circuits and Systems for Video Technology*, *16*(7), 896–903.
99. Xu, D., Yan, S., Zhang, L., Lin, S., Zhang, H., & Huang, T. S. (2008). Reconstruction and recogntition of tensor-based objects with concurrent subspaces analysis. *IEEE Transactions on Circuits and Systems for Video Technology*, *18*(1), 36–47.
100. Yin, M., Cai, S., & Gao, J. (2013). Robust face recognition via double low-rank matrix recovery for feature extraction. In *Proceedings of IEEE International Conference on Image Processing* (pp. 3770–3774). Melbourne, Australia.

101. Yin, M., Gao, J., & Lin, Z. (2016). Laplacian regularized low-rank representation and its applications. *IEEE Transactions on Pattern Analysis and Machine Intelligence*, *38*(3), 504–517.
102. Yokota, T., Zhao, Q., & Cichocki, A. (2016). Smooth PARAFAC decomposition for tensor completion. *IEEE Transactions on Signal Processing*, *64*(20), 5423–5436.
103. Zafeiriou, S. (2009). Discriminant nonnegative tensor factorization algorithms. *IEEE Transactions on Neural Networks*, *20*(2), 217–235.
104. Zhou, T., & Tao, D. (2011). GoDec: Randomized low-rank & sparse matrix decomposition in noisy case. In *Proceedings of the 28th International Conference on Machine Learning* (pp. 33–40). Bellevue, WA.
105. Zhou, Y., Wilkinson, D., Schreiber, R., & Pan, R. (2008). Large-scale parallel collaborative filtering for the netix prize. In *Proceedings of the 4th International Conference on Algorithmic Aspects in Information and Management* (pp. 337–348). Berlin: Springer.
106. Zhou, G., Cichocki, A., Zhao, Q., & Xie, S. (2015). Efficient nonnegative tucker decompositions: Algorithms and uniqueness. *IEEE Transactions on Image Processing*, *24*(12), 4990–5003.

Chapter 20
Kernel Methods

20.1 Introduction

The kernel method was originally invented in [2]. The key idea is to project the training set in a lower dimensional space into a high-dimensional kernel (feature) space by means of a set of nonlinear kernel functions. As stated by the Cover theorem, the data will be more likely linearly separable when they are nonlinearly mapped to a higher dimensional space. The kernel method is a powerful nonparametric modeling tool in machine learning and data analysis. Well-known examples include kernel density estimator (also called the Parzen window estimator) as well as the RBF network and SVM. Kernel-based methods have played an important role in many fields such as pattern recognition, approximation, modeling, and data mining.

The kernel method generates algorithms that, by replacing the inner product with an appropriate positive definite function, implicitly perform a nonlinear mapping of the input data into a high-dimensional feature space. Introduced with SVM, the kernel trick has attracted much attention because of its efficient and elegant way of modeling nonlinear patterns. The kernel trick has been applied to construct nonlinear equivalents of a wide range of classical linear statistical models. An important advantage of kernel models is that the parameters of the model are typically given by the solution of a convex optimization problem, with a single, global optimum.

Kernel methods have been known as *kernel machines*. A kernel function $k(\boldsymbol{x}, \boldsymbol{x}')$ is a transformation function that satisfies Mercer's theorem. A Mercer kernel, i.e., a continuous, symmetric, and positive definite function, indicates that the kernel matrix has to be semidefinite; that means it only has positive eigenvalues.

Reproducing kernel Hilbert spaces (RKHSs), defined by Aronszajn in [7], are now commonly used as hypothesis spaces in learning theory. Combined with regularization techniques, often they allow good generalization capabilities of the learned models and enforce desired smoothness properties of the solutions to the learning problems.

Another way for nonlinear feature generation is the kernels-as-features idea [9], where the kernel function is directly considered as features. Given a kernel

K.-L. Du and M. N. S. Swamy, *Neural Networks and Statistical Learning*,
https://doi.org/10.1007/978-1-4471-7452-3_20

function $k(\cdot)$ and l data $\{\boldsymbol{x}_1, \ldots, \boldsymbol{x}_l\}$ of the input space X, we can map each $\boldsymbol{x} \in X$ into an l-dimensional kernel feature space, called φ-space, by defining $\varphi(\boldsymbol{x}) = (k(\boldsymbol{x}, \boldsymbol{x}_1), \ldots, k(\boldsymbol{x}, \boldsymbol{x}_l))^T$, and then, certain algorithms can be performed in the φ-space instead of the input space to deal with nonlinearity.

Both the kernel trick and kernels-as-features ideas produce nonlinear feature spaces to perform certain linear algorithms for dealing with nonlinearity. However, the feature spaces produced by the two ideas are different: the former is implicit and can only be accessed by the kernel function as a black box program of inner product, whereas the latter is explicitly constructed using a set of data and a kernel function. An exact equivalence between the two kernel ideas applied to PCA and LDA is established in [72]. There is an equivalence up to different scalings on each feature between the kernel trick and kernels-as-features ideas applied to certain feature extraction algorithms, i.e., LDA, PCA, and CCA [123].

The notion of refinable kernels [121] leads to the introduction of wavelet-like reproducing kernels, yielding multiresolution analysis of RKHSs. Refinable kernels provide computational advantages for solving various learning problems. The dominant set of eigenvectors of the symmetrical kernel Gram matrix is used in many important kernel methods in machine learning. An efficient incremental approach is presented in [47] for fast calculation of the dominant kernel eigenbasis.

After the success of SVM, many linear learning methods have been formulated using kernels, producing the other kernel-based methods: kernel PCA [100, 107], kernel LDA [78, 125], kernel clustering [39], kernel BSS [76], kernel ICA [8], kernel CCA [60], and the minimax probability machine [61].

Kernel methods are extended from RKHSs to Krein spaces [70] and Banach spaces [104]. A class of reproducing kernel Banach spaces with the L_1 norm that satisfies the linear representer theorem can be applied in machine learning [104].

20.2 Kernel Functions and Representer Theorem

Definition 20.1 (*Kernel Function*) A kernel $k : \mathcal{X} \times \mathcal{X} \rightarrow \mathcal{R}$ can be expressed as an inner product operation in some high-dimensional feature space:

$$k(\boldsymbol{x}, \boldsymbol{x}') =< \phi(\boldsymbol{x}), \phi(\boldsymbol{x}') >, \quad \forall \boldsymbol{x}, \boldsymbol{x}' \in \mathcal{X} \tag{20.1}$$

where $\phi : \mathcal{I} \rightarrow \mathcal{F}$, i.e., $\phi(\boldsymbol{x})$ is the image of $\boldsymbol{x}$ from input space $\mathcal{I}$ to space $\mathcal{F}$.

There exists a Hilbert space $\mathcal{H}$ known as a feature space of k and $\phi : \mathcal{X} \rightarrow \mathcal{H}$ as a feature map of k. Notice that both $\mathcal{H}$ and ϕ are far from being unique. However, for a given kernel, there exists an RKHS. In kernel methods, data are represented as functions or elements in RKHSs, which are associated with positive definite kernels.

Equation (20.1) is commonly referred to as the *kernel trick*. For complex-valued signal processing, one can map the input data into a complex RKHS using pure complex kernels or real kernels (via the complexification trick) [14, 15].

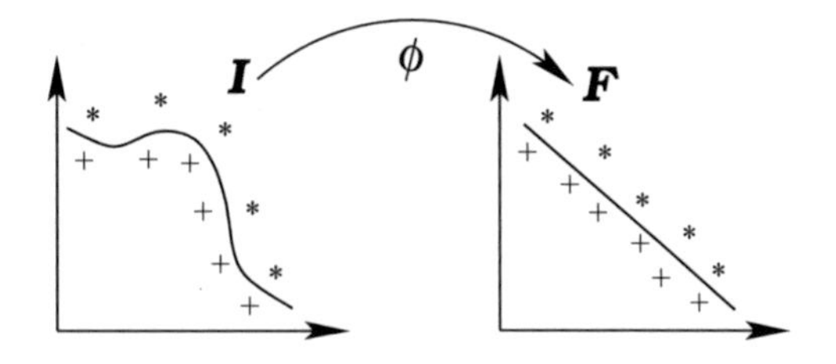

Fig. 20.1 Kernel-based transformation: from input space to feature space. Linear separation is produced in the feature space

Popular kernel functions used in the kernel method are the polynomial, Gaussian, and sigmoidal kernels, which are, respectively, given by [79]

$$k\left(\boldsymbol{x}_i, \boldsymbol{x}_j\right) = \left(\boldsymbol{x}_i^T \boldsymbol{x}_j + \theta\right)^m, \tag{20.2}$$

$$k\left(\boldsymbol{x}_i, \boldsymbol{x}_j\right) = \mathrm{e}^{-\frac{\|\boldsymbol{x}_i - \boldsymbol{x}_j\|^2}{2\sigma^2}}, \tag{20.3}$$

$$k\left(\boldsymbol{x}_i, \boldsymbol{x}_j\right) = \tanh\left(a\boldsymbol{x}_i^T \boldsymbol{x}_j + \theta\right), \tag{20.4}$$

where m is a positive integer, $\sigma > 0$, and $a, \theta \in R$. Notice that $\boldsymbol{x}_i^T \boldsymbol{x}_j =< \boldsymbol{x}_i, \boldsymbol{x}_j >$. Even if the exact form of $\phi(\cdot)$ does not exist, any symmetric function $k\left(\boldsymbol{x}_i, \boldsymbol{x}_j\right)$ satisfying Mercer's theorem can be used as a kernel function [79].

According to Mercer's work [77], a nonnegative linear combination of Mercer kernels is also a Mercer kernel, and the product of Mercer kernels is also a Mercer kernel. The performance of every kernel-based method depends on the kernel type selected. However, there are no general theories for choosing a kernel in a data-dependent way.

Nonlinear kernel functions are used to overcome the curse of dimensionality. The space of the input examples R^n is mapped onto a high-dimensional feature space so that the optimal separating hyperplane built on this space allows a good generalization capacity. By choosing an adequate mapping, the input examples become linearly or almost linearly separable in the high-dimensional space. This mapping transforms nonlinear separable data points in the input space into linear separable ones in the resulting high-dimensional space (see Fig. 20.1).

Definition 20.2 (*Kernel Definiteness*) Let the kernel matrix be $\mathbf{K} = \left[k(\boldsymbol{x}_i, \boldsymbol{x}_j)\right]_{n\times n}$. If for all the n data points and any vector $\boldsymbol{v} \in R^n$ the inequality $\boldsymbol{v}^T\mathbf{K}\boldsymbol{v} \geq 0$ holds, then $k(\cdot)$ is said to be *positive definite*. If this is only satisfied for those $\boldsymbol{v}$ with $\mathbf{1}_n^T \boldsymbol{v} = 0$, then $k(\cdot)$ is said to be *conditionally positive definite*. A kernel is *indefinite*, if for some $\mathbf{K}$ there exist vectors $\boldsymbol{v}$ and $\boldsymbol{v}'$ with $\boldsymbol{v}^T\mathbf{K}\boldsymbol{v} > 0$ and $\boldsymbol{v}'^T\mathbf{K}\boldsymbol{v}' < 0$.

The squared Euclidean distance has been generalized into a high-dimensional space $\mathcal{F}$ via the kernel trick

$$\|\phi(\boldsymbol{x}) - \phi(\boldsymbol{y})\|^2 = k(\boldsymbol{x}, \boldsymbol{x}) + k(\boldsymbol{y}, \boldsymbol{y}) - 2k(\boldsymbol{x}, \boldsymbol{y}). \tag{20.5}$$

This generalization becomes possible provided that the kernel is conditionally positive definite.

The representer theorem can be stated as: Any function defined in an RKHS can be represented as a linear combination of Mercer kernel functions.

Theorem 20.1 (Representer Theorem) *A mapping f can be written as a linear combination of kernel functions*

$$f(\boldsymbol{x}) = \sum_{i=1}^{N} \alpha_i k(\boldsymbol{x}_i, \boldsymbol{x}), \tag{20.6}$$

where $\alpha_i \in R$ are suitable coefficients, the kernel $k(\cdot)$ can be expressed as an inner product operation in some high-dimensional feature space.

Example 20.1 We revisit the XOR problem discussed in Example 11.1. Define the kernel $k(\boldsymbol{x}, \boldsymbol{x}_i) = (1 + \boldsymbol{x}^T \boldsymbol{x}_i)^2$, where $\boldsymbol{x} = (x_1, x_2)^T$ and $\boldsymbol{x}_i = (x_{i1}, x_{i2})^T$. The training samples $\boldsymbol{x}_1 = (-1, -1)$ and $\boldsymbol{x}_4 = (+1, +1)$ belong to class 0, and $\boldsymbol{x}_2 = (-1, +1)$, $\boldsymbol{x}_3 = (+1, -1)$ to class 1.

Expanding the kernel function, we have

$$k(\boldsymbol{x}, \boldsymbol{x}_i) = 1 + x_1^2 x_{i1}^2 + 2x_1 x_2 x_{i1} x_{i2} + x_2^2 x_{i2}^2 + 2x_1 x_{i1} + 2x_2 x_{i2} = \phi(\boldsymbol{x}) \cdot \phi(\boldsymbol{x}_i),$$

where

$$\phi(\boldsymbol{x}) = (1, x_1^2, \sqrt{2} x_1 x_2, x_2^2, \sqrt{2} x_1, \sqrt{2} x_2)^T,$$

$$\phi(\boldsymbol{x}_i) = (1, x_{i1}^2, \sqrt{2} x_{i1} x_{i2}, x_{i2}^2, \sqrt{2} x_{i1}, \sqrt{2} x_{i2})^T.$$

The feature space defined by $\phi(\boldsymbol{x})$ is six-dimensional. To discriminate the four examples in the feature space, we define the decision boundary in $x_1 x_2 = 0$. When $x_1 x_2 \geq 0$, an example is categorized into class 0, and otherwise into class 1.

20.3 Kernel PCA

Kernel PCA [100] introduces kernel functions into PCA. It first maps the original input data into a high-dimensional feature space using the kernel method and then calculates PCA in the high-dimensional feature space. Linear PCA in the high-dimensional feature space corresponds to a nonlinear PCA in the original input space. The decomposition of a Gram matrix is a particularly elegant method for extracting nonlinear features from multivariate data.

Given an input pattern set $\left\{\boldsymbol{x}_i \in R^{J_1} \middle| i = 1, \ldots, N\right\}$, $\phi : R^{J_1} \rightarrow R^{J_2}$ is a nonlinear map from the J_1-dimensional input to the J_2-dimensional feature space. A J_2-by-J_2 correlation matrix in the feature space is defined by

$$\mathbf{C}_1 = \frac{1}{N} \sum_{i=1}^{N} \phi(\boldsymbol{x}_i) \phi^T(\boldsymbol{x}_i). \tag{20.7}$$

Like PCA, the set of feature vectors is constrained to zero mean, $\frac{1}{N}\sum_{i=1}^{N} \phi(\boldsymbol{x}_i) = \mathbf{0}$. A procedure for selecting ϕ is given in [99].

The principal components are then computed by solving the eigenvalue problem [79, 100]

$$\lambda \boldsymbol{v} = \mathbf{C}_1 \boldsymbol{v} = \frac{1}{N} \sum_{j=1}^{N} \left(\phi(\boldsymbol{x}_j)^T \boldsymbol{v} \right) \phi(\boldsymbol{x}_j). \tag{20.8}$$

Thus, $\boldsymbol{v}$ must be in the span of the mapped data

$$\boldsymbol{v} = \sum_{i=1}^{N} \alpha_i \phi(\boldsymbol{x}_i). \tag{20.9}$$

After premultiplying both sides of (20.9) by $\phi(\boldsymbol{x}_j)$ and performing mathematical manipulations, the kernel PCA problem reduces to

$$\mathbf{K}\boldsymbol{\alpha} = \lambda \boldsymbol{\alpha}, \tag{20.10}$$

where λ and $\boldsymbol{\alpha} = (\alpha_1, \ldots, \alpha_N)^T$ are, respectively, the eigenvalues and the corresponding eigenvectors of $\mathbf{K}$, and $\mathbf{K}$ is an $N \times N$ kernel matrix with

$$K_{ij} = k(\boldsymbol{x}_i, \boldsymbol{x}_j) = \phi^T(\boldsymbol{x}_i) \phi(\boldsymbol{x}_j). \tag{20.11}$$

Arrange the eigenvalues in descending order, $\lambda_1 \geq \lambda_2 \geq \cdots \geq \lambda_{J_2} > 0$, and denote their corresponding eigenvectors as $\boldsymbol{\alpha}_1, \ldots, \boldsymbol{\alpha}_{J_2}$. The eigenvectors are further normalized as

$$\boldsymbol{\alpha}_k^T \boldsymbol{\alpha}_k = \frac{1}{\lambda_k}. \tag{20.12}$$

The nonlinear principal components of $\boldsymbol{x}$ can be extracted by projecting the mapped pattern $\phi(\boldsymbol{x})$ onto $\boldsymbol{v}_k$ [79, 100]

$$\boldsymbol{v}_k^T \phi(\boldsymbol{x}) = \sum_{j=1}^{N} \alpha_{k,j} k(\boldsymbol{x}_j, \boldsymbol{x}), \quad k = 1, 2, \ldots, J_2, \tag{20.13}$$

where $\alpha_{k,j}$ is the jth element of $\boldsymbol{\alpha}_k$.

Kernel PCA is much more complicated and may sometimes be caught more easily in local minima. The kernel PCA method via eigendecomposition involves a time complexity of $O(N^3)$, with N being the number of training vectors. Second, the resulting kernel principal components have to be defined implicitly by linear expansions of the training data; thus, all data must be saved after training.

PCA needs to deal with an eigenvalue problem of a $J_1 \times J_1$ matrix, while kernel PCA needs to solve an eigenvalue problem of an $N \times N$ matrix. Sparse approximation methods can be applied to reduce the computational cost [79]. An algorithm proposed in [16] enables us to recover the number of leading kernel PCA components relevant for good classification.

Traditionally, kernel methods require computation and storage of the entire kernel matrix and preparation of all training samples beforehand. This requirement can be eliminated by repeatedly cycling through the dataset, computing kernels on demand, as implemented in decomposition methods for SVM. This is done for kernel PCA by the kernel Hebbian algorithm as an online version of kernel PCA [56], which suffers from slow convergence. Kernel Hebbian algorithm, introduced by kernelizing GHA, has a scalar gain parameter that is either held constant or decreased according to a predetermined annealing schedule. Gain adaptation can improve convergence of kernel Hebbian algorithm by incorporating the reciprocal of the current estimated eigenvalues [43]. Subset kernel PCA uses a subset of samples for calculating the bases of nonlinear principal components. Online subset kernel PCA such as subset kernel Hebbian algorithm gradually adds and exchanges a sample in the basis set, thus it can be applied to time-varying patterns [116].

For kernel PCA, the L_2 loss function used is not robust, and outliers can skew the solution from the desired one. Kernel PCA lacks of sparseness because the principal components are expressed in terms of a dense expansion of kernels associated with every training data point. Some approaches introduce sparseness into kernel PCA, e.g., [101, 103]. Incremental kernel PCA in Krein space does not require the calculation of preimages and therefore is both efficient and exact [70].

The adaptive kernel PCA method in [31] has the flexibility to accurately track the kernel principal components. First, kernel principal components are recursively formulated from the recursive eigendecomposition of kernel covariance matrix. Kernel covariance matrix is then correctly updated to adapt to the changing characteristics of data. In this adaptive method, the kernel principal components are adaptively adjusted without re-eigendecomposing the kernel Gram matrix. The method not only maintains constant update speed and memory usage as the data size increases, but also alleviates suboptimality of the kernel PCA method for nonstationary data.

An LS-SVM approach to kernel PCA is given in [107]. Introducing the LS-SVM formulation to kernel PCA, kernel PCA is extended to a generalized form of kernel component analysis with a general loss function made explicit [4]. Robustness and sparseness are introduced into kernel component analysis by using an ϵ-insensitive robust loss function.

An incremental kernel SVD algorithm based on kernelizing incremental linear SVD is given in [25]. It does not require adaptive centering of the incremental data and the appropriate adjustment of the factorized subspace bases. Kernel PCA and kernel SVD return vastly different results if the dataset is not centered. In [25], reduced set expansions are constructed to compress the kernel SVD basis so as to achieve constant incremental update speed. By using a better compression strategy and adding a kernel subspace re-orthogonalization scheme, an incremental kernel PCA [24] has linear time complexity to maintain constant update speed and memory usage.

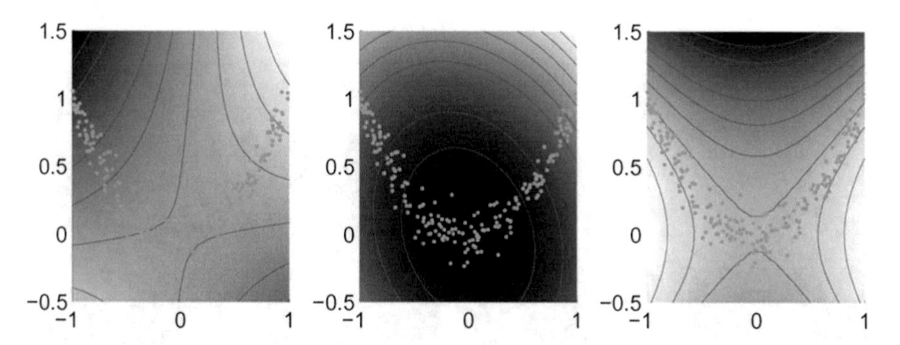

Fig. 20.2 The first three principal components for the two-dimensional dataset. ©IEEE, 2005 [56]

A kernel projection pursuit method [103] chooses sparse directions and orthogonalizes in the same way as PCA. Utilizing a similar methodology, a general framework for feature extraction is formalized based on an orthogonalization procedure, and two sparse kernel feature extraction methods are derived in [28]. These approaches have a training time $O(N)$.

A class of robust procedures for kernel PCA is proposed in [49] based on EVD of weighted covariance. The procedures place less weight on deviant patterns and thus is more resistant to data contamination and model deviation.

Kernel entropy component analysis [51] for data transformation and dimension reduction reveals structure relating to the Renyi entropy of the input space dataset, estimated via a kernel matrix using Parzen windowing. This is achieved by projections onto a subset of entropy-preserving kernel PCA axes.

A general widely linear complex kernel PCA framework defined in [86] efficiently performs widely linear PCA in small sample size problems.

Example 20.2 We replicate this example from [56]. For a two-dimensional dataset, with 150 data points generated from $y_i = -x_i^2 + \xi$, where x_i is generated from uniform distribution in $[-1, 1]$ and ξ is normal noise with standard deviation 0.2. Contour lines of constant value of the first three principal components for the dataset are obtained from kernel PCA with degree-2 polynomial kernel, as shown in Fig. 20.2.

Example 20.3 (*Image denoising*) A noisy image can be denoised by using kernel PCA. A large image is regarded as a composition of multiple patches. The multiple patches are used as training samples for learning the few leading eigenvectors by kernel PCA. The image is reconstructed by reconstructing each of the component patches and the noise is thus removed. This approach is similar to wavelet-based methods for image denoising in the sense that the objective is to find a good feature space in which the noise shows low power or is concentrated on a small subspace.

We replicate this example from [56]. Two different noisy images were constructed by adding white Gaussian noise (SNR 7.72 dB) and salt-and-pepper noise (SNR 4.94 dB) to the 256×256-sized Lena image. From each image, 12×12 overlapping

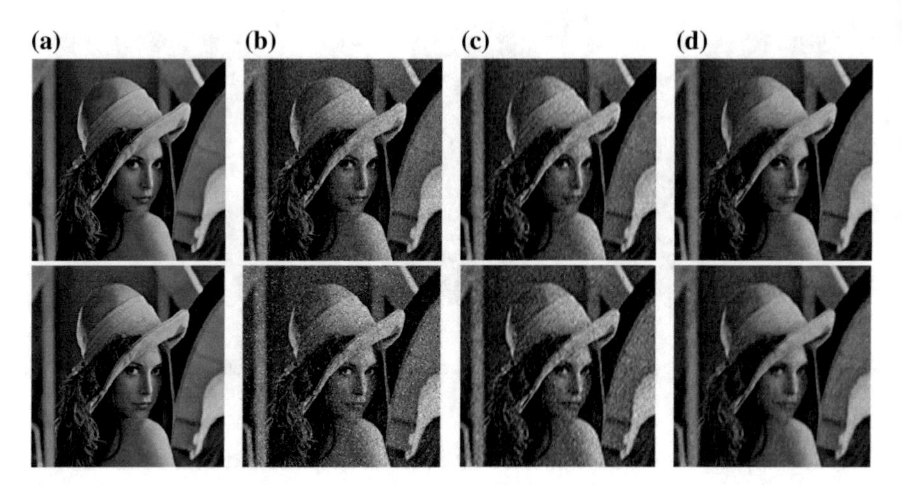

Fig. 20.3 Denoising Gaussian noise (Upper): **a** Original image, **b** input noisy image, **c** PCA ($r = 20$), and **d** KHA ($r = 40$). Denoising salt and pepper type noise (Lower): **a** original image, **b** input noisy image, **c** PCA ($r = 20$), and **d** KHA ($r = 20$). ©IEEE, 2005 [56]

image patches were sampled at a regular interval of two pixels. The kernel PCA model for the Gaussian kernel ($\sigma = 1$) was obtained by training the kernel Hebbian algorithm on each training set with a learning rate $\eta = 0.05$ for around 800 epochs through the dataset. The denoised images were then obtained by reconstructing the input image using the first r principal components from each kernel PCA model. The kernel Hebbian algorithm performs well for both noise types. Similar to linear PCA, kernel PCA is expected to work best if the noise characteristics are Gaussian in the feature space. The original, noisy, and denoised Lena images are shown in Fig. 20.3.

20.4 Kernel LDA

Kernel LDA consists of a two-stage procedure. The first step consists of embedding the data space X into a possibly infinite-dimensional RKHS F via a kernel function. The second step simply applies LDA in this new data space. Like LDA, kernel LDA always encounters the ill-posed problem, and some form of regularization needs to be included. Kernel LDA can be formulated for two classes. The inner product matrix can be made nonsingular by adding a scalar matrix [78]. By introducing kernel into linear $\boldsymbol{w}$, nonlinear discriminant analysis is obtained [78, 79, 119]. A multiple of the identity or the kernel matrix can be added to $\mathbf{S}_w$ (or its reformulated form $\widetilde{\mathbf{S}}_w$) after introducing the kernels to penalize $\|\boldsymbol{w}\|^2$ (or $\|\widetilde{\boldsymbol{w}}\|^2$) [78].

Hard margin linear SVM is equivalent to LDA when all the training points are support vectors [102]. SVM and kernel LDA usually have similar performance. In [95], the vector obtained by kernel QP feature selection is equivalent to the kernel

Fisher vector and therefore, an interpretation of kernel LDA is given which provides some computational advantages for highly unbalanced datasets.

Kernel LDA for multiple classes is formulated in [11]. The method employs QR decomposition to avoid singularity. Kernel LDA is much more effective than kernel PCA in face recognition [71]. Kernel direct discriminant analysis [71] generalizes direct LDA. Based on the kernel PCA plus LDA framework, complete kernel LDA [125] can be used to carry out discriminant analysis in double discriminant subspaces. It can make full use of regular and irregular discriminant information. Complete kernel LDA outperforms other kernel LDA algorithms. A kernelized one-class LDA is presented in [32].

Kernel quadratic discriminant [87] is based on the regularized kernel Mahalanobis distance in both complete and class-related subspaces. The method can be advantageous for data with unequal class spreads in the kernel-induced spaces.

The properties of kernel uncorrelated discriminant analysis and kernel regularized discriminant analysis are studied in [52]. Under a mild condition, both LDA and kernel uncorrelated discriminant analysis project samples in the same class to a common vector in the dimension-reduced space. Uncorrelated discriminant analysis may suffer from the overfitting problem if there are a large number of samples in each class. Regularization can be applied to overcome the overfitting problem.

Robust kernel fuzzy discriminant analysis [46] uses fuzzy memberships to reduce the effect of outliers and adopts kernel methods to accommodate nonlinearly separable cases. In [115], a scatter matrix-based class separability criterion is extended to a kernel space and developed as a feature selection criterion. Compared with the radius-margin bound, feature selection with this criterion is faster. The proposed criterion is more robust in the case of small sample set and is less vulnerable to noisy features. This criterion is proven to be a lower bound of the maximum value of kernel LDA's objective function.

The common vector method is a linear subspace classifier method which allows one to discriminate between classes of datasets. This method utilizes subspaces that represent classes during classification. The kernel discriminant common vector method [21] yields an optimal solution for maximizing a modified LDA criterion. A modified common vector method and its kernelized version are given in [22].

The kernel trick is applied to transform the linear-domain Foley–Sammon optimal, with respect to orthogonality constraints, discriminant vectors, resulting in a nonlinear LDA method [135]. The kernel Foley–Sammon method may suffer from the heavy computation problem due to the inverse of matrices, resulting in a cubic complexity for each discriminant vector. A fast algorithm for solving this kernelized model [137] is based on rank-one update of the eigensystems and QR decomposition of matrices, to incrementally establish the eigensystems for the discriminant vectors. It only requires a square complexity for each discriminant vector. To further reduce the complexity, a kernel Gram–Schmidt orthogonalization method [117] is adopted to replace kernel PCA in the preprocessing stage.

A criterion is derived in [129] for finding a kernel representation where the Bayes classifier becomes linear. It maps the original class (or subclass) distributions into a kernel space where these are best separated by a hyperplane. The approach aims to maximize the distance between the distributions of different classes, thus maximizing generalization. It is applied to LDA, nonparametric discriminant analysis and subclass discriminant analysis. A kernel version of subclass discriminant analysis yields the highest recognition rates [129].

Leave-one-out cross-validation for kernel LDA [18] is extended in [98] such that the leave-one-out error can be re-estimated following a change in the regularization parameter in kernel LDA. This reduces the computational complexity from $O(N^3)$ to $O(N^2)$ operations for N training patterns. The method is competitive with model selection based on k-fold cross-validation in terms of generalization, while being considerably faster.

20.5 Kernel Clustering

Kernel-based clustering first nonlinearly maps the patterns into an arbitrarily high-dimensional feature space, and then performs clustering in the feature space. Some examples are kernel C-means [100], kernel subtractive clustering [54], a kernel-based algorithm that minimizes the trace of the within-class scatter matrix [39], and support vector clustering. Support vector clustering can effectively deal with the outliers.

C-means can only find linearly separable clusters. Kernel C-means [100] identifies nonlinearly separable clusters; it monotonically converges if the kernel matrix is positive semidefinite. If the kernel matrix is not positive semidefinite, the convergence of the algorithm is not guaranteed. Kernel C-means requires $O(N^2\tau)$ scalar operations, where τ is the number of iterations until convergence is achieved. Performing C-means in the kernel PCA space is equivalent to kernel C-means [65]. Weighted kernel C-means [29], which assigns different weights to each data point, is closely related to graph partitioning as its objective becomes equivalent to many graphs cut criteria if the weights and kernel are set appropriately. Multilevel kernel C-means clustering [30] does not require to compute the whole kernel matrix on the training dataset, and thus, is extremely efficient.

Kernelized FCM [132] substitutes the Euclidean distance with kernel function, and it outperforms FCM. Kernel C-means and kernel FCM perform similarly in terms of classification quality when using a Gaussian kernel function and generally perform better than their standard counterparts [55]. Generic FCM and Gustafson–Kessel FCM are compared with two typical generalizations of kernel-based fuzzy clustering in [41]. The kernel-based FCM algorithms produce a marginal improvement over FCM and Gustafson–Kessel for most of the datasets. But they are in a number of cases highly sensitive to the selection of specific values of the kernel parameters.

A kernel-induced distance is used to replace the Euclidean distance in the potential function [54] of the subtractive clustering method. This enables clustering of the data that is linearly inseparable in the original space into homogeneous groups in the transformed high-dimensional space, where the data separability is increased.

A kernel version of SOM [73] is derived from kernelizing C-means with added neighborhood learning, based on the distance kernel trick. In the self-organizing mixture network [127], neurons in SOM are treated as Gaussian (or other) kernels, and the resulting map approximates a mixture of Gaussian (or other) distributions of the data. Kernel SOM can be derived naturally by minimizing an energy function, and the resulting kernel SOM unifies the approaches to kernelize SOM and can be performed entirely in the feature space [63]. SOM approximates the kernel method naturally, and further kernelizing SOM may not be necessary: there is no clear evidence showing that kernel SOMs are always superior to SOM [63].

The kernel-based maximum entropy learning rule (kMER) [112] is an approach that formulates an equiprobabilistic map using maximum entropy learning to avoid the underutilization problem for clustering. The kMER approach updates the prototype vectors and the corresponding radii of the kernels centered on these vectors to model the input density at convergence. kMER considers several winners at a time. This leads to computational inefficiency and a slow formation of the topographic map. SOM-kMER [110] integrates SOM and kMER. It allocates a kernel at each neuron which is a prototype of a cluster of input samples. Probabilistic SOM-kMER [111] utilizes SOM-kMER and the probabilistic neural network for classification problems. Instead of using all the training samples, the prototype vectors from each class of the trained SOM-kMER map are used for nonparametric density estimation.

The kernel version of neural gas [91] applies the soft rule for the update to the code vectors in feature space. An online version of kernel C-means can be found in [100]. Other methods are kernel possibilistic C-means [34] and kernel spectral clustering [5].

Self-adaptive kernel machine [13] is an online clustering algorithm for evolving clusters from nonstationary data, based on SVM and the kernel method. The algorithm designs an unsupervised feedforward network architecture with self-adaptive abilities.

20.6 Kernel Auto-associators, Kernel CCA, and Kernel ICA

Kernel Auto-associators

Kernel associative memory introduces the kernel approach to associative memory by nonlinearly mapping the data into some high-dimensional feature space through operating a kernel function with input space. Kernel auto-associators are generic one-class learning machines. They perform auto-association mapping via the kernel feature space. They can be applied to novelty detection and m-class classification problems. Kernel auto-associators have the same expression as that of the kernel associative memory [133]. Thus, kernel associative memories are a special form of kernel auto-associators. In [38], the update equation of the dynamic associative memory is interpreted as a classification step. A model of recurrent kernel associative memory [38] is analyzed in [89], and this model consists of a kernelization of RCAM [89].

Kernel CCA

If there is nonlinear correlation between two variables, CCA may not correctly correlate this relationship. Kernel CCA [8, 60] first maps the data into a high-dimensional feature space induced by a kernel and then performs CCA. In this way, nonlinear relationships can be found. Kernel CCA has been used in a kernel ICA algorithm [8].

Kernel CCA tackles the singularity problem of the Gram matrix by simply adding a regularization term to the Gram matrix. A mathematical proof of the statistical convergence of kernel CCA is given in [36]. The result also gives a sufficient condition for convergence on the regularization coefficient involved in kernel CCA. An improved kernel CCA algorithm based on EVD approach rather than the regularization method is given in [136].

LS weighted kernel reduced rank regression [26] is a unified framework for formulating many component analysis methods such as PCA, LDA, CCA, locality preserving projections, and spectral clustering, and their kernel and regularized extensions.

Kernel ICA

A class of kernel ICA algorithms [8] use contrast functions based on canonical correlations in an RKHS. In [3], the proposed kernel-based contrast function for ICA corresponds to a regularized correlation measure in high-dimensional feature spaces induced by kernels. The formulation is a multivariate extension of the LS-SVM formulation to kernel CCA. The kernel-based nonlinear BSS method [76] exploits second-order statistics in a kernel-induced feature space. This method extends a linear algorithm to the nonlinear domain using the kernel-based method of extracting nonlinear features applied in SVM and kernel PCA. kTDSEP [44] combines kernel feature spaces and second-order temporal decorrelation BSS using temporal information.

Kernel Classification

The kernel ridge regression classifier is provided by the generalized kernel machine toolbox [19]. The constrained covariance and the kernel mutual information [42] measure independence based on the covariance between functions of the random variables in RKHSs.

Kernel-based MAP classification [122] makes a Gaussian distribution assumption instead of a linearly separable assumption in the feature space. Robust methods are proposed to estimate the probability densities, and the kernel trick is utilized to calculate the model. The model can output probability or confidence for classification.

A sparsity-driven kernel classifier based on the minimization of a data-dependent generalization error bound is considered in [88]. The objective function consists of the usual hinge loss function penalizing training errors and a concave penalty function of the expansion coefficients. The problem of minimizing the nonconvex bound is addressed by a successive linearization approach, whereby the problem is transformed into a sequence of linear programs. The algorithm produces error rates comparable to SVM but significantly reduces the number of support vectors. The kernel classifier in [53] optimizes the L_2 or integrated squared error of a difference of densities. Like SVM, this classifier is sparse and results from solving a quadratic program. The method allows data-adaptive kernels, and does not require an independent sample.

Kernel logistic regression corresponds to the penalized logistic classification in an RKHS [50, 138]. It is easier to analyze than SVM because it has the logistic loss function and SVM has a hinge loss function. The computational cost for kernel logistic regression is much higher than that for SVMs with decomposition algorithms. A decomposition algorithm for the kernel logistic regression is introduced in [138]. In kernel logistic regression, the kernel expansion is nonsparse in the data.

Kernel matching pursuit [113] is an effective greedy discriminative sparse kernel classifier that extends the matching pursuit method.

20.7 Other Kernel Methods

Conventional online linear algorithms have been extended to their kernel versions: kernel LMS algorithm [68, 90], kernel adaline [35], kernel RLS [33], kernel Wiener filter [128], complex kernel LMS [14], kernel LS [97], and kernel online learning [58].

In the finite training data case, kernel LMS [68] is well-posed in RKHS without the addition of an regularization term. The extended kernel RLS algorithm [69] is the kernelized extended RLS algorithm. Kernel affine projection algorithms [67] inherit the simplicity and online nature of kernel LMS while reducing its gradient noise and boosting performance. Kernel affine projection provides a unifying model for kernel LMS, kernel adaline, sliding-window kernel RLS, and regularization networks. Kernel adaptive ARMA algorithm [66] is a kernel adaptive recurrent filtering algorithm based on the autoregressive moving-average (ARMA) model, which is trained with recurrent stochastic gradient descent in the RKHSs.

Kernel CG algorithm in online mode [134] converges as fast as the kernel RLS algorithm, but with only a quarter of the computational cost of the kernel RLS algorithm and without user-defined parameters. Kernelized partial LS [96] is equivalent to kernel CG, but with a much higher computational cost.

Kernel-based reinforcement learning [84] computes a decision policy which converges to a unique solution and is statistically consistent. Unfortunately, the model constructed grows with the number of sample transitions. Kernel-based stochastic factorization [10] compresses the information in the model of kernel-based reinforcement learning into an approximator of fixed size. The computational complexity is linear in the number of sample transitions. It allows for an incremental implementation that makes the amount of memory used independently of the number of sample transitions.

Regularized sparse kernel slow feature analysis generates an orthogonal basis in the unknown latent space for a given real-world time series by utilizing the kernel trick in combination with sparsification [12].

Two generalizations of NMF in kernel feature space are polynomial kernel NMF [17] and projected gradient kernel NMF (PGKNMF) [130]. An RKHS framework for spike trains is introduced in [85].

A common problem of kernel-based online algorithms is the amount of memory required to store the online hypothesis, which may increase without bound as the algorithm progresses. Furthermore, the computational load of such algorithms grows linearly with the amount of memory used to store the hypothesis.

Kernel density estimation is a nonparametric method that yields a pdf, given a set of observations. The resulting pdf is the sum of kernel functions centered in the data points. The computational complexity of kernel density estimation makes its application to data streams impossible. The cluster kernel approach [45] provides continuously computed kernel density estimators over streaming data.

Based on perceptron algorithm, projectron [83] projects the instances onto the space spanned by the previous online hypothesis. Projectron++ is deduced based on the notion of large margin. The performance of the projectron algorithm is slightly worse than, but very similar to, that of the perceptron algorithm, for a wide range of the learning rate. Projectron++ outperforms perceptron, projectron, forgetron [27] and randomized budget perceptron [20], with a similar hypothesis size. For a given target accuracy, the size of the support sets of projectron or projectron++ are much smaller than those of forgetron and randomized budget perceptron.

A reformulation of the sampling theorem using an RKHS [80] gives us a unified viewpoint for many generalizations and extensions of the sampling theorem. In [81], a framework of the optimal approximation, rather than a perfect reconstruction, of a function in the RKHS is introduced by using the orthogonal projection onto the linear subspace spanned by the given system of kernel functions corresponding to a finite number of sampling points. This framework is extended to infinite sampling points in [109].

In [37], learning kernels under the LASSO formulation is implemented via adopting a generative Bayesian learning and inference approach. A robust learning algorithm proposed produces a sparse kernel model with the capability of learning regularized parameters and kernel hyperparameters.

Kernelized version of low-rank representation and robust kernel low-rank representation approach are proposed in [118] to effectively cope with nonlinear data. An efficient optimization algorithm solves the robust kernel low-rank representation based on the alternating direction method.

Bounds on expected leave-one-out cross-validation errors for kernel methods are derived in [131], which lead to expected generalization bounds for various kernel algorithms. In addition, variance bounds for leave-one-out errors are obtained. In [74], prior knowledge over arbitrary general sets is incorporated into nonlinear kernel approximation problems in the form of linear constraints in a linear program.

20.7.1 Random Kitchen Sinks and Fastfood

The kernel method requires the computation of the kernel matrix as well as its operation. Nystrom method is a well-known sampling-based algorithm for approximating large kernel matrices and their eigensystems, which has been discussed in Sect. 14.4. Random kitchen sinks and fastfood are two approximation methods of kernel matrices.

Random kitchen sinks [92, 93] approximates the shift-invariant kernel based on its Fourier transform. It approximates a function f by means of multiplying the input with a Gaussian random matrix, followed by applying a nonlinearity. Sampling is used to approximate the kernel representation of the sum in an inner product in Mercer's theorem. For the expansion dimension n and the input dimension d, it requires $O(nd)$ time and memory to evaluate the function f. For large problems with sample size $m \gg n$, this is much faster than the kernel trick.

Fastfood [64] use random Fourier features to approximate the kernel function. It uses low-rank representation to approximate kernel matrix to speed up solving kernel SVM. Fastfood accelerates random kitchen sinks from $O(nd)$ to $O(n \log d)$ time. It relies on the fact that Hadamard matrices, when multiplied by diagonal Gaussian matrices, behave like Gaussian random matrices. Hadamard and diagonal matrices can be used in lieu of Gaussian matrices in random kitchen sinks. Fastfood approximation is unbiased, has low variance, and concentrates almost at the same rate as random kitchen sinks. Fastfood requires $O(n \log d)$ time and $O(n)$ storage to compute n nonlinear basis functions in d dimensions, without sacrificing accuracy.

Given the noisy measurements, the RKHS estimate represents the posterior mean (minimum variance estimate) of a Gaussian random field with covariance proportional to the kernel associated with the RKHS. A statistical interpretation is provided in [6] when more general losses are used. For any finite set of sampling locations, the maximum a posteriori estimate for the signal samples is given by the RKHS estimate evaluated at the sampling locations.

20.8 Multiple Kernel Learning

Multiple kernel learning (MKL) [62, 82] considers multiple kernels or the combination of kernels rather than a single fixed kernel. It tries to form an ensemble of kernels so as to fit for a given application. MKL can offer some needed flexibility and manipulate well the case that involves multiple, heterogeneous data sources. It has become a popular learning paradigm for combining multiple views, where base kernels are constructed from each view.

Any convex set of kernel matrices is a set of semidefinite programs. In [62], MKL was formulated as a semidefinite programming (SDP) problem. A convex quadratically constrained quadratic program (QCQP) is constructed by the conic combinations of multiple kernels $k = \sum_i \alpha_i k_i$ from a library of candidate kernels k_i [62]. In order to extend the method to large-scale problems, QCQP is reconstructed as a semi-infinite linear program that recycles the SVM implementations [105].

In [57], a sparsity-inducing multiple kernel LDA is introduced, where an L_1 norm is used to regularize the kernel weights. This optimal kernel selection problem can be reformulated as a tractable convex optimization problem which interior-point methods can solve globally and efficiently. For regularized kernel discriminant analysis, the optimal kernel matrix is obtained as a linear combination of prespecified kernel matrices [126]. The kernel learning problem can be formulated as a convex semidefinite program. A convex QCQP formulation is proposed for binary-class kernel learning in regularized kernel discriminant analysis, and the QCQP formulations are solved using the MOSEK interior-point optimizer for LP [126]. Multiclass regularized kernel discriminant analysis can be decomposed into a set of binary-class kernel learning problems which are constrained to share a common kernel; SDP formulations are then proposed, which lead naturally to QCQP and semi-infinite LP formulations.

Sparse MKL [106] generalizes group feature selection to kernel selection. It is capable of exploiting existing efficient single kernel algorithms while providing a sparser solution in terms of the number of kernels used as compared to the existing MKL framework.

Most MKL methods employ the L_1-norm simplex constraints on the kernel combination weights, which therefore involve a sparse but nonsmooth solution for the kernel weights. Despite the success of their efficiency, they tend to discard informative complementary or orthogonal base kernels and yield degenerated generalization performance. To tackle these problems, by introducing an elastic-net-type constraint on the kernel weights, a generalized MKL model has been proposed in [124], which enjoys the favorable sparsity property on the solution and also facilitates the grouping effect.

SimpleMKL [94] performs a reduced gradient descent on the kernel weights. HessianMKL [23] replaces the gradient descent update of SimpleMKL with a Newton update. At each iteration, HessianMKL solves a QP problem with the size of the number of kernels to obtain the Newton update direction. HessianMKL shows second-order convergence.

In a nonsparse scenario, L_p-norm MKL yields strictly better bounds than L_1-norm MKL and vice versa [1, 59]. The analytical solver for nonsparse MKL is compared with some L_1-norm MKL methods, namely, SimpleMKL, HessianMKL, SILP-based wrapper, and SILP-based chunking optimization [105]. SimpleMKL and the analytical solver become more efficient with increasing number of kernels, but the capacity remains limited due to memory restriction. HessianMKL is considerably faster than SimpleMKL but slower than the nonsparse interleaved methods and SILP. Overall, the interleaved analytic and cutting plane based optimization strategies [59] achieve a speedup of up to one and two orders of magnitude over HessianMKL and SimpleMKL, respectively.

SpicyMKL [108] is applicable to general convex loss functions and general types of regularization. By iteratively solving smooth minimization problems, there is no need of solving SVM, LP, or QP internally. SpicyMKL can be viewed as a proximal minimization method and converges super-linearly. The cost of inner minimization is roughly proportional to the number of active kernels.

In [120], a soft margin framework for MKL is proposed by introducing kernel slack variables. The commonly used hinge loss, square hinge loss, and square loss functions can be incorporated into this framework. Many MKL methods are special cases under the soft margin framework. The algorithms can efficiently achieve an effective yet sparse solution for MKL. L_1-norm MKL can be deemed as hard margin MKL.

SMO-MKL (http://manikvarma.org/) [114] implements linear MKL regularized with the p-norm squared, or with certain Bregman divergences, which is trained using SMO.

Bayesian efficient MKL (http://users.ics.aalto.fi/gonen/icml12.php) with sparse and nonsparse weights [40] allows to combine hundreds or thousands of kernels very efficiently based on a fully conjugate Bayesian formulation and a deterministic variational approximation.

The formulation of MKL is equivalent to maximum entropy discrimination with a noninformative prior over multiple views [75]. A hierarchical Bayesian model is presented to learn the proposed data-dependent prior and classification model simultaneously. Multiple kernel FCM [48] extends the FCM algorithm with a multiple kernel learning setting.

Problems

20.1 Show that Mercer kernels are positive definite.

20.2 Synthetically generate two-dimensional data which lie on a circle and are corrupted by the Gaussian noise. Use STPRtool (http://cmp.felk.cvut.cz/cmp/software/stprtool/) to find the first principal components of the data by using PCA and kernel PCA model with RBF kernel.

20.3 **Face image superresolution and denoising** [56]. Apply the kernel PCA method on a face image for image superresolution purpose and for denoising purpose [56]. The Yale Face Database B is used for the experiments. The database contains 5,760 images of 10 persons, down-sampled to 60×60 pixels. A total of 5,000 images are used for training and the remaining are used for testing. Kernel Hebbian algorithm is applied during the training and 16 eigenvectors are obtained. Ten test samples are randomly selected from the test set.

(a) Downsample the test samples to 20×20 resolution and then resize them to 60×60 resolution by mapping each pixel to a 3×3 block of identical pixel values. Project the processed test samples to the obtained eigenvectors. Reconstruct the test samples by finding the closest samples in the training set.

(b) Project the test samples directly to the obtained eigenvectors. Reconstruct the test samples from the projections.

20.4 **Superresolution of a natural image** [56]. Apply the kernel PCA method on a natural image of low resolution to get a superresolution image [56].

20.5 Consider an artificial 4×4 checkerboard data based on a uniform distribution. Use kernel LDA to separate the samples. Plot the separating boundary.

References

1. Aflalo, J., Ben-Tal, A., Bhattacharyya, C., Nath, J. S., & Raman, S. (2011). Variable sparsity kernel learning. *Journal of Machine Learning Research*, *12*, 565–592.
2. Aizerman, M., Braverman, E., & Rozonoer, L. (1964). Theoretical foundations of the potential function method in pattern recognition learning. *Automation and Remote Control*, *25*, 821–837.
3. Alzate, C., & Suykens, J. A. K. (2008). A regularized kernel CCA contrast function for ICA. *Neural Networks*, *21*, 170–181.
4. Alzate, C., & Suykens, J. A. K. (2008). Kernel component analysis using an epsilon-insensitive robust loss function. *IEEE Transactions on Neural Networks*, *19*(9), 1583–1598.
5. Alzate, C., & Suykens, J. A. K. (2010). Multiway spectral clustering with out-of-sample extensions through weighted kernel PCA. *IEEE Transactions on Pattern Analysis and Machine Intelligence*, *32*(2), 335–347.
6. Aravkin, A. Y., Bell, B. M., Burke, J. V., & Pillonetto, G. (2015). The connection between Bayesian estimation of a Gaussian random field and RKHS. *IEEE Transactions on Neural Networks and Learning Systems*, *26*(7), 1518–1524.
7. Aronszajn, N. (1950). Theory of reproducing kernels. *Transactions of the American Mathematical Society*, *68*, 337–404.
8. Bach, F. R., & Jordan, M. I. (2002). Kernel independent component analysis. *Journal of Machine Learning Research*, *3*, 1–48.
9. Balcan, M.-F., Blum, A., & Vempala, S. (2004). Kernels as features: On kernels, margins, and low-dimensional mappings. In *Proceedings of the 15th International Conference on Algorithmic Learning Theory* (pp. 194–205).
10. Barreto, A. M. S., Precup, D., & Pineau, J. (2016). Practical kernel-based reinforcement learning. *Journal of Machine Learning Research*, *17*, 1–70.
11. Baudat, G., & Anouar, F. (2000). Generalized discriminant analysis using a kernel approach. *Neural Computation*, *12*(10), 2385–2404.
12. Bohmer, W., Grunewalder, S., Nickisch, H., & Obermayer, K. (2012). Generating feature spaces for linear algorithms with regularized sparse kernel slow feature analysis. *Machine Learning*, *89*, 67–86.
13. Boubacar, H. A., Lecoeuche, S., & Maouche, S. (2008). SAKM: Self-adaptive kernel machine. A kernel-based algorithm for online clustering. *Neural Networks*, *21*, 1287–1301.
14. Bouboulis, P., & Theodoridis, S. (2011). Extension of Wirtinger's calculus to reproducing kernel Hilbert spaces and the complex kernel LMS. *IEEE Transactions on Signal Processing*, *59*(3), 964–978.
15. Bouboulis, P., Slavakis, K., & Theodoridis, S. (2012). Adaptive learning in complex reproducing kernel Hilbert spaces employing Wirtinger's subgradients. *IEEE Transactions on Neural Networks and Learning Systems*, *23*(3), 425–438.
16. Braun, M. L., Buhmann, J. M., & Muller, K.-R. (2008). On relevant dimensions in kernel feature spaces. *Journal of Machine Learning Research*, *9*, 1875–1908.
17. Buciu, I., Nikolaidis, N., & Pitas, I. (2008). Nonnegative matrix factorization in polynomial feature space. *IEEE Transactions on Neural Networks*, *19*(6), 1090–1100.
18. Cawley, G. C., & Talbot, N. L. C. (2003). Efficient leave-one-out cross-validation of kernel Fisher discriminant classifiers. *Pattern Recognition*, *36*(11), 2585–2592.

19. Cawley, G. C., Janacek, G. J., & Talbot, N. L. C. (2007). Generalised kernel machines. In *Proceedings of the IEEE/INNS International Joint Conference on Neural Networks, Orlando, FL* (pp. 1720–1725).
20. Cesa-Bianchi, N., Conconi, A., & Gentile, C. (2006). Tracking the best hyperplane with a simple budget Perceptron. In *Proceedings of the 19th International Conference on Learning Theory* (pp. 483–498).
21. Cevikalp, H., Neamtu, M., & Wilkes, M. (2006). Discriminative common vector method with kernels. *IEEE Transactions on Neural Networks*, *17*(6), 1550–1565.
22. Cevikalp, H., Neamtu, M., & Barkana, A. (2007). The kernel common vector method: A novel nonlinear subspace classifier for pattern recognition. *IEEE Transactions on Systems, Man, and Cybernetics Part B*, *37*(4), 937–951.
23. Chapelle, O., & Rakotomamonjy, A. (2008). Second order optimization of kernel parameters. In *NIPS Workshop on Kernel Learning: Automatic Selection of Optimal Kernels, Whistler, Canada*.
24. Chin, T.-J., & Suter, D. (2007). Incremental kernel principal component analysis. *IEEE Transactions on Image Processing*, *16*(6), 1662–1674.
25. Chin, T.-J., Schindler, K., & Suter, D. (2006). Incremental kernel SVD for face recognition with image sets. In *Proceedings of the 7th IEEE Conference on Automatic Face and Gesture Recognition* (pp. 461–466).
26. De la Torre, F. (2012). A least-squares framework for component analysis. *IEEE Transactions on Pattern Analysis and Machine Intelligence*, *34*(6), 1041–1055.
27. Dekel, O., Shalev-Shwartz, S., & Singer, Y. (2007). The Forgetron: A kernel-based perceptron on a budget. *SIAM Journal on Computing*, *37*(5), 1342–1372.
28. Dhanjal, C., Gunn, S. R., & Shawe-Taylor, J. (2009). Efficient sparse kernel feature extraction based on partial least squares. *IEEE Transactions on Pattern Analysis and Machine Intelligence*, *31*(8), 1347–1361.
29. Dhillon, I. S., Guan, Y., & Kulis, B. (2004). Kernel k-means, spectral clustering and normalized cuts. In *Proceedings of the 10th ACM SIGKDD International Conference on Knowledge Discovery and Data Mining* (pp. 551–556).
30. Dhillon, I. S., Guan, Y., & Kulis, B. (2007). Weighted graph cuts without eigenvectors: A multilevel approach. *IEEE Transactions on Pattern Analysis and Machine Intelligence*, *29*(11), 1944–1957.
31. Ding, M., Tian, Z., & Xu, H. (2010). Adaptive kernel principal component analysis. *Signal Processing*, *90*, 1542–1553.
32. Dufrenois, F. (2015). A one-class kernel Fisher criterion for outlier detection. *IEEE Transactions on Neural Networks and Learning Systems*, *26*(5), 982–994.
33. Engel, Y., Mannor, S., & Meir, R. (2004). The kernel recursive least-squares algorithm. *IEEE Transactions on Signal Processing*, *52*(8), 2275–2285.
34. Filippone, M., Masulli, F., & Rovetta, S. (2010). Applying the possibilistic c-means algorithm in kernel-induced spaces. *IEEE Transactions on Fuzzy Systems*, *18*(3), 572–584.
35. Frieb, T.-T., & Harrison, R. F. (1999). A kernel-based ADALINE. In *Proceedings of the European Symposium on Artificial Neural Networks, Bruges, Belgium* (pp. 245–250).
36. Fukumizu, K., Bach, F. R., & Gretton, A. (2007). Statistical consistency of kernel canonical correlation analysis. *Journal of Machine Learning Research*, *8*, 361–383.
37. Gao, J., Kwan, P. W., & Shi, D. (2010). Sparse kernel learning with LASSO and Bayesian inference algorithm. *Neural Networks*, *23*, 257–264.
38. Garcia, C., & Moreno, J. A. (2004). The Hopfield associative memory network: Improving performance with the kernel "trick". *Advances in artificial intelligence – IBERAMIA 2004*. LNCS (Vol. 3315, pp. 871–880). Berlin: Springer.
39. Girolami, M. (2002). Mercer kernel-based clustering in feature space. *IEEE Transactions on Neural Networks*, *13*(3), 780–784.
40. Gonen, M. (2012). Bayesian efficient multiple kernel learning. In *Proceedings of the 29th International Conference on Machine Learning, Edinburgh, UK* (Vol. 1, pp. 1–8).

41. Graves, D., & Pedrycz, W. (2010). Kernel-based fuzzy clustering and fuzzy clustering: A comparative experimental study. *Fuzzy Sets and Systems*, *161*, 522–543.
42. Gretton, A., Herbrich, R., Smola, A., Bousquet, O., & Scholkopf, B. (2005). Kernel methods for measuring independence. *Journal of Machine Learning Research*, *6*, 2075–2129.
43. Gunter, S., Schraudolph, N. N., & Vishwanathan, S. V. N. (2007). Fast iterative kernel principal component analysis. *Journal of Machine Learning Research*, *8*, 1893–1918.
44. Harmeling, S., Ziehe, A., Kawanabe, M., & Muller, K.-R. (2003). Kernel-based nonlinear blind source separation. *Neural Computation*, *15*, 1089–1124.
45. Heinz, C., & Seeger, B. (2008). Cluster kernels: Resource-aware kernel density estimators over streaming data. *IEEE Transactions on Knowledge and Data Engineering*, *20*(7), 880–893.
46. Heo, G., & Gader, P. (2011). Robust kernel discriminant analysis using fuzzy memberships. *Pattern Recognition*, *44*(3), 716–723.
47. Hoegaerts, L., De Lathauwer, L., Goethals, I., Suykens, J. A. K., Vandewalle, J., & De Moor, B. (2007). Efficiently updating and tracking the dominant kernel principal components. *Neural Networks*, *20*, 220–229.
48. Huang, H.-C., Chuang, Y.-Y., & Chen, C.-S. (2012). Multiple kernel fuzzy clustering. *IEEE Transactions on Fuzzy Systems*, *20*(1), 120–134.
49. Huang, S.-Y., Yeh, Y.-R., & Eguchi, S. (2009). Robust kernel principal component analysis. *Neural Computation*, *21*, 3179–3213.
50. Jaakkola, T., & Haussler, D. (1999). Probabilistic kernel regression models. In *Proceedings of the 7th International Workshop on Artificial Intelligence and Statistics*. San Francisco, CA: Morgan Kaufmann.
51. Jenssen, R. (2010). Kernel entropy component analysis. *IEEE Transactions on Pattern Analysis and Machine Intelligence*, *32*(5), 847–860.
52. Ji, S., & Ye, J. (2008). Kernel uncorrelated and regularized discriminant analysis: A theoretical and computational study. *IEEE Transactions on Knowledge and Data Engineering*, *20*(10), 1311–1321.
53. Kim, J., & Scott, C. D. (2010). L_2 kernel classification. *IEEE Transactions on Pattern Analysis and Machine Intelligence*, *32*(10), 1822–1831.
54. Kim, D. W., Lee, K. Y., Lee, D., & Lee, K. H. (2005). A kernel-based subtractive clustering method. *Pattern Recognition Letters*, *26*, 879–891.
55. Kim, D. W., Lee, K. Y., Lee, D., & Lee, K. H. (2005). Evaluation of the performance of clustering algorithms kernel-induced feature space. *Pattern Recognition*, *38*(4), 607–611.
56. Kim, K. I., Franz, M. O., & Scholkopf, B. (2005). Iterative kernel principal component analysis for image modeling. *IEEE Transactions on Pattern Analysis and Machine Intelligence*, *27*(9), 1351–1366.
57. Kim, S.-J., Magnani, A., & Boyd, S. (2006). Optimal kernel selection in kernel Fisher discriminant analysis. In *Proceedings of the International Conference on Machine Learning* (pp. 465–472).
58. Kivinen, J., Smola, A., & Williamson, R. C. (2004). Online learning with kernels. *IEEE Transactions on Signal Processing*, *52*(8), 2165–2176.
59. Kloft, M., Brefeld, U., Sonnenburg, S., & Zien, A. (2011). l_p-norm multiple kernel learning. *Journal of Machine Learning Research*, *12*, 953–997.
60. Lai, P. L., & Fyfe, C. (2000). Kernel and nonlinear canonical correlation analysis. *International Journal of Neural Systems*, *10*(5), 365–377.
61. Lanckriet, G. R. G., Ghaoui, L. E., Bhattacharyya, C., & Jordan, M. I. (2002). A robust minimax approach to classification. *Journal of Machine Learning Research*, *3*, 555–582.
62. Lanckriet, G. R. G., Cristianini, N., Bartlett, P., Ghaoui, L. E., & Jordan, M. I. (2004). Learning the kernel matrix with semidefinite programming. *Journal of Machine Learning Research*, *5*, 27–72.
63. Lau, K. W., Yin, H., & Hubbard, S. (2006). Kernel self-organising maps for classification. *Neurocomputing*, *69*, 2033–2040.
64. Le, Q., Sarlos, T., & Smola, A. (2013). Fastfood – Approximating kernel expansions in loglinear time. In *Proceedings of the 30th International Conference on Machine Learning, Atlanta, GA* (Vol. 28, pp. 244–252).

65. Li, J., Tao, D., Hu, W., & Li, X. (2005). Kernel principle component analysis in pixels clustering. In *Proceedings of the IEEE/WIC/ACM International Conference on Web Intelligence* (pp. 786–789).
66. Li, K., & Principe, J. C. (2016). The kernel adaptive autoregressive-moving-average algorithm. *IEEE Transactions on Neural Networks and Learning Systems, 27*(2), 334–346.
67. Liu, W., & Principe, J. C. (2008). Kernel affine projection algorithms. *EURASIP Journal on Advances in Signal Processing, 2008*, Article ID 784292, 12 pp.
68. Liu, W., Pokharel, P. P., & Principe, J. C. (2008). The kernel least-mean-square algorithm. *IEEE Transactions on Signal Processing, 56*(2), 543–554.
69. Liu, W., Park, I., Wang, Y., & Principe, J. C. (2009). Extended kernel recursive least squares algorithm. *IEEE Transactions on Signal Processing, 57*(10), 3801–3814.
70. Liwicki, S., Zafeiriou, S., Tzimiropoulos, G., & Pantic, M. (2012). Efficient online subspace learning with an indefinite kernel for visual tracking and recognition. *IEEE Transactions on Neural Networks and Learning Systems, 23*(10), 1624–1636.
71. Lu, J., Plataniotis, K. N., & Venetsanopoulos, A. N. (2003). Face recognition using kernel direct discriminant analysis algorithms. *IEEE Transactions on Neural Networks, 14*(1), 117–126.
72. Ma, J. (2003). Function replacement vs. kernel trick. *Neurocomputing, 50*, 479–483.
73. MacDonald, D., & Fyfe, C. (2000). The kernel self organising map. In *Proceedings of the 4th International Conference on Knowledge-Based Intelligence Engineering Systems and Allied Technologies* (Vol. 1, pp. 317–320).
74. Mangasarian, O. L., & Wild, E. W. (2007). Nonlinear knowledge in kernel approximation. *IEEE Transactions on Neural Networks, 18*(1), 300–306.
75. Mao, Q., Tsang, I. W., Gao, S., & Wang, L. (2015). Generalized multiple kernel learning with data-dependent priors. *IEEE Transactions on Neural Networks and Learning Systems, 26*(6), 1134–1148.
76. Martinez, D., & Bray, A. (2003). Nonlinear blind source separation using kernels. *IEEE Transactions on Neural Networks, 14*(1), 228–235.
77. Mercer, T. (1909). Functions of positive and negative type and their connection with the theory of integral equations. *Philosophical Transactions of the Royal Society of London Series A, 209*, 415–446.
78. Mika, S., Ratsch, G., Weston, J., Scholkopf, B., & Muller, K.-R. (1999). Fisher discriminant analysis with kernels. In *Proceedings of the IEEE Signal Processing Society Workshop on Neural Networks for Signal Processing* (pp. 41–48).
79. Muller, K. R., Mika, S., Ratsch, G., Tsuda, K., & Scholkopf, B. (2001). An introduction to kernel-based learning algorithms. *IEEE Transactions on Neural Networks, 12*(2), 181–201.
80. Nashed, M. Z., & Walter, G. G. (1991). General sampling theorem for functions in reproducing kernel Hilbert space. *Mathematics of Control Signals and Systems, 4*(4), 363–390.
81. Ogawa, H. (2009). What can we see behind sampling theorems? *IEICE Transactions on Fundamentals, E92-A*(3), 688–707.
82. Ong, C. S., Smola, A. J., & Williamson, R. C. (2005). Learning the kernel with hyperkernels. *Journal of Machine Learning Research, 6*, 1043–1071.
83. Orabona, F., Keshet, J., & Caputo, B. (2009). Bounded kernel-based online learning. *Journal of Machine Learning Research, 10*, 2643–2666.
84. Ormoneit, D., & Sen, S. (2002). Kernel-based reinforcement learning. *Machine Learning, 49*, 161–178.
85. Paiva, A. R. C., Park, I., & Principe, J. C. (2009). A reproducing kernel Hilbert space framework for spike train signal processing. *Neural Computation, 21*, 424–449.
86. Papaioannou, A., & Zafeiriou, S. (2014). Principal component analysis with complex kernel: The widely linear model. *IEEE Transactions on Neural Networks and Learning Systems, 25*(9), 1719–1726.
87. Pekalska, E., & Haasdonk, B. (2009). Kernel discriminant analysis for positive definite and indefinite kernels. *IEEE Transactions on Pattern Analysis and Machine Intelligence, 31*(6), 1017–1031.

88. Peleg, D., & Meir, R. (2009). A sparsity driven kernel machine based on minimizing a generalization error bound. *Pattern Recognition*, *42*, 2607–2614.
89. Perfetti, R., & Ricci, E. (2008). Recurrent correlation associative memories: A feature space perspective. *IEEE Transactions on Neural Networks*, *19*(2), 333–345.
90. Pokharel, P. P., Liu, W., & Principe, J. C. (2007). Kernel LMS. In *Proceedings of the IEEE International Conference on Acoustics, Speech, and Signal Processing (ICASSP), Honolulu, HI* (Vol. 3, pp. 1421–1424).
91. Qin, A. K., & Suganthan, P. N. (2004). Kernel neural gas algorithms with application to cluster analysis. In *Proceedings of the 17th International Conference on Pattern Recognition* (Vol. 4, pp. 617–620).
92. Rahimi, A., & Recht, B. (2007). Random features for large-scale kernel machines. In *Advances in Neural Information Processing Systems* (Vol. 20, pp. 1177–1184). Red Hook, NY: Curran & Associates Inc.
93. Rahimi, A., & Recht, B. (2008). Weighted sums of random kitchen sinks: Replacing minimization with randomization in learning. In *Advances in Neural Information Processing Systems* (Vol. 21, pp. 1313–1320). Red Hook, NY: Curran & Associates Inc.
94. Rakotomamonjy, A., Bach, F., Canu, S., & Grandvalet, Y. (2008). SimpleMKL. *Journal of Machine Learning Research*, *9*, 2491–2521.
95. Rodriguez-Lujan, I., Santa Cruz, C., & Huerta, R. (2011). On the equivalence of kernel Fisher discriminant analysis and kernel quadratic programming feature selection. *Pattern Recognition Letters*, *32*, 1567–1571.
96. Rosipal, R., & Trejo, L. J. (2001). Kernel partial least squares regression in reproducing kernel Hilbert spaces. *Journal of Machine Learning Research*, *2*, 97–123.
97. Ruiz, A., & Lopez-de-Teruel, P. E. (2001). Nonlinear kernel-based statistical pattern analysis. *IEEE Transactions on Neural Networks*, *12*(1), 16–32.
98. Saadi, K., Talbot, N. L. C., & Cawley, G. C. (2007). Optimally regularised kernel Fisher discriminant classification. *Neural Networks*, *20*, 832–841.
99. Scholkopf, B. (1997). *Support vector learning*. Munich, Germany: R Oldenbourg Verlag.
100. Scholkopf, B., Smola, A., & Muller, K.-R. (1998). Nonlinear component analysis as a kernel eigenvalue problem. *Neural Computation*, *10*, 1299–1319.
101. Scholkopf, B., Mika, S., Burges, C. J. C., Knirsch, P., Muller, K.-R., Scholz, M., et al. (1999). Input space versus feature space in kernel-based methods. *IEEE Transactions on Neural Networks*, *10*(5), 1000–1017.
102. Shashua, A. (1999). On the relationship between the support vector machine for classification and sparsified Fisher's linear discriminant. *Neural Processing Letters*, *9*(2), 129–139.
103. Smola, A. J., Mangasarian, O., & Scholkopf, B. (1999). *Sparse kernel feature analysis*. Technical report 99-03. Madison, WI: Data Mining Institute, University of Wisconsin.
104. Song, G., & Zhang, H. (2011). Reproducing kernel Banach spaces with the l_1 Norm II: Error analysis for regularized least square regression. *Neural Computation*, *23*, 2713–2729.
105. Sonnenburg, S., Ratsch, G., Schafer, C., & Scholkopf, B. (2006). Large scale multiple kernel learning. *Journal of Machine Learning Research*, *7*, 1531–1565.
106. Subrahmanya, N., & Shin, Y. C. (2010). Sparse multiple kernel learning for signal processing applications. *IEEE Transactions on Pattern Analysis and Machine Intelligence*, *32*(5), 788–798.
107. Suykens, J. A. K., Van Gestel, T., Vandewalle, J., & De Moor, B. (2003). A support vector machine formulation to PCA analysis and its kernel version. *IEEE Transactions on Neural Networks*, *14*(2), 447–450.
108. Suzuki, T., & Tomioka, R. (2011). SpicyMKL: A fast algorithm for multiple kernel learning with thousands of kernels. *Machine Learning*, *85*, 77–108.
109. Tanaka, A., Imai, H., & Miyakoshi, M. (2010). Kernel-induced sampling theorem. *IEEE Transactions on Signal Processing*, *58*(7), 3569–3577.
110. Teh, C. S., & Lim, C. P. (2006). Monitoring the formation of kernel-based topographic maps in a hybrid SOM-kMER model. *IEEE Transactions on Neural Networks*, *17*(5), 1336–1341.

111. Teh, C. S., & Lim, C. P. (2008). An artificial neural network classifier design based-on variable kernel and non-parametric density estimation. *Neural Processing Letters*, *27*, 137–151.
112. van Hulle, M. M. (1998). Kernel-based equiprobabilistic topographic map formation. *Neural Computation*, *10*(7), 1847–1871.
113. Vincent, P., & Bengio, Y. (2002). Kernel matching pursuit. *Machine Learning*, *48*, 165–187.
114. Vishwanathan, S. V. N., Sun, Z., Ampornpunt, N., & Varma, M. (2010). Multiple kernel learning and the SMO algorithm. *Advances in neural information processing systems*. Cambridge, MA: MIT Press.
115. Wang, L. (2008). Feature selection with kernel class separability. *IEEE Transactions on Pattern Analysis and Machine Intelligence*, *30*(9), 1534–1546.
116. Washizawa, Y. (2012). Adaptive subset kernel principal component analysis for time-varying patterns. *IEEE Transactions on Neural Networks and Learning Systems*, *23*(12), 1961–1973.
117. Wolf, L., & Shashua, A. (2003). Learning over sets using kernel principal angles. *Journal of Machine Learning Research*, *4*, 913–931.
118. Xiao, S., Tan, M., Xu, D., & Dong, Z. Y. (2016). Robust kernel low-rank representation. *IEEE Transactions on Neural Networks and Learning Systems*, *27*(11), 2268–2281.
119. Xiong, H., Swamy, M. N. S., & Ahmad, M. O. (2005). Optimizing the kernel in the empirical feature space. *IEEE Transactions on Neural Networks*, *16*(2), 460–474.
120. Xu, X., Tsang, I. W., & Xu, D. (2013). Soft margin multiple kernel learning. *IEEE Transactions on Neural Networks and Learning Systems*, *24*(5), 749–761.
121. Xu, Y., & Zhang, H. (2007). Refinable kernels. *Journal of Machine Learning Research*, *8*, 2083–2120.
122. Xu, Z., Huang, K., Zhu, J., King, I., & Lyua, M. R. (2009). A novel kernel-based maximum a posteriori classification method. *Neural Networks*, *22*, 977–987.
123. Yang, C., Wang, L., & Feng, J. (2008). On feature extraction via kernels. *IEEE Transactions on Systems, Man, and Cybernetics Part B*, *38*(2), 553–557.
124. Yang, H., Xu, Z., Ye, J., King, I., & Lyu, M. R. (2011). Efficient sparse generalized multiple kernel learning. *IEEE Transactions on Neural Networks*, *22*(3), 433–446.
125. Yang, J., Frangi, A. F., Yang, J.-Y., Zhang, D., & Jin, Z. (2005). KPCA plus LDA: A complete kernel Fisher discriminant framework for feature extraction and recognition. *IEEE Transactions on Pattern Analysis and Machine Intelligence*, *27*(2), 230–244.
126. Ye, J., Ji, S., & Chen, J. (2008). Multi-class discriminant kernel learning via convex programming. *Journal of Machine Learning Research*, *9*, 719–758.
127. Yin, H., & Allinson, N. (2001). Self-organising mixture networks for probability density estimation. *IEEE Transactions on Neural Networks*, *12*, 405–411.
128. Yoshino, H., Dong, C., Washizawa, Y., & Yamashita, Y. (2010). Kernel Wiener filter and its application to pattern recognition. *IEEE Transactions on Neural Networks*, *21*(11), 1719–1730.
129. You, D., Hamsici, O. C., & Martinez, A. M. (2011). Kernel optimization in discriminant analysis. *IEEE Transactions on Pattern Analysis and Machine Intelligence*, *33*(3), 631–638.
130. Zafeiriou, S., & Petrou, M. (2010). Nonlinear nonnegative component analysis algorithms. *IEEE Transactions on Image Processing*, *19*, 1050–1066.
131. Zhang, T. (2003). Leave-one-out bounds for kernel methods. *Neural Computation*, *15*, 1397–1437.
132. Zhang, D. Q., & Chen, S. C. (2003). Clustering incomplete data using kernel-based fuzzy C-means algorithm. *Neural Processing Letters*, *18*, 155–162.
133. Zhang, B., Zhang, H., & Ge, S. S. (2004). Face recognition by applying wavelet subband representation and kernel associative memory. *IEEE Transactions on Neural Networks*, *15*(1), 166–177.
134. Zhang, M., Wang, X., Chen, X., & Zhang, A. (2018). The kernel conjugate gradient algorithms. *IEEE Transactions on Signal Processing*, *66*(16), 4377–4387.
135. Zheng, W., Zhao, L., & Zou, C. (2005). Foley-Sammon optimal discriminant vectors using kernel approach. *IEEE Transactions on Neural Networks*, *16*(1), 1–9.

136. Zheng, W., Zhou, X., Zou, C., & Zhao, L. (2006). Facial expression recognition using kernel canonical correlation analysis (KCCA). *IEEE Transactions on Neural Networks*, *17*(1), 233–238.
137. Zheng, W., Lin, Z., & Tang, X. (2010). A rank-one update algorithm for fast solving kernel Foley-Sammon optimal discriminant vectors. *IEEE Transactions on Neural Networks*, *21*(3), 393–403.
138. Zhu, J., & Hastie, T. (2002). Kernel logistic regression and the import vector machine. *Advances in neural information processing systems* (Vol. 14). Cambridge, MA: MIT Press.

Chapter 21
Support Vector Machines

21.1 Introduction

SVM [14, 199] is a three-layer feedforward network. It implements the structural risk minimization (SRM) principle that minimizes the upper bound of the generalization error. This induction principle is based on the fact that the generalization error is bounded by the sum of a training error and a confidence-interval term that depends on the VC dimension. Generalization errors of SVMs are not related to the input dimensionality, but to the margin with which it separates the data. Instead of minimizing the training error, SVM purports to minimize an upper bound of the generalization error and maximizes the margin between a separating hyperplane and the training data.

The goal of SVM is to minimize the VC dimension by finding the optimal hyperplane between classes, with the maximal margin, where the margin is defined as the distance of the closest point in each class to the separating hyperplane. It has a general-purpose linear learning algorithm and a problem-specific kernel that computes the inner product of input data points in a feature space, where the projections of the training examples are always linearly separable in the feature space. The hippocampus, a brain region critical for learning and memory processes, has been reported to possess pattern separation function similar to SVM [6].

Similar to biological systems, an SVM ignores typical examples but pays attention to borderline cases and outliers. SVM is not obviously applicable to the brain. Bio-SVM [82] is a biologically feasible SVM. An unstable associative memory oscillates between support vectors and interacts with a feedforward classification pathway. Instant learning of surprising events and offline tuning of support vector weights train the system. Emotion-based learning, forgetting trivia, sleep, and brain oscillations are phenomena that agree with the Bio-SVM model, and a mapping to the olfactory system is suggested.

SVM is a universal approximator for various kernels [69]. It is popular for classification, regression, and clustering. One of the main features of SVM is the absence of local minima. SVM is defined in terms of a subset of the learning data, called *support*

K.-L. Du and M. N. S. Swamy, *Neural Networks and Statistical Learning*,
https://doi.org/10.1007/978-1-4471-7452-3_21

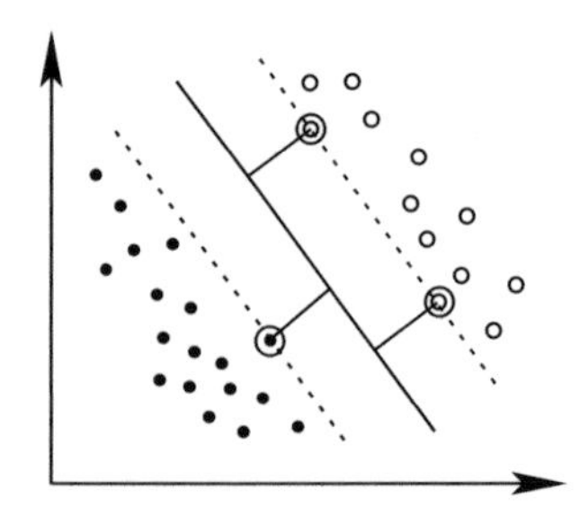

Fig. 21.1 An illustration of the hyperplane in the two-dimensional feature space of SVM: The margin is defined by the distance between the hyperplane and the nearest of the examples in the two classes. Those examples in circles are support vectors, against which the margin pushes

vectors. It is a sparse representation of the training data, and allows the extraction of a condensed dataset based on the support vectors.

Under mild assumptions, the solution to support vector regression (SVR) can be written as a linear combination of kernel functions. This is known as the representer theorem, given by Theorem 20.1. From (20.6) only samples $\boldsymbol{x}_i$ with $\alpha_i \neq 0$ have influence on $f(\boldsymbol{x})$; such samples are called *support vectors*. If $k(\cdot)$ is a polynomial kernel it is easy to see that the representation (20.6) is not unique whenever the sample size N is too large.

The representer theorem is generalized to differentiable loss functions [39] and even arbitrary monotonic ones [162]. A quantitative representer theorem has been proven in [183] without using the dual problem when convex loss functions are considered.

A classifier is called *universally consistent* if the risks of its decision functions converge to the Bayes risk in probability for all underpinning distributions. Lower (asymptotical) bounds on the number of support vectors is established in [183].

A relationship between SVM and a sparse approximation scheme that resembles the basis pursuit denoising algorithm is given in [59]. If the data are noiseless, the modified basis pursuit denoising method [59] is equivalent to SVM. The SVM technique can also be derived in the framework of regularization theory, establishing a connection between SVM, sparse approximation, and regularization theory [59].

21.2 SVM Model

SVM was originally proposed for binary classification. Let $f : \boldsymbol{x} \rightarrow \{-1, 1\}$ be an unknown function and $\mathcal{D} = \{(\boldsymbol{x}_i, y_i) | i = 1, \ldots, N\} \subset \mathcal{R}^n \times \{-1, 1\}$ be the training example set. SVM aims to find the function f with the optimal hyperplane that maximizes the margin between the examples of two different classes, as illustrated in Fig. 21.1.

The optimal hyperplane for linear SVM can be constructed by solving the following primal QP problem [199]:

$$\min E_0(\boldsymbol{w}, \boldsymbol{\xi}) = \frac{1}{2}\|\boldsymbol{w}\|^2 + C\sum_{i=1}^{N}\xi_p \tag{21.1}$$

subject to

$$y_p\left(\boldsymbol{w}^T\boldsymbol{x}_p + \theta\right) \geq 1 - \xi_p, \quad p = 1, \ldots, N, \tag{21.2}$$

$$\xi_p \geq 0, \quad p = 1, \ldots, N, \tag{21.3}$$

where $\boldsymbol{\xi} = (\xi_1, \cdots, \xi_N)^T$, ξ_p are slack variables, $\boldsymbol{w}$ and θ are the weight and bias parameters for determining the hyperplane, $y_p \in \{-1, +1\}$ is the described output of the classifier, and C is a regularization parameter that trades off wide margin with a small number of margin failures. Typically, C is optimized by employing statistical model selection procedures, e.g., cross-validation.

In the above SVM formulation, the slack variable ξ_p in (21.1) corresponds to a loss function using L_1-norm. L_1-SVM solves the following unconstrained optimization problem:

$$\min_{\boldsymbol{w}} J(\boldsymbol{w}) = \frac{1}{2}\boldsymbol{w}^T\boldsymbol{w} + C\sum_{j=1}^{l}\max(1 - y_j\boldsymbol{w}^T\boldsymbol{x}_j, 0). \tag{21.4}$$

L_2-SVM uses the sum of squared losses, and solves

$$\min_{\boldsymbol{w}} J(\boldsymbol{w}) = \frac{1}{2}\boldsymbol{w}^T\boldsymbol{w} + C\sum_{j=1}^{l}\left[\max(1 - y_j\boldsymbol{w}^T\boldsymbol{x}_j, 0)\right]^2. \tag{21.5}$$

L_1-SVMs are more popularly used than L_2-SVMs because they usually yield classifiers with a much less number of support vectors, thus leading to better classification speed. SVM is related to regularized logistic regression, which solves

$$\min_{\boldsymbol{w}} J(\boldsymbol{w}) = \frac{1}{2}\boldsymbol{w}^T\boldsymbol{w} + C\sum_{j=1}^{l}\log(1 + e^{-y_j\boldsymbol{w}^T\boldsymbol{x}_j}). \tag{21.6}$$

By applying the Lagrange multiplier method and replacing $\boldsymbol{x}_p^T\boldsymbol{x}$ by the kernel function $k\left(\boldsymbol{x}_p, \boldsymbol{x}\right)$, the ultimate objective of SVM learning is to find α_p, $p = 1, \ldots, N$, so as to minimize the dual quadratic form [199]

$$E_{\text{SVM}} = \frac{1}{2}\sum_{p=1}^{N}\sum_{i=1}^{N} y_p y_i k\left(\boldsymbol{x}_p, \boldsymbol{x}_i\right)\alpha_p\alpha_i - \sum_{p=1}^{N}\alpha_p \tag{21.7}$$

subject to

$$\sum_{p=1}^{N} y_p \alpha_p = 0, \tag{21.8}$$

$$0 \le \alpha_p \le C, \quad p = 1, \ldots, N, \tag{21.9}$$

where α_p is the weight for the kernel corresponding to the pth example. The kernel function $k\left(\boldsymbol{x}_p, \boldsymbol{x}\right) = \boldsymbol{\phi}^T\left(\boldsymbol{x}_p\right)\boldsymbol{\phi}\left(\boldsymbol{x}\right)$, where the form of $\boldsymbol{\phi}(\cdot)$ is implicitly defined by the choice of the kernel function and does not need to be given. When $k(\cdot)$ is a linear function, that is, $k\left(\boldsymbol{x}_p, \boldsymbol{x}\right) = \boldsymbol{x}_p^T\boldsymbol{x}$, SVM reduces to linear SVM [198]. The popular Gaussian and polynomial kernels are, respectively, given by (20.3) and (20.2).

The SVM output is given by

$$y(\boldsymbol{x}) = \text{sign}(\boldsymbol{w}^T\boldsymbol{\phi}(\boldsymbol{x}) + \theta), \tag{21.10}$$

where θ is a threshold. After manipulation, the SVM output gives a classification decision as

$$y(\boldsymbol{x}) = \text{sign}\left(\sum_{p=1}^{N} \alpha_p y_p k(\boldsymbol{x}_p, \boldsymbol{x}) + \theta\right). \tag{21.11}$$

The QP problem given by (21.7) through (21.9) will terminate when all of the KKT conditions are fulfilled

$$\begin{cases} y_p u_p \ge 1, \; \alpha_p = 0 \\ y_p u_p = 1, \; 0 < \alpha_p < C \\ y_p u_p \le 1, \; \alpha_p = C \end{cases}, \tag{21.12}$$

where u_p is the SVM output for the pth example. Those patterns with nonzero α_p are the support vectors, which lie on the margin.

In two-class L_2-SVM, all the slack variables ξ_i's receive the same penalty factor C. For imbalanced datasets, a common remedy is to use different C's for the two classes. In the more general case, each training pattern can have its own penalty factor C_i.

The generalization ability of SVM depends on the geometrical concept of span of support vectors [201]. The value of the span is proved to be always smaller than the diameter of the smallest sphere containing the support vectors, used in previous bounds [200]. The prediction of the test error given by the span is very accurate and has direct application in the choice of the optimal parameters of SVM. Bounds on the expectation of error from SVM is derived from the leave-one-out estimator, which is an unbiased estimate of the probability of test error [201]. An important approach for efficient SVM model selection is to use differentiable bounds of the leave-one-out error.

The VC dimension of hyperplanes with margin ρ is less than $D^2/4\rho^2$, where D is the diameter of the smallest sphere containing the training points [199]. SVM can have very large (even infinite) VC dimension by computing the VC dimension for homogeneous polynomial and Gaussian RBF kernels [18]. This means that an SVM has very strong classification/regression capacity.

21.2.1 SVM Versus Neural Networks

Generative machine learning approaches, such as the naive Bayes classifier, LDA, and neural networks, require computations of underlying conditional probability distributions over the data. Solutions are highly dependent on the initialization and termination criteria. The generative assumptions and parameter estimations simplify the learning process.

The discriminative approach directly optimizes the prediction accuracy instead of learning the underlying distribution. A discriminant classification function takes a data point $\boldsymbol{x}$ and assigns it to one of the different classes. From a geometric perspective, learning a classifier is equivalent to finding the equation for a multidimensional surface that best separates the different classes in the feature space.

SVM is a discriminative technique. It solves the convex optimization problem analytically, and always returns the same optimal hyperplane parameter. The separating hyperplane for two-class classification obtained by SVM is shown to be equivalent to the solution obtained by LDA on the set of support vectors [169].

Unlike neural networks, SVM does not suffer from the classical local minima, the curse of dimensionality and overfitting problems. By minimizing the structural risk rather than the empirical risk, as in the case of neural networks, SVM avoids overfitting. Neural networks control model complexity by limiting the feature set, while SVM automatically determines the model complexity by selecting the number of support vectors.

SVM has three attractive features:

- **SVM is a sparse technique**. Once the model parameters are identified, SVM prediction depends only on a subset of these training instances, called *support vectors*, which define the margins of the hyperplanes. The complexity of the classification task with SVM depends on the number of support vectors rather than the dimensionality of the input space.
- **SVM is a kernel technique**. The SVM optimization phase will entail learning only a linear discriminant surface in the mapped space.
- **SVM is a maximum-margin separator**. Beyond minimizing the error or a cost function based on the training datasets, the hyperplane needs to be situated such that it is at a maximum distance from the different classes.

21.3 Solving the Quadratic Programming Problem

L_1-norm soft-margin SVM, also called quadratic programming (QP) SVM, was introduced with polynomial kernels in [14] and with general kernels in [38]. Linear programming (LP) SVM is efficient and performs even better than QP-SVM for some purposes because of its linearity and flexibility for large datasets. An error analysis shows that the convergence behavior of LP-SVM is almost the same as that of QP-SVM [210]. By employing the L_1 or L_∞-norm in maximizing margins, SVMs result in an LP problem that requires a lower computational load compared to SVMs with L_2-norm.

The use of kernel mapping transforms SVM learning into a quadratic optimization problem, which has one global solution. Many general optimization tools such as CPLEX, LOQO, MATLAB linprog, and quadprog capable of solving linear and quadratic programs are derived from SVMs. However, given N training patterns, a naive implementation of the QP solver takes $O(N^3)$ training time and at least $O(N^2)$ space.

General convex QPs are typically solved by an interior-point method or an active-set method. If the Hessian $\mathbf{Q}$ of an objective function and/or the constraint matrix of the QP problem is large and sparse, then an interior-point method is usually selected. If the problem is of moderate size but the matrices are dense, then an active-set method is preferable. In SVM problems the $\mathbf{Q}$ matrix is typically dense. Thus, large SVM problems present a challenge for both the approaches. For some classes of SVMs, $\mathbf{Q}$ is dense but low-rank; in such a case, one can adapt an interior-point method to work very efficiently [51, 52]. However, if the rank of $\mathbf{Q}$ is high, an active-set method seems to be suitable.

The simplex method for LP problems is a traditional active-set method. It has very good practical performance. The idea is also used for solving QP problems. The active-set methods can benefit very well from warm starts, while the interior-point methods cannot. For instance, if some additional labeled training data become available, the old optimal solution is used as a starting point for the active-set algorithm and the new optimal solution is typically obtained within a few iterations. An active-set algorithm called SVM-QP [159] solves the convex QP problem by using the simplex method for convex quadratic problems. SVM-QP has an overall performance better than that of SVMlight, and has identical generalization properties. In addition, SVM-QP has better theoretical properties and it naturally, and almost without change, extends to incremental mode. This method fixes, at each iteration, all variables in the current dual active-set at their current values (0 or C), and then solves the reduced dual problem.

The most common approach to large SVM problems is to use a restricted active-set method, such as chunking [14] or decomposition [85, 146, 152], where at each iteration only a small number of variables are allowed to be varied. These methods tend to have slow convergence when getting closer to the optimal solution. Moreover, their performance is sensitive to the changes in the chunk size and there is no good way of predicting a good choice for the size of the chunks. A full active-set method avoids these disadvantages. Active-set methods for SVM were used in [72] for generating the entire regularization path of the cost parameter for standard L_2-SVM.

21.3.1 Chunking

The chunking technique [198, 200] breaks down a large QP problem into a series of smaller QP subproblems, whose ultimate goal is to identify all nonzero α_p, since training examples with $\alpha_p = 0$ do not change the solution. The chunking algorithm starts with an arbitrary subset (chunk of data, working set) which can fit in the memory and solves the optimization problem on it by the general optimizer, and trains an initial SVM. Support vectors remain in the chunk while other points are discarded and replaced by a new working set with gross violations of KKT conditions. Then, the SVM is retrained and the whole procedure is repeated. Chunking suffers from the problem that the entire set of support vectors that have been identified will still need to be trained at the end of the training process.

Chunking is based on the sparsity of SVM's solution, and support vectors actually take up a small fraction of the whole dataset. There may be many more active candidate support vectors during the optimization process than the final ones so that their size can go beyond the chunking space. However, the resultant kernel matrix may still be too large to fit into memory. The method of selecting a new working set by evaluating KKT conditions without efficient kernel caching may lead to a high computational cost. Standard projected conjugate gradient chunking algorithm scales somewhere between $O(N)$ and $O(N^3)$ in the training set size N [36, 85].

21.3.2 Decomposition

A large QP problem can be decomposed into smaller QP subproblems [146, 152, 202]. Each subproblem is initialized with the results of the previous subproblem. However, decomposition requires a numerical QP algorithm such as the projected conjugate gradient algorithm. For decomposition approach, the time complexity can be reduced from $O(N^3)$ to $O(N^2)$.

A basic strategy commonly used in decomposition methods is to execute two operations repeatedly until some optimality condition is satisfied; one is to select q variables among l and the other is to minimize the objective function by updating only the selected q variables. The set of q variables selected for updating at each step is called the working set. Only a fixed-size subset (working set) of the training data are optimized each time, while the variables corresponding to the other patterns are frozen [146]. Sequential minimal optimization (SMO) [152] selects only two variables for the working set in each iteration. SVMlight [85] sets the size of the working set to any even number q.

SMO selects at each iteration a working set of size exactly two. Each small QP problem involves only two α_p and is solved analytically. The first variable is chosen among points that violate the KKT conditions, while the second variable is chosen so as to have a large increase in the dual objective. This two-variable joint optimization process is repeated until the loose KKT conditions are fulfilled for all training patterns. This avoids using QP optimization as an inner loop. The amount of memory required for SMO is $O(N)$. SMO has a computational complexity of somewhere between $O(N)$ and $O(N^2)$, and it is faster than projected conjugate gradient chunking. The computational complexity of SMO is dominated by SVM evaluation, thus SMO is very fast for linear SVMs and sparse datasets [152].

The performance of SMO is enhanced in [98] by replacing one-thresholded parameters with two-thresholded parameters, since the pair of patterns chosen for optimization is theoretically determined by two-thresholded parameters. The two-parameter SMO algorithm performs significantly faster than SMO. Three-parameter SMO [118], as a natural generalization to the two-parameter SMO algorithm, jointly optimizes three chosen parameters in a manner similar to that of the two-parameter SMO. It outperforms two-parameter SMO significantly in both the executing time and the computational complexity for classification as well as regression benchmarks.

SMO algorithms have a strict decrease of the objective function if and only if the working set is a violating pair [78]. However, the use of generic violating pairs as working sets is not sufficient to guarantee convergence properties of the sequence generated by a decomposition algorithm. As each iteration only involves two variables in the optimization, SMO has slow convergence. Nevertheless, as each iteration is computationally simple, an overall speedup is often observed in practice. Decomposition methods with α-seeding are extremely useful for solving a sequence of linear SVMs with more data than attributes [91]. Analysis shows why α-seeding is much more effective for linear than nonlinear SVMs [91].

Working-set selection is an important step in decomposition methods for training SVMs. A popular way to select the working set is via the maximal-violating pair. An SMO-type algorithm using maximal-violating pairs as working sets is usually called maximal-violating-pair algorithm.

SMO is used in LIBSVM (http://www.csie.ntu.edu.tw/~cjlin/libsvm/) [20]. In LIBSVM ver. 2.8, a new working-set selection partially exploits the second-order information, thus increasing the convergence speed but also getting a moderate increase of the computational cost with respect to standard selections [49]. This achieves a theoretical property of linear convergence. A similar working-set selection strategy, called hybrid maximum-gain working-set selection [60], minimizes the number of kernel evaluations per iteration by the avoidance of cache misses in the decomposition algorithm. It reselects almost always one element of the previous working set, and at most one matrix row needs to be computed in every iteration. Hybrid maximum-gain working-set selection converges to an optimal solution. In contrast, for small problems LIBSVM ver. 2.8 is faster.

In SVMlight [85], a good working set is selected by finding the steepest feasible direction of descent with q nonzero elements. The q variables that correspond to these elements compose the working set. When q is equal to 2, the selected working set corresponds to the optimal pair in a modified SMO method [20]. SVMlight caches q rows of kernel matrix (row caching) to avoid kernel reevaluations and LRU (least recently used) is applied to update the rows in the cache. However, when the size of the training set is very large, the number of cached rows, which is dictated by the user, becomes small due to limited memory. The number of active variables is not large enough to achieve fast optimization. A generalized maximal-violating-pair policy for the working-set selection and a numerical solver for the inner QP subproblems are needed. For small working sets ($q = O(10)$), SVMlight often exhibits comparable performance with LIBSVM.

Sigmoidal kernels may lead to nonpositive-semi-definite kernel matrices which are required by the SVM framework to obtain the solution by means of QP techniques. SMO decomposition is used to solve non-convex dual problems, leading to the software LIBSVM, which is able to provide a solution with sigmoidal kernels. An improved SVM with a sigmoidal kernel, called support vector perceptron [138], provides very accurate results in many classification problems, providing maximal margin solutions when classes are separable, and also producing very compact architectures comparable to MLPs. In contrast, LIBSVM with sigmoidal kernel has a much larger architecture.

SimpleSVM [202] is a scale-up method. At each iteration, a point violating the KKT conditions is added to the working set by using rank-one update on the kernel matrix. However, storage is still a problem when SimpleSVM is applied to large dense kernel matrices. SimpleSVM divides the database into three groups: those groups for the nonbounded support vectors ($0 < \alpha_w < C$), for the bounded points—misclassified or in the margins ($\alpha_C = C$) and for the nonsupport vectors ($\alpha_0 = 0$). SimpleSVM solves an optimization problem such that the optimality condition leads to a linear system with α_w as unknown.

Many kernel methods can be equivalently formulated as minimum enclosing ball problems in computational geometry [194]. By adopting an efficient approximate minimum enclosing ball algorithm, one obtains provably approximate optimal solutions with the idea of core sets. The core set in core vector machine plays a similar role as the working set in other decomposition algorithms. Kernel methods (including the soft-margin one-class and two-class SVMs) are formulated as equivalent minimum enclosing ball problems, and then approximately optimal solutions are efficiently obtained by using core sets. Core vector machine [194] can be used with nonlinear kernels and has a time complexity $O(N)$ and a space complexity that is independent of N. Compared with SVM implementations, it is as accurate but is much faster and can handle much larger datasets. On relatively small datasets where $N \ll 2/\varepsilon$, SMO can be faster. Trade-off between efficiency and approximation quality is adjusted by ε, and $\varepsilon = 10^{-6}$ is acceptable for most tasks. The minimum enclosing ball is equivalent to the hard-margin support vector data description (SVDD) [187]. Core vector machine is similar to decomposition algorithms, but subset selection is much simpler. Moreover, while decomposition algorithms allow training patterns to join and leave the working set multiple times, patterns once recruited as core vectors by core vector machine will remain there during the whole training process. Core vector machine solves the QP on the core set only using SMO and thus obtains α. The stopping criterion is analogous to that for ν-SVM [22]. Core vector machine critically requires the kernel function $k(\boldsymbol{x}, \boldsymbol{x}) =$ constant for any $\boldsymbol{x}$. This condition is satisfied for the isotropic kernel (e.g., Gaussian kernel), the dot product kernel (e.g., polynomial kernel) with normalized inputs, and any normalized kernel.

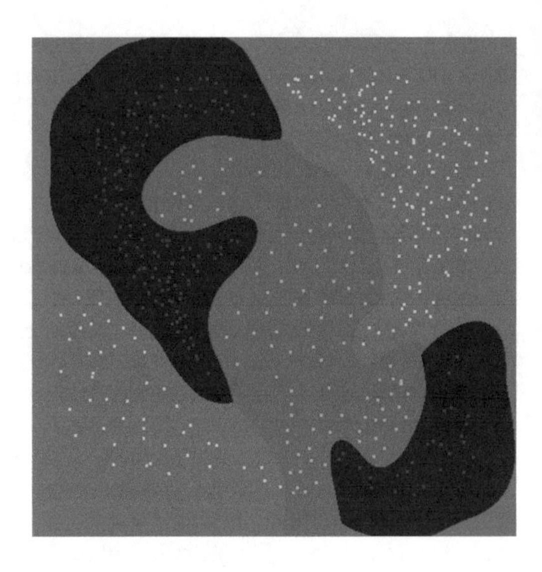

Fig. 21.2 SVM boundary for a three-class classification problem in two-dimensional space

Core vector machine does not converge toward the solution for all hyperparameters [119]. It also requires that the kernel methods do not have a linear term in their dual objectives so as to make them minimum enclosing ball problems; it has been shown in [194] that this holds for the one-class SVM [163] and two-class SVM, but not for SVR. Generalized core vector machine [195] introduces the center-constrained minimum enclosing ball problem to make SVR also a minimum enclosing ball problem. It can be used with any linear/nonlinear kernel and can also be applied to kernel methods such as SVR, the ranking SVM, and two-class SVM for imbalanced data. It has the same asymptotic time complexity and space complexity as those of core vector machine. It has good performance, but is faster and produces few support vectors on very large datasets.

Example 21.1 We use LIBSVM to classify 463 samples belonging to three classes in two-dimensional space. By selecting RBF with $e^{-\gamma\|\boldsymbol{u}-\boldsymbol{v}\|^2}$ with $\gamma = 200$ and $C = 20$, all the samples are correctly classified. The total number of supporting vectors is 172. The classification result is shown in Fig. 21.2.

Example 21.2 We revisit the Iris dataset. By setting every fifth sample of the dataset as a test sample and the remaining as training samples, we implement LIBSVM for classification. By selecting a polynomial $(\gamma \boldsymbol{u}^T \boldsymbol{v} + c_0)^d$, with $\gamma = 10$, $d = 3$, $c_0 = 0$, $C = 2$, we get a testing accuracy of 96.6667 % (29/30). The total number of supporting vectors is 13.

21.3.3 Convergence of Decomposition Methods

From the theoretical point of view, the policy for updating the working set plays a crucial role since it can guarantee a strict decrease of the objective function at each step [78]. The global convergence property of decomposition methods and SMO algorithms for classification has been clarified in [29, 96, 115, 116]. The convergence properties of SVMlight algorithm have been proved in [115, 116] under suitable convexity assumptions.

In case of working sets of minimal size 2, a proper selection via the maximal-violating-pair principle is sufficient to ensure asymptotic convergence of the decomposition scheme [29, 96, 116]. For larger working sets, convergence proofs are available under a further condition which ensures that the distance between two successive approximations tends to zero [115]. The generalized SMO algorithm has been proved to terminate within a finite number of iterations under a prespecified stopping condition and tolerance [96]. A simple asymptotic convergence proof of the linear convergence of SMO-type decomposition methods under a general and flexible way of choosing the two-element working set are given in [29]. The generalized SMO algorithm and SVMlight have been proved in [184] to have the global convergence property.

SVM is well understood when using conditionally positive-definite kernel functions. However, in practice, non-conditionally positive-definite kernels arise in SVMs. Using these kernels causes loss of convexity. LIBSVM software does converge for indefinite kernels such as the sigmoidal kernel. A geometric interpretation of SVMs with indefinite kernel functions is given in [67]. Such SVMs are shown to be optimal hyperplane classifiers not by margin maximization, but by minimization of distances between convex hulls in pseudo-Euclidean spaces. They are minimum distance classifiers with respect to certain points from the convex hulls of embedded training points.

21.4 Least Squares SVMs

Least squares SVM (LS-SVM) [178, 180] is a variant of SVM which simplifies the training process of SVM to a great extent. It is introduced as a reformulation to SVM by replacing the inequality constraints with equality ones. It obtains an analytical solution directly from solving a set of linear equations instead of a QP problem.

The unknown parameters in the decision function $f(\boldsymbol{x})$, namely α_p and θ, can be solved through the primal problem:

$$\min_{w,b,\xi} \frac{1}{2}\boldsymbol{w}^T\boldsymbol{w} + \frac{C}{2}\sum_{p=1}^{N}\xi_p^2 \tag{21.13}$$

subject to

$$y_p(\boldsymbol{w}^T\phi(\boldsymbol{x}_p)+\theta)=1-\xi_p, \quad p=1,\ldots,N, \tag{21.14}$$

where $\phi(\cdot)$ is a linear or nonlinear function which maps the input space into a higher dimensional feature space, $\boldsymbol{w}$ is a weight vector to be determined, C is a regularization constant and ξ_p's are slack variables. The Lagrangian is obtained as

$$L(\boldsymbol{w},b,\xi,\alpha)=\frac{1}{2}\boldsymbol{w}^T\boldsymbol{w}+\frac{C}{2}\sum_{p=1}^{N}\xi_p^2-\sum_{p=1}^{N}\alpha_p\left[y_p(\boldsymbol{w}^T\phi(\boldsymbol{x}_p)+\theta)-1+\xi_p\right], \tag{21.15}$$

where the Lagrange multipliers α_p are used as the same α_p in $f(\boldsymbol{x})$. The necessary conditions for the optimality are given by the KKT conditions:

$$\frac{\partial L}{\partial \boldsymbol{w}}=0 \Longrightarrow \boldsymbol{w}=\sum_{p=1}^{N}\alpha_p y_p\phi(\boldsymbol{x}_p), \tag{21.16}$$

$$\frac{\partial L}{\partial \theta}=0 \Longrightarrow \sum_{p=1}^{N}\alpha_p y_p=0, \tag{21.17}$$

$$\frac{\partial L}{\partial \xi_p}=0 \Longrightarrow \alpha_p=C\xi_p, \quad p=1,\ldots,N, \tag{21.18}$$

$$\frac{\partial L}{\partial \alpha_p}=0 \Longrightarrow y_p(\boldsymbol{w}^T\phi(\boldsymbol{x}_p)+\theta-1+\xi_p=0, \quad p=1,\ldots,N. \tag{21.19}$$

Eliminating $\boldsymbol{w}$ and ξ_p leads to

$$\begin{bmatrix} 0 & -\boldsymbol{Y}^T \\ \boldsymbol{Y} & \mathbf{Q}+\mathbf{C}^{-1}\mathbf{I} \end{bmatrix}\begin{bmatrix} \theta \\ \alpha \end{bmatrix}=\begin{bmatrix} 0 \\ \mathbf{1} \end{bmatrix}, \tag{21.20}$$

where $\boldsymbol{Y}=(y_1,\ldots,y_N)^T$, $q_{ij}=y_iy_j\phi(\boldsymbol{x}_i)^T\phi(\boldsymbol{x}_j)$, $\mathbf{1}=(1,\ldots,1)^T$, and $\boldsymbol{\alpha}=(\alpha_1,\ldots,\alpha_N)^T$. The function $\phi(\boldsymbol{x})$ satisfies Mercer's theorem: $\phi(\boldsymbol{x}_i)^T\phi(\boldsymbol{x}_j)=k(\boldsymbol{x}_i,\boldsymbol{x}_j)$.

The explicit solution to (21.20) is given by

$$\theta=\frac{\boldsymbol{Y}^T\mathbf{A}^{-1}\mathbf{1}}{\boldsymbol{Y}^T\mathbf{A}^{-1}\boldsymbol{Y}}, \quad \alpha=\mathbf{A}^{-1}(\mathbf{1}-\theta\boldsymbol{Y}), \tag{21.21}$$

where $\mathbf{A}=\mathbf{Q}+\mathbf{C}^{-1}\mathbf{I}$. The LS-SVM output for classification is given by (21.11).

LS-SVM obtains good performance on classification problems. However, the computational complexity of LS-SVM usually scales $O(N^3)$ for N samples, and the solution of LS-SVM lacks sparseness. There are some fast algorithms for LS-

SVM, such as CG algorithm [34, 179], SMO algorithm [97], and coordinate-descent algorithm [112]. These algorithms achieve low complexity, but their solutions are not sparse.

Fast sparse approximation for LS-SVM (http://homes.cs.washington.edu/~lfb/software/FSALSSVM.htm) [84] is a fast greedy algorithm. It iteratively builds the decision function by adding one basis function from a kernel-based dictionary at a time based on a flexible and stable ε-insensitive stopping criterion. A probabilistic version of the algorithm further improves its speed by employing a probabilistic speedup scheme.

Pruning LS-SVM

In [180], a sparse LS-SVM is constructed by deleting training examples associated with the smallest magnitude $\alpha(i)$ term that is proportional to the training error e_i. It recursively prunes the support vectors which are farthest from the decision boundary until the performance degrades. This algorithm is refined in [42]. Choosing the smallest $\alpha(i)$ does not necessarily result in the smallest change in the training error when the parameters of the algorithm are updated. A pruning method based on minimizing the output error is used. The algorithm uses a pruning method with no regularization ($\gamma = \infty$), leading to inversion of a singular matrix. A procedure of pruning with regularization (γ finite and nonzero) is implemented in [104] to make the data matrix nonsingular; it uses a selective window algorithm that is computationally more efficient, as it adds and deletes training examples.

Pruning in LS-SVM is investigated using an SMO-based pruning method [217]. It requires solving a set of linear equations for pruning each sample, causing a large computational cost.

The pruning methods [42, 104, 217] have to solve a set of linear equations many times, causing a huge computational cost.

Enlightened by incremental and decremental learning [19], an adaptive pruning algorithm for LS-SVM without solving primal non-sparse LS-SVM is developed in [215], based on a bottom-to-top strategy. Its training speed is much faster than SMO for the large-scale classification problems with no noises.

Primal LS-SVM

Primal space LS-SVM is derived as an equivalent LS-SVM model by the representer theorem. It still solves a set of linear equations $\mathbf{A}\boldsymbol{\alpha} = \boldsymbol{h}$. However, the symmetrical coefficient matrix $\mathbf{A}$ is always singular if the related kernel matrix is of low rank or can be approximated with a low-rank matrix. Primal space LS-SVM model solves an overdetermined system of linear equations in the primal space. The method may have multiple solutions, which include some sparse solutions.

Fixed-size LS-SVM [178] solves the LS problem in the primal space. It can rapidly find the sparse approximate solution of LS-SVM by using an explicit expression for the feature map using Nystrom method. In [221], two primal space LS-SVM algorithms are proposed for finding the sparse (approximate) solution by Cholesky factorization.

In [125], LS-SVM and fixed-size LS-SVM models (https://www.esat.kuleuven.be/sista/ADB/mall/softwareFS.php) are iteratively sparsified through L_0-norm-based reductions.

Weighted Least Squares SVM

Weighted LS-SVM [182] improves LS-SVM by adding weights on error variables to correct the biased estimation of LS-SVM and to obtain robust estimation from noisy data. It first trains the samples using LS-SVM, then calculates the weights for each sample according to its error variable, and finally retrains weighted LS-SVM.

For two-dimensional samples such as images, MatLSSVM [203] is a classifier design method based on matrix patterns, such that the method can not only directly operate on original matrix patterns, but also efficiently reduce memory for the weight vector from $d_1 d_2$ to $d_1 + d_2$. MatLSSVM inherits LS-SVM's existence of unclassifiable regions when extended to multiclass problems. A fuzzy version of MatLSSVM [203] removes unclassifiable regions effectively for multiclass problems.

The iterative reweighted LS procedure for solving SVM [151] solves a sequence of weighted LS problems that lead to the true SVM solution. Iterative reweighted LS is also applicable for solving regression problems. The C-loss kernel classifier with the Tikhonov regularization term [212] is equivalent to an iterative weighted LS-SVM. This explains the robustness of iterative weighted LS-SVM from the correntropy and density estimation perspectives.

21.5 SVM Training Methods

21.5.1 SVM Algorithms with Reduced Kernel Matrix

To reduce the time and space complexities, a popular technique is to obtain low-rank approximations on the kernel matrix, by using Nystrom method [209], greedy approximation [176], or matrix decomposition [52].

Approximations of the Gram matrix [45, 52] have been proposed to increase the training speed and reduce the memory requirements of SVM solvers. The Gram matrix is the $N \times N$ square matrix composed of the kernel products.

A method suggested in [176] approximates the dataset in the feature space by a set in a low-dimensional subspace. A small subset of data points is randomly selected to form the basis of the approximating subspace. All other data points are then approximated by linear combinations of the elements of the basis. The basis is built iteratively, each new candidate element is chosen by a greedy method to reduce the bound on the approximation error as much as possible. The QP subproblem solver used is loqo, which is an interior-point-method solver provided with the SVMlight package.

The reduced SVM formulation for binary classification is derived from generalized SVM [127] and smooth SVM [111]. Prior to training, reduced SVM [110] randomly selects a portion of the dataset so as to generate a thin rectangular kernel matrix, which is then used to replace the full kernel matrix in the nonlinear SVM formulation. Reduced SVM uses a nonstandard SVM cost function. No constraints are needed and a quadratically converging Newton algorithm can be used for training. The time complexity of the optimization routine is $O(N)$. Though it has higher training errors than SVM, reduced SVM has comparable, or sometimes slightly better, generalization ability [108]. On some small datasets, reduced SVM performs even better than SVM. The technique of using a reduced kernel matrix has been applied to other kernel-based learning algorithms, such as proximal SVM [56], ε-smooth SVR [108], Lagrangian SVM [130], active-set SVR [136], and LS-SVM [178].

A direct method [211] that is similar to reduced SVM is to build sparse kernel learning algorithms by adding one more constraint to the convex optimization problem, such that the sparseness of the resulting kernel machine is explicitly controlled while performance is kept as high as possible.

Lagrangian SVM [130] is a fast and extremely simple iterative algorithm, capable of classifying datasets with millions of points. The full algorithm is given in 11 lines of MATLAB code without any special optimization tools such as LP or QP solvers. For nonlinear kernel classification, Lagrangian SVM can handle any positive semi-definite kernel and is guaranteed to converge. For a positive semi-definite nonlinear kernel, an inversion of a single $N \times N$ Hessian matrix of the dual is required. Hence, Lagrangian SVM can handle only intermediate size problems.

Like LS-SVM, proximal SVM [56] replaces the inequality by equality in the defining constraint structure of the SVM framework and uses the LS concept. It replaces the absolute error measure by the squared error measure in defining the minimization problem.

Generalized eigenvalue proximal SVM [131] relaxes the parallelism condition on proximal SVM. It classifies points by assigning them to the closest of two nonparallel planes which are generated by their generalized eigenvalue problems. Each plane is closest to the points of its own class and farthest to the points of the other class.

Twin SVM [83] determines two nonparallel proximal hyperplanes by solving two smaller related SVM-type problems. It aims at generating two nonparallel hyperplanes such that each plane is closer to one of the two classes and is as far as possible from the other. This makes twin SVM almost four times faster than SVM. The twin SVM formulation is in the spirit of proximal SVMs via generalized eigenvalues. Twin SVM is not only fast, but compares favorably with SVM and generalized eigenvalue proximal SVM in terms of generalization. When twin SVMs are used with a nonlinear kernel, a classifier may be obtained very rapidly for unbalanced datasets. A coordinate-descent margin-based twin SVM [167] leads to very fast training. It handles one data point at a time, and can process very large datasets that need not reside in memory. Twin bounded SVM [168] tries to minimize the structural risk by adding a regularization term, with the idea of maximizing the margin.

21.5.2 ν-SVM

In ν-SVM [161] algorithms, parameter ν controls the number of support vectors, and it enables one of the other free parameters of the algorithm to be eliminated. Compared to SVM, the formulation of ν-SVM is more complicated. A decomposition method for ν-SVM [21] is competitive with existing methods for SVM.

In [80], the geometrical meaning of SVMs with L_p norm is investigated. The ν-SVM(p) solution $\boldsymbol{w}_p$ is closely related to the ν-SVM solution $\boldsymbol{w}_2$ and has little dependency on p, and the generalization error barely depends on p. These results are applicable to SVR, since it has a similar geometrical structure.

Par-ν-SVM [70] is a modification of ν-SVM, and the use of a parametric-insensitive/margin model with an arbitrary shape is demonstrated. As in ν-SVM, ν is used to control the number of errors and support vectors. By devising a parametric-insensitive loss function, par-ν-SVR automatically adjusts a flexible parametric-insensitive zone of arbitrary shape and minimal radius to include the given data.

Through an enhancement of the range of the parameter ν, extended ν-SVM is able to produce a wider variety of decision functions. ν-SVM can be interpreted as a nearest-point problem in reduced convex hulls. Extended ν-SVM can be reformulated as a geometrical problem that generalizes a nearest-point problem in reduced convex hulls [7]. RapMinos algorithm [7] is able to solve extended ν-SVM for any choice of regularization norm $L_{p\geq 1}$ seamlessly.

A common approach to classifier design is to optimize the expected misclassification (Bayes) cost. Often, this approach is impractical because either the prior class probabilities or the relative cost of false alarms and misses are unknown. Two alternatives to the Bayes cost for the training of SVM classifiers are the minimax and Neyman–Pearson criteria, which require no knowledge of prior class probabilities or misclassification costs [41]. Cost-sensitive extensions of SVM and ν-SVM are $2C$-SVM [147] and 2ν-SVM [31]. $2C$-SVM is proved equivalent to 2ν-SVM [41].

In twin SVM, the patterns of one class are at least a unit distance away from the hyperplane of the other class; this might increase the number of support vectors. ν-TWSVM [150] extended the concept of ν-SVM. The parameter ν in ν-TWSVM controls the bounds on the number of support vectors, similar to ν-SVM, and further the unit distance of twin SVM is modified to variable ρ, which is optimized in the primal problem involved.

Iν-TWSVM (Improvements on ν-twin SVM) and fast Iν-TWSVM [99] are two binary classifiers motivated by ν-TWSVM. Similar to ν-TWSVM, Iν-TWSVM determines two nonparallel hyperplanes such that they are closer to their respective classes and are at least ρ distance away from the other class. Iν-TWSVM solves one smaller-sized QP problem and one unconstrained minimization problem. Fast Iν-TWSVM avoids solving a smaller-sized QP problem by transforming it into a unimodal function, which can be solved using line-search methods. Iν-TWSVM is computationally faster than twin bounded SVM and ν-TWSVM but having comparable generalization ability.

21.5.3 Cutting-Plane Technique

For large-scale L_1-SVM, SVMperf [87] uses a cutting-plane technique to obtain the solution of (21.4). Cutting-plane algorithms solve the primal problem by successively tightening a piecewise linear approximation. The cutting-plane algorithm [190] is a general approach for solving problem (21.1). It is based on iterative approximation of the risk term by cutting planes. It solves a reduced problem obtained by substituting the cutting-plane approximation of the risk into the original problem (21.1). The cutting-plane model makes it straightforward to add basis vectors that are not in the training set. A closely related method [211] explores training SVMs with kernels that can represent the learned rule using arbitrary basis vectors, not just the support vectors from the training set [89].

It can be shown that cutting-plane methods converge to an ε-accurate solution of the regularized risk minimization problem in $O(1/\varepsilon\lambda)$ iterations, where λ is the trade-off parameter between the regularizer and the loss function [191].

An optimized cutting-plane algorithm [54] solves large-scale risk minimization problems by extending standard cutting-plane algorithm [190]. An efficient line-search procedure for the optimization of (21.1) is the only additional requirement of the optimized cutting-plane algorithm compared to standard cutting-plane algorithm. The number of iterations the optimized cutting-plane algorithm requires to converge to an ε-precise solution is approximately $O(N)$. An optimized cutting-plane algorithm-based linear binary SVM solver outperforms SVMlight, SVMperf and the cutting-plane algorithm, achieving a speedup factor of more than 1,200 over SVMlight on some datasets and a speedup factor of 29 over SVMperf, while obtaining the same precise support vector solution. A cutting-plane algorithm-based linear binary SVM solver often shows faster convergence than gradient descent and Pegasos, and its linear multiclass version achieves a speedup factor of up to 10 compared to multiclass SVM.

For an equivalent 1-slack reformulation of the linear SVM training problem, the cutting-plane method has time complexity $O(N)$ [88]. In particular, the number of iterations does not depend on N, and it is linear in the desired precision and the regularization parameter. The cutting-plane algorithm includes the training algorithm of SVMperf [87] for linear two-class SVMs as a special case. In [88], not only individual data points are considered as potential support vectors, but also linear combinations of those. This increased flexibility allows for solutions with far fewer nonzero dual variables, leading to the small cutting-plane models.

The cutting-plane subspace pursuit method [89], like basis pursuit methods, iteratively constructs the basis set. The method is efficient and modular. Its classification rules can be orders of magnitude sparser than the conventional support vector representation while providing comparable prediction accuracy. The algorithm produces sparse solutions that are superior to approximate solutions of Nystrom method [209], incomplete Cholesky factorization [52], core vector machine, ball vector machine [196], and LASVM with margin-based active selection and finishing [11]. Both Nystrom method and incomplete Cholesky factorization are implemented in SVMperf.

21.5.4 Gradient-Based Methods

Successive overrelaxation [129] is a derivative of coordinate-descent method. It updates only one variable at each iteration, without need to reside in memory. The method is used for solving symmetric linear complementarity problem and quadratic programs to train an SVM with very large datasets. The algorithm converges linearly to a solution. On smaller problems, the successive overrelaxation method is faster than SVMlight and comparable or faster than SMO. LIBLINEAR [48] applied successive overrelaxation method to dual coordinate-descent method for large-scale linear SVM.

oLBFGS algorithm [165] compares the derivatives $\boldsymbol{g}_{t-1}(\boldsymbol{w}_{t-1})$ and $\boldsymbol{g}_{t-1}(\boldsymbol{w}_t)$ for an example $(\boldsymbol{x}_{t-1}, y_{t-1})$. Compared to the first-order stochastic gradient descent, each iteration of oLBFGS computes the additional quantity $\boldsymbol{g}_{t-1}(\boldsymbol{w}_t)$ and updates the list of k rank-one updates. Setting the global learning gain is very difficult [13].

The stochastic gradient-descent quasi-Newton (SGD-QN) algorithm [12] together with corrected SGD-QN [13] is a stochastic gradient-descent algorithm for linear SVMs that makes use of second-order information and splits the parameter update into independently scheduled components. It estimates a diagonal rescaling matrix using a technique inspired by oLBFGS. SGD-QN iterates nearly as fast as a first-order gradient descent, but requires less iterations to achieve the same accuracy. Corrected SGD-QN [13] discovers sensible diagonal scaling coefficients. Similar speed improvements can be achieved by simple preconditioning techniques such as normalizing the means and the variances of each feature and normalizing the length of each example. Corrected SGD-QN can adapt automatically to skewed feature distributions or very sparse data.

To solve linear SVM in large-scale scenarios, modified Newton methods for training L_2-SVM are given in [95, 128]. For L_2-SVM, a single-variable piecewise quadratic function (21.5) is minimized, which is differentiable but not twice differentiable. To obtain the Newton direction, they use the generalized Hessian matrix. Trust-region Newton method [117] is a fast implementation for L_2-SVM. A trust-region Newton method for logistic regression is proposed in [117].

For L_2-SVM, a coordinate-descent method [218] updates one component of $\boldsymbol{w}$ at a time while fixing the other variables by solving a one-variable subproblem by applying a modified Newton method with the line-search technique similar to the trust-region method [24]. With a necessary condition of convexity, the coordinate-descent method maintains a strict decrease in the function value. In [24], the full Newton step is used if possible, thus leading to faster convergence, more efficient and stable than Pegasos and trust-region Newton method; the method is proved to globally converge to the unique minimum at the linear rate.

21.5.5 Training SVM in the Primal Formulation

The literature on SVM mainly concentrates on the dual optimization problem. Duality theory provides a convenient way to deal with the constraints. The dual optimization problem can be written in terms of dot products, thereby making it possible to use kernel functions. The primal QP problem can be prohibitively large while its Wolfe dual QP problem is considerably smaller. For solving the problem in the primal, the optimization problem is mainly written as an unconstrained one and the representer theorem is used. It is common to employ a two-stage training process where the first stage produces an approximate solution to the dual QP problem and the second stage maps this approximate dual solution to an approximate primal solution. In terms of both the solution and the time complexity, when it comes to approximate solution, primal optimization is superior because it is directly focused on minimizing the primal objective function [27]. Also, the corresponding implementation is very simple and does not require any optimization libraries.

A wide range of machine learning methods can be described as the unconstrained regularized risk minimization problem (21.1), where $\boldsymbol{w} \in R^n$ denotes the parameter vector to be learned, $\frac{1}{2}\|\boldsymbol{w}\|^2$ is a quadratic regularization term, $C > 0$ is a fixed regularization constant and the second term is a nonnegative convex risk function approximating the empirical risk. Using the primal formulation (21.1) is efficient when N is very large and the dimension of the input data is moderate or the inputs are sparse.

Primal optimization of linear SVMs has been studied in [95, 128]. The finite Newton method [128] is a direct primal algorithm for L_2-SVM that exploits the sparsity property. It is rather effective for linear SVM. In [95], the finite Newton method is modified by bringing CG techniques to implement the Newton iterations to obtain a very fast method for solving linear SVMs with L_2 loss function. The method is much faster than decomposition methods such as SVMlight and SMO when the number of examples is large. For linear SVMs, the primal optimization is definitely superior to the dual optimization [95].

Primal optimization of nonlinear SVMs has been implemented in smooth SVM [111]. On larger problems, smooth SVM is comparable or faster than SVMlight, successive overrelaxation [129] and SMO. The recursive finite Newton method [27] can efficiently solve the primal problem for both linear and nonlinear SVMs. Performing Newton optimization in the primal yields exactly the same computational complexity as optimizing the dual. In [79], algorithms that accept an accuracy ε_p of the primal QP problem is described as an input and they are guaranteed to produce an approximate solution that satisfies this accuracy in low-order polynomial time.

Another approach to the primal optimization is based on decomposing the kernel matrix and thus effectively linearizing the problem. Among the most efficient solvers are Pegasos [166] and stochastic gradient descent (http://leon.bottou.org/projects/sgd) [15]. Pegasos is a primal estimated sub-gradient solver for L_1-SVM which alternates between stochastic gradient-descent steps and projection steps. It is among the fastest linear SVM solvers, and it outperforms SVMperf.

A fast accelerated proximal gradient method with a theoretical convergence guarantee is developed for a unified formulation of various classification models including SVM [81]. Backtracking line-search and adaptive restarting strategy are designed in order to speed up the practical convergence. When solving SVMs with a linear kernel, it is substantially faster than LIBSVM which implements SMO and the MATLAB package SeDuMi, which implements an interior-point method. Moreover, it often runs faster than LIBLINEAR [48] especially for large-scale datasets with feature dimension $n > 2000$.

LapSVM for Semi-supervised Learning

Manifold regularization methods assume that all points are located in a low-dimensional manifold, and graph is used for an approximation of the underlying manifold. Neighboring point pairs connected by large weight edges tend to have the same labels and vice versa. In this way, the labels associated with data can be propagated throughout the graph. By employing the graph Laplacian, Laplacian SVM (LapSVM) [8] is a semi-supervised version for SVM. Like SVM, LapSVM also uses the hinge loss function.

PLapSVM-Newton and PLapSVM-PCG [134] are two strategies for solving the primal Laplacian SVM (PLapSVM) problem. In particular, training a Laplacian SVM in the primal can be efficiently performed with preconditioned CG [134]. Training is sped up by using an early stopping strategy based on the prediction on unlabeled data or, if available, on labeled validation examples. The computational complexity of the training algorithm is reduced from $O(N^3)$ to $O(kN^2)$, where N is the combined number of labeled and unlabeled examples and $k \ll N$.

Unlike LapSVM and PLapSVM, fast Laplacian SVM (FLapSVM) [154] has a dual problem having the same formulation as that of standard SVMs, and the kernel trick can be applied directly into the optimization model. FLapSVM can be solved efficiently by successive overrelaxation technique [129], which converges linearly to a solution. Combining the strategies of random scheduling of subproblem and two stopping conditions, the training time of FLapSVM is faster than LapSVM, and PLapSVM, with better average accuracy.

21.5.6 Clustering-Based SVM

Clustering-based SVM [216] maximizes the SVM performance for very large datasets given a limited amount of resource. It applies BIRCH algorithm to get finer descriptions at places close to the classification boundary and coarser descriptions at places far from the boundary. The training complexity is $O(n^2)$ when having n support vectors. Clustering-based SVM can be used to classify very large datasets of relatively low dimensions. It performs especially well when the important data occur infrequently or when the incoming data includes irregular patterns, resulting in different distributions between training and testing data.

Bit-reduction SVM [102] is a simple strategy to speed up the training and prediction procedures for an SVM. It groups similar examples together by reducing their resolution. Bit reduction reduces the resolution of the input data and groups similar data into one bin. A weight is assigned to each bin according to the number of examples from a particular class in it, and a weighted example is created. This data reduction and aggregation step is very fast and scales linearly with respect to the number of examples. Then, an SVM is built on a set of weighted examples which are the exemplars of their respective bins. Optimal compression parameters need only to be computed once and can be reused if data arrive incrementally. It is typically more accurate than random sampling when the data are not overcompressed.

Multi-prototype SVM [2] extends multiclass SVM to multiple prototypes per class. It allows to combine several vectors in a principled way to obtain large margin decision functions. This extension defines a non-convex problem. The algorithm reduces the overall problem into a series of simpler convex problems. The approach compares favorably versus LVQ.

In [5], SVM approach is combined with the fast nearest-neighbor condensation classification rule. On very large and multidimensional datasets, the training is one or two orders of magnitude faster than SVM, and the number of support vectors is more than halved with respect to SVM, at the expense of a little loss of accuracy.

21.5.7 Other SVM Methods

A fast SMO procedure [101] solves the dual optimization problem of potential SVM. It consists of a sequence of iteration steps in which the Lagrangian is optimized with respect to either one (single SMO) or two (dual SMO) of the Lagrange multipliers while keeping the other variables fixed. Potential SVM is applied using dual SMO, block optimization, and ε-annealing. In contrast to SVMs, potential SVM is applicable to arbitrary dyadic datasets. Dyadic data are based on relationships between objects. For problems that are also solvable by standard SVM methods, computation time of potential SVM is comparable to or somewhat higher than SVM. The number of support vectors found by potential SVM is usually much smaller for the same generalization performance.

LASVM algorithm [11] performs SMO during learning. It allows efficient online and active learning. LASVM algorithm uses active selection techniques to train SVMs on a subset of the training dataset. In the limit of arbitrarily many epochs, LASVM converges to the exact SVM solution [11]. LASVM use the L_1-norm of the slack variables. In [61], LASVM is considerably improved in learning speed, accuracy, and sparseness by replacing the working-set selection in the SMO steps. A second-order working-set selection strategy, which greedily maximizes the progress in each single step, is incorporated.

Decision tree SVM [25] uses a decision tree to decompose a given data space and train SVMs on the decomposed regions. For datasets whose size can be handled by standard kernel-based SVM training techniques, the proposed method speeds up the training by a factor of thousands, with comparable test accuracy.

SVM with automatic confidence (SVMAC) [220] calculates the label confidence value of each training sample. Thus, the label confidence values of all of the training samples can be considered in training SVMs. By incorporating the label confidence value of each training sample into learning, the corresponding QP problems is derived. The generalization performance of SVMAC is superior to that of traditional SVMs. In comparison with traditional SVMs, the main additional cost of training SVMACs is to construct a decision boundary γ for labeling the confidence value of each training sample.

SVMs with a hybrid kernel can be designed [186] by minimizing the upper bound of the VC dimension. This method realizes an SRM and utilizes a flexible kernel function such that a superior generalization over test data can be obtained. A hybrid kernel is developed using common Mercer kernels.

DirectSVM [158] is a very simple learning algorithm based on the proposition that the two closest training points of opposite class in a training set are support vectors, on the condition that the training points in the set are linearly independent. This condition is always satisfied for soft-margin SVMs with quadratic penalties. Other support vectors are found using the following conjecture: the training point that maximally violates the current hyperplane is also a support vector. DirectSVM converges to a maximal margin hyperplane in $M - 2$ iterations, if the number of support vectors is M. DirectSVM has a generalization performance similar to other SVM implementations, and is faster than a QP approach.

Time-adaptive SVM [62] generates adaptive classifiers, capable of learning concepts that change with time. It uses a sequence of classifiers, each appropriate for a small time window but learning all the hyperplanes in a global way. The addition of a new term in the cost function of the set of SVMs (that penalizes the diversity between consecutive classifiers) produces a coupling of the sequence that allows time-adaptive SVM to learn as a single adaptive classifier. Time-adaptive SVM needs to compute matrix inversion and matrix pseudo-inversion. Improved time-adaptive SVM [173] uses a common vector shared by all the SVM classifiers involved. It not only keeps an SVM formulation, but also avoids the computation of matrix inversion. Based on the equivalence with the minimum enclosing ball problem, a fast implementation of improved time-adaptive SVM by using core vector machine technique [194], that is, improved time-adaptive core vector machine, is developed, which has asymptotic linear time complexity for large nonstationary datasets.

Sparse support vector classification [76] leads to sparse solutions by automatically setting the irrelevant parameters exactly to zero. It adopts the L_0-norm regularization term and is trained by an iteratively reweighted learning algorithm. The approach contains a hierarchical-Bayes interpretation. Set covering machine [132] tries to find the sparsest classifier making few training errors.

Methods like core vector machine, ball vector machine [196], and LASVM with margin-based active selection and finishing [11] greedily select as to which basis vectors to include in the classification rule. They are limited to selecting basis vectors from the training set.

Max-min margin machine [75] is a general large margin classifier. It extends SVM by considering class structures into decision boundary determination via utilizing the Mahalanobis distance.

By minimizing the L_0-norm of the separating hyperplane, support feature machine [100] finds the smallest subspace (the least number of features) of a dataset such that within this subspace, two classes are linearly separable without error. Classification of unbalanced and nonseparable data is straightforward. These capabilities qualify support feature machine as a universal method for feature selection, especially for high-dimensional small-sample-size datasets.

Approximate extreme points SVM [137] relies on conducting the SVM optimization over a carefully selected subset, called the representative set, of the training dataset. A linear-time algorithm based on convex hulls and extreme points is used to compute the representative set in kernel space. The implementation trains much faster than LIBSVM, core vector machine [194], ball vector machine [196], LASVM [11], SVMperf [89], and the random features method [155] do, while its classification accuracy is similar to that of LIBSVM in all cases. The method also gives competitively fast classification time.

Conventional SVM with the hinge loss (C-SVM) $L_{hinge}(u) = \max(0, u), u \in R$ is sparse, but sensitive to feature noise. Pinball loss SVM (pin-SVM) [77] $L_\tau(u) = \max(u, -\tau u), u \in R, 0 \leq \tau \leq 1$ gives penalty on numerous rightly classified points, and in this way, pin-SVM approximates a model which maximizes the quantile distance between two classes. Pin-SVM is still a convex model and possesses many attractive theoretical properties, including noise insensitivity, robustness, and bounded misclassification error. Pin-SVM loses sparsity because the sub-gradient of the pinball loss is not equal to zero almost everywhere. Pinball loss with ϵ-insensitive zone [77] $L_\tau^\epsilon(u) = \max(u - \epsilon, 0, -\tau u - \epsilon), u \in R, 0 \leq \tau \leq 1, \epsilon \geq 0$ is similar to the ϵ-insensitive loss used for SVM regression. Since the horizontal part in loss function is related to the sparsity, the truncated pinball loss flattens the negative part of the pinball loss at an appropriate position. The corresponding truncated pinball loss SVM ($\overline{pin}$-SVM) [171] is trained by using CCCP to handle non-convexity.

Field-SVM [74] trains and predicts a group of patterns (i.e., a field pattern) simultaneously, when patterns occur as groups. Field-SVM classifier learns simultaneously both the classifier and the style normalization transformation for each group of data (called field). By appropriately exploring the style consistency in each field, field-SVM is able to significantly improve the classification accuracy.

An SVM framework of regression and quaternary classification for complex data [16] uses both complex kernels as well as complexified real ones. This problem is equivalent to solving two separate real SVM tasks employing an induced real kernel.

Support matrix machine is a matrix classification method that leverages the structure of the data matrices and has the grouping effect property. Support matrix machine [121] is defined as a hinge loss plus a so-called spectral elastic net penalty. An alternating direction method of multipliers (ADMM) is devised for solving the convex optimization problem.

21.6 Pruning SVMs

Traditional convex SVM solvers rely on the hinge loss to solve the QP problem. Hinge loss imposes no limit on the influence of the outliers. All misclassified training instances become support vectors. A theoretical result shows that the number n of support vectors grows in proportion to the number of training examples [183]. Predicting a new example involves a computational complexity of $O(n)$ for n support vectors. It is desirable to build SVMs with a small number of support vectors, maintaining the property that their hidden-layer weights are a subset of the data (the support vectors). ν-SVM [161] and sparse SVMs [94] maintain this property.

The non-convex ramp loss function can overcome the scalability problems of convex SVM solvers. It is amenable to constrained concave–convex procedure (CCCP) optimization since it can be decomposed into a difference of convex parts. By leveraging the ramp function to avoid the outliers to become support vectors, an online learning framework to generate LASVM variants [47] leads to a significant reduction in the number of wrongly discarded instances, and sparser models compared to LASVM [11], without sacrificing generalization performance.

In order to reduce the number of support vectors, some methods operate as a postprocessing step after standard SVM training. In [164] L_1 regularization is applied on the bias θ to obtain sparse approximation. A simple but effective idea of pruning SVMs based on linear dependence is given in [200], and is further developed in [44], which gives an exact algorithm to prune the support vector set after an SVM classifier is built. In the regression case, the initial support vectors are generated using SVMTorch [36] and then those support vectors that are identified as linearly dependent are eliminated.

The pruning algorithm is structured by building into a newly defined kernel row space $\mathcal{K}$ and is related to feature space $\mathcal{H}$ [114]. By analyzing the overlapped information of kernel outputs, a method of pruning SVMs to an architecture containing at most M support vectors in the M-dimensional space H is systematically developed in [114]. This results in a decrease in the upper bound for support vectors from $M + 1$ [153] to M while retaining the separating hyperplane. The method also circumvents the problem of explicitly discerning support vectors in feature space as the SVM formulation does. In [113], the method in [44, 114, 200] is generalized by relaxing linear dependence to orthogonal projection using an LS approximation in space $\mathcal{K}$. The support vectors are further pruned in batches through a clustering technique.

To overcome the problem of a large number of support vectors, a primal method devised in [94] decouples the idea of basis functions from the concept of support vectors; it greedily finds a set of kernel basis functions of a specified maximum size ($d_{\max}$) to approximate the SVM primal cost function well; it is efficient and roughly scales as $O(Nd_{\max}^2)$. The method incrementally finds basis functions (support vectors) to maximize accuracy, starting with an empty set of basis functions. In many cases, the method efficiently forms classifiers which have an order of magnitude smaller number of basis functions compared to SVM, while achieving nearly the same level of accuracy.

Discarding even a small proportion of the support vectors can lead to a severe reduction in generalization performance. There exist nontrivial cases where the reduced-set approximation is exact, showing that the support vector set delivered by SVM is not always minimal [17]. The solution is approximated using a reduced set of vectors that are generally not support vectors, but are computed from the original support vector set to provide the best approximation to the original decision surface. In [139], the reduction process iteratively selects two nearest support vectors belonging to the same class and replaces them by a newly constructed one.

A pattern selection algorithm based on neighborhood properties [174] selects only the patterns that are likely to be located near the decision boundary. A neighborhood property is that a pattern located near the decision boundary tends to have more heterogeneous neighbors in its class membership. A well-known entropy concept can be utilized for the measurement of heterogeneity of class labels among k nearest neighbors. And the measure will lead us to estimate the proximity accordingly.

Low-rank modifications to LS-SVM [142] are useful for fast and efficient variable selection. Recursive feature elimination (RFE) is used for variable selection. The method attempts to find the best subset r of input dimensions which lead to the largest margin of class separation, using an SVM classifier. Relevant variables are selected according to a closed form of the leave-one-out error estimator, which is obtained as a by-product of the low-rank modifications.

21.7 Multiclass SVMs

To solve a multiclass classification problem with SVM, many strategies can be adopted. For multiclass SVM, two types of approaches for training and classification are mainly applied: to consider all the classes in one big optimization problem, or to combine several binary classifiers.

By considering all the classes in one optimization problem, one creates multiclass versions of SVM using the single-machine approach [40, 208]. This approach generates a very large optimization problem. The multiclass categorization problem is cast as a constrained optimization problem with a quadratic objective function. An efficient fixed-point algorithm is described for solving this reduced optimization problem and its convergence is proved in [40]. In [111], multiclass smooth SVM is solved by using a fast Newton–Armijo algorithm, which is globally convergent to the unique solution with quadratic time.

Among the strategies to decompose a multiclass problem proposed are one-against-all (one-versus-rest) [157, 200, 208], one-against-one [103], all-against-all [103], and error-correcting output codes (ECOC) [4, 43]. Comparative studies of these methods can be found in [73, 157]. The one-against-one and one-against-all methods are often recommended because of their lower computational cost and conceptual simplicity. These strategies are introduced in Sect. 25.7.

SVMs with binary tree architecture [30] reduce the number of binary classifiers and achieve a fast decision. The method needs to train $m-1$ classifiers and test $\log_2 m$ times for the final decision. But to get a good classifier of one node, it has to evaluate 2^m grouping possibilities with m classes in this node. An architecture named binary tree of SVM [50] achieves high classification efficiency for multiclass problems. Binary tree of SVM and centered binary tree of SVM decrease the number of binary classifiers to the greatest extent. In the training phase, binary tree of SVM has $m-1$ binary classifiers in the best situation, while it has $\log_{4/3}((m+3)/4)$ binary tests on average when making a decision. Maintaining comparable accuracy, the average convergent efficiency of binary tree of SVM is $\log_2((m+3)/4)$; it is much faster

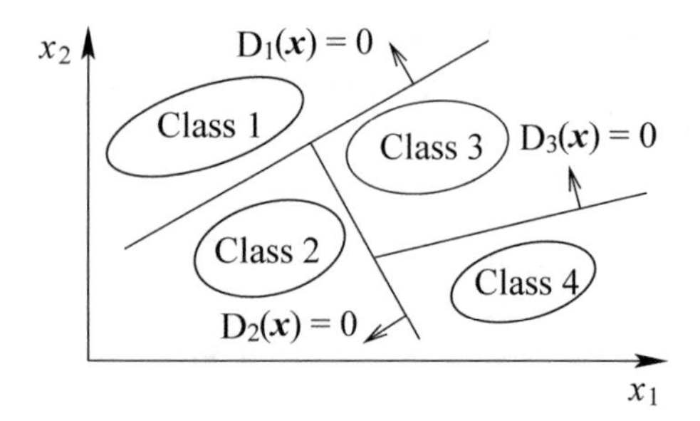

Fig. 21.3 A decision tree for classification. The discriminant functions $D_i(x)$ define class boundary

than directed acyclic graph SVM (DAG-SVM) [103] and ECOC in problems with big class number.

To resolve unclassifiable regions in one-against-all strategy, in decision tree-based SVMs, we train $m-1$ SVMs; the ith $(i = 1, \ldots, m-1)$ SVM is trained so that it separates data of the ith class from data belonging to one of classes $i+1, i+2, \ldots, m$. After training, classification is performed from the first to the $(m-1)$th SVM. If the ith SVM classifies a sample into class i, classification terminates; otherwise, classification is performed until the data sample is classified into the definite class.

Figure 21.3 shows an example of class boundaries for four classes, when linear kernels are used. The classes with smaller class numbers have larger class regions. Thus the processing order affects the generalization ability. In a usual decision tree, each node separates one set of classes from another set.

Example 21.3 By using STPRtool (http://cmp.felk.cvut.cz/cmp/software/stprtool/), we implement multiclass SVM classification with both the one-against-all and one-against-one strategies. The SMO binary solver is used to train the binary SVM subtasks. Gaussian kernel with $\sigma = 1$ is selected, and C is set to 50. The training data and the decision boundary are plotted in Fig. 21.4.

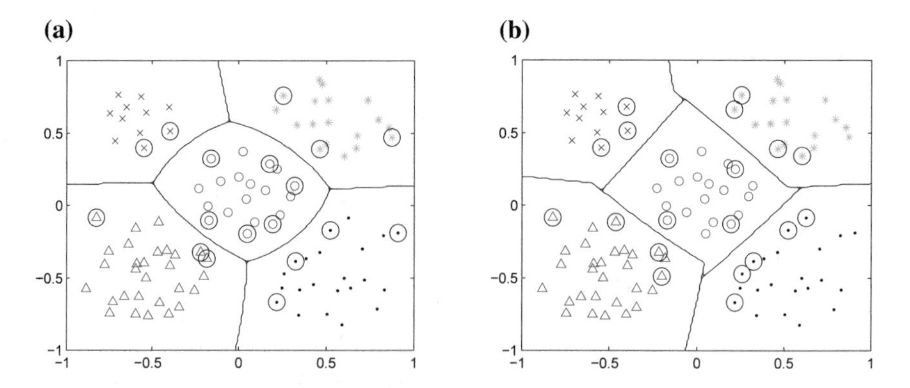

Fig. 21.4 Multiclass SVM classification using: **a** one-against-all strategy, **b** one-against-one strategy

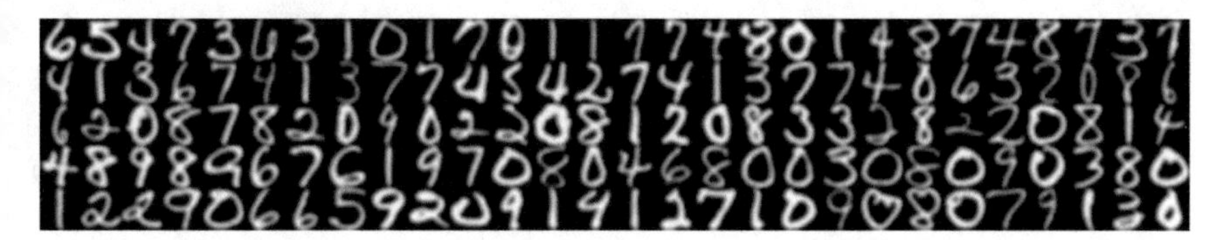

Fig. 21.5 Sample digits from the USPS database

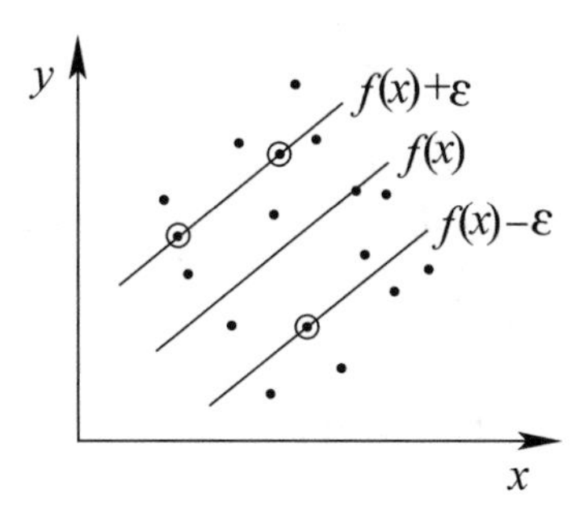

Fig. 21.6 Illustration of the hyperplanes in the two-dimensional feature space of SVR. The objective function penalizes examples whose y values are not within $(f(x)-\varepsilon, f(x)+\varepsilon)$. Those examples in circles are support vectors, against which the margin pushes

Example 21.4 The US-Postal Service (USPS) handwritten digit database contains 7291 training and 2007 testing images of 10 handwritten digits with size 16×16. The number of features is 256. Some samples of the digits are shown in Fig. 21.5.

By using LIBSVM, we select C-SVM with the polynomial $\gamma(\boldsymbol{u}^T\boldsymbol{v}+c_0)^d$ as kernel function, where $C=110$, $\gamma=1/n$, $n=256$ is the number of features, $d=3$, and $c_0=0$. A pairwise strategy is employed. The testing accuracy for classification is 95.4695% (1916/2007). The total number of supporting vectors for the trained SVM is 1445.

21.8 Support Vector Regression

SVM has been extended to regression. The basic idea of SVR is to map the data into a higher dimensional feature space via nonlinear mapping F and then to perform linear regression in this space. Regression approximation addresses the problem of estimating a function for a given dataset $\mathcal{D}=\{(\boldsymbol{x}_i, y_i)|i=1,\ldots,N\}$, where $\boldsymbol{x}_i \in R^n$ and $y_i \in R$. An illustration of SVR is given in Fig. 21.6.

The objective is to find a linear regression

$$f(\boldsymbol{x})=\boldsymbol{w}^T\boldsymbol{x}+\theta \tag{21.22}$$

such that the regularized risk function is minimized:

$$\min R(C) = \frac{1}{2}\|\boldsymbol{w}\|^2 + C\sum_{p=1}^{N} \left\| y_p - f\left(\boldsymbol{x}_p\right) \right\|_{\varepsilon}, \tag{21.23}$$

where the ε-insensitive loss function $\| \cdot \|_{\varepsilon}$ is used to define an empirical risk functional, defined by

$$\|\boldsymbol{x}\|_{\varepsilon} = \max\{0, \|\boldsymbol{x}\| - \varepsilon\}, \tag{21.24}$$

and $\varepsilon > 0$ and the regularization constant $C > 0$ are prespecified. Other robust statistics based loss functions such as Huber's function can also be used. If a data point $\boldsymbol{x}_p$ lies inside the insensitive zone called the ε-tube, i.e., $|y_p - f(\boldsymbol{x}_p)| \le \varepsilon$, then it will not incur any loss.

Introduction of slack variables ξ_p and ζ_p, the optimization problem (21.23) can be transformed into a QP problem [199]:

$$\min R(\boldsymbol{w}, \boldsymbol{\xi}, \boldsymbol{\zeta}) = \frac{1}{2}\|\boldsymbol{w}\|^2 + C\sum_{p=1}^{N} \left(\xi_p + \zeta_p\right) \tag{21.25}$$

subject to

$$\left(\boldsymbol{w}^T \boldsymbol{x}_p + \theta\right) - y_p \le \varepsilon + \xi_p, \quad p = 1, \ldots, N, \tag{21.26}$$

$$y_p - \left(\boldsymbol{w}^T \boldsymbol{x}_p + \theta\right) \le \varepsilon + \zeta_p, \quad p = 1, \ldots, N, \tag{21.27}$$

$$\xi_p \ge 0, \quad \zeta_p \ge 0, \quad p = 1, \ldots, N. \tag{21.28}$$

When the error is smaller than ε, the slack variables ξ_p and ζ_p take zero. The first term, $\frac{1}{2}\|\boldsymbol{w}\|^2$, is used as a measurement of function flatness.

By replacing $\boldsymbol{x}$ by $\phi(\boldsymbol{x})$, linear regression is generalized to kernel-based regression estimation, and regression is performed in the kernel space. Define the kernel function that satisfies Mercer's condition, $k(\boldsymbol{x}, \boldsymbol{y}) = \phi^T(\boldsymbol{x})\phi(\boldsymbol{y})$. Applying the Lagrange multiplier method, we get the following optimization problem [199, 200]:

$$\min L(\boldsymbol{\alpha}, \boldsymbol{\beta}) = \frac{1}{2}\sum_{p=1}^{N}\sum_{i=1}^{N} \left(\alpha_p - \beta_p\right)\left(\alpha_i - \beta_i\right) k\left(\boldsymbol{x}_p, \boldsymbol{x}_i\right) + \sum_{p=1}^{N} \left(\alpha_p - \beta_p\right) y_p + \sum_{p=1}^{N} \left(\alpha_p + \beta_p\right) \varepsilon \tag{21.29}$$

subject to

$$\sum_{p=1}^{N} \left(\alpha_p - \beta_p\right) = 0, \quad p = 1, \ldots, N, \tag{21.30}$$

$$0 \leq \alpha_p, \beta_p \leq C, \quad p = 1, \ldots, N, \tag{21.31}$$

where the Lagrange multipliers α_p and β_p, respectively, correspond to (21.26) and (21.27).

The SVM output generates the regression

$$u(\boldsymbol{x}) = f(\boldsymbol{x}) = \sum_{p=1}^{N} (\beta_p - \alpha_p) k(\boldsymbol{x}_p, \boldsymbol{x}) + \theta, \tag{21.32}$$

where θ can be solved using the boundary conditions. Those vectors with $\alpha_p - \beta_p \neq 0$ are called support vectors. SVR has the sparseness property.

The above ε-SVR model is formulated as a convex QP problem. Solving a QP problem needs $O(N^2)$ memory and time resources. The idea of LS-SVM has been extended for SVR [181].

SVR is a robust method due to the introduction of the ε-insensitive loss function. Varying ε influences the number of support vectors and thus controls the complexity of the model. The choice of ε reduces many of the weights $\alpha_p - \beta_p$ to zero, leading to a sparse solution in (21.31). Kernel selection is application-specific. For ε-SVR, ε is required to be set a priori. The performance of SVR is sensitive to the hyperparameters.

21.8.1 Solving Support Vector Regression

Bounded influence SVR [46] downweights the influence of outliers in all the regression variables. It adopts an adaptive weighting strategy, which is based on a robust adaptive scale estimator for large regression residuals.

In [23], various leave-one-out bounds for SVR are derived and the difference from those for classification is discussed. The proposed bounds are competitive with Bayesian SVR for parameter selection.

Regularization path algorithms [204] explore the path of possibly all solutions with respect to some regularization hyperparameter for model selection in an efficient way. An ε-path algorithm possesses the desirable piecewise linearity property. It possesses competitive advantages over the λ-path algorithm [66], which computes the entire solution path of SVR. The ε-path algorithm has a very simple initialization step and is efficient in finding a good regression function with the desirable sparseness property that can generalize well. It initializes the tube width ε to infinity, implying that it starts with no support vectors. It then reduces ε so that the number of support vectors increases gradually.

ε-smooth SVR [108] applies the smoothing technique for smooth SVM [111] to replace the ε-insensitive loss function by accurate smooth approximation so as to solve ε-SVR as an unconstrained minimization problem by using the Newton–Armijo method. For ε-smooth SVR, only a system of linear equations needs to be solved iteratively. In the linear case, ε-smooth SVR is much faster than SVR implemented by LIBSVM and SVMlight while with comparable correctness.

SVR is formulated as a convex QP problem with pairs of variables. Some SVR-oriented SMO algorithms make use of the close relationship between α_i and α_i^*. In [177], the method selects two pairs of variables (or four variables): α_i, α_i^*, α_j and α_j^* at each step according to a strategy similar to SMO, and solves the QP subproblem with respect to the selected variables analytically. The method for updating the bias is inefficient, and some improvements are made in [172] based on SMO for classification problems. Nodelib [53] includes some enhancements to SMO, where α_i and α_i^* are selected simultaneously. The QP problems with $2l$ variables can be transformed into nonsmooth optimization problems with l variables $\beta_i = \alpha_i - \hat{\alpha}_i, i = 1, 2, \ldots, l$, and an SMO algorithm solves these nonsmooth optimization problems.

SVMTorch [36] is a decomposition algorithm for regression problems, which is similar to SVMlight for classification problems. A convergence proof exists for SVMTorch [36]. SVMTorch selects α_i independently of their counterparts α_i^*. SVMTorch is usually many times faster than Nodelib [53], and training time generally scales slightly less than $O(N^2)$. Subproblems of size 2 is solved analytically, as is done in SMO. A cache-keeping part of the kernel matrix enables the program to solve large problems without keeping quadratic resources in memory and without recomputing every kernel evaluation.

The global convergence of a general SMO algorithm for SVR is given in [185] based on the formulation given in [53]. By using the same approach as in [96], the algorithm is proved to reach an optimal solution within a finite number of iterations if two conditions are satisfied [185].

In the spirit of twin SVM [83], twin SVR [148] aims at generating a pair of nonparallel ε-insensitive down- and up-bound functions for the unknown regressor. It solves two smaller-sized QP problems instead of the large one as in classical SVR, thus making the twin SVR work faster than SVR, with comparable generalization. By introducing a quadratic function to approximate its loss function, primal twin-SVR [149] directly optimizes the pair of QP problems of twin SVR in the primal space based on a series of sets of linear equations, thus obviously improves the learning speed of twin SVR without loss of the generalization.

To overcome the difficulty of selecting ε, ν-SVR [22, 161] is a modification of the ε-SVR algorithm, and it automatically minimizes ε. ν-SVR automatically adjusts the width of the tube so that at most a fraction ν of the data points lie outside the tube. ν-SVR is a batch learning algorithm. Through an approximation of the SVR model, parameter C can be dropped by considering its relation with other SVR hyperparameters (γ and ε). The decomposition algorithm for ν-SVR is similar to that for ν-SVM. The implementation is part of LIBSVM.

Pairing ν-SVR [71] combines the advantages of twin SVR and classical ε-SVR algorithms. In the spirit of twin SVR, pairing ν-SVR solves two QP problems of smaller size, and thus has a learning speed faster than that of ε-SVR. It improves twin SVR in the prediction speed and generalization ability by introducing the concepts of the insensitive zone and the regularization term. The parameter ν controls the bounds on fractions of support vectors and errors.

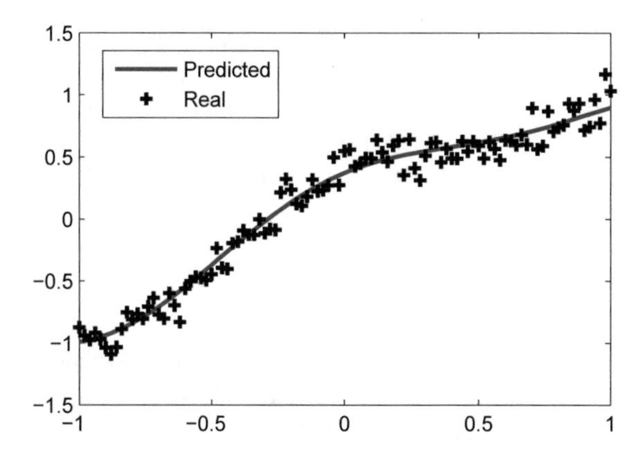

Fig. 21.7 Approximation of the samples

SVR model with parameterized lncosh loss [92] makes it possible to learn a loss function leading to a regression model which is maximum likelihood optimal for a specific input–output data. The SVR model with lncosh loss with ε-insensitiveness feature possesses the desirable characteristics of loss functions such as Vapnik's loss, the squared loss, and Huber's loss function for different parameters λ.

SVR lacks the flexibility to capture the local trend of data. Localized SVR [214] adapts the margin locally and flexibly, while the margin in SVR is fixed globally. It can be regarded as the regression extension of the max-min margin machine [75]. The associated optimization of localized SVR can be relaxed as a second-order cone programming problem, which can attain global optimal solution in polynomial time. Kernelization is applicable to the localized SVR model.

A recursive finite Newton method for nonlinear SVR in the primal is presented in [10] and it is comparable with dual optimizing methods like LIBSVM 2.82.

Example 21.5 Consider a function

$$y = x + 0.5\exp(-10x^2) + 0.1N, \quad x \in [-1, 1],$$

where N is added Gaussian noise with mean 0 and variance 0.2. We generate 101 training data from the equation. We use LIBSVM and select ε-SVR for regression. By selecting $C = 100$, $\gamma = 100$, and RBF kernel $\exp(-\gamma\|\boldsymbol{u} - \boldsymbol{v}\|^2)$, the obtained MSE is 0.0104 and the corresponding number of supporting vectors is 38. The result is shown in Fig. 21.7.

Support Vector Ordinal Regression

Support vector ordinal regression [170] is a generalization of the binary SVM to learning to rank or ordinal regression. It attempts to find an optimal mapping direction $\boldsymbol{w}$ and $r - 1$ thresholds, $b_1, \ldots, b_{r-1}$, which define $r - 1$ parallel discriminant hyperplanes for the r ranks accordingly. This formulation is improved in [35] by including the ordinal inequalities on the thresholds $b_1 \leq b_2 \leq \ldots \leq b_{r-1}$, and a good generalization performance is achieved with an SMO-type algorithm. The approaches proposed in [35] optimize multiple thresholds to define parallel discriminant hyperplanes for the ordinal scales. The size of these optimization problems is $O(N)$. Linear rankSVM [109] is a popular pairwise method for learning to rank.

21.9 Support Vector Clustering

Support vector clustering [9] uses a Gaussian kernel to transform the data points into a high-dimensional feature space. Clustering is conducted in the feature space and is then mapped back to the data space. The approach attempts to find in the feature space the smallest sphere of radius R that encloses all the data points in a set $\{\boldsymbol{x}_p\}$ of size N. It can be described by minimizing

$$E(R, \boldsymbol{c}, \boldsymbol{\xi}) = R^2 + C \sum_{p=1}^{N} \xi_p \tag{21.33}$$

subject to

$$\left\| \phi\left(\boldsymbol{x}_p\right) - \boldsymbol{c} \right\|^2 \leq R^2 + \xi_p, \quad p = 1, \ldots, N, \tag{21.34}$$

$$\xi_p \geq 0, \quad p = 1, \ldots, N, \tag{21.35}$$

where $\phi(\cdot)$ maps a pattern onto the feature space, ξ_p is a slack variable for the pth data point, $\boldsymbol{c}$ is the center of the enclosing sphere, and C is a penalty constant controlling the noise.

Based on the Lagrange multiplier method, the problem is transformed into

$$\min E_{\text{SVC}} = \sum_{p=1}^{N} \sum_{i=1}^{N} \alpha_p \alpha_i k\left(\boldsymbol{x}_p, \boldsymbol{x}_i\right) - \sum_{p=1}^{N} \alpha_p k\left(\boldsymbol{x}_p, \boldsymbol{x}_i\right) \tag{21.36}$$

subject to

$$\sum_{p=1}^{N} \alpha_p = 1, \tag{21.37}$$

$$0 \leq \alpha_p \leq C, \quad p = 1, \ldots, N, \tag{21.38}$$

where $k(\cdot)$ is selected as the Gaussian kernel and α_p is the Lagrange multiplier corresponding to the pth data point. The width σ of the Gaussian kernel controls the cluster scale while the soft margin ξ_p helps in coping with the outliers and overlapping clusters. By varying α_p and ξ_p, support vector clustering maintains a minimal number of support vectors so as to generate smooth cluster boundaries of arbitrary shape.

The distance between the mapping of an input pattern and the spherical center can be computed as

$$\begin{aligned} d^2(\boldsymbol{x}, \boldsymbol{c}) &= \|\phi(\boldsymbol{x}) - \boldsymbol{c}\|^2 \\ &= k(\boldsymbol{x}, \boldsymbol{x}) - 2\sum_{p=1}^{N} \alpha_p k\left(\boldsymbol{x}_p, \boldsymbol{x}\right) + \sum_{p=1}^{N}\sum_{i=1}^{N} \alpha_p \alpha_i k\left(\boldsymbol{x}_p, \boldsymbol{x}_i\right). \end{aligned} \quad (21.39)$$

Those data points that are on the boundary of the contours are support vectors.

A support function is defined as a positive scalar function $f : R^n \rightarrow R^+$, where a level set of f estimates a support of a data distribution. The level set of f can normally be decomposed into several disjoint-connected sets

$$L_f(r) = \{\boldsymbol{x} \in R^n : f(\boldsymbol{x}) \leq r\} = \mathcal{C}_1 \cup \cdots \cup \mathcal{C}_m, \quad (21.40)$$

where $\mathcal{C}_i$ are disjoint-connected sets corresponding to different clusters and m is the number of clusters determined by f. A support function is generated by the SVDD method (or one-class SVM) [163, 187]. SVDD maps data points to a high-dimensional feature space and finds a sphere with minimal radius that contains most of the mapped data points in the feature space. This sphere, when mapped back to the data space, can separate into several components, each enclosing a separate cluster of points.

Support vector clustering consists in general of two main steps: SVM training step to estimate a support function and cluster labeling step to assign each data point to its corresponding cluster. The time complexity of the cluster labeling step is $O(N^2m)$ for N data points and m ($\ll N$) sampling points on each edge. Support vector clustering has the ability to generate cluster boundaries of arbitrary shape and to deal with outliers. By utilizing the concept of dynamical consistency, labeling time can be reduced to $O(\log N)$ [90].

A heuristic rule is used to determine the width parameters of the Gaussian kernels and the soft-margin constant. Support vector clustering can be stopped when the fraction of support vectors and bounded support vectors exceeds a certain threshold (approximately 10% of the data points) [9]. Multisphere support vector clustering [32] creates multiple spheres to adaptively represent individual clusters. It is an adaptive cell-growing method, which essentially identifies dense regions in the data space by finding the corresponding spheres with minimal radius in the feature space. It can obtain cluster prototypes as well as cluster memberships.

Example 21.6 This example is taken from [9]. There are a dataset with 183 samples. The clusters can be separated by selecting $\sigma = 0.1443$ and $C = 1$. Support vector clustering gives the result in Fig. 21.8, where support vectors are indicated by small circles, and clusters are represented by different colors (gray scales) of the samples.

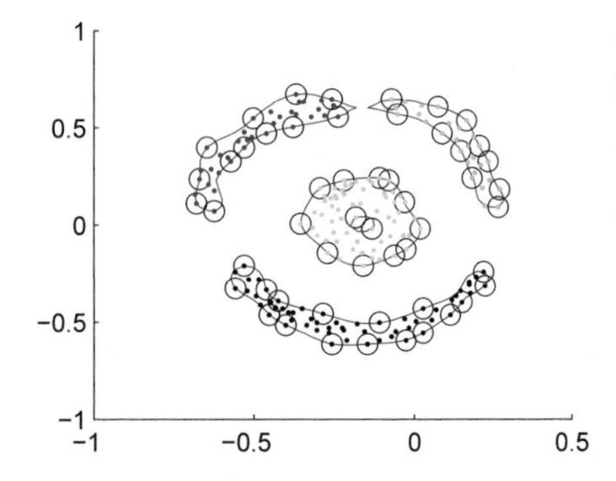

Fig. 21.8 Clustering of a dataset using support vector clustering with $C = 1$ and $\sigma = 0.1443$. ©Ben-Hur, Fig. 1d of [9]

Twin support vector clustering (http://www.optimal-group.org/Resource/TWSVC.html) [206] is a twin SVM-type plane-based clustering method. It determines k cluster center planes by solving a series of QP problems. An initialization algorithm based on the nearest-neighbor graph is suggested.

Maximum-margin clustering [213] performs clustering by simultaneously finding the largest margin separating hyperplane between clusters. It leads to a non-convex integer programming problem which can be reformulated and relaxed as a semi-definite programming problem. Alternating optimization can also be performed on the original non-convex problem [219]. A key step to avoid premature convergence in the resultant iterative procedure is to change the loss function from the hinge loss to the Laplacian/square loss to penalize overconfident predictions, leading to more accurate and two to four orders of magnitude faster. Generalized maximum-margin clustering [197] reduces the scale of the original semi-definite programming problem from $O(N^2)$ to $O(N)$.

A cutting-plane maximum-margin clustering algorithm [205] first decomposes the non-convex maximum-margin clustering problem into a series of convex subproblems by using CCCP and the cutting-plane algorithm for solving convex programs, then it adopts the cutting-plane algorithm to solve each subproblem. It outperforms maximum-margin clustering, both in efficiency and accuracy. The algorithm takes $O(sN)$ time to converge with guaranteed accuracy, for N samples in the dataset and the sparsity s of the dataset, i.e., the average number of nonzero features of the data samples. The multiclass algorithm is also derived in [205].

Maximum volume clustering [141] is a discriminative clustering model based on the large volume principle, and two approximation schemes, namely, a soft-label method using sequential QP and a hard-label method using semi-definite programming, are proposed to solve this model. The optimization problems of spectral clustering, relaxed C-means clustering, and information-maximization clustering are special limit cases of maximum volume clustering when its regularization parameter goes to infinity.

21.10 SVMs for One-Class Classification

Information retrieval using only positive examples for training is important in many applications. Consider trying to classify sites of interest to a web surfer where the only information available is the history of the user's activities. Novelty detection is the identification of novel patterns of which the learning system is trained with a few samples. This problem happens when novel or abnormal examples are expensive or difficult to obtain. Novelty detection is usually in the context of imbalanced datasets. With samples from both novel and normal patterns, novelty detection can be viewed as a usual binary classification problem. The purpose of data description, also called *one-class classification*, is to give a compact description of the target data that represents most of its characteristics.

One-class SVM [9, 163, 187] is a kernel method based on a support vector description of a dataset consisting of positive examples only. The two well-known approaches to one-class classification are the separation of data points from the origin [163] and spanning of data points with a sphere of minimum volume [187, 189]. The first approach is to extract a hyperplane in a kernel feature space such that a given fraction of training objects may reside beyond the hyperplane, while at the same time the hyperplane has maximal distance to the origin [163]. After transforming the feature via a kernel, they treat the origin as the only member of the second class using relaxation parameters, and standard two-class SVMs are then employed [163]. Both the approaches lead to similar, and in certain cases such as Gaussian kernel function, even identical formulations of dual optimization problems. If all data are inliers, one-class SVM computes the smallest sphere in feature space enclosing the image of the input data.

The hard- (soft-) margin SVDD then yields identical solution as the hard- (soft-) margin one-class SVM, and the weight $\boldsymbol{w}$ in the one-class SVM solution is equal to the center $\boldsymbol{c}$ in the SVDD solution [163]. Finding the soft-margin one-class SVM is essentially the same as fitting the minimum enclosing ball with outliers.

In SVDD [188, 189], the compact description of target data is given in a hyperspherical model, which is determined by support vectors as a hypersphere $(\boldsymbol{a}, R)$ with minimum volume containing most of the target data. SVDD has limitations to reflect overall characteristics of a target dataset with respect to its density distribution. In SVDD, support vectors fully determine the solution of target data description, whereas all of the nonsupport vectors have no influence on the solution of target description, regardless of the density distribution. The kernel trick is utilized to find a more flexible data description in a high-dimensional feature space [189]. To address the problem in SVDD, a density-induced SVDD [107] reflects the density distribution of a target dataset by introducing the notion of a relative density degree for each data point. By using density-induced distance measurements for both target data and negative data, density-induced SVDD can shift the center of hypersphere to the denser region based on the assumption that there are more data points in a denser region. When information of negative data is available, the method has a performance comparable to that of k-NN and SVM.

Outlier-SVM [126] is based on identifying outliers as representative of the second class. In the context of information retrieval, one-class SVM [163] and outlier-SVM [126] outperform the prototype, nearest neighbors, and naive Bayes methods. While one-class SVM is more robust with regard to smaller categories, a one-class neural network method based on bottleneck compression generated filters and outlier-SVM give good results by emphasizing success in the larger categories.

By reformulating a standard one-class SVM [163], LS one-class SVM [33] is derived with a reformulation very similar to that of LS-SVM. It extracts a hyperplane as an optimal description of training objects in a regularized LS sense. LS one-class SVM uses a quadratic loss function and equality constraints, and extracts a hyperplane with respect to which the distances from training objects are minimized in a regularized LS sense. Like LS-SVM, LS one-class SVM loses the sparseness property of standard one-class SVMs. One may overcome the loss of the sparseness by pruning the training samples.

21.11 Incremental SVMs

Some incremental learning techniques for SVM are given in [19, 55, 123, 133, 152, 194, 202]. Incremental SVMs are more efficient than batch SVMs in terms of computational cost.

$ALMA_p$ [58] is an incremental learning algorithm which approximates the maximal margin hyperplane with regard to norm $p \geq 2$ for a set of linearly separable data. By avoiding QP methods, $ALMA_p$ is as fast as the perceptron algorithm is. The accuracy levels achieved by $ALMA_2$ are superior to those achieved by incremental algorithms such as perceptron algorithm, but slightly inferior to that achieved by SVM. Compared to SVM, the $ALMA_2$ solution is significantly sparser. On the other hand, $ALMA_2$ is much faster and easier to implement than SVM training. When learning sparse target vectors (typical in text processing tasks), $ALMA_p$ with $p > 2$ largely outperforms perceptron-like algorithms such as $ALMA_2$. $ALMA_2$ operates directly on (an approximation to) the primal maximal margin problem. $ALMA_p$ is a large margin variant of the p-norm perceptron algorithm.

Exact incremental SVM learning (http://www.cpdiehl.org/code.html) [19] updates an optimal solution of an SVM training problem at each step only after a training example is added (or removed). It offers an advantage of immediate availability of the exact solution and reversibility, but has a large memory requirement, since the set of support vectors must be retained in memory during the entire learning process. Based on an analysis of convergence and algorithmic complexity of exact incremental SVM learning [19], a design using the gaxpy-type updates of the sensitivity vector speeds up the training of an incremental SVM by a factor of 5–20 [105].

Condensed SVM [140] involves integrating the vector combination for SVM simplification into an incremental framework for working-set selection in SVM training. The integration keeps the number of support vectors to the minimum.

Incremental asymmetric proximal SVM (IAPSVM) [156] employs a greedy search across the training data to select the basis vectors of the classifier, and tunes parameters automatically using the simultaneous perturbation stochastic approximation after incremental additions are made. The greedy search strategy substantially improves the accuracy of the resulting classifier compared to reduced-set methods introduced by proximal SVM. IAPSVM compares favorably with SVMTorch and core vector machine at reduced complexity levels.

Kernel Adatron [55] adapts Adatron to the problem of maximum-margin classification with kernels. An active-set approach to incremental SVM [175] uses a warm-start algorithm for training, which takes advantage of natural incremental properties of standard active- set approach to linearly constrained optimization problems. In an online algorithm for L_1-SVM [11], a close approximation of the exact solution is built online. This algorithm scales well to several hundred thousand examples, however, its online solution is not as accurate as the exact solution.

Core vector machine [194] is based on L_2-SVM and scales to several million of examples. It approximates a solution to L_2-SVM by a solution to the two-class minimum enclosing ball problem, for which several efficient online algorithms are available. While its scalability is very impressive, the method can lead to higher test errors.

In [93], the incremental training method uses one-class SVM. A hypersphere is generated for each class and data that exist near the boundary of the hypersphere are kept as candidates for support vectors while others are deleted.

Online independent SVM [145] approximately converges to the SVM solution each time new observations are added; the approximation is controlled via a user-defined parameter. The method employs a set of linearly independent observations and tries to project every new observation onto the set obtained so far, dramatically reducing time and space requirements at the price of a negligible loss in accuracy. Online-independent SVM produces a smaller model compared to that by standard SVM, with a training complexity of asymptotically $O(N^2)$. It uses the L_2-norm of the slack variables.

Accurate online ν-SVM algorithm (AONSVM) [65] extends exact incremental SVM learning [19] to ν-SVM. It can handle the conflict between a pair of equality constraints during the process of incremental learning.

Accurate incremental SVR algorithm [133] is obtained by extending exact incremental SVM learning [19] to SVR. Following exact incremental SVM learning [19], accurate online SVR [123] efficiently updates a trained SVR function whenever a sample is added to or removed from the training set. Accurate online SVR assumes that the new samples and the training samples are of the same characteristics. Accurate online SVR with varying parameters [143] uses varying SVR parameters rather than fixed ones and hence accounts for the variability that may exist in the samples. An incremental ν-SVR learning algorithm [64] is derived by combining a special procedure called initial adjustments with the two steps of AONSVM.

An incremental support vector ordinal regression algorithm [63] is obtained by extending AONSVM to a modified support vector ordinal regression formulation based on a sum-of-margins strategy. The algorithm can converge to the optimal solution in a finite number of steps.

21.12 SVMs for Active, Transductive, and Semi-supervised Learnings

21.12.1 SVMs for Active Learning

An active learning algorithm using SVM identifies positive examples in a dataset [207]. Selection of next point to be labeled is carried out in the algorithm using two heuristics that can be derived from an SVM classifier trained on points with known labels. The largest positive heuristic selects the point that has the largest classification score among all examples still unlabeled. The near boundary heuristic selects the point whose classification score has the smallest absolute value. In both cases, SVM has to be retrained after each selection. A better way is to apply incremental learning.

Given a set of labeled training data and a Mercer kernel, there is a set of hyperplanes that separate the data in the induced feature space. This set of consistent hypotheses are called the version space. In pool-based active learning [193], the learner has access to a pool of unlabeled instances and can request the labels for some of them. An algorithm for performing active learning with SVM chooses as to which instances should be requested next. The method significantly reduces the need for labeled training instances in both standard inductive and transductive settings.

21.12.2 SVMs for Transductive or Semi-supervised Learning

In addition to regular induction, SVM can also be used for transduction. SVM can perform transduction by finding the hyperplane that maximizes the margin relative to both the labeled and unlabeled data. See Fig. 21.9 for an example. Transductive SVM has been used for text classification, attaining improvements in precision/recall breakeven performance over regular inductive SVM [193].

Transduction utilizes the prior knowledge of the unlabeled test patterns [198]. It is an essentially easier task than first learning a general inductive rule and then applying it to the test examples. Transductive bounds address the performance of the trained system on these test patterns only. When the test and training data are not

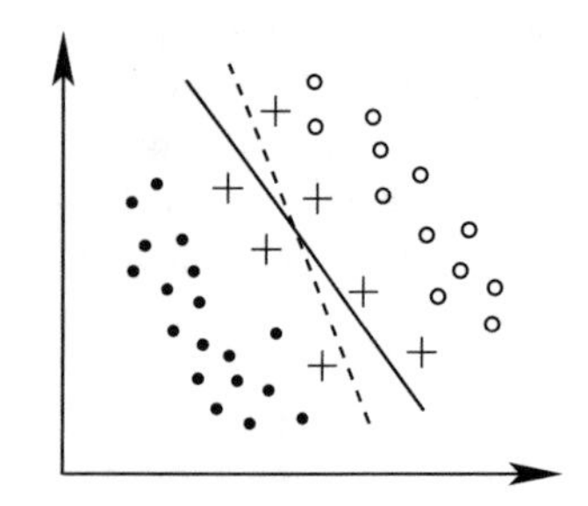

Fig. 21.9 SVM (solid line) and transductive SVM (dotted line). "+" signs represent unlabeled instances

identically distributed, the concept of transduction could be particularly worthwhile. Transductive SVM solves the SVM problem while treating the unknown labels as additional optimization variables. By maximizing the margin in the presence of unlabeled data, one learns a decision boundary that traverses through low data-density regions while respecting labels in the input space. In other words, this approach implements the cluster assumption for semi-supervised learning [26]. Transductive SVM learns an inductive rule defined over the entire input space.

Transductive SVM [199] is a method of improving the generalization accuracy of SVM by using unlabeled data. It learns a large margin hyperplane classifier using labeled training data, but simultaneously forces this hyperplane to be far away from the unlabeled data. Transduction (labeling a test set) is inherently easier than induction (learning a general rule) [199]. Transductive SVM can provide considerable improvement in generalization over SVM, if the number of labeled points is small and the number of unlabeled points is large.

The original transductive SVM problem is described as follows. The training set consists of L labeled examples $\{(\boldsymbol{x}_i, y_i)\}_{i=1}^{L}$, $y_i = \pm 1$, and U unlabeled examples $\{\boldsymbol{x}_i\}_{i=l+1}^{N}$, with $N = L + U$. Find among the possible binary vectors $\mathcal{Y} = \{(y_{L+1}, \ldots, y_{L+U})\}$ the one such that an SVM trained on $\mathcal{L} \cup (\mathcal{U} \times \mathcal{Y})$ yields the largest margin. This is a combinatorial problem, but one can approximate it as finding an SVM separating the training set under constraints which force the unlabeled examples to be as far as possible from the margin [199]. A primal method [26] scales as $(L + U)^3$, and stores the entire $(L + U) \times (L + U)$ kernel matrix in memory.

Following a formulation using an integer programming method, transductive linear SVM with an L_1-norm regularizer [28], where the corresponding loss function is decomposed as a sum of a linear function and a concave function, is algorithmically close to CS^{3}VM [57]. SVMLight-TSVM [86] is a combinatorial approach that is practical for a few thousand examples. ∇TSVM [26] is optimized by performing gradient descent in the primal space.

One problem with transductive SVM is that in high dimensions with few training examples, it is possible to classify all the unlabeled examples as belonging to only one of the classes with a very large margin, which leads to poor performance. One can constrain the solution by introducing a balancing constraint that ensures the unlabeled data are assigned to both classes. The fraction of positive and negatives assigned to the unlabeled data can be assumed to be the same fraction as found in the labeled data [86].

Transductive SVM learning provides a decision boundary in the entire input space and can be considered as inductive, rather than strictly transductive, semi-supervised learning. Semi-supervised SVM is based on applying the margin maximization principle to both labeled and unlabeled examples. Thus, semi-supervised SVMs are inductive semi-supervised methods and not strictly transductive [28]. Many techniques are available for solving the non-convex optimization problem associated with semi-supervised SVM or transductive SVM, for example, local combinatorial search [86], gradient descent [26], convex–concave procedures [37, 57], and semi-definite programming [213].

In linear semi-supervised SVM, the minimization problem is solved over both the hyperplane parameters $(\boldsymbol{w}, b)$ and the label vector $\boldsymbol{y}_U = (y_{L+1} \ldots y_N)^T$,

$$\min_{(\boldsymbol{w},b),\boldsymbol{y}_U} I(\boldsymbol{w}, b, \boldsymbol{y}_U) = \frac{1}{2}\|\boldsymbol{w}\|^2 + C\sum_{i=1}^{L} V(y_i, o_i) + C^* \sum_{i=L+1}^{N} V(y_i, o_i), \quad (21.41)$$

where $o_i = \boldsymbol{w}^T \boldsymbol{x}_i + b$ and the loss function V is usually selected as the hinge loss,

$$V(y_i, o_i) = [\max(0, 1 - y_i o_i)]^p . \quad (21.42)$$

It is common to select either $p = 1$ or $p = 2$. Nonlinear decision boundaries can be constructed using the kernel trick.

The first two terms in the objective function (21.41) define SVM. The third term incorporates unlabeled data. The loss over labeled and unlabeled examples is weighted by two hyperparameters, C and C^*, which reflect confidence in the labels and in the cluster assumption, respectively. The problem (21.41) is solved under the class-balancing constraint [86]:

$$\frac{1}{U}\sum_{i=L+1}^{N} \max(y_i, 0) = r, \quad \text{or equivalently,} \quad \frac{1}{U}\sum_{i=L+1}^{N} y_i = 2r - 1. \quad (21.43)$$

This constraint helps in avoiding unbalanced solutions by enforcing that a certain user-specified fraction, r, of the unlabeled data should be assigned to the positive class. r is estimated from the class ratio on the labeled set, or from prior knowledge of the classification problem.

CCCP is applied to semi-supervised SVM in [37, 57]. Using CCCP, a large-scale training method solves a series of SVM optimization problems with $L + 2U$ variables. It involves iterative solving of standard dual QP problems, and usually requires just a few iterations. This provides a highly scalable algorithm in the nonlinear case. Successive convex optimization is performed using an SMO implementation. CCCP-TSVM runs orders of magnitude faster than SVMlight-TSVM and ∇TSVM [37]. Both SVMlight-TSVM and ∇TSVM use an annealing heuristic for hyperparameter C^*.

An inductive semi-supervised learning method proposed in [106] first builds a trained Gaussian kernel support function that estimates a support of a data distribution via an SVDD procedure using both labeled and unlabeled data. Then, it partitions the whole data space into separate clustered regions. Finally, it classifies the decomposed regions utilizing the information of the labeled data and the topological structure of the clusters described by the constructed support function. Its formulation leads to a non-convex optimization problem.

S^{3}VMlight [86], the semi-supervised SVM algorithm implemented in SVMlight, is based on local combinatorial search guided by a label-switching procedure. ∇S^{3}VM [26] minimizes directly the objective function by gradient descent, with a

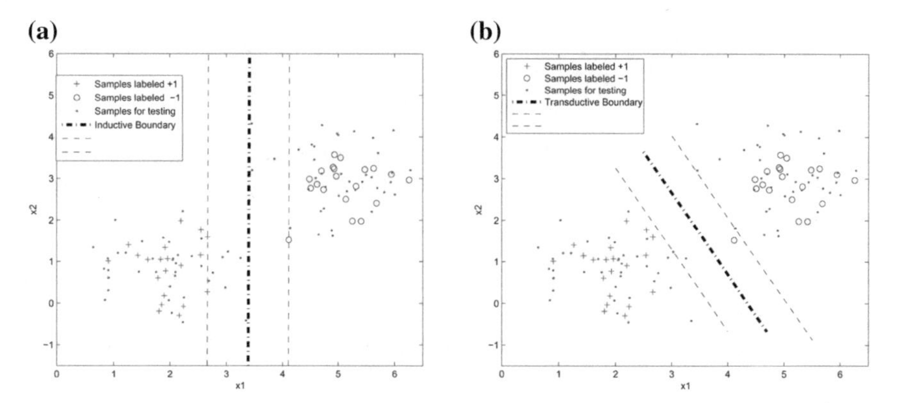

Fig. 21.10 Illustration of the transductive accuracy in a two-dimensional input space. **a** Boundary found by inductive learning with SVM classifier. Only labeled points (+ and o) are used for training. **b** Boundary found by transductive learning. All points, labeled and testing (unlabeled), are used for training. ©IEEE, 2009 [1]

complexity $O(N^3)$. In comparison, S^3VMlight typically scales as $O(n^3 + N^2)$ for n support vectors. A loss function for unlabeled points and an associated Newton semi-supervised SVM method are proposed in [28] (along the lines of [27]), bringing down the complexity of ∇S^3VM. Semi-supervised LS-SVM classifier [1] uses the transductive inference formulation, and different combinatorial optimization approaches are deduced from the transductive SVM idea to train the classifier. Fully weighted semi-supervised LS-SVM [124] formulates the optimization problem as a simple integer program. The derived algorithm is a generalization of the algorithms given in [1].

Example 21.7 The use of inductive inference for estimating the value of a function at given points involves two steps. The training points are first used to estimate a function for the entire input space, and the values of that function on separate test points are then computed based on the estimated parameters. In transductive inference, the end goal is to determine the values of the function at the predetermined test points. Figure 21.10 gives the classification boundaries of inductive and transductive learning with SVM classifier. Good classification is obtained on the testing samples for transductive learning.

21.13 Solving SVM with Indefinite Matrices

Using LP to solve the SVM problem, positive definiteness is not required. LP-SVM can be used for solving SVM with indefinite matrices [3]. LP-SVM is a kernel-as-features method, where the kernel entries are actually used as new features.

Once the kernel matrix is positive semi-definite, a convex optimization algorithm can be used to solve the resulting SVM. An indefinite kernel matrix $\mathbf{K}$ can be converted to a positive semi-definite matrix by one of the following methods [135]:

- Clip: the negative part is simply removed (i.e., negative eigenvalues cut to 0). IndefiniteSVM [122] solves SVM with indefinite kernel by doing some spectrum modification (clip).
- Shift: the complete spectrum is shifted until the least eigenvalue is 0.
- Flip: the absolute value of the spectrum is used (i.e., the negative part becomes positive). IKFD [68] performs LDA with the absolute value of the spectrum (flip).
- Square: eigenvalues are squared, which is equivalent to using $\mathbf{K}\mathbf{K}^T$ instead of $\mathbf{K}$. Potential SVM [101] and relevance vector machine [192] is based on the usage of $\mathbf{K}\mathbf{K}^T$, which is a direct way to consider the kernel rows as features.

Krein spaces are indefinite inner product spaces endowed with a Hilbertian topology. Reproducing kernel Krein spaces are introduced to allow the definition of SVMs in this framework as a projection. Positive semi-definiteness is not a requirement, and the representer theorem is also valid for reproducing kernel Krein space [144]. The SVM problem with indefinite kernels are defined in Krein spaces and the solution is the one which stabilizes the cost function.

Built on the work in [144], Krein-space SVM (http://gaelle.loosli.fr/research/) [120] is obtained by decomposition of Krein spaces into Hilbert spaces and transforming the dual SVM maximization problem with an appropriately converted kernel matrix into a convex minimization. The dual SVM maximization problem is equivalent to the primal stabilization problem in Krein space. Krein-space SVM has quadratic to cubic complexity and a non-sparse decision function.

Krein-space core vector machine (iCVM) [160] can be applied to indefinite kernels. A sparse model with linear runtime complexity can be obtained under a low rank assumption. iCVM has similar efficiency as Krein-space SVM but with substantially lower cost.

Problems

21.1 Acceptable kernels must satisfy Mercer's condition. Let k_1 and k_2 be kernels defined on $R^d \times R^d$. Show that the following are also kernels:
(a) $k(\boldsymbol{x}, z) = ak_1(\boldsymbol{x}, z) + bk_2(\boldsymbol{x}, z)$, with $a, b \in R^+$.
(b) $k(\boldsymbol{x}, z) = g(\boldsymbol{x})g(z)$, with $g(\cdot)$ being a real-valued function.

21.2 For Example 20.1, compute the kernel matrix $\mathbf{K} = [k_{ji}]$. Verify its positive definiteness.

21.3 Consider the mapping which maps two-dimensional points (p, q) to three-dimensional points $(p^2, q^2, \sqrt{2}pq)$. Specify the dot product of two vectors in three-dimensional space in terms of the dot products of the corresponding points in two-dimensional space.

21.4 For Example 20.1, establish the QP problem, given by (21.7)–(21.9), for SVM learning. Solve α_p, $p = 1, \ldots, 4$.

21.5 Derive the dual optimization problem for a soft-margin SVM.

21.6 The lncosh loss function [92] is a continuously differentiable convex loss function controlled by a single parameter λ. Its minimum is optimal in the maximum likelihood sense for a hyperbolic-secant distribution. It approaches ϵ-insensitive Laplace loss function as $\lambda \to \infty$, and approximates Vapnik's ϵ-insensitive loss function well as λ increases. It approaches ϵ-insensitive squared loss as $\lambda \to 0$. It becomes similar to ϵ-insensitive Huber's loss function for moderate values of λ.

The lncosh loss function is given by [92]

$$\text{lncosh}_\epsilon(\psi) = \begin{cases} 0 & \text{for } |\psi| < \epsilon \\ \frac{1}{\lambda}\ln(\cosh_\epsilon(\lambda\psi)) & \text{otherwise} \end{cases}$$

where $\lambda \in (0, +\infty)$, the determined tolerance $\epsilon > 0$, and $\cosh_\epsilon = \frac{e^{\lambda\psi}+e^{-\lambda\psi}}{2}$.

The influence model of lncosh loss is described by

$$\varphi(\psi) = \begin{cases} 0 & \text{for } |\psi| < \epsilon \\ \tanh_\epsilon(\lambda\psi) & \text{for } |\psi| \geq \epsilon. \end{cases}.$$

Plot the lncosh loss function and its influence function for different λ and obverve the properties of the function.

References

1. Adankon, M. M., Cheriet, M., & Biem, A. (2009). Semisupervised least squares support vector machine. *IEEE Transactions on Neural Networks*, *20*(12), 1858–1870.
2. Aiolli, F., & Sperduti, A. (2005). Multiclass classification with multi-prototype support vector machines. *Journal of Machine Learning Research*, *6*, 817–850.
3. Alabdulmohsin, I., Zhang, X., & Gao, X. (2014). Support vector machines with indefinite kernels. In *JMLR Workshop and Conference Proceedings: Asian Conference on Machine Learning* (Vol. 39, pp. 32–47).
4. Allwein, E. L., Schapire, R. E., & Singer, Y. (2000). Reducing multiclass to binary: A unifying approach for margin classifiers. *Journal of Machine Learning Research*, *1*, 113–141.
5. Angiulli, F., & Astorino, A. (2010). Scaling up support vector machines using nearest neighbor condensation. *IEEE Transactions on Neural Networks*, *21*(2), 351–357.
6. Baker, J. L. (2003). Is there a support vector machine hiding in the dentate gyrus? *Neurocomputing*, *52–54*, 199–207.
7. Barbero, A., Takeda, A., & Lopez, J. (2015). Geometric intuition and algorithms for Eν-SVM. *Journal of Machine Learning Research*, *16*, 323–369.
8. Belkin, M., Niyogi, P., & Sindhwani, V. (2006). Manifold regularization: A geometric framework for learning from examples. *Journal of Machine Learning Research*, *7*, 2399–2434.

9. Ben-Hur, A., Horn, D., Siegelmann, H., & Vapnik, V. (2001). Support vector clustering. *Journal of Machine Learning Research*, *2*, 125–137.
10. Bo, L., Wang, L., & Jiao, L. (2007). Recursive finite Newton algorithm for support vector regression in the primal. *Neural Computation*, *19*(4), 1082–1096.
11. Bordes, A., Ertekin, S., Wesdon, J., & Bottou, L. (2005). Fast kernel classifiers for online and active learning. *Journal of Machine Learning Research*, *6*, 1579–1619.
12. Bordes, A., Bottou, L., & Gallinari, P. (2009). SGD-QN: Careful quasi-Newton stochastic gradient descent. *Journal of Machine Learning Research*, *10*, 1737–1754.
13. Bordes, A., Bottou, L., Gallinari, P., Chang, J., & Smith, S. A. (2010). Erratum: SGDQN is less careful than expected. *Journal of Machine Learning Research*, *11*, 2229–2240.
14. Boser, B. E., Guyon, I. M., & Vapnik, V. N. (1992). A training algorithm for optimal margin classifiers. In *Proceedings of the 5th ACM Annals Workshop on Computational Learning Theory (COLT)* (pp. 144–152).
15. Bottou, L., & Bousquet, O. (2007). The tradeoffs of large scale learning. *Advances in neural information processing systems* (Vol. 20, pp. 161–168). Cambridge: MIT Press.
16. Bouboulis, P., Theodoridis, S., Mavroforakis, C., & Evaggelatou-Dalla, L. (2015). Complex support vector machines for regression and quaternary classification. *IEEE Transactions on Neural Networks and Learning Systems*, *26*(6), 1260–1274.
17. Burges, C. J. C. (1996). Simplified support vector decision rules. In *Proceedings of 13th International Conference on Machine Learning* (pp. 71–77). Bari, Italy.
18. Burges, C. J. C. (1998). A tutorial on support vector machines for pattern recognition. *Data Mining and Knowledge Discovery*, *2*, 121–167.
19. Cauwenberghs, G., & Poggio, T. (2001). Incremental and decremental support vector machine learning. In T. K. Leen, T. G. Dietterich, & V. Tresp (Eds.), *Advances in neural information processing systems* (Vol. 13, pp. 409–415). Cambridge: MIT Press.
20. Chang, C.-C., & Lin, C.-J. (2001). *LIBSVM: A library for support vector machines*. Technical Report, Department of Computer Science and Information Engineering, National Taiwan University.
21. Chang, C.-C., & Lin, C.-J. (2001). Training ν-support vector regression: Theory and algorithms. *Neural Computation*, *13*(9), 2119–2147.
22. Chang, C. C., & Lin, C. J. (2002). Training ν-support vector regression: Theory and algorithms. *Neural Computation*, *14*, 1959–1977.
23. Chang, M.-W., & Lin, C.-J. (2005). Leave-one-out bounds for support vector regression model selection. *Neural Computation*, *17*, 1188–1222.
24. Chang, K.-W., Hsieh, C.-J., & Lin, C.-J. (2008). Coordinate descent method for large-scale L2-loss linear support vector machines. *Journal of Machine Learning Research*, *9*, 1369–1398.
25. Chang, F., Guo, C.-Y., Lin, X.-R., & Lu, C.-J. (2010). Tree decomposition for large-scale SVM problems. *Journal of Machine Learning Research*, *11*, 2935–2972.
26. Chapelle, O., & Zien, A. (2005). Semi-supervised classification by low density separation. In *Proceedings of the 10th International Workshop on Artificial Intelligence Statistics* (pp. 57–64).
27. Chapelle, O. (2007). Training a support vector machine in the primal. *Neural Computation*, *19*(5), 1155–1178.
28. Chapelle, O., Sindhwani, V., & Keerthi, S. S. (2008). Optimization techniques for semi-supervised support vector machines. *Journal of Machine Learning Research*, *9*, 203–233.
29. Chen, P.-H., Fan, R.-E., & Lin, C.-J. (2006). A study on SMO-type decomposition methods for support vector machines. *IEEE Transactions on Neural Networks*, *17*(4), 893–908.
30. Cheong, S., Oh, S. H., & Lee, S.-Y. (2004). Support vector machines with binary tree architecture for multi-class classification. *Neural Information Processing - Letters and Reviews*, *2*(3), 47–51.
31. Chew, H. G., Bogner, R. E., & Lim, C. C. (2001). Dual-ν support vector machine with error rate and training size biasing. In *Proceedings of the IEEE International Conference on Acoustics, Speech, and Signal Processing (ICASSP)* (pp. 1269–1272).

32. Chiang, J.-H., & Hao, P.-Y. (2003). A new kernel-based fuzzy clustering approach: Support vector clustering with cell growing. *IEEE Transactions on Fuzzy Systems*, *11*(4), 518–527.
33. Choi, Y.-S. (2009). Least squares one-class support vector machine. *Pattern Recognition Letters*, *30*, 1236–1240.
34. Chu, W., Ong, C. J., & Keerthy, S. S. (2005). An improved conjugate gradient method scheme to the solution of least squares SVM. *IEEE Transactions on Neural Networks*, *16*(2), 498–501.
35. Chu, W., & Keerthi, S. S. (2007). Support vector ordinal regression. *Neural Computation*, *19*, 792–815.
36. Collobert, R., & Bengio, S. (2001). SVMTorch: Support vector machines for large-scale regression problems. *Journal of Machine Learning Research*, *1*, 143–160.
37. Collobert, R., Sinz, F., Weston, J., & Bottou, L. (2006). Large scale transductive SVMs. *Journal of Machine Learning Research*, *7*, 1687–1712.
38. Cortes, C., & Vapnik, V. (1995). Support vector networks. *Machine Learning*, *20*, 1–25.
39. Cox, D., & O'Sullivan, F. (1990). Asymptotic analysis of penalized likelihood and related estimators. *Annals of Statistics*, *18*, 1676–1695.
40. Crammer, K., & Singer, Y. (2001). On the algorithmic implementation of multiclass kernel-based vector machines. *Journal of Machine Learning Research*, *2*, 265–292.
41. Davenport, M. A., Baraniuk, R. G., & Scott, C. D. (2010). Tuning support vector machines for minimax and Neyman-Pearson classification. *IEEE Transactions on Pattern Analysis and Machine Intelligence*, *32*(10), 1888–1898.
42. de Kruif, B. J., & de Vries, T. J. A. (2003). Pruning error minimization in least squares support vector machines. *IEEE Transactions on Neural Networks*, *14*(3), 696–702.
43. Dietterich, T., & Bakiri, G. (1995). Solving multiclass learning problems via error-correcting output codes. *Journal of Artificial Intelligence Research*, *2*, 263–286.
44. Downs, T., Gates, K. E., & Masters, A. (2001). Exact simplification of support vector solutions. *Journal of Machine Learning Research*, *2*, 293–297.
45. Drineas, P., & Mahoney, M. W. (2005). On the Nystrom method for approximating a gram matrix for improved kernel-based learning. *Journal of Machine Learning Research*, *6*, 2153–2175.
46. Dufrenois, F., Colliez, J., & Hamad, D. (2009). Bounded influence support vector regression for robust single-model estimation. *IEEE Transactions on Neural Networks*, *20*(11), 1689–1706.
47. Ertekin, S., Bottou, L., & Giles, C. L. (2011). Nonconvex online support vector machines. *IEEE Transactions on Pattern Analysis and Machine Intelligence*, *33*(2), 368–381.
48. Fan, R.-E., Chang, K.-W., Hsieh, C.-J., Wang, X.-R., & Lin, C.-J. (2008). LIBLINEAR: A library for large linear classification. *Journal of Machine Learning Research*, *9*, 1871–1874.
49. Fan, R.-E., Chen, P.-H., & Lin, C.-J. (2005). Working set selection using second order information for training support vector machines. *Journal of Machine Learning Research*, *6*, 1889–1918.
50. Fei, B., & Liu, J. (2006). Binary tree of SVM: A new fast multiclass training and classification algorithm. *IEEE Transactions on Neural Networks*, *17*(3), 696–704.
51. Ferris, M. C., & Munson, T. S. (2000). *Interior point methods for massive support vector machines*. Technical Report 00-05. Madison, WI: Computer Sciences Department, University of Wisconsin.
52. Fine, S., & Scheinberg, K. (2001). Efficient SVM training using low-rank kernel representations. *Journal of Machine Learning Research*, *2*, 243–264.
53. Flake, G. W., & Lawrence, S. (2002). Efficient SVM regression training with SMO. *Machine Learning*, *46*, 271–290.
54. Franc, V., & Sonnenburg, S. (2009). Optimized cutting plane algorithm for large-scale risk minimization. *Journal of Machine Learning Research*, *10*, 2157–2192.
55. Friess, T., Cristianini, N., & Campbell, C. (1998). The kernel-adatron algorithm: A fast and simple learning procedure for support vector machines. In *Proceedings of the 15th International Conference on Machine Learning* (pp. 188–196). Madison, WI.

56. Fung, G., & Mangasarian, O. (2001). Proximal support vector machines. In *Proceedings of the 7th ACM SIGKDD International Conference on Knowledge Discovery and Data Mining* (pp. 77–86). San Francisco, CA.
57. Fung, G., & Mangasarian, O. (2001). Semi-supervised support vector machines for unlabeled data classification. *Optimization Methods and Software*, *15*, 29–44.
58. Gentile, C. (2001). A new approximate maximal margin classification algorithm. *Journal of Machine Learning Research*, *2*, 213–242.
59. Girosi, F. (1998). An equivalence between sparse approximation and support vector machines. *Neural Computation*, *10*, 1455–1480.
60. Glasmachers, T., & Igel, C. (2006). Maximum-gain working set selection for SVMs. *Journal of Machine Learning Research*, *7*, 1437–1466.
61. Glasmachers, T., & Igel, C. (2008). Second-order SMO improves SVM online and active learning. *Neural Computation*, *20*, 374–382.
62. Grinblat, G. L., Uzal, L. C., Ceccatto, H. A., & Granitto, P. M. (2011). Solving nonstationary classification problems with coupled support vector machines. *IEEE Transactions on Neural Networks*, *22*(1), 37–51.
63. Gu, B., Sheng, V. S., Tay, K. Y., Romano, W., & Li, S. (2015). Incremental support vector learning for ordinal regression. *IEEE Transactions on Neural Networks and Learning Systems*, *26*(7), 1403–1416.
64. Gu, B., Sheng, V. S., Wang, Z., Ho, D., Osman, S., & Li, S. (2015). Incremental learning for ν-support vector regression. *Neural Networks*, *67*, 140–150.
65. Gu, B., Wang, J.-D., Yu, Y.-C., Zheng, G.-S., Huang, Y. F., & Xu, T. (2012). Accurate on-line ν-support vector learning. *Neural Networks*, *27*, 51–59.
66. Gunter, L., & Zhu, J. (2007). Efficient computation and model selection for the support vector regression. *Neural Computation*, *19*, 1633–1655.
67. Haasdonk, B. (2005). Feature space interpretation of SVMs with indefinite kernels. *IEEE Transactions on Pattern Analysis and Machine Intelligence*, *27*(4), 482–92.
68. Haasdonk, B., & Pekalska, E. (2008). Indefinite kernel Fisher discriminant. In *Proceedings of the 19th International Conference on Pattern Recognition* (pp. 1–4). Tampa, FL.
69. Hammer, B., & Gersmann, K. (2003). A note on the universal approximation capability of support vector machines. *Neural Processing Letters*, *17*, 43–53.
70. Hao, P.-Y. (2010). New support vector algorithms with parametric insensitive/margin model. *Neural Networks*, *23*, 60–73.
71. Hao, P.-Y. (2017). Pair-ν-SVR: A novel and efficient pairing ν-support vector regression algorithm. *IEEE Transactions on Neural Networks and Learning Systems*, *28*(11), 2503–2515.
72. Hastie, T., Rosset, S., Tibshirani, R., & Zhu, J. (2004). The entire regularization path for the support vector machine. *Journal of Machine Learning Research*, *5*, 1391–1415.
73. Hsu, C.-W., & Lin, C.-J. (2002). A comparison of methods for multiclass support vector machines. *IEEE Transactions on Neural Networks*, *13*(2), 415–425.
74. Huang, K., Jiang, H., & Zhang, X.-Y. (2017). Field support vector machines. *IEEE Transactions on Emerging Topics in Computational Intelligence*, *1*(6), 454–463.
75. Huang, K., Yang, H., King, I., & Lyu, M. R. (2008). Maxi-min margin machine: learning large margin classifiers locally and globally. *IEEE Transactions on Neural Networks*, *19*(2), 260–272.
76. Huang, K., Zheng, D., Sun, J., Hotta, Y., Fujimoto, K., & Naoi, S. (2010). Sparse learning for support vector classification. *Pattern Recognition Letters*, *31*, 1944–1951.
77. Huang, X., Shi, L., & Suykens, J. A. K. (2014). Support vector machine classifier with pinball loss. *IEEE Transactions on Pattern Analysis and Machine Intelligence*, *36*(5), 984–997.
78. Hush, D., & Scovel, C. (2003). Polynomial-time decomposition algorithms for support vector machines. *Machine Learning*, *51*, 51–71.
79. Hush, D., Kelly, P., Scovel, C., & Steinwart, I. (2006). QP Algorithms with guaranteed accuracy and run time for support vector machines. *Journal of Machine Learning Research*, *7*, 733–769.

80. Ikeda, K., & Murata, N. (2005). Geometrical properties of Nu support vector machines with different norms. *Neural Computation, 17*, 2508–2529.
81. Ito, N., Takeda, A., & Toh, K.-C. (2017). A unified formulation and fast accelerated proximal gradient method for classification. *Journal of Machine Learning Research, 18*, 1–49.
82. Jandel, M. (2010). A neural support vector machine. *Neural Networks, 23*, 607–613.
83. Jayadeva, Khemchandani, R., & Chandra, S. (2007). Twin support vector machines for pattern classification. *IEEE Transactions on Pattern Analysis and Machine Intelligence, 29*(5), 905–910.
84. Jiao, L., Bo, L., & Wang, L. (2007). Fast sparse approximation for least squares support vector machine. *IEEE Transactions on Neural Networks, 18*(3), 685–697.
85. Joachims, T. (1999). Making large-scale SVM learning practical. In B. Scholkopf, C. J. C. Burges, & A. J. Smola (Eds.), *Advances in kernel methods - support vector learning* (pp. 169–184). Cambridge: MIT Press.
86. Joachims, T. (1999). Transductive inference for text classification using support vector machines. In *Proceedings of the 16th International Conference on Machine Learning* (pp. 200–209). San Mateo: Morgan Kaufmann.
87. Joachims, T. (2006). Training linear SVMs in linear time. In *Proceedings of the 12th ACM SIGKDD International Conference on Knowledge Discovery and Data Mining* (pp. 217–226).
88. Joachims, T., Finley, T., & Yu, C.-N. J. (2009). Cutting-plane training of structural SVMs. *Machine Learning, 77*, 27–59.
89. Joachims, T., & Yu, C.-N. J. (2009). Sparse kernel SVMs via cutting-plane training. *Machine Learning, 76*, 179–193.
90. Jung, K.-H., Lee, D., & Lee, J. (2010). Fast support-based clustering method for large-scale problems. *Pattern Recognition, 43*, 1975–1983.
91. Kao, W.-C., Chung, K.-M., Sun, C.-L., & Lin, C.-J. (2004). Decomposition methods for linear support vector machines. *Neural Computation, 16*, 1689–1704.
92. Karal, O. (2017). Maximum likelihood optimal and robust support vector regression with lncosh loss function. *Neural Networks, 94*, 1–12.
93. Katagiri, S., & Abe, S. (2006). Incremental training of support vector machines using hyperspheres. *Pattern Recognition Letters, 27*, 1495–1507.
94. Keerthi, S. S., Chapelle, O., & DeCoste, D. (2006). Building support vector machines with reduced classifier complexity. *Journal of Machine Learning Research, 7*, 1493–1515.
95. Keerthi, S. S., & DeCoste, D. (2005). A modified finite Newton method for fast solution of large scale linear SVMs. *Journal of Machine Learning Research, 6*, 341–361.
96. Keerthi, S. S., & Gilbert, E. G. (2002). Convergence of a generalized SMO algorithm for SVM classifier design. *Machine Learning, 46*, 351–360.
97. Keerthi, S. S., & Shevade, S. K. (2003). SMO for least squares SVM formulations. *Neural Computation, 15*, 487–507.
98. Keerthi, S. S., Shevade, S. K., Bhattacharyya, C., & Murthy, K. R. K. (2001). Improvements to Platt's SMO algorithm for SVM classifier design. *Neural Computation, 13*(3), 637–649.
99. Khemchandan, R., Saigal, P., & Chandra, S. (2016). Improvements on ν-twin support vector machine. *Neural Networks, 79*, 97–107.
100. Klement, S., Anders, S., & Martinetz, T. (2013). The support feature machine: Classification with the least number of features and application to neuroimaging data. *Neural Networks, 25*(6), 1548–1584.
101. Knebel, T., Hochreiter, S., & Obermayer, K. (2008). An SMO algorithm for the potential support vector machine. *Neural Computation, 20*, 271–287.
102. Kramer, K. A., Hall, L. O., Goldgof, D. B., Remsen, A., & Luo, T. (2009). Fast support vector machines for continuous data. *IEEE Transactions on Systems, Man, and Cybernetics Part B, 39*(4), 989–1001.
103. Kressel, U. H.-G. (1999). Pairwise classification and support vector machines. In B. Scholkopf, C. J. C. Burges, & A. J. Smola (Eds.), *Advances in kernel methods - support vector learning* (pp. 255–268). Cambridge: MIT Press.

104. Kuh, A., & De Wilde, P. (2007). Comments on pruning error minimization in least squares support vector machines. *IEEE Transactions on Neural Networks*, *18*(2), 606–609.
105. Laskov, P., Gehl, C., Kruger, S., & Muller, K.-R. (2006). Incremental support vector learning: Analysis, implementation and applications. *Journal of Machine Learning Research*, *7*, 1909–1936.
106. Lee, D., & Lee, J. (2007). Equilibrium-based support vector machine for semisupervised classification. *IEEE Transactions on Neural Networks*, *18*(2), 578–583.
107. Lee, K. Y., Kim, D.-W., Lee, K. H., & Lee, D. (2007). Density-induced support vector data description. *IEEE Transactions on Neural Networks*, *18*(1), 284–289.
108. Lee, Y.-J., Hsieh, W.-F., & Huang, C.-M. (2005). ε-SSVR: A smooth support vector machine for ε-insensitive regression. *IEEE Transactions on Knowledge and Data Engineering*, *17*(5), 678–685.
109. Lee, C.-P., & Lin, C.-J. (2014). Large-Scale Linear RankSVM. *Neural Computation*, *26*, 781–817.
110. Lee, Y. J., & Mangasarian, O. L. (2001). RSVM: Reduced support vector machines. In *Proceedings of the 1st SIAM International Conference on Data Mining* (pp. 1–17). Chicago, IL.
111. Lee, Y. J., & Mangasarian, O. L. (2001). SSVM: A smooth support vector machine. *Computational Optimization and Applications*, *20*(1), 5–22.
112. Li, B., Song, S., & Li, K. (2013). A fast iterative single data approach to training unconstrained least squares support vector machines. *Neurocomputing*, *115*, 31–38.
113. Liang, X. (2010). An effective method of pruning support vector machine classifiers. *IEEE Transactions on Neural Networks*, *21*(1), 26–38.
114. Liang, X., Chen, R.-C., & Guo, X. (2008). Pruning support vector machines without altering performances. *IEEE Transactions on Neural Networks*, *19*(10), 1792–1803.
115. Lin, C.-J. (2001). On the convergence of the decomposition method for support vector machines. *IEEE Transactions on Neural Networks*, *12*(6), 1288–1298.
116. Lin, C.-J. (2002). Asymptotic convergence of an SMO algorithm without any assumptions. *IEEE Transactions on Neural Networks*, *13*(1), 248–250.
117. Lin, C.-J., Weng, R. C., & Keerthi, S. S. (2008). Trust region Newton method for logistic regression. *Journal of Machine Learning Research*, *9*, 627–650.
118. Lin, Y.-L., Hsieh, J.-G., Wu, H.-K., & Jeng, J.-H. (2011). Three-parameter sequential minimal optimization for support vector machines. *Neurocomputing*, *74*, 3467–3475.
119. Loosli, G., & Canu, S. (2007). Comments on the core vector machines: Fast SVM training on very large data sets. *Journal of Machine Learning Research*, *8*, 291–301.
120. Loosli, G., Canu, S., & Ong, C. S. (2016). Learning SVM in Krein spaces. *IEEE Transactions on Pattern Analysis and Machine Intelligence*, *38*(6), 1204–1216.
121. Luo, L., Xie, Y., Zhang, Z., & Li, W.-J. (2015). Support matrix machines. In *Proceedings of the 32nd International Conference on Machine Learning*. Lille, France.
122. Luss, R., & d'Aspremont, A. (2007). Support vector machine classification with indefinite kernels. *Advances in Neural Information Processing Systems* (Vol. 20, pp. 953–960). Vancouver, Canada.
123. Ma, J., Theiler, J., & Perkins, S. (2003). Accurate online support vector regression. *Neural Computation*, *15*(11), 2683–2703.
124. Ma, Y., Liang, X., Kwok, J. T., Li, J., Zhou, X., & Zhang, H. (2018). Fast-solving quasi-optimal LS-S^3VM based on an extended candidate set. *IEEE Transactions on Neural Networks and Learning Systems*, *29*(4), 1120–1131.
125. Mall, R., & Suykens, J. A. K. (2015). Very sparse LSSVM reductions for large-scale data. *IEEE Transactions on Neural Networks and Learning Systems*, *26*(5), 1086–1097.
126. Manevitz, L. M., & Yousef, M. (2001). One-class SVMs for document classification. *Journal of Machine Learning Research*, *2*, 139–154.
127. Mangasarian, O. L. (2000). Generalized support vector machines. In A. Smola, P. Bartlett, B. Scholkopf, & D. Schuurmans (Eds.), *Advances in large margin classifiers* (pp. 135–146). Cambridge: MIT Press.

128. Mangasarian, O. L. (2002). A finite Newton method for classification. *Optimization Methods and Software*, *17*(5), 913–929.
129. Mangasarian, O. L., & Musicant, D. R. (1999). Successive overrelaxation for support vector machines. *IEEE Transactions on Neural Networks*, *10*(5), 1032–1037.
130. Mangasarian, O. L., & Musicant, D. R. (2001). Lagrangian support vector machines. *Journal of Machine Learning Research*, *1*, 161–177.
131. Mangasarian, O. L., & Wild, E. W. (2006). Multisurface proximal support vector classification via generalized eigenvalues. *IEEE Transactions on Pattern Analysis and Machine Intelligence*, *28*(1), 69–74.
132. Marchand, M., & Shawe-Taylor, J. (2002). The set covering machine. *Journal of Machine Learning Research*, *3*, 723–746.
133. Martin, M. (2002). On-line support vector machine regression. In *Proceedings of the 13th European Conference on Machine Learning, LNAI* (Vol. 2430, pp. 282–294). Berlin: Springer.
134. Melacci, S., & Belkin, M. (2011). Laplacian support vector machines trained in the primal. *Journal of Machine Learning Research*, *12*, 1149–1184.
135. Munoz, A., & de Diego, I. M. (2006). From indefinite to positive semidefinite matrices. In *Proceedings of the Joint IAPR International Workshop on Structural, Syntactic, and Statistical Pattern Recognition* (pp. 764–772).
136. Musicant, D. R., & Feinberg, A. (2004). Active set support vector regression. *IEEE Transactions on Neural Networks*, *15*(2), 268–275.
137. Nandan, M., Khargonekar, P. P., & Talathi, S. S. (2014). Fast SVM training using approximate extreme points. *Journal of Machine Learning Research*, *15*, 59–98.
138. Navia-Vazquez, A. (2007). Support vector perceptrons. *Neurocomputing*, *70*, 1089–1095.
139. Nguyen, D., & Ho, T. (2006). A bottom-up method for simplifying support vector solutions. *IEEE Transactions on Neural Networks*, *17*(3), 792–796.
140. Nguyen, D. D., Matsumoto, K., Takishima, Y., & Hashimoto, K. (2010). Condensed vector machines: Learning fast machine for large data. *IEEE Transactions on Neural Networks*, *21*(12), 1903–1914.
141. Niu, G., Dai, B., Shang, L., & Sugiyama, M. (2013). Maximum volume clustering: A new discriminative clustering approach. *Journal of Machine Learning Research*, *14*, 2641–2687.
142. Ojeda, F., Suykens, J. A. K., & Moor, B. D. (2008). Low rank updated LS-SVM classifiers for fast variable selection. *Neural Networks*, *21*, 437–449.
143. Omitaomu, O. A., Jeong, M. K., & Badiru, A. B. (2011). Online support vector regression with varying parameters for time-dependent data. *IEEE Transactions on Systems, Man, and Cybernetics Part A*, *41*(1), 191–197.
144. Ong, C. S., Mary, X., Canu, S., & Smola, A. J. (2004). Learning with nonpositive kernels. In *Proceedings of the 21th International Conference on Machine Learning* (pp. 639–646). Banff, Canada.
145. Orabona, F., Castellini, C., Caputo, B., Jie, L., & Sandini, G. (2010). On-line independent support vector machines. *Pattern Recognition*, *43*(4), 1402–1412.
146. Osuna, E., Freund, R., & Girosi, F. (1997). An improved training algorithm for support vector machines. In *Proceedings of IEEE Workshop on Neural Networks for Signal Processing* (pp. 276–285). New York.
147. Osuna, E., Freund, R., & Girosi, F. (1997). *Support vector machines: Training and applications*. Technical Report A.I. Memo No. 1602, MIT Artificial Intelligence Laboratory.
148. Peng, X. (2010). TSVR: An efficient twin support vector machine for regression. *Neural Networks*, *23*(3), 365–372.
149. Peng, X. (2010). Primal twin support vector regression and its sparse approximation. *Neurocomputing*, *73*, 2846–2858.
150. Peng, X. (2010). A ν-twin support vector machine (ν-TSVM) classifier and its geometric algorithms. *Information Sciences*, *180*(20), 3863–3875.
151. Perez-Cruz, F., Navia-Vazquez, A., Rojo-Alvarez, J. L., & Artes-Rodriguez, A. (1999). A new training algorithm for support vector machines. In *Proceedings of the 5th Bayona Workshop on Emerging Technologies in Telecommunications* (pp. 116–120). Baiona, Spain.

152. Platt, J. (1999). Fast training of support vector machines using sequential minimal optimization. In B. Scholkopf, C. Burges, & A. Smola (Eds.), *Advances in kernel methods - support vector learning* (pp. 185–208). Cambridge: MIT Press.
153. Pontil, M., & Verri, A. (1998). Properties of support vector machines. *Neural Computation, 10*, 955–974.
154. Qi, Z., Tian, Y., & Shi, Y. (2015). Successive overrelaxation for Laplacian support vector machine. *IEEE Transactions on Neural Networks and Learning Systems, 26*(4), 674–683.
155. Rahimi, A., & Recht, B. (2007). Random features for large-scale kernel machines. In *Advances in neural information processing systems* (pp. 1177–1184).
156. Renjifo, C., Barsic, D., Carmen, C., Norman, K., & Peacock, G. S. (2008). Improving radial basis function kernel classification through incremental learning and automatic parameter selection. *Neurocomputing, 72*, 3–14.
157. Rifkin, R., & Klautau, A. (2004). In defense of one-vs-all classification. *Journal of Machine Learning Research, 5*, 101–141.
158. Roobaert, D. (2002). DirectSVM: A simple support vector machine perceptron. *Journal of VLSI Signal Processing, 32*, 147–156.
159. Scheinberg, K. (2006). An efficient implementation of an active set method for SVMs. *Journal of Machine Learning Research, 7*, 2237–2257.
160. Schleif, F.-M., & Tino, P. (2017). Indefinite core vector machine. *Pattern Recognition, 71*, 187–195.
161. Scholkopf, B., Smola, A. J., Williamson, R. C., & Bartlett, P. L. (2000). New support vector algorithm. *Neural Computation, 12*(5), 1207–1245.
162. Scholkopf, B., Herbrich, R., & Smola, A. J. (2001). A generalized representer theorem. In *Proceedings of the 14th Annual Conference on Computational Learning Theory, LNCS* (Vol. 2111, pp. 416–426). Berlin: Springer.
163. Scholkopf, B., Platt, J., Shawe-Taylor, J., Smola, A., & Williamson, R. (2001). Estimating the support of a high-dimensional distribution. *Neural Computation, 13*(7), 1443–1471.
164. Scholkopf, B., Mika, S., Burges, C. J. C., Knirsch, P., Muller, K. R., Ratsch, G., et al. (1999). Input space versus feature space in kernel-based methods. *IEEE Transactions on Neural Networks, 5*(10), 1000–1017.
165. Schraudolph, N., Yu, J., & Gunter, S. (2007). A stochastic quasi-Newton method for online convex optimization. In *Proceedings of the 11th International Conference on Artificial Intelligence and Statistics (AIstats)* (pp. 433–440). Society for AIstats.
166. Shalev-Shwartz, S., Singer, Y., & Srebro, N. (2007). Pegasos: Primal estimated sub-gradient solver for SVM. In *Proceedings of the 24th International Conference on Machine Learning (ICML)* (pp. 807–814). New York: ACM Press.
167. Shao, Y.-H., & Deng, N.-Y. (2012). A coordinate descent margin based-twin support vector machine for classification. *Neural Networks, 25*, 114–121.
168. Shao, Y.-H., Zhang, C.-H., Wang, X.-B., & Deng, N.-Y. (2011). Improvements on twin support vector machines. *IEEE Transactions on Neural Networks, 22*(6), 962–968.
169. Shashua, A. A. (1999). On the equivalence between the support vector machine for classification and sparsified Fisher's linear discriminant. *Neural Processing Letters, 9*(2), 129–139.
170. Shashua, A., & Levin, A. (2002). Ranking with large margin principle: Two approaches. *Advances in neural information processing systems* (Vol. 15, pp. 937–944).
171. Shen, X., Niu, L., Qi, Z., & Tian, Y. (2017). Support vector machine classifier with truncated pinball loss. *Pattern Recognition, 68*, 199–210.
172. Shevade, S. K., Keerthi, S. S., Bhattacharyya, C., & Murthy, K. R. K. (2000). Improvements to the SMO algorithm for SVM regression. *IEEE Transactions on Neural Networks, 11*(5), 1188–1193.
173. Shi, Y., Chung, F.-L., & Wang, S. (2015). An improved TA-SVM method without matrix inversion and its fast implementation for nonstationary datasets. *IEEE Transactions on Neural Networks and Learning Systems, 26*(9), 2005–2018.
174. Shin, H., & Cho, S. (2007). Neighborhood property-based pattern selection for support vector machines. *Neural Computation, 19*, 816–855.

175. Shilton, A., Palamiswami, M., Ralph, D., & Tsoi, A. (2005). Incremental training of support vector machines. *IEEE Transactions on Neural Networks*, *16*, 114–131.
176. Smola, A. J., & Scholkopf, B. (2000). Sparse greedy matrix approximation for machine learning. In *Proceedings of the 17th International Conference on Machine Learning* (pp. 911–918). Stanford University, CA.
177. Smola, A. J., & Scholkopf, B. (2004). A tutorial on support vector regression. *Statistics and Computing*, *14*(3), 199–222.
178. Suykens, J. A. K., & Vandewalle, J. (1999). Least squares support vector machine classifiers. *Neural Processing Letters*, *9*, 293–300.
179. Suykens, J. A. K., Lukas, L., Van Dooren, P., De Moor, B., & Vandewalle, J. (1999). Least squares support vector machine classifiers: A large scale algorithm. In *Proceedings of European Conference on Circuit Theory and Design* (pp. 839–842).
180. Suykens, J. A. K., Lukas, L., & Vandewalle, J. (2000). Sparse approximation using least squares support vector machines. In *Proceedings of IEEE International Symposium on Circuits and Systems (ISCAS)* (Vol. 2, pp. 757–760). Genvea, Switzerland.
181. Suykens, J. A. K., Van Gestel, T., De Brabanter, J., De Moor, B., & Vandewalle, J. (2002). *Least squares support vector machines*. Singapore: World Scientific.
182. Suykens, J. A. K., De Brabanter, J., Lukas, L., & Vandewalle, J. (2002). Weighted least squares support vector machines: Robustness and sparse approximation. *Neurocomputing*, *48*, 85–105.
183. Steinwart, I. (2003). Sparseness of support vector machines. *Journal of Machine Learning Research*, *4*, 1071–1105.
184. Takahashi, N., & Nishi, T. (2006). Global convergence of decomposition learning methods for support vector machines. *IEEE Transactions on Neural Networks*, *17*(6), 1362–1369.
185. Takahashi, N., Guo, J., & Nishi, T. (2008). Global convergence of SMO algorithm for support vector regression. *IEEE Transactions on Neural Networks*, *19*(6), 971–982.
186. Tan, Y., & Wang, J. (2004). A support vector machine with a hybrid kernel and minimal Vapnik-Chervonenkis dimension. *IEEE Transactions on Knowledge and Data Engineering*, *16*(4), 385–395.
187. Tax, D. M. J., & Duin, R. P. W. (1999). Support vector domain description. *Pattern Recognition Letters*, *20*, 1191–1199.
188. Tax, D. M. J. (2001). *One-class classification: Concept-learning in the absence of counter-examples*. Ph.D. dissertation. Delft, The Netherlands: Electrical Engineering, Mathematics and Computer Science, Delft University of Technology.
189. Tax, D. M. J., & Duin, R. P. W. (2004). Support vector data description. *Machine Learning*, *54*, 45–66.
190. Teo, C. H., Smola, A., Vishwanathan, S. V., & Le, Q. V. (2007). A scalable modular convex solver for regularized risk minimization. In *Proceedings of ACM SIGKDD International Conference on Knowledge Discovery and Data Mining (KDD)* (pp. 727–736).
191. Teo, C. H., Vishwanthan, S. V. N., Smola, A., & Le, Q. (2010). Bundle methods for regularized risk minimization. *Journal of Machine Learning Research*, *11*, 311–365.
192. Tipping, M. E. (2001). Sparse Bayesian learning and the relevance vector machine. *Journal of Machine Learning Research*, *1*, 211–244.
193. Tong, S., & Koller, D. (2001). Support vector machine active learning with applications to text classification. *Journal of Machine Learning Research*, *2*, 45–66.
194. Tsang, I. W., Kwok, J. T., & Cheung, P.-M. (2005). Core vector machines: Fast SVM training on very large data sets. *Journal of Machine Learning Research*, *6*, 363–392.
195. Tsang, I. W.-H., Kwok, J. T.-Y., & Zurada, J. M. (2006). Generalized core vector machines. *IEEE Transactions on Neural Networks*, *17*(5), 1126–1140.
196. Tsang, I. W., Kocsor, A., & Kwok, J. T. (2007). Simpler core vector machines with enclosing balls. In *Proceedings of the 24th International Conference on Machine Learning* (pp. 911–918). Corvalis, OR.
197. Valizadegan, H., & Jin, R. (2007). Generalized maximum margin clustering and unsupervised kernel learning. *Advances in neural information processing systems* (Vol. 19, pp. 1417–1424). Cambridge: MIT Press.

198. Vapnik, V. N. (1982). *Estimation of dependences based on empirical data*. New York: Springer.
199. Vapnik, V. N. (1995). *The nature of statistical learning theory*. New York: Springer.
200. Vapnik, V. N. (1998). *Statistical learning theory*. New York: Wiley.
201. Vapnik, V., & Chapelle, O. (2000). Bounds on error expectation for support vector machines. *Neural Computation*, *12*, 2013–2036.
202. Vishwanathan, S. V. N., Smola, A. J., & Murty, M. N. (2003). SimpleSVM. In *Proceedings of the 20th International Conference on Machine Learning* (pp. 760–767). Washington, DC.
203. Wang, Z., & Chen, S. (2007). New least squares support vector machines based on matrix patterns. *Neural Processing Letters*, *26*, 41–56.
204. Wang, G., Yeung, D.-Y., & Lochovsky, F. H. (2008). A new solution path algorithm in support vector regression. *IEEE Transactions on Neural Networks*, *19*(10), 1753–1767.
205. Wang, F., Zhao, B., & Zhang, C. (2010). Linear time maximum margin clustering. *IEEE Transactions on Neural Networks*, *21*(2), 319–332.
206. Wang, Z., Shao, Y.-H., Bai, L., & Deng, N.-Y. (2015). Twin support vector machine for clustering. *IEEE Transactions on Neural Networks and Learning Systems*, *26*(10), 2583–2588.
207. Warmuth, M. K., Liao, J., Ratsch, G., Mathieson, M., Putta, S., & Lemmem, C. (2003). Support vector machines for active learning in the drug discovery process. *Journal of Chemical Information Sciences*, *43*(2), 667–673.
208. Weston, J., & Watkins, C. (1999). Multi-class support vector machines. In M. Verleysen (Ed.), *Proceedings of European Symposium on Artificial Neural Networks*. Brussels: D. Facto Press.
209. Williams, C. K. I., & Seeger, M. (2001). Using the Nystrom method to speed up kernel machines. In T. Leen, T. Dietterich, & V. Tresp (Eds.), *Advances in neural information processing systems* (Vol. 13, pp. 682–688). Cambridge: MIT Press.
210. Wu, Q., & Zhou, D.-X. (2005). SVM soft margin classifiers: Linear programming versus quadratic programming. *Neural Computation*, *17*, 1160–1187.
211. Wu, M., Scholkopf, B., & Bakir, G. (2006). A direct method for building sparse kernel learning algorithms. *Journal of Machine Learning Research*, *7*, 603–624.
212. Xu, G., Hu, B.-G., & Principe, J. C. (2018). Robust C-loss kernel classifiers. *IEEE Transactions on Neural Networks and Learning Systems*, *29*(3), 510–522.
213. Xu, L., Neufeld, J., Larson, B., & Schuurmans, D. (2004). Maximum margin clustering. *Advances in neural information processing systems* (Vol. 17). Cambridge: MIT Press.
214. Yang, H., Huang, K., King, I., & Lyu, M. R. (2009). Localized support vector regression for time series prediction. *Neurocomputing*, *72*, 2659–2669.
215. Yang, X., Lu, J., & Zhang, G. (2010). Adaptive pruning algorithm for least squares support vector machine classifier. *Soft Computing*, *14*, 667–680.
216. Yu, H., Yang, J., Han, J., & Li, X. (2005). Making SVMs scalable to large data sets using hierarchical cluster indexing. *Data Mining and Knowledge Discovery*, *11*, 295–321.
217. Zeng, X. Y., & Chen, X. W. (2005). SMO-based pruning methods for sparse least squares support vector machines. *IEEE Transactions on Neural Networks*, *16*(6), 1541–1546.
218. Zhang, T., & Oles, F. J. (2001). Text categorization based on regularized linear classification methods. *Information Retrieval*, *4*(1), 5–31.
219. Zhang, K., Tsang, I. W., & Kwok, J. T. (2009). Maximum margin clustering made practical. *IEEE Transactions on Neural Networks*, *20*(4), 583–596.
220. Zheng, J., & Lu, B.-L. (2011). A support vector machine classifier with automatic confidence and its application to gender classification. *Neurocomputing*, *74*, 1926–1935.
221. Zhou, S. (2016). Sparse LSSVM in primal ssing Cholesky factorization for large-scale problems. *IEEE Transactions on Neural Networks and Learning Systems*, *27*(4), 783–795.

Chapter 22
Probabilistic and Bayesian Networks

22.1 Introduction

The Bayesian network model was introduced by Pearl in 1985 [146]. It is the best-known family of graphical models in artificial intelligence (AI). Bayesian networks are a powerful tool of common knowledge representation and reasoning for partial beliefs under uncertainty. They are probabilistic models that combine probability theory and graph theory. The formalism is sometimes called a *causal probabilistic network* or a *probabilistic belief network*. It possesses the characteristic of being both a statistical and a knowledge representation formalism. The formalism is model-based, as domain knowledge can be structured by exploiting causal and other relationships between domain variables.

Bayesian inference is widely established as one of the principal foundations for machine learning. A Bayesian network is essentially an expert system. It can be used for causality relationship modeling, uncertain knowledge representation, probabilistic inference, and reply to probabilistic query. Probabilistic graphical models are particularly appealing due to their natural interpretation.

The Bayesian network has wide applications in bioinformatics and medicine, engineering, classification, data fusion, and decision support systems. A well-known application is in medical diagnosis systems [46], which give the diagnosis given symptoms. Bayesian networks are well suited to human–computer intelligent interaction tasks because they are easily mapped onto a comprehensible graphical network representation. Some well-known applications in Microsoft are technical support troubleshooters, such as the Office Assistant in Microsoft Office ("Clippy"), which observe some user actions and give helpful advice, Microsoft Windows help system, and automates fault diagnostics such as the printer fault-diagnostic system and software debugging [19].

Special cases of Bayesian networks were independently invented by many different communities [168], such as genetics (linkage analysis), speech recognition (hidden Markov models), tracking (Kalman filtering), data compression (density estimation) and coding (turbo codes). Inference is solved by forward–backward algorithm,

K.-L. Du and M. N. S. Swamy, *Neural Networks and Statistical Learning*,
https://doi.org/10.1007/978-1-4471-7452-3_22

and the maximum a posteriori (MAP) problem is handled by Viterbi algorithm [149, 168]. Forward–backward and Viterbi algorithms are directly equivalent to Pearl's algorithms [146], which are valid for any probability model that can be represented as a graphical model. Kalman filters and related linear models for dynamical systems are similar to hidden Markov models, but the hidden state variables are real-valued rather than discrete.

22.1.1 Classical Versus Bayesian Approach

There are two main opposing schools of statistical reasoning, namely frequentist and Bayesian approaches. The frequentist or classical approach has dominated scientific research, but Bayesianism (Thomas Bayes, 1702–1761) is changing the situation. Whereas, a classical probability of an event X is a true or physical probability, the Bayesian probability of an event X is a person's degree of belief in that event, thus known as a personal probability. Unlike physical probability, the measurement of the personal probability does not need repeated trials.

In the classical approach of learning from data, the parameter θ is fixed, and all datasets of size N are assumed to be generated by sampling from the distribution determined by θ. Each dataset $\mathcal{D}$ occurs with some probability $p(\mathcal{D}|\theta)$ and will produce an estimate $\theta^*(\mathcal{D})$. To evaluate an estimator, we obtain the expectation and variance of the estimate:

$$\mathrm{E}_{p(\mathcal{D}|\theta)}(\theta^*) = \sum_{\mathcal{D}} p(\mathcal{D}|\theta)\theta^*(\mathcal{D}), \tag{22.1}$$

$$\mathrm{var}_{p(\mathcal{D}|\theta)}(\theta^*) = \sum_{\mathcal{D}} p(\mathcal{D}|\theta)\left[\theta^*(\mathcal{D}) - \mathrm{E}_{p(\mathcal{D}|\theta)}(\theta^*)\right]^2. \tag{22.2}$$

An estimator that balances the bias $\theta - \mathrm{E}_{p(\mathcal{D}|\theta)}(\theta^*)$ and the variance can be chosen. A commonly used estimator is the ML estimator, which selects the value of θ that maximizes the likelihood $p(\mathcal{D}|\theta)$.

In the Bayesian approach, $\mathcal{D}$ is fixed, and all values of θ are assumed to be possibly generated. The estimate of θ is the expectation of θ with respect to our posterior beliefs about its value:

$$\mathrm{E}_{p(\theta|\mathcal{D},\xi)}(\theta) = \int \theta p(\theta|\mathcal{D}, \xi) d\theta, \tag{22.3}$$

where ξ is state of information (background knowledge).

The estimations given by (22.1) and (22.3) are different, and in many cases, leads to different estimates. This is due to their different definitions.

22.1.2 Bayes' Theorem and Bayesian Classifiers

Bayes' theorem is the origin and fundamental of the Bayesian approach. In this chapter, we use $P(\cdot)$ to denote a probability.

Definition 22.1 (*Conditional probability*) Let A and B be two events, the conditional probability of A conditioned on B is defined by

$$P(A|B) = \frac{P(A, B)}{P(B)}, \tag{22.4}$$

where $P(A, B)$ is the joint probability of both A and B happening.

This leads to the chain rule: $P(A, B) = P(A|B)P(B)$.

Given the occurrence of evidence E depending on a hypothesis H, i.e., $H \longrightarrow E$, the probability for both E and H to happen can be expressed as

$$P(H, E) = P(H)P(E|H). \tag{22.5}$$

By symmetry,

$$P(H, E) = P(H)P(E|H) = P(E, H) = P(E)P(H|E), \tag{22.6}$$

from which we obtain Bayes' theorem. Bayes' theorem forms the basis of the probability distributions between the nodes of the Bayesian network.

Theorem 22.1 (Bayes' theorem) *Given the occurrence of evidence E depending on hypothesis H, there is a relation between the probabilities:*

$$P(E|H) = \frac{P(E)P(H|E)}{P(H)}. \tag{22.7}$$

From a statistical point of view, $P(E|H)$ denotes the conditional probability of (belief in) evidence E caused by the hypothesis H. $P(H|E)$ is the a posteriori belief in H faced with evidence E; it means the probability value such that when evidence E is detected the degree of belief that hypothesis H has actually occurred. $P(H)$ denotes a priori belief in hypothesis H. $P(E)$ is the prior probability of evidence E.

A conventional statistical classifier is the parametric Bayesian classifier. Bayesian classifiers use Bayes' theorem to predict the class label c of a vector $\boldsymbol{x}$:

$$c = \arg \max_{i=1,\ldots,K} P(c_i|\boldsymbol{x})P(\boldsymbol{x}) = \arg \max_{i=1,\ldots,K} P(\boldsymbol{x}|c_i)P(c_i), \tag{22.8}$$

where $P(c_i|\boldsymbol{x})$ is the posterior probability that $\boldsymbol{x}$ belongs to class c_i, $P(c_i)$ is the prior probability of class c_i, and $P(\boldsymbol{x}|c_i)$ is the class conditional pdf, respectively. The class prior $P(c_i)$ can be estimated simply from the fractions of training data in each of the classes.

Different Bayesian classifiers are obtained by the way $P(\boldsymbol{x}|c_i)$, $i = 1, \ldots, K$, is approximated. Naive Bayes is the simplest Bayesian classifier which assumes independence of the attributes given the class label, i.e., $P(\boldsymbol{x}|c_i) = P(x_1|c_i)P(x_2|c_i)\ldots P(x_d|c_i)$, $i = 1, \ldots, K$. Naive Bayes classifier has continued to be one of the top 10 algorithms for classification.

22.1.3 Graphical Models

Graphical models are graphs in which nodes represent random variables. The arcs between nodes represent conditional dependence. Undirected graphical models are called Markov networks or Markov random fields. Markov networks are popular with the physics and vision communities [71]. They are also used for spatial data mining. Examples of undirected probabilistic independent networks are Markov networks [71] and Boltzmann machines.

Definition 22.2 (*Conditional Independence*) Two sets of nodes $\mathcal{A}$ and $\mathcal{B}$ are said to be *conditionally independent*, if all paths between the nodes in $\mathcal{A}$ and $\mathcal{B}$ are separated by a node in a third set $\mathcal{C}$. More specifically, two random variables X and Y are conditionally independent given another random variable Z if

$$P(X|Z) = P(X|Y, Z). \tag{22.9}$$

Definition 22.3 (*d-Separation*) Let $\mathcal{V}$ be the set of nodes. Two variables A and B in a Bayesian network are d-separated by $\mathcal{X} \subseteq \mathcal{V}$ if all paths between A and B are blocked by $\mathcal{X}$.

Directed graphical models without directed cycles are called *Bayesian networks*; that is, Bayesian networks are directed acyclic graphs (DAGs). Independence defined for Bayesian networks has to take into account the directionality of the arcs. An arc from node A to node B can be regarded as A *causes* B.

Dynamic Bayesian networks are directed graphical models of stochastic processes. They generalize hidden Markov models (HMMs) and linear dynamical systems by representing the hidden and observed state in terms of state variables.

Example 22.1 The Bayesian network $Z \to Y \to X$ is shown in Fig. 22.1. We have $P(X|Y, Z) = P(X|Y)$, since Y is the only parent of X and Z is not a descendant of X (that is, X is conditionally independent of Z). Prove that independence is symmetric, that is, $P(Z|Y, X) = P(Z|Y)$.

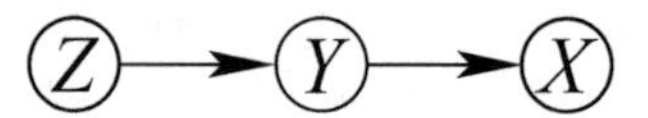

Fig. 22.1 A simple Bayesian network

Proof Since $P(X|Y, Z) = P(X|Y)$, we have

$$
\begin{aligned}
P(Z|X, Y) &= \frac{P(X, Y|Z)P(Z)}{P(X, Y)} && \text{(Bayes' rule)} \\
&= \frac{P(Y|Z)P(X|Y, Z)P(Z)}{P(X|Y)P(Y)} && \text{(Chain rule)} \\
&= \frac{P(Y|Z)P(X|Y)P(Z)}{P(X|Y)P(Y)} && \text{(By assumption)} \\
&= \frac{P(Y|Z)P(Z)}{P(Y)} = P(Z|Y) && \text{(Bayes' rule).}
\end{aligned}
$$

That completes the proof.

22.2 Bayesian Network Model

A Bayesian network encodes the joint probability distribution of a set of v variables in a problem domain, $\mathcal{X} = \{X_1, \ldots, X_v\}$, as a DAG $G = (V, E)$. Each node of this graph represents a random variable X_i in $\mathcal{X}$ and has a conditional probability table (CPT). Arcs stand for conditional dependence relationship among these nodes.

It is easy to identify the parent–child relationship or the probability dependency between two nodes. The parents of X_i are denoted by $\mathrm{pa}(X_i)$; the children of X_i are denoted by $\mathrm{ch}(X_i)$; and spouses of X_i (other parents of X_i's children) are denoted by $\mathrm{spo}(X_i)$. A CPT contains probabilities of the node being a specific value given its parents, that is, a CPT specifies the conditional probabilities $P(\boldsymbol{u}|\mathrm{pa}(X_i))$, where $\boldsymbol{u}$ is a configuration of the parents of X_i.

Definition 22.4 (*Markov Blanket*) The *Markov blanket* for node X_i is a set of all parents of X_i, children of X_i, and spouses of X_i.

In Bayesian networks, any variable X_i is independent of variables outside of Markov blanket of X_i, that is, $P(X_i|X_{\neq i}) = P(X_i|\text{Markov blanket}(X_i))$.

The Bayesian network model is based on the Markov condition: every variable is independent of its nondescendant nonparents given its parents. This leads to a unique joint probability density:

$$
p(\mathcal{X}) = \prod_{i=1}^{v} p(X_i|\mathrm{pa}(X_i)), \tag{22.10}
$$

where each X_i is associated with a conditional probability density $p(X_i|\mathrm{pa}(X_i))$. This holds as long as the network was designed such that $\mathrm{pa}(X_i) \in \{X_{i+1}, \ldots, X_n\}$. X_i is conditionally independent from all $X_j \in \mathrm{pa}(X_i)$.

A node can represent a discrete random variable that takes values from a finite set, and a numeric or continuous variable that takes values from a set of continuous numbers. Bayesian networks can thus be classified into discrete, continuous, and mixed Bayesian networks.

When a Bayesian network is used in conjunction with statistical techniques, the graphical model has several advantages for data analysis. As the model encodes dependencies among all variables, it readily handles situations where some data entries are missing. It is an ideal representation for combining prior knowledge and data.

There are some inherent limitations on Bayesian networks. Directly identifying the Bayesian network structures from input data $\mathcal{D}$ remains a challenge. While the resulting ability to describe the network can be performed in linear time, this process of network discovery is an NP-hard task for practically all model selection criteria such as AIC, BIC, and marginal likelihood [35, 46]. The second problem is concerned with the quality and extent of the prior beliefs used in Bayesian inference. A Bayesian network is useful only when this prior knowledge is reliable. Selecting a proper distribution model to describe the data has a notable effect on the quality of the resulting network.

Multiply sectioned Bayesian networks [199] relax the single Bayesian network paradigm. The framework allows a large domain to be modeled modularly and the inference to be performed distributively, while maintaining the coherence. Inference in a Bayesian network can be performed effectively using its junction tree representation. The multiply sectioned Bayesian network framework is an extension of these junction tree-based inference methods with the HUGIN method [93] the most relevant. Multiply sectioned Bayesian networks provide a coherent and flexible formalism for representing uncertain knowledge in large domains. Global consistency among subnets is achieved by communication.

Example 22.2 Probabilistic queries with respect to a Bayesian network are interpreted as queries with respect to the CPT the network specifies. Figure 22.2 depicts a simple Bayesian network with two of its CPTs:
The first CPT: $P(X = \mathrm{T}) = 0.3$, $P(X = \mathrm{F}) = 0.7$.
The second CPT: $P(Y = \mathrm{T}|X = \mathrm{T}) = 0$, $P(Y = \mathrm{F}|X = \mathrm{T}) = 1$, $P(Y = \mathrm{T}|X = \mathrm{F}) = 0.8$, $P(Y = \mathrm{F}|X = \mathrm{F}) = 0.2$.

Example 22.3 Assume the long-term experience with a specific kind of tumor is $P(\text{tumor}) = 0.01$ and $P(\text{no tumor}) = 0.99$. Tumor may cause positive testing result. The Bayesian network representation gives that the (causal) direction from tumor to positive is extracted from $P(\text{positive}|\text{tumor})$. The CPT for tumor $\rightarrow$ positive

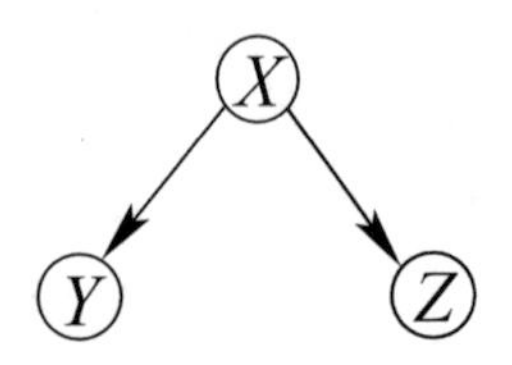

Fig. 22.2 A Bayesian network with two of its CPTs

is $P(\text{positive}|\text{tumor}) = 0.7$, $P(\text{negative}|\text{tumor}) = 0.3$, $P(\text{positive}|\text{no tumor}) = 0.2$, $P(\text{negative}|\text{no tumor}) = 0.8$. We now solve for $P(\text{tumor}|\text{positive})$.

Solution. We have

$$P(\text{positive}) = P(\text{positive}|\text{tumor})P(\text{tumor}) + P(\text{positive}|\text{no tumor}) \times P(\text{no tumor}) = 0.7 \times 0.01 + 0.2 \times 0.99 = 0.205.$$

From Bayes' theorem,

$$P(\text{tumor}|\text{positive}) = \frac{P(\text{positive}|\text{tumor})P(\text{tumor})}{P(\text{positive})} = \frac{0.7 \times 0.01}{0.205} = 0.0341.$$

After a positive test the probability of a tumor increases from 0.01 to 0.0341.

Probabilistic relational models [133] extend the standard attribute-based Bayesian network representation to incorporate a much richer relational structure. A probabilistic relational model, together with a particular database of objects and relations, defines a probability distribution over the attributes of the objects and the relations. A unified statistical framework for content and links [72] builds on probabilistic relational models. The standard approach for inference with a relational model is based on the generation of a propositional instance of the model in the form of a Bayesian network, and then applying classical algorithms, such as jointree [93], to compute answers to queries. A relational model describes a situation involving themes succinctly. This makes constructing a relational model much easier and less error-prone than constructing a Bayesian network. A relational model with a dozen or so general rules may correspond to a Bayesian network that involves hundreds of thousands of CPT parameters.

Cumulative distribution networks [88] is a class of graphical models for directly representing the joint cumulative distribution function of many random variables. In order to perform inference in such models, we describe the derivative-sum-product message-passing algorithm in which messages correspond to derivatives of the joint cumulative distribution function.

Datasets

Well-known benchmarks of Bayesian network learning algorithms include the Asia [108], Insurance [14], and Alarm [12] networks. The Asia network, shown in Fig. 22.3, is a small network that studies the effect of several parameters on the incidence of having lung cancer. The network has 8 nodes and 8 edges. The Insurance network was originally used for evaluating car insurance risks; it contains 27 variables and 52 edges. The Alarm network is used in the medical domain for potential anesthesia diagnosis in the operating room; it has 37 nodes, of which many have multiple values, and 46 directed edges.

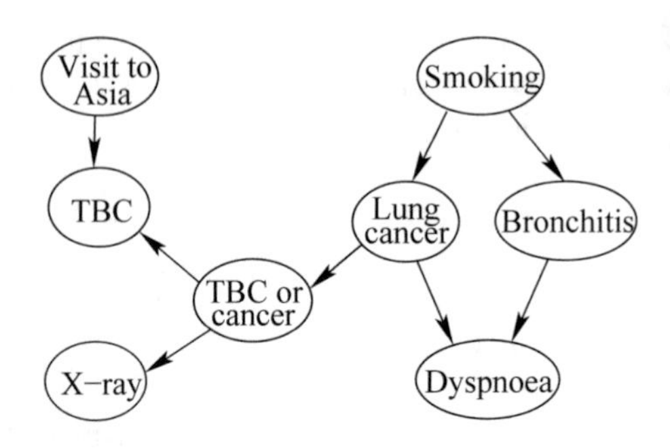

Fig. 22.3 The Asia network

22.3 Learning Bayesian Networks

Learning a Bayesian network from data requires the construction of the structure and CPTs from a given database of cases. It requires to learn the structure, the parameters for the structure (i.e., the conditional probabilities among variables), hidden variables, and missing values. Learning the structure is a much more challenging problem than estimating the parameters.

Given a training set $\mathcal{D} = \{\boldsymbol{x}_1, \ldots, \boldsymbol{x}_N\}$, the goal is to find a Bayesian network B that approximates the joint distribution $p(\boldsymbol{x})$. The network B can be found by maximizing the likelihood or the log-likelihood of the data:

$$L(B|\mathcal{D}) = \sum_{n=1}^{N} \log p_B(\boldsymbol{x}_n) = \sum_{n=1}^{N}\sum_{i=1}^{v} \log p_B(x_{n,i}|\text{pa}_{n,i}), \tag{22.11}$$

where $\text{pa}_{n,i}$ is the parent set of the ith variable for the data point $\boldsymbol{x}_n$.

The physical joint probability distribution for $\mathcal{D}$ is encoded in a network structure S. The problem of learning probabilities in a Bayesian network can now be stated: Given a random sample $\mathcal{D}$, compute the posterior distribution $p(\boldsymbol{\theta}_S|\mathcal{D}; S^h)$, where $\boldsymbol{\theta}_S$ is the vector of parameters $(\boldsymbol{\theta}_1; \ldots; \boldsymbol{\theta}_v)$, $\boldsymbol{\theta}_i$ being the vector of parameters for node i, and S^h denotes the event that the physical joint probability distribution can be factored according to the network structure S.

We have

$$p(\boldsymbol{x}|\boldsymbol{\theta}_S, S^h) = \prod_{i=1}^{v} p(x_i|\text{pa}_i, \boldsymbol{\theta}_i, S^h), \tag{22.12}$$

where $p(x_i|\text{pa}_i, \boldsymbol{\theta}_i, S^h)$ is a local distribution function. Thus, a Bayesian network can be viewed as a collection of probabilistic models, organized by conditional-independence relationships.

Popular Bayesian network software packages, such as Netica (http://www.norsys.com) and BNLearn [160], are designed for discrete variables. Some approaches extend Bayesian network learning algorithms to continuous variables. The conditional probability density of each continuous variable can be modeled using specific families of parametric distributions, and Bayesian network learning algorithms are then redesigned based on the parameterizations. Gaussian graphical model [195] is an example of such a method. Particle representations and Gaussian processes [91] are methods that use nonparametric distributions. Discretization approach [31] discretizes all continuous variables before learning a Bayesian network structure.

22.3.1 Learning the Structure

Structural learning of a Bayesian network can generally be divided into two classes of methods: independence analysis-based methods and score-based methods. Score-based approach [103] maps every structure of Bayesian network to a score and searches into the space of all structures for a good Bayesian network that fits the dataset best. Exhaustive search for the best network structure is NP-hard, even from a small sample dataset and when each node has at most two parents [35]. Identifying high-scoring DAGs from large datasets when using a consistent scoring criterion is also NP-hard [37]. A stochastic optimization method is usually used to search for the best network structure such as greedy search, iterated hill climbing, and simulated annealing. Score-based algorithms are more robust for small datasets, and it works with a wide range of probabilistic models.

Independence analysis-based algorithms do not require computation of the parameters of the model during the structure discovery process, and thus are efficient. They are generally more efficient than the score-based approach for sparse networks. However, most of these algorithms need an exponential number of conditional independence tests. Conditional independence tests with large condition sets may be unreliable unless the volume of the dataset is enormous [46]. The conditional independence approach is equivalent to minimizing Kullback–Leibler divergence using the score-based approach [47]. Hybrid methods take advantage of both the approaches [184].

An incremental structural learning method gradually modifies a Bayesian network structure to fit a sequential stream of observations, where the underlying distribution can change during the sampling of the database [134].

Independence Analysis-Based Approach

Exploiting the fact that the network structure represents the conditional independence, independence analysis-based methods find causal relations between the random variables, and deduce the structure of the graph [170]. They conduct a number of conditional independence tests on the data, successively constrain the number of possible structures consistent with the results of those tests to a singleton (if possible), and infer that the structure as the only possible one. The tests are usually done by using statistical [170] or information-theoretic [32] measures. The independence-based approach has been exemplified by the SGS [170], PC [170], GES [36], and grow-shrink [122] algorithms.

In an independence analysis-based method, search for separators of vertex pairs is a key issue for orientation of edges and for recovering DAG structures and causal relationships among variables. To recover structures of DAGs, the inductive causation algorithm [190] searches for a separator of two variables from all possible variable subsets such that the two variables are independent conditionally on the separator. A systematic way of searching for separators in increasing the order of cardinality is implemented in the PC algorithm [170].

When ignoring the directions of a DAG, one gets the skeleton of a DAG. The PC algorithm starts from a complete, undirected graph and deletes recursively edges based on conditional independence decisions. It runs in the worst case in exponential time with respect to the number of nodes, but if the true underlying DAG is sparse, which is often a reasonable assumption, this reduces to a polynomial runtime [95]. The PC algorithm is asymptotically consistent for the equivalence class of the DAG and its skeleton with corresponding very high-dimensional, sparse Gaussian distribution [95]. It is computationally feasible for such high-dimensional, sparse problems. The R-package CPDAG can be used to estimate from data the underlying skeleton or equivalence class of a DAG. For low-dimensional problems, there are a number of other implementations of the PC algorithm: Hugin, Murphy's Bayes Network Toolbox and Tetrad IV. An extensive comparative study of different algorithms is given in [184].

In [200], a recursive method for structural learning of DAGs is proposed, in which the problem of structural learning for a DAG is recursively decomposed into two problems of structural learning for two vertex subsets until no subset can be decomposed further. Search for separators of a pair of variables in a large DAG is localized to small subsets, and thus the approach can improve the efficiency of searches and the power of statistical tests for structural learning. These locally learned subgraphs are finally gradually combined into the entire DAG. Statistical test is used to determine a skeleton as in the inductive causation algorithm [190] and the PC algorithm [170].

Score-Based Approach

The score of a structure is generally based on Occam's razor principle. The score can be based on penalized log-likelihood such as AIC, BIC, and MDL criteria, or Bayesian scoring methods such as K2 [46] and BDeu [82]. The Bayesian score is equivalent to the marginal likelihood of the model given the data. For most criteria, Bayesian network structures are interpreted as independence constraints in some distribution from which the data was generated.

The K2 criterion applied to a DAG G evaluates the relative posterior probability that the generative distribution has the same independence constraints as those entailed by G. The criterion is not score-equivalent because the prior distributions over parameters corresponding to different structures are not always consistent. The (Bayesian) BDeu criterion [82] measures the relative posterior probability that the distribution from which the data were generated has the independence constraints given in the DAG. It uses a parameter prior that has uniform means, and requires both a prior equivalence sample size and a structure prior.

Although the BIC score decomposes into a sum of local terms, one per node, local search is still expensive, because we need to run EM at each step. An alternative iterative approach is to do the local search steps inside the M-step of EM: this is called structural EM, and provably converges to a local maximum of the BIC score [65]. Structural EM can adapt the structure in the presence of hidden variables, but usually performs poorly without prior knowledge about the cardinality and location of the hidden variables.

A general approach for learning Bayesian networks with hidden variables [60] builds on the information bottleneck framework and its multivariate extension. The approach is able to avoid some of the local maxima in which EM can get trapped when learning with hidden variables. The algorithmic framework allows learning of the parameters as well as the structure of a network.

The Bayesian BDeu criterion is very sensitive to the choice of prior hyperparameters [166]. AIC and BIC are derived through asymptotics and their behavior is suboptimal for small sample sizes. It is hard to set the parameters for the structures selected with AIC or BIC. Factorized normalized ML [166] is an effective scoring criterion, with no tunable parameters. The combination of the factorized normalized ML criterion and a sequential normalized ML-based parameter learning method yields a complete non-Bayesian method for learning Bayesian networks [167]. The approach is based on the minimax optimal normalized ML distribution, motivated by the MDL principle. Computationally, the method is parameter-free, robust, and as efficient as its Bayesian counterparts.

Algorithm B [21] is a greedy construction heuristic. It starts with an empty DAG and adds at each step, the arc with the maximum increase in the (decomposable) scoring metric such as BIC, but avoiding the inclusion of directed cycles in the graph. The algorithm ends when adding any additional valid arc does not increase the value of the metric.

An exact score-based structure discovery algorithm is given for Bayesian networks of a moderate size (say, 25 variables or less) by using dynamic programming in [100]. A parallel implementation of the score-based optimal structure search using dynamic programming has $O(n2^n)$ time and space complexity [173]. It is possible to learn the best Bayesian network structure with over 30 variables (http://b-course.hiit.fi/bene) [165].

A distributed algorithm for computing the MDL in learning Bayesian networks from data is presented in [104]. The algorithm exploits both properties of the MDL-based score metric and a distributed, asynchronous, adaptive search technique called nagging. Nagging is intrinsically fault-tolerant, has dynamic load balancing features, and scales well. The distributed algorithm can provide optimal solutions for larger problems as well as good solutions for Bayesian networks of up to 150 variables.

The problem of learning Bayesian network structures from data based on score functions that are decomposable is addressed in [54]. It describes properties that strongly reduce the time and memory costs of many known methods without losing global optimality guarantees. These properties are derived for different score criteria such as MDL (or BIC), AIC, and Bayesian Dirichlet criterion. A branch-and-bound algorithm integrates structural constraints with data in a way to guarantee global optimality. In [29], a conditional independence test-based approach is used to find node-ordering information, which is then fed to the K2 algorithm for structure learning. The performance of the algorithm is mainly dependent on the stage that identifies the order of the nodes, which avoids exponential complexity.

An asymptotic approximation of the marginal likelihood of data is presented in [157], given a naive Bayesian model with binary variables. It proves that the BIC score that penalizes the log-likelihood of a model by $\frac{d}{2}\ln N$ is incorrect for Bayesian networks with hidden variables and suggests an adjusted BIC score. Moreover, no uniform penalty term exists for such models in the sense that the penalty term depends on the averaged sufficient statistics. This claim stands in contrast to linear and curved exponential families, where the BIC score has been proven to provide a correct asymptotic approximation for the marginal likelihood.

The sparse Bayesian network structure learning algorithm proposed in [89] employs a formulation involving one L_1-norm penalty term to impose sparsity and another penalty term to ensure that the learned Bayesian network is a DAG. The sparse Bayesian network has a computational complexity that is linear in the sample size and quadratic in the number of variables. This makes the sparse Bayesian network more scalable and efficient than most of the existing algorithms.

K2 Algorithm

The basic idea of the Bayesian approach is to maximize the probability of the network structure given the data, that is, to maximize $P(B_S|\mathcal{D})$ over all possible network structures B_S given the cases of the dataset $\mathcal{D}$.

K2 algorithm [46] is a well-known greedy search algorithm for learning structure of Bayesian networks from the data. It uses a Bayesian scoring metric known as K2 metric, which measures the joint probability of a Bayesian network G and a dataset $\mathcal{D}$. K2 metric assumes that a complete dataset $\mathcal{D}$ of sample cases over a set of attributes that accurately models the network is given. K2 algorithm assumes that a prior ordering on the nodes is available and that all structures are equally likely. It searches, for every node, the set of parent nodes that maximizes the K2 metric. If there is a predefined ordering of variables, K2 algorithm can efficiently determine the structure.

A probabilistic network B over the set of variables $\mathcal{U}$ is a pair $B = (B_S, B_P)$ where the network structure B_S is a DAG with a node for every variable in $\mathcal{U}$ and B_P is a set of CPTs associated with B_S. For every variable $X_i \in \mathcal{U}$, the set B_P contains a CPT $P(X_i|\text{pa}_i)$ that enumerates the probabilities of all values of X_i given all combinations of values of the variables in its parent set pa_i in the network structure B_S. The network B represents the joint probability distribution represented $P(\mathcal{U})$ defined by $P(\mathcal{U}) = \prod_{i=1}^{n} P(X_i|\text{pa}_i)$.

K2 algorithm is a greedy heuristic algorithm for selecting a network structure that considers at most $O(n^3)$ different structures for n nodes. All nodes are considered independent of one another. For each node, a parent set is calculated by starting with the empty parent set and successively adding to the parent set the node that maximally improves $P(B_S, \mathcal{D})$ until no more node can be added such that $P(B_S, \mathcal{D})$ increases, for dataset $\mathcal{D}$ and all possible network structures B_S. A major drawback of K2 algorithm is that the ordering that is chosen on the nodes influences the resulting network structure and the quality of this structure. K2 algorithm is given in Algorithm 22.1.

Algorithm 22.1 (Algorithm K2)

1. *Let the variables of* $\mathcal{U}$ *be ordered* $X_1, \ldots, X_n$.
2. ***for*** $i = 1, \ldots, n$
   $\mathrm{pa}_i^{\mathrm{new}} \leftarrow \mathrm{pa}_i^{\mathrm{old}} \leftarrow \emptyset$.
3. ***for*** $i = 2, \ldots, n$
   Repeat until $(\mathrm{pa}_i^{\mathrm{new}} = \mathrm{pa}_i^{\mathrm{old}}$ ***or*** $|\mathrm{pa}_i^{\mathrm{new}}| = i - 1)$:
   a. $\mathrm{pa}_i^{\mathrm{old}} \leftarrow \mathrm{pa}_i^{\mathrm{new}}$.
   b. Let B_S *be defined by* $\mathrm{pa}_1^{\mathrm{old}}, \ldots, \mathrm{pa}_n^{\mathrm{old}}$.
   c.
   $$Z \leftarrow \arg\max_Y \left\{ \frac{P(B_{S_Y}, D)}{P(B_S, D)} | Y \in \{X_1, \ldots, X_{i-1}\right\} \setminus \mathrm{pa}_{i,old},$$
   where B_{S_Y} *is* B_S *but with* $\mathrm{pa}_i = \mathrm{pa}_i^{\mathrm{old}} \cup \{Z\}$.
4. *Output* B_S *defined by* $\mathrm{pa}_1^{\mathrm{new}} \ldots \mathrm{pa}_n^{\mathrm{new}}$.

The algorithm for learning discrete Bayesian networks from continuous data (https://github.com/sisl/LearnDiscreteBayesNets.jl) proposed in [31] alternates between K2 structure learning and discretization. It has quadratic complexity instead of the cubic complexity of other standard techniques.

K3 [15] is a modification of K2 where the Bayesian measure is replaced by the MDL measure and a uniform prior distribution over network structures is assumed. The MDL measure is approximately equal to the logarithm of the Bayesian measure. The results are comparable for K2 and K3, but K3 tends to be slightly faster, outputting network structures with fewer arcs than K2. A major drawback of both K2 and K3 is that their performance is highly dependent on the ordering on the variables taken as point of departure. The two measures have the same properties for infinite large databases. For smaller databases, the MDL measures assign equal quality to networks that represent the same set of independencies while the Bayesian measure does not.

22.3.2 Learning the Parameters

The goal of parametric learning is to find the parameters of each cumulative probability density that maximizes the likelihood of the training data. The ML method leads, with the classical decomposition of the joint probability in a product, to estimate separately each term of the product with the data. It asymptotically converges toward the true probability, if the proposed structure is exact. The Bayesian method tries to calculate the most probable parameters given the data, and this is equivalent to weight the parameters with an a priori law. The most used prior is the Dirichlet distribution. The VC dimension of the set of instanced Bayesian networks is upper bounded by the number of parameters of the set [70].

When learning the parameters of a Bayesian network with missing values or hidden variables, the common approach is to use the EM algorithm [56]. EM performs a greedy search of the likelihood surface and converges to a local stationary point. EM is faster and simpler than the gradient-descent method since it uses the natural gradient.

The concavity of the log-likelihood surface for logistic regression is a well-known result. The condition, under which Bayesian network models correspond to logistic regression with completely freely varying parameters, is given in [154]. Only then can we guarantee that there are no local maxima in the likelihood surface.

An inductive transfer learning method for Bayesian networks [118] induces a model for a target task from data of this task and of other related auxiliary tasks. The method includes both structure and parameter learning. The structure learning method is based on the PC algorithm, and it combines the dependency measures obtained from data in the target task, with those obtained from data in the auxiliary tasks. The parameter learning algorithm uses an aggregation process, combining the parameters estimated from the target task, with those estimated from the auxiliary data. A significant improvement is observed in terms of structure and parameters when knowledge is transferred from similar tasks [118].

Online Learning of the Parameters

Let Z_i be a node in the network that takes any value from the set $\{z_i^1, \ldots, z_i^{r_i}\}$. Let Pa_i be the set of parents of Z_i in the network that takes one of the configurations denoted by $\{\mathrm{pa}_i^1, \ldots, \mathrm{pa}_i^{q_i}\}$. An entry in the CPT of the variable Z_i is given by $\theta_{ijk} = P(Z_i = z_i^k | \Pi_i = \mathrm{pa}_i^j)$. We are given a dataset $\mathcal{D} = \{\boldsymbol{y}_1, \ldots, \boldsymbol{y}_t, \ldots\}$, and we have a current set of parameters $\bar{\boldsymbol{\theta}}$ that defines the network. The dataset is either complete or incomplete.

The network parameters are updated by

$$\tilde{\boldsymbol{\theta}} = \arg\max_{\boldsymbol{\theta}} [\eta L_{\mathcal{D}}(\boldsymbol{\theta}) - d(\boldsymbol{\theta}, \bar{\boldsymbol{\theta}})], \tag{22.13}$$

where $L_{\mathcal{D}}(\boldsymbol{\theta})$ is the normalized log-likelihood of the data given the network, $d(\boldsymbol{\theta}, \bar{\boldsymbol{\theta}})$ is the χ^2-distance between the two models and η is the learning rate.

Online learning of Bayesian network parameters has been discussed in [10, 169]. EM(η) [10] is derived by solving the maximization subject to the constraint that $\sum_k \theta_{ijk} = 1 \forall i, j$. Voting EM [42] is obtained by adapting EM(η) to online learning.

Voting EM takes a frequentist approach while the Spiegelhalter–Lauritzen algorithm [169] uses a Bayesian approach to parameter estimation.

Voting EM is given by

$$\theta_{ijk}(t) = \theta_{ijk}(t-1) + \eta \left[\frac{P(z_i^k, \mathrm{pa}_i^j | \boldsymbol{y}_t, \boldsymbol{\theta}(t-1))}{P(\mathrm{pa}_i^j | \boldsymbol{y}_t, \boldsymbol{\theta}(t-1))} - \theta_{ijk}(t-1) \right]$$

$$\text{if} \quad P(\mathrm{pa}_i^j | \boldsymbol{y}_t, \boldsymbol{\theta}(t-1)) \neq 0; \tag{22.14}$$

$$\theta_{ijk}(t) = \theta_{ijk}(t-1) \quad \text{otherwise.} \tag{22.15}$$

The learning rate $\eta \in (0, 1)$ controls the influence of the past; it can be selected by the Robbins–Monro conditions.

Mixtures of truncated exponentials is a model for dealing with discrete and continuous variables simultaneously in Bayesian networks without imposing any restriction on the network topology and avoiding rough approximations of methods based on the discretization of the continuous variables [130]. A method for inducing the structure of such network from data is proposed in [153].

22.3.3 Constraint-Handling

Constraints can be embedded within belief networks by modeling each constraint as a CPT. One approach is to add a new variable for each constraint that is perceived as its effect (child node) in the corresponding causal relationship and then to clamp its value to true [45, 146].

The use of several types of structural restrictions within algorithms for learning Bayesian networks is considered in [53]. These restrictions may codify expert knowledge in a given domain, in such a way that a Bayesian network representing this domain should satisfy them. Three types of restrictions are formally defined: existence of arc and/or edge, absence of arc and/or edge, and ordering restrictions. Two learning algorithms are investigated: a score-based local search algorithm, with the operators of arc addition, arc removal and arc reversal, and the PC algorithm.

A framework that combines deterministic and probabilistic networks [124] allows two distinct representations: causal relationships that are directional and normally quantified by CPTs and symmetrical deterministic constraints. In particular, a belief network has a set of variables instantiated (e.g., evidence) as a mixed network, by regarding the evidence set as a set of constraints.

Models that use different types of parameter sharing include dynamic Bayesian networks, HMMs, and Kalman filters. Parameter sharing methods constrain parameters to share the same value, but do not capture more complicated constraints among parameters such as inequality constraints or constraints on sums of parameter values.

22.4 Bayesian Network Inference

After constructing a Bayesian network, one can obtain various probabilities from the model. The computation of a probability from a model is known as probabilistic inference. Inference using a Bayesian network, also called belief updating, is based on Bayes' theorem as well as the CPTs. Given the joint probability distribution, all possible inference queries can be answered by marginalization (summing out over irrelevant variables). Bayesian network inference is to compute the inference probability $P(X = x|E = e)$, i.e., the probabilities of query nodes $(X = x)$ given the values of evidence nodes $(E = e)$.

Exact methods exploit the independence structure contained in the network to efficiently propagate uncertainty [146]. A message-passing scheme updates the probability distributions for each node in a Bayesian network in response to observations of one or more variables. The commonly used exact algorithm for discrete variables is the evidence propagation algorithm in [108], improved later in [93], which first transforms the Bayesian network into a tree where each node in the tree corresponds to a subset of variables in X and then exploits properties of this tree to perform probabilistic inference. This algorithm needs an ordering of the variables in order to make the triangulation of the moral graph associated with the original Bayesian network structure. A graph is triangulated if it has no cycles with a length greater than three without a cord. Obtaining the best triangulation for a Bayesian network is an NP-hard problem.

Although we use conditional independence to simplify probabilistic inference, exact inference in an arbitrary Bayesian network for discrete variables is NP-hard [45]. Even approximate inference is NP-hard [48], but approximate methods perform this operation in an acceptable amount of time. Approximating the inference probability in any sense, even for a single evidence node, is NP-hard [49]. Some popular approximate inference methods are sampling methods, variational Bayesian methods, and loopy belief propagation [206]. A randomized approximation algorithm, called the bounded-variance algorithm [49], is a variant of the likelihood-weighting algorithm.

The explaining Bayesian network inferences [205] procedure explains how variables interact to reach conclusions. The approach explains the value of a target node in terms of the influential nodes in the target's Markov blanket under specific contexts. Working back from the target node, the approach shows the derivation of each intermediate variable, and finally explains how missing and erroneous evidence values are compensated.

22.4.1 *Belief Propagation*

A complete graph is a graph with every pair of vertices joined by an edge. A clique is a complete subgraph—a set of vertices that are all adjacent to one another. In other words, the set of nodes in a clique is fully connected. A clique is maximal if no other vertices can be added to it so as to still yield a clique.

Belief propagation is a popular method of performing approximate inference on arbitrary graphical models. The belief propagation algorithm, also called the *sum-product algorithm*, is a popular means of solving inference problems exactly on a tree, but approximately in graphs with cycles [146]. It calculates the marginal pdfs for random variables by passing messages in a graphical model. Belief propagation has its optimality for tree-structured graphical models (with no loops). It is also widely applied to graphical models with cycles for approximate solution. Some additional justifications for loopy belief propagation have been developed in [177].

For tree-structured graphical models, belief propagation can be used to efficiently perform exact marginalization. However, as observed in [146], one may also apply belief propagation to arbitrary graphical models by following the same local message-passing rules at each node and ignoring the presence of cycles in the graph; this procedure is typically referred to as *loopy belief propagation*.

The goal of belief propagation is to compute the marginal distribution $p(X_t)$ at each node t. Belief propagation takes form of a message-passing algorithm between nodes, expressed in terms of an update to the outgoing message at iteration i from each node t to each neighbor s in terms of the previous iteration's incoming messages from node t's neighbors Γ_t. Typically, each message is normalized so as to integrate (sum) to unity.

The sum-product algorithm [102] is a generic message-passing algorithm that operates in a factor graph. It computes, either exactly or approximately, various marginal functions derived from the global function. Factor graphs are a straightforward generalization of the Tanner graphs obtained by applying them to functions. Bounds on the accumulation of errors in the system of approximate belief propagation message-passing are given in [92]. This analysis leads to convergence conditions for traditional belief propagation message-passing.

A wide variety of algorithms developed in AI, signal processing, and digital communications can be derived as specific instances of the sum-product algorithm, including the forward–backward algorithm, the Viterbi algorithm, the iterative turbo or LDPC decoding algorithm [58], Pearl's belief propagation algorithm [146] for Bayesian networks, the Kalman filter, and certain FFT algorithms [102]. The forward–backward algorithm, sometimes referred to as the BCJR or MAP algorithm in coding theory, is an application of the sum-product algorithm to HMM or to the trellises in which certain variables are observed at the output of a memoryless channel. The basic operations in the forward–backward recursions are therefore sums of products. When all codewords are a priori equally likely, MAP amounts to ML sequence detection. The Viterbi algorithm operates in the forward direction only; however, since memory of the best path is maintained and some sort of traceback is performed in making a decision, it might be viewed as being bidirectional.

Variational message-passing [197] applies variational inference to Bayesian networks. Like belief propagation, the method proceeds by sending messages between nodes in the network and updating posterior beliefs using local operations at each node. In contrast to belief propagation, it can be applied to a very general class of conjugate-exponential models. By introducing additional variational parameters, it can be applied to models containing non-conjugate distributions. The method is guaranteed to converge to a local minimum of the Kullback–Leibler divergence. It has been implemented in a general-purpose inference engine VIBES (http://vibes.sourceforge.net) which allows models to be specified graphically or in a text file containing XML. XMLBIF (XML for Bayesian networks interchange format) is an XML-based file format for representing Bayesian networks.

Loopy Belief Propagation

In loopy belief propagation [206], Pearl's belief propagation algorithm is applied to the original graph, even if it has loops (undirected cycles). Loopy belief propagation can be used to compute approximate marginals in Bayesian networks and Markov random fields. However, when applied to graphs with cycles, it does not always converge after any number of iterations. In practice, the procedure often arrives at a reasonable set of approximations to the correct marginal distributions. Double-loop algorithms guarantee convergence [207], but are an order of magnitude slower than standard loopy belief propagation.

The effect of n iterations of loopy belief propagation at any particular node s is equivalent to exact inference on a tree-structured unrolling of the graph from s. The computation tree with depth n consists of all length-n paths emanating from s in the original graph which do not immediately backtrack (though they may eventually repeat nodes).

The convergence and fixed points of loopy belief propagation may be considered in terms of a Gibbs measure on the graph's computation tree [177]. Loopy belief propagation is guaranteed to converge if the graph satisfies certain condition [177]. Fixed points of loopy belief propagation correspond to extrema of the so-called Bethe free energy [206]. Sufficient conditions for the uniqueness of loopy belief propagation fixed points are derived in [84].

Lazy propagation [107] is a scheme for modeling and exact belief update in the restricted class of mixture Bayesian networks known as conditional linear Gaussian Bayesian networks that is more general than Pearl's scheme. The basic idea of lazy propagation is to instantiate potentials to reflect evidence and to postpone the combination of potentials until it becomes mandatory by a variable elimination operation. Lazy propagation yields a reduction in potential domain sizes and a possibility of avoiding some of the postponed potential combinations. In traditional message-passing schemes, a message consists of a single potential over the variables shared by the sender and receiver cliques. In lazy propagation, a message consists of a set of potentials.

22.4.2 Factor Graphs and Belief Propagation Algorithm

A Bayesian network can easily be represented as a factor graph by introducing a factor for each variable, namely, the CPT of the variable given its parents in the Bayesian network. A Markov random field can be represented as a factor graph by taking the clique potentials as factors. Factor graphs naturally express the factorization structure of probability distributions; thus, they form a convenient representation for approximate inference algorithms that exploit this factorization. The exact solution to MAP inference in graphical models is well-known to be exponential in the size of the maximal cliques of the triangulated model, while approximate inference is typically exponential in the size of the model's factors.

A factor graph is an undirected graph consisting of nodes and edges. The factor nodes $\{F\}$ and the variable nodes $\{X\}$ are the two types of nodes in a factor graph. A factor node F represents a real-valued function and a variable node X represents a random variable. An edge E connecting a factor node F and a variable node X is an unordered pair $E = (F, X) = (X, F)$. There is one edge $E = (F, X)$ connecting a factor node F and a variable node X if and only if F is a function of X. There are no edges between any two-factor nodes or any two variable nodes. The factor node F represents a pdf and the product of all factors equals the joint pdf of the probability model.

Example 22.4 Consider an error-correcting code, expressed by a system of linear equations

$$X_1 + X_2 + X_3 = 0,$$
$$X_2 + X_4 + X_5 = 0,$$
$$X_1 + X_4 + X_6 = 0,$$

where the variables take value of 0 or 1, and addition is modulo-2. By denoting the operation + as boxes and the variables as circles, the factor graph is as shown in Fig. 22.4.

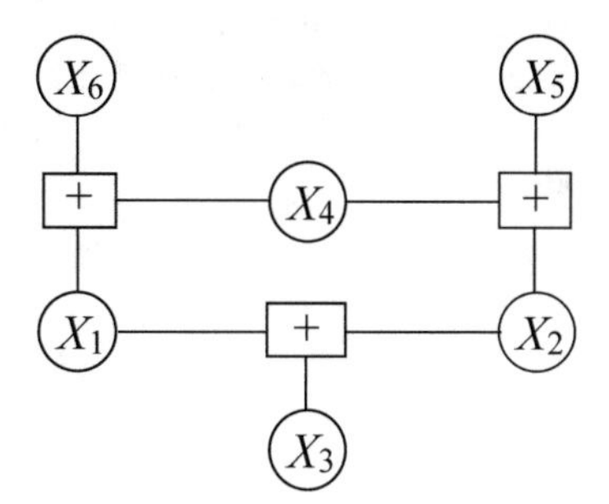

Fig. 22.4 A factor graph for the error-correcting code

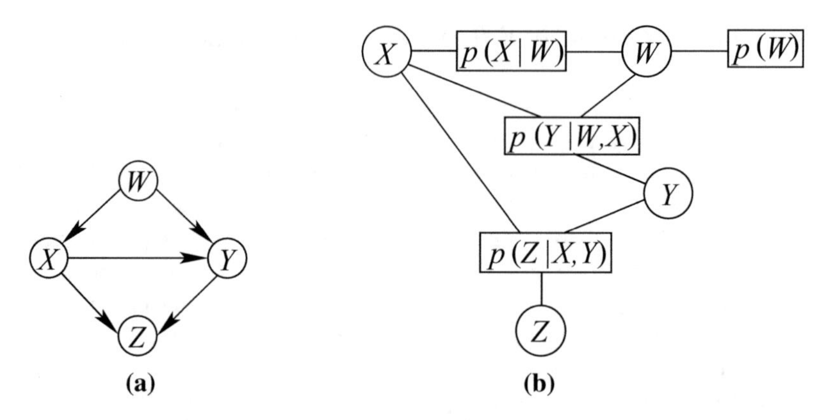

Fig. 22.5 A Bayesian network of four variables. **a** Bayesian network. **b** Factor graph

Example 22.5 Consider a Bayesian network with four variables X, Y, Z, W, as shown in Fig. 22.5a. The graph represents probability factorization:

$$p(X, Y, Z, W) = p(Z|X, Y)p(Y|X, W)p(X|W)p(W).$$

The corresponding factor graph is shown in Fig. 22.5b.

Any probability model can be represented by a factor graph by the following three steps:

- Factorize the joint pdf of the probability model as a product of pdfs and conditional pdfs;
- Associate each random variable with a variable node and each pdf with a factor node;
- Link one variable node X to one-factor node F if the pdf F is a function of the random variable X.

The belief propagation algorithm operates by passing messages along edges in a factor graph. Messages on an edge $E = (F, X)$ are functions of the variable node X. Let $\gamma_{X \to F}$ be the message from X to F and $\gamma_{F \to X}$ be the message from F to X. The message $\gamma_{X \to F}(X)$ (or $\gamma_{F \to X}(X)$) can be interpreted as an approximation of the pdf on all the information that the variable node X (or the factor node F) currently has except for that coming from the factor node F (or the variable node X). Let $n(\cdot)$ be the set of neighboring nodes of a node. The message update rules are summarized below:

- The message $\gamma_{X \to F}$ is updated by

$$\gamma_{X \to F}(X) = c_1 \prod_{G \in n(X) \setminus \{F\}} \gamma_{G \to X}(X), \tag{22.16}$$

where the constant c_1 makes $\gamma_{X \to F}(X)$ a pdf.

- The message $\gamma_{F\to X}$ is updated by

$$\gamma_{F\to X}(X) = c_2 \int_{\sim\{X\}} \left(F(\cdot) \prod_{Y\in n(f)\backslash\{X\}} \gamma_{Y\to F}(Y) \right) d(\sim\{X\}), \tag{22.17}$$

where the constant c_2 makes $\gamma_{F\to X}(X)$ a pdf, $F(\cdot)$ is the pdf associated with the factor node F, and $\int_{\sim\{X\}}$ is the integration over all arguments of the integrand except X.

The order of message propagation is not unique if messages are propagated through all edges in both directions in one iteration. The marginal pdf of X can be estimated by

$$\hat{F}(X) = c_3 \prod_{G\in n(X)} \gamma_{G\to X}(X), \tag{22.18}$$

where the constant c_3 makes $\hat{F}(X)$ a pdf.

When the factor graph contains no loop, the belief propagation algorithm produces exact marginal pdfs. However, when there are loops in the factor graph, belief propagation only performs approximate inference. Although it is difficult to prove its convergence, the loopy belief propagation algorithm is widely used in practice in view of its low computational complexity and giving good approximation to the marginal pdfs in many cases. Applications of loopy belief propagation include the iterative decoding of Turbo codes and LDPC codes [58].

The belief propagation-based sparse Bayesian learning algorithm discussed in [174] recovers sparse transform coefficients in large-scale compressed sensing problems. It is based on a hierarchical Bayesian model, which is turned into a factor graph. The algorithm has a computational complexity that is proportional to the number of transform coefficients.

Stochastic belief propagation [136] is an adaptively randomized version of belief propagation, in which each node passes randomly chosen information to each of its neighbors. This significantly reduces the computational complexity from $O(m^2)$ to $O(m)$ per iteration, and the communication complexity from $m-1$ real numbers for standard belief propagation message updates to only $\log_2 m$ bits per edge and iteration, where m is the state dimension.

AD3 (http://www.ark.cs.cmu.edu/AD3) [123] is an algorithm for approximate MAP inference on factor graphs, based on the alternating directions method of multipliers (ADMM). AD3 has a modular architecture, where local subproblems are solved independently, and their solutions are gathered to compute a global update. Each local subproblem has a quadratic regularizer, leading to faster convergence.

Sequential reweighted message-passing [101] is a family of message-passing techniques for MAP estimation in graphical models. It can be significantly slower than the popular max-product LP algorithm [74]. Special cases include a faster sequential tree-reweighted message-passing. A family of convex message-passing algorithms [80] are used as one of the building blocks. These algorithms all perform block-coordinate ascent on the objective function.

22.5 Sampling (Monte Carlo) Methods

Monte Carlo sampling is a standard tool for inference in statistical models. It represents a family of simulation-based algorithms that aim at generating samples from a target posterior distribution. The algorithms, referred to as *Markov chain Monte Carlo (MCMC) algorithms*, are based on constructing a Markov chain that has the desired distribution as its equilibrium distribution. The most prominent MCMC algorithms are the Metropolis–Hastings and Gibbs sampling algorithms.

MCMC is an efficient approach in high dimensions. It is a general class of algorithms used for optimization, search, and learning. The key idea is to build an ergodic Markov chain whose equilibrium distribution is the desired posterior distribution. MCMC is a sampling scheme for surveying a space S with a prescribed probability measure π. It has particular importance in Bayesian analysis, where $x \in S$ represents a vector of parameters and $\pi(x)$ is the posterior distribution of the parameters conditioned on the data. MCMC can also be used to solve the so-called missing data problem in frequentist statistics, where $x \in S$ represents the value of a latent or unobserved random variable, and $\pi(x)$ is its distribution conditioned on the data. In either case, MCMC serves as a tool for numerical computation of complex integrals and is often the only workable approach for problems involving a large space with a complex structure.

The numerical integration problem consists of calculating the value of the integral $\int f(x)\pi(x)dx$. MCMC approach consists in defining a Markov chain X_t with equilibrium distribution π, and to use the theorem that

$$\int f(x)\pi(x)dx \approx \frac{1}{t}\sum_{t=1}^{T} f(X_t) \tag{22.19}$$

for large T. In other words, for large number of samples the value of the integral is approximately equal to the average value of $f(X_t)$.

Probably the most widely used version of MCMC is the Metropolis–Hastings method [79]. Metropolis–Hastings algorithms are a class of Markov chains that are commonly used to perform large-scale calculations and simulations in physics and statistics. Two common problems which are approached by Metropolis–Hastings algorithms are simulation and numerical integration. The Metropolis–Hastings method gives a solution to the following problem: Construct an ergodic Markov chain with states $1, 2, \ldots, N$ and with a prescribed stationary distribution vector $\boldsymbol{\pi}$. Constructing a Markov chain amounts to defining its state transition probabilities.

Simulated annealing can be regarded as MCMC with a varying temperature. The approach consists in thinking of the function to be maximized as a probability density, and running a Markov chain with this density as the equilibrium distribution. At the same time the chain is run, the density is gradually raised to higher powers (i.e., lowering the temperature). In the limit as the temperature goes to zero, the successive densities turn into a collection of point masses located on the set of global maxima of the function. Thus when the temperature is very low, the corresponding Markov chain will spend most of its time near the global maxima.

The stochastic approximation Monte Carlo algorithm [114] has a self-adjusting mechanism. If a proposal is rejected, the weight of the subregion that the current sample belongs to is adjusted to a larger value, and thus the proposal of jumping out from the current subregion is less likely to be rejected in the next iteration. Annealing stochastic approximation Monte Carlo [113] can be regarded as a space annealing version of stochastic approximation Monte Carlo. Under mild conditions, it can converge weakly at a rate of $O(1/\sqrt{t})$ toward a neighboring set (in the space of energy) of the global minimizers.

Reversible-jump MCMC [76] is a framework for the construction of reversible Markov chain samplers that jump between parameter subspaces of differing dimensionality. It is essentially a random sweep Metropolis–Hastings method. This iterative algorithm does not depend on the initial state. At each step, a transition from the current state to a new state is accepted with a probability. This acceptance ratio is computed so that the detailed balance condition is satisfied, under which the algorithm converges to the measure of interest. A characteristic feature of the algorithm is that the proposition kernel can be decomposed into several kernels, each corresponding to a reversible move. In order to ensure the jump between different dimensions, the various moves used are the birth move, death move, split move, merge move, and perturb move, each selected with equal probability (0.2) [3]. In [3], simulated annealing with reversible-jump MCMC method is also proposed for optimization, with proved convergence.

22.5.1 Gibbs Sampling

Introduced in [71], Gibbs sampling can be used to approximate any function of an initial joint distribution $p(\boldsymbol{X})$ provided certain conditions are met. Given variables $\boldsymbol{X} = \{X_1, \ldots, X_K\}$ with some joint distribution $p(\boldsymbol{X})$, we can use a Gibbs sampler to approximate the expectation of a function $f(\boldsymbol{x})$ with respect to $p(\boldsymbol{X})$. The Gibbs sampler is an iterative adaptive scheme. First, the Gibbs sampler proceeds by generating a value for the conditional distribution for each component of $\boldsymbol{X}$, given the values of all other components of $\boldsymbol{X}$.

The Gibbs sampling implements iterations for a vector $\boldsymbol{x} = (x_1, x_2, \ldots, x_K)$ at each time t as: $x_1(t)$ is drawn from the distribution of X_1, given $x_2(t-1), x_3(t-1), \ldots, x_K(t-1)$, followed by implementations for $x_i, i = 2, \ldots, K$. Each component of the random vector is visited in the natural order. The new value of component $x_i, i = 1, 2, \ldots, K, i \neq k$ is immediately used for drawing a new value of X_k.

Then, we sample a state for X_i based on this probability distribution, and compute $f(\boldsymbol{X})$. The two steps are iterated, keeping track of the average value of $f(\boldsymbol{X})$. In the limit, as the number of cases approach infinity, this average is equal to $E_{p(X)}(f(\boldsymbol{X}))$ provided two conditions are met. First, the Gibbs sampler must be irreducible: $p(\boldsymbol{X})$ must be such that we can eventually sample any possible configuration of $\boldsymbol{X}$ given any possible initial configuration of $\boldsymbol{X}$. Second, each X_i must be chosen infinitely often. In practice, an algorithm for deterministically rotating through the variables is typically used. Gibbs sampling is a special case of the general MCMC method for approximate inference [132]. Like the Metropolis–Hastings algorithm, the Gibbs sampler generates a Markov chain with the Gibbs distribution as the equilibrium distribution. However, the transition probabilities associated with the Gibbs sampler are nonstationary [71].

Under mild conditions, the following three theorems hold for Gibbs sampling [69, 71].

Theorem 22.2 (Convergence Theorem) *The random variable $X_k(n)$ converges in distribution to the true probability distributions of X_k for $k = 1, 2, \ldots, K$ as n approaches infinity.*

Specifically, convergence of Gibbs sampling still holds under an arbitrary visiting scheme provided that this scheme does not depend on the values of the variables and that each component of $\boldsymbol{X}$ is visited on an infinitely-often basis [71].

Theorem 22.3 (Convergence of joint cumulative distribution) *Assuming that the components of $\boldsymbol{X}$ are visited in the natural order, the joint cumulative distribution of the random variables $X_1(n), X_2(n), \ldots, X_K(n)$ converges to the true joint cumulative distribution of $X_1, X_2, \ldots, X_K$ at a geometric rate in n.*

Theorem 22.4 (Ergodic Theorem) *For any measurable function g of the random variables $X_1, X_2, \ldots, X_K$ whose expectation exists, we have*

$$\lim_{n\to\infty} \frac{1}{n} \sum_{i=1}^{n} g(X_1(i), X_2(i), \ldots, X_K(i)) \to E[g(X_1, X_2, \ldots, X_K)],$$

with probability 1 (i.e., almost surely).

The ergodic theorem tells us how to use the output of the Gibbs sampler to obtain numerical estimations of the desired marginal densities.

Gibbs sampling is used in the Boltzmann machine to sample from distribution over hidden neurons. In the context of a stochastic machine using binary units (e.g., the Boltzmann machine), it is noteworthy that Gibbs sampler is exactly the same as a variant of Metropolis–Hastings algorithm.

Herded Gibbs [30] is a herding variant of Gibbs sampler, and is entirely deterministic. Herded Gibbs has an $O(1/T)$ convergence rate for models with independent variables and for fully connected probabilistic graphical models [30].

22.5.2 Importance Sampling

Importance sampling is a popular sampling tool used for Monte Carlo computing [152]. Together with MCMC methods, importance sampling has provided a foundation for simulation-based approaches to numerical integration since its introduction as a variance-reduction technique in statistical physics [78].

Importance sampling refers to a collection of Monte Carlo methods where a mathematical expectation with respect to a target distribution is approximated by a weighted average of random draws from another simple distribution. The weighting of the samples accounts for the mismatch between the target and the simple densities, and the desired inference is performed using the weighted samples. In importance sampling, we draw random samples $\boldsymbol{x}_i$ from the distribution on the hidden variables $p(\boldsymbol{X})$, and then weight the samples by their likelihood $p(y|\boldsymbol{x}_i)$, where y is the evidence.

Importance sampling methods are able to estimate normalizing constants of the target distribution, but this feature is not shared by MCMC methods. The performance of importance sampling methods depends on the selected densities [152]. During implementation, weight degeneracy may occur: only few of the importance sampling weights take relevant values, while the rest are negligible [152].

During implementation, the quality of the samples improves with iterations, and the inference becomes more accurate. This leads to adaptive importance sampling, which is able to learn from previously sampled values of the unknowns and are consequently more accurate.

Particle filters are based on importance sampling with a bootstrap resampling step, which aims at combating against weight degeneracy. Sequential importance sampling is a version of particle filters.

22.5.3 Particle Filtering

Particle filters [77], also known as *sequential Monte Carlo methods* [55], are used to model stochastic dynamical systems from a series of observations over a period of time, in which the posterior distribution of the target state is updated recursively as a new observation comes in.

Particle filters assume that the state transition is a first-order Markov process, and they use a number of particles to sequentially compute the expectation of any function of the state.

Particle filtering is a well-established Bayesian Monte Carlo technique for estimating the current state of a hidden Markov process using a fixed number of samples. Unlike Gaussian-based Kalman filters, particle filters are more suitable for nonlinear and non-Gaussian models and impose much weaker constraints on the underlying models, particularly with a sufficiently large number of samples.

Particle filtering builds the posterior density function of state variables using several random samples called *particles*. A large number of particles are required to obtain a small variance.

Particle filtering has a prediction phase and an updating phase. In the prediction phase, each particle is modified according to the existing model. In the updating phase, the weight of each particle is reevaluated using the last observation, and particles with lower weights are removed. Filtering gives an estimation of a weighted sum of several particles.

Let $\boldsymbol{y}(t)$ denote the state variable describing the location and shape of a target at time t. For all the t observations $\boldsymbol{x}(1:t) = \{\boldsymbol{x}(1), \ldots, \boldsymbol{x}(t)\}$, the distribution of the target is inferred by Bayes' rule as

$$p(\boldsymbol{y}(t)|\boldsymbol{x}(1:t-1)) = \int p(\boldsymbol{y}(t)|\boldsymbol{y}(t-1))p(\boldsymbol{y}(t-1)|\boldsymbol{x}(1:t-1))d\boldsymbol{y}(t-1), \tag{22.20}$$

$$p(\boldsymbol{y}(t)|\boldsymbol{x}(1:t)) \propto p(\boldsymbol{x}(t)|\boldsymbol{y}(t))p(\boldsymbol{y}(t)|\boldsymbol{x}(1:t-1)), \tag{22.21}$$

where $p(\boldsymbol{y}(t)|\boldsymbol{y}(t-1))$ is the state transition distribution and $p(\boldsymbol{x}(t)|\boldsymbol{y}(t))$ is the observation likelihood estimated by the appearance model.

The posterior probability is approximated by a finite set of n particles $\{\boldsymbol{y}^i(t)\}_{i=1}^n$ with importance weight $w^i(t)$. At each t, $w^i(t)$ is updated by the observation likelihood $p(\boldsymbol{x}(t)|\boldsymbol{y}^i(t))$ according to the bootstrap filter

$$w^i(t) \propto w^i(t-1)p(\boldsymbol{x}(t)|\boldsymbol{y}^i(t)). \tag{22.22}$$

Then, a set of n equally weighted particles are resampled according to the importance weight using $p(\boldsymbol{y}(t)|\boldsymbol{y}(t-1))$.

Particle MCMC [4] combines sequential Monte Carlo and MCMC. Particle Gibbs sampler uses sequential Monte Carlo to construct efficient, high-dimensional MCMC kernels. Particle Gibbs is similar to standard particle filter, but with the difference that one particle trajectory is specified a priori as a reference trajectory. Particle Gibbs with ancestor sampling [115] alleviates the path degeneracy problem by modifying the particle Gibbs kernel with an ancestor sampling step, thereby achieving the same effect as backward sampling.

22.6 Variational Bayesian Methods

The mean field methods such as variational Bayes, loopy belief propagation, expectation propagation, and expectation consistent have aroused much interest because of their potential as approximate Bayesian inference engines [94, 126, 139–141, 206]. They represent a very attractive compromise between accuracy and computational complexity. The mean field approximation exploits the law of large numbers to approximate large sums of random variables by their means. In many instances, the typical polynomial complexity makes the mean field methods the only available Bayesian inference option. An undesirable property of the mean field methods is that the approximation error is unattainable. A message-passing interpretation of the mean field approximation is given in [197].

Expectation consistent and expectation propagation are closely related to the adaptive Thouless–Anderson–Palmer mean field theory framework [139, 140]. Expectation consistent is a formalization of the same underlying idea aiming at giving an approximation to the marginal likelihood. A set of complementary variational distributions, which share sufficient statistics, arise naturally in the framework. Expectation propagation is an intuitively appealing scheme for tuning the variational distributions to achieve this expectation consistency.

Variational Bayesian, so-called *ensemble learning*, methods provide an approach for the design of approximate inference algorithms [85, 94, 127, 188]. Variational Bayesian learning is an approximation to Bayesian learning. They are very flexible general modeling tools. Variational Bayesian is developed to approximate posterior density, with the objective of minimizing the misfit between them [189]. It can be used for both parametric and variational approximation. In the parametric approximation, some parameters for the posterior pdf are optimized, while in variational approximation a whole function is optimized. Variational Bayesian learning can avoid overfitting, which is difficult to avoid if ML or MAP estimation is used.

Variational Bayesian treatments of statistical models present significant advantages over ML-based alternatives: ML approaches have the undesirable property of being ill-posed since the likelihood function is unbounded from above [203]. The adoption of a Bayesian model inference algorithm, providing posterior distributions over the model parameters instead of point estimates, would allow for the natural resolution of these issues [203]. Another central issue that ML treatments of generative models are confronted with is the selection of the optimal model size.

Variational Bayesian is an EM-like iterative procedure, and it is guaranteed to increase monotonically at each iteration. It alternately performs an E-step and an M-step. It requires only a modest amount of computational time, compared to that of EM. The BIC and MDL criteria for model selection are obtained from the variational Bayesian method in a large sample limit [6]. In this limit, variational Bayesian is equivalent to EM. Model selection in variational Bayesian is automatically accomplished by maximizing an estimation function. Variational Bayesian often shows a better generalization ability than EM when the number of data points is small. The online variational Bayesian algorithm [159] has proved convergence. It is a gradient method with the inverse of the Fisher information matrix for the posterior parameter distribution as a coefficient matrix [159], that is, the variational Bayesian method is a type of natural-gradient method.

The stochastic complexity in variational Bayesian learning, which is an important quantity for selecting models, is also called the variational free energy and corresponds to a lower bound for the marginal likelihood or the Bayesian evidence [159]. Variational Bayesian learning of Gaussian mixture models is discussed in [192] and upper and lower bounds of variational stochastic complexities are derived. Variational Bayesian learning of mixture models of exponential families that include mixtures of distributions such as Gaussian, binomial, and gamma is treated in [193]. A variational Bayesian algorithm for Student-t mixture models [5] is useful for constructing robust mixture models.

22.7 Hidden Markov Models

HMMs are probabilistic finite state machines used to find structures in sequential data. An HMM is defined by a set of states, the transition probabilities between states, and a table of emission probabilities associated with each state for all possible symbols that occur in the sequence. This allows domain information to be built into its structure while allowing fine details to be learned from the data by adjusting the transition and emission probabilities. At each step, the system transits to another state and emits an observable quantity to a state-specific probability distribution. HMMs are very successful for speech recognition and gene prediction.

Definition 22.5 An HMM is defined by
(1) A set of N possible states $\mathcal{Q} = \{q_1, q_2, \ldots, q_N\}$.
(2) A state transition matrix $\mathbf{A} = [a_{ij}]$, where $a_{ij} = P(x_t = j|x_{t-1} = i)$, $i, j = 1, \ldots, N$, denotes the probability of making a transition from state q_i to q_j, $\sum_{j=1}^{N} a_{ij} = 1 \forall i$, and x_t denotes the state at time t.
(3) A prior distribution over the state of the system at an initial time.
(4) A set of M possible outputs $\mathcal{O} = \{o_1, o_2, \ldots, o_M\}$.
(5) A state-conditioned probability distribution over observations $\mathbf{B} = [b_{jk}]$, where $b_{jk} = P(y_t = k|x_t = j) \geq 0, j = 1, \ldots, N, k = 1, \ldots, M$, denotes the observation probability for state q_j over observations o_k, $\sum_{k=1}^{M} b_{jk} = 1, \forall j$, and y_t denotes the output at time t.

First-order HMM is a simple probability model, as shown in Fig. 22.6. There exist efficient algorithms ($O(N)$) for solving the inference and MAP problems. An HMM has one discrete hidden node and one discrete or continuous observed node per slice. If the Bayesian network is acyclic, one can use a local message-passing algorithm, which is a generalization of the forward–backward algorithm for HMMs.

Denote a fixed length observation sequence by $\boldsymbol{y} = (y_1, y_2, \ldots, y_n)$ and the corresponding state sequence by $\boldsymbol{x} = (x_1, x_2, \ldots, x_n)$. HMM defines a joint probability distribution over observations as

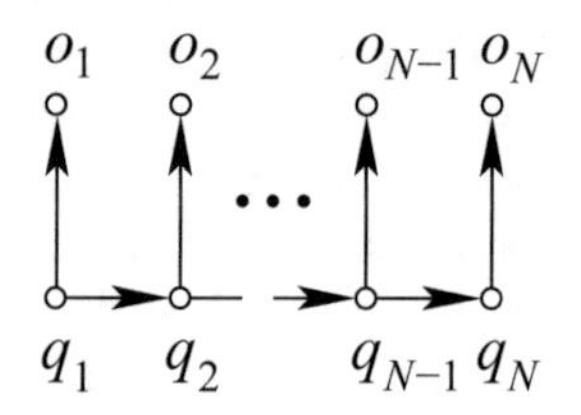

Fig. 22.6 DAG of a first-order HMM. q_i and o_i, $i = 1, \ldots, N$, are the hidden state variables and the observable variables, respectively

$$P(\mathbf{y}) = \sum_{\mathbf{x}} P(\mathbf{x})P(\mathbf{y}|\mathbf{x})$$
$$= \sum_{\mathbf{x}} P(x_1)P(x_2|x_1)\cdots P(x_n|x_{n-1})P(y_1|x_1)P(y_2|x_2)\cdots P(y_n|x_n)$$
$$= \sum_{\mathbf{x}} P(x_1)P(y_1|x_1)\prod_{i=2}^{n} P(x_i|x_{i-1})P(y_i|x_i), \tag{22.23}$$

where $P(x_i|x_{i-1})$ is specified by the state transition matrix, and $P(y_i|x_i)$ is specified by the model.

The standard method of estimating the parameters of HMM is the EM algorithm [56, 149]. Learning consists of adjusting the model parameters to maximize the likelihood of a given training sequence $\mathbf{y}_{0\to T}$. The EM procedure is guaranteed to converge to a local ML estimate for the model parameters under very general conditions. Unfortunately, EM is strongly dependent on the selection of the initial values of the model parameters. The simulated annealing version of EM [90] combines simulated annealing with EM by reformulating the HMM estimation process using a stochastic step between the EM steps and simulated annealing. In contrast to HMMs, Kalman filters are relevant when the hidden states are described by continuous variables. An online EM algorithm for HMMs, which does not require the storage of the inputs by rewriting the EM update in terms of sufficient statistics updated recursively, is given in [129]. This scheme is generalized to the case where the model parameters can change with time by introducing a discount factor into the recurrence relations. The resulting algorithm is equivalent to the batch EM algorithm, for appropriate discount factor and scheduling of parameters update [129].

For HMMs, the EM algorithm results in reestimating the model parameters according to the Baum–Welch formulas:

$$\hat{a}_{ij}^{(n+1)} = \frac{\sum_{t=1}^{T} P(x_{t-1} = i, x_t = j|\mathbf{y}_{0\to T}, \hat{\boldsymbol{\theta}}_n)}{\sum_{t=1}^{T} P(x_{t-1} = i|\mathbf{y}_{0\to T}, \hat{\boldsymbol{\theta}}_n)}, \tag{22.24}$$

$$\hat{b}_{jk}^{(n+1)} = \frac{\sum_{t=1}^{T} P(x_t = j, y_t = k|\mathbf{y}_{0\to T}, \hat{\boldsymbol{\theta}}_n)}{\sum_{t=1}^{T} P(x_t = j|\mathbf{y}_{0\to T}, \hat{\boldsymbol{\theta}}_n)}, \tag{22.25}$$

where the probabilities on the right-hand side are conditioned on the training sequence $\mathbf{y}_{0\to T}$ and on the current parameters' estimate $\hat{\boldsymbol{\theta}}_n = \left(\hat{a}_{ij}^{(n)}, \hat{b}_{jk}^{(n)}\right)$. They can be efficiently computed using the forward–backward procedure. This, however, requires storing the whole training sequence. Baum–Welch is an iterative EM technique specialized for batch learning of HMM parameters via ML estimation to best fit the observed data, used to estimate the transition matrix $\mathbf{P}$ of an HMM. Traditionally, the Baum–Welch algorithm is used to infer the state transition matrix of a Markov chain and symbol output probabilities associated to the states of the chain, given an initial Markov model and a sequence of symbolic output values [149].

The forward–backward algorithm [11, 25] is a dynamic programming technique that forms the basis for estimation of HMM parameters using the Baum–Welch technique. Given a finite sequence of training data, it efficiently evaluates the likelihood of this data given an HMM, and computes the smoothed conditional state probability densities for updating HMM parameters according to the Baum–Welch algorithm.

Despite suffering from numerical instability, the forward–backward algorithm is more famous than the forward filtering backward smoothing algorithm [62]. The latter algorithm [144, 150] is probabilistically more meaningful since it propagates probability densities in its forward and backward passes. An efficient version of the algorithm [97] reduces the memory complexity without the computational overhead. Efficient forward filtering backward smoothing generates the same results and the same time complexity $O(N^2T)$ for an HMM with N states and an observation sequence of length T, but reducing the memory complexity of $O(NT)$ for forward–backward to $O(N)$.

The Student-t HMM is a robust form of conventional continuous density HMMs, trained by means of the EM algorithm. A variational Bayesian inference algorithm [27] for this model yields the variational Bayesian Student-t HMM. The approach provides an efficient and more robust alternative to EM-based methods, tackling their singularity and overfitting proneness, while allowing for the automatic determination of the optimal model size without cross-validation.

The Viterbi algorithm is a dynamical programming method that uses the Viterbi path to discover the single most likely explanation for the observations. The evidence propagation algorithm [93, 108] is the inference algorithm for directed probabilistic independent networks. A closely related algorithm [52] solves the MAP identification problem with the same time complexity as the evidence propagation inference algorithm. The two algorithms are strict generalizations of the forward–backward and Viterbi algorithms for HMM, respectively [168].

Two common categories of algorithms for learning Markov network structure from data are score-based [148] and independence- or constraint-based [20, 170] algorithms. An efficient method for incrementally inducing features of Markov random fields is suggested in [148]. However, evaluation of these scores has been proved to be NP-hard for undirected models. A class of efficient algorithms for structure and parameter learning of factor graphs that subsume Markov and Bayesian networks are introduced in [1].

GSMN (grow-shrink Markov network) and GSIMN (grow-shrink inference Markov network) are two independence-based structure learning algorithms of Markov networks [20]. GSMN is an adaptation to Markov networks of the grow-shrink algorithm for learning the structure of Bayesian networks [122]. GSIMN is nearly optimal in terms of the number of tests it can infer, under a fixed ordering of the tests performed. Dynamic GSIMN learning [66] improves on GSIMN by dynamically selecting the locally optimal test that will increase the state of knowledge about the structure the most. Both GSIMN and dynamic GSIMN extend and improve GSMN by additionally exploiting Pearl's theorems on the properties of conditional independence relation to infer additional dependencies and independences from the set of already known ones resulting from statistical tests and previous inferences, thus avoiding the execution of these tests on data and therefore speeding up the structure learning process.

22.8 Dynamic Bayesian Networks

Dynamic Bayesian networks [99] are standard extensions of Bayesian networks to temporal processes. They generalize HMMs and Kalman filters. Dynamic Bayesian networks model a dynamic system by discretizing time and providing a Bayesian network fragment that represents the probabilistic transition of the state at time t to the state at time $t + 1$. The state of the system is represented at different points in time implicitly. It is very difficult to query a dynamic Bayesian network for a distribution over the time at which a particular event takes place. Moreover, since dynamic Bayesian networks slice time into fixed increments, one must always propagate the joint distribution over the variables at the same rate.

Dynamic Bayesian networks generalize HMMs by representing the hidden state (and observed states) in terms of state variables related in a DAG. This effectively reduces the number of parameters to be specified. A dynamic Bayesian network can be converted into an HMM. In a dynamic Bayesian network case, a probabilistic network models a system as it evolves over time. Dynamic Bayesian networks are also time-invariant since the topology of the network is a repeating structure and the CPTs do not change over time. Continuous-time Markov networks are the undirected counterparts of continuous-time Bayesian networks.

Since dynamic Bayesian networks are only a subclass of Bayesian networks, the structure-based algorithms developed for Bayesian networks can be immediately applied to reasoning with dynamic Bayesian networks. Constant-space algorithms for dynamic Bayesian networks [50] are efficient algorithms whose space complexity is independent of the time span T.

Continuous-time Bayesian networks [135] are based on the framework of continuous-time, finite state, and homogeneous Markov processes. Exact inference in a continuous-time Bayesian network can be performed by generating a single-joint intensity matrix over the entire state space of the continuous-time Bayesian network and running the forward–backward algorithm on the joint intensity matrix of the homogeneous Markov process. Inference in such models is intractable even in relatively simple structured networks. In a mean field variational approximation [43], a product of inhomogeneous Markov processes is used to approximate a joint distribution over trajectories. Additionally, it provides a lower bound on the probability of observations, thus making it attractive for learning tasks.

Importance sampling-based approximate inference [63] does not require computing the exact posterior distribution. It is extended to continuous-time particle filtering and smoothing algorithms. These three algorithms can estimate the expectation of any function of a trajectory, conditioned on any evidence set constraining the values of subsets of the variables over subsets of the timeline. Compared to approximate inference algorithms based on expectation propagation and Gibbs sampling, the importance sampling algorithm outperforms both in most of the experiments. In the situation of a highly deterministic system, Gibbs sampling performs better.

The continuous-time Bayesian network reasoning and learning engine (CTBN-RLE) [163] software provides C++ libraries and programs for most of the algorithms developed for continuous-time Bayesian networks. Exact inference as well as approximate inference methods including Gibbs sampling [59] and importance sampling [63] are implemented. In an MCMC procedure [59], a Gibbs sampler is used to generate samples from the posterior distribution given the evidence. The Gibbs sampling algorithm can handle any type of evidence.

22.9 Expectation–Maximization Method

The EM method [56] is the most popular optimization approach to the exact ML solution for the parameters given an incomplete data. EM splits a complex learning problem into a group of separate small-scale subproblems and solves each of the subproblems using a simple method, and is thus computationally efficient. It is a technique for finding a local ML or MAP. Although EM is generally considered to converge linearly, it has a significantly faster convergence rate than gradient descent. EM can be viewed as a deterministic version of Gibbs sampling, and can be used to search for the MAP estimate of model parameters. A message-passing interpretation for EM is given in [51].

For a large amount of data, $p(\boldsymbol{\theta}_S|\mathcal{D}, S^h) \propto p(\mathcal{D}|\boldsymbol{\theta}_S, S^h) \cdot p(\boldsymbol{\theta}_S|S^h)$ can be approximated as a multivariate-Gaussian distribution [81]. Define

$$\tilde{\boldsymbol{\theta}}_S = \arg\max_{\boldsymbol{\theta}_S} \log\{p(\mathcal{D}|\boldsymbol{\theta}_S, S^h) \cdot p(\boldsymbol{\theta}_S|S^h)\}. \tag{22.26}$$

Thus, $\tilde{\boldsymbol{\theta}}_S$ also maximizes $p(\boldsymbol{\theta}_S|\mathcal{D}, S^h)$, and the solution is known as the MAP solution of $\boldsymbol{\theta}_S$.

As the sample size increases, the effect of the prior $p(\boldsymbol{\theta}_S|S^h)$ diminishes, thus MAP reduces to the ML configuration

$$\hat{\boldsymbol{\theta}}_S = \arg\max_{\boldsymbol{\theta}_S}\{p(\mathcal{D}|\boldsymbol{\theta}_S, S^h)\}. \tag{22.27}$$

The explicit determination of ML estimates for the conditional probabilities of the nodes of a discrete Bayesian network using incomplete data is usually not possible, and iterative methods such as EM or Gibbs sampling are normally required.

The EM method alternates between performing the expectation step (E-step) and the maximization step (M-step), which, respectively, compute the expected values of the latent variables and the ML or MAP estimates of the parameters. $\boldsymbol{\theta}_S$ is initialized at random. In the E-step, the expected sufficient statistics for the missing entries of the database $\mathcal{D}$ are computed conditioned on the assigned configuration of $\boldsymbol{\theta}_S$ and the known data $\mathcal{D}$. In the M-step, the expected sufficient statistics are taken as though they were the actual sufficient statistics of a database $\mathcal{D}'$, and the mean of the parameters $\boldsymbol{\theta}_S$ is calculated such that the probability of observation of $\mathcal{D}'$ is maximized.

Under certain regularity conditions, iteration of the E- and M-steps will converge to a local maximum [56]. The EM algorithm is typically applied when sufficient statistics exist, i.e., when local distribution functions are in the exponential family. It is fast, but it does not provide a distribution over the parameters $\boldsymbol{\theta}$. A switch can be made to alternative algorithms when near a solution in order to overcome the slow convergence of EM when near local maxima [125].

BP and the generalized EM are both iterative gradient algorithms. BP is a special case of the generalized EM algorithm for iteratively maximizing a likelihood or log-likelihood [8]. The forward step of BP corresponds to the E-step, while its backward error-passing corresponds to the M-step. The hidden-layer parameters of a network corresponds to the EM latent variables. Minimization of squared error in BP is equivalent to maximization of a likelihood in the form of the exponential of the negative squared error in EM.

The asymptotic convergence rate of the EM algorithm for Gaussian mixtures locally around the true solution is given in [120]. The large sample local convergence rate for the EM algorithm tends to be asymptotically superlinear when the measure of the average overlap of Gaussians in the mixture tends to zero.

Singularities in the parameter spaces of hierarchical learning machines are known to be a main cause of slow convergence of gradient-descent learning. EM is a good alternative to overcome the slow learning speed of the gradient-descent method. The slow convergence of the EM method in the case of large component overlap is a widely known phenomenon.

The dynamics of EM for Gaussian mixtures around singularities is analyzed in [145]; there exists a slow manifold caused by a singular structure, which is closely related to the slow convergence of the EM algorithm. In the case of the mixture of densities from exponential families, the convergence speed depends on the separation of the component populations in the mixture. When the component populations are poorly separated, the convergence speed of EM becomes extraordinarily slow.

The noisy EM theorem states that a suitably noisy EM algorithm estimates the EM estimate in fewer steps on average than does the corresponding noiseless EM algorithm [143]. Noisy EM adds noise to the data at each EM iteration. The sufficient condition for a noise boost is an average positivity (non-negativity) condition [142]. Then such noise climbs a local hill of probability or log-likelihood. The injected noise decays with the iterations to ensure convergence to the optimal parameters of the original data model. Many centroid-based clustering algorithms including C-means benefits from noise because they are special cases of EM algorithm.

A Gibbs sampler called data augmentation [175] is quite similar to EM, but instead of calculating the expected values of the sufficient statistics, a value is drawn from a predictive distribution and imputed (I-step). Similarly, instead of calculating the ML estimates, a parameter value is drawn from the posterior distribution on the parameter space conditioned on the most recent fully imputed data sample (P-step). Based on the MCMC theory, this will, in the limit, return parameter realizations from the posterior parameter distribution conditioned on the observed data.

22.10 Mixture Models

Mixture models provide a rigorous framework for density estimation, and clustering. Mixtures of Gaussians are widely used throughout the fields of machine learning and statistics for data modeling.

The mixture model with K components is given by

$$p(\boldsymbol{x}, \boldsymbol{v}, \boldsymbol{\theta}) = \sum_{i=1}^{K} v_i \psi(\boldsymbol{x}, \theta_i), \tag{22.28}$$

where $\boldsymbol{v} = (v_1, v_2, \ldots, v_K)^T$ with v_i being a mixing coefficient for the ith component, $\boldsymbol{\theta} = (\theta_1, \theta_2, \ldots, \theta_K)^T$ with θ_i being a parameter associated with the ith component, and $\psi(\boldsymbol{x})$ is a basic pdf.

A latent variable model seeks to relate a n-dimensional measured vector $\boldsymbol{y}$ to a corresponding m-dimensional vector of latent variables $\boldsymbol{x}$ [182]:

$$\boldsymbol{y} = \boldsymbol{f}(\boldsymbol{x}, \boldsymbol{w}) + \boldsymbol{n}, \tag{22.29}$$

where $\boldsymbol{f}$ is a function of the latent variables $\boldsymbol{x}$ and parameters $\boldsymbol{w}$, and $\boldsymbol{n}$ is a noise process. Generally, one selects $m < n$ in order to ensure that the latent variables offer a more parsimonious description of the data. The model parameters can be determined by the ML method.

The task of estimating the parameters of a given mixture can be achieved with different approaches: ML, MAP, or Bayesian inference. ML estimates the parameters by means of the maximization of a likelihood function. One of the most common methods for fitting mixtures to data is EM. However, EM is unable to make model selection, i.e., to determine the appropriate number of model components in a density mixture. The same is true for the MAP estimation approach which tries to find the parameters that correspond to the location of the MAP density function.

Competitive EM [208], analogous to the SMEM algorithm [186], utilizes a heuristic split-and-merge mechanism to either split the model components in an underpopulated region or merge the components in an overpopulated region iteratively so as to avoid local solutions. It also exploits a heuristic component annihilation mechanism to determine the number of model components. In [13], the entropy of the pdf associated to each kernel is used to measure the quality of a given mixture model with a fixed number of kernels. Entropy-based EM starts with a unique kernel and performs only splitting by selecting the worst kernel in a global mixture entropy-based criterion called Gaussianity deficiency in order to find the optimum number of components of the mixture.

Another way is to implement parameter estimation and model selection jointly in a single paradigm by using MCMC or variational methods [73, 131]. Variational algorithms are guaranteed to provide a lower bound of the approximation error [73]. Reversible-jump MCMC has been applied to the Gaussian mixture model [151]. A Bayesian method for mixture model training [44] simultaneously treats feature selection and model selection. The algorithm follows the variational framework and can simultaneously optimize over the number of components, the saliency of the features, and the parameters of the mixture model.

The Gaussian mixture model can be thought of as a prototype method, similar in spirit to C-means and LVQ. Each cluster is described in terms of a Gaussian density, which has a centroid (as in C-means) and a covariance matrix. The two alternating steps of EM algorithm are very similar to the two steps in C-means. Gaussian mixture model is often referred to as a soft-clustering method, while C-means is hard.

When the data are not Gaussian, mixtures of generalized Dirichlet distributions may be adopted as a good alternative [16]. In fact, the generalized Dirichlet distribution is more appropriate for modeling data that are compactly supported, such as data originating from videos, images, or text. The conditional independence assumption among features commonly used in modeling high-dimensional data becomes a fact for generalized Dirichlet datasets without loss of accuracy. In [17], an unsupervised approach is presented for extraction of independent and non-Gaussian features in mixtures of generalized Dirichlet distributions. The proposed model is learned using EM algorithm by minimizing the message length of the dataset.

22.11 Bayesian and Probabilistic Approach to Machine Learning

A principled Bayesian learning approach to neural networks can lead to many improvements. In particular, by approximating the distributions of the weights with Gaussians and adopting smoothing priors, it is possible to obtain estimates of the weights and output variances as well as to set the regularization coefficients automatically [121]. Bayes learning rule is the optimal perceptron learning rule that gives rise to a lower bound on the generalization error [61].

An important approximation to Bayesian method is proposed and analyzed in [138] for online learning on perceptrons and relies on a projection of the posterior probabilities of the parameters to be estimated on a space of tractable distributions minimizing the Kullback–Leibler divergence between both. Probabilistic model-based approach to learning spiking neural networks is given in [57].

In a Bayesian approach to RBF networks [86], the model complexity can be adjusted according to the complexity of the problem using an MCMC sampler. A hierarchical full Bayesian model for RBF networks [3] treats the model size, model parameters, regularization parameters, and noise parameters as unknown random variables. Simulated annealing with reversible-jump MCMC method is used to perform the Bayesian computation.

The EM method has been applied to obtain ML estimates of the network parameters for feedforward network learning. In [119], training of three-layer MLP is first decomposed into a set of single neurons, and the individual neurons are then trained via a linearly weighted regression algorithm. The EM method has also been applied for RBF network learning [105, 109]. The EM algorithm for normalized Gaussian RBF network is derived in [202]. For classification using RBF network [204], Bayesian method is applied to explore the network structure, and an EM algorithm is used to estimate the weights.

A Bayesian interpretation of LDA is considered in [24]. With the use of a Gaussian process prior, the model is shown to be equivalent to a regularized kernel Fisher's discriminant.

A family of probabilistic latent variable models are used for analysis of nonnegative data [162]. The latent variable decompositions are numerically identical to the NMF algorithm that optimizes a Kullback–Leibler metric, where ML parameter estimation is performed via the EM algorithm. In [23], NMF with a Kullback–Leibler measure is described as a hierarchical generative model consisting of an observation and a prior component. Full Bayesian inference is developed via variational Bayes or Monte Carlo. Under the framework of MAP probability, two-dimensional NMF [67] is adaptively tuned using the variational approach. The method enables a generalized criterion for variable sparseness to be imposed onto the solution, and prior information to be explicitly incorporated into the basis features.

A Bayesian approach toward echo state networks, called the echo state Gaussian process, combines the merits of echo state networks and Gaussian processes to provide a more robust solution to reservoir computing while offering a measure of confidence on the generated predictions [26]. In [164], a variational Bayesian framework combined with automatic regularization and delay-and-sum readout adaptation is proposed for efficient training of echo state networks.

The kernel-based Bayesian network paradigm is introduced for supervised classification [147]. This paradigm is a Bayesian network which estimates the true density of the continuous variables using kernels. It uses a nonparametric kernel-based density estimation instead of a parametric Gaussian one.

Quasi-Newton algorithms can be interpreted as approximations to Bayesian linear regression under Gaussian and other priors [83]. This analysis gives rise to a class of Bayesian nonparametric quasi-Newton algorithms, which have a computational cost similar to its predecessors. These use a kernel model to learn from all observations in each line search, explicitly track uncertainty, and thus achieve faster convergence toward the true Hessian.

Sparse coding is modeled in a Bayesian network framework in [161], where sparsity-favoring Laplace prior is applied on the coefficients of the linear model and expectation propagation inference is employed. In [117], sparse coding is interpreted from a Bayesian perspective, which results in an objective function. Through MAP estimation, the obtained solution can have smaller reconstruction errors than that obtained by standard method using L_1 regularization.

In [210], a hierarchical Bayesian generative model is constructed for the L_1-norm low-rank matrix factorization problem and a mean field variational method is designed to automatically infer all the parameters involved in the model by closed-form equations. A probabilistic model for L_1-norm low-rank matrix factorization (http://winsty.net/prmf.html) [191] is solved by utilizing conditional EM algorithm. A probabilistic formulation for low-rank matrix factorization [128] utilizes MAP to estimate the factors. Probabilistic latent semantic analysis solves the problem of NMF with KL divergence [68].

To naturally deal with missing data, the probabilistic framework for tensor factorization was exploited in [40]. A generative model for robust tensor factorization in the presence of both missing data and outliers [211] explicitly infers the underlying low-CP-rank tensor capturing the global information and a sparse tensor capturing the local information (also considered as outliers). An efficient variational inference under a fully Bayesian treatment can effectively prevent the overfitting problem and scales linearly with data size. Bayesian multi-tensor factorization [96] is the Bayesian formulation for joint factorization of multiple matrices and tensors.

22.11.1 Probabilistic PCA

Probabilistic model-based PCA methods consider the linear generative model, and derive iterative algorithms by using EM within an ML framework for Gaussian data. These are batch algorithms that find principal subspace. A probabilistic formulation of PCA provides a good foundation for handling missing values. However, these methods conduct PSA rather than PCA. Two examples of probabilistic model-based PCA methods are probabilistic PCA [181] and EM-PCA [156].

The starting point for probabilistic PCA is a factor analysis style latent variable model (13.112). EM algorithm for PCA described in [156] is computationally very efficient in space and time; it does not require computing the sample covariance and has a complexity limited by $O(knp)$ operations for k leading eigenvectors to be learned, p dimensions and n data.

EM-ePCA [2] finds exact principal directions without rotational ambiguity by the minimization of an integrated-squared error measure for derivation of PCA. In addition, GHA can be derived using the gradient-descent method by minimizing the integrated-squared error. A number of Bayesian formulations of PCA have followed from the probabilistic formulation of PCA [181], with the necessary marginalization being approximated through both Laplace approximations and variational bounds.

Dual probabilistic PCA [110] is a probabilistic interpretation of PCA. It has an additional advantage that the linear mappings from the embedded space can easily be nonlinearized through Gaussian processes. This model is referred to as a Gaussian process latent variable model. The model is related to kernel PCA and multidimensional scaling.

Exponential PCA [194] reduces the dimension of the parameters of probability distributions using Kullback information as a distance between two distributions. It also provides a framework for dealing with various data types such as binary and integer for which the Gaussian assumption on the data distribution is inappropriate. A learning algorithm for those mixture models based on the variational Bayesian method is derived.

Bilinear probabilistic PCA is a probabilistic PCA model on 2D data [209]. A probabilistic model for GLRAM is formulated as a two-dimensional PCA method [201].

A Bayesian model family is developed as a unified probabilistic formulation of the latent variable models such as factor analysis and PCA [112]. It employs exponential family distributions to specify various types of factors and a Gibbs sampling procedure as a general computation routine. The EM approach to kernel PCA [155] is a computationally efficient method, especially for a large number of data points.

Variational autoencoders [98] are generative autoencoders for compressing and learning latent representations of the data through a variational approach to Bayesian inference.

22.11.2 Probabilistic Clustering

Model-based clustering approaches are usually based on mixture-likelihood and classification-likelihood. EM and classification EM [22] are the corresponding examples of these approaches. They are sensitive to the initial conditions of the model parameters.

EM clustering [18] represents each cluster using a probability distribution, typically a Gaussian distribution. Each cluster is represented by a mean and a $J_1 \times J_1$ covariance matrix. Each pattern belongs to all the clusters with the probabilities of membership determined by the distributions of the corresponding clusters. Thus, EM clustering can be treated as a fuzzy clustering technique. C-means is equivalent to classification EM corresponding to the uniform spherical Gaussian model [22]. C-means is similar to EM in two alternating steps.

The learning process of a probabilistic SOM is considered as a model-based data clustering procedure that preserves the topological relationships between data clusters [33]. A coupling-likelihood mixture model extends the reference vectors in SOM to multivariate-Gaussian distributions, and three EM-type learning algorithms are given in [33]. In [116], the probabilistic SOM is derived from a probabilistic mixture of multivariate Student-t components to improve the robustness of the map against outliers.

Variational Bayesian clustering algorithms utilize an evaluation function that can be described as the log-likelihood of given data minus the Kullback–Leibler divergence between the prior and the posterior of model parameters [172]. The update process of variational Bayesian clustering with finite mixture Student-t distribution is derived, taking the penalty term for the degree of freedom into account.

Maximum weighted likelihood [34] provides a general learning paradigm for density mixture model selection and learning, in which weight design is a key issue. A rival penalized EM algorithm for density mixture clustering makes the components in a density mixture compete with one another at each time step. Not only are the associated parameters of the winner updated to adapt to an input, but also all rivals' parameters are penalized with a strength proportional to the corresponding posterior density probabilities. Rival penalized EM can automatically select an appropriate number of densities by fading out the redundant densities during the learning process. Compared to RPCL and its variants, this method avoids the preselection of the delearning rate.

Affinity propagation [64] is an algorithm for exemplar-based clustering. It associates each data point with one exemplar, resulting in a partitioning of the whole dataset into clusters by minimizing the overall sum of similarities between data points and their exemplars. Real-valued messages are exchanged between data points until a high-quality set of exemplars and corresponding clusters gradually emerges. Affinity propagation has been derived as an instance of the max-product (belief propagation) algorithm in a loopy factor graph.

Using a probabilistic approach, the similarity measure for directional clustering given by (10.17) is proportional to the log-likelihood of the von Mieses–Fisher distribution. A von Mieses–Fisher mixture model has been used together with an EM algorithm for directional clustering in [9].

22.11.3 Probabilistic ICA

Bayesian ICA algorithms [7, 85, 127], offer accurate estimations for the linear model parameters. For instance, universal density approximation using a mixture of Gaussians may be used for the source distributions. In Bayesian ICA, however, the source distributions are modeled using a mixture of Gaussians model. Variational Bayesian nonlinear ICA [106, 188] uses MLP to model the nonlinear mixing transformation.

ICA and PCA are closely related to factor analysis. Factor analysis uses the Gaussian model and is not suitable for BSS, and the ML estimate of the mixing matrix is not unique. Independent factor analysis [7] recovers independent hidden sources from their observed mixtures. It generalizes and unifies factor analysis, PCA, and ICA. The source densities, mixing matrix and noise covariance are first estimated from the observed data by ML using an EM algorithm. Each source is described by a mixture of Gaussians; thus, all the probabilistic calculations can be performed analytically. The sources are then reconstructed from the observed data by an optimal nonlinear estimator. A variational approximation of this algorithm is derived for cases with a large number of sources. The complexity of independent factor analysis grows exponentially in the number of sources.

ICA can be performed using a constrained version of EM [196]. The source distributions are modeled as d one-dimensional mixtures of Gaussians. The observed data are modeled as linear mixtures of the sources with additive, isotropic noise. EM algorithm allows factoring the posterior density. This avoids an exponential increase of complexity with the number of sources and allows an exact treatment in the case of many sources.

Variational Bayesian applied to ICA can handle small datasets with high observed dimension. By extending the Gaussian analyzers mixture model to an ICA mixture model, variational Bayesian inference and structure determination are employed to construct an approach for modeling non-Gaussian, discontinuous manifolds; the method automatically determines the local dimensions of each manifold and uses variational inference to calculate the optimum number of ICA components [39].

In mean field approaches to probabilistic ICA [85], the sources are estimated from the mean of their posterior distribution and the mixing matrix is estimated by MAP. For mean field ICA methods [85], the flexibility with respect to the prior makes the inside of the black box rather complicated; convergence of EM-based learning is slow and no universal stopping criteria can be given. icaMF [198] solves these problems by using efficient optimization schemes of EM. The expectation consistent framework and expectation propagation message-passing algorithm are applied to the ICA model. The method is flexible with respect to choice of source prior, dimensionality and constraints of the mixing matrix (unconstrained or nonnegativity), and structure of the noise covariance matrix. The required expectations over the source posterior are estimated with mean field methods.

The Bayesian nonstationary source separation algorithm of [87] recovers nonstationary sources from noisy mixtures. In order to exploit the temporal structure of the data, a time-varying autoregressive process is used to model each source signal. Variational Bayesian learning is then adopted to integrate the source model with BSS in probabilistic form. The separation algorithm makes full use of temporally correlated prior information and avoids overfitting in the separation process. Variational EM steps are applied to derive approximate posteriors and a set of update rules for the parameters of these posteriors.

The convergence speed of FastICA is analyzed in [187]. The analysis suggests that the nonlinearity used in FastICA can be interpreted as denoising and taking Bayesian noise filtering as the nonlinearity resulted from fast Bayesian ICA. This denoising interpretation is generalized to a source separation framework called denoising source separation [158]. The algorithms differ in the denoising function while the other parts remain mostly the same. Some existing ICA algorithms are reinterpreted within the denoising source separation framework and some robust BSS algorithms are suggested.

The mixing system and source signals in ICA may be nonstationary. In [38], a separation procedure is established in the presence of nonstationary and temporally correlated mixing coefficients and source signals. In this procedure, the evolved statistics are captured from sequential signals according to online Bayesian learning for the Gaussian process. A variational Bayesian inference is developed to approximate the true posterior for estimating the nonstationary ICA parameters and for characterizing the activity of latent sources.

22.11.4 Probabilisitic Approach to SVM

In Bayesian techniques for support vector classification [41], Bayesian inference is used to implement model adaptation, while keeping the merits of support vector classifier, such as sparseness and convex programming. A differentiable loss function called trigonometric loss function has the desirable characteristic of natural normalization in the likelihood function.

Posterior probability SVM [176] modifies SVM to utilize class probabilities instead of using hard $-1/+1$ labels. It uses soft labels derived from estimated posterior probabilities so as to be more robust to noise and outliers. The method uses a window-based density estimator for the posterior probabilities. It achieves an accuracy similar to that of standard SVM by storing fewer support vectors. The decrease in error by posterior probability SVM is due to a decrease in bias rather than variance. The method is extended to the multiclass case in [75], where a neighbor-based density estimator is proposed and is also extended to the multiclass case.

Extended soft margin SVM (EC-SVM) can be cast as a robust optimization problem, where buffered probability of exceedance (bPOE) is minimized with data lying in a fixed uncertainty set [137]. bPOE is simply one minus the inverse of the superquantile. EC-SVM is simply a buffered way of minimizing the probability that misclassification errors exceed a threshold $-C$. Over the range of its free parameter (i.e., optimal objective value), EC-SVM has both a convex and a non-convex case, which are respectively associated with pessimistic and optimistic views of uncertainty.

C-SVM (soft-margin SVM) and EC-SVM, when formulated with any general norm and nonnegative parameter values, produce the same set of optimal hyperplanes. Thus, C-SVM is equivalent to minimization of bPOE [137]. C-SVM, formulated with any regularization norm, is equivalent to the convex EC-SVM. The optimal objective value of C-SVM, divided by sample size, equals a probability level.

Eν-SVM is equivalent to superquantile minimization, with its free parameter being equivalent to the free choice of probability level in superquantile minimization [171]. Eν-SVM is equivalent to EC-SVM over its entire parameter range, which includes both the convex and non-convex cases [137].

22.11.5 Relevance Vector Machines

Predictions are not probabilistic in the SVM outputs. Ideally, one prefers to estimate the conditional distribution $p(y|\boldsymbol{x})$ in order to capture uncertainty. It is necessary to estimate the error/margin trade-off parameter C (as well as the insensitivity parameter ε in regression); this generally entails a cross-validation procedure. Relevance vector machine [179] is a Bayesian treatment of the SVM prediction which does not suffer from any of these limitations. Relevance vector machine and informative vector machine [111] are sparse probabilistic kernel classifiers in a Bayesian setting. These non-SVM models can match the accuracy of SVM, while considerably bringing down the number of kernel functions as well as the training cost.

The basic idea of relevance vector machine is to assume a prior of the expansion coefficients which favors sparse solutions. Sparsity is obtained because the posterior distributions of many of the weights become sharply peaked around zero. Those training vectors associated with the remaining nonzero weights are termed *relevance vectors*. Each weight is assumed to be a Gaussian variable. Relevance vector machine acquires relevance vectors and weights by maximizing a marginal likelihood. Through a learning process, the data points with very small variances of the weights will be discarded. Derived by exploiting a probabilistic Bayesian learning framework, accurate prediction models typically utilize dramatically fewer kernel functions than a comparable SVM does while offering a number of additional advantages: probabilistic predictions, automatic estimation of parameters, and the facility to utilize arbitrary kernel functions (e.g., non-Mercer kernels). The learned model is dependent only on relevance vectors.

The primary disadvantage of the sparse Bayesian method is the computational complexity of the learning algorithm. Although the presented update rules are very simple in form, they require $O(N^2)$ memory and $O(N^3)$ computation for N training examples [180].

Analysis of relevance vector machine shows that adopting the same prior for different classes may lead to unstable solutions [28]. In order to tackle this problem, a signed and truncated Gaussian prior is adopted over every weight in probabilistic classification vector machine (PCVM) [28], where the sign of the prior is determined by the class label, i.e., $+1$ or -1. The truncated Gaussian prior not only restricts the sign of the weights but also leads to a sparse estimation of the weight vectors. The superiority of PCVM formulation is also discussed using MAP analysis and margin analysis in [28].

Using EM and variational approximation methods, variational Bayesian LS approach [178] offers a computationally efficient and statistically robust black box approach to generalized linear regression with high-dimensional inputs. Relevance vector machine is derived with variational Bayesian LS at its core. The iterative nature of variational Bayesian LS makes it most suitable for real-time incremental learning.

Incremental relevance vector machine [183] starts with an empty model, and at each iteration, a kernel function (centered at a training sample) might be added to or deleted from the model to maximize the marginal likelihood. Otherwise, the model hyperparameters are reestimated. Like relevance vector machine, an incremental Bayesian method for supervised learning given in [185] learns the parameters of the kernels during training. Specifically, different parameter values are learned for each kernel, resulting in a very flexible model. A sparsity-enforcing prior is used to control the effective number of model parameters.

Problems

22.1 Bayes' decision theory makes decision on two classes. A sample $\boldsymbol{x}$ belongs to class C_1 if $P(C_1|\boldsymbol{x}) > P(C_2|\boldsymbol{x})$, or it belongs to class C_2 if $P(C_2|\boldsymbol{x}) > P(C_1|\boldsymbol{x})$. Show that the rule can be represented by: Decide that $\boldsymbol{x}$ belongs to C_1 if $P(\boldsymbol{x}|C_1)P(C_1) > P(\boldsymbol{x}|C_2)P(C_2)$, and to C_2 otherwise.

22.2 **Monty Hall Problem**. In Monty Hall's television game show, there are three doors: behind one door is a car and behind the other two are two goats. You have chosen one door, say door A. The door remains closed for the time being. Monty Hall now has to open one of the two remaining doors (B or C). He selects door C, and there is a goat behind it. Then he asks you: "Do you want to switch to Door B?" What is your decision? Solve it using Bayes' theorem.

22.3 Let X_1 and X_2 be results of flips of two coins, and X_3 an identifier if X_1 and X_2 coincide. We have $P(X_1 = \text{head}) = P(X_1 = \text{tail}) = P(X_2 = \text{head}) = P(X_2 = \text{tail}) = 0.5$.
(1) Without evidence on X_3, are nodes X_1 and X_2 are marginally independent?
(2) Show that X_1 and X_1 are dependent if the value of X_3 is known: $P(X_1 = \text{head}, X_2 = \text{head}|X_3 = 1) \neq P(X_1 = \text{tail}|X_3 = 1)P(X_2 = \text{tail}|X_3 = 1)$.

22.4 Derive the likelihood ratio $\frac{p(x|C_1)}{p(x|C_2}$ in the case of Gaussian densities $p(x|C_1) \sim N(\mu_1, \sigma_1^2)$ and $p(x|C_2) \sim N(\mu_2, \sigma_2^2)$.

22.5 The likelihood ratio for a two-class problem is $\frac{p(x|C_1)}{p(x|C_2}$. Show that the discriminant function can be defined by $g(x) = p(x|C_1)P(C_1) - p(x|C_2)P(C_2)$: If $g(x) > 0$, we can choose class C_1, otherwise class C_2. Derive the discriminant function in terms of the likelihood ratio.

22.6 For Possion's distribution $p(x|\theta) = \theta^x \exp(-\theta)$, find θ using the maximum likelihood method [Hint: find θ that maximizes $p(x|\theta)$.]

22.7 A discrete random variable X has the binomial distribution with parameters n and p if its pdf is given by

$$p(x|n, p) = \binom{n}{x} p^x (1-p)^{n-x}, \quad x \in \{0, 1, \ldots, n\}, p \in (0, 1).$$

Four products are made. Assume that the probability of any product being defective is $p = 0.3$ and each product is independently produced.
(a) Calculate the probability of $x \in \{0, 1, 2, 3, 4\}$ products being defective.
(b) Calculate the probability of less than x products being defective.

22.8 I was at work. I felt that the office building had shaken suddenly. I thought that a heavy-duty truck had hit the building by accident. From the web, there were some messages that there had been an earthquake a moment before.

Sometimes a blast nearby also causes a building to shake. Was there an earthquake? Construct a Bayesian network, and tentatively give CPTs.

22.9 Download the Bayes Nets Toolbox for MATLAB (http://bnt.googlecode.com), and learn to use this software for structure learning of Bayesian networks.

22.10 A Bayesian network is shown in Fig. 22.7. Three CPTs are known. The first CPT is given by $P(C = \text{T}) = 0.5$, $P(C = \text{F}) = 0.5$. The second CPT is given by $P(S = \text{T}|C = \text{T}) = 0.1$, $P(S = \text{F}|C = \text{T}) = 0.9$, $P(S = \text{T}|C = \text{F}) = 0.5$, $P(S = \text{F}|C = \text{F}) = 0.5$. The third CPT is given by: $P(R = \text{T}|C = \text{T}) = 0.8$, $P(R = \text{F}|C = \text{T}) = 0.2$, $P(R = \text{T}|C = \text{F}) = 0.3$, $P(R = \text{F}|C = \text{F}) = 0.7$. Calculate the CPT for $P(W|S, R)$.

22.11 Develop a computer program for random simulations of state transitions in a Markov chain with discrete time.

22.12 Given the joint probability $P_{A,B}(a_i, b_j)$: $P_{A,B}(a_1, b_1) = 0.03$, $P_{A,B}(a_1, b_2) = 0.05$, $P_{A,B}(a_1, b_3) = 0.20$, $P_{A,B}(a_2, b_1) = 0.15$, $P_{A,B}(a_2, b_2) = 0.02$, $P_{A,B}(a_2, b_3) = 0.25$, $P_{A,B}(a_3, b_1) = 0.08$, $P_{A,B}(a_3, b_2) = 0.25$, $P_{A,B}(a_3, b_3) = 0.30$. Compute P_A, P_B, $P_{A|B}$, $P_{B|A}$.

22.13 Use the HUGIN software to make queries on the Bayesian network given by Fig. 22.7). Assume that we have CPT 1: $P(S = 0|C = 0) = 0.5$, $P(S = 1|C = 0) = 0.5$, $P(S = 0|C = 1) = 0.8$, $P(S = 0|C = 1) = 0.2$;
CPT 2: $P(R = 0|C = 0) = 0.9$, $P(R = 1|C = 0) = 0.1$, $P(R = 0|C = 1) = 0.1$, $P(R = 0|C = 1) = 0.9$;
CPT 3: $P(W = 1|R = 0, S = 0) = 0.05$, $P(W = 1|R = 0, S = 1) = 0.8$, $P(W = 1|R = 1, S = 0) = 0.8$, $P(W = 1|R = 1, S = 1) = 0.95$;
CPT 4: $P(C = 0) = P(C = 1) = 0.5$.
(a) Calculate $P(C = 1|W = 1)$, $P(C = 1|R = 1)$, $P(C = 1|S = 1)$.
(b) Draw the DAG. Verify the result using HUGIN.

22.14 Consider the probability model as $\phi(x_1, x_2, x_3) = p_1(x_1, x_2)p_2(x_2, x_3)$. Plot the factor graph of the model.

22.15 Discuss the similarities and differences between the Metropolis algorithm and the Gibbs sampler.

22.16 Given the expression for the area of a circle, $A = \pi r^2$, and using uniformly distributed random variates, devise a sampling approach for computing π.

22.17 Netica (http://www.norsys.com) is a popular Bayesian network development software, widely used by the world's leading companies and government agencies. Practise using the software for Bayesian network development.

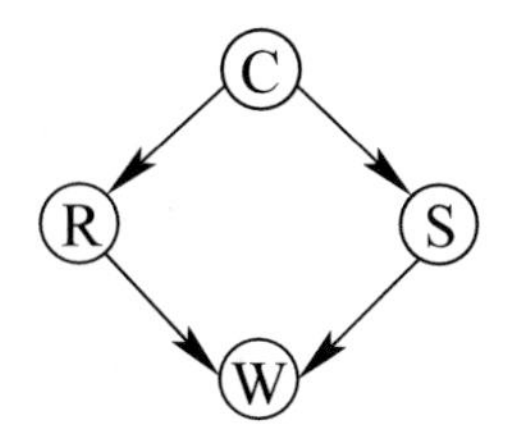

Fig. 22.7 The sprinkler, rain and wet pavement: C = cloudy, S = sprinkler, R = rain, W = wet grass

References

1. Abbeel, P., Koller, D., & Ng, A. Y. (2006). Learning factor graphs in polynomial time and sample complexity. *Journal of Machine Learning Research*, *7*, 1743–1788.
2. Ahn, J.-H., Oh, J.-H., & Choi, S. (2007). Learning principal directions: Integrated-squared-error minimization. *Neurocomputing*, *70*, 1372–1381.
3. Andrieu, C., de Freitas, N., & Doucet, A. (2001). Robust full Bayesian learning for radial basis networks. *Neural Computation*, *13*, 2359–2407.
4. Andrieu, C., Doucet, A., & Holenstein, R. (2010). Particle Markov chain Monte Carlo methods. *Journal of the Royal Statistical Society: Series B*, *72*(3), 269–342.
5. Archambeau, C., & Verleysen, M. (2007). Robust Bayesian clustering. *Neural Networks*, *20*, 129–138.
6. Attias, H. (1999). Inferring parameters and structure of latent variable models by variational Bayes. In *Proceedings of the 15th Annual Conference on Uncertainty in AI* (pp. 21–30).
7. Attias, H. (1999). Independent factor analysis. *Neural Computation*, *11*, 803–851.
8. Audhkhasi, K., Osoba, O., & Kosko, B. (2013). Noise benefits in backpropagation and deep bidirectional pre-training. In *Proccedings of International Joint Conference on Neural Networks (IJCNN)* (pp. 1–8). Dallas, TX.
9. Banerjee, A., Dhillon, I. S., Ghosh, J., & Sra, S. (2005). Clustering on the unit hypersphere using von Mises-Fisher distributions. *Journal of Machine Learning Research*, *6*, 1345–1382.
10. Bauer, E., Koller, D., & Singer, Y. (1997). Update rules for parameter estimation in Bayesian networks. In *Proceedings of Annual Conference on Uncertainty in AI* (pp. 3–13).
11. Baum, L. E., Petrie, T., Soules, G., & Weiss, N. (1970). A maximization technique occurring in the statistical analysis of probabilistic functions of Markov chains. *Annals of Mathematical Statistics*, *41*(1), 164–171.
12. Beinlich, I. A., Suermondt, H. J., Chavez, R. M., & Cooper, G. F. (1989). The ALARM monitoring system: A case study with two probabilistic inference techniques for Bayesian networks. In *Proceedings of the 2nd European Conference on Artificial Intelligence in Medicine* (pp. 247–256).
13. Benavent, A. P., Ruiz, F. E., & Saez, J. M. (2009). Learning Gaussian mixture models with entropy-based criteria. *IEEE Transactions on Neural Networks*, *20*(11), 1756–1771.
14. Binder, J., Koller, D., Russell, S., & Kanazawa, K. (1997). Adaptive probabilistic networks with hidden variables. *Machine Learning*, *29*, 213–244.
15. Bouchaert, R. R. (1994). *Probabilistic network construction using the minimum description length principle*. Technical Report UU-CS-1994-27, Utrecht University, Department of Computer Science, The Netherlands.
16. Bouguila, N., & Ziou, D. (2007). High-dimensional unsupervised selection and estimation of a finite generalized Dirichlet mixture model based on minimum message length. *IEEE Transactions on Pattern Analysis and Machine Intelligence*, *29*(10), 1716–1731.
17. Boutemedjet, S., Bouguila, N., & Ziou, D. (2009). A hybrid feature extraction selection approach for high-dimensional non-Gaussian data clustering. *IEEE Transactions on Pattern Analysis and Machine Intelligence*, *31*(8), 1429–1443.

18. Bradley, P. S., Fayyad, U. M., & Reina, C. A. (1998). *Scaling EM (expectation-maximization) clustering to large databases*. MSR-TR-98-35, Microsoft Research.
19. Breese, J. S., & Heckerman, D. (1996). Decision-theoretic troubleshooting: a framework for repair and experiment. In *Proceedings of the 12th Conference on Uncertainty in AI* (pp. 124–132). Portland, OR.
20. Bromberg, F., Margaritis, D., & Honavar, V. (2006). Effcient Markov network structure discovery using independence tests. In *Proceedings of the 6th SIAM International Conference on Data Mining* (pp. 141–152).
21. Buntine, W. (1991). Theory refinement of Bayesian networks. In B. D. D'Ambrosio, P. Smets, & P. P. Bonisone (Eds.), *Proceedings of the 7th Conference on Uncertainty in AI* (pp. 52–60). Burlington: Morgan Kaufmann.
22. Celeux, G., & Govaert, G. (1992). A classification EM algorithm for clustering and two stochastic versions. *Computational Statistics & Data Analysis*, *14*(3), 315–332.
23. Cemgil, A. T. (2009). Bayesian Inference for Nonnegative Matrix Factorisation Models. *Computational Intelligence and Neuroscience*, *2009*, 785152, 17 p.
24. Centeno, T. P., & Lawrence, N. D. (2006). Optimising kernel parameters and regularisation coefficients for non-linear discriminant analysis. *Journal of Machine Learning Research*, *7*, 455–491.
25. Chang, R., & Hancock, J. (1966). On receiver structures for channels having memory. *IEEE Transactions on Information Theory*, *12*(4), 463–468.
26. Chatzis, S. P., & Demiris, Y. (2011). Echo state Gaussian process. *IEEE Transactions on Neural Networks*, *22*(9), 1435–1445.
27. Chatzis, S. P., & Kosmopoulos, D. I. (2011). A variational Bayesian methodology for hidden Markov models utilizing Student's-t mixtures. *Pattern Recognition*, *44*(2), 295–306.
28. Chen, H., Tino, P., & Yao, X. (2009). Probabilistic classification vector machines. *IEEE Transactions on Neural Networks*, *20*(6), 901–914.
29. Chen, X.-W., Anantha, G., & Lin, X. (2008). Improving Bayesian network structure learning with mutual information-based node ordering in the K2 algorithm. *IEEE Transactions on Knowledge and Data Engineering*, *20*(5), 1–13.
30. Chen, Y., Bornn, L., de Freitas, N., Eskelin, M., Fang, J., & Welling, M. (2016). Herded Gibbs sampling. *Journal of Machine Learning Research*, *17*, 1–29.
31. Chen, Y.-C., Wheeler, T. A., & Kochenderfer, J. (2017). Learning discrete Bayesian networks from continuous data. *Journal of Artificial Intelligence Research*, *59*, 103–132.
32. Cheng, J., Greiner, R., Kelly, J., Bell, D., & Liu, W. (2002). Learning Bayesian networks from data: An information-theory based approach. *Artificial Intelligence*, *137*(1), 43–90.
33. Cheng, S.-S., Fu, H.-C., & Wang, H.-M. (2009). Model-based clustering by probabilistic self-organizing maps. *IEEE Transactions on Neural Networks*, *20*(5), 805–826.
34. Cheung, Y. M. (2005). Maximum weighted likelihood via rival penalized EM for density mixture clustering with automatic model selection. *IEEE Transactions on Knowledge and Data Engineering*, *17*(6), 750–761.
35. Chickering, D. M. (1996). Learning Bayesian networks is NP-complete. In D. Fisher & H. Lenz (Eds.), *Learning from Data: Artificial Intelligence and Statistics* (Vol. 5, pp. 121–130). Berlin: Springer.
36. Chickering, D. M. (2002). Optimal structure identification with greedy search. *Journal of Machine Learning Research*, *3*, 507–554.
37. Chickering, D. M., Heckerman, D., & Meek, C. (2004). Large-sample learning of Bayesian networks is NP-hard. *Journal of Machine Learning Research*, *5*, 1287–1330.
38. Chien, J.-T., & Hsieh, H.-L. (2013). Nonstationary source separation using sequential and variational Bayesian learning. *IEEE Transactions on Neural Networks and Learning Systems*, *24*(5), 681–694.
39. Choudrey, R. A., & Roberts, S. J. (2003). Variational mixture of Bayesian independent component analyzers. *Neural Computation*, *15*, 213–252.
40. Chu, W., & Ghahramani, Z. (2009). Probabilistic models for incomplete multi-dimensional arrays. In *Proceedings of the 12nd International Conference on Artificial Intelligence and Statistics (AISTATS)* (pp. 89–96). Clearwater Beach, FL.

41. Chu, W., Keerthi, S. S., & Ong, C. J. (2003). Bayesian trigonometric support vector classifier. *Neural Computation, 15*, 2227–2254.
42. Cohen, I., Bronstein, A., & Cozman, F. G. (2001). *Adaptive online learning of Bayesian network parameters*. HP Laboratories Palo Alto, HPL-2001-156.
43. Cohn, I., El-Hay, T., Friedman, N., & Kupferman, R. (2010). Mean field variational approximation for continuous-time Bayesian networks. *Journal of Machine Learning Research, 11*, 2745–2783.
44. Constantinopoulos, C., Titsias, M. K., & Likas, A. (2006). Bayesian feature and model selection for Gaussian mixture models. *IEEE Transactions on Pattern Analysis and Machine Intelligence, 28*(6), 1013–1018.
45. Cooper, G. F. (1990). The computational complexity of probabilistic inference using Bayesian Inference. *Artificial Intelligence, 42*, 393–405.
46. Cooper, G. F., & Herskovits, E. (1992). A Bayesian method for the induction of probabilistic networks from data. *Machine Learning, 9*, 309–347.
47. Cowell, R. (2001). Conditions under which conditional independence and scoring methods lead to identical selection of Bayesian network models. In J. Breese, & D. Koller (Eds.), *Proceedings of the 17th Conference on Uncertainty in AI* (pp. 91–97). Burlington: Morgan Kaufmann.
48. Dagum, P., & Luby, M. (1993). Approximating probabilistic inference in Bayesian belief networks is NP-hard. *Artificial Intelligence, 60*(1), 141–154.
49. Dagum, P., & Luby, M. (1997). An optimal approximation algorithm for Bayesian inference. *Artificial Intelligence, 93*, 1–27.
50. Darwiche, A. (2001). Constant-space reasoning in dynamic Bayesian networks. *International Journal of Approximate Reasoning, 26*(3), 161–178.
51. Dauwels, J., Korl, S., & Loeliger, H.-A. (2005). Expectation maximization as message passing. In *Proceedings of IEEE International Symposium on Information Theory* (1–4). Adelaide, Australia.
52. Dawid, A. P. (1992). Applications of a general propagation algorithm for probalilistic expert systems. *Statistics and Computing, 2*, 25–36.
53. de Campos, L. M., & Castellano, J. G. (2007). Bayesian network learning algorithms using structural restrictions. *International Journal of Approximate Reasoning, 45*, 233–254.
54. de Campos, C. P., & Ji, Q. (2011). Efficient structure learning of Bayesian networks using constraints. *Journal of Machine Learning Research, 12*, 663–689.
55. Del Moral, P., Doucet, A., & Jasra, A. (2006). Sequential Monte Carlo samplers. *Journal of the Royal Statistical Society: Series B, 68*(3), 411–436.
56. Dempster, A. P., Laird, N. M., & Rubin, D. B. (1977). Maximum Likelihood from Incomplete Data via the EM Algorithm. *Journal of the Royal Statistical Society, Series B, 39*(1), 1–38.
57. Deneve, S. (2008). Bayesian spiking neurons. I: Inference. *Neural Computation, 20*(1), 91–117.
58. Du, K.-L., & Swamy, M. N. S. (2010). *Wireless communication systems*. Cambridge: Cambridge University Press.
59. El-Hay, T., Friedman, N., & Kupferman, R. (2008). Gibbs sampling in factorized continuous-time Markov processes. In *Proceedings of the 24th Conference on Uncertainty in AI*.
60. Elidan, G., & Friedman, N. (2005). Learning hidden variable networks: The information bottleneck approach. *Journal of Machine Learning Research, 6*, 81–127.
61. Engel, A., & Van den Broeck, C. (2001). *Statistical mechanics of learning*. Cambridge: Cambridge University Press.
62. Ephraim, Y., & Merhav, N. (2002). Hidden markov processes. *IEEE Transactions on Information Theory, 48*(6), 1518–1569.
63. Fan, Y., Xu, J., & Shelton, C. R. (2010). Importance sampling for continuous time Bayesian networks. *Journal of Machine Learning Research, 11*, 2115–2140.
64. Frey, B. J., & Dueck, D. (2007). Clustering by passing messages between data points. *Science, 305*(5814), 972–976.

65. Friedman, N. (1997). Learning Bayesian networks in the presence of missing values and hidden variables. In D. Fisher (Ed.), *Proceedings of the 14th Conference on Uncertainty in AI* (pp. 125–133). San Francisco: Morgan Kaufmann.
66. Gandhi, P., Bromberg, F., & Margaritis, D. (2008). Learning Markov network structure using few independence tests. In *Proceedings of SIAM International Conference on Data Mining* (pp. 680–691).
67. Gao, B., Woo, W. L., & Dlay, S. S. (2012). Variational regularized 2-D nonnegative matrix factorization. *IEEE Transactions on Neural Networks and Learning Systems*, *23*(5), 703–716.
68. Gaussier, E., & Goutte, C. (2005). Relation between PLSA and NMF and implications. In *Proceedings of Annual ACMSIGIR Conference on Research and Development in Information Retrieval* (pp. 601–602).
69. Gelfand, A. E., & Smith, A. F. M. (1990). Sampling-based approaches to calculating marginal densities. *Journal of the American Statistical Association*, *85*, 398–409.
70. Gelly, S., & Teytaud, O. (2005). Bayesian networks: A better than frequentist approach for parametrization, and a more accurate structural complexity measure than the number of parameters. In *Proceedings of CAP*. Nice, France.
71. Geman, S., & Geman, D. (1984). Stochastic relaxation, Gibbs distributions, and the Bayesian restoration of images. *IEEE Transactions on Pattern Analysis and Machine Intelligence*, *6*(6), 721–741.
72. Getoor, L., Friedman, N., Koller, D., & Taskar, B. (2002). Learning probabilistic models of link structure. *Journal of Machine Learning Research*, *3*, 679–707.
73. Ghahramani, Z., & Beal, M. (1999). Variational inference for Bayesian mixture of factor analysers. *Advances in neural information processing systems* (Vol. 12). Cambridge: MIT Press.
74. Globerson, A., & Jaakkola, T. (2007). Fixing max-product: Convergent message passing algorithms for MAP LP-relaxations. In *Advances in neural information processing systems* (Vol. 20, pp. 553–560). Vancouver, Canada.
75. Gonen, M., Tanugur, A. G., & Alpaydin, E. (2008). Multiclass posterior probability support vector machines. *IEEE Transactions on Neural Networks*, *19*(1), 130–139.
76. Green, P. J. (1995). Reversible jump Markov chain Monte Carlo computation and Bayesian model determination. *Biometrika*, *82*, 711–732.
77. Handschin, J. E., & Mayne, D. Q. (1969). Monte Carlo techniques to estimate the conditional expectation in multi-stage non-linear filtering. *International Journal of Control*, *9*(5), 547–559.
78. Hammersely, J. M., & Morton, K. W. (1954). Poor man's Monte Carlo. *Journal of the Royal Statistical Society: Series B*, *16*, 23–38.
79. Hastings, W. K. (1970). Monte Carlo sampling methods using Markov chains and their applications. *Biometrika*, *57*, 97–109.
80. Hazan, T., & Shashua, A. (2010). Norm-product belief propagation: Primal-dual message-passing for approximate inference. *IEEE Transactions on Information Theory*, *56*(12), 6294–6316.
81. Heckerman, D. (1995). *A Tutorial on learning with Bayesian networks*. Microsoft Technical Report MSR-TR-95-06 (Revised Nov 1996).
82. Heckerman, D., Geiger, D., & Chickering, D. M. (1995). Learning Bayesian networks: The combination of knowledge and statistical data. *Machine Learning*, *20*(3), 197–243.
83. Hennig, P., & Kiefel, M. (2013). Quasi-Newton methods: A new direction. *Journal of Machine Learning Research*, *14*, 843–865.
84. Heskes, T. (2004). On the uniqueness of loopy belief propagation fixed points. *Neural Computation*, *16*, 2379–2413.
85. Hojen-Sorensen, P. A., & d. F. R., Winther, O., & Hansen, L. K., (2002). Mean-field approaches to independent component analysis. *Neural Computation*, *14*, 889–918.
86. Holmes, C. C., & Mallick, B. K. (1998). Bayesian radial basis functions of variable dimension. *Neural Computation*, *10*(5), 1217–1233.

87. Huang, Q., Yang, J., & Zhou, Y. (2008). Bayesian nonstationary source separation. *Neurocomputing*, *71*, 1714–1729.
88. Huang, J. C., & Frey, B. J. (2011). Cumulative distribution networks and the derivative-sum-product algorithm: Models and inference for cumulative distribution functions on graphs. *Journal of Machine Learning Research*, *12*, 301–348.
89. Huang, S., Li, J., Ye, J., Fleisher, A., Chen, K., Wu, T., et al. (2013). A sparse structure learning algorithm for Gaussian Bayesian network identification from high-dimensional data. *IEEE Transactions on Pattern Analysis and Machine Intelligence*, *35*(6), 1328–1342.
90. Huda, S., Yearwood, J., & Togneri, R. (2009). A stochastic version of expectation maximization algorithm for better estimation of hidden Markov model. *Pattern Recognition Letters*, *30*, 1301–1309.
91. Ickstadt, K., Bornkamp, B., Grzegorczyk, M., Wieczorek, J., Sheriff, M. R., Grecco, H. E., et al. (2010). Nonparametric Bayesian network. Bayesian. *Statistics*, *9*, 283–316.
92. Ihler, A. T., Fisher, J. W, I. I. I., & Willsky, A. S. (2005). Loopy belief propagation: Convergence and effects of message errors. *Journal of Machine Learning Research*, *6*, 905–936.
93. Jensen, F. V., Lauritzen, S. L., & Olesen, K. G. (1990). Bayesian updating in causal probabilistic networks by local computations. *Computational Statistics Quaterly*, *4*, 269–282.
94. Jordan, M. I., Ghahramani, Z., Jaakkola, T. S., & Saul, L. K. (1999). An introduction to variational methods for graphical models. *Machine Learning*, *37*, 183–233.
95. Kalisch, M., & Buhlmann, P. (2007). Estimating high-dimensional directed acyclic graphs with the PC-algorithm. *Journal of Machine Learning Research*, *8*, 613–636.
96. Khan, S. A., Leppaaho, E., & Kaski, S. (2016). Bayesian multi-tensor factorization. *Machine Learning*, *105*(2), 233–253.
97. Khreich, W., Granger, E., Miri, A., & Sabourin, R. (2010). On the memory complexity of the forward-backward algorithm. *Pattern Recognition Letters*, *31*, 91–99.
98. Kingma, D. P., & Welling, M. (2014). Auto-encoding variational Bayes. In *Proceedings of the 2nd International Conference on Learning Representations* (pp. 1–14). Banff, Canada.
99. Kjaerulff, U. (1995). dHugin: A computational system for dynamic time-sliced Bayesian networks. *International Journal of Forecasting*, *11*(1), 89–113.
100. Koivisto, M., & Sood, K. (2004). Exact Bayesian structure discovery in Bayesian networks. *Journal of Machine Learning Research*, *5*, 549–573.
101. Kolmogorov, V. (2015). A new look at reweighted message passing. *IEEE Transactions on Pattern Analysis and Machine Intelligence*, *37*(5), 919–930.
102. Kschischang, F. R., Frey, B. J., & Loeliger, H.-A. (2001). Factor graphs and the sum-product algorithm. *Transactions on Information Theory*, *47*(2), 498–519.
103. Lam, W., & Bacchus, F. (1994). Learning Bayesian belief networks: An approach based on the MDL principle. *Computational Intelligence*, *10*(3), 269–293.
104. Lam, W., & Segre, A. M. (2002). A distributed learning algorithm for Bayesian inference networks. *IEEE Transactions on Knowledge and Data Engineering*, *14*(1), 93–105.
105. Langari, R., Wang, L., & Yen, J. (1997). Radial basis function networks, regression weights, and the expectation-maximization algorithm. *IEEE Transactions on Systems, Man, and Cybernetics, Part A*, *27*(5), 613–623.
106. Lappalainen, H., & Honkela, A. (2000). Bayesian nonlinear independent component analysis by multilayer perceptron. In M. Girolami (Ed.), *Advances in independent component analysis* (pp. 93–121). Berlin: Springer.
107. Lauritzen, S. L. (1992). Propagation of probabilities, means and variances in mixed graphical association models. *Journal of the American Statistical Association*, *87*(420), 1098–1108.
108. Lauritzen, S. L., & Spiegelhalter, D. J. (1988). Local computations with probabilities on graphical structures and their application on expert systems. *Journal of the Royal Statistical Society, Series B*, *50*(2), 157–224.
109. Lazaro, M., Santamaria, I., & Pantaleon, C. (2003). A new EM-based training algorithm for RBF networks. *Neural Networks*, *16*, 69–77.
110. Lawrence, N. (2005). Probabilistic non-linear principal component analysis with Gaussian process latent variable models. *Journal of Machine Learning Research*, *6*, 1783–1816.

111. Lawrence, N., Seeger, M., & Herbrich, R. (2003). Fast sparse Gaussian process methods: The informative vector machine. In *Advances in Neural Information Processing Systems* (Vol. 15, pp. 609–616).
112. Li, J., & Tao, D. (2013). Exponential family factors for Bayesian factor analysis. *IEEE Transactions on Neural Networks and Learning Systems*, *24*(6), 964–976.
113. Liang, F. (2007). Annealing stochastic approximation Monte Carlo algorithm for neural network training. *Machine Learning*, *68*, 201–233.
114. Liang, F., Liu, C., & Carroll, R. J. (2007). Stochastic approximation in Monte Carlo computation. *Journal of the American Statistical Association*, *102*, 305–320.
115. Lindsten, F., Jordan, M. I., & Schon, T. B. (2014). Particle Gibbs with ancestor sampling. *Journal of Machine Learning Research*, *15*, 2145–2184.
116. Lopez-Rubio, E. (2009). Multivariate Student-t self-organizing maps. *Neural Networks*, *22*, 1432–1447.
117. Lu, X., Wang, Y., & Yuan, Y. (2013). Sparse coding from a Bayesian perspective. *IEEE Transactions on Neural Networks and Learning Systems*, *24*(6), 929–939.
118. Luis, R., Sucar, L. E., & Morales, E. F. (2010). Inductive transfer for learning Bayesian networks. *Machine Learning*, *79*, 227–255.
119. Ma, S., Ji, C., & Farmer, J. (1997). An efficient EM-based training algorithm for feedforward neural networks. *Neural Networks*, *10*, 243–256.
120. Ma, J., Xu, L., & Jordan, M. I. (2000). Asymptotic convergence rate of the EM algorithm for Gaussian mixtures. *Neural Computation*, *12*, 2881–2907.
121. Mackay, D. J. C. (1992). A practical Bayesian framework for backpropagation networks. *Neural Computation*, *4*(3), 448–472.
122. Margaritis, D., & Thrun, S. (2000). Bayesian network induction via local neighborhoods. In S. A. Solla, T. K. Leen, & K.-R. Muller (Eds.), *Advances in neural information processing systems* (Vol. 12, pp. 505–511). Cambridge: MIT Press.
123. Martins, A. F. T., Figueiredo, M. A. T., Aguiar, P. M. Q., Smith, N. A., & Xing, E. P. (2015). AD^3: Alternating directions dual decomposition for MAP inference in graphical models. *Journal of Machine Learning Research*, *16*, 495–545.
124. Mateescu, R., & Dechter, R. (2009). Mixed deterministic and probabilistic networks: A survey of recent results. *Annals of Mathematics and Artificial Intelligence*, *54*, 3–51.
125. Meilijson, I. (1989). A fast improvement to the EM algorithm on its own terms. *Journal of the Royal Statistical Society: Series B*, *51*(1), 127–138.
126. Minka, T. (2001). *Expectation propagation for approximate Bayesian inference*. Doctoral Dissertation, MIT Media Lab.
127. Miskin, J. W., & MacKay, D. J. C. (2001). Ensemble learning for blind source separation. In S. Roberts & R. Everson (Eds.), *Independent component analysis: Principles and practice* (pp. 209–233). Cambridge: Cambridge University Press.
128. Mnih, A., & Salakhutdinov, R. R. (2007). Probabilistic matrix factorization. In *Advances in neural information processing systems* (Vol. 20, pp. 1257–1264). Red Hook: Curran & Associates Inc.
129. Mongillo, G., & Deneve, S. (2008). Online learning with hidden Markov models. *Neural Computation*, *20*, 1706–1716.
130. Moral, S., Rumi, R., & Salmeron, A. (2001). Mixtures of truncated exponentials in hybrid Bayesian networks. In S. Benferhat, & P. Besnard (Eds.), *Proceedings of the 6th European Conference on Symbolic and Quantitative Approaches to Reasoning with Uncertainty, LNCS 2143* (pp. 156–167). Berlin: Springer.
131. Nasios, N., & Bors, A. (2006). Variational learning for Gaussian mixtures. *IEEE Transactions on Systems Man and Cybernetics, Part B*, *36*(4), 849–862.
132. Neal, R. M. (1993). *Probabilistic inference using Markov chain Monte Carlo methods*. Technical Report CRG-TR-93-1, Department of Computer Science, University of Toronto, Toronto.
133. Ngo, L., & Haddawy, P. (1995). Probabilistic logic programming and bayesian networks. In *Algorithms, Concurrency and Knowledge (Proceedings of Asian Computing Science Conference), LNCS* (Vol. 1023, pp. 286–300). Berlin: Springer.

134. Nielsen, S. H., & Nielsen, T. D. (2008). Adapting Bayes network structures to non-stationary domains. *International Journal of Approximate Reasoning*, *49*, 379–397.
135. Nodelman, U., Shelton, C. R., & Koller, D. (2002). Continuous time Bayesian networks. In *Proceedings of the 18th Conference on Uncertainty in Artificial Intelligence (UAI)* (pp. 378–387).
136. Noorshams, N., & Wainwright, M. J. (2013). Stochastic belief propagation: A low-complexity alternative to the sum-product algorithm. *IEEE Transactions on Information Theory*, *59*(4), 1981–2000.
137. Norton, M., Mafusalov, A., & Uryasev, S. (2017). Soft margin support vector classification as buffered probability minimization. *Journal of Machine Learning Research*, *18*, 1–43.
138. Opper, M. (1998). A Bayesian approach to online learning. In D. Saad (Ed.), *On-line learning in neural networks* (pp. 363–378). Cambridge: Cambridge University Press.
139. Opper, M., & Winther, O. (2000). Gaussian processes for classification: Mean field algorithms. *Neural Computation*, *12*, 2655–2684.
140. Opper, M., & Winther, O. (2001). Tractable approximations for probabilistic models: The adaptive Thouless-Anderson-Palmer mean field approach. *Physical Review Letters*, *86*, 3695–3699.
141. Opper, M., & Winther, O. (2005). Expectation consistent approximate inference. *Journal of Machine Learning Research*, *6*, 2177–2204.
142. Osoba, O., & Kosko, B. (2016). The noisy expectation-maximization algorithm for multiplicative noise injection. *Fluctuation and Noise Letters*, *15*(1), paper ID 1650007.
143. Osoba, O., Mitaim, S., & Kosko, B. (2011). Noise benefits in the expectation-maximization algorithm: NEM theorems and models. In *Proccedings of the International Joint Conference on Neural Networks (IJCNN)* (pp. 3178–3183). San Jose, CA.
144. Ott, G. (1967). Compact encoding of stationary markov sources. *IEEE Transactions on Information Theory*, *13*(1), 82–86.
145. Park, H., & Ozeki, T. (2009). Singularity and slow convergence of the EM algorithm for Gaussian mixtures. *Neural Processing Letters*, *29*, 45–59.
146. Pearl, J. (1988). *Probabilistic reasoning in intelligent systems: Networks of plausible inference*. San Mateo: Morgan Kaufmann.
147. Perez, A., Larranaga, P., & Inza, I. (2009). Bayesian classifiers based on kernel density estimation: Flexible classifiers. *International Journal of Approximate Reasoning*, *50*, 341–362.
148. Pietra, S. D., Pietra, V. D., & Lafferty, J. (1997). Inducing features of random fields. *IEEE Transactions on Pattern Analysis and Machine Intelligence*, *19*(4), 380–393.
149. Rabiner, L. R. (1989). A tutorial on hidden Markov models and selected applications in speech recognition. *Proceedings of the IEEE*, *77*(2), 257–286.
150. Raviv, J. (1967). Decision making in markov chains applied to the problem of pattern recognition. *IEEE Transactions on Information Theory*, *13*(4), 536–551.
151. Richardson, S., & Green, P. J. (1997). On Bayesian analysis of mixtures with an unknown number of components (with discussion). *Journal of the Royal Statistical Society: Series B*, *59*(4), 731–792.
152. Robert, C. P., & Casella, G. (2004). *Monte Carlo Statistical Methods*. New York: Springer.
153. Romero, V., Rumi, R., & Salmeron, A. (2006). Learning hybrid Bayesian networks using mixtures of truncated exponentials. *International Journal of Approximate Reasoning*, *42*, 54–68.
154. Roos, T., Grunwald, P., & Myllymaki, P. (2005). On discriminative Bayesian network classifiers and logistic regression. *Machine Learning*, *59*, 267–296.
155. Rosipal, R., & Girolami, M. (2001). An expectation-maximization approach to nonlinear component analysis. *Neural Computation*, *13*, 505–510.
156. Roweis, S. (1998). EM algorithms for PCA and SPCA. *Advances in neural information processing systems* (Vol. 10, pp. 626–632). Cambridge: MIT Press.
157. Rusakov, D., & Geiger, D. (2005). Asymptotic model selection for naive Bayesian networks. *Journal of Machine Learning Research*, *6*, 1–35.

158. Sarela, J., & Valpola, H. (2005). Denoising source separation. *Journal of Machine Learning Research*, *6*, 233–272.
159. Sato, M. (2001). Online model selection based on the variational Bayes. *Neural Computation*, *13*, 1649–1681.
160. Scutari, M. (2010). Learning Bayesian networks with the bnlearn R package. *Journal of Statistical Software*, *35*(3), 1–22.
161. Seeger, M. W. (2008). Bayesian inference and optimal design for the sparse linear model. *Journal of Machine Learning Research*, *9*, 759–813.
162. Shashanka, M., Raj, B., & Smaragdis, P. (2008). Probabilistic latent variable models as non-negative factorizations. *Computational Intelligence and Neuroscience*, *2008*, 947438, 9 p.
163. Shelton, C. R., Fan, Y., Lam, W., Lee, J., & Xu, J. (2010). Continuous time Bayesian network reasoning and learning engine. *Journal of Machine Learning Research*, *11*, 1137–1140.
164. Shutin, D., Zechner, C., Kulkarni, S. R., & Poor, H. V. (2012). Regularized variational Bayesian learning of echo state networks with delay & sum readout. *Neural Computation*, *24*, 967–995.
165. Silander, T., & Myllymaki, P. (2006). A simple approach for finding the globally optimal Bayesian network structure. In *Proceedings of the 22th Conference on Uncertainty in AI* (pp. 445–452).
166. Silander, T., Kontkanen, P., & Myllymaki, P. (2007). On sensitivity of the MAP Bayesian network structure to the equivalent sample size parameter. In R. Parr, & L. van der Gaag (Eds.), *Proceedings of the 23rd Conference on Uncertainty in AI* (pp. 360–367). AUAI Press.
167. Silander, T., Roos, T., & Myllymaki, P. (2009). Locally minimax optimal predictive modeling with Bayesian networks. In *Proceedings of the 12th International Conference on Artificial Intelligence and Statistics (AISTATS), JMLR Proceedings Track* (Vol. 5, 504–511). Clearwater Beach, FL.
168. Smyth, P., Hecherman, D., & Jordan, M. I. (1997). Probabilistic independent networks for hidden Markov probabilities models. *Neural Computation*, *9*(2), 227–269.
169. Spiegelhalter, D. J., & Lauritzen, S. L. (1990). Sequential updating of conditional probabilities on directed graphical structures. *Networks*, *20*(5), 579–605.
170. Spirtes, P., Glymour, C., & Scheines, R. (2000). *Causation, prediction, and search* (2nd ed.). Cambridge: MIT Press.
171. Takeda, A., & M. Sugiyama, (2008). ν-support vector machine as conditional value-at-risk minimization. In *Proceedings of the ACM 25th international conference on machine learning* (pp. 1056–1063).
172. Takekawa, T., & Fukai, T. (2009). A novel view of the variational Bayesian clustering. *Neurocomputing*, *72*, 3366–3369.
173. Tamada, Y., Imoto, S., & Miyano, S. (2011). Parallel algorithm for learning optimal Bayesian network structure. *Journal of Machine Learning Research*, *12*, 2437–2459.
174. Tan, X., & Li, J. (2010). Computationally efficient sparse Bayesian learning via belief propagation. *IEEE Transactions on Signal Processing*, *58*(4), 2010–2021.
175. Tanner, M., & Wong, W. (1987). The calculation of posterior distributions by data augmentation. *Journal of the American Statistical Association*, *82*(398), 528–540.
176. Tao, Q., Wu, G., Wang, F., & Wang, J. (2005). Posterior probability support vector machines for unbalanced data. *IEEE Transactions on Neural Networks*, *16*(6), 1561–1573.
177. Tatikonda, S., & Jordan, M. (2002). Loopy belief propagation and Gibbs measures. In *Proceedings of the 18th Conference on Uncertainty in Artificial Intelligence* (pp. 493–500). San Francisco, CA: Morgan Kaufmann.
178. Ting, J.-A., D'Souza, A., Vijayakumar, S., & Schaal, S. (2010). Efficient learning and feature selection in high-dimensional regression. *Neural Computation*, *22*, 831–886.
179. Tipping, M. E. (2000). The relevance vector machine. In *Advances in neural information processing systems* (Vol. 12, pp. 652–658).
180. Tipping, M. E. (2001). Sparse Bayesian learning and the relevance vector machine. *Journal of Machine Learning Research*, *1*, 211–244.
181. Tipping, M. E., & Bishop, C. M. (1999). Probabilistic principal component analysis. *Journal of the Royal Statistical Society: Series B*, *61*(3), 611–622.

182. Tipping, M. E., & Bishop, C. M. (1999). Mixtures of probabilistic principal component analyzers. *Neural Computation*, *11*, 443–482.
183. Tipping, M. E., & Faul, A. C. (2003). Fast marginal likelihood maximisation for sparse Bayesian models. In *Proceedings of the 9th International Workshop on Artificial Intelligence and Statistics* (pp. 1–13). Key West, FL.
184. Tsamardinos, I., Brown, L. E., & Aliferis, C. F. (2006). The max-min hill-climbing bayesian network structure learning algorithm. *Machine Learning*, *65*(1), 31–78.
185. Tzikas, D. G., Likas, A. C., & Galatsanos, N. P. (2009). Sparse Bayesian modeling with adaptive kernel learning. *IEEE Transactions on Neural Networks*, *20*(6), 926–937.
186. Ueda, N., Nakano, R., Ghahramani, Z., & Hinton, G. E. (2000). SMEM algorithm for mixture models. *Neural Computation*, *12*, 2109–2128.
187. Valpola, H., & Pajunen, P. (2000). Fast algorithms for Bayesian independent component analysis. In *Proceedings of the 2nd International Workshop on Independent Component Analysis and Signal Separation* (pp. 233–237). Helsinki, Finland.
188. Valpola, H. (2000). Nonlinear independent component analysis using ensemble learning: Theory. In *Proceedings of the 2nd International Workshop on Independent Component Analysis and Signal Separation* (pp. 251–256). Helsinki, Finland.
189. Valpola, H., & Karhunen, J. (2002). An unsupervised ensemble learning for nonlinear dynamic state-space models. *Neural Computation*, *141*(11), 2647–2692.
190. Verma, T., & Pearl, J. (1990). Equivalence and synthesis of causal models. In *Proceedings of the 6th Conference on Uncertainty in AI* (255–268). Cambridge, MA.
191. Wang, N., Yao, T., Wang, J., & Yeung, D.-Y. (2012). A probabilistic approach to robust matrix factorization. In *Proceedings of the 12th European Conference on Computer Vision* (pp. 126–139). Florence, Italy.
192. Watanabe, K., & Watanabe, S. (2006). Stochastic complexities of Gaussian mixtures in variational Bayesian approximation. *Journal of Machine Learning Research*, *7*(4), 625–644.
193. Watanabe, K., & Watanabe, S. (2007). Stochastic complexities of general mixture models in variational Bayesian learning. *Neural Networks*, *20*, 210–219.
194. Watanabe, K., Akaho, S., Omachi, S., & Okada, M. (2009). VB mixture model on a subspace of exponential family distributions. *IEEE Transactions on Neural Networks*, *20*(11), 1783–1796.
195. Weiss, Y., & Freeman, W. T. (2001). Correctness of belief propagation in Gaussian graphical models of arbitrary topology. *Neural Computation*, *13*(10), 2173–2200.
196. Welling, M., & Weber, M. (2001). A constrained EM algorithm for independent component analysis. *Neural Computation*, *13*, 677–689.
197. Winn, J., & Bishop, C. M. (2005). Variational message passing. *Journal of Machine Learning Research*, *6*, 661–694.
198. Winther, O., & Petersen, K. B. (2007). Flexible and efficient implementations of Bayesian independent component analysis. *Neurocomputing*, *71*, 221–233.
199. Xiang, Y. (2000). Belief updating in multiply sectioned Bayesian networks without repeated local propagations. *International Journal of Approximate Reasoning*, *23*, 1–21.
200. Xie, X., & Geng, Z. (2008). A recursive method for structural learning of directed acyclic graphs. *Journal of Machine Learning Research*, *9*, 459–483.
201. Xie, X., Yan, S., Kwok, J., & Huang, T. (2008). Matrix-variate factor analysis and its applications. *IEEE Transactions on Neural Networks*, *19*(10), 1821–1826.
202. Xu, L., Jordan, M. I., & Hinton, G. E. (1995). An alternative model for mixtures of experts. In G. Tesauro, D. S. Touretzky, & T. K. Leen (Eds.), *Advances in neural information processing systems* (Vol. 7, pp. 633–640). Cambridge: MIT Press.
203. Yamazaki, K., & Watanabe, S. (2003). Singularities in mixture models and upper bounds of stochastic complexity. *Neural Networks*, *16*, 1023–1038.
204. Yang, Z. R. (2006). A novel radial basis function neural network for discriminant analysis. *IEEE Transactions on Neural Networks*, *17*(3), 604–612.
205. Yap, G.-E., Tan, A.-H., & Pang, H.-H. (2008). Explaining inferences in Bayesian networks. *Applied Intelligence*, *29*, 263–278.

206. Yedidia, J. S., Freeman, W. T., & Weiss, Y. (2001). Generalized belief propagation. In T. K. Leen, T. G. Dietterich, & V. Tresp (Eds.), *Advances in neural information processing systems* (Vol. 13, pp. 689–695). Cambridge: MIT Press.
207. Yuille, A. (2002). CCCP algorithms to minimize the Bethe and Kikuchi free energies: Convergent alternatives to belief propagation. *Neural Computation*, *14*, 1691–1722.
208. Zhang, B., Zhang, C., & Yi, X. (2004). Competitive EM algorithm for finite mixture models. *Pattern Recognition*, *37*, 131–144.
209. Zhao, J., Yu, P. L. H., & Kwok, J. T. (2012). Bilinear probabilistic principal component analysis. *IEEE Transactions on Neural Networks and Learning Systems*, *23*(3), 492–503.
210. Zhao, Q., Meng, D., Xu, Z., Zuo, W., & Yan, Y. (2015). L_1-norm low-rank matrix factorization by variational Bayesian method. *IEEE Transactions on Neural Networks and Learning Systems*, *26*(4), 825–839.
211. Zhao, Q., Zhou, G., Zhang, L., Cichocki, A., & Amari, S.-I. (2016). Bayesian robust tensor factorization for incomplete multiway data. *IEEE Transactions on Neural Networks and Learning Systems*, *27*(4), 736–748.

Chapter 23
Boltzmann Machines

23.1 Boltzmann Machines

Boltzmann machine as well as some other stochastic models can be treated as generalizations of the Hopfield model. The Boltzmann machine integrates the global optimization capability of simulated annealing. By a combination of the concept of energy and the neural network topology, these models provide a method to deal with notorious COPs.

Boltzmann machine is a stochastic recurrent network based on physical systems [1, 28]. It has the same network architecture as that of Hopfield model, that is, it is highly recurrent with $w_{ij} = w_{ji}$ and $w_{ii} = 0$, $i, j = 1, \ldots, J$. Unlike the Hopfield network, Boltzmann machine can have hidden units. The Hopfield network operates in an unsupervised manner, while Boltzmann machine can be trained in an unsupervised or supervised manner. Boltzmann machine, operated in sequential or synchronous mode, is a universal approximator for arbitrary functions defined on finite sets [60].

Neurons of a Boltzmann machine are divided into visible and hidden units, as illustrated in Fig. 23.1. In Fig. 23.1a, the visible units are clamped onto specific states determined by the environment, while the hidden units always operate freely. By capturing high-order statistical correlations in the clamping vector, the hidden units simulate the underlying constraints contained in the input vectors. This type of Boltzmann machine uses unsupervised learning and can perform pattern completion. When the visible units are further divided into input and output neurons, as shown in Fig. 23.1b, this type of Boltzmann machine can be trained in a supervised manner. The recurrence eliminates the difference in input and output cells.

K.-L. Du and M. N. S. Swamy, *Neural Networks and Statistical Learning*,
https://doi.org/10.1007/978-1-4471-7452-3_23

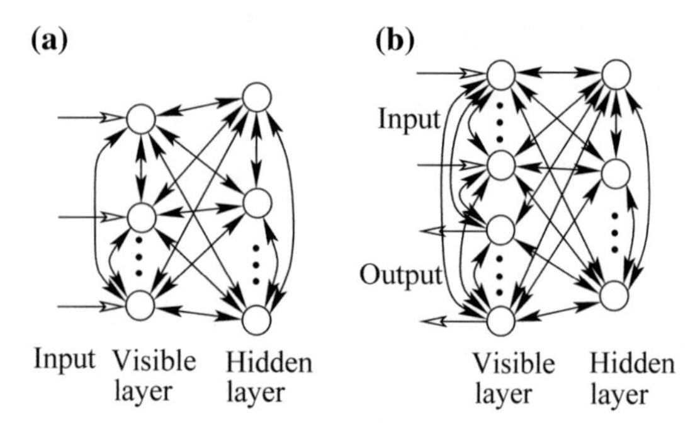

Fig. 23.1 Architecture of Boltzmann machine. **a** An architecture with visible and hidden neurons. **b** An architecture with input, output, and hidden neurons

Instead of using a sigmoidal function in the Hopfield network, the activation at each neuron takes either 0 or 1, depending on the probability of a temperature variable T

$$net_i = \sum_{j=1, j\neq i}^{J} w_{ji} x_j = \boldsymbol{w}_i \boldsymbol{x}, \tag{23.1}$$

$$x_i = \begin{cases} 1, & \text{with probability } P_i \\ 0, & \text{with probability } 1 - P_i \end{cases}, \tag{23.2}$$

$$P_i = \frac{1}{1 + \mathrm{e}^{-\frac{net_i}{T}}}. \tag{23.3}$$

When T is very large or net_i approaches 0, x_i takes either 1 or 0 with equal probability. For very small T, x_i is deterministically 1. The input and output states can be fixed or variable.

Search for all the possible states is performed at temperature T in the Boltzmann machine. At the steady state, the relative probability of two states in the Boltzmann machine is determined by the Boltzmann distribution of the energy difference between the two states

$$\frac{P_\alpha}{P_\beta} = \mathrm{e}^{-\frac{E_\alpha - E_\beta}{T}}, \tag{23.4}$$

where E_α and E_β are the corresponding energy levels of the two states. The energy can be computed by the same formula as for the Hopfield model

$$E(t) = -\frac{1}{2}\boldsymbol{x}^T(t)\mathbf{W}\boldsymbol{x}(t) = -\sum_{i<j} x_i(t) x_j(t) w_{i,j}. \tag{23.5}$$

In [4], a synchronous Boltzmann machines as well as its learning algorithm has been introduced to facilitate parallel implementations.

Like the complex-valued multistate Hopfield model, a multivalued Boltzmann machine proposed in [37] extends the binary Boltzmann machine. Each neuron of the multivalued Boltzmann machine can only take L discrete stable states, and the angle between two adjacent directions is given by $\varphi_0 = \frac{2\pi}{L}$. The probability of state change is determined by the Boltzmann distribution of the energy difference between the two states.

23.1.1 Boltzmann Learning Algorithm

When using the Boltzmann machine with hidden units, the generalized Hebbian rule cannot be used in an unsupervised manner. For supervised learning of Boltzmann machine, BP is not applicable due to the different network architecture. Simulated annealing is used by Boltzmann machines to learn weights corresponding to the global optimum. The learning process in the Boltzmann machine is computationally very expensive. The learning complexity of the exact Boltzmann machine is exponential in the number of neurons [31].

For constraint-satisfaction problems, some of the neurons are externally clamped to some input patterns, and we then find the global minimum for these particular input patterns. The integration of simulated annealing into the Boltzmann learning rule makes the Boltzmann machine especially suitable for constraint-satisfaction tasks involving a large number of weak constraints [28].

Boltzmann machines can be regarded as Markov random fields with binary random variables. For binary cases, they are equivalent to the Ising-spin model in statistical mechanics. Learning in Boltzmann machines is an NP-hard problem. The original Boltzmann learning algorithm [1, 28] is based on counting occurrences. The Boltzmann learning algorithm based on correlations [44] provides a better performance than the original algorithm. The correlation-based learning procedure for Boltzmann machine is given in Algorithm 23.1 [23, 44].

Algorithm 23.1 (Boltzmann Learning)

1. *Initialization.*
   a. *Initialize* w_{ji}*: Set* w_{ji} *as uniform random values in* $[-a_0, a_0]$, *typically* $a_0 = 0.5$ *or* 1.
   b. *Set the initial and the final temperatures:* T_0 *and* T_f.
2. *Clamping phase.*
   Present the patterns. For unsupervised learning, all the visible nodes are clamped to the patterns. For supervised learning, all the input and output nodes are clamped to the pattern pairs.
   a. *For each example, perform simulated annealing until* T_f *is reached.*
      i. *At each* T*, relax the network by the Boltzmann distribution for a length of time through updating the states of the unclamped (hidden) units*

$$x_i = \begin{cases} +1, & \text{with probability } P_i \\ -1, & \text{with probability } 1 - P_i \end{cases} \quad (23.6)$$

      where P_i *is calculated by (23.1) and (23.3).*
      ii. *Update* T *by the annealing schedule.*
   b. *At* T_f*, estimate the correlation in the clamped condition*

$$\rho_{ij}^{+} = E\left[x_i x_j\right], \quad i, j = 1, 2, \ldots, J; \quad i \neq j. \quad (23.7)$$

3. *Free-running phase.*
   a. *Repeat Step 2a. For unsupervised learning, all the visible neurons are now free-running. For supervised learning, only the input neurons are clamped and the output neurons are free-running.*
   b. *At* T_f*, estimate the correlation in the free-running condition*

$$\rho_{ij}^{-} = E\left[x_i x_j\right], \quad i, j = 1, 2, \ldots, J; \quad i \neq j. \quad (23.8)$$

4. *Weight update.*
   The weight update is performed as

$$\Delta w_{ij} = \eta\left(\rho_{ij}^{+} - \rho_{ij}^{-}\right), \quad i, j = 1, 2, \ldots, J; \quad i \neq j, \quad (23.9)$$

   where $\eta = \frac{\varepsilon}{T}$*, with* ε *being a small positive constant.*
5. *Repeat Steps 2 through 4 for next epoch until there is no change in* w_{ij}, $\forall i, j$.

Boltzmann learning is implemented in four steps: initialization, clamping phase, free-running phase, and weight update. The algorithm iterates from second to fourth steps for each epoch. Equation (23.9) is called the *Boltzmann learning rule*.

Boltzmann machine is suitable for modeling biological phenomena, since biological neurons are stochastic systems. This process is, however, too slow though it can find the global optimum.

Contrastive Hebbian learning [5] is a supervised learning algorithm based on Hebb's rule. Similar to Boltzmann learning, learning operates in a free-running phase and a clamping phase, based on the neural dynamics. All the neural activities in all the layers may be updated simultaneously without waiting for the convergence of previous or subsequent layers. In [58], the weights are updated after each phase has reached its fixed point; when the feedback gain is small (such as in the clamping phase), contrastive Hebbian learning is equivalent to BP [58].

Similar to feedback alignment [36], random contrastive Hebbian learning [16] replaces the transpose of the synaptic weights in contrastive Hebbian learning with fixed random matrices. It is biologically plausible by using randomly fixed instead of bidirectional synapses to transform the feedback signals during the clamping phase.

23.2 Restricted Boltzmann Machines

Boltzmann machines are two-layer neural network architectures composed of binary stochastic nodes connected in an interlayer and intralayer fashion. Restricted Boltzmann machines (RBMs) [49], as shown in Fig. 23.2, are Boltzmann machines with a single layer of feature-detecting units. The nodes in a visible layer are fully connected to the nodes in a hidden layer, and there is no intralayer connection between nodes within the same layer. This connectivity ensures that all hidden units (features) are statistically decoupled when visible units are clamped to the observed values. The restricted connectivity allows us to train an RBM efficiently on the basis of cheap inference and finite Gibbs sampling [25]. An RBM forms an autoassociative memory.

An RBM is a form of product-of-experts model [25], which is also a Boltzmann machine with a bipartite connectivity graph. It is a Markov random field with a bipartite structure. It is an undirected graphical model with stochastic binary units in both the visible layer and the hidden layer. The RBM maximizes the log-likelihood

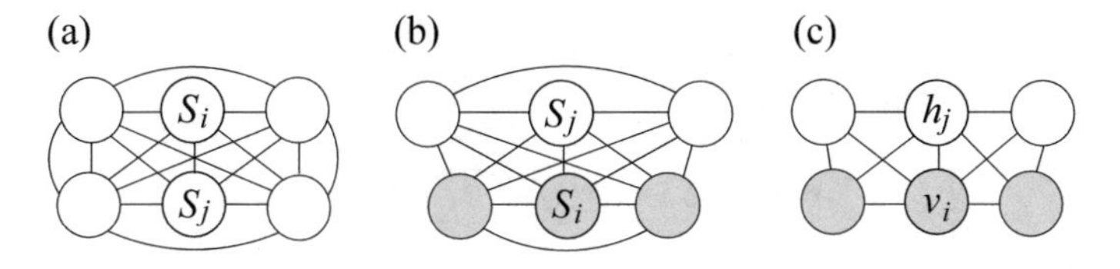

Fig. 23.2 Architecture: **a** A Boltzmann machine. **b** A Boltzmann machine partitioned into visible (shaded) and hidden units. **c** An RBM

for the joint distribution of all visible units, that is, the features and targets. It is trained in an unsupervised manner to model the distribution of the inputs.

A hybrid Boltzmann machine is an RBM in which the hidden units can take on continuous values, while the visible units have binary values [7]. When the functions are marginalized over the hidden units, the hybrid Boltzmann machine is thermodynamically equivalent to the Hopfield model [7].

An RBM could learn a probability distribution over its set of inputs. It defines Gibbs–Boltzmann probability distributions over the observable states of the network, depending on the interaction weights and biases. RBMs have been used as generative models of many different types of data. Typically, they are stacked in a hierarchy to build a deep network [8]. They have been successfully used as basic building blocks in deep neural networks for automatic features extraction, and unsupervised weights initialization, but also as standalone models in other types of applications such as density estimators. The pretraining step for RBMs aims to discover an unknown stationary distribution based on the sample data that has the lowest energy.

RBM can be trained in a supervised or unsupervised fashion. Its success is due to Hinton's contrastive divergence algorithm [25] for unsupervised training. Contrastive divergence algorithm approximates the log-likelihood gradient and requires fewer sampling steps than the MCMC algorithm does [25]. It performs k steps of Gibbs sampling and gradient descent to find the weight vector that maximizes the objective function, which is the product of probabilities. A typical value is $k = 1$.

Consider a model with input (visible) units $\boldsymbol{v} \in \{0, 1\}^D$, and hidden units $\boldsymbol{h} \in \{0, 1\}^M$. The visible and hidden units are connected with a weight matrix, $\mathbf{W} \in R^{D\times M}$ and have bias vectors $\boldsymbol{c} \in R^M$ and $\boldsymbol{b} \in R^D$, respectively. RBM can learn a joint probability distribution over its D visible units and M hidden feature units.

The energy function and the joint distribution for a visible vector and a hidden vector with the Bernoulli distribution of the logistic activation function are defined by

$$E(\boldsymbol{v}, \boldsymbol{h}|\boldsymbol{\theta}) = -\boldsymbol{h}^T \mathbf{W} \boldsymbol{v} - \boldsymbol{b}^T \boldsymbol{h} - \boldsymbol{c}^T \boldsymbol{v}, \tag{23.10}$$

$$p(\boldsymbol{v}, \boldsymbol{h}|\boldsymbol{\theta}) = \frac{e^{-E(\boldsymbol{v},\boldsymbol{h},\boldsymbol{\theta})}}{\sum_{\boldsymbol{v}} \sum_{\boldsymbol{h}} e^{-E(\boldsymbol{v},\boldsymbol{h},\boldsymbol{\theta})}}, \tag{23.11}$$

where the parameters $\boldsymbol{\theta} = \{\mathbf{W}, \boldsymbol{b}, \boldsymbol{c}\}$ denote connection weights and biases between cross-layer units.

For binary visible and hidden units, it is easy to obtain the conditional probability distributions:

$$p(h_j|\boldsymbol{v}, \boldsymbol{\theta}) = \sigma(b_j + \sum_i W_{ij} v_i), \tag{23.12}$$

$$p(v_i|\boldsymbol{h}, \boldsymbol{\theta}) = \sigma(c_i + \sum_j W_{ij} h_j), \tag{23.13}$$

where $\sigma(\cdot)$ is the activation function, typically the logistic function $\sigma(x) = 1/(1 + e^{-x})$.

Parameters of RBM and Gaussian RBM models are fitted by maximizing the likelihood function. In maximum likelihood, the learning phase actually minimizes the Kullback–Leiber (KL) measure between the input data distribution and the model approximation.

Training large RBMs by steepest ascent on the log-likelihood gradient is computationally intractable, because the gradient involves averages over an exponential number of terms. Therefore, the gradient is approximated by MCMC (e.g., methods based on Gibbs sampling [15, 25, 55]). For the learning algorithms, the bias of the approximation increases with increasing absolute values of the model parameters [9, 18].

RBM is usually used as a feature extractor, but not as a standalone classifier. It is popularly used as a building block of multilayer learning architectures such as deep belief networks [26] or deep autoencoders [27]. By introducing a discriminative component to RBM training, discriminative RBM [33] can model the joint distribution of inputs and the associated target labels.

23.2.1 Universal Approximation

Boltzmann machine and RBM are universal approximators for probability distributions on binary vectors [34, 60].

RBM is a universal approximator of discrete distributions, as given by Theorem 23.1 [34].

Theorem 23.1 (Universal Approximation of RBM, Le Roux and Bengio [34]) *Any distribution over* $\{0, 1\}^n$ *can be approximated arbitrarily well (in the sense of the Kullback–Leibler divergence) with an RBM with* $k + 1$ *hidden units where* k *is the number of input vectors whose probability is not* 0.

The universal approximation of conditional RBM follows immediately from that of Boltzmann machine [39]. There are exponentially many deterministic conditional distributions which can only be approximated arbitrarily well by a conditional RBM if the number of hidden units is exponential in the number of input units. It may be possible with a drastically smaller number of hidden units than RBM universal approximation, for the same number of visible units [39]. Universal approximation results for temporal RBM and recurrent temporal RBM are proved in [42].

In [45], a factored three-way restricted Botlzmann machine allows the factors or states of hidden units in restricted Botlzmann machine to modulate the pairwise interactions between visible units in natural images. The weights in the restricted Botlzmann machine are parameterized as a tensor.

23.2.2 Contrastive Divergence Algorithm

Contrastive divergence [25] is used as an approximate to the gradient of the log-likelihood of the data. The parameters $\boldsymbol{\theta}$ are trained to minimize the reconstruction error using contrastive divergence [25]. The method minimizes the negative log-likelihood of N data with respect to parameters $\boldsymbol{\theta}$:

$$L(\boldsymbol{\theta}) = -\sum_n \log p(\boldsymbol{x}_n|\boldsymbol{\theta}). \tag{23.14}$$

By maximizing the log-likelihood of the input data, the parameters can be derived by gradient descent:

$$\Delta W_{ij} = \eta \frac{\partial \log p(\boldsymbol{v})}{\partial W_{ij}} \approx \eta(< v_i h_j >_{data} - < v_i h_j >_{model}), \tag{23.15}$$

$$\Delta c_i = \eta \frac{\partial \log p(\boldsymbol{v})}{\partial c_i} \approx \eta(< v_i >_{data} - < v_i >_{model}), \tag{23.16}$$

$$\Delta b_j = \eta \frac{\partial \log p(\boldsymbol{v})}{\partial h_j} \approx \eta(< h_j >_{data} - < h_j >_{model}), \tag{23.17}$$

where η is the learning rate, and $< \cdot >$ is the operator of expectation with the subscript denoting the distribution.

Contrastive divergence algorithm was proposed as an efficient training method for product-of-expert models, of which RBMs are a special case [25]. It has been shown to scale well to large problems. The general procedure for training RBM using contrastive divergence is given by Algorithm 23.2. It can be trained using the open-source machine learning library Shark [29].

Algorithm 23.2 (Contrastive Divergence Algorithm)

Initialize the weights to 0.
For each sample from the training batch:

1. *Apply the sample to the network input.*
2. *For to* $k - 1$ *sampling steps,*
   a. for each hidden layer neuron from 1 to n, sample $h_i^{(t)} \; p(h_i|\boldsymbol{v}^{(t)})$;
   b. for each input layer neuron from 1 to m, sample $v_j^{(t)} \; p(v_j|\boldsymbol{h}^{(t)})$.
3. *For each input and hidden layer neuron, compute*
   a. $\Delta w_{ij} = \Delta w_{ij} + p(H_i = 1|v^{(0)})v_j^{(0)} - p(H_i = 1|v^{(k)})v_j^{(k)}$
   b. $\Delta b_j = \Delta b_j + v_j^{(0)} - v_j^{(k)}$
   c. $\Delta c_i = \Delta c_i + p(H_i = 1|v^{(0)} - p(H_i = 1|v^{(k)}$.

Contrastive divergence algorithm generally performs 1-step Gibbs sampling. This causes the reconstructed data to lie in the immediate neighborhood of the training data, and the reconstructed data are generally noisy versions of the training data. The issue still exists for k-step Gibbs sampling, since k is often a small number in practice. As such, contrastive divergence often has the issues of biased samples, poor mixing of Markov chains and high-mass probability modes.

Some efficient algorithms, such as persistent contrastive divergence [54], fast persistent contrastive divergence [55], parallel tempering [15], dissimilar contrastive divergence [48], and higher order mean field method [19], have been proposed to address these issues.

In persistent contrastive divergence [54] and fast persistent contrastive divergence [55], the Gibbs chain is not initialized by a training example but maintains its current value between approximation steps. Parallel tempering (also known as *replica exchange Monte Carlo sampling*) [15] introduces supplementary Gibbs chains that sample from more and more smoothed variants of the true probability distribution and allows samples to swap between chains. This leads to faster mixing but introduces computational overhead. The performance of these methods strongly depends on the mixing properties of the Gibbs chain. Dissimilar contrastive divergence [48] uses dissimilar data to allow the Markov chain to explore new modes in the probability space. This method can be used with all variants of contrastive divergence (or persistent contrastive divergence), and across all energy-based deep learning models. A deterministic iterative procedure based on a higher order mean field method known as the Thouless–Anderson–Palmer approach [19] provides performance equal to, and sometimes superior to, persistent contrastive divergence, while also providing a clear and easy to evaluate objective function.

Weighted contrastive divergence [46] modifies the negative phase of the gradients in the weights update rule in contrastive divergence as a weighted average over the members of the batch. It provides a significant improvement over contrastive divergence and persistent contrastive divergence at a small additional computational cost.

Contrastive divergence is biased [9] and it may not converge in some cases [11]. Contrastive divergence and variants, such as persistent contrastive divergence [54] and fast persistent contrastive divergence [55], can lead to a steady decrease of the log-likelihood of the training data during learning [15]. A simple reconstruction error is often used as a stopping criterion. However, in many cases, the evolution curve of the reconstruction error is monotonic, while the log-likelihood is not. An estimation of the log-likelihood based on annealed importance sampling works better than the reconstruction error in most cases, but it is considerably more expensive to compute.

A variant of STDP can approximate the contrastive divergence algorithm for training a spiking RBM [41].

23.2.3 Related Methods

A Metropolis-type MCMC algorithm [10] relies on a transition operator maximizing the probability of state changes. The transition operator can replace Gibbs sampling in RBM learning algorithms without producing computational overhead. This leads to faster mixing and more accurate learning. The operator is guaranteed to lead to an ergodic and thus properly converging Markov chain when using a periodic updating scheme.

Learning could be improved by incorporating some prior about the world or the representation itself. The prior can be introduced in the form of constraints or as a regularization term in the learning objective. For deep learning, the widely used regularization is weight decay.

RBMs and Gaussian RBMs are bipartite graphs which naturally have a small-world topology [38]. Scale-free networks are sparse [14]. By constraining RBMs and Gaussian RBMs to a scale-free topology, the number of weights that need to be computed is reduced by a few orders of magnitude, at virtually no loss in generative performance [38].

With enough hidden units, an RBM is able to represent a binary probability distribution fitting the training data as well as possible. However, as the number of hidden units increases, many learned features become strongly correlated [50], which increases the risk of overfitting. The weights of an infinite number of binary hidden units are conceptually tied in [40]. A similar effect can be achieved by dropout [50], which is equivalent to combining exponentially many different subnetworks, and also serves as regularization by an adaptive weight decay and sparse representation [6].

Infinite RBM [13] automatically adjusts the effective number of hidden units participating in the energy function during training according to the data. It mixes infinitely many RBMs with a different number of hidden units, and all the RBMs choose the hidden units from the same sequence starting from the first one. However, the convergence of learning is slow, due to the ordering of its hidden units. A training strategy [43] randomly regroups the hidden units before each gradient-descent step. It potentially achieves a mixing of infinitely many infinite RBMs with different permutations of the hidden units, which has a similar effect as dropout.

The efficient statistical modeling of RBMs has been combined with the dynamic properties of recurrent networks [26, 34, 52]. These include temporal RBM, recurrent temporal RBM, and conditional RBM.

A conditional RBM [52] is defined by clamping the states of an input subset of the visible units of an RBM. These networks define models of conditional probability distributions on the states of the output units given the states of the input units, parameterized by interaction weights and biases.

A discriminative learning approach [17] provides a self-contained RBM method for classification, inspired by free-energy-based function approximation. For classification, the free-energy RBM method computes the output for an input vector and a class vector by the negative free energy of an RBM. Learning is achieved by stochastic gradient descent using a mean-squared error. The learning performance can be further improved by computing the output by the negative expected energy [17].

Modeling Real-Valued Data

Gaussian RBM model [27] extends RBM model for applications with real-valued feature vectors. The binary units from the visible layer v are replaced by linear units with Gaussian noise. The hidden units $\boldsymbol{h}$ remain binary. The total energy function for a state $\{\boldsymbol{v}, \boldsymbol{h}\}$ of Gaussian RBMs is calculated in a manner similar to that in RBMs, but includes a slight change to take into consideration the Gaussian noise of the visible neurons. Gaussian–Bernoulli deep Boltzmann machines [12, 52] are deep networks based on Gaussian RBM.

Gaussian RBM extends binary RBM by replacing the Bernoulli distribution with the Gaussian distribution for the visible data [56]. The energy function of different configurations of visible units and hidden ones is defined by

$$E(\boldsymbol{v}, \boldsymbol{h}) = -\sum_i \frac{(v_i - c_i)^2}{2\sigma_i^2} - \sum_i \sum_j W_{ij} \frac{v_i}{\sigma_i} h_j - \sum_j b_j h_j, \qquad (23.18)$$

where w_{ij}, c_i, b_j and σ_i are the model parameters. In practice, the visible data is usually normalized to zero mean and unit variance. Exact maximum likelihood learning is also intractable, and thus, contrastive divergence algorithm is used to perform efficient learning.

Modeling Count Data

For the count data, we can use the replicated softmax RBM [47] to model the sparse bag of words feature vector. Given a document that contains D words, the energy function of a state $(\boldsymbol{v}, \boldsymbol{h})$ is defined by

$$E(\boldsymbol{v}, \boldsymbol{h}) = -\sum_i \sum_j w_{ij} v_i h_j - \sum_i c_i v_i - D \sum_j b_j h_j, \qquad (23.19)$$

where w_{ij}, c_i, and b_j are the model parameters. Efficient learning of this model is also performed using contrastive divergence.

23.3 Mean-Field-Theory Machine

Mean-field approximation is a well-known method in statistical physics [21, 53]. The mean-field annealing algorithm was proposed to accelerate the convergence of the Boltzmann machine [44]. The Boltzmann machine with such an algorithm is also termed the *mean-field-theory machine* or *deterministic Boltzmann machine*.

Mean-field annealing, which replaces all the states in the Boltzmann machine by their averages, can be treated as a deterministic form of Boltzmann learning. In [44], the correlations in the Boltzmann learning rule is replaced by the naive mean-field approximation. Instead of the stochastic binary neuron output for the Boltzmann machine, continuous neuron outputs, which are calculated as the average of the probability of the binary neuron variables at temperature T, are used. The average of state x_i is calculated for a specific value of activation net_i according to (23.6), (23.1) and (23.3)

$$\begin{aligned} E\left[x_i\right] &= (+1) P_i + (-1)\left(1 - P_i\right) = 2P_i - 1 \\ &= \tanh\left(\frac{net_i}{T}\right). \end{aligned} \tag{23.20}$$

The correlation in the Boltzmann learning rule is replaced by the mean-field approximation [44]

$$E\left[x_i x_j\right] \simeq E\left[x_i\right] E\left[x_j\right]. \tag{23.21}$$

The above approximation method is usually termed the *naive or zero-order mean-field approximation*. The mean-field-theory machine is one to two orders of magnitude faster than the Boltzmann machine [22, 44].

However, the validity of the naive mean-field algorithm is challenged in [20, 31]. By applying naive mean-field approximation to a finite system with nonrandom interactions, the true stochastic system is not faithfully represented in many situations. The independence assumption is shown to be unacceptably inaccurate in multiple-hidden-layer configurations. As a result, the mean-field-theory machine only works in supervised mode with a single hidden layer [20, 23]. The mean state is not a sufficient representation for the free-running probability distribution and thus, the mean-field method is ineffective for unsupervised learning. In [31], the naive mean-field approximation of the learning rules is shown to not converge in general; it leads to a converging gradient-descent algorithm only when (23.21) is satisfied for $i \neq j$.

In [32], the equivalence between the asynchronous mean-field-theory machine and the continuous Hopfield model is established in terms of the same fixed points for networks using the same Hopfield topology and energy function. The naive mean-field-theory machine performs the steepest descent on an appropriately defined cost function under certain circumstances, and has been empirically used to solve a variety of supervised learning problems [24].

An approximate mean-field algorithm for the Boltzmann machine [31] has a computational complexity of cubic in the number of neurons. In the absence of a hidden unit, the weights can be directly computed from the fixed-point equation of the learning rules, and thus a gradient-descent procedure is avoided. The solutions are close to the optimal ones; thus the method yields a significant improvement when correlations play a significant role.

The mean-field annealing algorithm can be derived by optimizing the Kullback–Leibler divergence between the factorial approximating distribution and the ideal joint distribution of the binary neural variables in terms of the mean activations. In [57], two interactive mean-field algorithms are derived by extending the internal representations to include both the mean activations and the mean correlations. The two algorithms, respectively, estimate the mean activations subject to the mean correlations, and the mean correlations subject to the mean activations by optimizing the objective quantified by a combination of the Kullback–Leibler divergence and the correlation strength between any two distinct variables. The interactive mean-field algorithms improve the mean-field approximation in both performance and relaxation efficiency.

In variational approaches, the posterior distributions are either approximated by factorized Gaussians, or integrals over the posteriors are evaluated by saddle-point approximations [3]. The resulting algorithm is an EM-like procedure with the four estimations performed sequentially. Practical learning algorithms for Boltzmann machines are proposed in [59] by using the belief propagation algorithm and the linear response approximation, which are often referred to as advanced mean-field methods.

23.4 Stochastic Hopfield Networks

Like the Hopfield network, both the Boltzmann machine and the mean-field-theory machine can be used as associative memory. The Boltzmann and mean-field-theory machines that use hidden units have a far higher capacity for storage and error-correcting retrieval of random patterns and improved basins of attraction than the Hopfield network does [22].

When the Boltzmann machine is trained as associative memory using an adaptive association rule [30], it does not suffer from spurious states. The association rule, which creates a sphere of influence around each stored pattern, is a generalization of the generalized Hebbian rule. Spurious fixed points, whose regions of attraction are not recognized by the rule, are skipped, due to the finite probability to escape from any state. The upper and lower bounds on retrieval probabilities of each stored pattern are also given in [30].

Due to the existence of the hidden units, neither the Boltzmann machine nor the mean-field-theory machine can be trained and retrieved in the same way as in the case of the Hopfield model. The retrieval process is as follows [22]. The visible neurons are clamped to a corrupted pattern, the whole network is annealed to a lower temperature, where the state of the hidden neurons approximates the learned internal representation of the stored pattern, and then the visible neurons are released. The annealing process continues until the whole network is settled.

The Gaussian machine [2] is a general framework that includes the Hopfield network, the Boltzmann machine, and also other stochastic networks. Stochastic distribution is realized by adding thermal noise, a stochastic external input ε, to each unit, and the network dynamics are the same as that of the Hopfield network. The stochastic term ε obeys a Gaussian distribution with zero mean and variance σ^2, where the deviation $\sigma = kT$, and T is the temperature. The stochastic term ε can occasionally bring the network to states with a higher energy. When $k = \sqrt{\frac{8}{\pi}}$, the distribution of the outputs has the same behavior as a Boltzmann machine. When employing noise obeying a logistic distribution rather than a Gaussian distribution in the original definition, we can obtain a Gaussian machine identical to a Boltzmann machine. When the noise in the Gaussian machine takes a Cauchy distribution with zero as the peak location and the half-width at the maximum $\sigma = T\sqrt{\frac{8}{\pi}}$, we get a Cauchy machine [51]. A similar idea was embodied in the stochastic network given in [35], where in addition to a cooling schedule for temperature T, gain annealing is also applied. The gain $\frac{1}{\beta}$ has to be decreased more slowly than T, and kept bounded away from zero.

Problem

23.1 Representation of the XOR problem requires to use hidden node in the MLP and the RBF network. This is also true for the Boltzmann machine. Solve the problem using the Boltzmann machine.

References

1. Ackley, D. H., Hinton, G. E., & Sejnowski, T. J. (1985). A learning algorithm for Boltzmann machines. *Cognitive Science*, *9*, 147–169.
2. Akiyama, Y., Yamashita, A., Kajiura, M., & Aiso, H. (1989). Combinatorial optimization with Gaussian machines. In *Proceedings of International Joint Conference on Neural Networks* (pp. 533–540). Washington, DC.
3. Attias, H. (1999). Inferring parameters and structure of latent variable models by variational Bayes. In *Proceedings of the 15th Annual Conference on Uncertainty in AI* (pp. 21–30).
4. Azencott, R., Doutriaux, A., & Younes, L. (1993). Synchronous Boltzmann machines and curve identification tasks. *Network*, *4*, 461–480.
5. Baldi, P., & Pineda, F. (1991). Contrastive learning and neural oscillations. *Neural Computation*, *3*(4), 526–545.
6. Baldi, P., & Sadowski, P. (2014). The dropout learning algorithm. *Artificial Intelligence*, *210*, 78–122.
7. Barra, A., Bernacchia, A., Santucci, E., & Contucci, P. (2012). On the equivalence of Hopfield networks and Boltzmann machines. *Neural Networks*, *34*, 1–9.
8. Bengio, Y. (2009). Learning deep architectures for AI. *Foundations and Trends in Machine Learning*, *2*(1), 1–127.
9. Bengio, Y., & Delalleau, O. (2009). Justifying and generalizing contrastive divergence. *Neural Computation*, *21*(6), 1601–1621.
10. Brugge, K., Fischer, A., & Igel, C. (2013). The flip-the-state transition operator for restricted Boltzmann machines. *Machine Learning*, *93*(1), 53–69.

11. Carreira-Perpinan, M. A., & Hinton, G. E. (2005). On contrastive divergence learning. In *Proceedings of the 10th International Workshop on Artificial Intelligence and Statistics* (pp. 59–66).
12. Cho, K. H., Raiko, T., & Ilin, A. (2013). Gaussian–Bernoulli deep Boltzmann machine. In *Proceedings of International Joint Conference on Neural Networks (IJCNN)* (pp. 1–7).
13. Cote, M. A., & Larochelle, H. (2016). An infinite restricted Boltzmann machine. *Neural Computation, 28*, 1265–1289.
14. Del Genio, C. I., Gross, T., & Bassler, K. E. (2011). All scale-free networks are sparse. *Physical Review Letters, 107*(19), Paper No. 178701.
15. Desjardins, G., Courville, A., Bengio, Y., Vincent, P., & Dellaleau, O. (2010). Parallel tempering for training of restricted Boltzmann machines. In *Proceedings of the 13th International Conference on Artificial Intelligence and Statistics (AISTATS'10)* (pp. 145–152).
16. Detorakis, G., Bartley, T., & Neftci, E. (2019). Contrastive Hebbian learning with random feedback weights. *Neural Networks, 114*, 1–14.
17. Elfwing, S., Uchibe, E., & Doya, K. (2015). Expected energy-based restricted Boltzmann machine for classification. *Neural Networks, 64*, 29–38.
18. Fischer, A., & Igel, C. (2011). Bounding the bias of contrastive divergence learning. *Neural Computation, 23*(3), 664–673.
19. Gabrie, M., Tramel, E. W., & Krzakala, F. (2015). Training restricted Boltzmann machine via the Thouless–Anderson–Palmer free energy. In *Advances in neural information processing systems* (pp. 640–648).
20. Galland, C. C. (1993). The limitations of deterministic Boltzmann machine learning. *Network, 4*, 355–380.
21. Glauber, R. J. (1963). Time-dependent statistics of the Ising model. *Journal of Mathematical Physics, 4*, 294–307.
22. Hartman, E. (1991). A high storage capacity neural network content-addressable memory. *Network, 2*, 315–334.
23. Haykin, S. (1999). *Neural networks: A comprehensive foundation* (2nd ed.). Upper Saddle River, NJ: Prentice Hall.
24. Hinton, G. E. (1989). Deterministic Boltzmann learning performs steepest descent in weight-space. *Neural Computation, 1*, 143–150.
25. Hinton, G. E. (2002). Training products of experts by minimizing contrastive divergence. *Neural Computation, 14*(8), 1771–1800.
26. Hinton, G. E., Osindero, S., & Teh, Y.-W. (2006). A fast learning algorithm for deep belief nets. *Neural Computation, 18*(7), 1527–1554.
27. Hinton, G. E., & Salakhutdinov, R. R. (2006). Reducing the dimensionality of data with neural networks. *Science, 313*(5786), 504–507.
28. Hinton, G. E., & Sejnowski, T. J. (1986). Learning and relearning in Boltzmann machines. In D. E. Rumelhart & J. L. McClelland (Eds.), *Parallel distributed processing: Explorations in microstructure of cognition* (Vol. 1, pp. 282–317). Cambridge, MA: MIT Press.
29. Igel, C., Glasmachers, T., & Heidrich-Meisner, V. (2008). Shark. *Journal of Machine Learning Research, 9*, 993–996.
30. Kam, M., & Cheng, R. (1989). Convergence and pattern stabilization in the Boltzmann machine. In D. S. Touretzky (Ed.), *Advances in neural information processing systems* (Vol. 1, pp. 511–518). San Mateo, CA: Morgan Kaufmann.
31. Kappen, H. J., & Rodriguez, F. B. (1998). Efficient learning in Boltzmann machine using linear response theory. *Neural Computation, 10*, 1137–1156.
32. Kurita, N., & Funahashi, K. I. (1996). On the Hopfield neural networks and mean field theory. *Neural Networks, 9*, 1531–1540.
33. Larochelle, H., & Bengio, Y. (2008). Classification using discriminative restricted Boltzmann machines. In *Proceedings of the 25th International Conference on Machine Learning* (pp. 536–543). Helsinki, Finlan.
34. Le Roux, N., & Bengio, Y. (2008). Representational power of restricted Boltzmann machines and deep belief networks. *Neural Computation, 20*(6), 1631–1649.

35. Levy, B. C., & Adams, M. B. (1987). Global optimization with stochastic neural networks. In *Proceedings of the 1st IEEE Conference on Neural Networks* (Vol. 3, pp. 681–689). San Diego, CA.
36. Lillicrap, T. P., Cownden, D., Tweed, D. B., & Akerman, C. J. (2016). Random synaptic feedback weights support error backpropagation for deep learning. *Nature Communications*, *7*, Paper No. 13276.
37. Lin, C. T., & Lee, C. S. G. (1995). A multi-valued Boltzmann machine. *IEEE Transactions on Systems Man and Cybernetics*, *25*(4), 660–669.
38. Mocanu, D. C., Mocanu, E., Nguyen, P. H., Gibescu, M., & Liotta, A. (2016). A topological insight into restricted Boltzmann machines. *Machine Learning*, *104*(2), 243–270.
39. Montufar, G., Ay, N., & Ghazi-Zahedi, K. (2015). Geometry and expressive power of conditional restricted Boltzmann machines. *Journal of Machine Learning Research*, *16*, 2405–2436.
40. Nair, V., & Hinton, G. E. (2010). Rectified linear units improve restricted Boltzmann machines. In *Proceedings of the International Conference on Machine Learning (ICML)* (pp. 807–814).
41. Neftci, E., Das, S., Pedroni, B., Kreutz-Delgado, K., & Cauwenberghs, G. (2014). Event-driven contrastive divergence for spiking neuromorphic systems. *Frontiers in Neuroscience*, *8*, 1–14.
42. Odense, S., & Edwards, R. (2016). Universal approximation results for the temporal restricted Boltzmann machine and the recurrent temporal restricted Boltzmann machine. *Journal of Machine Learning Research*, *17*, 1–21.
43. Peng, X., Gao, X., & Li, X. (2918). On better training the infinite restricted Boltzmann machines. *Machine Learning*, *107*(6), 943–968.
44. Peterson, C., & Anderson, J. R. (1987). A mean field learning algorithm for neural networks. *Complex Systems*, *1*(5), 995–1019.
45. Ranzato, M. A., Krizhevsky, A., & Hinton, G. E. (2010). Factored 3-way restricted Boltzmann machines for modeling natural images. In *Proceedings of the 13th International Conference on Artificial Intelligence and Statistics (AISTATS)* (pp. 621–628). Sardinia, Italy.
46. Romero, E., Mazzantib, F., Delgado, J., & Buchaca, D. (2019). Weighted contrastive divergence. *Neural Networks*, *114*, 147–156.
47. Salakhutdinov, R., & Hinton, G. (2009). Replicated softmax: An undirected topic model. In *Advances in neural information processing systems* (Vol. 22, pp. 1607–1614). Vancouver, Canada.
48. Sankar, A. R., & Balasubramanian, V. N. (2015). Similarity-based contrastive divergence methods for energy-based deep learning models. In *JMLR Workshop and Conference Proceedings* (Vol. 45, pp. 391–406).
49. Smolensky, P. (1986). Information processing in dynamical systems: Foundations of harmony theory. In D. E. Rumelhart, J. L. McClelland, & the PDP Research Group (Eds.), *Parallel distributed processing: Explorations in the microstructure of cognition* (Vol. 1, pp. 194–281). Cambridge, MA: MIT Press.
50. Srivastava, N., Hinton, G., Krizhevsky, A., Sutskever, I., & Salakhutdinov, R. (2014). Dropout: A simple way to prevent neural networks from overfitting. *Journal of Machine Learning Research*, *15*, 1929–1958.
51. Szu, H. H., & Hartley, R. L. (1987). Nonconvex optimization by fast simulated annealing. *Proceedings of the IEEE*, *75*, 1538–1540.
52. Taylor, G. W., Hinton, G. E., & Roweis, S. T. (2011). Two distributed-state models for generating high-dimensional time series. *Journal of Machine Learning Research*, *12*, 1025–1068.
53. Thouless, D. J., Anderson, P. W., & Palmer, R. G. (1977). Solution of "solvable model of a spin glass". *Philosophical Magazine*, *35*(3), 593–601.
54. Tieleman, T. (2008). Training restricted Boltzmann machines using approximations to the likelihood gradient. In W. W. Cohen, A. McCallum, & S. T. Roweis (Eds.), *Proceedings of the 25th International Conference on Machine Learning* (pp. 1064–1071). New York: ACM.
55. Tieleman, T., & Hinton, G. E. (2009). Using fast weights to improve persistent contrastive divergence. In A. P. Danyluk, L. Bottou, & M. L. Littman (Eds.), *Proceedings of the 26th Annual International Conference on Machine Learning* (pp. 1033–1040). New York: ACM.

56. Welling, M., Rosen-Zvi, M., & Hinton, G. (2004). Exponential family harmoniums with an application to information retrieval. In *Advances in neural information processing systems* (Vol. 17, pp. 1481–1488).
57. Wu, J. M. (2004). Annealing by two sets of interactive dynamics. *IEEE Transactions on Systems, Man, and Cybernetics Part B*, *34*(3), 1519–1525.
58. Xie, X., & Seung, H. S. (2003). Equivalence of backpropagation and contrastive Hebbian learning in a layered network. *Neural Computation*, *15*(2), 441–454.
59. Yasuda, M., & Tanaka, K. (2009). Approximate learning algorithm in Boltzmann machines. *Neural Computation*, *21*, 3130–3178.
60. Younes, L. (1996). Synchronous Boltzmann machines can be universal approximators. *Applied Mathematics Letters*, *9*(3), 109–113.

Chapter 24
Deep Learning

24.1 Introduction

In 1959, Hubel and Wiesel [39] found that cells in cat's visual cortex are responsible for detecting light in receptive fields. Neurons in primary visual cortex (V1), as the first cortical area in the visual hierarchy of the mammalian brain, detect primary visual features from input images [39]. Each V1 neuron is selective to a particular orientation. The human brain is a six-layered structure consisting of a very large number of neurons strongly connected via feedforward and feedback connections [23]. The neocortex has its structural and functional uniformity: all units in the network seem similar, and they perform the same basic operation. Connections between nonconsecutive layers are present in the human cortex.

Deep architectures adopt the hierarchical structure of the human neocortex, based on the evidence of a common computational algorithm in the brain that makes the brain deal with sensory information (i.e., visual, auditory, olfactory, and so on) in a similar way. Different regions in the brain connect in a hierarchy, such that information coalesces, building higher abstractions of the sensory stimuli at each successive level. This continuous information stream is by nature hierarchical, and a collection of smaller building blocks form a larger picture of the world. For instance, similar hierarchies exist for speech and text from sounds to phones, phonemes, syllables, words, and sentences. With digital imagery, pixels combine into edges, edges into contours, contours into shapes, and shapes into objects.

Cortical algorithms [23] are modeled after the human visual cortex, which stores sequences of patterns in an invariant form and recalls those patterns autoassociatively. Like this brain architecture, cortical algorithm architecture has minicolumns of varying thickness. A minicolumn is a group of neurons that share the same receptive field: neurons belonging to a minicolumn are associated with the same sensory input region. The network is trained in two stages: the first stage trains the columns to identify independent features from the patterns occurring in an unsupervised manner; the second stage relies on supervised learning to create invariant representations. An association of minicolumns is called a hypercolumn or layer.

K.-L. Du and M. N. S. Swamy, *Neural Networks and Statistical Learning*,
https://doi.org/10.1007/978-1-4471-7452-3_24

Inspired by the natural visual perception mechanism, Fukushima [29] proposed the neocognitron in 1980, which could be regarded as the predecessor of convolutional neural network. A convolutional neural network convolves the input data with a set of filters. This operation is a rough analogy to the use of receptive fields in the retina [39].

In 1990, Le Cun and collaborators [47] published the seminal paper establishing the modern framework of convolutional neural network, and later improved it in [48]. The developed multilayer neural network called *LeNet-5* could classify handwritten digits. LeNet-5 can be trained with BP algorithm. With LeNet-5, it is possible to recognize visual patterns directly from raw pixels with little-to-none preprocessing.

As the depth of a network increases, the performance of BP degrades due to the vanishing gradient problem [6, 9, 36]. The error propagated back in the network shrinks as it moves from layer to layer, becoming negligible in deep architectures and making it difficult to update the weights in the early layers.

In 1991, Schmidhuber [74] performed credit assignment across hundreds of nonlinear operators or neural layers, by using unsupervised pretraining for a hierarchy of recurrent networks. The algorithm trains a multilevel hierarchy of recurrent networks by using unsupervised pretraining on each layer and then fine-tuning the resulting weights via BP [74].

In 2006, Hinton and collaborators [37] introduced the deep belief network, which is a breakthrough of unsupervised learning algorithms. Ever since, deep learning has emerged as a new area of machine learning research [14, 37].

Deep learning methods are representation learning methods with multilevel representation. Deep neural networks can automatically find compact low-dimensional representations (features) of high-dimensional data. They are usually supervised, and typically consist of more than five processing layers. Falling hardware prices and the development of GPUs have contributed to the concept of deep learning.

Deep convolutional networks have achieved empirical success in processing images, video, speech, and audio, whereas deep recurrent networks are successful in sequential data such as text and speech. Deep learning is used in numerous products in speech recognition engines, object recognition (Google goggles), image and music information retrieval (Google image search, Google music), as well as computational advertising.

TensorFlow (https://www.tensorflow.org/) from Google is an open-source library in Python that focuses on deep learning. Tensorflow uses computational data-flow graphs to represent complicated neural network architecture. The nodes denote mathematical computations called ops (operations), whereas the edges denote the data tensors transferred between them. The relevant gradients are stored at each node, and they are combined to get the gradients with respect to each weight during backpropagation. Implementation using TensorFlow has three steps: define a network/model using a computational graph, compute the gradients using the automatic differentiation capabilities of TensorFlow, and fit the model using stochastic gradient descent. Tensorflow provides facilities to view computational graphs and other metrics/summaries while training a model via TensorBoard. It supports the training of large models in a multiple GPU, distributed setting.

24.2 Deep Neural Networks

A deep learning architecture is a multilayer stack of simple modules with nonlinear input–output mappings, which are subject to learning. Each module in the stack transforms its input to increase both the selectivity and the invariance of the representation. Deep neural networks are multilayer networks with many hidden layers.

Popular deep learning architectures are convolutional neural network [43], deep belief network [19], and deep stacking network [20]. Deep belief network is a deep network as a probabilistic generative structure. Multilayer convolutional neural networks [29, 47] have gained prominence due to excellent results on a number of vision benchmarks [46]. Deep MLPs initialized by unsupervised pretraining were successfully applied to speech recognition [62].

Without any training pattern deformations, a deep belief network fine-tuned by BP achieved 1.2% error rate [38] on the MNIST handwritten digits. In [69], a convolutional neural network [47] trained by BP set an MNIST record of 0.39%, using training pattern deformations but no unsupervised pretraining. An error rate of 0.35% was achieved on the MNIST database in [17].

An association between the structural complexity of Bayesian networks and their representational power is established in [54]. The maximum number of nodes' parents is used as the measure for the structural complexity of Bayesian networks, and the maximum number of XORs contained in a target function as the measure for the function complexity. Discrete Bayesian networks with each node having at most k parents cannot represent any function containing $(k + 1)$-XORs [54].

Generally, deep neural networks are trained in two phases [44]: pretraining layer at a time using unsupervised learning in a way that preserves information from the input and disentangles factors of variation; fine-tuning the whole network with respect to the ultimate criterion of interest.

Instead of learning the weights of millions of connections at once, this layerwise unsupervised pretraining scheme finds the optimal solution for a single layer at a time, which makes it a greedy algorithm. This is accomplished by tying all the weights of the following layers and learning only the weights of the current layer. This pretraining strategy helps the optimization by initializing weights in a region near a good local minimum, but also implicitly acts as a sort of regularization that brings better generalization and encourages internal distributed representations that are high-level abstractions of the input [25, 44]. It minimizes variance and introduces bias toward configurations of the parameter space that are useful for unsupervised learning. This type of regularization strategy is similar to the early stopping idea for training neural networks [25]. When training using gradient descent, the beneficial generalization effects due to pretraining do not appear to diminish as the number of labeled examples grows very large.

The unsupervised representation learning algorithms for pretraining can be used for restricted Boltzmann machine (RBM) [37], autoencoder [7], or a sparsifying form of autoencoder similar to sparse coding [69].

It was observed that on deep networks, even with the vanishing gradient problem, given enough epochs, BP algorithm can achieve results comparable to those of other, more complex training algorithms [17]. A deep network with an unsupervised greedy layerwise pretraining almost always outperforms the scenarios without the pretraining phase [25]. This phenomenon has been explained as the pretraining phase acting as a regularizer [6], and as an aid [7] for the supervised optimization problem.

However, deep neural networks have high computational cost and difficult to scale. Deep stacking network addresses the scalability problem of deep neural network, simple classifiers are stacked on top of each other in order to construct more complex classifier [85].

24.2.1 Deep Networks Versus Shallow Networks

Complexity theory of circuits strongly suggests that deep architectures can be much more efficient than their shallow counterparts, in terms of computational elements and parameters required to represent some functions. Theoretical results on circuit complexity theory have shown that shallow digital circuits can be exponentially less efficient than deeper ones [32]. An equivalent result has been proved for architectures whose computational elements are linear threshold units [33].

Functions that exhibit repeating patterns can be encoded much more efficiently in the deep representation, resulting in significant reduction in complexity. Depth is an effective encoder of repeating patterns in the data, in terms of parameter complexity and thus the VC-dimension [80]. This confirms the hypothesis that deep networks can generalize better by having a more compact representation [51]. Shallow architectures are shown to be more suitable for problems where data are clustered according to target label.

For sum-product networks, certain classes of polynomials have proved to be much more easily represented by deep networks than by shallow networks [18]. In [63] the number of linear regions of deep neural network is proved to grow exponentially in the number of hidden layers L and polynomially in the number of hidden neurons n, which is much faster than that of shallow neural networks with nL hidden neurons. In [16], it is proved that a three-layer MLP cannot provide localized approximation in an Euclidean space of dimension higher than one, while a four-layer MLP can.

Deep networks may be more expressive than shallow ones of comparable size. Deep rectified linear unit (ReLU) networks have proved to more efficiently approximate smooth functions than shallow ones [87]. To approximate a deep network with a shallower one, an exponential number of units may be required. For a ReLU network with L layers and $U = \Theta(L)$ units, any network approximating it with only $O(L^{1/3})$ layers must have $\Omega(2^{L^{1/3}})$ units [81]. Similarly, a high-dimensional 3-layer network cannot be approximated by a two-layer network except with an exponential blow-up in the number of nodes [24].

Deep networks are more effective than shallow networks with respect to computational complexity and generalization ability for a function approximation task with periodic characteristic over binary inputs [80]. It is proved that deep networks can approximate a set of compositional functions with the same accuracy as shallow networks, but with exponentially lower VC-dimensions [60].

Betti numbers characterize topological properties of functions represented by networks. Betti numbers of input–output functions of certain deep networks grow exponentially with the number of hidden units, whereas they grow merely polynomially for shallow networks with the same types of units [10].

A theoretical analysis of singular points of deep neural networks [65] yields deep neural network models having no critical points introduced by a hierarchical structure. Such deep neural network models have good nature for gradient-based optimization. The existence of critical points introduced by a hierarchical structure is determined by the rank and the regularity of weight matrices for a specific class of deep neural networks.

The deep belief network is a universal approximator even when each hidden layer is restricted to a relatively small number of hidden nodes [52, 78]. Deep but narrow generative networks do not require more parameters than shallow ones to achieve universal approximation [52]. Deep but narrow feedforward neural networks with sigmoidal units can represent any Boolean expression [52].

While most recent works advocate deep neural networks, shallower but wider networks may outperform a deep network. The success of residual network (ResNet) has been interpreted as ensembling exponentially many short networks, since paths in a ResNet do not strongly depend on one another, although they are trained jointly [82]. A wide 16-layer ResNet outperforms the original thin thousand-layer ResNet on CIFAR-10 and, and a 50-layer ResNet outperforms a 152-layer one on ImageNet [88]. Also, wide ResNets are several times faster to train. The performance of a ResNet is related to the number of trainable parameters [88].

Tight upper and lower bounds on the VC-dimension of deep ReLU networks are derived in [4]. For a model of W weights and L layers, the VC-dimension is proved to be $O(WL\log(W))$. In terms of U nonlinear units, a tight bound $\Theta(WU)$ on the VC-dimension is proved. These bounds generalize to arbitrary piecewise linear activation functions.

24.3 Deep Belief Networks

Deep Bayesian network, also known as *deep belief network*, is a generative neural network model with many layers of hidden causal factors [37, 38]. It is a sequential stack of RBMs, with no intralayer neuron connections. The first two layers of the network contain neurons with undirected symmetric connections, which form an associative memory, whereas the remaining hidden layers form a directed acyclic graph. Each RBM perceives pattern representations from the lower level and learns to encode them in an unsupervised fashion. Upper layers are supposed to represent more abstract concepts that explain the input observation x, whereas lower layers extract low-level features from x.

Deep belief network uses a layerwise unsupervised learning to pretraining the initial weights of the networks, and then with global supervised learning for fine-tuning [37]. The pretraining unsupervised step utilizes large amount of unlabeled training data for extracting structures and regularities in input features [38]. Training becomes feasible and fast, involving training RBM units independently before adjusting the weights, using an up-down algorithm to avoid underfitting [37].

The choice of input distribution in RBMs could be important for continuous-valued input and yields different types of filters at the first layer.

In theory, under certain assumptions, adding more layers improves a bound on the data's negative log probability (equivalent to the data's description length) [37]. Deep belief network has been applied to dimensionality reduction (image compression) [38].

Modular deep belief network trains different parts of the network separately, while adjusting the learning rate as training progresses [67]. This allows modular deep belief network to avoid forgetting features learned early in training. Sparse deep belief network learns sparse features by adding a penalty in the objective function for deviations from the expected activation of hidden units in the RBM formulation [49]. Convolutional deep belief network integrates translation invariance into image representations by sharing weights between locations in an image, allowing inference to be done when the image is scaled up by using convolution [50].

Deep Boltzmann machines are similar to, but have a more general deep architecture than deep belief networks. They are composed of Boltzmann machines stacked on top of one another [72]. The two-way edges let deep Boltzmann machines propagate input uncertainty better than deep belief networks. The greedy training algorithm of deep belief network is modified to achieve a more efficient training algorithm for deep Boltzmann machine by using an approximate inference algorithm. However, deep Boltzmann machine training is approximately three times slower than deep belief network training [73].

24.3.1 Training Deep Belief Networks

The algorithm for training a deep belief network consists of two phases. The first phase is a greedy layerwise training to learn the weights by tying the weights of the unlearned layers, and then applying contrastive divergence to learn the weights of the current layer. The second phase is an algorithm for fine-tuning the weights.

This process of tying, learning, and untying weights is repeated until all layers have been processed. A deep belief network with tied weights resembles an RBM, and the RBM is learned using contrastive divergence algorithm. Once the weights have been learned for each layer, a variant of the wake-sleep algorithm with the contrastive divergence weight update rule is used to fine-tune the learned parameters. This training algorithm is shown in Algorithm 24.1 [38].

Algorithm 24.1 (Two-Phase Algorithm)

1. *In the bottom-up pass:*
   a. *Compute positive phase probabilities.*
   b. *Sample states.*
   c. *Compute contrastive divergence statistics, using the positive phase probabilities.*
   d. *Perform Gibbs sampling for a predefined number of iterations, based on the associative memory part of the network.*
   e. *Compute negative phase contrastive divergence statistics, using information from step 1d.*
2. *In the top-down pass:*
   a. *Calculate negative phase probabilities.*
   b. *Sample states.*
   c. *Compute predictions.*
3. *Update generative parameters.*
4. *Update associative memory part of the network.*
5. *Update inference parameters.*

Note that Algorithm 24.1 can only be applied if the first two layers form an undirected graph, and the remaining hidden layers form a directed acyclic graph.

24.4 Deep Autoencoders

Autoencoders capture the input distribution by learning to reconstruct input configurations. An autoencoder is an unsupervised neural network-based feature extraction algorithm, which learns the best parameters required to reconstruct its output as close to its input as possible. It can provide a more powerful and nonlinear generalization compared to PCA. A basic linear autoencoder learns essentially the same representation as PCA. An autoencoder used in Kramer's nonlinear PCA is introduced in Sect. 13.6.1.

Like RBM, tied weight autoencoder [7] learns features by using tied weights. It consists of an encoder stage and a decoder stage. The tied weight concept means to use the same weights in the encoder stage and decoder stage of an autoencoder. This constrains the tied weight autoencoder to learn the same set of weights for dimension reduction and to reconstruct the data from the lower dimensional space.

A deep autoencoder is composed of two symmetrical deep belief networks, which typically have four or five shallow layers for encoding, and a second set of four or five layers for decoding. The output from each hidden layer is used as the input for a progressively higher level.

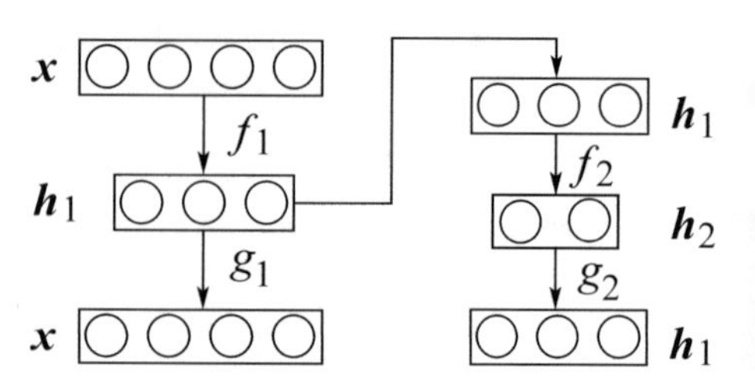

Fig. 24.1 A stacked autoencoder

A stacked autoencoder [38] consists of multiple layers of autoencoders in which the training of the deep network is performed by training each layer in turn. The structure of a stacked autoencoder is shown in Fig. 24.1.

24.5 Deep Convolutional Neural Networks

The revival of convolutional neural networks occurred until the success of AlexNet in the ImageNet competition in 2012 [71]. Convolutional networks are standard feedforward neural networks for large-scale image recognition. Convolutional networks are now the dominant approach for recognition and detection tasks. They have been explored for speech recognition [1], also showing improvements over deep neural networks.

Neocognitron [29] introduced convolutional neural networks, where the receptive field of a convolutional unit with a given weight vector is shifted step by step across a two-dimensional array of input values. Due to massive weight replication, relatively few parameters may be necessary to describe the behavior of such a convolutional layer. Neocognitron is similar to the architecture of supervised, feedforward deep learners with alternating convolutional and downsampling layers. However, the weights are set by local, WTA-based unsupervised learning rules, or by pre-wiring, not by BP. Spatial averaging is used for downsampling purposes instead of max pooling, currently a popular WTA mechanism.

In 1989, BP was applied to Neocognitron-like, weight-sharing, convolutional layers with adaptive connections [47]. Deep convolutional network [48] uses hierarchical layers of tiled convolutional filters to mimic the effects of receptive fields. LeNet-5 [47] consists of three types of layers, namely, convolutional, pooling, and fully connected layers. AlexNet [71] is similar to LeNet-5, but has a deeper structure. It consists of eight layers, totaling 60 million trainable parameters. It has significant improvements over previous methods for image classification. ZFNet [90], visual geometry group network (VGGNet) [77], GoogleNet [79], and ResNet [35] further improve the performance of AlexNet by getting deeper. ResNet, which won the championship of ILSVRC 2015, has 1,202 trainable layers; it is about 20 times deeper than AlexNet and 8 times deeper than VGGNet. Two residual deep networks, namely, CompNet [26] and SRSubBandNet [27], have been used for single image super-resolution.

24.5.1 Solving the Difficulties of Gradient Descent

In the ImageNet competition 2012, deep convolutional networks achieved the best result for object recognition with a large margin [43]. Hinton credited much of their success to the dropout training technique in a talk at NIPS 2012. The success is also contributed by the use of GPUs, ReLU function, and techniques for generating more training examples by deforming the existing ones.

A deeper network can better approximate the target function with increased nonlinearity and get better feature representations. However, the increased network complexity makes it more difficult to train and easier to get overfitting, according to the bias–variance trade-off. BP algorithm experiences three difficulties in training a deep network: vanishing gradient, overfitting, and computational load. Deep convolutional networks solve all these difficulties.

A representative solution to the vanishing gradient of BP algorithm is the use of ReLU function as the activation function. ReLU function $\sigma(x) = \max(0, x)$ is a nonsaturated activation function [64]. It is a piecewise linear function which prunes the negative part to zero and retains the positive part. The max operation allows us to compute much faster than sigmoid or tanh activation functions, and it also induces the sparsity in the hidden units. ReLU function better transmits the error than sigmoid function, since sigmoid function limits the node outputs to the unity. The derivative of ReLU function is given as $\varphi(x) = 1$ for $x > 0$, or 0 for $x \leq 0$. Thus, ReLU function has constant zero gradient whenever a unit is inactive. A unit that is inactive initially will never be active. Constant zero gradient may also slow down the training process. Leaky ReLU [56] compresses the negative part rather than mapping it to constant zero, and this allows for a small, nonzero gradient for inactive units.

Although ResNet also employs ReLU function, it was empirically found that almost all of the gradient updates in a 110-layer ResNet can only go through paths between 5 and 17 residual units [82]. Thus, ResNets do not resolve the vanishing gradient problem by preserving gradient flow throughout the entire depth of the network; rather, most of the gradient in a ResNet is interpreted as coming from by ensembling exponentially many short networks [82].

The most representative solution to overfitting is dropout, which trains only some of the randomly selected nodes rather than the entire network. The method is very effective, and its implementation is also not complex. Some nodes are randomly selected at a certain percentage and their outputs are set to be zero to deactivate the nodes. Dropout effectively prevents overfitting as it continuously alters the nodes and weights in the training process.

Deep neural networks use a lot of matrix multiplication, especially convolution, for both the forward pass and for backpropagation. GPUs are good at matrix-to-matrix multiplication. Backpropagating gradients through a convolutional network is as simple as through a regular deep network, allowing all the weights in all the filter banks to be trained. Convolutional networks are amenable to hardware implementations in chips or field-programmable gate arrays (FPGAs) [11, 28].

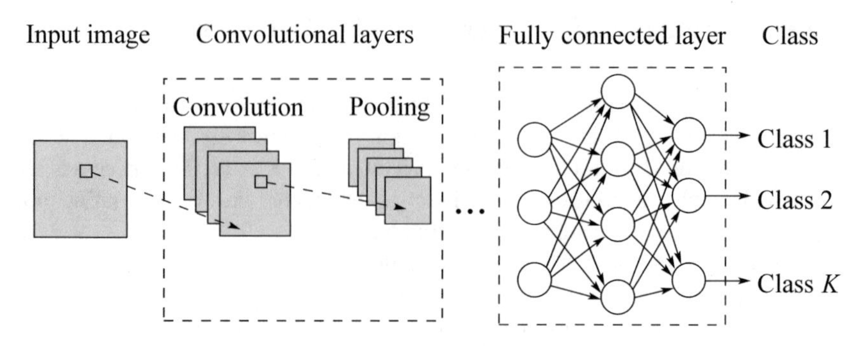

Fig. 24.2 Convolutional neural network for image classification

24.5.2 Implementing Deep Convolutional Neural Networks

Convolution is a technique for images and time series data. In a convolutional setting, the hidden units are not fully connected to the input but instead divided into locally connected segments. Convolution has been applied to both RBMs and autoencoders to create convolutional RBMs [50] and convolutional autoencoders [58]. A time-delay neural network exploits the temporal structure of the input by performing convolutions on overlapping windows. It was used for the recognition of phonemes and simple words [84].

Convolutional network, as illustrated in Fig. 24.2, consists of convolutional and pooling (or subsampling) layers, which are grouped into modules. Either one or more fully connected layers follow these modules. Modules are often stacked on top of each other to form a deep model. Convolutional networks commonly use mammalian primary visual cortex (V1)-like receptive field kernels in first few layers to extract features from stimuli by convolving the kernels over the input (e.g., image).

A convolutional layer detects local conjunctions of features from the previous layer. Neurons in a convolutional layer are organized in feature maps, within which each neuron is connected to local patches in the feature maps of the previous layer through a set of weights called a filter bank. Such a neighborhood is the neuron's receptive field in the previous layer. A convolution layer is composed of several convolution kernels, which are used to compute different feature maps.

Inputs are convolved with the learned weights in order to compute a new feature map, and the convolved results are sent through a nonlinear activation function. All neurons within a feature map have weights that are constrained to be equal; however, different feature maps within the same convolutional layer have different weights so that several features can be extracted at each location [46, 48].

The kth output feature map Y_k can be computed as

$$Y_k = \sigma(\mathbf{W}_k * \mathbf{X}), \tag{24.1}$$

where $\mathbf{X}$ stands for the input image, $\mathbf{W}_k$ for the convolutional filter related to the kth feature map, $*$ for the 2D convolutional operator, and $\sigma(\cdot)$ for the activation function, typically, sigmoid, hyperbolic tangent, or ReLU function.

A pooling layer is usually placed between two convolutional layers. Each feature map of a pooling layer is connected to its corresponding feature map of the preceding convolutional layer. The pooling layer merges similar features into one. Pooling combines nearby values in input or feature space through a max, average or histogram operator. The purpose is to achieve invariance to small local distortions and reduce the dimensionality of the feature space.

A typical pooling unit computes the maximum of a local patch of units in one feature map. Neighboring pooling units take input from patches that are shifted by more than one row or column. Probabilistic max pooling is introduced in the context of convolutional RBMs [50]. L_p pooling [12], a biologically inspired pooling process modeled on complex cells, provides better generalization than max pooling. Specially, when $p = 1$, it corresponds to average pooling, and when $p = \infty$, it reduces to max pooling. Stochastic pooling [89] is a dropout-inspired pooling method. It randomly picks the activations according to a multinomial distribution. Spectral pooling [70] performs dimensionality reduction by cropping the representation of input in frequency domain. Spatial pyramid pooling [34] can generate a fixed-length representation regardless of the input sizes. It pools input feature map in local spatial bins with sizes proportional to the image size, resulting in a fixed number of bins. Weight-sharing mechanism can drastically decrease the number of parameters.

Convolutional neural networks can be used to model spatial and temporal correlation, while reducing translational variance in signals. They capture translational invariance with far fewer parameters than fully connected deep neural networks by averaging the outputs of hidden units in different local time and frequency regions [45]. The space-time deep belief network [13] uses convolutional RBMs together with a spatial pooling layer and a temporal pooling layer to build invariant features from spatiotemporal data.

Careful noise injection speeds up the BP training of a convolutional network. Noisy convolutional neural network algorithm [2] speeds training on average because BP algorithm is a special case of generalized EM algorithm. The noise benefit is most pronounced for smaller datasets, since the largest EM hill-climbing gains tend to occur in the first few iterations. This noise effect can assist random sampling from large datasets because it allows a smaller random sample to give the same or better performance than a noiseless sample gives. Noisy convolutional neural network algorithm uses this noisy-EM result to produce a hyperplane in noise space that separates helpful noise from harmful noise [2].

Convolution filter in convolutional neural networks is a generalized linear model for the underlying local image patch. Network-in-network structure [53] replaces the linear filter of the convolutional layer by a micronetwork, e.g., yielding an MLP convolution layer, for better representation. The overall structure is the stacking of such micronetworks.

A tensor-factorized MLP with tensor inputs [15] is trained through the tensor-factorized error backpropagation procedure. The multiway information is preserved

Fig. 24.3 Samples of normalized digits from the testing set. From [47]

through layerwise factorization. This model is further extended to realize the convolutional tensor-factorized neural network by looking at small subtensors through the factorized convolution.

A deep learning method for single image super-resolution [22] directly learns an end-to-end mapping between the low/high-resolution images. The mapping is represented as a deep convolutional neural network that takes the low-resolution image as the input and outputs the high-resolution one. Traditional sparse-coding-based super-resolution methods can be viewed as a deep convolutional network [22].

Example 24.1 **Handwritten digit recognition** can be commercially used in postal services [47] or in banking services to recognize handwritten digits on envelopes or bank cheques. This is a real image recognition problem. The input consists of black or white pixels. The mapping is from the two-dimensional image space into 10 output classes.

The MNIST dataset contains handwritten digits. The training set is $9,298$ segmented numerals digitized from handwritten postal codes that appeared on real U.S. mail. Additional $3,349$ printed digits coming from 35 different fonts are also added to the training set. Around 79% of the training set is used for training and the remaining 21% for testing. The size of the characters is normalized to a 16×16 pixel image by using a linear transformation. Due to the linear transformation, the resulting image is not binary, but has multiple gray levels. The gray-leveled image is further scaled to the range of -1 to 1. Some samples of normalized digits from the testing set are shown in Fig. 24.3.

A large six-layer MLP (784-576-576-768-192-10) trained with BP is designed to recognize handwritten digits. Weight sharing and other heuristics are used among the hidden layers. The input layer uses a 28×28 plane instead of a 16×16 plane to avoid problems arising from a kernel overlapping a boundary. The network has 4,635 nodes, 98,442 connections, and 2,578 independent parameters. This architecture was optimized by using the OBD technique. The large network with weight sharing was shown to outperform a small network with the same number of free parameters.

An incremental BP algorithm based on an approximate Newton's method was employed, and the training time for 30 epochs through the training set plus test. The error rates on the training set and testing set were, respectively, 1.1% and 3.4%. When a rejection criterion was employed to the testing set, the result was 9% rejection rate for 1% error. These rejected samples may be due to fault segmentation, or ambiguous writing even to humans.

Digit Recognition

An MLP with a single hidden layer of 800 units achieved 0.70% error [76]. However, more complex methods listed on the MNIST web page always seemed to outperform MLPs. Convolutional neural networks achieved a record-breaking 0.40% error rate [76], using elastic training image deformations. Later, convolutional network training methods pretrain each hidden layer one by one in an unsupervised fashion, then use supervised learning to achieve 0.39% error rate [69].

24.6 Deep Reinforcement Learning

Integration of deep learning and reinforcement learning is in its infancy, but it has already outperformed passive vision systems for classification tasks [3] and produced impressive results in learning to play many video games [61]. Deep reinforcement learning revolutionizes the field of AI and represents a step toward building autonomous systems with a higher level understanding of the visual world.

Two recent breakthroughs in learning to play video games by Mnih and colleagues at Google DeepMind kickstarted the revolution in deep reinforcement learning [61, 75]. The workhorse of deep reinforcement learning remains backpropagation.

Value function-based deep reinforcement learning algorithms with the deep Q-network [61] could learn to play a range of Atari 2600 video games [5] at a level comparable to that of a professional human games tester across a set of 49 games, directly from image pixels. Solutions for the instability of function approximation techniques in reinforcement learning are provided. Reinforcement learning agents could be trained on raw, high-dimensional observations, solely based on a reward signal.

A deep Q-network (https://sites.google.com/a/deepmind.com/dqn) [61] can learn successful policies directly from high-dimensional sensory inputs using end-to-end reinforcement learning. It combines reinforcement learning with deep neural networks. A deep convolutional neural network [48] is used to approximate the optimal state-action value (i.e., Q value) function from high-dimensional input space. The agent selects actions in a fashion that maximizes cumulative future reward. The inputs to the deep Q-network are four gray scale frames of the game, concatenated over time, which are initially processed by several convolutional layers to extract spatiotemporal features. The final feature map from the convolutional layers is processed by several fully connected layers, which more implicitly encode the effects of actions. A biologically inspired mechanism termed experience replay [59] that randomizes over the data are used to remove correlations in the observation sequence and smooth over changes in the data distribution. RMSProp algorithm with mini batches of size 32 is used.

AlphaGo [75] was a hybrid deep reinforcement learning system that defeated a human champion in Go. It uses value networks to evaluate board positions and policy networks to select moves. AlphaGo comprises deep neural networks that were trained using supervised learning from human expert games and reinforcement learning from games of self-play, in combination with traditional Monte Carlo tree search programs that simulate thousands of random games of self-play.

24.7 Other Deep Neural Network Methods

Sum-product network [68] is a deep probabilistic model represented as a rooted DAG with weighted edges. It is a tractable density estimator with variables on the leaves of the graph. It can be seen as a feedforward deep neural network whose hidden nodes consist of sum and product operations, input neurons are pdfs, and edges represent computational units. This is in contrast to the classical probabilistic graphical models, such as Markov networks and Bayesian networks, where nodes represent random variables and edges the statistical dependencies among them.

Sum-product network is trained using BP and EM algorithms. It is more accurate, faster to train, and more tractable than classical feature extractors like the RBM, the deep belief network and the deep probabilistic autoencoders [42]. While classical probabilistic graphical models perform exact inference exponentially in the model tree-width, sum-product network computes many kinds of queries in linear time of the network size, due to structural constraints like decomposability and completeness. The DAG constrained topology, due to completeness and decomposability, determines sparse and local connections, similar to convolutional networks. Each hidden neuron acts as probabilistic part-based feature extractor. Sum-product network can be interpreted as sparse, labeled and generative MLPs whose layers are arranged in a DAG [83].

Deep embedded clustering [86] pretrains a neural network by means of an autoencoder and then fine-tunes by jointly optimizing cluster centroids in output space and the underlying feature representation. Soft assignments are first computed between the data and cluster centroids based on a Student's t-distribution; the parameters are then optimized by minimizing the Kullback–Leibler divergence, so as to match soft assignments to a target distribution. Deep divergence-based clustering network [40] leverages the discriminative power of divergence measures. It does not require pretraining steps. The divergence-based loss function incorporates geometric regularization constraints. The network scales well to large datasets.

Generative adversarial models [31] are an approach to unsupervised deep learning. The framework estimates generative models via an adversarial process, in which two models are simultaneously trained: a generative model for capturing the data distribution, and a discriminative model for estimating the probability that a sample came from the training data rather than from the generative model. The generative model is trained to maximize the probability of the discriminative model making a mistake. This framework corresponds to a minimax two-player game. In the space of arbitrary functions of generative model and discriminative model, a unique solution exists, with the generative model recovering the training data distribution and the discriminative model equal to 1/2 everywhere. When the two models are defined by MLPs, the entire system can be trained with BP.

Deep associative neural network [55] is an unsupervised deep associative memory model. The deep architecture consists of a perception layer and hierarchical propagation layers. Using a probability model very similar to that of RBM and an energy function inspired from Hebb learning rule, the model is optimized by a modified contrastive divergence algorithm with an iterated sampling process.

Deep cascade learning [57], motivated by the cascade-correlation algorithm for sequentially training MLPs, circumvents the vanishing gradient problem by ensuring that the output is always adjacent to the layer being trained. Deep Fisher discriminant analysis [21] is a straightforward nonlinear extension of LS-LDA.

DeepESN model [30] is characterized by a stacked hierarchy of reservoirs. DeepESN is characterized by multiple temporal representations, richness of reservoir states, and higher short-term memory capacity. Introducing a layered reservoir construction into the architectural design has the effect of reducing the number of nonzero recurrent connections. The readout layer is the only part of the network that is trained.

Deep spiking neural networks have been proposed for deep spiking MLP [8], deep spiking convolutional network [41], spiking deep Bayesian network [66]. These methods are based on STDP learning rules as the gradient descent learning rule.

Problems

24.1 Explain the principle of deep learning.

24.2 Plot the ReLU function $\sigma(x) = \max(0, x)$.

24.3 Consider a 4×4 pixel image

1	2	3	3
4	5	6	7
3	5	7	20
0	3	5	8

.

The convolution operation $(*)$ is the sum of the products of the elements located on the same positions of the two matrices. Calculate the 3×3 feature maps of this image using the two convolution filters:

$$\begin{bmatrix} 1 & 0 \\ 0 & 1 \end{bmatrix}, \quad \begin{bmatrix} 0 & -1 \\ -1 & 0 \end{bmatrix}.$$

24.4 DeepLearn MATLAB Toolbox (https://github.com/rasmusbergpalm/DeepLearnToolbox) for deep learning includes deep belief networks, stacked autoencoders, convolutional neural networks, convolutional autoencoders, and MLPs. Each method has examples to get you started. Practice the methods using a dataset.

References

1. Abdel-Hamid, O., Mohamed, A., Jiang, H., & Penn, G. (2012). Applying convolutional neural network concepts to hybrid NN-HMM model for speech recognition. In *Proceedings of the IEEE International Conference on Acoustics, Speech and Signal Processing (ICASSP)*. Kyoto, Japan.
2. Audhkhasi, K., Osoba, O., & Kosko, B. (2016). Noise-enhanced convolutional neural networks. *Neural Networks*, *78*, 15–23.
3. Ba, J., Mnih, V., & Kavukcuoglu, K. (2014). Multiple object recognition with visual attention. *Proceedings of International Conference on Learning Representations*.
4. Bartlett, P. L., Harvey, N., Liaw, C., & Mehrabian, A. (2019). Nearly-tight VC-dimension and pseudodimension bounds for piecewise linear neural networks. *Journal of Machine Learning Research*, *20*, 1–17.
5. Bellemare, M. G., Naddaf, Y., Veness, J., & Bowling, M. (2013). The arcade learning environment: An evaluation platform for general agents. *Journal of Artificial Intelligence Research*, *47*, 253–279.
6. Bengio, Y. (2009). Learning deep architectures for AI. *Foundations and Trends in Machine Learning*, *2*(1), 1–127.
7. Bengio, Y., Lamblin, P., Popovici, D., & Larochelle, H. (2006). Greedy layer-wise training of deep networks. In B. Schlkopf, J. Platt, & T. Hofmann (Eds.), *Advances in neural information processing systems* (Vol. 19, pp. 153–160). Cambridge, MA: MIT Press.
8. Bengio, Y., Lee, D.-H., Bornschein, J., Mesnard, T., & Lin, Z. (2015). Towards biologically plausible deep learning. arXiv:1502.04156, 1–10.
9. Bengio, Y., Simard, P., & Frasconi, P. (1994). Learning longterm dependencies with gradient descent is difficult. *IEEE Transactions on Neural Networks*, *5*(2), 157–166.
10. Bianchini, M., & Scarselli, F. (2014). On the complexity of neural network classifiers: A comparison between shallow and deep architectures. *IEEE Transactions on Neural Networks and Learning Systems*, *25*(8), 1553–1565.
11. Boser, B., Sackinger, E., Bromley, J., LeCun, Y., & Jackel, L. (1991). An analog neural network processor with programmable topology. *IEEE Journal of Solid-State Circuits*, *26*, 2017–2025.
12. Bruna, J., Szlam, A., & LeCun, Y. (2014). Signal recovery from pooling representations. In *Proceedings of the 31st International Conference on Machine Learning* (pp. 307–315).
13. Chen, B., Ting, J.-A., Marlin, B., & de Freitas, N. (2010). Deep learning of invariant spatio-temporal features from video. In *Proceedings of NIPS Workshop on Deep Learning and Unsupervised Feature Learning*.
14. Chen, X.-W., & Lin, X. (2014). Big data deep learning: Challenges and perspectives. *IEEE Access*, *2*, 514–525.
15. Chien, J.-T., & Bao, Y.-T. (2018). Tensor-factorized neural networks. *IEEE Transactions on Neural Networks and Learning Systems*, *29*(5), 1998–2011.
16. Chui, C. K., Li, X., & Mhaskar, H. N. (1994). Neural networks for localized approximation. *Mathematics of Computation*, *63*(208), 607–623.
17. Ciresan, D. C., Meier, U., Gambardella, L. M., & Schmidhuber, J. (2010). Deep, big, simple neural nets for handwritten digit recognition. *Neural Computation*, *22*(12), 3207–3220.

18. Delalleau, O., & Bengio, Y. (2011). Shallow vs. deep sum-product networks. *Advances in neural information processing systems* (pp. 666–674).
19. Dahl, G. E., Yu, D., Deng, L., & Acero, A. (2012). Context-dependent pretrained deep neural networks for large-vocabulary speech recognition. *IEEE/ACM Transactions on Audio, Speech, and Language Processing*, *20*(1), 30–42.
20. Deng, L., Yu, D., & Platt, J. (2012). Scalable stacking and learning for building deep architectures. In *Proceedings of IEEE International Conference on Acoustics, Speech, and Signal Processing (ICASSP)* (pp. 2133–2136).
21. Diaz-Vico, D., & Dorronsoro, J. R. (2019). Deep least squares fisher discriminant analysis. *IEEE Transactions on Neural Networks and Learning Systems*. https://doi.org/10.1109/TNNLS.2019.2906302.
22. Dong, C., Loy, C. C., He, K., & Tang, X. (2016). Image super-resolution using deep convolutional networks. *IEEE Transactions on Pattern Analysis and Machine Intelligence*, *38*(2), 295–307.
23. Edelman, G. M., & Mountcastle, V. B. (1978). *The mindful brain: Cortical organization and the group-selective theory of higher brain function*. Cambridge, MA: MIT Press.
24. Eldan, R., & Shamir, O. (2016). The power of depth for feedforward neural networks. In *Proceedings of the 29th Annual Conference on Learning Theory* (PMLR Vol. 49, pp. 907–940). New York, NY.
25. Erhan, D., Bengio, Y., Courville, A., Manzagol, P.-A., Vincent, P., & Bengio, S. (2010). Why does unsupervised pre-training help deep learning? *Journal of Machine Learning Research*, *11*, 625–660.
26. Esmaeilzehi, A., Ahmad, M. O., & Swamy, M. N. S. (2018). CompNet: A new scheme for single image super resolution based on deep convolutional neural network. *IEEE Access*, *6*, 59963–59974.
27. Esmaeilzehi, A., Ahmad, M. O., & Swamy, M. N. S. (2019). SRSubBandNet: A new deep learning scheme for single image super resolution based on subband reconstruction. In *Proceedings of the IEEE International Symposium on Circuits and Systems*. Sapporo, Japan.
28. Farabet, C., LeCun, Y., Kavukcuoglu, K., Culurciello, E., Martini, B., Akselrod, P., et al. (2011). Large-scale FPGA-based convolutional networks. In R. Bekkerman, M. Bilenko, & J. Langford (Eds.), *Machine learning on very large data sets* (pp. 399–419). Cambridge, UK: Cambridge University Press.
29. Fukushima, K. (1980). Neocognitron: A self-organizing neural network model for a mechanism of pattern recognition unaffected by shift in position. *Biological Cybernetics*, *36*, 193–202.
30. Gallicchio, C., Micheli, A., & Pedrelli, L. (2017). Deep reservoir computing: A critical experimental analysis. *Neurocomputing*, *268*, 87–99.
31. Goodfellow, I., Pouget-Abadie, J., Mirza, M., Xu, B., Warde-Farley, D., Ozair, S., Courville, A., & Bengio, Y. (2014). Generative adversarial nets. In *Advances in neural information processing systems* (Vol. 27, pp. 2672–2680).
32. Hastad, J. T. (1987). *Computational limitations for small depth circuits*. Cambridge, MA: MIT Press.
33. Hastad, J., & Goldmann, M. (1991). On the power of small-depth threshold circuits. *Computational Complexity*, *1*(2), 113–129.
34. He, K., Zhang, X., Ren, S., & Sun, J. (2015). Spatial pyramid pooling in deep convolutional networks for visual recognition. *IEEE Transactions on Pattern Analysis and Machine Intelligence*, *37*(9), 1904–1916.
35. He, K., Zhang, X., Ren, S., & Sun, J. (2016). Deep residual learning for image recognition. In *Proceedings of the IEEE Conference on Computer Vision and Pattern Recognition (CVPR)* (pp. 770–778).
36. Hinton, G. E. (2007). To recognize shapes, first learn to generate images. *Progress in Brain Research*, *165*, 535–547.
37. Hinton, G. E., Osindero, S., & Teh, Y.-W. (2006). A fast learning algorithm for deep belief nets. *Neural Computation*, *18*, 1527–1554.

38. Hinton, G. E., & Salakhutdinov, R. R. (2006). Reducing the dimensionality of data with neural networks. *Science*, *313*(5786), 504–507.
39. Hubel, D. H., & Wiesel, T. N. (1959). Receptive fields of single neurones in the cat's striate cortex. *Journal of Physiology*, *148*(3), 574–591.
40. Kampffmeyer, M., Lokse, S., Bianchi, F. M., Livi, L., Salberg, A.-B., & Jenssen, R. (2019). Deep divergence-based approach to clustering. *Neural Networks*, *113*, 91–101.
41. Kheradpisheh, S. R., Ganjtabesh, M., Thorpe, S. J., & Masquelier, T. (2018). STDP-based spiking deep convolutional neural networks for object recognition. *Neural Networks*, *99*, 56–67.
42. Kingma, D. P., & Welling, M. (2014). Auto-encoding variational Bayes. In *Proceedings of the 2nd International Conference on Learning Representations*. Banff, Canada.
43. Krizhevsky, A., Sutskever, I., & Hinton, G. E. (2012). ImageNet classification with deep convolutional neural networks. In *Advances in neural information processing systems* (Vol. 25, pp. 1090–1098).
44. Larochelle, H., Bengio, Y., Louradour, J., & Lamblin, P. (2009). Exploring strategies for training deep neural networks. *Journal of Machine Learning Research*, *1*, 1–40.
45. LeCun, Y., & Bengio, Y. (1995). Convolutional networks for images, speech, and timeseries. In *The handbook of brain theory and neural networks*. Cambridge, MA: MIT Press.
46. LeCun, Y., Bengio, Y., & Hinton, G. (2015). Deep learning. *Nature*, *521*(7553), 436–444.
47. Le Cun, Y., Boser, B., Denker, J. S., Henderson, D., Howard, R. E., Hubbard, W., et al. (1989). Handwritten digit recognition with a back-propagation network. In D. S. Touretzky (Ed.), *Advances in neural information processing systems* (Vol. 2, pp. 396–404). San Mateo, CA: Morgan Kaufmann.
48. LeCun, Y., Bottou, L., Bengio, Y., & Haffner, P. (1998). Gradient-based learning applied to document recognition. *Proceedings of the IEEE*, *86*(11), 2278–2324.
49. Lee, H., Ekanadham, C., & Ng, A. Y. (2007). Sparse deep belief net model for visual area V2. In J. C. Platt, D. Koller, Y. Singer, & S. T. Roweis (Eds.), *Advances in neural information processing systems* (Vol. 20, pp. 873–880).
50. Lee, H., Grosse, R., Ranganath, R., Ng, A. Y. (2009). Convolutional deep belief networks for scalable unsupervised learning of hierarchical representations. In L. Bottou, & M. Littman (Eds.), *Proceedings of the 26th Annual International Conference on Machine Learning* (pp. 609–616). New York: ACM.
51. Le Roux, N., & Bengio, Y. (2008). Representational power of restricted Boltzmann machines and deep belief networks. *Neural Computation*, *20*(6), 1631–1649.
52. Le Roux, N., & Bengio, Y. (2010). Deep belief networks are compact universal approximators. *Neural Computation*, *22*, 2192–2207.
53. Lin, M., Chen, Q., & Yan, S. (2014). Network in network. In *Proceedings of the 2nd International Conference on Learning Representations*. Banff, Canada.
54. Ling, C. X., & Zhang, H. (2002). The representational power of discrete Bayesian networks. *Journal of Machine Learning Research*, *3*, 709–721.
55. Liu, J., Gong, M., & He, H. (2019). Deep associative neural network for associative memory based on unsupervised representation learning. *Neural Networks*, *113*, 41–53.
56. Maas, A. L., Hannun, A. Y., & Ng, A. Y. (2013). Rectifier nonlinearities improve neural network acoustic models. In *Proceedings of the International Conference on Machine Learning* (Vol. 30).
57. Marquez, E. S., Hare, J. S., & Niranjan, M. (2018). Deep cascade learning. *IEEE Transactions on Neural Networks and Learning Systems*, *29*(11), 5475–5485.
58. Masci, J., Meier, U., Ciresan, D., & Schmidhuber, J. (2011). Stacked convolutional autoencoders for hierarchical feature extraction. In *Proceedings of the 21st International Conference on Artificial Neural Networks* (Vol. 1, pp. 52–59). Espoo, Finland.
59. McClelland, J. L., McNaughton, B. L., & O'Reilly, R. C. (1995). Why there are complementary learning systems in the hippocampus and neocortex: Insights from the successes and failures of connectionist models of learning and memory. *Psychological Review*, *102*, 419–457.

60. Mhaskar, H., Liao, Q., & Poggio, T. (2016). *Learning functions: When is deep better than shallow*. CBMM Memo No. 045. https://arxiv.org/pdf/1603.00988v4.pdf.
61. Mnih, V., Kavukcuoglu, K., Silver, D., Rusu, A. A., Veness, J., Bellemare, M. G., et al. (2015). Human-level control through deep reinforcement learning. *Nature*, *518*(7540), 529–533.
62. Mohamed, A., Dahl, G., & Hinton, G. (2009). Deep belief networks for phone recognition. In *Proceedings of NIPS Workshop on Deep Learning for Speech Recognition and Related Applications*.
63. Montufar, G. F., Pascanu, R., Cho, K., & Bengio, Y. (2014). On the number of linear regions of deep neural networks. In *Advances in neural information processing systems* (Vol. 27, pp. 2924–2932).
64. Nair, V., & Hinton, G. E. (2010). Rectified linear units improve restricted Boltzmann machines. In *Proceedings of the International Conference on Machine Learning (ICML)* (pp. 807–814).
65. Nitta, T. (2017). Resolution of singularities introduced by hierarchical structure in deep neural networks. *IEEE Transactions on Neural Networks and Learning Systems*, *28*(10), 2282–2293.
66. O'Connor, P., Neil, D., Liu, S.-C., Delbruck, T., & Pfeiffer, M. (2013). Real-time classification and sensor fusion with a spiking deep belief network. *Frontiers in Neuroscience*, *7*, 1–13.
67. Pape, L., Gomez, F., Ring, M., & Schmidhuber, J. (2011). Modular deep belief networks that do not forget. In *Proceedings of IEEE International Joint Conference on Neural Networks* (pp. 1191–1198).
68. Poon, H., & Domingos, P. (2011). Sum-product networks: A new deep architecture. In *Proceedings of the 27th Conference on Uncertainty in Artificial Intelligence* (pp. 337–346). Barcelona, Spain.
69. Ranzato, M. A., Poultney, C., Chopra, S., & LeCun, Y. (2006). Efficient learning of sparse representations with an energy-based model. In *Advances in neural information processing systems* (Vol. 19, 1137–1144).
70. Rippel, O., Snoek, J., & Adams, R. P. (2015). Spectral representations for convolutional neural networks. In *Advances in neural information processing systems* (Vol. 28, pp. 2449–2457).
71. Russakovsky, O., Deng, J., Su, H., Krause, J., Satheesh, S., Ma, S., et al. (2015). Imagenet large scale visual recognition challenge. *International Journal of Computer Vision*, *115*(3), 211–252.
72. Salakhutdinov, R., & Hinton, G. (2009). Deep Boltzmann machines. In D. van Dyk, & M. Welling (Eds.), *Proceedings of the 12th International Conference on Artificial Intelligence and Statistics* (PMLR Vol. 5, pp. 448–455).
73. Salakhutdinov, R., & Larochelle, H. (2010). Efficient learning of deep Boltzmann machines. In Y. W. Teh, & M. Titterington, (Eds.), *Proceedings of the 13th Annual International Conference on Artificial Intelligence and Statistics* (pp. 693–700).
74. Schmidhuber, J. (1992). Learning complex, extended sequences using the principle of history compression. *Neural Computation*, *4*, 234–242.
75. Silver, D., Huang, A., Maddison, C. J., Guez, A., Sifre, L., van den Driessche, G., et al. (2016). Mastering the game of go with deep neural networks and tree search. *Nature*, *529*(7587), 484–489.
76. Simard, P., Steinkraus, D., & Platt, J. C. (2003). Best practices for convolutional neural networks applied to visual document analysis. In *Proceedings of the 7th International Conference on Document Analysis and Recognition* (pp. 958–963).
77. Simonyan, K., & Zisserman, A. (2015). Very deep convolutional networks for large-scale image recognition. In *Proceedings of the International Conference on Learning Representations (ICLR)*.
78. Sutskever, I., & Hinton, G. E. (2008). Deep, narrow sigmoid belief networks are universal approximators. *Neural Computation*, *20*(11), 2629–2636.
79. Szegedy, C., Liu, W., Jia, Y., Sermanet, P., Reed, S., Anguelov, D., Erhan, D., Vanhoucke, V., & Rabinovich, A. (2015). Going deeper with convolutions. In *Proceedings of the IEEE Conference on Computer Vision and Pattern Recognition (CVPR)* (pp. 1–9).
80. Szymanski, L., & McCane, B. (2014). Deep networks are effective encoders of periodicity. *IEEE Transactions on Neural Networks and Learning Systems*, *25*(10), 1816–1827.

81. Telgarsky, M. (2016). Benefits of depth in neural networks. In *Proceedings of the 29th Annual Conference on Learning Theory* (PMLR Vol. 49, pp. 1517–1539). New York, NY.
82. Veit, A., Wilber, M., & Belongie, S. (2016). Residual networks behave like ensembles of relatively shallow networks. In *Advances in neural information processing systems* (Vol. 29, pp. 550–558).
83. Vergari, A., Di Mauro, N., & Esposito, F. (2019). Visualizing and understanding sum-product networks. *Machine Learning*, *108*, 551–573.
84. Waibel, A., Hanazawa, T., Hinton, G. E., Shikano, K., & Lang, K. (1989). Phoneme recognition using time-delay neural networks. *IEEE Transactions on Acoustics Speech and Signal Processing*, *37*, 328–339.
85. Wolpert, D. H. (1992). Stacked generalization. *Neural Networks*, *5*(2), 241–259.
86. Xie, J., Girshick, R., & Farhadi, A. (2016). Unsupervised deep embedding for clustering analysis. In *Proceedings of the 33rd International Conference on Machine Learning* (Vol. 48, pp. 478–487). New York, NY.
87. Yarotsky, D. (2017). Error bounds for approximations with deep ReLU networks. *Neural Networks*, *94*, 103–114.
88. Zagoruyko, S., & Komodakis, N. (2016). Wide residual networks. In *Proceedings of British Machine Vision Conference* (pp. 87.1–87.12). Newcastle, UK.
89. Zeiler, M. D., & Fergus, R. (2013). Stochastic pooling for regularization of deep convolutional neural networks. In *Proceedings of the 1st International Conference on Learning Representations*. Scottsdale, AZ.
90. Zeiler, M. D., & Fergus, R. (2014). Visualizing and understanding convolutional networks. In *Proceedings of the European Conference on Computer Vision (ECCV)* (pp. 818–833).

Chapter 25
Combining Multiple Learners: Data Fusion and Ensemble Learning

25.1 Introduction

Different learning algorithms have different accuracies. No-free-lunch theorem asserts that no single learning algorithm always achieves the best performance in any domain. They can be combined to attain higher accuracy. Averaging outputs of an infinite number of unbiased and independent classifiers may lead to the same response as the optimal Bayes classifier [94]. Data fusion is the process of fusing multiple records representing the same real-world object into a single, consistent, and clean representation. Fusion of data for improving prediction accuracy and reliability is an important problem in machine learning.

Fusion strategies can be implemented in different levels, namely, signal enhancement and sensor level (data level), feature level, classifier level, decision level, and semantic level. Evidence theory [85] falls within the theory of imprecise probabilities.

For classification, different classifiers can be generated by different initializations of a classifier, training a classifier with different training data, or training a classifier using different feature sets. Ensemble techniques [6, 79] build a number of different predictors (base learners), then combine them to form the composite predictor to classify the test set. This phenomenon is known as *diversity* [22]. Two classifiers are said to be *diverse* if they make different incorrect predictions on new data points. Classifier diversity plays a critical role in ensemble learning. The ensemble of predictors is often called a *committee machine* (or *mixture of experts*). Ensemble learning has its capability of improving the classification accuracy of any single classifier, given the same amount of training information.

For ensemble learning, each of the classifiers composing the ensemble can be constructed either independently (e.g., bagging [6]) or sequentially (e.g., boosting [79]). The majority voting scheme is the most popular classifier fusion method. In this method, the final class is determined by the maximum number of votes counted among all the classifiers fused. Averaging is a simple but effective method and is used in many classification problems; the final class is determined by the average of continuous outputs of all classifiers fused.

K.-L. Du and M. N. S. Swamy, *Neural Networks and Statistical Learning*,
https://doi.org/10.1007/978-1-4471-7452-3_25

25.1.1 Ensemble Learning Methods

Bagging [6] and boosting [79] are two popular committee machines. They run a learning algorithm on different distributions over the training data. Bagging builds training sets, called *bags*, of the same size of the original dataset by applying random sampling with replacement. Unlike bagging, boosting draws tuples randomly, according to a distribution, and tries to concentrate on harder examples by adaptively changing the distributions of the training set on the base of the performance of the previous classifiers. Bagging and random forests are ensemble methods for classification, where a committee of trees each cast a vote for the predicted class. In boosting, unlike random forests, the committee of weak learners evolves over time, and the members cast a weighted vote.

Subspace or multiview learning creates multiple classifiers from different feature spaces. In this way, different classifiers build their decision boundaries in different views of the feature space. Multiview learning utilizes the agreement among learners to improve the overall classification performance. Representative works in this area are the random subspace method [45], the random forest method [44], and the rotation forest [77].

The mixture of experts [46] is a divide-and-conquer algorithm that contains a gating network for soft partitioning the input space and expert networks modeling each of these partitions. The methodology provides a tool of classification when the set of classifiers are mixed according to a final gating mechanism. Both the classifiers and the gating mechanism are trained at the same time over a given dataset. The mixture of experts can be treated as an RBF network where the second-layer weights $\boldsymbol{w}$ are outputs of linear models, each taking the input, and these weights are called *experts*. Bounds for the VC dimension of the mixtures-of-experts architecture is derived in [47].

In Bayesian committee machine [93], the dataset is divided into M subsets of the same size and M models are derived from the individual sets. The predictions of the individual models are combined using a weight scheme which is derived from a Bayesian perspective in the context of Gaussian process regression. That is, the weight for each individual model is the inverse covariance of its prediction. Although it can be applied to a combination of any kind of estimators, the main foci are Gaussian process regression and related systems such as regularization networks and smoothing splines for which the degrees of freedom increase with the number of training data. The performance of Bayesian committee machine improves if several test points are queried at the same time and is optimal if the number of test points is at least as large as the degrees of freedom of the estimator. Bayesian model averaging [21] is also an ensemble learning method from a Bayesian learning perspective.

Stacking is an approach to combining the strengths of a number of fitted models. It replaces a simple average by a weighted average, where the weights take account of the complexity of the model or other aspects. Stacking is a non-Bayesian model averaging, where the estimated weights, corresponding to Bayesian priors that downweight complex models, are no longer posterior probabilities of models;

they are obtained by a technique based on cross-validation. In [13], Bayesian model averaging is compared to stacking. When the correct data generating model is on the list of models under consideration, Bayesian model averaging is never worse than stacking. Bayesian model averaging is more sensitive to model approximation error than stacking is, when the variabilities of the random quantities are roughly comparable. It is outperformed by stacking when the bias exceeds one term of size equal to the leading terms in the model or when the direction of deviation has a different functional form (with higher variability) that the model list cannot approximate well. Overall, stacking has better robustness properties than Bayesian model averaging in the most important settings.

Stack generalization [100] extends voting by combining the base learners through a combiner, which is another learner. Stacking estimates and corrects the biases of the base learners. The combiner should be trained on data that are not used for training the base learners.

Cascading is a multistate method where d_j (class j) is used only if all preceding learners, d_k, $k < j$ are not confident. Associated with each learner is a confident w_j such that d_j is confident of its output and can be used if $w_j > \theta_j$, where the confidence threshold satisfies $1/K < \theta_j \leq \theta_{j+1} < 1$, where K is the number of classes. For classification, the confidence function is set to the highest posterior: $w_j = \max_i d_{ji}$.

Neural networks and SVMs can also be regarded as an ensemble method. Bayesian methods for nonparametric regression can also be viewed as ensemble methods, where a large number of candidate models are averaged with respect to the posterior distribution of their parameter settings. A method designed for multiclass classification using error-correcting output codes (ECOCs) [19] is a learning ensemble. In fact, one could characterize any dictionary method, such as regression splines, as an ensemble method, with the basis functions serving the role of weak learners. A survey of tree-based ensemble methods is given in [18].

25.1.2 Aggregation

Aggregation operators combine data from several sources to improve the quality of information. In fuzzy logic, t-norm and t-conorm are two aggregation operators. The ordered weighted averaging operator [103] is a well-known aggregation operator for multicriteria decision-making. It provides a parameterized family of aggregation operators with the maximum, minimum and average as special cases.

The base algorithms can be different algorithms, or the same algorithm with different parameters, or the same algorithm using different features of the same input, or different base learners trained with a different subset of the training set, or cascaded training-based base learners. The main task can also be defined in terms of a number of subtasks to be implemented by the base learners, as is in the case of ECOCs. The combination can be in parallel or multistage implementation.

Voting is the simplest way of combining multiple classifiers. It takes a linear combination of the outputs of learners. The final output is computed by

$$y_i = \sum_{j=1}^{L} w_j d_{ji}, \tag{25.1}$$

subject to

$$\sum_{j=1}^{L} w_j = 1, \quad w_j \geq 0, \forall j, \tag{25.2}$$

where d_{ji}, $j = 1, \ldots, L$, is the vote of learner j for class C_i, and w_j is the weight of its vote.

When $w_j = 1/L$, we have *simple voting*. In the case of classification, this is called plurality voting where the class having the maximum number of votes is the winner. For two classes, this is *majority voting*, where the winner class gets more than half of the votes. Voting schemes can be viewed as a Bayesian framework with weights as prior model probabilities and model decisions as model-conditional likelihoods.

25.2 Majority Voting

Majority voting works well for the agnostic information about labeling quality of labelers, underlying class distributions and the difficulties of instances. Agnostic methods are more attractive in crowdsourcing systems. Majority voting works well under the assumptions that the overall labeling accuracy of most labelers is greater than 50% in binary labeling tasks, and the errors of each labeler are approximately uniformly distributed over all classes.

In a weighted majority vote, several classifiers (or voters) are assigned a specific weight. PAC-Bayesian theory [62] aims to provide PAC guarantees to Bayesian-like learning algorithms. PAC-Bayesian approach indirectly bounds the risk of a Q-weighted majority vote by bounding the risk of an associated Gibbs classifier. PAC-Bayesian theorem provides a risk bound for the "true" risk of Gibbs classifier, by considering the empirical risk of this Gibbs classifier on the training data and the Kullback–Leibler divergence between a posterior distribution Q and a prior distribution P. It is well known that the risk of the (deterministic) majority vote classifier is upper bounded by twice the risk of the associated Gibbs classifier [55, 61]. Unfortunately, this bound on the majority voting classifier is far from being tight, especially for weak voters, even if the PAC-Bayesian bound itself generally gives a tight bound on the risk of Gibbs classifier.

C-bound, originally presented in [54], is an accurate indicator of the risk of the majority vote. Minimizing C-bound allows to minimize the true risk of the weighted majority vote, and it reduces to a simple quadratic program. Justified by PAC-Bayesian theory, MinCq algorithm [56] optimizes the weights of a set of voters $\mathcal{H}$ by minimizing C-bound involving the first two statistical moments of the margin achieved on the training data. MinCq returns a posterior distribution Q on $\mathcal{H}$ that gives the weight of each voter.

The behavior of majority votes in binary classification is extensively analyzed in [40]. C-bound can be smaller than the risk of Gibbs classifier and can even be arbitrarily close to zero even if the risk of Gibbs classifier is close to $1/2$. MinCq achieves state-of-the-art performance, compared with both AdaBoost and SVM.

P-MinCq [4] extends MinCq by incorporating a priori knowledge in the form of a constraint over the distribution of the weights, along with general proofs of convergence that stand in the sample compression setting for data-dependent voters. P-MinCq is applied to a vote of k-NN classifiers with a specific modeling of the voters' performance. P-MinCq is quite robust to overfitting for high-dimensional problems.

Majority votes are central in the Bayesian approach [39]. In this setting, the majority vote is generally called Bayes classifier. Classifiers produced by kernel methods, such as SVM, can also be viewed as majority votes. Indeed, to classify an example $\boldsymbol{x}$, SVM classifier computes $\text{sgn}\left(\sum_{i=1}^{|S|} \alpha_i y_i k(\boldsymbol{x}_i, \boldsymbol{x})\right)$, where $k(\cdot, \cdot)$ is a kernel function, and the input–output pairs $(\boldsymbol{x}_i, y_i)$ are the examples in the training set $\mathcal{S}$. Thus, one can interpret each $y_i k(\boldsymbol{x}_i, \cdot)$ as a voter that chooses confidence level $|k(\boldsymbol{x}_i, \boldsymbol{x})|$ between "positive" and "negative", and α_i as the respective weight of this voter in the majority vote. Similarly, each neuron of the last layer of a neural network can be interpreted as a majority vote.

25.3 Bagging

Bagging, short for *Bootstrap AGGregatING*, is a popular technique for stabilizing statistical learners. Bagging works by training each classifier on a bootstrap sample [6]. The essential idea in bagging is to average many noisy but approximately unbiased models, hence reducing the prediction variance without affecting the prediction bias. It seems to work especially well for high-variance, low-bias procedures, such as trees. The effect of bagging on bias is uncertain, as a number of contradictory findings have been reported. The performance of bagging is generally worse than that of boosting.

Trees are ideal candidates for bagging, since they can capture complex interaction structures in the data and have relatively low bias if grown sufficiently deep. Since each tree generated in bagging is identically distributed, the expectation of an average of B such trees is the same as the expectation of any one of them. This means the bias of bagged trees is the same as that of the individual trees. This is in contrast to boosting, where the trees are grown in an adaptive way to remove bias [34], and hence are not identically distributed.

Bagging enhances the performance of a predictor by repeatedly evaluating the predictor on bootstrap samples and then forming an average over those samples. Bagging works well for unstable modeling procedures, i.e., those for which the conclusions are sensitive to small changes in the data [6]. Bagging is based on bootstrap samples of the same size of the training set. Each bootstrap sample is created by uniformly sampling instances from the whole training set with replacement, thus some examples may appear more than once, while others may not appear at all. Bagging then constructs a new learner from each, and averages the predictions. In boosting, predictions are averaged with different weights.

Bagging and boosting are non-Bayesian procedures that have some similarity to MCMC in a Bayesian model. The Bayesian approach fixes the data and perturbs the parameters, according to current estimate of the posterior distribution. Bagging perturbs the data in an i.i.d fashion and then re-estimates the model to give a new set of model parameters. Finally, a simple average of the model predictions from different bagged samples is computed. Bagging has the drawback of the behavior randomness, since its constituting classifiers are induced in total isolation from one another.

Boosting is similar to bagging, but fits a model that is additive in the models of each individual base learner, which are learned using non-i.i.d. samples.

We can write all of these models in the form

$$\hat{f}(\boldsymbol{x}_{\text{new}}) = \sum_{l=1}^{L} w_l E(y_{\text{new}}|\boldsymbol{x}_{\text{new}}, \hat{\boldsymbol{\theta}}_l), \tag{25.3}$$

where $\hat{\boldsymbol{\theta}}_l$ is a large collection of model parameters. For the Bayesian model, $w_l = 1/L$, and the average estimates the posterior mean (25.3) by sampling $\boldsymbol{\theta}_l$ from the posterior distribution. For bagging, $w_l = 1/L$ as well, and $\hat{\boldsymbol{\theta}}_l$ corresponds to the parameters refit to bootstrap resamples of the training data. For boosting, $w_l = 1$, but $\hat{\boldsymbol{\theta}}_l$ is typically chosen in a nonrandom sequential fashion to constantly improve the fit.

Online bagging [67] implements bagging sequentially. It asymptotically approximates the results of batch bagging, and it is not guaranteed to produce the same results as batch bagging. A variation is to replace the ordinary bootstrap with the Bayesian bootstrap. The online Bayesian version of bagging algorithm [57] is exactly equivalent to its batch Bayesian counterpart. The Bayesian approach produces a completely lossless bagging algorithm. It can lead to increased accuracy and decreased prediction variance for smaller datasets.

25.4 Boosting

Boosting, also known as ARCing (adaptive resampling and combining), was introduced in [79] for boosting the performance of any weak learning algorithm, i.e., an algorithm that generates classifiers which need only be a little bit better than random guessing. Schapire proved that the strong and weak PAC learnability are equivalent to each other [79], which is the theoretic basis for boosting. Boosting algorithms belong to a class of voting methods that produce a classifier as a linear combination of base or weak classifiers. Similar to boosting, multiple classifier systems [102] use a group of classifiers to compromise on a given task.

Boosting works by repeatedly running a given weak learning machine on different distributions of training examples and combining their outputs. Boosting is known as a gradient-descent algorithm over some classes of loss functions [34]. Boosting was believed to seldom overfit, which can arise when the number of classifiers is large [34]. It continues to decrease generalization error long after the sample training error becomes zero, by adding more weak classifiers to the linear combination of classifiers. Some studies have suggested that boosting might suffer from overfitting [60, 76], especially for noisy datasets.

The original boosting approach, known as boosting by filtering [79], was motivated by PAC learning theory. It requires a large number of training examples. This limitation is overcome by AdaBoost [29, 30]. In boosting by subsampling, a fixed sampling size and a set of training examples are used, and they are resampled according to a given probability distribution during training. In boosting by reweighting, all the training examples are used to train the weak learning machine, with weights assigned to each example. This technique is applicable only when the weak learning machine can handle the weighted examples.

For binary classification, the output of a strong classifier $H(\boldsymbol{x})$ is obtained from a weighted combination of all the weak hypotheses $h_t(\boldsymbol{x})$:

$$H(\boldsymbol{x}) = \text{sign}\left(f(\boldsymbol{x})\right) = \text{sign}\left(\sum_{t=1}^{T} \alpha_t h_t(\boldsymbol{x})\right), \tag{25.4}$$

where T is the number of iterations, and $f(\boldsymbol{x}) = \sum_{t=1}^{T} \alpha_t h_t(\boldsymbol{x})$ is a strong classifier. In order to minimize the learning error, one must seek to minimize h_t in each round of boosting, requiring the use of a specific confidence α_t.

Upper bounds on the risk of boosted classifiers are obtained, based on the fact that boosting tends to maximize the margin of the training examples [81]. Under some assumptions on the underlying distribution population, boosting converges to the Bayes risk as the number of iterations goes to infinity [9].

Boosting performs worse than bagging in the presence of noise [18], and it concentrates not only on the hard areas, but also on outliers and noise [3]. Boosting has relative resistance to overfitting. Boosting, when running for an arbitrary large number of steps, overfits, though it takes a very long time to do it. In boosting, unlike in bagging, the committee of weak learners evolves over time, and the members cast a weighted vote. Boosting appears to dominate bagging on most problems, and becomes the preferred choice. Learn++.NC [65] is a variant of boosting.

25.4.1 AdaBoost

Adaptive boosting (AdaBoost) algorithm [29, 30] is a popular approach to ensemble learning. Theoretically, AdaBoost can decrease the error of any weak learning algorithm. From a statistical view of boosting, AdaBoost is a stagewise optimization of an exponential loss function [34]. From a computer science view, generalization error guarantees are derived using VC bounds from PAC learning theory and margins [41]. AdaBoost classifier is better than its competitors such as CART, neural networks, logistic regression, and is substantively better than the technique of creating an ensemble using the bootstrap [6].

AdaBoost uses the whole dataset to train each classifier serially. It adaptively changes the distribution of the sample depending on how difficult each example is to classify. After each round, it gives more focus to difficult instances that were incorrectly classified during the current iteration. Hence, it gives more focus to examples that are harder to classify. After each iteration, the weights of misclassified instances are increased, while those of correctly classified instances are decreased. Each individual classifier is also weighted according to its overall accuracy; these weights are then used in the test phase. Finally, when a new instance is presented, each classifier gives a weighted vote, and the class label is obtained by majority voting.

Assume that a training set $\mathcal{S} = \{(\boldsymbol{x}_i, y_i), i = 1, \ldots, N\}$ with $y_i \in \{-1, +1\}$ is a sample of i.i.d. observations distributed as the random variable $(\boldsymbol{x}, y)$ over an unknown distribution P. AdaBoost works to find a strong classifier $f(\boldsymbol{x}) = \sum_{t=1}^{T} \alpha_t h_t(\boldsymbol{x})$ that minimizes the convex criterion, which is the average of the negative exponential of the margin $y(\boldsymbol{x}) f(\boldsymbol{x})$ of the sample $\mathcal{S}$

$$J = \frac{1}{N} \sum_{i=1}^{N} e^{-y_i f(\boldsymbol{x}_i)}. \tag{25.5}$$

AdaBoost allows to continue adding weak learners until a desired low training error has been achieved.

AdaBoost is shown in Algorithm 25.1, where p_t is the data distribution. The output is a boosted classifier $H(\boldsymbol{x})$. Note that in the updating equation of p_t, $-\alpha_t y_i h_t(\boldsymbol{x}_i) < 0$ when $y(i) = h_t(\boldsymbol{x}_i)$, and > 0 when $y(i) \neq h_t(\boldsymbol{x}_i)$. As a result, after selecting an optimal classifier h_t for p_t, the examples $\boldsymbol{x}_i$ identified correctly by the classifier are weighted less and those identified incorrectly are weighted more. When testing the classifiers on p_{t+1}, it selects a classifier that better identifies those examples missed by the previous classifier.

Algorithm 25.1 (AdaBoost)

1. $p_1(i) = 1/N, i = 1, \ldots, N.$
2. ***for*** $t = 1$ ***to*** T:
   a. *Find a weak learner* h_t *from* $\mathcal{S}, p_t$:
      $\epsilon_t = \sum_{i=1, y_i \neq h_t(\boldsymbol{x}_i)}^{N} p_t(i)$;
      $h_t = \arg\min_{h_j} \epsilon_t$.
   b. ***if*** $\epsilon_t \geq 0.5, T \leftarrow t - 1$;
      return
      end if
   c. $\alpha_t = \frac{1}{2} \ln \left(\frac{1-\epsilon_t}{\epsilon_t} \right)$.
   d. $p_{t+1}(i) = p_t(i) e^{-\alpha_t h_t(\boldsymbol{x}_i) y_i}, i = 1, \ldots, N;$
      $p_{t+1}(i) = \frac{p_{t+1}(i)}{\sum_{j=1}^{N} p_{t+1}(j)}, i = 1, \ldots, N.$

   end for
3. *Output:* $H(\boldsymbol{x}) = sign \left(\sum_{t=1}^{T} \alpha_t h_t(\boldsymbol{x}) \right)$.

AdaBoost can be viewed as a coordinate descent algorithm that iteratively minimizes an exponential function of the margin over the training set [34, 80, 81]. AdaBoost asymptotically converges to the minimum possible exponential loss [14]. AdaBoost provably achieves arbitrarily good bounds on its training and generalization errors provided that weak classifiers can perform slightly better than random guessing on every distribution over the training set [30].

It is demonstrated that a simple stopping strategy suffices for universal consistency [2]: the number of iterations is a fixed function of the sample size. Provided AdaBoost is stopped after $N^{1-\epsilon}$ iterations, for sample size N and $\epsilon \in (0, 1)$, the sequence of risks, or probabilities of error, of the classifiers it produces approaches the Bayes risk.

The convergence rate of AdaBoost to the minimum of the exponential loss is investigated in [66]. Within C/ϵ iterations, AdaBoost achieves a value of the exponential loss that is at most ϵ more than the best possible value, where C depends on the dataset. This dependence of the rate on ϵ is optimal up to constant factors, that is, at least $W(1/\epsilon)$ rounds are necessary to achieve within ϵ of the optimal exponential loss.

AdaBoost tends to overfit to the noisy data. This arises from the exponential loss function, which puts unrestricted penalties to the misclassified samples with very large margins. AdaBoost is highly affected by outliers.

AdaBoost finds a linear separator with a large margin [81]. However, it does not converge to the maximal margin solution [75]. If the weak learnability assumption holds, then the data are linearly separable [29]. AdaBoost with shrinkage asymptotically converges to an L_1-margin-maximizing solution [75]. AdaBoost* [75] converges to the maximal margin solution in $O(\log(N)/\epsilon^2)$ iterations. The family of algorithms proposed in [86] has the same convergence properties. These algorithms are effective when the data is linearly separable. Weak learnability is shown to be equivalent to linear separability with L_1 margin [86]. Efficient boosting algorithms for maximizing hard and soft versions of the L_1 margin are obtained [86].

FloatBoost [58] learns a boosted classifier for achieving the minimum error rate. FloatBoost learning uses a backtrack mechanism after each iteration of AdaBoost to minimize the error rate directly, rather than minimizing an exponential function of the margin as in AdaBoost. A stagewise approximation of the posterior probability is used for learning best weak classifiers. These techniques lead to a classifier which requires fewer weak classifiers than AdaBoost, yet achieves lower error rates in both training and testing. This is at the cost of longer training time [58]. MultBoost [36] is a parallel variant of AdaBoost that can achieve parallelization both in space and time. Unlike AdaBoost, LogitBoost [34] is based on additive logistic regression model.

AdaBoost.M1 and AdaBoost.M2 [30] extend AdaBoost from binary classification to the multiclass case. AdaBoost.M1, as a straightforward generalization, halts if the classification error rate of the weak classifier produced in any iterative step is $\geq 50\%$ [81]. To avoid the problem, AdaBoost.M2 attempts to minimize a more sophisticated error measure called *pseudoloss*. The boosting process continues as long as the weak classifier produced has pseudoloss slightly better than random guessing. In addition, the introduction of the mislabel distribution enhances the communication between the learner and the booster. AdaBoost.M2 can focus the learner not only on hard-to-classify examples, but on the incorrect labels. AdaBoost.MH [80] is an extension of AdaBoost to multiclass/multilabel classification problems based on Hamming loss. SAMME [108] is a multiclass extension of AdaBoost, which minimizes an exponential loss for multiclass classification. The considered weak learners have to solve weighted multiclass classification problems.

AdaBoost.R [30] further extends AdaBoost.M2 to boosting regression problems. It solves regression problems by reducing them to classification ones. AdaBoost.RT [88] is a boosting algorithm for regression problems. It filters out the examples with the relative estimation error that is higher than the preset threshold value, and then follows the AdaBoost procedure.

25.4.2 Other Boosting Algorithms

Gradient boosting [32, 34] is typically used with decision trees of a fixed size as base learners, and, in this setting, is called gradient tree boosting. Gradient tree boosting sequentially produces a model in the form of linear combinations of decision trees, by solving an infinite-dimensional optimization problem. It has exceptional performance. Accelerated gradient boosting [5] incorporates Nesterov's accelerated descent into the gradient boosting procedure [32]. It is less sensitive to the shrinkage parameter, and the output predictors are considerably more sparse in the number of trees.

For noisy data, overfitting effects can be avoided by regularizing boosting so as to limit the complexity of the function class. AdaBoostReg [76] and BrownBoost [28] are designed for this purpose. Using loss functions for robust boosting, robust eta-boost algorithm [48] is robust against both mislabels and outliers, especially for the estimation of conditional probability.

For inseparable datasets, LogLoss Boost algorithm [14] tries to minimize the cumulative logistic loss, which is less sensitive to noise. BrownBoost [28] improves AdaBoost performance by damping the influences of those samples that are hard to learn. It uses the error function (erf) as a margin-based loss function. BrownBoost works well in the inseparable case and is noise tolerant. SmoothBoost [84] builds on the idea of generating only smooth distributions by capping the maximal weight of a single example. It can tolerate relatively high rates of malicious noise. As a special case of SmoothBoost, a linear threshold learning algorithm obtained matches the sample complexity and malicious noise tolerance of the online perceptron algorithm.

AdaBoost selects weak learners by merely minimizing the training error rate, that is, the ratio of the number of misclassified samples to the total number of samples. Each misclassified sample is weighted equivalent, so training error can hardly represent the error degrees. However, the training error-based criterion does not work well in some cases, especially for small-sample-size problems, since more than one weak learner may give the same training error. Two key problems for AdaBoost are how to select the most discriminative weak learners and how to optimally combine them. To deal with these problems, error-degree-weighted training error [37] is defined based on error degree, which is related to the distances from the samples to the separating hyperplane. The most discriminative weak learners are first selected by these criteria; after getting the coefficients that are set empirically, the weak learners are optimally combined by tuning the coefficients using kernel-based perceptron.

SM-Boost [38] directly minimizes the expected binary loss, considering a stochastic decision rule parameterized by a softmax conditional distribution. This risk is minimized by performing a gradient descent in the function space linearly spanned by a base learner. SM-Boost is less sensitive to noise than AdaBoost.MH and SAMME.

A cascaded detector [97] is a sequence of detector stages to detect instances from a target class. Examples from this class are denoted positive while all others are denoted negative. Each detector is trained on the examples rejected by its predecessors. Each stage is a linear combination of weak learners. An example rejected, i.e., declared negative, by any stage is rejected by the cascade. Examples classified positive are propagated to subsequent stages. The cascade uses simple classifiers in the early stages and complex ones later on. The total number of stages and the target detection/false positive rate for each stage are selected first. A high detection rate is critical, while the false positive rate is less critical. The stages are designed with AdaBoost. Though fast and accurate, this detector is not optimal under any definition of cascade optimality.

Fast cascade boosting (FCBoost) [78] generalizes AdaBoost by minimizing a Lagrangian risk that jointly accounts for classification accuracy and speed. The concept of neutral predictors enables FCBoost to automatically determine the cascade configuration, i.e., number of stages and number of weak learners per stage, for the learned cascades, by minimization of the Lagrangian risk. The procedure is compatible with existing cost-sensitive extensions of boosting [98] that guarantee cascades of high detection rate.

L_2Boosting [11] can be interpreted as a stepwise additive learning scheme that minimizes the L_2 loss. L_2Boosting was proved to be consistent [2] and overfitting resistance [11]. However, its numerical convergence rate is slow, because the step size derived via linear search is usually not the most appropriate. ϵ-Boosting [43] specifies the step size as a fixed small positive number ϵ. RSBoosting [23] multiplies a small regularized factor to the step size deduced from linear search. RTBoosting [105] truncates the linear search in a small interval. RBoosting [59] rescales the ensemble estimator and implements linear search without any restriction on the step size in each iteration. Almost optimal numerical convergence rates of RBoosting with convex loss functions were derived [59].

25.5 Random Forests

Decision trees represent a classical induction method for classification and regression tasks. The method is simple to understand and interpret, and offers good performance. Any sequence of tests along the path from the root to a leaf represents an if-then rule. The most famous implementation is C4.5 [73]. It is hard to define a global objective to optimally learn decision trees. Trees can be learned based on Gini impurity [10] and information-theoretic considerations [73].

A C4.5 classifier inputs a collection of cases described by its n-dimensional attribute or feature vector, wherein each case (sample) is preclassified to one of the existing classes. C4.5 algorithms generate classifiers that are expressed as decision trees. Each node in the tree structure characterizes a feature, with branches representing values connecting features and leaves representing the class. An instance can be classified by tracing the path of nodes and branches to the terminating leaf. Given a set S of instances, C4.5 employs a divide-and-conquer method to grow an initial tree.

A classification and regression tree (CART) [10] is a decision tree that uses a binary recursive partitioning scheme starting with the root node, to classify a complete sample. CART algorithm is greedy because it searches for the best outcome based on the present split. The tree-growing process involves splitting among all the possible splits at each node, such that the resulting child nodes are the purest, that is, having lowest overall impurity (heterogeneity). The objective is to yield a sequence of nested pruned trees. A tree of proper size is identified by evaluating the predictive performance of every tree in the pruning sequence.

A forest is a graph where all its connected components are trees. Random forest is a popular ensemble method that extends the idea of bagged trees. In random forest method, multiple decision trees are systematically generated by randomly selecting subsets of feature spaces [44] or subsets of training instances [8]. Rotation forest method [77] uses K-axis rotations to form new features to train multiple classifiers.

A random forest [8] is a collection of identically distributed classification trees, each of which is constructed by some partitioning rule. It is formed by taking bootstrap samples from the training set. The idea in random forests is to improve the variance reduction of bagging by reducing the correlation between the trees, without introducing a significant increase in the bias. This is achieved in the tree-growing process through random selection of the input variables. As in bagging, the bias of a random forest is the same as the bias of any of the individual sampled trees. Prediction is performed by a voting mechanism among all trees. Hence, the improvements in prediction obtained by bagging or random forests are solely a result of variance reduction. Above a certain number of trees, adding more trees does not improve the performance [8].

Convergence of the generalization error is proved for random forests with growing number of trees, and there exists an upper bound for the generalization error [8].

For each bootstrap sample, a classification tree is formed, and there is no pruning—the tree grows until all terminal nodes are pure. After the tree is grown, one drops a new case down each of the trees. The classification that receives the majority vote is the one that is assigned. When used for classification, random forest obtains a class vote from each tree and then classifies using majority vote. When used for regression, the predictions from individual trees are simply averaged.

As a classifier, random forest is fully competitive with SVM. It generates an internal unbiased estimate of the generalization error. It handles missing data very well, and can maintain high levels of accuracy. It also provides estimates of the relative importance of each of the covariates in the classification rule.

For random forest, misclassification error is less sensitive to variance than MSE is. It has often been observed that boosting like random forest, does not overfit, or is slow to overfit. Random forest classifier is, in fact, a weighted version of k-NN classifier. On many problems, the performance of random forest is very similar to that of boosting, but it is simpler to train and tune. As a consequence, random forest method is popular, and is implemented in a variety of packages.

The variance estimation of bagged predictors and random forests only requires the bootstrap replicates that were used to form the bagged prediction itself, and so can be obtained with moderate computational overhead [99]. Bagged predictors are computed using a finite number B of bootstrap replicates. The variance estimation only requires $B = O(N)$ bootstrap replicates to converge, where N is the training set size [99].

Random forest algorithm can be summarized as Algorithm 25.2.

Algorithm 25.2 (Random Forest)

1. *To construct B trees, select n bootstrap samples from the dataset.*
2. *For each bootstrap sample:*
   a. *Grow a classification or regression tree.*
   b. *At each node of the tree:*
      - Select m predictor variables at random from all the predictor variables.
      - The predictor variable that provides the best split performs the binary split on that node.
      - The next node randomly selects another set of m variables from all predictor variables and performs the preceding step.
3. *Given a new dataset to be classified, take the majority vote of all the B trees.*

Random forests can be rewritten as kernel methods (called kernel based on random forests) which are more interpretable and easier to analyze [83].

25.5.1 AdaBoost Versus Random Forests

The statistics community was perplexed by two properties of AdaBoost: interpolation was achieved after relatively few iterations, and generalization error continues to decrease even after interpolation is achieved and maintained. The statistical view understands AdaBoost as a stagewise optimization of an exponential loss, which requires regularization of tree size and control on the number of iterations. the conventional wisdom requires regularization or early stopping and should be limited to low complexity classes of learners.

Random forest does not directly optimize any loss function across the entire ensemble. Each tree is grown independently, and it may optimize a criteria such as the Gini index. A random forest model is a self-averaging, weighted ensemble of interpolating classifiers by construction. A random forest seems to perform best with large trees and as many iterations as possible.

Breiman conjectured that AdaBoost emulates a random forest in its later stages [8]. AdaBoost is shown to be a weighted ensemble of interpolating classifiers [101]. These later iterations lead to an averaging effect, which causes AdaBoost to behave like a random forest [101]. Boosting should be used like random forests: with large decision trees, and without regularization or early stopping [101]. In such a view, AdaBoost is actually a "random" forest of forests. The trees in random forest and the forests in AdaBoost each interpolate the data without error [101].

Interpolation provides a kind of robustness to noise: if a classifier fits the data extremely locally, a noise point in one region will not affect the fit of the classifier at a nearby location. Fitting the training data in extremely local neighborhoods actually prevents overfitting in the presence of averaging [101]. Random forests and AdaBoost both achieve this desirable level of local interpolation by fitting deep trees. Thus reduces overfitting as the number of iterations increase.

25.6 Topics in Ensemble Learning

25.6.1 Ensemble Neural Networks

Boosting is used to combine a large number of SVMs, each trained on only a small data subsample [68]. Other parallel approaches to SVMs split the training data into subsets and distribute them among the processors. For a parallel mixture of SVMs [15], the model first trains many SVMs on small subsets and then combines their outputs using a gater such as linear hyperplane or MLP. The training time complexity can be driven down to $O(N)$. Surprisingly, this leads to a significant improvement in generalization [15].

An extended experimental analysis of bias–variance decomposition of the error in SVM is presented in [96], considering Gaussian, polynomial and dot product kernels. The bias–variance decomposition offers a rationale to develop ensemble methods using SVMs as base learners. The characterization of bias–variance decomposition of error for single SVMs also holds for bagged and random aggregating ensembles of SVMs: the main characteristics are maintained, with an overall reduction of the variance component [96].

A mixture model of linear SVMs [35] exploits a divide-and-conquer strategy by partitioning the feature space into subregions of linearly separable data points and learning a linear SVM for each of these regions. One can impose priors on the mixing coefficients and do implicit model selection in a top-down manner during the parameter estimation process. This guarantees the sparsity of the learned model. This is done by using the EM algorithm on a generative model, which permits the use of priors on the mixing coefficients.

Ensemble clustering aggregates the multiple clustering solutions into one solution that maximizes the agreement in the input ensemble [89]. Cluster ensemble is a more accurate alternative to individual clustering algorithms. The cluster ensembles considered in [53] are based on C-means clusterers. Each clusterer is assigned a random target number of clusters, k and is started from a random initialization. Vector quantization methods based on bagging and AdaBoost [87] can achieve a good performance in shorter learning times than conventional ones such as C-means and neural gas. Exact bagging of k-NN learners extends exact bagging methods from the conventional bootstrap sampling to bootstrap subsampling schemes [91].

For online learning of recurrent networks, the RTRL algorithm takes $O(n^4)$ computations for n neurons. Although EKF offers superior convergence properties to gradient descent, the computational complexity per time step for EKF is equivalent to RTRL and it also depends on RTRL derivatives. Through a sequential Bayesian filtering framework, the ensemble Kalman filter [64] is an MCMC method for estimating time evolution of the state distribution, along with an efficient algorithm for updating the state ensemble whenever a new measurement arrives. It avoids the computation of the derivatives. The ensemble Kalman filter has superior convergence properties to gradient-descent learning and EKF filtering. It reduces the computational complexity to $O(n^2)$.

25.6.2 Diversity Versus Ensemble Accuracy

The ensembles generated by existing techniques are sometimes unnecessarily large. The purpose of ensemble pruning is to search for a good subset of ensemble members that performs as well as, or better than, the original ensemble. A straightforward pruning method is to rank the classifiers according to their individual performance on a held-out test set and pick the best ones. This simple approach may work well but is theoretically unsound. For example, an ensemble of three identical classifiers with 90% accuracy is worse than an ensemble of three classifiers with 67% accuracy and least pairwise correlated error. Ensemble pruning can be viewed as a discrete version of weight-based ensemble optimization problem, which can be formulated as a quadratic integer programming problem to look for a subset of classifiers that has the optimal accuracy–diversity trade-off and SDP can be applied as a good approximate solution technique [106].

It is commonly admitted that large diversity between classifiers in a team is preferred. The most used diversity measure is certainly the Q-statistic [63]. Two statistically independent classifiers will have $Q = 0$. Q varies between -1 and 1, the lower the value the more diverse the classifiers. Classifiers that tend to recognize the same objects correctly will have positive values of Q, and those which commit errors on different objects will make Q negative.

Empirical studies show that the relationship between diversity measures and ensemble accuracy is somewhat confusing [52]. Theoretical insights show that the diversity measures are in general ineffective [92]. It has been proved that using diversity measures usually produces ensembles with large diversity, but not maximum diversity [92].

The diversity between classifiers and the individual accuracies of the classifiers clearly influence the performances of an ensemble of classifiers. An information-theoretic score is proposed in [63] to express a trade-off between individual accuracy and diversity. This technique can be directly used for selecting an optimal ensemble in a pool of classifiers. In the context of overproduction and selection of classifiers, the information-theoretic score-based selection outperforms diversity-based selection techniques.

25.6.3 Theoretical Analysis

The improved generalization capabilities of ensembles of learning machines can be interpreted in the framework of large margin classifiers [1], in the context of stochastic discrimination theory [50], and in the light of bias–variance analysis [7, 33]. Ensembles enlarge the margins, enhancing the generalization capabilities of learning algorithms [1, 81]. Ensembles can reduce variance [7] and also bias [51].

Historically, the bias–variance insight uses squared-loss as the loss function. For classification problems, where the $0/1$ loss is the main criterion, bias–variance decompositions related to the $0/1$ loss have been proposed in [7, 20, 31]. For classification problems, the $0/1$ loss function in a unified framework of bias–variance decomposition of the error is considered in [20]. Bias and variance are defined for an arbitrary loss function. Based on the unified bias–variance theory [20], methods and procedures have been proposed in [95] to evaluate and quantitatively measure the bias–variance decomposition of error in ensembles of learning machines. A bias–variance decomposition in the context of ECOC ensembles is described in [51].

The notion of margins [81] can be expressed in terms of bias and variance and vice versa [20], showing the equivalence of margin-based and bias–variance-based approaches. Bias and variance are not purely additive: Certain types of bias can be canceled by low variance to produce accurate classification [31]. This can dramatically mitigate the effect of the bias associated with some simple estimators like naive Bayes, and the bias induced by the curse of dimensionality on nearest-neighbor procedures. This explains why such simple methods are often competitive with and sometimes superior to more sophisticated ones for classification, and why bagging/aggregating classifiers can often improve accuracy [31].

25.6.4 Ensembles for Streams

Learn++ is a chunk-based ensemble approach to stationary streams [71]. The ensemble constructs new neural network models on each incoming chunk of data, and then combines their outputs using majority voting. This approach retains all previously learned classifiers, thus is inefficient for handling massive datasets. Bagging++ [107] improves on Learn++ by utilizing bagging to construct new models from incoming chunks of data. The approach gives results comparable to Learn++, but is significantly faster.

Online bagging and online boosting for stationary streams [67] alleviate the limitations of requiring the entire training set available beforehand for learning.

Chunk-based ensembles for nonstationary environments usually adapt to concept drifts by creating new base classifiers from new chunks of training examples. In general, base classifiers of the ensemble are constructed from chunks which correspond to different parts of the stream. Thus, the ensemble represents a mixture of different distributions (concepts) present in the data stream. Learning a new component from the most recent chunk is a natural way of adapting to drifts [109]. Learn++ for nonstationary environments (Learn++.NSE) [24] sets the weights of the training examples from a new data chunk based on the ensemble error on this chunk. The predictions by the ensemble are based on weighted majority voting.

Online ensembles for nonstationary streams are able to learn the data stream in one pass, potentially being faster and requiring less memory than chunk-based approaches.

25.7 Solving Multiclass Classification

A common way to model multiclass classification problems is to design a set of binary classifiers and then to combine them. ECOCs are a general framework to combine binary problems to address the multiclass problem [19].

25.7.1 *One-Against-All Strategy*

One-against-all strategy for K ($K > 2$) classes is the most simple and frequently used one. Each class is trained against the remaining $K - 1$ classes that have been collected together. For the ith two-class problem, the original K-class training data are labeled as belonging to or not belonging to class i and are used for training. Thus, a total of K binary classifiers are required. Each classifier needs to be trained on the whole training set, and there is no guarantee that good discrimination exists between one class and the remaining classes. This method also results in imbalanced data learning problems.

We determine K direct decision functions that separate one class from the remaining classes. Let the ith decision function, with the maximum margin that separates class i from the remaining classes, be $D_i(\boldsymbol{x})$. On the boundary, $D_i(\boldsymbol{x}) = 0$. To avoid the unclassifiable region, shown as shaded region in Fig. 25.1, data sample $\boldsymbol{x}$ is classified into the class with $i = \arg\max_{i=1,\ldots,K} D_i(\boldsymbol{x})$.

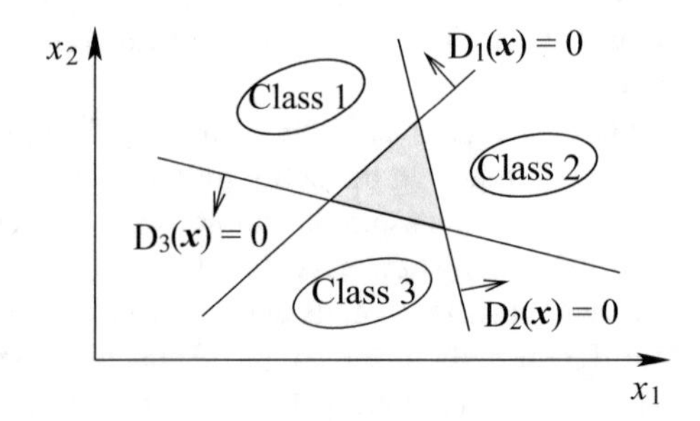

Fig. 25.1 Unclassifiable regions by the one-against-all formulation

25.7.2 One-Against-One Strategy

One-against-one (pairwise voting) strategy reduces the unclassifiable regions that occur for one-against-all strategy. But unclassifiable regions still exist. In one-against-one strategy, we determine the decision functions for all the combinations of class pairs. When determining a decision function for a class pair, we use the training data for the corresponding two classes. Thus, in each training session, the number of training data is reduced considerably compared to one-against-all strategy, which uses all the training data. One-against-one strategy needs to train $K(K-1)/2$ binary classifiers, compared to K for one-against-all strategy, and each classifier separating a pair of classes. Outputs of $K(K-1)/2$ times binary tests are required to make a final decision with majority voting. This approach is prohibitive for large K.

Let the decision function for class i against class j, with the maximum margin, be $D_{ij}(\boldsymbol{x})$. We have $D_{ij}(\boldsymbol{x}) = -D_{ji}(\boldsymbol{x})$. The regions

$$R_i = \{\boldsymbol{x} | D_{ij}(\boldsymbol{x}) > 0, j = 1, \ldots, K, j \neq i\}, \quad i = 1, \ldots, K \tag{25.6}$$

do not overlap. If $\boldsymbol{x} \in R_i$, $\boldsymbol{x}$ is considered to belong to class i. The problem that $\boldsymbol{x}$ may not be in any of R_i may occur. Therefore, we classify $\boldsymbol{x}$ by voting. By calculating

$$D_i(\boldsymbol{x}) = \sum_{j=1, j\neq i}^{K} \text{sign}(D_{ij}(\boldsymbol{x})), \tag{25.7}$$

we classify $\boldsymbol{x}$ into class $k = \arg\max_{i=1,\ldots,K} D_i(\boldsymbol{x})$.

If $\boldsymbol{x} \in R_i$, $D_i(\boldsymbol{x}) = K - 1$ and $D_j < K - 1$ for $j \neq i$. In this case, $\boldsymbol{x}$ is correctly classified. If any of $D_i(\boldsymbol{x}) \neq K - 1$, k may have multiple values. In this case, $\boldsymbol{x}$ is unclassifiable. In Fig. 25.2, the shaded region is unclassifiable, but it is much smaller than that for the one-against-all case.

Similar to the one-against-all formulation, the membership function is introduced to resolve unclassifiable regions while realizing the same classification results with those of the conventional one-against-one classification for the classifiable regions. The all-and-one approach [69] is based on the combination of the one-against-all and one-against-one methods and partially avoids their respective sources of failure.

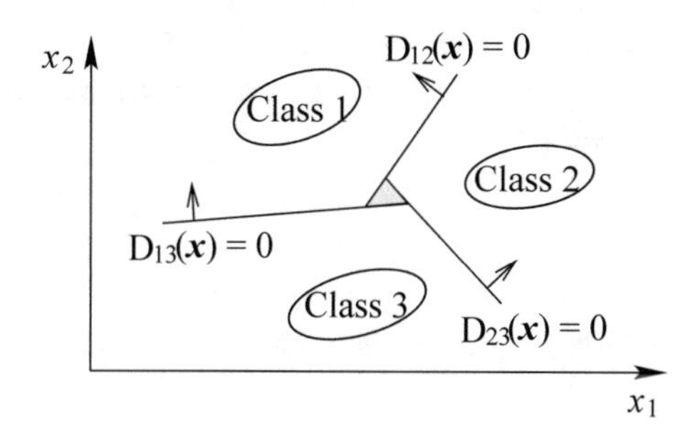

Fig. 25.2 Unclassifiable regions by the one-against-one formulation

A modification to one-against-one is made in directed acyclic graph SVM (DAGSVM) [70]. The training phase is same as that of the one-against-one method, however, in the testing phase, it uses a rooted binary directed acyclic graph with $K(K-1)/2$ internal nodes and K leaves. It requires to be evaluated only $K-1$ binary classifiers during the testing phase.

A one-versus-one-versus-rest scheme is implemented in multiclass smooth SVM [12] that decomposes the problem into $K(K-1)/2$ ternary classification subproblems based on the assumption of ternary voting games. The approach outperforms the one-against-one and one-against-rest for all datasets.

25.7.3 Error-Correcting Output Codes (ECOCs)

In the ECOC framework, at the coding step, a set of binary problems is defined based on the learning of different subpartitions of classes by means of a base classifier. Then, each of the partitions is embedded as a column of a coding matrix $\mathbf{Q}$. The rows of $\mathbf{Q}$ correspond to the codewords codifying each class. At the decoding step, a new data sample that arrives in the system is tested, and a codeword is formed as a result of the output of the binary problems. For a test sample, it looks for the most similar class codeword by using a distance metric such as the Hamming or the Euclidean distance. If the minimum Hamming distance between any pair of codewords is t, up to $\lfloor (t-1)/2 \rfloor$ single bit errors in $\mathbf{Q}$ can be corrected.

Unlike the voting procedure, the information provided by the ECOC dichotomizers is shared among classes in order to obtain a precise classification decision, being able to reduce errors caused by the variance and the bias produced by the learners [51].

In the binary ECOC framework, all positions from the coding matrix $\mathbf{Q}$ belong to the set $\{+1, -1\}$. This makes all classes to be considered by each dichotomizer as a member of one of the two possible partitions of classes that define each binary problem. In this case, the standard binary coding designs are the one-against-all strategy and the dense random strategy [19], which requires N and $10 \log_2 N$ dichotomizers, respectively [1].

In ECOCs, the main classification task is divided into a number of subtasks that are implemented by base learners. We then combine the simpler classifiers and get the final classification result. Base learners are binary classifiers with output $-1/+1$, and we have a coding matrix $\mathbf{Q}$ of $K \times L$, for K rows of binary codes of classes and L base learners d_j. ECOC can be given by a voting scheme where the entries q_{ij} are vote weights:

$$y_i = \sum_{j=1}^{L} q_{ij} d_j \tag{25.8}$$

and we choose the class with the highest y_i.

In ternary ECOCs [1], the positions of the coding matrix $\mathbf{Q}$ can be $+1$, -1 or 0. The zero symbol means that a given class is not considered in the learning process of a particular dichotomizer. The ternary framework contains a larger set of binary problems. The huge set of possible bipartitions of classes from the ternary ECOC framework has suggested the use of problem-dependent designs as well as new decoding strategies [25, 26, 72]. The coding designs in the same ternary ECOC framework are the one-against-one strategy [42], one-versus-all, and the sparse random strategy [1]. One-against-one strategy considers all possible pairs of classes. Thus, its codeword length is $K(K-1)/2$, corresponding to $K(K-1)/2$ binary classification problems. The sparse random strategy is similar to the dense random design, but it includes the third symbol 0 with another probability, and a sparse code of length $15 \log_2 K$ has been suggested in [1]. The discriminant ECOC approach [72] requires $N-1$ classifiers, where $N-1$ nodes of a binary tree structure is codified as dichotomizers of a ternary problem-dependent ECOC design.

In the ECOC framework, one-against-one coding tends to achieve higher performance than other coding designs in real multiclass problems [26]. A high percentage of the positions are coded by zero of the coding matrix, which implies a high sparseness degree. The zero symbol introduces two kinds of biases that require redefinition of the decoding design [26]. A type of decoding measure and two decoding strategies are defined. These decoding strategies avoid the bias produced by the zero symbol and all the codewords work in the same dynamic range, thus significant performance improvement is obtained on the ECOC designs [26]. A general extension of the ECOC framework to the online learning scenario is given in [27]. The final classifier handles the addition of new classes independently of the base classifier used. The online ECOC approaches tend to the results of the batch approach, and they provide a feasible and robust way for handling new classes using any base classifier.

In ECOC strategy, ECOCs are employed to improve the decision accuracy. Although the codewords generated by ECOCs have good error-correcting capabilities, some subproblems generated may be difficult to learn [1]. The simpler one-against-all and one-against-one strategies can present results comparable or superior to those produced by ECOC strategy in several applications. ECOC strategy at least needs K times the number of tests. A MATLAB ECOC library [26] contains both state-of-the-art coding and decoding designs.

The combination of several binary ones in ECOC strategy is typically done via a simple nearest-neighbor rule that finds the class that is closest to the outputs of the binary classifiers. For these nearest-neighbor ECOCs, existing bounds on the error rate of the multiclass classifier is improved given the average binary distance [49]. The results show as to why elimination and Hamming decoding often achieve the same accuracy. In addition to generalization improvement, ECOCs can be used to resolve unclassifiable regions.

For ECOCs, let g_{ij} be the target value of $D_{ij}(\boldsymbol{x})$, the jth decision function for class i:

$$g_{ij} = \begin{cases} 1, & \text{if } D_{ij}(\boldsymbol{x}) > 0, \\ -1, & \text{otherwise.} \end{cases}, \quad i = 1, \ldots, K. \tag{25.9}$$

The jth column vector $g_j = (g_{1j}, \ldots, g_{Kj})^T$ is the target vector for the jth decision function. If all the elements of a column are 1 or -1, classification is not performed by this decision function and two column vectors with $g_i = -g_j$ result in the same decision function. Thus the maximum number of distinct decision functions is $2^{K-1} - 1$.

The ith row vector $(g_{i1}, \ldots, g_{ik})$ corresponds to a codeword for class i, where k is the number of decision functions. In error-correcting codes, if the minimum Hamming distance between pairs of codewords is t, the code can correct at least $\lfloor (t-1)/2 \rfloor$-bit errors, where $\lfloor a \rfloor$ gives the maximum integer value that does not exceed a. For three-class problems, there are at most three decision functions as $\left[g_{ij}\right] = \begin{bmatrix} 1 & -1 & -1 \\ -1 & 1 & -1 \\ -1 & -1 & 1 \end{bmatrix}$, which is equivalent to one-against-all formulation, and there is no error-correcting capability. By introducing don't-care outputs 0, one-against-all, one-against-one and ECOC schemes are unified [1]. One-against-one classification for three classes can be shown as $\left[g_{ij}\right] = \begin{bmatrix} 1 & 0 & -1 \\ -1 & 1 & 0 \\ 0 & -1 & 1 \end{bmatrix}$.

25.8 Dempster–Shafer Theory of Evidence

Dempster–Shafer theory of evidence, which originates from the upper and lower probabilities, was first proposed by Dempster [16] and then further developed by Shafer [85]. It can be viewed as a generalization of the Bayesian probability calculus for combining probability judgements based on different bodies of evidence. The Dempster–Shafer method combines evidence regarding the truth of a hypothesis from different sources. It is the most frequently used fusion method at the decision-making level. The Dempster–Shafer method uses *belief* representing the extent to which the evidence supports a hypothesis and *plausibility* representing the extent to which the evidence fails to refute the hypothesis. These resemble necessity and possibility in fuzzy logic. Dempster–Shafer theory uses intervals for representing imprecise probabilities, which is in the same spirit as interval-valued fuzzy sets.

Definition 25.1 (*Frame of Discernment*) A frame of discernment Θ is defined as a finite and exhaustive set of N mutually exclusive singleton hypotheses that constitutes a source's scope of expertise

$$\Theta = \{\mathcal{A}_1, \mathcal{A}_2, \ldots, \mathcal{A}_N\}. \tag{25.10}$$

A hypothesis $\mathcal{A}_i$, referred to as a *singleton*, is the lowest level of discernible information.

Definition 25.2 (*Basic Probability Assignment Function*) The basic probability assignment function $m(\mathcal{A})$ of event (or proposition) $\mathcal{A}$ is defined such that that the mapping: 2^{Θ} (all events of Θ) $\rightarrow [0, 1]$ must satisfy $m(\emptyset) = 0$ and

$$\sum_{\mathcal{A}\subseteq\Theta} m(\mathcal{A}) = 1. \tag{25.11}$$

The power set of Θ, 2^Θ, contains all subsets of Θ. For example, if $\Theta = \{\mathcal{A}_1, \mathcal{A}_2, \mathcal{A}_3\}$, then $2^\Theta = \{\emptyset, \mathcal{A}_1, \mathcal{A}_2, \mathcal{A}_3, \mathcal{A}_1 \cup \mathcal{A}_2, \mathcal{A}_1 \cup \mathcal{A}_3, \mathcal{A}_2 \cup \mathcal{A}_3, \Theta\}$.

With any event $\mathcal{A}$ in the frame of discernment Θ, if the basic probability assignment $m(\mathcal{A}) > 0$, the event $\mathcal{A}$ is called a *focal element* of m. If m has a single focal element $\mathcal{A}$, it is said to be categorical and denoted as $m_\mathcal{A}$. If all focal elements of m are singletons, then m is said to be Bayesian.

Values of a basic probability assignment function are called *belief masses*. A basic probability assignment function with $m(\emptyset) = 0$ is said to be *normal*.

Definition 25.3 (*Belief Function*) With any hypothesis $\mathcal{A} \subseteq \Theta$, its belief function $Bel(\mathcal{A}) : 2^\Theta \to [0, 1]$ is defined as the sum of the corresponding basic probabilities of all its subsets, namely

$$Bel(\mathcal{A}) = \sum_{\mathcal{B}\subseteq\mathcal{A}} m(\mathcal{B}), \forall \mathcal{A} \subseteq \Theta. \tag{25.12}$$

The belief function, which is also called lower limit function, represents the minimal support for $\mathcal{A}$ and can be interpreted as a global measure of one's belief that hypothesis $\mathcal{A}$ is true. From the definition, we know $Bel(\emptyset) = 0$, $Bel(\Theta) = 1$. Note that the basic probability assignment and belief functions are in one-to-one correspondence.

Definition 25.4 (*Plausibility Function*) The plausibility function of $\mathcal{A}$,$Pl(\mathcal{A})$: $2^\Theta \to [0, 1]$, is the amount of belief not committed to the negation of $\mathcal{A}$

$$Pl(\mathcal{A}) = 1 - Bel(\bar{\mathcal{A}}) = \sum_{\mathcal{B}\cap\mathcal{A}\neq\emptyset} m(\mathcal{B}), \quad \forall \mathcal{A} \subseteq \Theta. \tag{25.13}$$

The plausibility function, which is also called upper limit function, expresses the greatest potential belief degree in $\mathcal{A}$. The plausibility function is a possibility measure, and the belief function is the dual necessity measure. It can be proved that [85]

$$Pl(\mathcal{A}) \geq Bel(\mathcal{A}), \quad \forall \mathcal{A} \subseteq \Theta. \tag{25.14}$$

If m is Bayesian, then function Bel is identical to Pl and it is a probability measure.

Definition 25.5 (*Uncertainty Function*) The uncertainty of $\mathcal{A}$ is defined by

$$U(\mathcal{A}) = Pl(\mathcal{A}) - Bel(\mathcal{A}). \tag{25.15}$$

Definition 25.6 (*Commonality Function*) The commonality function states how much basic probability assignment is committed to $\mathcal{A}$ and all of its supersets

$$Com(\mathcal{A}) = \sum_{\mathcal{B} \supseteq \mathcal{A}} m(\mathcal{B}). \tag{25.16}$$

Belief functions are widely used formalisms for uncertainty representation and processing. For a combination of beliefs, Dempster's rule of combination is used in Dempster–Shafer theory. Under strict probabilistic assumptions, its results are probabilistically correct and interpretable. Dempster's rule, also called *orthogonal summing rule*, produces a new belief (probability) represented by a basic probability assignment output by synthesizing the basic probability assignments from many sources. Assuming that m_1 and m_2 are basic probability assignment values from two different information sources in the same frame of discernment, we have [85]

$$m(\emptyset) = 0, \quad m(\mathcal{A}) = m_1 \oplus m_2 = \frac{1}{1-K} \sum_{\mathcal{B} \cap \mathcal{C} = \mathcal{A}} m_1(\mathcal{B}) m_2(\mathcal{C}), \tag{25.17}$$

where

$$K = \sum_{\mathcal{B} \cap \mathcal{C} = \emptyset} m_1(\mathcal{B}) m_2(\mathcal{C}) > 0 \tag{25.18}$$

measures the conflict between various evidence sources. If $K = 1$, the two pieces of evidence are logically contradictory and they cannot be combined. In order to apply Dempster's rule in the presence of highly conflicting beliefs, all conflicting belief masses can be allocated to a missing (empty) event based on the open-world assumption. The open-world assumption states that some possible event must have been overlooked and thus is missing in the frame of discernment [90].

In general, within the same frame of discernment Θ, the combining result of n basic probability assignment values $m_1, m_2, \ldots, m_n$ is given by

$$m(\mathcal{A}) = m_1 \oplus m_2 \oplus \cdots m_n = \frac{1}{1-K} \sum_{\cap \mathcal{A}_i = \mathcal{A}} \left(\prod_{i=1}^{n} m(\mathcal{A}_i) \right), \tag{25.19}$$

$$K = \sum_{\cap \mathcal{A}_i = \emptyset} \left(\prod_{i=1}^{n} m(\mathcal{A}_i) \right). \tag{25.20}$$

The advantage of the Dempster–Shafer over the Bayesian method is that the Dempster–Shaferr method does not require prior probabilities; it combines current evidence. The Dempster–Shafer method fails for fuzzy systems, since it requires the hypotheses to be mutually exclusive.

Dempster's rule assumes the classifiers to be independent. For combining nonindependent classifiers, the cautious rule and, more generally, t-norm-based rules with behavior ranging between Dempster's rule and the cautious rule can be used [74]. An optimal combination scheme can be learned based on a parameterized family of t-norms.

Dempster's rule considers all the sources equally reliable. Counterintuitive results are obtained in some cases, especially when there is high conflict among bodies of evidence [104]. A degree of falsity is defined based on the conflict coefficient obtained using Dempster's rule [82]. Based on the degree of falsity, discounting factors for evidence combination can be generated.

When implementing the decision-making level fusion with Dempster–Shafer theory, it is required to set up corresponding basic probability assignments. To avoid the trouble of establishing basic probability assignments, MLP can be used as an aid. The independent diagnosis of each category of feature data is first conducted using MLP. The MLP outputs are processed by normalization and then are taken as the basic probability assignments of tremor types.

Logistic regression and multilayer feedforward networks can be viewed as converting input or higher level features into mass functions and aggregating them by Dempster's rule of combination [17]. The probabilistic outputs of these classifiers are the normalized plausibilities corresponding to the combined mass function.

Problems

25.1 An ordered weighted averaging operator of dimension n is a mapping OWA: $R^n \rightarrow R$ that has an associated weighting vector $\boldsymbol{w}$ of dimension n with $\sum_{j=1}^{n} w_j = 1$ and $w_j \in [0, 1]$, such that

$$\text{OWA}\,(a_1, a_2, \ldots, a_n) = \sum_{j=1}^{n} w_j b_j,$$

where b_j is the jth largest of the a_i.

(a) Show that it has the maximum, the minimum and the average as special cases.

(b) Show that the operator is commutative, monotonic, bounded, and idempotent.

(c) Show that the median operator can be expressed an OWA function with the weighting vector $\boldsymbol{w} = (0, \ldots, 0, 1, 0, \ldots, 0)^T$ for odd n and $\boldsymbol{w} = (0, \ldots, 0, \frac{1}{2},$
$\frac{1}{2}, 0, \ldots, 0)^T$ for even n.

(d) Show that for the voting scheme of removing the maximum and minimum votes and averaging over the remaining votes, the weighting vector is given by $\boldsymbol{w} = (0, \frac{1}{n-2}, \ldots, \frac{1}{n-2}, 0)$.

25.2 Describe the difference between bagging and boosting.

25.3 Consider the dataset of two-dimensional patterns: (1, 1, 1), (1, 2, 1), (2, 1, 1), (2, 2, 1), (4, 3, 2), (4, 2, 2), (5, 2, 2), (5, 3, 2), (4, 4, 3), (5, 4, 3), (5, 5, 3),

(4, 5, 3), where each pattern is represented by two features and the class label. Implement bagging on the dataset and classify the test pattern (3.5, 2.5) by the following steps:

(a) Select two patterns from each class at random and use the 1-NN algorithm to classify the test pattern.
(b) Perform this procedure five times and classify the test pattern according to majority voting.

25.4 In the cascading method, it is required that $\theta_{j+1} \geq \theta_j$. Give an explanation as to why this is required.

25.5 Consider the dataset of patterns: (1, 1, 1), (1, 2, 1), (2, 1, 1), (2, 2, 1), (3.5, 2, 2), (4, 1.5, 2), (4, 2, 2), (5, 1, 2), where each pattern is represented by two features and the class label. Classify a pattern with two features (3.1, 1) by using AdaBoost with the following weak classifiers:

(a) If $x_1 \leq 2$ then the pattern belongs to class 1; else to class 2.
(b) If $x_1 \leq 3$ then the pattern belongs to class 1; else to class 2.
(c) If $x_1 + x_2 \leq 4$ then the pattern belongs to class 1; else to class 2.

25.6 Write a program implementing AdaBoost.

25.7 Show that in case of AdaBoost, for a given performance level of a strong learner, the more discriminative each weak learner is, the less the number of weak learners needed and the shorter the training time consumed.

References

1. Allwein, E. L., Schapire, R. E., & Singer, Y. (2000). Reducing multiclass to binary: A unifying approach for margin classifiers. *Journal of Machine Learning Research, 1*, 113–141.
2. Bartlett, P. L., & Traskin, M. (2007). AdaBoost is consistent. *Journal of Machine Learning Research, 8*(1), 2347–2368.
3. Bauer, E., & Kohavi, R. (1999). An empirical comparison of voting classification algorithms: Bagging, boosting, and variants. *Machine Learning, 36*, 105–139.
4. Bellet, A., Habrard, A., Morvant, E., & Sebban, M. (2014). Learning a priori constrained weighted majority votes. *Machine Learning, 97*, 129–154.
5. Biau, G., Cadre, B., & Rouviere, L. (2019). Accelerated gradient boosting. *Machine Learning, 108*, (6), 971–992.
6. Breiman, L. (1996). Bagging predictors. *Machine Learning, 24*(2), 123–140.
7. Breiman, L. (1996). *Bias variance and arcing classifiers* (Technical report TR 460). Berkeley, CA: Statistics Department, University of California.
8. Breiman, L. (2001). Random forests. *Machine Learning, 45*(1), 5–32.
9. Breiman, L. (2004). Population theory for predictor ensembles. *Annals of Statistics, 32*(1), 1–11.
10. Breiman, L., Friedman, J. H., Olshen, R. A., & Stone, C. J. (1984). *Classification and regression trees*. London: Chapman & Hall/CRC.
11. Buhlmann, P., & Yu, B. (2003). Boosting with the L_2 loss: Regression and classification. *Journal of the American Statistical Association, 98*(462), 324–339.

12. Chang, C.-C., Chien, L.-J., & Lee, Y.-J. (2011). A novel framework for multi-class classification via ternary smooth support vector machine. *Pattern Recognition*, *44*, 1235–1244.
13. Clarke, B. (2003). Comparing Bayes model averaging and stacking when model approximation error cannot be ignored. *Journal of Machine Learning Research*, *4*, 683–712.
14. Collins, M., Schapire, R. E., & Singer, Y. (2002). Logistic regression, AdaBoost and Bregman distances. *Machine Learning*, *47*, 253–285.
15. Collobert, R., Bengio, S., & Bengio, Y. (2002). A parallel mixture of SVMs for very large scale problems. *Neural Computation*, *14*, 1105–1114.
16. Dempster, A. P. (1967). Upper and lower probabilities induced by multivalued mappings. *Annals of Mathematics and Statistics*, *38*, 325–339.
17. Denoeux, T. (2019). Logistic regression, neural networks and Dempster-Shafer theory: A new perspective. *Knowledge-Based Systems*, *176*, 54–67.
18. Dietterich, T. G. (2000). An experimental comparison of three methods for constructing ensembles of decision trees: Bagging, boosting, and randomization. *Machine Learning*, *40*(2), 139–158.
19. Dietterich, T. G., & Bakiri, G. (1995). Solving multiclass learning problems via error-correcting output codes. *Journal of Artificial Intelligence Research*, *2*, 263–286.
20. Domingos, P. (2000). A unified bias-variance decomposition for zero-one and squared loss. In *Proceedings of the 17th National Conference on Artificial Intelligence* (pp. 564–569). Austin, TX.
21. Domingos, P. (2000). Bayesian averaging of classifiers and the overfitting problem. In *Proceedings of the 17th International Conference on Machine Learning* (pp. 223–230). San Mateo, CA: Morgan Kaufmann.
22. Du, K.-L., & Swamy, M. N. S. (2010). *Wireless communication systems*. Cambridge, UK: Cambridge University Press.
23. Ehrlinger, J., & Ishwaran, H. (2012). Characterizing L2 boosting. *Annals of Statistics*, *40*(2), 1074–1101.
24. Elwell, R., & Polikar, R. (2011). Incremental learning of concept drift in nonstationary environments. *IEEE Transactions on Neural Networks*, *22*(10), 1517–1531.
25. Escalera, S., Tax, D., Pujol, O., Radeva, P., & Duin, R. (2008). Subclass problem dependent design of error-correcting output codes. *IEEE Transactions on Pattern Analysis and Machine Intelligence*, *30*, 1041–1054.
26. Escalera, S., Pujol, O., & Radeva, P. (2010). On the decoding process in ternary error-correcting output codes. *IEEE Transactions on Pattern Analysis and Machine Intelligence*, *32*(1), 120–134.
27. Escalera, S., Masip, D., Puertas, E., Radeva, P., & Pujol, O. (2011). Online error correcting output codes. *Pattern Recognition Letters*, *32*, 458–467.
28. Freund, Y. (2001). An adaptive version of the boost by majority algorithm. *Machine Learning*, *43*, 293–318.
29. Freund, Y., & Schapire, R. E. (1996). Experiments with a new boosting algorithm. In *Proceedings of the 13th International Conference on Machine Learning* (pp. 148–156). San Mateo, CA: Morgan Kaufmann.
30. Freund, Y., & Schapire, R. E. (1997). A decision-theoretic generalization of on-line learning and an application to boosting. *Journal of Computer and System Sciences*, *55*(1), 119–139.
31. Friedman, J. H. (1997). On bias, variance, 0/1-loss, and the curse-of-dimensionality. *Data Mining and Knowledge Discovery*, *1*, 55–77.
32. Friedman, J. H. (2001). Greedy function approximation: A gradient boosting machine. *Annals of Statistics*, *29*, 1189–1232.
33. Friedman, J., & Hall, P. (2000). *On bagging and nonlinear estimation* (Technical report). Stanford, CA: Statistics Department, Stanford University.
34. Friedman, J., Hastie, T., & Tibshirani, R. (2000). Additive logistic regression: A statistical view of boosting. *Annals of Statistics*, *28*(2), 337–407.
35. Fu, Z., Robles-Kelly, A., & Zhou, J. (2010). Mixing linear SVMs for nonlinear classification. *IEEE Transactions on Neural Networks*, *21*(12), 1963–1975.

36. Gambs, S., Kegl, B., & Aimeur, E. (2007). Privacy-preserving boosting. *Data Mining and Knowledge Discovery*, *14*, 131–170.
37. Gao, C., Sang, N., & Tang, Q. (2010). On selection and combination of weak learners in AdaBoost. *Pattern Recognition Letters*, *31*, 991–1001.
38. Geist, M. (2015). Soft-max boosting. *Machine Learning*, *100*, 305–332.
39. Gelman, A., Carlin, J. B., Stern, H. S., & Rubin, D. B. (2004). *Bayesian data analysis*. London: Chapman & Hall/CRC.
40. Germain, P., Lacasse, A., Laviolette, F., Marchand, M., & Roy, J.-F. (2015). Risk bounds for the majority vote: From a PAC-Bayesian analysis to a learning algorithm. *Journal of Machine Learning Research*, *16*, 787–860.
41. Guestrin, C. (2006). PAC-learning, VC dimension and margin-based bounds. *Machine Learning - 10701/15781*, Carnegie Mellon University.
42. Hastie, T., & Tibshirani, R. (1998). Classification by pairwise grouping. In *Advances in neural information processing systems* (Vol. 11, pp. 451–471). Cambridge, MA: MIT Press.
43. Hastie, T., Taylor, J., Tibshirani, R., & Walther, G. (2007). Forward stagewise regression and the monotone lasso. *Electronic Journal of Statistics*, *1*, 1–29.
44. Ho, T. K. (1995). Random decision forests. In *Proceedings of the International Conference on Document Analysis and Recognition* (pp. 278–282). Washington, DC.
45. Ho, T. K. (1998). The random subspace method for constructing decision forests. *IEEE Transactions on Pattern Analysis and Machine Intelligence*, *20*(8), 832–844.
46. Jacobs, R. A., Jordan, M. I., Nowlan, S. J., & Hinton, G. E. (1991). Adaptive mixtures of local experts. *Neural Computation*, *3*(1), 79–87.
47. Jiang, W. (2000). The VC dimension for mixtures of binary classifiers. *Neural Computation*, *12*, 1293–1301.
48. Kanamori, T., Takenouchi, T., Eguchi, S., & Murata, N. (2007). Robust loss functions for boosting. *Neural Computation*, *19*, 2183–2244.
49. Klautau, A., Jevtic, N., & Orlitsky, A. (2003). On nearest-neighbor error-correcting output codes with application to all-pairs multiclass support vector machines. *Journal of Machine Learning Research*, *4*, 1–15.
50. Kleinberg, E. (2000). On the algorithmic implementation of stochastic discrimination. *IEEE Transactions on Pattern Analysis and Machine Intelligence*, *22*(5), 473–490.
51. Kong, E., & Dietterich, T. G. (1995). Error-correcting output coding correct bias and variance. In *Proceedings of the 12th International Conference on Machine Learning* (pp. 313–321). San Francisco, CA: Morgan Kauffmanm.
52. Kuncheva, L. I., & Whitaker, C. J. (2003). Measures of diversity in classifier ensembles and their relationship with the ensemble accuracy. *Machine Learning*, *51*(2), 181–207.
53. Kuncheva, L. I., & Vetrov, D. P. (2006). Evaluation of stability of k-means cluster ensembles with respect to random initialization. *IEEE Transactions on Pattern Analysis and Machine Intelligence*, *28*(11), 1798–1808.
54. Lacasse, A., Laviolette, F., Marchand, M., Germain, P., & Usunier, N. (2006). PAC-Bayes bounds for the risk of the majority vote and the variance of the Gibbs classifier. In *Advances in neural information processing systems* (Vol. 19, pp. 769–776).
55. Langford, J., & Shawe-Taylor, J. (2002). PAC-Bayes & margins. In *Advances in neural information processing systems* (Vol. 15, pp. 423–430).
56. Laviolette, F., Marchand, M., & Roy, J.-F. (2011). From PAC-Bayes bounds to quadratic programs for majority votes. In *Proceedings of the 28th International Conference on Machine Learning* (pp. 649–656). Bellevue, WA.
57. Lee, H. K. H., & Clyde, M. A. (2004). Lossless online Bayesian bagging. *Journal of Machine Learning Research*, *5*, 143–151.
58. Li, S. Z., & Zhang, Z. (2004). FloatBoost learning and statistical face detection. *IEEE Transactions on Pattern Analysis and Machine Intelligence*, *26*(9), 1112–1123.
59. Lin, S., Wang, Y., & Xu, L. (2015). Re-scale boosting for regression and classification. arXiv:1505.01371.

60. Mease, D., & Wyner, A. (2008). Evidence contrary to the statistical view of boosting. *Journal of Machine Learning Research*, *9*, 131–156.
61. McAllester, D. (2003). Simplified PAC-Bayesian margin bounds. In *Computational learning theory and kernel machines*. LNCS (Vol. 2777, pp. 203–215).
62. McAllester, D. A. (1999). PAC-Bayesian model averaging. In *Proceedings of the 12th ACM Annual Conference on Computational Learning Theory* (pp. 164–170).
63. Meynet, J., & Thiran, J.-P. (2010). Information theoretic combination of pattern classifiers. *Pattern Recognition*, *43*, 3412–3421.
64. Mirikitani, D. T., & Nikolaev, N. (2010). Efficient online recurrent connectionist learning with the ensemble Kalman filter. *Neurocomputing*, *73*, 1024–1030.
65. Muhlbaier, M. D., Topalis, A., & Polikar, R. (2009). Learn++.NC: Combining ensemble of classifiers with dynamically weighted consult-and-vote for efficient incremental learning of new classes. *IEEE Transactions on Neural Networks*, *20*(1), 152–168.
66. Mukherjee, I., Rudin, C., & Schapire, R. E. (2013). The rate of convergence of AdaBoost. *Journal of Machine Learning Research*, *14*, 2315–2347.
67. Oza, N. C., & Russell, S. (2001). Online bagging and boosting. In T. Richardson & T. Jaakkola (Eds.), *Proceedings of the 18th International Workshop on Artificial Intelligence and Statistics (AISTATS)* (pp. 105–112). Key West, FL. San Mateo, CA: Morgan Kaufmann.
68. Pavlov, D., Mao, J., & Dom, B. (2000). Scaling-up support vector machines using boosting algorithm. In *Proceedings of the 15th International Conference on Pattern Recognition* (pp. 2219–2222). Barcelona, Spain.
69. Pedrajas, N. G., & Boyer, D. O. (2006). Improving multiclass pattern recognition by the combination of two strategies. *IEEE Transactions on Pattern Analysis and Machine Intelligence*, *28*(6), 1001–1006.
70. Platt, J. C., Christiani, N., & Shawe-Taylor, J. (1999). Large margin DAGs for multiclass classification. In S. A. Solla, T. K. Leen, & K. R. Muller (Eds.), *Advances in neural information processing systems* (Vol. 12, pp. 547–553). Cambridge, MA: MIT Press.
71. Polikar, R., Upda, L., Upda, S. S., & Honavar, V. (2001). Learn++: An incremental learning algorithm for supervised neural networks. *IEEE Transactions on Systems Man and Cybernetics Part C*, *31*(4), 497–508.
72. Pujol, O., Radeva, P., & Vitria, J. (2006). Discriminant ECOC: A heuristic method for application dependent design of error correcting output codes. *IEEE Transactions on Pattern Analysis and Machine Intelligence*, *28*, 1001–1007.
73. Quinlan, J. R. (1986). Induction of decision trees. *Machine Learning*, *1*, 81–106.
74. Quost, B., Masson, M.-H., & Denoeux, T. (2011). Classifier fusion in the Dempster-Shafer framework using optimized t-norm based combination rules. *International Journal of Approximate Reasoning*, *52*, 353–374.
75. Ratsch, G., & Warmuth, M. K. (2005). Efficient margin maximizing with boosting. *Journal of Machine Learning Research*, *6*, 2153–2175.
76. Ratsch, G., Onoda, T., & Muller, K.-R. (2001). Soft margins for AdaBoost. *Machine Learning*, *43*(3), 287–320.
77. Rodriguez, J. J., Kuncheva, L. I., & Alonso, C. J. (2006). Rotation forest: A new classifier ensemble method. *IEEE Transactions on Pattern Analysis and Machine Intelligence*, *28*(10), 1619–1630.
78. Saberian, M., & Vasconcelos, N. (2014). Boosting algorithms for detector cascade learning. *Journal of Machine Learning Research*, *15*, 2569–2605.
79. Schapire, R. E. (1990). The strength of weak learnability. *Machine Learning*, *5*, 197–227.
80. Schapire, R. E., & Singer, Y. (1999). Improved boosting algorithms using confidence-rated predictions. *Machine Learning*, *37*(3), 297–336.
81. Schapire, R. E., Freund, Y., Bartlett, P. L., & Lee, W. S. (1998). Boosting the margin: A new explanation for the effectiveness of voting methods. *Annals of Statistics*, *26*(5), 1651–1686.
82. Schubert, J. (2011). Conflict management in Dempster-Shafer theory using the degree of falsity. *International Journal of Approximate Reasoning*, *52*(3), 449–460.

83. Scornet, E. (2016). Random forests and kernel methods. *IEEE Transactions on Information Theory*, *62*(3), 1485–1500.
84. Servedio, R. A. (2003). Smooth boosting and learning with malicious noise. *Journal of Machine Learning Research*, *4*, 633–648.
85. Shafer, G. (1976). *A mathematical theory of evidence*. Princeton, NJ: Princeton University Press.
86. Shalev-Shwartz, S., & Singer, Y. (2010). On the equivalence of weak learnability and linear separability: New relaxations and efficient boosting algorithms. *Machine Learning*, *80*, 141–163.
87. Shigei, N., Miyajima, H., Maeda, M., & Ma, L. (2009). Bagging and AdaBoost algorithms for vector quantization. *Neurocomputing*, *73*, 106–114.
88. Shrestha, D. L., & Solomatine, D. P. (2006). Experiments with AdaBoost.RT, an improved boosting scheme for regression. *Neural Computation*, *18*, 1678–1710.
89. Singh, V., Mukherjee, L., Peng, J., & Xu, J. (2010). Ensemble clustering using semidefinite programming with applications. *Machine Learning*, *79*, 177–200.
90. Smets, P. (1990). The combination of evidence in the transferable belief model. *IEEE Transactions on Pattern Analysis and Machine Intelligence*, *12*(5), 447–458.
91. Steele, B. M. (2009). Exact bootstrap k-nearest neighbor learners. *Machine Learning*, *74*, 235–255.
92. Tang, E. K., Suganthan, P. N., & Yao, X. (2006). An analysis of diversity measures. *Machine Learning*, *65*(1), 247–271.
93. Tresp, V. (2000). A Bayesian committee machine. *Neural Computation*, *12*, 2719–2741.
94. Tumer, K., & Ghosh, J. (1996). Analysis of decision boundaries in linearly combined neural classifiers. *Pattern Recognition*, *29*(2), 341–348.
95. Valentini, G. (2005). An experimental bias-variance analysis of SVM ensembles based on resampling techniques. *IEEE Transactions on Systems, Man, and Cybernetics, Part B*, *35*(6), 1252–1271.
96. Valentini, G., & Dietterich, T. G. (2004). Bias-variance analysis of support vector machines for the development of SVM-based ensemble methods. *Journal of Machine Learning Research*, *5*, 725–775.
97. Viola, P., & Jones, M. (2001). Robust real-time object detection. *International Journal of Computer Vision*, *57*(2), 137–154.
98. Viola, P., & Jones, M. (2002). Fast and robust classification using asymmetric AdaBoost and a detector cascade. In *Advances in neural information processing systems* (Vol. 14, pp. 1311–1318).
99. Wager, S., Hastie, T., & Efron, B. (2014). Confidence intervals for random forests: The Jackknife and the infinitesimal Jackknife. *Journal of Machine Learning Research*, *15*, 1625–1651.
100. Wolpert, D. H. (1992). Stacked generalization. *Neural Networks*, *5*, 241–259.
101. Wyner, A. J., Olson, M., Bleich, J., & Mease, D. (2017). Explaining the success of AdaBoost and random forests as interpolating classifiers. *Journal of Machine Learning Research*, *18*, 1–33.
102. Xu, L., Krzyzak, A., & Suen, C. Y. (1992). Methods of combining multiple classifiers and their applications to handwriting recognition. *IEEE Transactions on Systems, Man, and Cybernetics*, *22*, 418–435.
103. Yager, R. R. (1988). On ordered weighted averaging aggregation operators in multicriteria decision-making. *IEEE Transactions on Systems, Man, and Cybernetics*, *18*(1), 183–190.
104. Zadeh, L. A. (1986). A simple view of the Dempster-Shafer theory of evidence and its implication for the rule of combination. *AI Magazine*, *2*, 85–90.
105. Zhang, T., & Yu, B. (2005). Boosting with early stopping: Convergence and consistency. *Annals of Statistics*, *33*(4), 1538–1579.
106. Zhang, Y., Burer, S., & Street, W. N. (2006). Ensemble pruning via semi-definite programming. *Journal of Machine Learning Research*, *7*, 1315–1338.

107. Zhao, Q., Jiang, Y., & Xu, M. (2010). Incremental learning by heterogeneous bagging ensemble. In *Proceedings of the 6th International Conference on Advanced Data Mining and Applications* (Vol. 2, pp. 1–12). Chongqing, China.
108. Zhu, J., Zou, H., Rosset, S., & Hastie, T. (2009). Multi-class AdaBoost. *Statistics and Its Interface*, *2*, 249–360.
109. Zliobaite, I. (2010). *Adaptive training set formation*. Ph.D. thesis, Vilnius University.

Chapter 26
Introduction to Fuzzy Sets and Logic

26.1 Introduction

The concept of fuzzy sets was first proposed by Zadeh [44]. The theory of fuzzy sets and logic, as a mathematical extension to classical theory of sets and binary logic, has become a general mathematical tool for data analysis. Fuzzy sets serve as information granules quantifying a given input or output variable, and fuzzy logic is a means of knowledge representation. In fuzzy logic, the knowledge of experts is modeled by linguistic IF-THEN rules, which build up a fuzzy inference system. Some fuzzy inference systems have universal function approximation capability; they can be used in many areas where neural networks are applicable. An exact model is not needed for model design. Fuzzy logic has also been widely applied in control, data analysis, regression, and signal and image processing.

Knowledge-based systems represent a different perspective on the human brain's epigenetic process. Unlike neural networks, there is no attempt to model the physical structure; an inference engine fulfills that role. Learning is modeled by the construction of rules that are produced under the guidance of domain experts and held in a knowledge base. The abstract process of reasoning occurs when the inference engine fires rules as a result of data input. This explicit knowledge representation and reasoning offer the advantage that knowledge can be updated dynamically. A knowledge-based system can provide a rationale behind its decisions.

Fuzzy logic uses the notion of membership. It is most suitable for the representation of vague data and concepts on an intuitive basis such as human linguistic description, e.g., the expressions *approximately*, *good*, *strong*. The conventional or *crisp* set can be treated as a special case of a fuzzy set. A fuzzy set is uniquely determined by its membership function, and it is also associated with a linguistically meaningful term.

Fuzzy logic provides a systematic framework to incorporate human experience. It is based on three core concepts, namely fuzzy sets, linguistic variables, and possibility distributions. A fuzzy set is an effective means to represent linguistic variables. A linguistic variable is a variable, whose value can be described qualitatively using a

K.-L. Du and M. N. S. Swamy, *Neural Networks and Statistical Learning*,
https://doi.org/10.1007/978-1-4471-7452-3_26

linguistic expression and quantitatively using a membership function [45]. Linguistic expressions are useful for communicating concepts and knowledge with human beings, whereas membership functions are useful for processing numeric input data. When a fuzzy set is assigned to a linguistic variable, it imposes an elastic constraint, called a *possibility distribution*, on the possible values of the variable.

Fuzzy logic is a rigorous mathematical discipline. Fuzzy reasoning is a straightforward formalism for encoding human knowledge or common sense in a numerical framework, and fuzzy inference systems can approximate arbitrarily well any continuous function on a compact domain [17, 41]. Fuzzy inference systems and feedforward networks can approximate each other to any degree of accuracy [5]. In [15], the Mamdani model and feedforward networks are shown to be able to approximate each other to an arbitrary accuracy. Gaussian-based Mamdani systems have the ability of approximating any sufficiently smooth function and reproducing its derivatives up to any order [10]. The functional equivalence between a multilayer feedforward network and a zero-order Takagi–Sugeno–Kang (TSK) fuzzy system is proven in [21]. The TSK model is proved to be equivalent to the RBF network under certain conditions [13]. Fuzzy systems with Gaussian membership functions are proved to be universal approximators for a smooth function and its derivatives [19]. In [42], the fuzzy system with nth-order B-spline membership functions and the CMAC network with nth-order B-spline basis functions are proved to be universal approximators for a smooth function and its derivatives up to the $(n-2)$th order.

26.2 Definitions and Terminologies

In this section, we give some definitions and terminologies used in the fuzzy logic literature.

Definition 26.1 (*Universe of Discourse*) The universal set $\mathcal{X} : \mathcal{X} \rightarrow [0, 1]$ is called the *universe of discourse*, or simply the *universe*. The implication $\mathcal{X} \rightarrow [0, 1]$ is the abbreviation for the IF-THEN rule: "IF x is in $\mathcal{X}$, THEN its membership function $\mu_{\mathcal{X}}(x)$ is in $[0, 1]$."

The universe $\mathcal{X}$ may contain either discrete or continuous values.

Definition 26.2 (*Linguistic Variable*) A linguistic variable is a variable, whose value is linguistic terms in a natural or artificial language.

For example, the size of an object is a linguistic variable, whose value can be *small*, *medium* and *large*.

Definition 26.3 (*Fuzzy Set*) A *fuzzy set* $\mathcal{A}$ in $\mathcal{X}$ is defined by

$$\mathcal{A} = \{(x, \mu_{\mathcal{A}}(x)) | x \in \mathcal{X}\}, \tag{26.1}$$

where $\mu_{\mathcal{A}}(x) \in [0, 1]$ is the membership function of x in $\mathcal{A}$. For $\mu_{\mathcal{A}}(x)$, the value 1 stands for complete membership of the set $\mathcal{A}$, while 0 represents that x does not belong to the set at all.

A fuzzy set can also be represented by

$$\mathcal{A} = \begin{cases} \sum_{x_i \in \mathcal{X}} \frac{\mu_{\mathcal{A}}(x_i)}{x_i}, & \text{if } \mathcal{X} \text{ is discrete} \\ \int_{\mathcal{X}} \frac{\mu_{\mathcal{A}}(x)}{x}, & \text{if } \mathcal{X} \text{ is continuous} \end{cases}. \tag{26.2}$$

The summation, integral, and division signs syntactically denote the union of $(x, \mu_{\mathcal{A}}(x))$ pairs.

Definition 26.4 (*Support*) The elements on a fuzzy set $\mathcal{A}$, whose membership is larger than zero are called the *support* of the fuzzy set

$$\mathrm{supp}(\mathcal{A}) = \left\{x \in \mathcal{A} \middle| \mu_{\mathcal{A}}(x) > 0\right\}. \tag{26.3}$$

Definition 26.5 (*Height*) The *height* of a fuzzy set $\mathcal{A}$ is defined by

$$h(\mathcal{A}) = \sup \{\mu_{\mathcal{A}}(x) | x \in \mathcal{X}\}. \tag{26.4}$$

Definition 26.6 (*Normal Fuzzy Set*) If $h(\mathcal{A}) = 1$, then a fuzzy set $\mathcal{A}$ is said to be *normal*.

Definition 26.7 (*Non-normal Fuzzy Set*) If $0 < h(\mathcal{A}) < 1$, a fuzzy set $\mathcal{A}$ is said to be *non-normal*. It can be normalized by dividing it by its height

$$\overline{\overline{\mu}}_{\mathcal{A}}(x) = \frac{\mu_{\mathcal{A}}(x)}{h(\mathcal{A})}. \tag{26.5}$$

Definition 26.8 (*Fuzzy Partition*) For a linguistic variable, a number of fuzzy subsets are enumerated as the value of the variable. This collection of fuzzy subsets is called a *fuzzy partition*. Each fuzzy subset has a membership function. For a finite fuzzy partition $\{\mathcal{A}_1, \mathcal{A}_2, \ldots, \mathcal{A}_n\}$ of a set $\mathcal{A}$, the membership function for each $x \in \mathcal{A}$ satisfies

$$\sum_{i=1}^{n} \mu_{\mathcal{A}_i}(x) = 1 \tag{26.6}$$

and $\mathcal{A}_i$ is normal, that is, the height of $\mathcal{A}_i$ is unity.

A fuzzy partition is illustrated in Fig. 26.1. The fuzzy set for representing the linguistic variable *human age* is partitioned into three fuzzy subsets, namely *young*, *middle age*, and *old*. Each fuzzy subset is characterized by a membership function.

Definition 26.9 (*Empty Set*) The subset of $\mathcal{X}$ having no element is called the *empty set*, denoted by $\emptyset$.

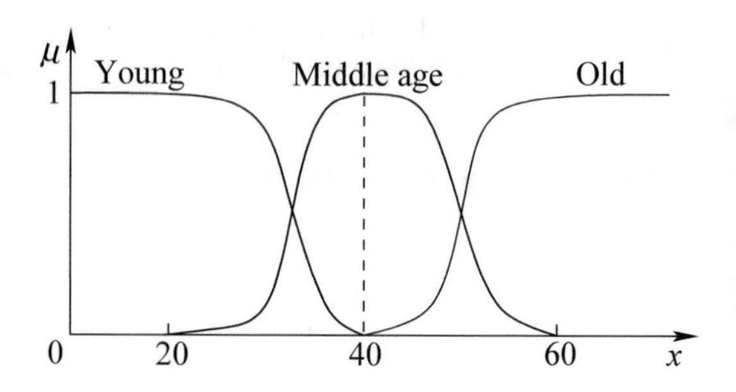

Fig. 26.1 A fuzzy partition of human age

Definition 26.10 (*Complement*) The *complement* of $\mathcal{A}$, written as $\overline{\mathcal{A}}$, $\neg\mathcal{A}$ or NOT $\mathcal{A}$, is defined as $\mu_{\overline{\mathcal{A}}}(x) = 1 - \mu_{\mathcal{A}}(x)$. Thus, $\overline{\mathcal{X}} = \emptyset$ and $\overline{\emptyset} = \mathcal{X}$.

Definition 26.11 (α-*Cut*) The α-*cut* or α-*level set* of a fuzzy set $\mathcal{A}$, written as $\mu_{\mathcal{A}}[\alpha]$, is defined as the set of all elements in $\mathcal{A}$ whose degree of membership is not less than α

$$\mu_{\mathcal{A}}[\alpha] = \{x \in \mathcal{A} | \mu_{\mathcal{A}}(x) \ge \alpha\}, \quad \alpha \in [0, 1]. \tag{26.7}$$

A fuzzy set $\mathcal{A}$ is usually represented by its membership function $\mu(x)$, $x \in \mathcal{A}$. The inverse of $\mu(x)$ can be represented by $x = \mu^{-1}(\alpha)$, $\alpha \in [0, 1]$, where each value of α may correspond to one or more values of x. A fuzzy set is usually represented by a finite number of its membership values.

The resolution principle uses α-cuts to represent membership to a fuzzy set

$$\mu_{\mathcal{A}} = \bigvee_{0<\alpha\le 1} \left[\alpha \cdot \mu_{\mathcal{A}_\alpha}(x)\right], \tag{26.8}$$

where the maximum is taken over all values of α.

Definition 26.12 (*Kernel or Core*) All the elements in a fuzzy set $\mathcal{A}$ with membership degree 1 constitute a subset called the *kernel* or *core* of the fuzzy set, written as $\ker(\mathcal{A}) = \mu_{\mathcal{A}}[1]$.

The support, kernel, and α-cut of a fuzzy set are shown in Fig. 26.2 for a trapezoid membership function, where a, b, c, and d are shape parameters. The α-cut shown is represented by $\mu_{\mathcal{A}}[\alpha] = [a_1, a_2]$.

Definition 26.13 (*Convex Fuzzy Set*) A fuzzy set $\mathcal{A}$ is said to be *convex* if and only if

$$\mu_{\mathcal{A}}(\lambda x_1 + (1-\lambda)x_2) \ge \mu_{\mathcal{A}}(x_1) \wedge \mu_{\mathcal{A}}(x_2), \quad \forall \lambda \in [0, 1], x_1, x_2 \in \mathcal{X}, \tag{26.9}$$

where $\wedge$ denotes the minimum operation.

Any α-cut set of a convex fuzzy set is a closed interval.

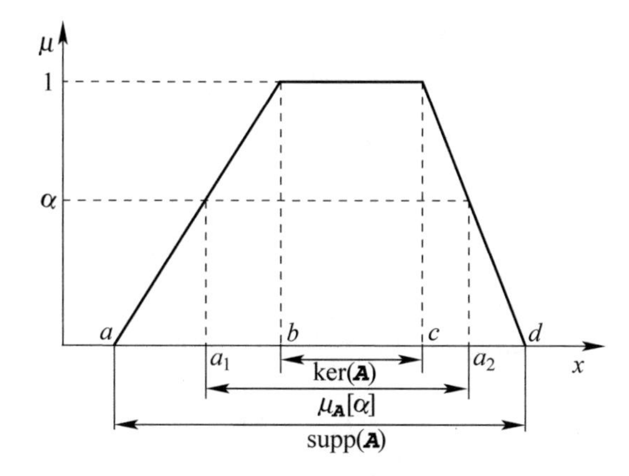

Fig. 26.2 Support, kernel, and α-cut of a fuzzy set $\mathcal{A}$

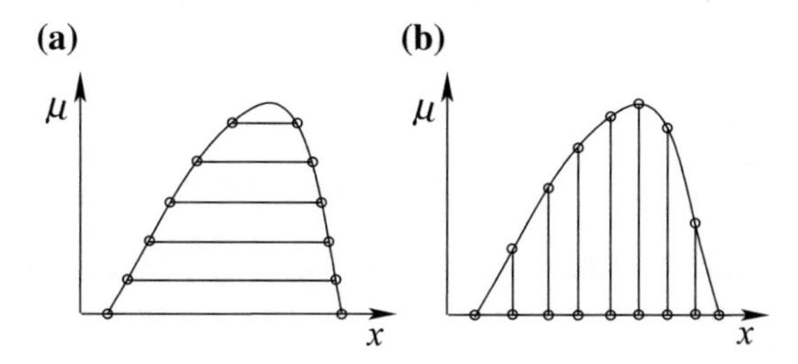

Fig. 26.3 Representations of a fuzzy number. **a** α-level sets. **b** Discretized membership function

Definition 26.14 (*Concave Fuzzy Set*) A fuzzy set $\mathcal{A}$ is said to be *concave* if and only if

$$\mu_{\mathcal{A}}(\lambda x_1 + (1-\lambda)x_2) \le \mu_{\mathcal{A}}(x_1) \vee \mu_{\mathcal{A}}(x_2), \quad \forall \lambda \in [0, 1], x_1, x_2 \in \mathcal{X}, \quad (26.10)$$

where $\vee$ denotes the maximum operation.

Definition 26.15 (*Fuzzy Number*) A fuzzy number $\mathcal{A}$ is a fuzzy set of the real line with a normal, convex and continuous membership function of bounded support.

Fuzzy numbers are fuzzified versions of classical crisp intervals. Thus, the theory of fuzzy numbers and their arithmetic should be a fuzzified version of interval analysis. A fuzzy number is usually represented by a family of α-level sets or by a discretized membership function, as illustrated in Fig. 26.3.

Definition 26.16 (*Fuzzy Singleton*) A fuzzy set $\mathcal{A} = \left\{(x, \mu_{\mathcal{A}}(x)) \,\middle|\, x \in \mathcal{X}\right\}$ is said to be a *fuzzy singleton* if $\mu_{\mathcal{A}}(x) = 1$ for $x \in \mathcal{X}$ and $\mu_{\mathcal{A}}(x') = 0\ \forall x' \in \mathcal{X}$ with $x' \neq x$.

Definition 26.17 (*Cardinality*) Given a fuzzy set $\mathcal{A}$ defined in a finite or countable universe $\mathcal{X}$, its cardinality, denoted $\mathrm{card}(A)$, is defined by

$$\mathrm{card}(\mathcal{A}) = |\mathcal{A}| = \sum_{x \in \mathcal{X}} \mu_{\mathcal{A}}(x). \quad (26.11)$$

The cardinality of a fuzzy set is derived by summing up the membership degrees. Cardinality is used to measure the magnitude of a fuzzy set. It is associated with the concept of granularity of information granules. Relative cardinality is obtained by dividing the magnitude of fuzzy set $\mathcal{A}$ by that of universal set $\mathcal{X}$

$$\|\mathcal{A}\| = \frac{|\mathcal{A}|}{|\mathcal{X}|}. \tag{26.12}$$

Definition 26.18 (*Equality*) Fuzzy sets $\mathcal{A}$ and $\mathcal{B}$ defined in the same universe $\mathcal{X}$ are said to be equal, $\mathcal{A} = \mathcal{B}$, if and only if

$$\mu_{\mathcal{A}}(x) = \mu_{\mathcal{B}}(x), \quad \forall x \in \mathcal{X}. \tag{26.13}$$

Definition 26.19 (*Fuzzy Subset and Inclusion*) A fuzzy set $\mathcal{A} = \{(x, \mu_{\mathcal{A}}(x)) | x \in \mathcal{X}\}$ is said to be a *fuzzy subset* of $\mathcal{B} = \{(x, \mu_{\mathcal{B}}(x)) | x \in \mathcal{X}\}$, denoted $\mathcal{A} \subseteq \mathcal{B}$, where $\subseteq$ is the inclusion operator, if and only if every element of $\mathcal{A}$ is also an element of $\mathcal{B}$, that is,

$$\mu_{\mathcal{A}}(x) \leq \mu_{\mathcal{B}}(x), \quad \forall x \in \mathcal{X}. \tag{26.14}$$

Definition 26.20 (*Product of Fuzzy Sets*) The product of fuzzy sets $\mathcal{A}$ and $\mathcal{B}$, defined on the same universe of discourse $\mathcal{X}$, denoted $\mathcal{A} \cdot \mathcal{B}$, is also a fuzzy set, whose membership function is given by

$$\mu_{\mathcal{A} \cdot \mathcal{B}}(x) = \mu_{\mathcal{A}}(x) \mu_{\mathcal{B}}(x). \tag{26.15}$$

Definition 26.21 (*Hedge*) A hedge transforms a fuzzy set into a new fuzzy set. Hedges are modifiers, adjectives, or adverbs, which change truth values.

Hedges are used to intensify or dilute the characteristic of a fuzzy set such as *very* and *quite*, or to approximate a fuzzy set or convert a scalar to a fuzzy set such as *about*, *nearly*, *roughly*. The use of hedges enables dynamical creation of fuzzy sets and this also helps to reduce the complexity of rules. For example, for a fuzzy set *good* with membership degree $\mu_{\mathcal{A}}(x)$, *very good* can be described using membership degree $\mu^2_{\mathcal{A}}(x)$, while *quite good* can be described using membership degree $\mu^{\frac{1}{2}}_{\mathcal{A}}(x)$. An illustration of hedge operations is given in Fig. 26.4. The membership function $\mu_1(x)$ is a hedge operation that transforms a scalar 5 into a fuzzy number *close to* 5, and $\mu_2(x) = \mu_1^2(x)$ and $\mu_3(x) = \mu_1^{\frac{1}{2}}(x)$ are, respectively, hedge operators that realize *very close to* 5 and *quite close to* 5.

Definition 26.22 (*Power of a Fuzzy Set*) The power of a fuzzy set $\mathcal{A}$ is a new fuzzy set, $\mathcal{A}^\alpha$, whose membership function is given by

$$\mu_{\mathcal{A}^\alpha}(x) = [\mu_{\mathcal{A}}(x)]^\alpha. \tag{26.16}$$

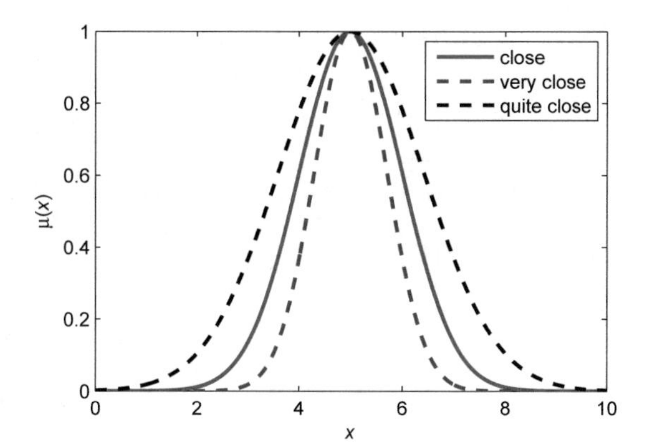

Fig. 26.4 An illustration of hedge operations

This actually define the hedge functions. *Concentration* raises a fuzzy set $\mathcal{A}$ to power 2 and obtains a fuzzy set $\mathcal{A}^2$. *Dilation* of $\mathcal{A}$ generates a fuzzy set $\mathcal{A}^{1/2}$.

In traditional fuzzy systems, the structure is characterized by using type-1 fuzzy sets. Type-1 fuzzy sets, defined on a universe of discourse, maps an element of the universe of discourse onto a precise number in [0, 1]. A type-2 fuzzy set can be informally defined as a fuzzy set that is characterized by a fuzzy membership function in the computational level. More generally,

Definition 26.23 (*Type-n Fuzzy Set*) A type-n fuzzy set is a fuzzy set whose membership values are type-$(n-1)$, $n > 1$, fuzzy sets on [0, 1].

A type-2 fuzzy logic has more computational complexity. Type-2 fuzzy logic demonstrates improved performance and robustness relative to type-1 fuzzy logic when confronted with various sources of data uncertainties [16].

The representations of α-planes [23] and zSlices [40] offer a viable framework for representing and computing with the general type-2 fuzzy sets. The α-planes and the zSlices representation theorems allow us to treat the general type-2 fuzzy sets as a composition of multiple interval type-2 fuzzy sets, each raised to the respective level of either α or z.

Definition 26.24 (*Fuzzy Transform and Inverse Fuzzy Transform*) A direct fuzzy transform [31] uses a fuzzy partition of an interval $[a, b]$ to convert a continuous function assigned on $[a, b]$ in a suitable n-dimensional vector. Then an inverse fuzzy transform converts this n-dimensional vector into another continuous function which approximates the original function up to an arbitrary accuracy.

Fuzzy transform explains modeling with fuzzy IF-THEN rules as a specific transformation. Fuzzy transform can be regarded as a specific type of Takagi–Sugeno fuzzy system, with several additional properties [2].

26.3 Membership Function

A fuzzy set $\mathcal{A}$ over the universe of discourse $\mathcal{X}$, $\mathcal{A} \subseteq \mathcal{X} \to [0, 1]$, is described by the degree of membership $\mu_{\mathcal{A}}(x) \in [0, 1]$ for each $x \in \mathcal{X}$. Unimodality and normality are two important aspects of the membership functions.

Piecewise-linear functions such as triangles and trapezoids are often used as membership functions in applications. The triangular membership function can be defined by

$$\mu(x; a, b, c) = \begin{cases} \frac{x-a}{b-a}, & a \leq x \leq b \\ \frac{c-x}{c-b}, & b < x \leq c \\ 0, & \text{otherwise} \end{cases}, \tag{26.17}$$

where the shape parameters satisfy $a < b < c$ and $b \in \mathcal{X}$. The triangular membership function is useful for modeling linguistic terms such as "The value is close to 10".

The trapezoid membership function can be defined by

$$\mu(x; a, b, c, d) = \begin{cases} 0, & x \leq a \text{ or } x \geq d \\ \frac{x-a}{b-a}, & a < x < b \\ 1, & b \leq x \leq c \\ \frac{d-x}{d-c}, & c < x < d \end{cases}, \tag{26.18}$$

where the shape parameters satisfy $a < b < c < d$. This function is shown in Fig. 26.2. It is suitable for modeling such linguistic terms as "He is in his twenties".

The Gaussian and bell-shaped functions have continuous derivatives, and are usually used to replace the triangular membership function when shape parameters are adapted using a gradient-descent procedure. The Gaussian function is given by

$$\mu(x; c, \sigma) = \mathrm{e}^{-\frac{(x-c)^2}{2\sigma^2}}, \tag{26.19}$$

and the bell-shaped function is defined by

$$\mu(x; c, a, b) = \frac{1}{1 + \left[\left(\frac{x-c}{a}\right)^2\right]^b}. \tag{26.20}$$

In (26.19) and (26.20), c is the center of the curves, and a, b and σ are their shape parameters.

Another popular membership function is the sigmoidal function of the form

$$\mu(x; c, \beta) = \frac{1}{1 + \mathrm{e}^{-\beta(x-c)}}, \tag{26.21}$$

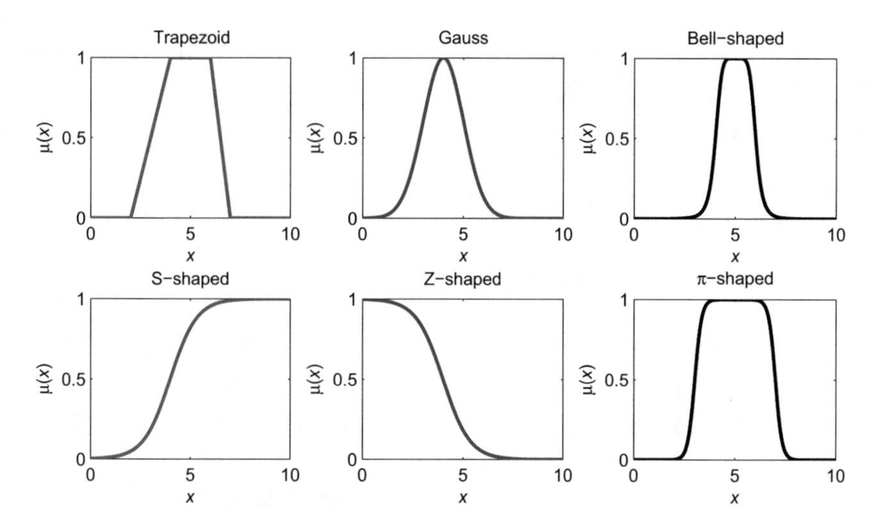

Fig. 26.5 Shapes of some popular membership functions. The parameters for each membership function are selected as: **a** Trapezoid, $a = 2$, $b = 4$, $c = 6$, $d = 7$. **b** Gaussian, $\sigma = 1$, $c = 4$. **c** Bell-shaped, $a = 1$, $b = 3$, $c = 5$. **d** S-shaped, $\beta = 1.5$, $c = 5$. **e** Z-shaped, $\beta = -1.5$, $c = 5$. **f** π-shaped, $\beta_1 = 6$, $c_1 = 3$, $\beta_2 = 6$, $c_2 = 7$

where c shifts the function to the left or to the right, and β controls the shape of the function. When $\beta > 1$ it is an S-shaped function, and when $\beta < -1$ it is a Z-shaped function.

When an S-shaped function is multiplied by a Z-shaped function, a π-shaped function is obtained:

$$\mu(x; c_1, \beta_1, c_2, \beta_2) = \frac{1}{1 + \mathrm{e}^{-\beta_1(x-c_1)}} \cdot \frac{1}{1 + \mathrm{e}^{-\beta_2(x-c_2)}} \tag{26.22}$$

where $\beta_1 > 1$, $\beta_2 < -1$, and $c_1 < c_2$. π-shaped membership functions can be used in situations where trapezoid membership functions are used. These popular membership functions are illustrated in Fig. 26.5.

26.4 Intersection, Union and Negation

The set operations *intersection* and *union* correspond to the logic operations *conjunction* (*AND*) and *disjunction* (*OR*), respectively. *Intersection* is described by the so-called *triangular norm (t-norm)*, denoted by $T(x, y)$, whereas *union* is described by the so-called *triangular conorm (t-conorm)*, denoted by $C(x, y)$.

If $\mathcal{A}$ and $\mathcal{B}$ are fuzzy subsets of $\mathcal{X}$, then intersection $\mathcal{I} = \mathcal{A} \cap \mathcal{B}$ is defined by

$$\mu_{\mathcal{I}}(x) = T(\mu_{\mathcal{A}}(x), \mu_{\mathcal{B}}(x)). \tag{26.23}$$

Definition 26.25 (*t-Norm*) A mapping $T : [0, 1] \times [0, 1] \to [0, 1]$ with the following four properties is called *t-norm*. For all $x, y, z \in [0, 1]$,

- Commutativity: $T(x, y) = T(y, x)$.
- Monotonicity: $T(x, y) \le T(x, z)$, if $y \le z$.
- Associativity: $T(x, T(y, z)) = T(T(x, y), z)$.
- Linearity: $T(x, 1) = x$.

Basic t-norms are listed below [6]:

$$T_m(x, y) = \min(x, y) \quad \text{(standard intersection)}, \tag{26.24}$$

$$T_b(x, y) = \max(0, x + y - 1) \quad \text{(bounded sum)}, \tag{26.25}$$

$$T_p(x, y) = xy \quad \text{(algebraic product)}, \tag{26.26}$$

$$T^*(x, y) = \begin{cases} x, & \text{if } y = 1 \\ y, & \text{if } x = 1 \\ 0, & \text{otherwise} \end{cases} \quad \text{(drastic intersection)}. \tag{26.27}$$

There is relation

$$T^*(x, y) \le T_b(x, y) \le T_p(x, y) \le T_m(x, y), \quad \forall x, y \in [0, 1], \tag{26.28}$$

where $T^*(x, y)$ and $T_m(x, y)$ are, respectively, the lower and upper bounds of any t-norm.

Similarly, union $\mathcal{U} = \mathcal{A} \cup \mathcal{B}$ is defined by

$$\mu_{\mathcal{U}}(x) = C\left(\mu_{\mathcal{A}}(x), \mu_{\mathcal{B}}(x)\right). \tag{26.29}$$

Definition 26.26 (*t-Conorm*) A mapping $C : [0, 1] \times [0, 1] \to [0, 1]$ having the following four properties is called *t-conorm*. For all $x, y, z \in [0, 1]$,

- Commutativity: $C(x, y) = C(y, x)$.
- Monotonicity: $C(x, y) \le C(x, z)$, if $y \le z$.
- Associativity: $C(x, C(y, z)) = C(C(x, y), z)$.
- Linearity: $C(x, 0) = x$.

Basic t-conorms are defined by [6]

$$C_m(x, y) = \max(x, y) \quad \text{(standard union)}, \tag{26.30}$$

$$C_b(x, y) = \min(1, x + y) \quad \text{(bounded sum)}, \tag{26.31}$$

$$C_p(x, y) = x + y - xy \quad \text{(algebraic sum)}, \tag{26.32}$$

$$C^*(x, y) = \begin{cases} x, & \text{if } y = 0 \\ y, & \text{if } x = 0 \\ 1, & \text{otherwise} \end{cases} \quad \text{(drastic union)}. \tag{26.33}$$

Accordingly,

$$C_m(x, y) \le C_p(x, y) \le C_b(x, y) \le C^*(x, y), \quad \forall x, y \in [0, 1], \tag{26.34}$$

where $C_m(x, y)$ and $C^*(x, y)$ are, respectively, the lower and upper bounds of any t-contorm.

When t-norm and t-conorm satisfy

$$1 - T(x, y) = C(1 - x, 1 - y), \tag{26.35}$$

T and C are said to be *dual*. This makes De Morgan's laws $\overline{\mathcal{A} \cap \mathcal{B}} = \overline{\mathcal{A}} \cup \overline{\mathcal{B}}$ and $\overline{\mathcal{A} \cup \mathcal{B}} = \overline{\mathcal{A}} \cap \overline{\mathcal{B}}$ still hold in fuzzy set theory. The above basic t-norms and t-conorms with the same subscripts are dual. To satisfy the principle of duality, they are usually used in pairs.

Definition 26.27 (*Negation*) A function $N : [0, 1] \to [0, 1]$ is called a (fuzzy) negation ($\neg$) if it is monotonically nonincreasing, continuous, $N(0) = 1$ and $N(1) = 0$. A negation N is said to be *strict* if it is strictly decreasing and continuous, or *strong* if it is an involution, that is, $N(N(x)) = x, \forall x \in [0, 1]$.

Negation generalizes the set notion of complement.

26.5 Fuzzy Relation and Aggregation

Definition 26.28 (*Extension Principle*) Given mapping $f : \mathcal{X} \to \mathcal{Y}$, if we have a fuzzy set $\mathcal{A} = \{(x, \mu_{\mathcal{A}}(x)) | x \in \mathcal{X}\}$, $\mu_{\mathcal{A}}(x) \in [0, 1]$, the *extension principle* is defined by

$$f(\mathcal{A}) = f(\{(x, \mu_{\mathcal{A}}(x)) | x \in \mathcal{X}\}) = \{(f(x), \mu_{\mathcal{A}}(x)) | x \in \mathcal{X}\}. \tag{26.36}$$

Application of the extension principle transforms x into $f(x)$, but does not affect the membership function $\mu_{\mathcal{A}}(x)$.

Definition 26.29 (*Cartesian Product*) If $\mathcal{X}$ and $\mathcal{Y}$ are two universal sets, then $\mathcal{X} \times \mathcal{Y}$ is the set of all ordered pairs $\{(x, y) | x \in \mathcal{X}, y \in \mathcal{Y}\}$. Let $\mathcal{A}$ be a fuzzy set of $\mathcal{X}$ and $\mathcal{B}$ a fuzzy set of $\mathcal{Y}$. The Cartesian product is defined by

$$\mathcal{A} \times \mathcal{B} = \{(z, \mu_{\mathcal{A}\times\mathcal{B}}(z)) | z = (x, y) \in \mathcal{Z}, \mathcal{Z} = \mathcal{X} \times \mathcal{Y}\}, \tag{26.37}$$

where $\mu_{\mathcal{A}\times\mathcal{B}}(z) = \mu_{\mathcal{A}}(x) \wedge \mu_{\mathcal{B}}(y)$, and $\wedge$ denotes t-norm operator.

Definition 26.30 (*Fuzzy Relation*) If $\mathcal{R}$ is a subset of $\mathcal{X} \times \mathcal{Y}$, then $\mathcal{R}$ is said to be a relation between $\mathcal{X}$ and $\mathcal{Y}$, or a relation on $\mathcal{X} \times \mathcal{Y}$. Mathematically,

$$\mathcal{R}(x, y) = \{((x, y), \mu_{\mathcal{R}}(x, y)) |\, (x, y) \in \mathcal{X} \times \mathcal{Y}, \mu_{\mathcal{R}}(x, y) \in [0, 1]\}, \tag{26.38}$$

where $\mu_{\mathcal{R}}(x, y)$ is the degree of membership for association between x and y.

A fuzzy relation is a fuzzy set defined on Cartesian products. Fuzzy relation is used to describe the association between two things.

The height of the fuzzy relation $\mathcal{R}(x, y)$ is given by

$$h(\mathcal{R}(x, y)) = \max_{y \in \mathcal{Y}} \max_{x \in \mathcal{X}} \mu_{\mathcal{R}}(x, y). \tag{26.39}$$

Definition 26.31 (*Fuzzy Matrix*) Given finite, discrete fuzzy sets $\mathcal{X} = \{x_1, x_2, \ldots, x_m\}$ and $\mathcal{Y} = \{y_1, y_2, \ldots, y_n\}$, a fuzzy relation on $\mathcal{X} \times \mathcal{Y}$ can be represented by an $m \times n$ matrix called a *fuzzy matrix*, $\mathbf{R} = \left[\mu_{\mathcal{R}}\left(x_i, y_j\right)\right]$.

A fuzzy relation can also be represented by

$$\mathcal{R}(x, y) = \sum_{i,j} \mu_{\mathcal{R}}(x, y)/(x_i, y_j). \tag{26.40}$$

The inverse relation of $\mathcal{R}(x, y)$, denoted by $\mathcal{R}^{-1}(x, y)$, can be represented as the transpose of a membership matrix, and $(\mathcal{R}^{-1}(x, y))^{-1} = \mathcal{R}(x, y)$.

Definition 26.32 (*Fuzzy Graph*) A fuzzy relation $\mathcal{R}(x, y)$ can be represented by a fuzzy graph. In a fuzzy graph, all x_i and y_j are vertices, and the grade $\mu_{\mathcal{R}}\left(x_i, y_j\right)$ is added to the connection from x_i and y_j.

Definition 26.33 (*Aggregation of Fuzzy Relations*) Consider two fuzzy relations, $\mathcal{R}_1$ on $\mathcal{X} \times \mathcal{Y}$ and $\mathcal{R}_2$ on $\mathcal{Y} \times \mathcal{Z}$,

$$\begin{aligned} \mathcal{R}_1(x, y) &= \left\{\left((x, y), \mu_{\mathcal{R}_1}(x, y)\right) \middle|\, (x, y) \in \mathcal{X} \times \mathcal{Y}, \mu_{\mathcal{R}_1}(x, y) \in [0, 1]\right\}, \\ \mathcal{R}_2(y, z) &= \left\{\left((y, z), \mu_{\mathcal{R}_2}(y, z)\right) \middle|\, (y, z) \in \mathcal{Y} \times \mathcal{Z}, \mu_{\mathcal{R}_2}(y, z) \in [0, 1]\right\}. \end{aligned} \tag{26.41}$$

The *max–min composition* of $\mathcal{R}_1$ and $\mathcal{R}_2$, denoted by $\mathcal{R}_1 \circ \mathcal{R}_2$ with membership function $\mu_{\mathcal{R}_1 \circ \mathcal{R}_2}$, is given by a fuzzy relation on $\mathcal{X} \times \mathcal{Z}$

$$\mathcal{R}_1 \circ \mathcal{R}_2 = \left\{\left((x, z), \max_y \left\{\min\left(\mu_{\mathcal{R}_1}(x, y), \mu_{\mathcal{R}_2}(y, z)\right)\right\}\right) |(x, z) \in \mathcal{X} \times Z, y \in \mathcal{Y}\right\}. \tag{26.42}$$

Aggregation or composition operations on fuzzy sets provide a means for combining several sets in order to produce a single fuzzy set. In the definition, the aggregation operators max and min correspond to t-conorm and t-norm, respectively.

There are some other composition operations, such as the min-max composition, denoted by $\mathcal{R}_1 \diamond \mathcal{R}_2$, with the difference that the role of max and min is interchanged. The two compositions are related by $\overline{\mathcal{R}_1 \diamond \mathcal{R}_2} = \overline{\mathcal{R}_1} \circ \overline{\mathcal{R}_2}$.

26.6 Fuzzy Implication

Classical *modus ponens* is "If A then B," which can be read "If proposition A is true, then infer that proposition B is true." *modus ponens* itself is a proposition, sometimes written as "A `implies` B" or "$A \to B$," where `implies` is a logical operator with A and B as operands.

The implication operator is indispensable in the inference mechanisms of any logic, like *modus ponens*, *modus tollens*, and hypothetical syllogism in classical logic. Fuzzy implication is one of the key operations in fuzzy logic and approximate reasoning. It is used for the management of fuzzy conditionals of the type "If $\mathcal{A}$ then $\mathcal{B}$." Fuzzy implication operators are usually functionally expressed through numerical functions $I : [0, 1] \times [0, 1] \to [0, 1]$ called implication functions or simply implications.

Fuzzy implication $\mathcal{A} \to \mathcal{B}$ interprets the fuzzy rule: "IF x is $\mathcal{A}$ THEN y is $\mathcal{B}$". It is a mapping I of an input fuzzy region $\mathcal{A}$ onto an output fuzzy region $\mathcal{B}$ according to the defined fuzzy relation $\mathcal{R}$ on $\mathcal{A} \times \mathcal{B}$: $\mu_{\mathcal{R}}(x, y) = I(x, y)$. For a fuzzy rule expressed as a fuzzy implication using the defined fuzzy relation $\mathcal{R}$, the output linguistic variable $\mathcal{B}$ is denoted by

$$\mathcal{B} = \mathcal{A} \circ \mathcal{R}, \tag{26.43}$$

which is characterized by $\mu_{\mathcal{B}}(y) = \vee_x \left(\mu_{\mathcal{A}}(x) \wedge \mu_{\mathcal{R}}(x, y)\right)$.

Definition 26.34 A fuzzy implication is a function $I : [0, 1]^2 \to [0, 1]$ that satisfies the following properties:

- Monotonicity: $x_1 \le x_2 \Rightarrow I(x_1, y) \ge I(x_2, y)$.
- Monotonicity: $y_1 \le y_2 \Rightarrow I(x, y_1) \le I(x, y_2)$.
- Dominance of falsity: $I(0, y) = 1$.
- Neutrality of truth: $I(1, y) = y$.
- Exchange: $I(x_1, I(x_2, y)) = I(x_2, I(x_1, y))$.

It is obvious that $I(0, 0) = I(0, 1) = I(1, 1) = 1$, $I(1, 0) = 0$, and $I(a, a) = 1$, which satisfy the properties of classical implication. A fuzzy rule "IF x is $\mathcal{A}$ THEN y is $\mathcal{B}$" is expressed as $I(a, b)$, where a and b are the membership grades of $\mathcal{A}$ and $\mathcal{B}$, respectively.

Since conjunctions, disjunctions, and negations are usually performed by t-norms, t-conorms, and strong negations, the majority of the known implication functions are directly derived from these operators. Several well-known implication operators so defined even do not satisfy the definition.

The four most usual definitions for implications are S-, R-, QL-, and D-implications [22]. They are equivalent in any Boolean algebra and consequently in classical logic. However, in fuzzy logic they yield distinct classes of fuzzy implications. The two classes most commonly used are R- and S-implications.

S-implications are defined by

$$I(x, y) = C(N(x), y), \quad x, y \in [0, 1], \tag{26.44}$$

where C is t-conorm and N is strong negation. They appear as an immediate generalization of the classical boolean implication $p \to q \equiv \neg p \vee q$.

R-implications are defined by

$$I(x, y) = \sup\{z \in [0, 1] | T(x, z) \leq y\}, \quad x, y \in [0, 1], \tag{26.45}$$

where T is a left-continuous t-norm. They come from residuated lattices based on the residuation property that in the case of t-norms can be written as

$$T(x, y) \leq z \Longleftrightarrow I(x, z) \geq y \quad \forall x, y, z \in [0, 1], \tag{26.46}$$

The majority of known implications not only satisfy the definition but also belong to some of the four types. For instance, the Lukasiewicz implication ($I(x, y) = \min(1, 1 - x + y)$) belongs to the four types. The Kleene-Dienes implication ($I(x, y) = \max(1 - x, y)$) is an S-implication derived from the t-conorm maximum and the negation $N(x) = 1 - x$.

The implication functions are used not only to represent IF-THEN statements but also to perform forward and backward inferences in fuzzy systems, with the two main classical inference rules (deduction rules) of *modus ponens* and *modus tollens*, respectively. The choice of fuzzy implication cannot be made independently of the inference rule that is going to be applied.

QL-implication, known as *propositional calculus*, is based on the classical logic form $\neg(a \wedge \neg(a \wedge b)) = \neg a \vee (a \wedge b)$ and logical operators are substituted by fuzzy operators. S-implication, called *material implication*, derives from the classical logic form $a \to b = \neg a \vee b$. R-implication and D-implication reflect a partial ordering on propositions and are based on a generalization of modus ponens and modus tollens, respectively.

26.7 Reasoning and Fuzzy Reasoning

Reasoning processes are categorized into deductive and reductive types. Deductive reasoning proceeds with inference from premises to a conclusion: If premises P and $P \to Q$, then the conclusion Q. Reductive reasoning carries inference from conclusions to a set of plausible premises: If $P \to Q$ and Q, then P. Whereas deductive reasoning is exact, reductive reasoning is more intuitive. Inductive reasoning, as a special type of reductive reasoning, generalizes evidence to a hypothesis for a population. Inductive reasoning has no more justification than random guessing, and this can be drawn from the no free lunch theorem.

Logics as bases for reasoning can be distinguished essentially by three items: truth values, vocabulary (operators), and reasoning procedures (tautologies, syllogisms). A formal logical system largely consists of an axiom system (knowledge base) and an inference system. The axiom system consists of a set of axiom schemes, and the inference system consists of a set of rules of inference. Two popular inference rules in mathematical logic, mathematics, and AI are *modus ponens* and *modus tollens*. Modus ponens and modus tollens belong to rules of deduction.

Fuzzy reasoning, also called *approximate reasoning*, is an inference procedure for deriving conclusions from a set of fuzzy rules and one or more conditions. Fuzzy reasoning employs the generalized fuzzy modus ponens. The compositional rule of inference is the essential rational behind fuzzy reasoning.

26.7.1 Modus Ponens *and* Modus Tollens

Modus Ponens

Modus ponens is an important tool in classical logic for inferring one proposition from another, and has been used for roughly 2000 years.

Consider *modus ponens* as one tautology:

$$(P \wedge (P \to Q)) \to Q. \tag{26.47}$$

Modus ponens has a general form

$$\begin{array}{ll} P \to Q & \\ \underline{P \qquad\quad} & \\ \therefore \quad\quad Q & \end{array} \qquad \begin{array}{ll} \text{Implication: If } P \text{ then } Q. & \\ \underline{\text{Premise: } \quad P \text{ is true.}\quad} & \\ \text{Conclusion: } \quad Q \text{ is true.} & \end{array}$$

Thus, if P is true and $P \to Q$ is true, then the conclusion Q is also true.

Modus Tollens

Modus tollens is given by

$$\begin{array}{l} P \quad \to Q \\ \underline{\neg Q \qquad\qquad} \\ \therefore \qquad\quad \neg P \end{array}$$

Thus, if Q is false and $P \to Q$ is true, then the conclusion $\neg P$ is true. From an experience-based reasoning viewpoint, this is a formalized summary of experience.

26.7.2 *Generalized* Modus Ponens

The truth value of a proposition is a measure in the interval [0, 1] of how sure we are that the proposition is true. Data have truth values, a measure of the extent to which the values of the data are valid; rules have truth values, a measure of the extent to which the rule itself is valid. In general, the truth value of something is a measure in [0, 1] of its validity.

A basic truth-functional operator used in fuzzy logic is logical implication written $\mathcal{P} \rightarrow \mathcal{Q}$. It may be expressed in terms of `or` and `not` as

$$(\mathcal{P} \rightarrow \mathcal{Q}) \equiv ((\neg\mathcal{P}) \vee \mathcal{Q}). \tag{26.48}$$

Generalized *modus ponens* is given by

$$\begin{array}{ll} \mathcal{P} & \rightarrow \mathcal{Q} \\ \mathcal{P}' & \\ \hline \therefore & \mathcal{Q}' \end{array}$$

where $\mathcal{P}'$ and $\mathcal{Q}'$ correspond to the two compound propositions $\mathcal{P}$ and $\mathcal{Q}$.

For example,

Implication:	If a tomato is red, then the tomato is ripe.
Premise:	This tomato is very red.
Conclusion:	This tomato is very ripe.

For composition of propositions, one uses fuzzy connectives including negation ($\neg$), conjunction ($\wedge$), disjunction ($\vee$), implication ($\rightarrow$), and equivalence ($\Leftrightarrow$). Let $\mathcal{P}(x)$ and $\mathcal{Q}(y)$ be two fuzzy propositions which have the truth degree $\mu_{\mathcal{P}}(x)$ and $\mu_{\mathcal{Q}}(y)$, respectively, with $x \in R_{\mathcal{P}}$ and $y \in R_{\mathcal{Q}}$. The degrees of truth yielded by these fuzzy connectives are defined by

$$\mu_{\mathcal{P}\rightarrow\mathcal{Q}}(x, y) = \max((\mu_{\mathcal{P}}(x) \leq \mu_{\mathcal{Q}}(y)), \mu_{\mathcal{Q}}(y)) \quad \text{(Implication)}, \tag{26.49}$$

$$\mu_{\mathcal{P}\Leftrightarrow\mathcal{Q}}(x, y) = \max((\mu_{\mathcal{P}}(x) == \mu_{\mathcal{Q}}(y)), \min(\mu_{\mathcal{P}}(x), \mu_{\mathcal{Q}}(y))) \quad \text{(Equivalence)}. \tag{26.50}$$

In fuzzy logic, after interpreting algebraically in terms of true values, generalized *modus ponens* becomes: "The true value of $\mathcal{P} \wedge (\mathcal{P} \rightarrow \mathcal{Q})$ must be less than or equal to the true value of $\mathcal{Q}$," which can be expressed as

$$T_1(x, I(x, y)) \leq y, \quad \forall x, y \in [0, 1], \tag{26.51}$$

where T_1 is a t-norm performing conjunction and I is an implication function performing the conditional relation. When satisfying (26.51) with respect to t-norm T_1, it is said that I is T_1-conditional. It is known that the R-implication derived from a left-continuous t-norm T_1 is always T_1-conditional (in fact, it is the greatest T_1-conditional).

26.7.3 Fuzzy Reasoning Methods

Generally, fuzzy systems can be divided into three categories:

- Takagi–Sugeno reasoning—consequents are functions of inputs.
- Mamdani-type reasoning—consequents and antecedents are related by the min operator or generally by t-norm.
- Logical-type reasoning—consequents and antecedents are related by fuzzy implications.

Consider a fuzzy set $\mathcal{A} = \left\{(x, \mu_{\mathcal{A}}(x)) \,\middle|\, x \in \mathcal{X}\right\}$ and a fuzzy relation $\mathcal{R}$ on $\mathcal{A} \times \mathcal{B}$, $\mathcal{R}(x, y) = \left\{((x, y), \mu_{\mathcal{R}}(x, y)) \,\middle|\, (x, y) \in \mathcal{X} \times \mathcal{Y}\right\}$. Fuzzy set $\mathcal{B}$ can be inferred from $\mathcal{A}$ and $\mathcal{R}$ according to the max–min composition

$$\mathcal{B} = \mathcal{A} \circ \mathcal{R} = \left\{\left(y, \max_x \{\min\left(\mu_{\mathcal{A}}(x), \mu_{\mathcal{R}}(x, y)\right)\}\right) \,\middle|\, x \in \mathcal{X}, y \in \mathcal{Y}\right\}. \tag{26.52}$$

Generalized *modus ponens* can be formulated as: "If x is $\mathcal{A}$ then y is $\mathcal{B}$. From $x = \mathcal{A}'$, infer that $y = \mathcal{B}'$." Here $\mathcal{A}$ and $\mathcal{A}'$ are fuzzy sets defined on the same universe, and $\mathcal{B}$ and $\mathcal{B}'$ are also fuzzy sets defined on the same universe, which may be different from the universe on which $\mathcal{A}$ and $\mathcal{A}'$ are defined. By computing the fuzzy conclusion $\mathcal{B}'$ using the compositional rule of inference $\mathcal{B}' = \mathcal{A}' \circ \mathcal{R}$, we have

$$\mathcal{B}' = \mathcal{A}' \circ (\mathcal{A} \to \mathcal{B}). \tag{26.53}$$

The inverse problem of approximate reasoning is to conclude $\mathcal{A}'$ from $\mathcal{B}'$ and $\mathcal{A} \to \mathcal{B}$. By using the law of contrapositive symmetry, similarity-based inverse approximate reasoning [26] provides a solution to the problem.

Example 26.1 Assume that

$$\mathcal{A} = \left\{\frac{0.3}{x_1}, \frac{0.7}{x_2}, \frac{1.0}{x_3}\right\}, \quad \mathcal{B} = \left\{\frac{0.5}{y_1}, \frac{1.0}{y_2}, \frac{0.6}{y_3}\right\}, \quad \mathcal{A}' = \left\{\frac{1.0}{x_1}, \frac{0.6}{x_2}, \frac{0.3}{x_3}\right\}.$$

By choosing the Lukasiewicz implication operator $I(x, y) = \min(1, 1 - x + y)$, we obtain $\mathcal{R}(x, y)$ of $\mathcal{A} \to \mathcal{B}$ as

$$\mathcal{R} = \begin{bmatrix} 1 & 1 & 0.5 \\ 0.8 & 1 & 1 \\ 0.6 & 1 & 0.6 \end{bmatrix}.$$

We are now ready to obtain $\mathcal{B}'$ from $\mathcal{A}' \circ \mathcal{R}$. Using $T = T_m$, we get

$$\mathcal{B}'(y_1) = \max(\min(1, 1), \min(0.6, 0.8), \min(0.3, 0.6)) = 1,$$

$$\mathcal{B}'(y_2) = \max(\min(1, 1), \min(0.6, 1), \min(0.3, 1)) = 1,$$

$$\mathcal{B}'(y_3) = \max(\min(1, 0.5), \min(0.6, 1), \min(0.3, 0.6)) = 0.6,$$

and

$$\mathcal{B}' = \left\{ \frac{1.0}{y_1}, \frac{1.0}{y_2}, \frac{0.6}{y_3} \right\}.$$

26.8 Fuzzy Inference Systems

In control systems, the inputs to the systems are the error and the change in the error of the feedback loop, while the output is the control action. Fuzzy logic based controllers are popular control systems. The general architecture of a fuzzy controller is depicted in Fig. 26.6. Fuzzy controllers are knowledge-based, where knowledge is defined by fuzzy IF-THEN rules. The core of a fuzzy controller is a fuzzy inference system, in which the data flow involves fuzzification, knowledge-base evaluation, and defuzzification. A fuzzy inference system is also termed a *fuzzy expert system* or a *fuzzy model.*

In a fuzzy inference system, the knowledge base is comprised of the fuzzy rule base and the database. The database contains the linguistic term sets considered in the linguistic rules and the membership functions defining the semantics of the linguistic variables, and information about domains. The rule base contains a collection of linguistic rules that are joined by the `also` operator. An expert provides his knowledge in the form of linguistic rules. The fuzzification process collects the inputs and then converts them into linguistic values or fuzzy sets. The decision logic, called *fuzzy*

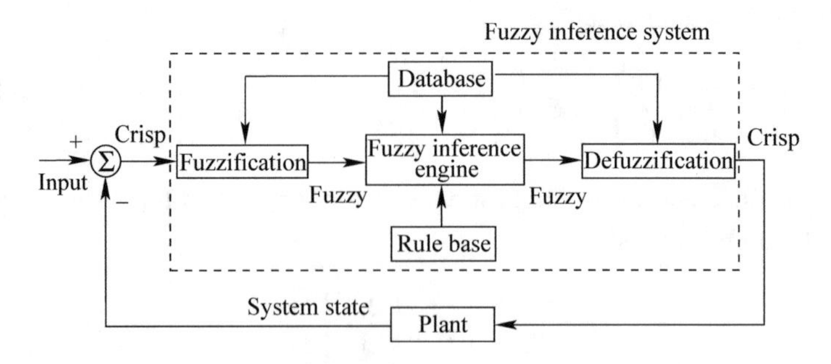

Fig. 26.6 The architecture of a fuzzy controller

inference engine, generates output from the input, and finally the defuzzification process produces a crisp output for control action.

Fuzzy inference systems are universal approximators capable of performing nonlinear mappings between inputs and outputs. The interpretations of a certain rule and the rule base depend on the fuzzy system model. The Mamdani [20] and TSK [36] models are two popular fuzzy inference systems. The Mamdani model is a non-additive fuzzy model that aggregates the output of fuzzy rules using the maximum operator, while the TSK model is an additive fuzzy model that aggregates the output of rules using the addition operator. Kosko's standard additive model [18] is another additive fuzzy model. All these models can be derived from fuzzy graph [43], and are universal approximators [4, 5, 17, 41].

Both neural networks and fuzzy logic can be used to approximate an unknown control function. Neural networks achieve a solution using the learning process, while fuzzy inference systems apply a vague interpolation technique. Fuzzy inference systems are appropriate for modeling nonlinear systems, whose mathematical models are not available. Unlike neural networks and other numerical models, fuzzy models operate at a level of information granules—fuzzy sets.

26.8.1 Fuzzy Rules and Fuzzy Interference

There are two types of fuzzy rules, namely *fuzzy mapping rules* and *fuzzy implication rules* [43]. A fuzzy mapping rule describes a functional mapping relationship between inputs and an output using linguistic terms, while a fuzzy implication rule describes a generalized logic implication relationship between two logic formulas involving linguistic variables. Fuzzy implication rules generalize set-to-set implications, whereas fuzzy mapping rules generalize set-to-set associations. The former was motivated to allow intelligent systems to draw plausible conclusions in a way similar to human reasoning, while the latter was motivated to approximate complex relationships such as nonlinear functions in a cost-effective and easily comprehensible way. The foundation of fuzzy mapping rule is fuzzy graph, while the foundation of fuzzy implication rule is a generalization to two-valued logic.

A rule base consists of a number of rules in IF-THEN logic: "IF *condition*, THEN *action*." The condition, also called *premise*, is made up of a number of *antecedents* that are negated or combined by different operators such as `and` or `or` computed with t-norms or t-conorms. In a fuzzy rule system, some variables are linguistic variables and the determination of the membership function for each fuzzy subset is critical. Membership functions can be selected according to human intuition, or by learning from training data.

A fuzzy inference is made up of several rules with the same output variables. Given a set of fuzzy rules, the inference result is a combination of the fuzzy values of the conditions and the corresponding actions. For example, we have a set of N_r rules

$$R_i : \text{IF } (condition = \mathcal{C}_i) \text{ THEN } (action = \mathcal{A}_i), \quad i = 1, \ldots, N_r,$$

where $\mathcal{C}_i$ and $\mathcal{A}_i$ are fuzzy sets. Assuming that a condition has a membership degree of μ_i associated with $\mathcal{C}_i$. The condition is first converted into a fuzzy category using a syntactical representation

$$condition = \sum_{i=1}^{N_r} \frac{\mathcal{C}_i}{\mu_i} = \frac{\mathcal{C}_1}{\mu_1} + \frac{\mathcal{C}_2}{\mu_2} + \cdots + \frac{\mathcal{C}_{N_r}}{\mu_{N_r}}. \tag{26.54}$$

Notice the difference from the definition of a finite fuzzy set in (26.2). We can see that each rule is valid to a certain extent. A fuzzy inference is the combination of all the possible consequences. The action coming from a fuzzy inference is also a fuzzy category, which can be syntactically represented by $action = \sum_{i=1}^{N_r} \frac{\mathcal{A}_i}{\mu_i}$. The inference procedure depends on fuzzy reasoning. This result can be further processed or transformed into a crisp value.

26.8.2 Fuzzification and Defuzzification

Fuzzification is to transform crisp inputs into fuzzy subsets. Given crisp inputs x_i, $i = 1, \ldots, n$, fuzzification is to construct the same number of fuzzy sets $\mathcal{A}^i$,

$$\mathcal{A}^i = \text{fuzz}\,(x_i)\,, \tag{26.55}$$

where fuzz$(\cdot)$ is a fuzzification operator. Fuzzification is determined according to the defined membership functions.

Defuzzification maps fuzzy subsets of real numbers into real numbers. It is applied after aggregation. Defuzzification is necessary in fuzzy controllers, since the machines cannot understand control signals in the form of a complete fuzzy set. Popular defuzzification methods include the centroid defuzzifier [20] and the mean-of-maxima defuzzifier [20].

The best-known centroid defuzzifier finds the centroid of the area surrounded by the membership function and the horizontal axis. A discrete centroid defuzzifier is given by

$$\text{defuzz}(\mathcal{B}) = \frac{\sum_{i=1}^{K} \mu_{\mathcal{B}}\,(y_i)\, y_i}{\sum_{i=1}^{K} \mu_{\mathcal{B}}\,(y_i)}, \tag{26.56}$$

where K is the number of quantization steps by which the universe of discourse $\mathcal{Y}$ of the membership function $\mu_{\mathcal{B}}(y)$ is discretized.

Aggregation and defuzzification can be combined into a single phase, such as the weighted-mean method [11]

$$\text{defuzz}(\mathcal{B}) = \frac{\sum_{i=1}^{N_r} \mu_i b_i}{\sum_{i=1}^{N_r} \mu_i}, \tag{26.57}$$

where N_r is the number of rules, μ_i is the degree of activation of the ith rule, and b_i is a numerical value associated with the consequent of the ith rule, $\mathcal{B}_i$. The parameter b_i can be selected as the mean value of the α-level set when $\alpha = \mu_i$ [11].

26.9 Fuzzy Models

Given a set of N examples $\{(\boldsymbol{x}_p, \boldsymbol{y}_p) \,|\boldsymbol{x}_p \in R^n, \boldsymbol{y}_p \in R^m\}$, the underlying system can be identified by using some fuzzy models. Two popular fuzzy inference system models are the Mamdani and TSK models.

26.9.1 Mamdani Model

For the Mamdani model with N_r rules, the ith rule is given by

$$R_i : \text{ IF } \boldsymbol{x} \text{ is } \mathcal{A}_i, \text{ THEN } \boldsymbol{y} \text{ is } \mathcal{B}_i, \quad i = 1, \ldots, N_r,$$

where $\mathcal{A}_i = \{A_i^1, A_i^2, \ldots, A_i^n\}$, $\mathcal{B}_i = \{B_i^1, B_i^2, \ldots, B_i^m\}$, and A_i^j, B_i^k are, respectively, fuzzy sets that define an input and output space partitioning.

For an n-tuple input in the form of "$\boldsymbol{x}$ is $\mathcal{A}'$," the output "$\boldsymbol{y}$ is $\mathcal{B}'$" is characterized by combining the rules according to

$$\mu_{\mathcal{B}'}(\boldsymbol{y}) = \bigvee_{i=1}^{N_r} \left(\mu_{\mathcal{A}_i'}(\boldsymbol{x}) \wedge \mu_{\mathcal{B}_i}(\boldsymbol{y})\right), \tag{26.58}$$

where $\mathcal{A}' = \{A'^1, A'^2, \ldots, A'^n\}, \mathcal{B}' = \{B'^1, B'^2, \ldots, B'^m\}$, and A'^j, B'^k are, respectively, fuzzy sets that define an input and output space partitioning,

$$\mu_{\mathcal{A}_i'}(\boldsymbol{x}) = \mu_{\mathcal{A}'}(\boldsymbol{x}) \wedge \mu_{\mathcal{A}_i}(\boldsymbol{x}) = \bigwedge_{j=1}^{n} \left(\mu_{A'^j} \wedge \mu_{A_i^j}\right), \tag{26.59}$$

$\mu_{\mathcal{A}'}(\boldsymbol{x}) = \bigwedge_{j=1}^n \mu_{A'^j}$ and $\mu_{\mathcal{A}_i}(\boldsymbol{x}) = \bigwedge_{j=1}^n \mu_{A_i^j}$ being, respectively, the membership degrees of $\boldsymbol{x}$ to the fuzzy sets $\mathcal{A}'$ and $\mathcal{A}_i$, $\mu_{\mathcal{B}_i}(\boldsymbol{y}) = \bigwedge_{k=1}^m \mu_{B_i^k}$ is the membership degree of $\boldsymbol{y}$ to the fuzzy set $\mathcal{B}_i$, $\mu_{A_i'^j}$ is the association between the jth input of $\mathcal{A}'$ and the ith rule, $\mu_{B_i^k}$ is the association between the kth input of $\mathcal{B}$ and the ith rule, $\wedge$ is the intersection operator, and $\vee$ is the union operator.

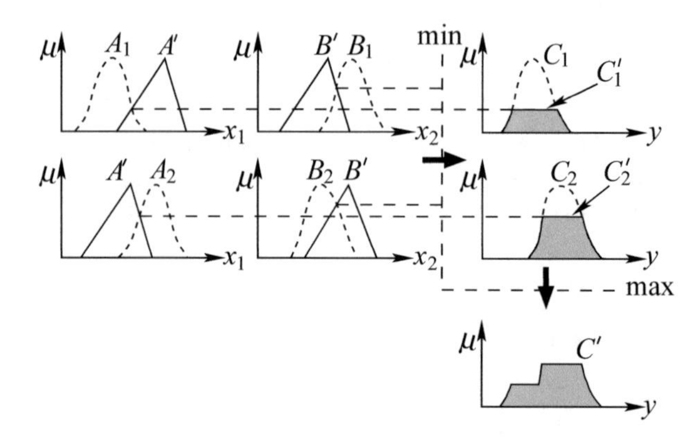

Fig. 26.7 The inference procedure of the Mamdani model with the max–min composition and fuzzy inputs

Minimum and product are the most common intersection operators. When minimum and maximum are, respectively, used as the intersection and union operators, the Mamdani model is called a *max–min model*. Kosko's standard additive model [18] has the same rule form, but it uses the product operator for the fuzzy intersection operation, and sup-product as well as addition as the composition operators.

We now illustrate the inference procedure for the Mamdani model. Assume that we have a two-rule Mamdani system with the rules of the form

$$R_i : \text{IF } x_1 \text{ is } \mathcal{A}_i \text{ and } x_2 \text{ is } \mathcal{B}_i, \text{ THEN } y \text{ is } \mathcal{C}_i, \text{ for } i = 1, 2.$$

When the max–min composition is employed, for the inputs "x_1 is $\mathcal{A}'$" and "x_2 is $\mathcal{B}'$", the fuzzy reasoning procedure for the output y is illustrated in Fig. 26.7. When two crisp inputs x_1' and x_2' are fed, the derivation of the output y' is illustrated in Fig. 26.8. As a comparison with Figs. 26.7 and 26.9 illustrates the result when the max-product composition is used to replace the max–min composition. A defuzzification strategy is needed to get a crisp output value.

The Mamdani model offers a high semantic level and a good generalization capability. It contains fuzzy rules built from expert knowledge. However, fuzzy inference systems based only on expert knowledge may result in insufficient accuracy. For accurate numerical approximation, the TSK model can usually generate better performance.

26.9.2 *Takagi–Sugeno–Kang Model*

In the TSK model [36], for the same set of examples $\left\{\left(\boldsymbol{x}_p, \boldsymbol{y}_p\right)\right\}$, fuzzy rules are given in the form

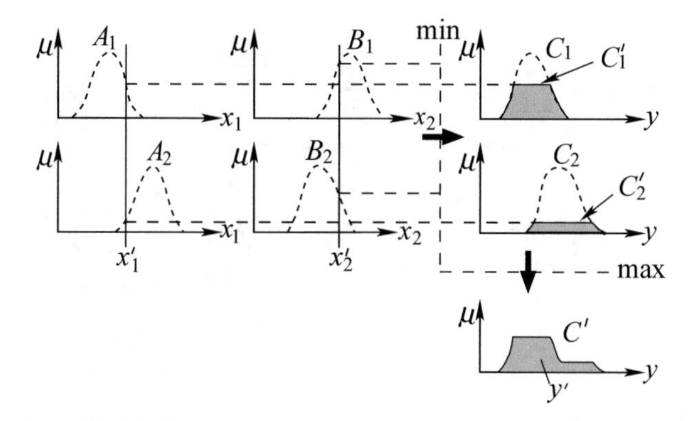

Fig. 26.8 The inference procedure of the Mamdani model with the max–min composition and crisp inputs

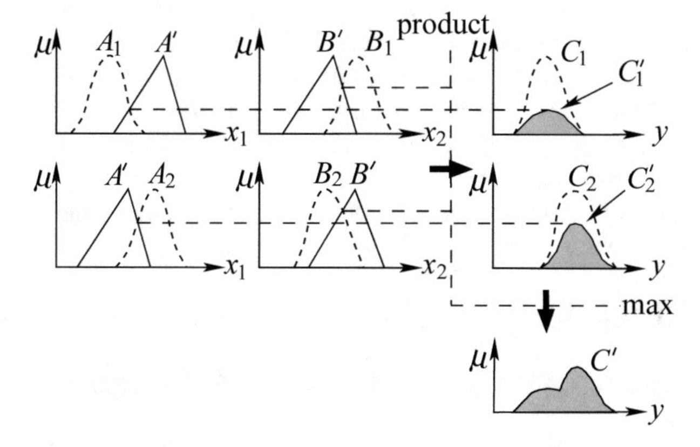

Fig. 26.9 The inference procedure of the Mamdani model with the max-product composition and fuzzy inputs

$$R_i : \text{IF } \boldsymbol{x} \text{ is } \mathcal{A}_i, \text{ THEN } \boldsymbol{y} = \boldsymbol{f}_i(\boldsymbol{x}), \quad i = 1, 2, \ldots, N_r,$$

where $\boldsymbol{f}_i(\boldsymbol{x}) = \left(f_i^1(\boldsymbol{x}), f_i^2(\boldsymbol{x}), \ldots, f_i^m(\boldsymbol{x})\right)^T$ is a crisp vector function of $\boldsymbol{x}$, and $f_i^j(\boldsymbol{x})$ is typically selected as a linear relation of $\boldsymbol{x}$

$$f_i^j(\boldsymbol{x}) = a_{i,0}^j + a_{i,1}^j x_1 + \cdots + a_{i,n}^j x_n, \tag{26.60}$$

with $a_{i,k}^j$, $k = 0, 1, \ldots, n$, being adjustable parameters.

For an n-tuple input in the form of "$\boldsymbol{x}$ is $\mathcal{A}'$," the output $\boldsymbol{y}'$ is obtained by combining the rules according to

$$\boldsymbol{y}' = \frac{\sum_{i=1}^{N_r} \mu_{\mathcal{A}_i'}(\boldsymbol{x}) \boldsymbol{f}_i(\boldsymbol{x})}{\sum_{i=1}^{N_r} \mu_{\mathcal{A}_i'}(\boldsymbol{x})}, \tag{26.61}$$

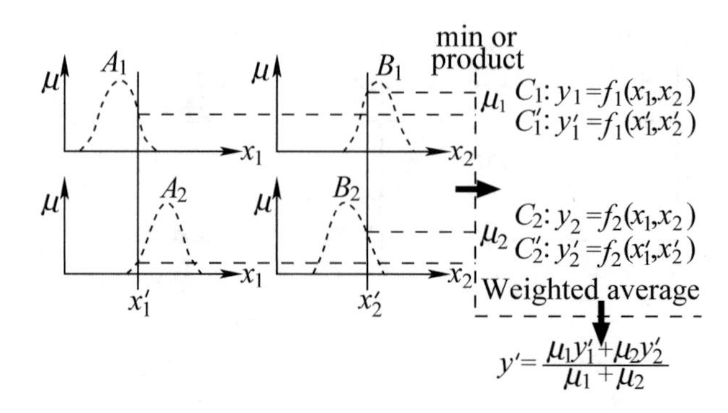

Fig. 26.10 The inference procedure for the TSK model with the min or product operator

where $\mu_{A_i'}(\boldsymbol{x})$ is defined by (26.59). This model produces a real-valued function, and it is essentially a model-based fuzzy control method. The stability analysis of the TSK model is given in [38]. When $f_i^j(\cdot)$ are first-order polynomials, the model is termed the *first-order TSK model*, which is the typical form of the TSK model. When $f_i^j(\cdot)$ are constants, it is called the *zero-order TSK model*, which can be viewed as a special case of the Mamdani model.

Similarly, we illustrate the inference procedure of the TSK model. Given a two-rule TSK fuzzy inference system with the rules of the form

$$R_i : \text{ IF } x_1 \text{ is } \mathcal{A}_i \text{ and } x_2 \text{ is } \mathcal{B}_i, \text{ THEN } y = f(x_1, x_2), \text{ for } i = 1, 2.$$

When two crisp inputs x_1' and x_2' are fed, the inference for the output y' is as illustrated in Fig. 26.10.

In comparison with the Mamdani model, the TSK model, which is based on automatic learning from the data, can accurately approximate a function using fewer rules. It has a stronger and more flexible representation capability than the Mamdani model. In the TSK model, rules are extracted from the data, but the generated rules may be meaningless to experts. The TSK model has found more successful applications in building fuzzy systems.

26.10 Complex Fuzzy Logic

Complex fuzzy sets and logic are mathematical extensions of fuzzy sets and logic from the real domain to the complex domain [34, 35]. A complex fuzzy set $\mathcal{S}$ is characterized by a complex-valued membership function, and the membership degree of any element x in $\mathcal{S}$ is given by a complex value of the form

$$\mu_{\mathcal{S}}(x) = r_{\mathcal{S}}(x) \mathrm{e}^{j\varphi_{\mathcal{S}}(x)}, \tag{26.62}$$

where the amplitude $r_S(x) \in [0, 1]$, and φ_S is the phase, that is, $\mu_S(x)$ is within a unit circle in the complex plane.

In [34, 35], basic set operators for fuzzy logic have been extended for the complex fuzzy logic, and some additional operators, such as the vector aggregation, set rotation, and set reflection, are also defined. The operations of intersection, union, and complement for complex fuzzy sets are defined on the modulus of the complex membership degree without consideration of its phase information. In [8], the complex fuzzy logic is extended to a logic of vectors in the plane, rather than scalars.

Complex fuzzy sets are superior to the Cartesian products of two fuzzy sets. Complex fuzzy logic maintains both the advantages of fuzzy logic and the properties of complex fuzzy sets. In complex fuzzy logic, rules constructed are strongly related and a relation manifested in the phase term is associated with complex fuzzy implications. In a complex fuzzy inference system, the output of each rule is a complex fuzzy set, and phase terms are necessary when combining multiple rules so as to generate the final output.

The fuzzy complex number [3] is a different concept from the complex fuzzy set [35]. The fuzzy complex number was introduced by incorporating the complex number into the support of the fuzzy set. A fuzzy complex number is a fuzzy set of complex numbers, which have real-valued membership degree in the range [0, 1]. The operations of addition, subtraction, multiplication, and division for fuzzy complex numbers are derived using the extension principle, and closure of the set of fuzzy complex numbers is proved under each of these operators. To sum up, a fuzzy complex number is a fuzzy set in one dimension, while a complex fuzzy set or number is a fuzzy set in two dimensions.

26.11 Possibility Theory

Fuzzy logic can be treated as a possibility theory. A possibility of 0.5 for element x may imply a 0.5 degree of evidence or belief that x belongs to a certain set. In possibility theory, possibility distribution is analogous to the notion of probability distribution in probability theory.

Let V be a variable that takes values on some element x of $\mathcal{X}$ and $\mathcal{A}$ be a fuzzy set, defined by $\mu_{\mathcal{A}}(x)$. The possibility of V being assigned x, $\pi_V(x)$, is $\mu_{\mathcal{A}}(x)$. Possibility distributions are fuzzy sets.

Both fuzzy theory and probability theory express uncertainty, and use variables having values in the range of [0, 1]. The membership function $\mu_{\mathcal{A}}(x)$ is defined as a possibility distribution function for set $\mathcal{A}$ on the universal set $\mathcal{X}$. Possibility measures are softer than probability measures, and their interpretations are different. Probability has a frequentistic interpretation: it quantifies the frequency of occurrence of an event. It is defined on a sample space $\mathcal{S}$ and must sum up to one. Possibility has a context-dependent interpretation: it quantifies the meaning of an event. It is defined on a universal set $\mathcal{X}$ but there is no limit for the sum. Probability is an objectivistic measure, while possibility or fuzzy measures are subjectivistic measures.

The probability/possibility consistency principle [46] states that possibility is an upper bound for probability. That is, the possibility $\mu(\mathcal{A})$ and probability $P(\mathcal{A})$ of an event $\mathcal{A}$ have the relation: $\mu(\mathcal{A}) \geq P(\mathcal{A})$. If an event is not possible, it is not probable.

Example 26.2 Suppose the proposition: "Cynric has x siblings, $x \in \mathcal{N} = \{1, 2, 3, 4, \ldots 8\}$." Both probability distribution and possibility distribution can be used to define x in $\mathcal{N}$. The probability and possibility of having x siblings are denoted by $P(x)$ and $\mu_{\mathcal{N}}(x)$, respectively. The set $\mathcal{N}$ is considered as a sample space in the probability distribution and as a universal set in the possibility distribution.

x	1	2	3	4	5	6	7	8
$P(x)$	0.5	0.3	0.15	0.05	0	0	0	0
$\mu(x)$	1.0	1.0	0.8	0.3	0.1	0	0	0

From the above table, we see that the sum of the probabilities is 1 but that of the possibilities is greater than 1. We can see that higher possibility does not always mean higher probability, but lower possibility leads to lower probability.

Definition 26.35 (*Crisp Probability of a Fuzzy Event*) Let event $\mathcal{A}$ be a fuzzy event or a fuzzy set considered in space R^n, $\mathcal{A} = \{(x, \mu_{\mathcal{A}}(x))|x \in R^n\}$. The probability for $\mathcal{A}$ is defined by

$$P(\mathcal{A}) = \int_{\mathcal{A}} \mu_{\mathcal{A}} dP, \tag{26.63}$$

and alternatively,

$$P(\mathcal{A}) = \sum_{x \in \mathcal{A}} \mu_{\mathcal{A}}(x) P(x). \tag{26.64}$$

Given a fuzzy event in the sample space $\mathcal{S}$, $\mathcal{A} = \{(x, \mu_{\mathcal{A}}(x))|x \in \mathcal{S}\}$. The α-cut set of the event (set) $\mathcal{A}$ is given as $\mathcal{A}_\alpha = \{x|\mu_{\mathcal{A}}(x) \geq \alpha\}$. The probability of the α-cut event is given by

$$P(\mathcal{A}_\alpha) = \sum_{x \in \mathcal{A}_\alpha} P(x). \tag{26.65}$$

Here, $\mathcal{A}_\alpha$ is the union of mutually exclusive events. The probability of $\mathcal{A}_\alpha$ is the sum of the probability of each event in $\mathcal{A}_\alpha$. For the probability of the α-cut event, we can say that the possibility of the probability of $\mathcal{A}_\alpha$ being $P(\mathcal{A}_\alpha)$ is α.

Definition 26.36 (*Fuzzy probability of a fuzzy event*) Let fuzzy event $\mathcal{A}$, its α-cut event $\mathcal{A}_\alpha$ and the probability $P(\mathcal{A}_\alpha)$ be defined from the above procedure. The fuzzy probability of $\mathcal{A}$ is defined by

$$P(\mathcal{A}) = \{(P(\mathcal{A}_\alpha), \alpha)|\alpha \in [0, 1]\}. \tag{26.66}$$

26.12 Case-Based Reasoning

Experience-based reasoning is a widely used reasoning paradigm based on logical arguments. It models human reasoning. Case-based reasoning is only a method of experience-based reasoning. It relies on using encapsulated prior experiences as a basis for dealing with similar new situations, i.e., a case-based reasoner solves new problems by adapting solutions that were used to solve old problems. Case-based reasoning is a kind of similarity-based reasoning from a logical viewpoint. The system consists of case bases, each of which consists of the previously encountered problem and its solution, and experience-based reasoning.

There are five different types of case-based reasoning systems, and although they share similar features, each of them is more appropriate for a particular type of problem [1]: exemplar-based, instance-based, memory-based, analogy-based, and typical case-based reasoning. Case-based reasoning systems are normally used in problems for which it is difficult to define rules.

The traditional process models of case-based reasoning are the R^4 model [1] and the problem space model [14]. Both models basically describe the major process stages for performing case-based reasoning, i.e., case retrieval, case reuse, and case adaptation. In addition, the R^4 model stresses the cyclic feature of case-based reasoning; problem space model emphasizes the solution obtained based on two different types of similarity in the problem space and the solution space.

Case retrieval is a process of finding and retrieving a case or a set of cases in the case base that is considered to be similar to the current problem. Case adaptation is a process in which the solutions of previous similar cases with successful outcomes are modified to suit the current case, bearing in mind the lessons from previous similar cases with unsuccessful solutions. Case updating is a process of revising cases or insertion of new cases in the case base.

Fuzzy set-based models in case-based reasoning and a logical formalization of the basic case-based reasoning inference are proposed in [9]. A logical approach to case-based reasoning using fuzzy similarity relations is proposed in [33], based on the graded consequence relations named approximation entailment and proximity entailment. A unified logical foundation for the case-based reasoning cycle is based on an integration of traditional mathematical logic, fuzzy logic, and similarity-based reasoning [12].

26.13 Granular Computing and Ontology

Humans tend to think on granular, abstract levels rather than on the level of detailed and precise data. Too much information may cause an information overload that reduces the quality of human decisions and actions [7]. Granular computing approximates detailed machine-like information by a coarser presentation on a human-like level. Within granular computing, continuous variables are mapped into intervals for the formulation of linguistic variables. Important approaches to granular computing are fuzzy sets, interval regression [37] and granular box regression [32], rough sets [27, 28], soft sets [25], and shadowed sets [29].

The theory of rough sets [27, 28] was proposed by Pawlak in 1982 as a mathematical tool for managing ambiguity, vagueness, and general uncertainty that arise from granularity in the universe of discourse. It can be approached as an extension to the classical theory of sets. It is a framework for the construction of approximations of concepts when only incomplete information is available. The objects of the universe of discourse $\mathcal{U}$, called *rough sets*, can be identified only within the limits determined by the knowledge represented by a given indiscernibility relation, which defines a partition in $\mathcal{U}$. A rough set is an imprecise representation of a concept (set) in terms of a pair of subsets, namely a lower approximation and an upper approximation. The approximations themselves can be crisp, imprecise, or fuzzy. The lower approximation is the set of objects definitely belonging to the vague concept, whereas the upper approximation is the set of objects possibly belonging to the vague concept. These approximations are used to define the notions of discernibility matrices, discernibility functions, reducts, and dependency factors, all of which are necessary for reduction of knowledge [24]. Hybridizations for rule generation and exploiting the characteristics of rough sets with neural, fuzzy, and evolutionary approaches are available in the literature [24]. Fuzzy rough sets [39] are a generalization of rough sets to deal with both fuzziness and vagueness in data.

Fuzzy sets tend to capture vagueness exclusively through membership values. This poses a dilemma of excessive precision in describing the imprecise phenomenon. The notion of shadowed sets [29] tries to solve this problem of selecting the optimum level of resolution in precision. Shadowed sets can be sought as a symbolic representation of numeric fuzzy sets [30]. Three quantification levels that are elements of the set $\{0, 1, [0, 1]\}$ are utilized to simplify the relevant fuzzy sets in shadowed set theory. Conceptually, shadowed sets are close to rough sets. The concepts of negative region, lower bound, and boundary region in rough set theory correspond to three-logical values 0, 1 and [0, 1] in shadowed sets, namely excluded, included and uncertain, respectively. In this sense, shadowed sets bridges fuzzy and rough sets.

Ontology is defined as that branch of metaphysics concerned with the "nature of being." It typically has a more restricted definition, namely "a working model of entities and interactions." Ontologies, as sets of concepts and their interrelations in a specific domain, are a useful tool in the areas of digital libraries, the semantic web, and personalized information management. Many different kinds of semantic or linguistic relations can be defined between terms or concepts, such as synonymy, hypernymy (is-a), meronymy (part-of) relations. These relations and their representations are more formal than in a taxonomy, since ontologies are generally used to model complex knowledge about the real world and to infer additional knowledge.

Problems

26.1 Describe the state of an environment by quantifying temperature as very cold, cold, cool, comfortable, warm, hot, and very hot. Define an approximate universal of discourse. Represent state values using: (a) sets and (b) fuzzy sets.

26.2 Show that tv($\mathcal{P}$ or $\mathcal{Q}$) $= C(\text{tv}(\mathcal{P}), \text{tv}(\mathcal{Q}))$ for any t-conorm, where tv denotes true value.

26.3 Prove the relations given by (26.28) and (26.34).

26.4 List the truth table for the logical operator $\mathcal{P} \rightarrow \mathcal{Q}$.

26.5 Consider fuzzy sets $\mathcal{A}$ and $\mathcal{B}$, defined by membership functions

$$\mu_{\mathcal{A}}(x) = \frac{1}{1 + 0.2(x - 6)^2}, \quad \mu_{\mathcal{B}} = \frac{1}{1 + (x/4)^3}.$$

(a) Calculate the union and intersection of $\mathcal{A}$ and $\mathcal{B}$, and the complements of $\mathcal{A}$ and $\mathcal{B}$. Plot their membership functions.
(b) Plot the membership function of the product of $\mathcal{A}$ and $\mathcal{B}$.

26.6 Fuzzy set $\mathcal{A}$ describes "Temperature is higher than 35 °C" by using

$$\mu_{\mathcal{A}}(x) = \begin{cases} \frac{1}{1+(x-35)^{-2}}, & x > 35 \\ 0, & x \leq 35 \end{cases}$$

and fuzzy set $\mathcal{B}$ describes "Temperature is approximately 38 °C" by using

$$\mu_{\mathcal{B}}(x) = \frac{1}{1 + (x - 38)^5}.$$

(a) Define a fuzzy set $\mathcal{C}$: "Temperature is higher than 35 and approximately equal to 38."
(b) Draw the membership function for $\mathcal{C}$, by using three t-norms for representing and.
(c) Do the same when and is changed to or.

26.7 Consider the following fuzzy sets defined on the finite universe of discourse $\mathcal{X} = \{1, 2, 3, 4, 5\}$:

$$\mathcal{A} = \{0, 0.1, 0.3, 0.4, 1\}, \mathcal{B} = \{0, 0.1, 0.2, 0.3, 1\}, C = \{0.1, 0.2, 0.4, 0.6, 0\}.$$

(a) Verify the relations $\mathcal{B} \subseteq \mathcal{A}$ and $\mathcal{C} \subseteq \mathcal{B}$.
(b) Give the cardinalities of $\mathcal{A}$, $\mathcal{B}$, and $\mathcal{C}$.

26.8 Show that the Yager negation function $N(x) = (1 - x^w)^{1/w}$, $w \in (0, \infty)$, satisfies the definition for negation.

26.9 Develop a rule-based model to approximate $f(x) = x^3 + x + 4$ in the interval $x \in [-1, 3]$.

26.10 In a classical example given by Zadeh [46], the proposition "Hans ate V eggs for breakfast," where $V \in \{1, 2, \ldots\}$. A possibility distribution $\pi_V(x)$ and a probability distribution $p_V(x)$ are associated with V.

(a) For each x, propose your distributions.
(b) What is the difference between the two distributions?

26.11 The membership matrices of the relations $\mathcal{R}_1$ on $\mathcal{X} \times \mathcal{Y}$ and $\mathcal{R}_2$ on $\mathcal{Y} \times \mathcal{Z}$ are

$$\mathcal{R}_1 = \begin{bmatrix} 1.0 & 0.4 & 0.8 & 0.0 \\ 0.4 & 1.0 & 0.6 & 1.0 \\ 0.8 & 0.6 & 1.0 & 0.7 \\ 0.0 & 1.0 & 0.7 & 1.0 \end{bmatrix}, \quad \mathcal{R}_2 = \begin{bmatrix} 1.0 & 0.9 & 0.8 \\ 1.0 & 0.1 & 0.6 \\ 0.5 & 0.4 & 0.0 \\ 0.1 & 0.3 & 0.2 \end{bmatrix}.$$

Calculate the max–min composition $\mathcal{R}_1 \circ \mathcal{R}_2$.

26.12 What type of fuzzy controller/model do you prefer, Mamdani-type or TSK-type? Justify your choice. Build some Mamdani fuzzy rules and TSK fuzzy rules of your own.

26.13 Show that the following two characteristics of crisp set operators do not hold for fuzzy sets:

(a) $\mathcal{A} \cap \overline{\mathcal{A}} = \emptyset$.
(b) $\mathcal{A} \cup \overline{\mathcal{A}} = \mathcal{X}$.

26.14 A property unique to fuzzy sets (not for crisp sets) is $\mathcal{A} \cap \emptyset = \emptyset$. Justify this statement.

26.15 Show that the two fuzzy sets satisfy De Morgan's law:

$$\mu_{\mathcal{A}}(x) = \frac{1}{1 + (x - 8)}, \quad \mu_{\mathcal{B}}(x) = \frac{1}{1 + 2x^2}.$$

26.16 Consider the definition of fuzzy implication.

(a) Show that $I(x, 1) = 1, \forall x \in [0, 1]$.
(b) Show that the restriction of I to $\{0, 1\}^2$ coincides with the classical material implication.

26.17 Show that the following two implication functions satisfy the definition of fuzzy implication.

(a) The Lukasiewicz implication: $I(x, y) = \min(1, 1 - x + y)$.
(b) The Zadeh implication: $I(x, y) = \max(1 - x, \min(x, y))$.
(c) Plot these functions.

26.18 Show that the Godel and Goguen (also known as product) implications, given by

$$I(x, y) = \begin{cases} 1, & \text{if } x \le y \\ y, & \text{if } x > y \end{cases}, \quad I(x, y) = \begin{cases} 1, & \text{if } x \le y \\ y/x, & \text{if } x > y \end{cases},$$

are R-implications derived from the t-norms minimum and product, respectively.

26.19 Using the definition of fuzzy composition of two fuzzy relations, verify that

$$(R_1(x, y) \circ R_2(y, z))^{-1} = R_2^{-1}(z, y) \circ R_1^{-1}(y, x),$$

$$(R_1(X, Y) \circ R_2(Y, Z)) \circ R(Z, W) = R_1(X, Y) \circ (R_2(Y, Z) \circ R(Z, W))$$

are similar to crisp binary relations.

26.20 Assume that

$$\mathcal{R}_1(\mathcal{X}, \mathcal{Y}) = \begin{pmatrix} 0.3\ 1.0\ 0.9\ 0.2 \\ 1.0\ 0.2\ 0.0\ 0.4 \\ 0.5\ 0.8\ 0.6\ 1.0 \end{pmatrix}, \quad \mathcal{R}_2(\mathcal{Y}, \mathcal{Z}) = \begin{pmatrix} 0.9\ 0.4\ 0.8\ 0.2 \\ 0.4\ 0.8\ 0.7\ 0.4 \\ 0.1\ 0.4\ 0.6\ 0.1 \\ 0.1\ 0.6\ 0.7\ 1.0 \end{pmatrix}.$$

Compute $\mathcal{R}_1 \circ \mathcal{R}_2$.

26.21 Consider a probability distribution $P(x)$, and two events $\mathcal{A}$ and $\mathcal{B}$: $P(a) = 0.2$, $P(b) = 0.4$, $P(c) = 0.3$, $P(d) = 0.1$; $\mathcal{A} = \{a, b, c\}$; $\mathcal{B} = \{(a, 0.4),$ $(b, 0.8), (c, 0.9), (d, 0.2)\}$.

(a) Find the probability of event $\mathcal{A}$.
(b) Find the crisp probability of fuzzy event $\mathcal{B}$.
(c) Find the fuzzy probability of fuzzy event $\mathcal{B}$.

References

1. Aamodt, A., & Plaza, E. (1994). Case-based reasoning: Foundational issues, methodological variations, and system approaches. *AI Communications*, *7*(1), 39–59.
2. Bede, B., & Rudas, I. J. (2011). Approximation properties of fuzzy transforms. *Fuzzy Sets and Systems*, *180*(1), 20–40.
3. Buckley, J. J. (1989). Fuzzy complex numbers. *Fuzzy Sets and Systems*, *33*, 333–345.
4. Buckley, J. J. (1993). Sugeno type controllers are universal controllers. *Fuzzy Sets and Systems*, *53*, 299–304.
5. Buckley, J. J., Hayashi, Y., & Czogala, E. (1993). On the equivalence of neural nets and fuzzy expert systems. *Fuzzy Sets and Systems*, *53*, 129–134.
6. Buckley, J. J., & Eslami, E. (2002). *An introduction to fuzzy logic and fuzzy sets*. Heidelberg: Physica-Verlag.
7. Davis, J. G., & Ganeshan, S. (2009). Aversion to loss and information overload: An experimental investigation. In *Proceedings of the International Conference on Information Systems* (Paper no. 11). Phoenix, AZ.
8. Dick, S. (2005). Toward complex fuzzy logic. *IEEE Transactions on Fuzzy Systems*, *13*(3), 405–414.
9. Dubois, D., Esteva, F., Garcia, P., Godo, L., de Mantaras, R. L., & Prade, H. (1997). Fuzzy modelling of case-based reasoning and decision. In D. B. Leake & E. Plaza (Eds.), *Case-based reasoning research and development* (Vol. 1266, pp. 599–610). LNAI. Berlin: Springer.

10. Ferrari-Trecate, G., & Rovatti, R. (2002). Fuzzy systems with overlapping Gaussian concepts: Approximation properties in Sobolev norms. *Fuzzy Sets and Systems, 130*, 137–145.
11. Figueiredo, M., Gomides, F., Rocha, A., & Yager, R. (1993). Comparison of Yager's level set method for fuzzy logic control with Mamdani and Larsen methods. *IEEE Transactions on Fuzzy Systems, 2*, 156–159.
12. Finnie, G., & Sun, Z. (2003). A logical foundation for the case-based reasoning cycle. *International Journal of Intelligent Systems, 18*, 367–382.
13. Jang, J. S. R., & Sun, C. I. (1993). Functional equivalence between radial basis function networks and fuzzy inference systems. *IEEE Transactions on Neural Networks, 4*(1), 156–159.
14. Leake, D. (1996). *Case-based reasoning: Experiences, lessons, and future direction* (p. 420). Menlo Park: AAAI Press/MIT Press.
15. Li, H. X., & Chen, C. L. P. (2000). The equivalence between fuzzy logic systems and feedforward neural networks. *IEEE Transactions on Neural Networks, 11*(2), 356–365.
16. Karnik, N. N., & Mendel, J. M. (1999). Type-2 fuzzy logic systems. *IEEE Transactions on Fuzzy Systems, 7*(6), 643–658.
17. Kosko, B. (1992). Fuzzy system as universal approximators. In *Proceedings of IEEE International Conference on Fuzzy Systems* (pp. 1153–1162). San Diego, CA.
18. Kosko, B. (1997). *Fuzzy engineering*. Englewood Cliffs: Prentice Hall.
19. Kreinovich, V., Nguyen, H. T., & Yam, Y. (2000). Fuzzy systems are universal approximators for a smooth function and its derivatives. *International Journal of Intelligent Systems, 15*, 565–574.
20. Mamdani, E. H. (1974). Application of fuzzy algorithms for control of a simple dynamic plant. *Proceedings of the IEEE, 12*(1), 1585–1588.
21. Mantas, C. J., & Puche, J. M. (2008). Artificial neural networks are zero-order TSK fuzzy systems. *IEEE Transactions on Fuzzy Systems, 16*(3), 630–643.
22. Mas, M., Monserrat, M., Torrens, J., & Trillas, E. (2007). A survey on fuzzy implication functions. *IEEE Transactions on Fuzzy Systems, 15*(6), 1107–1121.
23. Mendel, J. M., Liu, F., & Zhai, D. (2009). α-plane representation for type-2 fuzzy sets: Theory and applications. *IEEE Transactions on Fuzzy Systems, 17*(5), 1189–1207.
24. Mitra, S., & Hayashi, Y. (2000). Neuro-fuzzy rule generation: Survey in soft computing framework. *IEEE Transactions on Neural Networks, 11*(3), 748–768.
25. Molodtsov, D. (1999). Soft set theory–first results. *Computers & Mathematics with Applications, 37*, 19–31.
26. Mondal, B., & Raha, S. (2011). Similarity-based inverse approximate reasoning. *IEEE Transactions on Fuzzy Systems, 19*(6), 1058–1071.
27. Pawlak, Z. (1982). Rough sets. *International Journal of Computer and Information Sciences, 11*, 341–356.
28. Pawlak, Z. (1991). *Rough sets–Theoretical aspects of reasoning about data*. Dordrecht: Kluwer.
29. Pedrycz, W. (1998). Shadowed sets: Representing and processing fuzzy sets. *IEEE Transactions on Systems, Man, and Cybernetics, Part B, 28*, 103–109.
30. Pedrycz, W. (2009). From fuzzy sets to shadowed sets: Interpretation and computing. *International Journal of Intelligent Systems, 24*, 48–61.
31. Perfilieva, I. (2006). Fuzzy transforms: Theory and applications. *Fuzzy Sets and Systems, 157*, 993–1023.
32. Peters, G. (2011). Granular box regression. *IEEE Transactions on Fuzzy Systems, 19*(6), 1141–1152.
33. Plaza, E., Esteva, F., Garcia, P., Godo, L., & de Mantaras, R. L. (1996). A logical approach to case-based reasoning using fuzzy similarity relations. *Information Sciences, 106*, 105–122.
34. Ramot, D., Friedman, M., Langholz, G., & Kandel, A. (2003). Complex fuzzy logic. *IEEE Transactions on Fuzzy Systems, 11*(4), 450–461.
35. Ramot, D., Milo, R., Friedman, M., & Kandel, A. (2002). Complex fuzzy sets. *IEEE Transactions on Fuzzy Systems, 10*(2), 171–186.

36. Takagi, T., & Sugeno, M. (1985). Fuzzy identification of systems and its applications to modelling and control. *IEEE Transactions on Systems Man and Cybernetics*, *15*(1), 116–132.
37. Tanaka, H. (1987). Fuzzy data analysis by possibilistic linear models. *Fuzzy Sets and Systems*, *24*, 363–375.
38. Tanaka, K., & Sugeno, M. (1992). Stability analysis and design of fuzzy control systems. *Fuzzy Sets and Systems*, *45*, 135–150.
39. Tsang, E. C. C., Chen, D., Yeung, D. S., Wang, X.-Z., & Lee, J. W. T. (2008). Attributes reduction using fuzzy rough sets. *IEEE Transactions on Fuzzy Systems*, *16*(5), 1130–1141.
40. Wagner, C., & Hagras, H. (2010). Toward general type-2 fuzzy logic systems based on *z*Slices. *IEEE Transactions on Fuzzy Systems*, *18*(4), 637–660.
41. Wang, L. X. (1992). Fuzzy systems are universal approximators. In *Proceedings of IEEE International Conference on Fuzzy Systems* (pp. 1163–1170). San Diego, CA.
42. Wang, S., & Lu, H. (2003). Fuzzy system and CMAC network with B-spline membership/basis functions are smooth approximators. *Soft Computing*, *7*, 566–573.
43. Yen, J. (1999). Fuzzy logic–A modern perspective. *IEEE Transactions on Knowledge and Data Engineering*, *11*(1), 153–165.
44. Zadeh, L. A. (1965). Fuzzy sets. *Information and Control*, *8*, 338–353.
45. Zadeh, L. A. (1975). The concept of a linguistic variable and its application to approximate reasoning–I, II, III. *Information Sciences*, *8*, 199–249, 301–357; *9*, 43–80.
46. Zadeh, L. A. (1978). Fuzzy sets as a basis for theory of possibility. *Fuzzy Sets and Systems*, *1*, 3–28.

Chapter 27
Neurofuzzy Systems

27.1 Introduction

The neurofuzzy system is inspired by the biological–cognitive synergism in human intelligence. It is the synergism between the neuronal transduction/processing of sensory signals, and the corresponding cognitive, perceptual, and linguistic functions of the brain.

As an example, we describe how the human is aware of the ambient temperature. The human skin has two kinds of temperature receptors to sense the temperature: one for *warm* and the other for *cold* [57]. Neural fibers from numerous temperature receptors enter the spinal cord to form synapses. These neural signals are passed forward to medial and ventrobasal thalamus in the lower part of the brain, and then further carried to the cerebral cortex. The process can be modeled by a neural network. On the cerebral cortex, the temperature outputs are fused, and expressed linguistically as *cold*, *warm*, or *hot*. This part can be modeled by a fuzzy system. Based on this knowledge, one can make a decision on whether to put on more clothes or turn off the air conditioner.

Existing neurofuzzy systems are mainly customized for clustering, classification, and regression. The learning capability of neural networks is exploited to adapt the knowledge base from a given data, and this work is traditionally conducted by human experts. The application of fuzzy logic endows neural networks with the capability of explaining their actions. Neurofuzzy models usually achieve a faster convergence speed with smaller network size, compared to neural networks. Interpretability and accuracy are contradictory requirements: While interpretability is the capability to express the behavior of the real system in an understandable way, accuracy is the capability to represent faithfully the real system. A trade-off between the two edges must be achieved.

Both neural networks and fuzzy systems are dynamic, parallel distributed processing systems that estimate functions. Many neural networks and fuzzy systems are universal approximators. They estimate a function without any mathematical model and learn from experience with sample data. From the point of view of an

K.-L. Du and M. N. S. Swamy, *Neural Networks and Statistical Learning*,
https://doi.org/10.1007/978-1-4471-7452-3_27

expert system, fuzzy systems and neural networks are similar to inference systems. An inference system involves knowledge representation, reasoning, and knowledge acquisition:

- A trained neural network represents knowledge using connection weights and neurons in a distributed manner. In a fuzzy system, knowledge is represented using IF-THEN rules.
- When an input is presented to a neural network, an output is generated. This is a reasoning process. In a fuzzy system, reasoning is logic based.
- Knowledge acquisition is via learning in a neural network, while in a fuzzy system knowledge is encoded by a human expert.

Fuzzy systems can be applied to problems with knowledge represented in the form of IF-THEN rules. Problem-specific a priori knowledge can be integrated into the systems. Training pattern set and system modeling are not needed, and only heuristics are used. During the tuning process, one needs to add, remove, or change a rule, or even change the weight of a rule, using knowledge of experts. On the other hand, neural networks are useful when we have a training pattern set. A trained neural network is a black box that represents knowledge in its distributed structure. However, any prior knowledge of the problem cannot be incorporated into the learning process. It is difficult for human beings to understand the internal logic of the system. By extracting rules from neural networks, users can understand what neural networks have learned and how they predict.

27.1.1 Interpretability

A motivation for using fuzzy systems is due to their interpretability. Interpretability helps to check the plausibility of a system, leading to easy maintenance of the system. It can also be used to acquire knowledge from a problem characterized by numerical examples. An improvement in interpretability can enhance the performance of generalization when the data set is small.

The interpretability of a rule base is usually related to continuity, consistency, and completeness [25]. Continuity guarantees that small variations of the input do not induce large variations in the output. Consistency means that if two or more rules are simultaneously fired, their conclusions are coherent. Completeness means that for any possible input vector, at least one rule is fired and there is no inference breaking. Two neighboring fuzzy subsets in a fuzzy partition overlap.

When neurofuzzy systems are used to model nonlinear functions described by training sets, the approximation accuracy can be optimized by the learning procedure. However, since learning is accuracy oriented, it usually causes a reduction in the interpretability of the generated fuzzy system. The loss of interpretability can be due to [39]: incompleteness of fuzzy partitions, indistinguishability of fuzzy partitions (subsets), inconsistency of fuzzy rules, too fuzzy or too crisp fuzzy subsets, and non-compactness of the fuzzy system.

To improve the interpretability of neurofuzzy systems, one can add to the cost function, regularization terms that apply constraints on the parameters of fuzzy membership functions. For example, the order of the centers of all the fuzzy subsets $\mathcal{A}^i$, which are partitions of the fuzzy set $\mathcal{A}$, should be specified and remain unchanged during learning. Similar membership functions should be merged to improve the distinguishability of fuzzy partitions and to reduce the number of fuzzy subsets. One can also reduce the number of free parameters in defining fuzzy subsets. To increase the interpretability of the designed fuzzy system, the same linguistic term should be represented by the same membership function. This results in weight sharing [56]. For the TSK model, one practice for good interpretability is to keep the number of fuzzy subsets much smaller than the number of fuzzy rules N_r, especially when N_r is large.

27.2 Rule Extraction from Trained Neural Networks

There are many techniques for extracting rules from trained neural networks [6, 8, 10, 32, 37, 45]. This leads to the functional equivalence between neural networks and fuzzy rule-based systems.

27.2.1 Fuzzy Rules and Multilayer Perceptrons

For a three-layer MLP with $\phi^{(1)}(\cdot)$ as the logistic function and $\phi^{(2)}(\cdot)$ as the linear function, there always exists a fuzzy additive system that calculates the same function as the network does [6]. In [6], a fuzzy logic operator, called *interactive-or (i-or)*, is defined by applying the concept of f-duality to the logistic function. The use of i-or operator explains clearly the acquired knowledge of a trained MLP. The i-or operator is defined by [6]

$$a \otimes b = \frac{a \cdot b}{(1 - a) \cdot (1 - b) + a \cdot b}. \tag{27.1}$$

The i-or operator works on $(0, 1)$. It is a hybrid between both a t-norm and a t-conorm. Based on the i-or operator, the equality between MLPs and fuzzy inference systems has been established [6]. The equality proof also yields an automated procedure for knowledge acquisition. An extension of the method has been presented in [8].

In [20], relations between input uncertainties and fuzzy rules have been established. Sets of crisp logic rules applied to uncertain inputs have been shown to be equivalent to fuzzy rules with sigmoidal membership functions applied to crisp inputs. Crisp logic and fuzzy rule systems have been shown to be, respectively, equivalent to the logical network and the three-layer MLP. Keeping fuzziness on the input side enables an easier understanding of the networks or the rule systems.

In [10, 69], MLPs are interpreted by fuzzy rules in such a way that the sigmoidal activation function is decomposed into three TSK fuzzy rules with one TSK fuzzy rule for each partition. An algorithm for rule extraction given in [10] extracts $O(N)$ rules for N examples. Rule generation from a trained neural network can be done by analyzing the saturated zones of the fuzzy activation functions [69].

27.2.2 Fuzzy Rules and RBF Networks

The normalized RBF network is found functionally equivalent to a class of TSK systems [36]. For the convenience of presentation, we reproduce the output of the J_1-J_2-J_3 normalized RBF network given by (11.50):

$$y_j = \frac{\sum_{i=1}^{J_2} w_{ij}\phi\left(\|\boldsymbol{x} - \boldsymbol{c}_i\|\right)}{\sum_{i=1}^{J_2} \phi\left(\|\boldsymbol{x} - \boldsymbol{c}_i\|\right)}, \quad j = 1, \ldots, J_3. \tag{27.2}$$

When the t-norm in the TSK model is selected as algebraic product and the membership functions are selected as RBFs, the two models are mathematically equivalent [36]. Note that each hidden unit corresponds to a fuzzy rule.

To have a perfect match between $\phi\left(\|\boldsymbol{x} - \boldsymbol{c}_i\|\right)$ in (27.2) and $\mu_{\mathcal{A}_i'}(\boldsymbol{x})$ in (26.61), one is required to select factorizable $\phi\left(\|\boldsymbol{x} - \boldsymbol{c}_i\|\right)$ such that

$$\mu_{\mathcal{A}_i'}(\boldsymbol{x}) = \prod_{j=1}^{J_1} \mu_{A_i'^j}\left(x_j\right) \Longleftrightarrow \phi\left(\|\boldsymbol{x} - \boldsymbol{c}_i\|\right) = \prod_{j=1}^{J_1} \phi\left(\left|x_j - c_{i,j}\right|\right). \tag{27.3}$$

Each component $\phi\left(\left|x_j - c_{i,j}\right|\right)$ corresponds to a membership function $\mu_{A_i'^j}$. Note that the Gaussian RBF is the only strictly factorizable function.

In the normalized RBF network, w_{ij}s typically take constant values and the normalized RBF network corresponds to the zero-order TSK model. When the RBF weights are linear regression functions of the input variables, the model is functionally equivalent to the first-order TSK model.

In a practical implementation of the TSK model, one can select some $\mu_{A_i'^j} = 1$ or some $\mu_{A_i'^j} = \mu_{A_k'^j}$ in order to increase the distinguishability of the fuzzy partitions. Correspondingly, one should share some component RBFs or set some component RBFs to unity. This considerably reduces the effective number of free parameters in the RBF network. When implementing component RBF or membership function sharing, a Euclidean-like distance measure is used to describe the similarity between two-component RBFs. A gradient-descent procedure is conducted so as to extract interpretable fuzzy rules from a trained RBF network [39].

A fuzzy system can be first constructed according to heuristic knowledge and existing data, and then converted into an RBF network. This is followed by a refinement of the RBF network using a learning algorithm. Due to this learning procedure, the interpretability of the original fuzzy system may be lost. The RBF network is then again converted into interpretable fuzzy system, and knowledge is extracted from the network. This process refines the original fuzzy system design.

The fuzzy basis function network [79] has a structure similar to that of the RBF network. It is also based on the TSK model. It can readily adopt various learning algorithms developed for the RBF network.

27.2.3 Rule Extraction from SVMs

Rules can be extracted from trained SVMs. By using support vectors from a trained SVM, it is possible to use any RBF network learning technique for rule extraction, while avoiding the overlapping problem between classes [60]. Merging node centers and support vectors explanation rules can be obtained in the form of ellipsoids and hyperrectangles.

In decompositional approach, rejoins are formed in the input space, utilizing the SVM decision functions and the support vectors, which are then mapped to rules. Three types of rejoins are formed: ellipsoids [59], hyperrectangles [59], and hypercubes [22].

SVM+ prototype method [59] utilizes a clustering algorithm to determine prototype vectors for each class, which are then used together with the support vectors to define ellipsoid and hyperrectangle regions in the input space. Ellipsoids are then mapped to IF-THEN rules. This iterative procedure first trains an SVM model, which divides the training data into two subsets: those with positive predicted class and those with negative predicted class. For each of these subsets, clusters are generated. Based on the cluster prototype and the farthest support vector, interval or ellipsoid rules can be created. The rules extracted by this method are of high accuracy and fidelity; however, it produces a relatively large number of rules.

Rule extraction from linear SVMs or from any of the hyperplane-based linear classifiers is approached based on an LP formulation of SVMs with linear kernels [22]. Each rule extracted defines a hypercube, which is a subset of one of the bounded regions and must have one vertex that lies on the separating hyperplane for the rules to be disjoint. The method is decompositional as it is only applicable when the underlying model provides a linear decision boundary.

In [84], parsimonious L_2-SVM-based fuzzy classifiers are constructed considering model selection and feature ranking performed simultaneously, in which fuzzy rules are generated from data by L_2-SVM learning. As a prototype-based classifier, the L_2-SVM fuzzy classifier has the number of support vectors that equals the number of induced fuzzy rules.

An exact representation of SVMs as TSK fuzzy systems is given for every used kernel function in [9]. The behavior of SVMs is explained by means of fuzzy logic and the interpretability of the system is improved by introducing the λ-fuzzy rule-based system (λ-FRBS). λ-FRBS exactly approximates the SVM's decision boundary, and its rules and membership functions are very simple, aggregating the antecedents with uninorms as compensation operators. The rules of λ-FRBS are limited to two, and the number of fuzzy propositions in each rule only depends on the cardinality of the set of support vectors. Hence, λ-FRBS overcomes the curse of dimensionality.

27.2.4 Rule Generation from Other Neural Networks

Rule generation encompasses both rule extraction and rule refinement. Rule extraction is to extract knowledge from trained neural networks using the network parameters, while rule refinement is to refine the rules that are extracted from neural networks and initialized with crude domain knowledge.

Rule extraction from trained neural networks can be categorized into three main families: learning-based, decompositional, and eclectic. Learning-based approaches treat the model as a black box describing only the relationship between the inputs and the outputs. Decompositional approaches open the model, look into its individual components, and then attempt to extract rules at the level of these components. The eclectic approach lies in between the two families.

Feedforward networks generally do not have the capability to represent recursive rules when the depth of the recursion is not known a priori. Recurrent networks have the ability to store information over indefinite periods of time, to develop hidden states through learning, and thus to conveniently represent recursive linguistic rules [54]. They are particularly well suited for problem domains, where incomplete or contradictory prior knowledge is available. In such cases, knowledge revision or refinement is also possible.

Discrete-time recurrent networks have been used to correctly classify strings of a regular language [61]. Recurrent networks are suitable for crisp/fuzzy grammatical inference. For rule extraction from recurrent networks, the recurrent network is transformed into an equivalent deterministic finite-state automata by applying clustering algorithms in the output space of neurons. An augmented recurrent network that encodes fuzzy finite-state automata and recognizes a given fuzzy regular language with an arbitrary accuracy has been constructed in [62]. The granularity within both extraction techniques is at the level of ensemble of neurons, and thus, the approaches are not strictly decompositional. Rule extraction from recurrent networks aims to find models of a recurrent network, typically in the form of finite-state machines. This is carried out using four steps [33]: quantization of the continuous state space of the recurrent network, resulting in a discrete set of states; state and output generation by feeding the recurrent network with input patterns; construction of the corresponding deterministic finite-state automaton, based on the observed transitions; and minimization of the deterministic finite-state automaton.

In the all-permutations fuzzy rule base method [45], the input–output mapping of a specific fuzzy rule base is a linear sum of sigmoidal functions. This Mamdani-type fuzzy model is shown to be mathematically equivalent to the standard feedforward network. It was used to extract and insert symbolic information into feedforward networks [45]. The method is also used to extract symbolic knowledge from recurrent networks [46].

Rule extraction has also been carried out on Kohonen networks [75]. A comprehensive survey on rule generation from trained neural networks has been provided in [54], where the optimization capability of evolutionary algorithms are emphasized for rule refinement. An overview of rule extraction from recurrent networks is given in [33].

27.3 Extracting Rules from Numerical Data

Fuzzy inference systems can be designed directly from expert knowledge and data. The design process is usually decomposed into two phases, namely, rule generation and system optimization [25]. Rule generation leads to a basic system with a given space partitioning and the corresponding set of rules, while system optimization can be the optimization of membership parameters and rule base. Design of fuzzy rules can be conducted in one of three ways, namely, all possible combinations of fuzzy partitions, one rule for each data pair, or dynamically choosing the number of fuzzy sets.

For good interpretability, a suitable selection of variables and the reduction of the rule base are necessary. During the system optimization phase, merging techniques such as cluster merging and fuzzy-set merging are usually used for interpretability purposes. Fuzzy-set merging leads to higher interpretability than cluster merging. The reduction of a set of rules results in a loss of numerical performance on the training data set, but a more compact rule base has a better generalization capability and is also easier for human understanding.

Methods for designing fuzzy inference systems from data are analyzed and surveyed in [25], with emphasis on clustering methods for rule generation and evolutionary algorithms on system optimization. They are grouped into several families and compared based on rule interpretability.

27.3.1 Rule Generation Based on Fuzzy Partitioning

For rule generation, fuzzy partitioning is used for structure identification for fuzzy inference systems, and a learning algorithm is then used for parameter identification. There are usually three methods for partitioning the input space, namely grid partitioning, tree partitioning and scatter partitioning. These partitioning methods in the two-dimensional input space are illustrated in Fig. 27.1.

The grid structure has easy interpretability and is most widely used for generating fuzzy rules. Fuzzy sets of each variable are shared by all the rules. However, the number of fuzzy rules grows exponentially with input dimension, that is, the problem of *curse of dimensionality*. For n input variables, each being partitioned into m_i fuzzy

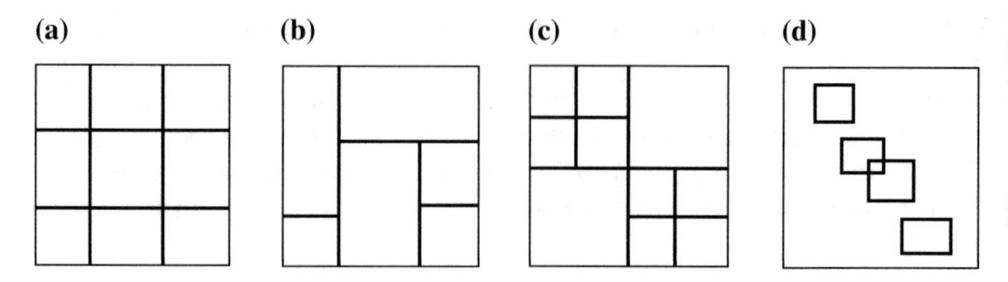

Fig. 27.1 Partitioning of the two-dimensional input space. **a** Grid partitioning. **b** *k-d* tree partitioning. **c** Multilevel grid partitioning. **d** Scatter partitioning

sets, a total of $\prod_{i=1}^{n} m_i$ rules are needed to cover the whole input space. Since each rule has a few parameters to adjust, there are too many parameters to adapt during the learning process. This reduces the interpretability of the fuzzy system. The grid structure is a static structure, and is appropriate for a low-dimensional data set with good coverage. The performance of the resultant model depends entirely on the initial definition of these grids. Thus, a training procedure can be applied to optimize the grid structure and the rule consequences. The grid structure is illustrated in Fig. 27.1a.

k-d tree and multilevel grid structures are two hierarchical partitioning techniques [71]. The input space is first partitioned roughly, and a subspace is recursively divided until a desired approximation performance is achieved. The *k-d* tree results from a series of *guillotine cuts* that is entirely across the subspace to be partitioned. After the ith guillotine cut, the entire space is partitioned into $i + 1$ regions. A *k-d* tree partitioning is illustrated in Fig. 27.1b. For the multilevel grid structure, the top-level grid coarsely partitions the whole space into equal-sized and evenly spaced fuzzy boxes, which are recursively partitioned into finer grids until a criterion is met. Hence, a multilevel grid structure is also called a *box tree*. The criterion can be that the resulting boxes have a similar number of training examples or that an application-specific evaluation in each grid is below a threshold. A multilevel grid partitioning is illustrated in Fig. 27.1c. A multilevel grid in the two-dimensional space is called a *quad tree*. Tree partitioning significantly relieves the problem of rule explosion, but it needs some heuristics to extract rules.

Scatter partitioning uses multidimensional antecedent fuzzy sets. It usually generates fewer fuzzy regions than the grid and tree partitioning techniques owing to the natural clustering property of training patterns. Fuzzy clustering algorithms form a family of rule-generation techniques. The training examples are gathered into homogeneous groups and a rule is associated with each group. The fuzzy sets are not shared by the rules, but each of them is tailored for one particular rule. Thus, the resulting fuzzy sets are usually difficult to interpret [25]. Scatter partitioning of high-dimensional feature spaces is difficult, and some learning or evolutionary procedures may be necessary. A scatter partitioning is illustrated in Fig. 27.1d. Some clustering-based methods for extracting fuzzy rule for function approximation have been proposed based on the TSK model. Clustering can be used for identification of the antecedent part of the model such as determination of the number of rules and initial rule parameters. Each cluster center corresponds to a fuzzy rule. The consequent part of the model can be estimated by the LS method. Based on the Mamdani model, a clustering-based method for function approximation is also given in [80].

27.3.2 Other Methods

Hierarchical structure for fuzzy rule systems can also effectively solve the rule-explosion problem [51, 77]. A hierarchical fuzzy system is comprised of a number of low-dimensional fuzzy systems connected in a hierarchical fashion. The low-dimensional fuzzy systems can be TSK systems, each constituting a level in the hierarchical fuzzy system. The total number of rules increases only linearly with the number of input variables. For a hierarchical fuzzy system comprised of $n-1$ two-input TSK systems, the n input variables are x_i, $i=1,\ldots,n$, the output is denoted by y, and y_i is the output of the ith TSK system:

$$y_i = f_i\left(y_{i-1}, x_{i+1}\right), \quad i = 1, \ldots, n-1, \tag{27.4}$$

where f_i is the nonlinear relation described by the ith TSK system, y_i is the output of the ith TSK system, and $y_0 = x_1$. The final output $y = y_{n-1}$ is easily obtained by a recursive procedure.

Hierarchical TSK systems [77] and generalized hierarchical TSK systems [51] have been shown to be universal approximators of any continuous function defined on a compact set. If there are n variables each of which is partitioned into m_i fuzzy subsets, the total number of rules is only $\sum_{i=1}^{n-1} m_i m_{i+1}$. However, the curse of dimensionality is inherent in the system. In a standard fuzzy system, the degree of freedom is unevenly distributed over the IF and THEN parts of the rules, with a comprehensive IF part to cover the whole domain and a simple THEN part. The hierarchical fuzzy system, on the other hand, provides with an incomplete IF part but a more complex THEN part. Generally, conventional fuzzy systems achieve universal approximation using piecewise-linear functions, while the hierarchical fuzzy system achieves it through piecewise-polynomial functions [51, 77].

Designing fuzzy systems from pattern pairs is a nonlinear regression problem. In a simple look-up-table technique [78], each pattern pair generates one fuzzy rule and then a selection process determines the important rules, which are used to construct the final fuzzy system. The input membership functions do not change with the sampling data, thus the designed fuzzy system uniformly covers the domain of interest. The input and output spaces are first divided into fuzzy regions, then a fuzzy rule is generated from a given pattern pair, and finally a degree is assigned to each rule to resolve rule conflicts and reduce the number of rules. When a new pattern pair becomes available, a rule is created for this pattern pair and the fuzzy rule base is updated. The look-up-table technique is implemented in five steps in [78, 80]. The fuzzy system thus constructed is proved to be a universal approximator by using the Stone–Weierstrass theorem [78]. The approach is a simple and fast one-pass procedure. This algorithm produces an enormous number of rules. The problem of contradictory rules also arises, and noisy data in the training examples will affect the consequence of a rule. In a similar grid partitioning based method, each datum generates one rule [1].

Many other general methods can be used to automatically extract fuzzy rules from a set of numerical examples and to build a fuzzy system for function approximation, such as heuristics-based approaches [73] and hybrid neural-fuzzy approaches [34]. A function approximation problem can be first converted into a pattern-classification problem, and then solved by using a fuzzy system [73]. The universe of discourse of the output variable is divided into multiple intervals, each regarded as a class, and then a class is assigned to each of the training data according to the desired value of the output variable. The data of each class are then partitioned in the input space to achieve a higher accuracy in the approximation of the class regions until a termination criterion is satisfied.

27.4 Synergy of Fuzzy Logic and Neural Networks

While neural networks have strong learning capabilities at the numerical level, it is difficult for the users to understand them at the logic level. Fuzzy logic, on the other hand, has a good capability of interpretability and can also integrate expert's knowledge. The synergy of both paradigms yields the capabilities of learning, good interpretation and incorporating prior knowledge.

The combination can be in different forms. The simplest form may be the concurrent neurofuzzy model, where a fuzzy system and a neural network work separately. The output of one system can be fed as the input to the other system. The cooperative neurofuzzy model corresponds to the case, where one system is used to adapt the parameters of the other system. Neural networks can be used to learn the membership values for fuzzy systems, to construct IF-THEN rules [24], or to construct a decision logic.

The true synergy of the two paradigms is a hybrid neural–fuzzy system, which captures the merits of both the systems. It can be in the form of either a fuzzy neural network or a neurofuzzy system. A hybrid neural–fuzzy system does not use multiplication, addition, or the sigmoidal function. Alternatively, fuzzy logic operations such as t-norm and t-conorm are used.

A fuzzy neural network is a neural network equipped with the capability of handling fuzzy information, where the input signals, activation functions, weights, and/or the operators are based on fuzzy-set theory. Thus, symbolic structure is incorporated [63]. The network can be represented in an equivalent rule-based format, where the premise is the concatenation of fuzzy AND and OR logic, and the consequence is the network output. The fuzzy AND and OR neurons are defined by

$$y_{\mathrm{AND}} = \wedge\left(\vee\left(w_1, x_1\right), \vee\left(w_2, x_2\right)\right) = T\left(C\left(w_1, x_1\right), C\left(w_2, x_2\right)\right), \tag{27.5}$$

$$y_{\mathrm{OR}} = \vee\left(\wedge\left(w_1, x_1\right), \wedge\left(w_2, x_2\right)\right) = C\left(T\left(w_1, x_1\right), T\left(w_2, x_2\right)\right). \tag{27.6}$$

Weights always have values in [0, 1], and negative weight is achieved by using the NOT operator. The weights of the fuzzy neural network can be interpreted as calibration factors of the conditions and rules. In [64], fuzzy logic networks for logic-based data analysis are treated. The networks are homogeneous architectures comprising OR/AND neurons. The developed network realizes a logic approximation of multidimensional mappings between unit hypercubes, that is, transformations from $[0, 1]^n$ to $[0, 1]^m$.

A neurofuzzy system is a fuzzy system whose parameters are learned by a learning algorithm obtained from neural networks. It can always be interpreted as a system of fuzzy rules. Learning is used to adaptively adjust the rules in the rule base, and to produce or optimize the membership functions of a fuzzy system. A neurofuzzy system has a neural-network architecture constructed from fuzzy reasoning. Structured knowledge is codified as fuzzy rules, while the adapting and learning capabilities of neural networks are retained. Expert knowledge can increase learning speed and estimation accuracy.

Both fuzzy neural networks and neurofuzzy systems can be treated as neural networks, where the units employ t-norm or t-conorm operator instead of an activation function. The weights are fuzzy sets, and the neurons apply t-norm or t-conorm operations. The hidden layers are usually used as rule layers. The layers before the rule layers perform as premise layers, while those after perform as consequent layers. As there is no distinct borderline between a neurofuzzy system or a fuzzy neural network, we call both types of synergisms as neurofuzzy systems. When only the input is fuzzy, it is a type-I neurofuzzy system. When everything except the input is fuzzy, we get a type-II model. A type-III model is defined as one where the inputs, weights, and shift terms are all fuzzy.

The functions realizing the inference process are usually nondifferentiable and thus, the popular gradient-descent or BP algorithm cannot always be applied for training neurofuzzy systems. To make use of gradient-based algorithms, one has to select differential functions. For nondifferentiable inference functions, training can be performed by using evolutionary algorithms. The shape of the membership functions, the number of fuzzy partitions, and the rule base can all be evolved by using evolutionary algorithms. Roughly speaking, the neurofuzzy method is superior to the neural-network method in terms of the convergence speed and compactness of the structure.

27.5 ANFIS Model

ANFIS is a well-known neurofuzzy model [34, 36, 37]. ANFIS model, shown in Fig. 27.2, has a six-layer (n-nK-K-K-K-1) architecture, and is a graphical representation of TSK model. The symbol N in the circles denotes the normalization operator, and $\boldsymbol{x} = (x_1, x_2, \ldots, x_n)^T$.

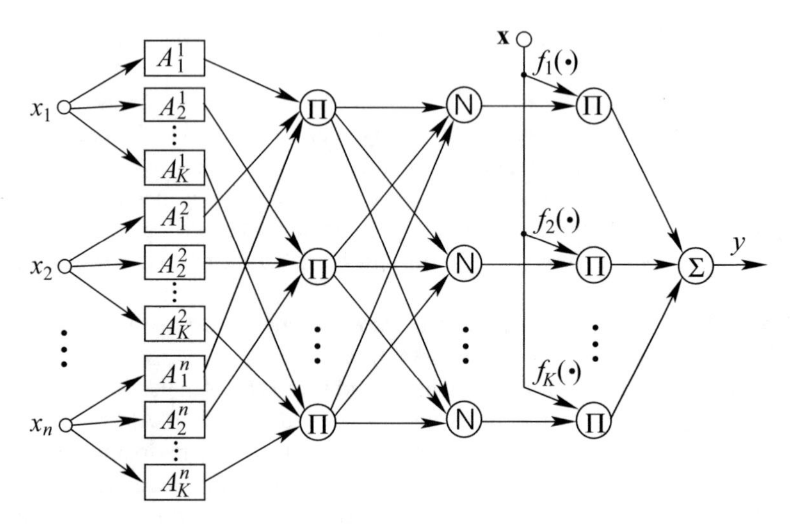

Fig. 27.2 ANFIS: graphical representation of TSK model

Layer 1 is the input layer with n nodes. Layer 2 has nK nodes, each outputting the membership value of the ith antecedent of the jth rule

$$o_{ij}^{(2)} = \mu_{\mathcal{A}_j^i}(x_i), \quad i = 1, \ldots, n, j = 1, \ldots, K, \tag{27.7}$$

where $\mathcal{A}_j^i$ defines a partition of the space of x_i, and $\mu_{\mathcal{A}_j^i}(x_i)$ is typically selected as a generalized bell membership function defined by (26.20), $\mu_{\mathcal{A}_j^i}(x_i) = \mu\left(x_i; c_j^i, a_j^i, b_j^i\right)$. The parameters c_j^i, a_j^i and b_j^i are referred to as *premise parameters*.

Layer 3 has K fuzzy neurons with the product t-norm as the aggregation operator. Each node corresponds to a rule, and the output of the jth neuron determines the degree of fulfillment of the jth rule

$$o_j^{(3)} = \prod_{i=1}^{n} \mu_{\mathcal{A}_j^i}(x_i), \quad j = 1, \ldots, K. \tag{27.8}$$

Each neuron in layer 4 performs normalization, and the outputs are called *normalized firing strengths*

$$o_j^{(4)} = \frac{o_j^{(3)}}{\sum_{k=1}^{K} o_k^{(3)}}, \quad j = 1, \ldots, K. \tag{27.9}$$

The output of each node in layer 5 is defined by

$$o_j^{(5)} = o_j^{(4)} f_j(\boldsymbol{x}), \quad j = 1, \ldots, K, \tag{27.10}$$

where $f_j(\cdot)$ is given for the jth node in layer 5. Parameters in $f_j(\boldsymbol{x})$ are referred to as *consequent parameters*.

The outputs of layer 5 are summed and the output of the network, $o^{(6)} = \sum_{j=1}^{K} o_j^{(5)}$, gives the TSK model (26.61).

In the ANFIS model, functions used at all the nodes are differentiable, thus BP can be used to train the network. Each membership function $\mu_{\mathcal{A}_i^j}$ is specified by a predefined shape and its corresponding shape parameters. The shape parameters are adjusted by a learning algorithm using a sample set of size N, $\{(\boldsymbol{x}_p, y_p)\}$. For nonlinear modeling, the effectiveness of the model is dependent on the membership functions used.

TSK fuzzy rules are employed in ANFIS model

$$R_i\text{: IF } \boldsymbol{x} \text{ is } \mathcal{A}_i, \text{ THEN } y = f_i(\boldsymbol{x}) = \sum_{j=1}^{n} a_{i,j} x_j + a_{i,0}, \quad i = 1, \ldots, K,$$

where $\mathcal{A}_i = \{\mathcal{A}_i^1, \mathcal{A}_i^2, \ldots, \mathcal{A}_i^n\}$ are fuzzy sets and $a_{i,j}$, $j = 0, 1, \ldots, n$, are consequent parameters. The output of the network for pattern p is thus given by

$$\hat{y}_p = \frac{\sum_{i=1}^{K} \mu_{\mathcal{A}_i}(\boldsymbol{x}_p) f_i(\boldsymbol{x})}{\sum_{i=1}^{K} \mu_{\mathcal{A}_i}(\boldsymbol{x}_p)}, \tag{27.11}$$

where $\mu_{\mathcal{A}_i}(\boldsymbol{x}_p) = \bigwedge_{j=1}^{n} \mu_{\mathcal{A}_i^j}(x_{p,j}) = \prod_{j=1}^{n} \mu_{\mathcal{A}_j^i}(x_{p,j})$. Accordingly, the error measure for pattern p is defined by

$$E_p = \frac{1}{2}(\hat{y}_p - y_p)^2. \tag{27.12}$$

After the rule base is specified, ANFIS adjusts only the membership functions of the antecedents and the consequent parameters. BP algorithm can be used to train both the premise and consequent parameters. A more efficient procedure is to learn the premise parameters by BP, but to learn the linear consequent parameters $a_{i,j}$ by the RLS method [34]. The learning rate η can be adaptively adjusted by using a heuristic used for MLP learning. This hybrid learning method provides better results than the MLP trained with BP and the cascade-correlation network [34]. Second-order methods are also applied for training ANFIS. Compared to the hybrid method, the LM method for ANFIS training [35] achieves a better precision, but the interpretability of the final membership functions is quite weak. In [12], RProp and RLS are used to learn the premise parameters and the consequent parameters, respectively.

ANFIS is attractive for applications in view of its network structure and the standard learning algorithm. However, it is computationally expensive due to the curse-of-dimensionality problem arising from grid partitioning. Constraints on membership functions and initialization using prior knowledge cannot be provided to ANFIS model due to the learning procedure. The learning results may be difficult to interpret. In order to preserve the plausibility of ANFIS, one can add some regularization terms to the cost function so that some constraints on interpretability are considered [37].

Coactive ANFIS [55] is a generalization of ANFIS obtained by introducing nonlinearity into the TSK rules. Generalized ANFIS [3] is based on a generalization of the TSK model and a generalized Gaussian RBF network. The generalized fuzzy model is trained by using the generalized RBF network model, based on the functional equivalence between the two models. In sigmoid-ANFIS [83], only sigmoidal membership functions are employed. Sigmoid-ANFIS is a combination of the additive TSK-type MLP and the additive TSK-type fuzzy inference system. It adopts the interactive-or operator as its fuzzy connectives.

Unfolding-in-time is a method to transform a recurrent network into a feedforward network so that BP algorithm can be used. ANFIS-unfolded-in-time [68] is a method that duplicates ANFIS T times to integrate temporal information, where T is the number of time intervals needed in the specific problem. ANFIS-unfolded-in-time is designed for prediction of time series data.

While the ANFIS model uses a fixed space partitioning with adaptive fuzzy rule parameters, adaptive parsimonious neurofuzzy systems can be achieved by using a constructive approach and a simultaneous adaptation of space partitioning and fuzzy rule parameters. The dynamic fuzzy neural network [81] is an online, constructive implementation of the TSK fuzzy system based on an extended RBF network and its learning algorithm. The extended RBF network has five layers and no bias, and the weights may be a linear regression of the input.

Example 27.1 We use ANFIS model to solve the IRIS classification problem. For the 120 patterns, the ranges of the input and output variables are $x_1 \in [4.3, 7.9]$, $x_2 \in [2.0, 4.4]$, $x_3 \in [1.0, 6.9]$, $x_4 \in [0.1, 2.5]$, $y \in [1, 3]$.

An initial TSK fuzzy inference system is first generated by using grid partitioning. Each of the variables is partitioned into 3 subsets. The Gaussian membership function is selected. The maximum epochs is 100. The fuzzy partitioning for the input space as well as the training error is illustrated in Fig. 27.3. The classification error rate is 0. The ANFIS model generates 193 nodes, 405 linear parameters, 24 nonlinear parameters, and 81 fuzzy rules. The training time is 53.70 s. The classification error for the training set is 0.

Example 27.2 We solve the IRIS problem using the ANFIS with scatter partitioning. Clustering the input space is a desired method for generating fuzzy rules. This can significantly reduce the total number of fuzzy rules, hence offer a better generalization capability. Subtractive clustering is used for rule extraction so as to find an initial fuzzy inference system for ANFIS training. Radius $r \in [0, 1]$ specifies the range of influence of the cluster center for each input or output dimension. The training error can be controlled by adjusting r. Specifying a smaller cluster radius usually yields more, smaller clusters in the data, and hence more rules.

Since the range of the input space is very small compared to that of the output space, we select $r = 0.8$ for all the input dimensions and the output space. The training time is 1.4203 s for 200 epochs. After training the MSE testing error is 0.0126. The ANFIS model has 37 nodes, 15 linear parameters, 24 nonlinear parameters, and 3 fuzzy rules. The classification error is 1.33%. The scatter partitioning is shown in Fig. 27.4a, b, and the training and testing errors are illustrated in Fig. 27.4c.

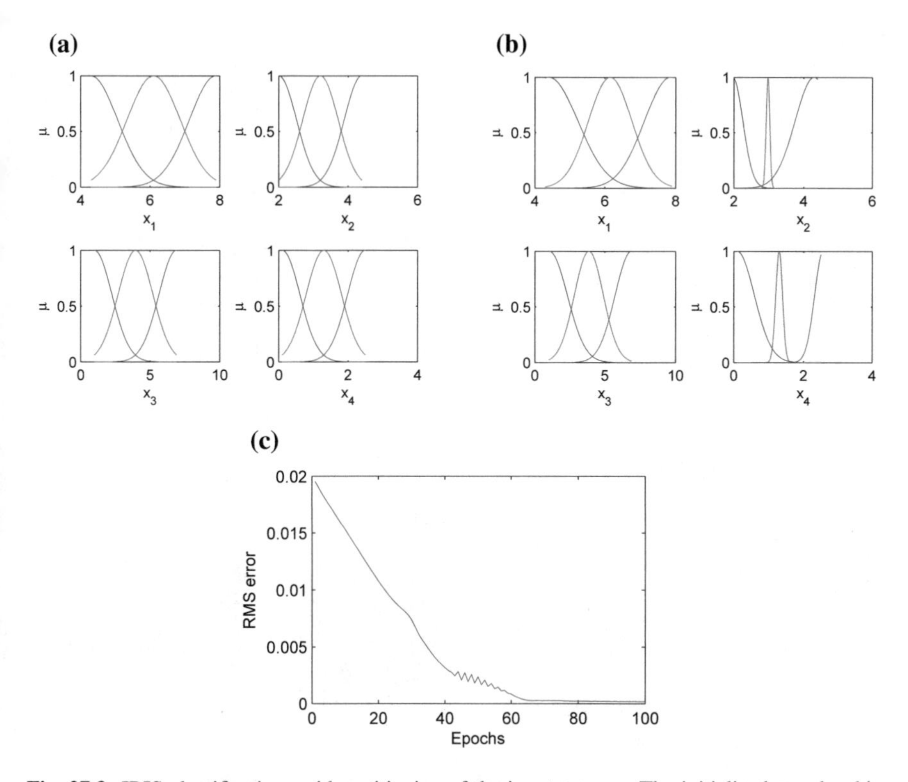

Fig. 27.3 IRIS classification: grid partitioning of the input space. **a** The initialized membership functions. **b** The learned membership functions. **c** The training RMS error

In order to further increase the training accuracy, we can select $r = 0.3$ for all the input dimensions and the output space to get finer clustering. Then we can get more rules. The ANFIS model has 107 nodes, 50 linear parameters, 80 nonlinear parameters, and 10 fuzzy rules. The training time is 3.3866 s for 200 epochs. After training, the MSE testing error is 1.5634×10^{-5}. The classification error is 0. The result is shown in Fig. 27.5.

For the 10 rules generated, each rule has its own membership function for each input variable. For example, the ith rule is given by

$$R_i\text{: IF } x_1 \text{ is } \mu_{i,1} \text{ AND } x_2 \text{ is } \mu_{i,2} \text{ AND } x_3 \text{ is } \mu_{i,3} \text{ AND } x_4 \text{ is } \mu_{i,4} \text{ THEN } y \text{ is } \mu_{i,y}$$

where $\mu_{i,k}$, $k = 1, \ldots, 4$, and $\mu_{i,y}$ are membership functions. Each row of plots in Fig. 27.5d corresponds to one rule, and each column corresponds to either an input variable x_i or the output variable y.

Example 27.3 Data is generated from the Mackey-Glass time-delay differential equation defined by

$$\frac{dx(t)}{dt} = \frac{0.2x(t-\tau)}{1 + x(t-\tau)^{10}} - 0.1x(t).$$

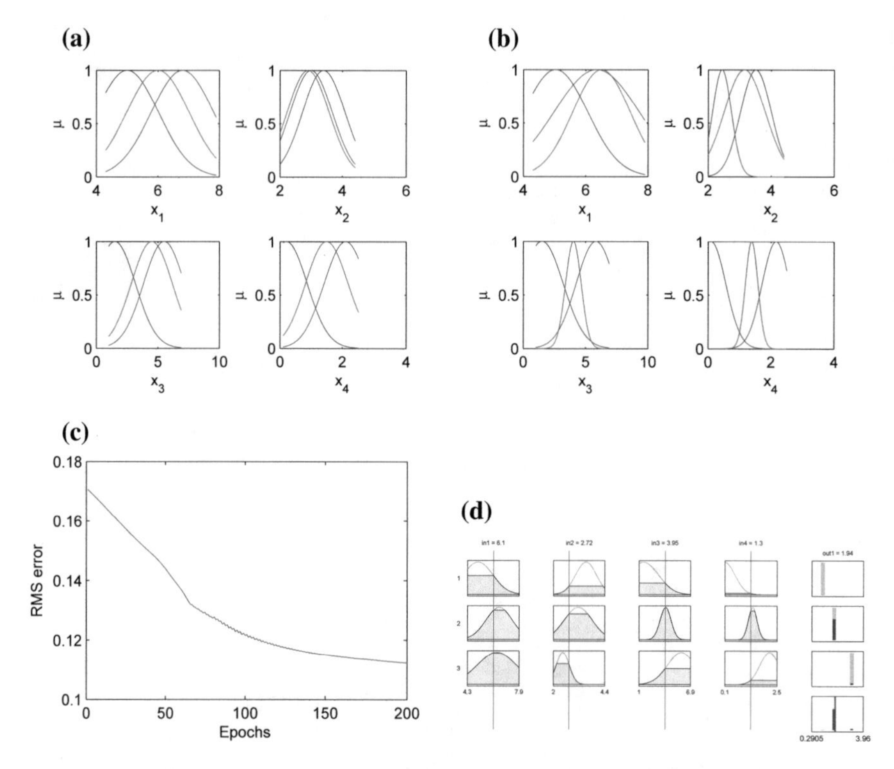

Fig. 27.4 IRIS classification: scatter partitioning of the input space. **a** The initialized membership functions. **b** The learned membership functions. **c** The training RMS error. **d** The 3 fuzzy rules generated. Note that some membership functions coincide. $\boldsymbol{r} = [0.8, 0.8, 0.8, 0.8, 0.8]$

When $x(0) = 1.2$ and $\tau = 17$, we have a non-periodic and non-convergent time series that is very sensitive to initial conditions. We assume $x(t) = 0$ when $t < 0$.

We build an ANFIS that can predict $x(t + 6)$ from the past values of this time series, that is, $x(t - 18)$, $x(t - 12)$, $x(t - 6)$, and $x(t)$. Therefore the training data format is $[x(t - 18), x(t - 12), x(t - 6), x(t); x(t + 6]$. From $t = 118$ to 1117, we collect 1000 data pairs of the above format. The first 500 are used for training while the others are used for checking.

We first generate an initial fuzzy inference system employing 2 membership functions using the generalized bell function and a grid partition using the training data, and then applying ANFIS. The number of training epochs is 10. The first 100 data points are ignored to avoid the transient portion of the data. The result is shown in Fig. 27.6.

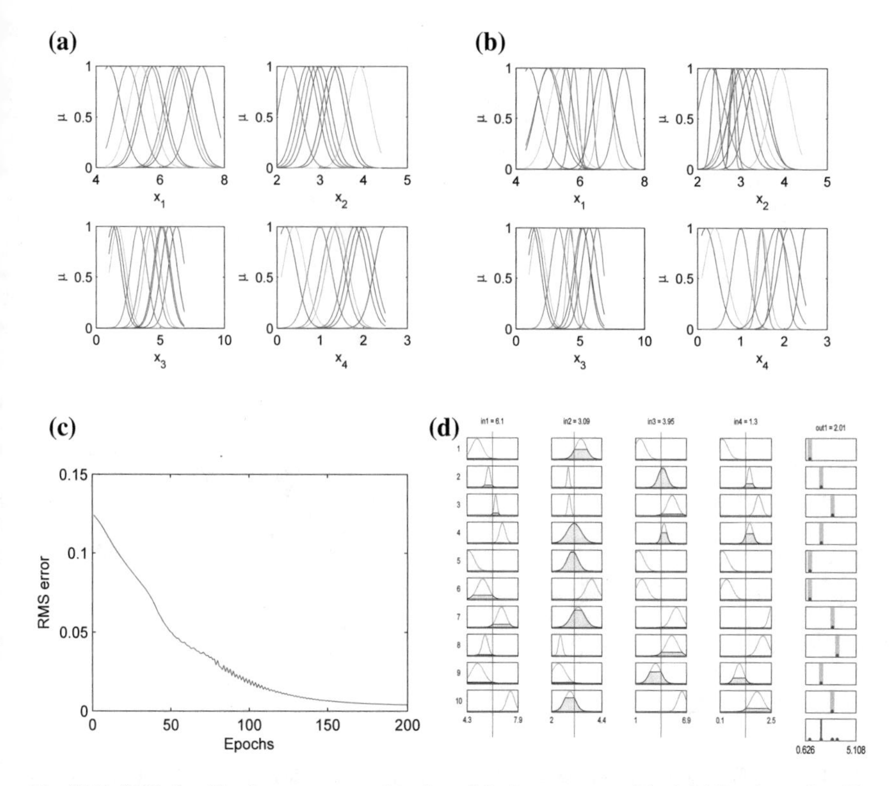

Fig. 27.5 IRIS classification: scatter partitioning of the input space. **a** The initialized membership functions. **b** The learned membership functions. **c** The training RMS error. **d** The 10 fuzzy rules generated. Note that some membership functions coincide. $\boldsymbol{r} = [0.3, 0.3, 0.3, 0.3, 0.3]$

27.6 Generic Fuzzy Perceptron

Generic fuzzy perceptron [56] has a structure similar to that of the three-layer MLP. The network inputs and the weights are modeled as fuzzy sets, and t-norm or t-conorm is used as the activation at each unit. The hidden layer acts as the rule layer. The output units usually use a defuzzification function. Generic fuzzy perceptron can interpret its structure in the form of linguistic rules and the structure of generic fuzzy perceptron can be treated as a linguistic rule base, where the weights between the input and hidden (rule) layers are called *fuzzy antecedent weights* and the weights between the hidden (rule) and output layers, *fuzzy consequent weights*. The generic fuzzy perceptron model is based on the Mamdani model. Due to the use of nondifferentiable t-norm and t-conorm, the gradient-descent method cannot be applied. A set of linguistic rules are used for describing the performance of the models. This knowledge-based fuzzy error is independent of the range of the output value. Based on the generic fuzzy perceptron model, there are three fuzzy models [56], namely neurofuzzy controller (NEFCON), neurofuzzy classification (NEFCLASS),

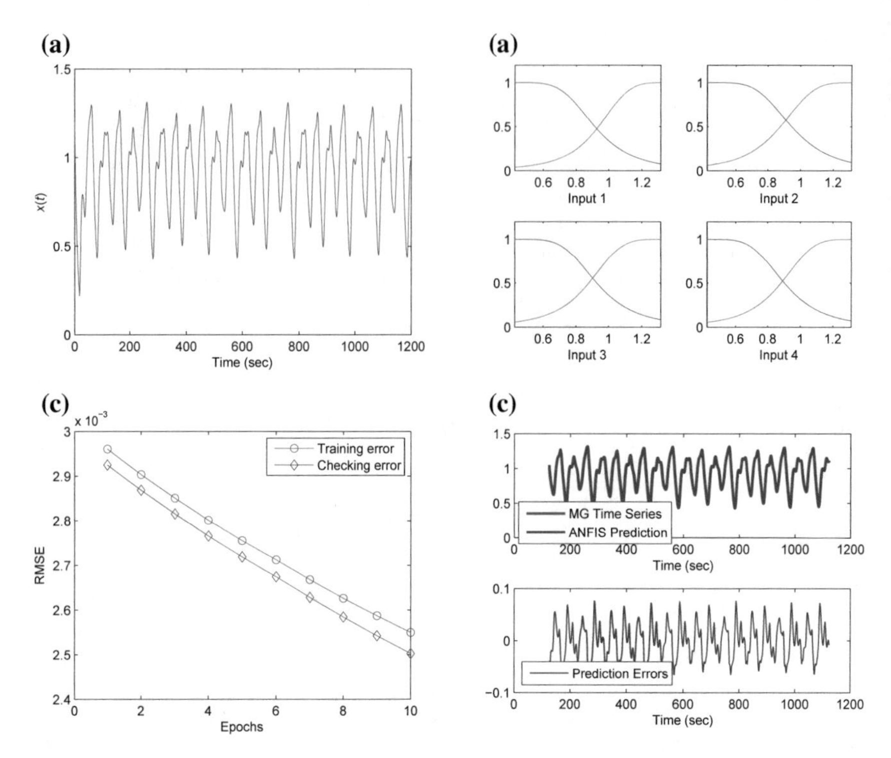

Fig. 27.6 ANFIS regression. **a** Data from Mackey-Glass chaotic time series. **b** The learned membership functions. **c** The training and testing RMS errors. **d** The prediction error

and neuronfuzzy function approximation (NEFPROX). Learning algorithms for all these models are derived from the fuzzy error using simple heuristics.

Initial fuzzy partitions are needed to be specified for each input variable. Some connections with identical linguistic values are forced to have the same weights so as to keep the interpretability. Prior knowledge can be integrated into the form of fuzzy rules to initialize the neurofuzzy systems, and the remaining rules are obtained by learning. NEFCON has a single output node, and is used for control. A reinforcement learning algorithm is used for online learning. NEFCLASS and NEFPROX can learn rules by using supervised learning. NEFCLASS does not use membership functions in the rules' consequents. NETPROX is more general. The architecture of NETPROX is shown in Fig. 27.7. When there is only a single output, NEFPROX has the same architecture as NEFCON, and when no membership functions are used in the consequent parts, NEFPROX has the same architecture as NEFCLASS. The hidden layer is the rule layer with each node corresponding to a rule, and the output layer is the defuzzification layer. All $\mu_i^{(1)}$ and $\mu_i^{(2)}$ are, respectively, the membership functions used in the premise and consequent parts. NEFPROX is an order of magnitude faster than ANFIS, but with a higher approximation error [56]. NEFPROX represents a Mamdani system with too many rules.

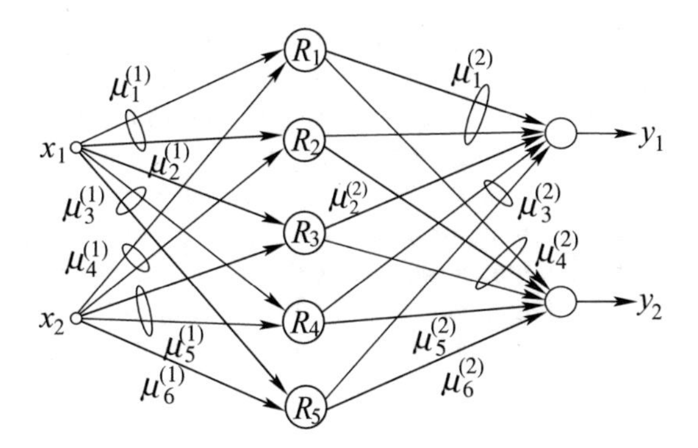

Fig. 27.7 NEFPROX as an example of the generic fuzzy perceptron model

27.7 Fuzzy SVMs

In the support-vector based fuzzy neural network for classification [48], the initial number of rules is equal to the number of Gaussian-kernel support vectors. A learning algorithm is then used to remove irrelevant fuzzy rules. The consequent part of each rule is of the fuzzy singleton type. A learning algorithm consists of three learning phases. First, the fuzzy rules and membership functions are determined by clustering. Then, the parameters of fuzzy neural network are calculated by the SVM with an adaptive fuzzy kernel function. Finally, the relevant fuzzy rules are selected.

The self-organizing TSK-type fuzzy network with support vector learning [42] is a fuzzy system constructed by the hybridization of fuzzy clustering and SVM. The antecedent part is generated via fuzzy clustering of the input data, and then SVM is used to tune the consequent part parameters.

Local SVR can model local behavior of models, but it still has the problem of boundary effects, which may generate a large bias on the boundary and also need more time to calculate. Fuzzy weighted SVR with fuzzy partition [16] eliminates the boundary effects. It first employs FCM clustering to split the training data into several training subsets. Then, local-regression models are independently obtained by SVR for each training subset. Those local-regression models are then combined by a fuzzy weighted mechanism to form the output. The approach needs less computational time and can have more accurate results than the local SVR does.

In a fuzzy modeling network based on SVM for regression and classification [15], the fuzzy basis function is regarded as the kernel function in an SVM, and fuzzy rules are generated based on the extracted support vectors. In [15], the rule base of a fuzzy system is extracted from learned SVM. A zeroth-order TSK fuzzy system is obtained; the number of fuzzy rules is equal to the number of support vectors.

TSK-based SVR [41] is motivated by TSK-type fuzzy rules and its parameters are learned by a combination of fuzzy clustering and linear SVR. In TSK-based SVR, a one-pass clustering algorithm clusters the input training data; a TSK-kernel, which corresponds to a TSK-type fuzzy rule, is then constructed by the product of a cluster output and a linear combination of input variables. The output is a linear weighted sum of the TSK-kernels.

Support vector interval regression network [38] utilizes the ϵ-SVR for interval regression analysis. A two-step approach constructs two independent RBF networks, identifying the lower and upper bounds of the data interval, respectively, after determining the initial structures and parameters of support vector interval regression networks through SVR mechanism. BP learning is employed to adjust the RBF networks. Support vector interval regression machine [30] evaluates the interval regression model combining the possibility estimation formulation and the property of central tendency with the principle of ϵ-SVR. It performs better for the data set with outliers and is conceptually simpler than the support vector interval regression network [38].

SVM is very sensitive to outliers or noises. This may result in the overfitting problem. Fuzzy SVM [47] deals with the overfitting problem. In fuzzy SVMs, training examples are assigned different fuzzy membership values based on their importance, and these membership values are incorporated into the SVM learning algorithm to make it less sensitive to outliers and noise. In [5], fuzzy SVM is improved for class imbalance learning in the presence of outliers and noise.

As there exist problems of finite samples and uncertain data in the estimation, in [82], the input and output variables are described as fuzzy numbers, and fuzzy ν-SVM is proposed by combining the fuzzy theory with ν-SVM. A fuzzy version of LS-SVM in [74] removes the unclassifiable regions for multiclass problems. Fuzzy one-class SVM [26] incorporates the concept of fuzzy-set theory into the one-class SVM model. It treats the training data points with different importance in the training process.

Total margin-based adaptive fuzzy SVM (TAF-SVM) [52] solves the class-boundary-skew problem due to the very imbalanced data sets and the overfitting problem resulted from outliers. By introducing the total margin algorithm to replace the conventional soft-margin algorithm, it achieves a lower generalization error bound.

27.8 Other Neurofuzzy Models

Neurofuzzy systems can employ network topologies similar to those of layered feedforward networks [27], RBF network [81], SOM model [76], and recurrent networks [49], mainly used for function approximation. They typically have a layered feedforward network architecture and are based on TSK-type fuzzy inference systems. Neurofuzzy systems are usually trained by using gradient descent, which in this case is sometimes termed as *fuzzy BP*. CG for training a neurofuzzy systems is also known as fuzzy CG algorithm [50].

Fuzzy clustering is primarily based on competitive learning networks such as Kohonen network and ART models. Based on the fuzzification of the linear autoassociative neural networks, fuzzy PCA [17] can extract a number of relevant features from high-dimensional fuzzy data. A fuzzy wavelet network has been introduced for approximating arbitrary nonlinear functions in [28].

Fuzzy perceptron network [11] is a type-I neurofuzzy system. The input to the network can be either fuzzy IF-THEN rules or numerical data. The learning scheme is derived based on the α-cut concept, which extends perceptron learning to fuzzy input vectors. Moreover, the fuzzy pocket algorithm is derived in [11] and then further incorporated into fuzzy perceptron learning scheme to tackle inseparable cases. Fuzzy BP [70] shows considerably greater convergence speed than BP does and can easily escape from local minima. For aggregation of input values (forward propagation), Sugeno fuzzy integral is employed. For weight learning, error backpropagation takes place. QuickFBP is a modification of fuzzy BP, where the modified computation of the net function is significantly faster [58]. Fuzzy BP is proved to be of exponential complexity in the case of large-sized networks with a large number of inputs, while QuickFBP is of polynomial complexity.

Hybrid neural fuzzy inference system (HyFIS) [44] is a five-layer neurofuzzy model based on the Mamdani fuzzy inference system. Layer 1 is the input layer with crisp values. Expert knowledge can be used for the initialization of these membership functions. HyFIS employs a hybrid learning scheme comprising two phases, namely, rule generation from data (structure learning), and rule tuning using BP learning (parameter learning). HyFIS first extracts fuzzy rules from data by using the look-up-table technique [78]. This is used as the initial structure so that the learning process can be fast, reliable and highly intuitive. Gradient descent is then applied to tune the membership functions of input/output linguistic variables and the network weights. Only a few training iterations are needed for the model to converge, since the initial structure and weights of the model are set properly. HyFIS is comparable in performance with that of ANFIS.

Fuzzy min–max neural networks are a class of neurofuzzy models using min–max hyperboxes for clustering, classification, and regression [23, 66, 67]. Max–min fuzzy Hopfield network [49] is a fuzzy recurrent network for fuzzy associative memory. The manipulations of the hyperboxes involve mainly comparison, addition and subtraction operations, thus learning is extremely efficient. An implicative fuzzy associative memory [72] consists of a network of completely interconnected Pedrycz logic neurons with threshold whose connection weights are determined by the minimum of implications of presynaptic and postsynaptic activations.

Fuzzy k-NN [43] replaces k-NN rule by associating each sample with a membership value expressing how closely the pattern belongs to a given class. In fuzzy k-NN, the importance of a neighbor is determined based on the relative distance between the neighbor and the test pattern. The algorithm possesses the possibilistic classification capability. It considers all training patterns as the neighbors with different degrees. It avoids the problem of choosing the optimal value of K.

To reduce the effect of outliers, fuzzy memberships are introduced to robustly estimate the scatter matrices, leading to fuzzy LDA [13]. Fuzzy LDA still cannot accommodate nonlinearly separable cases.

A fuzzy inference system is adopted to determine the learning rate for BSS [53], yielding good signal separation. Fuzzy FastICA [29] can handle fuzziness in the iterative algorithm by using clustering as preprocessing.

In continuous fuzzy reinforcement learning, fuzzy inference system is used to obtain an approximate model for the value function in continuous state space and to generate continuous actions [7]. A critic-only based fuzzy reinforcement learning uses only a fuzzy system for approximating action-value function, and generates action with probability proportional to this function. Fuzzy Q-learning [40] is a Q-learning method with linear function approximators using fuzzy systems, is a critic-only fuzzy reinforcement learning algorithm, and is applied to some problems with continuous state and action spaces. Fuzzy Sarsa learning [18] is based on linear Sarsa, and the existence of stationary points is established for it. It is an extension of Sarsa for continuous state and action spaces using fuzzy system as function approximator. It achieves higher learning speed and action quality compared to that of fuzzy Q-learning. A fuzzy balance management scheme between exploration and exploitation can be implemented in any critic-only fuzzy reinforcement learning method [19]. An enhanced fuzzy Sarsa learning [19] integrates an adaptive learning rate and a fuzzy balancer into fuzzy Sarsa learning.

Fuzzy RBM [14] extends RBM by replacing all the real-valued parameters with fuzzy numbers. It has better representation capability, and better robustness to noisy data, compared with RBM. In the learning process, the fuzzy free energy function is defuzzified before the probability is defined [14]. The fuzzy deep belief network [21] is devised based on fuzzy RBMs.

Interval-valued data provides a way of representing the available information in complex problems with uncertainty, inaccuracy or variability. When a neural network has at least one of its input, output or weight sets being interval valued, it is an interval neural network. The interval neural network for fuzzy regression analysis [31] has all its weights, biases and output being interval valued, but input data being crisp. This kind of interval neural network is proved to be a universal approximator [4]. Interval MLP [65] approximates nonlinear interval functions with interval-valued inputs and outputs, single-valued weights and biases, and a transfer function operating with interval-valued inputs and outputs. Interval-SVM [2] directly incorporates an interval approach into SVM by inserting information into SVM in the form of intervals.

Problems

27.1 The fuzzy AND and OR neurons are intrinsically excitatory, since higher input implies higher output. To generate inhibitory behaviors, one can negate the input x, that is, $1 - x$. A nonlinear activation function can also be applied to the AND or OR neuron output. These neurons can constitute a layered fuzzy neural network. Consider how a bias can be incorporated into the definition of the AND and OR neuron.

27.2 Train an ANFIS model to identify the nonlinear dynamic system given by

$$y(n) = \frac{y(n-1)y(n-2)(y(n-1)-1)}{1+y^2(n-1)+y^2(n-2)} + u(n-1).$$

The input $u(n)$ is uniformly selected in $[-1, 1]$ and the test input $u(n) = 0.5\sin(2\pi n/29)$. Produce 20,000 and 200 observation data for training and testing, respectively.

References

1. Abe, S., & Lan, M. S. (1995). Fuzzy rules extraction directly from numerical data for function approximation. *IEEE Transactions on Systems, Man, and Cybernetics*, *25*(1), 119–129.
2. Angulo, C., Anguita, D., Gonzalez-Abril, L., & Ortega, J. A. (2008). Support vector machines for interval discriminant analysis. *Neurocomputing*, *71*, 1220–1229.
3. Azeem, M. F., Hanmandlu, M., & Ahmad, N. (2000). Generalization of adaptive neuro-fuzzy inference systems. *IEEE Transactions on Neural Networks*, *11*(6), 1332–1346.
4. Baker, M. R., & Patil, R. B. (1998). Universal approximation theorem for interval neural networks. *Reliable Computing*, *4*, 235–239.
5. Batuwita, R., & Palade, V. (2010). FSVM-CIL: Fuzzy support vector machines for class imbalance learning. *IEEE Transactions on Fuzzy Systems*, *18*(3), 558–571.
6. Benitez, J. M., Castro, J. L., & Requena, I. (1997). Are artificial neural networks black boxes? *IEEE Transactions on Neural Networks*, *8*(5), 1156–1164.
7. Berenji, H. R., & Vengerov, D. (2003). A convergent actor–critic-based FRL algorithm with application to power management of wireless transmitters. *IEEE Transactions on Fuzzy Systems*, *11*(4), 478–485.
8. Castro, J. L., Mantas, C. J., & Benitez, J. M. (2002). Interpretation of artificial neural networks by means of fuzzy rules. *IEEE Transactions on Neural Networks*, *13*(1), 101–116.
9. Castro, J. L., Flores-Hidalgo, L. D., Mantas, C. J., & Puche, J. M. (2007). Extraction of fuzzy rules from support vector machines. *Fuzzy Sets and Systems*, *158*, 2057–2077.
10. Cechin, A., Epperlein, U., Koppenhoefer, B., & Rosenstiel, W. (1996). The extraction of Sugeno fuzzy rules from neural networks. In M. Verleysen (Ed.), *Proceedings of the European Symposium on Artificial Neural Networks* (pp. 49–54). Bruges, Belgium.
11. Chen, J. L., & Chang, J. Y. (2000). Fuzzy perceptron neural networks for classifiers with numerical data and linguistic rules as inputs. *IEEE Transactions on Fuzzy Systems*, *8*(6), 730–745.
12. Chen, M. S., & Liou, R. J. (1999). An efficient learning method of fuzzy inference system. In *Proceedings of IEEE International Fuzzy Systems* (pp. 634–638). Seoul, Korea.
13. Chen, Z.-P., Jiang, J.-H., Li, Y., Liang, Y.-Z., & Yu, R.-Q. (1999). Fuzzy linear discriminant analysis for chemical datasets. *Chemometrics and Intelligent Laboratory Systems*, *45*, 295–302.
14. Chen, C. L. P., Zhang, C.-Y., Chen, L., & Gan, M. (2015). Fuzzy restricted Boltzmann machine for the enhancement of deep learning. *IEEE Transactions on Fuzzy Systems*, *23*(6), 2163–2173.
15. Chiang, J. H., & Hao, P. Y. (2004). Support vector learning mechanism for fuzzy rule-based modeling: A new approach. *IEEE Transactions on Fuzzy Systems*, *12*(1), 1–12.
16. Chuang, C.-C. (2007). Fuzzy weighted support vector regression with a fuzzy partition. *IEEE Transactions on Systems, Man, and Cybernetics, Part B*, *37*(3), 630–640.
17. Denoeux, T., & Masson, M. H. (2004). Principal component analysis of fuzzy data using autoassociative neural networks. *IEEE Transactions on Fuzzy Systems*, *12*(3), 336–349.
18. Derhami, V., Majd, V. J., & Ahmadabadi, M. N. (2008). Fuzzy Sarsa learning and the proof of existence of its stationary points. *Asian Journal of Control*, *10*(5), 535–549.
19. Derhami, V., Majd, V. J., & Ahmadabadi, M. N. (2010). Exploration and exploitation balance management in fuzzy reinforcement learning. *Fuzzy Sets and Systems*, *161*, 578–595.

20. Duch, W. (2005). Uncertainty of data, fuzzy membership functions, and multilayer perceptrons. *IEEE Transactions on Neural Networks*, *16*(1), 10–23.
21. Feng, S., Chen, C. L. P., & Zhang, C.-Y. (2019). A fuzzy deep model based on fuzzy restricted boltzmann machines for high-dimensional data classification. *IEEE Transactions on Fuzzy Systems*. https://doi.org/10.1109/TFUZZ.2019.2902111 (in press).
22. Fung, G., Sandilya, S., & Rao, R. (2005). Rule extraction from linear support vector machines. In *Proceedings of the 11th ACM SIGKDD International Conference on Knowledge Discovery in Data Mining (KDD)* (pp. 32–40).
23. Gabrays, B., & Bargiela, A. (2000). General fuzzy min–max neural networks for clustering and classification. *IEEE Transactions on Neural Networks*, *11*(3), 769–783.
24. Gallant, S. I. (1988). Connectionist expert systems. *Communications of the ACM*, *31*(2), 152–169.
25. Guillaume, S. (2001). Designing fuzzy inference systems from data: An interpretability-oriented review. *IEEE Transactions on Fuzzy Systems*, *9*(3), 426–443.
26. Hao, P.-Y. (2008). Fuzzy one-class support vector machines. *Fuzzy Sets and Systems*, *159*, 2317–2336.
27. Hayashi, Y., Buckley, J. J., & Czogala, E. (1993). Fuzzy neural network with fuzzy signals and weights. *International Journal of Intelligent Systems*, *8*(4), 527–537.
28. Ho, D. W. C., Zhang, P. A., & Xu, J. (2001). Fuzzy wavelet networks for function learning. *IEEE Transactions on Fuzzy Systems*, *9*(1), 200–211.
29. Honda, K., Ichihashi, H., Ohue, M., & Kitaguchi, K. (2000). Extraction of local independent components using fuzzy clustering. In *Proceedings of the 6th International Conference on Soft Computing (IIZUKA2000)* (pp. 837–842).
30. Hwang, C., Hong, D. H., & Seok, K. H. (2006). Support vector interval regression machine for crisp input and output data. *Fuzzy Sets and Systems*, *157*, 1114–1125.
31. Ishibuchi, H., Tanaka, H., & Okada, H. (1993). An architecture of neural networks with interval weights and its application to fuzzy regression analysis. *Fuzzy Sets and Systems*, *57*, 27–39.
32. Ishikawa, M. (2000). Rule extraction by successive regularization. *Neural Networks*, *13*(10), 1171–1183.
33. Jacobsson, H. (2005). Rule extraction from recurrent neural networks: A taxonomy and review. *Neural Computation*, *17*(6), 1223–1263.
34. Jang, J. S. R. (1993). ANFIS: Adaptive-network-based fuzzy inference systems. *IEEE Transactions on Systems, Man, and Cybernetics*, *23*(3), 665–685.
35. Jang, J. S. R., & Mizutani, E. (1996). Levenberg–Marquardt method for ANFIS learning. In *Proceedings of Biennial International North American Fuzzy Information Processing (NAFIPS)* (pp. 87–91). Berkeley, CA.
36. Jang, J. S. R., & Sun, C. I. (1993). Functional equivalence between radial basis function Networks and fuzzy inference systems. *IEEE Transactions on Neural Networks*, *4*(1), 156–159.
37. Jang, J. S. R., & Sun, C. I. (1995). Neuro-fuzzy modeling and control. *Proceedings of the IEEE*, *83*(3), 378–406.
38. Jeng, J.-T., Chuang, C.-C., & Su, S.-F. (2003). Support vector interval regression networks for interval regression analysis. *Fuzzy Sets and Systems*, *138*, 283–300.
39. Jin, Y. (2003). *Advanced fuzzy systems design and applications*. Heidelberg, Germany: Physica-Verlag.
40. Jouffe, L. (1998). Fuzzy inference system learning by reinforcement methods. *IEEE Transactions on Systems, Man, and Cybernetics, Part C*, *28*(3), 338–355.
41. Juang, C.-F., & Hsieh, C.-D. (2009). TS-fuzzy system-based support vector regression. *Fuzzy Sets and Systems*, *160*, 2486–2504.
42. Juang, C.-F., Chiu, S.-H., & Chang, S.-W. (2007). A self-organizing TS-type fuzzy network with support vector learning and its application to classification problems. *IEEE Transactions on Fuzzy Systems*, *15*(5), 998–1008.
43. Keller, J. M., Gray, M. R., & Givens, J. A, Jr. (1985). A fuzzy K-nearest neighbor algorithm. *IEEE Transactions on Systems, Man, and Cybernetics*, *15*(4), 580–585.

44. Kim, J., & Kasabov, N. (1999). HyFIS: Adaptive neuro-fuzzy inference systems and their application to nonlinear dynamical systems. *Neural Networks*, *12*, 1301–1319.
45. Kolman, E., & Margaliot, M. (2005). Are artificial neural networks white boxes? *IEEE Transactions on Neural Networks*, *16*(4), 844–852.
46. Kolman, E., & Margaliot, M. (2009). Extracting symbolic knowledge from recurrent neural networks—A fuzzy logic approach. *Fuzzy Sets and Systems*, *160*, 145–161.
47. Lin, C.-F., & Wang, S.-D. (2002). Fuzzy support vector machines. *IEEE Transactions on Neural Networks*, *13*(2), 464–471.
48. Lin, C.-T., Yeh, C.-M., Liang, S.-F., Chung, J.-F., & Kumar, N. (2006). Support-vector-based fuzzy neural network for pattern classification. *IEEE Transactions on Fuzzy Systems*, *14*(1), 31–41.
49. Liu, P. (2000). Max–min fuzzy Hopfield neural networks and an efficient learning algorithm. *Fuzzy Sets and Systems*, *112*, 41–49.
50. Liu, P., & Li, H. (2004). Efficient learning algorithms for three-layer regular feedforward fuzzy neural networks. *IEEE Transactions on Neural Networks*, *15*(3), 545–558.
51. Liu, P., & Li, H. (2005). Hierarchical TS fuzzy system and its universal approximation. *Information Sciences*, *169*, 279–303.
52. Liu, Y.-H., & Chen, Y.-T. (2007). Face recognition using total margin-based adaptive fuzzy support vector machines. *IEEE Transactions on Neural Networks*, *18*(1), 178–192.
53. Lou, S. T., & Zhang, X. D. (2003). Fuzzy-based learning rate determination for blind source separation. *IEEE Transactions on Fuzzy Systems*, *11*(3), 375–383.
54. Mitra, S., & Hayashi, Y. (2000). Neuro-fuzzy rule generation: Survey in soft computing framework. *IEEE Transactions on Neural Networks*, *11*(3), 748–768.
55. Mizutani, E., & Jang, J. S. (1995). Coactive neural fuzzy modeling. In *Proceedings of IEEE International Conference on Neural Networks* (Vol. 2, pp. 760–765). Perth, Australia.
56. Nauck, D., Klawonn, F., & Kruse, R. (1997). *Foundations of neuro-fuzzy systems*. New York: Wiley.
57. Nicholls, J. G., Martin, A. R., & Wallace, B. G. (1992). *From neuron to brain: A cellular and molecular approach to the function of the nervous system* (3rd ed.). Sunderland, MA: Sinauer Associates.
58. Nikov, A., & Stoeva, S. (2001). Quick fuzzy backpropagation algorithm. *Neural Networks*, *14*, 231–244.
59. Nunez, H., Angulo, C., & Catala, A. (2002). Rule extraction from support vector machines. In *Proceedings of European Symposium on Artificial Neural Networks* (pp. 107–112).
60. Nunez, H., Angulo, C., & Catala, A. (2006). Rule-based learning systems for support vector machines. *Neural Processing Letters*, *24*, 1–18.
61. Omlin, C. W., & Giles, C. L. (1996). Extraction of rules from discrete-time recurrent neural networks. *Neural Networks*, *9*, 41–52.
62. Omlin, C. W., Thornber, K. K., & Giles, C. L. (1998). Fuzzy finite-state automata can be deterministically encoded into recurrent neural networks. *IEEE Transactions on Fuzzy Systems*, *6*, 76–89.
63. Pedrycz, W., & Rocha, A. F. (1993). Fuzzy-set based models of neurons and knowledge-based networks. *IEEE Transactions on Fuzzy Systems*, *1*(4), 254–266.
64. Pedrycz, W., Reformat, M., & Li, K. (2006). OR/AND neurons and the development of interpretable logic models. *IEEE Transactions on Neural Networks*, *17*(3), 636–658.
65. Roque, A. M. S., Mate, C., Arroyo, J., & Sarabia, A. (2007). iMLP: Applying multi-layer perceptrons to interval-valued data. *Neural Processing Letters*, *25*, 157–169.
66. Simpson, P. K. (1992). Fuzzy min–max neural networks—Part I: Classification. *IEEE Transactions on Neural Networks*, *3*, 776–786.
67. Simpson, P. K. (1993). Fuzzy min–max neural networks—Part II: Clustering. *IEEE Transactions on Fuzzy Systems*, *1*(1), 32–45.
68. Sisman-Yilmaz, N. A., Alpaslan, F. N., & Jain, L. (2004). ANFIS-unfolded-in-time for multivariate time series forecasting. *Neurocomputing*, *61*, 139–168.

69. Soria-Olivas, E., Martin-Guerrero, J. D., Camps-Valls, G., Serrano-Lopez, A. J., Calpe-Maravilla, J., & Gomez-Chova, L. (2003). A low-complexity fuzzy activation function for artificial neural networks. *IEEE Transactions on Neural Networks*, *14*(6), 1576–1579.
70. Stoeva, S., & Nikov, A. (2000). A fuzzy backpropagation algorithm. *Fuzzy Sets and Systems*, *112*, 27–39.
71. Sun, C. T. (1994). Rule-base structure identification in an adaptive-network-based inference system. *IEEE Transactions on Fuzzy Systems*, *2*(1), 64–79.
72. Sussner, P., & Valle, M. E. (2006). Implicative fuzzy associative memories. *IEEE Transactions on Fuzzy Systems*, *14*(6), 793–807.
73. Thawonmas, R., & Abe, S. (1999). Function approximation based on fuzzy rules extracted from partitioned numerical data. *IEEE Transactions on Systems, Man, and Cybernetics, Part B*, *29*(4), 525–534.
74. Tsujinishi, D., & Abe, S. (2003). Fuzzy least squares support vector machines for multiclass problems. *Neural Networks*, *16*, 785–792.
75. Ultsch, A., Mantyk, R., & Halmans, G. (1993). Connectionist knowledge acquisition tool: CONKAT. In D. J. Hand (Ed.), *Artificial intelligence frontiers in statistics: AI and statistics* (Vol. 3, pp. 256–263). London: Chapman & Hall.
76. Vuorimaa, P. (1994). Fuzzy self-organizing map. *Fuzzy Sets and Systems*, *66*(2), 223–231.
77. Wang, L. X. (1999). Analysis and design of hierarchical fuzzy systems. *IEEE Transactions on Fuzzy Systems*, *7*(5), 617–624.
78. Wang, L. X., & Mendel, J. M. (1992). Generating fuzzy rules by learning from examples. *IEEE Transactions on Systems, Man, and Cybernetics*, *22*(6), 1414–1427.
79. Wang, L. X., & Mendel, J. M. (1992). Fuzzy basis functions, universal approximation, and orthogonal least-squares learning. *IEEE Transactions on Neural Networks*, *3*(5), 807–814.
80. Wang, L. X., & Wei, C. (2000). Approximation accuracy of some neuro-fuzzy approaches. *IEEE Transactions on Fuzzy Systems*, *8*(4), 470–478.
81. Wu, S., & Er, M. J. (2000). Dynamic fuzzy neural networks—A novel approach to function approximation. *IEEE Transactions on Systems, Man, and Cybernetics, Part B*, *30*(2), 358–364.
82. Yan, H.-S., & Xu, D. (2007). An approach to estimating product design time based on fuzzy ν-support vector machine. *IEEE Transactions on Neural Networks*, *18*(3), 721–731.
83. Zhang, D., Bai, X. L., & Cai, K. Y. (2004). Extended neuro-fuzzy models of multilayer perceptrons. *Fuzzy Sets and Systems*, *142*, 221–242.
84. Zhou, S.-M., & Gan, J. Q. (2007). Constructing L2-SVM-based fuzzy classifiers in high-dimensional space with automatic model selection and fuzzy rule ranking. *IEEE Transactions on Fuzzy Systems*, *15*(3), 398–409.

Chapter 28
Neural Network Circuits and Parallel Implementations

28.1 Introduction

Analog hardware for implementing neural networks results in low-cost parallelism, low power, high sample rate or bandwidth, and small size. Design of analog hardware requires good theoretical knowledge of transistor physics as well as experience. Weights in a neural network can be coded by one single analog element (e.g., a resistor). Very simple rules such as Kirchoff's laws can be used to carry out the addition of input signals. As an example, Boltzmann machines can be easily implemented by amplifying the natural noise present in analog devices.

However, connection storage is usually volatile, and inaccurate circuit parameters affect the computational accuracy. Analog VLSI circuits are sensitive to device mismatches, circuit layout, sensitivity to ambient noise and to temperature, and parasitic elements; consequently, design automation in analog circuits is still quite primitive. In contrast, components in neural networks do not have to be of high precision or fast switching. The learning capability of neural networks can compensate initial device mismatches and long-term drift of the device characteristics. The analog standard-cell method is especially suitable for VLSI neural designs [52].

Digital systems, though subject to discretization, power consumption, and circuit size, are preferred for high accuracy, high repeatability, low noise sensitivity, good testability, high flexibility, and compatibility with other types of preprocessing. Digital systems can be designed more easily using computer-aided design tools. Nevertheless, digital designs are slow in computing. The majority of the available digital chips use CMOS technology.

Many neuro-chips have been designed and built. A major limitation of VLSI implementation for general-purpose neural networks is the large number of interconnections and synapses. The number of synapses increases quadratically with that of neurons, and thus silicon area is mainly occupied by the synaptic cells and the interconnection channels. A single integrated circuit is planar with limited possibility for crossover connections. A possible solution would be using optical interconnections. Hardware implementations of neural networks are commonly based on build-

K.-L. Du and M. N. S. Swamy, *Neural Networks and Statistical Learning*,
https://doi.org/10.1007/978-1-4471-7452-3_28

ing blocks and thus allow for the inherent parallelism of neural networks. For highly localized networks such as cellular networks, the VLSI implementation is relatively simple.

Field programmable gate arrays (FPGAs) provide an excellent, quick, general-purpose development platform for digital systems. FPGA implementation is a cheap alternative to VLSI for research or low production quantities. FPGAs have availability of IP (intellectual property) cores. They can be digitally configured (programmed) to implement virtually any digital system, with accessibility, reprogrammability, and low costs. An FPGA platform supports development of fast, compact solutions, providing powerful integration of hardware design with the software-programming paradigm. Specifically, this integration is made possible with the use of a general-purpose hardware description language such as VHDL or Verilog, for different FPGA chips or ASICs. Although FPGAs do not achieve the power, clock rate, or gate density of custom chips, they provide a speedup of several orders of magnitude compared to software simulation.

A solution for overcoming the problem of limited FPGA density is to implement separate parts of the same system by time-multiplexing a single FPGA chip through runtime reconfiguration. This technique has been used in BP algorithm, dividing the algorithm into three sequential stages: forward, backward, and update stages. When the computations of one stage are completed, the FPGA is reconfigured for the next stage [32]. The efficiency of this approach depends on the reconfiguration time relative to the computation time. Finite precision errors are introduced due to the quantization of both the signals and the parameters.

Another solution is the use of pulse-stream arithmetic. The signals are stochastically coded in pulse sequences and therefore can be summed and multiplied using simple logic gates. This is a full digital implementation using analog circuitry. The pulse-width modulation is a hybrid pulse-stream technique that combines the advantages of both analog and digital VLSI implementations. In a pulse-stream-based architecture, signals are encoded by using pulse amplitude, width, density, frequency, or phase. Pulse-mode architecture has a number of advantages over analog and conventional digital implementations. For instance, signal multiplication can be realized by using a very simple digital circuit like an AND gate, and nonlinear activation function can also be easily implemented. An FPGA prototyping implementation of an on-chip BP algorithm presented in [46] uses parallel stochastic bitstreams. Pulse-stream-based architectures for neural networks have been implemented in [40].

Universal Turing machine [94] is the conceptual underpinning of all our modern digital computers. Computational complexity theory is based on this concept. A universal Turing machine is practically realized using von Neumann architecture, which can be viewed as a device with a central processing unit (CPU) that is physically separate from the memory. Implementation of learning algorithms on von Neumann architecture is highly inefficient, due to the locality of the learning rules. This locality, however, allows highly efficient parallel computation.

Memcomputing paradigm [28], inspired by our brain, puts all the computation burden directly into the memory. Universal memcomputing machines [92] are a class of general-purpose computing machines based on systems with memory, whereby processing and storing of information occur on the same physical location. They have universal computing power, intrinsic parallelism, functional polymorphism, and information overhead. A universal memcomputing machine is a nondeterministic Turing machine, namely, it can solve NP-complete problems in polynomial time, with an amount of memory cells (memprocessors) that grows polynomially with the problem size.

Two types of parallel general-purpose computers are single instruction multiple data (SIMD) and multiple instruction multiple data (MIMD). SIMD consists of a number of processors which execute the same instructions but on different data, whereas MIMD has a separate program for each processor. Fine-grained computers are usually SIMD, while coarse-grained computers tend to be MIMD. Systolic arrays take the advantage of laying out algorithms in two dimensions. The name systolic is derived from the analogy of pumping blood through a heart and feeding data through a systolic array.

28.2 Hardware/Software Codesign

Restrictive design specifications such as high-performance, reduced size, or low-power consumption are difficult to fulfill with a software approach. However, both the research activity and commercial interest in neural/fuzzy hardware have been decreasing due to the important increase in speed of software solutions based on general-purpose microprocessors or digital signal processors (DSPs). Software approaches are characterized by their high versatility, whereas dedicated hardware implementations provide a suitable solution only when extreme requirements, in terms of speed, power consumption, or size, are needed.

For implementations of neural and fuzzy systems, the use of hardware/software codesign is concluded as a means of exploiting the best from both the hardware and software techniques, as it allows a fast design of complex systems with the highest performance–cost ratio [78]. Heterogeneous hardware/software technologies have emerged as an optimal solution for many systems. This approach proposes the partition of the system into hardware and software parts by exploiting the advantages of both the hardware and software intrinsic characteristics.

In [26], ANFIS model is modified for efficient hardware/software implementation. The piecewise multilinear ANFIS exhibits approximation capabilities and learning abilities comparable to those of generic ANFIS. Two different on-chip design approaches are presented: a high-performance parallel architecture for offline training and a pipelined architecture suitable for online parameter adaptation. The device contains an ARM embedded-processor core and a large FPGA. The processor provides flexibility and high precision to implement the learning algorithms, while the FPGA allows the development of high-speed inference architectures for real-time embedded applications. The internal architecture is shown in Fig. 28.1. The processor subsystem contains a 32-bit ARM922T hard processor core, a memory subsystem, external memory interfaces, and standard peripherals, while the FPGA block consists of an APEX 20KE-like architecture with resources for integration.

Fig. 28.1 Internal architecture of Altera's Excalibur family used for implementation of the piecewise multilinear ANFIS

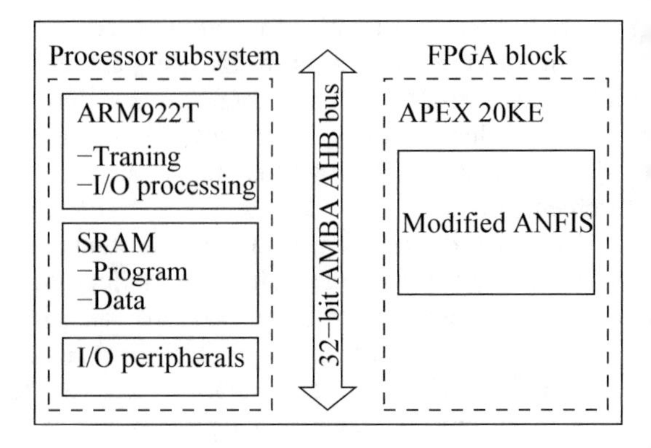

28.3 Topics in Digital Circuit Designs

Multiplication-free architectures are attractive, since digital multiplication operations in each neuron are very demanding in terms of time or chip area and create a bottleneck. In binary representation, multiplication between two integers can be substituted by a shift, if one of the integers is a powers-of-two.

The family of coordinate rotation digital computer (CORDIC) algorithms exploits the geometric properties of two-dimensional and three-dimensional vector rotations for the fast computation of transcendental functions through additions and shifts [2]. One version of these algorithms can be used for computing the exponential function e^x by performing two-dimensional rotations in a hyperbolic coordinate system and making use of the relation $e^x = \sinh(x) + \cosh(x)$.

Look-Up Tables

When the logistic or hyperbolic tangent function is used in digital designs, an exponential function needs to be calculated. The value of an exponential function is usually computed by using a Taylor series expansion, which requires many floating-point operations. In view of the piecewise-linear approximation of the sigmoidal and its derivative functions, we need to use two look-up tables to store many input–output associations. The output of a unit can be approximated by linear interpolation of the points in a table. Since the activation function usually has output in the interval $(0, 1)$ or $(-1, 1)$, it would be possible to adopt a fixed-point representation. Most neural chips integrate look-up tables and fixed-point representation of the activation function in order to simplify the logic design and increase the processing speed.

This method needs an external memory to store the look-up table, and thus simultaneous evaluation of the activation function for multiple neurons is not possible. In practical implementations, all the neurons are typically assumed to have the same activation functions. Depending on the type of processor, the calculation of nonlinear activation functions such as the logistic or hyperbolic tangent function may consume considerable time. Piecewise-linear approximation to the sigmoidal functions along with look-up tables works very well [40].

Quantization Effects

In embedded systems design, the digital hardware constraints can be quite severe and the use of fixed-point arithmetic requires both a careful algorithm implementation and a thorough analysis of quantization effects [3]. Typical examples are the devices for sensor networks [25], where the minimization of both area and power consumption is a strategic issue for any successful application. One of the possible solutions is to avoid hardware multipliers, which are quite demanding in terms of resources, especially if compared to that required by the adders. In most applications, a representation precision with 16-bit coding for weights and 8-bit coding for the outputs is sufficient for the convergence of a learning algorithm [9].

There are some cases for which the quantization effect is beneficial and the error rate becomes even lower with respect to the floating-point case. This is suggested by the large variation of the results for small k and is an already reported effect [4], which can be partially explained by the fact that the precision reduction acts as a pruning of the less important parameters of SVM. There are two byproducts of the quantization process. The first one is the increase of the sparsity of an SVM or a neural network, because some parameters are negligible with respect to the least significant bit and therefore, their values are rounded to zero. This increases the generalization ability. The second consequence is the reduction of the number of bits required to describe the network: This improves the generalization ability of a learning machine in the MDL framework.

28.4 Circuits for Neural Networks

The sigmoidal function can be generated by utilizing the current–voltage characteristics of a differential pair operating in the subthreshold regime (see Fig. 28.2a). By cascading two differential pairs, the basic Gaussian circuit has been developed, which has the problem of asymmetry. By attaching one more differential pair, we get Gilbert Gaussian circuit [44]. Figure 28.2b can be regarded as a circuit made by connecting input terminals in a Gilbert multiplier. It is often used as squaring circuit [100]. Because of the nonlinearity, the usage as a squarer suffers from narrow input range. However, by focusing on the single output of the differential current, it becomes a symmetric Gaussian circuit having a wide input range, yielding a Gilbert Gaussian circuit. A simple and area-efficient differential Gaussian circuit consists of only four transistors [98]. The circuit is more accurate, less susceptible to the process variations, and requires less on-chip area when compared to Gilbert Gaussian circuit.

28.4.1 Memristor

The memristor was predicted by Chua in 1971 as the missing fourth fundamental passive circuit element based on a nonlinear relationship between charge and flux [21]. A memristor made of titanium dioxide was manufactured in 2008 [89].

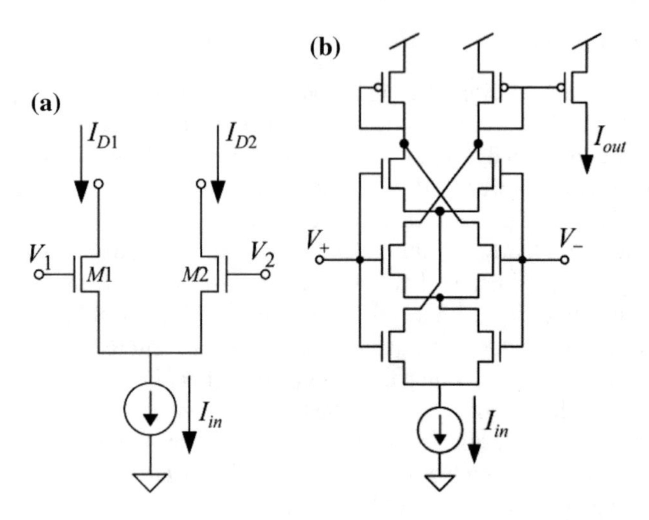

Fig. 28.2 Circuits for generating activation functions. **a** Sigmoidal circuit. **b** Gilbert Gaussian circuit. ©IEEE, 2010 [44]

A memristor is a resistor whose resistance changes according to time integral of the current through the device, or alternatively the integrated voltage upon the device. Memristors exhibit the pinched hysteresis loop fingerprint and properties of non-volatility, good scalability, low power, and nonlinearity. The activation-dependent dynamics of resistance makes memristors ideal for updating of synaptic weights.

The conductance of a memristor, G, depends directly on the time integral of the voltage upon the device, referred to as flux. Formally, a memristor obeys the following dynamics:

$$i(t) = G(s(t))v(t), \tag{28.1}$$

$$\dot{s}(t) = v(t), \tag{28.2}$$

where $s(t)$ is the state variable of the memristor, which is related to an integral of the voltage, and $G(s(t))$ its conductivity.

Synapses among neurons in biological neural networks are long-term memories, but resistors have no function of memory. A memristor has memory and behavior more like a biological synapse. Memristors are an electronic analogy of the Hodgkin–Huxley ion channels not only in terms of threshold switching effect as well as stochastic behavior. Switching effect in memristive devices is thermodynamically driven, which is stochastic in nature. The neuristor [74], built using two nanoscale Mott memristors, is an electronic device with properties similar to the Hodgkin–Huxley axon. The memristive synapse proposed in [33] is composed of a single memristor, connected to a shared terminal of two MOSFET transistors (p-type and n-type).

28.4.2 Circuits for MLPs

The computational power of the MLP using integer weights in a very restricted range has been analyzed in [30]. For classification problems, an existence result is derived for calculating a weight range that is able to guarantee the existence of a solution as a function of the minimum distance between patterns of different classes.

A parallel VLSI implementation of MLP with BP [66] uses only shift-and-add and rounding operations instead of multiplications. In the forward phase, weights are restricted to powers-of-two, and the activation function is computed through look-up table. In the backward phase, the derivative of the activation function is computed through look-up table, some internal terms are rounded to the nearest powers-of-two, and external terms like η are selected as powers-of-two terms. Decomposition of binary integers into powers-of-two terms can be accomplished very quickly and with a limited amount of circuitry. The gain with respect to multiplication is more than one order of magnitude both in speed and in chip area, and thus the overhead due to the decomposition operation is negligible. The rounding operations introduce randomness that helps BP to escape from local minima. This randomness also helps to improve the generalization performance of the network.

MLP with BP is implemented as a full digital system using pulse-mode neurons in [40]. A piecewise-linear activation function is used. BP algorithm is simplified to make the hardware implementation easier. The derivative of the activation function is generated by a pulse differentiator. A random pulse sequence is injected to the pulse differentiator output to improve the learning capability.

The circuit complexity of a sigmoidal MLP can be examined in the framework of classical Boolean and threshold gate circuit complexity by converting a sigmoidal MLP into an equivalent threshold gate circuit [12]. Sigmoidal MLPs can be implemented in polynomial-size Boolean circuits with a small constant fan-in at the expense of an increase in the number of layers by a logarithmic factor.

A method for parallel hardware implementation of neural networks using digital techniques is presented in [72]. Signals are represented using uniformly weighted single-bit stream, which can be obtained by using the front-end of a sigma-delta modulator. This single-bit representation offers significant advantages over multi-bit representations, since they mitigate the fan-in and fan-out issues which are typical to distributed systems. To process these bitstreams using neural network concepts, functional elements which perform summing, scaling, and squashing have been implemented. These elements are modular and are designed to be easily interconnected. Using these functional elements, an MLP can be easily constructed.

A CORDIC-like algorithm for computing the feedforward phase of an MLP in fixed-point arithmetic [65] uses only shift-and-add operations and avoiding multiplications. Digital BP learning [69] is implemented for three-layered neural networks with nondifferentiable binary units. A neural network using digital BP learning is fast and easy to implement in hardware. In [10], a real-time reconfigurable perceptron circuit element is presented using subthreshold operation. The circuit performs competitively with standard static CMOS implementation.

A general scheme to design hardware is presented in [88] for online gradient-descent learning rules. The scheme uses a memristor to store the synaptic weight, and two CMOS transistors to control the circuit. Temporal encoding is used as a mechanism to perform a multiplication operation. The design requires 2–8% of the area and static power of the CMOS-only circuits.

28.4.3 Circuits for RBF Networks

Most hardware implementations for the RBF network are developed for the Gaussian RBF network. The properties of the MOS transistor are desirable for analog designs of the Gaussian RBF network. In the subthreshold or weak-inversion region, the drain current of the MOS transistor has an exponential dependence on the gate bias and dissipates very low power. This exponential characteristic of the MOS devices is usually exploited for designing the Gaussian function [16, 57, 100]. In [16], a compact analog Gaussian cell is designed whose core takes only eight transistors and can be supplied with a high number of input pairs. On the other hand, the MOS transistor has a square-law dependence on the bias voltages in its strong-inversion or saturation region, based on which a compact programmable analog Gaussian synapse cell has been designed in [20].

Similarity measure, typically the Euclidean distance measure, is essential in many neural network models such as the clustering networks and the RBF network. The circuits for the Euclidean distance measure are usually based on the square-law property of the strong-inversion region [22, 57, 67, 105]. The generalized measure of similarity between two voltage inputs can be implemented by using the Gaussian-type or bump circuit and by using the bump–antibump circuit. These circuits are based on the concept of the current correlator developed for weak-inversion operation [27]. Based on the current correlator, an analog Gaussian/square function computation circuit is given in [57]. This circuit exhibits independent programmability for the center, width, and peak amplitude of the dc transfer curve. When operating in the strong-inversion region, it calculates the squared difference, whereas in the weak-inversion region it realizes a Gaussian-like function. In [22], circuits for calculating Euclidean distance and computing programmable Gaussian units with tunable center and variance are also designed based on the square-law of the strong-inversion region. A pulsed VLSI RBF network chip is fabricated in [67] where a collection of pulse-width-modulation analog circuits are combined on a single RBF network chip. The distance metric is based on the square-law property of the strong-inversion region, and a Gaussian-like RBF is produced using two MOS transistors.

In an analog VLSI circuit implementation of conic-sectional function network neurons [105], the circuit computes both the weighted sum for the MLP and the Euclidean distance for the RBF. The two propagation rules are then aggregated as the design of synapse and neuron of the conic-section function network. A hybrid VLSI/digital design of the RBF network integrates a custom analog VLSI circuit and a commercially available digital signal processor [100].

28.4.4 Circuits for Clustering

Many WTA models can be achieved based on the continuous-time Hopfield network topology [63, 90], or based on the cellular network model with linear circuit complexity [85]. There are also some circuits for realizing the WTA function such as a series of compact CMOS integrated circuits [51].

k-WTA networks are usually based on the continuous-time Hopfield network [63]. The k-WTA circuit devised in [51] has infinite resolution and is implemented using the Hopfield network based on the penalty method. In [96], a k-WTA circuit with $O(n)$ interconnect complexity extends the WTA circuits given in [51], where n is the number of neurons. k-WTA is formulated as a mathematical programming problem solved by direct analog implementation of the Lagrange multiplier method. The circuit has merits of real-time responses and short wire length, although it has a finite resolution. A discrete-time mathematical model of k-WTA neural circuit that can quickly identify the k winning nodes from n neurons is given and analyzed in [95]. For n competitors, the circuit is composed of n feedforward and one feedback of hard-limiting neurons that are used to determine the dynamic shift of input signals. In a k-WTA network based on a one-neuron recurrent network [56], the k-WTA operation is first converted equivalently into a linear programming problem. Finite-time convergence of the network is proved using the Lyapunov method.

In [79], improved neural gas and its analog VLSI subcircuitry are developed based on partial sorting. The VLSI architecture includes two chips, one for the Euclidean distance computation and the other for the programmable sorting of code vectors. The latter is based on the WTA structure [51]. The approach is empirically shown to reduce the training time by up to two orders of magnitude, without reducing the performance quality. A fully analog integrated circuit of SOM has been designed in [64].

28.4.5 Circuits for SVMs

SVM learning can be formulated as a dynamic problem, which can then be solved using a two-layer recurrent network. The neural network is suitable for analog hardware implementation [8, 91]. In [5], a digital architecture for SVM learning is modeled as a dynamical system and is implemented on an FPGA.

A one-layer recurrent network for SVM learning is presented in [103] for pattern classification and regression. The network is guaranteed to obtain the optimal solution of SVM and SVR. Compared with the two-layer neural network for SVM classification, the proposed network has a low complexity for implementation. Moreover, it can converge exponentially to the optimal solution of SVM learning. An analog neural network for SVM learning is proposed in [73], based on a partially dual formulation of the QP problem. The performance is substantially equivalent to that of [103], in terms of both the settling time and the steady-state solutions.

A CORDIC-like algorithm for computing the feedforward phase of an SVM in fixed-point arithmetic is proposed in [7], using only shift-and-add operations and avoiding multiplications. This result is obtained, thanks to a hardware-friendly kernel,

$$k(\boldsymbol{x}_i, \boldsymbol{x}_j) = 2^{-\gamma \|\boldsymbol{x}_i - \boldsymbol{x}_j\|_1}, \tag{28.3}$$

where the hyperparameter is an integer powers-of-two, i.e., $\gamma = 2^{\pm p}$ for integer p, and $\| \cdot \|_1$ is the L_1 norm. This kernel is proved to be an admissible Mercer's kernel. It greatly simplifies the SVM feedforward phase computation.

SVM with integer parameters [6] is based on a branch-and-bound procedure, derived from modern mixed-integer QP solvers, and is useful for implementing the feedforward phase of SVM in fixed-point arithmetic. This allows the implementation of the SVM algorithm on resource-limited hardware for building sensor networks, where floating-point units are rarely available.

An analog circuit architecture of Gaussian-kernel SVMs having on-chip training capability has been developed in [44]. It has a scalable array processor configuration, and the circuit size increases only in proportion to the number of learning samples. The learning function is realized by attaching a small additional circuitry to the SVM hardware composed of an array of Gaussian circuits. Although the system is inherently analog, the input and output signals including training results are all available in digital format. Therefore, the learned parameters are easily stored and reused after training sessions.

28.4.6 Circuits for Other Neural Network Models

An analog implementation of Hodgkin–Huxley model with 30 adjustable parameters is described in [77], which requires 4 mm^2 area for a single neuron. A circuit for generating spiking activity is designed in [83]. The circuit integrates charge on a capacitor such that when the voltage on the capacitor reaches a certain threshold, two consecutive feedback cycles generate a voltage spike and then bring the capacitor back to its resting voltage.

CAVIAR [86] is a massively parallel hardware implementation of a spike-based sensing–processing–learning–actuating system inspired by the physiology of the nervous system. It uses the asynchronous address-event representation communication framework. It performs 12 G synaptic operations per second and achieves millisecond object recognition and tracking latencies.

A signal processing circuit for a continuous-time recurrent network is implemented in [15] using subthreshold analog VLSI in mixed-mode (current and voltage) approach, where state variables are represented by voltages, while neural signals are conveyed as currents. The use of current allows for the accuracy of the neural signal stobe maintained over long distances, making this architecture relatively robust and scalable.

In [41], layer multiplexing technique is used to implement multilayer feedforward networks into a Xilinx FPGA. The suggested layer multiplexing involves implementing only the layer having the largest number of neurons. A separate control block is designed, which appropriately selects the neurons from this layer to emulate the behavior of any other layer and assigns the appropriate inputs, weights, biases, and excitation functions for every neuron of the layer that is currently being emulated in parallel. Each single neuron is implemented as a look-up table.

In [58], a hardware architecture for GHA operates the principal component computation and weight vector updating of GHA in parallel. An FPGA implementation of ICA algorithm for BSS and adaptive noise canceling is proposed in [45]. In [87], FastICA is implemented on an FPGA. There are some examples of hardware implementations of neural SVD algorithms [23, 61].

The mean-field annealing algorithm can be simulated by RC circuits, coupled with the local nature of the Boltzmann machine, which makes mean-field-theory machine suitable for massively parallel VLSI implementation [53, 84].

Restricted Boltzmann machine can be mapped to a high-performance hardware architecture on FPGA platforms [54]. A method is presented to partition large restricted Boltzmann machines into smaller congruent components, allowing the distribution of one restricted Boltzmann machine across multiple FPGA resources.

A digital liquid state machine for low-power VLSI-based applications [108] employs an online spike-based learning algorithm. The liquid state machine extracts information from input patterns on the fly without intermediate data storage. The learning rule is local such that each synaptic weight update is based only upon the firing activities of the corresponding presynaptic and postsynaptic neurons.

28.4.7 Circuits for Fuzzy Neural Models

Generally, fuzzy systems can be easily implemented in digital form, which can be either general-purpose microcontrollers running fuzzy inference and defuzzification programs, or dedicated fuzzy coprocessors, or RISC processors with specialized fuzzy support, or fuzzy ASICs.

Fuzzy coprocessors work in conjunction with a host processor. They are general-purpose hardware, and thus have a lower performance compared to a custom fuzzy hardware. There are many commercially available fuzzy coprocessors [80]. RISC processors with specialized fuzzy support are also available [24, 80]. A fuzzy-specific extension to the instruction set is defined and implemented using hardware/software codesign techniques. The fuzzy-specific instructions significantly speed up fuzzy computation with no increase in the processor cycle time and with only a minor increase in the chip area.

A common approach to general-purpose fuzzy hardware is to use a software design tool to generate the program code for a target microcontroller. Examples include the Motorola-Aptronix fuzzy inference development language and Togai InfraLogic's MicroFPL system [42]. Compared to dedicated fuzzy processors and ASICs, this approach leads to rapid design and testing at the cost of low performance.

The tool TROUT [42] automates fuzzy neural ASIC design. It produces a specification for small, customized, application-specific circuits called *smart parts*. A smart part is a dedicated compact-size circuit customized to a single function. A designer can package a smart part in a variety of ways. The model library of TROUT includes fuzzy or neural network models for implementation as circuits. To synthesize a circuit, TROUT takes as its input an application dataset, optionally augmented with user-supplied hints. It delivers, as output, technology-independent VHDL code, which describes a circuit implementing a specific fuzzy or neural network model optimized for the input dataset. As an example, TROUT has been used for the synthesis of the fuzzy min–max classification network.

On a versatile neurofuzzy platform with a topology strongly influenced by theories of fuzzy modeling [38], various critical hardware design issues are identified. With a hybrid learning scheme involving structural and parametric optimization, fuzzy neural networks are well suited in forming the adaptive logic processing core of this platform. A resolution of at most 4 or 5 bits is necessary to achieve performance on par with that of a continuous representation, with good results seen for as little as 2 bits of precision [38].

There are also many analog [49, 55] and mixed-signal [11, 13] fuzzy circuits. Analog circuits usually operate in current mode and are fabricated using CMOS technology, and this leads to the advantages of high-speed, small-circuit area, high-performance, and low power dissipation. A design methodology for fuzzy ASICs and general-purpose fuzzy processors is given in [49], based on the left–right fuzzy implication cells and the left–right fuzzy arithmetic cells. In [11, 13], the fabrication of mixed-signal CMOS chips for fuzzy controllers is considered; in these circuits, the computing power is provided by the analog part while the digital part is used for programmability.

28.5 Graphic Processing Unit (GPU) Implementation

GPUs were originally designed only to render triangles, lines, and points. Graphics hardware is now used for general-purpose computations for performance-critical algorithms that can be efficiently expressed in the streaming nature of GPU architecture. Graphics hardware has become competitive in terms of speed and programmability.

GPUs are specialized stream processors. Stream processors are capable of taking large batches of fragments that can be thought of as pixels in image processing applications, and computing similar independent calculations in parallel. Each calculation is with respect to a program, often called a kernel, which is an operation applied to every fragment in the stream.

The computational power of GPUs is increasing significantly faster than CPUs. GPUs have instructions to handle many linear algebra operations, such as the dot product, vector, and matrix multiplication, and computing the determinant of a matrix, given their vector processing architecture. GPUs are capable of executing more floating-point operations per second (flops) than CPUs do. NVIDIA GeForce 8800 GTX has 128 stream processors, a core clock of 575 MHz, and a shader clock of 1350 MHz; it is capable of over 350 Gflops, and the peak bandwidth to the graphics memory is 86.4 GB/s. The key concept required for converting a CPU program to a GPU program involves the idea that arrays are equivalent to textures. Data used in a GPU program are passed from the CPU as a texture. Due to the limited floating-point representation of current GPUs, the accuracy of GPU implementations is lower compared to that in the CPU case, but it is adequate for a large range of applications. However, the floating-point representation of the graphic cards has been improved rapidly.

Compute unified device architecture (CUDA) is a general-purpose parallel computing architecture developed by NVidia. It provides an application programming interface (API) to a GPU that enables designers to easily create GPU code. This model consists of three major abstractions: a hierarchy of thread groups, shared memories, and barrier synchronization. Using these CUDA features, developers can write C-like functions called kernels. Each kernel invocation is associated with an ordered set of thread groups in a grid executing the kernel in parallel on the GPU. CUDA supports implementing parallel algorithms with topological processing, running the same code segment on all nodes, which is similar to the cellular network architecture. On top of CUDA, a CUBLAS library provides basic linear algebra subroutines, especially matrix multiplication. The library is self-contained at the API level. The basic approach to use the CUBLAS library is to allocate memory space on the GPU memory, transfer data from the CPU to the GPU memory, call a sequence of CUBLAS, and transfer data from the GPU back to the CPU memory.

GPUmat (free, http://gp-you.org/), Jacket (commercial), and the Parallel Computing Toolbox of MATLAB (commercial) are MATLAB toolboxes for GPU-based processing on the market. Computing using GPUmat usually only involves recasting variables and requires minor changes to MATLAB scripts or functions.

A GPU implementation of the multiplication between the weights and the input vectors in each layer in the working phase of an MLP is proposed in [70]. A GPU implementation of the entire learning process of an RBF network [14] reduces the computational cost by about two orders of magnitude with respect to its CPU implementation. A generalized method for offloading fuzzy clustering to a GPU (http://cirl.missouri.edu/gpu/) [1] leads to a speed increase of over two orders of magnitude for particular clustering configurations and platforms. In [93], EKF algorithm for training of recurrent networks is implemented, where most computationally intensive tasks are performed on the GPU.

CURRENNT [101] is an open-source CUDA toolkit for recurrent networks. The recurrent network implemented is based on long short-term memory (LSTM) units, and feedforward network training is supported. Besides simple regression, CURRENNT includes logistic and softmax output layers for training of classification. A reference CPU implementation of LSTM-RNNs is available as open-source C++ code (http://sourceforge.net/projects/rnnl/). CURRENNT provides a C++ class library for deep LSTM-RNN modeling. The network architecture can be specified by the user in JavaScript Object Notation (JSON), and trained parameters are saved in the same format.

GPUSVM [18] is an open-source CUDA implementation of SMO. Optimized hierarchical decomposition SVM (https://github.com/OrcusCZ/OHD-SVM) [97] uses a hierarchical decomposition iterative algorithm that fits better to actual GPU architecture. The method excels significantly over five publicly available C++ SVM training GPU implementations in all datasets. It is the only one that can handle dense as well as sparse datasets.

Restricted Boltzmann machines can be accelerated using a GPU [76]. The implementation was written in CUDA and tested on an NVIDIA GeForce GTX 280. An implementation of the MPI developed specifically for embedded FPGA designs, called time-multiplexed differential data-transfer MPI [81], extends the MPI protocol to hardware. The hardware implementation is controlled entirely with MPI software code, using messages to abstract the hardware compute engines as computational processes, called ranks. In addition to ease of use, this feature also provides portability and versatility, since each compute engine is compartmentalized into message-passing modules that can be inserted or removed based on available resources and desired functionality.

28.6 Implementation Using Systolic Algorithms

A systolic array of processors is a parallel digital hardware system. It is an array of data processing units which are connected to a small number of nearest neighbor data processing units in a mesh-like topology. Data processing units perform a sequence of operations on data that flows between them.

A number of systolic algorithms are available for matrix-vector multiplication, the basic computation involved in the operation of a neural network. The systolic architecture uses a locally communicative interconnection structure for distributed computing. Systolic arrays can be readily implemented in programmable logic devices.

QR decomposition is a key step in many DSP applications. Divide and square root operations in Givens rotation algorithm can be avoided by using special operations such as CORDIC or special number systems such as the logarithmic number system. A two-dimensional systolic array QR decomposition is implemented [99] on an FPGA using Givens rotation algorithm. This design uses straightforward floating-point divide and square root operations, which make it easier to be used within a larger system. The input matrix size can be configured at compile time to many different sizes.

In a unified systolic architecture for implementing neural networks [48], proper ordering of the elements of the weight matrix makes it possible to design a cascaded dependency graph for consecutive matrix-vector multiplication, which requires the directions of data movement at both the input and output of the dependency graph to be identical. Using this cascaded dependency graph, the computations in both the recall and learning iterations of BP have been mapped onto a ring systolic array. The same mapping strategy has been used in [43] for mapping the HMM and the recursive BP network onto the ring systolic array. The main drawback of these implementations is the presence of spiral (global) communication links that damage the local property.

MLP with BP learning has been implemented on systolic arrays [36, 37, 62]. In [62], dependency graphs are derived for implementing operations in both the recall and learning phases of BP. These dependency graphs are mapped onto a linear bidirectional systolic array, and algorithms have been presented for executing both the recall and learning phases efficiently. In [36], BP is implemented online by using a pipelined adaptation, where a systolic array is implemented on an FPGA. Parallelism is better exploited because both forward and backward phases can be performed simultaneously. In [37], a pipelined modification of online BP better exploits the parallelism because both the forward and backward phases can be performed simultaneously.

28.7 Implementation on Parallel Computers

An increasing number of databases (such as weather, oceanographic, remote sensing, financial) are becoming online and distributed. Distributed processing naturally emerges when data are acquired in many places with different owners, and data privacy arises.

Multiple-core computers are becoming the norm, and datasets are becoming ever larger. Most of the previous approaches have considered the parallel computer system as a cluster of independent processors, communicating through a message-passing scheme such as MPI (Message-Passing Interface, http://www.mpi-forum.org). MPI is a library of functions. It allows one to easily implement an algorithm in parallel by running multiple CPU processors for improving efficiency. MPI provides a straightforward software–hardware interface. The message-passing paradigm is widely used in high-performance computing.

The SIMD mode, where different processors execute the same program but different data, is generally used in MPI for developing parallel programs. Advances in technology have resulted in systems where several processing cores have access to a single memory space, and such symmetric multiprocessing architectures are becoming prevalent. OpenMP API (http://www.openmp.org) works effectively on shared memory systems, while MPI can be used for message passing between nodes. Most high-performance computing systems are now clusters of symmetric multiprocessing nodes. On such hybrid systems, a combination of message-passing between symmetric multiprocessing nodes and shared memory techniques inside each node could potentially offer the best parallelization performance from the architecture. A standard approach to combining the two schemes involves OpenMP parallelization inside each MPI process, while communication between the MPI processes is made only outside of the OpenMP regions [75].

IBM RS/6000 SP system (http://www.rs6000.ibm.com/hardware/largescale) is a scalable distributed-memory multiprocessor consisting of up to 512 processing nodes connected by a high-speed switch. Each processing node is a specially packaged RS/6000 workstation CPU with local memory, local disk(s), and an interface to the high-performance switch. The SP2 parallel environment supports MPI for the development of message-passing applications.

Network-partitioned parallel methods for SOM algorithm, written in the SIMD programming model, preserve the recursive weight update, and hence produce exact agreement with the serial algorithm. A data-partitioned parallel method for SOM [50] is based on the batch SOM formulation in which the neural weights are updated at the end of each pass over the training data. The underlying serial algorithm is enhanced to take advantage of the sparseness often encountered in these datasets. A parallel version of DBSCAN algorithm [104] uses the shared-nothing architecture with multiple computers interconnected through a network. A fundamental component of a shared-nothing system is its distributed data structure. The dR*-tree, a distributed spatial index structure, is introduced, in which the data is spread among multiple computers and the indexes of the data are replicated on every computer.

Probabilistic inference is examined through parallel computation on real multiprocessors in [47]. Experiments are performed on a 32-processor Stanford DASH multiprocessor, a cache-coherent shared-address-space machine with physically distributed main memory.

Parallel implementations using cloud computing are mainly based on MapReduce framework. Based on MapReduce computing model and cascading model, a parallelized BP network [59] supplies advanced features including fault tolerance, data replication, and load balancing. AdaBoost.PL and LogitBoost.PL are two parallel boosting algorithms, and they are implemented in MapReduce framework [71]. Cloud computing is introduced in Chap. 31.

Differential privacy measures privacy risk by a parameter that bounds the log-likelihood ratio of output of an algorithm under two databases differing in a single individual [31]. It has strong semantic guarantees and is resistant to many attacks that succeed against alternative definitions of privacy. Differentially private PCA [19] approximations to PCA guarantee differential privacy. It uses the exponential mechanism to choose a k-dimensional subspace biased toward those which capture more of "energy" of the matrix. The method is implemented using an MCMC procedure. The sample complexity scales as $O(d)$. The problem of differentially private low-rank matrix reconstruction for sparse matrices is considered in [39].

An asynchronous and decentralized PCA algorithm of data spread over a network [34] is based on the integration of a dimension reduction step into a gossip consensus protocol. A straightforward dual formulation makes it suitable when observed dimensions are distributed. It is theoretically equivalent with a centralized PCA under a low-rank assumption on training data.

A distributed semi-supervised algorithm for low-rank matrix completion is proposed in [35] based on the framework of diffusion adaptation. It is efficient and scalable and can preserve privacy by the inclusion of flexible privacy-preserving mechanisms for similarity computation.

28.7.1 Distributed and Parallel SVMs

Distributed SVMs and parallel SVMs emphasize global optimality. A distributed SVM algorithm assumes training data to come from the same distribution and are locally stored in different locations with processing capabilities (nodes). A reasonably small amount of information is interchanged among nodes to obtain an SVM solution. A distributed SVM algorithm can find support vectors locally and process them altogether in a central processing center. The solution is not globally optimal. This method can be improved by allowing the data processing center to send support vectors back to the distributed data source and iteratively achieve the global optimum. This model is slow due to extensive data accumulation in each site.

Two distributed schemes are analyzed in [68]: a naive distributed chunking approach, where support vectors are communicated, and the distributed semiparametric SVM, which further reduces the total amount of information passed between nodes. The naive distributed chunking approach is simple to implement and has a performance slightly better than that of distributed semiparametric SVM. In the distributed semiparametric SVM, no raw data are exchanged. By selecting centroids as training patterns plus noise, privacy can be preserved.

A parallel implementation of SVMlight [106] splits the QP problem into smaller subproblems, which are then solved by a variable projection method. However, these methods need centralized access to the training data, and therefore, cannot be used in distributed classification applications. A parallel gradient projection-based decomposition technique [107] is implemented based on both the gradient projection QP solvers and the selection rules for large working sets. The software implements an iterative decomposition technique and exploits both the storage and computing resources available on multiprocessor systems.

HeroSVM [29] uses a block-diagonal approximation of the kernel matrix to derive hundreds of independent small SVMs and filter out the examples which are estimated to be non-support vectors; then a new serial SVM is trained on the collected support vectors. A parallel optimization step is introduced to quickly remove most of the non-support vectors. In addition, some effective strategies such as kernel caching and efficient computation of kernel matrix are integrated to speed up the training process. The algorithm complexity grows linearly with the number of classes and size of the dataset.

A parallel implementation of linear SVM training [102] uses a combination of MPI and OpenMP. Using an interior-point method for optimization and a reformulation that avoids the dense Hessian matrix, the structure of the augmented system matrix is exploited to partition data, and computations among parallel processors efficiently. Parallel SMO [17] is developed based on MPI. It first partitions the entire training dataset into smaller subsets, and then simultaneously runs multiple CPU processors to deal with each of the partitioned datasets. The efficiency of parallel SMO decreases with increasing number of processors, as there is more communication time for using more processors.

Distributed parallel SVM [60] is implemented for distributed data classification in a general network configuration, namely, strongly connected network. Support vectors carry all the classification information of the local dataset. Each site within a strongly connected network classifies subsets of training data locally via SVM, passes the calculated support vectors to its descendant sites, receives support vectors from its ancestor sites, recalculates the support vectors, passes them to its descendant sites, and so on. SVMlight is used as the local solver. Distributed parallel SVM is able to work on multiple arbitrarily partitioned working sets and achieve close to linear scalability if the size of the network is not too large. The algorithm is proved to converge to a globally optimal classifier (at every site) for arbitrarily distributed data over a strongly connected network in finite steps.

In [82], a semi-supervised SVM is trained for the samples distributed over a network of interconnected agents. A distributed training protocol over networks is designed, where communication is restricted only to neighboring agents. The semi-supervised SVM training problem is formulated as the distributed minimization of a non-convex social cost function using a standard relaxation. A (stationary) solution is searched in a distributed manner by employing a distributed gradient-descent algorithm and a framework for in-network non-convex optimization.

Problem

28.1 Design a systolic array for multiplication of a vector and a matrix.

References

1. Anderson, D. T., Luke, R. H., & Keller, J. M. (2008). Speedup of fuzzy clustering through stream processing on graphics processing units. *IEEE Transactions on Fuzzy Systems*, *16*(4), 1101–1106.
2. Andraka, R. (1998). A survey of CORDIC algorithms for FPGA based computers. In *Proceedings of ACM/SIGDA International Symposium on Field Programmable Gate Arrays* (pp. 191–200). Monterey, CA.
3. Anguita, D., & Boni, A. (2003). Neural network learning for analog VLSI implementations of support vector machines: A survey. *Neurocomputing*, *55*, 265–283.
4. Anguita, D., Boni, A., & Ridella, S. (1999). Learning algorithm for nonlinear support vector machines suited for digital VLSI. *Electronics Letters*, *35*(16), 1349–1350.
5. Anguita, D., Boni, A., & Ridella, S. (2003). A digital architecture for support vector machines: Theory, algorithm and FPGA implementation. *IEEE Transactions on Neural Networks*, *14*(5), 993–1009.
6. Anguita, D., Ghio, A., Pischiutta, S., & Ridella, S. (2008). A support vector machine with integer parameters. *Neurocomputing*, *72*, 480–489.
7. Anguita, D., Pischiutta, S., Ridella, S., & Sterpi, D. (2006). Feed-forward support vector machine without multipliers. *IEEE Transactions on Neural Networks*, *17*(5), 1328–1331.
8. Anguita, D., Ridella, S., & Rovetta, S. (1998). Circuital implementation of support vector machines. *Electronics Letters*, *34*(16), 1596–1597.

9. Asanovic, K., & Morgan, N. (1991). Experimental determination of precision requirements for back-propagation training of artificial neural networks. *Proceedings of the 2nd International Conference on Microelectronics for Neural Networks* (pp. 9–15). Munich, Germany.
10. Aunet, S., Oelmann, B., Norseng, P. A., & Berg, Y. (2008). Real-time reconfigurable subthreshold CMOS perceptron. *IEEE Transactions on Neural Networks, 19*(4), 645–657.
11. Baturone, I., Sanchez-Solano, S., Barriga, A., & Huertas, J. L. (1997). Implementation of CMOS fuzzy controllers as mixed-signal integrated circuits. *IEEE Transactions on Fuzzy Systems, 5*(1), 1–19.
12. Beiu, V., & Taylor, J. G. (1996). On the circuit complexity of sigmoid feedforward neural networks. *Neural Networks, 9*(7), 1155–1171.
13. Bouras, S., Kotronakis, M., Suyama, K., & Tsividis, Y. (1998). Mixed analog-digital fuzzy logic controller with continuous-amplitude fuzzy inferences and defuzzification. *IEEE Transactions on Fuzzy Systems, 6*(2), 205–215.
14. Brandstetter, A., & Artusi, A. (2008). Radial basis function networks GPU-based implementation. *IEEE Transactions on Neural Networks, 19*(12), 2150–2154.
15. Brown, B., Yu, X., & Garverick, S. (2004). A mixed-mode analog VLSI continuous-time recurrent neural network. In *Proceedings of the 2nd IASTED International Conference on Circuits, Signals and Systems* (pp. 104–108). Clearwater Beach, FL.
16. Cancelo, G., & Mayosky, M. (1998). A parallel analog signal processing unit based on radial basis function networks. *IEEE Transactions on Nuclear Science, 45*(3), 792–797.
17. Cao, L. J., Keerthi, S. S., Ong, C.-J., Zhang, J. Q., Periyathamby, U., Fu, X. J., et al. (2006). Parallel sequential minimal optimization for the training of support vector machines. *IEEE Transactions on Neural Networks, 17*(4), 1039–1049.
18. Catanzaro, B., Sundaram, N., & Keutzer, K. (2008). Fast support vector machine training and classification on graphics processors. In *Proceedings of the 25th ACM International Conference on Machine Learning* (pp. 104–111).
19. Chaudhuri, K., Sarwate, A. D., & Sinha, K. (2013). A near-optimal algorithm for differentially-private principal components. *Journal of Machine Learning Research, 14*, 2905–2943.
20. Choi, J., Sheu, B. J., & Chang, J. C. F. (1994). A Gaussian synapse circuit for analog VLSI neural networks. *IEEE Transactions on Very Large Scale Integration (VLSI) Systems, 2*(1), 129–133.
21. Chua, L. O. (1971). Memristor-the missing circuit element. *IEEE Transactions on Circuit Theory, 18*(5), 507–519.
22. Churcher, S., Murray, A. F., & Reekie, H. M. (1993). Programmable analogue VLSI for radial basis function networks. *Electronics Letters, 29*(18), 1603–1605.
23. Cichocki, A. (1992). Neural network for singular value decomposition. *Electronics Letters, 28*(8), 784–786.
24. Costa, A., De Gloria, A., Farabosch, P., Pagni, A., & Rizzotto, G. (1995). Hardware solutions of fuzzy control. *Proceedings of the IEEE, 83*(3), 422–434.
25. Culler, D., Estrin, D., & Srivastava, M. (2004). Overview of sensor networks. IEEE. *Computer, 37*(8), 41–49.
26. del Campo, I., Echanobe, J., Bosque, G., & Tarela, J. M. (2008). Efficient hardware/software implementation of an adaptive neuro-fuzzy system. *IEEE Transactions on Fuzzy Systems, 16*(3), 761–778.
27. Delbruck, T. (1991). 'Bump' circuits for computing similarity and dissimilarity of analog voltage. In *Proceedings of IEEE International Joint Conference on Neural Networks* (Vol. 1, pp. 475–479). Seattle, WA.
28. Di Ventra, M., & Pershin, Y. V. (2013). The parallel approach. *Nature Physics, 9*, 200–202.
29. Dong, J.-X., Krzyzak, A., & Suen, C. Y. (2005). Fast SVM training algorithm with decomposition on very large data sets. *IEEE Transactions on Pattern Analysis and Machine Intelligence, 27*(4), 603–618.
30. Draghici, S. (2002). On the capabilities of neural networks using limited precision weights. *Neural Networks, 15*, 395–414.

31. Dwork, C., McSherry, F., Nissim, K., & Smith, A. (2006). Calibrating noise to sensitivity in private data analysis. In S. Halevi & T. Rabin (Eds.), *Theory of cryptography, LNCS* (Vol. 3876, pp. 265–284). Berlin: Springer.
32. Elredge, J. G., & Hutchings, B. L. (1994). RRANN: A hardware implementation of the backpropagation algorithm using reconfigurable FPGAs. In *Proceedings of IEEE International Conference on Neural Networks* (pp. 77–80). Orlando, FL.
33. Feali1, M. S., & Ahmadi, A., (2017). Realistic Hodgkin-Huxley axons using stochastic behavior of memristors. *Neural Processing Letters*, *45*(1), 1–14.
34. Fellus, J., Picard, D., & Gosselin, P.-H. (2015). Asynchronous gossip principal components analysis. *Neurocomputing*, *169*, 262–271.
35. Fierimonte, R., Scardapane, S., Uncini, A., Panella, M. (2017). Fully decentralized semi-supervised learning via privacy-preserving matrix completion. *IEEE Transactions on Neural Networks and Learning Systems*, *28*(11), 2699–2711.
36. Gadea, R., Cerda, J., Ballester, F., & Mocholi, A. (2000). Artificial neural network implementation on a single FPGA of a pipelined on-line backprogation. In *Proceedings of the 13th International Symposium on System Synthesis* (pp. 225–230). Madrid, Spain.
37. Girones, R. G., Palero, R. C., & Boluda, J. C. (2005). FPGA implementation of a pipelined on-line backpropagation. *Journal of VLSI Signal Processing*, *40*, 189–213.
38. Gobi, A. F., & Pedrycz, W. (2006). The potential of fuzzy neural networks in the realization of approximate reasoning engines. *Fuzzy Sets and Systems*, *157*, 2954–2973.
39. Hardt, M., & Roth, A. (2012). Beating randomized response on incoherent matrices. In *Proceedings of the 44th Annual ACM Symposium on Theory of Computing* (pp. 1255–1268). New York, NY.
40. Hikawa, H. (2003). A digital hardware pulse-mode neuron with piecewise linear activation function. *IEEE Transactions on Neural Networks*, *14*(5), 1028–1037.
41. Himavathi, S., Anitha, D., & Muthuramalingam, A. (2007). Feedforward neural network implementation in FPGA using layer multiplexing for effective resource utilization. *IEEE Transactions on Neural Networks*, *18*(3), 880–888.
42. Hurdle, J. F. (1997). The synthesis of compact fuzzy neural circuits. *IEEE Transactions on Fuzzy Systems*, *5*(1), 44–55.
43. Hwang, J. N., Vlontzos, J. A., & Kung, S. Y. (1989). A systolic neural network architecture for hidden Markov models. *IEEE Transactions on Acoustics, Speech, and Signal Processing*, *32*(12), 1967–1979.
44. Kang, K., & Shibata, T. (2010). An on-chip-trainable Gaussian-kernel analog support vector machine. *IEEE Transactions on Circuits and Systems I*, *57*(7), 1513–1524.
45. Kim, C. M., Park, H. M., Kim, T., Choi, Y. K., & Lee, S. Y. (2003). FPGA implementation of ICA algorithm for blind signal separation and adaptive noise canceling. *IEEE Transactions on Neural Networks*, *14*(5), 1038–1046.
46. Kollmann, K., Riemschneider, K., & Zeider, H. C. (1996). On-chip backpropagation training using parallel stochastic bit streams. In *Proceedings of the 5th International Conference on Microelectronics for Neural Networks and Fuzzy Systems* (pp. 149–156). Lausanne, Switzerland.
47. Kozlov, A. V., & Singh, J. P. (1994). A parallel Lauritzen-Spiegelhalter algorithm for probabilistic inference. In *Proceedings of ACM/IEEE conference on Supercomputing* (pp. 320–329). Washington, DC.
48. Kung, S. Y., & Hwang, J. N. (1989). A unified systolic architecture for artificial neural networks. *Journal of Parallel and Distributed Computing*, *6*, 358–387.
49. Kuo, Y. H., & Chen, C. L. (1998). Generic LR fuzzy cells for fuzzy hardware synthesis. *IEEE Transactions on Fuzzy Systems*, *6*(2), 266–285.
50. Lawrence, R. D., Almasi, G. S., & Rushmeier, H. E. (1999). A scalable parallel algorithm for self-organizing maps with applications to sparse data mining problems. *Data Mining and Knowledge Discovery*, *3*, 171–195.
51. Lazzaro, J., Lyckebusch, S., Mahowald, M. A., & Mead, C. A. (1989). Winner-take-all networks of $O(n)$ complexity. In D. S. Touretzky (Ed.), *Advances in neural information processing systems* (Vol. 1, pp. 703–711). San Mateo, CA: Morgan Kaufmann.

52. Lee, B. W., & Shen, B. J. (1992). Design and analysis of analog VLSI neural networks. In B. Kosko (Ed.), *Neural networks for signal processing* (pp. 229–284). Englewood Cliffs, NJ: Prentice-Hall.
53. Lee, B. W., & Shen, B. J. (1993). Parallel hardware annealing for optimal solutions on electronic neural networks. *IEEE Transactions on Neural Networks*, *4*(4), 588–599.
54. Le Ly, D., & Chow, P. (2010). High-performance reconfigurable hardware architecture for restricted Boltzmann machines. *IEEE Transactions on Neural Networks*, *21*(11), 1780–1792.
55. Lemaitre, L., Patyra, M., & Mlynek, D. (1994). Analysis and design of CMOS fuzzy logic controller in current mode. *IEEE Journal of Solid-State Circuits*, *29*(3), 317–322.
56. Liu, Q., Dang, C., & Cao, J. (2010). A novel recurrent neural network with one neuron and finite-time convergence for k-winners-take-all operation. *IEEE Transactions on Neural Networks*, *21*(7), 1140–1148.
57. Lin, S. Y., Huang, R. J., & Chiueh, T. D. (1998). A tunable Gaussian/square function computation circuit for analog neural networks. *IEEE Transactions on Circuits and Systems II*, *45*(3), 441–446.
58. Lin, S.-J., Hung, Y.-T., & Hwang, W.-J. (2011). Efficient hardware architecture based on generalized Hebbian algorithm for texture classification. *Neurocomputing*, *74*, 3248–3256.
59. Liu, Y., Jing, W., & Xu, L. (2016). Parallelizing backpropagation neural network using MapReduce and cascading model. *Computational Intelligence and Neuroscience, 2016*, Article ID 2842780, 11 pages.
60. Lu, Y., Roychowdhury, V., & Vandenberghe, L. (2008). Distributed parallel support vector machines in strongly connected networks. *IEEE Transactions on Neural Networks*, *19*(7), 1167–1178.
61. Luo, F.-L., Unbehauen, R., & Li, Y.-D. (1997). Real-time computation of singular vectors. *Applied Mathematics and Computation*, *86*, 197–214.
62. Mahapatra, S., & Mahapatra, R. N. (2000). Mapping of neural network models onto systolic arrays. *Journal of Parallel and Distributed Computing*, *60*, 677–689.
63. Majani, E., Erlanson, R., & Abu-Mostafa, Y. (1989). On the k-winners-take-all network. In D. S. Touretzky (Ed.), *Advances in neural information processing systems 1* (pp. 634–642). San Mateo, CA: Morgan Kaufmann.
64. Mann, J. R., & Gilbert, S. (1989). An analog self-organizing neural network chip. In D. S. Touretzky (Ed.), *Advances in neural information processing systems 1* (pp. 739–747). San Mateo, CA: Morgan Kaufmann.
65. Marchesi, M., Orlandi, G., Piazza, F., & Uncini, A. (1993). Fast neural networks without multipliers. *IEEE Transactions on Neural Networks*, *4*(1), 53–62.
66. Marchesi, M. L., Piazza, F., & Uncini, A. (1996). Backpropagation without multiplier for multilayer neural networks. *IEE Proceedings—Circuits, Devices and Systems*, *143*(4), 229–232.
67. Mayes, D. J., Murray, A. F., & Reekie, H. M. (1996). Pulsed VLSI for RBF neural networks. In *Proceedings of the 5th IEEE International Conference on Microelectronics for Neural Networks* (pp. 177–184). Lausanne, Switzerland.
68. Navia-Vazquez, A., Gutierrez-Gonzalez, D., Parrado-Hernandez, E., & Navarro-Abellan, J. J. (2006). Distributed support vector machines. *IEEE Transactions on Neural Networks*, *17*(4), 1091–1097.
69. Oohori, T., & Naganuma, H. (2007). A new backpropagation learning algorithm for layered neural networks with nondifferentiable units. *Neural Computation*, *19*, 1422–1435.
70. Oh, K.-S., & Jung, K. (2004). GPU implementation of neural networks. *Pattern Recognition*, *37*(6), 1311–1314.
71. Palit, I., & Reddy, C. K. (2012). Scalable and parallel boosting with MapReduce. *IEEE Transactions on Knowledge and Data Engineering*, *24*(10), 1904–1916.
72. Patel, N. D., Nguang, S. K., & Coghill, G. G. (2007). Neural network implementation using bit streams. *IEEE Transactions on Neural Networks*, *18*(5), 1488–1503.
73. Perfetti, R., & Ricci, E. (2006). Analog neural network for support vector machine learning. *IEEE Transactions on Neural Networks*, *17*(4), 1085–1091.

74. Pickett, M. D., Medeiros-Ribeiro, G., & Williams, R. S. (2013). A scalable neuristor built with Mott memristors. *Nature Materials*, *12*(2), 114–117.
75. Rabenseifner, R., & Wellein, G. (2003). Comparison of parallel programming models on clusters of SMP nodes. In H. G. Bock, E. Kostina, H. X. Phu, & R. Rannacher (Eds.), *Modeling, simulation and optimization of complex processes* (pp. 409–426). Berlin: Springer.
76. Raina, R., Madhavan, A., & Ng, A. Y. (2009). Large-scale deep unsupervised learning using graphics processors. In *Proceedings of ACM International Conference on Machine Learning* (pp. 873–880).
77. Rasche, C., & Douglas, R. (2000). An improved silicon neuron. *Analog Integrated Circuits and Signal Processing*, *23*(3), 227–236.
78. Reyneri, L. M. (2003). Implementation issues of neuro-fuzzy hardware: Going toward HW/SW codesign. *IEEE Transactions on Neural Networks*, *14*(1), 176–194.
79. Rovetta, S., & Zunino, R. (1999). Efficient training of neural gas vector quantizers with analog circuit implementation. *IEEE Transactions on Circuits and Systems II*, *46*(6), 688–698.
80. Salapura, V. (2000). A fuzzy RISC processor. *IEEE Transactions on Fuzzy Systems*, *8*(6), 781–790.
81. Saldana, M., Patel, A., Madill, C., Nunes, D., Wang, D., Styles, H., Putnam, A., Wittig, R., & Chow, P. (2008). MPI as an abstraction for software-hardware interaction for HPRCs. In *Proceedings of the 2nd International Workshop on High-Performance Reconfigurable Computing Technology and Applications* (pp. 1–10). Austin, TX.
82. Scardapane, S., Fierimonte, R., Di Lorenzo, P., & Panella, M. (2016). A. Uncini. Distributed semi-supervised support vector machines. *Neural Networks*, *80*, 43–52.
83. Schaik, A. (2001). Building blocks for electronic spiking neural networks. *Neural Networks*, *14*, 617–628.
84. Schneider, R. S., & Card, H. C. (1998). Analog hardware implementation issues in deterministic Boltzmann machines. *IEEE Transactions on Circuits and Systems II*, *45*(3), 352–360.
85. Seiler, G., & Nossek, J. (1993). Winner-take-all cellular neural networks. *IEEE Transactions on Circuits and Systems II*, *40*(3), 184–190.
86. Serrano-Gotarredona, R., Oster, M., Lichtsteiner, P., & 15 colleagues,. (2009). CAVIAR: A 45k neuron, 5M synapse, 12G connects/s AER hardware sensory-processing-learning-actuating system for high-speed visual object recognition and tracking. *IEEE Transactions on Neural Networks*, *20*(9), 1417–1438.
87. Shyu, K.-K., Lee, M.-H., Wu, Y.-T., & Lee, P.-L. (2008). Implementation of pipelined FastICA on FPGA for real-time blind source separation. *IEEE Transactions on Neural Networks*, *19*(6), 958–970.
88. Soudry, D., Di Castro, D., Gal, A., Kolodny, A., & Kvatinsky, S. (2015). Memristor-based multilayer neural networks with online gradient descent training. *IEEE Transactions on Neural Networks and Learning Systems*, *26*(10), 2408–2421.
89. Strukov, D. B., Snider, G. S., Stewart, D. R., & Williams, R. S. (2008). The missing memristor found. *Nature*, *453*(7191), 80–83.
90. Sum, J. P. F., Leung, C. S., Tam, P. K. S., Young, G. H., Kan, W. K., & Chan, L. W. (1999). Analysis for a class of winner-take-all model. *IEEE Transactions on Neural Networks*, *10*(1), 64–71.
91. Tan, Y., Xia, Y., & Wang, J. (2000). Neural network realization of support vector methods for pattern classification. In *Proceedings of IEEE International Joint Conference on Neural Networks* (Vol. 6, pp. 411–416). Como, Italy.
92. Traversa, F. L., & Di Ventra, M. (2015). Universal Memcomputing Machines. *IEEE Transactions on Neural Networks and Learning Systems*, *26*(11), 2702–2715.
93. Trebaticky, P., & Pospichal, J. (2008). Neural network training with extended Kalman filter using graphics processing unit. In *Proceedings of the 18th International Conference Artificial Neural Networks (ICANN)* (Vol. 2, pp. 198–207). Berlin: Springer.
94. Turing, A. M. (1936). On computational numbers, with an application to the entscheidungsproblem. *Proceedings of the London Mathematical Society*, *42*(2), 230–265.

95. Tymoshchuk, P. V. (2009). A discrete-time dynamic K-winners-take-all neural circuit. *Neurocomputing, 72*, 3191–3202.
96. Urahama, K., & Nagao, T. (1995). K-winners-take-all circuit with $O(N)$ complexity. *IEEE Transactions on Neural Networks, 6*, 776–778.
97. Vanek, J., Michalek, J., & Psutka, J. (2017). A GPU-Architecture Optimized Hierarchical Decomposition Algorithm for Support Vector Machine Training. *IEEE Transactions on Parallel and Distributed Systems, 28*(12), 3330–3343.
98. Vrtaric, D., Ceperic, V., & Baric, A. (2013). Area-efficient differential Gaussian circuit for dedicated hardware implementations of Gaussian function based machine learning algorithms. *Neurocomputing, 118*, 329–333.
99. Wang, X., & Leeser, M. (2009). A truly two-dimensional systolic array FPGA implementation of QR decomposition. *ACM Transactions on Embedded Computing Systems Article, 9*(1), Article 3, 1–17.
100. Watkins, S. S., & Chau, P. M. (1992). A radial basis function neurocomputer implemented with analog VLSI circuits. In *Proceedings of International Joint Conference on Neural Networks* (Vol. 2, pp. 607–612). Baltimore, MD.
101. Weninger, F., Bergmann, J., & Schuller, B. (2015). Introducing CURRENNT: The Munich open-source CUDA RecurREnt Neural Network Toolkit. *Journal of Machine Learning Research, 16*, 547–551.
102. Woodsend, K., & Gondzio, J. (2009). Hybrid MPI/OpenMP parallel linear support vector machine training. *Journal of Machine Learning Research, 10*, 1937–1953.
103. Xia, Y., & Wang, J. (2004). A one-layer recurrent neural network for support vector machine learning. *IEEE Transactions on Systems, Man, and Cybernetics, Part B, 34*(2), 1261–1269.
104. Xu, X., & Jager, J. (1999). A fast parallel clustering algorithm for large spatial databases. *Data Mining and Knowledge Discovery, 3*, 263–290.
105. Yildirim, T., & Marsland, J. S. (1996). A conic section function network synapse and neuron implementation in VLSI hardware. In *Proceedings of IEEE International Conference on Neural Networks* (Vol. 2, pp. 974–979). Washington, DC.
106. Zanghirati, G., & Zanni, L. (2003). A parallel solver for large quadratic programs in training support vector machines. *Parallel Computing, 29*, 535–551.
107. Zanni, L., Serafini, T., & Zanghirati, G. (2006). Parallel software for training large scale support vector machines on multiprocessor systems. *Journal of Machine Learning Research, 7*, 1467–1492.
108. Zhang, Y., Li, P., Jin, Y., & Choe, Y. (2015). A digital liquid state machine with biologically inspired learning and its application to speech recognition. *IEEE Transactions on Neural Networks and Learning Systems, 26*(11), 2635–2649.

Chapter 29
Pattern Recognition for Biometrics and Bioinformatics

29.1 Biometrics

Biometrics are the personal or physical characteristics of a person. These biometric identities are usually used for identification or verification. Biometric recognition systems are increasingly being deployed as a more natural, more secure, and more efficient means than the conventional password-based method for the recognition of people. Many biometric verification systems have been developed for global security.

A biometric system may operate in either the verification or identification mode. The verification mode authenticates an individual's identity by comparing the individual with his/her own template(s) (Am I whom I claim I am?). It conducts one-to-one comparison. The identification mode recognizes an individual by searching the entire template database for a match (Who am I?). It conducts one-to-many comparisons.

Biometrics are usually classified into physiological biometrics and behavioral biometrics. Physiological biometrics use biometric characteristics that do not change with time. Some examples of these biometrics are fingerprint, face, facial thermogram, eye, eye's iris, eye's retina scan, ear, palmprint, footprint, palm, palm vein, hand vein, hand geometry, and DNA. Signature is also known to be unique to every individual. Behavioral biometrics are dynamic characteristics that change over time. For recognition purpose, one has to record at a certain time duration, depending on the Nyquist theorem. Examples of such biometrics are speech, keystroke, signature, gesture, and gait. Both types of biometrics can be fused for some complex systems.

Biometric cues such as fingerprints, voice, face, and signature are specific to an individual and characterizes that individual. Verification using fingerprints is the most widely used, as the fingerprint of an individual is unique [29]. The simplest, most pervasive in society, and least obtrusive biometric measure is that of human speech. Speech is unique for each individual. Typically, the biometric identifiers are scanned and processed in an appropriate algorithm to extract a feature vector, which is stored as a template in registration.

Several companies such as Identix sell high-accuracy face recognition software with databases of more than 1,000 people. Face recognition is fast but not extremely

K.-L. Du and M. N. S. Swamy, *Neural Networks and Statistical Learning*,
https://doi.org/10.1007/978-1-4471-7452-3_29

reliable, while fingerprint verification is reliable but inconvenient in fingerprint sampling. Faces and fingerprints can be integrated into a biometric system. Recently, the reliability of face recognition has improved rapidly due to the progress in deep learning.

29.1.1 Physiological Biometrics and Recognition

Human faces convey a significant amount of nonverbal information to facilitate the real-world human-to-human communication. Modern intelligent systems are expected to have the capability to accurately recognize and interpret human faces in real time. Facial attributes, such as identity, age, gender, expression, and ethnic origin, play a crucial role in applications of real facial image analysis including multimedia communication, human–computer interaction, and security. There are seven facial expressions: anger, disgust, happiness, fear, sadness, surprise, and neutral expression. Face recognition has a wide range of commercial, security, surveillance, and law enforcement applications, or even healthcare for helping patients with Alzheimer's disease. Face recognition is not as unobtrusive as other recognition methods such as fingerprint or other biometric recognition methods.

Fingerprint and iris technologies currently offer greater accuracy than face recognition, but require explicit cooperation from the user. A fingerprint is the pattern of ridges and furrows on the surface of a fingertip. The uniqueness of a fingerprint is exclusively determined by the local ridge characteristics and their relationships.

The eye's iris is the colored region between the pupil and the white region (sclera) of the eye. The primary role of the iris is to dilate and constrict the size of the pupil. Iris is unique even for twins. The human iris is shown is in Fig. 29.1a.

The eye's retina, as shown in Fig. 29.1b, is essentially a sensory tissue, which consists of multiple layers. It is located toward the back of the eye. Because of its internal location within the eye, the retina is not exposed to the external environment, and thus it possesses a very stable biometric. Retinal scanning devices can be purchased from EyeDentify, Inc. Uniqueness of retina comes from the uniqueness of the pattern distribution of the blood vessels at the top of the retina.

Palmprint verification [42] recognizes a person based on unique features in his palm, such as the principal lines, wrinkles, ridges, minutiae points, singular points, and texture. Palmprint is a promising biometric feature for use in access control and forensic applications. The human palmprint is shown in Fig. 29.1c.

The texture pattern produced by bending the finger knuckle is a highly distinctive, biometric authentication system using finger knuckle print imaging. The human knuckle print is shown in Fig. 29.1d.

Hand vein, shown in Fig. 29.1e, is the subcutaneous vascular pattern/network appearing on the back of hand. Vein patterns are quite stable in the age group of 20–50. Some commercial products that authenticate individuals from hand vein images are also available. Finger vein or hand vein using finger vein patterns are extracted from an infrared finger image.

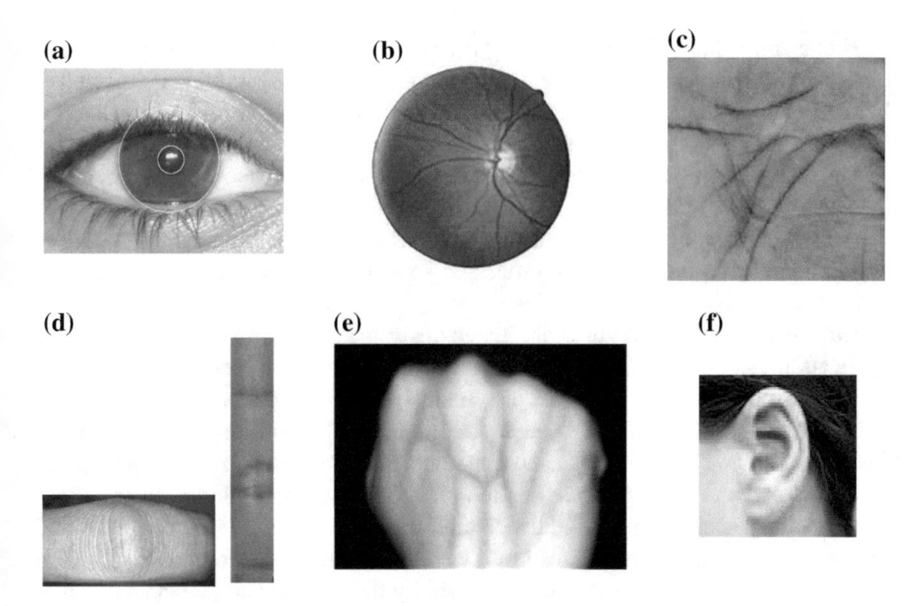

Fig. 29.1 **a** Human iris: the image showing both limbic and pupil boundaries. **b** Retina image. **c** Palmprint. **d** Knuckle print. **e** Human hand vein. **f** Ear

A human ear, as shown in Fig. 29.1f, remains consistent and fixed against a predictable background. Unlike faces, they do not change shape with different expressions or age. Human ear recognition is based on their unique ear shape. NEC company uses the resonation of sound determined by the unique shape of human ear cavities to identify individuals.

Hand image-based recognition is a viable secure access control scheme. The silhouettes of hand images are first registered to a fixed pose, which involves both rotation and translation of the hand and, of the individual fingers. Two feature sets, namely Hausdorff distance of the hand contours and independent component features of the hand silhouette images, have been assessed in [41].

The tongue is a unique organ which can be stuck out of the mouth for inspection. Tongue-print recognition [43] is a technology for noninvasive biometric assessment. The tongue can present both static features and dynamic features for authentication.

Machine recognition of biometrics from images and videos involves several disciplines such as pattern recognition, image processing, computer vision, and neural networks. Although any feature extraction technique such as DCT or FFT can be used, neural techniques such as PCA, LDA, and ICA are local methods and thus, are more attractive and widely used. After applying these transforms to the images, some of the coefficients are selected to construct feature vectors. For classification problems, MLP, RBF network, and SVM can be used for supervised classification, while Hopfield network and clustering are used for unsupervised classification.

Human retinal identification system [10] can be composed of three principal modules including blood vessel segmentation, feature generation, and feature matching. DCT is an appropriate feature extraction method for fingerprint recognition and for face recognition. Fingerprint registration is a critical step in fingerprint matching.

For face recognition, the system first detects the faces in the images, segments the faces from the cluttered scene, extracts features from a face region, and identifies the face according to a stored database of faces. The face region can be found by clustering. The region is then preprocessed so as to extract prominent features for further recognition.

Human age estimation via facial image analysis has lots of potential real-world applications, such as human–computer interactions. The aging process is determined by not only the person's gene, but also many external factors, such as health, living style, living location, and weather conditions. Males and females also age differently. An age-specific human–computer interaction system may be developed for secure network/system access control, electronic customer relationship management, security control, and surveillance monitoring, biometrics, entertainment, and cosmetology. The system ensures that kids have no access to Internet pages with adult materials. A vending machine, secured by the age-specific human–computer interaction system, can refuse selling alcohol or cigarettes to the underage people. Ad-agency can find out what kind of scroll advertisements can attract the passengers in what age ranges using a latent computer vision system. Computer-based age synthesis and estimation via faces have become particularly prevalent topics, such as forensic art.

Generally, human age progression is reflected by facial attributes that include smoothness, face shape, face acne, wrinkles, and bags under eyes. Each attribute at different ages obeys a relative ordering relationship. The face aging process is a typical ordinal procedure. Ages can be classified into age groups [19], or directly estimated by the ranking-based classification [40]. Age synthesis [12] is defined to rerender a face image esthetically with natural aging and rejuvenating effects on the individual face. Age estimation is defined to label a face image automatically with the exact age or the age group of the individual face.

Gender and race identifications from face images have many applications [24]: improving search engine retrieval accuracy, demographic data collection, and human–computer interfaces (adjusting the software behavior with respect to the user gender). Moreover, in a biometric recognition framework, gender identification can help by requiring a search of only half the subject database. Gender identification may aid shopkeepers to present targeted advertisements. One way to enhance gender identification techniques is to combine different cues such as face, gait, and voice.

Individuals can be authenticated by using triangulation of hand vein images and simultaneous extraction of knuckle shape information [18]. The method employs palm dorsal hand vein images acquired from the low-cost, near-infrared, and contactless imaging.

29.1.2 Behavioral Biometrics and Recognition

From speech, one can identify a speaker, his/her mental status and gender, or the content of the speech. Speech features should be constant for each person when measured many times and in multiple samples, and should discriminate other people as much as possible. Emotion recognition from speech signals can also be conducted. Common features extracted for speech are maximum amplitude obtained by cepstral analysis, peak and average power spectrum density, number of zero crossings, formant frequency, pitch and pitch amplitude, and time duration.

Automatic language identification is to quickly and accurately identify the language being spoken (e.g., English, Spanish, etc.). Language identification has numerous applications in a wide range of multi-lingual services. The language identification system can be used to route an incoming telephone call to a human operator fluent in the corresponding language.

Speaker recognition systems utilize human speech to recognize, identify, or verify an individual. It is important to extract features from each frame which can capture the speaker-specific characteristics. The same features adopted in speech recognition are equally successful when applied to speaker recognition. Mel-frequency cepstral coefficients (MFCCs) have been most commonly used in both speech recognition and speaker recognition systems. Linear prediction coefficients have received special attention in this respect. For each utterance contiguous 25 ms speech samples can be extracted and MFCC features obtained. Speaker segmentation aims at finding speaker change points in an audio stream, whereas speaker clustering aims at grouping speech segments based on speaker characteristics.

The task of text-independent speaker verification entails verifying a particular speaker from possible imposters without any knowledge of the text spoken. A conventional approach uses a classifier to generate scores based on individual speech frames. The scores are integrated over the duration of the utterance and compared against a threshold to accept or reject the speaker.

Speech emotion recognition can be implemented using speaking rate [17]. The emotions are anger, disgust, fear, happiness, neutral, sadness, sarcasm, and surprise. At the first stage, based on speaking rate, the eight emotions are categorized into three broad groups namely active (fast), normal and passive (slow). In the second stage, these three broad groups are further classified into individual emotions using vocal tract characteristics.

Visual speech recognition [5] uses the visual information of the speaker's face in performing speech recognition. A speaker produces speech using these articulatory organs together with the muscles that generate facial expressions. Because some of the articulators, such as the tongue, the teeth, and the lips are visible, there is an inherent relationship between the acoustic and visible speech. Visual speech recognition is also referred to as lipreading.

HMM-based speech synthesis first extracts parametric representations of speech including spectral (filter) and excitation (source) parameters from a speech database and then models them by using a set of subword HMMs. One of the key tasks of spoken-dialog systems is classification.

Gait is an efficient biometric feature for human identification at a distance. Gait should be analyzed within complete walking cycle(s) because it is a kind of periodic action. In a window of complete walking cycle(s), gait (motion) energy image is constructed on spatial domain as gait feature. These methods can be mainly classified into three categories: model-based, appearance-based, and spatiotemporal-based. For a walking person, there are two kinds of information in his/her gait signature: static and dynamic. This is also applicable for human action analysis. After the motion features are computed by obtaining sparse decomposition of the scale invariant feature transform (SIFT) features [39], an action is often represented by a collection of codewords in a predefined codebook; this is the bag-of-words model.

Hand gesture recognition is an important topic for human–computer interaction, sign language interpretation, and visual surveillance. Different hand gesture recognition methods are based on particular features, e.g., gesture trajectories and acceleration signals. Hand gestures can be used to exchange information with other people in a virtual space, to guide robots to perform certain tasks in a hostile environment, or to interact with computers. Hand gestures can be divided into two main categories: static and dynamic.

A dynamic hand gesture recognition technique based on the 2D skeleton representation of the hand is given in [15]. For each gesture, the hand skeletons of each posture are superposed providing a single image, which is the dynamic signature of the gesture. The recognition is performed by comparing this signature with the ones from a gesture alphabet, using Baddeley's distance as a measure of dissimilarities between model parameters.

Handwriting-based biometric recognition [27] can be handwriting recognition, forensic verification, and user authentication. Techniques for handwritten character recognition are reviewed and compared in [20]. Typically, handwriting-based biometric verification and identification use signatures. Signature as proof of authenticity is a socially well-accepted transaction, especially for legal document management and financial transactions. Online signature verification provides a reliable authentication. Signature verification techniques are based on the dynamics of a person's signature, namely, time series of pen-tip coordinates, writing forces, or inclination angles of a pen. However, signatures of a person may be very variable. Feature extraction of signatures can be based on Hough transform, which detects straight lines from the presented images.

29.2 Face Detection and Recognition

Research efforts in face processing are focused on face detection, face recognition, face tracking, pose estimation, and expression recognition. Face recognition is sufficiently mature and can be ported to real-time system. It is not robust in natural environments, where there is noise and illuminations and pose problems, and in real-time and video environment. Some kinds of information fusions are necessary for reliable recognition.

29.2.1 Face Detection

Face detection [37] is the first step for automatic face recognition. It is a difficult task in view of pose, facial expression, presence of some facial components (such as beards, mustaches, and glasses), and their variability texture (such as size, shape, color, occlusion, and imaging orientation and lighting). Human skin color is an effective feature in many applications from face detection to hand tracking. For different skin color, the major difference lies largely between their intensity rather than their chrominance.

Face detection is a typical two-class problem to distinguish face class from nonface class: faces and images not containing faces. The boundary between the face and nonface patterns is highly nonlinear because the face manifold due to variations in facial appearance, lighting, head pose, and expression is highly complex. The learning-based approach has so far been the most effective one for constructing face/nonface classifiers.

PCA can be used for the localization of a face region. Due to the fact that color is the most discriminating feature of a facial region, the first step can be a pixel-based color segmentation to detect skin-colored regions.

Human face regions can be rapidly detected in MPEG video sequences [34]. The underlying algorithm takes the inverse quantized DCT coefficients of MPEG video as the input, and outputs the locations of the detected face regions. The algorithm consists of three stages, where chrominance, shape, and frequency information are, respectively, used. By detecting faces directly in the compressed domain, there is no need to carry out inverse DCT. The algorithm can be applied to JPEG unconstrained images or motion JPEG video as well. A robust face tracking system presented in [44] extracts multiple face sequences from MPEG video. Specifically, a view-based DCT-domain face detection algorithm is first applied periodically to capture mostly frontal and slight slanting faces of variable sizes and locations. The face tracker then searches the target faces in local areas across frames in both the forward and backward directions. The tracking combines color histogram matching and skin color adaptation to provide robust tracking.

In [11], a real-time face detection system for color image sequences is presented. The system applies three different face detection methods and integrates the results so obtained to achieve a greater location accuracy. The groups of connected pixels (blobs) extracted from moving objects are then subject to outline analysis, skin color detection, and PCA. Outline analysis is applied to localize the human head. A skin color method is applied to the blobs to find skin regions. PCA is trained for frontal view faces only, and is used to classify if a particular skin region is a face or a nonface. Finally, the obtained face locations are fused to increase the detection reliability and to avoid false detections due to occlusions or unfavorable human poses.

The accuracy of face alignment affects the performance of a face recognition system. Since face alignment is usually conducted using eye positions, an accurate eye localization algorithm is essential for accurate face recognition. An automatic technique for eye detection is introduced in [35]. It has an overall 94.5% eye detection rate using FRGC 1.0 database, with the detected eyes very close to the manually provided eye positions. In [16], color, edge, and binary information are used to detect eye pair candidate regions from input image, and then face candidate region with the detected eye pair is extracted. The approach shows excellent face detection performance over 99.2%.

29.2.2 Face Recognition

Face recognition approaches could be categorized into feature-based approaches and holistic approaches. Holistic matching methods use the whole face region as the raw input to a recognition system. In feature-based (structural) matching methods, local features such as the eyes, nose, and mouth are first extracted and their locations and local statistics (geometric and/or appearance) are fed into a structural classifier.

PCA is applied on the training set of faces, yielding the eigenfaces approach [33]. It assumes that the set of all possible face images occupies a low-dimensional subspace, derived from the original high-dimensional input image space. Eigenfaces algorithm represents a face with 50–100 coefficients and uses a global representation of the face, so that the algorithm is faster, with a global encoding of the face.

A multimodal-part face recognition method based on PCA [32] combines multimodals: Five kinds of parts are dealt with by PCA to obtain eigenfaces, eigeneyebrows, eigeneyes, eigennoses, and eigenmouths. Thirty-one kinds of different face recognition models are created by using combinations of different parts. This method has a higher recognition rate and more flexibility compared to the eigenfaces method.

Fisherfaces approach first uses PCA for dimension reduction and then apply LDA. While eigenfaces algorithm derives the most expressive features, Fisherfaces derives the most discriminative features. Eigenfaces can typically achieve a classification rate of 90% for the ORL database, while Fisherfaces can achieves 95%. Neural network approach using MLP, RBF network and LVQ can outperform Fisherfaces. All these methods belong to the holistic approach. Eigenfaces usually needs many (usually >5) images for each person in the gallery. Either PCA or its derived methods are sensitive to variations caused by illumination and rotation.

Example 29.1 (*Face recognition using Fisherfaces*) In Example 14.2, we have given an example of face recognition using eigenfaces. For the recognition of $C = 30$ people by using $N = 60$ training samples, the method achieves a classification rate of 100%. In this example, we rework the same problem by using the Fisherfaces approach. Fisherfaces first applies PCA to project the samples onto a $(N - C)$-dimensional linear subspace, and then applies LDA to project the resulted projections onto $(C - 1)$-dimensional linear subspace so as to determine the most discriminating features between faces. The combined weights are called Fisherfaces. When a new sample is presented, the projection on the weights is derived. The projection of the presented sample is compared with those of all the training samples, and the training sample that has the minimum difference from the test sample is classified as the correct class. Experiment on the set of 30 testing samples shows a classification rate of 86.7% (26/30).

ICA representations are superior to PCA representations for recognizing faces across days and changes in expression [2]. Associative memory is another kind of neural network for face recognition. Associative memory-based classification learns how to perform recognition by categorizing positive examples of a subject.

An illumination normalization approach is presented for face recognition under varying lighting conditions in [6]. DCT is employed to compensate for illumination variations in logarithmic domain. Since illumination variations mainly lie in the low-frequency band, an appropriate number of DCT coefficients are truncated to minimize variations under different lighting conditions.

Wavelets can be used to extract global as well as local features, such as the nose and eye regions of a face. Face representation based on Gabor features has achieved great success in face recognition area. Gabor features are derived from the convolution of a Gabor filter and an image. Gabor wavelets model the receptive field profiles of cortical simple cells quite well. They exhibit desirable characteristics of spatial frequency, spatial locality, and orientation selectivity to cope with the variations due to illumination and facial expression changes. Moreover, to extract more local features from the original images, a series of Gabor filters with various scales and orientations (called Gabor filter bank) are needed in most cases of biometrics.

Gabor–Fisher classifier for face recognition [21] is robust to changes in illumination and facial expression. It applies enhanced LDA to an augmented Gabor feature vector derived from the Gabor wavelet representation of face images. The Gabor–Fisher classifier achieves 100% accuracy on face recognition using only 62 features, tested on face recognition using 600 FERET frontal face images corresponding to 200 subjects.

Much like recognition from visible imagery is affected by illumination, recognition with thermal face imagery is affected by a number of exogenous and endogenous factors such as weather conditions, environment change, and subject's metabolism. And while the appearance of some features may change, their underlying shape remains the same and continues to hold useful information for recognition [31].

Laplacianfaces approach [14] uses locality-preserving projections to map face images into a face subspace for analysis. Locality-preserving projections find an embedding that preserves local information, and obtains a face subspace that best detects the essential face manifold structure. The Laplacianfaces are the optimal linear approximations to the eigenfunctions of the Laplace–Beltrami operator on the face manifold. The unwanted variations resulting from changes in lighting, facial expression and pose may be eliminated or reduced. The method provides a better representation and achieves lower error rates in face recognition than eigenfaces and Fisherfaces do.

Video-based face recognition method provides more information in a video sequence than in a single image. A multiple classifiers fusion-based video face recognition algorithm can be used.

A 3D face image records the exact geometry of the subject, invariant to illumination and orientation changes. 3D model-based face recognition is robust against pose and lighting variations. In 3D face recognition, registration is a key preprocessing step. 3D face recognition can be implemented using reconstructed 3D models from a set of 2D images [13]. The reconstructed 3D model can be used to obtain the 2D projection images that are matched with probe images. In [22], 3D models are used to recognize 2.5D face scans, provided by commercial 3-D sensors, such as Minolta Vivid series. A 2.5D scan is a simplified 3D (x, y, z) surface representation that contains at most one depth value (z direction) for every point in the (x, y) plane, associated with a registered texture image.

29.3 Bioinformatics

Statistical and computational problems in biology and medicine have created *bioinformatics*. Typical bioinformatics problems are protein structure prediction from amino acid sequences, fold pattern recognition, homology modeling, multiple alignment, distant homology, motif finding, protein folding, and phylogeny, resulting in a large number of NP-hard optimization problems. Prognostic prediction for the recurrence of a disease or the death of a patient is also a classification or a regression problem.

The genetic information that defines the characteristics of living cells within an organism is encoded in the form of a moderately simple molecule, deoxyribonucleic nucleic acid (DNA). DNA is often represented as a string composed of four nucleotide bases with the chemical formulae: adenine (A, $C_5H_5N_5$), cytosine (C, $C_4H_5N_3O$), guanine (G, $C_5H_5N_5O$) and thymine (T, $C_5H_5N_2O_2$). DNA has the ability to perform two basic functions: replicating itself and storing information on the linear composition of the amino acids in proteins.

Ribonucleic acid (RNA) is a polymer of repeating units, namely ribonucleotides, with a structure analogous to single-stranded DNA. Sugar deoxyribose which appears in DNA is replaced in RNA by another sugar called ribose, and the base thymine (T) that appears in DNA is replaced in RNA by another organic base called uracil (U). Compared with DNA, RNA molecules are less stable and exhibit more variability in their three-dimensional structure. Although RNA and protein polymers seem to exhibit a much more variable functional spatial structure than DNA does, their linear content is always a copy of a short fragment of the genome.

Proteins are basic constructional blocks and functional elements of living organisms. In an organism, proteins are responsible for carrying out many different functions in the life cycle of the organism. Proteins are molecules with complicated three-dimensional structures, but they always have an underlying linear chain of amino acids as their primary structure. Each protein is a chain of 20 different amino acids in a specific order and it has unique functions. The length of protein is between 50 and 3000, with an average length of 200. The order of amino acids is determined by the DNA sequences in the gene which codes for a specific protein.

The production of a viable protein from a gene is called *gene expression*, and the regulation of gene expression is a fundamental process necessary to maintain the viability of an organism. Gene expression is a two-step process. To produce a specific protein, a specific segment on the genome codes a gene, the gene is first transcribed from DNA into a messenger RNA (mRNA), which is then converted into a protein via translation. The process of gene expression is regulated by certain proteins known as transcription factors. In both replication (creating a new copy of DNA) and transcription (creating an RNA sequence complementary to a DNA fragment), protein enzymes called polymerases slide along the template DNA strand. It is very difficult to measure the protein level directly because there are simply too many of them in a cell. Therefore, the levels of mRNA are used to identify how much a specific protein is presented in a sample, i.e., it gives an indication of the levels of gene expression.

The genome is the ensemble of genes in an organism, and genomics is the study of the genome. The major goal of genomics is to determine the function of each gene in the genome (i.e., to annotate the sequence). The genome is the complete DNA sequence of an organism. Genomics is concerned with the analysis of gene sequences, including the comparison of gene sequences, and analysis of the succession of symbols in sequences.

The international Human Genome Project had the primary goal of determining the sequence of 3 billions of the senucleotide bases that make up DNA (i.e., DNA sequencing) and to identify the genes of the human genome from both a physical and functional standpoint. The project began in 1990 and was initially headed by James D. Watson at the U.S. National Institutes of Health. The complete genome was released in 2003. Approximate numbers of different objects in the human body are 30,000 genes within the human genome, 10^5 mRNA, 3×10^5 proteins, 10^3–10^4 expressed proteins, 250 cell types, and 10^{13}–10^{14} cells [28]. The sequencing of entire genomes of various organisms has become one of the basic tools of biology.

One of the greatest challenges facing molecular biology is the understanding of the complex mechanisms regulating gene expression. Identification of the motif sequences is the first step toward unraveling the complex genetic networks composed of multiple interacting gene regulation. The goal of gene identification is for automatic annotation. For genome sequencing and annotation, sequence retrieval and comparison, a natural toolbox for problems of this nature is provided by HMMs and the Viterbi algorithm based on dynamic programming.

The proteome is the set of proteins of an organism. It is the vocabulary of the genome. Via the proteome, genetic regulatory networks can be elucidated. Proteomics combines the census, distribution, interactions, dynamics, and expression patterns of the proteins in living systems. The primary task is to correlate the pattern of gene expression with the state of the organism. For any given cell, typically only 10% of the genes are actually translated into proteins under a given set of conditions and at a particular epoch in the cell's life. On the other hand, a given gene sequence can give rise to tens of different proteins, by varying the arrangements of the exons and by post-translational modification. As proteins are the primary vehicle of phenotype, proteomics constitutes a bridge between genotype and phenotype. Comparison

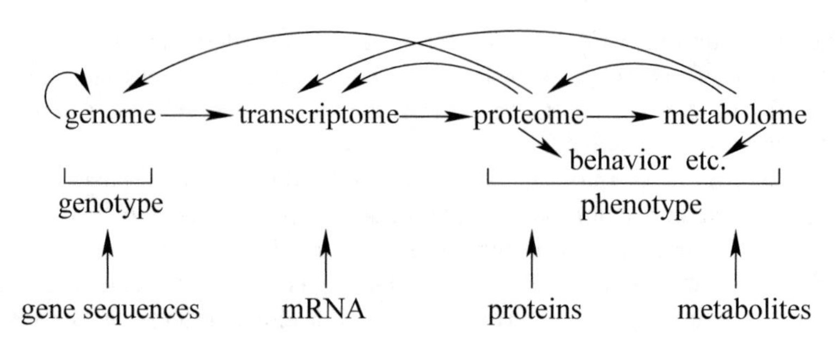

Fig. 29.2 The relation among genes, mRNA, proteins, and metabolites. The curved arrows denote regulatory processes [28]

between the proteomes of diseased and healthy organisms forms the foundation of the molecular diagnosis of disease.

Figure 29.2 highlights the principle objects of investigation of bioinformatics. The three main biological processes are DNA sequence determining protein sequence, protein sequence determining protein structure and protein structure determining protein function.

Example 29.2 The consensus sequence for the human mitochondrial genome has the GenBank accession number NC_001807. The nucleotide sequence for the human mitochondrial genome has 16571 elements in form of a character array:

```
gatcacaggtctatcaccctattaaccactcacgggagctctccatgcat
ttggtattttcgtctgggggtgtgcacgcgatagcattgcgagacgctg
gagccggagcaccctatgtcgcagtatctgtctttgattcctgcctcatt
ctat...
```

By analyzing a DNA sequence, sections of a sequence with a high percent of A+T nucleotides usually indicate intergenic parts of the sequence, while low A+T and higher G+C nucleotide percentages indicate possible genes. Often, high CG dinucleotide content is located before a gene. Figure 29.3 gives the sequence statistics of the human mitochondrial genome.

29.3.1 Microarray Technology

The idea of measuring the level of mRNA as a surrogate measure of the level of gene expression dates back to 1970s. The methods allow only a few genes to be studied at a time. Microarrays allow to measure mRNA levels in thousands of genes in a single experiment and check whether those genes are active, hyperactive or silent. A microarray is typically a small glass slide or silicon wafer, upon which genes or gene fragments are deposited or synthesized in a high-density manner.

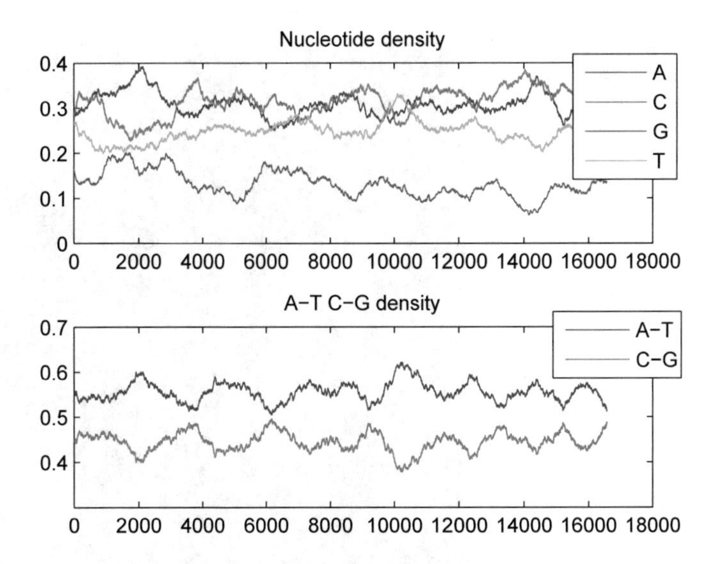

Fig. 29.3 The nucleotide density of the human mitochondrial genome

DNA microarrays evaluate the behavior and inter-relationship between genes on a genomic scale. This is done by quantifying the amount of mRNA for that gene which is contained within the cell. DNA microarray technology permits a systematic study of the correlation of the expression of thousands of genes. It is very useful for drug development and test, gene function annotation, and cancer diagnosis.

To make a microarray, the first stage is probe selection. It determines the genetic materials to be deposited or synthesized on the array. The genetic materials deposited on the array serve as probes to detect the level of expressions for various genes in the sample. Each gene is represented by a single probe. A probe is normally single-stranded (denatured) DNA, so the genetic material from the sample can bind with the probe. Once the probes are selected, each type of probe will be deposited or synthesized on a predetermined spot on the array. Each spot will have thousands of probes of the same type, so the level of intensity detected at each spot can be traced back to the corresponding probe. A sample is prepared after the microarray is made. The mRNA in the sample is first extracted and purified, then is reverse transcribed into single-stranded DNA and a fluorescent marker is attached to each transcript. The single-stranded DNA transcript will only bind with the probe that is complementary with the transcript, that is, binding will only occur if the DNA transcript from the sample is coming from the same gene as the probe. By measuring the amount of fluorescence in each spot using a scanner, the level of expression of each gene can be measured. The result of a DNA microarray experiment is shown in Fig. 29.4.

A microarray is a small chip onto which a large number of DNA molecules (probes) are attached in fixed grids. The chip is made of chemically coated glass, nylon, membrane, or silicon. Each grid cell of a microarray chip corresponds to a DNA sequence. For cDNA (complementary DNA) microarray experiment, the first

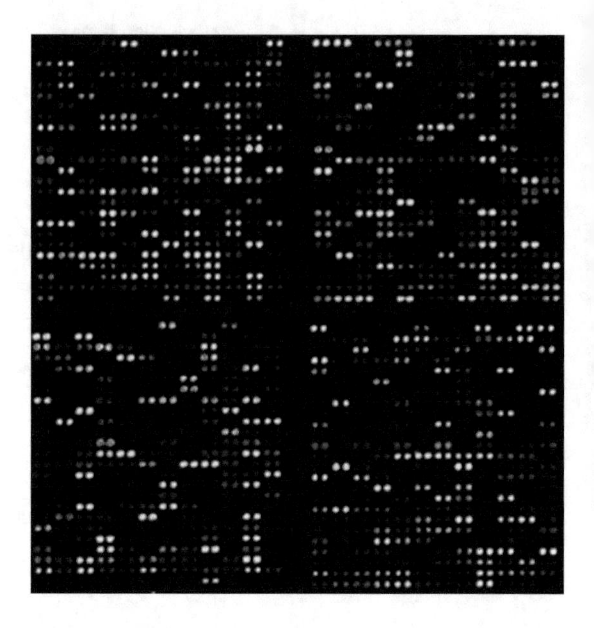

Fig. 29.4 The result of a DNA microarray experiment. The fluorescence intensity level at each spot indicates the corresponding gene's expression level

step is to extract RNA from a tissue sample and amplification of RNA. Thereafter two mRNA samples are reverse transcribed into cDNA labeled using red-fluorescent dye Cy5 and green-fluorescent dye Cy3. The cDNA binds to the specific oligonucleotides on the array. In the subsequent stage, the dye is excited by a laser so that the amount of cDNA can be quantified by measuring the fluorescence intensities. The log ratio of two intensities of each dye is used as the gene expression profiles: gene expression level $= \log_2 \frac{\text{intensity}(Cy5)}{\text{intensity}(Cy3)}$. A microarray gene expression data matrix is a real-valued $n \times d$ matrix, where the rows correspond to n genes and the columns correspond to d conditions (or time points).

Protein microarrays allow a simultaneous assessment of expression levels for thousands of genes across various treatment conditions and time. The main difference compared with nucleic acid arrays is the difficulty and expense of placing thousands of protein capture agents on the array.

Labeling exhaustively all the experimental samples can be simply impossible. For extracting knowledge from such huge volume of microarray gene expression data, computational analysis is required. Clustering is one of the primary approaches to analyze such large amount of data to discover the groups of coexpressed genes. In [25], an attempt has been made in order to improve the performance of fuzzy clustering by combining it with SVM classifier. Genetic linkage analysis is a statistical method for mapping genes onto chromosomes, and determining the distance between them. A tool called RC_Link [1] can model genetic linkage analysis problems as Bayesian networks and perform inference efficiently.

The regulation of gene expression is a dynamic process. There are inherent difficulties in discovering regulatory networks using microarray data [9]. Genetic networks provide a concise representation of the interaction between multiple genes at the system level. The inference of genetic regulatory networks based on the time series gene expression data from microarray experiments has become an important and effective way to achieve this goal. Genetic regulatory network inference is critically important for revealing fundamental cellular processes, investigating gene functions, and understanding their relations. Pharmaceutical companies can test how cells will react to new drug treatments by observing gene expression patterns pre- and post-treatment. Genetic regulatory networks can be inferred by using Boolean networks, Bayesian networks, continuous models, and fuzzy logic models. Genetic regulatory networks can be inferred from time series gene expression data by using a recurrent network and particle swarm optimization, where gene interaction is explained through a connection weight matrix [36].

29.3.2 Motif Discovery, Sequence Alignment, Protein Folding, and Coclustering

The motif is often defined more formally along the lines of a sequence of amino acids that defines a substructure in a protein, which can be connected in some way to protein function or structural stability. Motifs are defined as transcription binding sites in DNA sequences. Identification of motifs is a difficult task since they are very short (generally 6–20 nucleotides long) and may have variations as a result of mutations, insertions, and deletions [8]. Moreover, these genomic patterns reside on very long DNA sequences, which make the task irresolvable for traditional computational methods. In general, many algorithms either adopt probabilistic methods such as Gibbs sampling, or based on word-enumerative exhaustive search methods. Various motif discovery tools are evaluated and compared with respect to their performances in [8]. DNA motif discovery helps to better understand the regulation of the transcription in the protein synthesis process.

In molecular biology, DNA or protein sequences store genetic information. A typical bioinformatic problem is provided by polymorphisms in the human genome. Any two human DNA sequences differ at random points located, on average, several hundred nucleotides apart. The human genome sequence is based on only a few individuals. Sequence alignment is used to arrange sequences of DNA, RNA, or proteins in order to identify regions of similarity as these regions might be a consequence of functional, structural, or evolutionary relationships between the sequences. BLAST and ClustalW are well-known tools for sequence alignment by calculating the statistical significance of matches to sequence databases. GenBank and EMBL-Bank are sequence databases, which provide annotated DNA sequences.

Computational approaches to sequence alignment are generally dynamic programming-based optimal methods or heuristic methods. The Smith–Waterman algorithm [30] is a dynamic programming technique for global pairwise alignment of two DNA sequences. The most widely used methods for multiple sequence alignment include scalar product-based alignment of groups of sequences. Scalar product-based alignment algorithms can be significantly speeded up by general-purpose GPUs [3]. The local alignment problem is stated as follows. Given a set of unaligned biologi-

Fig. 29.5 Sequence alignment of the human and mouse amino acid sequences. The color for highlighting the matches is red (|) and the color for highlighting similar residues that are not exact matches is magenta (:)

cal (DNA or protein) sequences, locate some recurrent short sequence elements or motifs that are shared among the set of sequences.

Example 29.3 Sequence alignment functions can be used to find similarities between two nucleotide sequences. Alignment functions return more biologically meaningful results when you are using amino acid sequences. Nucleotide sequences are converted to amino acid sequences and the open reading frames are identified. For the human and mouse amino acid sequences, the alignment is very good between amino acid position 69 and 599. Realigning the trimmed sections with high similarity, the percent identity is 84% (446/530). A segment of the sequence alignment is shown in Fig. 29.5.

The protein folding problem is to predict a protein's three-dimensional structure from its one-dimensional amino acid sequence. If this problem is solved, rapid progress will be made in the field of protein engineering and rational drug design. NMF in combination with three nearest-neighbor classifiers is explored for protein fold recognition in [26].

Extracting biologically relevant information from DNA microarrays is a very important task for drug development and test, function annotation, and cancer diagnosis. When analyzing the large and heterogeneous collections of gene expression data, conventional clustering algorithms often cannot produce a satisfactory solution. Coclustering refers to the simultaneous clustering of both rows and columns of a data matrix [23]. The goal is to find submatrices, that is, subgroups of genes and subgroups of conditions, where the genes exhibit highly correlated activities for every condition. Minimum sum-squared residue coclustering algorithm [7] is a residue-based coclustering algorithm that simultaneously identify coclusters with coherent values in both rows and columns via an alternating C-means-like iterative algorithm, resulting in a checkerboard structure. Specific strategies are proposed to enable the algorithm to escape poor local minima and resolve the degeneracy problem in partitional clustering algorithms. For microarray data analysis, matrix decomposition methods such as SVD [38] and NMF [4] are used to detect biclusters in gene expression profiles.

References

1. Allen, D., & Darwiche, A. (2008). RC_Link: Genetic linkage analysis using Bayesian networks. *International Journal of Approximate Reasoning*, *48*, 499–525.
2. Bartlett, M. S., Movellan, J. R., & Sejnowski, T. J. (2002). Face recognition by independent component analysis. *IEEE Transactions on Neural Networks*, *13*(6), 1450–1464.
3. Bassoy, C. S., Torgasin, S., Yang, M., & Zimmermann, K.-H. (2010). Accelerating scalar-product based sequence alignment using graphics processor units. *Journal of Signal Processing Systems*, *61*, 117–125.
4. Carmona-Saez, P., Pascual-Marqui, R. D., Tirado, F., Carazo, J. M., & Pascual-Montano, A. (2006). Biclustering of gene expression data by non-smooth non-negative matrix factorization. *BMC Bioinformatics*, *7*, Article No. 78.
5. Chen, T. (2001). Audiovisual speech processing. *IEEE Signal Processing Magazine*, *18*(1), 9–21.
6. Chen, W., Er, M. J., & Wu, S. (2006). Illumination compensation and normalization for robust face recognition using discrete cosine transform in logarithm domain. *IEEE Transactions on Systems, Man, and Cybernetics, Part B*, *36*(2), 458–466.
7. Cho, H., & Dhillon, I. S. (2008). Coclustering of human cancer microarrays using minimum sum-squared residue coclustering. *IEEE/ACM Transactions on Computational Biology and Bioinformatics*, *5*(3), 385–400.
8. Das, M. K., & Dai, H. K. (2007). A survey of DNA motif finding algorithms. *BMC Bioinformatics*, *8*, S21.
9. ElBakry, O., Ahmad, M. O., & Swamy, M. N. S. (2012). Identification of differentially-expressed genes for time-course microarray data based on modified RM ANOVA. *IEEE/ACM Transactions on Computational Biology and Bioinformatics*, *9*, 451–466.
10. Farzin, H., Abrishami-Moghaddam, H., & Moin, M.-S. (2008). A novel retinal identification system. *EURASIP Journal on Advances in Signal Processing*, *2008*, Article ID 280635, 10 pp.
11. Foresti, G. L., Micheloni, C., Snidaro, L., & Marchiol, C. (2003). Face detection for visual surveillance. In *Proceedings of the 12th International Conference on Image Analysis and Processing* (pp. 115–120). Mantova, Italy.
12. Fu, Y., Guo, G., & Huang, T. S. (2010). Age synthesis and estimation via faces: A survey. *IEEE Transactions on Pattern Analysis and Machine Intelligence*, *32*(11), 1955–1976.
13. Georghiades, A. S., Belhumeur, P. N., & Kriegman, D. J. (2001). From few to many: Illumination cone models for face recognition under variable lighting and pose. *IEEE Transactions on Pattern Analysis and Machine Intelligence*, *23*(6), 643–660.
14. He, X., Yan, S., Hu, Y., Niyogi, P., & Zhang, H.-J. (2005). Face recognition using Laplacianfaces. *IEEE Transactions on Pattern Analysis and Machine Intelligence*, *27*(3), 328–340.
15. Ionescu, B., Coquin, D., Lambert, P., & Buzuloiu, V. (2005). Dynamic hand gesture recognition using the skeleton of the hand. *EURASIP Journal on Advances in Signal Processing*, *2005*(13), 2101–2109.
16. Jee, H., Lee, K., & Pan, S. (2004). Eye and face detection using SVM. In *Proceedings of Intelligent Sensors, Sensor Networks and Information Processing Conference* (pp. 577–580). Melbourne, Australia.
17. Koolagudi, S. G., & Krothapalli, R. S. (2011). Two stage emotion recognition based on speaking rate. *International Journal of Speech Technology*, *14*, 35–48.
18. Kumar, A., & Prathyusha, K. V. (2009). Personal authentication using hand vein triangulation and knuckle shape. *IEEE Transactions on Image Processing*, *18*(9), 2127–2136.
19. Kwon, Y., & Lobo, N. (1999). Age classification from facial images. *Computer Vision and Image Understanding*, *74*(1), 1–21.
20. Le Cun, Y., Bottou, L., Bengio, Y., & Haffner, P. (1998). Gradient-based learning applied to document recognition. *Proceedings of the IEEE*, *86*(11), 2278–2324.
21. Liu, C., & Wechsler, H. (2002). Gabor feature based classification using the enhanced Fisher linear discriminant model for face recognition. *IEEE Transactions on Image Processing*, *11*(4), 467–476.

22. Lu, X., Jain, A. K., & Colbry, D. (2006). Matching 2.5D face scans to 3D models. *IEEE Transactions on Pattern Analysis and Machine Intelligence*, *28*(1), 31–43.
23. Madeira, S. C., & Oliveira, A. L. (2004). Biclustering algorithms for biological data analysis: A survey. *IEEE/ACM Transactions on Computational Biology and Bioinformatics*, *1*(1), 24–45.
24. Makinen, E., & Raisamo, R. (2008). Evaluation of gender classification methods with automatically detected and aligned faces. *IEEE Transactions on Pattern Analysis and Machine Intelligence*, *30*(3), 541–547.
25. Mukhopadhyay, A., & Maulik, U. (2009). Towards improving fuzzy clustering using support vector machine: Application to gene expression data. *Pattern Recognition*, *42*, 2744–2763.
26. Okun, O., & Priisalu, H. (2006). Fast nonnegative matrix factorization and its application for protein fold recognition. *EURASIP Journal on Advances in Signal Processing*, *2006*, Article ID 71817, 8 pp.
27. Plamondon, R., & Srihari, S. N. (2000). On-line and off-line handwriting recognition: A comprehensive survey. *IEEE Transactions on Pattern Analysis and Machine Intelligence*, *22*(1), 63–84.
28. Ramsden, J. J. (2009). *Bioinformatics: An introduction* (2nd ed.). London: Springer.
29. Shalaby, M. A. W., & Ahmad, M. O. (2013). A multilevel structural technique for fingerprint representation and matching. *Signal Processing*, *93*(1), 56–69.
30. Smith, T., & Waterman, M. (1981). Identification of common molecular subsequences. *Journal of Molecular Biology*, *147*, 195–197.
31. Socolinsky, D. A., & Selinger, A. (2004). Thermal face recognition in an operational scenario. In *Proceedings of IEEE Conference on Computer Vision and Pattern Recognition (CVPR)* (Vol. 2, pp. 1012–1019).
32. Su, G., Zhang, C., Ding, R., & Du, C. (2002). MMP-PCA face recognition method. *Electronics Letters*, *38*(25), 1654–1656.
33. Turk, M., & Pentland, A. (1991). Eigenfaces for recognition. *Journal of Cognitive Neuroscience*, *3*, 71–86.
34. Wang, H., & Chang, S.-F. (1997). A highly efficient system for automatic face region detection in MPEG video. *IEEE Transactions on Circuits and Systems for Video Technology*, *7*(4), 615–628.
35. Wang, P., Green, M. B., Ji, Q., & Wayman, J. (2005). Automatic eye detection and its validation. In *Proceedings of IEEE Conference on Computer Vision and Pattern Recognition (CVPR)* (pp. 164–165). San Diego, CA.
36. Xu, R., Wunsch, D. C, I. I., & Frank, R. L. (2007). Inference of genetic regulatory networks with recurrent neural network models using particle swarm optimization. *IEEE/ACM Transactions on Computational Biology and Bioinformatics*, *4*(4), 681–692.
37. Yang, M.-H., Kriegman, D. J., & Ahuja, N. (2002). Detecting faces in images: A survey. *IEEE Transactions on Pattern Analysis and Machine Intelligence*, *24*(1), 34–58.
38. Yang, W. H., Dai, D. Q., & Yan, H. (2007). Biclustering of microarray data based on singular value decomposition. In *Emerging technologies in knowledge discovery and data mining, LNCS* (Vol. 4819, pp. 194–205). Berlin: Springer.
39. Yang, J., Yu, K., Gong, Y., & Huang, T. (2009). Linear spatial pyramid matching using sparse coding for image classification. In *Proceedings of IEEE Conference on Computer Vision and Pattern Recognition* (pp. 1794–1801). Miami, FL.
40. Yang, P., Zhong, L., & Metaxas, D. (2010). Ranking model for facial age estimation. In *Proceedings of IEEE Conference on Pattern Recognition* (pp. 3404–3407). Istanbul, Turkey.
41. Yoruk, E., Konukoglu, E., Sankur, B., & Darbon, J. (2006). Shape-based hand recognition. *IEEE Transactions on Image Processing*, *15*(7), 1803–1815.
42. Zhang, D., Kong, W. K., You, J., & Wong, M. (2003). Online palmprint identification. *IEEE Transactions on Pattern Analysis and Machine Intelligence*, *25*(9), 1041–1050.
43. Zhang, D., Liu, Z., & Yan, J.-Q. (2010). Dynamic tongue print: A novel biometric identifier. *Pattern Recognition*, *43*, 1071–1082.
44. Zhao, Y., & Chua, T.-S. (2003). Automatic tracking of face sequences in MPEG video. In *Proceedings of Computer Graphics International* (pp. 170–175). Tokyo, Japan.

Chapter 30
Data Mining

30.1 Introduction

The Web is the world's largest source of information. It records the real world from many aspects at every moment. This success is somewhat, thanks to XML-based technology, which provides a means of information interchange between applications, as well as a semi-structured data model for integrating information and knowledge. Information retrieval has enabled the development of useful web search engines. Relevance criteria based on both textual contents and link structure are very useful for effectively retrieving text-rich documents.

Data mining refers to a variety of techniques in the fields of databases, machine learning, and pattern recognition. The objective is to uncover useful patterns and associations from large databases. Data mining is to automatically search large stores of data for consistent patterns and/or relationships between variables so as to predict future behavior. The process of data mining consists of three phases, namely, data preprocessing and exploration, model selection and validation, as well as final deployment. Structured databases have well-defined features, and data mining can easily succeed with good results. Web mining is more difficult since the World Wide Web is a less structured database.

There are three types of web mining in general: web structure mining, web usage mining (context mining), and web content mining. Content mining unveils useful information about the relationships of web pages based on their content. In a similar way, context mining unveils useful information about the relationship of web pages based on past visitor activity. Context mining is usually applied to the access logs of the website. Some of the most common data items found in access logs are the IP address of the visitor, the date and time of the access, the time zone of the visitor, the size of the data transferred, the URL accessed, the protocol used, and the access method. The data stored in access logs is configurable at the web server with the items mentioned above appearing in most access logs.

Machine learning provides the technical basis of data mining. Data mining needs first to discover the structural features in a database, and exploratory techniques

K.-L. Du and M. N. S. Swamy, *Neural Networks and Statistical Learning*,
https://doi.org/10.1007/978-1-4471-7452-3_30

through self-organization such as clustering are particularly promising. Neurofuzzy systems are ideal tools for knowledge representation. Bayesian networks provide a consistent framework to model the probabilistic dependencies among variables. Classification is also a fundamental method in data mining.

Raw data contained in databases typically contain obsolete or redundant fields, outliers, and values not allowed. Data cleaning and data transformation may be required for data mining. A graphical method for identifying outliers for numeric variables is to examine a histogram of the variable. Outlier mining can be used in telecom or credit card frauds to detect the atypical usage of telecom services or credit cards, in medical analysis to test abnormal reactions to new medical therapies, and in marketing and customer segmentations to identify customers spending much more or much less than the average customer.

30.2 Document Representations for Text Categorization

Document classification requires first to transform text data into numerical data. This is the vectorization step. The widely used vector space model [107] for document representation is commonly referred to as the bag-of-words model. Documents are represented as a feature vector of word frequencies (real numbers) of the terms (words) that appear in all the document set. The vector space model represents a document by a weighted vector of terms. A weight assigned to a term represents the relative importance of that term. One common approach for term weighting uses the frequency of occurrence of a particular word in the document to represent the vector components. Each document can be represented as term vector $\boldsymbol{a} = (a_1, a_2, \ldots, a_n)$ in term space, where each term a_i has an associated weight w_i that denotes the normalized frequency of the word in the vector space, and n is the number of term dimensions.

The most widely used weighting approach for term weights is the combination of term frequency and inverse document frequency (tf-idf) [108]. The inverse document frequency is the inverse of the number of documents in which a word is present in the training dataset. Thus, less weight is given to words which occur in larger number of documents, ensuring that the commonly occurring words (like "the" or "for") are not given undue importance. The weight of term i in document j is defined as

$$w_{ji} = tf_{ji} \times idf = tf_{ji} \times \log_2(N/df), \tag{30.1}$$

where tf_{ji} is the number of occurrences of term i in the document j, df is the total term frequency in a dataset, and N is the number of documents. tf-idf takes into account the distribution of the words in the documents, or its variant tfc [61] also considers the different lengths of the documents. The document score is based on the occurrence of the query terms in the document. This representation can use Boolean features indicating whether a specific word occurs in a document or not. It can use the absolute frequency of a word (tf), as used in [23].

Techniques such as tf-idf vectorize the data easily. However, since each vectorized word is used to represent the document, this leads to the number of dimensions being equal to the number of words. Feature reduction can be performed by removing the stop words (e.g., articles, prepositions, conjunctions, and pronouns), and by mapping words with the same meaning to one morphology, which is known as stemming. Porter algorithm [23] strips common terminating strings (suffixes) from words in order to reduce them to their roots or stems. Stemming removes suffixes and prefixes to avoid duplicate entries and remove basic stop words.

List all the words (after stemming) in all the training documents sorted by the document frequency (i.e., the number of documents it appears in). Choose the top m such words (called keywords) according to this dictionary frequency. For binary representation of a specific document, choose the m-dimensional binary vector where the ith entry is 1 if the ith keyword appears in the document and 0 if it does not. For frequency representation, choose an m-dimensional real-valued vector, where the ith entry is the normalized frequency of appearance of the ith keyword in the specific document. For tf-idf representation, choose the m-dimensional real-valued vector.

Proximity methods in text retrieval provide higher scores to documents that contain query terms A and B separated by x terms compared to documents that contain query terms A and B separated by y terms, where $x < y$. Vector space model would give the same score to both documents, but it is much faster. Phrase-based analysis means that the similarity between documents should be based on matching phrases rather than on single words only.

A well-known normalization technique is the cosine normalization, in which the weight w_i of term i is computed as

$$w_i = \frac{tf_i \cdot idf_i}{\sqrt{\sum_{i=1}^{n} (tf_i \cdot idf_i)^2}}, \tag{30.2}$$

where tf_i denotes the term frequency of a_i, and idf_i denotes the inverse document frequency.

Similarity between documents includes the cosine measure. The cosine similarity between two documents with weight vectors $\boldsymbol{u} = (u_1, \ldots, u_n)$ and $\boldsymbol{v} = (v_1, \ldots, v_n)$ is given by [107]

$$\text{cosine}(\boldsymbol{u}, \boldsymbol{v}) = \frac{\sum_{i=1}^{n} f(u_i) \cdot f(v_i)}{\sqrt{\sum_{i=1}^{n} f(u_i)^2} \cdot \sqrt{\sum_{i=1}^{n} f(v_i)^2}}, \tag{30.3}$$

where $f(\cdot)$ is a damping function such as the square root or the logarithmic function.

The basic assumption is that the similarity between two documents is based on the ratio of how much they overlap to their union, all in terms of phrases. The term-based similarity measure is given as [74]

$$\text{sim}_t(\boldsymbol{d}_1, \boldsymbol{d}_2)^T = \cos(\boldsymbol{d}_1, \boldsymbol{d}_2) = \frac{\boldsymbol{d}_1 \cdot \boldsymbol{d}_2}{\|\boldsymbol{d}_1\| \|\boldsymbol{d}_2\|}, \tag{30.4}$$

where the vectors $\boldsymbol{d}_1$ and $\boldsymbol{d}_2$ represent term weights calculated using tf-idf weighting scheme.

Spectral document ranking methods employing DFT [98] use the term spectra in the document score calculation. They provide higher precision than other text retrieval methods, such as vector space proximity retrieval methods. Spectral document ranking using DFT relies on much longer query times. Spectral document ranking can be based on DCT [99]. By taking advantage of the properties of DCT and by employing vector space model, queries can be processed as fast as vector space model does and a much higher precision is achieved [99].

The simple-minded independent bag-of-words representation remains very popular. Other more sophisticated techniques for document representation are those that are based on higher order word statistics, string kernels [79].

30.3 Neural Network Approach to Data Mining

30.3.1 Classification-Based Data Mining

Text classification is a supervised learning task for assigning text documents to predefined classes of documents. It is used to find valuable information from a huge collection of text documents available in digital libraries, knowledge databases, and the web. Several characteristics have been observed in vector space-based methods for text classification [107], including the high dimensionality of the input space, sparsity of document vectors, linear separability in most text classification problems, and the belief that few features are irrelevant.

Centroid-based classification is one of the simplest classification methods. A test document is assigned to a class that has the most similar centroid. Using the cosine similarity measure, the original form of centroid-based classification finds the nearest centroid and assigns the corresponding class as the predicted class.

Text categorization based on the word-cluster representation [6] can outperform categorization based on the bag-of-words representation, although the performance may depend on the chosen dataset. Information bottleneck clustering [6] is applied for generating document representation in a word-cluster space, where each cluster is a distribution over document classes. When the contribution of low-frequency words to text categorization is significant, SVM with word-cluster representation significantly outperforms SVM with the bag-of-words representation in terms of categorization accuracy or representation efficiency. On two other datasets, the word-based representation slightly outperforms the word-cluster representation.

SVM using standard word frequencies as features yield good performance on a number of benchmark problems [61]. String kernels [79] are used to compute document similarity based on matching nonconsecutive subsequences of characters. In string kernels, the features are the extent to which all possible ordered subsequences of characters are represented in the document. Text categorization using string kernels operating at the character level yields performance comparable to kernels based on the bag-of-words representation. Furthermore, as gaps within the sequence are allowed, string kernels could also pick up stems of consecutive words. Examples of the sequence kernels are string kernel, syllable kernel, and word-sequence kernel. The problem of categorizing documents using kernel-based methods such as SVMs is addressed in [13]. This technique is used with sequences of words rather than characters. It is computationally more efficient, and it ties in closely with standard linguistic preprocessing techniques. A kernel method commonly used in the field of information retrieval is the ranking SVM [62].

In a method for spam filtering [125], instead of using keywords, the spamming behaviors are analyzed and the representative ones are extracted as features for describing the characteristics of e-mails. Spamming behaviors are identified according to the information recorded in headers and syslogs of e-mails. An enhanced MLP model is considered for two-pass classification: determining the spamming behaviors of incoming e-mails and identifying spam according to the behavior-based features. Since spamming behaviors are infrequently changed, compared with the change frequency of keywords used in spams, behavior-based features are more robust with respect to the change of time; thus the behavior-based filtering mechanism outperforms keyword-based filtering.

The notion of privacy-preserving data mining addresses the problem of performing data analysis on distributed data sources with privacy constraints. Necessary information is exchanged between several parties to compute aggregate results without sharing the private content with one another. A solution is to add noise to the source data. BiBoost (bipartite boosting) and MultiBoost (multiparty boosting) [45] allow two or more participants to construct a boosting classifier without explicitly sharing their datasets. The algorithms inherit the excellent generalization performance of AdaBoost, and independently of the number of participants, they perform close to AdaBoost executed using the entire dataset.

Since support vectors are intact tuples taken from the training dataset, releasing an SVM classifier for public use or to clients will disclose the private content of support vectors. Privacy-preserving SVM classifier [75] is designed to not disclose the private content of support vectors for the Gaussian kernel function. It is robust against adversarial attacks.

The iJADE web miner for Internet shopping [71] is based on the integration of neurofuzzy-based web mining technology, human face identification and recognition for user authentication, and interactive and mobile agent-based product search from a large database.

30.3.2 Clustering-Based Data Mining

Clusters provide a structure for organizing a large number of information sources for efficient browsing, searching, and retrieval. Document clustering is one of the most important text mining methods developed to help users effectively navigate, summarize, and organize text documents. The clustering techniques exploit naturally the graph formed by hyperlinks connecting documents to one another. Document clustering can be used to browse a collection of documents or to organize the results returned by a search engine in response to a user's query. For document clustering, the features are words and the samples are documents. Some data mining approaches that use clustering are database segmentation, predictive modeling, and visualization of large databases. The topology-preserving property for SOM makes it particularly suitable for web information processing.

Latent topic models, such as latent semantic indexing [32], probabilistic latent semantic analysis [117], and latent Dirichlet allocation [7], provide a statistical approach to semantically summarize and analyze large-scale document collections based on the bag-of-words assumption. Word clusters are called topics. Latent semantic indexing [32] is a spectral clustering method, where each document is represented by a histogram of word counts over a vocabulary of fixed size. The problems of polysemy and synonymy are considered. In probabilistic latent semantic analysis [117], the number of parameters grows linearly with the size of the training data, subject to overfitting. To address the problem, the latent Dirichlet allocation model [7] introduces priors over the parameters into the probabilistic latent semantic analysis model. Latent Dirichlet allocation [7] assigns the hidden topic labels (variables) to explain the observed words in document–word matrix. The latent topic models are extended to semantically learn the underlying structure of long time series based on the bag-of-words representation [123]. Each time series is treated as a text document, and a set of local patterns from the sequence is extracted as words by sliding a short temporal window along the sequence.

An efficient disk-based implementation of C-means [95] is designed to work inside a relational database management system. It can cluster large datasets having very high dimensionality. In general, it only requires three scans over the dataset and one additional one-time run to compute the global mean and covariance. It is optimized to perform heavy disk I/O, and its memory requirements are low. Its parameters are easy to set.

Current techniques for weblog mining utilize the content or the context [90] of the website. The WEBSOM system [67] utilizes SOM to cluster web pages based on their content. An SOM-based method [101] utilizes both content and context mining clustering techniques to help visitors identify relevant information quicker. The input of the content mining is the set of web pages of the website, whereas the source of the context mining is the access logs of the website. A scalable parallel implementation of batch SOM suitable for data mining applications [69] demonstrates essentially linear speedup on an SP2 parallel computer.

Applying FCM, fuzzy ART, fuzzy max–min, and the Kohonen network to document clustering with the bibliographic database LISA, the best results were found with Kohonen algorithm which also organizes the clusters topologically [50].

Correspondence analysis is a technique for analyzing the relations existing between the modalities of all the qualitative variables by completing a simultaneous projection of the modalities. SOM can be viewed as an extension of PCA due to its topology-preserving property. For qualitative variables, SOM has been generalized for multiple correspondence analysis [24].

SOM is not suitable for nonvectorial data analysis such as the structured data analysis. Examples of structured data are temporal sequences such as the time series, language and words, spatial sequences like the DNA chains, and tree- or graph-structured data arising from natural language parsing and from chemistry. Some unsupervised self-organizing models for nonvectorial data are temporal Kohonen map, recurrent SOM, recursive SOM, SOM for structured data, and merge SOM [52]. All these models introduce recurrence into SOM.

Co-clustering is an approach in which both words and documents are clustered at the same time [34]. It is assumed that all documents belonging to a particular cluster or category refer to a certain common topic. The topics of genuine categories are naturally best described by the titles given to these categories by human experts.

Proximity FCM is an extension of FCM incorporating a measure of similarity or dissimilarity as user's feedback on the clusters during web navigation [80]. The algorithm consists of two main phases that are realized in an interleaved manner. The first phase is FCM applied to the patterns. The second concerns an accommodation of the proximity-based hints and involves some gradient-oriented learning. Proximity FCM offers a relatively simple way of improving the web page classification according to the user interaction with the search engine. The merits of building text categorization systems by using supervised clustering techniques are discussed in [1].

Clustering problems have been studied for a data stream environment in [51]. Clustering continuous data streams allows for the observation of the changes in group behavior. It is assumed that at each time instant, data points from individual streams arrive simultaneously, and the data points are highly correlative to previous ones in the same stream. Clustering on-demand framework [27] dynamically clusters multiple data streams. It realizes online collection of statistics in a single data scan as well as compact multiresolution approximations. The framework consists of two phases, namely, the online maintenance phase and the offline clustering phase. The online maintenance phase provides an efficient mechanism to maintain summary hierarchies of data streams with multiple resolutions in time linear in both the number of streams and the number of data points in each stream. An adaptive clustering algorithm is devised for the offline phase to retrieve approximations of desired substreams from summary hierarchies according to clustering queries.

In data-intensive peer-to-peer environments, distributed data mining algorithms that avoid large-scale synchronization or data centralization offer an alternate choice. In distributed C-means clustering [29], the data and computing resources are distributed over a large peer-to-peer network. Approximate C-means clustering without centralizing the data [29] can approximate the results of a centralized clustering at reasonable communication cost. In distributed data mining, adopting a flat node distribution model can affect scalability. A hierarchically distributed peer-to-peer architecture and a clustering algorithm address the problem of modularity, flexibility, and scalability [53].

Integrating data mining algorithms with a relational DBMS is an important problem. Three SQL implementations of the C-means clustering are introduced to integrate it with a relational DBMS in [94]. SQL overhead is significant for small datasets but relatively low for large datasets, whereas export time for running clustering outside the DBMS becomes a bottleneck for C++, making SQL a more efficient choice.

Document clustering performance can be improved significantly in lower dimensional linear subspaces. NMF [126] and concept factorization [127] have been applied to document clustering with impressive results. Locally consistent concept factorization [11] is an approach to extract the document concepts which are consistent with the manifold geometry such that each concept corresponds to a connected component. By using the graph Laplacian to smooth the document-to-concept mapping, it can extract concepts with respect to the intrinsic manifold structure and thus documents associated with the same concept can be well clustered.

WordNet [87] is a widely used ontology. This English ontology contains general terms organized in synsets (sets of synonymous terms) related using semantic relations. It comprises a core ontology and a lexicon. It is an online lexical reference system which organizes nouns, verbs, adjectives, and adverbs into synonym sets (synsets). ANNIE is an information extraction component of GATE [26]. An unsupervised method uses ANNIE and WordNet lexical categories and WordNet ontology in order to create a well-structured document vector space whose low dimensionality allows common clustering algorithms to perform well [104].

30.3.3 Bayesian Network-Based Data Mining

Bayesian network models for information retrieval were first introduced in [119], where index terms, documents, and user queries are seen as events and are represented as nodes in a Bayesian network. Bayesian networks have also been applied to other information retrieval problems besides ranking as, for example, assigning structure to database queries [12], and document clustering and classification [40].

Bayesian networks provide an effective and flexible framework for modeling distinct sources of evidence in support of a ranking. They can be used to represent vector space model, and this representation can be extended to naturally incorporate new evidence from distinct information sources [31].

Latent Dirichlet allocation [7] is a hierarchical Bayesian model that can infer probabilistic topics from the document–word matrix by using variational Bayes method. It organizes and summarizes large corpora of text documents by discovering the semantic themes, or topics, within the data. When training a latent Dirichlet allocation model on a collection of documents, the distributions over word types for topics and the distributions over topics for documents are estimated from samples of the model's latent variables. Collapsed Gibbs sampler (CGS) (http://mallet.cs.umass.edu) [49] remains a popular choice for topic model estimation, due to its simplicity, along with the availability of efficient implementations leveraging sparsity and parallel architectures. In [131], the latent Dirichlet allocation model is represented as a factor graph, and the loopy belief propagation algorithm is used for approximate inference and parameter estimation.

Fast online EM [132] infers the topic distribution from the previously unseen documents incrementally with constant memory requirements. Within the stochastic approximation framework, it converges to the local stationary point of the Latent Dirichlet allocation's likelihood function. By dynamic scheduling for fast speed and parameter streaming for low memory usage, fast online EM is designed to process infinite documents with infinite vocabulary words for some lifelong topic modeling tasks. It is compared with five online latent Dirichlet allocation algorithms such as OGS (http://mallet.cs.umass.edu/) [130], OVB (http://www.cs.princeton.edu/_blei/topicmodeling.html) [56], RVB, SOI [88] (http://mallet.cs.umass.edu/), and SCVB [43].

Automatic subject indexing from a controlled vocabulary [47] and hierarchical text classification [110] are difficult. As each descriptor in the thesaurus represents a different class/category and a document may be associated with several classes, it is a multi-label problem of high dimensionality; there are explicit (hierarchical) relationships between the class labels; the training data can be quite unbalanced to each class. In [30], given a document to be classified, a method is described that automatically generates an ordered set of appropriate descriptors extracted from a thesaurus. The method creates a Bayesian network to model the thesaurus and uses probabilistic inference to select the set of descriptors having high posterior probability of being relevant given the available evidence (the document to be classified).

An enhanced hybrid classification method is implemented in [58] through the utilization of the naive Bayes approach and SVM. Bayes formula was used to vectorize a document according to a probability distribution reflecting the probable categories that the document may belong to. The dimensions are reduced from thousands (equal to the number of words in the document when using tf-idf) to typically less than 20 (number of categories the document may be classified to) through the use of the Bayes formula, and then this probability distribution is fed to SVM for training and classification purposes.

30.4 XML Format

Extensible markup language (XML) is a standard data representation for interoperability over the Internet. Data in XML documents are self-describing. A huge amount of information is formatted in XML documents. Decomposing the XML documents and storing them in relational tables is a popular practice. Similar to the popular hypertext markup language (HTML), XML is flexible in organizing data based on so-called nested tags. Tags in HTML associated with data express the presentation style of data, while tags in XML describe the semantics of data. The hierarchy formed by nested tags structures the content of XML documents. The role of nested tags in XML is similar to that of schemas in relational databases. At the same time, the nested XML model is far more flexible than the flat relational model. The basic data model of XML is a labeled and ordered tree.

XML queries concern not only the content but also the structure of XML data. Basically, the queries can be formed using twig patterns, in which nodes represent the content part and edges the structural part of the queries. XML queries are categorized into two classes [48]: database-style queries and information retrieval style queries. Database-style queries return all query results that precisely match the queries. Commercial database systems are mainly relational database management systems (RDBMSs), and examples include IBM DB2, Microsoft SQL Server, and Oracle DB. Information retrieval style queries allow imprecise or fuzzy query results, which are ranked based on their relevance to the queries. Only the top-ranked results are returned to users, which is similar to the semantics of keyword search queries in the traditional information retrieval context.

XML Path Language (XPath) and XQuery are mainstream (database-style) XML query languages. Twig patterns play a very important role in XPath and XQuery. In XML, each document is associated with a document-type description, which contains information describing the document structure. Large websites are becoming repositories of structured information that can benefit from being viewed and queried as relational databases.

One important problem in XML query processing is twig pattern matching, that is, finding all matches in an XML data tree that satisfy a specified twig (or path) query pattern. Major techniques for twig pattern matching are reviewed, classified, and compared in [48]. The relational approach and the native approach are two classes of major XML query processing techniques. A good trade-off between the two approaches would be storing XML data in the form of inverted lists by using existing relational databases, coupled with integrating efficient native join algorithms for XML twig queries into existing relational query optimizers [48].

XML documents can be compared as to their structural similarity, in order to group them into clusters so that different storage, retrieval, and processing techniques can be effectively exploited. Compared to standard methods based on graph-matching algorithms, the technique presented in [42] for detecting structural similarity between XML documents is based on the idea of representing an XML document as a time series of a numerical sequence. Thus the structural similarity between two documents can be computed by exploiting the DFT of the associated signals, allowing a significant reduction of the required computation costs.

The semantic web adds metadata and ontology information to web pages to make the web easier to be exploited by both humans and especially by programs. The paradigm of the semantic web helps use metadata as a largely untapped source in order to enhance activities of intelligent information management. Resource description format (RDF) has become the standard language for representing any semantic web. It describes a semantic web using statements which are triples of the form (subject, property, and object). Subjects are resources which are uniquely identified by a uniform resource identifier. In terms of semantic richness for multimedia information, the MPEG-7 structured annotations description scheme is among the most comprehensive and powerful, and has been applied to image annotation.

30.5 Association Mining

Associations are affinities between items. Association rule mining has been applied to analyze market baskets, helping managers realize which items are likely to be bought at the same time [76]. A well-known technique for discovering association rules from databases is the Apriori algorithm [3]. While association rules ignore ordering among the items, an Apriori variation respecting (temporal) ordering emerged under the name sequence mining [4]. The link analysis technique mines relationships and discovers knowledge. Association discovery algorithms find combinations where the presence of one item suggests the presence of another. Some algorithms can find association rules from nominal data [3].

Given a user keyword query, current web search engines return a list of individual web pages ranked by their goodness with respect to the query. Thus, the basic unit for search and retrieval is an individual page, even though information on a topic is often spread across multiple pages. This degrades the quality of search results, especially for long or uncorrelated (multitopic) queries, where a single page is unlikely to satisfy the user's information need. Given a keyword query, composed pages, which contain all query keywords, can be on the fly generated by extracting and stitching together relevant pieces from hyperlinked web pages and retaining links to the original web pages [120]. To rank the composed pages, both the hyperlink structure of the original pages and the associations between the keywords within each page are considered.

Given a time-stamped transaction database and a user-defined reference sequence of interest over time, similarity-profiled temporal association mining discovers all associated item sets whose prevalence variations over time are similar to the reference sequence. The similar temporal association patterns can reveal interesting relationships of data items which co-occur with a particular event over time. Most works in temporal association mining have focused on capturing special temporal regulation patterns such as cyclic patterns and calendar scheme-based patterns.

30.5.1 Affective Computing

Affective computing [44] aims to recognize, feel, infer, and interpret human emotions. It is the set of techniques aimed at performing affect recognition from data, in different modalities and at different granularity scales. Sentiment analysis and emotion recognition are two topics in the field of affective computing. Sentiment analysis performs coarse-grained affect recognition, while emotion recognition performs fine-grained affect recognition. With the proliferation of videos posted online for product reviews, movie reviews, political views, and more, affective computing research has increasingly evolved into complex forms of multimodal analysis.

A key application of text mining is comment mining, with focus on sentiment analysis or opinion mining. Sentiment analysis or opinion mining is the computational study of people's opinions, appraisals, attitudes, and emotions toward entities such as products, services, organizations, individuals, events, topics, and their different aspects. The goal is achieved by content extraction and classification. Sentiment class (negative/positive) serves to collect a set of relevant comments. A common practice is to use the tf-idf weighting for text preprocessing.

Collaborative filtering is used by recommender systems, whose goal is to forecast the user's interest in a given item, based on collective user experience. The main objective is to match people with similar interests to generate personalized recommendations.

Emotional Brain-Based Neural Networks

Emotional brain refers to the portions of the human brain that process external emotional stimuli such as reward and punishment received from the outside world. Due to the existence of shorter signal propagation paths in the emotional brain, emotional stimuli are processed much faster than normal stimuli [70]. An emotion is a personal appraisal of person–environment relationship.

Brain emotional learning-based neural networks [81] and emotional BP-based neural networks [64] are each associated with emotional learning. Brain emotional learning-based networks have been created via anatomical computational models and emotional BP-based networks have been made via appraisal computational models of emotion. By applying the emotional states in the learning process of MLP, emotional BP learning algorithm [64] has additional emotional weights that are updated using two emotional parameters: anxiety and confidence. A limbic-based artificial emotional neural network [82] models emotional situations such as anxiety and confidence in the learning process, the short paths, the forgetting processes, and inhibitory mechanisms of the emotional brain. It shows a higher accuracy than brain emotional learning and emotional BP-based networks do.

30.6 Web Usage Mining

Web usage mining consists in adapting data mining methods to access log file records. These files collect data such as the IP address of the connected machine, the requested URL, the date, and other information regarding the navigation of the user. Web usage mining techniques provide knowledge about the behavior of users in order to extract relationships from the recorded data. Sequential patterns are particularly suited to the study of logs. The access log file is first sorted by address and by transaction. Then all uninteresting data are pruned out from the file. During the sorting process, URLs and clients can be mapped to integers. Each time and date is also translated into a relative time with respect to the earliest time in the log file. The structure of a log file is close to the client–time–item structure used by sequential pattern algorithms.

In [128], clustering is used to segment user sessions into clusters or profiles that can later form the basis for personalization. Web utilization miner [114] discovers navigation patterns with user-specified characteristics over an aggregated materialized view of the weblog, consisting of a tree of sequences of web views. Popular clustering approaches build clusters based on usage patterns derived from users' page preferences. For the need to discover similarities in users' accessing behavior with respect to the time locality of their navigational acts, two time-aware clustering approaches define clusters with users that show similar visiting behavior at the same time period, by varying the priority given to page or time visiting [100].

In [92], a complete framework and findings in mining web usage patterns are presented from weblog files of a real web site that has all the challenging aspects of real-life web usage mining, including evolving user profiles and external data describing an ontology of the web content, as well as an approach for discovering and tracking evolving user profiles. The discovered user profiles can be enriched with explicit information need that is inferred from search queries extracted from weblog data [92].

A specific data mining process is proposed in [86] in order to reveal the densest periods automatically. The approach is able to extract both frequent sequential patterns and the associated dense periods.

Caching is a strategy for improving the performance of web-based systems. The heart of a caching system is its page replacement policy, which selects the pages to be replaced in a cache when a request arrives. A weblog mining method [129] caches web objects and a prediction algorithm predicts future web requests to improve the system performance.

30.7 Ranking Search Results

Information retrieval aims to retrieve objects from a database given the user's information need. Given a collection of objects $\mathcal{D} = \{d_1, d_2, \ldots, d_n\}$ and a query object q, the objects in $\mathcal{D}$ are ranked according to their relevance to q and a subset of the most relevant objects are selected.

In a traditional search engine like Google, a query is specified by giving a set of keywords, possibly linked through logic operators and enriched with additional constraints (i.e., document type, language, etc.). Traditional search engines do not have the necessary infrastructure for exploiting relation-based information that belongs to the semantic annotations for a web page. PageRank [9, 96] and hyperlink induced topics search (HITS) [65] are widely applied to analyze the structure of the web.

A simple counting of the number of links to a page does not take into account the fact that not all the citations have the same authority. PageRank used in Google web search engine very effectively ranks the results. The authority of a page is computed recursively as a function of the authorities of the pages that link the target page. At query time, these importance scores are used in conjunction with query-specific information retrieval scores to rank the query results. PageRank has a clear efficiency advantage over HITS algorithm, as the query time cost of incorporating the precomputed PageRank importance score for a page is low. Furthermore, as PageRank is generated using the entire web graph, rather than a small subset, it is less susceptible to localized link spam.

HITS estimates the authority and hub values of hyperlinked pages on the web, while PageRank merely ranks pages. For HITS, the authority is a measure of the page relevance as information source, and the hubness refers to the quality of a page as a link to authoritative resources. Documents with high authority scores have many links pointing to them. Documents with high hub scores point to many authoritative sites. HITS is query-dependent. User queries are issued to a search engine in order to create a set of seed pages. Crawling the web forward and backward from that seed is performed to mirror the web portion containing the information which is likely to be useful. A ranking criterion based on topological analyses can be applied to the pages belonging to the selected web portion.

In the semantic web, each page possesses semantic metadata that record additional details concerning the web page itself. Annotations are based on classes of concepts and relations among them. The vocabulary for the annotation is usually expressed by means of an ontology that provides a common understanding of terms within a given domain. Semantic search engines are capable of exploiting concepts (and relations) hidden behind each keyword together with natural language interpretation techniques to further refine the result set. A relation-based PageRank algorithm is used in conjunction with semantic web search engines that simply rely on information that could be extracted from user queries and on annotated resources [68]. Relevance is measured as the probability that a retrieved resource actually contains those relations whose existence was assumed by the user at the time of query definition.

Normalized discounted cumulative gain is a popular metric for evaluating ranking performance of a search engine, especially in case of more than two relevance degrees [59]. Google distance computes the distance between two concepts derived from the number of hits returned by Google search engine when querying both concepts. Two concepts with the same or similar meanings in a natural language sense tend to have small Google distance. Normalized Google distance [20] can ideally measure the textual conceptual relationship, i.e., the same and similarity, when the pairwise concepts frequently occur in the same web page.

30.7.1 Surfer Models

A surfer model models a surfer who browses the Internet. There are a variety of surfer models, such as random surfer [9], HITS [65], directed surfer [105], and topic-sensitive PageRank [55]. Random surfer model assumes that a surfer is browsing web pages at random by either following a link from the current page chosen uniformly at random or by typing its URL. Directed surfer model assumes that, when a surfer is at any page, he jumps to only one of those pages that is relevant to the context, the probability which is proportional to the relevance of each outlink. Both models guarantee the convergence of this stochastic process to a stationary distribution under mild assumptions like the irreducibility of the transition probability matrix. The computation of PageRank can be modeled by a single-surfer random walk by choosing a surfer model based only on two actions: the surfer jumps to a new random page with probability $1 - d$ or follows one link from the current page with probability d.

Directed surfer model encompasses the models which allow only forward walks. A surfer probabilistically chooses the next page to be visited depending on the content of the page and the query terms he is looking for [105]. A general probabilistic framework is proposed for web page scoring systems in [35]. The general web page scoring model extends both PageRank and HITS. A methodology that simultaneously performs page ranking and context extraction is dealt with in [97] based on the principle of surfer models. A scalable and convergent iterative procedure is provided for its implementation.

PageRank algorithm for improving the ranking of search query results computes a single vector, using the link structure of the web, to capture the relative importance of web pages, independent of any particular search query. To yield more accurate search results, a set of PageRank vectors, biased using a set of representative topics, are computed to capture more accurately the notion of importance with respect to a particular topic [55]. By using linear combinations of these (precomputed) biased PageRank vectors to generate context-specific importance scores for pages at query time, more accurate rankings can be generated than using a single, generic PageRank vector.

Based on HITS, an entropy-based analysis mechanism is proposed in [63] for analyzing the entropy of anchor texts and links to eliminate the redundancy of the hyperlinked structure so that the complex structure of a website can be distilled. However, to increase the value and the accessibility of pages, most of the content sites tend to publish their pages with intrasite redundant information, such as navigation panels, advertisements, copy announcements, etc. To further eliminate such redundancy, another mechanism InfoDiscoverer [63] applies the distilled structure to identify sets of article pages. InfoDiscoverer employs the entropy information to analyze the information measures of article sets and to extract informative content blocks from these sets. On average, the augmented entropy-based analysis leads to prominent performance improvement.

30.7.2 PageRank Algorithm

PageRank algorithm used by the Google search engine is an unsupervised learning method. The main idea is to determine the importance of a web page in terms of the importance assigned to the pages hyperlinking to it. The web is viewed as a directed graph of pages connected by hyperlinks. A random surfer starts from an arbitrary page and keeps clicking on successive links at random, visiting from page to page.

The PageRank value of a page corresponds to the relative frequency the random surfer visits that page, assuming that the surfer goes on infinitely. The more time spent by the random surfer on a page, the higher the PageRank importance of the page. PageRank algorithm considers a web page to be important if many other web pages point to it. It takes into account both the importance (PageRank) of the linking pages and the number of outgoing links that they have. Linking pages with higher PageRank are given more weight, while pages with more outgoing links are given less weight. The algorithm precomputes a rank vector that provides a priori importance estimates for all of the pages on the web. This vector is computed once, offline, and is independent of the search query. At the query time, lookup is implemented to find the value, and this is integrated with other strategies to rank the pages.

In network science, PageRank metric is a topological-based ranking criterion. It has been used for sorting of the graph nodes. It produces a ranking based on how many neighbors point to a specific node.

Let $l_{ij} = 1$ if page j points to page i, and $l_{ij} = 0$ otherwise. The number of pages pointed to by page j (number of outlinks) is denoted by $c_j = \sum_{i=1}^{N} l_{ij}$, where N is the number of pages. The PageRanks p_i are defined by the recursive relationship

$$p_i = (1 - \alpha) + \alpha \sum_{j=1}^{N} \frac{l_{ij}}{c_j} p_j, \tag{30.5}$$

where the damping factor α is a positive constant (0.85). The importance of page i is the sum of the importance of pages that point to that page. The sums are weighted by $1/c_j$, that is, each page distributes a total vote of 1 to other pages. The constant α ensures that each page gets a PageRank of at least $1 - \alpha$. The first term is to prevent the presence of pages with no forward links. The random surfer escapes from the dangling page by jumping to a randomly chosen page. The surfer can avoid getting trapped into a bucket of the web graph, which is a reachable strongly connected component without outgoing edges toward the rest of the graph.

In matrix form

$$\boldsymbol{p} = (1 - \alpha)\mathbf{1} + \alpha \mathbf{L}\mathbf{D}_c^{-1} \boldsymbol{p}, \tag{30.6}$$

where $\mathbf{1}$ is a vector of N ones and $\mathbf{D}_c = \text{diag}(\boldsymbol{c})$ is a diagonal matrix with diagonal elements c_j. Assuming that the average PageRank is 1, $\mathbf{1}^T \boldsymbol{p} = N$, we have

$$\boldsymbol{p} = [(1 - \alpha)\mathbf{1}\mathbf{1}^T/N + \alpha \mathbf{L}\mathbf{D}_c^{-1}]\boldsymbol{p} = \mathbf{A}\boldsymbol{p}, \tag{30.7}$$

where the matrix $\mathbf{A}$ is the expression in square braces.

The PageRank vector can be found by using the power method. Viewing PageRank as a Markov chain, $\mathbf{A}$ has a real eigenvalue equal to unity, and this is its largest eigenvalue. We can find the desired PageRanks $\hat{\boldsymbol{p}}$ by the power method

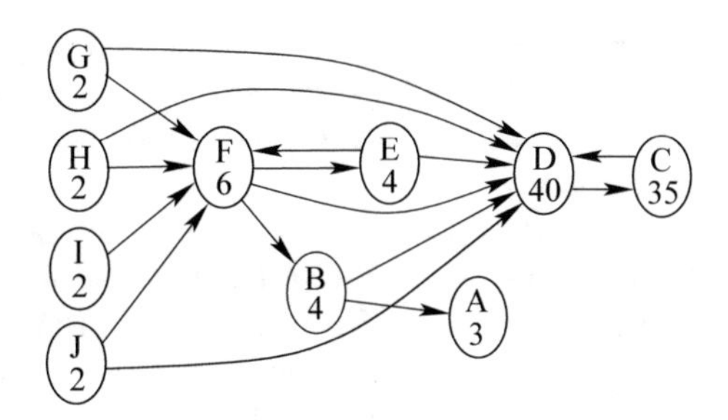

Fig. 30.1 A PageRank instance with solution

$$
\begin{aligned}
\boldsymbol{p}_k &\leftarrow \mathbf{A}\boldsymbol{p}_{k-1}, \\
\boldsymbol{p}_k &\leftarrow N\frac{\boldsymbol{p}_k}{\mathbf{1}^T\boldsymbol{p}_k},
\end{aligned}
\tag{30.8}
$$

starting with $\boldsymbol{p} = \boldsymbol{p}_0$. Since **A** has positive entries with each column summing to unity, Markov chain theory tells us that it has a unique eigenvector with eigenvalue one, corresponding to the stationary distribution of the chain.

The system is interpreted as that, with probability α the random surfer moves forward by following the links, and, with the complementary probability $1-\alpha$ the surfer gets bored of following the links and enters a new destination in the browser's URL line, possibly unrelated to the current page. Setting $\alpha = 0.85$ implies that after about five link clicks the random surfer chooses a random page.

Example 30.1 An example is shown in Fig. 30.1. Each node is labeled with its PageRank score. Scores are normalized to sum to 100. $\alpha = 0.85$. Page A is a dangling node, while pages C and D form a bucket. Page C receives only one link but from the most important page D, and its importance is high. Page F receives many more links, but from anonymous pages, and its importance is low compared to that of page C. Pages G, H, I, and J do not receive endorsements, and thus their scores correspond to the minimum amount of status of each page.

PageRank does not consider the lengths of time that the web surfer spends on the pages during the browsing process. The staying time information can be used as good indicators of the importance of the pages. BrowseRank [78] leverages the user staying time information on web pages for page importance computation. It collects the user behavior data in web surfing and builds a user browsing graph, which contains both user transition information and user staying time information. A continuous-time Markov process is employed in BrowseRank to model the browsing behaviors of a web surfer, and the stationary distribution of the process is regarded as the page importance scores.

Under the general Markov framework [46], a web Markov skeleton process is used to model the random walk conducted by the web surfer on a given graph. Page importance is then defined as the product of two factors: page reachability, the average possibility that the surfer arrives at the page, and page utility, the average value that the page gives to the surfer in a single visit. These two factors can be computed as the stationary probability distribution of the corresponding embedded Markov chain and the mean staying time on each page of the web Markov skeleton process, respectively. This general framework covers PageRank and BrowseRank as special cases.

PageRank favors pages with many in-links. Older pages may have accumulated many in-links. It is thus difficult to find the latest information on the web using PageRank. Time-sensitive ranking (TS-Rank) [73] extends PageRank by using a function of time in the damping factor d in PageRank.

As PageRank is an application of the random walker, the arc weights need to be normalized. This normalization introduces a series of unwanted side effects. Black-hole metric [10] generalizes PageRank to mitigate the problem. Black-hole metric globally preserves the proportions among the arc weights, and ensures at the same time that the out-strength is equal to 1 for each node. This is compatible with the random walker model. The transformation only requires the knowledge of the maximum, and the minimum value each weight can assume.

Some PageRank-inspired bibliometric indicators to evaluate the importance of journals using the academic citation network have been proposed and extensively tested: journal PageRank [8], Eigenfactor (http://www.eigenfactor.org), and SCImago (http://www.scimagojr.com). GeneRank [91] is a modification of PageRank for using connectivity data to produce a prioritization of the genes in a microarray experiment.

30.7.3 Hypertext-Induced Topic Search (HITS)

Let $\mathbf{L} = \left(l_{i,j}\right)$ be the adjacency matrix of the web graph, i.e., $l_{i,j} = 1$ if page i links to page j and $l_{i,j} = 0$ otherwise. HITS defines a pair of recursive equations:

$$\boldsymbol{x}^{(k)} = \mathbf{L}^T \boldsymbol{y}^{(k-1)}, \quad \boldsymbol{y}^{(k)} = \mathbf{L}\boldsymbol{x}^{(k)}, \tag{30.9}$$

where $\boldsymbol{x}$ is the authority vector containing the authority scores and $\boldsymbol{y}$ is the hub vector containing the hub scores, $k \geq 1$ and $\boldsymbol{y}^{(0)} = \mathbf{1}$, the vector of all ones. The first equation tells us that authoritative pages are those pointed to by good hub pages, while the second equation claims that good hubs are pages that point to authoritative pages.

Notice that (30.9) is equivalent to

$$\boldsymbol{x}^{(k)} = \mathbf{L}^T\mathbf{L}\boldsymbol{x}^{(k-1)}, \quad \boldsymbol{y}^{(k)} = \mathbf{L}\mathbf{L}^T \boldsymbol{y}^{(k-1)}. \tag{30.10}$$

It follows that the authority vector $\boldsymbol{x}$ is the dominant right eigenvector of the authority matrix $\mathbf{A} = \mathbf{L}^T\mathbf{L}$, and the hub vector $\boldsymbol{y}$ is the dominant right eigenvector of the hub matrix $\mathbf{H} = \mathbf{L}\mathbf{L}^T$. This is very similar to the PageRank method. To compute the dominant eigenpairs, the power method can be exploited. While the convergence of the power method is guaranteed, the computed solution is not necessarily unique, since the authority and hub matrices are not necessarily irreducible. A modification similar to the teleportation trick used for the PageRank method can be applied to recover the uniqueness of the solution [133]. HITS is related to SVD [37]. It follows that the HITS authority and hub vectors correspond, respectively, to the right- and left-singular vectors associated with the highest singular value of the adjacency matrix **L**.

An advantage of HITS with respect to PageRank is that it provides two rankings: the most authoritative pages and the most hubby pages. HITS has a higher susceptibility to spamming: while it is difficult to add incoming links to a favorite page, the addition of outgoing links is much easier. This leads to the possibility of purposely inflating the hub score of a page, indirectly influencing also the authority scores of the pointed pages.

30.8 Personalized Search

Web search engines are built to serve all users. Personalization of web search is to carry out retrieval for each user incorporating his/her interests. One approach of personalized search is to filter or rerank search results by checking content similarity between returned web pages and user profiles. User profiles store approximations of user interests. User profiles are either specified by users themselves [17, 103] or are automatically learnt from a user's historical activities [113]. A user profile is usually structured as a concept/topic hierarchy. User profiles are built by two groups of works: topical categories [17, 103] and keyword lists (bags of words) [115].

Several approaches represent user interests by using topical categories. User-issued queries and user-selected snippets/documents are categorized into concept hierarchies that are accumulated to generate a user profile. When the user issues a query, each of the returned snippets/documents is also classified. The documents are reranked based on how well the document categories match user interest profiles. In topical-interest-based personalization strategies [38], user profiles are automatically learned from users' past queries and clickthroughs in search engine logs. Other personalized search approaches use lists of keywords to represent user interests. In [115], user preferences are built as vectors of distinct terms and are constructed by aggregating past preferences, including both long-term and short-term preferences. In [17], keywords are associated with categories, and thus, user profiles are represented by a hierarchical category tree based on keyword categories.

Test queries can be divided into three types: clear queries, semiambiguous queries, and ambiguous queries. Personalization significantly increases output quality for ambiguous and semiambiguous queries; but for clear queries, a common web search and current web search ranking might be sufficient, and thus, personalization is unnecessary. Queries can also be divided into fresh queries and recurring queries. The recent history tends to be much more useful than the remote history, especially for fresh queries, whereas the entire history is helpful for improving the search accuracy of recurring queries.

Personalized PageRank is a modification of PageRank algorithm for personalized web search [96]. Multiple personalized PageRank scores, one for each main topic of the Open Directory Project (ODP) category hierarchy, are used to enable topic-sensitive web search [54]. HITS algorithm is extended in [116] by artificially increasing the authority and hub scores of pages marked relevant by the user in previous searches. In the personalized search approaches based on user group, search histories of users who have similar interests with a test user are used to refine the search. Collaborative filtering is used to construct user profiles and is a typical group-based personalization method used in personalized search [115].

Personalization may be ineffective for queries that show less variation among individuals. Click entropy is a simple measurement on whether a query should be personalized [38]. Click entropy measures the variation in user information needs for a query q as follows. If all users click only one identical page on query q, we have ClickEntroy$(q) = 0$. A smaller click entropy means that the majority of users agree with one another on a small number of web pages. In such cases, there is no need to do any personalization. A large click entropy indicates that many web pages were clicked for the query. In this case, personalization can help to filter the pages that are more relevant to users by making use of historical selections. In this case, personalization can be used to provide different web pages to different users. A method that incorporates click histories of a group of users with similar topical affinities to personalize web search is given in [38]. A large-scale evaluation framework is presented for personalized search based on query logs, and then five personalized search algorithms (including two click-based ones and three topical-interest-based ones) are evaluated using 12-day query logs of Windows Live Search. No personalization algorithm is found to outperform the others for all queries [38].

Web search ranking requires labeled data. The labels usually come in the form of relevance assessments made by editors. The relevance labels of the training examples could change over time. Click logs embed important information about user satisfaction with a search engine and can provide a highly valuable source of relevance information. Compared to editorial labels, clicks are much cheaper to obtain and always reflect current relevance. Click logs can provide an important source of implicit feedback and can be used as a cheap proxy for editorial labels. Clicks have been used in multiple ways by a search engine: to tune search parameters, to evaluate different ranking functions [14, 62], or as signals to directly influence ranking [62]. However, clicks are known to be biased, by the presentation order, the appearance (e.g., title and abstract) of the documents, and the reputation of individual sites. In [14], the relationship between clicks and relevance is modeled so that clicks can be used to unbiasedly evaluate search engine when lack of editorial relevance judgment.

The collective feedback of the users of an information retrieval system provides semantic information that can be useful in web mining tasks. A richer data structure is used to preserve most of the information available in the weblogs [36]. This data structure consists of three groups of entities, namely, users, documents, and queries, which are connected in a network of relations. Query refinements correspond to separate transitions between the corresponding query nodes in the graph, while users are linked to the queries they have issued and to the documents they have selected. The classical query/document transitions, which connect a query to the documents selected by the users in the returned result page, are also considered.

A query expansion method [25] is proposed based on user interactions recorded in the clickthrough data. The method focuses on mining correlations between query terms and document terms by analyzing user's clickthroughs. Document terms that are strongly related to the input query are used together to narrow down the search. In [72], clickthrough data are used to estimate user's conceptual preferences and personalized query suggestions are then provided for each individual user according to his/her conceptual needs. The motivation is that queries submitted to a search engine may have multiple meanings. To resolve the disadvantage of keyword-based clustering methods, clickthrough data have been used to cluster queries based on common clicks on URLs. One major problem with the clickthrough-based method is that the number of common clicks on URLs for different queries is limited. Thus, the chance for the users to see the same results would be small.

In [103], users' profiles are learned from their surfing histories and documents returned by a metasearch engine are reranked/filtered based on the profiles. In [77], a user profile and a general profile are learned from the user's search history and a category hierarchy, respectively. The two profiles are combined to map a user query into a set of categories which represent the user's search intention and serve as a context to disambiguate the words in the user's query. Web pages are retrieved by merging multiple lists of web pages from multiple query submissions. Web search is conducted based on both the user query and the set of categories.

With the proliferation of geo-positioned mobile devices, the geographical web (GeoWeb) is prospering. In GeoWeb, each object (e.g., web page) is associated with a geographical location and a textual description, which enables location-based services such as local search, advertisements, and map services. The reverse nearest neighbor query [66] finds the set of objects that have q as their nearest neighbor given a query point q. However, this category does not consider textual similarity at all. Another reverse query processing suggests weights to objects for the top-k results [121]. A spatio-textual query takes a user location and a keyword set as inputs, and returns the most spatially and textually relevant objects. It retrieves the best top-k objects given a user location and a keyword set [33].

30.9 Data Warehousing

Data warehousing is a paradigm specifically intended to provide vital strategic information [102]. A data warehouse is a repository that integrates information from multiple data sources, which may or may not be heterogeneous and makes them available for decision support querying and analysis. Materialized views collect data from databases into the warehouse but without copying each database into the warehouse. Queries on the warehouse can then be answered using the views instead of accessing the remote databases. When modification of data occurs on remote databases, they are transmitted to the warehouse. Architecture of a typical data warehouse is shown in Fig. 30.2.

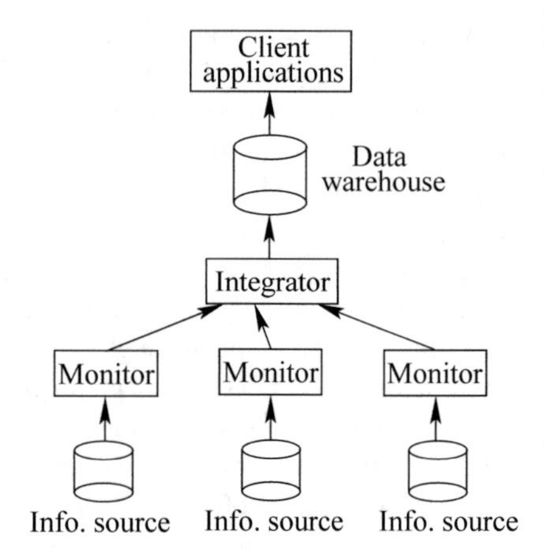

Fig. 30.2 Architecture of a typical data warehouse

Business intelligence, in the context of the data warehouse, is the ability of an enterprise to study past behaviors and actions in order to understand its past, determine its current situation, and predict or change what will happen in the future. Data marts are a subset of data warehouse data and are where most of the analytical activities in the business intelligence environment take place. The data in each data mart are usually tailored for a particular function, such as product profitability analysis and customer demographic analysis. Data mining fits well and plays a significant role in the data warehouse environment.

The data warehouse provides the best opportunity for analysis, and online analytical processing (OLAP) is the vehicle for carrying out involved analysis. OLAP helps the user to analyze the past and gain insights, while data mining helps the user to predict the future. OLAP tools conceptually model the information as multidimensional cubes. The cubes in a data warehouse can be stored by following either a relational OLAP and/or a multidimensional OLAP approach. In relational OLAP, the data are stored in relational tables.

Structured query language (SQL) is the interface for retrieving and manipulating data from relational databases. These methods are used in data warehouse environments. A data warehouse stores materialized views of data from one or more sources, for the purpose of implementing decision support or OLAP queries.

Database integrity is a very important motivation for studying temporal dependencies. Temporal constraints can take different forms. They can be expressed using first-order temporal logic [19], or using temporal dependencies [124], which are restricted classes of first-order temporal logic formulas. A temporal dependency, called trend dependency, captures a significant family of data evolution regularities (constraints). An example of such regularity is "Salaries of employees generally do not decrease." Trend dependencies compare attributes over time using operators of $\{<, =, >, \leq, \geq, \neq\}$, allowing us to express meaningful trends. Trend dependency mining is investigated in [124] for a temporal database.

Entities may have two or more representations in databases. We distinguish between two types of data heterogeneity: structural and lexical. Structural heterogeneity occurs when the fields of the tuples in the database are structured differently in different databases. Lexical heterogeneity occurs when the tuples have identically structured fields across databases, but the data use different representations to refer to the same real-world object. Duplicate records do not share a common key and/or they contain errors that make duplicate matching a difficult task. Duplicate record detection is reviewed in [41], and multiple techniques are described for improving the efficiency and scalability of approximate duplicate detection algorithms.

In uncertain data management, data records are typically represented by probability distributions rather than deterministic values. A survey of uncertain data mining and management applications is provided in [2].

30.10 Content-Based Image Retrieval

The retrieval of images is either content-based or text-based retrieval. Early approaches to image retrieval were text-based approaches, and most search engines return images solely based on the text of the pages from which the images are linked. In text-based retrieval, some form of textual description of the image contents is assumed to be stored with the image itself by some form of HTML tagged text. However, since images on the web are poorly labeled, standard keyword-based image searching techniques frequently yield poor results. Image-based features provide either alternative or additional signals for image retrieval.

In content-based image retrieval (CBIR) [15, 106, 112], image characteristics, such as color, shape, or texture, are used for indexing and searching. CBIR systems adopt query by example as the primary query model where the user query is specified through an example image or features of the example image, which is then compared to the images in the database. The images retrieved are ranked according to a distance metric from the query. Multiple exemplars can be used for capturing user's retrieval needs, and ranked results of individual queries using each image separately can be combined.

The low-level visual features (color, texture, shape, etc.) are automatically extracted to represent the images. Most CBIR systems are region-based. The regions correspond to objects or parts of objects in the image. Global feature-based retrieval is comparatively simpler. However, the low-level features may not accurately characterize the high-level semantic concepts. In CBIR, understanding the user's needs is a challenging task. Relevance feedback is an effective tool for taking the user's judgement into account [106]. With the user-provided negative and positive feedbacks, image retrieval can then be thought of as a classification problem.

On the Internet are many digital documents in scanned image format. Although the technology of document image processing may be utilized to automatically convert the digital images of these documents to the machine-readable text format using optical character recognition (OCR) technology, it is not a cost-effective and practical way to process a huge number of paper documents. A document image retrieval system [89] provides an answer of yes or no with respect to the user's query, rather than the exact recognition of a character/word as in document image processing. Words, rather than characters, are the basic units of meaning in information retrieval. Therefore, directly matching word images in a document image is an alternative way to retrieve information from the document. An approach with the capability of matching partial word images [84] addresses word spotting and similarity measurement between documents.

Approaches to CBIR typically extract a single signature from each image based on color, texture, or shape features. The images returned as the query result are the ones whose signatures are closest to the signature of the query image. While efficient for simple images, such methods do not work well for complex scenes since they fail to retrieve images that match the query only partially, that is, only certain regions of the image match. WAveLet-based Retrieval of User-specified Scenes (WALRUS) [93] is a similarity retrieval algorithm that is robust to scaling and translation of objects within an image. It employs a similarity model in which each image is first decomposed into its regions and the similarity measure between a pair of images is then defined to be the fraction of the area of the two images covered by matching regions from the images. In order to extract regions for an image, WALRUS considers sliding windows of varying sizes and then clusters them based on the proximity of their signatures. WALRUS builds a set of a variable number of signatures for an image, one signature per image region.

The aim of image auto-annotation is to find a group of keywords $\boldsymbol{w}^*$ that maximizes the conditional distributions $p(\boldsymbol{w}|I_q)$, where I_q is the uncaptioned query image and $\boldsymbol{w}$ are terms or phrases in the vocabulary. An attempt at model-free image annotation is a data-driven approach that annotates images by mining their search results [122]. The search process first discovers visually and semantically similar search results, the mining process then identifies salient terms from textual descriptions of the search results, and the annotation rejection process filters out noisy terms yielded. Since no training dataset is required, the approach enables annotating with unlimited vocabulary and is highly scalable and robust to outliers.

In [22], multiple sources of evidence related to the images are considered. To allow combining these distinct sources of evidence, an image retrieval model is introduced based on Bayesian belief networks. Retrieval using an image surrounding text passages is as effective as standard retrieval based on HTML tags.

In [60], the image-ranking problem is cast into the task of identifying authority nodes on an infrared visual similarity graph, and VisualRank is proposed to analyze the visual link structures among images. The images found to be authorities are chosen as those that answer the image-queries well. VisualRank is an end-to-end system to improve Google image search results with emphasis on robust and efficient computation of image similarities applicable to a large number of queries and images. VisualRank employs the random walk intuition to rank images based on the visual hyperlinks among the images. It incorporates the advances made in using link and network analysis for web document search into image search.

An image search reranking technique aims at refining text-based search results by mining images' visual content. The idea of one-class classification has been used in CBIR and image search reranking.

Bag-of-words (BoW) principle is a common practice in natural language processing. Word histograms are generated within a text document, and the frequencies of the words from a dictionary are counted. The resulting term frequency (tf) vector is then input to a classifier. BoW has been adopted by the visual computing community as bag-of-visual-words [111], where local image features are extracted and their general distribution is modeled by a histogram. The principle has also been employed in audio classification, under the name bag-of-audio-words. Many approaches utilize SIFT feature [83] to represent images and leverage BoW model to index large-scale image dataset for scalable retrieval. Vector quantization divides a large set of training SIFT features into nonoverlapped groups by clustering, with the clustering centers called visual word.

SIFT-based image retrieval usually uses Hessian–Affine local detector and SIFT descriptor. Since 2012, the popularity of SIFT-based models is overtaken by the convolutional neural network. Convolutional network-based retrieval methods can usually build compact fixed-length representations. Competitive performance compared to BoW models has been reported, even with short feature vectors. For SIFT-based methods, the cropped regions are usually used as query, whereas convolutional network-based methods may employ the full-sized query images.

Content-Based Music Retrieval

Music is one of the most popular types of online information on the web. Some of the huge music collections available have posed a major challenge for searching, retrieving, and organizing music content. Music search uses content-based methods.

Low-level audio features are measurements of audio signals that contain information about a musical work and music performance. In general, low-level audio features are segmented in three different ways: frame-based segmentations (periodic sampling at 10–1000 ms intervals), beat-synchronous segmentations (features are aligned to musical beat boundaries), and statistical measures that construct probability distributions out of features (bag of features models). Many low-level audio features are based on the short-time spectrum of the audio signal.

Music genre is probably the most popular description of music content. Most of the music genre classification algorithms resort to BoW approach [109], which models the audio signals by the long-term statistical distribution of their short-time spectral features. These features can be roughly classified into three classes (i.e., timbral texture features, rhythmic features, and pitch content features). Genres are next classified from the feature vectors extracted.

Music identification is based on audio fingerprinting techniques, which aim at describing digital recordings with a compact set of acoustic features. An alternative approach to music identification is audio watermarking. In this case, research on psychoacoustics is exploited to embed a watermark in a digital recording without altering sound perception. Similarly to fingerprints, audio watermarks should be robust to distortions, additional noise, D/A and A/D conversions, and lossy compression.

30.11 E-mail Anti-spamming

Fraud detection, intrusion detection, and medical diagnosis are recognized as anomaly detection problems. Anomaly detection is to find objects that are different from most other objects. The class imbalance problem is thus intrinsic to the anomaly detection applications.

The e-mail spam problem continues to grow drastically. Various methods of near-duplicate spam detection have been developed [28]. Based on an analysis of e-mail content text, this problem is modeled as a binary text classification task. Representatives of this category are the naive Bayes [57] and SVM [39] methods.

To achieve small storage size and efficient matching, prior works mainly represent each e-mail by a succinct abstraction derived from e-mail content text. Moreover, hash-based text representation is applied extensively. A common attack to this type of representation is to insert a random normal paragraph without any suspicious keywords into unobvious position of an e-mail. In such a context, if the whole e-mail content is utilized for hash-based representation, the near-duplicate part of spams cannot be captured. Hash-based text representation is not suitable for all languages. As important clues to spam detection, images and hyperlinks, however, are unable to be included in hash-based text representation.

Noncontent information such as e-mail header, e-mail social network [18], and e-mail traffic [21] is exploited to filter spams. Collecting notorious and innocent sender addresses (or IP addresses) from e-mail header to create black list and white list is a commonly applied method initially. MailRank [18] examines the feasibility of rating sender addresses with the PageRank algorithm in the e-mail social network. However, e-mail header can be altered by spammers to conceal the identity.

Regarding collaborative spam filtering with near-duplicate similarity matching scheme, peer-to-peer-based architecture [28] and centralized server-based system are generally employed. The primary idea of the similarity matching scheme for spam detection is to maintain a known spam database, formed by user feedback, to block subsequent near-duplicate spams. On purpose of achieving efficient similarity matching and reducing storage utilization, prior works mainly represent each e-mail by a succinct abstraction derived from e-mail content text. An e-mail abstraction scheme proposed in [118] considers e-mail layout structure to represent e-mails. The designed complete spam detection system Cosdes possesses an efficient near-duplicate matching scheme and a progressive update scheme.

Detection of malicious websites from the lexical and host-based features of their URLs is explored in [85]. Online algorithms not only process large numbers of URLs more efficiently than batch algorithms, but they also adapt more quickly to new features in the continuously evolving distribution of malicious URLs. A real-time system for gathering URL features is developed and is paired with a real-time feed of labeled URLs from a large web mail provider.

Problems

30.1 How is data mining different from OLAP? Explain briefly.

30.2 As a data mining consultant, you are hired by a large commercial bank that provides many financial services. The bank has a data warehouse. The management wants to find the existing customers who are most likely to respond to a marketing campaign offering new services. Outline the knowledge discovery process.

30.3 Consider a table of linked web pages:

Page	Link to page
A	B D E F
B	C D E F
C	B E F
D	A B F
E	A B C
F	A C

(a) Find the authorities and hubs of HITS by using two iterations.
(b) Find the PageRank scores for each page after one iteration using 0.25 as the dampening factor.

30.4 Consider the PageRank algorithm.
(a) Show that from definition (30.5) the sum of the PageRanks is N, the number of web pages.
(b) Write a program to compute the PageRank solutions by the power method using formulation (30.8). Apply it to the network of Fig. 30.1.

30.5 Explain why PageRank can effectively fight spam.

References

1. Aggarwal, C. C., Gates, S. C., & Yu, P. S. (2004). On using partial supervision for text categorization. *IEEE Transactions on Knowledge and Data Engineering*, *16*(2), 245–255.
2. Aggarwal, C. C., & Yu, P. S. (2009). A survey of uncertain data algorithms and applications. *IEEE Transactions on Knowledge and Data Engineering*, *21*(5), 609–623.
3. Agrawal, R., & Srikant, R. (1994). Fast algorithms for mining association rules. In *Proceedings of the 20th International Conference on Very Large Data Bases* (pp. 487–499). Santiago, Chile.
4. Agrawal, R., & Srikant, R. (1995). Mining sequential patterns. In *Proceedings of the 11th International Conference on Data Engineering* (pp. 3–14). Tapei, Taiwan.
5. Allen, D., & Darwiche, A. (2008). RC_Link: Genetic linkage analysis using Bayesian networks. *International Journal of Approximate Reasoning*, *48*, 499–525.
6. Bekkerman, R., El-Yaniv, R., Tishby, N., & Winter, Y. (2003). Distributional word clusters vs. words for text categorization. *Journal of Machine Learning Research*, *3*, 1183–1208.

7. Blei, D. M., Ng, A. Y., & Jordan, M. I. (2003). Latent Dirichlet allocation. *Journal of Machine Learning Research*, *3*, 993–1022.
8. Bollen, J., Rodriguez, M. A., & de Sompel, H. V. (2006). Journal status. *Scientometrics*, *69*(3), 669–687.
9. Brin, S., & Page, L. (1998). The anatomy of a large-scale hypertextual web search engine. In *Proceedings of the 7th International World Wide Web Conference (WWW)* (pp. 107–117).
10. Buzzanca, M., Carchiolo, V., Longheu, A., Malgeri, M., & Mangioni, G. (2018). Black hole metric: Overcoming the pagerank normalization problem. *Information Sciences*, *438*, 58–72.
11. Cai, D., He, X., & Han, J. (2011). Locally consistent concept factorization for document clustering. *IEEE Transactions on Knowledge and Data Engineering*, *23*(6), 902–913.
12. Calado, P., da Silva, A. S., Vieira, R. C., Laender, A. H. F., & Ribeiro-Neto, B. A. (2002). Searching web databases by structuring keyword-based queries. In *Proceedings of the 11th ACM International Conference on Information and Knowledge Management* (pp. 26–33). McLean, VA.
13. Cancedda, N., Gaussier, E., Goutte, C., & Renders, J.-M. (2003). Word-sequence kernels. *Journal of Machine Learning Research*, *3*, 1059–1082.
14. Carterette, B., & Jones, R. (2008). Evaluating search engines by modeling the relationship between relevance and clicks. In J. Platt, D. Koller, Y. Singer, & S. Roweis (Eds.), *Advances in neural information processing systems 20* (pp. 217–224). MIT Press.
15. Chang, E., Goh, K., Sychay, G., & Wu, G. (2003). CBSA: Content-based soft annotation for multimodal image retrieval using Bayes point machines. *IEEE Transactions on Circuits and Systems for Video Technology*, *13*(1), 26–38.
16. Chen, H.-L., Chuang, K.-T., & Chen, M.-S. (2008). On data labeling for clustering categorical data. *IEEE Transactions on Knowledge and Data Engineering*, *20*(11), 1458–1471.
17. Chirita, P.-A., Nejdl, W., Paiu, R., & Kohlschutter, C. (2005). Using ODP metadata to personalize search. In *Proceedings of the 28th Annual International ACM SIGIR Conference on Research and Development in Information Retrieval* (pp. 178–185). Salvador, Brazil.
18. Chirita, P.-A., Diederich, J., & Nejdl, W. (2005). Mailrank: Using ranking for spam detection. In *Proceedings of the 14th ACM International Conference on Information and Knowledge Management* (pp. 373–380). Bremen, Germany.
19. Chomicki, J. (1995). Efficient checking of temporal integrity constraints using bounded history encoding. *ACM Transactions on Database Systems*, *20*(2), 148–186.
20. Cilibrasi, R., & Vitanyi, P. (2005). Clustering by compression. *IEEE Transactions on Information Theory*, *51*(4), 1523–1545.
21. Clayton, R. (2007). Email Traffic: A quantitative snapshot. In *Proceedings of the 4th Conference on Email and Anti-Spam*. Mountain View, CA.
22. Coelho, T. A. S., Calado, P. P., Souza, L. V., Ribeiro-Neto, B., & Muntz, R. (2004). Image retrieval using multiple evidence ranking. *IEEE Transactions on Knowledge and Data Engineering*, *16*(4), 408–417.
23. Combarro, E. F., Montanes, E., Diaz, I., Ranilla, J., & Mones, R. (2005). Introducing a family of linear measures for feature selection in text categorization. *IEEE Transactions on Knowledge and Data Engineering*, *17*(9), 1223–1232.
24. Cottrell, M., Ibbou, S., & Letremy, P. (2004). SOM-based algorithms for qualitative variables. *Neural Networks*, *17*, 1149–1167.
25. Cui, H., Wen, J., Nie, J., & Ma, W. (2003). Query expansion by mining user logs. *IEEE Transactions on Knowledge and Data Engineering*, *15*(4), 829–839.
26. Cunningham, H., Maynard, D., Bontcheva, K., & Tablan, V. (2002). Gate: A framework and graphical development environment for robust NLP tools and applications. In *Proceedings of the 40th Annual Meeting of the Association for Computational Linguistics* (pp. 168–175). Philadelphia, PA.
27. Dai, B.-R., Huang, J.-W., Yeh, M.-Y., & Chen, M.-S. (2006). Adaptive clustering for multiple evolving streams. *IEEE Transactions on Knowledge and Data Engineering*, *18*(9), 1166–1180.

28. Damiani, E., di Vimercati, S. D. C., Paraboschi, S., & Samarati, P. (2004). P2P-based collaborative spam detection and filtering. In *Proceedings of the 4th IEEE International Conference on Peer-to-Peer Computing* (pp. 176–183). Zurich, Switzerland.
29. Datta, S., Giannella, C. R., & Kargupta, H. (2009). Approximate distributed K-means clustering over a peer-to-peer network. *IEEE Transactions on Knowledge and Data Engineering, 21*(10), 1372–1388.
30. de Campos, L. M., & Romero, A. E. (2009). Bayesian network models for hierarchical text classification from a thesaurus. *International Journal of Approximate Reasoning, 50*, 932–944.
31. de Cristo, M. A. P., Calado, P. P., & de Lourdes da Silveira, M., Silva, I., Muntz, R., & Ribeiro-Neto, B., (2003). Bayesian belief networks for IR. *International Journal of Approximate Reasoning, 34*, 163–179.
32. Deerwester, S. C., Dumais, S. T., Furnas, G. W., Landauer, T. K., & Harshman, R. A. (1990). Indexing by latent semantic analysis. *Journal of the American Society for Information Science, 41*, 391–407.
33. De Felipe, I., Hristidis, V., & Rishe, N. (2008). Keyword search on spatial databases. In *Proceedings of 24th Int. Conf. Data Eng.* (pp. 656–665).
34. Dhillon, I. S. (2001). Co-clustering documents and words using bipartite spectral graph partitioning. In *Proceedings of ACM SIGKDD International Conference on Knowledge Discovery and Data Mining* (pp. 269–274). San Francisco, CA.
35. Diligenti, M., Gori, M., & Maggini, M. (2004). A unified probabilistic framework for web page scoring systems. *IEEE Transactions on Knowledge and Data Engineering, 16*(1), 4–16.
36. Diligenti, M., Gori, M., & Maggini, M. (2011). A unified representation of web logs for mining applications. *Information Retrieval, 14*, 215–236.
37. Ding, C. H. Q., Zha, H., He, X., Husbands, P., & Simon, H. D. (2004). Link analysis: Hubs and authorities on the World Wide Web. *SIAM Review, 46*(2), 256–268.
38. Dou, Z., Song, R., Wen, J.-R., & Yuan, X. (2009). Evaluating the effectiveness of personalized web search. *IEEE Transactions on Knowledge and Data Engineering, 21*(8), 1178–1190.
39. Drucker, H., Wu, D., & Vapnik, V. N. (1999). Support Vector Machines for spam categorization. *IEEE Transactions on Neural Networks, 10*(5), 1048–1054.
40. Dumais, S. T., Platt, J., Heckerman, D., & Sahami, M. (1998). Inductive learning algorithms and representations for text categorization. In *Proceedings of the 7th ACM International Conference on Information and Knowledge Management* (pp. 148–155.). Bethesda, MA.
41. Elmagarmid, A. K., Ipeirotis, P. G., & Verykios, V. S. (2007). Duplicate record detection: A survey. *IEEE Transactions on Knowledge and Data Engineering, 19*(1), 1–16.
42. Flesca, S., Manco, G., Masciari, E., Pontieri, L., & Pugliese, A. (2005). Fast detection of XML structural similarity. *IEEE Transactions on Knowledge and Data Engineering, 17*(2), 160–175.
43. Foulds, J. R., Boyles, L., DuBois, C., Smyth, P., & Welling, M. (2013). Stochastic collapsed variational Bayesian inference for latent Dirichlet allocation. In *Proceedings of the 19th ACM SIGKDD International Conference on Knowledge Discovery and Data Mining* (pp. 446–454). Chicago, IL.
44. Fragopanagos, N., & Taylor, J. G. (2005). Emotion recognition in human-computer interaction. *Neural Networks, 18*(4), 389–406.
45. Gambs, S., Kegl, B., & Aimeur, E. (2007). Privacy-preserving boosting. *Data Mining and Knowledge Discovery, 14*, 131–170.
46. Gao, B., Liu, T.-Y., Liu, Y., Wang, T., Ma, Z.-M., & Li, H. (2011). Page importance computation based on Markov processes. *Information Retrieval, 14*(5), 488–514.
47. Golub, K. (2006). Automated subject classification of textual web documents. *Journal of Documentation, 62*(3), 350–371.
48. Gou, G., & Chirkova, R. (2007). Efficiently querying large XML data repositories: A survey. *IEEE Transactions on Knowledge and Data Engineering, 19*(10), 1381–1403.
49. Griffiths, T. L., & Steyvers, M. (2004). Finding scientific topics. *Proceedings of the National Academy of Sciences of the USA, 101*(Suppl. 1), 5228–5235.

50. Guerrero-Bote, V. P., Lopez-Pujalte, C., de Moya-Anegon, F., & Herrero-Solana, V. (2003). Comparison of neural models for document clustering. *International Journal of Approximate Reasoning*, *34*, 287–305.
51. Guha, S., Meyerson, A., Mishra, N., Motwani, R., & O'Callaghan, L. (2003). Clustering data streams: Theory and practice. *IEEE Transactions on Knowledge and Data Engineering*, *15*(3), 515–528.
52. Hammer, B., Micheli, A., Sperduti, A., & Strickert, M. (2004). Recursive self-organizing network models. *Neural Networks*, *17*, 1061–1085.
53. Hammouda, K. M., & Kamel, M. S. (2009). Hierarchically distributed peer-to-peer document clustering and cluster summarization. *IEEE Transactions on Knowledge and Data Engineering*, *21*(5), 681–698.
54. Haveliwala, T. H. (2002). Topic-sensitive pagerank. In *Proceedings of the 11th International World Wide Web Conference (WWW)* (pp. 517–526). New York: ACM Press.
55. Haveliwala, T. H. (2003). Topic-sensitive PageRank: A context-sensitive ranking algorithm for web search. *IEEE Transactions on Pattern Analysis and Machine Intelligence*, *15*(4), 784–796.
56. Hoffman, M., Blei, D., & Bach, F. (2010). Online learning for latent Dirichlet allocation. In *Advances in neural information processing systems* (Vol. 23, pp. 856–864).
57. Hovold, J. (2005). Naive Bayes spam filtering using word-position-based attributes. In *Proceedings of the 2nd Conference on Email and Anti-Spam*. Palo Alto, CA.
58. Isa, D., Lee, L. H., Kallimani, V. P., & RajKumar, R. (2008). Text document preprocessing with the Bayes formula for classification using the support vector machine. *IEEE Transactions on Knowledge and Data Engineering*, *20*(9), 1264–1272.
59. Jarvelin, K., & Kekalainen, J. (2000). IR evaluation methods for retrieving highly relevant documents. In *Proceedings of the 23rd Annual International ACM SIGIR Conference on Research and Development in Information Retrieval (SIGIR '00)* (pp. 41–48). Athens, Greece.
60. Jing, Y., & Baluja, S. (2008). VisualRank: Applying PageRank to large-scale image search. *IEEE Transactions on Pattern Analysis and Machine Intelligence*, *30*(11), 1877–1890.
61. Joachims, T. (1998). Text categorization with support vector machines: Learning with many relevant features. In *Proceedings of European Conference on Machine Learning, LNCS* (Vol. 1398, pp. 137–142). Berlin: Springer Verlag.
62. Joachims, T. (2002). Optimizing search engines using clickthrough data. In *Proceedings of 8th ACM SIGKDD International Conference on Knowledge Discovery and Data Mining* (pp. 133–142). Edmonton, Canada.
63. Kao, H.-Y., Lin, S.-H., Ho, J.-M., & Chen, M.-S. (2004). Mining web informative structures and contents based on entropy analysis. *IEEE Transactions on Pattern Analysis and Machine Intelligence*, *16*(1), 41–55.
64. Khashman, A. (2008). A modified back propagation learning algorithm with added emotional coefficients. *IEEE Transactions on Neural Networks*, *19*(11), 1896–1909.
65. Kleinberg, J. (1999). Authoritative sources in a hyperlinked environment. *Journal of the ACM*, *46*(5), 604–632.
66. Korn, F., & Muthukrishnan, S. (2000). Influence sets based on reverse nearest neighbor queries. In *Proceedings of ACM SIGMOD International Conference on Management of Data* (pp. 201–212). Dallas, TX.
67. Lagus, K., Kaski, S., & Kohonen, T. (2004). Mining massive document collections by the WEBSOM method. *Information Sciences*, *163*, 135–156.
68. Lamberti, F., Sanna, A., & Demartini, C. (2009). A relation-based page rank algorithm for semantic web search engines. *IEEE Transactions on Knowledge and Data Engineering*, *21*(1), 123–136.
69. Lawrence, R. D., Almasi, G. S., & Rushmeier, H. E. (1999). A scalable parallel algorithm for self-organizing maps with applications to sparse data mining problems. *Data Mining and Knowledge Discovery*, *3*, 171–195.
70. LeDoux, J. (1996). *The emotional brain*. New York: Simon and Schuster.

71. Lee, R. S. T., & Liu, J. N. K. (2004). iJADE Web-Miner: An intelligent agent framework for Internet shopping. *IEEE Transactions on Knowledge and Data Engineering*, *16*(4), 461–473.
72. Leung, K. W.-T., Ng, W., & Lee, D. L. (2008). Personalized concept-based clustering of search engine queries. *IEEE Transactions on Knowledge and Data Engineering*, *20*(11), 1505–1518.
73. Li, X., Liu, B., & Yu, P. (2008). Time sensitive ranking with application to publication search. In *Proceedings of the 8th IEEE International Conference on Data Mining* (pp. 893–898). Pisa, Italy.
74. Lin, D. (1998). An information-theoretic definition of similarity. In *Proceedings of the 15th International Conference on Machine Learning* (pp. 296–304). San Francisco, CA: Morgan Kaufmann.
75. Lin, K.-P., & Chen, M.-S. (2011). On the design and analysis of the privacy-preserving SVM classifier. *IEEE Transactions on Knowledge and Data Engineering*, *23*(11), 1704–1717.
76. Lin, Q. Y., Chen, Y. L., Chen, J. S., & Chen, Y. C. (2003). Mining inter-organizational retailing knowledge for an alliance formed by competitive firms. *Information Management*, *40*(5), 431–442.
77. Liu, F., Yu, C., & Meng, W. (2004). Personalized web search for improving retrieval effectiveness. *IEEE Transactions on Knowledge and Data Engineering*, *16*(1), 28–40.
78. Liu, Y., Gao, B., Liu, T., Zhang, Y., Ma, Z., He S., & Li, H. (2008). BrowseRank: Letting users vote for page importance. In *Proceedings of the 31st Annual International ACM SIGIR Conference* (pp. 451–458). Singpore.
79. Lodhi, H., Saunders, C., Shawe-Taylor, J., Cristianini, N., & Watkins, C. (2002). Text classification using string kernels. *Journal of Machine Learning Research*, *2*, 419–444.
80. Loia, V., Pedrycz, W., & Senatore, S. (2003). P-FCM: A proximity-based fuzzy clustering for user-centered web applications. *International Journal of Approximate Reasoning*, *34*, 121–144.
81. Lotfi, E., & Akbarzadeh-T., M. R., (2013a). Brain emotional learning-based pattern recognizer. *Cybernetics and Systems*, *44*(5), 402–421.
82. Lotfi, E., & Akbarzadeh-T., M.-R., (2014). Practical emotional neural networks. *Neural Networks*, *59*, 61–72.
83. Lowe, D. G. (2004). Distinctive image features from scale-invariant keypoints. *International Journal of Computer Vision*, *60*(2), 91–110.
84. Lu, Y., & Tan, C. L. (2004). Information retrieval in document image databases. *IEEE Transactions on Knowledge and Data Engineering*, *16*(11), 1398–1410.
85. Ma, J., Saul, L. K., Savage, S., & Voelker, G. M. (2011). Learning to detect malicious URLs. *ACM Transactions on Intelligent Systems and Technology, 2*(3), Article No. 30, 24 pages.
86. Masseglia, F., Poncelet, P., Teisseire, M., & Marascu, A. (2008). Web usage mining: Extracting unexpected periods from web logs. *Data Mining and Knowledge Discovery*, *16*, 39–65.
87. Miller, G., Beckwith, R., Fellbaum, C., Gross, D., & Miller, K. (1990). Introduction to wordnet: An on-line lexical database. *International Journal of Lexicography*, *3*, 235–244.
88. Mimno, D., Hoffman, M. D., & Blei, D. M. (2012). Sparse stochastic inference for latent Dirichlet allocation. In *Proceedings of the 29th International Conference on Machine Learning* (pp. 1599–1606). Edinburgh, UK.
89. Mitra, M., & Chaudhuri, B. B. (2000). Information retrieval from documents: A survey. *Information Retrieval*, *2*, 141–163.
90. Mobasher, B., Cooley, R., & Srivastava, J. (1999). Creating adaptive web sites through usage-based clustering of URLs. In *Proceedings of Workshop on Knowledge and Data Engineering Exchange* (pp. 19–25). Chicago, IL.
91. Morrison, J. L., Breitling, R., Higham, D. J., & Gilbert, D. R. (2005). GeneRank: using search engine technology for the analysis of microarray experiments. *BMC Bioinformatics*, *6*, 233–246.
92. Nasraoui, O., Soliman, M., Saka, E., Badia, A., & Germain, R. (2008). A web usage mining framework for mining evolving user profiles in dynamic web sites. *IEEE Transactions on Knowledge and Data Engineering*, *20*(2), 202–215.

93. Natsev, A., Rastogi, R., & Shim, K. (2004). WALRUS: A similarity retrieval algorithm for image databases. *IEEE Transactions on Knowledge and Data Engineering*, *16*(3), 301–316.
94. Ordonez, C. (2006). Integrating K-means clustering with a relational DBMS using SQL. *IEEE Transactions on Knowledge and Data Engineering*, *18*(2), 188–201.
95. Ordonez, C., & Omiecinski, E. (2004). Efficient disk-based K-means clustering for relational databases. *IEEE Transactions on Knowledge and Data Engineering*, *16*(8), 909–921.
96. Page, L., Brin, S., Motwani, R., & Winograd, T. (1999). *The PageRank citation ranking: Bringing order to the Web*. Technical Report 1999–66, Computer Science Department, Stanford University.
97. Pal, S. K., Narayan, B. L., & Dutta, S. (2005). A web surfer model incorporating topic continuity. *IEEE Transactions on Knowledge and Data Engineering*, *17*(5), 726–729.
98. Park, L. A. F., Ramamohanarao, K., & Palaniswami, M. (2004). Fourier domain scoring: A novel document ranking method. *IEEE Transactions on Knowledge and Data Engineering*, *16*(5), 529–539.
99. Park, L. A. F., Palaniswami, M., & Ramamohanarao, K. (2005). A novel document ranking method using the discrete cosine transform. *IEEE Transactions on Pattern Analysis and Machine Intelligence*, *27*(1), 130–135.
100. Petridou, S. G., Koutsonikola, V. A., Vakali, A. I., & Papadimitriou, G. I. (2008). Time aware web users clustering. *IEEE Transactions on Knowledge and Data Engineering*, *20*(5), 653–667.
101. Petrilis, D., & Halatsis, C. (2008). Two-level clustering of web sites using self-organizing maps. *Neural Processing Letters*, *27*, 85–95.
102. Ponniah, P. (2001). *Data warehousing fundamentals*. New York: John Wiley & Sons.
103. Pretschner, A., & Gauch, S. (1999). Ontology based personalized search. In *Proceedings of 11th 11th IEEE International Conference on Tools with Artificial Intelligence* (pp. 391–398).
104. Recupero, D. R. (2007). A new unsupervised method for document clustering by using WordNet lexical and conceptual relations. *Information Retrieval*, *10*, 563–579.
105. Richardson, M., & Domingos, P. (2002). The intelligent surfer: Probabilistic combination of link and content information in Pagerank. In *Advances in neural information processing systems 14* (pp. 1441–1448). MIT Press.
106. Rui, Y., Huang, T. S., Ortega, M., & Mehrotra, S. (1998). Relevance feedback: A power tool for interactive content-based image retrieval. *IEEE Transactions on Circuits and Systems for Video Technology*, *8*(5), 644–655.
107. Salton, G., & McGill, M. J. (1983). *Introduction to modern information retrieval*. New York: McGraw-Hill.
108. Salton, G., & Yang, C.-S. (1973). On the specification of term values in automatic indexing. *Journal of Documentation*, *29*(4), 351–372.
109. Scaringella, N., Zoia, G., & Mlynek, D. (2006). Automatic genre classification of music content: A survey. *IEEE Signal Processing Magazine*, *23*(2), 133–141.
110. Sebastiani, F. (2002). Machine learning in automated text categorization. *ACM Computing Surveys*, *34*, 1–47.
111. Sivic, J., & Zisserman, A. (2003). Video Google: A text retrieval approach to object matching in videos. In *Proceedings of the 9th IEEE International Conference on Computer Vision* (pp. 1470–1477). Nice, France.
112. Smeulders, A. W., Worring, M., Santini, S., Gupta, A., & Jain, R. (2000). Content-based image retrieval at the end of the early years. *IEEE Transactions on Pattern Analysis and Machine Intelligence*, *22*(12), 1349–1380.
113. Speretta, M., & Gauch, S. (2005). Personalized search based on user search histories. In *Proceedings of IEEE/WIC/ACM International Conference on Web Intelligence* (pp. 622–628). Compiegne, France.
114. Spiliopoulou, M., & Faulstich, L. C. (1998). WUM: A web utilization miner. In *Proceedings of International Workshop on The World Wide Web and Databases* (pp. 109–115). Valencia, Spain.

115. Sugiyama, K., Hatano, K., & Yoshikawa, M. (2004). Adaptive Web search based on user profile constructed without any effort from users. In *Proceedings of the 13th International World Wide Web Conference (WWW)* (pp. 675–684).
116. Tanudjaja, F., & Mui, L. (2002). Persona: A contextualized and personalized web search. In *Proceedings of the 35th Annual Hawaii International Conference on System Sciences* (pp. 1232–1240). Big Island, HI.
117. Thomas, H. (2001). Unsupervised learning by probabilistic latent semantic analysis. *Machine Learning*, *42*, 177–196.
118. Tseng, C.-Y., Sung, P.-C., & Chen, M.-S. (2011). Cosdes: A collaborative spam detection system with a novel e-mail abstraction scheme. *IEEE Transactions on Knowledge and Data Engineering*, *23*(5), 669–682.
119. Turtle, H. R., & Croft, W. B. (1990). Inference networks for document retrieval. In J.-L. Vidick (Ed.), *Proceedings of the 13th Annual International ACM SIGIR Conference on Research and Development in Information Retrieval* (pp. 1–24). Brussels, Belgium.
120. Varadarajan, R., Hristidis, V., & Li, T. (2008). Beyond single-page web search results. *IEEE Transactions on Knowledge and Data Engineering*, *20*(3), 411–424.
121. Vlachou, A., Doulkeridis, C., Kotidis, Y., & Norvag, K. (2010). Reverse top-*k* queries. In *Proceedings of IEEE 26th International Conference on Data Engineering* (pp. 365–376). Long Beach, CA.
122. Wang, X.-J., Zhang, L., Li, X., & Ma, W.-Y. (2008). Annotating images by mining image search results. *IEEE Transactions on Pattern Analysis and Machine Intelligence*, *30*(11), 1919–1932.
123. Wang, J., Sun, X., She, M. F. H., Kouzani, A., & Nahavandi, S. (2013). Unsupervised mining of long time series based on latent topic model. *Neurocomputing*, *103*, 93–103.
124. Wijsen, J. (2001). Trends in databases: Reasoning and mining. *IEEE Transactions on Knowledge and Data Engineering*, *13*(3), 426–438.
125. Wu, C.-H., & Tsai, C.-H. (2009). Robust classification for spam filtering by back-propagation neural networks using behavior-based features. *Applied Intelligence*, *31*, 107–121.
126. Xu, W., Liu, X., & Gong, Y. (2003). Document clustering based on non-negative matrix factorization. In *Proceedings of the 26th Annual International ACM SIGIR Conference on Research and Development in Information Retrieval* (pp. 267–273). Toronto, Canada.
127. Xu, W., & Gong, Y. (2004). Document clustering by concept factorization. In *Proceedings of the 26th Annual International ACM SIGIR Conference on Research and Development in Information Retrieval* (pp. 202–209). Sheffield, UK.
128. Yan, T., Jacobsen, M., Garcia-Molina, H., & Dayal, U. (1996). From user access patterns to dynamic hypertext linking. In *Proceedings of the 5th International World Wide Web Conference* (pp. 1007–1014). Paris, France.
129. Yang, Q., & Zhang, H. H. (2003). Web-log mining for predictive web caching. *IEEE Transactions on Pattern Analysis and Machine Intelligence*, *15*(4), 1050–1053.
130. Yao, L., Mimno, D., & McCallum, A. (2009). Efficient methods for topic model inference on streaming document collections. In *Proceedings of the 15th ACM SIGKDD International Conference on Knowledge Discovery and Data Mining* (pp. 937–946). Paris, France.
131. Zeng, J., Cheung, W. K., & Liu, J. (2013). Learning topic models by belief propagation. *IEEE Transactions on Pattern Analysis and Machine Intelligence*, *35*(5), 1121–1134.
132. Zeng, J., Liu, Z.-Q., & Cao, X.-Q. (2016). Fast online EM for big topic modeling. *IEEE Transactions on Knowledge and Data Engineering*, *28*(3), 675–688.
133. Zheng, A. X., Ng, A. Y., & Jordan, M. I. (2001). Stable algorithms for link analysis. In *Proceedings of the 24th Annual International ACM SIGIR Conference on Research and Development in Information Retrieval* (pp. 258–266). New Orleans, LA.

Chapter 31
Big Data, Cloud Computing, and Internet of Things

31.1 Big Data

31.1.1 Introduction to Big Data

Big data concerns large-volume, complex, growing datasets with multiple, autonomous sources. The data source can be generated by humans or by machines. Big data takes many forms, including web and social media data, machine-to-machine data, business transaction data, biometric data, and human-generated data from Internet search and social media. Big data is not just a massive database, but it is also often unstructured. Big data is now rapidly expanding in all domains.

Big data is characterized by the five Vs [21]: volume (large amounts of data in terabyte or petabyte), velocity (constantly generating, updating, or processing data), variety (the range of data types and sources), veracity (whether the data conform to the inherent and important characteristics of raw data), and value. The first four Vs are concerned about data collection, preprocessing, transmission, and storage. The fifth V focuses on extracting value from the data using statistical and analytical methods. Within each of the five Vs, the data are likely to be uneven, suggesting implications for privacy. The volume of information is a critical privacy parameter.

The real value of big data is to produce small data. Small data is usually actionable insight designed to answer a specific question or serve a particular goal. A big data project should be designed for simplicity. There are occasions when redundancy is necessary. However, redundancy makes the system much more complex. Nature takes a middle-of-the-road approach to redundancy.

Big data infrastructure comprises a big data repository, data analytics software, and the data scientists. Big data performs many computational tasks, some of which are not directly involved in analytics, e.g., retrieving data and organizing the retrieved data by a ranking system.

Big data systems address the challenges of capturing, storing, managing, analyzing, and visualizing big data. Big data analysis discovers new insights or knowledge from the processed data through statistical and analytical methods. Visualization

K.-L. Du and M. N. S. Swamy, *Neural Networks and Statistical Learning*,
https://doi.org/10.1007/978-1-4471-7452-3_31

tools can help to aid human interpretation and decision-making. Useful visualization tools include Tableau (http://www.tableau.com/) for map/location-based data and Cytoscape (http://www.cytoscape.org/) for network-specific visualization.

Big data analysis is a multistep process whereby data are extracted, filtered, and transformed. Data analysis is grouped into statistical analysis, modeling, and predictive analysis. Statistical analysis is associated with hypothesis testing. Modeling is to describe the behavior of a system or its objects. Predictive analysis is to guess how an individual, group, or data object will behave based on past outcomes or on the observed behaviors of similar individuals or groups. Predictive analytics includes the algorithms for recommenders, classifiers, and clustering. Recommender techniques measure the distance between one data object and other data items. The data objects that are closest to one another will tend to have the same preferences and can serve as recommenders.

Traditional relational database management systems are not efficient, nor effective in handling big data due to its unstructured nature and high volume. Apache Hadoop and NoSQL databases make up the infrastructure to maintain and process big data.

Big data processing technologies include

- Scalable computing infrastructures, such as high-performance and elastic data center cloud resources;
- Data ingestion frameworks, such as Apache Kafka (http://kafka.apache.org) and Amazon Kinesis (https://aws.amazon.com/kinesis);
- Data storage frameworks, such as MongoDB (http://www.mongodb.org), BigTable, MySQL, and Apache Cassandra (Facebook, http://cassandra.apache.org);
- Parallel programming frameworks, such as Apache Hadoop (http://hadoop.apache.org) and Apache Storm (http://storm.incubator.apache.org);
- Scalable data mining frameworks, such as Apache Mahout (http://mahout.apache.org), GraphLab, and MLBase.

Big data framework benchmarks include BigDataBench, BigBench, Hibench, PigMix, CloudSuite, and GridMix. Hibench has Sort, WordCount, TeraSort, PageRank, C-means, and Bayes classification workloads for loading Hadoop and Hive frameworks. Among these frameworks, BigDataBench is most comprehensive as it constitutes workload models for NoSQL, DBMS, SPEs, and batch processing frameworks; it targets search engine, social network, and e-commerce application domains. WEKA is a popular workbench for machine learning and statistical analysis. It comprises a very wide range of tools that are suitable for big data analysis.

31.1.2 MapReduce

MapReduce [10, 11], a distributed programming model pioneered by Google, has become an industry standard for storing and processing massive data due to its simplicity, scalability, fault tolerance, and flexibility. MapReduce libraries have been

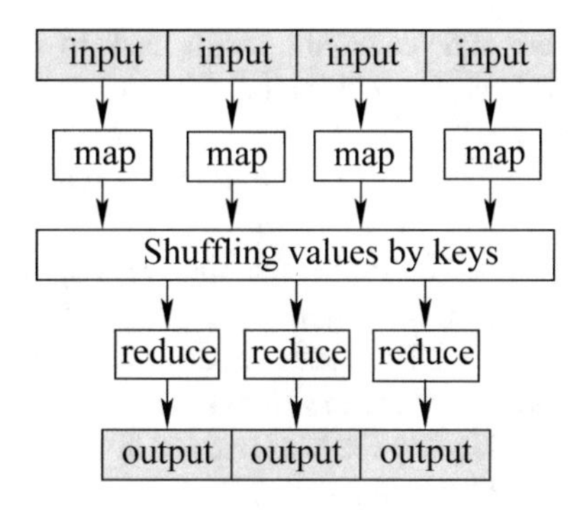

Fig. 31.1 The MapReduce framework: the map function is applied to all input records, while the generated immediate results are aggregated by the reduce function

written in Java, Python, and C++. It is mainly used to process large-scale (TB-level) data files. Programming is very simple in a cloud computing environment.

MapReduce is complementary to DBMS. All MapReduce tasks can be written as equivalent DBMS tasks. The goal of DBMS is efficiency, while MapReduce aims at scalability and fault tolerance.

MapReduce automatically parallelizes large-scale data processing. It enables distributed computation over massive data on thousands of servers or clusters of servers. Data are initially partitioned across the nodes (computers) of a cluster and stored in a distributed file system (DFS). Redundant data are stored in multiple areas across the cluster. The programming model resolves failures automatically by running portions of the program on various servers in the cluster.

MapReduce model is shown in Fig. 31.1. The input data are first processed by the map function, generating intermediate results as the input of the reduce function. A map task processes a block of data and produces $< key, value >$ output pairs, which are partitioned according to the range of key. A reduce task aggregates and operates on $< key, value >$ pairs of map output that fall under the assigned key range partition. Users need only to specify the computation in terms of a map and a reduce function, and the runtime system automatically parallelizes the computation across large-scale clusters of machines, handles machine failures, and schedules inter-machine communication. User can set the number of map functions to be used in the cloud. Map tasks are processed in parallel by the nodes in the cluster without sharing data with any other node. After all the map functions have completed their tasks, the outputs are transferred to reduce function(s). The reduce function produces a (possibly) smaller set of values.

The $< key, value >$ pair is the basic data structure in MapReduce, and all intermediate outputs are $< key, value >$ pairs. For each input $< key, value >$ pair, a map function is invoked once and some intermediate pairs are generated. These intermediate pairs are then shuffled by merging, grouping, and sorting values by keys. A reduce function is invoked once for each shuffled intermediate pair and generate output pairs as final results.

The master node segments the input dataset into independent blocks and distributes them to the worker nodes. The worker node processes the smaller problem and passes the result back to its master mode. The map function receives a $< key, value >$ pairs and generates a set of intermediate $< key, value >$ pairs as output:

$$map(key_1, value_1) \rightarrow list(key_2, value_2).$$

The MapReduction library groups all intermediate values associated with the same intermediate key and transforms them to speed up the computation in the reduce function. The master node collects the results of all the subproblems and combines them to form the final output. The reduce function accepts the intermediate key provided by the MapReduce library and generates as final results the corresponding pair of key and value:

$$reduce(key_2, list(value_2)) \rightarrow list(key_3, value_3).$$

Since the map function only takes a single record, all map operations are independent of one another and fully parallelizable. Reduce function can be executed in parallel on each set of intermediate pairs with the same key.

Hadoop (http://hadoop.apache.org) is the most popular open-source implementation of MapReduce. Hadoop is written in Java and is a top-level Apache project that started in 2006. Hadoop can be easily deployed on commodity hardware. It stores the intermediate computational results in local disks and then informs the appropriate workers to retrieve them for further processing. This introduces a considerable communication overhead. Moreover, Hadoops MapReduce API does not support configuring a map task over multiple iterations. CGL-MapReduce [15] is another open-source implementation of MapReduce. It utilizes NaradaBrokering, a streaming-based content dissemination network, for all the communications. This eliminates the overheads associated with communicating via a file system. IBM Parallel Machine Learning Toolbox is similar to MapReduce model by providing APIs for easy parallel implementation. Unlike MapReduce, it implements iterative learning, which requires multiple passes over data.

Hadoop MapReduce is made up of an execution runtime and a distributed file system. The execution runtime is responsible for job scheduling and execution. It is composed of one master node called JobTracker and multiple slave (worker) nodes called TaskTrackers. Map slot value is the maximum number of map tasks that run concurrently on a TaskTracker node. A JobTracker process monitors the job progress, and a set of TaskTracker processes perform the actual map and reduce tasks. When the JobTracker receives a job, it first splits the job into a number of map and reduce tasks, and then allocates them to the TaskTrackers. The distributed file system HDFS is used to manage task and data across nodes. A MapReduce system usually reads the input data and writes the final results to a distributed file system (e.g., HDFS), which divides a file into equal-sized blocks and stores the blocks across a cluster of machines.

When a flock of jobs is submitted to a MapReduce cluster, they compete for the shared resources and the overall job response times. Hadoop uses a default first-in-first-out (FIFO) scheduler at the job level. Each job will use the whole cluster and execute in the order of submission. Fair scheduler [48] improves the average job response times in shared Hadoop clusters by assigning to all jobs, on average, an equal share of resources over time. FIFO, fair, and capacity schedulers are adopted as the standard scheduling disciplines in MapReduce frameworks.

Dynamic priority task scheduling for heterogeneous Hadoop clusters [38] allows capacity distribution across concurrent users to change dynamically based on user preferences. PRISM, a fine-grained resource-aware MapReduce scheduler [49], divides tasks into phases, each having a constant resource usage profile, and performs scheduling at the phase level. A Hadoop scheduler called LsPS [47] leverages the knowledge of the workload patterns to reduce the average job response times by dynamically tuning the resource shares among users and the scheduling algorithms for each user. A lightweight information collector first tracks statistic information of recently finished jobs from each user.

Incremental MapReduce

Batch processing frameworks like Hadoop MapReduce are difficult to scale to multiple clouds due to latencies involved in intercloud data transfer and synchronization overheads during shuffle phase. This inhibits the MapReduce framework from guaranteeing performance at variable load surges without over-provisioning in the internal cloud.

HaLoop [7], a modified Hadoop, improves the efficiency of iterative computation by making the task scheduler loop-aware and by employing caching. It uses an extra MapReduce job to match structure and state data in each iteration. Incoop [5] extends MapReduce to support incremental processing. However, Incoop supports only task-level incremental processing. That is, it saves and reuses states at the granularity of individual map and reduce tasks. If Incoop detects any data changes in the input of a task, it will rerun the entire task. Incoop treats each iteration as a separate MapReduce job.

iMapReduce [51] creates the same number of map and reduce tasks, and connects every reduce task to a map task with a local connection to transfer the state data output from a reduce task to the corresponding map task. This approach assumes one-to-one dependency for join operation. It efficiently supports iterative computation. i^{2}MapReduce [50] is an incremental processing extension to MapReduce. Compared with Incoop, i^{2}MapReduce performs $< key, value >$-pair level incremental processing rather than task-level re-computation, and incorporates techniques to reduce I/O overhead for accessing preserved fine-grained computation states. i^{2}MapReduce provides general-purpose support, including one-to-one, one-to-many, many-to-one, and many-to-many correspondence. PageRank sees one-to-one dependency: every vertex is associated with both an out-neighbor set and a ranking score. In C-means, the map instance of every point requires the set of all centroids, showing many-to-one dependency.

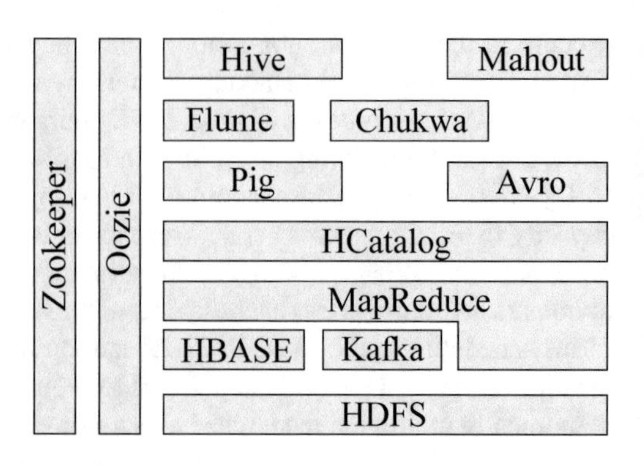

Fig. 31.2 Hadoop software stack

Pregel [29] follows the bulk synchronous processing model. This model can efficiently run a large number of iterative graph algorithms. Open-source Pregel implementations include Giraph (http://giraph.apache.org/), Hama (https://hama.apache.org/), and Pregelix (http://pregelix.ics.uci.edu/).

31.1.3 Hadoop Software Stack

Hadoop combines open-source projects and programming frameworks across a distributed system. It is a batch-oriented tool for processing a large volume of unstructured data. Hadoop software stack, illustrated in Fig. 31.2, can access data by three different sets of tools in particular layers. The middle layer Hadoop MapReduce system server serves for batch analytics. HBase is a key–value layer, i.e., NoSQL database. High-level languages HiveQL (Facebook), PigLatin (Yahoo!), and Jaql (IBM) are for some users at the outermost layer. Hadoop distributed file system (HDFS) and Google file system are distributed file systems. HDFS is good for sequential data access, while HBase provides random real-time read/write access to data.

Cluster contains two types of nodes. A name node acts as a master node. A data node acts as a slave node. The data nodes come in multiples. HDFS can also have secondary name node. HDFS stores files in blocks with the default block size 64 MB. All HDFS files are replicated in multiples to facilitate the parallel processing of large amounts of data.

HBase is an open-source distributed management system that is based on BigTable of Google. This system is column-based rather than row-based, which accelerates the performance of operations over similar values across large datasets. HBase is accessible through APIs. By default, HBase depends completely on a ZooKeeper instance.

ZooKeeper maintains, configures, and names large amounts of data. It also provides distributed synchronization and group services. This instance enables distributed processes to manage and contribute to one another through a namespace of data registers that is shared and hierarchical, such as a file system.

HCatalog manages HDFS. It stores metadata and generates tables for large amounts of data. HCatalog depends on Hive metastore and integrates it with other services, using a common data model.

Hadoop gains limited standard query language (SQL) support and lacks basic SQL functions. HDFS was built for efficiency, where data are replicated in multiples. Hadoop is much inferior to DBMSs in terms of structured data processing. Apache Hive and its SQL-like language HiveQL (Facebook) have become an open-source SQL interface for Hadoop. Hive support SQL-like analysis to the data in HDFS. Hive is primarily based on three related data structures: tables, partitions, and buckets. Tables correspond to HDFS directories and can be distributed in various partitions and, eventually, buckets.

PIG (Yahoo!) is similar to Hive. PIG consists of an open-source Perl-like language that allows for query execution over data stored on a Hadoop cluster, instead of a SQL-like language for Hive. PIG operates a runtime platform that enables to execute MapReduce on Hadoop.

Apache Mahout [22] is an open-source library for a wide range of machine learning and data mining algorithms that runs on top of a Hadoop system via the MapReduce framework.

Oozie is a module that coordinates, executes, and manages job flow. It combines actions and arranges Hadoop tasks using a directed acyclic graph (DAG).

Avro serializes data, conducts remote procedure calls, and passes data from one program or language to another.

Chukwa is a framework for data collection and analysis that is related to MapReduce and HDFS.

Flume is used to aggregate and transfer large amounts of data in and out of Hadoop. It utilizes two channels, namely, sources and sinks. Sources include Avro, files, and system logs, whereas sinks are HDFS and HBase.

Apache Kafka is a high-throughput distributed messaging system. A simple instance of large-scale data stream processing service consists of Apache Kafka for data ingestion, Apache Storm for data analytics, and Apache Cassandra for data storage.

WibiData, built on top of HBase, is a combination of web analytics with Hadoop.

Platfora is a platform that turns user's queries into Hadoop jobs.

31.1.4 Other Big Data Tools

Big data tools (platforms) fall into three categories: batch processing of large volumes of on-disk data with no time constraints, streaming processing of in-memory data in real time, and interactive analysis tools.

Most batch processing tools are based on Apache Hadoop infrastructure, such as MapReduce, Mahout, and Dryad. Dryad [23], which encompasses MapReduce and the relational algebra, is another programming model for implementing parallel and distributed programs that can scale up capability of processing from a small cluster to a large cluster. It is based on dataflow graph processing.

Spark (http://spark-project.org/), a cluster computing system, can be treated as a replacement of MapReduce and Hive. Spark uses a programming model that is optimized for memory-resident read-only objects. Spark will produce a large amount of intermediate data in memory during iterative computation. For small input, Spark exhibits much better performance than Hadoop does because of in-memory processing.

Streaming data analytic platforms include Storm (http://storm-project.net/), SQL-stream (http://www.sqlstream.com/products/server/), StreamCloud [18], Splunk, Apache Kafka (http://kafka.apache.org), and SAP Hana. Storm is an open-source distributed and fault-tolerant real-time computation system for processing limitless streaming data. A Storm cluster is similar to a Hadoop cluster, whereas on Storm users run different topologies for different Storm tasks. Distributed message-queuing frameworks, such as Amazon Kinesis and Apache Kafka, provide a reliable, high-throughput, low-latency system of queuing real-time data streams.

Google's Dremel [31] and Apache Drill are big data platforms based on interactive analysis. Dremel is a scalable, interactive ad hoc query system for analysis of read-only nested data. It acts as a complement of map/reduce-based computations. Apache Drill is an open-source version of Dremel. Drill supports a variety of query languages, data formats, and data sources. Drill and Dremel use HDFS for storage and map/reduce to perform batch analysis. By searching data either stored in columnar form or within a distributed file system, it is possible to scan over petabytes of data in seconds, as response to ad hoc queries.

Presto (Facebook) (http://prestodb.io) is an interactive distributed SQL query engine that runs fast on a Hadoop cluster. Presto is optimized for ad hoc analysis at interactive speed by avoiding MapReduce and supports various file formats. It performs ten times better than Hive/MapReduce in terms of CPU efficiency and latency for most queries.

Cloudera Impala, an open-source SQL-on-Hadoop system, uses its own processing framework to execute queries, bypassing MapReduce model. It has a speedup over Hive.

Shark [44] is a large-scale data warehouse system built on top of Spark, designed to be compatible with Apache Hive. Spark provides the fine granular lineage-based fault tolerance. Shark can answer HiveQL queries much faster than Hive can without modification to the existing data or queries. Shark run SQL queries up to 100X faster than Hive.

H2O is an open-source platform for machine learning and big data/big math. It interacts directly with Python, R, Scala, Spark, REST/JSON, and a JavaScript-based web browser.

31.1.5 NoSQL Databases

Many tools and techniques are available for data management, including Google BigTable, Simple DB, Not Only SQL (NoSQL), and Data Stream Management System. The mainstream big data platforms adopt NoSQL to break and transcend the rigidity of normalized DBMS schemas.

Traditional row-oriented databases fall short on query performance as the data volumes grow and as data become unstructured. Column-oriented databases store data with a focus on columns, allowing for huge data compression and very fast query times. They generally only allow batch updates, having a much slower update time than traditional models. NoSQL databases, such as key–value stores and document stores, focus on the storage and retrieval of large volumes of unstructured, semi-structured, or even structured data. Schema-free databases enable to quickly modify the structure of data without rewriting tables.

Data warehouses and data marts are two popular approaches for managing large-scale datasets in a structured way. A data warehouse is a relational database system that is used to store and analyze data, and also to report the results to users. A data mart is based on a data warehouse and facilitates the access and analysis of the data warehouse. Data warehouses and marts are SQL-based database systems.

NoSQL database frameworks allow data access based on predefined access primitives such as key–value pairs. A key uniquely identifies a value. This data access pattern results in better scalability and performance predictability that is suitable for storing and indexing real-time streams of big datasets, as NoSQL databases do not require fixed table schemas or support expensive join operations. NoSQL does not avoid SQL.

The value for a given key (or row-ID) can be a collection of couples composed from a name and a value attached to this name. Data access operations, typically CRUD (create, read, update, and delete) operations, have only a key as the address argument. The key–value approach enables efficient lookups.

In a more complex case, NoSQL database stores name–value couples collected into collections, i.e., rows addressed by a key. For column NoSQL databases, new columns can be added to these collections. There is a level of structure called super-columns, where a column contains nested (sub)columns. The most general models are document-oriented NoSQL databases such as MongoDB. JSON (JavaScript object notation) format is usually used to represent such data structures. Document stores allow arbitrarily complex documents, e.g., subdocuments within subdocuments and lists with documents, whereas column stores only allow a fixed format. NoSQL are of more categories, e.g., object-oriented, XML, and graph databases.

Some popular NoSQL databases are Apache Cassandra, Hbase, Apache CouchDB, MongoDB, HyperTable (http://hypertable.org), Amazon Dynamo, and Neo4J. However, many big data analytic platforms, like SQLstream and Cloudera Impala, still use SQL in their database systems.

31.2 Cloud Computing

Cloud computing is a revolutionary paradigm for infrastructure, platforms, and software consumption in which users consume from a shared pool of resources. Users pay for what they use. The cloud consists of a collection of interconnected and virtualized computers dynamically provisioned as unified computing resources. Resources are referred to collectively as *the cloud*. We have computing power *on tap*.

Amazon launched the core parts of its Amazon web services, namely, elastic compute cloud (EC2) and simple storage service (S3), in 2006. Apache Hadoop project added support for running Hadoop on EC2 and S3 in the same year. In 2008, Amazon launched its elastic MapReduce service for large-scale data processing. Google launched BigQuery, a web service for querying massive datasets, in 2010, and launched compute engine, an infrastructure as a service (IaaS), in 2012. In 2013, Microsoft launched Azure IaaS and a cloud-based Hadoop service called HDInsight. These are public clouds. One can build his own private cloud by using software offerings, such as Openstack and VMWare.

Commercial and public data centers provide computing, storage, and software resources as cloud services, which are enabled by virtualized software/middleware stacks. The ubiquity of cloud applications requires high quality-of-service (QoS), low costs, and low CO_2 emissions.

Three open-source cloud management platforms are Eucalyptus, Open Nebula, and Nimbus. Among them, Open Nebula provides the highest level of customization that allows users to switch almost every component from the virtual machine monitor to the front end. Nimbus also provides a high level of customization, but the major portion of customization is available to the administrator. Eucalyptus mimics Amazon EC2 and is an open-source implementation of Amazon web service API. Its low customization level makes it appropriate for a private company. Apache Hadoop on Demand (HOD) provides virtual Hadoop clusters over a large physical cluster based on Torque.

Mobile cloud computing improves the computational capabilities of resource-constrained mobile devices. The mobile users demand a certain level of QoS provisioning while they use services from the cloud, even if the interfacing gateway changes due to the mobility of the users.

31.2.1 Services Models, Pricing, and Standards

In the cloud computing context, network-accessible resources are defined as services. The cloud computing services are typically delivered via one of the following service models.

Infrastructure as a Service

Infrastructure as a service (IaaS) is a cloud computing model based on the principle that the entire infrastructure is deployed in an on-demand model. IaaS offers storage, computation, and network capabilities to service subscribers through virtual machines. IaaS model provides just the hardware and network; the customer installs or develops its own operating systems. IaaS is often considered utility computing because it treats compute resources much like utilities (such as electricity, telephony) are treated. The consumer has control over operating systems, storage, and deployed applications and possibly limited control of select networking components (e.g., host firewalls).

Platform as a Service

Platform as a service (PaaS) provides an environment for software application development and hosts a client's applications in a cloud computing infrastructure. In PaaS, an operating system, hardware, and network are provided, and the customer provides his own software and applications. PaaS provides the platform for developing, running, and managing applications, such as Google App Engine. The customer can deploy onto the cloud infrastructure applications created using programming languages, libraries, services, and tools supported by the provider. The customer has control over the deployed applications and possibly configuration settings for the application-hosting environment.

Software as a Service

Software as a service (SaaS) delivers on-demand software services via a computer network, eliminating the cost of software purchase and maintenance. A pre-made application, along with any required software, operating system, hardware, and network, is provided. SaaS provider provides a user the access to applications and resources that are stored and run on virtual servers in the cloud. The user does not manage the cloud infrastructure but with limited user-specific application configuration settings.

Workflows are submitted and executed in the cloud, and each workflow is usually associated with a deadline as performance guarantee. Workflow as a service (WaaS) is a SaaS model for hosting workflows in the cloud. Users are charged based on the execution of their workflows and the QoS requirements.

Many cloud infrastructure providers have deployed the MapReduce framework as an infrastructure service. Some cloud service providers also offer their own MapReduce as a service (MRaaS), which is typically set up as a kind of SaaS on the owned or provisioned MapReduce clusters of cloud instances.

Storage as a Service

Data storage is the most popular cloud service. Storage as a service (StaaS) allows users to store their data at remote disks and access them anytime from any place using Internet. In order to preserve the privacy of data owners, data are often stored in an encrypted form. However, encrypted data introduce new challenges for data deduplication, which is crucial for big data storage and processing in the cloud.

Encrypted data deduplication cannot flexibly support data access control and revocation. A scheme to deduplicate encrypted data stored in the cloud proposed in [46] is based on data ownership challenge and proxy re-encryption.

Database as a Service

Database as a service (DBaaS) [20] can support several Internet-based applications. DBaaS poses several challenges in terms of security and cost evaluation from a tenant's point of view. The cloud database must be able to execute SQL operations directly over encrypted data without accessing any decryption key. An initial solution is based on data aggregation techniques that associate plaintext metadata to sets of encrypted data [20]. However, plaintext metadata may leak sensitive information and data aggregation introduces unnecessary network overheads.

Pricing

Pricing is a critical component of cloud computing because it affects providers' revenue and customers' budget. Static pricing is the dominant pricing strategy. The challenge is to design a dynamic pricing policy so that the expected long-term revenue is maximized.

Amazon sells virtual machines as instances, where different types of instances are allocated different amounts of resources. In Amazon EC2, the smallest pricing time unit of an on-demand instance is one hour. Some IaaS providers offer optional fine-grained pricing schemes. CloudSigma (http://www.cloudsigma.com/) offers a burst pricing scheme that changes every 5 min upon busy status. Google compute engine offers a 10-minute-based pricing scheme.

The pricing schemes adopted in IaaS cloud market can be categorized into three types: pay-as-you-go offer, subscription option, and spot market. Under the pay-as-you-go scheme, users pay a fixed rate for cloud resource consumption per billing cycle (e.g., an hour) with no commitment. On-demand instances are often used to run short jobs. In the subscription scheme (Amazon-Reserved Instances), users need to pay an upfront fee to reserve resources for a certain period of time and in turn receive a significant price discount. For the spot scheme (Amazon EC2 Spot Instances), users simply bid on spare instances and run them whenever their bid prices exceed the current spot price. Spot Instances are suitable for time-flexible, interruption-tolerant tasks. The instances can be terminated by the provider at any time. The hourly spot price traces for Amazon EC2 Spot Instances are collected by using the ec2-describe-spot-price-history API call provided by the EC2 API tools.

In an optimized fine-grained and fair pricing scheme [25], an optimal price can be derived in an acceptable price range that satisfies both customers and providers simultaneously, but also find a best-fit billing cycle to maximize social welfare (i.e., the sum of the cost reductions for all customers and the revenue gained by the provider).

Service-level agreements regulate the costs that the cloud customers have to pay for the provided QoS [33]. In a large IaaS cloud, component failures are quite common. Service availability is usually specified in service-level agreements as downtime in minutes per year or as the percentage of time the service will be up in a year, average uptime per month, maximum response time, average response time per hour, or maximum time to detect and block an intrusion.

Standards

Various commercial and open-source solutions are available to resolve interoperability and portability issues. Openstack (http://www.openstack.org) is an IaaS cloud platform positioned as a *de facto* standard for cloud interoperability. OpenNebula (http://opennebula.org) and Deltacloud (http://deltacloud.apache.org) address IaaS interoperability by allowing to use APIs from different cloud vendors. Cloud Data Management Interface (http://www.snia.org/cdmi) from Storage Networking Industry Association addresses IaaS offers, focusing on infrastructure, services, and data storage management. Other cloud interoperability standards include Open Cloud Computing Interface (OCCI, http://occi-wg.org) from Open Grid Forum, and Cloud Infrastructure Management Interface [9]. OCCI focuses mostly on IaaS but, given its flexibility, could be applied to other service layers.

Cloud Foundry (http://cloudfoundry.org) and Red Hat Openshift (http://www.openshift.com) are open-source PaaS initiatives that support application portability for PaaS architectures. Cloud Application Management for Platforms from Organization for the Advancement of Structured Information Standards (OASIS) relates to PaaS offers, defining a set of basic APIs. Other open-source PaaS alternatives are Engine Yard (http://www.engineyard.com), Google App Engine (https://developers.google.com/appengine), Salesforce Heroku (http://www.heroku.com), AppFog (http://www.appfog.com), and ActiveState Stackato (www.activestate.com/stackato).

The IEEE P2301 group aims to develop a guide for Cloud Portability and Interoperability Profiles (CPIP) as an aid to vendors and users in developing, building, and using standard-based cloud computing products and services. The P2302 project, known as IEEE Intercloud Working Group, is developing Standard for Intercloud Interoperability and Federation (SIIF) and has the goal of operating an intercloud testbed (http://www.intercloudtestbed.org) in support of this work.

31.2.2 Virtual Machines, Data Centers, and Intercloud Connections

Virtual Machines

Virtualization can be defined as the process of abstracting the physical structure of innumerable technologies, such as hardware platform, operating system, a storage device, or other network resources. Every physical machine in a cloud can host several virtual machines, which is equivalent to a fully functional physical machine from a user's perspective. Moreover, virtual machines can start and stop anytime without any change to the underlying hardware. Migration of virtual machines between the physical machines is also possible without much disruption.

VMware virtual center, platform orchestration, and enomalism provide automatic monitoring and deployment of virtual machines in resource pools [41]. Several open-source virtual machine-based cloud management platforms have been launched, such as Eucalyptus [35], oVirt (http://www.ovirt.org/Home), and Enomaly Elastic Compute Platform (ECP) (http://www.enomaly.com/).

Two issues in deploying and provisioning virtual machine instances over IaaS environment are refined resource allocation and precise pricing for resource renting. Xen and KVM are the two most widely used open-source hypervisors by IaaS providers [40].

Virtual machines can be provisioned on-demand to crunch data after uploading into the virtual machines. Moreover, the elapsed time comes at a cost of provisioning virtual machines in the cloud and keeping them waiting to load the data. A big data provisioning service presented in [42] incorporates hierarchical and peer-to-peer data distribution techniques to speed up data loading into the virtual machines. Placing the right data on each virtual machine is an NP-hard problem [3].

Network facilities are also virtualized by the virtual machine monitor. Network I/O processing correlates highly with the resource allocation policy of virtual machine monitor scheduler. Single-root I/O virtualization (SR-IOV) [13] has become the de facto standard of network virtualization in cloud infrastructure. In this model, packets are sent from network interface card to the virtual machine without the data copy in the para-virtualized driver. SR-IOV offloads almost all the driver domain's overhead to network interface card hardware and achieves high I/O performance.

Virtual machines provided by current cloud infrastructures do not exhibit a stable performance in terms of execution times. An overall CPU performance variability of 24% on Amazon's EC2 cloud is reported in [39].

Data Centers

Data centers constitute the foundation of cloud computing platforms. A virtual data center is a collection of virtual machines, switches, and routers that are interconnected through virtual links. Each virtual link is characterized by its bandwidth capacity and its propagation delay. Compared to virtual machines, virtual data centers are able to provide better isolation of network resources. A data center network constitutes the communicational backbone of a data center. It needs to be robust to failures and uncertainties to deliver the required QoS level and satisfy service-level agreement. Scalability, high cross-section bandwidth, QoS concerns, energy efficiency, and service-level agreement assurance are the major challenges faced by cloud data center architectures.

VMware vSphere is a data center virtualization platform that enables infrastructure provisioning and virtual machine lifecycle management. It features distributed resource scheduler and distributed power management. Infrastructure resource allocation software solutions, such as Microsoft PRO, VMware DRS, and Citrix XenServer, currently only support dynamic allocation for a set of 32 or fewer hosts.

Multiple tenants with diverse resource and QoS requirements often share the same physical infrastructure offered by a cloud provider [6]. The virtualization of server, network, and storage resources adds challenges to managing data center infrastructures. Cloud providers must guarantee reliability and robustness in case of workload perturbations, hardware failures, and malicious attacks [6], and deliver the agreed services and QoS.

The huge number of interconnected servers in a data center leads to scalability as a major issue. Tree-structured data center architectures, such as ThreeTier, VL2, and FatTree, offer low scalability. FatTree architecture, capped by the number of network switch ports, delivers high bisection bandwidth and a 1:1 oversubscription ratio. Server-centric architectures (such as DCell and FiConn) and freely/randomly connected architectures (such as JellyFish and Scafida) deliver high scalability, but at the cost of low performance and high packet delays for high network loads.

Infrastructure providers should have the ability to provision requested virtual data centers across their distributed infrastructure to achieve multiple goals including revenue maximization, operational costs reduction, energy efficiency, and green IT, or to simply satisfy geographic location constraints of the virtual data centers. Greenhead [2] is a holistic resource management framework for embedding virtual data centers across geographically distributed data centers connected through a backbone network.

Intercloud Connections

In standard cloud application deployment, an application is deployed and managed over a single data center. Interconnecting multiple cloud-based data centers allows to improve overall QoS, reliability, and flexibility of applications.

Cloud-to-cloud data transfer can be either intracloud or intercloud. Intracloud data transfer performance is directly related to the size of the pipe within the cloud service, as well as to cache services. Typically, intracloud data transfer is between tenants, virtual machines, or applications and data stores. Intercloud data transfer is more complex, exchanging data with cloud services that might not like each other. The Internet is the typical mode of transport.

Cloud interoperability is primarily driven by two factors: customer demands and standards. Owners typically implement an application provisioner software program (such as RightScale https://www.rightcale.com or CloudSwitch https://home.cloudswitch.com), which distributes application components across multiple resource providers to meet the service-level agreements in an optimal way.

Cloud-of-clouds paradigm [4] denotes the integration of different clouds. It postulates integration of multiple data centers, cloud abstractions, and applications.

By abstracting from hardware addresses and lower level communication, the publish/subscribe abstraction supports communication across clouds, thus application interoperability and portability. Content-based publish/subscribe systems [1, 16] provide an expressive abstraction that matches well with the key–value pair model of many cloud storage and computing systems, and their decentralized overlay-based implementations scale up well. They typically employ an overlay network of brokers, with filtering happening downstream from publishers to subscribers based on upstream aggregation of subscriptions. However, they perform poorly at small scale, e.g., one-to-one or one-to-many communication. Overlay-based multi-hop routing will impose increased latency compared to a direct multi-send via UDP or TCP. Topic-based publish/subscribe systems support communication at small scale more efficiently but still route messages over multiple hops and also lack the flexibility of content-based publish/subscribe systems. Atmosphere [24] is a middleware solution that supports the expressive content-based publish/subscribe abstraction across data centers and clouds in a way that is effective for a wide range of communication patterns, and to elastically scale both up and down between these cases. Atmosphere allows to implement the popular HDFS and ZooKeeper systems which operate efficiently across data centers.

31.2.3 Cloud Infrastructure Requirements

Cloud infrastructure requirements include elasticity, scalability, data security, reliability, operational efficiency, data retrieval, and latency.

Cloud management deals with the operations of cloud infrastructures and cloud services, and the enforcement of service-level agreements. Management functions are traditionally broken down into five areas: fault management, configuration management, accounting management, performance management, and security management, known as FCAPS.

High energy consumption causes low system reliability and negative impacts on environment. Virtualization is an efficient approach to increase resource utilization and in turn reduce energy consumption. In virtualized data centers, consolidation of virtual machines on the minimum number of physical servers is a very efficient approach. The optimal assignment of virtual machines to the servers of a data center is analogous to the NP-hard bin packing problem, which assigns a given set of items of variable size to the minimum number of bins taken from a given set.

Elastic resource provisioning is a feature of IaaS cloud system. This increases the flexibility for cloud consumers according to changing requirement of the customer demands and service-level agreements. However, elasticity may lead to poor resource utilization. Admission control mechanisms are needed to increase the number of services accepted, raising the utilization without affecting services performance. Overbooking is a way of mitigating these resource utilization problems. However, overbooking always has a risk of resource congestion upon unexpected situations and consequently to service-level agreement violations.

The variability of users' demands increases when it comes to their requests for data-intensive applications. The cloud providers need to reshape their business structures and seek to improve their dynamic resource scaling capabilities. Federated clouds offer a practical platform for addressing this service management issue. A cloud federation is a collection of cloud providers that cooperate in order to provide the resources requested by users [37]. A cloud federation formation game considers the cooperation of the cloud providers in offering cloud IaaS services [30].

Security and Privacy

Security is one of the major concerns of the cloud users. Cloud computing brings in new concerns and challenges on security and privacy. In public clouds, monitored data need to be transferred from the monitored entity to the analytics engine. Cloud computing model is prone to denial of service (DoS) and distributed DoS, which aim at reducing the service availability and performance by exhausting the resources of the service's host system.

A key-policy attribute-based encryption with time-specified attributes [45] is a secure data self-destructing scheme in cloud computing. Every ciphertext is labeled with a time interval, while private key is associated with a time instant. The ciphertext can only be decrypted if both the time instant is in the allowed time interval and the attributes associated with the ciphertext satisfy the key's access structure. The sensitive data will be securely self-destructed after a user-specified expiration time.

Software-defined network is the software enablement of infrastructures. It is the next-generation networking architecture that is ideal for the high-bandwidth, dynamic applications. It is part of the software-defined data center concept that detaches the hardware from the software and replaces complex hardware devices with changeable layers of software. Software-defined networking consists of three layers—infrastructure, control, and application. The paradigm separates the control plane and data plane such that switches become simple data forwarding devices and network management is controlled by logically centralized servers [26].

Software-defined network provides better and more dynamic security and governance. The solution can dynamically update itself to protect against those threats. Network governance is much easier, since policies can be dynamically placed on network functions, and even integrated with other governance layers, such as services/APIs and data. However, the centralized control plane introduced imposes a great challenge for the network security. An attack may be made in a cooperative manner to multiple switches if they are distributed in different regions. An intuitive solution is to employ multiple controllers for each switch [27]. Unfortunately, other attacks require a more secure mechanism to protect controllers.

31.3 Internet of Things

The emerging Internet/web/network of things technologies enable the collection of data from an increasing volume and variety of networked sensors for analysis. The Internet of everything is an intelligent connection of people, process, data, and things into one cohesive whole. These applications have resulted in a hyper-world consisting of the social, cyber, and physical worlds, with data as a bridge.

The Internet of things (IoT) paradigm covers a diverse range of technologies, including sensing, networking, computing, information processing, and intelligent control technologies. IoT has four fundamentals: sensing, actuation, computing, and communication. The large-scale, complexity, and highly heterogeneous nature of IoT is the main challenge facing IoT technologies.

Data captured from the physical world through sensors tend to be noisy, incomplete, and unreliable. The inconsistency among multiple sensor measurements serves as an indicator for the data quality. For wireless sensor networks, the volume of data is reduced for processing, and aggregation also reduces the transmission requirements and increases the energy efficiency of battery-powered sensor nodes. Sensor-based applications depend on the functional (only statistical summary values) or recoverable (the full dataset) information that is required at the sink [43].

Big sensor data systems involve identifying and generating the required data from sensor farms and other sources. A sparse sensor farm deployment with a low sampling rate would produce a challenge. The primary goal is to infer the values of the missing data points from the sensor points which are available. Compressive sensing is effective to reduce the node energy consumption [8, 43].

The cloud of things paradigm integrates cloud computing with the IoT. Sensing and actuation features offered in the IoT can be abstracted and virtualized into cloud resources [32]. IoT capabilities and data are provided in the form of services.

31.3.1 Architecture of IoT

The architecture of IoT is typically divided into three layers: application layer, network layer, and perception layer. Service-oriented architecture is a component-based model, which can be designed to connect different functional units (also known as services) of an application via interfaces and protocols. Service-oriented architecture focuses on designing the workflow of coordinated services and enables the reuse of software and hardware components. Thus, it can be easily integrated into IoT architecture, in which data services provided by the network layer and the application layer can be extracted and form the service layer (also known as the interface layer or middleware layer). Middleware provides an abstraction. Thus, in a service-oriented architecture-based IoT architecture, the four layers are the perception layer, network layer, service layer, and application layer [28].

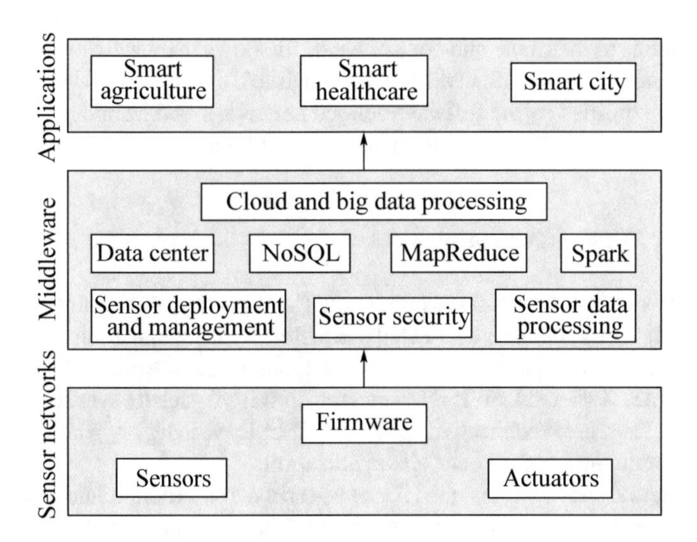

Fig. 31.3 A conceptual architecture for a sensor-based big data cyber-infrastructure

Figure 31.3 shows a conceptual architecture for a sensor-based big data cyber-infrastructure. The sensor network layer at the bottom senses and collects large sets of data in real or semi-real time. The middleware layer includes the engines for processing the collected big datasets. The top application layer includes a variety of applications.

Perception layer, also known as the sensor layer in IoT architecture, interacts with physical devices and components through smart devices (RFID, sensors, actuators, etc.). It connects things into the IoT network, and to measure, collect, and process the state information associated with these things via deployed smart devices, transmitting the processed information into upper layer via layer interfaces. The perception layer identifies and tracks objects. RFID is used to identify and track objects without contact. It supports data exchange via radio signals over a short distance. Barcode (1-D code) stores information in several black lines and white spacings, and 2-D code records information by using black and white pixels on the plane. Wireless sensor networks bridge the real world and the cyber-world.

Network layer, also known as the transmission layer in IoT architecture, is used to determine routes and provide data transmission support via integrated heterogeneous networks. Various devices (hub, switching, gateway, cloud computing perform, etc.) and various communication technologies (Bluetooth, Wi-Fi, long-term evolution, etc.) are integrated into this layer.

Service layer is located between network layer and application layer, providing services to support the application layer. It involves interface technology, service management technology, middleware technology, and resource management and sharing technology.

Application layer, also known as the business layer, receives the data transmitted from network layer and uses the data to provide required services or operations.

IoT requires to distribute and control the traffic flows in the network for load balancing and minimization of network delay. Such IoT network management requirements can be fulfilled by the software-defined network-based technology, as it leverages the global view of the network in a centralized manner.

Low-Power Wireless Networks

Low-power wireless personal area networks (LoWPANs) are organized by a large number of low-cost devices connected via wireless communications. LoWPAN is characterized by small packet sizes, low power, and low bandwidth. 6LoWPAN protocol combines IPv6 and LoWPAN. It can transmit IPv6 packets over IEEE 802.15.4 networks. It has a great connectivity and compatibility with legacy architectures, low energy consumption, and ad hoc self-organization.

ZigBee and Z-Wave are designed for short-term wireless communication with low cost, low energy consumption, and reliability. They operate in different frequency bands. ZigBee network, based on IEEE 802.15.4, has low data rate, low complexity, and security. It supports multiple topologies, including star, tree, and mesh topologies. ZigBee network supports end devices (slaves) up to 65,000, while Z-wave network only supports 232 end devices (slaves) and is simple in implementation.

Narrowband IoT (NB-IoT) and long range (LoRa) are two leading low-power wide-area technologies for the IoT market. Unlicensed LoRa has advantages in terms of battery lifetime, capacity, and cost. Meanwhile, licensed NB-IoT offers benefits in terms of QoS, latency, reliability, and range.

NB-IoT is set up by 3GPP as a part of Release 13. It aims to support massive number of user equipment operated in an enhanced coverage area. It is kept as simple as possible in order to reduce device costs and minimize battery consumption. It uses the same licensed frequency bands of LTE and uses multicarrier technology. LoRa is a proprietary spread spectrum modulation scheme that is a derivative of chirp spread spectrum modulation and has a fixed channel bandwidth. LoRa network applies an adaptive modulation technique with multichannel multi-modem transceiver. The data rate is directly proportional to the spreading factor. LoRa operates in a non-licensed band below 1 GHz.

31.3.2 Cyber-Physical System Versus IoT

Both the cyber-physical system and IoT aim to achieve the interactions between the cyber-world and the physical world [28]. In a cyber-physical system, *cyber* means using modern sensing, computing, and communication technologies to effectively monitor and control the physical components, while *physical* means the physical components in real world. There are similarities between a cyber-physical system and IoT.

A cyber-physical system consists of multiple heterogeneous distributed subsystems. It measures the state information of physical devices and ensures secure, efficient, and intelligent operation on physical devices. It is the integration of physical components, sensors, actuators, communication networks, and control centers.

A cyber-physical system has the sensor/actuator layer, communication layer, and application (control) layer. The sensor/actuator layer is used to collect real-time data and execute commands, the communication layer is used to deliver data to upper layer and commands to lower layer, and the application (control) layer is used to analyze data and make decisions. Figure 31.4 illustrates the three layers in a cyber-physical system. This figure indicates that cyber-physical system is a vertical architecture.

IoT is a networking infrastructure that connects a massive number of devices and monitors/controls devices by using technologies in cyber-space. IoT interconnects various networks so that data collection, resource sharing, analysis, and management can be implemented across heterogeneous networks. Thus, IoT is a horizontal architecture, which integrates communication layers of all cyber-physical systems for interconnection, as shown in Fig. 31.4. The control plane (interfaces, middleware, protocols, etc.) should ensure that data can be efficiently delivered across different networks and shared.

The requirements for both cyber-physical system and IoT are real-time, reliable, and secure data transmission. The primary goal for cyber-physical system is effective, reliable, accurate, real-time control, while IoT focuses on many services, including resource sharing and management, data sharing and management, interface among different networks, massive-scale/big data collection and storage, data mining, data aggregation and information extraction, and high quality of network QoS.

Smart Cities

One of the most representative applications that integrate cyber-physical system and IoT is smart cities. Smart cities are considered a complex IoT paradigm, which aims to manage public affairs of cities via information and communication technology (ICT) solutions. The focuses include citizen well-being, infrastructure, industry,

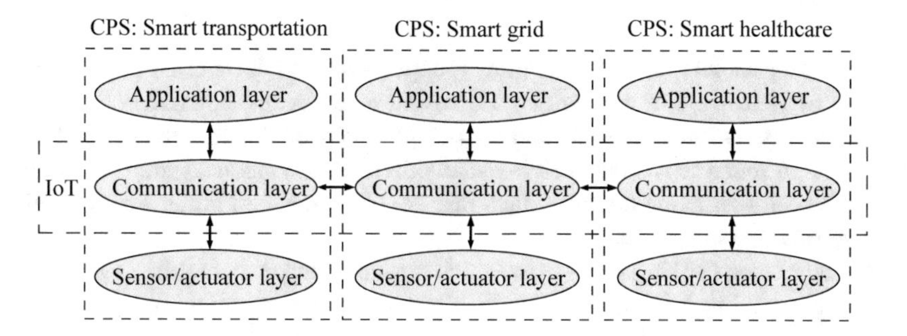

Fig. 31.4 Cyber-physical system versus IoT [28]

and government. For smart cities, several cyber-physical systems operate simultaneously, including smart grid, smart transportation, and smart health care, as shown in Fig. 31.4.

ISO 37120 is the first international standard on city data published by International Organization for Standardization in 2014. It defines and establishes definitions and methodologies for a set of 100 indicators to steer and measure the performance of city services and quality of life. The indicators (also referred to as ISO4City indicators) are categorized under 17 themes related to city services and quality of life: economy, education, energy, environment, finance, fire and emergency response, governance, health, recreation, safety, shelter, solid waste, telecommunication and innovation, transportation, urban planning, wastewater, and water and sanitation. The low-resolution indicators are static or slow changing quantities that merely show the city's average performance. Increasing the resolution of existing indicators and introducing new (related) indicators lead to a 3D discrete space with the following axes and discrete values: temporal, spatial (by area), and human (by population).

Tactile Internet and Industry 4.0

Tactile Internet is a communication network that is capable of delivering real-time control, touch, and sensing/actuation information through reliable, responsive, and intelligent connectivity. IEEE 1918.1 standards define tactile Internet that achieves latency as low as 1-ms round-trip for machine–client cases over distances longer than 150 km (or 100 km in fiber). This can be used for the scenarios such as industrial control and the human–client case of teleoperation and remote surgery. Tactile Internet will likely operate as an overlay on 5G ultra-reliable low-latency communications (URLLC) networks.

Industry 4.0, referred to as the fourth industrial revolution, is based on cyber-physical system and IoT technology. The Industry 4.0 framework comprises smart machines, storage systems, and production facilities, leading to efficient, flexible, and self-organized factories.

Social IoT

The emerging social IoT paradigm allows people and smart devices to interact within a social framework. Through unique addressing schemes and standard communication protocols, users can interact and cooperate with people within their social relationships. In social IoT, a network is usually divided into distinct groups to control the degree of interaction among things that are friends. Smart cities are a complex cyber-physical social system.

Social sensing [14] is a process by which physical sensors in mobile devices such as GPS are used to infer social relationships and human activities. A social sensor or human-based sensor is an agent that provides information on a social network after interaction with other agents. Inspired by participatory sensing and ubiquitous computing [17], mobile crowdsensing is a dynamic sensing paradigm that allows people to contribute data sensed or generated from their mobile devices, aggregates, and fuses the data in the cloud for crowd intelligence extraction and people-centric service delivery [19]. A mobile crowdsensing participant collects the information about a fire, and then he is treated as a fire sensor. Social sensors, social sensor receiver platform (e.g., Weibo and Twitter), and mobile crowdsensing paradigm compose a process by which physical sensors present in mobile devices are used to infer social relationships and human activities.

31.4 Fog/Edge Computing

Cloud computing paradigm can hardly satisfy low latency, mobility support, and location awareness. Fog computing concept was created by Cisco Systems in 2011 to bring cloud services to the edge of an enterprise network. Fog computing, also known as edge computing, aims at extending the cloud by bringing computational power, storage, and communication capabilities to the edge of the network, in support of IoT. A mobile device can offload its data or computationally expensive tasks to a distributed fog node within its proximity, instead of distant cloud, to decrease latency and improve QoS for IoT applications. Fog supports heterogeneity, which is a distinguished characteristic of IoT. Fog computing-based IoT becomes future infrastructure on IoT.

A fog computing node can be any network device with the capability of storage, computing, and network connectivity (routers, switches, video surveillance cameras, servers, etc.). In fog computing, routers themselves may become the virtualized infrastructure and services can be hosted at end devices or access points. IoT applications in a fog computing framework consist of the cloud, fog node, and things [28].

When fog nodes are allocated to provide services, different requirements need to be considered, including service availability, energy consumption, and even revenue. In addition, security and privacy issues in fog computing infrastructures integrating IoT remain challenging.

Fog might be specified in terms of functionality as fog edge nodes, fog server, and foglet, where fog edge nodes and fog server are hardware nodes, and foglet is the middleware in charge of data exchange. When fog is employed as a platform for IoT, a fog edge node accommodates its adjacent smart objects for network access and edge computing. A fog server is focused on the interplay between fog edge nodes and cloud. It controls, manages, and coordinates fog edge nodes at their one-hop proximity. Foglet offers the cross-platform capability of monitoring, liaising, and organizing of fog resources.

Mobile edge computing extends cloud computing services to the edge of networks leveraging mobile stations. It empowers mobile cloud computing by deploying cloud resources, e.g., storage and processing capacity, to the edge within the radio access network. Mobile edge computing was standardized by European Telecommunications Standards Institute (ETSI) as a prime technology for 5G networks. Since 2017, the ETSI mobile edge computing industry group has renamed mobile edge computing to multi-access edge computing.

Mobile edge computing connects the user directly to the nearest cloud service-enabled edge network. It is characterized by [36]:

- On-premises: Mobile edge computing platforms can run isolated from the rest of the network; they have access to local resources.
- Proximity and lower latency: Mobile edge computing services are deployed at the nearest location to user devices. User experience is high quality with ultralow latency and high bandwidth.
- Location awareness: Mobile edge computing receives information from edge devices within the local access network to discover the location of devices.
- Network context information: Applications providing network information and services of real-time network data can benefit businesses and events by implementing mobile edge computing in their business model.

The term cloudlet was coined at Carnegie Mellon University as part of a research project Elijah (http://elijah.cs.cmu.edu/). Cloudlets can be considered as a local small-box data center to enable localized cloud services; it offers high performance and faster access to cloud resources by multiple users simultaneously. Cloudlet is normally deployed at one wireless hop away from mobile devices, such as hospital, shopping center, and office building.

The open fog consortium defined a system-level horizontal architecture that distributes resources and services of computing, storage, control, and networking anywhere from cloud to things. Open-Edge Computing Initiative has offered cloudlets open-source code APIs as an extension to OpenStack to promote cloudlet as an enabling technology to OpenStack++. Cloudlet/open-edge computing (http://openedgecomputing.org/) has been motivated by the Internet community to optimize Internet demanding applications over resource-constrained mobile devices.

31.5 Blockchain

Blockchain concept was introduced when bitcoin was launched in 2008 [34]. Blockchain maintains a distributed, decentralized, authenticated, and synchronized ledger of transactions. Blockchain is a special data structure which stores historical states and transactions. In blockchain data structure, each block is linked to its predecessor via a cryptographic pointer, all the way back to the first block. Blockchains are distributed ledgers that enable parties (nodes) who do not fully trust one another to maintain a set of global states. The parties (nodes) agree on the existence, values, and histories of the states.

A ledger is a data structure that consists of an ordered list of transactions. In blockchains, the ledger is replicated over all the nodes. Transactions are grouped into blocks which are then chained together. Thus, the distributed ledger is essentially a replicated append-only data structure. A blockchain starts with some initial states, and the ledger records entire history of update operations made to the states. Being replicated, updates to the ledger must be agreed on by all parties. That is, multiple parties must come to a consensus. Network nodes devote their processing power to proofread transactions. The authenticated transactions are written into the public ledger in the form of blocks.

A blockchain can be either a public or a consortium (private) system. In a public blockchain, any node can join and leave, and thus the blockchain is fully decentralized, resembling a peer-to-peer system. A private blockchain enforces strict membership and access control. Every node is authenticated, and its identity is known to the other nodes.

Bitcoin is the most well-known public blockchain. In Bitcoin, the states are digital coins (cryptocurrencies), and a transaction moves coins from one set of addresses to another. Each node broadcasts a set of transactions it wants to perform. Network participants (miners) collect transactions into blocks, check for validity, and start a consensus protocol to append the blocks onto the blockchain. Bitcoin uses proof-of-work for consensus: only a miner which has successfully solved a computationally hard puzzle can append to the blockchain. Proof-of-work works well in a public setting, because it guards against Sybil attacks. However, it is unsuitable for banking and finance applications, which require deterministic and fast computation.

Hyperledger (https://www.hyperledger.org) is a popular private blockchain. Since node identities are known in the private settings, most blockchains adopt one protocol on distributed consensus. Deterministic consensus and support of smart contracts make Hyperledger particularly desirable in business and financial systems.

Blockchains use cryptographic techniques to ensure integrity of the ledgers. Integrity refers to the ability to detect tampering of the blockchain data. Network participants (miners) contribute their processing power to verify information correctness and integrity in blockchain. Accountability function is benefited by using block lookup, which helps to timely revoke the cryptographic materials of malicious users. Transactions are used to convey information among the distributed network using peer-to-peer mode. Security and privacy of the network are based on the proof-of-work. The security model assumes public-key cryptography.

A smart contract is a stored procedure invoked upon a transaction. The inputs, outputs, and states affected by the smart contract execution are agreed on by every node. All blockchains have built-in smart contracts that implement their transaction logics.

It is shown that current blockchains have limited performance, much lower than that of a state-of-the-art database system, based on a comprehensive evaluation of three major blockchain systems, namely, Ethereum (https://www.ethereum.org/), Parity (https://ethcore.io/parity.html), and Hyperledger (https://www.hyperledger.org) on BlockBench [12].

Problems

31.1 Discuss the duty and skills of a data scientist.

31.2 Consider an application scenario that contains most of the concepts treated in this chapter. Draw a conceptual architecture of the application.

References

1. Aguilera, M. K., Strom, R. E., Sturman, D. C., Astley, M., & Chandra, T. D. (1999). Matching events in a content-based subscription system. In *Proceedings of the 18th Annual ACM Symposium on Principles of Distributed Computing* (pp. 53–61). Atlanta, GA.
2. Amokrane, A., Zhani, M. F., Langar, R., Boutaba, R., & Pujolle, G. (2013). Greenhead: Virtual data center embedding across distributed infrastructures. *IEEE Transactions on Cloud Computing*, *1*(1), 36–49.
3. Andreev, K., & Racke, H. (2004). Balanced graph partitioning. In *Proceedings of the 16th Annual ACM Symposium on Parallelism in Algorithms and Architectures* (pp. 120–124). Barcelona, Spain.
4. Bessani, A., Correia, M., Quaresma, B., Andre, F., & Sousa, P. (2011). DepSky: Dependable and secure storage in a cloud-of-clouds. In *Proceedings of the 6th European Conference on Computer Systems* (pp. 31–46).
5. Bhatotia, P., Wieder, A., Rodrigues, R., Acar, U. A., & Pasquin, R. (2011). Incoop: Mapreduce for incremental computations. In *Proceedings of the 2nd ACM Symposium on Cloud Computing* (Article No. 7, 14 pp.). Cascais, Portugal.
6. Bilal, K., Manzano, M., Khan, S. U., Calle, E., Li, K., & Zomaya, A. Y. (2013). On the characterization of the structural robustness of data center networks. *IEEE Transactions on Cloud Computing*, *1*(1), 64–77.
7. Bu, Y., Howe, B., Balazinska, M., & Ernst, M. D. (2010). Haloop: Efficient iterative data processing on large clusters. *Proceedings of the VLDB Endowment*, *3*(1), 285–296.
8. Chen, W., & Wassell, I. J. (2011). Energy efficient signal acquisition in wireless sensor networks: A compressive sensing framework. In *Proceedings of the 6th International Symposium on Wireless and Pervasive Computing* (pp. 1–6). Hong Kong, China.
9. Davis, D., Pilz, G., & Zhang, A. (Eds.). (2012). *Cloud Infrastructure Management Interface (CIMI) Primer*, DSP2027, v. 1.0.1. Distributed Management Task Force.
10. Dean, J., & Ghemawat, S. (2004). MapReduce: Simplified data processing on large clusters. In *Proceedings of the 6th Symposium on Operating System Design and Implementation* (pp. 137–150). San Francisco, CA.
11. Dean, J., & Ghemawat, S. (2008). MapReduce: Simplified data processing on large clusters. *Communications of the ACM*, *51*(1), 107–113.
12. Dinh, T. T. A., Liu, R., Zhang, M., Chen, G., Ooi, B. C., & Wang, J. (2018). Untangling blockchain: A data processing view of blockchain systems. *IEEE Transactions on Knowledge and Data Engineering*, *30*(7), 1366–1385.
13. Dong, Y., Yang, X., Li, X., Li, J., Tian, K., & Guan, H. (2010). High performance network virtualization with SR-IOV. In *Proceedings of the 16th International Conference on High-Performance Computer Architecture* (pp. 1–10). Bangalore, India.
14. Eagle, N., & Pentland, A. (2006). Reality mining: Sensing complex social systems. *Personal and Ubiquitous Computing*, *10*(4), 255–268.
15. Ekanayake, J., Pallickara, S., & Fox, G. (2008). MapReduce for data intensive scientific analyses. In *Proceedings of the IEEE 4th International Conference on eScience* (pp. 277–284). Indianapolis, IN.

16. Fiege, L., Gartner, F. C., Kasten, O., & Zeidler, A. (2003). Supporting mobility in content-based publish/subscribe middleware. In *Proceedings of the ACM/IFIP/USENIX 2003 International Conference on Middleware* (pp. 103–122). Alzburg, Austria.
17. Ganti, R., Ye, F., & Lei, H. (2011). Mobile crowdsensing: Current state and future challenges. *IEEE Communications Magazine*, *49*(11), 32–39.
18. Gulisano, V., Jimenez-Peris, R., Patino-Martinez, M., Soriente, C., & Valduriez, P. (2012). Streamcloud: An elastic and scalable data streaming system. *IEEE Transactions on Parallel and Distributed Systems*, *23*(12), 2351–2365.
19. Guo, B., Chen, C., Zhang, D., Yu, Z., & Chin, A. (2016). Mobile crowd sensing and computing: When participatory sensing meets participatory social media. *IEEE Communications Magazine*, *54*(2), 131–137.
20. Hacigumus, H., Iyer, B., Li, C., & Mehrotra, S. (2002). Executing SQL over encrypted data in the database-service-provider model. In *Proceedings of ACM SIGMOD International Conference on Management of Data* (pp. 216–227). Madison, WI.
21. Huang, T., Lan, L., Fang, X., An, P., Min, J., & Wang, F. (2015). Promises and challenges of big data computing in health sciences. *Big Data Research*, *2*(1), 2–11.
22. Ingersoll, G. (2009). *Introducing apache mahout: Scalable, commercial-friendly machine learning for building intelligent applications*. IBM Corporation.
23. Isard, M., Budiu, M., Yu, Y., Birrell, A., & Fetterly, D. (2007). Dryad: Distributed data-parallel programs from sequential building blocks. In *Proceedings of the 2nd ACM SIGOPS/EuroSys European Conference on Computer Systems* (pp. 59-72). Lisbon, Portugal.
24. Jayalath, C., Stephen, J., & Eugster, P. (2014). Universal cross-cloud communication. *IEEE Transactions on Cloud Computing*, *2*(2), 103–116.
25. Jin, H., Wang, X., Wu, S., Di, S., & Shi, X. (2015). Towards optimized fine-grained pricing of IaaS cloud platform. *IEEE Transactions on Cloud Computing*, *3*(4), 436–448.
26. Koponen, T., Casado, M., Gude, N., Stribling, J., Poutievski, L., Zhu, M., et al. (2010). Onix: A distributed control platform for large-scale production networks. In *Proceedings of the 9th USENIX Symposium on Operating Systems Design and Implementation* (pp. 1–6). Vancouver, Canada.
27. Kreutz, D., Ramos, F. M., & Verissimo, P. (2013). Towards secure and dependable software-defined networks. In *Proceedings of the 2nd ACM SIGCOMM Workshop on Hot Topics in Software Defined Networking* (pp. 55–60). Hong Kong, China.
28. Lin, J., Yu, W., Zhang, N., Yang, X., Zhang, H., & Zhao, W. (2017). A Survey on Internet of things: Architecture, enabling technologies, security and privacy, and applications. *IEEE Internet of Things Journal*, *4*(5), 1125–1142.
29. Malewicz, G., Austern, M. H., Bik, A. J., Dehnert, J. C., Horn, I., Leiser, N., et al. (2010). Pregel: A system for large-scale graph processing. In *Proceedings of ACM SIGMOD International Conference on Management of Data* (pp. 135–146). Indianapolis, IN.
30. Mashayekhy, L., Nejad, M. M., & Grosu, D. (2015). Cloud federations in the sky: Formation game and mechanism. *IEEE Transactions on Cloud Computing*, *3*(1), 14–27.
31. Melnik, S., Gubarev, A., Long, J., Romer, G., Shivakumar, S., Tolton, M., et al. (2010). Dremel: Interactive analysis of web-scale datasets. In *Proceedings of the 36th International Conference on Very Large Data Bases* (pp. 330–339).
32. Mitton, N., Papavassiliou, S., Puliafito, A., & Trivedi, K. S. (2012). Combining cloud and sensors in a smart city environment. *EURASIP Journal on Wireless Communications and Networking*, *2012*(247), 1–10.
33. Mont, M. C., McCorry, K., Papanikolaou, N., & Pearson, S. (2012). Security and privacy governance in cloud computing via SLAS and a policy orchestration service. In *Proceedings of the 2nd International Conference on Cloud Computing and Services Science* (pp. 670–674). Porto, Portugal.
34. Nakamoto, S. (2008). *Bitcoin: A peer-to-peer electronic cash system*. https://bitcoin.org/bitcoin.pdf.
35. Nurmi, D., Wolski, R., Grzegorczyk, C., Obertelli, G., Soman, S., Youseff, L., et al. (2009). The Eucalyptus open-source cloud computing system. *Proceedings of the 9th IEEE/ACM International Symposium on Cluster Computing and the Grid* (pp. 124–131). Shanghai, China.

36. Patel, M., Hu, Y., Hédé, P., Joubert, J., Thornton, C., Naughton, B., et al. (2014). *Mobile-edge computing—Introductory technical white paper*. White paper, Mobile-Edge Computing (MEC) Industry Initiative.
37. Rochwerger, B., Breitgand, D., Epstein, A., Hadas, D., Loy, I., Nagin, K., et al. (2011). Reservoir—When one cloud is not enough. *Computer*, *44*(3), 44–51.
38. Sandholm, T., & Lai, K. (2010). Dynamic proportional share scheduling in Hadoop. In *Proceedings of the 15th International Workshop on Job Scheduling Strategies for Parallel Processing, LNCS* (Vol. 6253, pp. 110–131). Atlanta, GA. Berlin: Springer.
39. Schad, J., Dittrich, J., & Quiane-Ruiz, J.-A. (2010). Runtime measurements in the cloud: Observing, analyzing, and reducing variance. *Proceedings of the VLDB Endowment*, *3*, 460–471.
40. Sempolinski, P., & Thain, D. (2010). A comparison and critique of Eucalyptus, OpenNebula and Nimbus. In *Proceedings of IEEE 2nd International Conference on Cloud Computing Technology and Science* (pp. 417–426). Indianapolis, IN.
41. Sotomayor, B., Montero, R. S., Llorente, I. M., & Foster, I. (2008). Capacity leasing in cloud systems using the OpenNebula engine. In *Proceedings of Workshop on Cloud Computing and its Applications*. Chicago, IL.
42. Vaquero, L. M., Celorio, A., Cuadrado, F., & Cuevas, R. (2015). Deploying large-scale datasets on-demand in the cloud: Treats and tricks on data distribution. *IEEE Transactions on Cloud Computing*, *3*(2), 132–144.
43. Xiang, L., Luo, J., & Rosenberg, C. (2013). Compressed data aggregation: Energy-efficient and high-fidelity data collection. *IEEE/ACM Transactions on Networking*, *21*(6), 1722–1735.
44. Xin, R. S., Rosen, J., Zaharia, M., Franklin, M. J., Shenker, S., & Stoica, I. (2013). Shark: SQL and rich analytics at scale. In *Proceedings of ACM SIGMOD International Conference on Management of Data* (pp. 13–24). New York.
45. Xiong, J., Liu, X., Yao, Z., Ma, J., Li, Q., Geng, K., et al. (2014). A secure data self-destructing scheme in cloud computing. *IEEE Transactions on Cloud Computing*, *2*(4), 448–458.
46. Yan, Z., Ding, W., Yu, X., Zhu, H., & Deng, R. H. (2016). Deduplication on encrypted big data in cloud. *IEEE Transactions on Big Data*, *2*(2), 138–150.
47. Yao, Y., Tai, J., Sheng, B., & Mi, N. (2015). LsPS: A job size-based scheduler for efficient task assignments in Hadoop. *IEEE Transactions on Cloud Computing*, *3*(4), 411–424.
48. Zaharia, M., Borthakur, D., Sarma, J. S., Elmeleegy, K., Shenker, S., & Stoica, I. (2009). *Job scheduling for multi-user mapreduce clusters*. Technical Report UCB/EECS-2009-55, University of California, Berkeley.
49. Zhang, Q., Zhani, M. F., Yang, Y., Boutaba, R., & Wong, B. (2015). PRISM: Fine-grained resource-aware scheduling for MapReduce. *IEEE Transactions on Cloud Computing*, *3*(2), 182–194.
50. Zhang, Y., Chen, S., Wang, Q., & Yu, G. (2015). i^2MapReduce: Incremental MapReduce for mining evolving big data. *IEEE Transactions on Knowledge and Data Engineering*, *27*(7), 1906–1919.
51. Zhang, Y., Gao, Q., Gao, L., & Wang, C. (2012). iMapReduce: A distributed computing framework for iterative computation. *Journal of Grid Computing*, *10*(1), 47–68.

Appendix A
Mathematical Preliminaries

In this appendix, mathematical preliminaries on linear algebra, stability theory, probability and stochastic processes, and optimization are introduced. Some classification measures are also defined.

A.1 Linear Algebra

Pseudoinverse

Definition A.1 (*Pseudoinverse*) Pseudoinverse $\mathbf{A}^{\dagger}$, also called *Moore–Penrose generalized inverse*, of a matrix $\mathbf{A} \in R^{m \times n}$ is unique, which satisfies

$$\mathbf{A}\mathbf{A}^{\dagger}\mathbf{A} = \mathbf{A}, \tag{A.1}$$

$$\mathbf{A}^{\dagger}\mathbf{A}\mathbf{A}^{\dagger} = \mathbf{A}^{\dagger}, \tag{A.2}$$

$$\left(\mathbf{A}\mathbf{A}^{\dagger}\right)^{T} = \mathbf{A}\mathbf{A}^{\dagger}, \tag{A.3}$$

$$\left(\mathbf{A}^{\dagger}\mathbf{A}\right)^{T} = \mathbf{A}^{\dagger}\mathbf{A}. \tag{A.4}$$

$\mathbf{A}^{\dagger}$ can be calculated by

$$\mathbf{A}^{\dagger} = \left(\mathbf{A}^{T}\mathbf{A}\right)^{-1}\mathbf{A}^{T} \tag{A.5}$$

if $\mathbf{A}^{T}\mathbf{A}$ is nonsingular, and

$$\mathbf{A}^{\dagger} = \mathbf{A}^{T}\left(\mathbf{A}\mathbf{A}^{T}\right)^{-1} \tag{A.6}$$

if $\mathbf{A}\mathbf{A}^{T}$ is nonsingular. Pseudoinverse is directly associated with the linear LS problem.

K.-L. Du and M. N. S. Swamy, *Neural Networks and Statistical Learning*,
https://doi.org/10.1007/978-1-4471-7452-3

When $\mathbf{A}$ is a square nonsingular matrix, pseudoinverse $\mathbf{A}^{\dagger}$ reduces to its inverse $\mathbf{A}^{-1}$. For a scalar α, $\alpha^{\dagger} = \alpha^{-1}$ for $\alpha \neq 0$, and $\alpha^{\dagger} = 0$ for $\alpha = 0$.

For $n \times n$ identity matrix $\mathbf{I}$ and $n \times n$ singular matrix $\mathbf{J}$, namely, $\det(\mathbf{J}) = 0$, for $a \neq 0$ and $a + nb \neq 0$, we have [14]

$$(a\mathbf{I} + b\mathbf{J})^{-1} = \frac{1}{a}\left(\mathbf{I} - \frac{b}{a+nb}\mathbf{J}\right). \tag{A.7}$$

Linear Least Squares Problems

The linear LS or L_2-norm problem is basic to many signal processing techniques. It tries to solve a set of linear equations, written in matrix form

$$\mathbf{A}\boldsymbol{x} = \boldsymbol{b}, \tag{A.8}$$

where $\mathbf{A} \in R^{m\times n}$, $\boldsymbol{x} \in R^n$, and $\boldsymbol{b} \in R^m$.

This problem can be converted into the minimization of the squared error function

$$E(\boldsymbol{x}) = \frac{1}{2}\|\mathbf{A}\boldsymbol{x} - \boldsymbol{b}\|_2^2 = \frac{1}{2}(\mathbf{A}\boldsymbol{x} - \boldsymbol{b})^T(\mathbf{A}\boldsymbol{x} - \boldsymbol{b}). \tag{A.9}$$

The solution corresponds to one of the following three situations [7]:

- $\operatorname{rank}(\mathbf{A}) = n = m$. We get a unique exact solution

$$\boldsymbol{x}^* = \mathbf{A}^{-1}\boldsymbol{b} \tag{A.10}$$

  and $E(\boldsymbol{x}^*) = 0$.
- $\operatorname{rank}(\mathbf{A}) = n < m$. The system is overdetermined and has no exact solution. There is a unique solution in the least squares error sense

$$\boldsymbol{x}^* = \mathbf{A}^{\dagger}\boldsymbol{b}, \tag{A.11}$$

  where $\mathbf{A}^{\dagger} = \left(\mathbf{A}^T\mathbf{A}\right)^{-1}\mathbf{A}^T$. In this case,

$$E\left(\boldsymbol{x}^*\right) = \boldsymbol{b}^T\left(\mathbf{I} - \mathbf{A}\mathbf{A}^{\dagger}\right)\boldsymbol{b} \geq 0. \tag{A.12}$$

- $\operatorname{rank}(\mathbf{A}) = m < n$. The system is underdetermined, and the solution is not unique. But the solution with the minimum L_2-norm $\|\boldsymbol{x}\|_2^2$ is unique

$$\boldsymbol{x}^* = \mathbf{A}^{\dagger}\boldsymbol{b}. \tag{A.13}$$

  Here $\mathbf{A}^{\dagger} = \mathbf{A}^T\left(\mathbf{A}\mathbf{A}^T\right)^{-1}$. We have $E(\boldsymbol{x}^*) = 0$ and $\|\boldsymbol{x}^*\|_2^2 = \boldsymbol{b}^T\left(\mathbf{A}\mathbf{A}^T\right)^{-1}\boldsymbol{b}$.

Vector Norms

Definition A.2 (*Vector Norms*) A norm acts as a measure of distance. A vector norm on R^n is a mapping $f : R^n \to R$ that satisfies such properties: For any $\boldsymbol{x}, \boldsymbol{y} \in R^n, a \in R$,

- $f(\boldsymbol{x}) \geq 0$, and $f(\boldsymbol{x}) = 0$ iff $\boldsymbol{x} = 0$.
- $f(\boldsymbol{x} + \boldsymbol{y}) \leq f(\boldsymbol{x}) + f(\boldsymbol{y})$.
- $f(a\boldsymbol{x}) = |a| f(\boldsymbol{x})$.

The mapping is denoted as $f(\boldsymbol{x}) = \|\boldsymbol{x}\|$.

The p-norm or L_p-norm is a popular class of vector norms

$$\|\boldsymbol{x}\|_p = \left(\sum_{i=1}^{n} |x_i|^p \right)^{\frac{1}{p}} \tag{A.14}$$

with $p \geq 1$. The L_1, L_2, and L_∞ norms are more useful:

$$\|\boldsymbol{x}\|_1 = \sum_{i=1}^{n} |x_i| , \tag{A.15}$$

$$\|\boldsymbol{x}\|_2 = \sum_{i=1}^{n} \left(x_i^2\right)^{\frac{1}{2}} = \left(\boldsymbol{x}^T \boldsymbol{x}\right)^{\frac{1}{2}} , \tag{A.16}$$

$$\|\boldsymbol{x}\|_\infty = \max_{1 \leq i \leq n} |x_i| . \tag{A.17}$$

The L_2-norm is the popular Euclidean norm.

A matrix $\mathbf{Q} \in R^{m \times m}$ is called an *orthogonal matrix* or *unitary matrix* if $\mathbf{Q}^T \mathbf{Q} = \mathbf{I}$. The L_2-norm is invariant under orthogonal transforms, that is, for all orthogonal $\mathbf{Q}$ of appropriate dimensions

$$\|\mathbf{Q}\boldsymbol{x}\|_2 = \|\boldsymbol{x}\|_2. \tag{A.18}$$

Matrix Norms

A matrix norm is a generalization of the vector norm by extending from R^n to $R^{m \times n}$. For a matrix $\mathbf{A} = \left[a_{ij}\right]_{m \times n}$, the most frequently used matrix norms are the Frobenius norm

$$\|\mathbf{A}\|_F = \left(\sum_{i=1}^{m} \sum_{j=1}^{n} a_{ij}^2 \right)^{\frac{1}{2}} , \tag{A.19}$$

and the matrix p-norm

$$\|\mathbf{A}\|_p = \sup_{x \neq 0} \frac{\|\mathbf{A}\boldsymbol{x}\|_p}{\|\boldsymbol{x}\|_p} = \max_{\|x\|_p = 1} \|\mathbf{A}\boldsymbol{x}\|_p, \tag{A.20}$$

where sup is the supreme operation.

The matrix 2-norm and the Frobenius norm are invariant with respect to orthogonal transforms, that is, for all orthogonal $\mathbf{Q}_1$ and $\mathbf{Q}_2$ of appropriate dimensions

$$\|\mathbf{Q}_1\mathbf{A}\mathbf{Q}_2\|_F = \|\mathbf{A}\|_F, \tag{A.21}$$

$$\|\mathbf{Q}_1\mathbf{A}\mathbf{Q}_2\|_2 = \|\mathbf{A}\|_2. \tag{A.22}$$

Eigenvalue Decomposition

Definition A.3 (*Eigenvalue Decomposition*) Given a square matrix $\mathbf{A} \in R^{n \times n}$, if there exists a scalar λ and a nonzero vector $\boldsymbol{v}$ such that

$$\mathbf{A}\boldsymbol{v} = \lambda\boldsymbol{v}, \tag{A.23}$$

then λ and $\boldsymbol{v}$ are, respectively, called an *eigenvalue* of $\mathbf{A}$ and its corresponding *eigenvector*. All the eigenvalues $\lambda_i, i = 1, \ldots, n$, can be obtained by solving the characteristic equation

$$\det(\mathbf{A} - \lambda\mathbf{I}) = 0, \tag{A.24}$$

where $\mathbf{I}$ is an $n \times n$ identity matrix. The set of all the eigenvalues is called the *spectrum* of $\mathbf{A}$.

If $\mathbf{A}$ is nonsingular, $\lambda_i \neq 0$. If $\mathbf{A}$ is symmetric, then all λ_i are real. The maximum and minimum eigenvalues satisfy the Rayleigh quotient

$$\lambda_{\max}(\mathbf{A}) = \max_{v \neq 0} \frac{\boldsymbol{v}^T\mathbf{A}\boldsymbol{v}}{\boldsymbol{v}^T\boldsymbol{v}}, \qquad \lambda_{\min}(\mathbf{A}) = \min_{v \neq 0} \frac{\boldsymbol{v}^T\mathbf{A}\boldsymbol{v}}{\boldsymbol{v}^T\boldsymbol{v}}. \tag{A.25}$$

The trace of a matrix is equal to the sum of all its eigenvalues and the determinant of a matrix is equal to the product of its eigenvalues

$$\mathrm{tr}(\mathbf{A}) = \sum_{i=1}^{n} \lambda_i, \tag{A.26}$$

$$|\mathbf{A}| = \prod_{i=1}^{n} \lambda_i. \tag{A.27}$$

Singular Value Decomposition

Definition A.4 (*Singular Value Decomposition*) For a matrix $\mathbf{A} \in R^{m \times n}$, there exist real unitary matrices $\mathbf{U} = [\boldsymbol{u}_1, \boldsymbol{u}_2, \ldots, \boldsymbol{u}_m] \in R^{m \times m}$ and $\mathbf{V} = [\boldsymbol{v}_1, \boldsymbol{v}_2, \ldots, \boldsymbol{v}_n] \in R^{n \times n}$ such that

$$\mathbf{U}^T \mathbf{A} \mathbf{V} = \boldsymbol{\Sigma}, \tag{A.28}$$

where $\boldsymbol{\Sigma} \in R^{m \times n}$ is a real pseudodiagonal $m \times n$ matrix with σ_i, $i = 1, \ldots, p$, $p = \min(m, n)$, $\sigma_1 \geq \sigma_2 \geq \ldots \geq \sigma_p \geq 0$, on the diagonal and zeros off the diagonal. σ_i's are called the *singular values* of $\mathbf{A}$, and $\boldsymbol{u}_i$ and $\boldsymbol{v}_i$ are, respectively, called the *left singular vector* and *right singular vector* for σ_i. They satisfy the relations

$$\mathbf{A}\boldsymbol{v}_i = \sigma_i \boldsymbol{u}_i, \qquad \mathbf{A}^T \boldsymbol{u}_i = \sigma_i \boldsymbol{v}_i. \tag{A.29}$$

Accordingly, $\mathbf{A}$ can be written as

$$\mathbf{A} = \mathbf{U} \Sigma \mathbf{V}^T = \sum_{i=1}^{r} \lambda_i \boldsymbol{u}_i \boldsymbol{v}_i^T, \tag{A.30}$$

where r is the cardinality of the smallest nonzero singular value. In the special case when $\mathbf{A}$ is a symmetric nonnegative definite matrix, $\boldsymbol{\Sigma} = \operatorname{diag}\left(\lambda_1^{\frac{1}{2}}, \ldots, \lambda_p^{\frac{1}{2}}\right)$, where $\lambda_1 \geq \lambda_2 \geq \ldots \lambda_p \geq 0$ are the real eigenvalues of $\mathbf{A}$, $\boldsymbol{v}_i$ being the corresponding eigenvectors.

SVD is useful in many situations. The rank of $\mathbf{A}$ can be determined by the number of nonzero singular values. The power of $\mathbf{A}$ can be easily calculated by

$$\mathbf{A}^k = \mathbf{U} \boldsymbol{\Sigma}^k \mathbf{V}^T, \tag{A.31}$$

where k is a positive integer. SVD is extensively applied in linear inverse problems. The pseudoinverse of $\mathbf{A}$ can then be described by

$$\mathbf{A}^\dagger = \mathbf{V}_r \boldsymbol{\Sigma}_r^{-1} \mathbf{U}_r^T, \tag{A.32}$$

where $\mathbf{V}_r$, $\boldsymbol{\Sigma}_r$, and $\mathbf{U}_r$ are the matrix partitions corresponding to the r nonzero singular values.

The Frobenius norm can thus be calculated as

$$\|\mathbf{A}\|_F = \left(\sum_{i=1}^{p} \sigma_i^2 \right)^{\frac{1}{2}}, \tag{A.33}$$

and the matrix 2-norm is calculated by

$$\|\mathbf{A}\|_2 = \sigma_1. \tag{A.34}$$

SVD requires a time complexity of $O(mn \min\{m, n\})$ for a dense $m \times n$ matrix. Common methods for computing the SVD of a matrix are standard eigensolvers such as QR iteration and Arnoldi/Lanczos iteration.

QR Decomposition

For the full-rank or overdetermined linear LS case, $m \geq n$, (A.8) can also be solved by using QR decomposition procedure.

A is first factorized as

$$\mathbf{A} = \mathbf{QR}, \tag{A.35}$$

where $\mathbf{Q}$ is an $m \times m$ orthogonal matrix, that is, $\mathbf{Q}^T\mathbf{Q} = \mathbf{I}$, and $\mathbf{R} = \begin{bmatrix} \overline{\mathbf{R}} \\ \mathbf{0} \end{bmatrix}$ is an $m \times n$ upper triangular matrix with $\overline{\mathbf{R}} \in R^{n \times n}$.

Inserting (A.35) into (A.8) and premultiplying by $\mathbf{Q}^T$, we have

$$\mathbf{R}\boldsymbol{x} = \mathbf{Q}^T\boldsymbol{b}. \tag{A.36}$$

Denoting $\mathbf{Q}^T\boldsymbol{b} = \begin{bmatrix} \overline{\boldsymbol{b}} \\ \widetilde{\boldsymbol{b}} \end{bmatrix}$, where $\overline{\boldsymbol{b}} \in R^n$ and $\widetilde{\boldsymbol{b}} \in R^{m-n}$, we have

$$\overline{\mathbf{R}}\boldsymbol{x} = \overline{\boldsymbol{b}}. \tag{A.37}$$

Since $\overline{\mathbf{R}}$ is a triangular matrix, $\boldsymbol{x}$ can be easily solved using backward substitution. This is the procedure used in the GSO procedure.

When $\mathrm{rank}(\mathbf{A}) < n$, the rank-deficient LS problem has an infinite number of solutions, QR decomposition does not necessarily produce an orthonormal basis for $\mathrm{range}(\mathbf{A}) = \{\boldsymbol{y} \in R^m : \boldsymbol{y} = \mathbf{A}\boldsymbol{x} \text{ for some } \boldsymbol{x} \in R^n\}$. QR-cp can be applied to produce an orthonormal basis for $\mathrm{range}(\mathbf{A})$.

As a basic method for computing SVD, QR decomposition itself can be computed by means of the Givens rotation, the Householder transform, or GSO.

Condition Numbers

Definition A.5 (*Condition Number*) The condition number of a matrix $\mathbf{A} \in R^{m \times n}$ is defined by

$$\mathrm{cond}_p(\mathbf{A}) = \|\mathbf{A}\|_p \left\|\mathbf{A}^{\dagger}\right\|_p, \tag{A.38}$$

where p can be selected as 1, 2, ∞, Frobenius, or any other norm.

The relation, $\mathrm{cond}(\mathbf{A}) \geq 1$, always holds. Matrices with small condition numbers are well conditioned, while matrices with large condition number are poorly conditioned or ill-conditioned. The condition number is especially useful in numerical computation, where ill-conditioned matrices are sensitive to rounding errors.

For the L_2-norm,

$$\mathrm{cond}_2(\mathbf{A}) = \frac{\sigma_1}{\sigma_p}, \tag{A.39}$$

where $p = \min(m, n)$.

Householder Reflections and Givens Rotations

Orthogonal transforms play an important role in the matrix computation such as EVD, SVD, and QR decomposition. The Householder reflection, also termed the *Householder transform*, and Givens rotations, also called the *Givens transform*, are two basic operations in the orthogonalization process. These operations are easily constructed, and they introduce zeros in a vector so as to simplify matrix computations. The Householder reflection is exceedingly efficient for annihilating all but the first entry of a vector, while the Givens rotation is more effective to transform a specified entry of a vector into zero.

Let $\boldsymbol{v} \in R^n$ be nonzero. The Householder reflection is defined as a rank-one modification to the identity matrix

$$\mathbf{P} = \mathbf{I} - 2\frac{\boldsymbol{v}\boldsymbol{v}^T}{\boldsymbol{v}^T\boldsymbol{v}}. \tag{A.40}$$

The Householder matrix $\mathbf{P} \in R^{n\times n}$ is symmetric and orthogonal. $\boldsymbol{v}$ is called a *Householder vector*. The Householder transform of a matrix $\mathbf{A}$ is given by $\mathbf{PA}$. By specifying the form of the transformed matrix, one can find a suitable Householder vector $\boldsymbol{v}$. For example, one can define a Householder vector as $\boldsymbol{v} = \boldsymbol{u} - \alpha\boldsymbol{e}_1$, where $\boldsymbol{u} \in R^m$ is an arbitrary vector of length $|\alpha|$ and $\boldsymbol{e}_1 \in R^m$, wherein only the first entry is unity, all the other entries being zero. In this case, $\mathbf{P}\boldsymbol{x}$ becomes a vector with only the first entry nonzero, where $\boldsymbol{x} \in R^n$ is a nonzero vector.

The Givens rotation $\mathbf{G}(i, k, \theta)$ is a rank-two correction to the identity matrix $\mathbf{I}$. It modifies $\mathbf{I}$ by setting the (i, i)th entry as $\cos\theta$, the (i, k)th entry as $\sin\theta$, the (k, i)th entry as $-\sin\theta$, and the (k, k)th entry as $\cos\theta$. The Givens transform $\mathbf{G}(i, k, \theta)\boldsymbol{x}$ applies a counterwise rotation of θ radians in the (i, k) coordinate plane. One can specify an entry in a vector to zero by applying the Givens rotation and then calculate the rotation angle θ.

Matrix Inversion Lemma

The matrix inversion lemma is also called the *Sherman–Morrison–Woodbury formula*. It is useful in deriving many iterative algorithms. Assume that the relationship between the matrix $\mathbf{A} \in R^{n\times n}$ at iterations t and $t + 1$ is given as

$$\mathbf{A}(t+1) = \mathbf{A}(t) + \Delta\mathbf{A}(t). \tag{A.41}$$

If $\Delta\mathbf{A}(t)$ can be expressed as $\mathbf{U}\mathbf{V}^T$, where $\mathbf{U} \in R^{n\times m}$ and $\mathbf{V} \in R^{m\times n}$, it is referred to as a *rank-m update*. The matrix inversion lemma gives [7]

$$\begin{aligned}\mathbf{A}^{-1}(t+1) &= \mathbf{A}^{-1}(t) - \Delta\mathbf{A}^{-1}(t) \\ &= \mathbf{A}^{-1}(t) - \mathbf{A}^{-1}(t)\mathbf{U}\left(\mathbf{I} + \mathbf{V}^T\mathbf{A}^{-1}(t)\mathbf{U}\right)^{-1}\mathbf{V}^T\mathbf{A}^{-1}(t),\end{aligned} \tag{A.42}$$

where both $\mathbf{A}(t)$ and $\left(\mathbf{I} + \mathbf{V}^T\mathbf{A}^{-1}(t)\mathbf{U}\right)$ are assumed to be nonsingular. Thus, a rank-m correction to a matrix results in a rank-m correction to its inverse.

Some modifications to the formula are available, and one popular update is given here. If $\mathbf{A}$ and $\mathbf{B}$ are two positive-definite matrices, which have the relation

$$\mathbf{A} = \mathbf{B}^{-1} + \mathbf{C}\mathbf{D}\mathbf{C}^T, \tag{A.43}$$

where $\mathbf{C}$ and $\mathbf{D}$ are also matrices. The matrix inversion lemma gives the inverse of $\mathbf{A}$ as

$$\mathbf{A}^{-1} = \mathbf{B} - \mathbf{B}\mathbf{C}(\mathbf{D} + \mathbf{C}^T\mathbf{B}\mathbf{C})^{-1}\mathbf{C}^T\mathbf{B}. \tag{A.44}$$

Partial Least Squares Regression

Partial LS regression [17] is a statistical method for modeling a linear relationship between two datasets $\mathcal{X}$ and $\mathcal{Y}$. It finds projection vectors by maximizing the linear association between two latent components which are the projection of two deflation datasets.

It is a robust, iterative method that avoids matrix inversion for underconstrained datasets by decomposing the multivariate regression problem into successive univariate regressions. Partial LS iteratively chooses its projection directions according to the direction of maximum correlation between the (current residual) input and the output. Computation of each projection direction is $O(d)$ for d dimensions of the data. Successive iterations create orthogonal projection directions by removing the subspace of the input data used in the previous projection. The number of projection directions found by partial LS is bound only by the dimensionality of the data, with each univariate regression on successive projection components further reducing the residual error. Using all d projections leads to ordinary LS regression. If the distribution of the input data is spherical, then partial LS requires only a single projection to optimally reconstruct the output. Partial LS in statistics is equivalent to the CG method [11].

A.2 Data Preprocessing

Linear Scaling and Data Whitening

By linear normalization, all the raw data can be brought in the vicinity of an average value. For a one-dimensional dataset, $\{x_i | i = 1, \ldots, N\}$, the mean $\hat{\mu}$ and variance $\hat{\sigma}^2$ are estimated by

$$\hat{\mu} = \frac{1}{N} \sum_{i=1}^{N} x_i, \tag{A.45}$$

$$\hat{\sigma}^2 = \frac{1}{N-1} \sum_{i=1}^{N} (x_i - \hat{\mu})^2 . \tag{A.46}$$

The transformed data are now defined by

$$\widetilde{x}_i = \frac{x_i - \hat{\mu}}{\hat{\sigma}}. \tag{A.47}$$

The transformed dataset $\{\widetilde{x}_i | i = 1, \ldots, N\}$ has zero mean and unit standard deviation.

When the raw dataset $\{\boldsymbol{x}_i | i = 1, \ldots, N\}$ is composed of vectors, accordingly, the mean vector $\boldsymbol{\mu}$ and covariance matrix $\boldsymbol{\Sigma}$ are calculated by

$$\hat{\boldsymbol{\mu}} = \frac{1}{N} \sum_{i=1}^{N} \boldsymbol{x}_i, \tag{A.48}$$

$$\widehat{\boldsymbol{\Sigma}} = \frac{1}{N-1} \sum_{i=1}^{N} (\boldsymbol{x}_i - \hat{\boldsymbol{\mu}}) (\boldsymbol{x}_i - \hat{\boldsymbol{\mu}})^T . \tag{A.49}$$

Equations (A.46) and (A.49) are, respectively, the unbiased estimates of the variance and the covariance matrix. When the factor $\frac{1}{N-1}$ is replaced by $\frac{1}{N}$, the estimates for $\boldsymbol{\mu}$ and $\boldsymbol{\Sigma}$ are the ML estimates. The ML estimates for variance and covariance are biased.

New input vectors can be defined by the linear transformation

$$\widetilde{\boldsymbol{x}}_i = \Lambda^{-\frac{1}{2}} \mathbf{U}^T (\boldsymbol{x}_i - \hat{\boldsymbol{\mu}}), \tag{A.50}$$

where $\mathbf{U} = [\boldsymbol{u}_1, \ldots, \boldsymbol{u}_M]$, $\Lambda = \mathrm{diag}(\lambda_1, \ldots, \lambda_M)$, M is the dimension of data vectors, and λ_i and $\boldsymbol{u}_i$ are the eigenvalues and the corresponding eigenvectors of $\boldsymbol{\Sigma}$, which satisfy

$$\boldsymbol{\Sigma} \boldsymbol{u}_i = \lambda_i \boldsymbol{u}_i. \tag{A.51}$$

The new dataset $\{\widetilde{\boldsymbol{x}}_i\}$ has zero mean, and its covariance matrix is the identity matrix [1]. The above process is also called *data whitening*.

Gram–Schmidt Orthonormalization Transform

Ill-conditioning is usually measured for a data matrix $\mathbf{A}$ by its condition number ρ, defined as $\rho(\mathbf{A}) = \frac{\sigma_{\max}}{\sigma_{\min}}$, where $\sigma_{\max}$ and $\sigma_{\min}$ are, respectively, the maximum and minimum singular values of $\mathbf{A}$. In the batch LS algorithm, the information matrix $\mathbf{A}^T\mathbf{A}$ needs to be manipulated. Since $\rho\left(\mathbf{A}^T\mathbf{A}\right) = \rho(\mathbf{A})^2$, the effect of ill-conditioning on parameter estimation will be more severe. Orthogonal decomposition is a well-known technique to eliminate ill-conditioning.

The GSO procedure starts with QR decomposition of the full feature matrix. Denote

$$\mathbf{X} = [\boldsymbol{x}_1, \boldsymbol{x}_2, \ldots, \boldsymbol{x}_N], \tag{A.52}$$

where the ith pattern $\boldsymbol{x}_i = \left(x_{i,1}, x_{i,2}, \ldots, x_{i,J}\right)^T$, $x_{i,j}$ denotes the jth component of $\boldsymbol{x}_i$, and J is the dimensions of the raw data. We then represent $\mathbf{X}^T$ by

$$\mathbf{X}^T = \left[\boldsymbol{x}^1, \boldsymbol{x}^2, \ldots, \boldsymbol{x}^J\right], \tag{A.53}$$

where $\boldsymbol{x}^j = \left(x_{1,j}, x_{2,j}, \ldots, x_{N,j}\right)^T$.

QR decomposition is performed on $\mathbf{X}^T$

$$\mathbf{X}^T = \mathbf{QR}, \tag{A.54}$$

where $\mathbf{Q}$ is an orthonormal matrix, that is, $\mathbf{Q}^T\mathbf{Q} = \mathbf{I}_J$, $\mathbf{Q} = \left[\boldsymbol{q}_1, \boldsymbol{q}_2, \ldots, \boldsymbol{q}_J\right]$, $\boldsymbol{q}_i = \left(q_{i,1}, q_{i,2}, \ldots, q_{i,N}\right)^T$, $q_{i,j}$ denoting the jth component of $\boldsymbol{q}_i$, and $\mathbf{R}$ is an upper triangular matrix. QR decomposition can be performed by the Householder transform or Givens rotation [7], which is suitable for hardware implementation.

The GSO procedure is given as

$$\boldsymbol{q}_1 = \boldsymbol{x}^1, \tag{A.55}$$

$$\boldsymbol{q}_k = \boldsymbol{x}^k - \sum_{i=1}^{k-1} \alpha_{ik}\boldsymbol{q}_i, \tag{A.56}$$

$$\alpha_{ik} = \begin{cases} \frac{(\boldsymbol{x}^k)^T\boldsymbol{q}_i}{\boldsymbol{q}_i^T\boldsymbol{q}_i}, & \text{for} \quad i = 1, 2, \ldots, k-1 \\ 1, & \text{for} \quad i = k \\ 0, & \text{for} \quad i > k \end{cases}. \tag{A.57}$$

Thus $\boldsymbol{q}_k$ is a linear combination of $\boldsymbol{x}^1, \ldots, \boldsymbol{x}^k$, and the Gram–Schmidt features $\boldsymbol{q}_1, \ldots, \boldsymbol{q}_k$ and the vectors $\boldsymbol{x}^1, \ldots, \boldsymbol{x}^k$ are one-to-one mappings, for $1 \le k \le J$. GSO transform can be used for feature subset selection; it inherits the compactness of the orthogonal representation and at the same time provides features retaining their original meaning.

A.3 Stability of Dynamic Systems

For a dynamic system described by a set of ordinary differential equations, the stability of the system can be examined by Lyapunov's second theorem or the Lipschitz condition.

Lyapunov theorem is a sufficient but not a necessary tool for proving the stability of an equilibrium of a dynamic system. The method is dependent on finding a Lyapunov function for the equilibrium. It is especially important for analyzing the stability of recurrent networks and ordinary differential equations.

Theorem A.1 (Lyapunov Theorem) *Consider a function $L(\boldsymbol{x})$. Define a region Ω, where any point $\boldsymbol{x} \in \Omega$ satisfies $L(\boldsymbol{x}) < c$ for a constant c, with the boundary of Ω given by $L(\boldsymbol{x}) = c$, such that*

- $\frac{dL(x)}{dt} < 0, \forall \boldsymbol{x}, \boldsymbol{x}^* \in \Omega, \boldsymbol{x} \neq \boldsymbol{x}^*$.
- $\frac{dL(x^*)}{dt} = 0$.

Then, the equilibrium point $\boldsymbol{x} = \boldsymbol{x}^*$ is asymptotically stable, with a domain of attraction Ω.

Theorem A.2 (Lyapunov's Second Theorem) *For a dynamic system described by a set of differential equations*

$$\frac{d\boldsymbol{x}}{dt} = \boldsymbol{f}(\boldsymbol{x}), \tag{A.58}$$

where $\boldsymbol{x} = (x_1(t), x_2(t), \ldots, x_n(t))^T$ and $\boldsymbol{f} = (f_1, f_2, \ldots, f_n)^T$, if there exists a positive-definite function $E = E(\boldsymbol{x})$, called a Lyapunov function *or* energy function, *such that*

$$\frac{dE}{dt} = \sum_{j=1}^{n} \frac{\partial E}{\partial x_j} \frac{dx_j}{dt} \leq 0 \tag{A.59}$$

with $\frac{dE}{dt} = 0$ only for $\frac{dx}{dt} = \boldsymbol{0}$, then the system is stable, and the trajectories $\boldsymbol{x}$ will asymptotically converge to stationary points as $t \to \infty$.

The stationary points are also known as *equilibrium points* and *attractors*. The crucial step in applying the Lyapunov's second theorem is to find a suitable energy function.

Theorem A.3 (Lipschitz Condition) *For a dynamic system described by (A.58), a sufficient condition that guarantees the existence and uniqueness of the solution is given by the Lipschitz condition*

$$\|\boldsymbol{f}(\boldsymbol{x}_1) - \boldsymbol{f}(\boldsymbol{x}_2)\| \leq \gamma \|\boldsymbol{x}_1 - \boldsymbol{x}_2\|, \tag{A.60}$$

where γ is any positive constant, called Lipschitz constant, *and $\boldsymbol{x}_1$, $\boldsymbol{x}_2$ are any two variables in the domain of the function vector $\boldsymbol{f}$. $\boldsymbol{f}(\boldsymbol{x})$ is said to be Lipschitz continuous.*

If $\boldsymbol{x}_1$ and $\boldsymbol{x}_2$ are in some neighborhood of $\boldsymbol{x}$, then they are said to satisfy the Lipschitz condition locally and will reach a unique solution in the neighborhood of $\boldsymbol{x}$. The unique solution is a trajectory that will converge to an attractor asymptotically and reach it only at $t \to \infty$.

A.4 Probability Theory and Stochastic Processes

Conditional Probability

For two statements (or propositions) A and B, one writes $A|B$ to denote the situation that A is true subject to the condition that B is true. The probability of $A|B$, called *conditional probability*, is denoted by $P(A|B)$. This gives a measure for the plausibility of the statement $A|B$.

Gaussian Distribution

The Gaussian distribution, known as the *normal distribution*, is the most common assumption for error distribution. The pdf of the normal distribution is defined as

$$p(x) = \frac{1}{\sigma\sqrt{2\pi}} \mathrm{e}^{-\frac{(x-\mu)^2}{2\sigma^2}}, \quad x \in R, \tag{A.61}$$

where μ is the mean and $\sigma > 0$ is the standard deviation. For the Gaussian distribution, 99.73 % of the data are within the range of $[\mu - 3\sigma, \mu + 3\sigma]$. The Gaussian distribution has its first-order moment as μ, second-order moment as σ^2, and higher order moments as zero. If $\mu = 0$ and $\sigma = 1$, the distribution is called the *standard normal distribution*. The pdf is also known as the *likelihood function*. An ML estimator is a set of values (μ, σ) that maximizes the likelihood function for a fixed value of x.

The cumulative distribution function (cdf) is defined as the probability that a random variable is less than or equal to a value x, that is,

$$F(x) = \int_{-\infty}^{x} p(t)dt. \tag{A.62}$$

The standard normal cdf, conventionally denoted Φ, is given by setting $\mu = 0$ and $\sigma = 1$. The standard normal cdf is usually expressed by

$$\Phi(x) = \frac{1}{2}\left[1 + \mathrm{erf}\left(\frac{x}{\sqrt{2}}\right)\right], \tag{A.63}$$

where the error function $\mathrm{erf}(x)$ is a nonelementary function, which is defined by

$$\mathrm{erf}(x) = \frac{2}{\sqrt{\pi}} \int_0^x \mathrm{e}^{-t^2} dt. \tag{A.64}$$

When vector $\boldsymbol{x} \in R^n$, the pdf of the normal distribution is then defined by

$$p(\boldsymbol{x}) = \frac{1}{(2\pi)^{\frac{n}{2}} |\boldsymbol{\Sigma}|} \mathrm{e}^{-\frac{1}{2}(\boldsymbol{x}-\boldsymbol{\mu})^T \boldsymbol{\Sigma}^{-1} (\boldsymbol{x}-\boldsymbol{\mu})}, \tag{A.65}$$

where $\boldsymbol{\mu}$ is the mean vector and $\boldsymbol{\Sigma}$ is the covariance matrix.

The Gaussian distribution is only one of the canonical exponential distributions, and it is suitable for describing real-value data. In the case of binary-valued, integer-valued, or nonnegative data, the Gaussian assumption is inappropriate, and a family of exponential distributions can be used. For example, Poisson's distribution is better suited for integer data and the Bernoulli distribution to binary data, and an exponential distribution to nonnegative data.

Cauchy Distribution

The Cauchy distribution, also known as the *Cauchy–Lorentzian distribution*, is another popular data distribution model. The density of the Cauchy distribution is defined as

$$p(x) = \frac{1}{\pi\sigma\left[1 + \left(\frac{x-\mu}{\sigma}\right)^2\right]}, \quad x \in R, \tag{A.66}$$

where μ specifies the location of the peak and σ specifies the half-width at the half-maximum. When $\mu = 0$ and $\sigma = 1$, the distribution is called the *standard Cauchy distribution*.

Accordingly, the cdf of the Cauchy distribution is calculated by

$$F(x) = \frac{1}{\pi} \arctan\left(\frac{x - \mu}{\sigma}\right) + \frac{1}{2}. \tag{A.67}$$

None of the moments is defined for the Cauchy distribution. The median of the distribution is equal to μ. Compared to the Gaussian distribution, the Cauchy distribution has a longer tail; this makes it more valuable in stochastic search algorithms by searching larger subspaces in the data space.

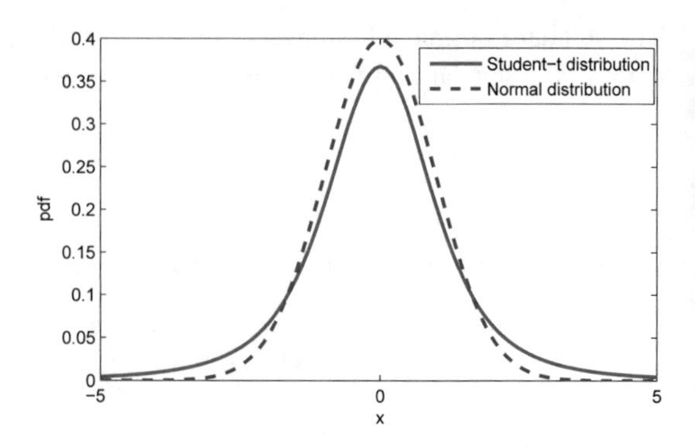

Fig. A.1 The Student-t distribution with $\nu = 4$ and standard normal distribution

Student-*t* Models

The Student-t pdf is given by

$$p(x) = \frac{\Gamma\left(\frac{\nu+1}{2}\right)}{\sqrt{\pi\nu}\,\Gamma\left(\frac{\nu}{2}\right)} \frac{1}{\left(1 + \frac{x^2}{\nu}\right)^{\frac{\nu+1}{2}}}, \tag{A.68}$$

where $\Gamma(\cdot)$ is the Gamma function, and ν is the degrees of freedom. The Gaussian distribution is a particular t distribution with $\nu = \infty$.

For a random sample of size n from a normal distribution with mean μ, we get the statistic

$$t = \frac{\overline{x} - \mu}{\sigma/\sqrt{n}}, \tag{A.69}$$

where $\overline{x}$ is the sample mean and σ is the sample standard deviation. t has a Student-t distribution with $n-1$ degrees of freedom.

The Student-t distribution has a longer tail than the Gaussian distribution. The pdfs of the Student-t distribution and the normal distribution are plotted in Fig. A.1.

Kullback–Leibler Divergence

Mutual information between two signals $\boldsymbol{x}$ and $\boldsymbol{y}$ is characterized by calculating the cross-entropy, known as *Kullback–Leibler divergence*, between the joint pdf $p(\boldsymbol{x}, \boldsymbol{y})$ of $\boldsymbol{x}$ and $\boldsymbol{y}$ and the product of the marginal pdfs $p(\boldsymbol{x})$ and $p(\boldsymbol{y})$

$$I(\boldsymbol{x}; \boldsymbol{y}) = \int p(\boldsymbol{x}, \boldsymbol{y}) \ln \frac{p(\boldsymbol{x}, \boldsymbol{y})}{p(\boldsymbol{x})p(\boldsymbol{y})} d\boldsymbol{x} d\boldsymbol{y}. \tag{A.70}$$

This may be implemented by estimating the pdfs in terms of the cumulants of the signals. This approach requires the numerical estimation of the joint and marginal densities.

Cumulants

For random variables $X_1, \ldots, X_4$, second-order cumulants are defined as $\mathrm{cum}(X_1, X_2) = E[\overline{X}_1 \overline{X}_2]$, where $\overline{X}_i = X_i - E[X_i]$, and the fourth-order cumulants are [4]

$$\begin{aligned}\mathrm{cum}(X_1, X_2, X_3, X_4) &= E\left[\overline{X}_1 \overline{X}_2 \overline{X}_3 \overline{X}_4\right] - E\left[\overline{X}_1 \overline{X}_2\right] E\left[\overline{X}_3 \overline{X}_4\right] \\ &\quad - E\left[\overline{X}_1 \overline{X}_3\right] E\left[\overline{X}_2 \overline{X}_4\right] - E\left[\overline{X}_1 \overline{X}_4\right] E\left[\overline{X}_2 \overline{X}_3\right]. \end{aligned} \quad (A.71)$$

The variance and kurtosis of a real random variable X are defined by

$$\mathrm{var}(X) = \sigma^2(X) = \mathrm{cum}(X, X) = E\left[\overline{X}^2\right], \quad (A.72)$$

$$\mathrm{kurt}(X) = \mathrm{cum}(X, X, X, X) = E\left[\overline{X}^4\right] - 3E^2\left[\overline{X}^2\right]. \quad (A.73)$$

They are the second- and fourth-order *autocumulants*. A cumulant having at least two different variables is called a *cross-cumulant*.

Markov Processes, Markov Chains and Markov-Chain Analysis

Markov processes constitute the best-known class of stochastic processes. A Markov process has a limited memory. Assume a stochastic process $\{X(t) : t \in \mathcal{T}\}$, where t is time, $X(t)$ is a state in the state space $\mathcal{S}$. A Markov process is defined as a stochastic process that satisfies the relation characterized by the conditional distribution

$$\begin{aligned} P\left[X(t_0 + t_1) \leq x \middle| X(t_0) = x_0, X(\tau) = x_\tau, -\infty < \tau < t_0\right] \\ = P\left[X(t_0 + t_1) \leq x \middle| X(t_0) = x_0\right] \end{aligned} \quad (A.74)$$

for any value of t_0 and for $t_1 > 0$. The future distribution of the process is determined by the present value of $X(t_0)$ only. This latter property is known as the *Markov property*.

When $\mathcal{T}$ and $\mathcal{S}$ are discrete, a Markov process is called a *Markov chain*. Conventionally, time is indexed using integers, and a Markov chain is a set of random variables that satisfy

$$\begin{aligned} P\left[X_n = x_n \middle| X_{n-1} = x_{n-1}, X_{n-2} = x_{n-2}, \ldots\right] \\ = P\left[X_n = x_n \middle| X_{n-1} = x_{n-1}\right]. \end{aligned} \quad (A.75)$$

This definition can be extended for multistep Markov chains, where a chain state has conditional dependency on only a finite number of its previous states.

For a Markov chain, $P\left[X_n = j | X_{n-1} = i\right]$ is the transition probability of state i to j at time $n-1$. If

$$P\left[X_n = j | X_{n-1} = i\right] = P\left[X_{n+m} = j | X_{n+m-1} = i\right], \quad m \geq 0, \; i, j \in \mathcal{S}, \tag{A.76}$$

the chain is said to be *time homogeneous*. In this case, one can denote

$$P_{i,j} = P\left[X_n = j | X_{n-1} = i\right] \tag{A.77}$$

and the transition probabilities can be represented by a matrix, called the *transition matrix*, $\mathbf{P} = \left[P_{i,j}\right]$, where $i, j = 0, 1, \ldots$. For finite $\mathcal{S}$, $\mathbf{P}$ has a finite dimension. An important property of Markov chains is their time homogeneity, which means that their transition probabilities p_{ij} do not depend on time.

In Markov chain analysis, the transition probability after k step transitions is $\mathbf{P}^k$. The *stationary distribution* or *steady-state distribution* is a vector that satisfies

$$\mathbf{P}^T \pi^* = \pi^*. \tag{A.78}$$

That is, π^* is the left eigenvector of $\mathbf{P}$ corresponding to the eigenvalue 1.

If $\mathbf{P}$ is irreducible and aperiodic, that is, every state is accessible from every other state and in the process none of the states repeats itself periodically, then $\mathbf{P}^k$ converges elementwise to a matrix each row of which is the unique stationary distribution π^*, with

$$\lim_{k \to \infty} \left(\mathbf{P}^k\right)^T \pi = \pi^*. \tag{A.79}$$

Many modeling applications are Markovian, and Markov chain analysis is widely used for convergence analysis for algorithms.

Example A.1 The transition probability matrix corresponding to the graph in Fig. A.2 is given by

$$\mathbf{P} = \begin{bmatrix} 0.3 & 0.5 & 0.2 & 0 \\ 0 & 0 & 0.2 & 0.8 \\ 0 & 0 & 0.3 & 0.7 \\ 0 & 0 & 0 & 1 \end{bmatrix}.$$

Probabilities of transitions from state i to all other states add up to one, i.e., $\sum_{j=1}^{N} P_{ij} = 1$.

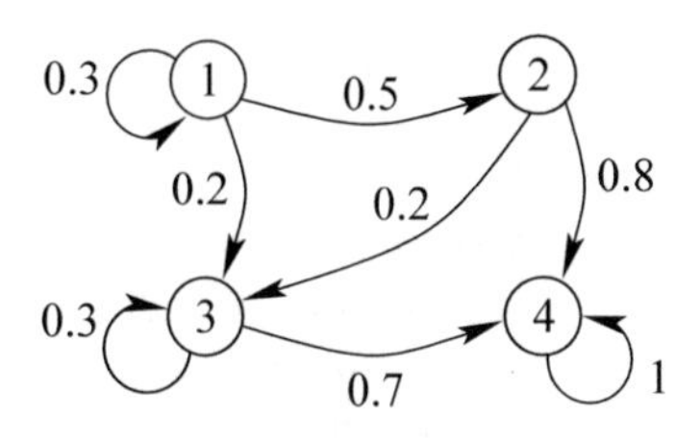

Fig. A.2 State diagram of a Markov chain

A.5 Numerical Optimization Techniques

Although optimization problems can be solved analytically in some cases, numerical optimization techniques are usually more powerful and are also indispensible for all disciplines in science and engineering. Optimization problems discussed in this book are mainly unconstrained continuous optimization problems, COPs, and quadratic programming problems. To deal with constraints, the KKT theorem, as a generalization to the Lagrange multiplier method, introduces a slack variable into each inequality constraint before applying the Lagrange multiplier method. The conditions derived from the procedure are known as the *KKT conditions* [5].

A Brief Taxonomy

Optimization techniques can generally be divided into derivative methods and non-derivative methods, depending on whether or not derivatives of the objective function are required for the calculation of the optimum. Derivative methods can be either gradient-search methods or second-order methods. Gradient-search methods include the gradient-descent method, CG methods, and the natural-gradient method. The gradient descent is also known as *steepest descent*. It searches for a local minimum by taking steps along the negative direction of the gradient of the function. If the steps are along the positive direction of the gradient, the method is known as *gradient ascent* or *steepest ascent*. The gradient-descent method is credited to Cauchy. Examples of second-order methods are Newton's method, the Gauss–Newton method, quasi-Newton methods, the trust-region method, and the LM method. CG methods can also be viewed as a reduced form of the quasi-Newton method, with systematic reinitializations of $\mathbf{H}_t$ to the identity matrix.

Derivative methods can also be classified into *model-based* and *metric-based* methods. Model-based methods improve the current point by a local approximating model. Newton and quasi-Newton methods are model-based methods. Metric-based methods perform a transformation of the variables and then apply a gradient-search method to improve the point. The steepest descent method, quasi-Newton methods, and CG methods belong to this latter category.

The vanilla stochastic gradient-descent method converges slower than the gradient-descent method does. A damped step size is used to guarantee convergence, due to the large variance of stochastic gradient. Nesterov's accelerated gradient descent [9] uses a proper momentum to accelerate the convergence rate of gradient descent to optimal. For smooth convex optimization, the sequence recovers the minimum at a quadratic convergence rate, with the same computational complexity as the vanilla gradient-descent method.

Coordinate-descent method [6] provides an efficient general solver. The method cyclically chooses one variable at a time and performs a simple analytical update. Stochastic coordinate-descent methods update a single randomly chosen coordinate at a time by moving in the direction of the negative partial derivative. One way to update more coordinates at each iteration is via partitioning the coordinates into blocks and operating on a single randomly chosen block at a time [10]. Theory was developed for methods that update a random subset of blocks of coordinates at a time [12]. Block coordinate-descent method may cycle and stagnate when being applied to solve non-convex problems.

Among quasi-Newton methods, limited memory BFGS reduces the computational complexity in each iteration to the same order as gradient descent. Stochastic quasi-Newton methods such as online BFGS [3, 13] and online limited memory BFGS [13] extend BFGS by using stochastic gradients both as descent directions and constituents of Hessian estimates. A sublinear convergence of online limited memory BFGS for solving optimization problems with stochastic objectives is established in [8].

Typical nonderivative methods for multivariable functions are random-restart hill climbing, simulated annealing, evolutionary algorithms, random search, many heuristic methods, and their hybrids. Hill climbing attempts to optimize a discrete or continuous function for a local optimum. When operating on continuous space, it is called *gradient ascent*. Other nonderivative search methods include univariant search parallel to an axis, sequential simplex method, and acceleration methods in direct search such as Hooke–Jeeves method, Powell's method, and Rosenbrock's method. Hooke–Jeeves method accelerates in distance, Powell's method accelerates in direction, and Rosenbrock's method accelerates in both direction and distance. Interior-point methods represent state-of-the-art techniques for solving linear, quadratic, and nonlinear optimization programs. Standard LP solvers are the simplex algorithm and the interior-point algorithm. LP can be solved using IBM ILOG CPLEX Optimizer (https://www.ibm.com/analytics/data-science/prescriptive-analytics/cplex-optimizer).

Lagrange Multiplier Method

The Lagrange multiplier method can be used to analytically solve continuous function optimization subject to equality constraints [5]. Let $f(\boldsymbol{x})$ be the objective function and $h_i(\boldsymbol{x}) = 0, i = 1, \ldots, m$, be the constraints. The Lagrange function can be constructed as

$$L(\boldsymbol{x}; \lambda_1, \ldots, \lambda_m) = f(\boldsymbol{x}) + \sum_{i=1}^{m} \lambda_i h_i(\boldsymbol{x}), \tag{A.80}$$

where λ_i, $i = 1, \ldots, m$, are called the *Lagrange multipliers*.

The constrained optimization problem is converted into an unconstrained optimization problem: Optimize $L(\boldsymbol{x}; \lambda_1, \ldots, \lambda_m)$. By setting

$$\frac{\partial}{\partial \boldsymbol{x}} L(\boldsymbol{x}; \lambda_1, \ldots, \lambda_m) = 0, \tag{A.81}$$

$$\frac{\partial}{\partial \lambda_i} L(\boldsymbol{x}; \lambda_1, \ldots, \lambda_m) = 0, \quad i = 1, \ldots, m, \tag{A.82}$$

and solving the resulting set of equations, we can obtain the $\boldsymbol{x}$ position at the extremum of $f(\boldsymbol{x})$ under the constraints.

Line Search

The popular quasi-Newton and CG methods implement a line search at each iteration. The efficiency of the line search method significantly affects the performance of these methods.

Bracketing and sectioning are two elementary operations for any line search method. A bracket is an interval (α_1, α_2) that contains an optimal value of α. Any three values of α that satisfy $\alpha_1 < \alpha_2 < \alpha_3$ form a bracket when the values of the function $f(\alpha)$ satisfies $f(\alpha_2) \leq \min(f(\alpha_1), f(\alpha_3))$. Sectioning is applied to reduce the size of the bracket at a uniform rate. Once a bracket is identified, it can be contracted by using sectioning or interpolation techniques or their combinations. Popular sectioning techniques are the gloden-section search, the Fibonacci search, the secant method, Brent's quadratic approximation, and Powell's quadratic convergence search without derivatives. The Newton–Raphson search is an analytical line search technique based on the gradient of the objective function. Wolfe's conditions are two inequality conditions for performing inexact line search. Wolfe's conditions enable an efficient selection of the step size without minimizing $f(\alpha)$.

Semidefinite Programming

For a convex optimization problem, a local solution is the global optimal solution. The semidefinite programming (SDP) problem is a convex optimization problem with a linear objective, and linear matrix inequality and affine equality constraints. It optimizes convex cost functions over the convex cone of positive semidefinite matrices. There exist interior-point algorithms to solve SDP problems with good theoretical and practical computational efficiency. One very useful tool to reduce a problem to an SDP problem is the so-called Schur complement lemma. The SDP problem can be efficiently solved using standard SDP solvers such as a C library for semidefinite programming [2], and the MATLAB packages CVX (http://cvxr.com/cvx/), SeDuMi (http://sedumi.ie.lehigh.edu/) [15], and SDPT3 [16].

Many stability or constrained optimization problems including the SDP problem can be converted into a quasi-convex optimization problem in the form of a linear matrix inequality (LMI)-based optimization problem. The LMI-based optimization problem can be efficiently solved by interior-point methods by using MATLAB LMI Control Toolbox. For verifying the stability of delayed neural networks, a Lyapunov function is usually constructed based on the LMI approach.

The constrained concave–convex procedure (CCCP) of [18] is used for solving the non-convex optimization problem. CCCP essentially decomposes a non-convex function into a convex component and a concave component. At each iteration, the concave part is replaced by a linear function (namely, the tangential approximation at the current point) and the sum of this linear function and the convex part is minimized to get the next iteration.

A.6 Classification Measures

When dealing with imbalanced datasets, overall accuracy is a biased measure of classifier goodness. Instead, the confusion matrix, and the true positive (TP) and false positive (FP) are better indications of classifier performance. Referred to as matching matrix in unsupervised learning, a confusion matrix provides a visual representation of actual versus predicted class accuracies.

Accuracy is the number of data points correctly classified:

$$Accuracy = \frac{TP + TN}{TP + TN + FN + FP}, \tag{A.83}$$

where the positive class is the class that is important and usually is the minority class, true positive (TP) is the number of data points from the positive class that are correctly classified, false positive (FP) is the number of data points from the negative class that are predicted to be in the positive class, true negative (TN) is the number of data points from the negative class that are correctly classified, and false negative (FN) is the number of data points from the positive class that are predicted to be in the negative class.

Sensitivity, true positive rate (TPR) or recall rate (RR) measures how well a classifier classifies data points in the positive class:

$$Sensitivity = \frac{TP}{TP + FN}. \tag{A.84}$$

Specificity or true negative rate (TNR) measures how well a classifier classifies data points in the negative class:

$$Specificity = \frac{TN}{TN + FP}. \tag{A.85}$$

Precision shows how exactly the classifier predicts the positive data among actual positive data:

$$Precision = \frac{TP}{TP + FP}. \tag{A.86}$$

Fall-out or false positive rate (FPR) measures the rate of false alarm.

$$FPR = \frac{TN}{TN + FP}. \tag{A.87}$$

Miss rate or false negative rate (FNR) measures the rate of missed samples.

$$FNR = \frac{FN}{FN + TP}. \tag{A.88}$$

F_1-measure combines precision and sensitivity as

$$F_1 = \frac{2 \times (Precision \times Sensitivity)}{Precision + Sensitivity}. \tag{A.89}$$

Accuracy measure is widely used in clustering. It calculates the largest rate of correct assignments by matching each data to the right cluster. For an imbalanced dataset, accuracy is usually high, while precision and recall are relatively low since the classifier tends to predict all data as majority class.

The receiver operating characteristic (ROC) curve offers another useful graphical representation for a binary classifier. It is a plot of the true positive rate (TPR) against the false positive rate (FPR). A higher area under the curve (AUC) indicates a better classifier.

Confidence

Consider a normal distribution Z with mean m and standard error σ. For a confidence level α, the confidence interval is given by $m \pm Z^* \frac{\sigma}{\sqrt{n}}$, where n is the size of each testing. This interval contains the true population mean with a probability of α; that is, $\Pr(-Z^* < Z < Z^*) = \alpha$. For a confidence level of 95%, $Z^* = 1.96$.

For a classification accuracy or a proportion, p, its standard error is calculated as $\sigma = \sqrt{\frac{p(1-p)}{n}}$. For a confidence level α, a normal distribution assumption leads to a confidence interval $p \pm Z^* \sqrt{\frac{p(1-p)}{n}}$. The normal distribution assumption requires $np \geq 10$ and $n(1-p) \geq 10$.

Problems

A.1 For nonbinary data, show that $\|\boldsymbol{x}\|_1 > \|\boldsymbol{x}\|_2 > \|\boldsymbol{x}\|_\infty$.

A.2 Draw the Student-t and Gaussian distributions.

A.3 Consider the function $f(x) = 10x^3 + 4x^2 + 3x + 12$.
(a) Compute its gradient.
(b) Find all its local and global maxima/minima.

A.4 Verify the Lipschitzness of the functions:
(a) $f(x) = |x|$.
(b) $f(x) = x^2$.

A.5 For a classification test, we have the following counts of true labels and output labels:

		Output	
		pos	neg
True	pos	70	90
	neg	50	900

Calculate the values of precision, recall, sensitivity, and specificity.

A.6 Suggest a field where:
(a) Precision is more important than recall.
(b) Recall is more important than precision.

A.7 Assum that a classification accuracy p satisfies the normal distribution and that $p = 0.9$.

(a) Specify a suitable size of the testing set.
(b) Calculate the standard error of the classification accuracy, for testing sets of size $n = 200$.
(c) Give the confidence interval for the 95% confidence level.
(d) Explain the influence of the size of the testing set on the classification confidence.

References

1. Bishop, C. M. (1995). *Neural networks for pattern recognition*. New York: Oxford Press.
2. Borchers, B. (1999). CSDP: A C library for semidefinite programming. *Optimization Methods and Software, 11*, 613–623.
3. Bordes, A., Bottou, L., & Gallinari, P. (2009). SGD-QN: Careful quasi-newton stochastic gradient descent. *Journal of Machine Learning Research, 10*, 1737–1754.
4. Cardoso, J.-F. (1999). High-order contrasts for independent component analysis. *Neural Computation, 11*, 157–192.

5. Fletcher, R. (1991). *Practical methods of optimization*. New York: Wiley.
6. Friedman, J., Hastie, T., Hofling, H., & Tibshirani, R. (2007). Pathwise coordinate optimization. *Annals of Applied Statistics, 1*(2), 302–332.
7. Golub, G. H., & van Loan, C. F. (1989). *Matrix computation* (2nd Edn.). Baltimore, MD: Johns Hopkins University Press.
8. Mokhtari, A., & Ribeiro, A. (2015). Global convergence of online limited memory BFGS. *Journal of Machine Learning Research, 16* 3151–3181.
9. Nesterov, Y. (1983). A method of solving a convex programming problem with convergence rate $O(1/k^2)$. *Soviet Mathematics Doklady, 27*, 372–376.
10. Nesterov, Y. (2012). Efficiency of coordinate descent methods on huge-scale optimization problems. *SIAM Journal on Optimization, 22*(2), 341–362.
11. Phatak, A., & de Hoog, F. (2002). Exploiting the connection between PLS, Lanczos methods and conjugate gradients: Alternative proofs of some properties of PLS. *Journal of Chemometrics, 16*(7), 361–367.
12. Richtarik, P., & Takac, M. (2015). Parallel coordinate descent methods for big data optimization. *Mathematical Programming*, 1–52.
13. Schraudolph, N. N., Yu, J., & Gunter, S. (2007). A stochastic quasi-newton method for online convex optimization. In *Proceedings of the 11th International Conference on Artificial Intelligence and Statistics* (pp. 436–443). San Juan, Puerto Rico.
14. Searle, S. R. (1982). *Matrix algebra useful for statistics*. New York: Wiley-Interscience.
15. Sturm, J. F. (1999). Using Sedumi 1.02: a MATLAB toolbox for optimization over symmetric cones. *Optimization Methods and Software, 11*, 625–653.
16. Toh, K. C., Todd, M. J., & Tutuncu, R. H. (1999). SDPT3 – A MATLAB software package for semidefinite programming. *Optimization Methods and Software, 11*, 545–581.
17. Wold, H. (1975). Soft modeling by latent variables: The nonlinear iterative partial least squares approach. In J. Gani (Ed.), *Perspectives in probability and statistics: Papers in honor of M. S. Bartlett*. London: Academic Press.
18. Yuille, A. L., & Rangarajan, A. (2003). The concave-convex procedure. *Neural Computation, 15*, 915–936.

Appendix B
Benchmarks and Resources

In this appendix, we provide some benchmarks and resources for machine learning, pattern recognition, and data mining.

B.1 Face Databases

The face image data have been standardized as ISO/IEC JTC 1/SC 37 N 506 (Biometric Data Interchange Formats, Part 5: Face Image Data).

AT&T Olivetti Research Laboratory (ORL) face recognition database (http://www.cl.cam.ac.uk/research/dtg/attarchive/facedatabase.html) includes 400 images from 40 individuals. Each individual has 10 images: five for training and five for testing. The face images in ORL only contain pose variation and are perfectly centralized/localized. All the images are taken against a dark homogeneous background but vary in sampling time, illuminations, facial expressions, facial details (glasses/no glasses), scale, and tilt. Each image with 256 gray scales is in the size of 92×112.

California Institute of Technology (CIT) Face Database (http://www.vision.caltech.edu/Image_Datasets/faces/) has 450 color images, the size of each being 320×240 pixels, and contains 27 different people and a variety of lighting, backgrounds, and facial expressions.

MIT CBCL Face Database (http://cbcl.mit.edu/cbcl/software-datasets/FaceData2.html) has 6, 977 training images (with 2,429 faces and 4,548 nonfaces) and 24,045 test images (472 faces and 23,573 nonfaces). All images are captured in grayscale at a resolution of 19×19 pixels, but rather than use pixel values as features.

Face Recognition Grand Challenge (FRGC) Database [7] consists of 1,920 images, corresponding to 80 individuals selected from the original collection. Each individual has 24 controlled or uncontrolled color images. The faces are automatically detected and normalized through a face detection method and an extraction method. FRGC dataset provides high-resolution images and 3D face data.

K.-L. Du and M. N. S. Swamy, *Neural Networks and Statistical Learning*,
https://doi.org/10.1007/978-1-4471-7452-3

FRGC v2 Database is the largest available database of 3D face images composed of 4,007 images with different facial expressions from 466 subjects with different facial expressions. All images have resolution of 640 × 480, acquired by a Minolta Vivid 910 laser scanner. The face images have frontal pose and several types of facial expression: neutral, happy, sad, disgusting, surprised, and puffy cheek. Moreover, some images present artifacts: stretched or distorted images, nose absence, holes around nose, or waves around mouth.

Carnegie Mellon University (CMU) Multi-PIE (pose, illumination, and expression) Face Database (http://www.flintbox.com/public/project/4742/) contains 337 subjects with more than 750,000 face images under various viewpoints, illuminations, and expressions. The face images are of size 32 × 32 pixels, captured as under 13 poses, 43 illumination conditions, and 4 expressions.

Yale Face Database (https://www.cvc.yale.edu/projects/yalefaces/yalefaces.html) contains 165 grayscale images in GIF format of 15 individuals. There are 11 images of 64 × 64 pixels per subject, one per different facial expressions or configurations: center light, w/glasses, happy, left light, w/no glasses, normal, right light, sad, sleepy, surprised, and wink.

Yale Face Database B (https://www.cvc.yale.edu/projects/yalefacesB/yalefacesB.html) contains 2,414 frontal face images of 38 persons in 9 poses/facial expressions and under 64 illumination conditions, with approximately 64 images for each person. The images were cropped to 192 × 168 pixels.

AR Face Database (http://www2.ece.ohio-state.edu/~aleix/ARdatabase.html) contains over 4,000 color images corresponding to 126 people's faces (70 men and 56 women). Images feature frontal view faces with different facial expressions (anger, smiling and screaming), illumination conditions (left and/or right light on), and occlusions (sunglasses and scarf).

Oulu Physics-Based Face Database (www.ee.oulu.fi/research/imag/color/pbfd.html) contains faces of 125 different individuals, each in 16 different camera calibrations and illumination conditions, an additional 16 if the person has glasses. Faces are in frontal position, captured under horizon, incandescent, fluorescent, and daylight illuminant. The database includes three spectral reflectances of skin per person measured from both cheeks and forehead, and contains RGB spectral response of camera used and spectral power distribution of illuminants.

Sheffield (previously UMIST) Face Database (https://www.sheffield.ac.uk/eee/research/iel/research/face) consists of 575 images of 20 individuals (mixed race/gender/appearance). Each individual is shown in a range of poses from profile to frontal views. The files are all in PGM format, approximately 220 × 220 pixels with 256-bit grayscale.

University of Notre Dame 3D Face Database (http://www.nd.edu/~cvrl/CVRL/Data_Sets.html) includes a total of 275 subjects, among which 200 subjects participated in both a gallery acquisition and a probe acquisition. The time lapse between the acquisitions of the probe image and the gallery image for any subject ranges between 1 and 13 weeks. The 3D scans in the database were acquired using a Minolta Vivid 900 range scanner. All subjects were asked to display a neutral facial expression and to look directly at the camera. The result is a 640 × 480 array of range data.

FG-NET Aging Database (http://www.fgnet.rsunit.com) contains 1,002 high-resolution color or grayscale face images of 82 subjects at different ages, with the minimum age being 0 and the maximum age being 69, with large variation of lighting, pose, and expression. The ages (0–69) are divided into six ranges: 0–9, 10–19, 20–29, 30–39, 40–49, and 50+.

MORPH Data Corpus (http://www.faceaginggroup.com/projects-morph.html) is a face aging dataset. It has two separate databases: Album1 and Album2. Album1 contains 1,690 images from 625 different subjects. Album2 contains more than 20,000 images from more than 4,000 subjects whose metadata (age, sex, ancestry, height, and weight) are also recorded.

Iranian Face Aging Database (http://kiau.ac.ir/bastanfard/IFDB_index.htm) contains digital images of people from 1 to 85 years of age. It is a large database that can support studies of the age classification systems. It contains over 3,600 color images.

Bosphorus Database (http://bosphorus.ee.boun.edu.tr/Home.aspx) is intended for research on 3D and 2D human face processing tasks including expression recognition, facial action unit detection, facial action unit intensity estimation, face recognition under adverse conditions, deformable face modeling, and 3D face reconstruction. There are 105 subjects and 4,666 faces in the database.

XM2VTS Face Video Database (http://www.ee.surrey.ac.uk/CVSSP/xm2vtsdb/) contains four recordings of 295 subjects taken over a period of 4 months. The BioID face detection database is available at http://support.bioid.com/downloads/facedb/index.php. Some other face databases are given at http://www.face-rec.org/databases/.

Yahoo! News Face Dataset was constructed from about half a million captioned news images collected from the Yahoo! News website by crawling from the web [1]. It consists of a large number of photographs taken in real-life conditions. As a result, there are a large variety of poses, illuminations, expressions, and environmental conditions. There are 1,940 images, corresponding to 97 largest face clusters, in which each individual cluster has 20 images. Faces are cropped from the selected images using the face detection and extraction methods.

CUHK Face Sketch FERET (CUFSF) Dataset (http://mmlab.ie.cuhk.edu.hk/archive/cufsf/) is used to evaluate photo-sketch face recognition. It contains 1,194 subjects with lighting variations, where examples in this dataset come from photo and sketch.

Databases for Facial Recognition in the Wild

WIDER FACE Dataset (http://mmlab.ie.cuhk.edu.hk/projects/WIDERFace/) is a face detection benchmark, whose images are selected from the publicly available WIDER dataset.

MS-Celeb-1M (https://www.msceleb.org/celeb1m/dataset) is a public face recognition dataset, containing more than 10 million labeled face images of the top 100,000 distinct identities from the 1 million celebrity list with significant pose, illumination, occlusion, and other variations.

CelebA Dataset (http://mmlab.ie.cuhk.edu.hk/projects/CelebA.html) is annotated with 40 face attributes and 5 keypoints by a professional labeling company for 202,599 face images of over 10,000 subjects.

MegaFace Dataset (http://megaface.cs.washington.edu/) is used to test the robustness of face recognition algorithms in the open-set setting with 1 million distractors. The dataset has two parts: the first part allows the use of any external training datasets and the other provides 4.7 million face images of 672,000 subjects.

YouTube Faces Dataset (https://www.cs.tau.ac.il/~wolf/ytfaces/) contains 3,425 videos of 1,595 different subjects and is the standard dataset used to evaluate video-face recognition algorithms.

Databases for Facial Expression Recognition

Japanese Female Facial Expression (JAFFE) Database (http://www.kasrl.org/jaffe.html) contains 213 images of seven facial expressions (six basic facial expressions + one neutral) posed by 10 Japanese female models. Each image has been rated on 6 emotion adjectives by 60 Japanese subjects.

Binghamton University BU-3DFE Database is a database of annotated 3D facial expressions [10]. There are a total of 100 subjects in the database, 56 females and 44 males. A neutral scan was captured for each subject, and then they were asked to perform six expressions: happiness, anger, fear, disgust, sad, and surprise. The expressions vary according to four levels of intensity (low, middle, high, and highest). Thus, there are 25 3D facial expression models per subject. A set of 83 manually annotated facial landmarks is associated to each model. These landmarks are used to define the regions of the face that undergo specific deformations due to muscle movements when conveying facial expression.

Audio Visual Emotion Challenge (AVEC 2011) Database (https://avec-db.sspnet.eu/), which is based on **SEMAINE Database**, contains spontaneous emotional states in naturalistic situations. It consists of 95 videos recorded at 49.979 frames per second. Binary labels along the four affective dimensions (activation, expectation, power, and valence) are provided for each video frame. **SEMAINE Database** allows to study interpersonal dynamics between speakers and listeners.

EmotiW Database [11] is a collection of short video clips collected from some popular movies, where the actor is expressing one of seven emotions (anger, disgust, fear, happy, neutral, sad, and surprise) under near real-world conditions. It contains realistic challenges like pose variations, various illumination conditions, occlusions, and spontaneous emotions. EmotiW consists of 380 training, 396 validation, and 312 testing video clips, respectively.

B.2 UCI Machine Learning Repository

Some popular datasets from UCI machine learning repository (http://archive.ics.uci.edu/ml/) are listed below.

- **HouseVotes Dataset** contains the 1,984 congressional voting records for 435 representatives voting on 17 issues. Votes are all three-valued: yes, no, or unknown. For each representative, the political party is given; this dataset is typically used in a classification setting to predict the political party of the representative based on the voting record.
- **Mushroom Dataset** contains physical characteristics of 8,124 mushrooms, as well as whether each mushroom is poisonous or edible. There are 22 physical characteristics for each mushroom, all of which are discrete.
- **Adult Dataset** has 48,842 patterns of 15 attributes, including eight categorical attributes, six numerical attributes, and one class attribute. The class attribute indicates whether the salary is over 50,000. In the dataset, 76% of the patterns have the value of $\leq$50,000. The goal is to predict whether a household has an income greater than $50,000.
- **Iris Dataset** has 150 data samples from three classes (setosa, versicolor, and virginica) with four measurements (sepal length, sepal width, petal length, and petal width).
- **Wisconsin Diagnostic Breast Cancer Data (WDBC)** contains 569 samples, each with 30 features. The samples are grouped into two clusters: 357 samples for benign and 212 for malignant.
- **Boston Housing Dataset** consists of 516 instances with 12 input variables (including a binary one) and an output variable representing the median housing values in suburbs of Boston.
- **Microsoft Web Training Data (MSWeb)** contains 32,711 instances of users visiting the www.microsoft.com website on one day in 1996. For each user, the data contain a variable indicating whether or not that user visited each of the 292 areas of the site.
- **Image Segmentation Database** consists of samples randomly drawn from a database of seven outdoor images. The images were hand segmented to create a classification for every pixel. Each sample has a 3×3 region and 19 attributes. There are a total of 7 classes, each having 330 samples. The attributes were normalized to lie in $[-1, 1]$.
- **Internet Advertisement Dataset** from UCI machine learning repository consists of 3,279 examples including 459 ads images (positive examples) and 2,820 non-ads images (negative examples). The first view describes the image itself (words in the image's URL, alt text and caption), while the other view contains all other features (words from the URLs of the pages that contain the image and the image points to).
- **Zoo Dataset** covers 101 animals with 17 Boolean-valued attributes, where the attributes contain hair, feathers, eggs, milk, legs, tail, etc. These animal data are grouped into seven classes.

- **Wine Quality Dataset** can be used for ordinal regression.
- **Isolet Dataset** contains acoustic features of isolated spoken letters from "A" to "Z".

B.3 Some Machine Learning Databases

KEEL Dataset Repository (http://www.keel.es/dataset.php) provides imbalanced datasets, multi-instance datasets, and multi-label datasets for evaluating algorithms.

CaliforniaDBpedia Dataset is a dataset for spatio-textual query. It is a synthesized dataset which combines the spatial data in California and a real collection of article categories from DBpedia.

MovieLens 10M (http://www.grouplens.org/) and **Netflix Prize Dataset** (http://www.netflixprize.com/) are two large publicly available collaborative filtering datasets. The $1M Netflix prize competition dataset is a training dataset of 100,480, 507 ratings that 480,189 users gave to 17,770 movies. Each training rating is a quadruplet <user, movie, date of grade, grade>. The user and movie fields are integer IDs, while grades are from 1 to 5 (integral) stars.

AMI Meeting Corpus is a dataset for understanding human multimodal behaviors during social interactions. It contains 100 h of video recordings of meetings, all fully transcribed and annotated.

EEG Datasets

EEGLAB (http://sccn.ucsd.edu/eeglab/) is an interactive MATLAB toolbox for processing continuous and event-related EEG, MEG and other electrophysiological data incorporating ICA, time/frequency analysis, artifact rejection, event-related statistics, and several useful modes of visualization of the averaged and single-trial data. The Sleep-EDF database gives sleep recordings and hypnograms in European data format (EDF).

The EEG Dataset From BCI Competition 2003 (http://www.bbci.de/competition/) has 28 channel input recorded from a single subject performing a self-paced key typing, that is, pressing with the index and little fingers corresponding keys in a self-chosen order and timing. **BCI Competition IV-2a** provides data from 9 subjects performing 4 imagery movement tasks, namely, right hand (RH), left hand (LH), both feet (F), and tongue (T), during 2 days of recordings. Each day the subjects performed 72 trials of each task (3 seconds per trial), and the EEG data were recorded using 20 electrodes. **Physiobank** (http://www.physionet.org/physiobank/database) contains EEG data from 109 subjects performing various combinations of real and imagined movements in one day of recordings.

Image Databases

Columbia Object Image Library (COIL-20) (http://www.cs.columbia.edu/CAVE/software/softlib/coil-20.php) contains the images of 20 different three-dimensional objects. The objects represent cups, toys, drugs, and cosmetics. For each object 72 training samples are available. The size of the images is 32 × 32 grayscale images viewed from varying angles.

CIFAR-10 and CIFAR-100 Datasets (https://www.cs.toronto.edu/~kriz/cifar.html) are labeled subsets of the 80 million tiny images dataset. CIFAR-10 dataset consists of 60,000 32 × 32 color images in 10 classes, with 6000 images per class: 50000 for training and 10000 for testing. CIFAR-100 dataset is just like CIFAR-10, except it has 100 classes containing 600 images each: 500 for training and 100 for testing per class. The 100 classes in the CIFAR-100 are grouped into 20 superclasses.

Microsoft Research Asia Internet Multimedia Dataset 1.0 (MSRA-MM 1.0) explored the query log of Microsoft Bing Image Search and selected a set of representative ones. The ground truth of queries is given. MSRA-MM 2.0 image dataset adds 1, 097 frequently used queries. These queries are manually classified into nine categories, i.e., Animal, Cartoon, Event, Object, NamedPerson, PeopleRelated, Scene, Time, and Misc. The total image number is 1,011,738. Each concept has approximately 500–1,000 images. Seven low-level features were extracted for each image.

Corel Images Dataset from UCI repository consists of 34 categories, each with 100 JPEG images of 384 × 256 or 256 × 384 resolution.

Corel and MSRA image data are very representative and frequently used in many tasks of multiple-instance learning.

NUS-WIDE Dataset (http://lms.comp.nus.edu.sg/research/NUS-WIDE.htm) is also a large-scale annotated web image dataset publicly available to researchers. It consists of 269,648 images collected from the website of Flickr and their ground truth annotations for 81 concepts.

Caltech-101 Dataset (http://www.vision.caltech.edu/Image_Datasets/Caltech101/) contains 101 categories of object images. It contains 8,677 images from 101 categories and 467 images from an additional background category.

Pascal Visual Object Classes (VOC) Challenge Dataset (http://host.robots.ox.ac.uk/pascal/VOC/) is a benchmark for image classification, detection, and segmentation. There are large variations in pose, view, scale, appearance, and clustered background. Pascal VOC 2012 contains 22,534 images including a trainval set of 11,540 images and a test set of 10,994 images. 5,717 images of the trainval set are used for training and the remaining 5,823 images for validation. Annotations of the test set are not publicly available.

ImageNet (http://image-net.org/) is a real-world image database containing roughly 15 million images organized according to the WordNet hierarchy. Currently, over 20 thousand noun synsets in WordNet are indexed and each synset has over 500 images on average. **ILSVRC 1000 (ImageNet Large Scale Visual Recognition Challenge 2010) Dataset** has 1,000 categories and 1,261,406 images. **ILSVRC 2013 Object Detection Set** is constructed following the style of PASCAL VOC but contains more images and categories: 200 basic-level categories, 395, 909 images for training, 20,121 images for validation, and 40,152 images for testing.

KITTI Vision Benchmark Suite (http://www.cvlibs.net/datasets/kitti/) provides datasets for developing automatic driving systems. The object detection dataset in the suite can be used for detecting cars in images taken "in the wild." It has three levels of difficulty: easy, moderate, and hard. There are 7,481 images for training and 7,518 for testing. The testing images have no ground truth.

Richly Annotated Pedestrian (RAP) Dataset (http://rap.idealtest.org/) contains 84,928 images with 72 types of attributes and additional tags of viewpoint, occlusion, body parts, and 2,589 person identities. It was collected from a high-definition (1,280 × 720) surveillance network at an indoor shopping mall. Other examples of datasets for pedestrian detection are **MIT Pedestrian Dataset** (http://cbcl.mit.edu/software-datasets/PedestrianData.html) and **INRIA Person Dataset** (http://pascal.inrialpes.fr/data/human/).

Some benchmark image databases are **Berkeley Image Segmentation Database** (http://www.eecs.berkeley.edu/Research/Projects/CS/vision/bsds/), brain MRI images from **BrainWeb Database** (http://www.bic.mni.mcgill.ca/brainweb/), and 3D shape objects from **NORB Dataset** (https://cs.nyu.edu/~ylclab/data/norb-v1.0/index.html). In the Berkeley segmentation database, a natural color image (Training Image #124084) is a flower, which contains four dominant colors. A collection of datasets for the annotations of the video sequence is available at http://www.vision.ee.ethz.ch/~bleibe/data/datasets.html.

There are several standard benchmarks for single image super-resolution (i.e., **Set5**, **Set14** and **BSD200** Datasets) and for super-resolution (i.e., **Videoset4 Dataset**) [2].

Image Motion Detection Databases

CDnet Dataset (http://www.changedetection.net/) is a real-world region-level motion detection benchmark. The 2012 dataset contains 31 video sequences that are divided into 6 video categories: baseline, dynamic background, intermittent object motion, thermal, camera jitter, and shadows, with 4 to 6 video sequences in each category. The resolution of the videos also varies from 320 × 240 to 480 × 720 with hundreds to thousands of frames. The 2014 dataset contains 11 video categories with 4 to 6 video sequences in each category.

Hopkins-155 Dataset (http://www.vision.jhu.edu/data/hopkins155/) is a benchmark for motion segmentation. It contains 120 two-motion and 35 three-motion videos.

Image Retrieval Databases

INRIA Holidays Dataset (http://lear.inrialpes.fr/~jegou/data.php#holidays) is collected from personal holiday albums. It has 1, 491 images composed of 500 groups of similar images. Each image group has 1 query, totaling 500 query images.

Ukbench Dataset consists of 10,200 images of various contents, such as objects, scenes, and CD covers. The images are divided into 2,550 groups, each having 4 images of the same object/scene, under various angles, illuminations, and translations. Each image is taken as a query, thus 10,200 queries in total.

Oxford Buildings Dataset (http://www.robots.ox.ac.uk/~vgg/data/oxbuildings/) consists of 5,062 images collected from Flickr by searching for particular Oxford landmarks using the names of 11 different landmarks in Oxford. The dataset defines 5 queries for each landmark by hand-drawn bounding boxes, totaling 55 query regions of interest (ROI). Each database image is assigned one of the four labels, good, OK, junk, or bad.

Flickr 100k Dataset (http://www.robots.ox.ac.uk/~vgg/data/oxbuildings/flickr 100k.html) contains 100, 071 high-resolution images crawled from Flickr's 145 most popular tags.

Paris Dataset (http://www.robots.ox.ac.uk/~vgg/data/parisbuildings/) is featured by 6,412 images collected from Flickr by searching from 11 queries on particular Paris landmarks. Each landmark has 5 queries, so there are also 55 queries with bounding boxes. The database images are annotated with the same four types of labels as Oxford Buildings Dataset.

Pittsburgh Dataset (http://www.ok.sc.e.titech.ac.jp/~torii/project/repttile/) is a geotagged image database for visual place recognition. It is formed by 254,064 perspective images, generated from 10,586 Google Street View panoramas of the Pittsburgh area.

Places Dataset (http://places2.csail.mit.edu/) contains more than 10 million images comprising 400+ unique scene categories. The dataset features 5,000–30,000 training images per class, consistent with real-world frequencies of occurrence.

FM2 Dataset (http://www.zemris.fer.hr/~ssegvic/datasets/unizg-fer-fm2.zip) for traffic scene recognition contains 6,237 images from eight classes: highway, road, tunnel, tunnel exit, settlement, overpass, toll booth, and dense traffic.

Cityscapes Dataset (https://www.cityscapes-dataset.com/) focuses on semantic understanding of urban street scenes. It consists of a diverse set of street scene photos taken from 50 different cities by car-carried cameras: 5,000 photos with high-quality pixel-level annotations and 20,000 photos with coarse annotations. The pictures belong to 30 semantic classes, such as road, car, pedestrian, and bicycle, which are grouped into eight categories, i.e., flat, nature, object, sky, construction, human, and vehicle, and void.

Biometric Databases

University of Notre Dame Iris Image Dataset (https://sites.google.com/a/nd.edu/public-cvrl/data-sets) contains 64,980 iris images obtained from 356 subjects (712 unique irises) between January 2004 and May 2005.

ATVS-FIr Database (http://atvs.ii.uam.es/fir_db.html) is an iris database from ATVS Biometric Recognition Group. The samples are taken from 50 random users of Biosec Baseline Iris Subcorpus. It contains iris samples of both eyes. Four samples of each iris were captured in 2 acquisition sessions. Fake samples were also acquired from high-quality printed images of the original samples. There are 800 real and 800 fake image samples.

CASIA-IrisV2 and CASIA-IrisV4 Databases (http://biometrics.idealtest.org/) are provided by Institute of Automation of the Chinese Academy of Sciences (CASIA). CASIA-IrisV2 includes two subsets, each including 1,200 images from 60 classes. CASIA-IrisV4 comprises six subsets.

IIITD Contact Lens Iris Database (http://www.iab-rubric.org/resources.html) is provided by Image Analysis and Biometrics Lab of IIIT, Delhi, India. It is composed of 6,570 iris images coming from 101 subjects. Both left and right iris images of each subject were captured, and therefore there are 202 iris classes.

Hong Kong Polytechnic University (PolyU) Palmprint Database (https://www4.comp.polyu.edu.hk/~biometrics/) includes 600 palmprint images with the size of 128 × 128 from 100 individuals, with six images from each.

AMI Ear Dataset, which was collected at the University of Las Palmas, consists of 700 images of a total of 100 distinct subjects in the age group of 19–65 years.

Annotated Web Ears (AWE) Dataset contains images collected from the web and is a dataset for ear recognition gathered in the wild. AWE MATLAB toolbox (http://awe.fri.uni-lj.si) contains tools for generating performance metrics and graphs, and for research in ear recognition. The dataset contains 1,000 ear images of 100 subjects. Each image in the dataset was annotated according to gender, ethnicity, accessories, occlusion, head pitch, head roll, head yaw, and head side.

USTB (University of Science and technology in Beijing) Ear Image Databases (http://www1.ustb.edu.cn/resb/en/visit/visit.htm) offers four collections of 2D ear and face profile images, and **UND (University of Notre Dame) Databases** (https://cvrl.nd.edu/projects/data/) offers five databases of 2D ear images.

Some other biometric databases are **PolyU Finger-Knuckle–Print Databases** (http://www4.comp.polyu.edu.hk/~biometrics/FKP.htm) and **CASIA Gait Database** (http://www.cbsr.ia.ac.cn/english/Gait%20Databases.asp).

Cambridge Gesture Dataset (https://labicvl.github.io/ges_db.htm) consists of 900 image sequences of nine hand gesture classes, which are divided into three primitive hand shapes and three primitive motions. Each class contains 100 image sequences including five illumination backgrounds, and each of the sequences was recorded in front of a fixed camera which roughly isolated gestures in space and time.

For human action recognition, **Weizmann Action Dataset** and **Ballet Dataset** (http://www.cs.sfu.ca/research/groups/VML/semilatent/) are video sequences of different actions of many subjects.

Datasets for One-Class Classification

Datasets for One-Class Classification

Intrusion Detection Dataset (http://kdd.ics.uci.edu/databases/kddcup99/kddcup99.html) consists in binary TCP dump data from 7 weeks of network traffic. Each original pattern has 34 continuous features and seven symbolic features. The training set contains 4,898,431 connection records, which are processed from about four gigabytes of compressed binary TCP dump data from 7 weeks of network traffic. Another 2 weeks of data produced the test data with 311,029 patterns. The dataset includes a wide variety of intrusions simulated in a military network environment. There are a total of 24 training attack types, and additional 14 types that appear in the test data only.

Promoter Database (from UCI Repository) consists of 106 samples, 53 for promoters, while the others for nonpromoters.

Datasets for Handwriting Recognition

The well-known real-world OCR benchmarks are the USPS dataset, the MNIST dataset, and the UCI Letter dataset (from UCI Repository).

MNIST handwritten Digits Database (http://yann.lecun.com/exdb/mnist/) consists of 60,000 training samples from approximately 250 writers and 10,000 test samples from a disjoint set of 250 other writers. It contains 784-dimensional nonbinary sparse vectors which resembles 28×28 pixel gray-level images of the handwritten digits.

US Postal Service (USPS) handwritten digit database (http://www.cs.nyu.edu/~roweis/data.html) contains 7,291 training and 2,007 images of handwritten digits, size 16×16.

Pendigits Dataset (from UCI Repository) contains 7,494 training digits and 3,498 testing digits represented as vectors in 16-dimensional space. The digit database collects 250 samples from 44 writers. The samples written by 30 writers are used for training, and the digits written by the other 14 are used for testing.

B.4 Datasets for Data Mining

Reuters-21578 Corpus (http://www.daviddlewis.com/resources/testcollections/reuters21578/) is a set of 21,578 economic news published by Reuters in 1987. Each article is typically designated into one or more semantic categories such as "earn", "trade", and "corn", where the total number of categories is 114. The commonly used ModApte split filters out duplicate articles and those without a labeled topic, and then uses earlier articles as the training set and later articles as the test set.

20 Newsgroups Dataset (http://people.csail.mit.edu/jrennie/20Newsgroups/) is a collection of approximately 20,000 newsgroup documents, partitioned (nearly) evenly across 20 different newsgroups. This corpus contains 26,214 distinct terms after stemming and stop word removal. Each document is then represented as a term frequency vector and normalized to one.

CMU WebKB Knowledge Base (http://www.cs.cmu.edu/afs/cs/project/theo-11/www/wwkb/) is a collection of 8,282 web pages obtained from 4 academic domains. The web pages in the WebKB set are labeled using two different polychotomies. The first is according to topic, and the second is according to web domain. The first polychotomy consists of 7 categories: course, department, faculty, project, staff, student, and other.

OHSUMED Dataset (http://ir.ohsu.edu/ohsumed/ohsumed.html) is a clinically oriented MEDLINE subset formed by 348,566 references of 270 medical journals published between 1987 and 1991. It consists of 348,566 references and 106 queries with their respective ranked results. The relevance degrees of references with regard to the queries are assessed by humans, on three levels: definitely, possibly, or not relevant. Totally, there are 16,140 query–document pairs with relevance judgments.

tr41 Dataset is derived from the TREC-5, TREC-6, and TREC-7 collections (http://trec.nist.gov). It includes 210 documents belonging to seven different classes. The dimension of this dataset is 7,454.

Spam Dataset (from UCI Repository) contains 4,601 examples of e-mails, roughly 39% of which are classified as spam. There are 57 attributes for each example, most of which represent how frequently certain words or characters appear in the e-mail.

B.5 Databases and Tools for Speech Recognition and Audio Classification

YOHO Speaker Verification Database consists of sets of four combination lock phrases spoken by 168 speakers. This database can be purchased from Linguistic Data Consortium as LDC94S16.

Isolet Spoken Letter Recognition Database (from the UCI Repository) contains 150 subjects who spoke the name of each letter of the alphabet twice. The speakers are grouped into sets of 30 speakers each and are referred to as isolets 1 through 5.

TIMIT Acoustic-Phonetic Continuous Speech Corpus contains a total of 6,300 sentences, 10 sentences spoken by 630 speakers selected from eight major dialect regions of the United States. 70% of the speakers are male, and 30% are female. It can be purchased from Linguistic Data Consortium as LDC93S1. The speech was labeled at both phonetic and lexical levels.

Oregon Graduate Institute Telephone Speech (OGI-TS) Corpus is a multilingual speech corpus for LID experiments. The OGI-TS speech corpus contains the speech from 11 languages. It includes recorded utterances from about 2,052 speakers.

CALLFRIEND Telephone Speech Corpus (http://www.ldc.upenn.edu/Catalog/) is a collection of unscripted conversations for 12 languages recorded over telephone lines. It is used in the NIST language recognition evaluations (http://www.itl.nist.gov/iad/mig/tests/lang/) tasks, which are performed as language detection: Given a segment of speech and a language hypothesis, the task is to decide whether that target language was spoken in the given segment. OGI-TS corpus and CALLFRIEND corpus are widely used in language identification evaluation.

HMM Tool Kit (http://htk.eng.cam.ac.uk/) is a de facto standard toolkit in C for training and manipulating HMMs in speech research. The HMM-based speech synthesis system (HTS) (http://hts-engine.sourceforge.net/) adds to HMM Tool Kit various functionalities in C for HMM-based speech synthesis. Some speech synthesis systems are Festival (http://www.cstr.ed.ac.uk/projects/festival/), Flite (Festival-lite) (http://www.speech.cs.cmu.edu/flite/), and MARY text-to-speech system (http://mary.dfki.de/).

CMU_ARCTIC Databases (http://festvox.org/cmu_arctic/) are phonetically balanced, U.S. English, single-speaker databases designed for speech synthesis research. The HTS recipes for building speaker-dependent and speaker-adaptive HTS voices use these databases.

Some open-source speech processing systems are Speech Signal Processing Toolkit (http://sp-tk.sourceforge.net/), STRAIGHT and STRAIGHTtrial (http://www.wakayama-u.ac.jp/~kawahara/STRAIGHTadv/index_e.html), and Edinburgh Speech Tools (http://www.cstr.ed.ac.uk/projects/speech_tools/).

auDeep (https://github.com/auDeep/auDeep) is a Python toolkit for deep unsupervised learning from acoustic data. It is based on a recurrent sequence to sequence autoencoder approach to learn representations of time series data. It provides a command-line interface. auDeep can be used for audio classification tasks, such as acoustic scene classification, environmental sound classification, and music genre classiffication.

Benchmarks for audio classification tasks are **TUT Acoustic Scenes 2017 Dataset** (http://www.cs.tut.fi/sgn/arg/dcase2017/challenge/task-acoustic-scene-classification) for acoustic scene classification, **ESC-50 Dataset** (https://github.com/karoldvl/ESC-50) for environmental sound classiffication (ESC), and **GTZAN Dataset** (http://opihi.cs.uvic.ca/sound/genres.tar.gz) for music genre classiffication.

B.6 Datasets for Microarray and for Genome Analysis

Yeast Sporulation Dataset (http://cmgm.stanford.edu/pbrown/sporulation) is a microarray dataset on the transcriptional program of sporulation in budding yeast. A DNA microarray containing 97 % of the known and predicted genes is used. The total number of genes is 6, 118. During the sporulation process, the mRNA levels were obtained at seven time points 0, 0.5, 2, 5, 7, 9, and 11.5 h. The ratio of each gene's mRNA level (expression) to its mRNA level in vegetative cells before transfer to the sporulation medium is measured.

Human Fibroblasts Serum Dataset (http://www.sciencemag.org/feature/data/984559.shl) contains the expression levels of 8,613 human genes. It has 13 dimensions. A subset of 517 genes whose expression levels changed substantially across the time points has been chosen.

Rat Central Nervous System Dataset (http://faculty.washington.edu/kayee/cluster) examines the expression levels of a set of 112 genes during rat central nervous system development over nine time points.

Yeast Cell Cycle Dataset (http://faculty.washington.edu/kayee/cluster) was extracted from a dataset that shows the fluctuation of expression levels of approximately 6,000 genes over two cell cycles (17 time points). Out of these 6,000 genes, 384 genes have been selected to be cell cycle regulated.

ELVIRA Biomedical Dataset Repository (http://leo.ugr.es/elvira/DBCRepository/index.html) includes high-dimensional biomedical datasets, including gene expression data, protein profiling data, and genomic sequence data that are related to classification. The colon cancer dataset consists of 62 samples of colon epithelial cells from colon cancer patients. The samples consist of tumor biopsies collected from tumors (40 samples), and normal biopsies collected from healthy part of the colons (22 samples) of the same patient. The number of genes in the dataset is 2,000.

Global Cancer Map (http://www.broadinstitute.org/cgi-bin/cancer/datasets.cgi) is a gene expression dataset consisting of 198 human tumor samples spanning 14 different cancer types.

General Databases

GenBank (http://www.ncbi.nlm.nih.gov/Genbank/index.html) is the NIH genomic database, an annotated collection of all publicly available DNA sequences. It contains all annotated nucleic acid and amino acid sequences. Apart from presenting and annotating sequences, these databases offer many functions related to searching and browsing sequences.

Rfam Database (http://rfam.sanger.ac.uk/) is a collection of RNA families, each represented by multiple sequence alignments, consensus secondary structures, and covariance models.

EMBL Nucleotide Sequence Database (EMBL-Bank) (http://www.ebi.ac.uk/embl/) constitutes Europe's primary nucleotide sequence resource.

Stanford Microarray Database (http://genome-www5.stanford.edu/) and gene expression omnibus are the two most famous and abundant gene expression databases in the world. Gene expression omnibus is a database including links to microarray-based experiments measuring mRNA, genomic DNA, and protein abundances, as well as non-array techniques such as serial analysis of gene expression and mass spectrometric proteomic data.

Analysis Tools

Some websites for genome analysis are Human Genome Project (http://www.ornl.gov/sci/techresources/Human_Genome/home.shtml), Ensembl Genome Browser (http://www.ensembl.org/index.html), and UCSC Genome Browser (http://genome.ucsc.edu/).

MeV (http://www.tm4.org/mev.html) is a versatile microarray tool, incorporating sophisticated algorithms for clustering, visualization, classification, statistical analysis, and biological theme discovery.

For sequence analysis, BLAST (http://blast.ncbi.nlm.nih.gov/Blast.cgi) finds regions of similarity between biological sequences, and ClustalW2 (http://www.ebi.ac.uk/Tools/clustalw/) is a general-purpose multiple sequence alignment program for DNA or proteins.

SignatureClust (http://infos.korea.ac.kr/sigclust.php) is a tool for landmark gene-guided clustering that enables biologists to get multiple views of the microarray data.

B.7 Software

Stuttgart Neural Network Simulator (http://www.ra.cs.uni-tuebingen.de/SNNS/) is a software simulator for neural networks on Unix systems. The simulator kernel is written in C, and it provides X graphical user interface. The simulator supports the following network architectures and learning procedures that are discussed in this book: online BP, BP with momentum term and flat spot elimination, batch BP, Quickprop, Rprop, generalized RBF network, ART 1, ART 2, ARTMAP, cascade correlation, dynamic LVQ, BPTT, Quickprop through time, SOM, TDNN with BP, Jordan networks, Elman networks, and associative memory.

SHOGUN (http://www.shogun-toolbox.org) is an open-source toolbox in C++ that runs on UNIX/Linux platforms and interfaces to MATLAB. It provides a generic interface to 15 SVM implementations (among them are SVMlight, LibSVM, GPDT, SVMLin, LibLinear, SVM SGD, SVMPegasos and OCAS, kernel ridge regression, SVR), multiple kernel learning, Naive Bayes classifier, k-NN, LDA, HMMs, C-means, and hierarchical clustering. SVMs can be combined with more than 35 different kernel functions. One of the SHOGUN's key features is the combined kernel to construct weighted linear combinations of multiple kernels that may even be defined on different input domains.

Dlib-ml (http://dclib.sourceforge.net) provides a similarly rich environment for developing machine learning software in C++. It contains an extensible linear algebra toolkit with built-in BLAS support. It also houses implementations of algorithms for performing inference in Bayesian networks and kernel-based methods for classification, regression, clustering, anomaly detection, and feature ranking. MLPACK (http://www.mlpack.org) is a scalable, multi-platform C++ machine learning library offering a simple, consistent API, high performance, and flexibility.

LRSLibrary (https://github.com/andrewssobral/lrslibrary) provides a collection of low-rank and sparse decomposition algorithms in MATLAB. It was designed for motion segmentation in videos but can be used for other computer vision problems. LRSLibrary offers more than 100 algorithms based on matrix and tensor methods, including robust PCA, subspace tracking, matrix completion, low-rank recovery, three-term decomposition, NMF, nonnegative tensor factorization, and tensor decomposition.

Netlab (http://www1.aston.ac.uk/eas/research/groups/ncrg/resources/netlab/) is another neural network simulator implemented in MATLAB.

ThunderSVM (https://github.com/zeyiwen/thundersvm) is an open-source SVM software toolkit which exploits the high performance of GPUs and multi-core CPUs. ThunderSVM supports all the functionalities of LibSVM. It is generally an order of magnitude faster than LibSVM while producing identical SVMs.

DOGMA (http://dogma.sourceforge.net) is a MATLAB toolbox for discriminative online learning. The library focuses on linear and kernel online algorithms, mainly developed in the relative mistake bound framework. Examples are perceptron, passive-aggressive, ALMA, NORMA, SILK, projectron, RBP, and Banditron.

Some resources for implementing RBF networks are: ELM (http://www.ntu.edu.sg/home/egbhuang/), optimally pruned ELM (https://research.cs.aalto.fi/aml/software/OPELM.zip), and the improved Levenberg–Marquardt algorithm for RBF networks (http://www.eng.auburn.edu/~wilambm/nnt/index.htm).

A MATLAB toolbox for implementing several PCA techniques is available at http://research.ics.tkk.fi/bayes/software/index.shtml. Some NMF tools are NMFPack (MATLAB,http://www.cs.helsinki.fi/u/phoyer/software.html), NMF package (C++, http://nmf.r-forge.r-project.org), and bioNMF (MATLAB, C, http://bionmf.cnb.csic.es).

Some resources for implementing ICA are JADE (http://www.tsi.enst.fr/icacentral/Algos/cardoso/), FastICA (http://www.cis.hut.fi/projects/ica/fastica/), efficient FastICA (http://itakura.kes.tul.cz/zbynek/downloads.htm), RADICAL (http://www.eecs.berkeley.edu/~egmil/ICA), and denoising source separation (http://www.cis.hut.fi/projects/dss/).

Some resources for implementing clustering are SOM_PAK and LVQ_PAK (http://www.cis.hut.fi/~hynde/lvq/), Java applets for TSP based on SOM and Kohonen network (http://sydney.edu.au/engineering/it/~irena/ai01/nn/tsp.html, http://www.sund.de/netze/applets/som/som2/), Java applet implementing several competitive learning-based clustering algorithms (http://www.sund.de/netze/applets/gng/full/GNG-U_0.html), C++ code for minimum sum-squared residue co-clustering algorithm (http://www.cs.utexas.edu/users/dml/Software/cocluster.html), and C++ code for single-pass fuzzy C-means and online fuzzy C-means (http://www.csee.usf.edu/~hall/scalable).

Some resources for implementing LDA are uncorrelated LDA and orthogonal LDA (http://www-users.cs.umn.edu/~jieping/UOLDA/), neighborhood component analysis (http://www.cs.berkeley.edu/~fowlkes/software/nca/), local LDA (http://sugiyama-www.cs.titech.ac.jp/~sugi/software/LFDA/), and semi-supervised local

Fisher discriminant analysis (http://sugiyama-www.cs.titech.ac.jp/~sugi/software/SELF).

Some resources for implementing SVMs are Lagrangian SVM (http://www.cs.wisc.edu/dmi/lsvm), potential SVM (http://ni.cs.tu-berlin.de/software/psvm), LASVM (http://leon.bottou.com/projects/lasvm, http://www.neuroinformatik.rub.de/PEOPLE/igel/solasvm), LS-SVM (http://www.esat.kuleuven.ac.be/sista/lssvmlab/), 2ν-SVM (dsp.rice.edu/software), Laplacian SVM in the primal (http://sourceforge.net/projects/lapsvmp/), SimpleSVM (http://sourceforge.net/projects/simplesvm/), decision-tree SVM (http://ocrwks11.iis.sinica.edu.tw/dar/Download/WebPages/DTSVM.htm), core vector machine (http://c2inet.sce.ntu.edu.sg/ivor/cvm.html), OCAS and OCAM in LIBOCAS (http://cmp.felk.cvut.cz/~xfrancv/ocas/html/) and as a part of the SHOGUN toolbox, Pegasos (http://ttic.uchicago.edu/~shai/code), EnsembleSVM (http://homes.esat.kuleuven.be/~claesenm/ensemblesvm/), MSVMpack in C (http://www.loria.fr/~lauer/MSVMpack/), and BMRM in C++ (http://users.cecs.anu.edu.au/~chteo/BMRM.html).

Some resources for implementing kernel methods are regularized kernel discriminant analysis (http://www.public.asu.edu/~jye02/Software/DKL/), L_p-norm multiple kernel learning (http://doc.ml.tu-berlin.de/nonsparse_mkl/, implemented within the SHOGUN toolbox), TRON and TRON-LR in LIBLINEAR (http://www.csie.ntu.edu.tw/~cjlin/liblinear), FaLKM-lib (http://disi.unitn.it/~segata/FaLKM-lib), SimpleMKL (http://asi.insa-rouen.fr/enseignants/~arakotom/code/mklindex.html), HessianMKL (http://olivier.chapelle.cc/ams/), LevelMKL (http://appsrv.cse.cuhk.edu.hk/~zlxu/toolbox/level_mkl.html, SpicyMKL (http://www.simplex.t.u-tokyo.ac.jp/~s-taiji/software/SpicyMKL), generalized kernel machine toolbox (http://theoval.cmp.uea.ac.uk/~gcc/projects/gkm), and JKernel-Machines (in Java https://github.com/davidpicard/jkernelmachines) for SVM and learning with kernels.

GPML (http://www.gaussianprocess.org/gpml/code/matlab/doc/) and GPstuff (http://research.cs.aalto.fi/pml/software/gpstuff/) are MATLAB toolboxes for Gaussian processes, which are Bayesian nonparametric models using a prior on functions. GPML toolbox implements approximate inference algorithms for Gaussian processes for a wide class of likelihood functions for both regression and classification. GPstuff toolbox is a versatile collection of Gaussian process models and computational tools required for inference. GPflow (http://github.com/GPflow/GPflow) is a Gaussian process library that uses TensorFlow for core computations and Python for its front end. It uses variational inference as the primary approximation method, provides concise code through using automatic differentiation, and is able to exploit GPU hardware.

A selected collection of tutorials, publications, computer codes for Gaussian processes, mathematical programming, SVM, and kernel methods can be found at http://www.kernel-machines.org.

For Bayesian Networks

XMLBIF (XML-based BayesNets Interchange Format) is an XML-based format that is very simple to understand and yet can represent DAGs with probabilistic relations, decision variables, and utility values. The XMLBIF format is implemented in the JavaBayes (http://www.cs.cmu.edu/~javabayes/) and GeNie (http://genie.sis.pitt.edu/) systems.

FastInf (http://compbio.cs.huji.ac.il/FastInf) is a C++ library for propagation-based approximate inference methods in large-scale discrete undirected graphical models. Various message-scheduling schemes that improve on the standard synchronous or asynchronous approaches are included. FastInf includes exact inference by the junction-tree algorithm [3], loopy belief propagation, generalized belief propagation [9], tree re-weighted belief propagation [8], propagation based on convexification of the Bethe free energy [4], variational Bayesian, and Gibbs sampling. All methods can be applied to both sum and max product propagation schemes, with or without damping of messages.

libDAI (http://www.libdai.org) is an open-source C++ library that provides implementations of various exact and approximate inference methods for graphical models with discrete-valued variables. libDAI uses factor graphs. Apart from exact inference by brute force enumeration and the junction-tree method, libDAI offers the following approximate inference methods for calculating partition sums, marginals, and MAP states: mean field, (loopy) belief propagation, tree expectation propagation [5], generalized belief propagation [9], loop-corrected belief propagation [6], a Gibbs sampler, and several other methods. In addition, libDAI supports parameter learning of conditional probability tables by ML or EM (in case of missing data).

Some resources for implementing Bayesian and probabilistic networks are: Murphy's Bayes Network Toolbox (in MATLAB, http://code.google.com/p/bnt/), Probabilistic Networks Library (http://sourceforge.net/projects/openpnl), GRMM (http://mallet.cs.umass.edu/grmm), Factorie (http://code.google.com/p/factorie), Hugin (http://www.hugin.com), and an applet showcasing common Markov chain algorithms (http://www.lbreyer.com/classic.html).

For Reinforcement Learning

RL-Glue (http://glue.rl-community.org) is a language-independent software for reinforcement learning experiments; it provides a common interface for a number of software and hardware projects in the reinforcement learning community. RL-Glue has been ported to a number of languages including C/C++/Java/MATLAB via sockets.

Libpgrl (http://code.google.com/p/libpgrl/) implements both model-free reinforcement learning and policy search algorithms, though not any model-based learning. Libpgrl is efficient in a distributed reinforcement learning environment. Libpgrl is a fast C++ implementation that has abstract classes to model a subset of reinforcement learning.

MATLAB Markov Decision Process Toolbox (http://www.inra.fr/mia/T/MDPtoolbox/) implements only a few basic algorithms such as tabular Q-learning, SARSA, and dynamic programming. Some resources on reinforcement learning are available at http://www-all.cs.umass.edu/rlr/.

IoT Platforms

ThingSpeak (https://thingspeak.com) is an open cloud platform that connects things and people. It includes real-time data collection and storage, MATLAB analytics and visualizations, alerts, scheduling, device communication, open API, and geolocation data.

NIMBITS (https://www.nimbits.com/index.jsp) is an open-source IoT platform for connecting people, sensors, and devices on the cloud.

EVRYTHNG (https://evrythng.com) manages billions of intelligent IoT identities on the cloud, giving each a persistent, addressable web presence. SensorCloud (https://www.sensorcloud.com) is a sensor data storage, visualization, and remote management platform based on the cloud.

Xively (https://xively.com) offers a PAAS that allows IoT devices to connect to the cloud.

Other Resources

CVX (http://cvxr.com/cvx/) is a MATLAB-based modeling system for convex optimization.

SDPT3 (http://www.math.nus.edu.sg/~mattohkc/sdpt3.html) is an SDP solver. The MATLAB function `fmincon` is an SQP solver with a quasi-Newton approximation to the Hessian of the Lagrangian using the BFGS method.

SparseLab(http://sparselab.stanford.edu/) is a MATLAB software package for sparse solutions to systems of linear equations.

Resources on random forests are available at http://www.math.usu.edu/~adele/forests/. WEKA machine learning archive (http://www.cs.waikato.ac.nz/ml/weka/) offers a Java implementation of random forests. The classification results for bagging and boosting can be obtained using WEKA on identical training and test sets.

MultiBoost package (http://www.multiboost.org/) provides a fast C++ implementation of multiclass/multi-label/multitask boosting algorithms.

C5.0 (http://www.rulequest.com/see5-info.html) is a sophisticated data mining tool in C for discovering patterns that delineate categories, assembling them into classifiers, and using them to make predictions.

Resources on GPU can be found at http://www.nvidia.com, http://www.gpgpu.org/.

SIFT descriptors for an image can be generated by using open-source C libraries such as openSIFT library (http://robwhess.github.io/opensift/) or ezSIFT (https://github.com/robertwgh/ezSIFT).

Pylearn2 (http://deeplearning.net/software/pylearn2) is a python machine learning library.

Megaman (https://github.com/mmp2/megaman) is a Python package for scalable manifold learning.

MLweb (http://mlweb.loria.fr/lalolab/) is an open-source JavaScript software toolkit for machine learning on the web. All computations are performed on the client side without the need to send data to a third-party server.

SAMOA (scalable advanced massive online analysis, https://github.com/abifet/samoa) is an open-source Java platform for mining big data streams. It provides a collection of distributed streaming algorithms for the most common data mining and machine learning tasks. Its pluggable architecture allows it to run on distributed stream processing engines such as Storm, S4, and Samza.

Gesture recognition toolkit (https://github.com/nickgillian/grt) is a cross-platform open-source C++ library designed to make real-time machine learning and gesture recognition.

SPMF (http://www.philippe-fournier-viger.com/spmf/) is an open-source Java library of more than 55 data mining algorithms. It is specialized for discovering patterns in transaction and sequence databases such as frequent itemsets, association rules, and sequential patterns.

MEKA project (http://waikato.github.io/meka/) provides an open-source Java implementation of dozens of methods for multi-label learning and evaluation. MEKA is based on WEKA machine learning toolkit.

Apache Spark is a popular open-source platform for large-scale data processing that is well suited for iterative machine learning tasks. MLlib is Spark's open-source distributed machine learning library.

scikit-learn (https://scikit-learn.org/stable/) is a popular open-source machine learning library in Python. scikit-multilearn (http://scikit.ml) is a Python library for performing multi-label classification.

Related to the WEKA project, MOA (https://moa.cms.waikato.ac.nz/, in Java) is a popular open-source machine learning framework for data stream mining. Built on scikit-learn, MOA and MEKA, scikit-multiflow (https://github.com/scikit-multiflow, in Python) is a framework for learning from data streams and multi-output learning.

TensorLy (https://github.com/tensorly) is a Python library that provides a high-level API for tensor methods and deep tensorized neural networks. They can be scaled on multiple CPU or GPU machines.

Imbalanced-learn (https://github.com/scikit-learn-contrib/imbalanced-learn) is an open-source Python toolbox providing a wide range of methods to cope with the problem of imbalanced datasets.

Tensor Toolbox for MATLAB (http://www.tensortoolbox.org/) provides functions for manipulating dense, sparse, and structured tensors.

OpenXBOW (https://github.com/openXBOW/) is an open-source Java toolkit for generating bag-of-words (BoW) representations from multimodal input.

SnapVX (http://snap.stanford.edu/snapvx) is a fast and scalable python solver for large convex optimization problems defined on networks. It is based on the alternating direction method of multipliers (ADMM).

BlockBench (https://github.com/ooibc88/blockbench) is a benchmarking framework for quantitatively evaluating private blockchains.

References

1. Berg, T. L., Berg, A. C., Edwards J., Maire M., White, R., Teh, Y.-W., Learned-Miller, E., & Forsyth, D. A. (2004). Names and faces in the news. In *Proceedings of IEEE Conference on Computer Vision and Pattern Recognition* (Vol. 2, pp. 848–854).
2. Dhall, A., Goecke, R., Joshi, J., Wagner, M., & Gedeon, T. (2013). Emotion recognition in the wild challenge 2013. In *Proceedings of the 15th ACM on International Conference on Multimodal Interaction* (pp. 509–516). Sydney, Australia.
3. Hayat, K. (2018). Multimedia super-resolution via deep learning: A survey. *Digital Signal-Processing, 81*, 198–217.
4. Lauritzen, S. L., & Spiegelhalter, D. J. (1988). Local computations with probabilities on graphical structures and their application on expert systems. *Journal of the Royal Statistical Society Series B, 50*(2), 157–224.
5. Meshi, O., Jaimovich, A., Globerzon, A., & Friedman, N. (2009). Convexifying the bethe free energy. In *Proceedings of the 25th Conference on Uncertainty in Artificial Intelligence (UAI)*. Montreal, Canada.
6. Minka, T. (2001). *Expectation propagation for approximate Bayesian inference*. Doctoral dissertation, MIT Media Lab.
7. Mooij, J., & Kappen, H. (2007). Sufficient conditions for convergence of the sum-product algorithm. *IEEE Transactions on Information Theory, 53*, 4422–4437.
8. Phillips, P. J., Flynn, P. J., Scruggs, T., Bowyer, K. W., Chang, J., Hoffman, K., Marques, J., Min, J., & Worek, W. (2005). Overview of the face recognition grand challenge. In *Proceedings of IEEE Conference on Computer Vision and Pattern Recognition* (Vol. 1, pp. 947–954).
9. Wainwright, M.J., Jaakkola, T. S., & Willsky, A.S. (2005). A new class of upper bounds on the log partition function. *IEEE Transactions on Information Theory, 51*(7), 2313–2335.
10. Yedidia, J. S., Freeman, W. T., & Weiss, Y. (2005). Constructing free energy approximations and generalized belief propagation algorithms. *IEEE Transactions on Information Theory, 51*, 2282–2312.
11. Yin, L., Wei, X., Sun, Y., Wang, J., & Rosato, M. J. (2006). A 3D facial expression database for facial behavior research. In *Proceedings of the 7th International Conference on Automatic Face and Gesture Recognition* (pp. 211–216).

Index

K.-L. Du and M. N. S. Swamy, *Neural Networks and Statistical Learning*,
https://doi.org/10.1007/978-1-4471-7452-3

S

Printed in the United States
By Bookmasters